CONTENTS

FOREWORD

The development of efficient methods for use in organic synthesis, particularly applicable to polyfunctionalized molecules and heterocycles, remains one of the top challenges in organic chemistry. Oftentimes older reactions are rediscovered and improved, sometimes substantially, in order to solve modern problems. A prime example of this is copper-catalyzed methodology. This topic, studied at the beginning of the twentieth century by Ullmann, Goldberg, and others has experienced a renaissance in the past two decades. A significant reason for this is the development of new and improved ligands to facilitate the transformations of interest.

While the most obvious difference between catalysts based on palladium and copper is the lower cost of the latter, this is only a small part of what makes copper catalysts so attractive. For most chemists in academia or discovery chemistry, the cost is not a major issue. In addition, one must take in to account many factors, such as volumetric productivity (reactor time, labor cost), substrate cost, yield, and substrate scope, which affect the total cost when comparing any two processes. Among the major advantages of using copper catalysts (in addition to cost) are that they tend to show excellent functional group compatibility and, in fact, are often at their best in promoting reactions of substrates that contain coordinating functional groups. A prime example, as delineated in Evano and Blanchard's preface to this book, is seen in Ullmann's seminal work: he was able to accomplish the coupling of aniline with 2-chlorobenzoic acid well before the same transformation could be carried out with palladium catalysts. In fact, the chemistry based on these two metals is really quite complementary. In support of this notion, today these are used interchangeably for many synthetic operations depending on the nature of the substrate combination to be examined.

The chapters in the book cover a range of synthetically and mechanistically important topics. Many aspects of carbon-heteroatom and carbon–carbon, bond-forming reactions are described. These have become an everyday part of the practicing synthetic organic chemist's arsenal. In addition, topics such as the formation of trifluoromethylated organic molecules and the application of copper-catalyzed methods to the total synthesis of natural products and pharmaceuticals are also covered. Taken together, the chapters in this book provide an up-to-date overview of the state of the field, providing the reader with the knowledge of both what has been accomplished and what new areas remain for further investigation.

Massachusetts Institute of Technology STEPHEN L. BUCHWALD

PREFACE

Copper Catalysis from a Historical Perspective: A Legacy from the Past

GWILHERM EVANO AND NICOLAS BLANCHARD

THE HISTORICAL AND REMARKABLE FIRST DISCOVERIES BY FRITZ ULLMANN AND IRMA GOLDBERG

It all started in 1901 in Geneva when Fritz Ullmann, born in 1875 in Fürth, where his father owned a fabric... of metallic powders,[1] reported, in a remarkable publication that we highly recommend, that "Erhitzt man *o*-Bromnitrobenzol mit fein vertheiltem Kupferpulver, so bemerkt man, dass letzteres seinen Glanz verliert und in eine matte graue Masse verwandelt wird. Bei Aufarbeitung des Reactionsproductes zeigte sich nun, dass das Kupfer zum grössten Theil in Cuprobromid und das Bromnitrobenzol in eine bromfreie Substanz verwandelt worden ist, welche sich bei näherer Untersuchung identisch, mit der von Tauber auf andere Weise dargestellten 2.2′-Dinitrobiphenyl, erwies." (If one heats *o*-bromonitrobenzene with finely divided copper powder, so one recognizes that the last one is losing its shine and turns into a matt grey mass. After purification of the reaction products, it appears that copper has turned into copper bromide and that the bromonitrobenzene has turned into a bromine-free substance, which, on a closer look, turns out to be identical with the 2,2-dinitrobiphenyl synthesized in a different way by Tauber) (Fig. P.1).*,[2] The reaction can also be performed from aryl iodides and, which is just astonishing, even from aryl chlorides (Scheme P.1). Two molecules of aryl

* The editors are truly indebted to Dr. Théophile Tschamber from the University of Mulhouse (France) for his expertise in translating the numerous original German research papers and patents that are cited in the Preface.

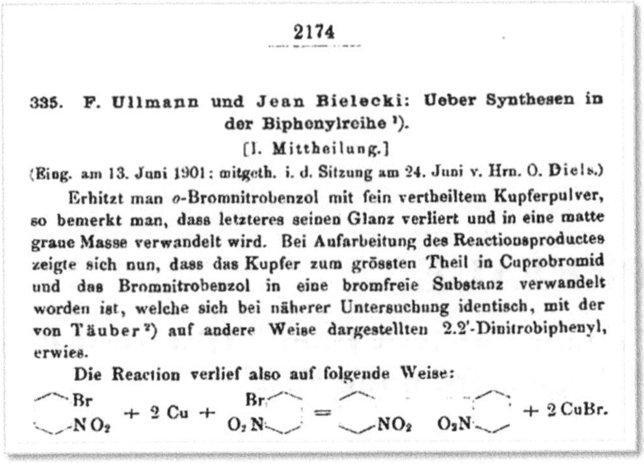

2174

335. F. Ullmann und Jean Bielecki: Ueber Synthesen in der Biphenylreihe [1]).

[I. Mittheilung.]

(Eing. am 13. Juni 1901; mitgeth. i. d. Sitzung am 24. Juni v. Hrn. O. Diels.)

Erhitzt man *o*-Bromnitrobenzol mit fein vertheiltem Kupferpulver, so bemerkt man, dass letzteres seinen Glanz verliert und in eine matte graue Masse verwandelt wird. Bei Aufarbeitung des Reactionsproductes zeigte sich nun, dass das Kupfer zum grössten Theil in Cuprobromid und das Bromnitrobenzol in eine bromfreie Substanz verwandelt worden ist, welche sich bei näherer Untersuchung identisch, mit der von Täuber [2]) auf andere Weise dargestellten 2.2'-Dinitrobiphenyl, erwies.

Die Reaction verlief also auf folgende Weise:

Figure P.1 Fritz Ullmann (1875–1939) and an extract from his original publication.

halides can be coupled together in the presence of stoichiometric amounts of metallic copper: The Ullmann reaction had just been discovered.

Obviously fascinated with the power of the metallic copper, Ullmann next reported in 1903 an equally fascinating reaction in a publication entitled "Über eine neue Bildungsweise von Diphenylaminderivaten" (*On a New Method for the Formation of Diphenylamine Derivatives*). By mixing aniline (or other aniline derivatives) and *ortho*-chlorobenzoic acid in the presence of metallic copper at reflux, the aniline was smoothly arylated to the corresponding diarylamine with a remarkable efficiency (Scheme P.2).[3] Two years after his initial discovery of the biaryl synthesis, Ullmann continued to set the foundations of copper-mediated cross-coupling reactions and the arylation of amines that would later on be named after him was invented. Also remarkable is that the aryl halide substrate chosen for this study, *ortho*-chlorobenzoic acid, already possessed a non-innocent chelating group: the *ortho*-effect, which still has a deep impact in copper catalysis today and which was crucial to the success of many copper-mediated transformations including in Nicolaou's synthesis of vancomycin (see further), was already touched upon in 1903.

Following these two reports, Ullmann's assistant in Geneva, Irma Goldberg, reported 3 years later that a related reaction could be performed using only catalytic amounts of copper. By reacting anthranilic acid and bromobenzene in the presence of potassium carbonate and catalytic amounts of "Naturkupfer C" in refluxing nitrobenzene for 3 hours, the reaction proceeded with a remarkable efficiency, the corresponding condensation product being isolated in 99% yield (Scheme P.3).[4] The arylation could also be efficiently performed with *para*-nitro-bromobenzene and, even better, the arylation of amides (benzamide in this case), known nowadays as the Goldberg condensation reaction, could also be catalyzed by traces of metallic copper. These three remarkable

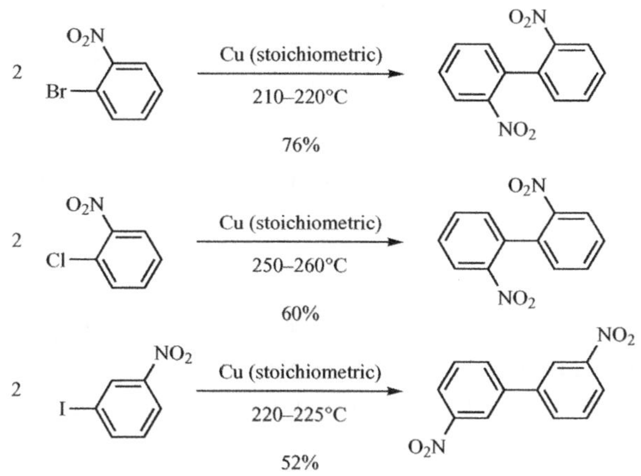

Scheme P.1 1901: the discovery of the Ullmann reaction.

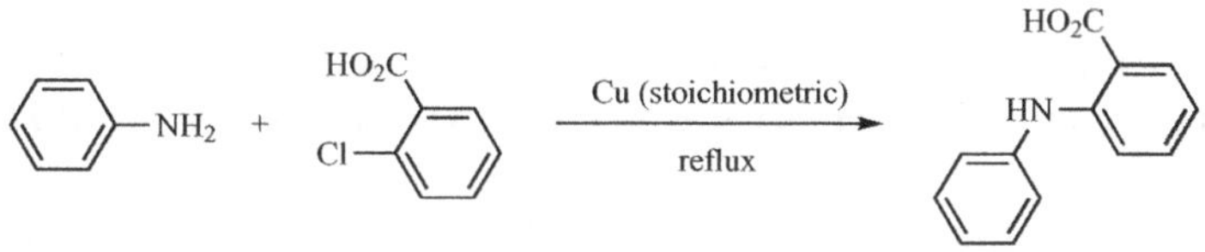

Scheme P.2 1903: the discovery of the Ullmann condensation reaction for the synthesis of diarylamines.

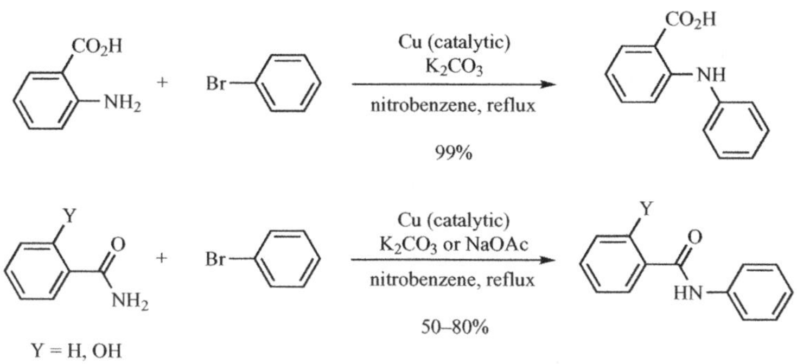

Scheme P.3 1906: the discovery of the copper-catalyzed arylation of anilines and amides by Irma Goldberg.

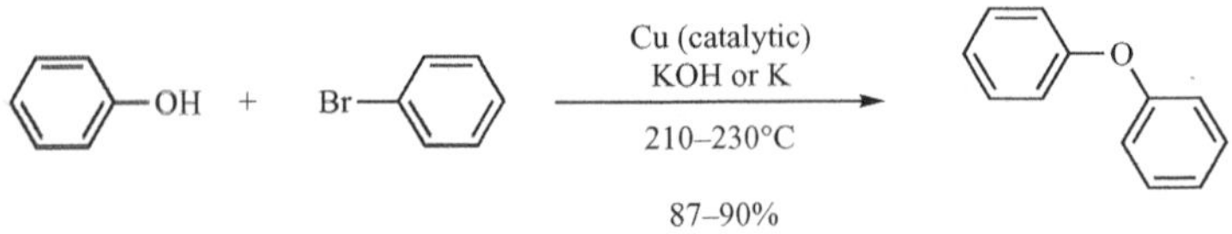

Scheme P.4 1905: the discovery of the copper-catalyzed arylation of phenols by Fritz Ullmann.

reactions published between 1901 and 1906 mark the beginning of cross-coupling reactions and clearly paved the way for all recent developments.

Last but not least, Ullmann also studied the cross coupling involving phenols and reported in 1905 that "if you try to react potassium phenoxide with bromobenzene, the yield of the biphenyl ether is 0.9%. If you however add small quantities of copper to the reaction mixture, the yield goes to 90%" (Scheme P.4).[5] It is therefore possible to catalyze the cross coupling between phenols and aryl bromides with small amounts of copper, which obviously have a dramatic effect on the reaction rate and yields.

FURTHER HISTORICAL DEVELOPMENTS: THE CYANATION OF ARYL HALIDES BY ROSENMUND AND VON BRAUN AND THE ARYLATION OF DIKETONES AND MALONATES BY HURTLEY

In a series of just four publications, Fritz Ullmann and Irma Goldberg, who eventually got married in 1910, clearly set the foundations of modern copper catalysis. This pioneering work was followed by other key reports in the next 30 years or so, which further demonstrated the potential of copper-mediated cross-coupling reactions.

The next major development in this field was stimulated by the interest in the copper-mediated cyanation of aryl halides, which appeared closely after the seminal publications of Ullmann and Goldberg. Although this transformation is known as the "Rosenmund–von Braun" reaction, many academic and industrial research groups contributed to the development of this very useful reaction.[6–8] Indeed, the first report of such a reaction appeared in the patent literature in 1913 by the Meister, Lucius & Brüning Company who found that chloro- and bromoanthraquinones reacted with copper(I) cyanide in pyridine at 150°C to give the corresponding nitriles (Fig. P.2 and Scheme P.5).[8]

Karl W. Rosenmund and his PhD student Erich Struck described next in 1919 that potassium cyanide in combination with copper(I) cyanide converted aryl halides to the corresponding benzoic acids in water at 200°C under pressure. The intermediate nitriles were not isolated, but the central role of the copper(I) species was put forward (Scheme P.6, Eq. 1).[9a] Rosenmund disclosed a year later[9b] that stoichiometric copper(I) thiocyanate in pyridine

KAISERLICHES PATENTAMT.

AUSGEGEBEN DEN 18. MÄRZ 1914.

PATENTSCHRIFT

— № 271790 —

KLASSE 12 *o*. GRUPPE 10.

FARBWERKE VORM. MEISTER LUCIUS & BRÜNING IN HÖCHST A. M.

Verfahren zur Darstellung von Anthrachinon-α-nitrilen.

Patentiert im Deutschen Reiche vom 18. Februar 1913 ab.

Es wurde gefunden, daß Anthrachinon-α-nitrile erhalten werden, wenn man α-Halogenanthrachinone mit Kupfercyanür in Gegenwart von indifferenten Mitteln erhitzt, während eine ähnliche Umsetzung mittels Cyankaliums nicht gelingt.

Dies war nicht vorauszusehen, da in verschiedenen organischen Halogenverbindungen mit leicht beweglichem Halogen sich dieses überhaupt nicht durch eine Cyangruppe ersetzen läßt (vgl. Windisch, Dissertation, Berlin 1887; V. v. Richter, Ber. IV, 1871, Seite 462).

H. Dienel hat bereits das α-Cyananthrachinon sowohl durch Oxydation von α-Cyananthracen wie durch trockne Destillation von anthrachinonsulfosaurem Kalium mit Cyankalium erhalten (Ber. XXXIX, 1906, Seite 932). Ullmann und van der Schalk haben dann nachgewiesen, daß man beim Nacharbeiten des letzteren Verfahrens an α-Cyananthrachinon im günstigsten Falle 1 Prozent der Theorie erhält (Ann. 388, 205), und gezeigt, daß α-Cyananthrachinon im allgemeinen aus den α-Aminoanthrachinonen über die Diazoniumverbindungen mittels der Sandmeyerschen Reaktion gewonnen werden (Ann. 388, 203); Patent 243788.

Den bisher bekannten Verfahren gegenüber bedeutet das vorliegende schon deshalb einen großen technischen Fortschritt, weil es verschiedene neue, auf anderem Wege nicht erhältliche Derivate des Anthrachinon-α-nitrils darzustellen gestattet.

Beispiele.

1. Anthrachinon-1-nitril: 24 Teile 1-Chloranthrachinon werden mit 9 Teilen Kupfercyanür und 100 Teilen Pyridin während einiger Stunden auf 150° erhitzt. Das Reaktionsprodukt wird mit Ammoniak verdünnt und das abgeschiedene rohe Nitril abgesaugt. Zur Reinigung wird es z. B. aus Chlorbenzol umkristallisiert und so in Form brauner Kristalle erhalten, die bei 247° schmelzen und eine reine blaue Küpe geben.

2. 2-Amino-3-bromanthrachinon-1-nitril: 20 Teile 1·3-Dibrom-2-aminoanthrachinon werden mit 5 Teilen Kupfercyanür und 100 Teilen Pyridin etwa 6 Stunden am Rückflußkühler erhitzt. Nach dem Erkalten wird das grünlichgelbe, schön kristallisierte Nitril abgesaugt und gewaschen. Aus Chlorbenzol kristallisiert es in gelblichen Nadeln, die bei 297 bis 300° schmelzen. Die Lösung in konzentrierter Schwefelsäure ist rotorange und wird beim Erwärmen rot. Gießt man die schwefelsaure Lösung des Nitrils in Wasser, so fällt die 2-Amino-3-bromanthrachinon-1-carbonsäure in gelben Flocken aus.

3. 2-Oxy-3-bromanthrachinon-1-nitril: 20 Teile 1·3-Dibrom-2-oxyanthrachinon werden mit 5 Teilen Kupfercyanür und 200 Teilen Pyridin 6 Stunden am Rückflußkühler erhitzt. Nach dem Erkalten wird das rote, in Pyridin schwer lösliche Pyridinsalz des 2-Oxy-3-bromanthrachinon-1-nitrils abgesaugt. Aus der Lö-

Figure P.2 The 1913 Meister, Lucius & Brüning Company patent.

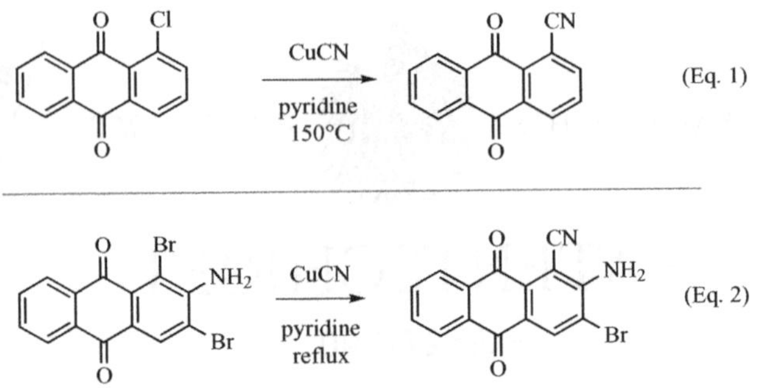

Scheme P.5 1913: The first copper(I)-mediated cyanations of haloanthraquinones by the Meister, Lucius & Brüning Company.

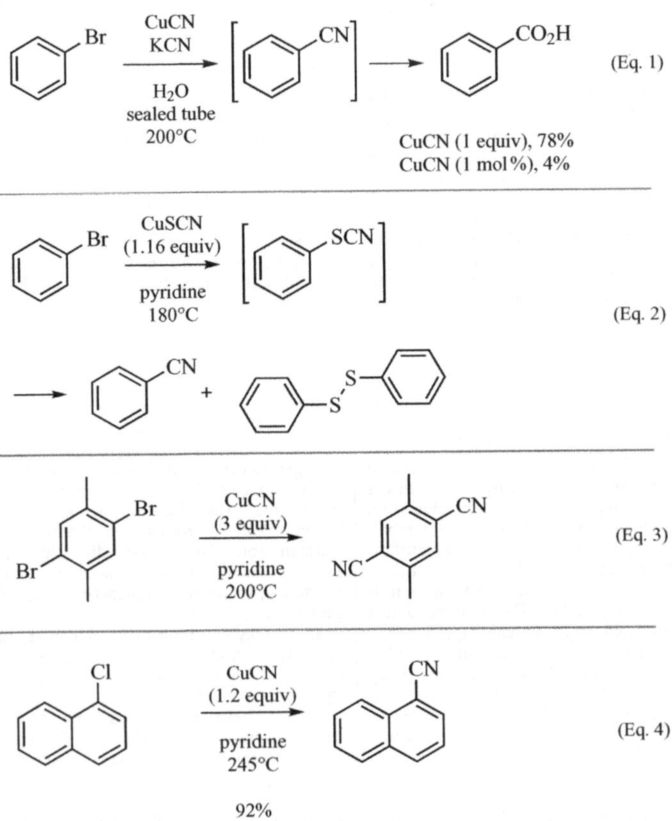

Scheme P.6 1919–1923: Karl W. Rosenmund and Henri de Diesbach early reports on the cyanation of aryl halides.

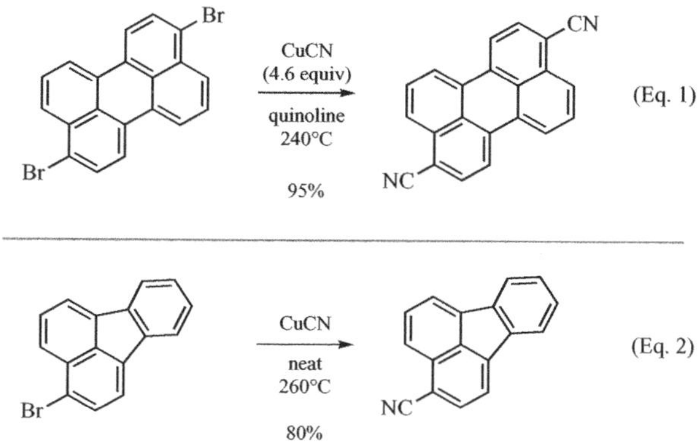

Scheme P.7 1927–1931: Alfred Pongratz and Julius von Braun cyanation of aryl halides.

converted aryl bromides to intermediate aryl thiocyanates, which spontaneously gave the corresponding nitriles at 180°C in the presence of the copper species,[6] the nitriles being characterized as carboxylic acids after aqueous acidic treatment (Scheme P.6, Eq. 2). Shortly after, Henri de Diesbach reported in two back-to-back articles in 1923 that dibromoxylene derivatives could be converted to the corresponding dinitriles in excellent yield using an excess of copper(I) cyanide in pyridine at 200°C (Scheme P.6, Eq. 3),[10] and even aryl chlorides could be transformed to the corresponding nitriles, as seen in Scheme P.6, Eq. 4.[11]

In 1927, Alfred Pongratz described that copper(I) cyanide in refluxing quinoline was efficient for the cyanation of perylene-3,9-dibromide (Scheme P.7, Eq. 1) and that the corresponding chloride could be converted in the absence of solvent at 300°C, although no yield was given in the latter case.[12] Finally, during investigation on the chemistry of fluoranthene derivatives, Julius von Braun, who gave his name to this cyanation reaction, demonstrated that 4-bromo-fluoranthene could be transformed to the corresponding nitrile in 80% yield when reacted with copper(I) cyanide at 260°C (Scheme P.7, Eq. 2).[13]

The pioneering investigations of Ullmann, Goldberg and Rosenmund[8b] were a source of inspiration for William R. H. Hurtley, who recognized their contributions in his seminal 1929 report, stating that "under the catalytic influence of copper, the halogen in *o*-bromo-benzoic acid is much more reactive than is commonly realized."[14] Hurtley showed that copper bronze or copper acetate can promote the C-arylation of some families of CH-acids in reaction with *o*-bromobenzoic acid in the presence of sodium ethoxide as a base (Scheme P.8). Since then, this transformation, commonly known as the "Hurtley reaction", has triggered mechanistic investigations as well as very elegant applications in organic synthesis. Mild conditions are now available thanks to

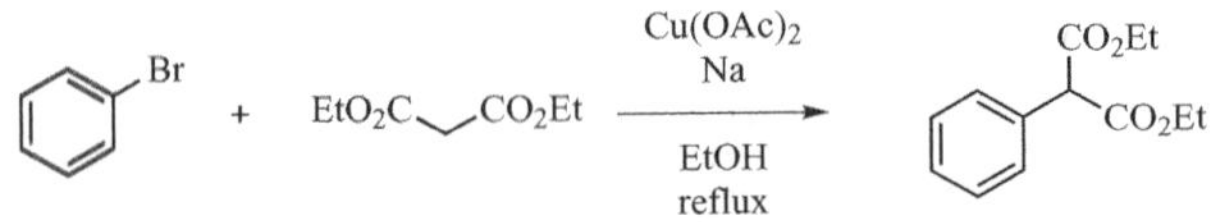

Scheme P.8 1929: the discovery of the copper-mediated arylation of CH-acid derivatives by William R. H. Hurtley.

new classes of ligands that even allowed the development of an asymmetric version of the Hurtley reaction.

FROM THE HISTORICAL DISCOVERIES TO THE DEVELOPMENT OF MODERN COPPER-MEDIATED CROSS-COUPLING REACTIONS

While remarkable, these reactions, however, have found only a limited number of applications, especially when compared to their impressive potential. Indeed, drawbacks associated with the use of harsh reaction conditions (strong bases, stoichiometric amounts of copper in some cases, and high reaction temperatures) have clearly hampered the development of these copper-mediated cross-coupling reactions. From the first discoveries in the early 1900s up to the beginning of the twenty-first century, no real breakthrough was made to significantly improve the original systems and the strategy was to tailor the substrate, not the catalyst, to obtain systems that would perform under milder conditions. Nicolaou's remarkable total synthesis of vancomycin is one of the most representative example of such a strategy. Capitalizing on the well-known *ortho*-effect in copper-mediated transformations, a mild copper-mediated arylation reaction was designed based on the ingenious incorporation of a triazene unit in the starting material strategically placed *ortho* to the bromine: this internal helper auxiliary served both as a potential "electron sink" and to coordinate the intermediate copper species. The triazene indeed considerably helped the reaction, which enabled the formation of the C–O–D and D–O–E ring systems under remarkably mild conditions since the reaction could be performed in refluxing acetonitrile (Scheme P.9).[15]

Besides these strategies based on substrate modifications, the inherent shortcomings of the original copper-mediated cross-coupling reactions were still not overcome at the end of the twentieth century and palladium was in most cases preferred to copper. Despite remarkable work done with palladium catalysis for the construction of an impressive and ever-increasing number of C–C and C–heteroatom bonds, limitations still exist and the high cost of palladium and ligands has forced chemists to consider alternative metals. This evolution eventually led to the renaissance of copper catalysis, which has been extremely revisited over the past decade. Following extensive and remarkable studies of the Ullmann condensation by Payne,[16] who showed that the active

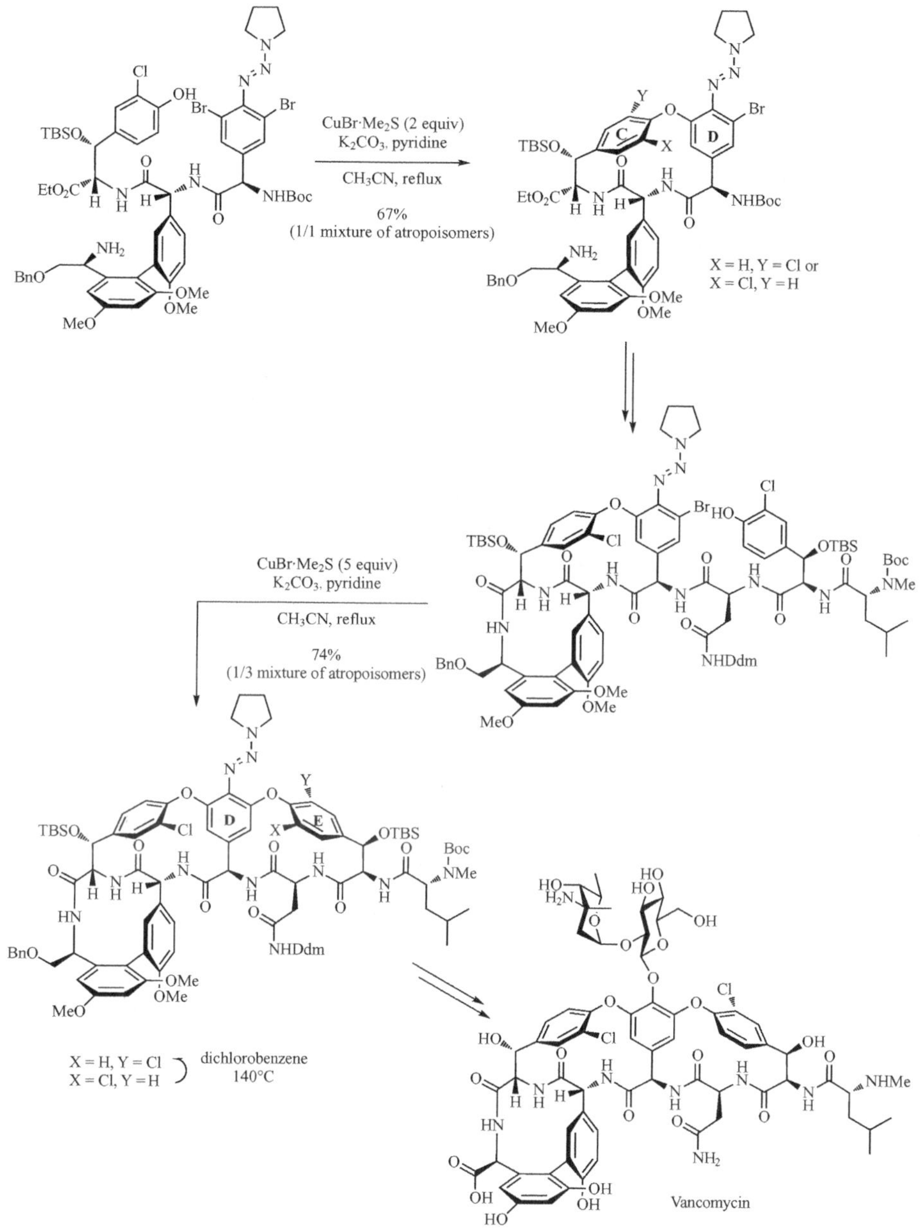

Scheme P.9 The Ullmann condensation reaction under milder conditions: redesigning the substrate to improve the efficiency (Nicolaou's total synthesis of vancomycin).

catalytic species are soluble cuprous ions and pioneering work by Bryant,[17] Capdevielle,[18] and Goodbrand,[19] who investigated the idea of a "ligated copper catalysis", the key breakthrough was the introduction of chelating ligands for copper.

Remarkable and pioneering studies in this area from, among others, the Buchwald, Ma, and Taillefer groups had a strong impact in the renaissance of copper catalysis, and the Ullmann, Goldberg, Rosenmund–von Braun, Hurtley, and other related copper-mediated processes can now be conducted under much milder conditions together with dramatically enhanced yield. Recent advances in the arylation of N-, O-, and S- as well as P-nucleophiles with aryl halides will be described in Chapters 1, 2, and 3, respectively.

In addition to the development of improved protocols for the arylation of various heteronucleophiles, the introduction of chelating ligands also considerably expanded the scope of the original procedures and allowed for the use of new reaction partners for the introduction of vinyl, alkyne, and allene functional groups. These copper-mediated syntheses of heterosubstituted alkenes, alkynes, and allenes will be overviewed in Chapter 5. Copper-catalyzed aromatic versions of the Finkelstein reaction have also been recently reported: they proceed with remarkable efficiencies and will be covered in Chapter 6.

These recent developments also had a strong impact on the development of new and improved protocols for the formation of C–C bonds, and a variety of robust, mild, and reliable procedures have been designed with the assistance of copper complexes.

The classical Hurtley-type transformation as well as the tremendous progresses that have been reported recently in this central C-arylation reaction of CH-acid derivatives will be presented in Chapter 8, with an emphasis on the importance of the different types of ligands, mechanisms, and synthetic applications. The copper-catalyzed synthesis of nitriles from aryl halides is then covered in Chapter 9, including discussions on the different catalytic systems and sources of cyanide that have been greatly improved in the past two decades.

A central reaction in the realm of copper-catalyzed and copper-mediated transformations is of course the aryl–aryl coupling that has continuously evolved into an efficient and practical tool with multiple applications. Chapter 10 is devoted to this vibrant field of research that has recently witnessed a shift of paradigm with even more direct reactions such as the copper-catalyzed direct C–H arylations. The next three chapters, Chapters 11, 12, and 13, have captured relevant advances in some of the most practical and efficient copper-catalyzed transformations that have clearly modified our conception of carbon–carbon bond disconnections: the alkenylation and alkynylation of aryl, alkynyl, and vinyl derivatives.

To close this section dedicated to the formation of carbon–carbon bonds, the central role of copper catalysts in methods aiming at the introduction of the trifluoromethyl moiety will be exposed. Indeed, the occurrence of the trifluoromethyl motif in pharmaceutical and agrochemical active ingredients is well recognized and has triggered the development of new methodologies for its selective introduction. The copper-catalyzed trifluoromethylation reaction and copper-mediated oxidative processes are efficient strategies in this context and are developed in detail in Chapter 14.

Besides substrate modifications and the design of more efficient catalytic systems, the drawbacks associated with the harsh conditions of the classical Ullmann–Goldberg and related transformations also motivated the search for alternative reagents to aryl halides that would enable the arylation of heteronucleophiles under milder reaction conditions. In 1998, independent back-to-back publications by Chan,[20] Evans,[21] and Lam[22] revolutionized the arylation of heteronucleophiles as they reported generally applicable protocols for the copper-mediated oxidative cross coupling with arylboronic acids at room temperature. This reaction is now widely applied and other reagents such as organobismuth, -lead, -stannanes, -siloxanes, or hypervalent iodonium salts have also been used. They are sometimes more reactive than aryl halides, allow overcoming some limitations met with these reagents, or are just more readily available in some cases. Advances in copper-mediated cross coupling using these reagents are covered in Chapter 4 as well as the heteroarylation involving masked nucleophiles and the direct introduction of a heteronucleophile from a C–H bond. The use of these alternative reagents for the alkenylation/alkynylation of heteronucleophiles will be overviewed in Chapter 5.

If the modern copper-catalyzed Ullmann and Goldberg cross-coupling reactions have recently evolved as reliable and efficient methods for the construction of products relevant to both industry and academic settings, their mechanisms have been studied only recently. Mechanistic insights into these reactions, which will be crucial for the development of even more efficient catalytic systems, will be described in Chapter 7.

MODERN COPPER-MEDIATED CROSS-COUPLING REACTIONS AND THEIR IMPACT IN ORGANIC CHEMISTRY

We have all heard that we have arrived at a point where it is possible to make almost any conceivable chemical structure by a rational approach, using the large toolbox of synthetic methods available today and that there was basically nothing left to do in organic synthesis. The impact that had the renaissance of copper catalysis over the past 10 years is one of the most striking examples that this is just as stupid as it seems. Many long-standing synthetic problems have been solved using copper-mediated cross-coupling technologies, which had an important impact on the way chemists design their synthetic routes. These transformations have found remarkable applications in heterocyclic synthesis, which are detailed in Chapter 15, or in natural product synthesis, as shown in Chapters 16–18, an area where chemical transformations are clearly challenged and pushed way beyond their limits. Copper-mediated cross-coupling reactions have also been readily integrated in process chemistry where remarkably efficient procedures have been developed on the basis of copper catalysis as shown by examples collected in Chapter 19.

Copper is cheap, at least for the moment, but high catalyst loading is still required in most cases and catalyst recovery is therefore a quite important issue. Reusable catalysts for copper-mediated cross-coupling reactions under heterogeneous conditions have also been developed recently and will be overviewed in Chapter 20.

As a concluding remark, all advances reported in the last decade in copper-mediated cross-coupling reactions are truly remarkable and had strong impact, in a relatively short period of time, in organic synthesis. Many of the transformations presented in this book are already classics in organic synthesis and, in many areas, copper really showed the way.

REFERENCES

[1] Meyer, K. H. *Helv. Chim. Acta* **1940**, *23*, 93–100.

[2] Ullmann, F.; Bielecki, J. *Ber. Dtsch. Chem. Ges.* **1901**, *34*, 2174–2185.

[3] Ullmann, F. *Ber. Dtsch. Chem. Ges.* **1903**, *36*, 2382–2384.

[4] Goldberg, I. *Ber. Dtsch. Chem. Ges.* **1906**, *39*, 1691–1692.

[5] Ullmann, F.; Sponagel, P. *Ber. Dtsch. Chem. Ges.* **1905**, *38*, 2211–2212.

[6] Mowry, D. T. *Chem. Rev.* **1948**, *42*, 189–283.

[7] Ellis, G. P.; Romney-Alexander, T. M. *Chem. Rev.* **1987**, *87*, 779–794.

[8] Farbwerke vormals Meister, Lucius & Brüning, assignee. Verfahren zur Darstellung von Anthrachinon-α-nitrilen. German patent 271,790. 1913 February 18.

[9] (a) Rosenmund, K. W.; Struck, E. *Ber. Dtsch. Chem. Ges.* **1919**, *52*, 1749–1746. (b) Rosenmund, K. W.; Harms, H. *Ber. Dtsch. Chem. Ges.* **1920**, *53*, 2226–2240.

[10] (a) de Diesbach, H. *Helv. Chim. Acta* **1923**, *6*, 539–548. (b) de Diesbach, H.; Schmidt, V.; Decker, E. *Helv. Chim. Acta* **1923**, *6*, 548–549.

[11] Newman, M. S. *J. Am. Chem. Soc.* **1937**, *59*, 2472.

[12] Pongratz, A. *Monatsh. Chem.* **1927**, *48*, 585–591.

[13] von Braun, J.; Manz, G. *Liebigs Ann. Chem.* **1931**, *488*, 111–126.

[14] Hurtley, W. R. H. *J. Chem. Soc.* **1929**, 1870–1873.

[15] (a) Nicolaou, K. C.; Boddy, C. N. C.; Natarajan, S.; Yue, T.-Y.; Li, H.; Bräse, S.; Ramanjulu, J. M. *J. Am. Chem. Soc.* **1997**, *119*, 3421–3422. (b) Nicolaou, K. C.; Li, H.; Boddy, C. N. C.; Ramanjulu, J. M.; Yue, T.-Y.; Natarajan, S.; Chu, X.-J.; Bräse, S.; Rübsam, F. *Chem. Eur. J.* **1999**, *5*, 2584–2601. (c) Nicolaou, K. C.; Boddy, C. N. C.; Li, H.; Koumbis, A. E.; Hughues, R.; Natarajan, S.; Jain, N. F.; Ramanjulu, J. M.; Bräse, S.; Solomon, M. E. *Chem. Eur. J.* **1999**, *5*, 2602–2621. (d) Nicolaou, K. C.; Koumbis, A. E.; Takayanagi, M.; Natarajan, S.; Jain, N. F.; Bando, T.; Li, H.; Hughues, R. *Chem. Eur. J.* **1999**, *5*, 2622–2647. (e) Nicolaou, K. C.; Mitchell, H. J.; Jain, N. F.; Bando, T.; Hughes, R.; Winssinger, N.; Natarajan, S.; Koumbis, A. E. *Chem. Eur. J.*

1999, *5*, 2648–2667. (f) Nicolaou, K. C.; Boddy, C. N. C. *J. Am. Chem. Soc.* **2002**, *124*, 10451–10455.

[16] Paine, A. J. *J. Am. Chem. Soc.* **1987**, *109*, 1496–1502.

[17] Bryant, R. J., inventor; Sterwin A.-G., assignee. Substitution of aromatic organic compounds. G. B. patent 2,089,672. 1982 June 30.

[18] Capdevielle, P.; Maumy, M. *Tetrahedron Lett.* **1993**, *34*, 1007–1010.

[19] Goodbrand, H. B.; Hu, N.-X. *J. Org. Chem.* **1999**, *64*, 670–674.

[20] Chan, D. M. T.; Monaco, K. L.; Wang, R.-P.; Winter, M. P. *Tetrahedron Lett.* **1998**, *39*, 2933–2936.

[21] Evans, D. A.; Katz, J. L.; West, T. R. *Tetrahedron Lett.* **1998**, *39*, 2937–2940.

[22] Lam, P. Y. S.; Clark, C. G.; Saubern, S.; Adams, J.; Winters, M. P.; Chan, D. M. T.; Combs, A. *Tetrahedron Lett.* **1998**, *39*, 2941–2944.

CONTRIBUTORS

Carole Alayrac, Laboratoire de Chimie Moléculaire et Thioorganique, UMR CNRS 6507, INC3M, FR3038, ENSICAEN & Université de Caen Basse-Normandie, 6 boulevard du Maréchal Juin, 14050 Caen, France

Irina P. Beletskaya, Moscow M.V. Lomonosov State University, Chemical Faculty, Department of Organic Chemistry, Leninskie gory 1, Moscow 119991, Russian Federation

Matthias Beller, Leibniz-Institut für Katalyse e.V. an der Universität Rostock, Albert-Einstein-Strasse 29a, 18059 Rostock, Germany

Nicolas Blanchard, Université de Strasbourg, Ecole Européenne de Chimie, Polymères et Matériaux, Laboratoire de Chimie Moléculaire Associé au CNRS, 25 rue Becquerel 67087 Strasbourg, France

Alicia Casitas, QBIS Group, Departament de Química, Universitat de Girona, Campus de Montilivi, 17071 Girona, Spain

Morgan Donnard, Université de Strasbourg, Faculté de Pharmacie, Laboratoire d'Innovation Thérapeutique Associé au CNRS, 74 route du Rhin—CS60024, 67401 Illkirch Cedex, France

Gwilherm Evano, Laboratoire de Chimie Organique, Service de Chimie et PhysicoChimie Organiques, Université Libre de Bruxelles, Avenue F. D. Roosevelt 50, CP160/06, B-1050 Brussels, Belgium

Alexey Yu. Fedorov, N.I. Lobachevky Nizhny Novgorod State University, Chemical Faculty, Department of Organic Chemistry, Gagarin avenue 23, 603950 Nizhny Novgorod, Russian Federation

Hua Fu, Key Laboratory of Bio-organic Phosphorus Chemistry and Chemical Biology, Department of Chemistry, Tsinghua University, Beijing 100084, P. R. China

Annie-Claude Gaumont, Laboratoire de Chimie Moléculaire et Thioorganique, UMR CNRS 6507, INC3M, FR3038, ENSICAEN & Université de Caen Basse-Normandie, 6 boulevard du Maréchal Juin, 14050 Caen, France

Céline Guissart, Laboratoire de Chimie Organique, Service de Chimie et PhysicoChimie Organiques, Université Libre de Bruxelles, Avenue F. D. Roosevelt 50, CP160/06, B-1050 Brussels, Belgium

Ruimao Hua, Department of Chemistry, Tsinghua University, Beijing 100084, P. R. China

Yongwen Jiang, State Key Laboratory of Bio-organic and Natural Products Chemistry, Shanghai Institute of Organic Chemistry, Chinese Academy of Sciences, 354 Fenglin Lu, Shanghai 200032, P. R. China

Kévin Jouvin, Institut Lavoisier de Versailles, Université de Versailles Saint-Quentin-en-Yvelines, 45 avenue des Etats Unis, 78035 Versailles Cedex, France

Klaus Kunz, Bayer Crop Science AG, Alfred-Nobel-Strasse 50, 40789 Monheim, Germany

Jihoon Lee, Department of Chemistry and Center for Chemical, Methodology and Library Development, Metcalf Center for Science and Engineering, 590 Commonwealth Avenue, Boston University, Boston, MA 02215, USA

Hao Li, Department of Chemistry, Atwood Hall, 1515 Dickey Drive, Emory University, Atlanta, GA, 30322-2210, USA

Jin-Heng Li, State Key Laboratory of Chemo/Biosensing and Chemometrics, College of Chemistry and Chemical Engineering, Hunan University, Changsha 410082, P. R. China

Lanny S. Liebeskind, Department of Chemistry, Atwood Hall, 1515 Dickey Drive, Emory University, Atlanta, GA, 30322-2210, USA

Songbai Liu, Department of Food Science and Nutrition, Zhejiang University, 388 Yuhangtang Road, Hangzhou, Zhejiang 310058, P. R. China

Norbert Lui, Bayer Crop Science AG, Alfred-Nobel-Strasse 50, 40789 Monheim, Germany

Dawei Ma, State Key Laboratory of Bio-organic and Natural Products Chemistry, Shanghai Institute of Organic Chemistry, Chinese Academy of Sciences, 354 Fenglin Lu, Shanghai 200032, P. R. China

Luc Neuville, Centre de Recherche de Gif, Laboratoire International Associé, Institut de Chimie des Substances Naturelles, CNRS, 91198 Gif-sur-Yvette Cedex, France

James S. Panek, Department of Chemistry and Center for Chemical, Methodology and Library Development, Metcalf Center for Science and

Engineering, 590 Commonwealth Avenue, Boston University, Boston, MA, 02215, USA

Doron Pappo, Department of Chemistry, Ben-Gurion University, P.O. Box 653, Be'er-Sheva 84105, Israel

Xavi Ribas, QBIS Group, Departament de Química, Universitat de Girona, Campus de Montilivi, 17071 Girona, Spain

Thomas Schareina, Leibniz-Institut für Katalyse e.V. an der Universität Rostock, Albert-Einstein-Strasse 29a, 18059 Rostock, Germany

Ren-Jie Song, State Key Laboratory of Chemo/Biosensing and Chemometrics, College of Chemistry and Chemical Engineering, Hunan University, Changsha 410082, P. R. China

Marc Taillefer, Institut Charles Gerhardt Montpellier, ENSCM, 8 rue de l'Ecole Normale, 34296 Montpellier Cedex 5, France

Cédric Theunissen, Laboratoire de Chimie Organique, Service de Chimie et PhysicoChimie Organiques, Université Libre de Bruxelles, Avenue F. D. Roosevelt 50, CP160/06, B-1050 Brussels, Belgium

Anis Tlili, Institut Charles Gerhardt Montpellier, ENSCM, 8 rue de l'Ecole Normale, 34296 Montpellier Cedex 5, France

Changfeng Wan, Hefei National Laboratory for Physical Sciences at Microscale, CAS Key Laboratory of Soft Matter Chemistry and Department of Chemistry, University of Science and Technology of China, Hefei, Anhui, 230026, P. R. China

Ye Wang, Hefei National Laboratory for Physical Sciences at Microscale, CAS Key Laboratory of Soft Matter Chemistry and Department of Chemistry, University of Science and Technology of China, Hefei, Anhui, 230026, P. R. China

Zhiyong Wang, Hefei National Laboratory for Physical Sciences at Microscale, CAS Key Laboratory of Soft Matter Chemistry and Department of Chemistry, University of Science and Technology of China, Hefei, Anhui, 230026, P. R. China

Yoshihiko Yamamoto, Department of Basic Medicinal Sciences, Graduate School of Pharmaceutical Sciences, Nagoya University, Chikusa, Nagoya 464-8601, Japan

Daoshan Yang, Key Laboratory of Bio-organic Phosphorus Chemistry and Chemical Biology, Department of Chemistry, Tsinghua University, Beijing 100084, P. R. China

PART I

FORMATION OF C–HETEROATOM BONDS

1

MODERN ULLMANN–GOLDBERG CHEMISTRY: ARYLATION OF N-NUCLEOPHILES WITH ARYL HALIDES

YONGWEN JIANG AND DAWEI MA

1.1 INTRODUCTION

The copper-catalyzed arylation of N-nucleophiles with aryl halides, named after their inventors (the "Ullmann reaction"[1] for the cross coupling with amines or the "Goldberg reaction"[2] for the cross coupling with amides), is a valuable transformation that has been widely applied in the preparation of an ever-growing number of pharmaceutically important compounds and in materials sciences as well. For quite a long time, however, these two reactions suffered from the requirement of high reaction temperatures, strong bases, and, in most cases, stoichiometric amounts of copper. Since the late 1990s, a great number of ligands, especially *N,N*-, *N,O*-, and *O,O*-bidentate ligands, have been demonstrated to have the ability to efficiently promote these copper-catalyzed arylation reactions.[3] With the assistance of these ligands, the copper-catalyzed arylation of most N-nucleophiles now proceeds smoothly and under relatively mild conditions, and the amounts of copper salt can be greatly reduced. These advantages not only make these two old reactions become real catalytic ones but also broaden their scope because more substrates—especially those that are sensitive to high reaction temperatures or basic conditions—can now be used in these arylation reactions.

Copper-Mediated Cross-Coupling Reactions, First Edition. Edited by Gwilherm Evano and Nicolas Blanchard.

In this chapter, we will present the developments in this field with emphasis on the most general procedures. This chapter has been divided into four sections, according to the nature of the N-nucleophile: amines, amides/imides/carbamates and related nucleophiles, conjugated N-heterocycles, and ammonia or synthetic equivalents. The arylation of these nucleophiles will be overviewed starting with the arylation of amines, a reaction that yields highly valuable, polysubstituted anilines from readily available starting materials. An impressive number of systems have been reported for each type of nucleophile. Instead of a fully exhaustive list of catalytic systems, this chapter will focus on the most versatile ones and on the ones that rely on the use of mild conditions. For example, systems that require the use of a strong base when others allow performing the same reaction in the presence of a milder one have been omitted for clarity.

1.2 ARYLATION OF AMINES

A major step forward toward the design of efficient catalytic systems that would allow performing the arylation of amines under mild/milder conditions was reported in the late 1990s based on the use of amino acids as ligands for copper. Indeed, amino acids have been demonstrated to have strong accelerating effects on the Ullmann arylation of amines. They could serve as both promoters and coupling partners, affording the corresponding *N*-aryl amino acids at 90°C, a temperature that is remarkably low for such a cross-coupling reaction (Scheme 1.1, Eq. 1).[4] Besides being arylated under mild conditions, amino acids can also serve efficiently as accelerating ligands to promote the arylation of various amines, provided they possess either a secondary (such as proline) or a tertiary (such as *N*,*N*-dimethylglycine) amine that will prohibit their competitive arylation. For the arylation of simple amines, L-proline was identified as the best ligand, although *N*-methylglycine and *N*,*N*-dimethylglycine gave similar results in some cases (Scheme 1.1, Eq. 2).[5,6] Under the assistance of L-proline, the reaction of cyclic secondary amines with aryl (or heteroaryl) halides proceeded well, while the use of more bulky acyclic secondary amines as coupling partners resulted in low yields of the corresponding coupling products (Scheme 1.1, Eq. 3).[5,6] When switching to less-reactive anilines as nucleophiles, L-proline promoted the copper(I) iodide-catalyzed arylation at 90°C; but only electron-rich anilines gave complete conversion, while electron-deficient anilines were more sluggish reaction partners and resulted in low yields of the arylated products (Scheme 1.1, Eq. 4).[6] Considering the aryl halide coupling partner, the cross coupling of aryl iodides and amines typically proceeds at a lower temperature than that of aryl bromides. The use of aryl chlorides, which are generally poor coupling partners in the Ullmann-type amination reaction, was found to be more sluggish and resulted in very low yields of the cross-coupling products, even with aryl chlorides bearing strongly electron-withdrawing groups. The general trend for reactivity, which

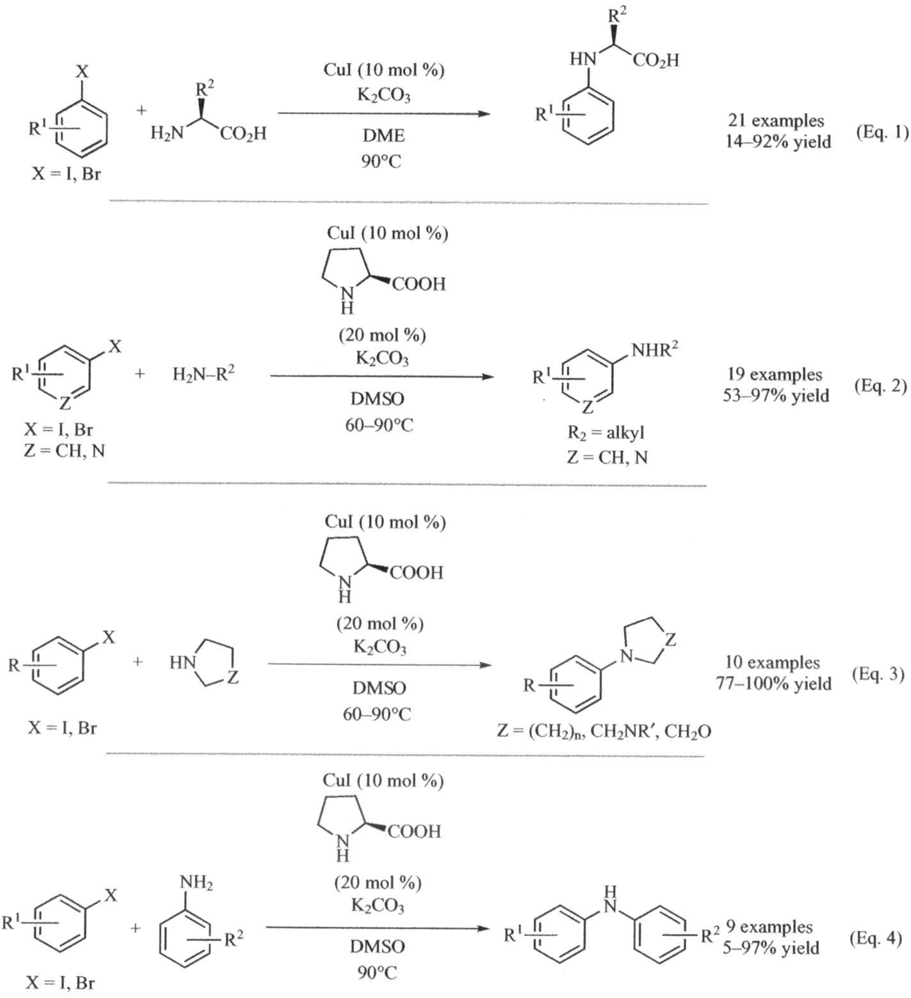

Scheme 1.1 Copper(I) iodide/L-proline-catalyzed cross coupling of aryl halides with amino acids, primary alkylamines, cyclic secondary amines, and anilines.

is typical for most Ullmann-type cross-coupling reactions, is the following: I > Br >> Cl.

With the assistance of amino acids and carefully chosen organic ionic bases such as tetrabutylammonium adipate (TBAA), tetrabutylphosphonium acetate (TBPE), or tetrabutylphosphonium malonate (TBPM), Fu and Liu were able to perform the amination of various aryl halides at room temperature, which is rarely attained except with activated substrates such as those bearing an *ortho*-chelating group (Scheme 1.2).[7] The amination of aryl iodides even proceeded at 0°C when copper(I) iodide and *N*,*N*-dimethylglycine were used

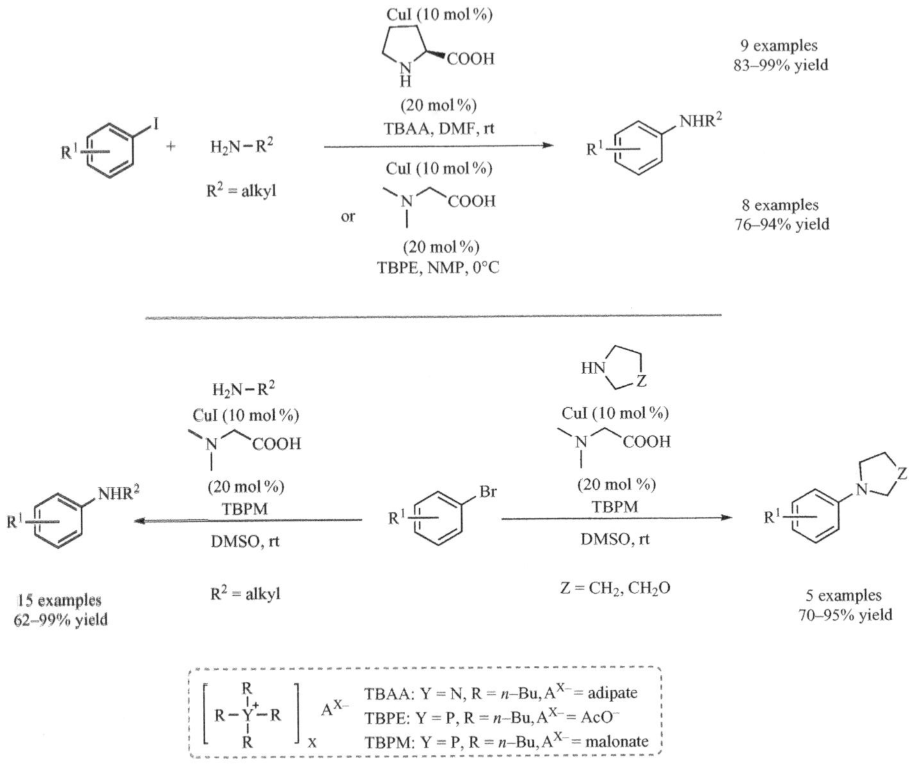

Scheme 1.2 Copper(I) iodide/amino acid-catalyzed cross coupling of aryl halides and amines under mild conditions with the assistance of soluble organic ionic bases.

together with TBPE in *N*-methyl-pyrrolidinone (NMP). It was proposed that the good ionization ability of these organic bases as well as their solubility in organic solvents might be responsible for their excellent performances.

More complex aromatic halides can be used as partners for the arylation of amines, as shown by Ishikawa and coworkers, who found that *N*-tosyl halo-indoles could be efficiently used, affording the corresponding *N*-indolyl amino acids with moderate yields after reaction with the corresponding α-amino acids and by using copper(I) iodide as the catalyst, and cesium carbonate as the base (Scheme 1.3).[8] These products are useful building blocks for synthesizing indolactam V and analogues, and no racemization at the sensitive amino acid residue is observed.

Based on these observations and on the use of proline as a ligand for copper salts in Ullmann-type arylation reactions, the efficiency of other amino acids was evaluated. Pipecolinic acid was demonstrated by Fu and coworkers to be an efficient ligand for the arylation of amines. It was found that most amines,

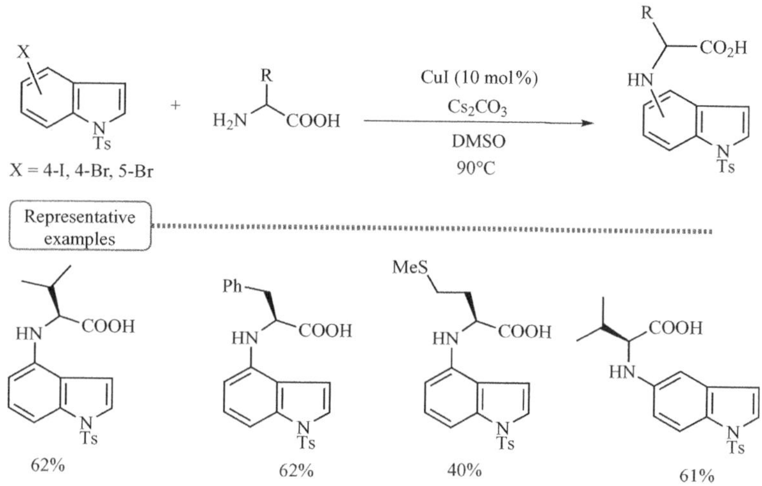

Scheme 1.3 Copper(I) iodide-catalyzed synthesis of *N*-indolyl amino acids.

and especially anilines, could be coupled with aryl iodides and bromides under the catalysis of copper(I) iodide/pipecolinic acid at 110°C, yielding the corresponding aryl amines with good yields (Scheme 1.4).[9] Attempts with aryl chlorides as substrate, however, failed to give the corresponding products in satisfactory yields.

Besides amino acids, various other bidentate ligands have been shown to efficiently promote the arylation of N-nucleophiles. After screening a series of diols as readily available promoters for the copper-catalyzed arylation of amines, Buchwald and coworkers found that ethylene glycol could be used as an ideal ligand in combination with copper(I) iodide for the amination of aryl iodides.[10] Using 5 mol % of copper(I) iodide and 200 mol % of ethylene glycol, the coupling of aryl iodides with amines proceeded smoothly at 80°C (Scheme 1.5, Eq. 1). This catalytic system was, however, not suitable for the amination of aryl bromides, and further screening of ligands, including several substituted phenols, revealed that diethylsalicylamide, bis-alkylated Goldberg's original chelating substrate, enabled a clean copper(I) iodide-catalyzed cross coupling of various amines with aryl bromides at 90°C in *N,N*-dimethylformamide (DMF). This reaction could also be performed at a relatively low loading of ligand (5 mol %) provided that the reaction temperature was increased to 100°C and that the reaction was performed under solvent-free conditions, which might be a limitation in some instances (Scheme 1.5, Eq. 2).[11]

The use of 1,3-diketones as ligands for copper salts was first reported to promote the copper-catalyzed formation of biaryl ethers.[12] In line with these studies, Buchwald and Shafir found that some diketones, particularly 2-isobutyrylcyclohexanone, could serve as powerful ligands for the copper-catalyzed amination of aryl halides. This catalytic system enabled the coupling

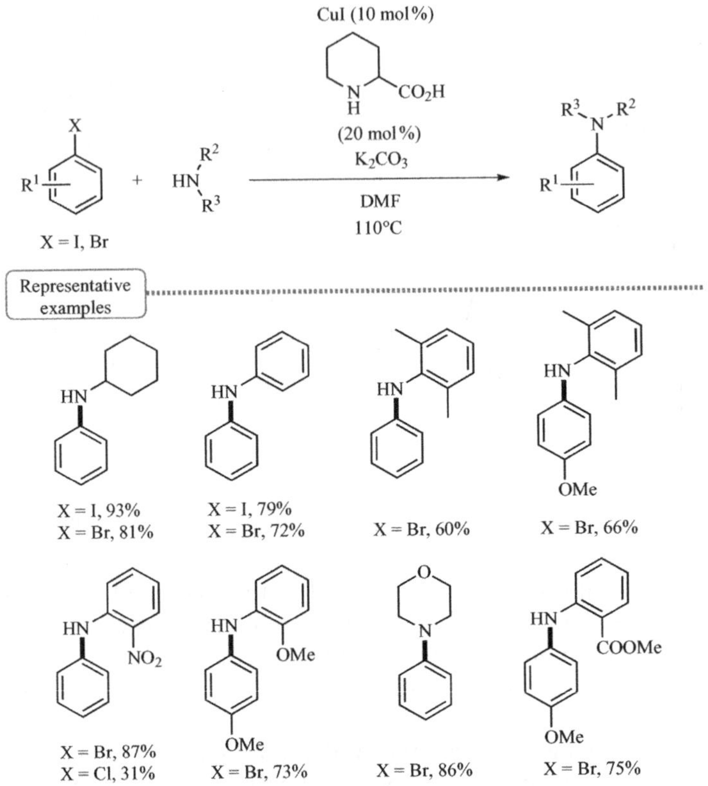

Scheme 1.4 Copper(I) iodide/pipecolinic acid-catalyzed cross coupling of aryl halides and amines (C-N bond formed shown in bold).

of aryl iodides and amines to occur at room temperature, while the coupling of aryl bromides and amines proceeded at 90°C (Scheme 1.6).[13]

The issue of chemoselectivity arises when amino alcohols are used as substrates, since competing N- and O-arylation might occur. Interestingly, clean and selective N- or O-arylation could be achieved by using different ligands (Scheme 1.7).[14] While the use of a combination of copper(I) iodide and 2-isobutyrylcyclohexanone provides the corresponding *N*-aryl amino alcohols, switching to 3,4,7,8-tetramethyl-1,10-phenanthroline as the ligand allowed for a complete reversal of selectivity and only O-arylation products were obtained. A tentative explanation for the observed selectivities was proposed by Buchwald and coworkers on the basis of density functional theory (DFT) calculations.[15a] The selectivities observed experimentally would not be derived from the initial formation of the Cu(I)(nucleophile) complex but from the subsequent steps involving aryl halide activation. This activation may occur via single-electron transfer (SET) or iodine atom transfer (IAT), depending on the electron-donating abilities of the ligand and nucleophile, while other mechanisms involving either oxidative addition/reductive elimination or

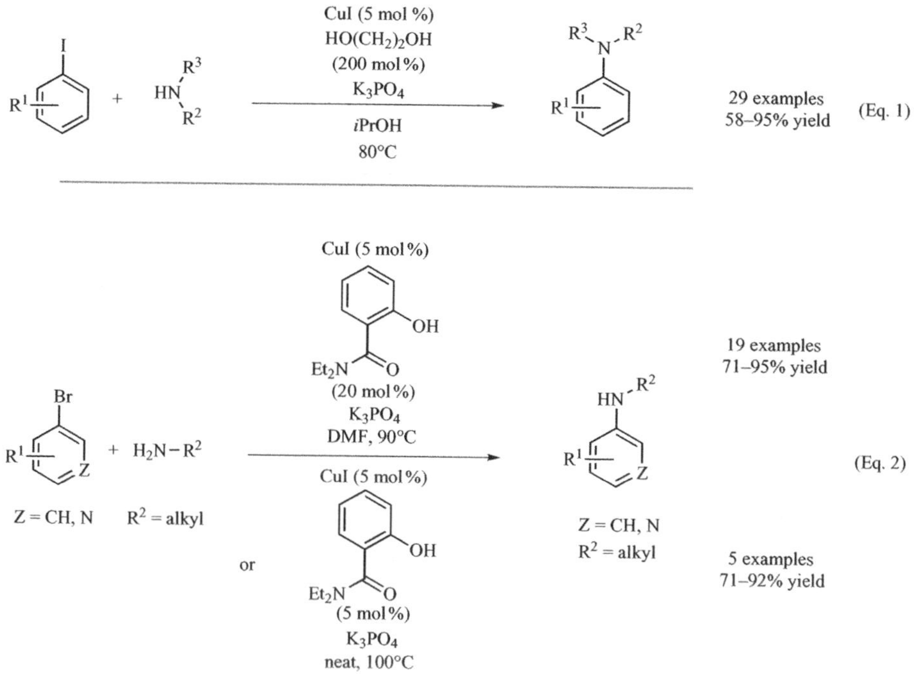

Scheme 1.5 Ethylene glycol and diethylsalicylamide as ligands for the copper(I) iodide-promoted cross coupling of aryl halides and amines.

σ-bond metathesis are disfavored. SET mechanisms are favored in reactions promoted by the β-diketone ligand, and the N-arylation is favored in these cases due to the faster reaction involving the Cu(I) amido complex compared to the one involving the Cu(I) methoxide one. In contrast, the phenanthroline ligand promotes O-arylation reactions via IAT mechanisms in preference to N-arylation reactions, which occur via SET mechanisms. The combined rate for the formation of the (phen)Cu(I)(alkoxide) complex and the concomitant IAT is faster than the combined rate for formation of the (phen)Cu(I)(amido) complex and SET, leading to the selective formation of O-arylated products. An alternative mechanistic explanation for this ligand-dependent selectivity has been proposed recently and is discussed in detail in Chapter 7 (Scheme 7.13).[15b]

For the cross coupling of aryl halides with anilines, which provides an interesting and straightforward entry to diarylamines, Buchwald and coworkers discovered that pyrrole 2-carboxylic acid was an efficient ligand.[16] Under the assistance of this ligand, both electron-rich and electron-deficient anilines could react with aryl halides and a variety of useful functional groups were tolerated. For aryl iodides, the reaction was complete at 70–90°C, while a slight increase in the reaction temperature was required in the case of aryl bromides (Scheme 1.8).

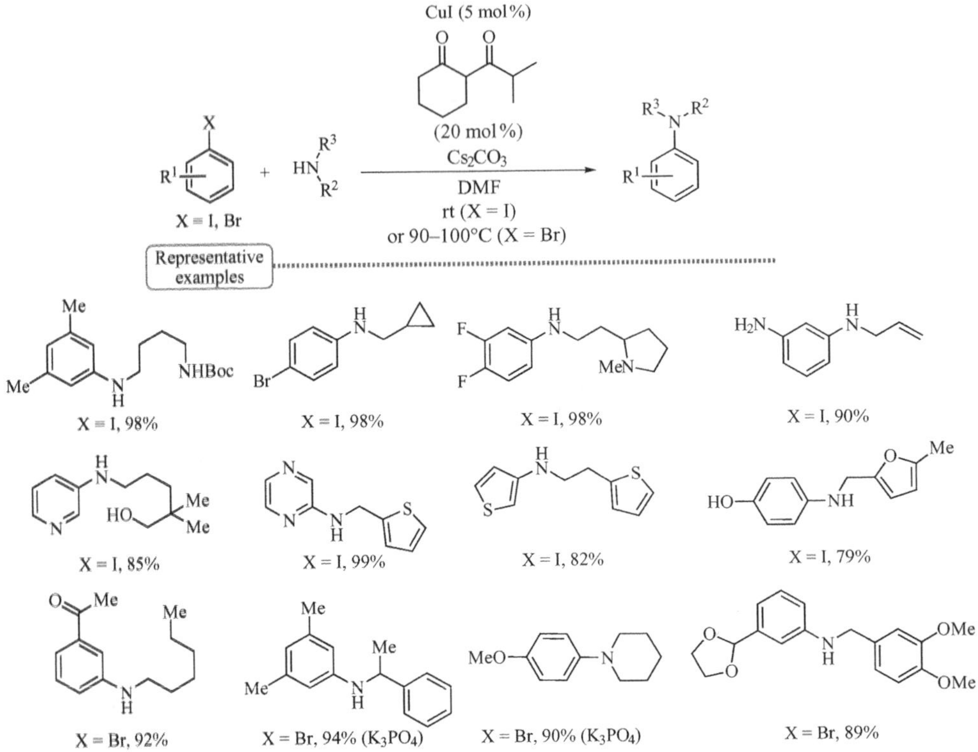

Scheme 1.6 Copper(I) iodide/2-isobutyrylcyclohexanone-catalyzed cross coupling of aryl halides and amines.

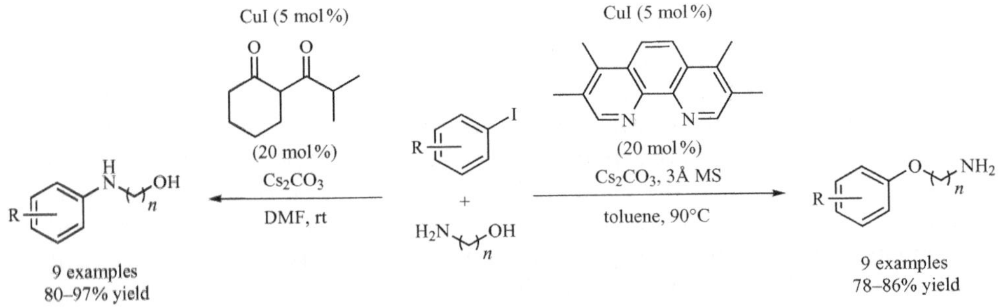

Scheme 1.7 Copper(I) iodide/ligand-catalyzed chemoselective N- or O-arylation of amino alcohols.

rac-BINOL was initially reported by Wan and coworkers as a possible ligand for the copper-catalyzed arylation of amines using rather harsh conditions since the reaction was carried out at 90–125°C (Scheme 1.9, Eq. 1).[17] A careful reexamination of this system by Fu and coworkers revealed that a combination of copper(I) bromide and *rac*-BINOL was in fact able to catalyze the amination of aryl iodides at room temperature (Scheme 1.9, Eq. 2).[18a] The

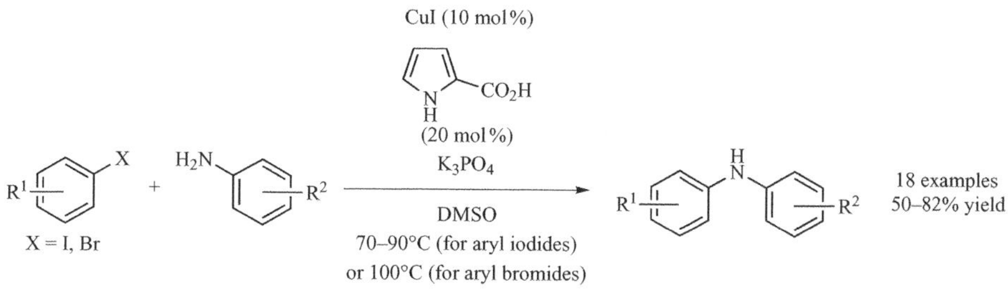

Scheme 1.8 Copper(I) iodide/pyrrole 2-carboxylic acid-catalyzed cross coupling of aryl halides and anilines.

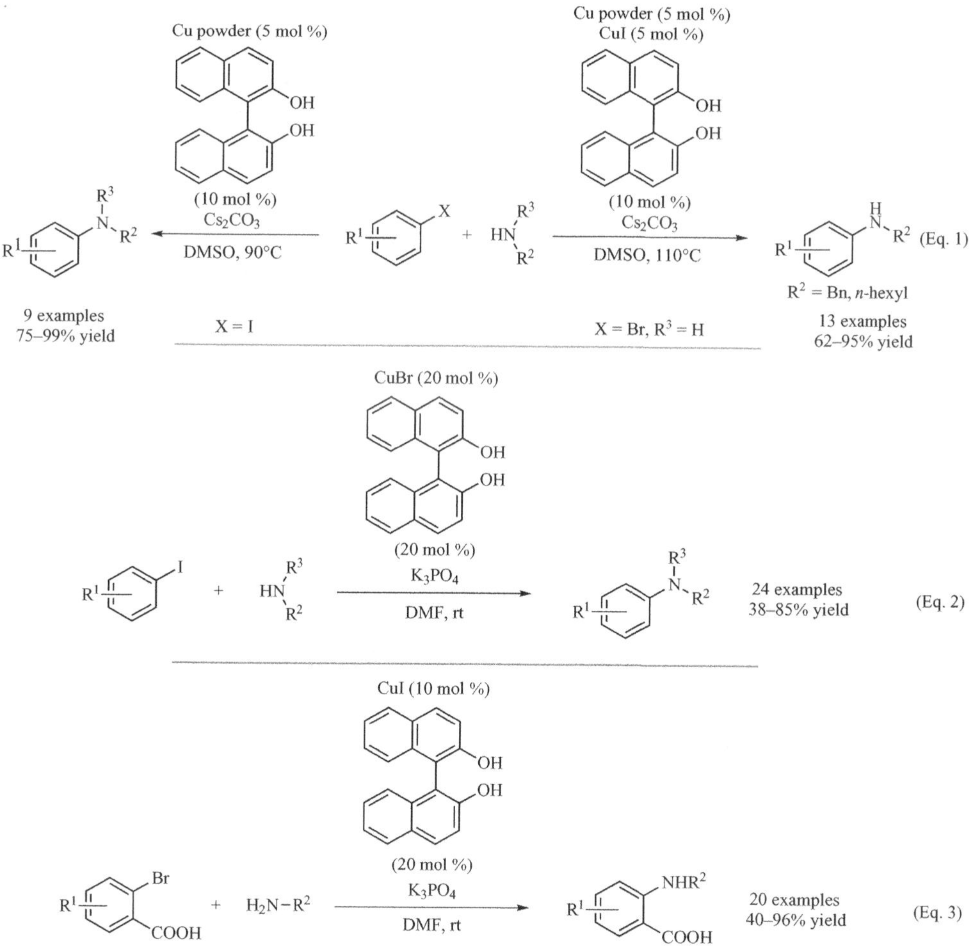

Scheme 1.9 Copper/*rac*-BINOL-catalyzed cross coupling of aryl halides and amines.

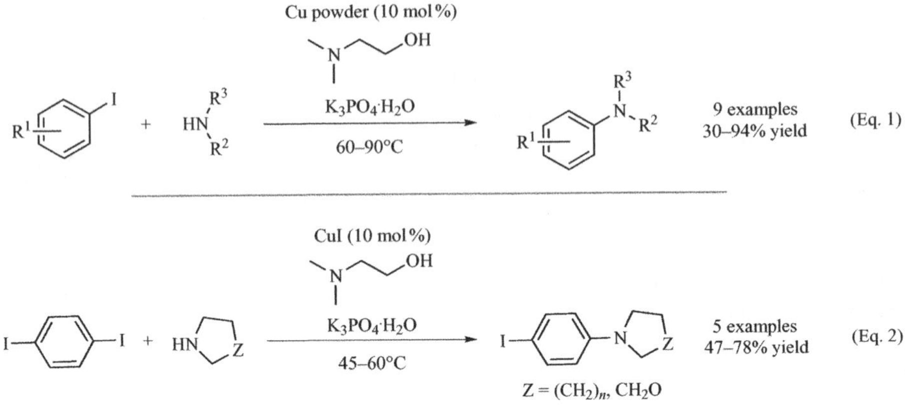

Scheme 1.10 Copper-catalyzed cross coupling of aryl iodides and amines in 2-dimethylaminoethanol.

copper source was found to have a marked influence in this coupling reaction because the combination of copper(I) iodide and *rac*-BINOL gave relatively low yields. Additionally, with the assistance of a carboxylic acid in the *ortho* position, the coupling of aryl bromides and amines could be carried out at room temperature (Scheme 1.9, Eq. 3).[18b] The use of BINOL derivatives provides excellent opportunities for the development of asymmetric arylation reactions starting from prochiral amines, as nicely demonstrated recently by Yu and coworkers who designed an elegant enantioselective intramolecular Ullmann reaction.[18c] Details on this reaction can be found in Chapter 15 (Scheme 15.1, Eq. 4).

The Twieg group described that 2-dimethylaminoethanol could be used as a ligand for the copper-catalyzed arylation of amines, although a high amount of the ligand—used as the solvent or co-solvent—was required.[19] The cross coupling of aryl iodides with simple primary aliphatic amines and cyclic secondary amines proceeded in 2-dimethylaminoethanol to afford the corresponding arylated products in good yields (Scheme 1.10, Eq. 1),[19a] while the combination of 2-dimethylaminoethanol and water was found more suitable when amino acids, amino alcohols, and even some peptides were used as nucleophiles, which is quite remarkable since the arylation of water could be a serious side reaction that was, however, not observed (Scheme 1.11).[19b] Bis iodides were selectively monoaminated even in the presence of 2.4 equiv of the starting amine, which provides interesting platforms for further functionalization (Scheme 1.10, Eq. 2). Using 2-dimethylaminoethanol as the solvent, a number of halothiophenes could be coupled with amines under the catalysis of copper and copper(I) iodide, yielding 2- and 3-amino-thiophenes with high efficiency (Scheme 1.12).[19c]

Besides *N,O* and *O,O*-bidentate ligands, *N,N*-ligands have also been used for the development of efficient catalytic systems for the arylation of amines. Indeed, 1,10-phenanthroline was found to be an especially suitable ligand for

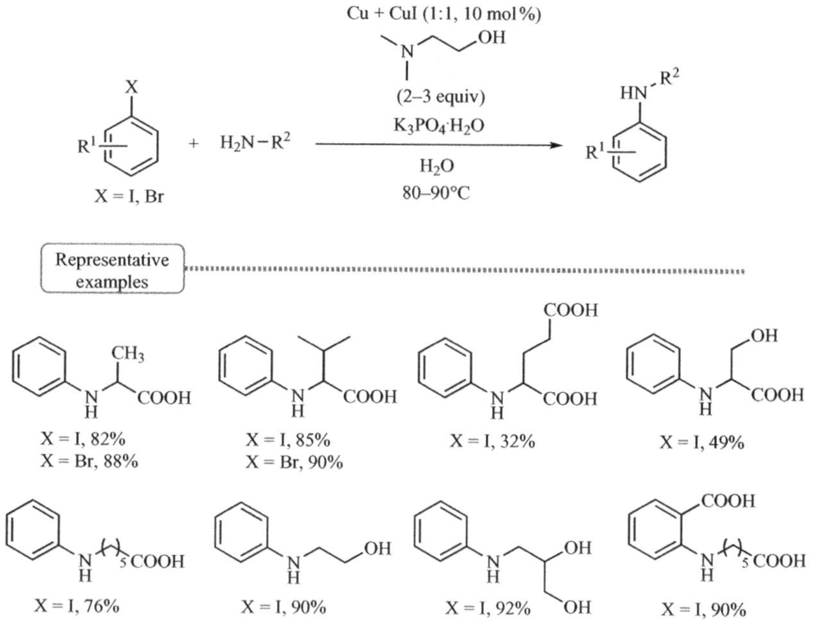

Scheme 1.11 Copper-catalyzed cross coupling of aryl iodides with amino acids and amino alcohols in 2-dimethylaminoethanol and water.

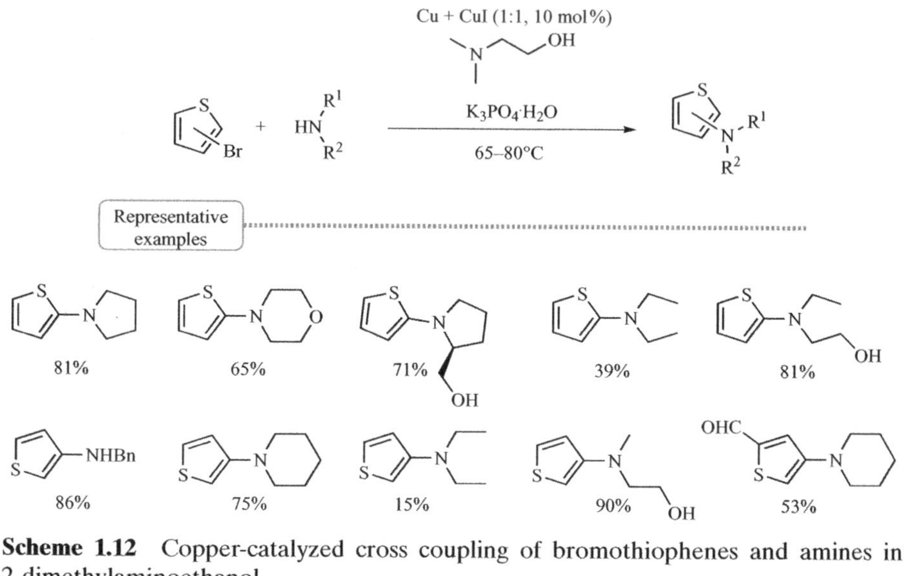

Scheme 1.12 Copper-catalyzed cross coupling of bromothiophenes and amines in 2-dimethylaminoethanol.

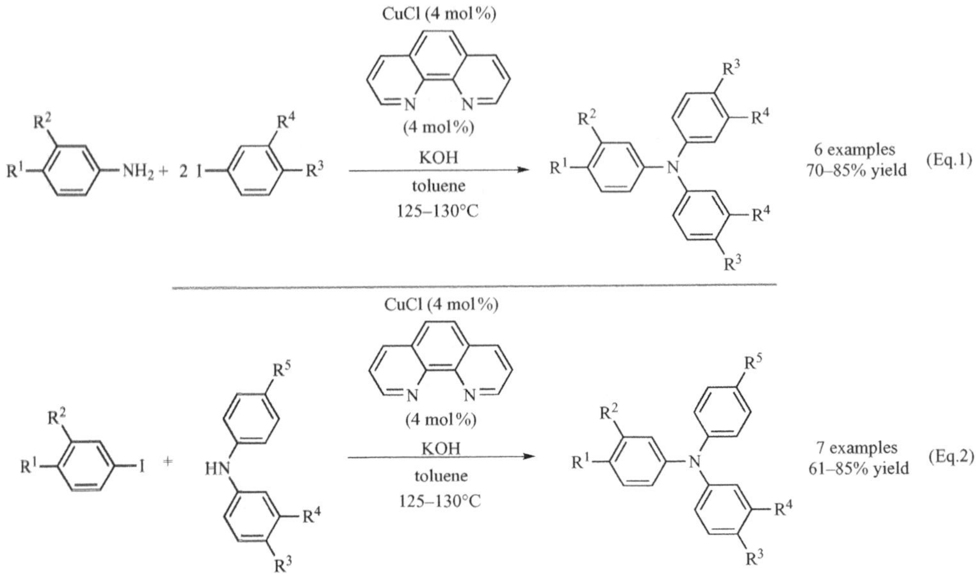

Scheme 1.13 Copper(I) chloride/1,10-phenanthroline-catalyzed synthesis of triarylamines.

the synthesis of triarylamines via a copper-catalyzed double arylation reaction starting from aryl halides and anilines (Scheme 1.13, Eq. 1).[20] Using the same conditions, the cross coupling between aryl iodides and diarylamines provides a straightforward and efficient entry to unsymmetrical triarylamines (Scheme 1.13, Eq. 2). As a note, related preformed neocuproine-derived copper complexes were found to have similar activity.[21] An alternative and straightforward entry to triarylamines relying on the direct arylation of ammonia might be an especially attractive option to consider: This point will be described later on (see Section 1.5).

Some other notable ligands are shown in Figure 1.1: They all contain bidentate moieties. Among them, diphenyl pyrrolidine-2-phosphonate **1**,[22] pyrrolidine-2-phosphonic acid phenyl monoester **2**,[23] 2-hydroxybenzaldehyde *N*-phenylhydrazone **3**,[24] 1-(5,6,7,8-tetrahydroquinolin-8-yl)ethanone **4**,[25] and 3-acetylcoumarin **5**[26] were found to be suitable for amination of both aryl iodides and aryl bromides, while ligand **6** was only applied to the amination of aryl iodides.[27] When glyoxal-derived dioxime **7** was used in combination with copper(I) chloride, the coupling of aryl halides with aliphatic amines could be performed in water,[28] thereby giving an alternative method for aryl amination.

Before closing this section, it should be mentioned that there is still no universal system for the arylation of amines, even if tremendous progresses have been made in this area and if copper catalysis is an excellent alternative to the Buchwald–Hartwig cross coupling. The added bonus with copper-based systems is that they do not require complex ligands or expensive metal sources, which renders them less costly, in addition to being operationally simpler. The

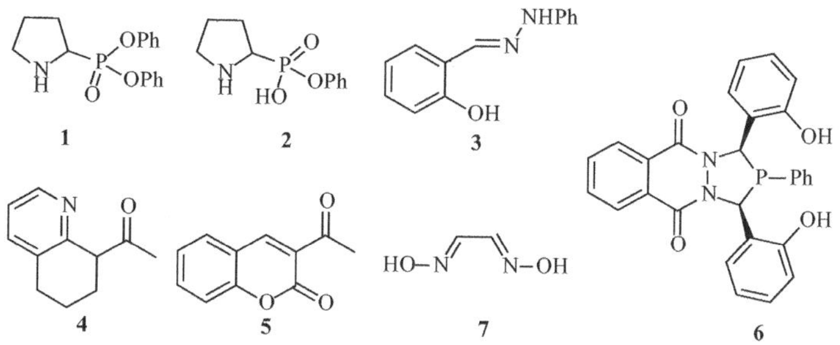

Figure 1.1 Some other efficient ligands for the cross coupling of aryl halides and amines (non-exhaustive list).

reader should keep in mind that other cross-coupling partners might be considered for the arylation of amines, such as boronic acids or other surrogates for aryl halides. The use of these alternative reagents is extensively described in Chapter 4.

Due to its high synthetic utility, the arylation of amides and related nucleophiles has also received a great deal of attention over the past 10 years. As with the arylation of amines, the introduction of simple, cheap, and readily available chelating ligands allowed the design of remarkably efficient catalytic systems and dramatic softening of the reaction conditions compared to the original procedure first described by Goldberg. Advances in this area will be covered in the next section.

1.3 ARYLATION OF AMIDES, IMIDES, AND CARBAMATES

From a synthetic point of view, the arylation of amides, that is, the Goldberg reaction, is a remarkable reaction that nicely complements the classical synthesis of anilides by acylation of anilines. The use of the former rather than the latter allows for completely different bond disconnections and has been extensively used in organic synthesis, notably for the total synthesis of complex natural products (see Chapter 16 for details), which clearly demonstrated its efficiency and reliability.

A key discovery in this area was reported in 2001 by Buchwald and coworkers, who demonstrated that various 1,2-diamines were remarkably efficient ligands for promoting the copper-catalyzed amidation of aryl halides.[29,30] These ligands, shown in Figure 1.2, include ethylenediamine (EDA), *N,N′*-dimethylethylenediamine (DMEDA), *trans*-cyclohexane-1,2-diamine (CyDA), and *trans-N,N′*-dimethylcyclohexane-1,2-diamine (DMCyDA), the second one probably being one of the most frequently used. Due to the commercial availability of these ligands, they have been widely used in synthesis and are certainly among the first ligands to evaluate for a planned transformation.[31] There are, however, no universal ligands, and they nicely complement each other since they display slight differences in efficiency with different reaction partners (i.e., N-nucleophile and aryl halide).

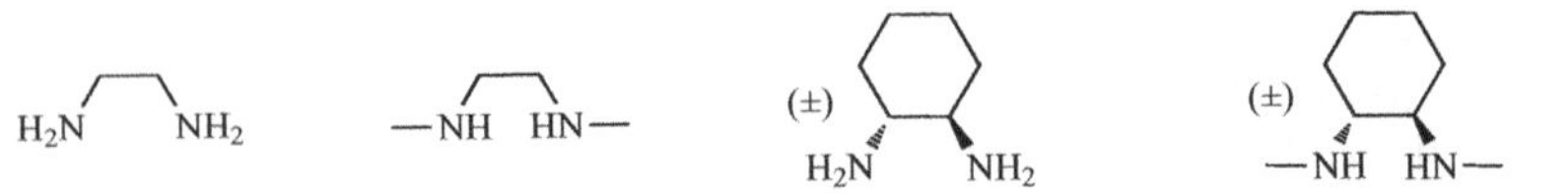

ethylenediamine (**EDA**) | *N,N′*-dimethyl-ethylenediamine (**DMEDA**) | *trans*-cyclo-hexane-1,2-diamine (**CyDA**) | *trans-N,N′*-dimethyl-cyclohexane-1,2-diamine (**DMCyDA**)

Figure 1.2 Four typical 1,2-diamine ligands for the copper-catalyzed amidation of aryl halides.

When CyDA was used as the ligand, primary amides, secondary amides as well as lactams could be coupled with aryl iodides using potassium phosphate or carbonate as the base and dioxane or toluene as the solvent at 110°C, affording the corresponding amidation products in good yields (Table 1.1, entry 1). For primary amides, lowering the catalyst loading to 1 mol % still gave good results, while an increase to 10 mol % was required for the amidation of aryl bromides. Toluene was in general the most efficient solvent when the ligand was changed to DMEDA, EDA, or DMCyDA (Table 1.1, entries 2, 3 and 4). In these cases, the reaction of aryl iodides and primary amides could be carried out at relatively low temperatures (60–80°C). In addition, aryl chlorides could also react with benzamide or β-lactam by using *trans-N,N′*-dimethyl-cyclohexane-1,2-diamine (DMCyDA) as the ligand provided that the reaction temperature was increased to 110–130°C (Table 1.1, entry 4). The major limitation in this case is that aryl chlorides have to serve as both substrate and solvent.

The Padwa group has extended this methodology to the preparation of 2 and 3-substituted amidofurans and amidothiophenes, useful building blocks for alkaloid synthesis.[32] In their studies, DMEDA and DMCyDA were found to be the most efficient ligands. Generally, good yields were observed when primary amides, lactams, and 2-oxazolidinones were used as coupling partners, while low yields were obtained in case of secondary amides and carbamates as substrates (Scheme 1.14).

By using CyDA as the ligand, Trehan and coworkers developed a highly efficient copper(I) iodide-catalyzed coupling of aryl bromides and 2-oxazolidinones (Scheme 1.15).[33] Various aryl bromides and substituted oxazolidinones were used to evaluate the utility of this procedure, which was found to be rather general, except in the case of an aromatic ring substituted with a formyl group or in the presence of an unprotected alcohol. Some pharmaceutically important *N*-aryl-oxazolidinones such as linezolid and toloxatone were conveniently synthesized using these conditions.

The major drawback of the procedures overviewed earlier is the use of heating, which can be problematic with some substrates. Recently, researchers at Novartis reported that for the copper(I) iodide/DMEDA-catalyzed amidation of aryl iodides, cesium fluoride could be used to replace the stronger bases like potassium carbonate and potassium phosphate that were used in the previous studies. In this case, most aryl iodides were readily coupled in high yields at room temperature (Scheme 1.16).[34] This noticeable improvement

TABLE 1.1 Copper/1,2-Diamines-Catalyzed Amidation of Aryl Halides

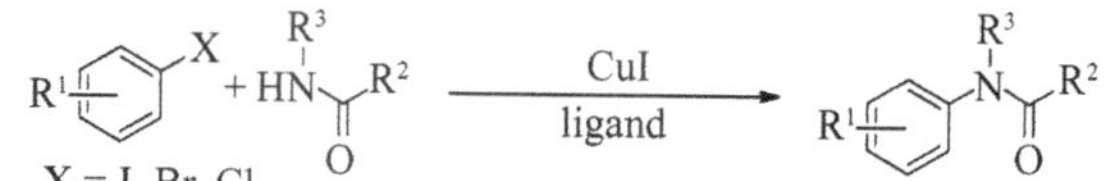

X = I, Br, Cl

Entry	N-Nucleophiles	Aryl Halides	Catalytic Systems		Conditions	Number of Examples	Yields (%)	Ref.
1	Primary amides	Aryl iodides	CuI (1 mol %)	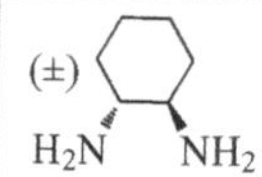 (10 mol %)	K_3PO_4, dioxane or toluene, 90–110°C	5	69–98	[29,30]
	Secondary amides Lactams	Aryl iodides	CuI (0.2–5 mol %)	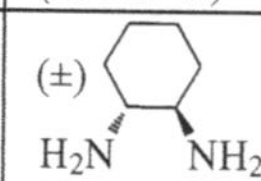 (5–20 mol %)	K_3PO_4, dioxane or toluene, 110°C	12	63–99	
	Amides Lactams	Aryl bromides	CuI (1–10 mol %)	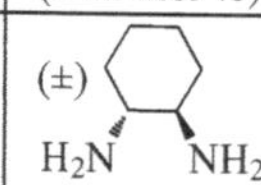 (10 mol %)	K_3PO_4 or K_2CO_3, dioxane, 110°C	6	81–97	
2	Primary amides	Aryl iodides	CuI (5 mol %)	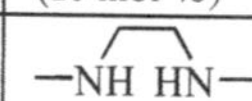 (10 mol %)	K_3PO_4, toluene or DMF, 60–80°C	3	88–95	[29,30]
	Secondary amides Lactams	Aryl iodides	CuI (5 mol %)	 (10–20 mol %)	K_3PO_4, toluene, 80–110°C	3	87–95	
	Amides Lactams	Aryl bromides	CuI (2–10 mol %)	 (10–20 mol %)	K_2CO_3, dioxane or toluene or diglyme, 110°C	7	49–98	
3	Secondary amides Lactams	Aryl iodides	CuI (1 mol %)	 (10 mol %)	K_3PO_4, toluene, 80°C	1	93	[29,30]
4	Amides Lactams	Aryl bromides	CuI (5 mol %)	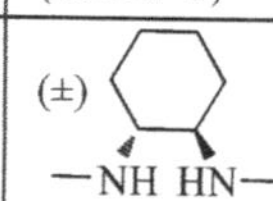 (10–11 mol %)	K_3PO_4, toluene, 110°C	4	91–99	[29,30]
	Primary amides β-Lactam	Aryl chlorides			K_2CO_3, neat, 110–130°C	4	51–95	

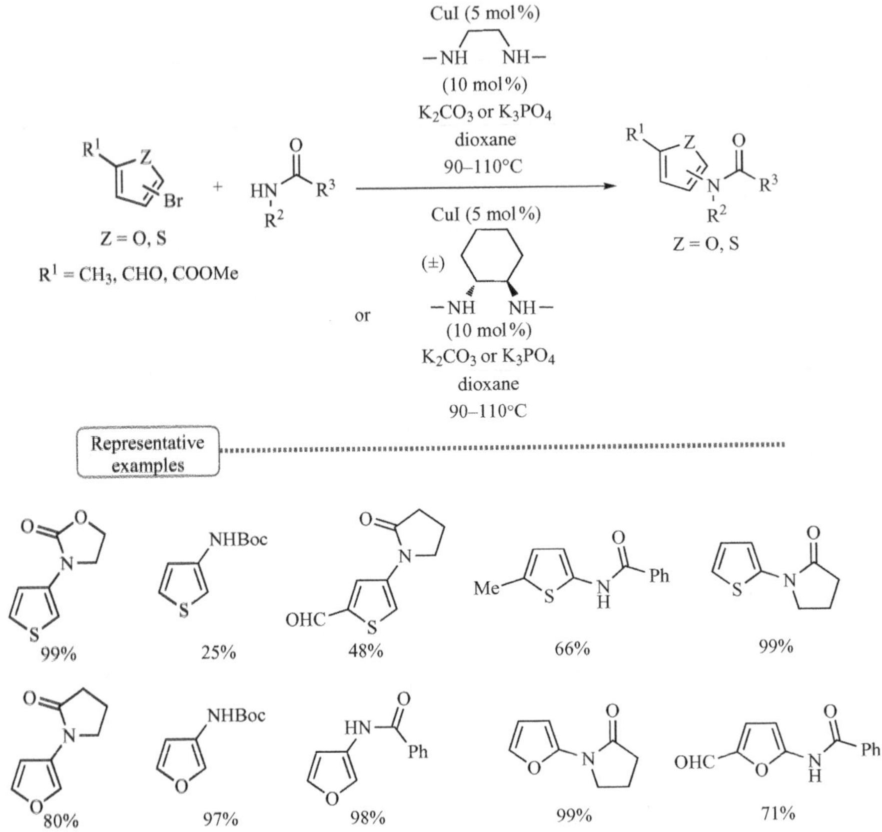

Scheme 1.14 Copper(I) iodide/*N*,*N*′-dimethylethylenediamine, *trans*-*N*,*N*′-dimethylcyclohexane-1,2-diamine-catalyzed amidation of 2- or 3-bromothiophenes and furans.

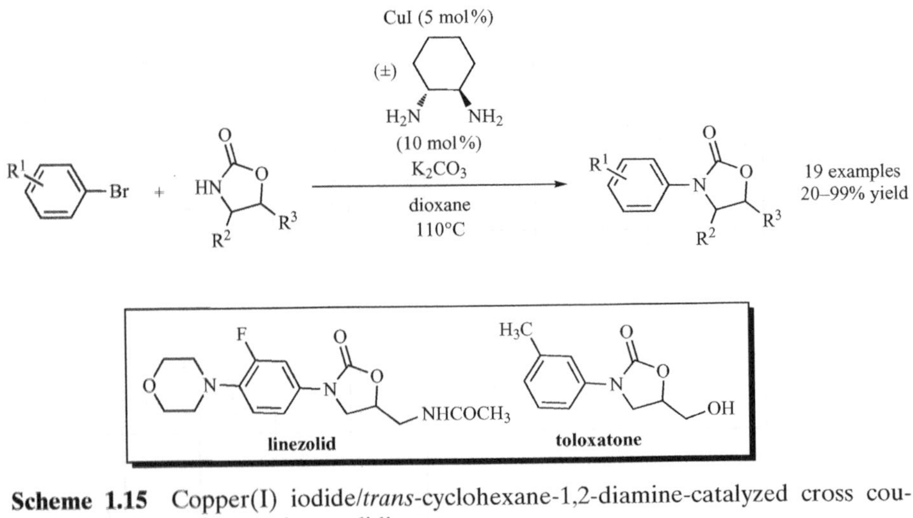

Scheme 1.15 Copper(I) iodide/*trans*-cyclohexane-1,2-diamine-catalyzed cross coupling of aryl bromides and oxazolidinones.

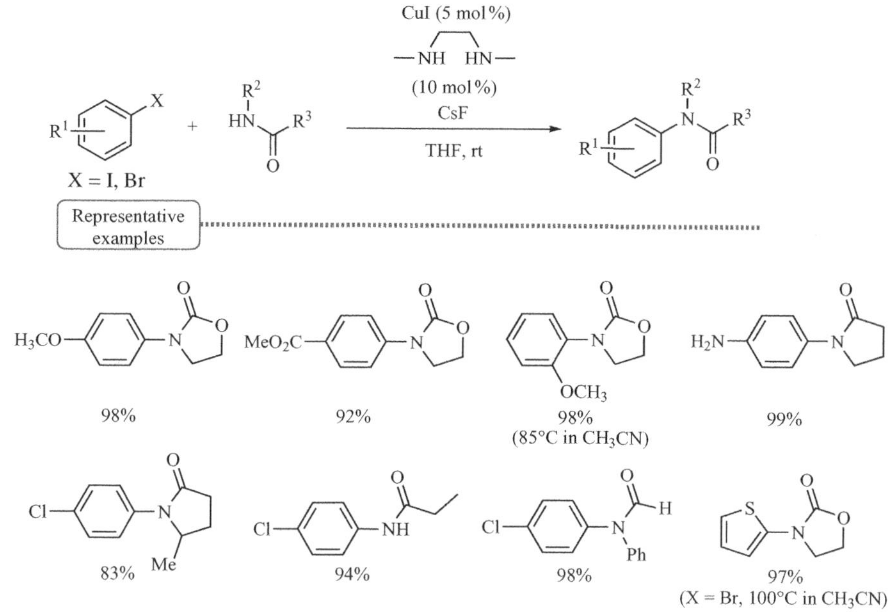

Scheme 1.16 Room temperature, copper(I) iodide/*N,N′*-dimethylethylenediamine-catalyzed cross coupling of aryl iodides and oxazolidinones and amides in the presence of cesium fluoride.

will be of benefit to cross-coupling reactions using substrates that are incompatible with high temperature and/or strongly basic conditions.

Substitution around the reacting centers can be detrimental in some cases and the arylation of derivatives of pyroglutamic acid can be problematic. Rigo and coworkers reported a systematic study of this specific reaction and found that the copper(I) iodide-promoted coupling of aryl bromides and DL-methyl pyroglutamate worked well by using DMEDA as the ligand.[35a] They were able to carry out the reaction at 60°C, or even room temperature in some cases, at the expense of high catalyst loading since 50 mol % of copper(I) iodide and 100 mol % of DMEDA were required (Scheme 1.17, Eq. 1). Finally, DMEDA was also found to be an efficient ligand in combination with copper(I) oxide for the arylation of hydroxamates with aryl iodides (Scheme 1.17, Eq. 2).[35b]

Besides 1,2-diamines, which have been extensively used as promoters for the copper-catalyzed arylation of amides and related nucleophiles, several other efficient chelating ligands have been shown to efficiently promote the copper-catalyzed amidation of aryl halides; the most efficient ones are summarized in Table 1.2. Among those, amino acids have been demonstrated to be another class of especially useful ligands for the amidation of a wide range of N-nucleophiles with various aryl halides. For example, glycine[36] and *N,N*-dimethylglycine[37] have been reported to promote the copper(I) iodide-catalyzed cross coupling of aryl iodides with a number of amides and lactams (Table 1.2, entry 1). Interestingly, high levels of chemoselectivity between aryl

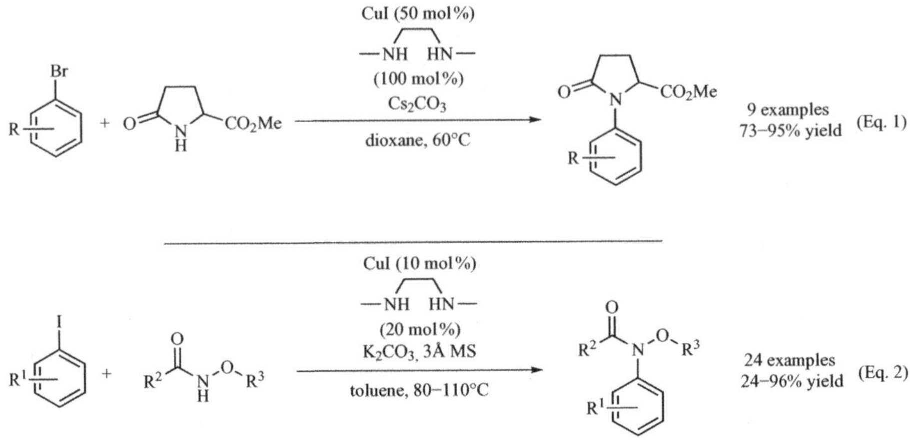

Scheme 1.17 Copper(I)/*N*,*N*′-dimethylethylenediamine-catalyzed cross coupling of aryl iodides and DL-methyl pyroglutamate and hydroxamates.

iodides and aryl bromides could be obtained, as evidenced by the selective amidation of 1-bromo-4-iodobenzene yielding to 4-bromo-anilide derivatives only (Table 1.2, entry 2).[37a] As a note, the use of 1-bromo-4-iodobenzene can result in a formal substitution of the bromine atom resulting from a Ullmann/Finkelstein sequence (see Chapter 6, Schemes 6.5 and 6.6 for details).[37b]

1,10-Phenanthroline, another ligand commonly used in copper catalysis which was proven over the years to be an especially efficient one, notably for the arylation of alcohols, has also been proven to be an effective promoter for the copper-catalyzed amidation of aryl iodides (Table 1.2, entry 3).[38] In this case, potassium fluoride on aluminum oxide was found to be the best base, which led to complete conversion within 1.5 hours in refluxing toluene. 4,7-Dimethoxy-1,10-phenanthroline has also been used as a ligand for related systems and was found to be especially efficient for the selective N-arylation of 2- and 4-hydroxypyridine derivatives. Compared to other systems, the use of this electron-rich phenanthroline enables the cross coupling with N-containing heteroaryl and *ortho*-substituted aryl halides (Table 1.2, entry 4).[39]

O,*O*-Bidentate ligands have also been reported to be useful ligands for the copper-catalyzed aryl amidation. For example, the combination of copper(I) bromide and ethyl 2-oxocyclohexane carboxylate efficiently catalyzes the cross-coupling reaction of aryl halides with lactams and primary amides, affording the corresponding amidation products in good yields (Table 1.2, entry 5).[40] The reaction rate seems to be highly dependent on the electronic nature of the aryl halides, as evidenced from the amidation of 1-bromo-4-iodobenzene and 1-chloro-4-iodobenzene, which readily proceed at room temperature, while an increased reaction temperature is required for electronic-rich aryl iodides. Recently, the Taillefer group found that another *O*,*O*-bidentate ligand, 2,2,6,6-tetramethyl-3,5-heptadione (TMHD), could drastically improve the coupling of aryl iodides and acyclic secondary aryl amides, a reaction that

TABLE 1.2 Other Efficient Catalytic Systems for the Amidation of Aryl Halides

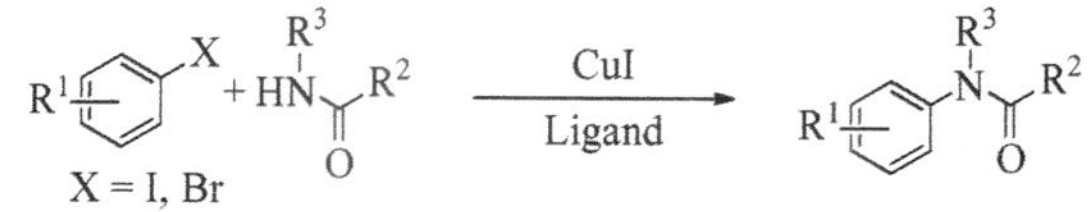

Entry	Amides	Aryl Halides	Catalytic Systems	Conditions	Number of Examples	Yield (%)	Ref.
1	Primary and secondary amides Lactams Sulfonamides Hydrazide	Aryl iodides	CuI (5 mol %), 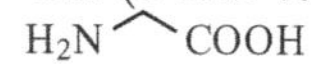 (10 mol %)	K_3PO_4, dioxane, 100°C	29	53–98	[36]
2	Primary and secondary amides Lactams	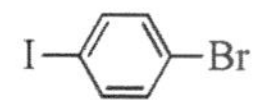	CuI (5 mol %), 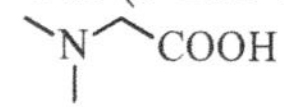 (20 mol %)	K_3PO_4, DMSO, 80–100°C	21	45–87	[37]
3	Primary and secondary amides Lactams	Aryl iodides	CuI (5 mol %), 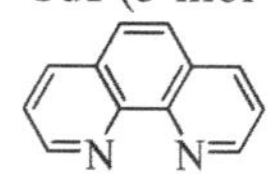 (10 mol %)	KF/Al_2O_3, toluene, reflux	19	67–99	[38]
4	2-Hydroxypyridines 4-Hydroxypyridines	Aryl iodides Aryl bromides	CuI (2–10 mol %), 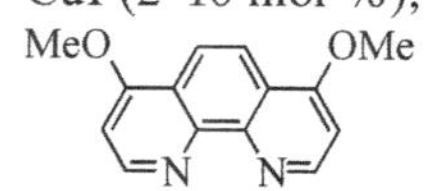 (7.5–40 mol %)	K_2CO_3, DMSO, 85–120°C	15	47–95	[39]
5	Primary amides β-Lactam	Aryl iodides Aryl bromides	CuBr (10 mol %), 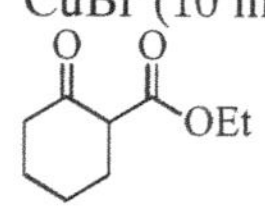 (20 mol %)	Cs_2CO_3, DMSO, rt–75°C	12	76–96	[40]
6	Acyclic aromatic secondary amides	Aryl iodides	CuI (5 mol %), 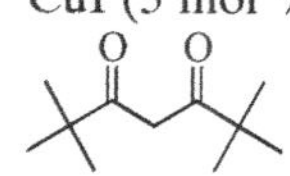 (10 mol %)	K_3PO_4, toluene, 130°C	24	23–95	[41]

is typically one of the most challenging to perform because of the low reactivity of acyclic secondary amides, affording tertiary acyclic amides with good yields (Table 1.2, entry 6).[41] The major drawback with all these ligands is that they are not reported to efficiently catalyze the amidation of aryl chlorides, a reaction for which diamine ligands are clearly much more efficient (Table 1.1, entry 4).

Due to the importance of *N*-arysulfonamides in pharmaceutical science, two groups have directed their attention to systematic studies of the ligand-promoted, copper-catalyzed cross coupling of aryl halides with sulfonamides (Scheme 1.18). It was reported by Guo and coworkers that copper(I) iodide/

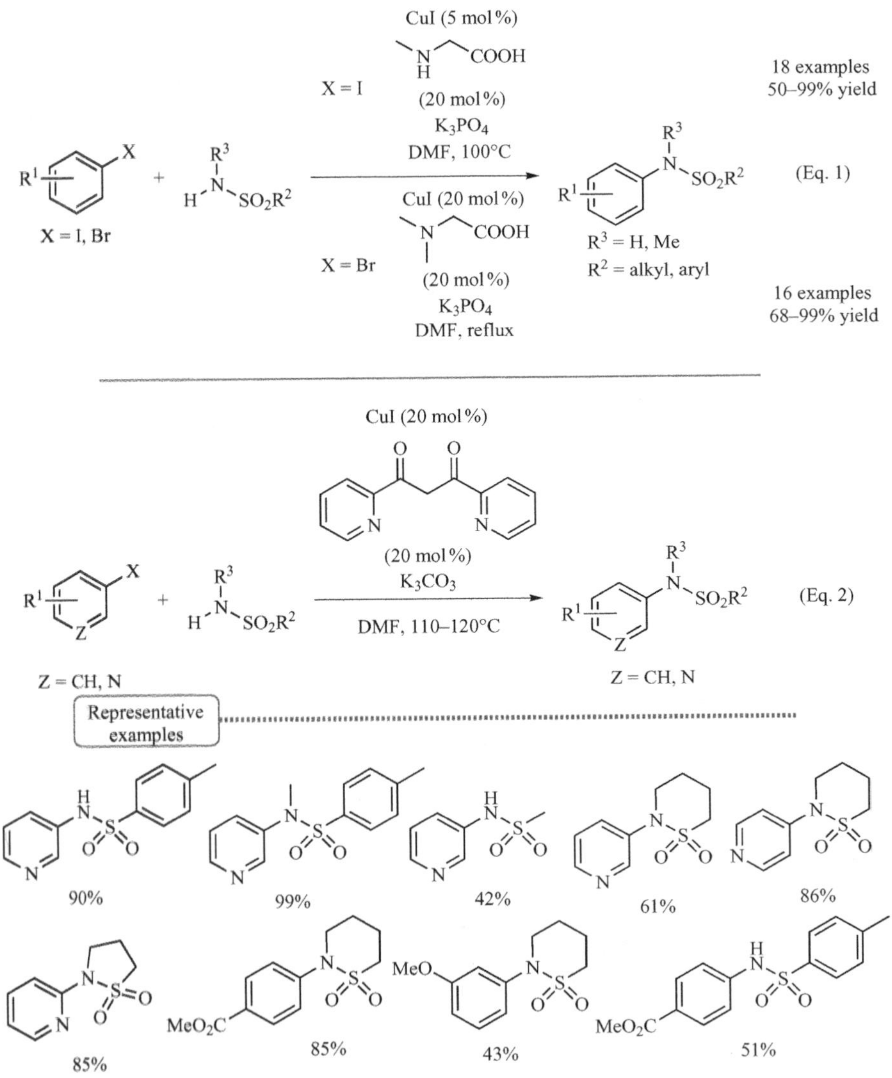

Scheme 1.18 Copper-catalyzed cross coupling of aryl halides and sulfonamides.

N-methylglycine-catalyzed the coupling reaction of aryl iodides and sulfonamides at 100°C in DMF. However, for aryl bromides, increasing reaction temperature and switching ligand to *N*,*N*-dimethylglycine were required to ensure complete conversion (Scheme 1.18, Eq. 1).[42] Recently, Han, from Bristol-Myers Squibb, demonstrated that 1,3-di(bipyridin-2-yl)propane-1,3-dione was also effective for this coupling. It seemed that electron-deficient aryl bromides, particularly 3-bromopyridine, could give satisfactory coupling yields, while electron-rich aryl bromides gave moderate or low yields of products (Scheme 1.18, Eq. 2).[43] Noteworthy is that in these two reports, both primary and secondary alkyl and aryl sulfonamides are suitable coupling partners, which allows the assembly of highly substituted *N*-aryl sulfonamides with great diversity.

These developments on the copper-catalyzed arylation of amides, lactams, oxazolidinones, and sulfonamides led to the study of the reactivity of other N-nucleophiles. In 2001, the Buchwald group reported that catalytic amounts of copper(I) iodide together with 1,10-phenanthroline as ligand could efficiently catalyze the cross coupling between aryl iodides and *N*-Boc-hydrazine, a reaction that could regioselectively provide N-arylated *N*-Boc hydrazines in DMF at 80°C (Scheme 1.19, Eq. 1).[44] A few years later, the Kwong group found that 2-picolinic acid was also efficient for this transformation (Scheme 1.19, Eq. 2),[45] while Ma and coworkers discovered that changing the solvent from DMF to DMSO could allow performing the copper(I) iodide-catalyzed arylation of *N*-Boc hydrazine with aryl iodides at 50°C without the assistance of any ligands (Scheme 1.19, Eq. 3).[46] More importantly, they described that when *trans*-4-hydroxy-L-proline was used as the ligand, aryl bromides could be successfully employed in this transformation, therefore increasing its scope and synthetic utility (Scheme 1.19, Eq. 3).[46]

The cross-coupling reactions of aryl halides with other N-nucleophiles are outlined in Scheme 1.20. Hosseinzadeh and coworkers developed a copper(I) iodide/*N*,*N′*-dibenzylethylenediamine-catalyzed arylation of phenylurea for assembling symmetric and unsymmetric *N*,*N′*-diarylureas (Scheme 1.20, Eq. 1).[47] Simultaneously, the Jones group demonstrated that copper(I) iodide/1,10-phenanthroline efficiently catalyzed the coupling reaction of aryl iodides and *N*-Boc-protected, O-functionalized hydroxylamines at 80°C to afford the corresponding N-arylation products (Scheme 1.20, Eq. 2).[48] More recently, Antilla and coworkers reported that combinations of copper(I) iodide and *N*,*N*-diethylsalicylamide catalyzed the guanidinylation of aryl iodides at 80°C, providing symmetrical, *N*,*N′*-disubstituted guanidines with moderate to good yields (Scheme 1.20, Eq. 3).[49a] Finally, 2-imidazolines can also be arylated with aryl iodides in the presence of copper(I) bromide, although the reaction needs heating at 150°C in NMP (Scheme 1.20, Eq. 4).[49b]

As evidenced with all results overviewed in this section, a wide range of N-nucleophiles can be efficiently arylated under relatively mild conditions and with high efficiency. The introduction of chelating ligands for copper also enabled the development of efficient procedures for the arylation of another most useful class of N-nucleophiles: conjugated N-heterocycles. Advances in this area will be overviewed in the next section of this chapter.

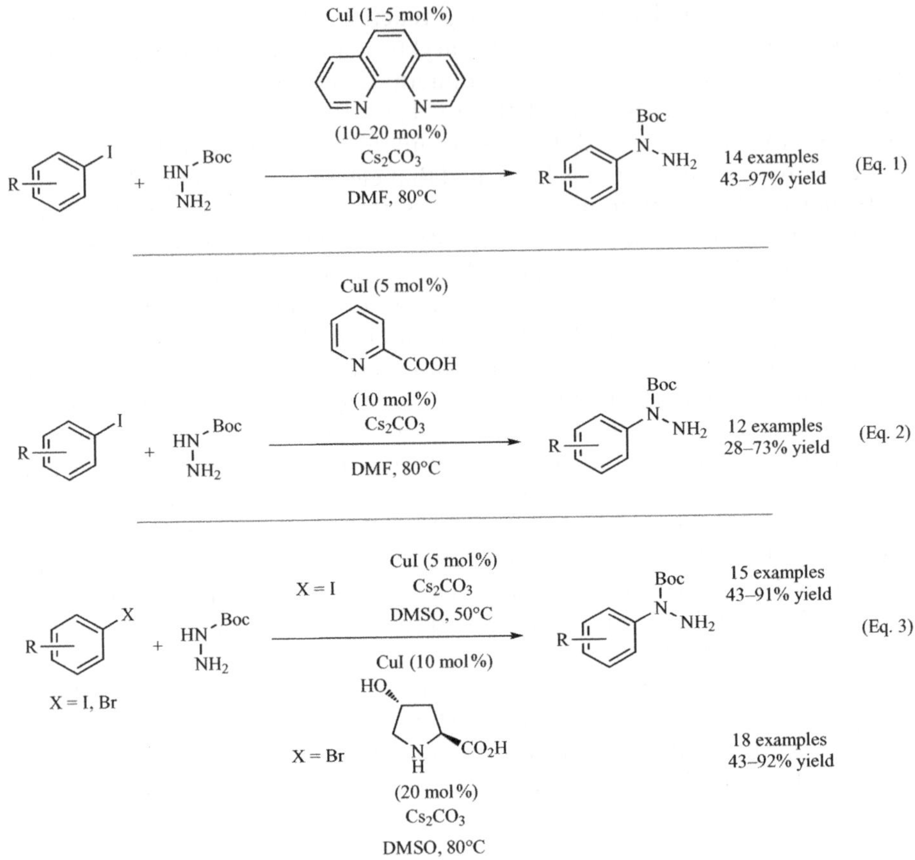

Scheme 1.19 Copper(I) iodide/ligand-catalyzed cross coupling of aryl halides and *N*-Boc-hydrazine.

1.4 ARYLATION OF CONJUGATED N-HETEROCYCLES

N-Aryl heterocycles are frequently found in naturally occurring and/or bioactive molecules and are useful precursors for materials science. Therefore, the development of reliable, efficient, and general procedures for the synthesis of *N*-aryl heterocycles has become an especially active area in organic synthesis, notably for the direct arylation of conjugated N-heterocycles with aryl halides or synthetic equivalents (details concerning the use of other reagents such as aryl boronic acids, potassium aryltrifluoroborates . . . can be found in Chapter 4 and will not be detailed in this chapter). Over the years, the copper-catalyzed cross coupling of aryl halides with N-heterocycles has been proven to be one of the most powerful approaches for assembling these compounds and is now commonly used in both academic and industrial research.

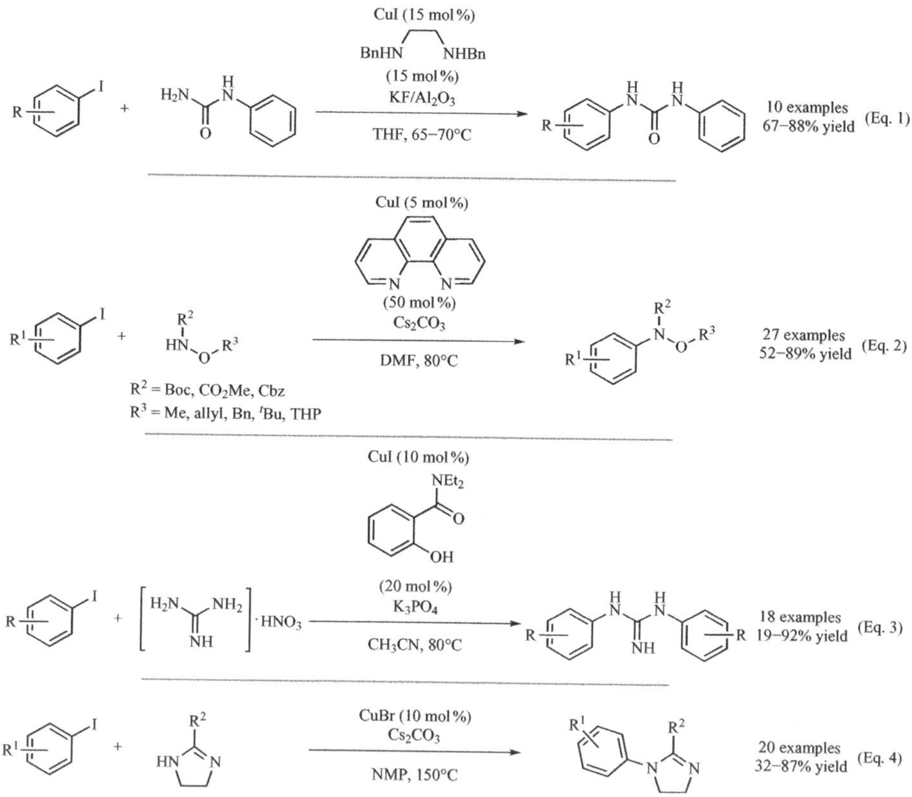

Scheme 1.20 Copper(I) iodide/ligand-catalyzed cross coupling of aryl iodides with ureas, substituted hydroxylamines, guanidine and 2-imidazolines.

A major advance to overcome the limitations of classical Ullmann-type procedures which involve elevated temperatures, stoichiometric amounts of copper, and the use of polar solvents was reported in 1999 by the Buchwald group, who found that 1,10-phenanthroline could efficiently promote the copper-catalyzed arylation of imidazoles with aryl iodides and bromides (Scheme 1.21, Eq. 1).[50] In the presence of this ligand as well as *trans,trans*-dibenzylideneacetone (dba) as an additive, catalytic amounts of $(CuOTf)_2{\cdot}C_6H_6$ was found to promote the coupling of aryl halides with imidazoles in xylenes at temperatures ranging from 110 to 125°C, which is still quite elevated. In this case a stoichiometric quantity of 1,10-phenanthroline was required for the success of the process. A few years later, the same group reported that the more electron-rich 4,7-dimethoxy-1,10-phenanthroline was a much more effective ligand than 1,10-phenanthroline (Scheme 1.21, Eq. 2).[51,52] Indeed, under the assistance of this electron-rich ligand, the copper(II) oxide-catalyzed arylation of imidazoles could be carried out at 80–120°C, and for the arylation of imidazole with iodobenzene, complete conversion could be achieved with

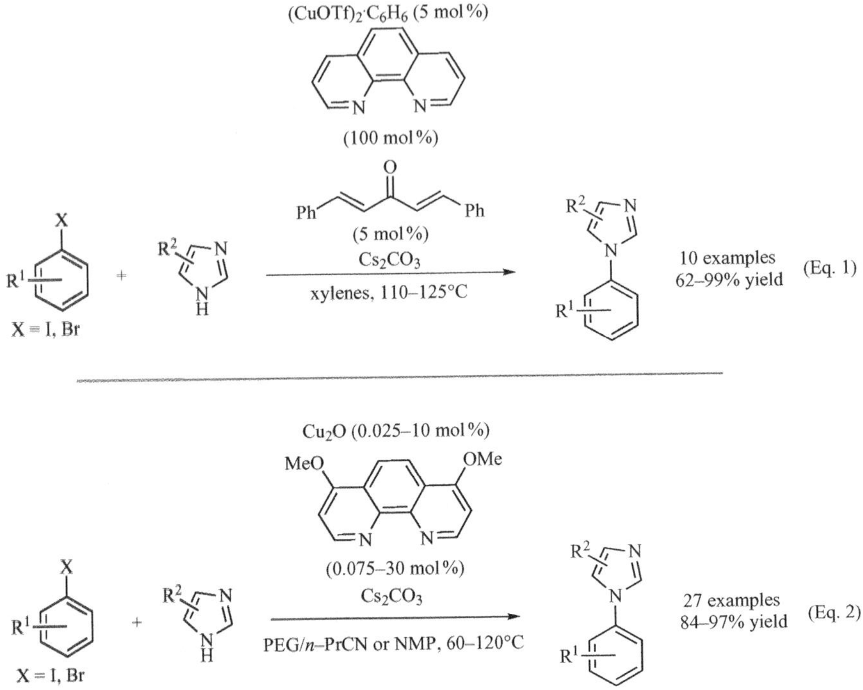

Scheme 1.21 Copper(I) salt/1,10-phenanthroline or 4,7-dimethoxy-1,10-phenanthroline-catalyzed cross coupling of aryl halides and imidazoles.

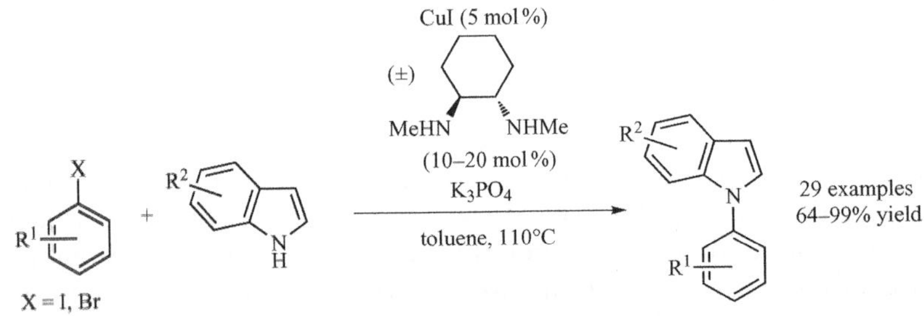

Scheme 1.22 Copper(I) iodide/*trans-N,N'*-dimethylcyclohexane-1,2-diamine-catalyzed cross coupling of aryl halides and indoles.

only 0.05 mol % copper(I) iodide and 0.075 mol % ligand, which is still quite rare for copper-based catalytic systems.

In order to further increase the scope of the copper(I) iodide/diamine system, and after screening several diamine ligands for the copper(I) iodide-catalyzed coupling of aryl halides with indoles, Buchwald and coworkers discovered that DMCyDA gave the best results, leading to a reaction that occurred in refluxing toluene to give *N*-aryl indoles with good to excellent yields (Scheme 1.22).[53]

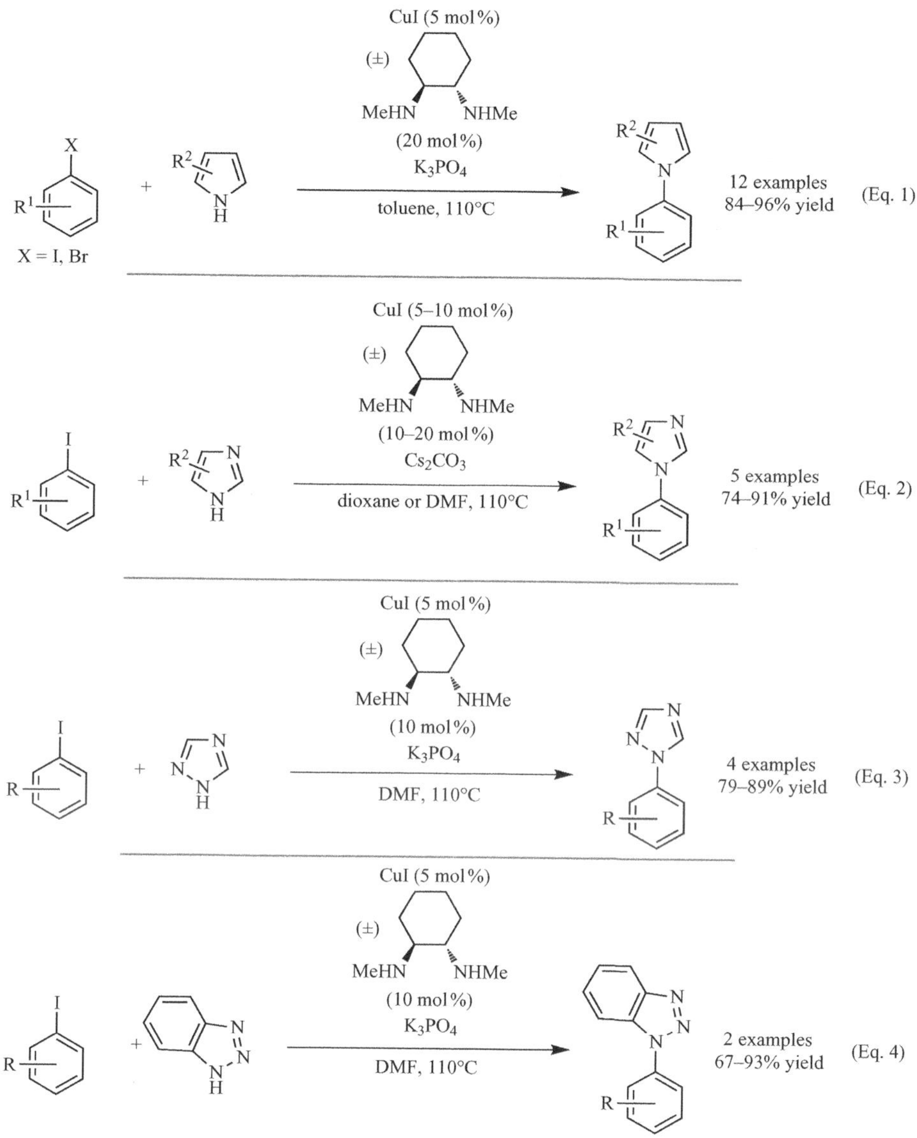

Scheme 1.23 Copper(I) iodide/*trans-N,N'*-dimethylcyclohexane-1,2-diamine-catalyzed arylation of pyrroles, imidazoles, triazole, and benzotriazole.

Further investigations indicated that the copper/DMCyDA system was also suitable for the arylation of a wide number of other N-heterocycles (Scheme 1.23 and Scheme 1.24),[54] with slightly different optimized conditions, however. For example, the use of potassium phosphate as the base in refluxing toluene was optimal for the arylation of pyrroles and indazoles, while triazoles and pyrazoles required changing the solvent to DMF and the base to potassium carbonate, respectively, and imidazoles needed both a different base (cesium

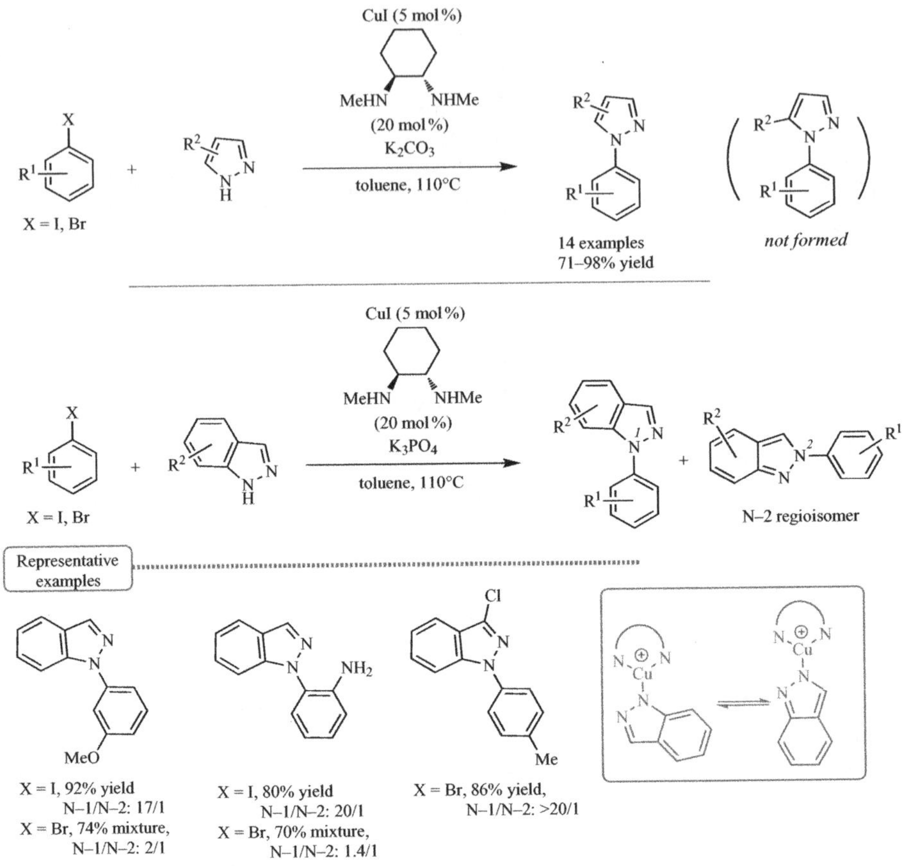

Scheme 1.24 Regioselectivity trends for the cross coupling of aryl halides and unsymmetrical N-heterocycles.

carbonate) and solvent (DMF or dioxane). Noteworthy is that for the coupling reactions involving imidazoles and triazoles, only aryl iodides were compatible with these conditions.

Some interesting trends are apparent for the cross coupling of aryl halides and substituted imidazoles, pyrazoles, as well as indazoles. For unsymmetrical pyrazoles and imidazoles, the N-arylation occurs at the less hindered nitrogen, while indazoles are selectively N-arylated at N-1 (Scheme 1.24), provided that aryl iodides are used for the arylation, the corresponding aryl bromides yielding to lower selectivities.[52,54] This dramatic change in regioselectivity was attributed to a faster oxidative addition with aryl iodides, which would avoid equilibration of the (diamine)Cu(indazole) intermediate formed in the initial step of the catalytic cycle.

Besides 1,2-diamines, the Ma group showed that amino acids were also able to promote the arylation of various N-heterocycles. For example, the use of

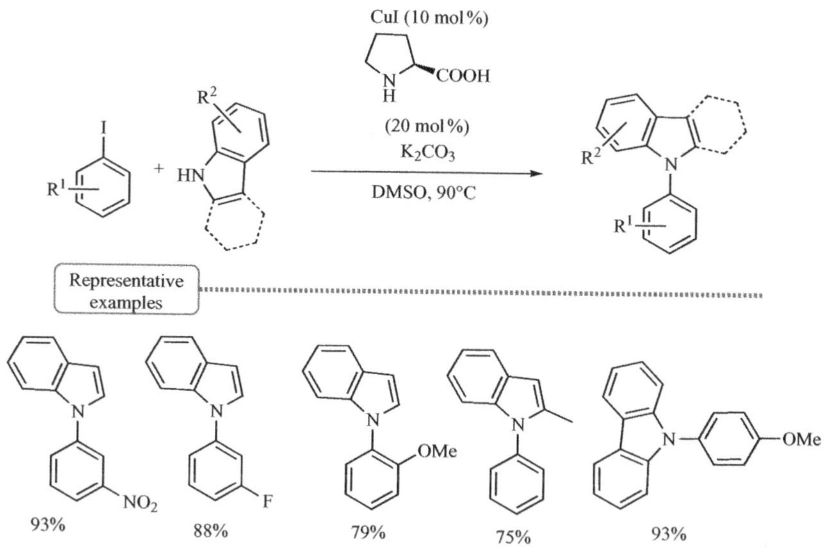

Scheme 1.25 Copper(I) iodide/L-proline-catalyzed cross coupling of aryl iodides and indoles and carbazole.

L-proline as a promoter allowed to efficiently perform the copper(I) iodide-catalyzed cross-coupling reaction of aryl iodides with indoles and carbazole in DMSO at 90°C to afford the corresponding *N*-aryl heterocycles with good to excellent yields (Scheme 1.25).[6,55] The copper(I) iodide/L-proline system was also suitable to catalyze the arylation of pyrroles, imidazoles, and pyrazoles with aryl iodides and electron-deficient aryl bromides at 60–90°C. However, the ligand had to be changed to *N,N*-dimethylglycine and the temperature raised to 110°C for the cross coupling involving electron-rich aryl bromides and pyrroles, imidazoles, or pyrazoles. In these cases, incomplete conversions were observed when L-proline was utilized, which might result from a fast competitive coupling between this ligand and aryl bromides at 110°C (Scheme 1.26).[6,55]

Interestingly, the copper(I) iodide/*N,N*-dimethylglycine catalytic system has been successfully applied for the large-scale preparation of GW876008, a novel corticotrophin releasing factor (CRF) antagonist developed by GSK that has entered phase II clinical trials for treatment of stress-related illnesses like anxiety and depression (Scheme 1.27).[56] In this case, the copper(I) iodide/*N,N*-dimethylglycine-catalyzed cross-coupling reaction of aryl iodide **8** with pyrazolyl compound **9** readily proceeded in DMSO to give the corresponding arylation product in good yield, which was further converted into GW876008 via methanolysis. This synthetic route worked well at the manufacturing stage (>100 kg), clearly demonstrating the efficiency of this process.

During the past years, tons of catalytic systems have been reported for these transformations, which renders a basic literature search quite tricky. Indeed, the copper-catalyzed arylation of N-heterocycles has become a real testing

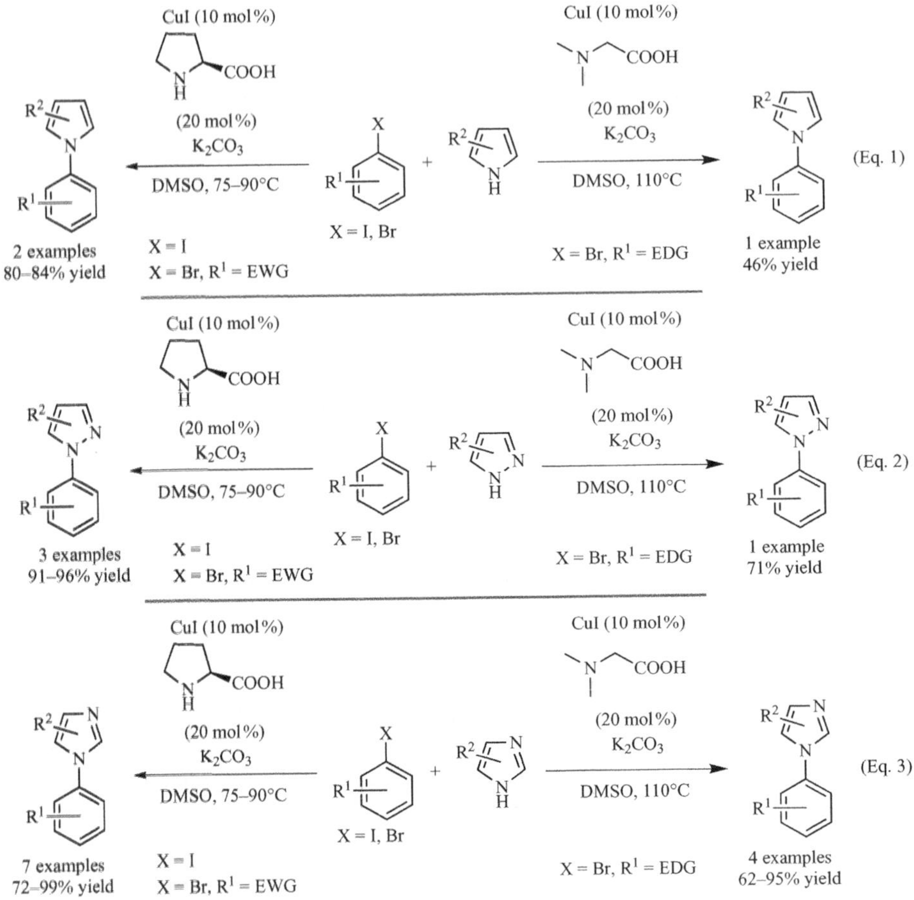

Scheme 1.26 Copper(I) iodide/amino acid-catalyzed cross coupling of aryl halides and pyrroles, pyrazoles, and imidazoles.

tube for ligand development, and a great number of new ligands for this transformation have been reported. In order to focus on the most efficient ones, they will not be all developed and systematically described. For example, some ligands that are suitable for copper-catalyzed aryl amination, which include *rac*-BINOL,[17,18a] pyrrolidine-2-phosphonic acid phenyl monoester **2**,[23] ethyl 2-oxocyclohexanecarboxylate,[40] 1-(5,6,7,8-tetrahydroquinolin-8-yl)ethanone **5**,[25] and 3-acetylcoumarin **6**,[26] were found to be also applicable for the coupling of aryl halides with N-heterocycles. The structures of some new ligands are outlined in Figure 1.3, which include salicylaldoxime **10**[57] and 8-hydroxyquinoline **11**,[58] which were recently shown to promote the chemoselective arylation of the endocyclic amine of 2-aminobenzimidazoles (while palladium catalysis allows for a selective arylation of the exocyclic amine)[58b] and its hydrogenation product **12**,[59] 2-aminopyrimidine-4,6-diol **13**,[60]

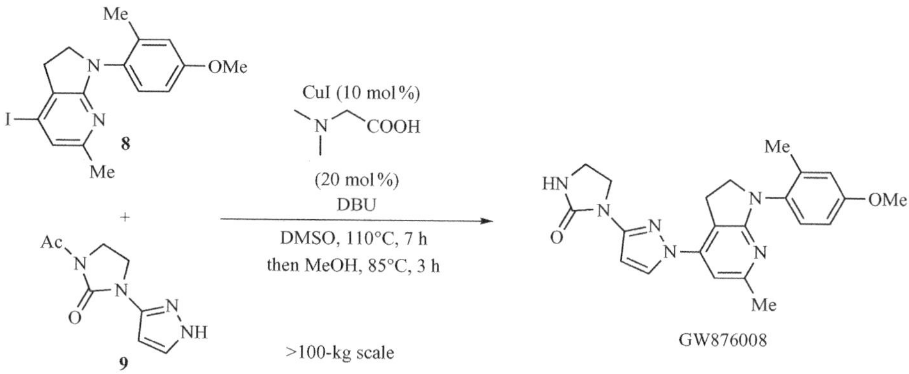

Scheme 1.27 Synthesis of GW876008 via copper(I) iodide/*N*,*N*-dimethylglycine-catalyzed cross coupling.

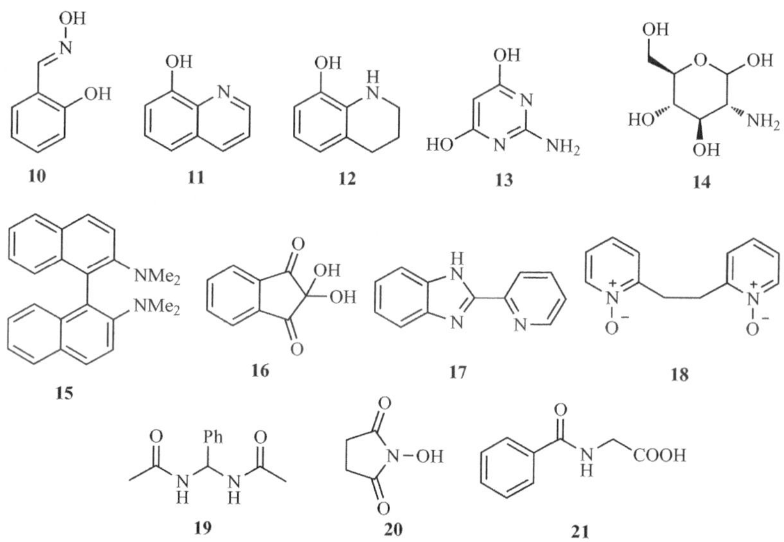

Figure 1.3 Structures of other ligands for the copper-catalyzed arylation of conjugated N-heterocycles (non-exhaustive list).

D-glucosamine **14**,[61] *N*,*N*,*N*′,*N*′-tetramethyl-1,1′-binaphthyl-2,2′-diamine **15**,[62] ninhydrin **16**,[63] 2-(2′-pyridyl)benzimidazole **17**,[64] bis(pyridine *N*-oxide) **18**,[65] *N*,*N*′-(phenylmethylene)diacetamide **19**,[66] *N*-hydroxysuccinimide **20**,[67] and hippuric acid **21**.[68] Although the efficiency of these ligands is similar, when not inferior, to diamine and amino acid ligands, these molecules may serve as new leads for further ligand design and can be evaluated when the most general ligands fail to provide an expected arylation product.

Before closing this section, a seminal report by Bolm and coworkers should be mentioned. A major drawback of most catalytic systems we have

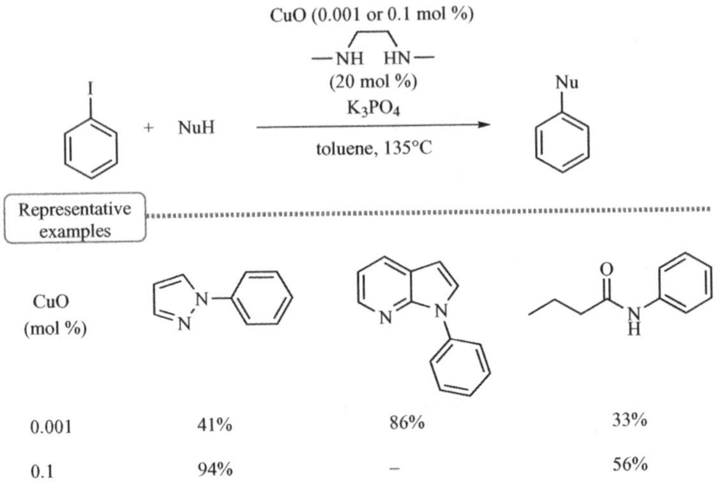

Scheme 1.28 Copper-catalyzed cross coupling with part-per-million catalyst loading.

overviewed here is that they require especially high catalyst loading, which is mostly attributed to the low solubility of copper salts. In fact, the development of copper-based catalytic systems do not frequently involve lowering the catalyst loading since most copper sources and ligands are rather cheap. However, the use of percents of the copper source is not systematically required, as shown by Bolm and coworkers, who found, upon realizing that most iron-mediated cross-coupling reactions were actually due to contamination of the iron salts with traces of copper, that some cross-coupling reactions did only require part-per-million catalyst loading (Scheme 1.28).[69] Two factors appear to be quite important in this case: the ligand quantity, since the reactions occurred at high ligand-to-metal ratio, and reaction temperature, providing the corresponding N-arylated heterocycles, as well as other products resulting from C–N bond formation.

As evidence by all results collected at this point of this chapter, most N-nucleophiles can be arylated, with high levels of efficiency in most cases provided that the appropriate combination of copper source (mostly copper(I) iodide), ligand, base, and solvent is chosen. A last nucleophile to be considered, whose arylation will be described in the last section of this chapter, is ammonia, for which a direct cross coupling with aryl halides provides straightforward, highly efficient synthesis of anilines.

1.5 SYNTHESIS OF ANILINES BY COUPLING WITH AMMONIA OR SYNTHETIC EQUIVALENTS

The conversion of aryl halides to primary aryl amines is a useful transformation in organic synthesis. In an early example, researchers at Merck have successfully

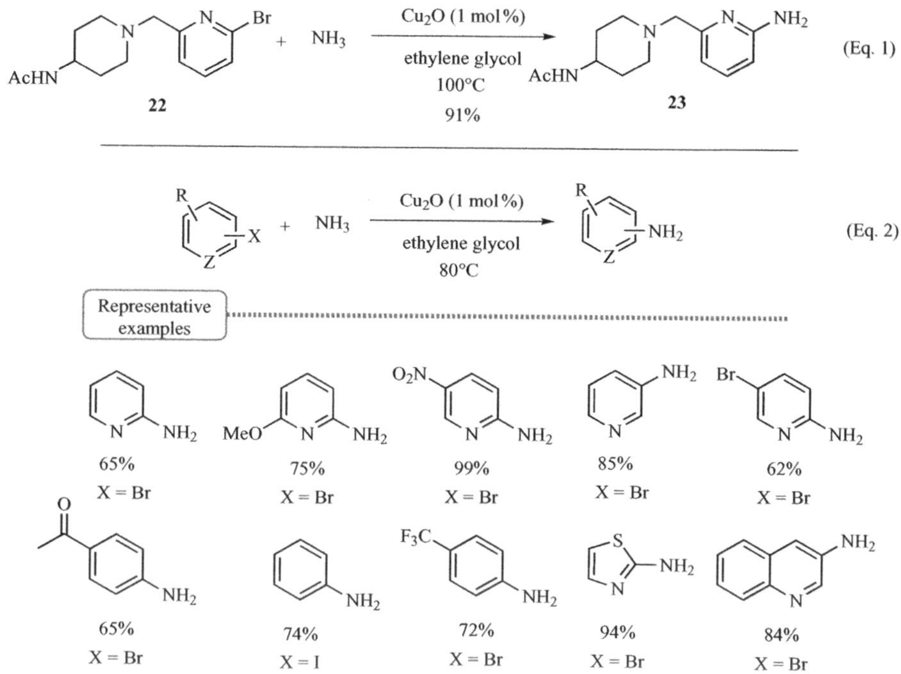

Scheme 1.29 Copper(I) oxide-catalyzed synthesis of primary heteroaryl amines using ammonia in ethylene glycol.

applied the copper-catalyzed aryl amination to transform aryl bromide **22** into 2-aminopyridine **23**, a key intermediate for synthesizing a potent muscarinic antagonist (Scheme 1.29, Eq. 1).[70] In this case, ethylene glycol may serve as both ligand and solvent, leading to the coupling reaction with ammonia at 100°C. These reaction conditions have also been shown to be applicable for the direct amination of a variety of bromopyridines and aryl halides (Scheme 1.29, Eq. 2).

After screening several ligands such as ethylene glycol, amino acids, and diamines, Kim and Chang reported an alternative catalytic system based on the combination of L-proline and copper(I) iodide, which enabled the coupling reaction of aryl iodides with ammonia, or ammonium chloride, at room temperature (Scheme 1.30, Eq. 1).[71] Increasing the reaction temperature to 80°C also led to complete conversion with electron-deficient aryl bromides. However, incomplete coupling was observed in the case of electron-rich aryl bromides as substrates. To solve this problem, another ligand, *trans*-4-hydroxy-L-proline, was tested and this promoter was found to be applicable for both electron-deficient and electron-rich aryl bromides, yielding to complete conversions at 50°C (Scheme 1.30, Eq. 2).[46]

Another especially efficient system was reported by Taillefer and coworkers who demonstrated that 1,3-diketones are especially powerful ligands for the copper-catalyzed coupling of aryl halides with ammonia. The optimized system

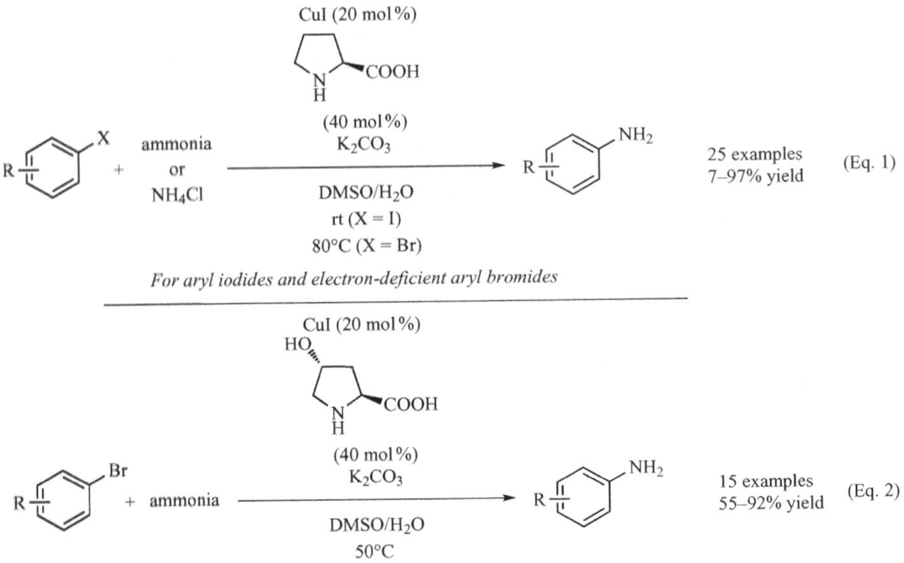

Scheme 1.30 Copper(I) iodide/amino acid-catalyzed synthesis of anilines from aryl halides and ammonia.

relies on a combination of copper(II) acetylacetonate and acetylacetone, enabling this transformation to work at 60–90°C in DMF (Scheme 1.31, Eq. 1).[72a] This system is rather general since both aryl iodides and aryl bromides are suitable coupling partners. This reaction can also be performed directly in liquid ammonia using a combination of copper(I) iodide and ascorbic acid as demonstrated by Page and coworkers (Scheme 1.31, Eq. 2). Aryl iodides can be smoothly converted to the corresponding anilines at room temperature with this system while aryl bromides and chlorides (substituted with nitro groups only) required heating the reaction mixture at 100°C.[72b] It also should be mentioned that this direct arylation of ammonia was extended to the use of heteroaryl bromides (Scheme 1.31, Eq. 3)[72c] and aryl dibromides (Scheme 1.31, Eq. 4),[72d] and that a selective one-pot synthesis of symmetrical and unsymmetrical di- and triarylamines by direct double or triple arylation of lithium amide was reported.[72e]

Other notable ligands that were shown to be able to promote this reaction include 1-(5,6,7,8-tetrahydroquinolin-8-yl)-2-methylpropane-1-one **24**,[73] D-glucosamine **14**,[74] and N^2,$N^{2\prime}$-diisopropyloxalohydrazide **26** (Fig. 1.4).[75]

Besides ammonia, whose use might not be compatible with all substrates, various surrogates can be employed to introduce a primary amine from the corresponding aryl halide. Amidines can be used for such a transformation: Indeed, during the studies on the copper(I) iodide/L-proline-catalyzed coupling of aryl halides with amidine hydrochlorides, Fu and coworkers discovered that primary anilines were directly isolated, most certainly through a coupling/hydrolysis sequence as indicated in Scheme 1.32.[76]

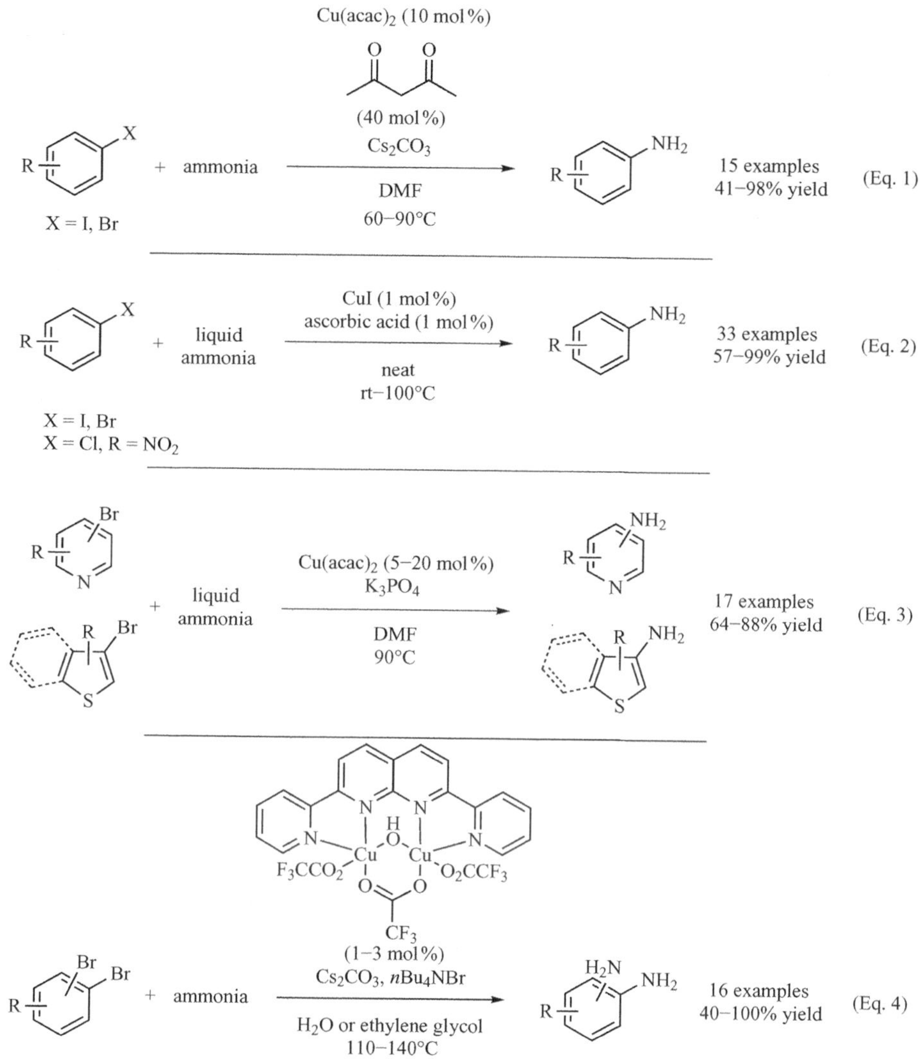

Scheme 1.31 Copper/ligand-catalyzed direct arylation of ammonia.

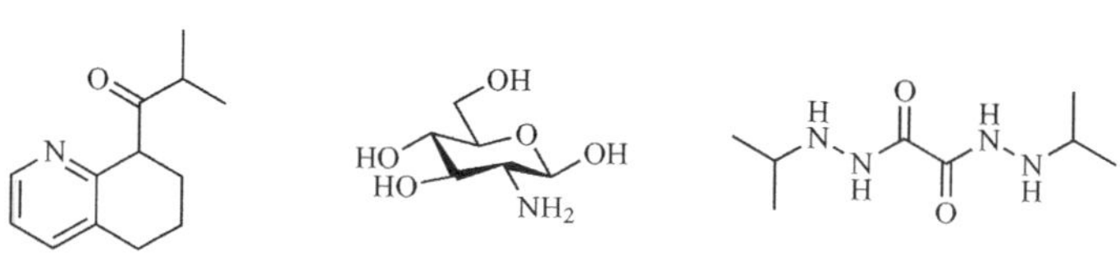

24 **14** **26**

Figure 1.4 Other reported ligands for the copper-catalyzed synthesis of anilines by direct arylation of ammonia (non-exhaustive list).

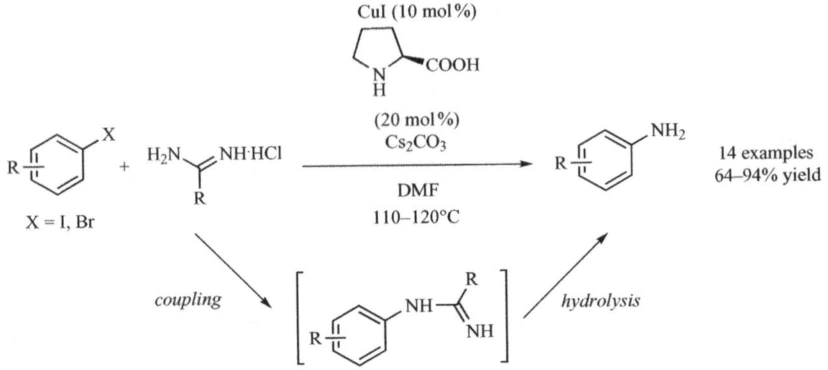

Scheme 1.32 Copper(I) iodide-catalyzed synthesis of anilines from aryl halides and amidines.

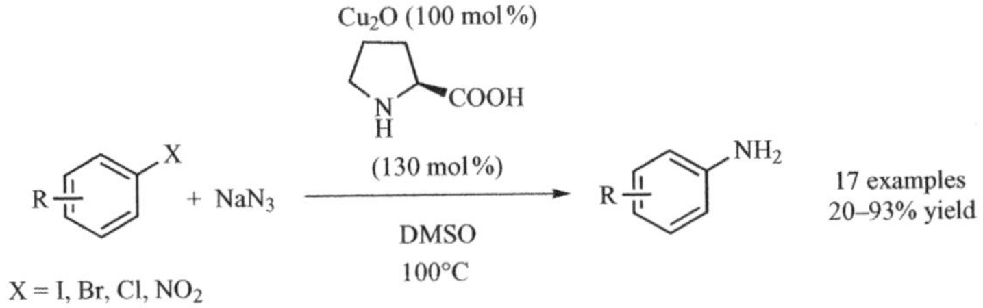

Scheme 1.33 Copper(I) oxide/L-proline-promoted synthesis of anilines from aryl halides and sodium azide.

It had been reported that the copper(I) iodide/L-proline or copper(I) iodide/DMEDA-catalyzed coupling of aryl halides with sodium azide logically provided the corresponding aryl azides.[77] However, when the Helquist group applied these reaction conditions for the synthesis of bioactive molecules, they were able to isolate various amounts of primary anilines as by-products. After carefully exploring the reaction conditions, they showed that these primary anilines could be produced exclusively if the reaction was carried out by using 100 mol % copper(I) oxide, 130 mol % L-proline, and two equivalents of sodium azide (Scheme 1.33).[78] Although the detailed mechanism is not clear yet, this reaction provides an alternative approach for assembling primary anilines from aryl halides, at the cost of high catalyst loading, however.

Guo, Liu, and coworkers reported another route for obtaining primary anilines from aryl halides relying on a copper(I) iodide/DMEDA-catalyzed cross coupling of trifluoroacetamide with aryl halides, a reaction that was found to proceed smoothly in DMF (for aryl iodides) or dioxane (for aryl bromides) to deliver the corresponding amidation products, which could be easily converted into primary amines via hydrolysis in a one-pot manner by simply adding methanol and water after completion of the reaction (Scheme 1.34).[79]

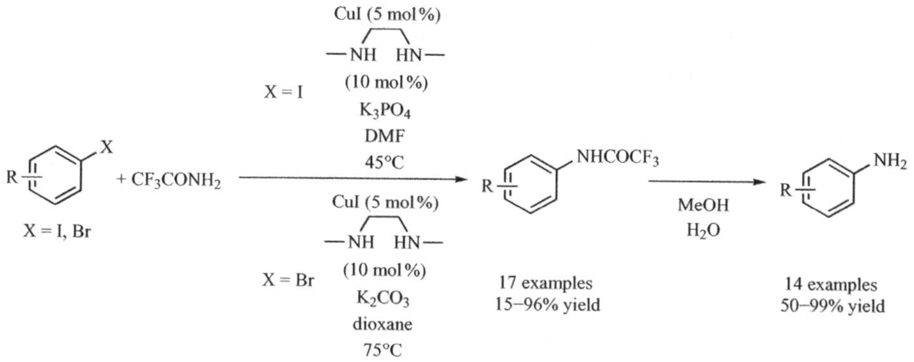

Scheme 1.34 Copper(I) iodide/*N*,*N*′-dimethylethylenediamine-catalyzed one-pot synthesis of anilines from aryl halides and 2,2,2-trifluoroacetamide.

1.6 CONCLUSION AND FUTURE PROSPECTS

In summary, we have collected in this chapter a number of examples showing that various ligands have remarkable accelerating effects on the copper-catalyzed Ullmann and Goldberg cross-coupling reactions, the most efficient and general ones certainly being diamines, amino acids, phenanthrolines, and 1,3-dicarbonyl compounds. The use of these readily available ligands has allowed the design of remarkably efficient and reliable catalytic systems that have found numerous applications in the synthesis of natural products, designed bioactive compounds (see Chapter 16 for details), or compounds for materials sciences. However, further investigations in this area are still required because the possible mechanisms for these transformations are still unclear, and both the reaction scope and catalytic efficiency still need to be improved, the most obvious example certainly being the absence of reactivity of aryl chlorides in most cases and the low one of deactivated aryl bromides. There is therefore still room for optimization and one approach will clearly involve developing new ligands to solve these problems and reduce the catalyst loading. The use of photoactivation will also be interesting to follow in the near future since it was recently demonstrated by Peters, Fu, and coworkers that photoinduced Ullmann-Goldberg-type cross-coupling reactions could proceed under remarkably mild conditions.[80]

REFERENCES

[1] Ullmann, F. *Justus Liebigs Ann. Chem.* **1904**, *332*, 38–81.
[2] Goldberg, I. *Ber. Dtsch. Chem. Ges.* **1906**, *39*, 1691–1692.
[3] For reviews, see: (a) Monnier, F.; Taillefer, M. *Angew. Chem. Int. Ed.* **2009**, *48*, 6954–6971. (b) Monnier, F.; Taillefer, M. *Angew. Chem. Int. Ed.* **2008**, *47*, 3096–3099. (c) Evano, G.; Blanchard, N.; Toumi, M. *Chem. Rev.* **2008**, *108*, 3054–3131. (d) Ma,

D.; Cai, Q. *Acc. Chem. Res.* **2008**, *41*, 1450–1460. (e) Beletskaya, I. P.; Cheprakov, A. V. *Coord. Chem. Rev.* **2004**, *248*, 2337–2364. (f) Ley, S. V.; Thomas, A. W. *Angew. Chem. Int. Ed.* **2003**, *42*, 5400–5449.

[4] (a) Ma, D.; Zhang, Y.; Yao, J.; Wu, S.; Tao, F. *J. Am. Chem. Soc.* **1998**, *120*, 12459–12467. (b) Ma, D.; Xia, C. *Org. Lett.* **2001**, *3*, 2583–2586.

[5] Ma, D.; Cai, Q.; Zhang, H. *Org. Lett.* **2003**, *5*, 2453–2455.

[6] Zhang, H.; Cai, Q.; Ma, D. *J. Org. Chem.* **2005**, *70*, 5164–5173.

[7] Yang, C.-T.; Fu, Y.; Huang, Y.-B.; Yi, J.; Guo, Q.-X.; Liu, L. *Angew. Chem. Int. Ed.* **2009**, *48*, 7398–7401.

[8] Kurokawa, M.; Nakanishi, W.; Ishikawa, T. *Heterocycles* **2007**, *71*, 847–854.

[9] Guo, X.; Rao, H.; Fu, H.; Jiang, Y.; Zhao, Y. *Adv. Synth. Catal.* **2006**, *348*, 2197–2202.

[10] Kwong, F. Y.; Klapars, A.; Buchwald, S. L. *Org. Lett.* **2002**, *4*, 581–584.

[11] Kwong, F. Y.; Buchwald, S. L. *Org. Lett.* **2003**, *5*, 793–796.

[12] Buck, E.; Song, Z. J.; Tschaen, D.; Dormer, P. G.; Volante, R. P.; Reider, P. J. *Org. Lett.* **2002**, *4*, 1623–1626.

[13] Shafir, A.; Buchwald, S. L. *J. Am. Chem. Soc.* **2006**, *128*, 8742–8743.

[14] Shafir, A.; Lichtor, P. A.; Buchwald, S. L. *J. Am. Chem. Soc.* **2007**, *129*, 3490–3491.

[15] (a) Jones, G. O.; Liu, P.; Houk, K. N.; Buchwal, S. L. *J. Am. Chem. Soc.* **2010**, *132*, 6205–6213. (b) Yu, H.-Z.; Jiang, Y.-Y.; Fu, Y.; Liu, L. *J. Am. Chem. Soc.* **2010**, *132*, 18078–18091.

[16] Altman, R. A.; Anderson, K. W.; Buchwald, S. L. *J. Org. Chem.* **2008**, *73*, 5167–5169.

[17] Zhu, D.; Wang, R.; Mao, J.; Xu, L.; Wu, F.; Wan, B. *J. Mol. Catal. A: Chem.* **2006**, *256*, 256–260.

[18] (a) Jiang, D.; Fu, H.; Jiang, Y.; Zhao, Y. *J. Org. Chem.* **2007**, *72*, 672–674. (b) Zeng, L.; Fu, H.; Qiao, R.; Jiang, Y.; Zhao, Y. *Adv. Synth. Catal.* **2009**, *351*, 1671–1676. (c) Zhou, F.; Guo, J.; Liu, J.; Ding, K.; Yu, S.; Cai, C. *J. Am. Chem. Soc.* **2012**, *134*, 14326–14329.

[19] (a) Lu, Z.; Twieg, R. J.; Huang, S. D. *Tetrahedron Lett.* **2003**, *44*, 6289–6292. (b) Lu, Z.; Twieg, R. J. *Tetrahedron Lett.* **2005**, *46*, 2997–3001. (c) Lu, Z.; Twieg, R. J. *Tetrahedron* **2005**, *61*, 903–918.

[20] Goodbrand, H. B.; Hu, N.-X. *J. Org. Chem.* **1999**, *64*, 670–674.

[21] Gujadhur, R. K.; Bates, C. G.; Venkataraman, D. *Org. Lett.* **2001**, *3*, 4315–4317.

[22] Rao, H.; Fu, H.; Jiang, Y.; Zhao, Y. *J. Org. Chem.* **2005**, *70*, 8107–8109.

[23] Rao, H.; Jin, Y.; Fu, H.; Jiang, Y.; Zhao, Y. *Chem. Eur. J.* **2006**, *12*, 3636–3646.

[24] Jiang, Q.; Jiang, D.; Jiang, Y.; Fu, H.; Zhao, Y. *Synlett* **2007**, 1836–1842.

[25] Wang, D.; Ding, K. *Chem. Commun.* **2009**, 1891–1893.

[26] Tao, C.-Z.; Liu, W.-W.; Sun, J.-Y.; Cao, Z.-L.; Li, H.; Zhang, Y.-F. *Synthesis* **2010**, 1280–1284.

[27] Yang, M.; Liu, F. *J. Org. Chem.* **2007**, *72*, 8969–8971.

[28] Li, X.; Yang, D.; Jiang, Y.; Fu, H. *Green Chem.* **2010**, *12*, 1097–1105.

[29] Klapars, A.; Antilla, J. C.; Huang, X.; Buchwald, S. L. *J. Am. Chem. Soc.* **2001**, *123*, 7727–7729.

[30] Klapars, A.; Huang, X.; Buchwald, S. L. *J. Am. Chem. Soc.* **2002**, *124*, 7421–7428.

[31] Surry, D. S.; Buchwald, S. L. *Chem. Sci.* **2010**, *1*, 13–31.

[32] (a) Crawford, K. R.; Padwa, A. *Tetrahedron Lett.* **2002**, *43*, 7365–7368. (b) Padwa, A.; Crawford, K. R.; Rashatasakhon, P.; Rose, M. *J. Org. Chem.* **2003**, *68*, 2609–2617.

[33] Mallesham, B.; Rajesh, B. M.; Reddy, P. R.; Srinivas, D.; Trehan, S. *Org. Lett.* **2003**, *5*, 963–965.

[34] Phillips, D. P.; Zhu, X.-F.; Lau, T. L.; He, X.; Yang, K.; Liu, H. *Tetrahedron Lett.* **2009**, *50*, 7293–7296.

[35] (a) Ghinet, A.; Oudir, S.; Henichart, J.-P.; Rigo, B.; Pommery, N.; Gautret, P. *Tetrahedron* **2010**, *66*, 215–221. (b) Kukosha, T.; Trufilkina, N.; Belyakov, S.; Katkevics, M. *Synlett* **2012**, *44*, 2413–2423.

[36] Deng, W.; Wang, Y.-F.; Zou, Y.; Liu, L.; Guo, Q.-X. *Tetrahedron Lett.* **2004**, *45*, 2311–2315.

[37] (a) Chen, W.; Li, J.; Fang, D.; Feng, C.; Zhang, C. *Org. Lett.* **2008**, *10*, 4565–4568. (b) Toto, P.; Gesquière, J.-C.; Deprez, B.; Willand, N. *Tetrahedron Lett.* **2006**, *47*, 1181–1186.

[38] Hosseinzadeh, R.; Tajbakhsh, M.; Mohadjerani, M.; Mehdinejad, H. *Synlett* **2004**, 1517–1520.

[39] Altman, R. A.; Buchwald, S. L. *Org. Lett.* **2007**, *9*, 643–646.

[40] Lv, X.; Bao, W. *J. Org. Chem.* **2007**, *72*, 3863–3867.

[41] Racine, E.; Monnier, F.; Vors, J.-P.; Taillefer, M. *Org. Lett.* **2011**, *13*, 2818–2821.

[42] Deng, W.; Liu, L.; Zhang, C.; Liu, M.; Guo, Q.-X. *Tetrahedron Lett.* **2005**, *46*, 7295–7298.

[43] Han, X. *Tetrahedron Lett.* **2010**, *51*, 360–362.

[44] Wolter, M.; Klapars, A.; Buchwald, S. L. *Org. Lett.* **2001**, *3*, 3803–3805.

[45] Lam, M. S.; Lee, H. W.; Chan, A. S. C.; Kwong, F. Y. *Tetrahedron Lett.* **2008**, *49*, 6192–6194.

[46] Jiang, L.; Lu, X.; Zhang, H.; Jiang, Y.; Ma, D. *J. Org. Chem.* **2009**, *74*, 4542–4546.

[47] Hosseinzadeh, R.; Sarrafi, Y.; Mohadjerani, M.; Mohammadpourmir, F. *Tetrahedron Lett.* **2008**, *49*, 840–843.

[48] Jones, K. L.; Porzelle, A.; Hall, A.; Woodrow, M. D.; Tomkinson, N. C. O. *Org. Lett.* **2008**, *10*, 797–800.

[49] (a) Cortes-Salva, M.; Nguyen, B.-L.; Cuevas, J.; Pennypacker, K. R.; Antilla, J. C. *Org. Lett.* **2010**, *12*, 1316–1319. (b) Davis, O. A.; Hughes, M.; Bull, J. A. *J. Org. Chem.* **2013**, *78*, 3470–3475.

[50] Kiyomori, A.; Marcoux, J.-F.; Buchwald, S. L. *Tetrahedron Lett.* **1999**, *40*, 2657–2660.

[51] Altman, R. A.; Buchwald, S. L. *Org. Lett.* **2006**, *8*, 2779–2782.

[52] Altman, R. A.; Koval, E. D.; Buchwald, S. L. *J. Org. Chem.* **2007**, *72*, 6190–6199.

[53] Antilla, J. C.; Klapars, A.; Buchwald, S. L. *J. Am. Chem. Soc.* **2002**, *124*, 11684–11688.

[54] Antilla, J. C.; Baskin, J. M.; Barder, T. E.; Buchwald, S. L. *J. Org. Chem.* **2004**, *69*, 5578–5587.

[55] Ma, D.; Cai, Q. *Synlett* **2004**, 128–130.

[56] Ribecai, A.; Bacchi, S.; Delpogetto, M.; Guelfi, S.; Manzo, A. M.; Perboni, A.; Stabile, P.; Westerduin, P.; Hourdin, M.; Rossi, S.; Provera, S.; Turco, L. *Org. Process Res. Dev.* **2010**, *14*, 895–901.

[57] (a) Cristau, H.-J.; Cellier, P. P.; Hamada, S.; Spindler, J.-F.; Taillefer, M. *Chem. Eur. J.* **2004**, *10*, 5607–5622. (b) Cristau, H.-J.; Cellier, P. P.; Hamada, S.; Spindler, J.-F.; Taillefer, M. *Eur. J. Org. Chem.* **2004**, 695–709.

[58] (a) Liu, L.; Frohn, M.; Xi, N.; Dominguez, C.; Hungate, R.; Reider, P. J. *J. Org. Chem.* **2005**, *70*, 10135–10138. (b) Ueda, S.; Buchwald, S. L. *Angew. Chem. Int. Ed.* **2012**, *51*, 10364–10367.

[59] Wang, H.; Li, Y.; Sun, F.; Feng, Y.; Jin, K.; Wang, X. *J. Org. Chem.* **2008**, *73*, 8639–8642.

[60] Xie, Y.-X.; Pi, S.-F.; Wang, J.; Yin, D.-L.; Lin, J.-H. *J. Org. Chem.* **2006**, *71*, 8324–8327.

[61] Cheng, D.; Gan, F.; Qian, W.; Bao, W. *Green Chem.* **2008**, *10*, 171–173.

[62] Rao, R. K.; Naidu, A. B.; Haseer, E. A.; Sekar, G. *Tetrahedron* **2009**, *65*, 4619–4624.

[63] Huang, Y.-Z.; Gao, J.; Ma, H.; Miao, H.; Xu, J. *Tetrahedron Lett.* **2008**, *49*, 948–951.

[64] Haneda, S.; Adachi, Y.; Hayashi, M. *Tetrahedron* **2009**, *65*, 10459–10462.

[65] Liang, L.; Li, Z.; Zhou, X. *Org. Lett.* **2009**, *11*, 3294–3297.

[66] Wang, J.-P.; Chai, Y.-F.; Wu, J.-M.; Pan, Y.-J. *Synlett* **2008**, 3068–3072.

[67] Ma, H.-C.; Jiang, X.-Z. *J. Org. Chem.* **2007**, *72*, 8943–8946.

[68] Mao, J.; Guo, J.; Song, H.; Ji, S. *Tetrahedron* **2008**, *64*, 1383–1387.

[69] Larsson, P.-F.; Correa, A.; Carril, M.; Norrby, P.-O.; Bolm, C. *Angew. Chem. Int. Ed.* **2009**, *48*, 5691–5693.

[70] Lang, F.; Zewge, D.; Houpis, I. N.; Volante, R. P. *Tetrahedron Lett.* **2001**, *42*, 3251–3254.

[71] Kim, J.; Chang, S. *Chem. Commun.* **2008**, 3052–3054.

[72] (a) Kaddouri, H.; Vicente, V.; Ouali, A.; Ouazzani, F.; Taillefer, M. *Angew. Chem. Int. Ed.* **2008**, *48*, 337–339. (b) Ji, P.; Atherton, J. H.; Page, M. I. *J. Org. Chem.* **2012**, *77*, 7471–7478. (c) Fantasia, S.; Windisch, J.; Scalone, M. *Adv. Synth. Catal.* **2013**, *355*, 627–631. (d) Liao, B.-S.; Liu, S.-T. *J. Org. Chem.* **2012**, *77*, 6653–6656. (e) Tlili, A.; Monnier, F.; Taillefer, M. *Chem. Commun.* **2012**, *48*, 6408–6410.

[73] Wang, D.; Cai, Q.; Ding, K. *Adv. Synth. Catal.* **2009**, *351*, 1722–1726.

[74] Thakur, K. G.; Ganapathy, D.; Sekar, G. *Chem. Commun.* **2011**, *47*, 5076–5078.

[75] Meng, F.; Zhu, X.; Li, Y.; Xie, J.; Wang, B.; Yao, J.; Wan, Y. *Eur. J. Org. Chem.* **2010**, *10*, 6149–6152.

[76] Gao, X.; Fu, H.; Qiao, R.; Jiang, Y.; Zhao, Y. *J. Org. Chem.* **2008**, *73*, 6864–6866.

[77] (a) Zhu, W.; Ma, D. *Chem. Commun.* **2004**, 888–889. (b) Andersen, J.; Madsen, U.; Bjorkling, F.; Liang, X. F. *Synlett* **2005**, 2209–2213.

[78] Markiewicz, J.; Wiest, O.; Helquist, P. *J. Org. Chem.* **2010**, *75*, 4887–4890.

[79] Tao, C.-Z.; Li, J.; Fu, Y.; Liu, L.; Guo, Q.-X. *Tetrahedron Lett.* **2008**, *49*, 70–75.

[80] (a) Creutz, S. E.; Lotito, K. J.; Fu, G. C.; Peters, J. C. *Science* **2012**, *338*, 647–651. (b) Bissember, A. C.; Lundgren, R. J.; Creutz, S. E.; Peters, J. C.; Fu, G. C. *Angew. Chem. Int. Ed.* **2013**, *52*, 5129–5133. (c) Uyeda, C.; Tan, Y.; Fu, G. C.; Peters, J. C. *J. Am. Chem. Soc.* **2013**, *135*, 9548–9552.

2

ULLMANN CONDENSATION TODAY: ARYLATION OF ALCOHOLS AND THIOLS WITH ARYL HALIDES

ANIS TLILI AND MARC TAILLEFER

2.1 INTRODUCTION

More than a century ago, Ullmann reported the arylation of a nucleophile with an aryl halide in the presence of a stoichiometric amount of copper (see the Preface for a historical perspective). This marked the beginning of a pioneering research program aimed at providing chemists with methods to perform the arylation of nitrogen, oxygen, and carbon nucleophiles. Before the discovery of palladium catalysis, the chemistry developed by Ullmann was used largely to form C(aryl)–N bonds (arylation of amines, amides, and carbamates: Ullmann–Goldberg condensation), C(aryl)–O bonds (arylation of phenols: Ullmann condensation), and C(aryl)–C bonds (arylation of active methylene compounds: Ullmann–Hurtley condensation).[1,2]

Despite their high value in life sciences and material industries, Ullmann-type coupling reactions had not been exploited to their full potential until the beginning of the twenty-first century due to a number of restraints, including harsh reaction conditions, restricted range of substrates, and the moderate yields obtained. Indeed, these reactions are known to be rather unpredictable. In addition, the synthetic scope of these transformations was often limited because of the use of high-boiling-point polar solvents (>200°C), stoichiometric amounts of copper reagents, and the requirement of activated or *o*-carboxyaryl halides.[2]

Copper-Mediated Cross-Coupling Reactions, First Edition. Edited by Gwilherm Evano and Nicolas Blanchard.

Over time, a couple of studies had shown a peculiar rate enhancement when arylations were performed in the presence of particular additives or substrates bearing chelating groups.[3] These compounds were thought to improve the solubility and stability of the copper catalyst or to prevent aggregation of the metal and these observations eventually evolved into a general tendency that gave rise to a renaissance of Ullmann-type cross-coupling reactions. In 2001, two academic groups[4] independently discovered versatile, general, and very efficient new copper/ligand systems for C–C, C–N, C–S, C–P, and C–O bond formations, which enabled the use of only catalytic amounts of copper under milder conditions. Since these important breakthroughs, which were based on a unique system for cross coupling involving all types of nucleophiles, a large number of methods that show an extraordinary functional group tolerance have been described in the literature. Nowadays, modified copper-catalyzed cross-coupling reactions of aryl halides with nucleophiles are serious rivals for the more expensive palladium-catalyzed procedures, which also suffer from toxicity issues.

This chapter will survey C(aryl)–O and C(aryl)–S bond-forming reactions via copper-catalyzed couplings of oxygen (water, alcohols, phenols) and sulfur (thiols, thiophenols) nucleophiles with aryl halides. Among the increasing number of publications related to this topic, this chapter focuses mainly on the most significant results and important breakthroughs that have been reported in the last 10 years or so. Its scope highlights the most efficient homogeneous and heterogeneous catalytic systems based on the association of a copper source with or without an original ligand. The use of other coupling partners, such as arylmetals or alkenyl halides, is outside the scope of this chapter and will be covered in Chapters 4 and 5, respectively.

2.2 FORMATION OF C–O BONDS VIA COPPER-CATALYZED CROSS-COUPLING REACTIONS WITH ARYL HALIDES

2.2.1 Coupling of Aryl Halides with Hydroxide Sources: Synthesis of Phenols

Efficient methods for the synthesis of phenol are of prime importance to the chemical industry as around 10 million tons of this important synthetic intermediate are nowadays produced per year. More than 90% of the worldwide phenol production relies on the Hock process, which involves peroxidation of cumene, itself obtained from benzene propylation.[5] This industrial process, which is not very efficient in terms of yield (5% overall) and energy use, also produces large amounts of acetone and α-methylstyrene as coproducts. Almost 50% of the worldwide production of phenol is used for the synthesis of bisphenol A (BPA), a molecule of extreme importance, in particular, for the stabilization of many polymers, although it has been recently questioned for reasons of public health (Fig. 2.1).

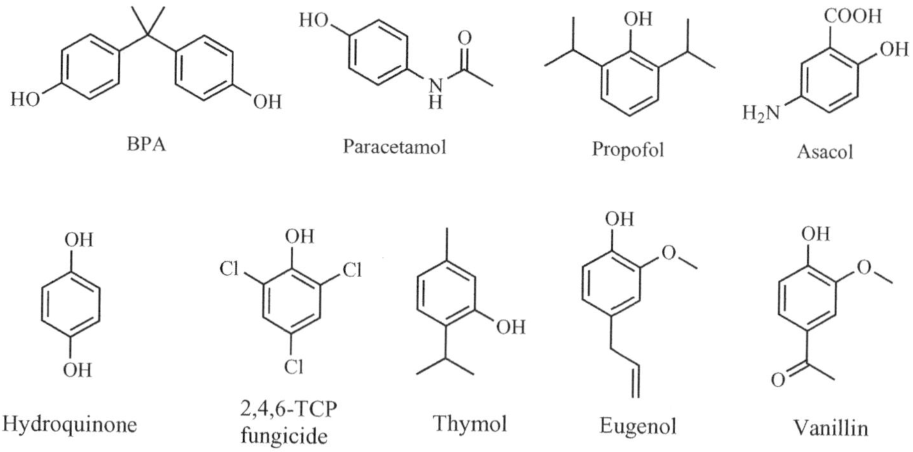

Figure 2.1 Representative phenol-derived drugs, agrochemicals, and perfumery.

Phenol derivatives, which are produced for the most part on incomparably smaller scales, possess much higher added value than phenol itself. These compounds are important intermediates and targets in the chemical, pharmaceutical, and materials industries.[6] Notable examples shown in Figure 2.1 include paracetamol, one of the world's best-selling drugs, propofol (an anesthetic), asacol, also known as mesalazine (an antiinflammatory drug used to treat inflammation of the digestive tract), and hydroquinone, a most useful reducing agent. Phenols have also been successfully exploited in agrochemicals, with, for example, the fungicide 2,4,6-TCP, and are also of great interest for perfumery and cosmetics (thymol, eugenol, and vanillin, among others) (Fig. 2.1).

In general, phenol derivatives are obtained either by nucleophilic aromatic substitution of activated aryl halides, by reactions with diazoarenes (via Sandmeyer-type reactions), or protocols involving benzyne intermediates.[7] These different routes are limited procedures for the synthesis of phenols and derivatives since even if a good selectivity is obtained, harsh reaction conditions are usually required. More recently, phenols and derivatives have been accessed by new transition-metal-catalyzed systems involving efficient and selective arylations of oxygen sources with palladium catalysts and sophisticated monophosphine-type ligands.[8] However, because of the cost and toxicity of the catalytic systems used, the development of efficient alternatives, which are at the same time general, cost-effective, environmentally friendly, and display low toxicity, therefore represents an important challenge for synthetic chemists and industry. In this field, advances in copper-catalyzed direct synthesis of phenol derivatives allowing the use of catalytic amounts of copper salts along with oxygen sources under mild temperature conditions (in contrast to the conditions used in the traditional Ullmann reaction) recently appeared in the literature. This emerging area will be the focus of this subchapter.

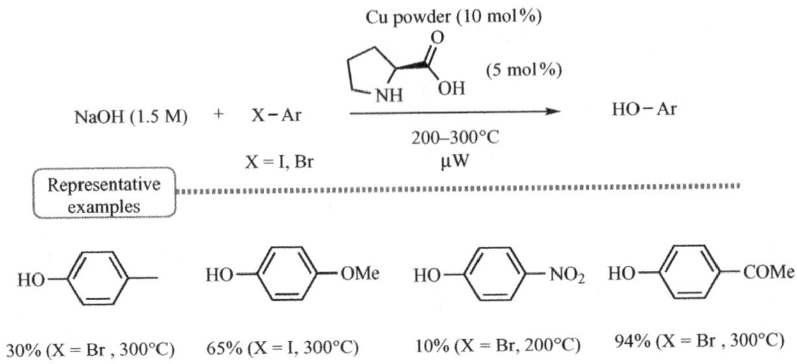

Scheme 2.1 Cu powder/L-proline hydroxylation of aryl halides under microwave heating developed by Leadbeater.

Though outside of the scope of this chapter because of the harsh experimental conditions used, the original work of Leadbeater describing the hydroxylation of aryl halides using catalytic quantities of copper powder (10 mol %), 5 mol % of L-proline, and a 1.5 M solution of sodium hydroxide in water deserves to be reported (Scheme 2.1).[9] The reaction was performed under microwave heating in heavy-walled quartz reaction vessels at very high temperatures (200°C) or in near critical water (80 bar, 300°C). Except in some particular cases, the conversion of aryl iodides and bromides to the corresponding phenols was observed with low-to-moderate yields. Thus, the need for high temperatures and the narrow scope of the reaction limit the interest of the method.

In 2009, the groups of Taillefer[10] and You[11] independently reported two new efficient, softer, and general methods for the direct hydroxylation of aryl halides.

These two methods involve the use of hydroxide salts (potassium or cesium hydroxide) in the presence of 10 mol % of copper(I) iodide and either 20 mol % of 1,10-phenanthroline[11] or 50 mol % of dibenzoylmethane[10] as ligands. The same ratio of DMSO/H_2O (1/1) was used at 110–130°C and the phenols were obtained in good-to-excellent yields from aryl iodides (substituted with both electron-withdrawing and electron-donating groups) and aryl bromides (substituted with electron-withdrawing groups) (Scheme 2.2, Eq. 1 and Eq. 2). Taillefer and coworkers demonstrated that the use of 50 mol % of *N,N′*-dimethylethylenediamine in the presence of sodium iodide (2 equiv) and an equimolar ratio of 1,4-dioxane/H_2O allows the extension of the reaction to all aryl bromides, including nonactivated ones (Scheme 2.2, Eq. 3). The reaction was, in addition, found to be very selective, as no trace of diaryl ethers as potential by-products has ever been observed.

Several other methods have emerged since these first discoveries. They will be overviewed in this chapter on the basis of their catalytic efficiency and the nature of the oxygen source used to obtain the corresponding phenols.

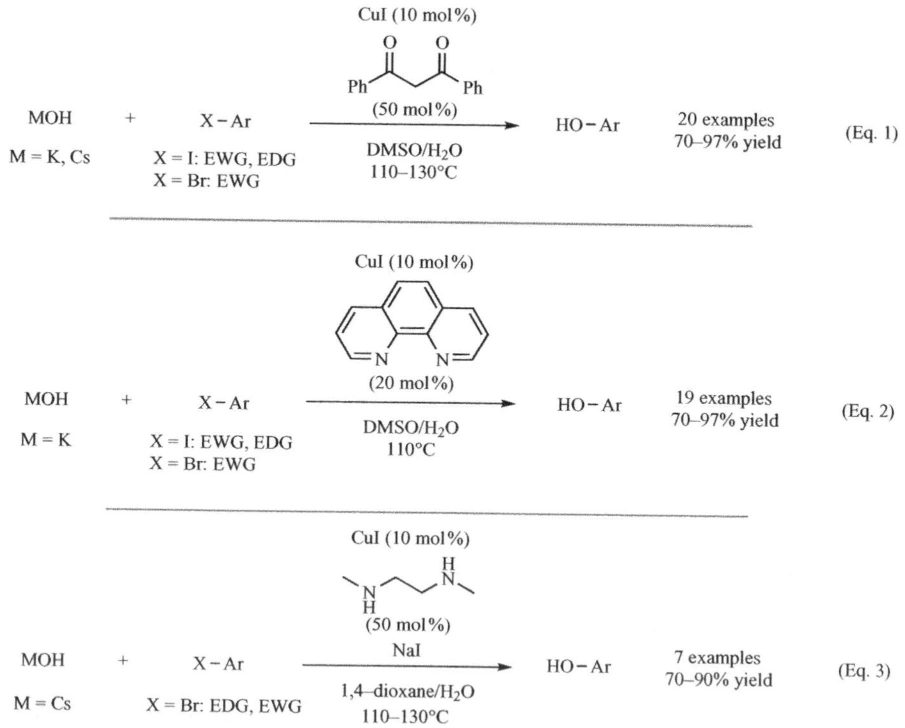

Scheme 2.2 Copper(I) iodide/ligand-catalyzed hydroxylation of aryl halides with hydroxide salts developed independently by You and Taillefer. Note that for all schemes in this chapter, EWG and EDG correspond to the substituents (electron-withdrawing and electron-donating groups) of the aromatic ring denoted “Ar.”

Tao and coworkers[12] and Mehmood and Leadbeater[13] have developed alternative processes for the synthesis of phenols by using different oxygen sources, 4-methoxybenzyl alcohol (the phenol is obtained after deprotection with TFA of the corresponding benzylic ether intermediate) and water, respectively, to avoid the use of strongly nucleophilic bases (MOH) which could lead to the formation of highly nucleophilic phenoxide anions (Scheme 2.3, Eq. 1 and Eq. 2). Despite the efficiency of these methods which use catalytic amounts of copper(I) iodide (5–10 mol %), the use of harsh reaction conditions (180°C, Scheme 2.3, Eq. 2) and the important loss in molecular weight (120 $g.mol^{-1}$ for each phenol; Scheme 2.3, Eq. 1) limit their interest. In both cases, some particular examples involving activated aryl chlorides are reported, but only low-to-fair yields have been obtained with these substrates.

The formation of phenol has also been undertaken by Yang and Fu[14] and Zhou and coworkers.[15] The reactions were performed in water in the presence of hydroxide salts (cesium or sodium hydroxide) and a phase transfer catalyst (20 mol %), TBAB ($n Bu_4 NBr$) or TBAF ($n Bu_4 NF$). The use of copper(I) oxide (5 mol %) along with pyridine-2-aldoxime (20 mol %) was efficient for

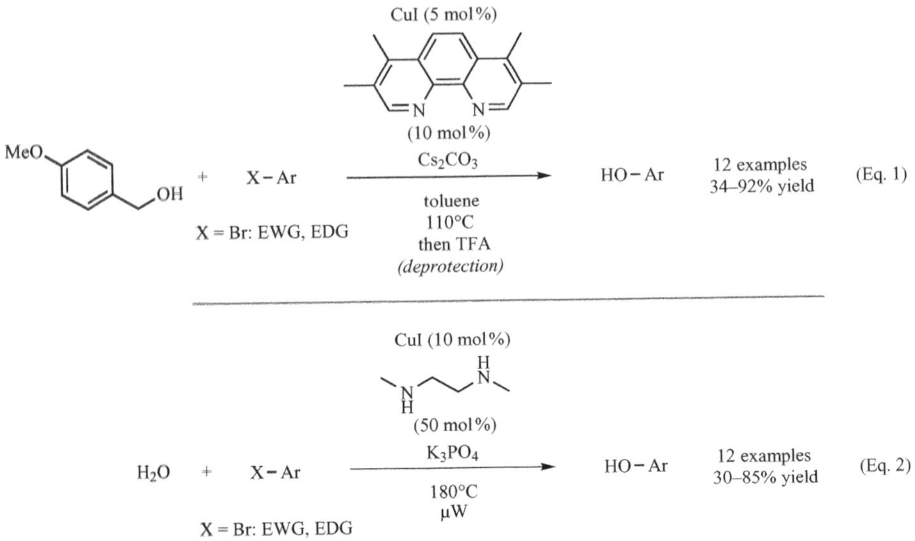

Scheme 2.3 Copper(I) iodide/ligand-catalyzed hydroxylation of aryl halides with different oxygen sources developed by Tao and Leadbeater.

the hydroxylation of aryl iodides substituted by electron-withdrawing and electron-donating groups. For aryl bromides, only those substituted by electron-withdrawing groups are reactive under these conditions at 110°C (Scheme 2.4, Eq. 1). The second system based on the use of copper(I) iodide (10 mol %) in conjunction with lithium pipecolinate (20 mol %) demonstrated a good reactivity for aryl iodides and bromides substituted either by electron-withdrawing and electron-donating groups, as good-to-excellent yields were obtained at 130°C (Scheme 2.4, Eq. 2). Finally, two particular examples of aryl chlorides activated by a carboxylic acid (in *ortho*-position – 66% yield) or strongly activated by two nitro groups were reported by the authors.

In addition, the arylation of potassium hydroxide was also studied by Ma and coworkers[16a] and Thakur and Sekar.[17] Good-to-excellent yields were obtained from aryl iodides when 8-hydroxyquinoline (20 mol %) or D-glucose (5 mol %) were used as ligands in association with copper(I) iodide (10 mol %) or copper(II) acetate (5 mol %), respectively. A three-solvent mixture (^{t}BuOH/DMSO/H_2O) is essential for high conversion rates at 100°C in the system developed by Ma, while heating at a more elevated temperature of 120°C in a DMSO/H_2O solvent mixture is required in the system of Sekar and coworkers (Scheme 2.5). These systems were also efficient for the hydroxylation of activated aryl bromides, and the system developed by Ma was recently used to develop a new synthesis of roflumilast, a selective phosphodiesterase type 4 inhibitor developed for the treatment of chronic obstructive pulmonary diseases launched as Daxas® on the European market in 2010, based on a copper-catalyzed hydroxylation as the key step.[16b]

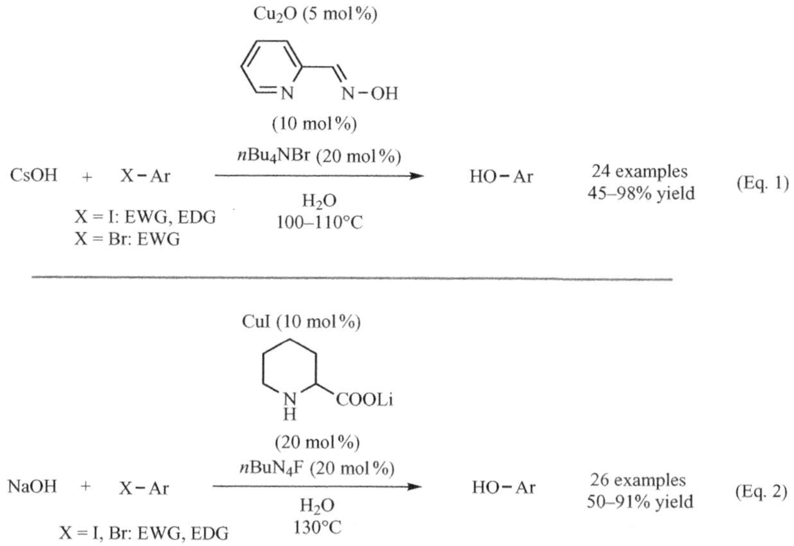

Scheme 2.4 Copper(I)/ligand-catalyzed hydroxylation of aryl halides with hydroxide bases developed by Fu and Zhou.

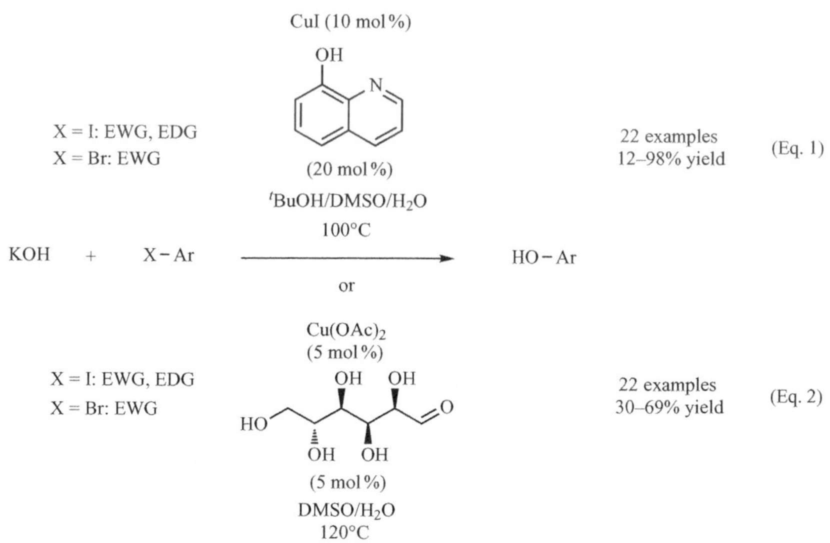

Scheme 2.5 Copper(I) or copper(II)/ligand-catalyzed hydroxylation of aryl halides with potassium hydroxide developed by Ma and Sekar.

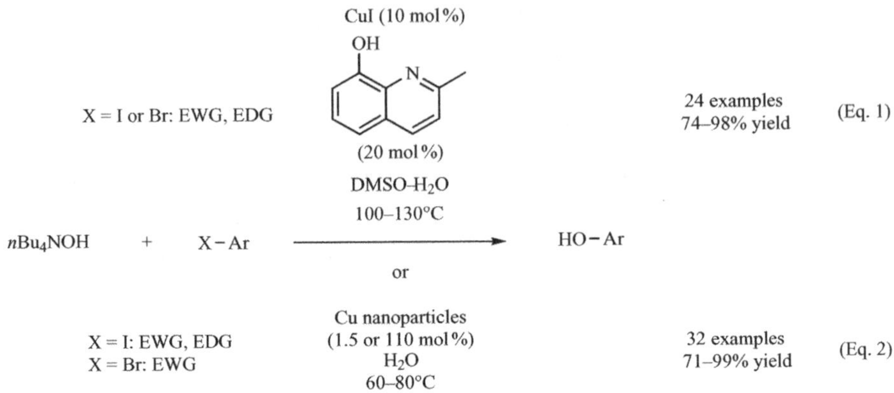

Scheme 2.6 Copper-catalyzed hydroxylation of aryl halides with tetrabutylammonium hydroxide developed by Punniyamurthy and Xu and Feng.

Another hydroxide source, tetrabutylammonium hydroxide (nBu$_4$NOH), was employed by the groups of Punniyamurthy (Scheme 2.6, Eq. 1)[18] and Xu and Feng (Scheme 2.6, Eq. 2).[19] When copper(I) iodide (10 mol %) and 8-hydroxyquinalidine (20 mol %) were employed, very good yields were obtained from aryl iodides and bromides at 100–130°C in a 2:3 mixture of DMSO/H_2O. Nanoparticles of copper(I) iodide, which combine features of heterogeneous and homogeneous catalyses, were also shown to be effective in the absence of additional ligands, as hydroxylated arenes can be obtained in good yields at lower temperatures for aryl iodides (60°C) and bromides (80°C).

Finally, Jiang and coworkers[20] developed more recently a new ligand, 8-hydroxyquinoline-*N*-oxide, that could promote a similar reaction, when 40 mol % was used, in combination with copper(I) iodide (10 mol %) and cesium hydroxide (3 equiv) in a 1:1 mixture of DMSO and water. This method provides an access to the direct hydroxylation of aryl iodides and bromides at 100–110°C (Scheme 2.7). From activated aryl chlorides, good yields are obtained at a temperature between 130 and 150°C (with one example from the deactivated *m*-chloroanisole). Under the conditions used, the strict necessity of the copper system does not, however, seem to have been checked, and a traditional S_NAr reaction could also account for the results obtained.

In summary, the recent results presented in this section show the great potential of copper catalysis for the synthesis of phenol and its derivatives. The use of copper-based catalysts, efficient with aryl iodides or bromides, is thus an interesting alternative to palladium-based catalytic systems. Generalizing this reaction to aryl chlorides remains an important challenge since only activated ones have been used in some rare examples for which an aromatic nucleophilic substitution mechanism cannot be excluded.

Besides significant advances in the copper-catalyzed hydroxylation of aryl halides, major breakthroughs have also been recently reported in the copper-mediated cross coupling involving aliphatic alcohols. They will be overviewed in the next section.

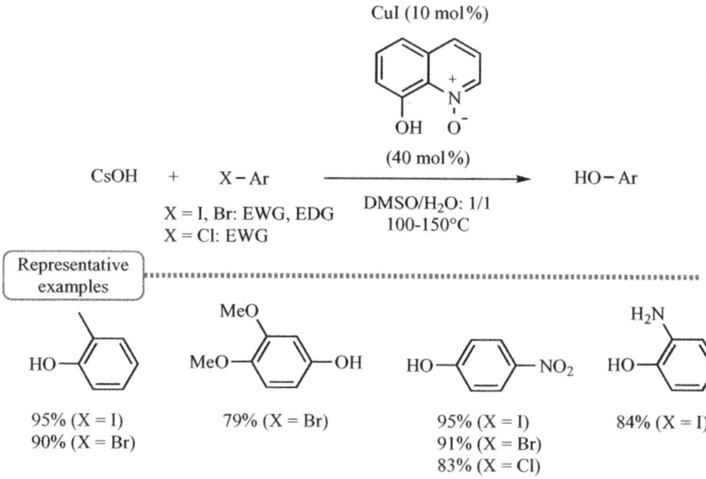

Scheme 2.7 Copper(I) iodide/8-hydroxyquinoline-*N*-oxide-catalyzed hydroxylation of aryl halides with cesium hydroxide developed by Jiang.

2.2.2 Coupling of Aryl Halides with Aliphatic Alcohols: Synthesis of Alkyl Aryl Ethers

Alkyl aryl ethers constitute important building blocks for the synthesis of medicinal compounds and several of them are important pharmaceuticals. Notable examples comprise butoxycain (vasoprotective—topical agent for the treatment of hemorrhoids), bufexamac (which displays antiinflammatory properties and is also an analgesic and an antipyretic), octabenzone (used as ultraviolet filter), bunamide (an antiarrhythmic), pramoxine (a topical anesthetic used as an antipruritic), and, finally, cinchocaine (a compound used as a local anesthetic) (Fig. 2.2). In addition, the aryl alkyl ether moiety is found in an impressive number of natural products.

The reaction of aliphatic alcohols with aryl halides mediated by copper (Ullmann condensation) is clearly a classical transformation that has been known for a long time for the synthesis of alkyl aryl ethers.[21] Despite the fact

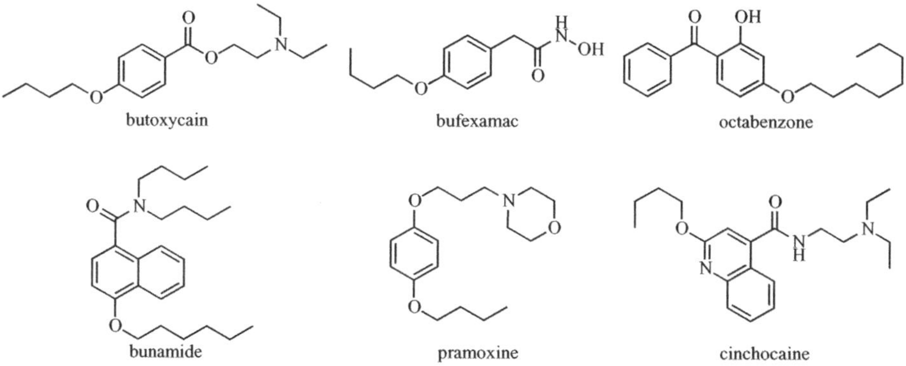

Figure 2.2 Representative biologically important alkyl aryl ethers.

that catalytic amounts of copper have been used for several years, the classical Ullmann condensation suffers from harsh reaction conditions such as, for example, the use of strong bases (such as alkoxides or sodium hydride), large amounts of alkoxides, or high reaction temperatures.[22]

In 2002, Buchwald and coworkers[23] developed a remarkably efficient system for the arylation of aliphatic alcohols using cesium carbonate as the base—a key parameter in this transformation—instead of the stronger alkoxide bases classically used. In the presence of copper(I) iodide (10 mol %), 1,10-phenanthroline (20 mol %), and cesium carbonate, the reaction of various alcohols, used as solvents, with aryl iodides (often substituted by donating groups, NH_2 or OH) proceeded to give the corresponding alkyl aryl ethers in good-to-excellent yields (Scheme 2.8). In some cases, and particularly for more precious alcohols, toluene was used as solvent in the presence of the corresponding alcohols, and the cross coupling of an unactivated aryl iodide with an enantiomerically pure benzylic alcohol led to the formation of the corresponding ether with complete retention of configuration.

Several months later, the same group[24] developed a selective system for the O-arylation of β-amino alcohols with aryl iodides catalyzed by copper(I) iodide (5 mol %) under ligandless conditions as shown in Scheme 2.9 (in fact, the β-amino alcohol could act as a supporting ligand). Cesium carbonate was found to be the most efficient base under these conditions, and the reactions were carried out in butyronitrile at 125°C. By an elegant tuning of the conditions, the N-arylated β-amino alcohols could be also selectively obtained. It is worth noting that, more recently, the authors improved again the selectivity and the scope of their system in the presence of additional ligands.[25] (See Chapter 1, Scheme 1.7 for details on the N-arylation of amino alcohols

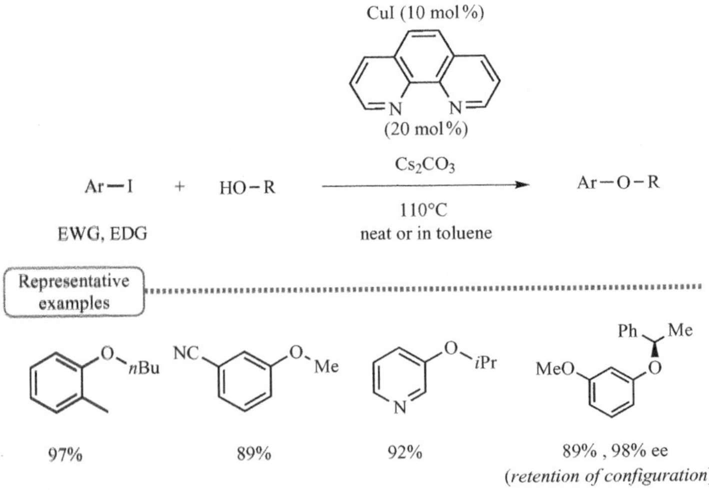

Scheme 2.8 Copper(I) iodide/1,10-phenanthroline-catalyzed reaction of alcohols with aryl halides developed by Buchwald.

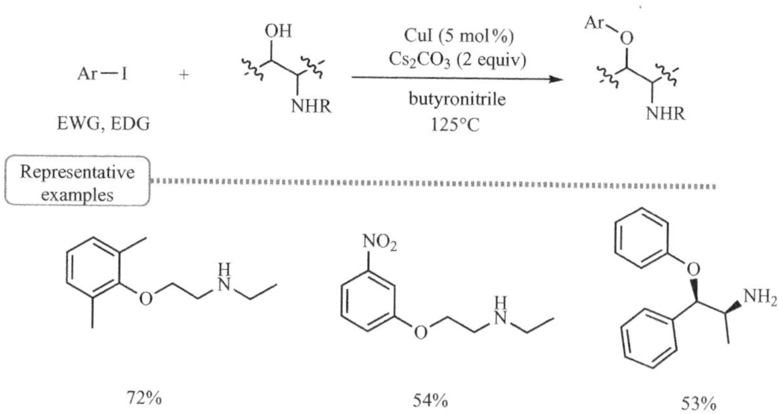

Scheme 2.9 Copper(I) iodide-catalyzed reactions of β-amino alcohols with aryl halides developed by Buchwald.

and Chapter 7, Schemes 7.12 and 7.13 for details on the origin of the chemoselectivity).

Hosseizadech, Tajbakhsh, and coworkers[26] used a copper(I) iodide/1,10-phenanthroline system, in the presence of KF/Al_2O_3 (potassium fluoride supported on alumina) as the base and a large excess of aliphatic alcohols (used as solvents) as nucleophiles, for the cross coupling with aromatic iodides. Good conversions in coupling products were obtained, except when alcohols with a low boiling point were employed.

Stockland and coworkers[27] have developed octanuclear copper clusters as precatalysts for the cross coupling of aryl iodides with aliphatic alcohols in the presence of cesium carbonate at 110°C or under microwave heating (300 W). The reactions could be carried under a stream of nitrogen or in air, slightly lower conversions being observed in this last case.

The usefulness of these reactions is nicely illustrated by the coupling of a highly functionalized cyclic alcohol with an aryl iodide to form a pyrrolidinyl aryl ether, as successfully reported by Evano and coworkers[28] in the first elegant total synthesis of the sedative cyclopeptide alkaloid paliurine F. The method made use of copper(I) iodide (10 mol %), 1,10-phenanthroline (20 mol %), and cesium carbonate, the reaction being carried out in toluene at 125°C (Scheme 2.10). The same authors demonstrated that this copper-catalyzed system was the only one that provided the desired product, as palladium-mediated couplings and nucleophilic aromatic substitution reactions were tested, with no trace of the products being observed in these cases.[29] Subsequent examples of such strategies in natural product synthesis are described in Chapter 17.

Other nitrogen-containing ligands have been applied in the arylation of aliphatic alcohols. Buchwald and coworkers[30] have used copper(I) iodide (5 mol %), 3,4,7,8-tetramethyl-1,10-phenanthroline (10 mol %), and cesium carbonate in toluene at 80–130°C: Using these improved conditions, the

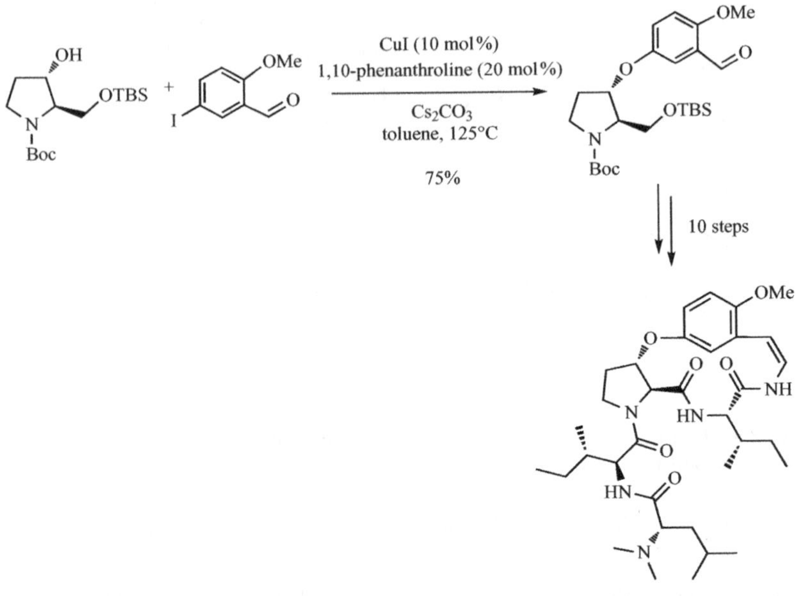

paliurine F

Scheme 2.10 Synthesis of paliurine F via a copper-catalyzed coupling of a functionalized cyclic alcohol with an aryl iodide by Evano.

nucleophilic alcohol partner is no more used in large excess, and this procedure is one of the most efficient ones to date (Scheme 2.11, Eq. 1). Naidu and Sekar[31] have employed (1,1′-binaphthalene)-2,2′-diamine (BINAM, 20 mol %) as the ligand for the coupling reaction between aliphatic alcohols (used here as solvent) in the presence of a quite high amount of copper(I) iodide (20 mol %) as shown in Scheme 2.11 (Eq. 2). In general, good-to-excellent yields were obtained from aryl iodides, but a limited number of examples are given with aryl bromides for these two catalytic systems.

Other types of ligands have been studied in the copper-catalyzed arylation reactions of alcohols with aryl halides. In 2009, Sekar and coworkers[32] demonstrated that diols are also possible ligands for this transformation. Aryl iodides were transformed to the corresponding aryl alkyl ethers in moderate-to-good yields at 82°C in the presence of cesium carbonate and an equal mixture of copper(I) iodide (20 mol %) and the racemic, anthracene-based diol, *trans*-9,10-dihydro-9,10-ethanoanthracene-11,12-dimethanol (20 mol %). As with most methods, aliphatic alcohols were used as solvent (Scheme 2.12, Eq. 1).

Lower catalytic charges of copper(I) iodide (2–5 mol %) and 8-hydroxyquinoline (4–10 mol %) were employed by Hu, Xu, and coworkers in the presence of potassium phosphate as the base.[33] Good-to-excellent yields were obtained using activated or deactivated aryl bromides for the intra- or intermolecular C–O bond-forming reactions (formation of linear or cyclic aryl ethers: Scheme 2.12, Eq. 2). Note that the alcohol was also the reaction solvent.

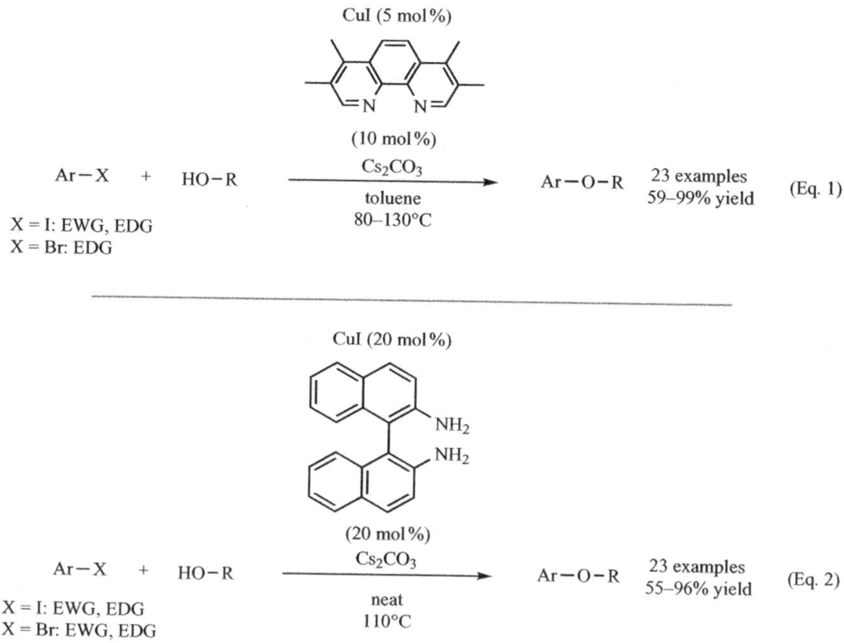

Scheme 2.11 Copper(I) iodide-catalyzed reactions of aliphatic alcohols with aryl halides developed by Buchwald and Sekar.

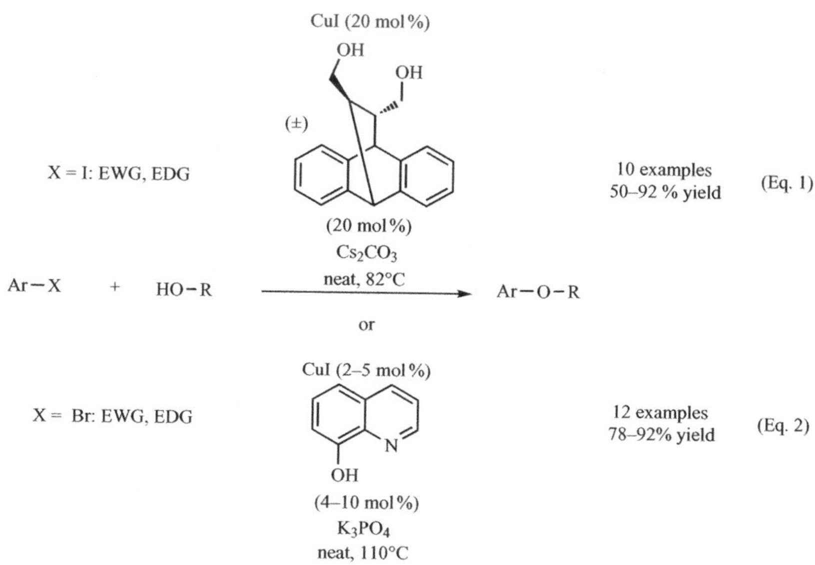

Scheme 2.12 Copper(I) iodide-catalyzed reactions of aliphatic alcohols with aryl halides developed by Sekar and Hu and Xu.

Two reports have appeared in 2011 for the development of "ligand-free" methods.[34,35] These systems, however, require the use of strong bases (LiO*t*Bu or NaO*t*Bu) for the *in situ* preparation of alkoxides, which strongly limit their potential use.[22]

Finally, a rare example of the use of a heterogeneous copper-based system for the arylation of aliphatic alcohols has been described by Nageswar and coworkers.[36y] The catalyst, based on alumina-supported copper(II) oxide (13%), allows the coupling between aliphatic alcohols and iodobenzene in good yields in the presence of potassium hydroxide in DMSO at 100°C.

As seen with examples described in this section, tremendous progress has been made in the development of efficient copper-based catalytic systems for the arylation of alcohols. These systems are among the most efficient to date to perform the cross coupling between alcohols and aryl halides, even if the Chan–Lam–Evans reaction, and related reactions overviewed in Chapter 4, might still represent a useful alternative that one should consider. These systems are, however, far from ideal, and advances are still needed for the development of improved catalysts that would allow performing the reaction under milder conditions, reducing the catalyst loading, or using readily available and cheap aryl chlorides as reaction partners. Nevertheless, the formation of aryl ether via a copper-catalyzed transformation complements well the traditional Williamson reaction or the classical nucleophilic aromatic substitutions of activated aryl chlorides and fluorides.

Remarkable advances have also been made with another class of O-nucleophiles, phenols: They will be overviewed in the next section.

2.2.3 Coupling Reactions of Aryl Halides with Phenols: Synthesis of Biaryl Ethers

The diaryl ether linkage constitutes an important structural motif that can be found in a variety of agrochemicals, as well as in important molecules from the pharmaceutical industry and polymer sciences (Fig. 2.3).

The synthesis of diaryl ethers derivatives has been known for more than a century since the discovery of the Ullmann condensation.[37] As a consequence of the recent renaissance of copper catalysis, an impressive number of publications have appeared on this type of reaction[38]: We have selected those that result in real improvement in terms of overall efficiency (the presentation is based on the nature of the catalytic systems [Cu]/ligand). Because of this selection, many other systems, though they represent interesting contributions, cannot be included in this section.[36]

Buchwald and coworkers reported in 1997[3e] that the arylation of phenols with aryl iodides and bromides could be performed using a copper(I) triflate-based catalytic system, in conjunction with cesium carbonate as a base, the presence of 5 mol % of ethyl acetate rendering the system most effective. The use of an additional stoichiometric amount of 1-naphthoic acid is recommended in the reactions of unactivated aryl halides with phenols substituted by electron-withdrawing substituents and with less soluble phenols as well (Scheme 2.13).

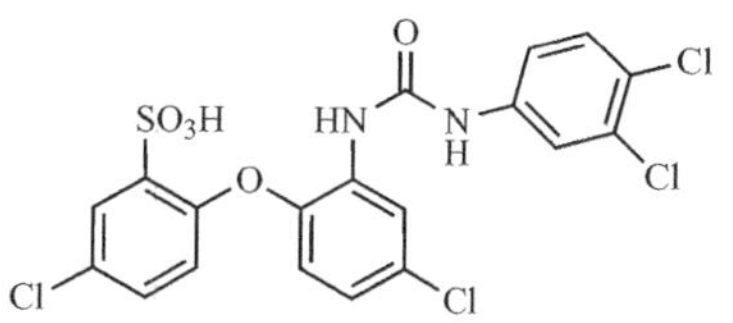

sulcofuron
insecticide, Syngenta

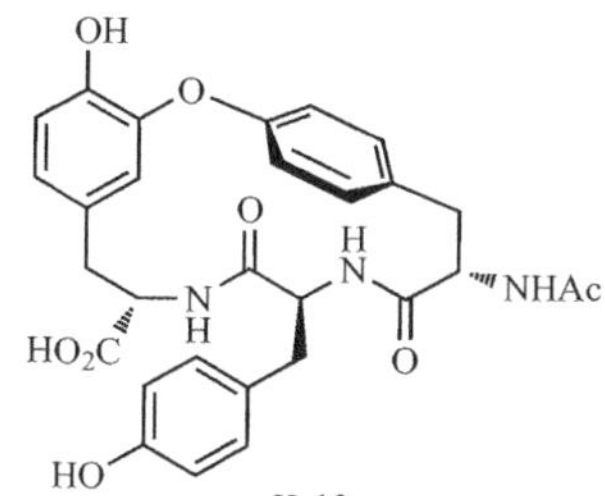

K-13
natural inhibitor of the angiotensin
converting enzyme (ACE)

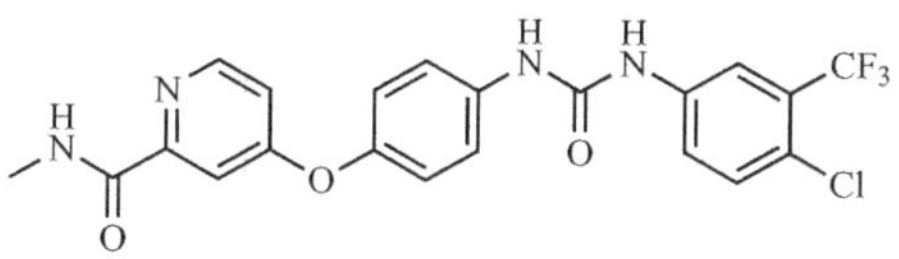

kinase insert domain receptor
(KDR) inhibitor

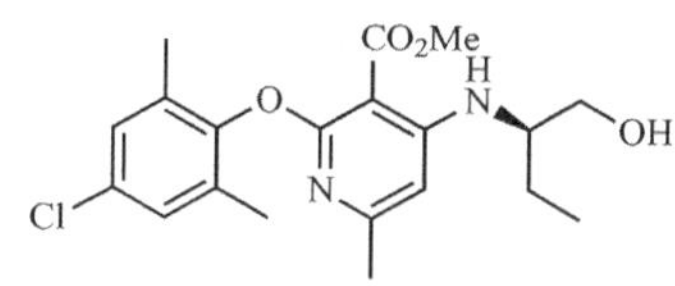

corticotropin-releasing
factor (CRF) antagonist

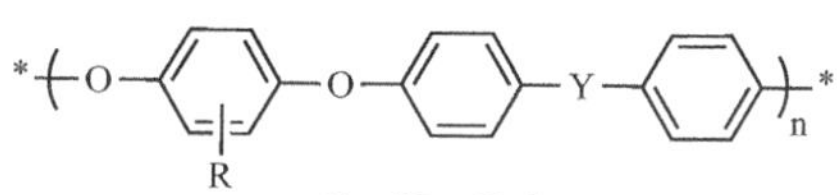

$Y = SO_2$, C=O

thermoplastic polymers

Figure 2.3 Representative drugs, agrochemicals, and materials featuring a diaryl ether moiety.

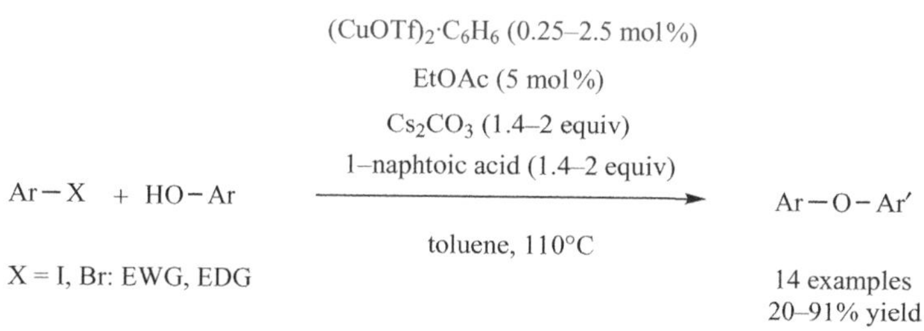

Scheme 2.13 Copper(I) triflate-catalyzed formation of diaryl ethers in the presence of additives developed by Buchwald.

In 1998, Palomo and coworkers[3g] also demonstrated that the copper-mediated (20–200 mol % of copper(II) bromide) arylation of phenols with aryl iodides and activated aryl bromides can take place in the presence of a strong Schwesinger phosphazene base, P_4-^{t}Bu, in refluxing toluene or 1,4-dioxane. The reaction is not general when the copper salt is employed under catalytic conditions (20 mol %).

Two years later, the synthesis of 2-phenoxy-4,6-dimethylaniline via the coupling of preformed potassium phenoxide with 2-bromo-4,6-dimethylaniline was performed by researchers from the DuPont Company. The desired biaryl ether could be efficiently obtained by using a catalytic amount of copper(I)

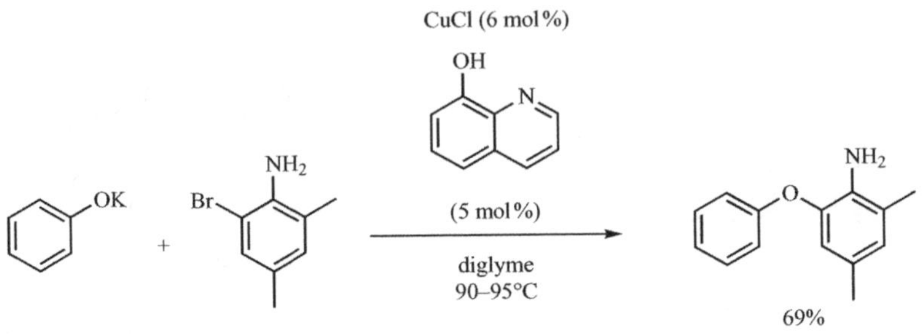

Scheme 2.14 Copper(I) chloride/8-hydroxyquinoleine for the arylation of potassium phenoxide developed by researchers from the DuPont Company.

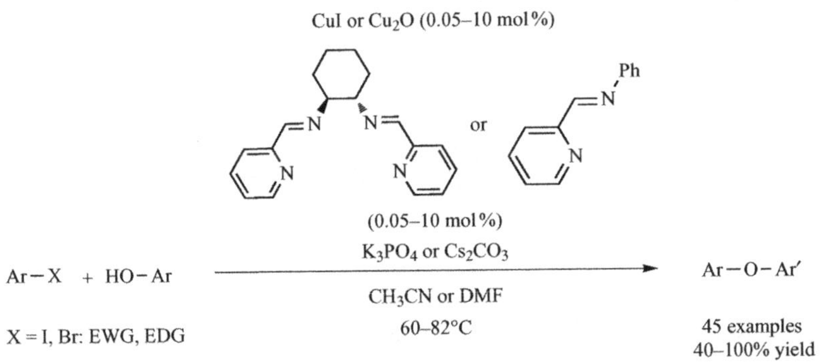

Scheme 2.15 Copper(I)/Schiff base ligands for the arylation of phenols with aryl bromides and iodides.

chloride (6 mol %) and an additional ligand (8-hydroxyquinoleine, 5 mol %) in diglyme at 90°C (Scheme 2.14).[3j]

In 2001, the Taillefer group reported a significant breakthrough with the discovery of versatile and efficient new copper-based catalytic systems for the arylation of various nucleophiles, including phenols (this system is also efficient for C–N and C–C bond-forming reactions).[4] The system requires the presence of Schiff base ligands combined with a catalytic amount of copper(I) iodide or copper(I) oxide, and cesium carbonate or potassium phosphate as bases (the latter, inexpensive, is more interesting from an industrial point of view). The method, general for substrates possessing both electron-withdrawing and electron-donating groups, enabled the cross coupling of a large range of aryl bromides or iodides with phenols under mild conditions (60–82°C). As demonstrated with couple examples, the reaction can be performed using very low metal and ligand loadings (Scheme 2.15).[4a,39]

Venkataraman and coworkers[40,41] reported in 2001 a protocol for the arylation of phenols based on the use of well-defined complexes $Cu(PPh_3)_3Br$ or $Cu(neocuproine)(PPh_3)Br$ (more soluble in organic solvents), respectively,

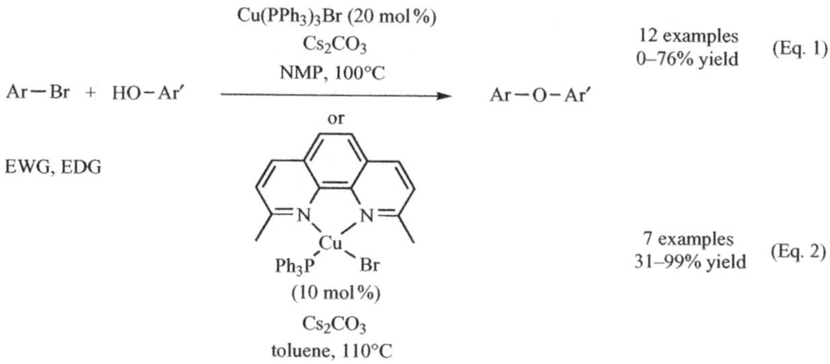

Scheme 2.16 Well-defined copper(I)/ligand complexes developed by Venkataraman for the synthesis of biaryl ethers.

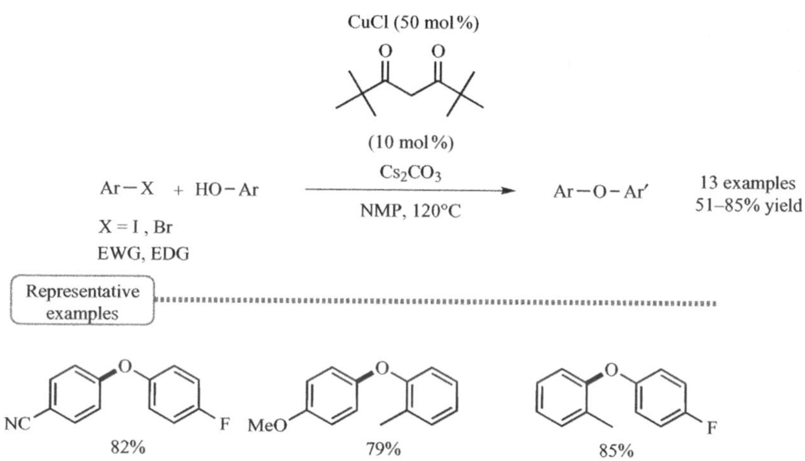

Scheme 2.17 Copper(I) chloride/TMHD in the arylation of phenols developed by Merck researchers (C–O bond formed shown in bold).

in *N*-methylpyrrolidinone (NMP) at 100°C and in toluene at 110°C. It is worth noting that when phenol derivatives are substituted with electron-withdrawing groups (*p*-CN, *o*-$COOCH_3$), the coupling products are not observed. The yields are also lower for aryl bromides bearing *ortho*-substituents (Scheme 2.16).

In 2002, researchers at Merck[42] have shown that the introduction of 2,2,6,6-tetramethyl-3,5-heptanedione (TMHD) in the presence of copper(I) chloride (50 mol %) accelerated the cross coupling between phenols and aryl bromides in the presence of cesium carbonate in NMP at 120°C. Note that these conditions can be hardly considered as catalytic (less than two cycles) and the synthesis of diaryl ethers from activated phenols is not possible with this procedure (Scheme 2.17).

Ren and coworkers[43] conducted in 2003 the arylation of phenols with aryl iodides in an ionic liquid medium (1-*n*-butyl-3-methylimidazolium iodide [bmiI]) in the presence of copper iodide and a strong base, sodium *tert*-butoxide (^{t}BuONa). The copper salt, which is introduced in catalytic amounts, is recycled and the products are easily isolated by extraction. The extrapolation of this method on a large scale, however, does not seem to be very competitive because of the cost of the ionic liquid which is used as the solvent.

The arylation of phenols with aryl bromides has also been described by Ma and Cai[44] in the presence of copper(I) iodide (10 mol %) and *N,N*-dimethylglycine (30–100 mol %) as the ligand. Good-to-excellent yields of diaryl ethers were obtained at 90–105°C with cesium carbonate employed as a base in 1,4-dioxane (Scheme 2.18, Eq. 1). The same group, in continuation of their work on the synthesis of naturally occurring peptides possessing a diaryl ether linkage, reported further development of this system, which was found to be remarkably efficient for the cross coupling between L-tyrosine derivatives and aromatic bromides substituted in the *ortho*-position with a chelating amide group NHCOR (Scheme 2.18, Eq. 2).[45] The main advantage

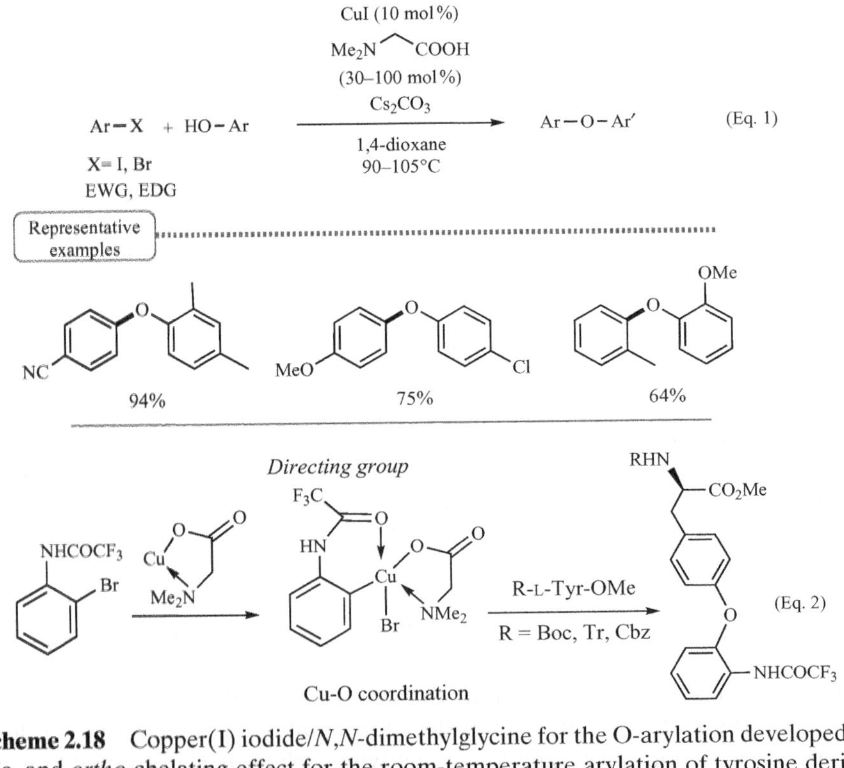

Scheme 2.18 Copper(I) iodide/*N,N*-dimethylglycine for the O-arylation developed by Ma, and *ortho*-chelating effect for the room-temperature arylation of tyrosine derivatives without racemization (C–O bond formed shown in bold).

in this case is that the reactions occur at ambient temperature without racemization of the tyrosine stereocenter while the drawback remains the high *N,N*-dimethylglycine ligand loading (100%). The presence of the *ortho*-amide substituent as a directing group able to stabilize the copper complex intermediate via O-coordination is a key factor for the success of this reaction (Scheme 2.18, Eq. 2).

In 2007, the research groups of Buchwald, Bao, and Taillefer showed that diketones could be efficient ligands for the copper-catalyzed arylation of phenols. Buchwald and Altman have reported the cross coupling of hydroxypyridines and nonconjugated hydroxyquinolines with aryl iodides and bromides in the presence of TMHD in 1,4-dioxane at 80–120°C (Scheme 2.19, Eq. 1).[46] Lv and Bao have reported the cross coupling of phenols with aryl iodides and bromides (bearing electron-donating substituents) in the presence of a β-ketoester (ethyl 2-oxocyclohexanecarboxylate) as the supporting ligand, yielding the corresponding diaryl ethers at mild temperatures (60–80°C) (Scheme 2.19, Eq. 2).[47]

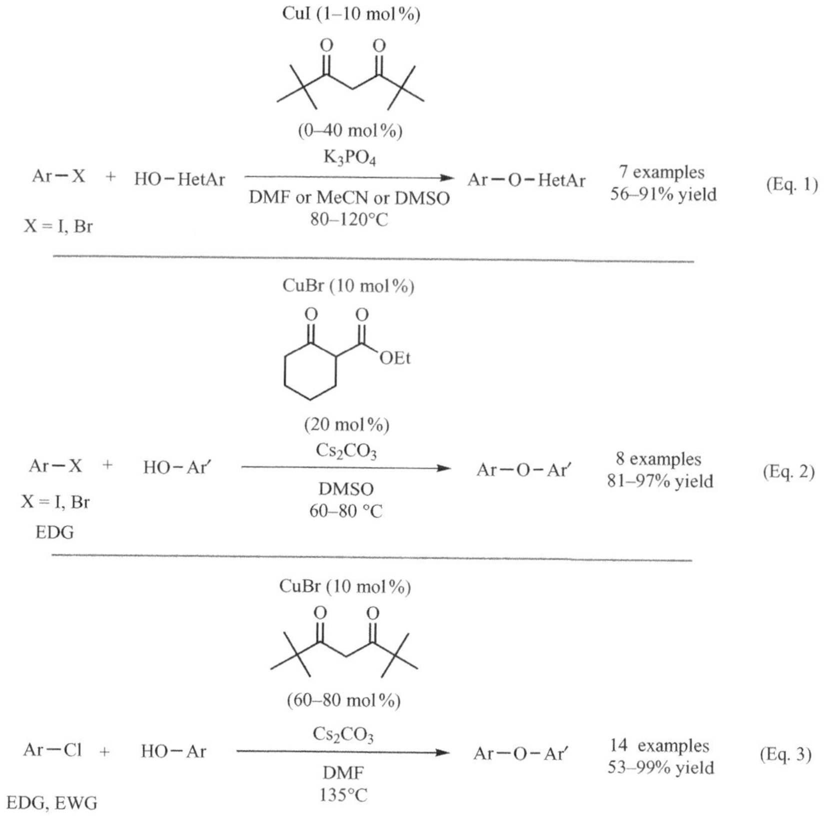

Scheme 2.19 Copper(I)/diketone-based catalytic systems developed by Buchwald, Bao, and Taillefer.

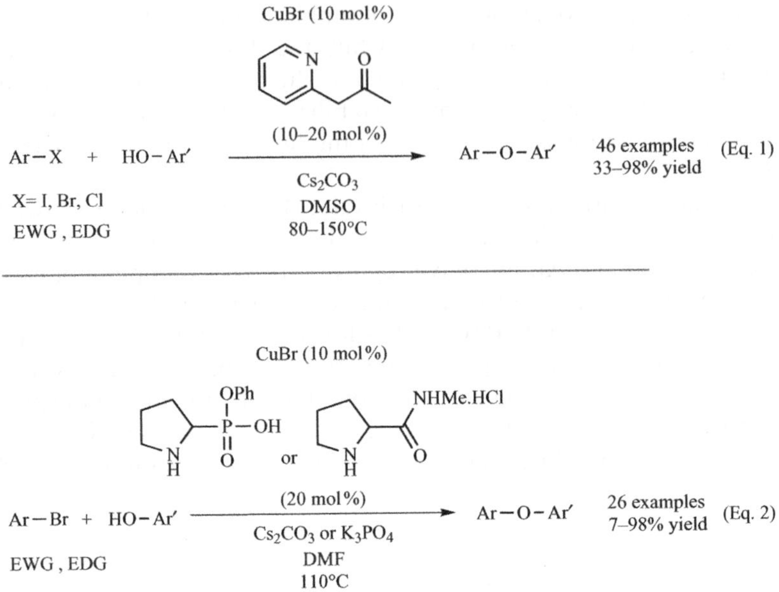

Scheme 2.20 Copper(I) bromide/ligand systems developed by Ding and Fu.

The first efficient method promoting the arylation of phenols with inexpensive and challenging activated and deactivated aryl chlorides was developed by Taillefer and coworkers.[48] The key parameter for the success of this reaction was found to be the combination of copper(I) bromide (10 mol %) with TMHD (Scheme 2.19, Eq. 3). Although relatively harsh conditions are used (60–80% of TMHD), this procedure has real economic advantages in that inexpensive materials are used.

Two years later, Ding and coworkers introduced the 2-pyridylacetone ligand (10–20 mol %) associated with copper(I) bromide to promote the O-arylation of aryl chlorides. The reaction was carried out at 120°C for activated ones, while low conversions were observed for nonactivated aryl chlorides (electron-donating group on the aromatic cycle) despite a much higher temperature (150°C). The cross coupling involving aryl bromides and iodides was promoted efficiently at a lower temperature (80°C) (Scheme 2.20, Eq. 1).[49]

Two analogues of L-proline were also employed as ligands in the presence of copper(I) bromide by Fu and coworkers.[50,51] Indeed, pyrrolidine-2-phosphonic acid phenyl monoester (PPAPM) in the presence of cesium carbonate was found to efficiently promote the cross coupling of aryl iodides and bromides in *N,N*-dimethylformamide (DMF) at 110°C (Scheme 2.20, Eq. 2), and the *N*-methylpyrrolidine-2-carboxamide ligand used in combination with the cheaper base potassium phosphate was demonstrated to be slightly more effective (Scheme 2.20, Eq. 2). In both cases, the copper loading was, however, quite high. In another publication, Fu and coworkers reported on the use of

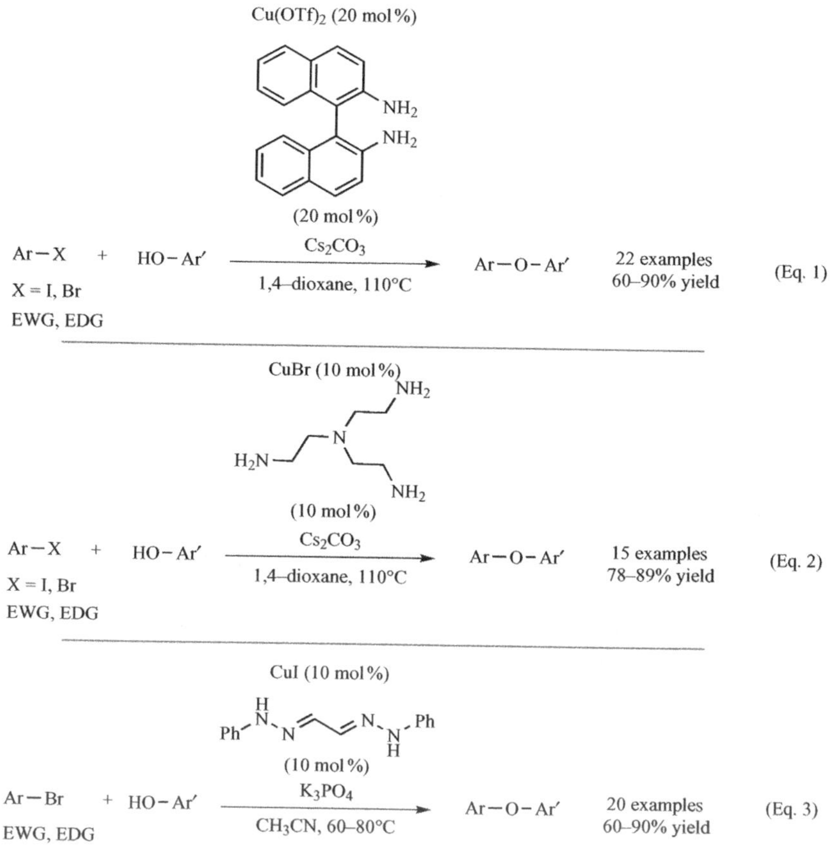

Scheme 2.21 Copper/*N*-ligand systems developed by Sekar, Shingare, and Yang.

dimethylaminomethylphosphonic acid derivatives as new ligands associated with copper(I) iodide: Efficient at 110°C for both aryl bromides and iodides, this method, however, also requires quite high metal (20 mol %) and ligand (30 mol %) loadings.[52]

Other N-ligands have also been reported for the copper-catalyzed arylation of phenols. They include the *N*,*N*′-dimethyl-1,1′-binaphthyl-2,2′-diamine (BINAM, 20 mol %) employed by Sekar (Scheme 2.21, Eq. 1) in the presence of copper(II) triflate (20 mol %),[53] and tris-(2-aminoethyl)amine, a tripodal ligand introduced by Shingare, which was used in combination with copper(I) bromide (Scheme 2.21, Eq. 2).[54] Similar reaction conditions (cesium carbonate as the base, 1,4-dioxane as the solvent, 110°C) allowed the synthesis of diaryl ethers in good-to-excellent yields from aryl iodides and bromides.

Much lower temperatures are required when glyoxalbis(phenylhydrazone) is used as the ligand (10 mol %) in the presence of copper(I) iodide (10 mol %). This system, developed by Yang and coworkers,[55] allows the reaction to be

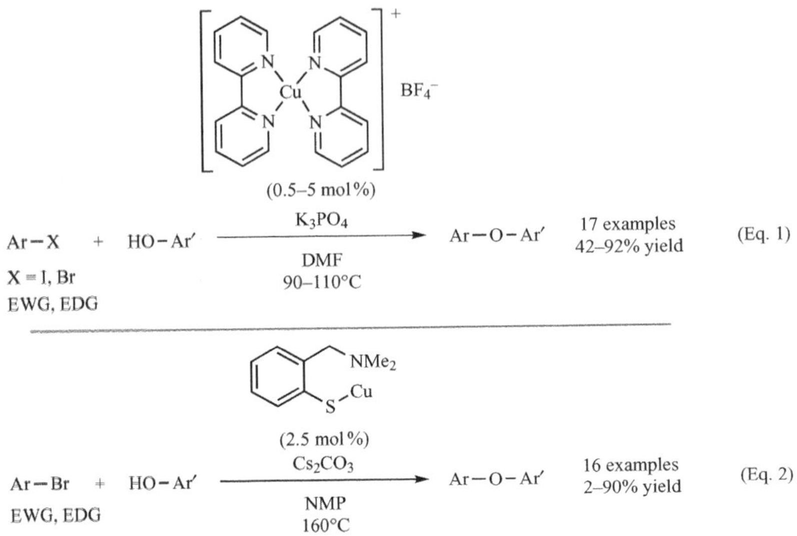

Scheme 2.22 Preformed copper complexes developed by Hu and van Koten.

carried out in acetonitrile at 60–80°C, using the less expensive potassium phosphate as a base (Scheme 2.21, Eq. 3).

Two well-defined copper complexes were also tested for the synthesis of biaryl ethers by arylation of phenol derivatives. The cross coupling of phenols with aryl iodides and bromides was realized in the presence of an air-stable bipyridyl copper(I) complex developed by Hu and coworkers: At low loadings, this catalyst is able to promote the coupling of hindered substrates in DMF at reasonable temperatures (Scheme 2.22, Eq. 1).[56] A much more higher temperature (160°C in NMP) was required when 2-aminomethylarene thiolato-copper(I) complexes, prepared by van Koten and coworkers, were employed in the presence of cesium carbonate as the base for the arylation of phenols (Scheme 2.22, Eq. 2).[57]

Sekar and coworkers showed that a quite sophisticated diol ligand (*trans*-9,10-dihydro-9,10-ethanoanthracene-11,12-dimethanol) could also promote the O-arylation of phenols with aryl iodides and bromides in the presence of copper(I) iodide (10 mol %) and cesium carbonate in refluxing acetonitrile, although the benefits of this complex ligand are not clear (Scheme 2.23).[32]

Buchwald and coworkers developed a new and highly efficient method for the chemoselective O-arylation of aminophenols with aryl iodides and bromides. Excellent yields and high levels of chemoselectivity were observed in the presence of picolinic acid as the ligand (10–20 mol %), copper(I) iodide (5–10 mol %), and potassium phosphate in dimethylsulfoxide (Scheme 2.24, Eq. 1). After this success, the authors reported alternative conditions for the selective arylation of 4-aminophenols based on the use of *trans-N,N′*-dimethylcyclohexane-1,2-diamine (20 mol %), copper(I) iodide (10 mol %), and potassium carbonate as the base in butyronitrile at 70°C (Scheme 2.24, Eq. 3). The resulting O-arylated aminophenols constitute a structural motif found in

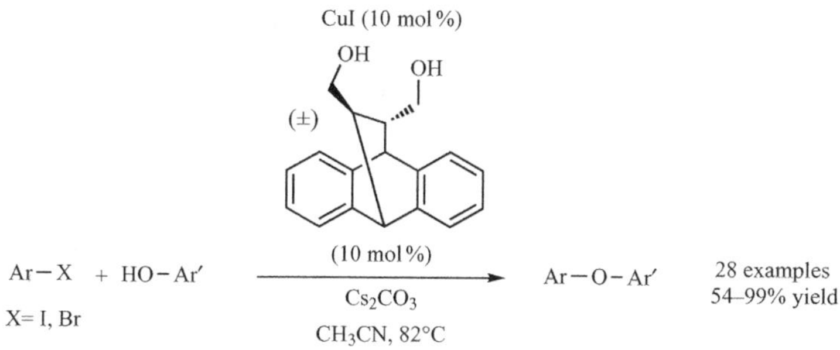

Scheme 2.23 Copper(I) iodide/racemic anthracene-based diol system developed by Sekar.

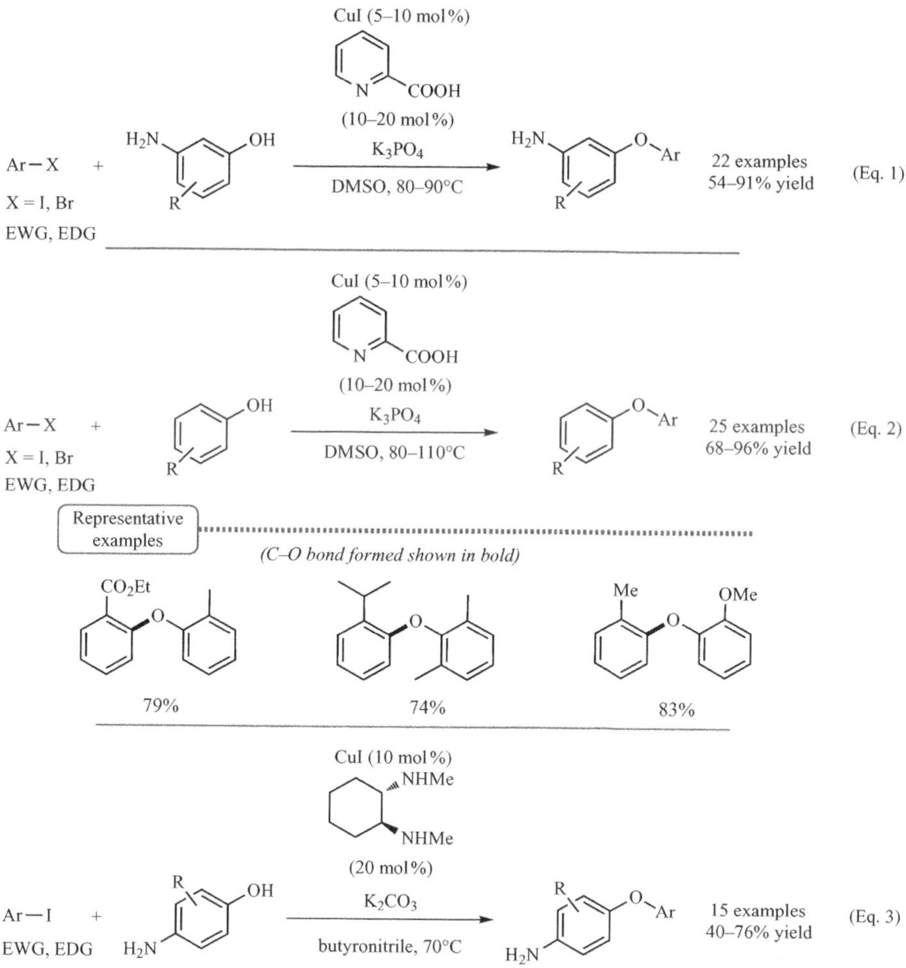

Scheme 2.24 Copper(I) iodide/ligand-catalyzed chemoselective O-arylation of 3-aminophenols and 4-aminophenols, and copper(I) iodide/picolinic acid-catalyzed synthesis of sterically hindered diaryl ethers developed by Buchwald.

numerous therapeutic agents, drug candidates, and in materials, which is also the case for the related N-arylated aminophenols, whose elegant chemoselective synthesis via a C–N bond formation was also described by the authors in the same publication by using an orthogonal palladium-based catalytic system.[58] Buchwald and Maiti have also shown that the catalytic system developed earlier based on picolinic acid as the ligand (Scheme 2.24, Eq. 1) can efficiently promote the cross coupling yielding to sterically hindered and heteroaryl diaryl ethers from the corresponding aryl iodides or bromides (Scheme 2.24, Eq. 2).[59] The yields obtained with sterically hindered substrates are particularly remarkable.

Taillefer, Monnier, and coworkers have recently developed a new strategy that affords a selective one-pot access to symmetrical and unsymmetrical diaryl ethers from aryl halides and water.[60] The former are obtained by a double arylation of water as an oxygen source in the presence of copper(I) iodide (10 mol %) and TMHD (30 mol %), potassium phosphate being the base of choice for this transformation performed with various aryl iodides or bromides in a 1:1 mixture of ethanol and water (Scheme 2.25, Eq. 1). The synthesis of unsymmetrical diaryl ethers was performed with a similar catalytic system in the presence of cesium hydroxide as the oxygen source: The use of cesium carbonate in DMSO with a quite high loading of the TMHD ligand (50 mol %) was found to be the most efficient system for obtaining diaryl ethers in good-to-excellent yields from aryl iodides, bromides, but also from challenging aryl chlorides (Scheme 2.25, Eq. 2).

Besides these systems, some recyclable catalytic systems for the condensation of aryl halides with phenols have been also reported (see Chapter 20 for

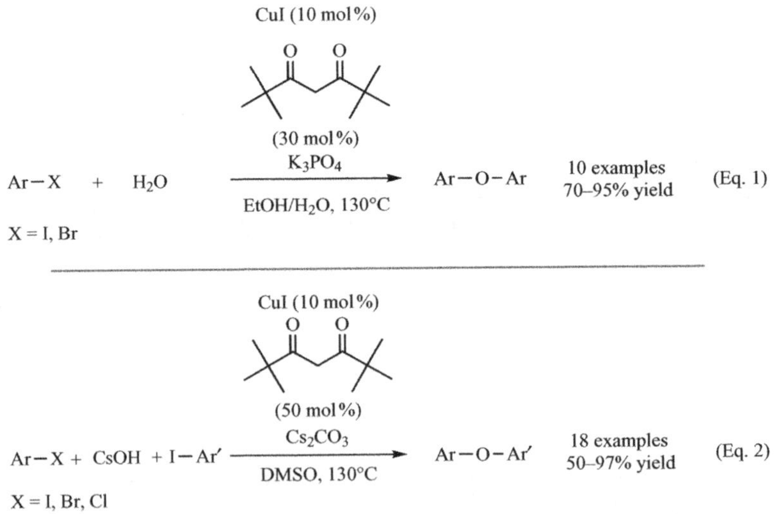

Scheme 2.25 Copper(I) iodide/2,2,6,6-tetramethyl-3,5-heptanedione-catalyzed double arylation of water or cesium hydroxide for the synthesis of symmetrical and unsymmetrical diaryl ethers.

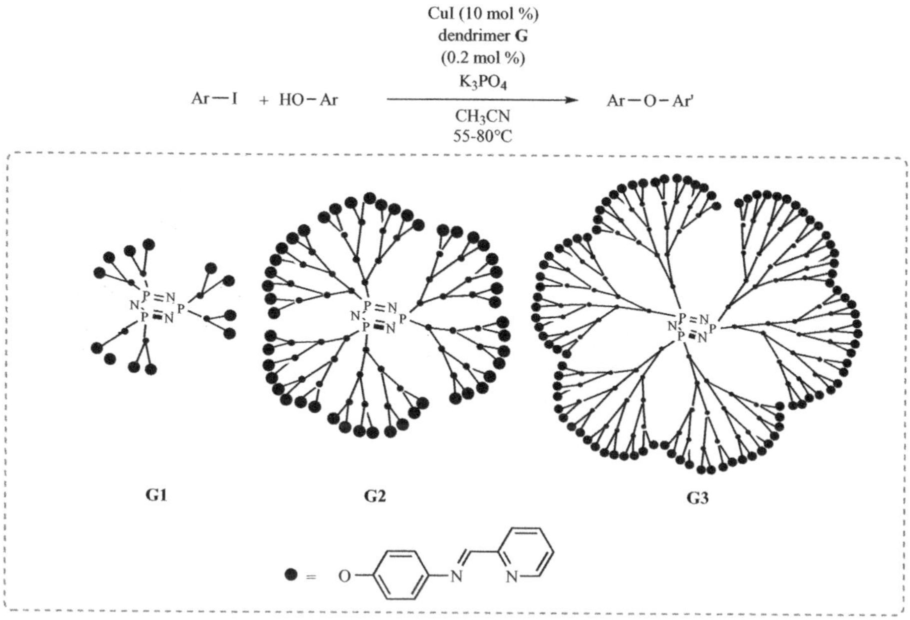

Scheme 2.26 Copper(I) iodide/dendrimer-bound iminopyridine ligand-catalyzed arylation of phenols developed by Taillefer and Majoral.

an extensive coverage of reusable copper catalyts). Among several hundred publications on the catalytic behavior of metallodendrimers, only a few report a strong positive catalytic effect of dendrimers compared to that of monomeric species. One of the rare examples was reported by Taillefer, Majoral, and coworkers, who demonstrated a strong dendritic effect for the copper-catalyzed arylation or vinylation of heteronucleophiles, including phenols. Copper(I) iodide (10 mol %) was used as a precatalyst and optimum catalytic activity was obtained from aryl iodides under mild temperature conditions (55–80°C) when using **G3** (0.2 mol %) dendrimer-bound iminopyridine as the ligand (Scheme 2.26).[61] The stability of the support enables this system to be reused.

In the context of reusable copper-based catalytic systems, Wang and Miao developed a 3-(2-aminoethylamino)propyl-functionalized silica gel-immobilized copper catalyst able to promote the arylation of phenols using all types of aryl iodides as well as aryl bromides and chlorides activated with electron-withdrawing groups as well. The best results were obtained in DMSO at 130°C and potassium fluoride was found to be the most effective base. The silica-supported copper catalyst could be recycled and used for 10 consecutive trials without loss of its catalytic efficiency (Scheme 2.27, Eq. 1).[62]

In 2008, Taillefer, Monnier, and Wong Chi Man synthesized a new hybrid silica with urea-based bypyridyl bridging units by a sol-gel process. The corresponding copper-complexed hybrid silica is an efficient heterogeneous catalyst for the Ullmann coupling (at 110°C in DMF) of phenols with aryl iodides and

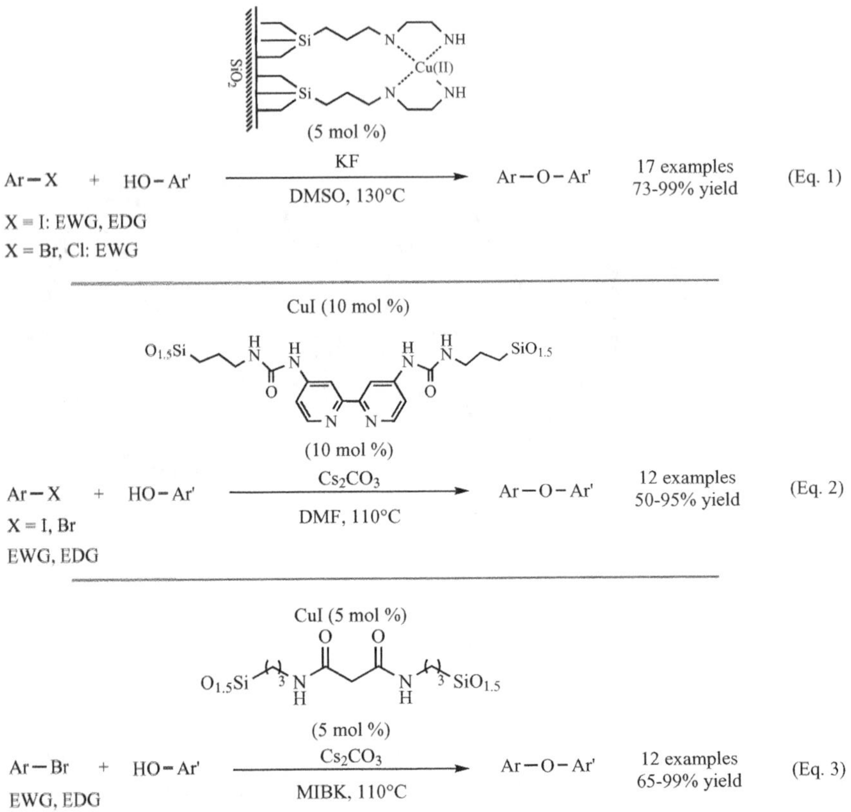

Scheme 2.27 Heterogeneous and reusable copper/ligand systems for the arylation of phenols.

bromides. For the latter, the reaction proceeds in the presence of potassium iodide (50 mol %), which allows for the *in situ* formation of an intermediate aryl iodide by a copper-catalyzed halogen exchange reaction before its subsequent cross coupling with phenols (see Chapter 6 for details on the aromatic Finkelstein reaction). The hybrid is easily recovered from the reaction media by filtration and proved to be a versatile catalyst for subsequent reactions with several substrates, yielding products with the same efficiency even after 10 reuses. This represents a promising route that may contribute to the reduction of waste while maintaining economic viability (Scheme 2.27, Eq. 2).[63]

Couple months later, the same groups developed a simple β-diamide ligand immobilized on hybrid silica for the copper-mediated O-arylation of phenols. This new heterogeneous catalytic system was found to be efficient to promote the cross coupling at relatively low loadings (5 mol % of copper(I) iodide and ligand were used) from aryl iodides and bromides without the need for additives. The reaction was carried out in methyl isobutylketone (MIBK), an industrial eco-friendly solvent with low toxicity. High reactivity was observed after more than five cycles, and negligible copper leaching (20 ppm) occurred after reaction in solution from the supported catalyst (Scheme 2.27, Eq. 3).[64]

As in the case of copper-catalyzed C–N bond formation, some reports appeared in the literature for ligand-free methods. In this context, we have shown in 2003 that the arylation of phenols with aryl iodides was possible in acetonitrile at 82°C by using a copper precursor without additional supporting ligands.[65] Two others examples were also reported, but very high reaction temperatures are needed (150–160°C), given that aryl iodides and bromides are employed.[66,67]

Nanoparticles of copper, which contain high surface areas, can also be involved as catalysts in copper-mediated C–O bond formation processes. In a first report, Kidwai and coworkers reported a mild heterogeneous system based on copper nanoparticles (Cu nano/18 nm, 10%) for the coupling of mainly iodo- and bromobenzene with phenols. The reactions take place at 50–60°C without any addition of ligand and the system is reusable, although a decrease in yield is observed for each subsequent run.[68a] This nanoparticle-based methodology was next further extended by other groups.[68] Interestingly, Sreedhar and coworkers showed that a catalytic amount of CuI nano (1.25 mol %) was able to promote the cross coupling from cheap, nonactivated aryl chlorides, in DMF at 110°C using potassium carbonate as the base.[68c] This was not the case in the work reported by Punniyamurthy and coworkers, who did not observe the arylation of the phenol from chlorobenezene, using CuO nanoparticles (2.5 mol %) at 110°C in DMSO with potassium hydroxide as the base.[68d] Both methods do not require the presence of additional ligands, and because of the advantages inherent to the use of nanoparticles (recyclable catalysts, simplicity of the systems, easy workup), interesting developments in this field should appear in the near future, notably for the copper-catalyzed arylation of phenols and other nucleophiles.

Since the renaissance of copper-catalyzed cross-coupling reactions, less work has been devoted to the development of copper-based systems for the formation of C(aryl)–S bonds than for the formation of related C(aryl)–O or C(aryl)–N bonds. However, the resulting thiophenols and aryl alkyl or diaryl thioethers are also at the core of numerous pharmaceuticals, polymers, and natural products. The next chapter deals with the most recent and significant advances in this field.

2.3 FORMATION OF C–S BONDS VIA COPPER-CATALYZED CROSS-COUPLING REACTIONS WITH ARYL HALIDES

2.3.1 Coupling of Aryl Halides with Sulfur Sources: Synthesis of Thiophenols

Compared to the number of methods developed for the synthesis of phenols by direct copper-mediated hydroxylation of aryl halides, only very few

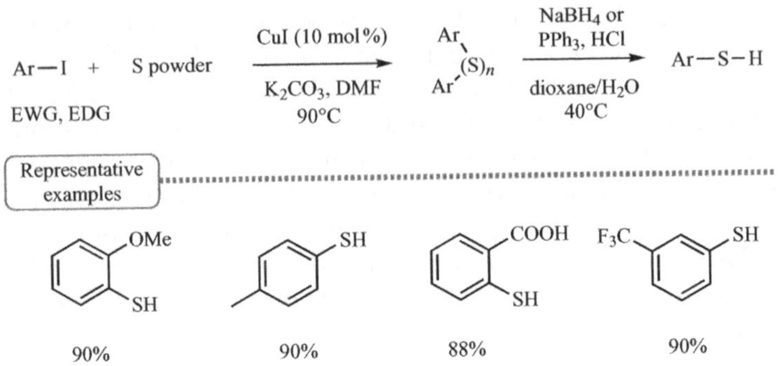

Scheme 2.28 Copper(I) iodide-catalyzed formation of thiophenols from aryl iodides developed by Ma and Jiang.

publications dealing with the related synthesis of thiophenols (or aryl thiols) have appeared in the literature. These compounds are, however, also important building blocks in the synthesis of pharmaceuticals and sulfur-containing natural products.

An interesting system was reported by Ma, Jiang, and coworkers in 2009.[69] Copper(I) iodide (10 mol %) promotes the coupling reaction between aryl iodides bearing electron-withdrawing or electron-donating groups with sulfur powder in the presence of potassium carbonate in DMF at 90°C. Subsequently, the reduction of the formed mixture of diarylpolysulfides is realized either by sodium borohydride or by a combination of triphenylphosphine and hydrochloric acid to give the corresponding thiophenols in good-to-excellent yields. It is worth noting that this catalytic system does not require the presence of additional ligands (Scheme 2.28). The authors also reported that for some aryl iodides bearing a strong electron-withdrawing group (such as the *para*-nitro substituent), the reaction could take place without the use of copper(I) iodide.

In 2010, Junsheng and coworkers developed a simple and efficient alternative procedure based on the arylation of thiourea, a cheap basic feedstock of chemistry, to obtain thiophenols starting from aryl iodides.[70] The system is based on the use of copper(I) iodide (10 mol %) and L-proline as the ligand (20 mol %). A strong base such as sodium *tert*-butoxide was required to obtain good-to-excellent yields and the reactions were carried out in DMSO at 90°C. The formation of thiophenols was explained by hydrolysis of intermediate S-aryl-isothiouronium iodides upon acidic workup (Scheme 2.29).

A copper nanoparticle-catalyzed procedure for the C–S bond-forming reaction to give thiophenols has been developed by Feng, Xu, and coworkers in 2010.[19] A low catalytic charge of CuI nanoparticles (1.5–3.0 mol %) allowed the arylation of sulfur powder in water, and a transfer phase agent (tetrabutylammonium hydroxide) was necessary to promote the reaction. The aryl disulfides obtained were reduced in this case by zinc dust and hydrochloric acid. Aryl iodides substituted with both electron-withdrawing or electron-donating groups gave thiophenols in good-to-excellent yields. In addition, a

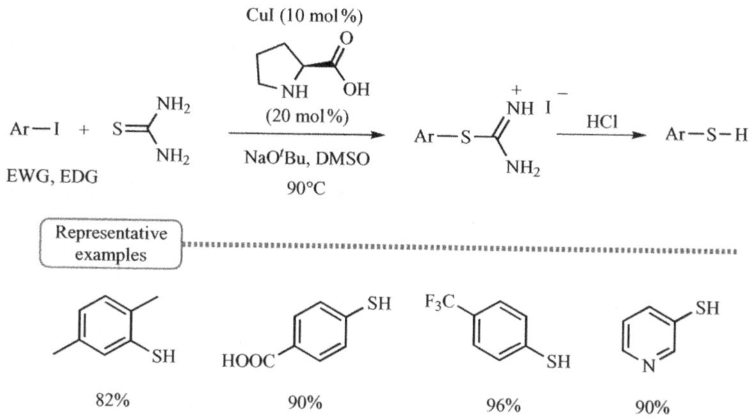

Scheme 2.29 Copper(I) iodide-catalyzed arylation of thiourea developed by Junsheng.

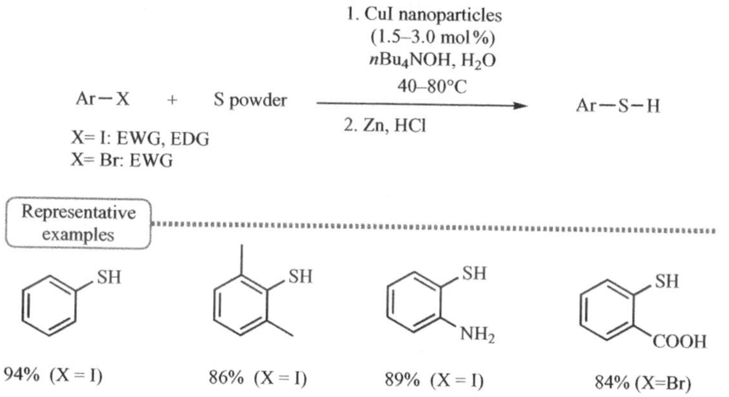

Scheme 2.30 Copper(I) iodide nanoparticle-catalyzed direct arylation of sulfur developed by Feng and Xu.

couple of tests have been realized starting from aryl bromides, and only those substituted by an electron-withdrawing group (a carboxyl substituent in *ortho*- or *para*-position) were reactive (Scheme 2.30).

Sawada and coworkers reported in 2006 a method allowing the coupling of aryl iodides and thiobenzoic acid in the presence of catalytic amounts of copper iodide (10 mol %) and 1,10-phenanthroline (20 mol %). Various thiophenols protected as *S*-aryl thiobenzoates were obtained in excellent yields from both electron-deficient and electron-rich aryl iodides with this procedure.[71] The deprotection, performed in mild conditions, was however illustrated by a single example (from *S*-phenyl thiobenzoate to quantitatively give the corresponding thiophenol).

Besides significant advances in the copper-catalyzed thiolation of aryl halides, other breakthroughs have also been recently reported in the copper-mediated cross coupling involving aliphatic thiols. They will be overviewed in the next section.

2.3.2 Coupling of Aryl Halides with Aliphatic Thiols: Synthesis of Alkyl Aryl Thioethers (or Alkyl Aryl Sulfides)

The importance of organosulfur compounds such as aryl thiols, previously described, or alkyl aryl thioethers (or alkyl aryl sulfides), presented in this section, stems from the prevalence of aryl–sulfur bonds in an impressive number of molecules of importance in pharmaceutical, biological, or material sciences.

The first study of the copper-catalyzed cross coupling of aryl iodides with alkyl thiols was undertaken by Venkataraman and coworkers in 2002.[72] The association of copper(I) iodide (10 mol %) and neocuproine (2,9-dimethyl-1,10-phenanthroline, 10 mol %) in the presence of a strong base, sodium *tert*-butoxide, in refluxing toluene allowed the formation of aryl alkyl sulfides in good-to-excellent yields. Behind that good reactivity, which has also been observed for aryl iodides bearing *ortho*-substituents, the authors showed particularly interesting examples of aryl iodides substituted by donating groups (Scheme 2.31, Eq. 1). In the same period, Kwong and Buchwald[73] developed a selective system for the arylation of aliphatic or branched alkyl thiols with aryl iodides. Relatively milder reaction conditions are used compared to Venkataraman's system since the reaction can be performed with lower catalyst loading (5 mol % of copper(I) iodide), with potassium carbonate as the base and with isopropyl alcohol as the solvent. The main drawbacks of this procedure are the use of overstoichiometric amounts of ethylene glycol as an additive and the use of aryl iodides substituted by electron-donating groups as in the previous procedure (Scheme 2.31, Eq. 2). The same catalytic system is also

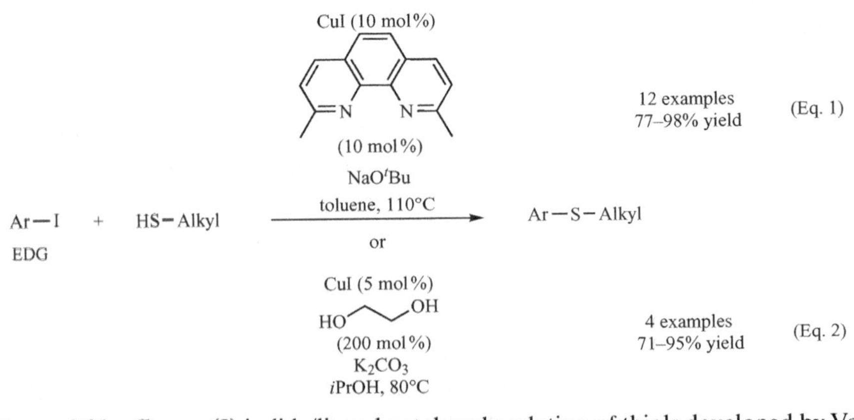

Scheme 2.31 Copper(I) iodide/ligand-catalyzed arylation of thiols developed by Venkataraman and Buchwald.

able to promote the coupling reaction of aryl iodides and thiophenols, which will be detailed in the next section.

Since these early developments by Venkataraman and Buchwald, several other procedures have been developed using homogeneous- or heterogeneous-based copper catalysts. Selected catalytic systems that have appeared in the literature will be highlighted in the following paragraphs, according to the nature of the catalyst and its efficiency.

The first examples involving other ligands were reported some years later. Indeed, Liu, Guo, and coworkers[74] reported in 2004 the use of an amino acid (*N*-methylglycine, 20 mol %) in the presence of copper(I) iodide (10 mol %). Cheap potassium hydroxide was employed as the base with this catalytic system for the arylation of 1-octanethiol, and the reactions proceeded efficiently in dioxane at 100°C to give good-to-excellent yields of the corresponding cross-coupled products (Scheme 2.32, Eq. 1). It is worth noting that a tuning of the reaction conditions (20 mol % of copper(I) iodide and 20 mol % of *N*,*N*-dimethylglycine, DMF instead of dioxane, potassium phosphate as the

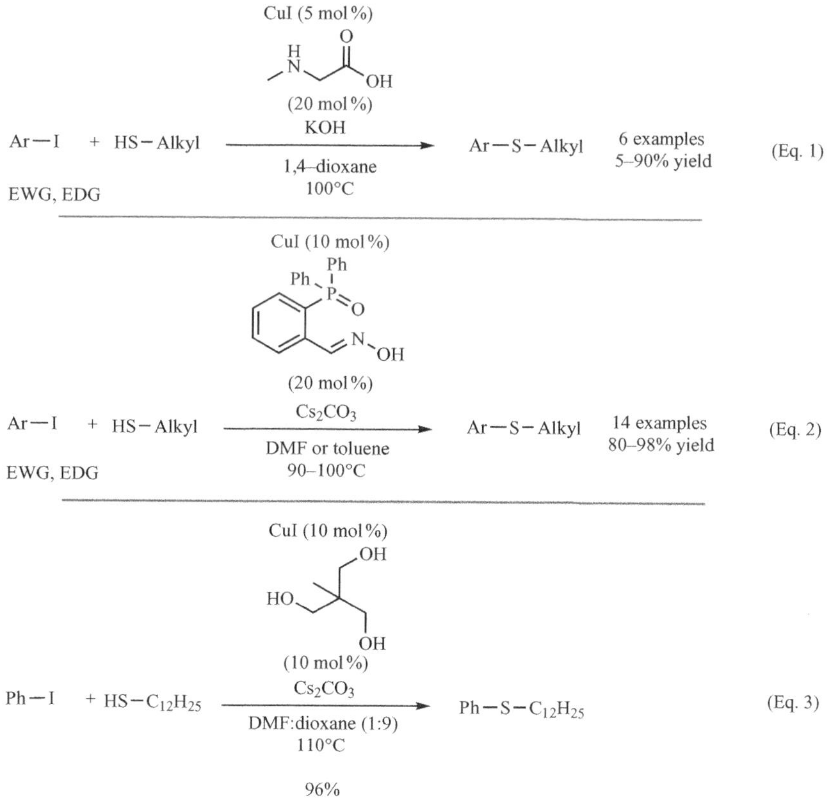

Scheme 2.32 Copper(I) iodide/ligand-catalyzed arylation of thiols developed by Guo, Liu, Wan, and Chen.

base, 120°C as optimal temperature) allowed starting from aryl bromides, although only a few examples were presented. The same copper salt (copper(I) iodide, 10 mol %) was used in the system developed by Wan and coworkers[75] in combination with a quite sophisticated oxime-phosphine oxide (20 mol %). The catalytic system efficiently promotes the cross-coupling reaction in DMF at 90°C using aryl iodides as starting materials (Scheme 2.32, Eq. 2). Finally, a tripod oxygen-type ligand (1,1,1-tris(hydroxymethyl)ethane, 10 mol %) associated with copper(I) iodide in a mixture of DMF and dioxane (1:9) showed good reactivity in the single example developed by Chen and Chen (Scheme 2.32, Eq. 3).[36h]

The study of the association of copper(I) oxide (5 mol %) and 1,10-phenanthroline (10 mol %) for the arylation of alkyl thiols with iodobenzene was undertaken by Xu and coworkers in 2009 and was found to be rather efficient.[76] The reactions were carried out in DMSO at 80°C in the presence of cesium carbonate to give good-to-excellent yields of the corresponding phenyl alkyl thioethers (Scheme 2.33).

The use of copper nanoparticles was introduced in 2007 by the groups of Punniyamurthy[77] and Ranu[78] for the arylation of various alkylthiols with iodobenzene. Under ligand-free conditions and with only 1.26 mol % of CuO nanoparticles, the reactions using Punniyamurthy's system were readily performed in DMSO at 80°C and in the presence of a cheap base, potassium hydroxide, to give good-to-excellent yields of the arylated products (Scheme 2.34, Eq. 1; Ar = Ph). As a note, both thiol groups of butane-1,4-dithiol underwent the arylation in the presence of 2 equiv of iodobenzene. Much harsher reaction conditions were necessary in the system developed by Ranu, which relied on the use of 20 mol % of copper nanoparticles in the presence of potassium carbonate (Scheme 2.34, Eq. 2). The cross coupling was performed in DMF under microwave irradiation at 120°C, yielding to the corresponding aryl alkyl thioethers in moderate-to-good yields (three examples were reported).

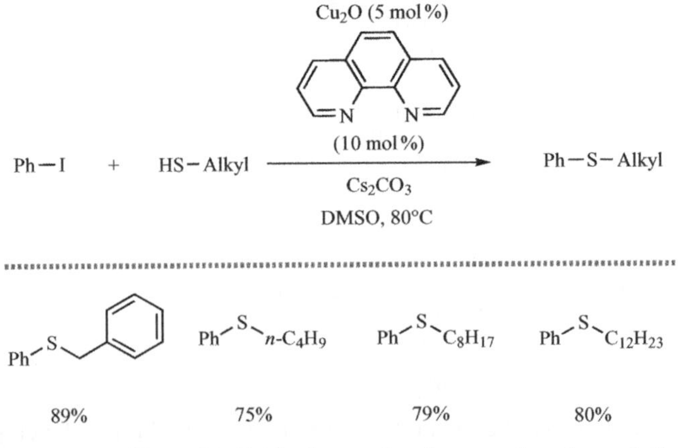

Scheme 2.33 Copper(I) oxide/1,10-phenanthroline-catalyzed arylation of thiols developed by Xu.

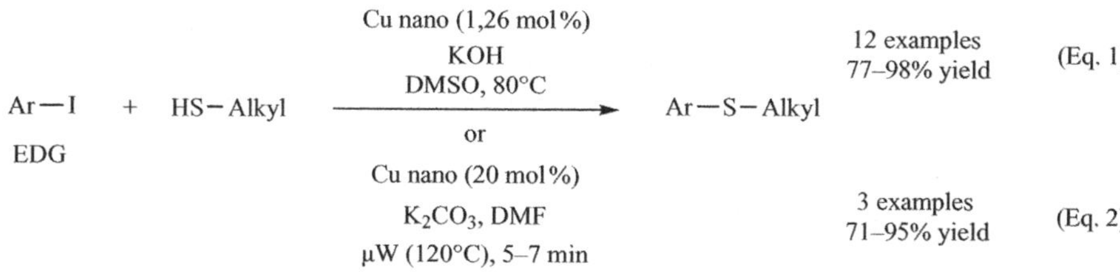

Scheme 2.34 Copper nanoparticles-catalyzed arylation of thiols developed by Punniyamurthy and Ranu.

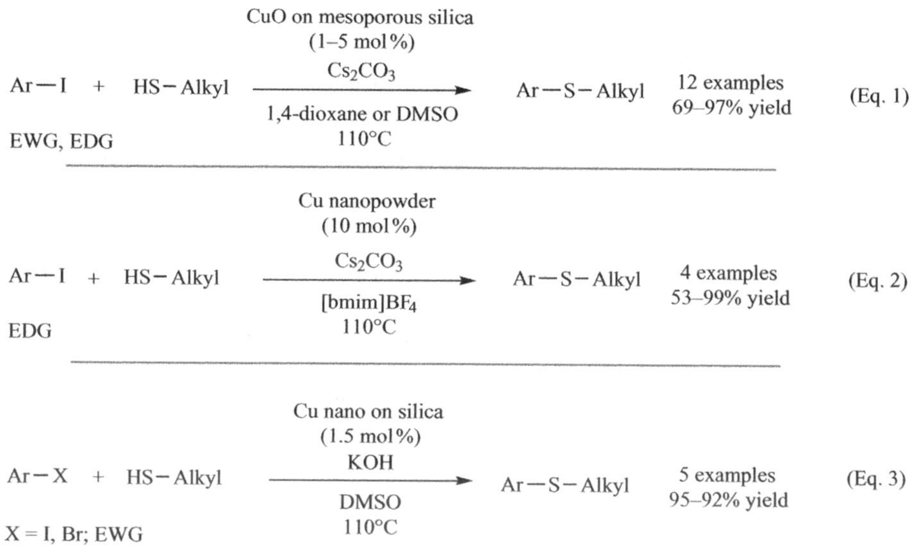

Scheme 2.35 Recyclable copper catalysts for the arylation of thiols.

To the best of our knowledge, only very few other heterogeneous copper-based systems have been developed for the arylation of alkyl thiols. Recently, Lee and coworkers[79] synthesized a copper(II) oxide supported on mesoporous silica, via a convenient sol-gel method. With its well-dispersed CuO sites, this material has been successfully applied to the arylation of various alkyl thiols with low catalyst loadings (1–5 mol %) using cesium carbonate as the base, the reactions being carried out in DMSO or dioxane. It is worth noting that the recyclability of this copper(II) oxide on mesoporous silica was successfully carried out without loss of reactivity even after four runs (Scheme 2.35, Eq. 1). The use of nanocopper oxide in ionic liquid was the key to the success for the system developed by Braga and coworkers.[80] Higher catalytic loadings were required (10 mol %) in the presence of cesium carbonate and the reactions were carried out in 1-butyl-2,3-dimethylimidazolium tetrafluoroborate ([bmim]BF_4) under ligandless conditions (Scheme 2.35, Eq. 2). Interestingly, biologically active sulfides such as captropil and *N*-Boc-protected L-penicillanime were successfully coupled with 4-methoxy iodobenzene with this catalytic system. The authors demonstrated the recyclability of both ionic liquid and the nanocopper only in the case of the synthesis of diaryl sulfides (see next section for details). Finally, Rajagobal and coworkers[81] used lower catalytic loadings (1.5 mol % of nanocopper supported on silica) in the presence of potassium hydroxide to successfully promote the arylation of alkyl thiols with iodobenzene and activated iodo- or bromoacetophenones (Scheme 2.35, Eq. 3).

Several other publications have appeared based on ligand-free, copper-based catalytic systems. In general, the studies have been focused on the arylation of thiophenols, with one or a couple of examples of aliphatic thiols. These publications will be detailed in the next chapter.[69,82]

Remarkable advances have also been made with another class of S-nucleophiles, thiophenols, and they will be overviewed in the next section.

2.3.3 Coupling of Aryl Halides with Aryl Thiols: Synthesis of Diaryl Thioethers (or Diaryl Sulfides)

The synthesis of diaryl thioethers and related derivatives by the classic Ullmann reaction has been known for a long time. As it is the case with other types of heteronucleophiles, it often requires harsh reaction conditions, but the recent renaissance of copper catalysis partially allowed overcoming this drawback. Fewer publications have appeared for the copper-catalyzed formation of C–S bonds than for the related C–O or C–N cross-coupling reactions, despite diaryl thioethers (or diaryl sulfides) being of great significance to the pharmaceutical industry, their framework being present in numerous pharmaceuticals, biologically active molecules, and also in materials.[83]

One of the first uses of a copper catalyst for the arylation of thiophenols was reported by Palomo and coworkers in 2000.[84] The procedure, already described for the synthesis of diaryl ethers[3g] in Section 2.2.3, could also be successfully extended to the formation of C–S bonds in the presence of 20 mol % of copper(I) bromide (although 200 mol % was required in some cases) and 2 equiv of an expensive Schwesinger phosphazene base, P_2–Et. Very good yields were obtained for the preparation of biarylthioethers from aryl iodides as starting materials with this system, but only two examples of cross coupling involving aryl bromides substituted with electron-withdrawing groups have been described (Scheme 2.36).

In 2002, the groups of Venkataraman[72] and Buchwald[73,85] independently reported two new catalytic copper-based systems for the arylation of thiophenols with aryl iodides. In the system developed by the former, the association of copper(I) iodide (10 mol %) and neocuproine (2,9-dimethyl-1,10-phenanthroline, 10 mol %) in the presence of a strong base, sodium *tert*-butoxide, in refluxing toluene allowed the formation of various biaryl sulfides in good-to-excellent yields (Scheme 2.37, Eq. 1). Efficient cross-coupling reactions were also observed with aryl iodides bearing *ortho*-substituents. Lower amounts of the same copper salt were shown to be sufficient in the system

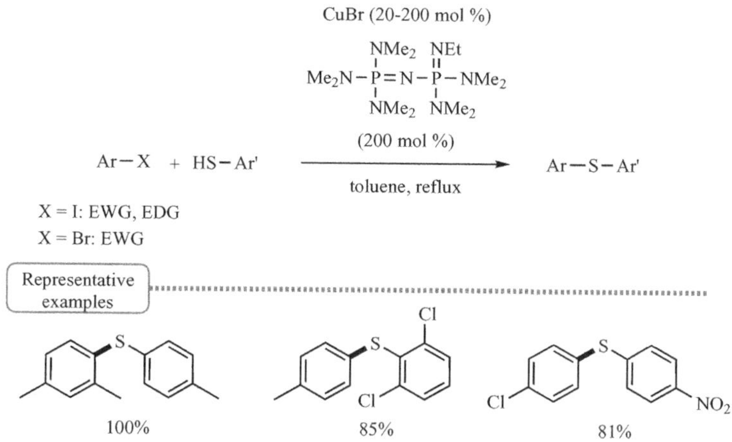

Scheme 2.36 Copper(I) bromide-catalyzed arylation of thiophenols developed by Palomo in the presence of Schwesinger's phosphazene P_2–Et base (C–S bond shown in bold).

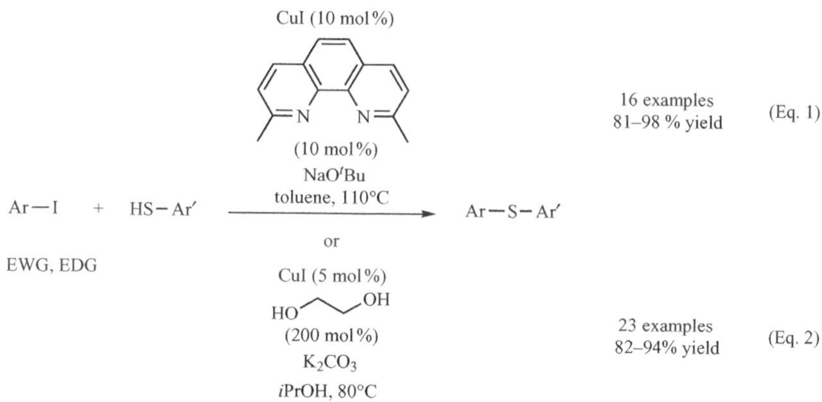

Scheme 2.37 Copper(I) iodide/ligand-catalyzed arylation of thiophenols developed by Venkataraman and Buchwald.

developed by Buchwald (Scheme 2.37, Eq. 2), although 2 equiv of ethylene glycol are required, in the presence of potassium carbonate as the base in isopropanol as the solvent. A good functional group tolerance is one of the interesting features of this system, which additionally is particularly efficient for *ortho*-substituted aryl iodides. In a general point of view, the method is very tolerant of steric hindrance.

In 2003, Wu and He[86] developed a microwave-assisted, copper-catalyzed cross-coupling reaction of aryl iodides and bromides with thiophenols under ligandless conditions. Copper(I) iodide was employed as a precursor (10 mol %) in the presence of cesium carbonate and the reaction was performed in NMP, providing good-to-excellent yields of the corresponding

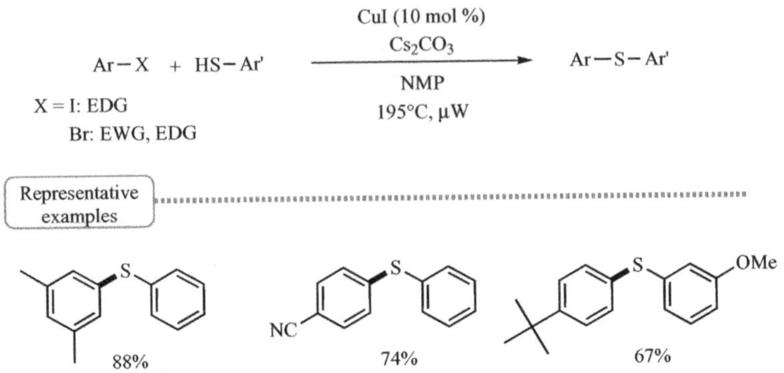

Scheme 2.38 Copper(I) iodide-catalyzed arylation of thiophenols under microwave heating developed by Wu (C–S bond formed shown in bold).

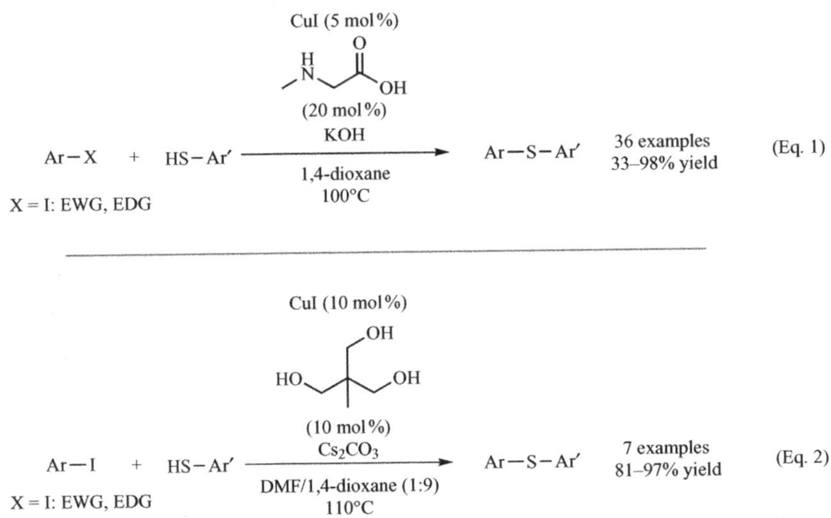

Scheme 2.39 Copper(I) iodide/ligand-catalyzed arylation of thiophenols developed by Guo and Liu, Wan, and Chen.

thioethers under microwave heating (195°C for 2 hours), regardless of the electronic properties of the starting aryl halides. Not surprisingly, only poor conversion was obtained with chlorobenzene as the substrate (Scheme 2.38).

Following these first developments, several other procedures have been developed using homogeneous or heterogeneous copper-based catalysts.

Liu, Guo, and coworkers[74] reported the use of an amino acid (*N*-methylglycine, 20 mol %) as the ligand in the presence of copper(I) iodide (5 mol %) and potassium hydroxide. The arylation of thiophenols with aryl iodides proceeded efficiently at 100°C to give very good yields of the corresponding cross-coupled products (Scheme 2.39, Eq. 1). The use of slightly

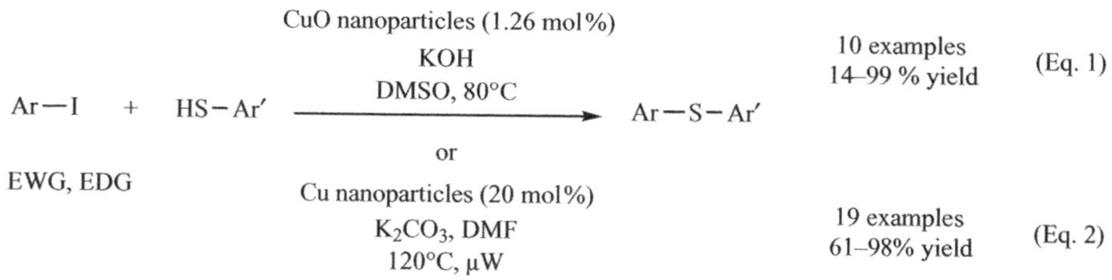

Scheme 2.40 Copper nanoparticles-catalyzed arylation of thiophenols developed by Punniyamurthy and Ranu.

different reaction conditions (20 mol % of the copper salt and 20 mol % of the *N*,*N*-dimethylglycine, DMF instead of dioxane, potassium phosphate as the base, 120°C as the optimal temperature) allowed the extension of this procedure to aryl bromides, although only a few examples were presented. Two years later, the Chen group[36h] showed that a tripod oxygen-type ligand (1,1,1-tris(hydroxymethyl)ethane, 10 mol %) associated with copper(I) iodide in a mixture of DMF and dioxane (1:9) at 110°C also allowed the synthesis of some diaryl sulfides from thiophenols and aryl iodides (Scheme 2.39, Eq. 2).

In 2007, the groups of Punniyamurthy[77] and Ranu[78] used copper nanoparticles to promote the cross coupling between aryl iodides and thiophenols. Under ligand-free conditions and with only 1.26 mol % of CuO nanoparticles, the reactions using Punniyamurthy's procedure were easily performed in DMSO at 80°C and in the presence of a cheap base, potassium hydroxide (Scheme 2.40, Eq. 1). Good-to-excellent yields of the arylated products were obtained when iodobenzene and 1-iodo-4-nitrobenzene were used in the presence of a variety of aromatic thiols, but only very poor conversion has been observed in reactions involving 4-iodoanisole as substrate (14% yield). The catalyst was easily recovered by centrifugation and was reused up to three times with only a slight decrease in catalytic activity. Harsher reaction conditions and higher copper nanoparticle catalyst loading (20 mol %) were used in the system developed by Ranu (Scheme 2.40, Eq. 2). The cross coupling yielding to various diaryl sulfides was performed under microwave irradiation at 120°C in DMF in the presence of potassium carbonate. However, good results have been obtained for aryl iodides substituted, this time, with electron-donating or electron-withdrawing groups, and it is worth noting that the system is also efficient with *ortho*-substituted aryl iodides.

The arylation of thiophenols in water was developed by the groups of Dominguez[87] and Punniyamurthy.[82b] Copper(I) chloride (8.5 mol %) associated with a large excess of the *trans*-cyclohexane-1,2-diamine ligand (3.9 equiv) was used in the system of Dominguez, which allows performing the cross coupling of thiophenols with aryl iodides or activated aryl bromides bearing electron-withdrawing substituents at 120°C (Scheme 2.41, Eq. 1). Copper(I) iodide (5 mol %) in combination with 1 equiv of a phase transfer catalyst, tetrabutylammonium bromide, and potassium hydroxide (1.5 equiv) is

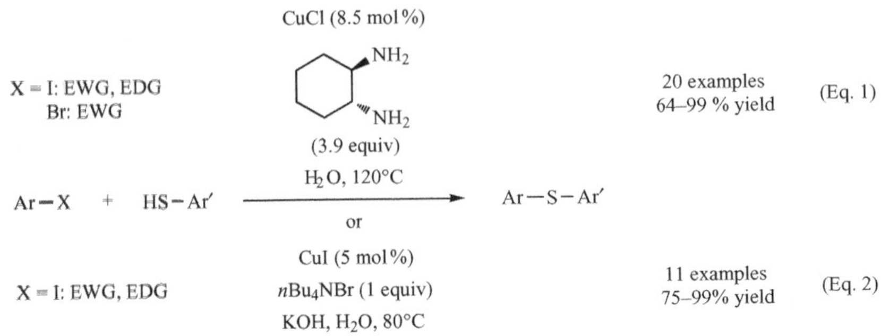

Scheme 2.41 Copper(I) halide-catalyzed arylation of thiophenols in water developed by Punniyamurthy and Dominguez.

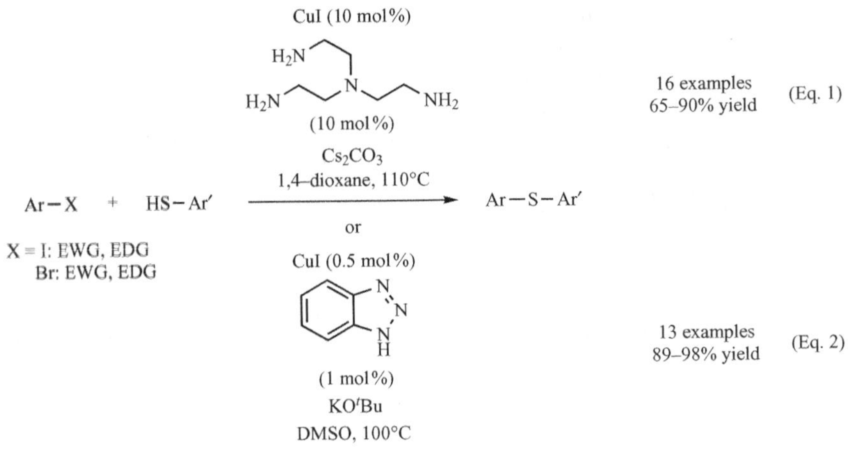

Scheme 2.42 Copper(I) iodide/ligand-catalyzed arylation of thiophenols developed by Verma and Shingare.

employed in the system of Punniyamurthy, which performs well at lower reaction temperature (80°C) (Scheme 2.41, Eq. 2). In this case, aryl iodides are the substrates of choice, while poor yields are obtained when aryl bromides or chlorides are used as starting coupling partners. Interestingly, these reactions do not require an inert atmosphere.

Two other families of nitrogen-based ligands have been employed by the groups of Verma[88] and Shingare.[89] In the procedure developed by the latter, thiophenols were arylated with aryl iodides and bromides substituted with electron-withdrawing and electron-donating groups in the presence of 10 mol % of copper(I) iodide and the tris-(2-aminoethyl)amine tripodal ligand. Cesium carbonate was used as the base and the reaction was carried out in dioxane at 110°C (Scheme 2.42, Eq. 1). Lower catalyst loading was employed in the procedure of Verma (Scheme 2.42, Eq. 2) since the use of only 0.5 mol %

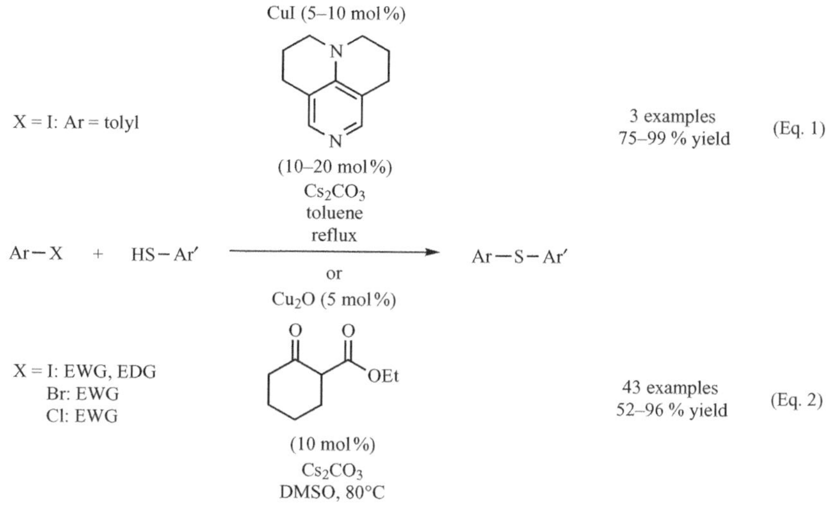

Scheme 2.43 Copper(I)/ligand-catalyzed arylation of thiophenols developed by Wong, Feng, and Fu.

of copper(I) iodide and 1 mol % of benzotriazole as the ligand in the presence of a strong base, potassium *tert*-butoxide, enabled the synthesis of diaryl sulfides in excellent yields from various aryl bromides when the reaction was carried out in DMSO at 100°C.

The development of a new ligand, 9-azajulolidine (obtained in four steps starting from 3,5-dibromo-4-aminopyridine), was undertaken by Wong.[36m] Associated with copper(I) iodide (5–10 mol %) and cesium carbonate as the base, this ligand allowed the synthesis of diaryl sulfides from iodotoluene (Scheme 2.43, Eq. 1) while 10 mol % of a commercially available ketoester, ethyl 2-oxocyclohexane-1-carboxylate, was used in combination with 5 mol % of copper(I) oxide in the method developed by Feng and coworkers (Scheme 2.43, Eq. 2).[90] Thiophenols were readily arylated with aryl iodides substituted with electron-withdrawing and electron-donating groups in good-to-excellent yields in the presence of cesium carbonate, and the reaction was carried out in DMSO at 80°C. Some examples using activated aryl bromides and chlorides substituted with electron-withdrawing groups have been reported, the corresponding diaryl sulfides being obtained in moderate to excellent yields. The presence of the ligand and the copper seems to be necessary in these cases since the blank reaction performed from 1-bromo-4-nitrobenzene and thiophenol yields only 8% of the product. A similar blank experiment was not performed for aryl chlorides.

In 2008, the groups of van Koten,[82a] Chan,[91] and Feng[82c] independently developed three different ligand-free methods for the arylation of thiophenols with aryl iodides substituted by either electron-withdrawing or electron-donating groups in good-to-excellent yields. Low catalyst loading has been

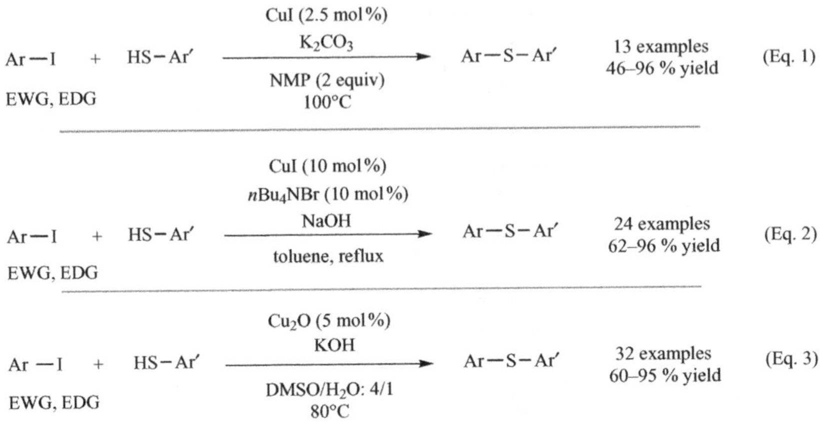

Scheme 2.44 Copper(I)/ligand-free-catalyzed arylation of thiophenols developed by Van Koten, Chan, Fu, and Feng.

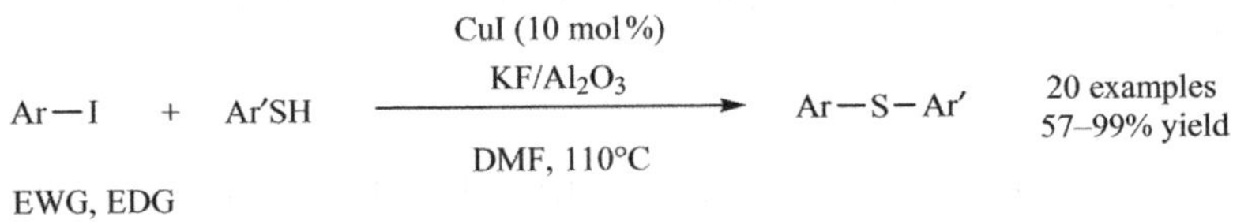

Scheme 2.45 Copper(I) iodide-catalyzed arylation of thiophenols developed by Xu.

used in the system of van Koten (2.5 mol % of copper(I) iodide) in the presence of potassium carbonate as the base and in only 2 equiv of NMP at 100°C (Scheme 2.44, Eq. 1). A phase transfer catalyst, tetrabutylammonium bromide, (10 mol %) was used in the system developed by Chan, which requires higher catalyst loading (10 mol % of copper(I) iodide) and sodium hydroxide as the base in refluxing toluene (Scheme 2.44, Eq. 2). A lower temperature of 80°C could be used when a combination of DMSO and water (4/1) was employed as the solvent mixture in the procedure presented by Fu and Feng using potassium hydroxide as the base and 5 mol % of the cheap copper(I) oxide as the precatalyst (Scheme 2.44, Eq. 3).

The main feature of the ligand-free system reported by Xu and coworkers[82e] lies in the use of potassium fluoride associated with aluminum oxide as solid and low moisture-sensitive base. The reactions could be carried out in the presence of copper(I) iodide (10 mol %) in DMF at 110°C and the diaryl sulfides have been obtained in good-to-excellent yields starting from aryl iodides substituted with electron-withdrawing and electron-donating groups (Scheme 2.45).

Recently, Prasad and Sekar[92] showed that *trans*-9,10-dihydro-9,10-ethanoanthracene-11,12-dimethanol, a quite sophisticated diol ligand already used for the copper-catalyzed O-arylation of phenols, was also able to promote, with a high loading of copper(I) bromide (20 mol %), the arylation of thiophenol

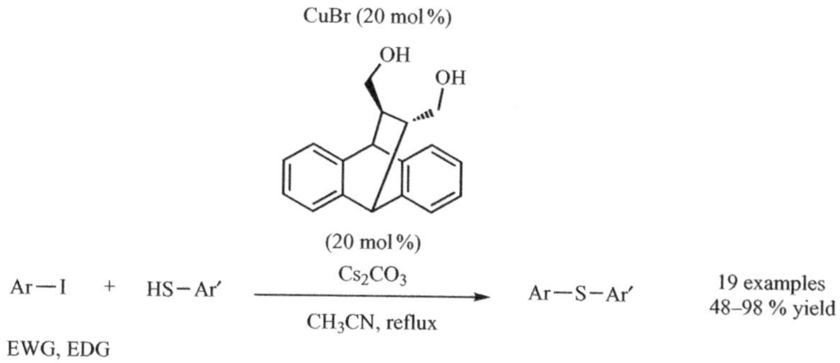

Scheme 2.46 Copper(I) iodide/racemic anthracene-based diol system for the arylation of thiophenols developed by Sekar.

derivatives starting from various aryl iodides, including sterically hindered *ortho*-substituted ones (Scheme 2.46).

More recently, Zhou and coworkers[93] reported a copper-based system for the synthesis of symmetrical and unsymmetrical diaryl sulfides using potassium thiocyanate (KSCN) as the source of sulfur. The procedure is based on the use of copper(II) chloride (10 mol %) associated with 1,10-phenanthroline (10 mol %), and the reactions were carried out in water in the presence of a phase transfer agent, tetrabutylammonium fluoride. Symmetrical diaryl sulfides were obtained in moderate-to-good yields starting from both aryl iodides and bromides bearing electron-withdrawing or electron-donating groups (Scheme 2.47, Eq. 1). In the case of the synthesis of unsymmetrical diaryl sulfides, two different aryl iodides were used as coupling partners (Scheme 2.47, Eq. 2). The general reaction pathway proposed on the base of control experiments first involves the intermediate formation of an aryl thiocyanate, which is further hydrolyzed to generate a thiolate ion. The latter then reacts with a second equivalent of aryl halide to afford the C–S coupling product.

Copper nanoparticles, whose use for C–S bond formation have been discussed earlier (Scheme 2.40),[77,78] have also been used for the synthesis of diaryl sulfides by the groups of Punniyamurthy[68d] and Nageswar.[94] In both procedures, copper(II) oxide nanoparticles (CuO nano) were employed as catalysts with different approaches and reaction conditions. In the system of Punniyamurthy, 2.5–5 mol % of CuO nano were used in the presence of potassium hydroxide and the reaction was carried out in DMSO at 80°C, allowing the cross coupling of thiophenols with aryl iodides bearing electron-withdrawing or electron-donating groups in moderate to very good yields (Scheme 2.48, Eq. 1). In the approach published by Nageswar, copper(II) oxide nanoparticles (5 mol %) were used as catalysts for the formation of symmetrical diaryl sulfides from inexpensive thiourea as a source of sulfur. The products were obtained in good-to-excellent yields, especially from aryl iodides bearing electron-withdrawing or electron-donating groups (as a note, two heterocyclic bromides were also used in the scope and limitations studies). The reaction

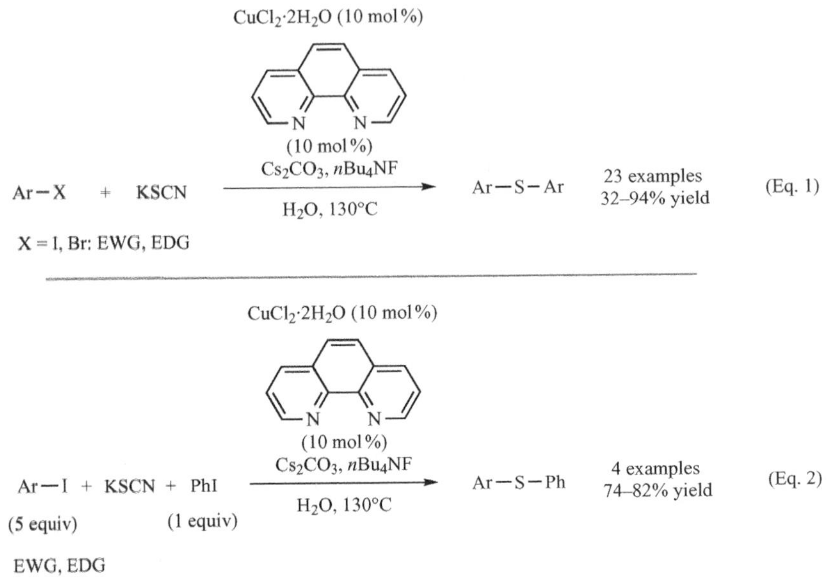

Scheme 2.47 Copper(II) chloride/phenanthroline-catalyzed double arylation of potassium thiocyanate to diaryl sulfides developed by Zhou.

was carried out in DMSO as in the system of Punniyamurthy, but the more expensive cesium carbonate was used at higher temperature (110°C). Excellent yields were obtained after the recovery and reuse of the catalyst even after four runs. The first step of the reaction probably corresponds to the formation of a C_{Ar}–S bond to give a S-aryl-isothiouronium iodide intermediate. Hydrolysis of the latter and formation of the corresponding thiophenoxide would allow for a second cross coupling with the remaining equivalent of aryl halide to give the symmetrical diaryl thioether (Scheme 2.48, Eq. 2). In the procedure of Akkilagunta and Kakulapati,[95] (Scheme 2.48, Eq. 3) another sulfur surrogate, potassium ethyl xanthogenate, was used as the source of sulfur. In the presence of recyclable copper(II) oxide nanoparticles (7 mol %) as catalyst in DMSO at 85°C, the synthesis of unsymmetrical sulfides starting from aryl iodides bearing both electron-donating and electron-withdrawing groups was performed in good yields. In the first step, the aryl iodide forms a C_{Ar}–S bond with the xanthogenate, and hydrolysis generates the corresponding thiophenate, which undergoes cross coupling with the second aryl iodide, added sequentially, thus affording the unsymmetrical diaryl sulfide.

Recently, Ranu developed an alumina-supported copper(II) sulfate as a heterogeneous catalyst for the cross coupling of thiophenols with aryl iodides and bromides substituted with both electron-withdrawing or electron-donating groups.[96] The reaction, undertaken with a copper/alumina charge of 5 mol % in the presence of potassium carbonate in DMF at 110°C, gave the diaryl sulfides in good-to-excellent yields. This heterogeneous catalyst was reused

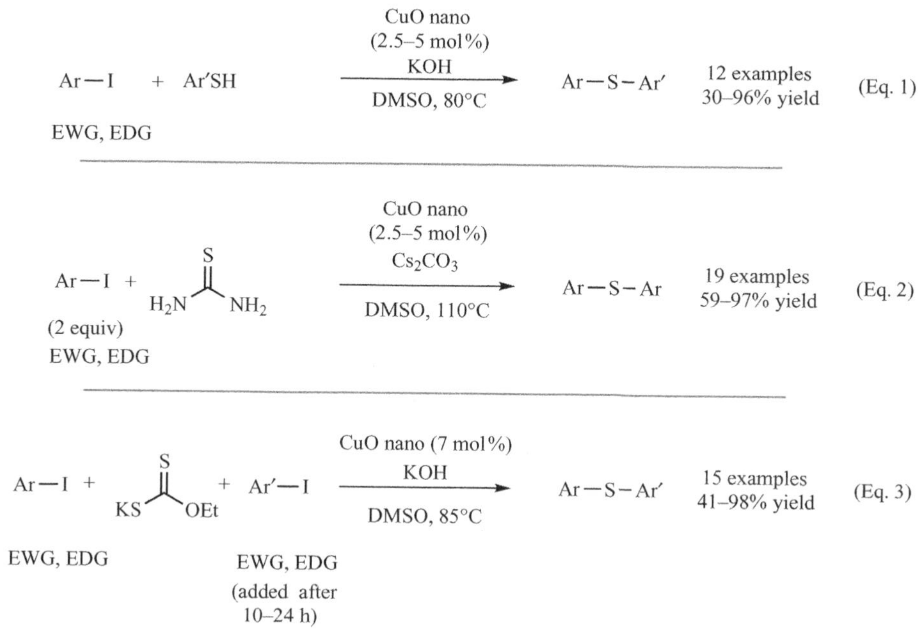

Scheme 2.48 Copper(II) oxide nanoparticles-catalyzed synthesis of diarylsulfides developed by Punniyamurthy, Nageswar, and Kakulapati.

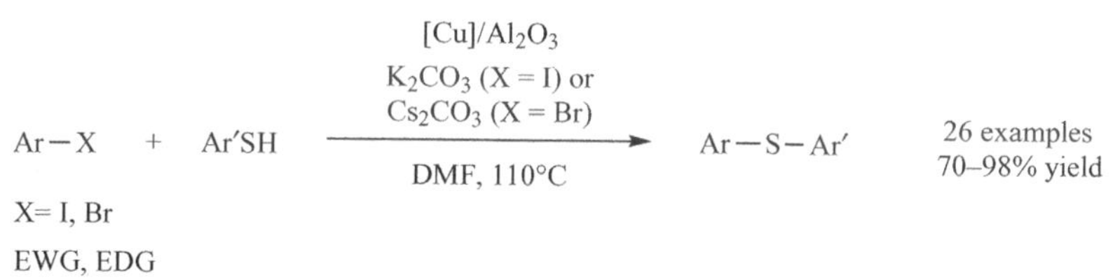

Scheme 2.49 Heterogeneous copper-catalyzed arylation of thiophenols developed by Ranu.

eight times without any loss of efficiency and no appreciable copper leaching was detected as measured by inductively coupled plasma mass spectrometry (ICP-MS) (Scheme 2.49).

Finally, Rajagobal designed a new dispersed silica-supported nanocopper-catalytic system using a sol-gel technique and applied it for the cross coupling between thiophenols and aryl iodides substituted with electron-withdrawing or electron-donating groups.[81] Low catalyst loading was employed (1.5 mol %) in this reaction undertaken in the presence of potassium hydroxide and carried out in DMSO at 110°C. The system, which showed negligible copper leaching, could be successfully reused several times without loss of catalytic activity (Scheme 2.50).

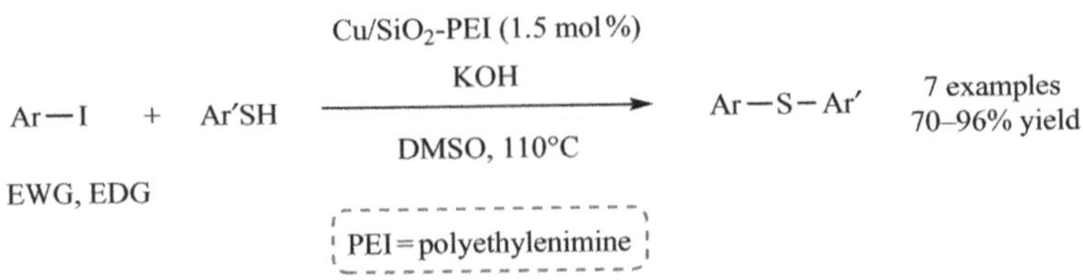

Scheme 2.50 Heterogeneous copper-catalyzed arylation of thiophenols developed by Rajagobal.

Note that other systems, though they represent interesting contributions, cannot be included within the scope of this book chapter, which has been limited to the most efficient systems for clarity.[19,57,97]

2.4 CONCLUSION

Since the renaissance of copper-catalyzed cross-coupling reactions at the end of the last century, very important achievements have already been made, not only for the formation of C_{Ar}–O and C_{Ar}–S bonds, detailed in this chapter, but also for all types of related couplings involving the formation of C_{Ar}–N and C_{Ar}–C bonds as evidenced in other chapters of this book. Nevertheless, in a general point of view, significant challenges, listed further, remain to be overcome whatever the type of the coupling considered.

The first one could be the discovery of new ligands and catalytic systems efficient at lower copper loading (≤1 mol %) both under milder temperature conditions and with higher turnover frequencies. This would allow reduction of the amount of metal waste and would improve the competitiveness of copper-catalyzed procedures. Along these lines, the development of more efficient recoverable and reusable catalysts (copper nanoparticles, hybrid silica prepared by sol-gel processes, copper powder, polymers . . .) could also offer interesting catalytic systems with improved efficiencies than those described in this chapter. For some of these systems, an advantage would lie in the efficient trapping of copper traces, an important point for the preparation of pharmaceuticals.

Another important goal is the discovery of general and efficient procedures that would enable the cross coupling of nonactivated aryl chlorides (substituted on the aromatic ring by electron-donating groups), which are less expensive but also, unfortunately, less reactive than the corresponding aryl iodides or bromides. Note that some systems, often based on heterogeneous catalysts, seem to provide interesting solutions for activated chlorides (substituted by electron-withdrawing groups). However, because the blank experiments (the tests without the copper/ligand systems) are often missing, it is difficult to discard a possible S_NAr-type coupling mechanism. Other interesting starting reagents that could be used for the arylation of sulfur nucleophiles (and others) would be aryl sulfonates, which would greatly expand the copper catalysis toolbox. Indeed, if these substrates are quite expensive and sensitive, their use as reaction partners would allow the introduction of phenol derivatives in the Ullmann-type cross-coupling chemistry, which would definitely offer new possibilities.

The question of industrial applications of the Ullmann reaction in its modern catalytic version is an important point as highlighted with representative and selected examples in Chapter 19. The method will probably find in the next future an important place and some companies, among which, for example, Rhodia or the Shasun Chemical Company, produce (or have produced) important building blocks using copper catalysis on a ton scale.

Finally, a better understanding of the mechanisms involved in copper-mediated cross-coupling reactions could also enable the development of more efficient copper-based systems that could bring useful solutions both to industrial and academic challenges. In this perspective, a major step forward for the formation of C–S bonds under much milder conditions than the one previously available was reported in 2013 by the groups of Fu and Peters. In continuation of their studies of photoinduced copper-catalyzed cross coupling reactions,[98a] they recently reported a versatile photoinduced, copper-catalyzed arylation of thiophenols with aryl iodides and bromides. The use of 10 mol % of copper(I) iodide and sodium *tert*-butoxide enabled the efficient coupling of a broad range of thiophenols with various aryl halides, including hindered, electron-rich and electron-poor ones, at 0°C with a thiophenol/aryl halide/base ratio of 1:1:1 upon irradiation with a 100-watt-Hg lamp (Scheme 2.51). While activation of C(aryl)-X bonds is usually believed to occur via oxidative addition at copper, the success of this transformation is based on a shift in the reaction pathway, a single-electron transfer (SET) mechanism being operative in this case.[98b] Further developments will definitively open new avenues in copper catalysis.

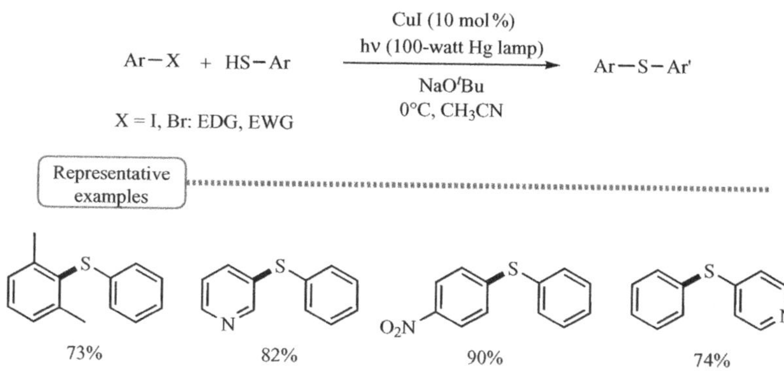

Scheme 2.51 Photoinduced, copper-catalyzed cross coupling of thiophenols with aryl halides at 0°C via SET (C–S bond formed shown in bold).

REFERENCES

[1] (a) Ley, S. V.; Thomas, A. W. *Angew. Chem. Int. Ed.* **2003**, *42*, 5400–5449. (b) Kunz, K.; Scholz, U.; Ganzer, D. *Synlett* **2003**, *15*, 2428–2439. (c) Beletskaya, I. P.; Cheprakov, A. V. *Coord. Chem. Rev.* **2004**, *248*, 2337–2364. (d) Frlan, R.; Kikelj, D. *Synthesis* **2006**, 2271–2285. (e) Corbet, J. P.; Mignani, G. *Chem. Rev.* **2006**, *106*, 2651–2710. (f) Kienle, M.; Dubakka, S. R.; Brade, K.; Knochel, P. *Eur. J. Org. Chem.* **2007**, 4166–4176. (g) Evano, G.; Blanchard, N.; Toumi, M. *Chem. Rev.* **2008**, *108*, 3054–3131. (h) Monnier, F.; Taillefer, M. *Angew. Chem. Int. Ed.* **2008**, *47*, 3096–3099. (i) Monnier, F.; Taillefer, M. *Angew. Chem. Int. Ed.* **2009**, *48*, 6954–6971.

[2] (a) Pozharskii, A. F.; Martsokha, B. K.; Simonov, A. M. *J. Gen. Chem. USSR* **1963**, *33*, 994–995. (b) Sitkina, L. M.; Simonov, A. M. *Chem. Heterocyclic Compounds USSR* **1968**, *2*, 103–105. (c) Khan, M. A.; Polya, J. B. *J. Chem. Soc. C* **1970**, 85–91. (d) Cirigottis, K. A.; Ritchie, E.; Taylor, W. C. *Aust. J. Chem.* **1974**, *27*, 2209–2228. (e) Bacon, R. G. R.; Murray, J. C. F. *J. Chem. Soc. Perkin Trans. 1* **1975**, 1267–1272. (f) Bruggink, A.; McKillop, A. *Tetrahedron* **1975**, *31*, 2607–2619. (g) Yamamoto, T.; Kurata, Y. *Can. J. Chem.* **1983**, *61*, 86–91. (h) Tironi, C.; Fruttero, R.; Garrone, A. *IlFarmaco* **1990**, *45*, 473–478. (i) Sugaya, T.; Mimura, Y.; Kato, N.; Ikuta, M.; Mimura, T.; Kasai, M.; Tomioka, S. *Synthesis* **1994**, 73–76. (j) Sugahara, M.; Ukita, T. *Chem. Pharm. Bull.* **1997**, *45*, 719–721. (k) Quallich, G. J.; Makowski, T. W.; Sanders, A. F.; Urban, F. J.; Vazquez, E. *J. Org. Chem.* **1998**, *63*, 4116–4119.

[3] (a) Weingarten, H. *J. Org. Chem.* **1964**, *29*, 3624–3626. (b) Couture, C.; Paine, A. J. *Can. J. Chem.* **1985**, *63*, 111–120. (c) Soula, G. *J. Org. Chem.* **1985**, *50*, 3717–3721. (d) Oi, R.; Shimakawa, S.; Takenaka, S. *Chem. Lett.* **1988**, 899–900. (e) Marcoux, J. F.; Doye, S.; Buchwald, S. L. *J. Am. Chem. Soc.* **1997**, *119*, 10539–10540. (f) Ma, D.; Zhang, Y.; Yao, J.; Wu, S.; Tao, F. *J. Am. Chem. Soc.* **1998**, *120*, 12459–12467. (g) Palomo, C.; Oiarbide, M.; Lopez, R.; Bengoa, E.-G. *Chem. Commun.* **1998**, 2091–2092. (h) Kiyomori, F.; Marcoux, J. F.; Buchwald, S. L. *Tetrahedron Lett.* **1999**, *40*, 2657–2660. (i) Goodbrand, H. B.; Hu, N. X. *J. Org. Chem.* **1999**, *64*, 670–674. (j) Fagan, P. J.; Hauptman, E.; Shapiro, R.; Casalnuovo, A. *J. Am. Chem. Soc.* **2000**, *122*, 5043–5051.

[4] Two important breakthroughs were reported in 2001 for copper-catalyzed C–C, C–N or C–O coupling by the groups of Taillefer and Buchwald: (a) Taillefer, M.; Cristau, H. J.; Cellier, P. P.; Spindler, J. F., inventors; Rhodia Chimie, assignee. Method for forming a carbon-carbon or carbon-heteroatom linkage. FR 01 16547, 2001 December 20; PCT FR2002/04499. (b) Taillefer, M.; Cristau, H. J.; Cellier, P. P.; Spindler, J. F.; Ouali, A., inventors; Rhodia Chimie, assignee. Method for forming a carbon-carbon or carbon-heteroatom linkage. FR 026717, 2002 May 31; PCT FR2003/01647. (c) Buchwald, S. L.; Klapars, A.; Antilla, J. C.; Job, G. E.; Wolter, M.; Kwong, F. Y.; Nordmann, G.; Hennessy, E. J., inventors; Massachusetts Institute of Technology, assignee. Copper-catalyzed formation of carbon-heteroatom and carbon-carbon bonds. US 60/286,268, 2001 April 24; PCT US2002/12785.

[5] Arpe, H. J. *Industrial Organic Chemistry*, 5th edition. Wiley-VCH: Weinheim, **1997**.

[6] Rappoport, Z. (ed.), *In Patai Series: The Chemistry of Functional Groups. The Chemistry of Phenols*. John Wiley & Sons Ltd: Jerusalem, **2003**.

[7] (a) Fyfe, C. A. *In Patai Series, The Chemistry of the Hydroxyl Group*, Vol. 1. Wiley Interscience: New York, **1971**, pp. 83–127. (b) Hanson, P.; Jones, J. R.; Taylor, A. B.;

Walton, P. H.; Timms, A. W. *J. Chem. Soc., Perkin Trans. 2* **2002**, 1135–1150. (c) George, T.; Mabon, R.; Sweeney, G.; Sweeney, J. B.; Tavassoli, A. *J. Chem. Soc. Perkin Trans. 1* **2000**, 2529–2574.

[8] (a) Alonso, A. D.; Najera, C.; Pastor, M. I.; Yus, M. *Chem. Eur. J.* **2010**, *16*, 5274–5284. (b) Enthaler, S.; Company, A. *Chem. Soc. Rev.* **2011**, *40*, 4912–4924.

[9] Kormos, C. M.; Leadbeater, N. E. *Tetrahedron* **2006**, *62*, 4728–4732.

[10] (a) Tlili, A.; Xia, N.; Monnier, F.; Taillefer, M. *Angew. Chem. Int. Chem* **2009**, *48*, 8725–8728. (b) Taillefer, M.; Tlili, A.; Monnier, F.; Xia, N., inventors; Centre Nationale de la Recherche Scientifique, Ecole Nationale Supérieure de Chimie de Montpellier, assignee. Method for the hydroxylation of halogenated aryl compounds. FR 09 02767, 2009 June 8; PCT FR2010/051140.

[11] Zhao, D.; Wu, N.; Zhang, S.; Xi, P.; Su, X.; Lan, J.; You, J. *Angew. Chem. Int. Ed.* **2009**, *48*, 8729–8732.

[12] Tao, C. Z.; Liu, W. W.; Sun, J. Y. *Chin. Chem. Lett.* **2009**, *20*, 1170–1174.

[13] Mehmood, A.; Leadbeater, N. E. *Catal. Comm.* **2010**, *12*, 64–66.

[14] Yang, D.; Fu, H. *Chem. Eur. J.* **2010**, *16*, 2366–2370.

[15] Jing, L.; Wei, J.; Zhou, L.; Huang, Z.; Li, Z.; Zhou, X. *Chem. Commun.* **2010**, *46*, 4767–4769.

[16] (a) Maurer, S.; Liu, W.; Zhang, X.; Jiang, Y.; Ma, D. *Synlett* **2010**, 976–978. (b) Ni, F.; Li, J. *Synthesis* **2012**, *44*, 3598.

[17] Thakur, K. G.; Sekar, G. *Chem. Commun.* **2011**, *47*, 6692–6694.

[18] Paul, R.; Ali, M.; Punniyamurthy, T. *Synthesis* **2010**, 4268–4272.

[19] Xu, H. J.; Liang, Y. F.; Cai, Z. Y.; Qi, H. X.; Yang, C. Y.; Feng, Y. S. *J. Org. Chem.* **2011**, *76*, 2296–2300.

[20] Kai, K.; Li, Z.; Wang, Z.; Yao, Z.; Jiang, S. *Org. Lett.* **2011**, *13*, 4340–4344.

[21] For a review, see: Lindley, J. *Tetrahedron* **1984**, *40*, 1433–1456.

[22] (a) Alten, H. L.; Van Koten, G.; Grove, D. M.; Kuilman, T.; Piestra, O. G.; Hulshof, L. A.; Sheldon, R. A. *Tetrahedron* **1989**, *45*, 5565–5578. (b) Keegstra, M. A.; Peters, T. H. A.; Brandsma, L. *Tetrahedron* **1992**, *48*, 3633–3652. (c) Capdevielle, P.; Maumy, M. *Tetrahedron Lett.* **1993**, *34*, 1007–1010. (d) Morishima, Y.; Fujita, J.; Ikeda, T.; Kamachi, M. *Chem. Lett.* **1994**, 557–560. (e) Zhu, J.; Price, B. A.; Zhao, S. X.; Skonezny, P. M. *Tetrahedron Lett.* **2000**, *41*, 4011–4041.

[23] Wolter, M.; Nordmann, G.; Job, E.; Buchwald, S. L. *Org. Lett.* **2002**, *4*, 973–976.

[24] Job, G. E.; Buchwald, S. L. *Org. Lett.* **2002**, *4*, 3703–3706.

[25] Shafir, A.; Lichtor, P. A.; Buchwald, S. L. *J. Am. Chem. Soc.* **2007**, *129*, 3490–3491.

[26] Hosseizadech, R.; Tajbakhsh, M.; Mohadjerani, M.; Alikarami, M. *Synlett* **2005**, 1101–1104.

[27] Manbeck, G. F.; Lipman, A. J.; Stockland, R. A.; Freidl, A. L.; Hasler, A. F.; Stone, J. J.; Guzei, I. *J. Org. Chem.* **2005**, *70*, 244–250.

[28] Toumi, M.; Couty, F.; Evano, G. *Angew. Chem. Int. Ed.* **2007**, *46*, 572–575.

[29] Toumi, M.; Rincheval, V.; Young, A.; Gergeres, D.; Turos, E.; Couty, F.; Mignotte, B.; Evano, G. *Eur. J. Org. Chem.* **2009**, 3368–3386.

[30] Altman, R. A.; Shafir, A.; Choi, A.; Lichtor, P. A.; Buchwald, S. L. *J. Org. Chem.* **2008**, *73*, 284–286.

[31] Naidu, A. B.; Sekar, G. *Tetrahedron Lett.* **2008**, *49*, 3147–3151.

[32] Naidu, A. B.; Jaseer, E. A.; Sekar, G. *J. Org. Chem.* **2009**, *74*, 3675–3679.

[33] Niu, J.; Guo, P.; Kang, J.; Li, Z.; Xu, J.; Hu, S. *J. Org. Chem.* **2009**, *74*, 5075–5078.

[34] Maiti, D. *Chem. Commun.* **2011**, *47*, 8340–8342.

[35] Huang, J.; Chen, Y.; Chan, J.; Ronk, M. L.; Larsen, R. D.; Faul, M. M. *Synlett* **2011**, 1419–1422.

[36] (a) He, H.; Wu, Y.-J. *Tetrahedron Lett.* **2003**, *44*, 3445–3446. (b) Ghosh, R.; Samuelson, A. G. *New J. Chem.* **2004**, *28*, 1390–1393. (c) Heiden, M. R.; Frey, G. D.; Plenio, H. *Organometallics* **2004**, *23*, 3548–3551. (d) Lohmann, S.; Andrews, S. P.; Burke, B. J.; Smith, M. D.; Attfield, J. P.; Tanaka, H.; Kaneko, K.; Ley, S. V. *Synlett* **2005**, 1291–1295. (e) Yang, F.; Yan, L.; Ma, K.; Yang, L.; Li, J.; Chen, L.; You, J. *Eur. J. Org. Chem.* **2006**, 1109–1112. (f) Cai, Q.; He, G.; Ma, D. *J. Org. Chem.* **2006**, *71*, 5268–5273. (g) Takeuchi, D.; Asano, I.; Osakada, K. *J. Org. Chem.* **2006**, *71*, 8614–8617. (h) Chen, Y.-J.; Chen, H.-H. *Org. Lett.* **2006**, *8*, 5609–5612. (i) D'Angelo, N. D.; Peterson, J. J.; Booker, S. K.; Fellows, I.; Dominguez, C.; Hungate, R.; Reidera, P. J.; Kim, T.-S. *Tetrahedron Lett.* **2006**, *47*, 5045–5048. (j) Wang, B.-A.; Zeng, R.-S.; Wei, H.-Q.; Jia, A.-Q.; Zou, J.-P. *Chin. J. Chem.* **2006**, *24*, 1062–1065. (k) Zhu, X.-H.; Chen, G.; Ma, Y.; Song, H.-C.; Xu, Z.-L.; Wan, Y.-Q. *Chin. J. Chem.* **2007**, *25*, 546–552. (l) Lipshutz, B. H.; Unger, J. B.; Taft, B. R. *Org. Lett.* **2007**, *9*, 1089–1092. (m) Wong, K.-T.; Ku, S.-Y.; Yenb, F.-W. *Tetrahedron Lett.* **2007**, *48*, 5051–5054. (n) Chen, W.; Li, J.; Fang, D.; Feng, C.; Zhang, C. *Org. Lett.* **2008**, *10*, 4565–4568. (o) Tubaro, C.; Biffis, A.; Scattolin, E.; Basato, M. *Tetrahedron* **2008**, *64*, 4187–4195. (p) Schareina, T.; Zapf, A.; Cotté, A.; Müller, N.; Beller, M. *Tetrahedron Lett.* **2008**, *49*, 1851–1855. (q) Ouali, A.; Oshovsky, G. V.; Duhayon, C.; Taillefer, M.; Majoral, J. P. *Organometallics* **2008**, *27*, 5733–5736. (r) Gershoni-Poranne, R.; Pappo, D.; Solel, E.; Keinan, E. *Org. Lett.* **2009**, *11*, 5146–5149. (s) Sreedhar, B.; Arundhathi, R.; Reddy, M. A.; Kanatm, M. L. *Synthesis* **2009**, 483–487. (t) Liu, Y.; Li, G.; Yang, L. *Chin. J. Chem.* **2009**, *27*, 423–427. (u) Kaddouri, H.; Vicente, V.; Ouali, A.; Ouazzani, F.; Taillefer, M. *Angew. Chem. Int. Ed.* **2009**, *48*, 333–336. (v) Ellul, C. E.; Reed, G.; Mahon, M. F.; Pascu, S. I.; Whittlesey, M. K. *Organometallics* **2010**, *29*, 4097–4104. (w) Khalilzadeh, M. A.; Hosseini, A.; Pilevar, A. *Eur. J. Org. Chem.* **2011**, 1587–1592. (x) Yang, H.; Xi, C.; Miao, Z.; Chen, R. *Eur. J. Org. Chem.* **2011**, 3353–3360. (y) Swapna, K.; Murthy, S. N.; Jyothi, M. T.; Nageswar, Y. V. D. *Org. Biomol. Chem.* **2011**, *9*, 5978–5988. (z) Wan, J.-P.; Wang, C.; Liu, Y. *Org. Biomol. Chem.* **2011**, *9*, 6481–6483. (aa) Islam, S. K. M.; Mondal, S.; Mondal, P.; Roy, A. S.; Tuhina, K.; Salam, N.; Mobarak, M. J. *Organomet. Chem.* **2012**, *696*, 4264–4274. (ab) Verma, S.; Kuma, N.; Jain, S. L. *Tetrahedron Lett.* **2012**, *53*, 4665–4668. (ac) Cheng, A.-Y.; Hsieh, J.-C. *Tetrahedron Lett.* **2012**, *53*, 71–75.

[37] (a) Ullmann, F. *Ber. Dtsch. Chem. Ges.* **1904**, *37*, 853–854. (b) Ullmann, F.; Sponagel, P. *Ber. Dtsch. Chem. Ges.* **1905**, *38*, 2211–2212.

[38] For reviews on the synthesis of diaryl ethers, see: (a) Frlan, R.; Kikelj, D. *Synthesis* **2006**, 2271–2285. (b) Pitsinos, E. N.; Vidali, V. P.; Couladouros, E. A. *Eur. J. Org. Chem.* **2011**, 1207–1222.

[39] (a) Cristau, H.-J.; Cellier, P. P.; Hamada, S.; Spindler, J.-F.; Taillefer, M. *Org. Lett.* **2004**, *6*, 913–916. (b) Ouali, A.; Spindler, J.-F.; Cristau, H.-J.; Taillefer, M. *Adv. Synth. Cat.* **2006**, *348*, 499–505. (c) Ouali, A.; Spindler, J.-F.; Jutand, A.; Taillefer, M. *Adv. Synth. Cat.* **2007**, *349*, 1906–1916. (d) Ouali, A.; Taillefer, M.; Spindler, J.-F.; Jutand, A. *Organometallics* **2007**, *26*, 65–74.

[40] Gujadhur, R.; Venkataraman, D. *Synth. Commun.* **2001**, *31*, 139–153.

[41] Gujadhur, R. K.; Bates, C. G.; Venkataraman, D. *Org. Lett.* **2001**, *3*, 4315–4317.

[42] Buck, E.; Song, Z. J.; Tschaen, D.; Dormer, P. G.; Volante, R. P.; Reider, P. J. *Org. Lett.* **2002**, *9*, 1623–1626.

[43] Luo, Y.; Wu, J. X.; Ren, R. X. *Synlett* **2003**, 1734–1736.

[44] Ma, D.; Cai, Q. *Org. Lett.* **2003**, *5*, 3799–3802.

[45] Cai, Q.; Zou, B.; Ma, D. *Angew. Chem. Int. Ed.* **2006**, *45*, 1276–1279.

[46] Altman, R. A.; Buchwald, S. L. *Org. Lett.* **2007**, *9*, 643–646.

[47] Lv, X.; Bao, W. *J. Org. Chem.* **2007**, *72*, 3863–3867.

[48] (a) Taillefer, M.; Xia, N.; Monnier, F.; Tlili, A., inventors; Centre National de la Recherche Scientifique, assignee. New catalytic systems for cross-coupling reactions. US60 996830, 2007 Dec 6; PCT IB08/03715. (b) Xia, N.; Taillefer, M. *Chem. Eur. J.* **2008**, *14*, 6037–6039.

[49] Zhang, Q.; Wang, D.; Wang, X.; Ding, K. *J. Org. Chem.* **2009**, *74*, 7187–7190.

[50] Rao, H.; Jin, Y.; Fu, H.; Jiang, Y.; Zhao, Y. *Chem. Eur. J.* **2006**, *12*, 3636–3646.

[51] Liu, X.; Fu, H.; Jiang, Y.; Zhao, Y. *Synlett* **2008**, 221–224.

[52] Jin, Y.; Liu, J.; Yin, Y.; Fu, H.; Jiang, Y.; Zhao, Y. *Synlett* **2006**, 1564–1568.

[53] Naidu, A. B.; Raghunath, O. R.; Prasad, D. J. C.; Sekar, G. *Tetrahedron Lett.* **2008**, *49*, 1057–1061.

[54] Jogdand, N. R.; Shingate, B. B.; Shingare, M. S. *Tetrahedron Lett.* **2009**, *50*, 4019–4021.

[55] Liu, Y. H.; Li, G.; Yang, L.-M. *Tetrahedron Lett.* **2009**, *50*, 343–346.

[56] Niu, J.; Zhou, H.; Li, Z.; Xu, J.; Hu, S. *J. Org. Chem.* **2008**, *73*, 7814–7817.

[57] Sperotto, E.; van Klink, G. P. M.; de Vries, J. G.; van Koten, G. *Tetrahedron* **2010**, *66*, 9009–9020.

[58] Maiti, D.; Buchwald, S. L. *J. Am. Chem. Soc.* **2009**, *131*, 17423–17429.

[59] Maiti, D.; Buchwald, S. L. *J. Org. Chem.* **2010**, *75*, 1791–1794.

[60] Tlili, A.; Monnier, F.; Taillefer, M. *Chem. Eur. J.* **2010**, *16*, 12299–12302.

[61] Ouali, A.; Laurent, R.; Caminade, A. M.; Majoral, J. P.; Taillefer, M. *J. Am. Chem. Soc.* **2006**, *128*, 15990–15991.

[62] Miao, T.; Wang, L. *Tetrahedron Lett.* **2007**, *48*, 95–99.

[63] Benyahya, S.; Monnier, F.; Taillefer, M.; Wong Chi Man, M.; Bied, C.; Ouazzani, F. *Adv. Synth. Catal.* **2008**, *350*, 2205–2208.

[64] Benyahya, S.; Monnier, F.; Wong.Chi. Man, M.; Bied, C.; Ouazzani, F.; Taillefer, M. *Green Chem.* **2009**, *11*, 1121–1123.

[65] Taillefer, M.; Cristau, H. J.; Cellier, P. P.; Ouali, A.; Spindler, J.-F., inventors; Rhodia Chimie, assignee. Method for forming a carbon-carbon or carbon-heteroatom bond. FR 03 10253, 2003 August 28; PCT FR2004/002021.

[66] Sperotto, E.; de Vries, S. J. G.; Van Klink, G. P. M.; Van Koten, G. *Tetrahedron Lett.* **2007**, *48*, 7366–7370.

[67] Chang, J. W. W.; Chee, S.; Mak, S.; Buranaprasertsuk, P.; Chavasiri, W.; Chan, W. H. P. *Tetrahedron Lett.* **2008**, *49*, 2018–2022.

[68] (a) Kidwai, M.; Mishra, N. K.; Bansal, V.; Kumar, A.; Mozumdar, S. *Tetrahedron Lett.* **2007**, *48*, 8883–8887. (b) Zhang, J.; Zhang, Z.; Wang, Y.; Zheng, X.; Wang, Z.

Eur. J. Org. Chem. **2008**, 5112–5116. (c) Sreedhar, B.; Arundhathi, R.; Reddy, P. L.; Kantam, M. L. *J. Org. Chem.* **2009**, *74*, 7951–7954. (d) Jammi, S.; Sakthivel, S.; Rout, L.; Mukherjee, T.; Mandal, S.; Mitra, R.; Saha, P.; Punniyamurthy, T. *J. Org. Chem.* **2009**, *74*, 1971–1976. (e) Hu, T.; Chen, X.; Wang, J.; Huang, J. *Chem. Cat. Chem.* **2011**, 661–665.

[69] Jiang, Y.; Qin, Y.; Xie, S.; Zhang, X.; Dong, J.; Ma, D. *Org. Lett.* **2009**, *11*, 5250–5253.

[70] Shu, Q.; Kun, X.; Junsheng, Q. *Chin. J. Chem.* **2010**, *28*, 1441–1443.

[71] Sawada, N.; Itoh, T.; Yasuda, N. *Tetrahedron Lett.* **2006**, *47*, 6595–6597.

[72] Bates, C. G.; Gujadhur, R. K.; Venkataraman, D. *Org. Lett.* **2002**, *4*, 2803–2806.

[73] Kwong, F. Y.; Buchwald, S. L. *Org. Lett.* **2002**, *4*, 3517–3520.

[74] Deng, W.; Zou, Y.; Wang, Y.-F.; Liu, L.; Guo, Q.-X. *Synlett* **2004**, 1254–1258.

[75] Zhu, D.; Xu, L.; Wu, F.; Wan, B. *Tetrahedron Lett.* **2006**, *47*, 5781–5784.

[76] Feng, Y.; Zhao, X.; Wang, J.; Zheng, F.; Xu, H. *Chin. J. Chem.* **2009**, *27*, 2423–2425.

[77] Rout, L.; Sen, T. K.; Punniyamurthy, T. *Angew. Chem. Int. Ed.* **2007**, *46*, 5583–5586.

[78] Ranu, B. C.; Saha, A.; Jana, R. *Adv. Synth. Catal.* **2007**, *349*, 2690–2696.

[79] Chen, C.-K.; Chen, Y. W.; Lin, C.-H.; Lin, H.-P.; Lee, C.-F. *Chem. Commun.* **2010**, *46*, 282–284.

[80] Schwab, R. S.; Singh, D.; Alberto, E. E.; Piquini, P.; Rodrigues, O. E. D.; Braga, A. L. *Catal. Sci. Technol.* **2011**, *1*, 569–573.

[81] Veerakumar, P.; Velayudham, M.; Lu, K.-L.; Rajagobal, S. *Catal. Sci. Technol.* **2011**, *1*, 1512–1525.

[82] (a) Sperotto, E.; van Klink, G. P. M.; de Vries, J. G.; van Koten, G. *J. Org. Chem.* **2008**, *73*, 5625–5628. (b) Rout, L.; Saha, P.; Jammi, S.; Punniyamurthy, T. *Eur. J. Org. Chem.* **2008**, 640–643. (c) Xu, H.-J.; Zhao, X.-Y.; Fu, Y.; Feng, Y.-S. *Synlett* **2008**, 3063–3067. (d) She, J.; Jiang, Z.; Wang, Y. *Tetrahedron Lett.* **2009**, *50*, 593–596. (e) Feng, Y.-S.; Li, Y.-Y.; Tang, L.; Wu, W.; Xu, H.-J. *Tetrahedron Lett.* **2010**, *51*, 2489–2492.

[83] (a) Cremlyn, R. J. *An Introduction to Organosulfur Chemistry*. John Wiley & Sons Ltd: Chichester, **1996**. (b) Nicolaou, K. C.; Boddy, C. N. C.; Bräse, S.; Winssinger, N. *Angew. Chem. Int. Ed.* **1999**, *38*, 2096–2152. (c) Theil, F. *Angew. Chem. Int. Ed.* **1999**, *38*, 2345–2347. (d) Alcaraz, M.-L.; Atkinson, S.; Cornwall, P.; Foster, A. C.; Gill, D. M.; Humphries, L. A.; Keegan, P. S.; Kemp, R.; Meriefield, E.; Nixon, R. A.; Noble, A. J.; O'Beirne, D.; Patel, Z. M.; Perkins, J.; Rowan, P.; Sadler, P.; Singleton, J. T.; Tornos, J.; Watts, A. J.; Woodland, I. A. *Org. Process Res. Dev.* **2005**, *9*, 555–569.

[84] Palomo, C.; Oiarbide, M.; Lopez, R.; Gomez-Bengoa, E. *Tetrahedron Lett.* **2000**, *41*, 1283–3517.

[85] Enguehard-Gueiffier, C.; Thery, I.; Gueiffier, A.; Buchwald, S. L. *Tetrahedron* **2006**, *62*, 6042–6049.

[86] Wu, Y. J.; He, H. *Synlett* **2003**, 1789–1790.

[87] Carril, M.; SanMartin, R.; Dominguez, E.; Tellitu, I. *Chem. Eur. J.* **2007**, *13*, 5100–5105.

[88] Verma, A. K.; Singh, J.; Chaudhary, R. *Tetrahedron Lett.* **2007**, *48*, 7199–7202.

[89] Jogdand, N. R.; Shingate, B. B.; Shingare, M. S. *Tetrahedron Lett.* **2009**, *50*, 6092–6094.

[90] Xu, H.-J.; Zhao, X.-Y.; Deng, J.; Fu, Y.; Feng, Y.-S. *Tetrahedron Lett.* **2009**, *50*, 434–437.

[91] Buranaprasertsuk, P.; Chang, J. W. W.; Chavasiri, W.; Chan, P. W. H. *Tetrahedron Lett.* **2008**, *49*, 2023–2025.

[92] Prasad, D. J. C.; Sekar, G. *Synthesis* **2010**, 79–84.

[93] Ke, F.; Qu, Y.; Jiang, Z.; Li, Z.; Wu, D.; Zhou, X. *Org. Lett.* **2011**, *13*, 454–457.

[94] Reddy, K. H. V.; Reddy, V. P.; Shankar, J.; Madhav, B.; Kumar, B. S. P. A.; Nageswar, Y. V. D. *Tetrahedron Lett.* **2011**, *52*, 2679–2682.

[95] Akkilagunta, V. K.; Kakulapati, R. R. *J. Org. Chem.* **2011**, *76*, 6819–6824.

[96] Bhadra, S.; Sreedhar, B.; Ranu, B. C. *Adv. Synth. Catal.* **2009**, *351*, 2369–2378.

[97] (a) Zeni, G. *Tetrahedron Lett.* **2005**, *46*, 2647–2651. (b) Zhang, S.; Qian, P.; Zhang, M.; Hu, M.; Cheng, J. *J. Org. Chem.* **2010**, *75*, 6732–6735. (c) Li, Y.; Li, X.; Wang, H.; Chen, T.; Xie, Y. *Synthesis* **2010**, 3602–3608. (d) Tao, C.; Lv, A.; Zhao, N.; Yang, S.; Liu, X.; Zhou, J.; Liu, W.; Zhao, J. *Synlett* **2011**, 134–138. (e) Kovacs, S.; Novak, Z. *Org. Biomol. Chem.* **2011**, *9*, 711–716. (f) Mondal, J.; Modak, A.; Dutta, A.; Bhaumik, A. *Dalton Trans.* **2011**, *40*, 5228–5235. (g) Cabrero-Antonino, J. R.; Garcia, T.; Rubio-Marqués, P.; Vidal-Moya, J. A.; Leyva-Peírez, A.; Al-Deyab, S. S.; Al-Resayes, S. I.; Díaz, U.; Corma, A. *ACS Catal.* **2011**, *1*, 147–158. (h) Mallick, S.; Rana, S.; Parida, K. *Dalton Trans.* **2011**, *40*, 9169–9175.

[98] (a) Creutz, S. E.; Lotito, K. J.; Fu, G. C.; Peters, J. C. *Science* **2012**, *338*, 647–651. (b) Uyeda, C.; Tan, Y.; Fu, G. C.; Peters, J. C. *J. Am. Chem. Soc.* **2013**, *135*, 9548–9552.

3

COPPER-CATALYZED FORMATION OF C–P BONDS WITH ARYL HALIDES

CAROLE ALAYRAC AND ANNIE-CLAUDE GAUMONT

3.1 INTRODUCTION

Copper-catalyzed P-arylation reactions have recently emerged as useful, cheaper, and less-toxic alternatives to the analogous palladium-catalyzed transformations. Although promising, this research area is still in its infancy,[1a] and mechanistic studies are so far quite limited.[1b] Early studies related to Ullmann-type cross-coupling reactions with P-nucleophiles involved stoichiometric quantities of copper in the absence of ligand, harsh reaction conditions and were limited to phosphites or H-phosphonates as reaction partners. Since 2003, the combination of a set of ligands—mostly diamines and amino acids—with various copper salts has allowed the development of catalytic P-arylation reactions under rather mild conditions with a broad variety of phosphorus derivatives: phosphines, phosphine oxides, H-phosphonates, H-phosphinates, and ammonium hypophosphite (Scheme 3.1). These reactions will be overviewed in this chapter and have been classified according to the nature of the P-nucleophile. Related transformations involving the cross coupling with alkenyl/alkynyl halides or with terminal alkynes will be overviewed in Chapter 5, Sections 5.2.4 and 5.3.4.

This chapter will be devoted mostly to cross-coupling reactions involving aryl halides, which are the reaction partners most frequently encountered. Alternative reagents such as boronic acids are emerging as alternative reaction partners (see Chapter 4 for a detailed survey) and will be also briefly described.

Copper-Mediated Cross-Coupling Reactions, First Edition. Edited by Gwilherm Evano and Nicolas Blanchard.

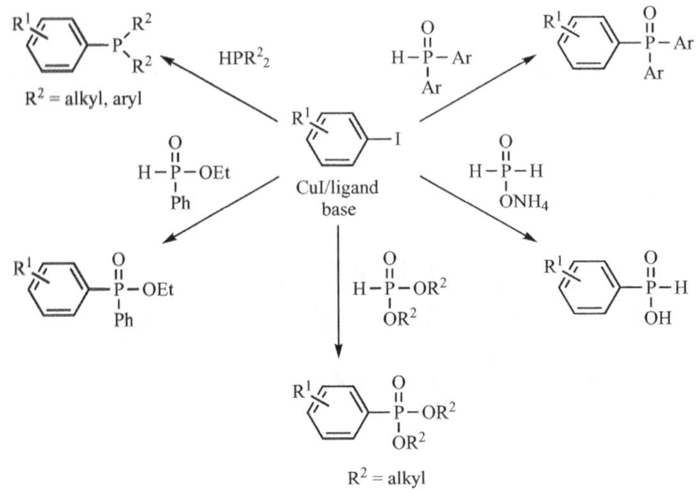

Scheme 3.1 Copper-catalyzed C–P cross-coupling reactions with aryl halides—an overview of the possibilities.

3.2 ARYLATION OF PHOSPHINES

3.2.1 Arylation of Diarylphosphines

The first examples of copper-catalyzed arylation of secondary phosphines were reported independently by the groups of Venkataraman[2] and Buchwald[3a] in 2003. The group of Venkataraman showed that the cross-coupling reaction between diphenylphosphine and various substituted aryl iodides (1.2 equiv) readily proceeded, using a catalytic amount of copper(I) iodide (10 mol %) in the presence of cesium carbonate in toluene at 110°C (Scheme 3.2).

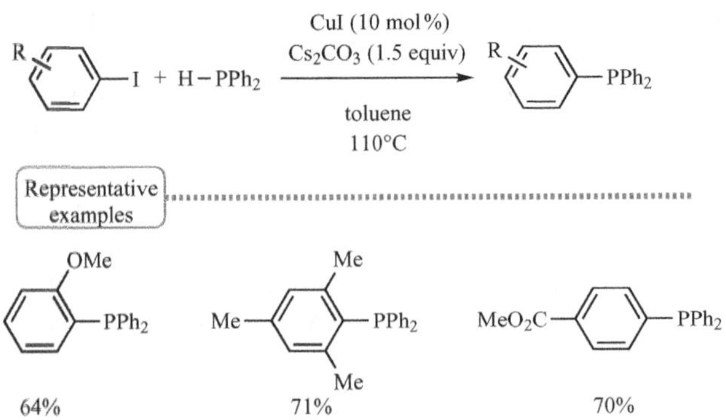

Scheme 3.2 Copper(I) iodide-catalyzed arylation of diphenylphosphine.

The reaction scope encompassed electron-rich to electron-poor substrates, and *ortho*-substituted aryl iodides were coupled with similar efficiency. Interestingly, no rate acceleration was observed in the presence of a ligand such as 1,10-phenanthroline, or with $Cu(PPh_3)_3Br$ as the catalyst, suggesting that diphenylphosphine itself might contribute as a ligand to the catalytic process. When applied to bromobenzene, the optimized reaction conditions, however, furnished the corresponding cross-coupled product (i.e. triphenylphosphine) in only 10% yield, a limitation that was nicely turned into an advantage by Buchwald and coworkers to design a new entry to (2-bromophenyl)diphenylphosphine from 2-bromo-1-iodo-benzene in 84% yield. In addition to the poor reactivity of aryl bromides, the same authors observed as further limitations of the ligand-free protocol: (i) the occurrence of halide reduction as a side reaction, in particular with *ortho*-substituted aryl iodides and (ii) an incomplete conversion when a 1:1 ratio of aryl iodides and phosphines was used. The use of an equimolar amount of the two coupling partners aimed at improving the reaction yields by simplifying the purification step. The group of Buchwald demonstrated that some limits encountered with the ligand-free reaction conditions could be circumvented when *N,N′*-dimethylethylenediamine (DMEDA) as ligand (35 mol %) was combined with copper(I) iodide (5 mol %) (Scheme 3.3, Eq. 1).

Using this protocol, sterically hindered substrates, aryl iodides bearing unusual *ortho*-substituents such as a free amine, as well as aryl bromides activated by an electron-withdrawing *para*-substituent (CO_2Me, 70% yield) were successfully coupled in good-to-excellent yields.

Quite recently, the Buchwald protocol was successfully combined to *in situ* catalytic P=O reduction for the direct conversion of secondary phosphine oxides into tertiary phosphines (Scheme 3.3, Eq. 2). The reduction of diphenylphosphine oxide or di-(3,5-xylyl)phosphine oxide was achieved using 2 equiv of tetramethyldisiloxane (TMDS) in toluene at 100°C with copper(II) triflate (10–15 mol %) as the catalyst. The latter was able to catalyze also the P-arylation of the *in situ* generated diarylphosphines in the presence of 20 mol % of DMEDA and 2 equiv of cesium carbonate at 110°C (13 examples, 58–89%).[3b]

3.2.2 Arylation of Dialkylphosphines

The cross-coupling reaction involving the more challenging dialkylphosphines was next investigated. Buchwald and coworkers successfully coupled di(isobutyl)phosphine and di(cyclohexyl)phosphine with aryl iodides under reaction conditions similar to those optimized for the cross coupling with diarylphosphines, using copper(I) iodide (5 mol %) and DMEDA (20–35 mol %) as a catalytic system in toluene at 110°C (Scheme 3.4).

Dialkylphosphines proved to be less reactive than diarylphosphines toward aryl halides under copper catalysis. Indeed, a slight excess of either dialkylphosphines or aryl iodides was usually found necessary to reach a complete

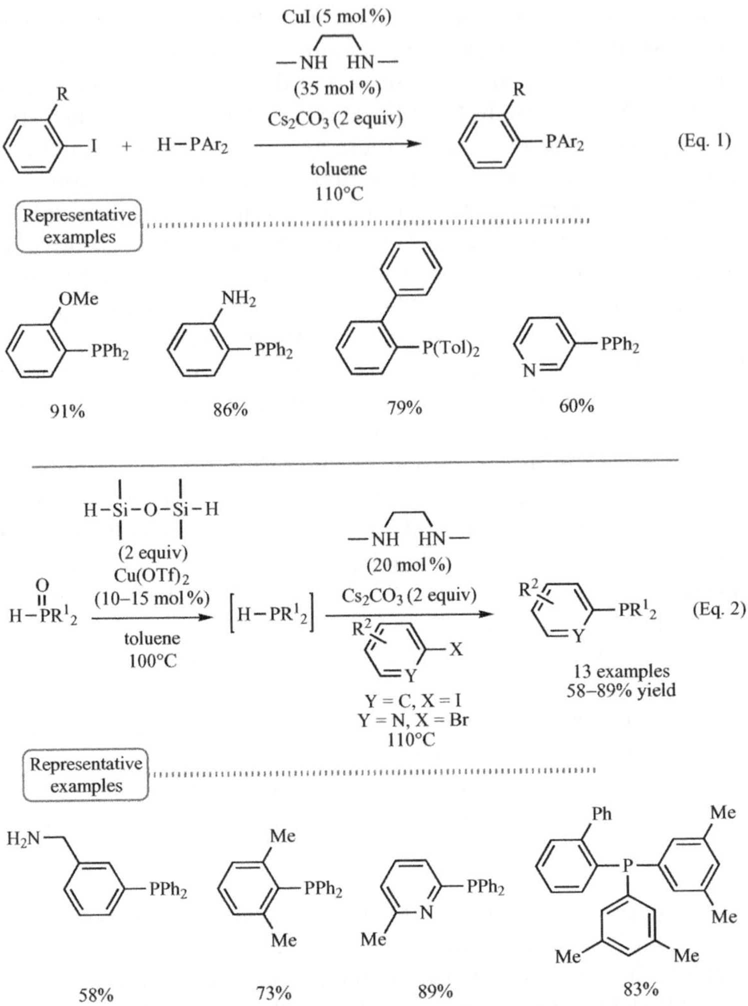

Scheme 3.3 Copper-catalyzed cross coupling of diarylphosphines with aryl iodides.

conversion. No cross coupling with aryl bromides was observed, even in the presence of the ligand. Not only the electronic effect but also the steric demand of the dialkylphosphine has a strong influence, as shown with di(*tert*-butyl) phosphine which failed to undergo the arylation under the defined reaction conditions. It is noteworthy that the targeted phenyl-di-*tert*-butylphosphine was obtained in 56% yield when the copper(II)triflate-catalyzed domino reduction—arylation sequence developed by Beller and coworkers (see Section 3.2.1) was applied to di-*tert*-butylphosphine oxide and iodobenzene, suggesting a positive influence of the excess of silane used in the reduction step.[3b]

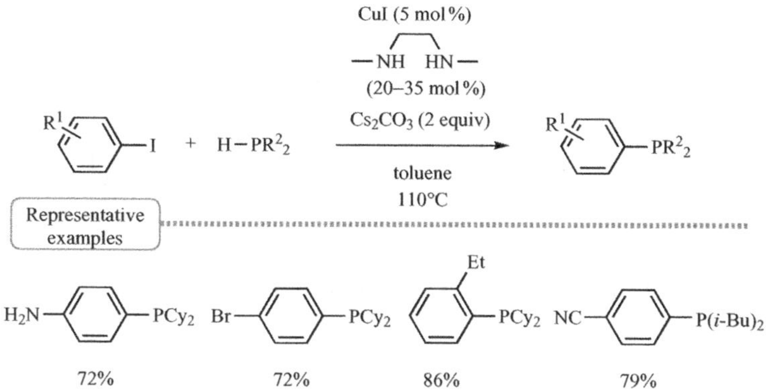

Scheme 3.4 Copper(I) iodide/diamine-catalyzed cross coupling of dialkylphosphines with aryl iodides.

3.2.3 Application to the Synthesis of Bulky Tertiary Phosphine Ligands

The main feature of the Buchwald procedure is its tolerance toward sterically hindered aryl halides, which allows access to a broad diversity of bulky phosphine ligands. It is, for example, applicable to the cross-coupling reaction between *ortho*-substituted aryl iodides and dialkylphosphines (Scheme 3.5) and might therefore offer a straightforward entry to biaryldialkylphosphines, most useful ligands for transition metal-catalyzed transformations.[4]

Stoltz and coworkers[5] reported a facile and modular synthesis of an important class of *P,N*-ligands, the phosphinooxazoline (PHOX) ligands,[6] based on the Buchwald protocol. The latter was slightly modified, mostly regarding the amount of catalyst and reagents, in order to efficiently couple various aryl bromides bearing an oxazoline moiety in the *ortho*-position with diarylphosphines (Scheme 3.6).

It is noteworthy that good yields, ranging from 53% to 89%, were obtained regardless of the bulkiness of the oxazoline moiety and the electronic nature of the aryl bromides used in the cross coupling. Moreover, electron-rich (Ar = *p*-Tol) and electron-poor (Ar = *p*-$CF_3C_6H_4$) diarylphosphines displayed a similar reactivity. The authors were also able to extend the coupling reaction to dialkylphosphines, which are known to be less reactive than diarylphosphines (Scheme 3.7). Di(isobutyl)phosphine and di(cyclohexyl) phosphine were introduced as their tetrafluoroboronic acid salts[7] to facilitate their handling. Increasing the amount of base (5 equiv instead of 3.75 equiv) allowed the *in situ* generation of the free secondary dialkylphosphines and their conversion to the corresponding aryldialkylphosphines after a reaction time of 24–27 hours (vs. 2–4 hours in the case of diarylphosphines). The coupling products were isolated as their borane adducts after treatment with borane-tetrahydrofuran complex (BH_3·THF).

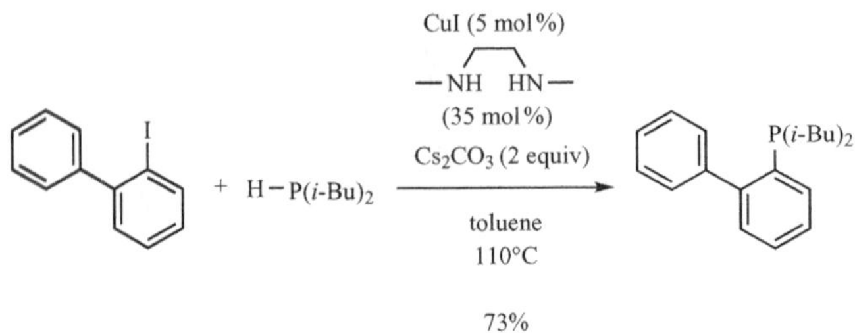

Scheme 3.5 Copper-catalyzed synthesis of biaryldialkylphosphines.

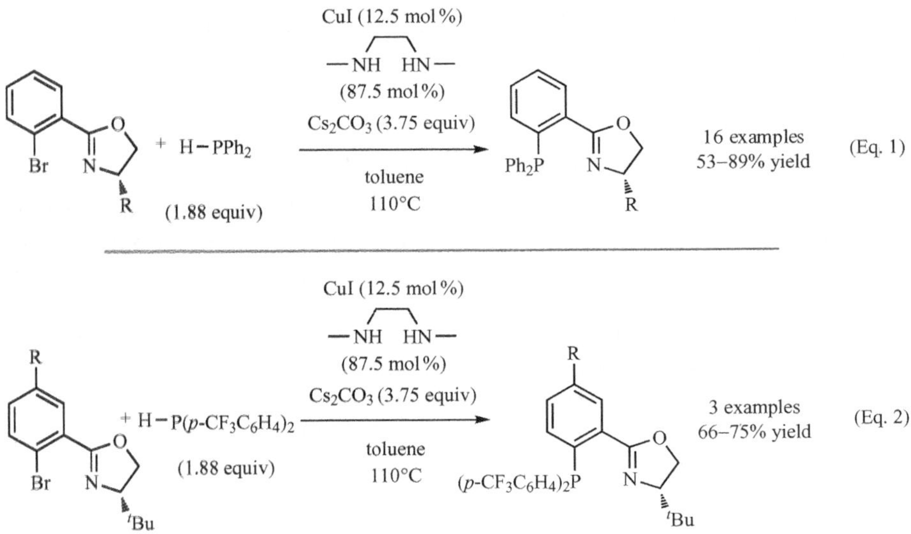

Scheme 3.6 Copper-catalyzed synthesis of diaryl phosphinooxazoline (PHOX) ligands.

Decomplexation of these tertiary phosphine boranes can be readily achieved in near quantitative yields by using the Hiyama conditions (4Å MS, THF/MeOH).[8]

3.3 ARYLATION OF PHOSPHINE OXIDES AND PHOSPHITES

3.3.1 Arylation of Phosphine Oxides

The few examples of copper-catalyzed P-arylation of phosphine oxides reported in the literature are mostly focusing on diarylphosphine oxides as cross-coupling partners. The group of Stoltz showed that the reaction

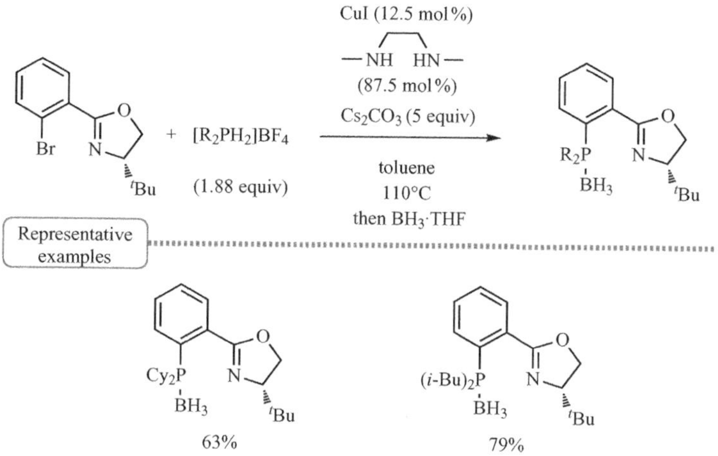

Scheme 3.7 Copper-catalyzed synthesis of dialkyl phosphinooxazoline (PHOX) ligands.

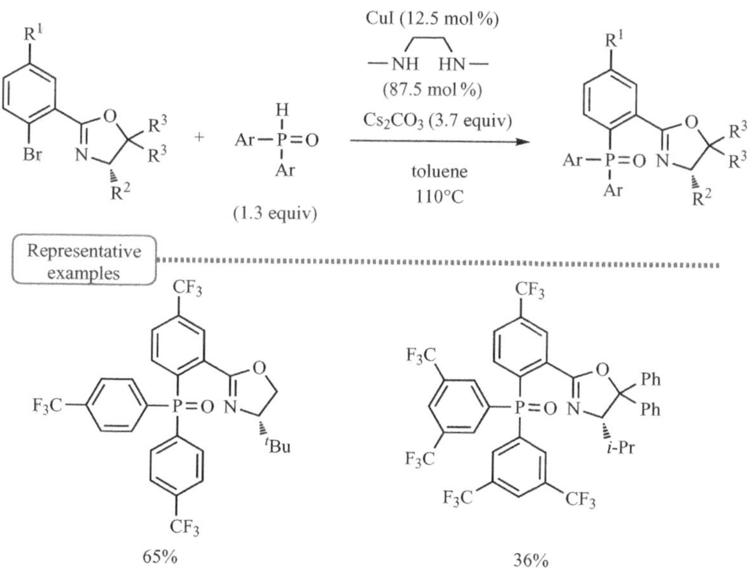

Scheme 3.8 Copper-catalyzed synthesis of electron-deficient triarylphosphine oxides.

conditions defined for the copper-catalyzed cross-coupling reaction between oxazoline-containing aryl bromides and secondary phosphines (see Section 3.2.3) were applicable to electron-deficient secondary phosphine oxides bearing trifluoromethylaryl substituents, provided that the reaction time was extended to 38–42 hours (Scheme 3.8).[9] The catalytic system, consisting of copper(I) iodide (12.5 mol %) and DMEDA (87.5 mol %), promoted the cross

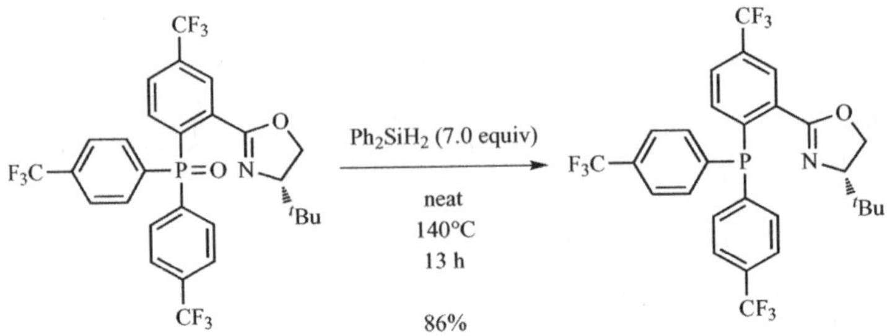

Scheme 3.9 Synthesis of (S)-p-$(CF_3)_3$-^{t}BuPHOX via silane reduction of the corresponding triarylphosphine oxide.

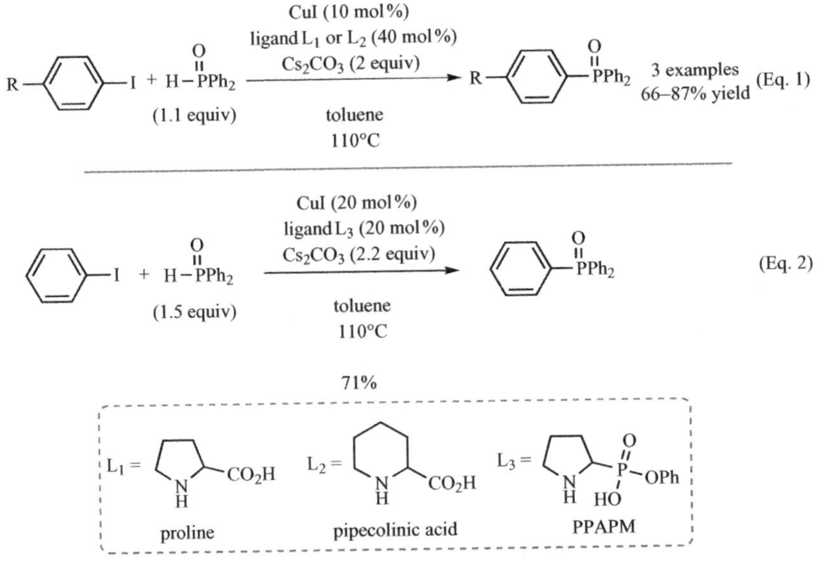

Scheme 3.10 Copper-catalyzed arylation of diphenylphosphine oxide using N,O-ligands.

coupling to triarylphosphine oxides in yields ranging from 36% to 46% and from 55% to 85% starting from bis- and mono-trifluoromethyl-substituted diaryl phosphine oxides, respectively.

The reduction of the resulting tertiary phosphine oxides with diphenylsilane provides a variety of new electron-deficient PHOX ligands as illustrated with the example shown in Scheme 3.9.

As an alternative to DMEDA, the group of Fu showed that proline[10] and related bidentate ligands such as pipecolinic acid and pyrrolidine-2-phosphonic acid phenyl monoester (PPAPM),[11] in combination with copper(I) iodide, were efficient promoters for the copper-catalyzed P-arylation of diphenylphosphine oxide (Scheme 3.10). The cross-coupling reaction, which was only tested on three examples of aryl iodides, was performed in toluene at 110°C for 36 hours. The catalyst loading was 10 mol % of copper(I) iodide combined with 40 mol % of ligand in the case of proline and pipecolinic acid, whereas a higher catalyst loading of 20 mol % of copper(I) iodide combined with 20 mol % of ligand was used in the case of PPAPM. The resulting triarylphosphine oxides were isolated in good to high yields ranging from 66% to 87%.

The same group also reported that the copper-catalyzed coupling of 2-halotrifluoroacetanilides with diphenylphosphine oxide proceeded under especially mild conditions (50°C), which is rather unusual for a copper-catalyzed P-arylation reaction (Scheme 3.11, Eq. 1).[12a] The ligand of choice for this transformation was found to be *N*-methylpyrrolidine-2-carboxamide, and the resulting aryl phosphine oxides were isolated in 78–84% yields. Nevertheless, this method displays some limits: (i) the high catalyst loading (30 mol % of copper(I) iodide and 100 mol % of ligand) and (ii) the narrow scope of this reaction, which seems to be driven by an *ortho*-effect of the trifluoroacetamide group. Indeed, the P-arylation of the analogous *meta*- or *para*-halotrifluoroacetanilides did not proceed under the exact same reaction conditions. Worthy of note is the first example of copper-catalyzed C–P bond-forming reaction involving an aryl chloride, which was achieved in 50% yield provided that the reaction temperature was raised to 110°C (Scheme 3.11, Eq. 2).

According to a recent report from the group of Stankevič, (*S*)-α-methylbenzylamine seems to be superior to previously employed ligands to promote the P-arylation of secondary phosphine oxides (Scheme 3.11, Eq. 3).[12b] Indeed the amount of monodentate ligand required in combination with copper(I) iodide (10 mol %) is moderate (20 mol %). The defined catalytic system proved to be highly efficient in terms of yields and scope, and it was successfully applied to the cross-coupling reaction of diaryl-, aryl-alkyl, and dialkyl-phosphine oxides (69–99% yield). It should be mentioned that all obtained P-stereogenic cross-coupling products were racemic.

3.3.2 Synthesis of Aryl Phosphonates by Arylation of Trialkylphosphites and H-Phosphonates

The first examples of copper-mediated synthesis of aryl phosphonates were related to Ullmann-type coupling reactions with trialkyl phosphites, H-phosphonates, or their anions. A stoichiometric amount of copper in the absence of ligand was used under harsh conditions (very high temperatures ranging from 120 to 160°C, toxic solvent hexamethylphosphoric triamide [HMPT]). Since 2003, the use of ligands in combination with copper salts has allowed development of much milder reaction conditions. A variety of ligands combined with copper(I) iodide as copper source is indeed effective in the coupling of aryl halides with H-phosphonates.[3,10,11,13] Very recently, a new procedure was developed based on a Chan–Lam–Evans type aerobic coupling of dialkyl H-phosphonates with aryl boronic acids, using a catalytic system

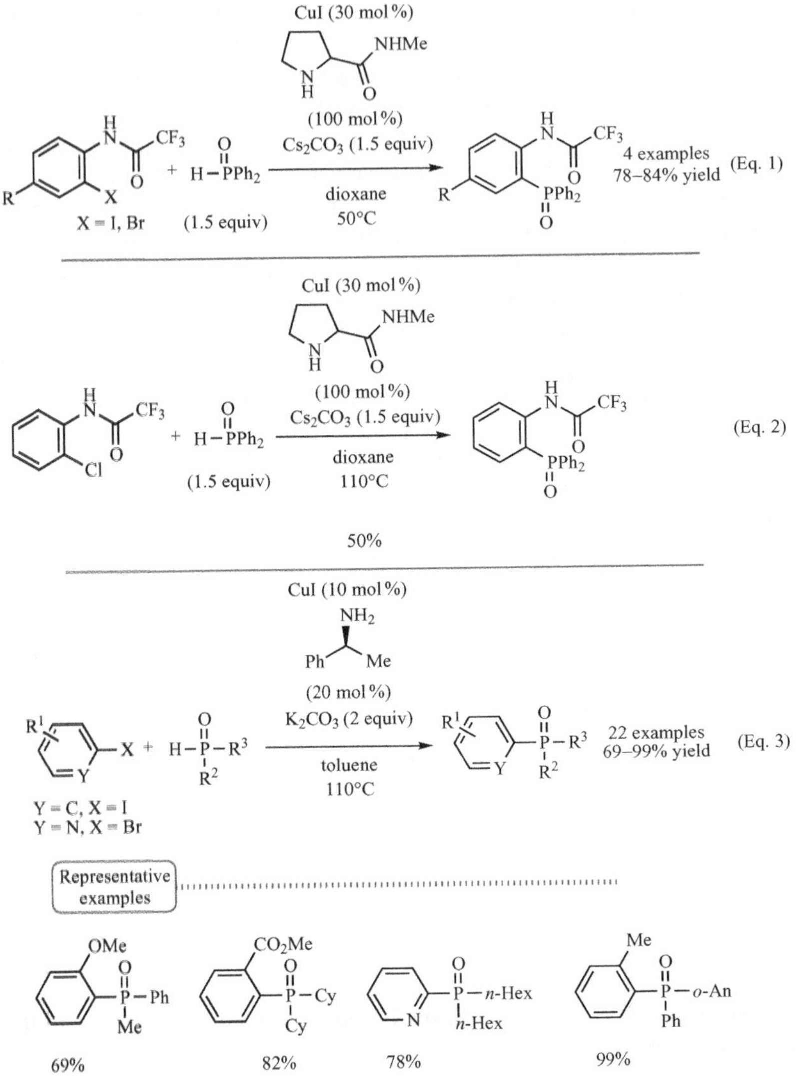

Scheme 3.11 *Ortho* effect in the copper-catalyzed cross coupling of diphenylphosphine oxide with 2-halo-trifluoroacetanilides and copper-catalyzed cross-coupling of disubstituted phosphine oxides using (*S*)-α-methylbenzylamine as a ligand.

based on copper(I) oxide and 1,10-phenanthroline (see Chapter 4 for more details on the Chan–Lam–Evans cross-coupling reaction).[14]

3.3.2.1 Early Developments: Stoichiometric Conditions. In 1967, Tavs and Korte reported the copper-mediated synthesis of aryl phosphonates from trialkyl phosphites (Scheme 3.12, Eq. 1).[15] This extension of the Arbuzov

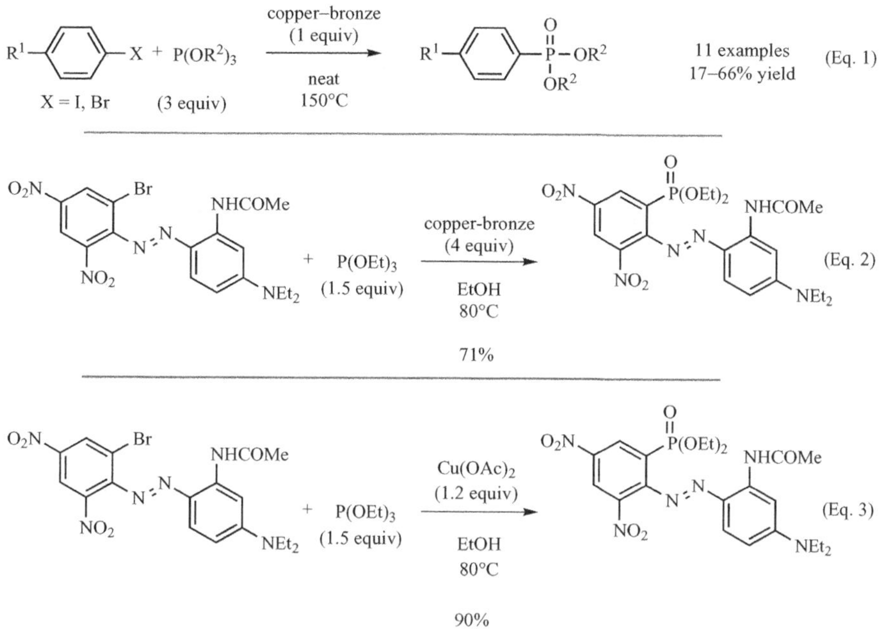

Scheme 3.12 Synthesis of aryl phosphonates via a copper-mediated cross coupling of aryl halides with trialkylphosphites.

reaction, usually performed on alkyl halides,[16] is made possible by the use of a stoichiometric amount of copper–bronze. The reaction proceeded under harsh conditions since it was run at 150°C for 1 day without solvent. The aryl phosphonates were obtained in low-to-moderate yields of 17–25% and 53–66% from aryl bromides and iodides, respectively, whereas no P-arylation reaction was observed with aryl chlorides. About 10 years later, this procedure was adapted to sensitive *ortho*-bromodiarylazo-compounds by Hall and Price.[17] These particular activated aryl bromides underwent the coupling reaction with trialkyl phosphites under milder conditions (80°C, 15 minutes, EtOH as solvent) (Scheme 3.12, Eq. 2), which might be attributed to a strong chelating effect of the diazo group. Other copper sources such as copper(II) acetate and copper(I) acetate were also able to promote this transformation (Scheme 3.12, Eq. 3).

The same authors showed that these functionalized aryl phosphonates were also available through the use of H-phosphonates as coupling partners with *ortho*-bromodiarylazo-compounds, in the presence of copper(I) iodide and sodium acetate in refluxing ethanol.[18] Ogawa and coworkers reported another example of copper(I) iodide promoted P-arylation of diethyl H-phosphonate with unactivated bromobenzene (Scheme 3.13),[19] a cross coupling that required harsher reaction conditions (tetramethylurea [TMU], 120–140°C, 24 hours) due to the absence of an *ortho*-effect.

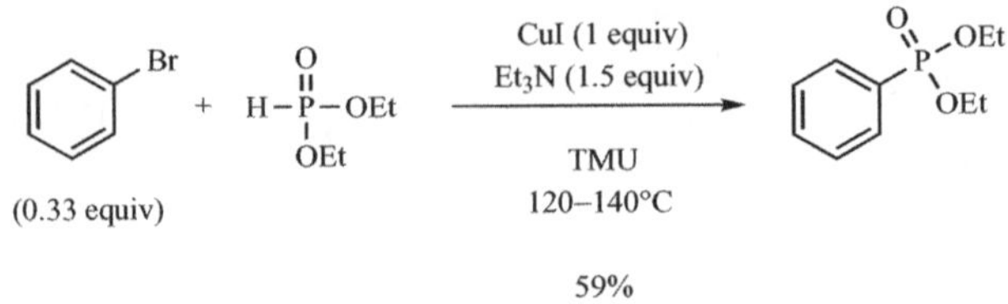

Scheme 3.13 Copper(I) iodide-mediated arylation of diethyl H-phosphonate with bromobenzene.

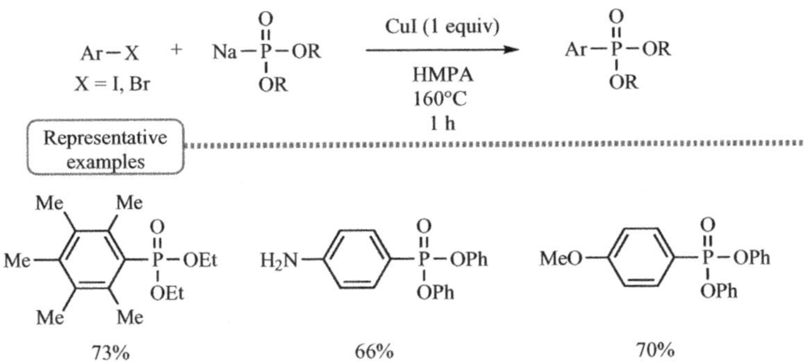

Scheme 3.14 Copper(I) iodide-mediated synthesis of aryl phosphonates from sodium phosphonates and aryl halides.

Osuka and coworkers developed an interesting entry to aryl phosphonates based on the arylation of sodium phosphonates in the presence of a stoichiometric amount of copper(I) iodide.[20] The aryl phosphonates were obtained in good yields (54–81%) from either sodium diphenylphosphonate or sodium diethylphosphonate. The reaction scope is broad, and bulky substrates as well as functional groups such as a free amino group, which are rather uncommon in copper-mediated couplings, are tolerated (Scheme 3.14). However, the reaction conditions are harsh (160°C) and a toxic solvent, hexamethylphosphoramide (HMPA), is required. The reaction was also applied to aryl chlorides but required an even higher reaction temperature (170–180°C) and provided the corresponding aryl phosphonates in low yields (20–30%).

3.3.2.2 The Emergence of Catalytic Conditions. While quite efficient for the preparation of simple derivatives, these early procedures for the arylation of phosphonates were clearly suffering from the harshness of the conditions and the use of stoichiometric amounts of copper. These limitations could be circumvented by the introduction of ligands for copper salts, which allowed the development of catalytic and rather mild procedures by solubilizing the catalyst, protecting it, and/or rendering the reaction a lot faster. In a pioneering work, Buchwald and coworkers reported that dibutyl H-phosphonate readily underwent copper-catalyzed P-arylation under the reaction conditions defined

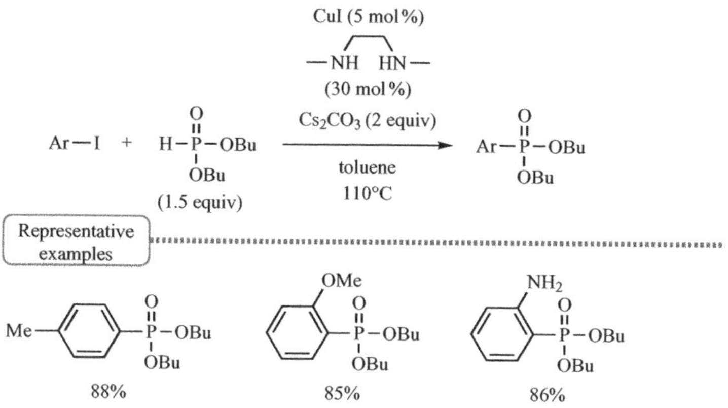

Scheme 3.15 Copper(I) iodide/diamine-catalyzed arylation of dibutyl H-phosphonate with aryl iodides.

for phosphines (see Section 3.2.1).[3a] The cross coupling of dibutyl H-phosphonate was achieved with three examples of *para-* or *ortho*-substituted aryl iodides, using a combination of copper(I) iodide (5 mol %) and DMEDA (30 mol %) in toluene at 110°C (Scheme 3.15). The resulting aryl phosphonates were obtained in high yields ranging from 85% to 88%.

The group of Fu introduced additional ligands capable of promoting the copper-catalyzed P-arylation of diakyl H-phosphonates: commercially available, bidentate proline and pipecolinic acid as well as PPAPM. These three ligands are efficient in combination with copper(I) iodide (Scheme 3.16).[10,11] Worthy of note is, however, that the catalyst loading is twice to four times higher than the one used with DMEDA as ligand in the Buchwald protocol. As with the latter, cesium carbonate was found the best base for this transformation. Aryl iodides showed significantly higher reactivity than aryl bromides, therefore enabling the selective monocoupling of 4-bromo-1-idobenzene with diethyl H-phosphonate, affording 4-bromophenylphosphonate in high yield (Scheme 3.16, Eq. 3). The addition of 2 equiv of potassium iodide was necessary to perform efficient P-arylation of H-phosphonates from aryl bromides, through *in situ* conversion of aryl bromides into the corresponding aryl iodides, via an aromatic Finkelstein reaction (see Chapter 6 for details on the Finkelstein reaction).

The same group reported that the copper-catalyzed cross coupling of dialkyl H-phosphonates with 2-iodo- as well as 2-bromo-trifluoroacetanilides proceeded under very mild conditions (45°C, 1.5–4.5 hours) and high yields (90–93%) using *N*-methylpyrrolidine-2-carboxamide as ligand in combination with copper(I) iodide (Scheme 3.17).[12a] The activating *ortho*-effect of the trifluoroacetamide group was demonstrated by the fact that the reaction of 2,4-dibromotrifluoroacetanilide with diisopropyl H-phosphonate only provided the monocoupling product. The same observation was made with diphenylphosphine oxide as coupling partner (see Section 3.3.1).

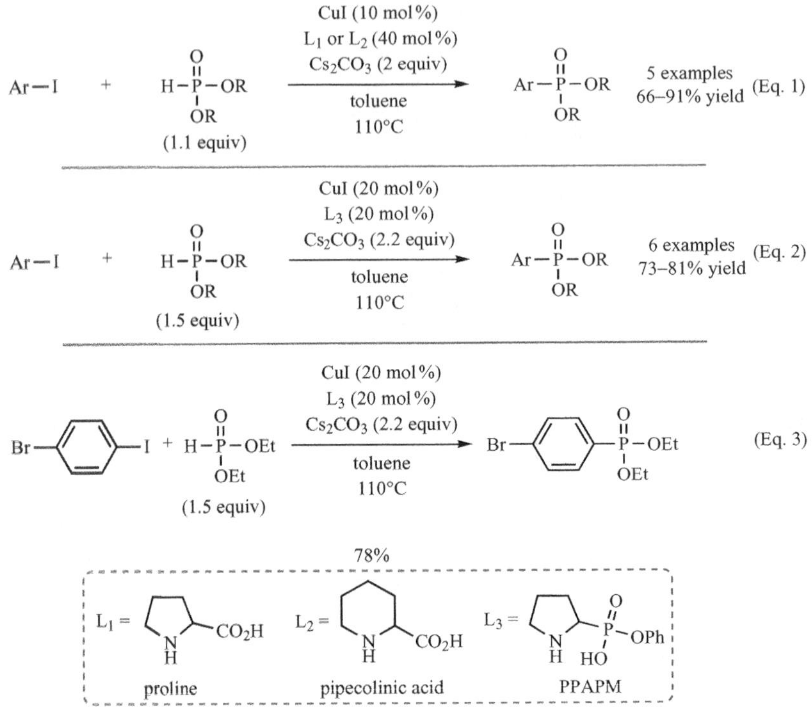

Scheme 3.16 Copper-catalyzed arylation of dialkyl H-phosphonates using *N,O*-ligands.

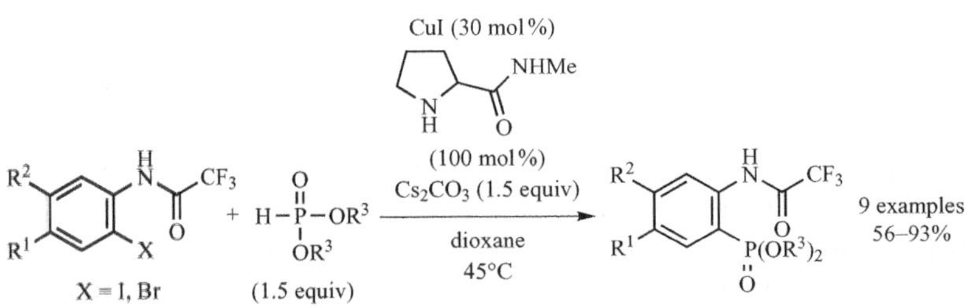

Scheme 3.17 *Ortho* effect in the copper-catalyzed cross coupling of dialkyl H-phosphonates with 2-halo-trifluoroacetanilides.

The few catalytic procedures for copper-catalyzed P-arylation described so far usually required high amounts of ligand (30–100 mol %). Recent studies by Karlstedt and Beletskaya dealing with copper-catalyzed cross coupling of diethyl phosphonate with aryl iodides demonstrated that the amount of ligand could be reduced to 10 mol % when 1,10-phenanthroline was used in combination with copper(I) iodide (Scheme 3.18).[13] Interestingly, higher yields were obtained when cesium carbonate was used instead of potassium carbonate as

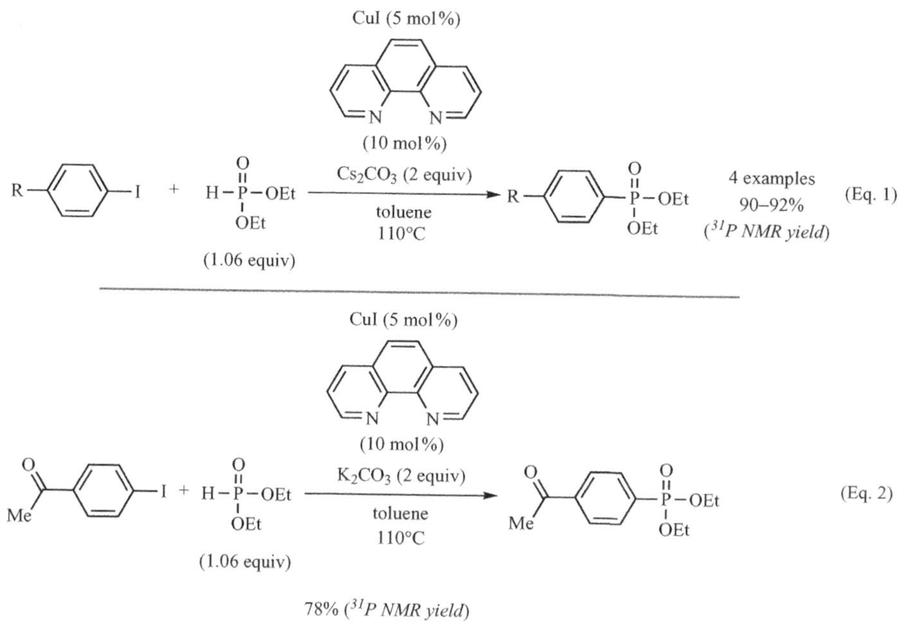

Scheme 3.18 Copper(I) iodide/1,10-phenanthroline-catalyzed cross coupling of diethyl phosphonate with aryl iodides.

the base, with the exception of the arylation of diethyl phosphonate with *p*-acetylphenyl iodide (Scheme 3.18, Eq. 2). In this case, the use of potassium carbonate was preferred to minimize the formation of an α-hydroxy phosphonate resulting from a side Abramov reaction (nucleophilic attack of the phosphorus atom onto the carbonyl function).[21]

Very recently, an original protocol was reported by Zhao and coworkers.[14] It contrasts with the previously known copper-catalyzed syntheses of aryl phosphonates by the fact that the coupling substrates are not aryl halides anymore but aryl boronic acids (see Chapter 4 for a detailed account on the use of aryl boronic acids in copper-promoted cross-coupling reactions). The optimal catalytic system was found to be a combination of copper(I) oxide and 1,10-phenanthroline. The reaction conditions are very mild (room temperature, under air, acetonitrile, 24 hours): The reaction tolerates a broad variety of functional groups (ketone, ester, vinyl, halide) and both electron-rich and electron-poor aryl boronic acids are suitable (Scheme 3.19). Interestingly, aryl halides did not undergo C–P bond formation under the defined reaction conditions, which might prove to be especially useful for chemoselective transformations. Only a trace amount of coupling product was obtained when the reaction was performed under inert atmosphere, demonstrating that oxygen is essential. The authors suggest that oxygen could oxidize the intermediate copper(II) complex generated after the transmetallation step of a copper(II)

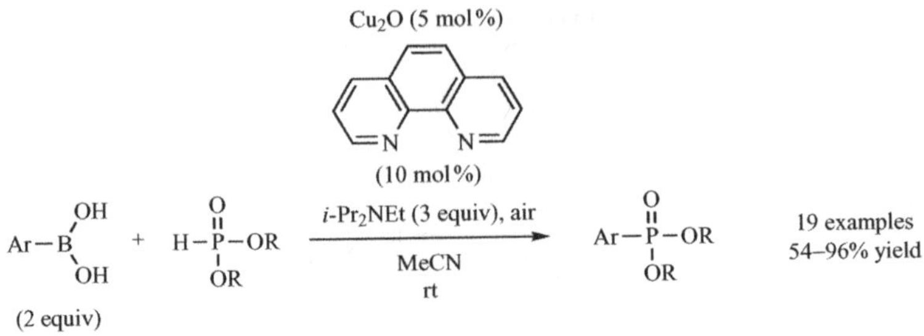

Scheme 3.19 Copper-catalyzed aerobic cross coupling of aryl boronic acids with dialkyl H-phosphonates.

active species to a copper(III) complex, therefore facilitating the reductive elimination step (see Chapter 7 for details on the mechanism of the Chan–Lam–Evans cross-coupling reaction). In addition a competing side reaction, the formation of tetraethyl hypodiphosphate, through homocoupling of H-phosphonates,[22] took place when the reaction was performed under pure oxygen and significantly lowered the yield. The reaction outlined on Scheme 3.19 is attractive in view of its mild conditions (room temperature) and its broad scope regarding the aryl boronic acids that can be used as reaction partners.

3.3.3 Arylation of Other Phosphorus Derivatives

The group of Fu successfully extended the procedure involving combinations of copper(I) iodide and proline-related ligands for the arylation of less common phosphorus derivatives such as ethyl phenylphosphinate and ammonium hypophosphite.[10,11] These are the only examples of copper-catalyzed synthesis of bis-aryl phosphinates and mono- or bis-aryl phosphinic acid.

3.3.3.1 Arylation of Ethyl Phenylphosphinate. Fu and coworkers reported that copper(I) iodide, combined with proline, pipecolinic acid, or PPAPM as ligands, was the optimal catalytic system to achieve the P-arylation of ethyl phenylphosphinate with aryl iodides (Scheme 3.20).[10,11] The reaction conditions (toluene, 110°C and cesium carbonate as base) are identical to those defined in the case of diphenylphosphine oxide and dialkyl H-phosphonates (see Sections 3.3.1 and 3.3.2.2). This method offers a straightforward entry to symmetrical and unsymmetrical bis-aryl phosphinates in good to high yields (67–87%).

3.3.3.2 Arylation of Ammonium Hypophosphite. The same authors showed that the copper-based catalytic system that they developed for the P-arylation of phosphine oxides and H-phosphonates was also applicable to ammonium hypophosphite using iodobenzene as the coupling partner.[10,11] Noteworthy is that some modifications of the reaction conditions were required, regarding

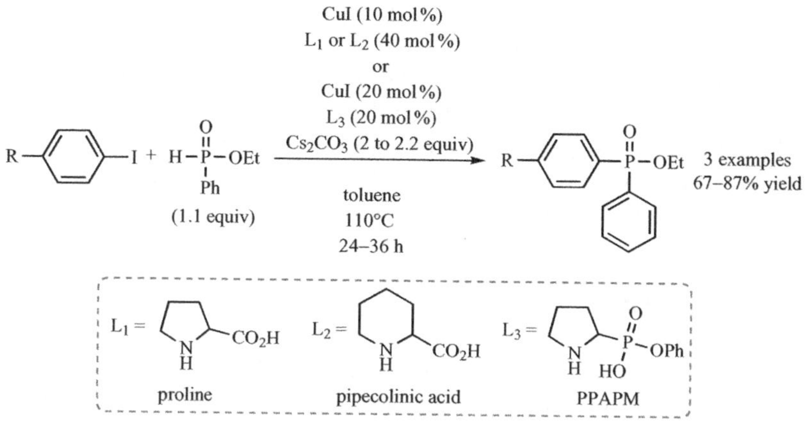

Scheme 3.20 Copper-catalyzed arylation of ethyl phenylphosphinate with aryl iodides.

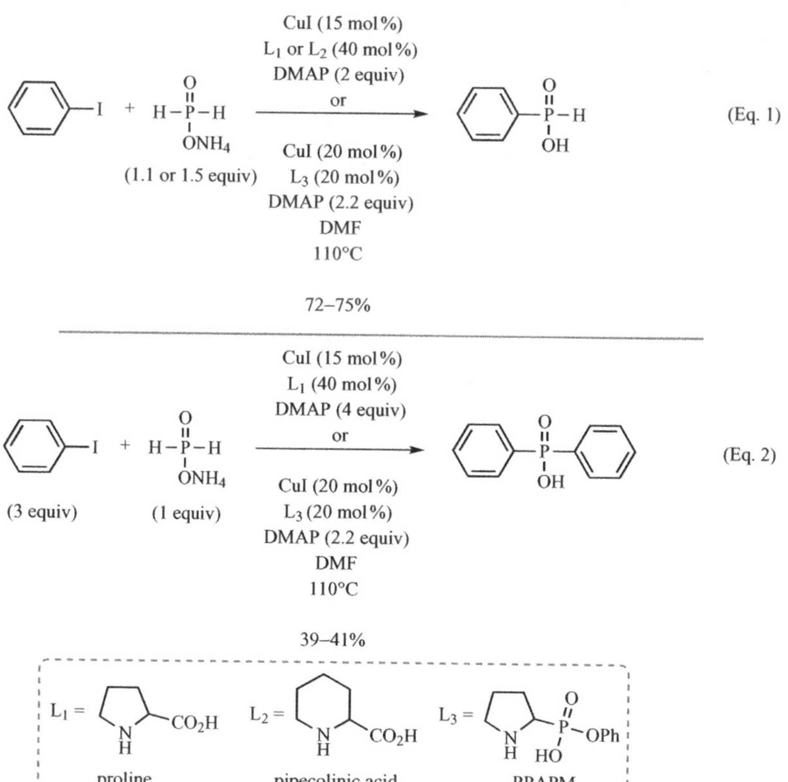

Scheme 3.21 Copper-catalyzed cross coupling of ammonium hypophosphite with iodobenzene.

the solvent (*N*,*N*-dimethylformamide [DMF] instead of toluene), the base (4-dimethylaminopyridine [DMAP] instead of cesium carbonate), and the reaction time, which was prolonged to 48 hours. Depending on the number of equivalents of iodobenzene used, either the mono- (less than 1 equiv) or the bis-aryl phosphinic acid (about 3 equiv) could be obtained in about 73% and 40% yield, respectively (Scheme 3.21). The scope of this reaction was not further explored with substituted aryl iodides.

3.4 CONCLUSION

Within the last 10 years, various copper-catalyzed P-arylation reactions have emerged as straightforward and efficient approaches toward a variety of aryl phosphorus derivatives. Copper iodide and aryl halides (mostly aryl iodides) are usually used as copper source and substrates, respectively. The set of ligands employed so far includes DMEDA, proline derivatives, and phenanthroline-type ligands. The scope of phosphorus coupling partners up to now comprises diaryl and dialkyl phosphines, or phosphine oxides, dialkyl H-phosphonates, ethyl phenylphosphinate, and ammonium hypophosphite. It is noteworthy that the electronic effects and the steric demand of the aryl halides have little influence on these couplings. Consequently, the copper-catalyzed P-arylation of sterically hindered aryl halides, notably those bearing bulky substituents *ortho*-located to the halide, proceeds with high yields with minor arene reduction as side reaction. This makes copper-catalyzed couplings a useful alternative to palladium-catalyzed transformations to access bulky aryl phosphorus derivatives, beyond the criteria related to costs and toxicity. Some of the drawbacks of these copper-catalyzed cross couplings to be mentioned are (i) a high catalyst loading (up to 30 mol % of copper(I) iodide), (ii) a high amount of ligand (up to 1 equiv), and (iii) a rather high reaction temperature (typically 110°C). Finding new catalytic systems that would allow milder reaction conditions is an important issue, notably for future extension to the asymmetric series. Quite recently, a new catalytic synthesis of aryl phosphonates proceeding at room temperature was reported based on the aerobic coupling of dialkyl H-phosphonates with aryl boronic acids. It is likely that other mild protocols will be developed soon and that the scope of copper C–P couplings will be broadened in the near future.

REFERENCES

[1] (a) Tappe, F. M. J.; Trepohl, V. T.; Oestreich, M. *Synthesis* **2010**, 3037–3062. (b) Font, M.; Parella, T.; Costas, M.; Ribas, X. *Organometallics* **2012**, *31*, 7976–7982.

[2] Van Allen, D.; Venkataraman, D. *J. Org. Chem.* **2003**, *68*, 4590–4593.

[3] (a) Gelman, D.; Jiang, L.; Buchwald, S. L. *Org. Lett.* **2003**, *5*, 2315–2318. (b) Li, Y.; Das, S.; Zhou, S.; Junge, K.; Beller, M. *J. Am. Chem. Soc.* **2012**, *134*, 9727–9732.

[4] Surry, D. S.; Buchwald, S. L. *Angew. Chem. Int. Ed.* **2008**, *47*, 6338–6361.

[5] Tani, K.; Behenna, D. C.; McFadden, R. M.; Stoltz, B. M. *Org. Lett.* **2007**, *9*, 2529–2531.

[6] Helmchen, G.; Pfaltz, A. *Acc. Chem. Res.* **2000**, *33*, 336–345.

[7] Netherton, M. R.; Fu, G. C. *Org. Lett.* **2001**, *3*, 4295–4298.

[8] Schröder, M.; Nozaki, K.; Hiyama, T. *Bull. Chem. Soc. Jpn* **2004**, *77*, 1931–1932.

[9] McDougal, N. T.; Streuff, J.; Mukherjee, H.; Virgil, S. C.; Stoltz, B. M. *Tetrahedron Lett.* **2010**, *51*, 5550–5554.

[10] Huang, C.; Tang, X.; Fu, H.; Jiang, Y.; Zhao, Y. *J. Org. Chem.* **2006**, *71*, 5020–5022.

[11] Rao, H.; Jin, Y.; Fu, H.; Jiang, Y.; Zhao, Y. *Chem. Eur. J.* **2006**, *12*, 3636–3646.

[12] (a) Jiang, D.; Jiang, Q.; Fu, H.; Jiang, Y.; Zhao, Y. *Synthesis* **2008**, 3473–3477. (b) Stankevič, M.; Włodarczyk, A. *Tetrahedron* **2013**, *69*, 73–81.

[13] Karlstedt, N. B.; Beletskaya, I. P. *Russ. J. Org. Chem.* **2011**, *47*, 1011–1014.

[14] Zhuang, R.; Xu, J.; Cai, Z.; Tang, G.; Fang, M.; Zhao, Y. *Org. Lett.* **2011**, *13*, 2110–2113.

[15] Tavs, P.; Korte, F. *Tetrahedron* **1967**, *23*, 4677–4679.

[16] Bhattacharya, A. K.; Thyagarajan, G. *Chem. Rev.* **1981**, *81*, 415–430.

[17] Hall, N.; Price, R. *J. Chem. Soc. Perkin Trans. 1* **1979**, 2634–2641.

[18] Hall, N.; Price, R. *J. Chem. Soc. Perkin Trans. 1* **1979**, 2873–2877.

[19] Ogawa, T.; Usuki, N.; Ono, N. *J. Chem. Soc. Perkin Trans. 1* **1998**, 2953–2958.

[20] Osuka, A.; Ohmasa, N.; Yoshida, Y.; Suzuki, H. *Synthesis* **1983**, 69–71.

[21] Engel, R. *Org. React.* **1988**, *36*, 175–248.

[22] Zhou, Y.; Yin, S.; Gao, Y.; Zhao, Y.; Goto, M.; Han, L.-B. *Angew. Chem. Int. Ed.* **2010**, *49*, 6852–6855.

4

ALTERNATIVE AND EMERGING REAGENTS FOR THE ARYLATION OF HETERONUCLEOPHILES

LUC NEUVILLE

4.1 INTRODUCTION

Functional groups such as aryl ethers, aryl amines, or thioethers are ubiquitous units found in a number of important natural or biologically relevant compounds, agrochemicals, or materials. Copper has been playing a central role for the construction of such bonds since the discovery by Fritz Ullmann and Irma Goldberg of a reaction that would be named after them, the copper-promoted Ullmann–Goldberg condensation.[1] However, the harsh conditions associated with the original protocol seriously limited the scope of the method,[2] and a useful alternative was found through the rapid development of palladium-catalyzed cross-coupling reactions for the formation of carbon-heteroatom bonds.[3] This approach is, however, not without drawbacks, requiring elevated temperature, strong bases, and expensive ligands and palladium complexes. The limitations associated with cost issues have drawn the chemical community's attention back to the "old" copper processes. Tremendous progress has recently been achieved in the classical Ullman synthesis as shown in various chapters of this book.[4] Driven by the need for milder conditions in order to build carbon-heteroatom bonds on increasingly more complex targets, alternative copper-promoted approaches have emerged. Two excellent and comprehensive reviews covering these methods appeared in the literature in 2002 and 2003.[5]

Copper-Mediated Cross-Coupling Reactions, First Edition. Edited by Gwilherm Evano and Nicolas Blanchard.

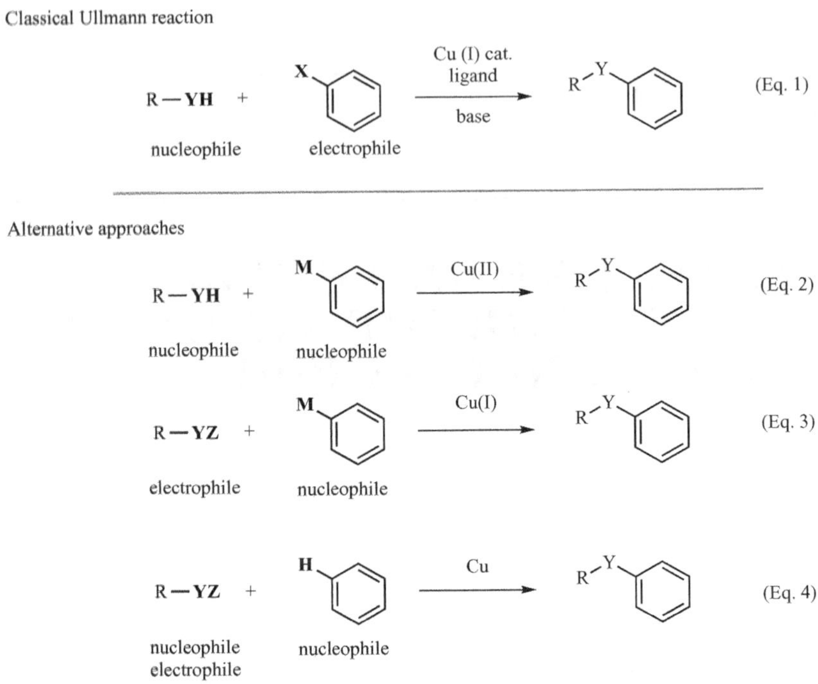

Figure 4.1 Classical versus alternative approaches of copper-promoted heteroarylation.

These alternative approaches are illustrated in Figure 4.1 and can be divided in three categories based on the polarity of the reagents. Schematically, the classical Ullmann condensation involves a copper(I)-catalyzed reaction of a heteronucleophile with an electrophilic arene, usually an aryl halide (Figure 4.1, Eq. 1). The use of (pseudo)-metallated arenes as arylating agents (Figure 4.1, Eq. 2), is certainly the most investigated alternative approach. Illustrative of such an approach is the Chan–Lam–Evans coupling involving boronic acids, which already belongs to the classical toolbox of chemists. Studies closely related to the classical Ullman condensation, but using reverse electrophilicity/nucleophilicity of both reacting partners, have also led to the development of effective heteroarylation methods. In such processes, an oxidized form of the heteroatom is reacted with a metallated arene (Figure 4.1, Eq. 3). Finally, recent research in the field of C–H functionalization has opened the way for copper-promoted direct C–H functionalization of arenes or heteroarenes (Figure 4.1, Eq. 4).

All these approaches will be covered in this chapter, which is organized as follows: The first part is devoted to oxidative copper-promoted reactions involving organoboron species, while the use of alternative metallated aryl derivatives is covered in the second section. Heteroarylation involving masked nucleophiles is discussed in the third part, followed by a section dealing with reactions involving direct introduction of a heteronucleophile from a C–H bond.

4.2 CHAN–LAM–EVANS COUPLING: COPPER(II)-PROMOTED OXIDATIVE ARYL TRANSFER FROM ARYLBORON DERIVATIVES

In 1998, three back-to-back publications established the basis of a process, now commonly known as the Chan–Lam–Evans cross-coupling reaction, that strongly impacted the field of the copper-mediated arylation of heteronucleophiles.[6] This reaction and its development have been recently reviewed. Readers are encouraged to consider these excellent publications.[7]

4.2.1 Initial Discovery of the Chan–Lam–Evans Coupling

The initial discovery was established by Chan in 1997 and his first publication appeared in 1998.[6a] In this paper, Chan reported a new N- and O-arylation method based on a copper-promoted cross coupling of arylboronic acids with heteronucleophiles (Scheme 4.1). A wide range of N–H-containing substrates, including primary and secondary amines, amides, ureas, sulfonamides, carbamates, and various cyclic structures such as isatines, phthalimides, and benzimidazolones, were found to be conveniently arylated using a combination of 1 equiv of copper(II) acetate and 2 equiv of triethylamine or pyridine in dichloromethane, under especially mild conditions since the reaction was performed at room temperature, which represents a dramatic improvement compared to the conditions classically needed with the classical Ullmann–Goldberg reactions. Additional conjugated N-heterocycles proved to be compatible with the newly reported process, as demonstrated in the following publication by Lam and coworkers.[6c] Interestingly, in the case of poly-aza-heterocycles, where multiple reaction sites are present, formation of a single (or at least a major) isomer corresponding to the arylation of the nitrogen shown in bold in Scheme 4.1 was always observed.

Simultaneously, Evans reported a similar process for the preparation of biaryl ethers using a copper-promoted coupling between phenols and boronic acids under the exact same conditions: stoichiometric amounts of copper(II) acetate in the presence of triethylamine or pyridine in dichloromethane at room temperature (Scheme 4.2).[6b] Electron-rich phenols were found to provide better yields of the corresponding biaryl ethers, but both electron-rich and electron-poor boronic acids were still found to be suitable partners for the reaction. *Ortho*-substitution was best tolerated on the phenolic unit, and the reaction was found to be compatible with a broad range of functional groups such as alkyl groups, esters, ethers, halogens (Cl, F, and I), silyl ethers, allyloxy substituents, carbamates, and a nitro group.

For both C–N and C–O bond-forming reactions, conditions were established as follows: boronic acid (2 equiv), nitrogen-containing compound or

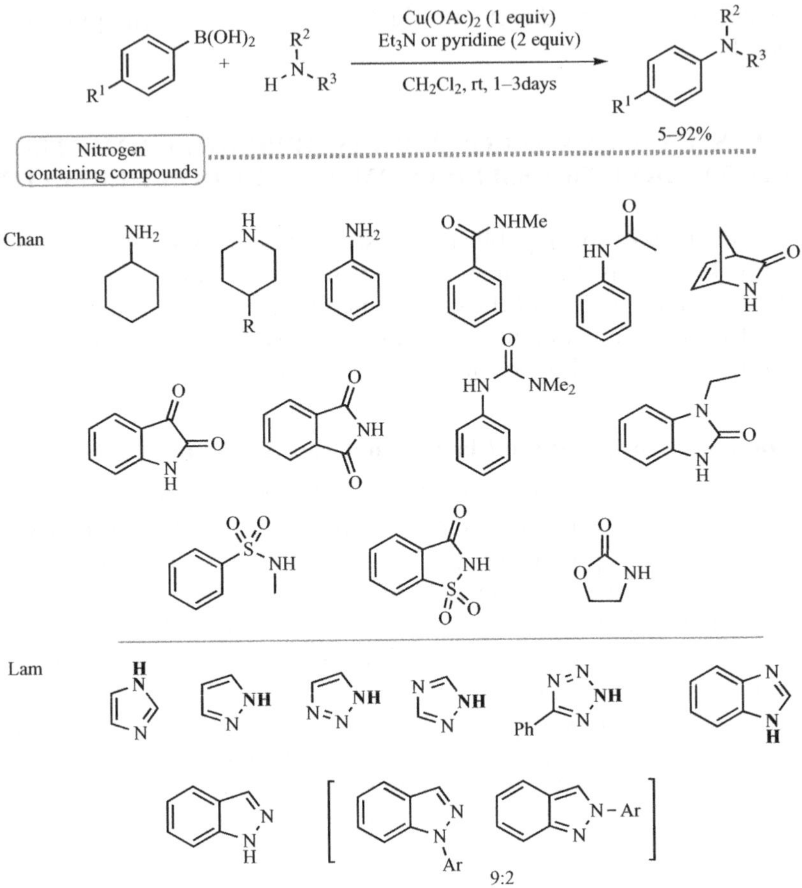

Scheme 4.1 First and historical developments in the copper-promoted arylation of nitrogen nucleophiles with boronic acids by Chan and Lam.

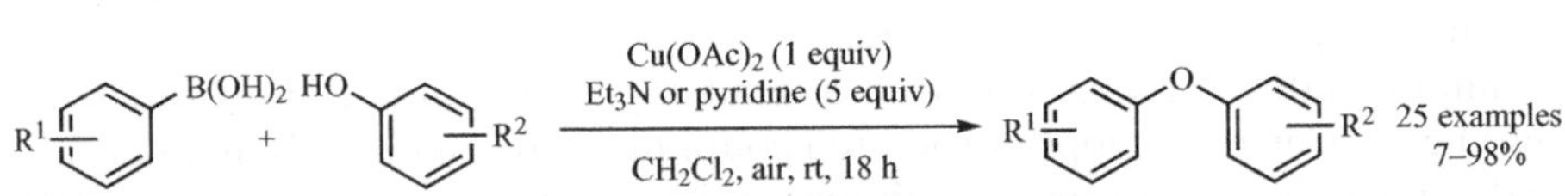

Scheme 4.2 Initial development in the copper-promoted biaryl ether synthesis with boronic acids by Evans.

phenol (1 equiv), copper(II) acetate (1–2 equiv), and triethylamine or pyridine (2–5 equiv) under air at room temperature for 18–72 hours. The reaction was best performed in dichloromethane, though solvents ranging from apolar (toluene) to polar (dimethylformamide [DMF]) were also found to be suitable. As noted by Evans, the addition of 4Å molecular sieves often greatly improved the yield, even if its precise role is still a matter of debate. No clear correlation between base and substrate could be drawn, even if pyridine was generally found to be better with nitrogen heterocycles.

The remarkable simplicity and mildness of the reaction conditions, along with the commercial availability of most boronic acids (thanks to the development of the Suzuki cross-coupling reaction) contributed to the immediate popularity of the process. It should be noted that the original protocol is still used in many applications, and notably in natural product synthesis (see Chapters 16 and 17 for details). Such a process is, however, not without its drawbacks, and limitations appeared in terms of substrate scope. For example, pyrroles, indoles, or aliphatic alcohols were not effectively arylated under such reaction conditions, which led later on to the development of alternative procedures in an attempt to address some of these limitations and shortcomings.

4.2.2 Developments in the Oxidative Copper-Mediated Arylation of Heteronucleophiles with Boronic Acids

4.2.2.1 Catalytic Reactions. Compared to other metals, copper is relatively inexpensive, and the development of conditions that would allow reduction of the amount of copper used in the Chan–Lam–Evans cross coupling might have appeared to be of limited interest, which might account for the limited efforts that have been devoted to this task. However, the development of a catalytic reaction would result in an overall improvement of the system, as well as allowing the evaluation of ligands that could play a critical role for further improvement. In fact, Evans already noted in his seminal paper that a substoichiometric amount of copper still promoted the reaction, albeit in modest yield.[6b] A major step forward was reported by Collman, who described the first catalytic system allowing for the arylation of imidazole with arylboronic acids (Scheme 4.3, Eq. 1).[8] A single catalyst, $[Cu(OH)Cl{\cdot}TMEDA]_2Cl_2$, was investigated in this study, and the optimal loading was established as being 10 mol %. The use of 2 equiv of the arylboronic acid was necessary to achieve good yields, and molecular oxygen proved to be more efficient than air in the reaction. It should be noted that no additional base was required in the reaction and that 3-mono-substituted imidazoles were found to lead to a mixture of regioisomeric products, the major one being the least hindered one – a regioselectivity that could be further improved by lowering the amount of boronic acid to 1 equiv, however, at the expense of the yield.

Subsequently, the same authors evaluated a series of sp^2-*N* as well as sp^3-*N* bidentate ligands in the same reaction (Scheme 4.3, Eq. 2).[9] If all ligands complexed to copper(II) displayed catalytic activity, *N*,*N*,*N*′,*N*′-tetramethylethylenediamine (TMEDA) was found to be superior to all others. The study also revealed that steric hindrance next to the chelating nitrogen, as for ligand **L5**, **L6**, or **L7**, seriously decreased the catalyst efficiency and lowered the yield.

Soon after, it was shown that a μ-hydroxo copper(II) complex was not required in order to perform a catalytic reaction. Buchwald reported a general copper(II) acetate-promoted catalytic N-arylation procedure involving a stoichiometric amount of 2,6-lutidine together with 10–40 mol % of myristic acid

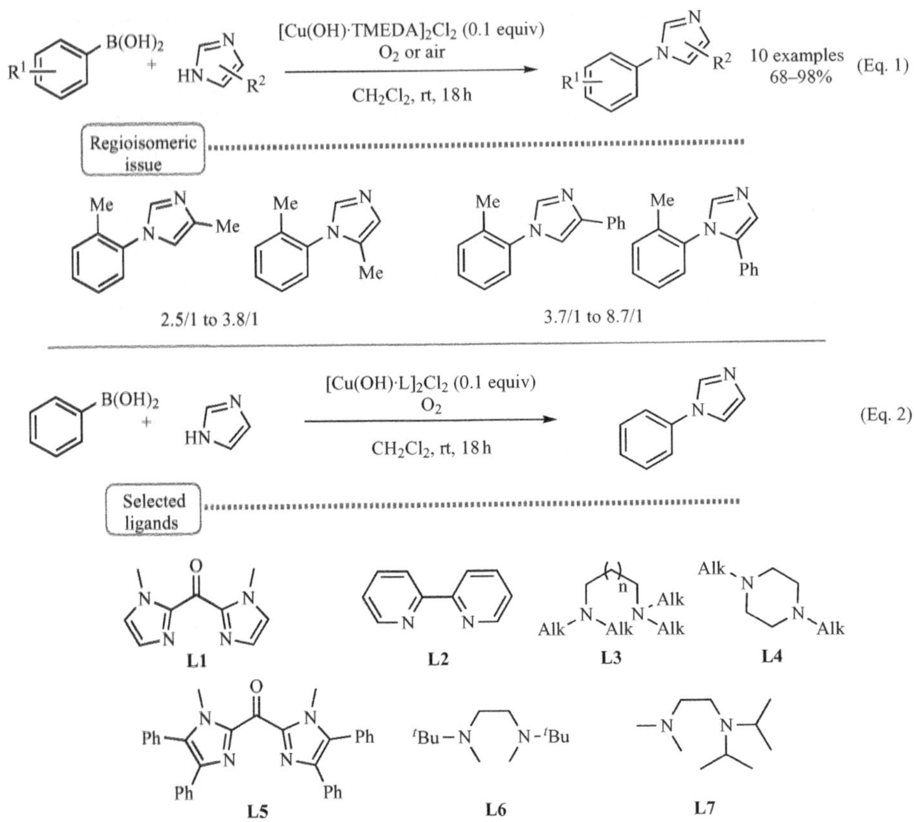

Scheme 4.3 Catalytic copper-promoted N-arylation of imidazoles according to Collman.

as an additive, which is believed to increase the solubility of the catalyst (Scheme 4.4, Eq. 1).[10] Vigorous stirring and a large reaction vessel allowed performing the reaction under air instead of pure oxygen. Copper(II) acetate was also reported to be catalytically active under ligandless and base-free conditions according to Batey (Scheme 4.4, Eq. 2).[11] Whereas aliphatic amines gave excellent results under ligand-free conditions, Buchwald's protocol proved to be superior for anilines. It should be noted that potassium trifluoroboronates could be used as surrogates for boronic acids under the reaction conditions described in Scheme 4.4, Equation 2 (the use of other boron derivatives will be covered in Section 4.2.2.5).

In 2004, studies revealed that ligand- and base-free catalytic conditions could be efficiently performed in protic solvent such as methanol at reflux (Scheme 4.5, Eq. 1).[12] Whereas copper(I) chloride or copper(II) acetate performed best in the reaction, a number of other copper sources were also found to be efficient (copper(I) iodide, copper(II) chloride, copper(I) bromide, copper(I) perchlorate, copper(II) nitrate, or copper(I) oxyde). Compared with

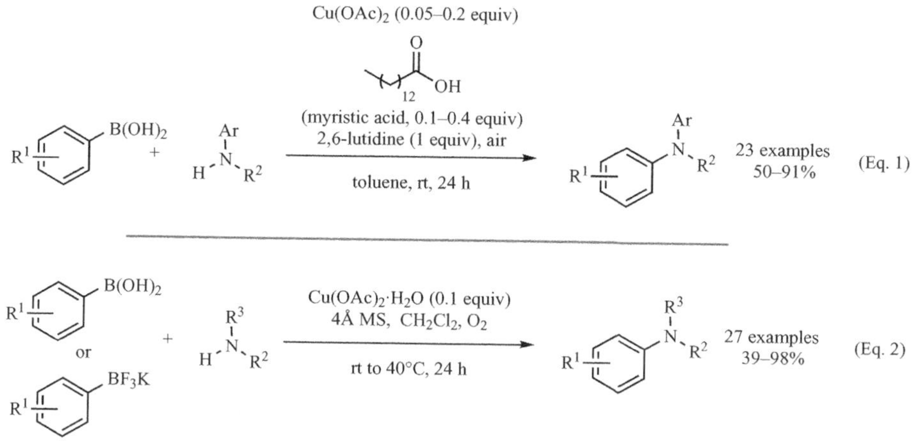

Scheme 4.4 Copper(II) acetate-catalyzed N-arylations with boronic acids or surrogates.

previous conditions, the reaction was much faster, allowing the arylation of imidazoles in 3–6 hours (Scheme 4.5, Eq. 2). Imides, anilines, primary amides, and suflonamides were also efficiently arylated under such reaction conditions. The reaction was, however, strongly impacted by steric factors since *ortho*-substituted phenylboronic acids were not tolerated, a problem that could be efficiently solved by performing the reaction in the presence of a ligand (Scheme 4.5, Eq. 3). The simple addition of 1 equiv of TMEDA relative to the amount of copper salt clearly improved the reaction efficiency, allowing performing the N-arylation of substituted imidazoles at room temperature with reasonable-to-good yields (Scheme 4.5, Eq. 3).[13]

4.2.2.2 Solvent. A notable feature of the Chan–Lam–Evans reaction is its extreme tolerance toward the reaction medium. Toluene, dichloromethane, tetrahydrofuran (THF), 1,4-dioxane, 1,2-dimethoxyethane (DME), acetonitrile (CH_3CN), *N*,*N*-dimethylformamide (DMF), *N*-methylpyrrolidinone (NMP), dimethyl sulfoxide (DMSO), and ethyl acetate (EtOAc) have been used for various reactions, with dichloromethane being the most commonly used one. Based on a case study involving the N-arylation of morpholine, Combs and coworkers reported the following order for various solvents: 1,4-dioxane – CH_2Cl_2 >> DMF > EtOAc > THF > toluene > DMSO.[14] It should, however, be noted that this is not a fully general trend since the preferred solvent appears to be substrate-dependent, and mixed solvent systems can sometimes be beneficial. Finally, the N-arylation of a range of nitrogen-containing functional groups was also achieved in an ionic liquid, [bmim][BF_4] (Scheme 4.6) and the one-to-one stoichiometry of the nucleophile relative to the boronic acid is noteworthy, as an excess of the latter is usually required for such transformations.[15]

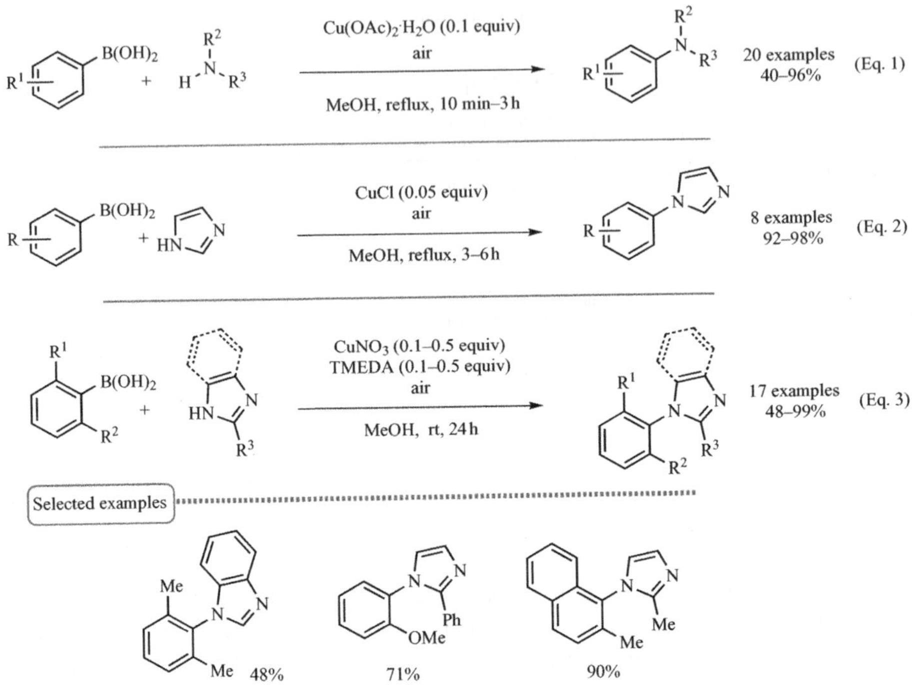

Scheme 4.5 Catalytic N-arylations with arylboronic acids in methanol.

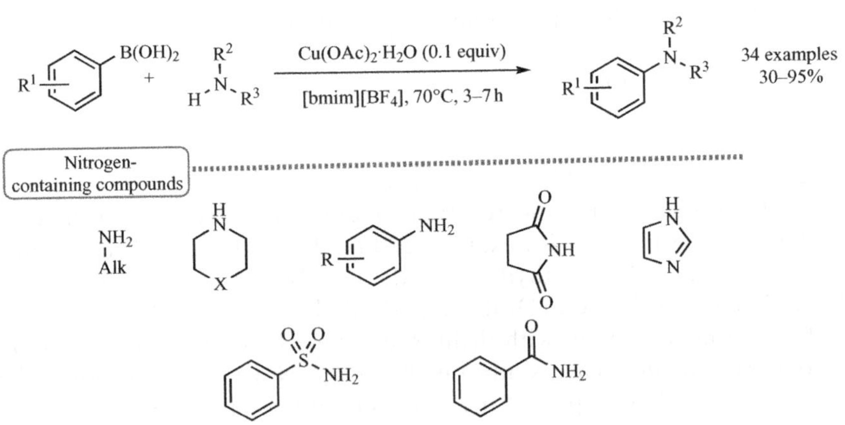

Scheme 4.6 N-Arylation with arylboronic acids performed in an ionic liquid.

As previously mentioned, methanol can be conveniently used as solvent in various catalytic N-arylation reactions with boronic acids.[12] It has also been shown that it can have an especially beneficial role as an additive, as highlighted with the installation of the key biaryl ether of chloropeptin I using an intramolecular Chan–Lam–Evans reaction (Scheme 4.7).[16] In this step, 10 equiv of methanol relative to the substrate was found to be crucial for the

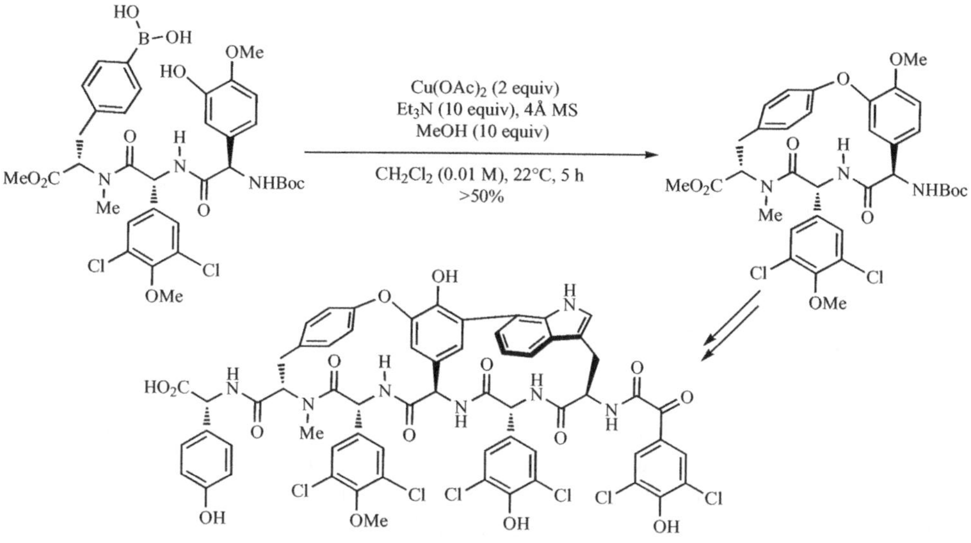

Scheme 4.7 Intramolecular O-arylation during the synthesis of chloropeptin I: dramatic effect of methanol as an additive.

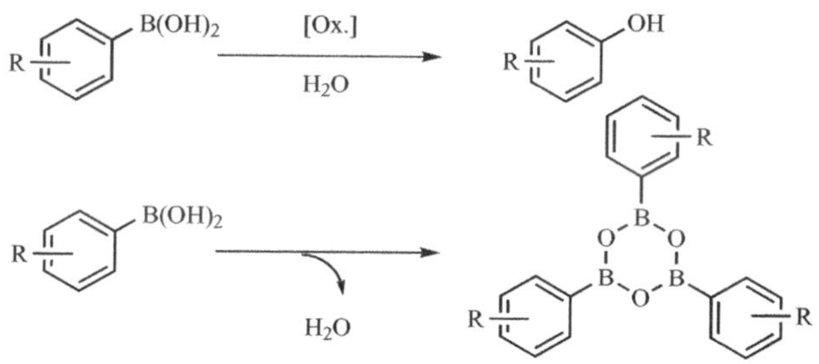

Scheme 4.8 Competitive reactions of boronic acids involving water.

efficient formation of the biaryl ether, which was attributed to an increased solubility of the catalyst or the *in situ* formation of the corresponding dimethyl boronate. It was also observed during this study that methanol did not compete at all with the phenol residue for the O-arylation.

Like methanol, water has early on been considered as deleterious for the reaction since it could possibly compete with the nucleophile. Lam showed, using oxygen isotopic labeling, that arylation of water with boronic acids yielding to the formation of phenols could be a competitive reaction, which would account for the typical use of more than 1 equiv of boronic acid.[17] Even under anhydrous conditions, water can be produced *in situ*, as it is formed during the trimerization of boronic acids to the corresponding boroxines (Scheme 4.8).[18] Therefore, the addition of 4Å molecular sieves frequently improves the reactions, even if the exact role of this additive is still unclear.

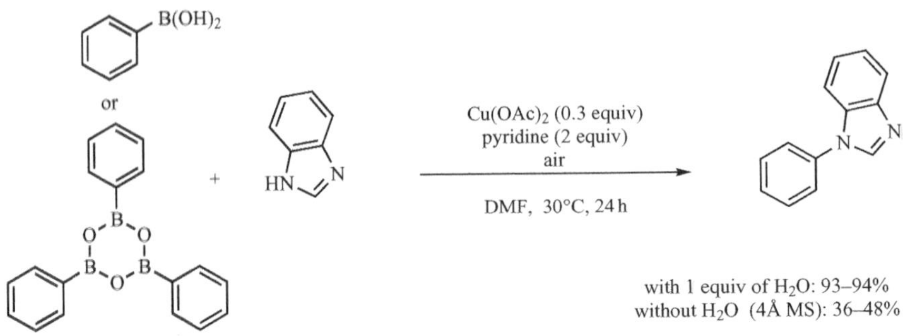

Scheme 4.9 Beneficial effect of water in the arylation of benzimidazole.

The negative impact of water is not always observed as the arylation of imidazoles in methanol/water or even pure water has been reported.[12,19] During a work aimed at the large-scale arylation of benzimidazole, an interesting study actually revealed a beneficial accelerating effect of water (Scheme 4.9).[20] In addition, it was found that the competitive formation of phenols was not dependent on the amount of water in the reaction.

4.2.2.3 Oxidant. From the seminal paper of Evans, it was recognized that this reaction performed best in the presence of an oxidant. It was hypothesized that such oxidants would facilitate the formation of a copper(III) complex from which reductive elimination is facilitated. For stoichiometric reactions, the use of an additional oxidant is not required. Copper itself might play the role of the oxidant, which accounts for the typical use of an excess of the copper(II) salt. For catalytic reactions, the presence of an oxidant is not only beneficial but also simply a necessity. The presence of an oxidant in the Chan–Lam–Evans coupling is, however, one of the drawbacks of this reaction. Besides the boronic acids, which can be oxidized to phenols, heteronucleophiles might also be sensitive to oxidants. This is particularly the case with phenols, which can lead to quinones,[21] with aliphatic amines for which oxidative C–N bond cleavage is known, or with thiols, which easily dimerize to form disulfides (Scheme 4.10).[22]

If air is usually preferred as a cheap and convenient oxidant, alternative oxidants were disclosed by Lam and coworkers in 2001 (Scheme 4.11).[23] The best systems explored for C–N bond formation involved $Cu(OAc)_2$ (0.2 equiv) and a combination of pyridine *N*-oxide or 2,2,6,6-tetramethyl-1-piperidinyloxy (TEMPO) (1.1 equiv) and air, but oxygen alone performed sometimes as efficiently. Among the co-oxidants evaluated for the *p*-tolylation of *N*-ethyl-benzimidazolone, the following beneficial effect was established: *N*-methylmorpholine *N*-oxide (NMO) > di-*tert*-butylnitroxide > (1*R*)-(10-camphorsulfonyl)oxaziridine > sodium perborate ($NaBO_3{\cdot}H_2O$) > potassium ferricyanide ($K_3Fe(CN)_6$) > *meta*-chloroperoxybenzoic acid (*m*-CPBA). For C–O bond formation, oxygen alone was by far the best oxidant. It should be noted that in this case, a gentle heating at 50°C was applied.

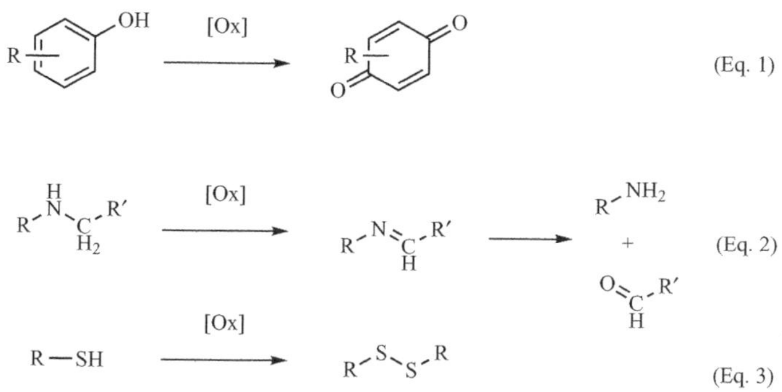

Scheme 4.10 General competitive oxidative reactions of phenols, amines, and thiols.

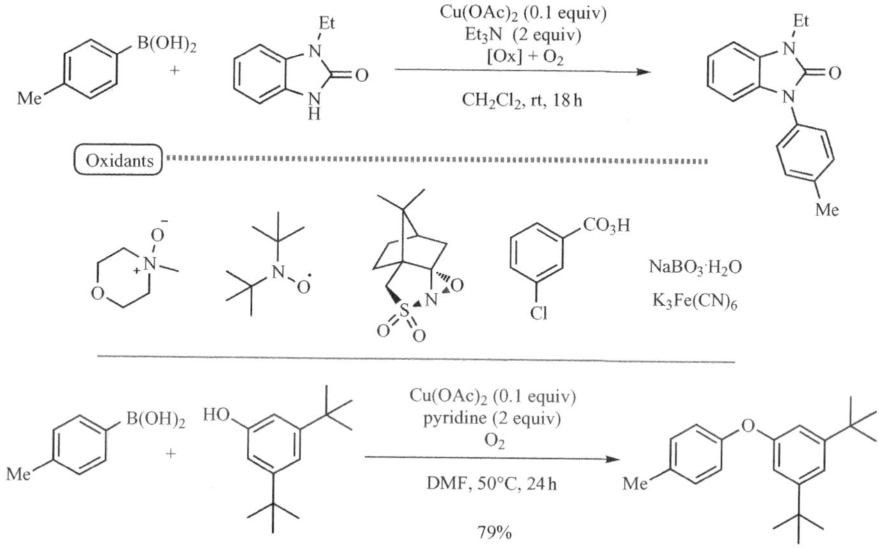

Scheme 4.11 Alternative oxidants to air in the Chan–Lam–Evans coupling.

An interesting symmetrical biaryl ether formation was disclosed by Prakash and Petasis in 2001 (Scheme 4.12).[24] Taking into account that boronic acids are easily oxidized to phenols, they devised a one-pot sequential oxidation/C–O bond formation from arylboronic acids. In this sequence, hydrogen peroxide was first used to oxidize part of the arylboronic acid to the corresponding phenol, and subsequent addition of copper(II) acetate and triethylamine allowed the C–O bond formation to proceed, yielding to symmetrical biaryl ethers in good yields. Addition of 4Å molecular sieves before the second step was clearly beneficial to trap the water formed in the process.

4.2.2.4 Bases. The stoichiometric Chan–Lam–Evans coupling requires the presence of a base. The exact role of this base is unclear, but two roles

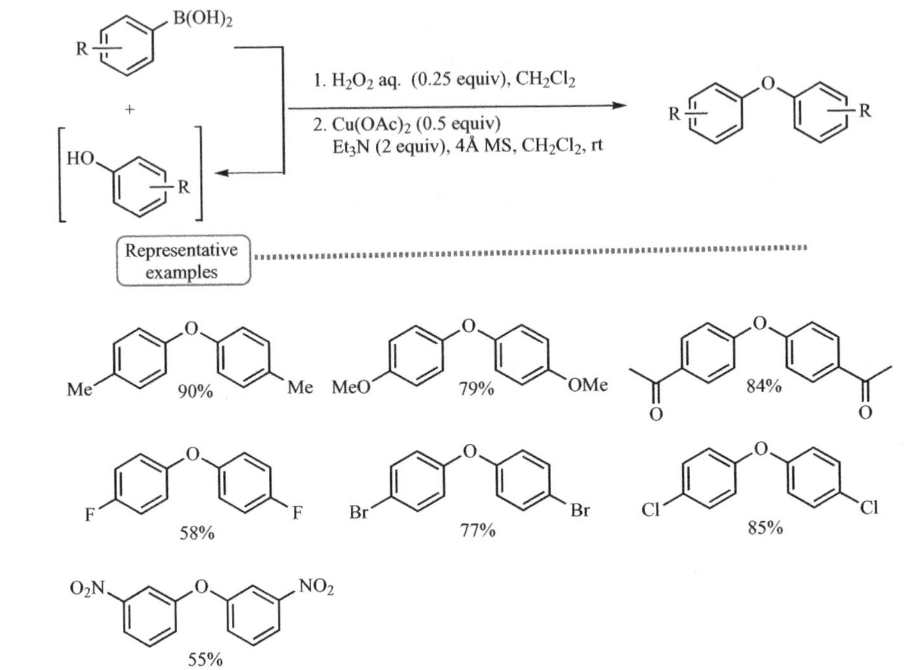

Scheme 4.12 Direct conversion of arylboronic acids to symmetrical diaryl ethers.

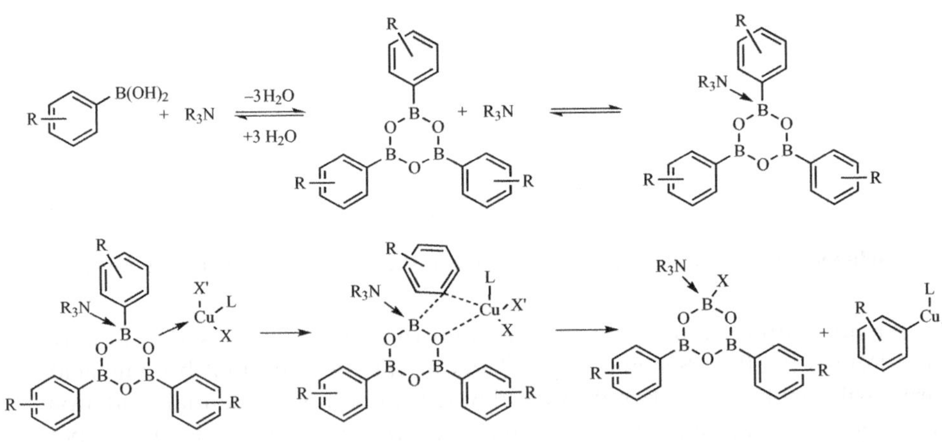

Scheme 4.13 Alternative role of the base in the Chan–Lam–Evans coupling.

have been advanced: It could serve as a ligand for any copper species present during the reaction and/or act as proton scavenger. McKinley and O'Shea reported an interesting study in which they proposed an additional role for the organic base (Scheme 4.13).[25] First, the base would catalyze the formation of the boroxine from the starting boronic acid. Second, the base would activate the boroxine by the formation of a tetracoordinate species, and the

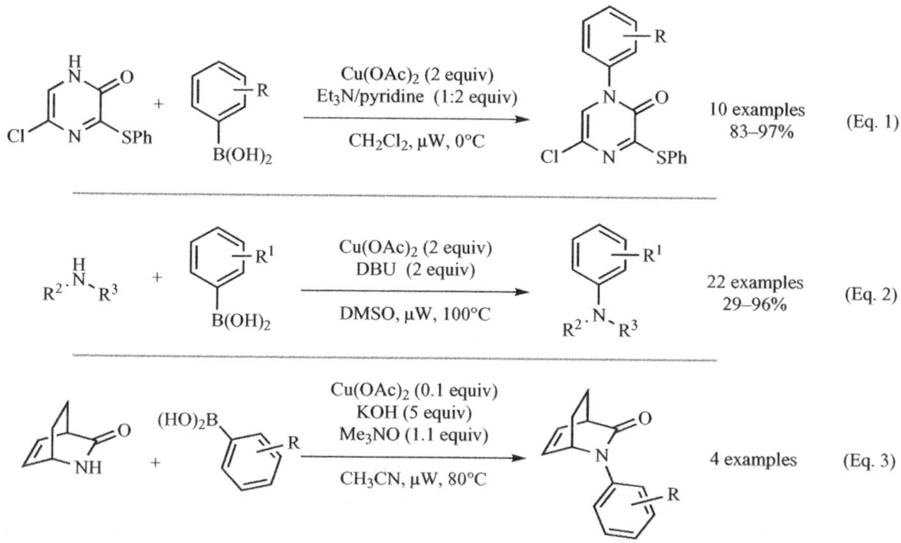

Scheme 4.14 Microwave promoted Chan–Lam–Evans N-arylation involving boronic acid and various bases.

transmetallation from this activated species would next be facilitated, as in related Suzuki–Miyaura cross-coupling reactions.[3]

There is, however, no clear tendency for the optimal base to use. In early studies, triethylamine and pyridine have been used: Pyridine was found to be superior for the N-arylation of most nitrogen-containing heterocycles, whereas triethylamine was preferred for amines and anilines. In the case of C–O bond formation, both bases were widely used. In selected cases, mixed base systems can also afford optimal results, as for the N-arylation of pyrazinone (Scheme 4.14, Eq. 1).[26] During this study, it was also demonstrated that the coupling was greatly improved when performed under microwave irradiation with simultaneous cooling at 0°C.

The bases for the Chan–Lam–Evans coupling are not limited to triethylamine or pyridine. N-Arylation of amines and anilines were, for example, achieved under microwave irradiation at 100°C using 1,8-diazabicyclo[5.4.0]undec-7-ene (DBU) as the base (Scheme 4.14, Eq. 2). *N*,*N*-Diisopropylethylamine was also found to be crucial for the N-arylation of indoles lacking a chelating electron-withdrawing group at the 2-position, as no product was obtained with triethylamine or pyridine (Scheme 4.15).[27] A combination of inorganic base (potassium hydroxide) and organic co-oxidant (trimethylamine-*N*-oxide) allowed for the copper-catalyzed N-arylation of a bicyclic bridged lactam using microwave heating (Scheme 4.14, Eq. 3).[28] Finally, stoichiometric 4-dimethylaminopyridine and 1,10-phenanthroline have also been used in combination with triethylamine for the N-arylation of various heterocyclic compounds.[29]

While the proper choice of the stoichiometry, solvent, oxidant, and base can have a deep effect on the outcome of the Chan–Lam–Evans cross coupling,

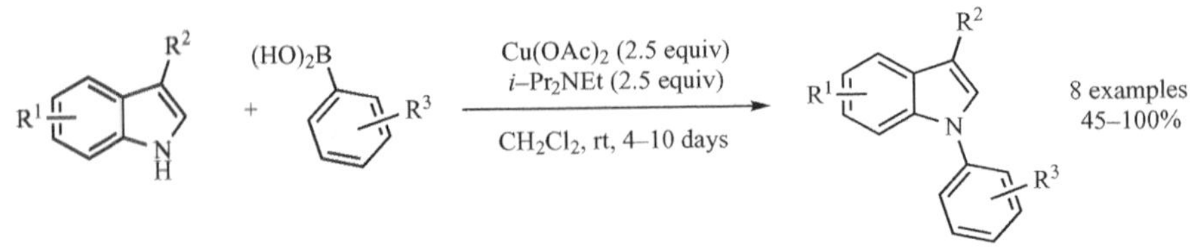

Scheme 4.15 N-Arylation of indoles with boronic acids under copper-mediated conditions and using Hunig's base.

the nature of the boron derivative itself can also have a dramatic effect, and various surrogates can be efficiently used.

4.2.2.5 Boron Derivatives. From a general point of view, electronically and structurally diverse aryl-substituted boron derivatives are well tolerated. Heteroaryl and electron-poor boronic acids tend to furnish lower yields than the corresponding electron-rich derivatives. The susceptibility toward protodeboronation of electron-deficient (hetero)arylboronic donors explains this tendency and justifies the stoichiometry of the reaction, requiring the use of an excess of the aryl donor. In addition to electronic, the reaction is also sensitive to steric factors, and *ortho,ortho'*-disubstitued arylboronic acids tend to be problematic coupling partners, which can also be the case with aryl halides, and is still one of the problems to be solved in most copper-mediated transformations.

Whereas boronic acids have been extensively used, alternative boron derivatives have been reported in such arylations. Lam performed a comparative study of various boronic esters for the arylation of three prototypical heteronucleophiles: a phenol, a secondary amine, and a cyclic urea (Scheme 4.16).[30]

Importantly, it was established that each boron derivative was a competent aryl donor, with different reactivities, however. Sterically hindered pinacolate-derived boronic ester was the least efficient donor furnishing only a trace amount of product for O- and N(urea)-arylations, but was shown to be comparable to phenylboronic acid for the N-arylation of the secondary amine. Such esters have been barely used, but in selected cases, they provide good results as described for the synthesis of halo-oxybispyridines.[31] Catechol-derived phenylboronic ester was comparable to phenylboronic acid itself, which is understandable considering the limited stability of such an ester that is easily hydrolyzed to the corresponding boronic acid. However, it should be kept in mind that the release of the catechol, which is easily oxidized, could be problematic. More interestingly, it was shown that the boroxine performed better than the parent acid. It has therefore been suggested that it is not the boronic acid but the cyclic boroxine which is the reactive aryl donor, which would also explain why the addition of molecular sieves often improves the reaction efficiency, as removal of water favors the formation of the cyclic anhydride. Other boronic esters were found to be more efficient than the free acid but less or equally potent than the boroxine.

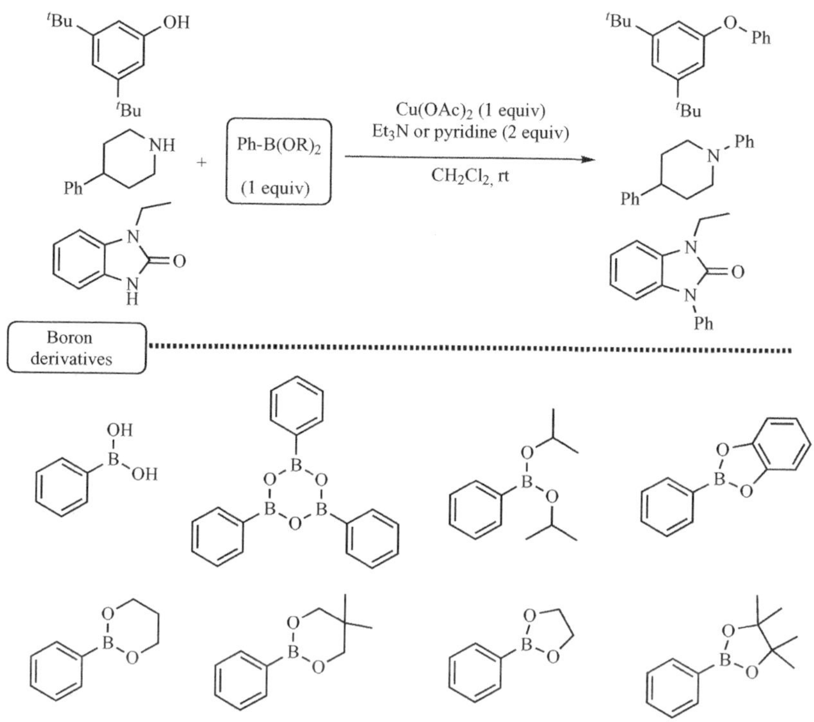

Scheme 4.16 Compared reactivity of various boron species.

Other especially attractive surrogates for boronic acids are potassium organotrifluoroborates, highly crystalline solids that are air and moisture tolerant and easily prepared. These compounds have been used as surrogates to boronic acids in the palladium-catalyzed Suzuki–Miyaura cross-coupling reaction and other processes,[32] and Batey and Quach initially investigated the reactivity of these salts in copper-mediated O-arylation reactions (this point will be discussed in Section 4.2.4.1, Scheme 4.20).[33] They demonstrated that the fluoroborate salts were not only effective but provided superior results when compared to boronic acids, and their efficiency in N-arylation reactions was also established (see Scheme 4.4).[11] These reagents are mostly soluble in polar solvents, but reactions could be performed in DCE. Taking advantage of the solubility issues, N-arylation of various heterocycles could be performed in water, as recently reported by Vaultier and coworkers (Scheme 4.17).[19b] For this catalytic reaction, various ligands such as amine- or keto-derivatives were evaluated, and ethyl acetoacetate stood as the most efficient one. Among the trifluoroborates evaluated, potassium trifluoro(4-nitrophenyl)borate was completely ineffective, mostly leading to a homocoupling product, whereas other electron-rich or electron-deficient aryl units were well tolerated.

Yamamoto and Miyaura recently introduced potassium triolborates as an alternative to potassium trifluoroborates (Scheme 4.18). These compounds

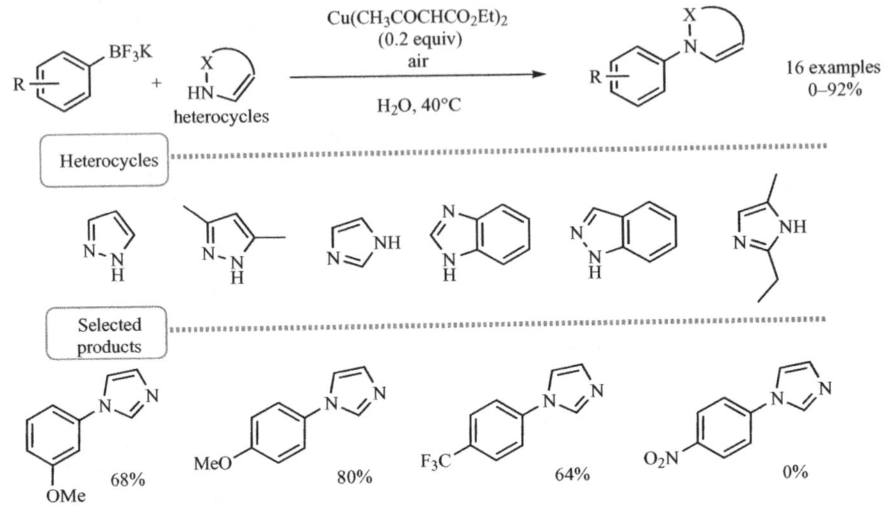

Scheme 4.17 Potassium aryltrifluoroborates for the copper-catalyzed N-arylation of azoles in water.

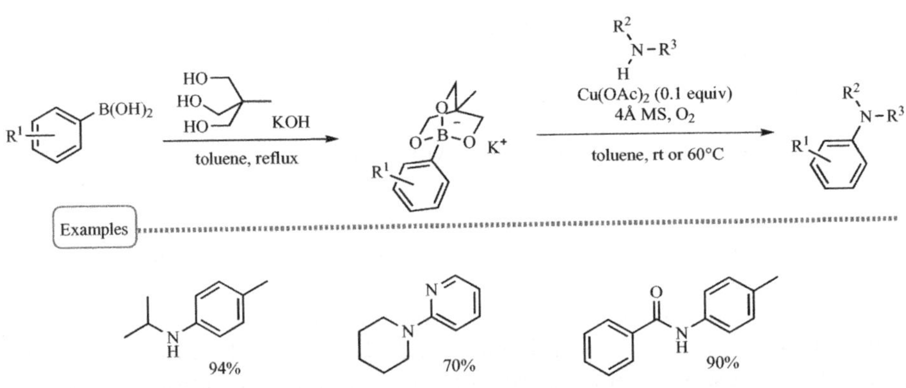

Scheme 4.18 Copper-catalyzed N-arylation with potassium triolarylborates.

share the characteristic features of potassium trifluoroborates, being air and water stable, but in addition, they proved to be soluble in most organic solvents.[34] They are easily prepared by simply refluxing the corresponding boronic acids and triols with potassium hydroxide in toluene. A catalytic arylation of selected N-containing compounds demonstrated their potential in oxidative copper-promoted C–N bond formation: Isopropylamine, piperidine, and benzamide were effectively arylated using 0.1 equiv of copper(II) acetate in toluene under an oxygen atmosphere, and the addition of 4Å molecular sieves avoided the formation of undesired phenol side products. It is interesting to note that pyridine-2-yl- and pyridine-3-yl triolborates are efficiently coupled in high yields to nitrogen nucleophiles (piperidine was used in the study), a reaction that is especially challenging to perform with the corresponding boronic acids, which are especially sensitive to hydrolysis and are poorly reactive toward transmetallation. The scope of this reaction was further evaluated, and it was established that trimethylamine *N*-oxide could advantageously replace oxygen for the N-arylation of anilines and imidazoles.[35]

4.2.3 Mechanism of the Coupling

Evans postulated the mechanism of the copper-promoted C–O bond formation in his seminal contribution (Fig. 4.2). Following Barton's hypothesis on the related copper-mediated transformation from triarylbismuth diacetates [$BiAr_3(OAc)_2$], the following elementary steps were proposed: (i) transmetallation of Cu(II)X_2 with the arylboronic acid, (ii) ligand exchange from ArCu(II)-X to ArCu(II)-OAr, and (iii) reductive elimination to afford the product. In order to account for the beneficial effect of air or oxygen in the process, the authors speculated that the copper(II) species could be oxidized to copper(III) prior to reductive elimination, such reductive elimination being faster from copper(III) species than from copper(II) ones. Even if not mentioned in this paper, the catalytic cycle can be easily drawn from this scheme by adding an additional reoxidation step of the copper(I) species to the copper(II) species.

Stahl and coworkers reported a solid mechanistic study of the copper-catalyzed aerobic oxidative coupling between arylboronic esters and methanol in 2009. The proposed mechanism is depicted in Figure 4.3.[36] This study established that the mechanism was initiated by a transmetallation step and supported the copper(III) intermediate. However, contrary to the initial speculation, it was shown that oxygen was only a secondary oxidant. A disproportionation reaction of Ar–Cu(II) yields to the formation of copper(III) and

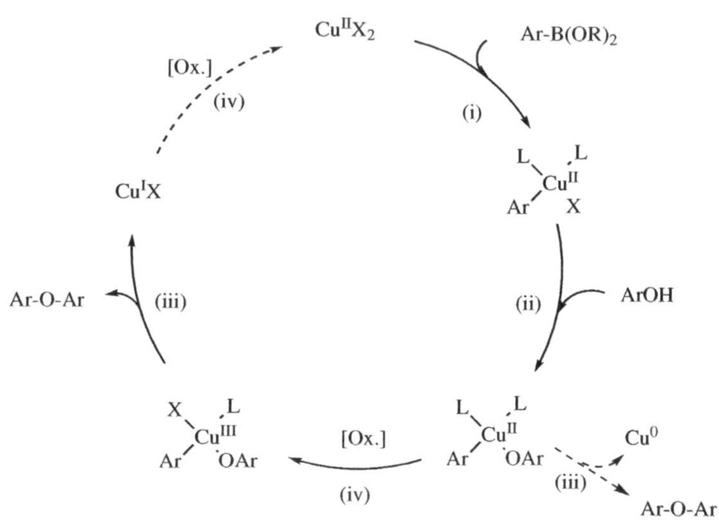

Figure 4.2 Initially postulated mechanism of the cross coupling between boronic acids and alcohols. (i) Transmetallation, (ii) ligand exchange, (iii) reductive elimination, and (iv) oxidation.

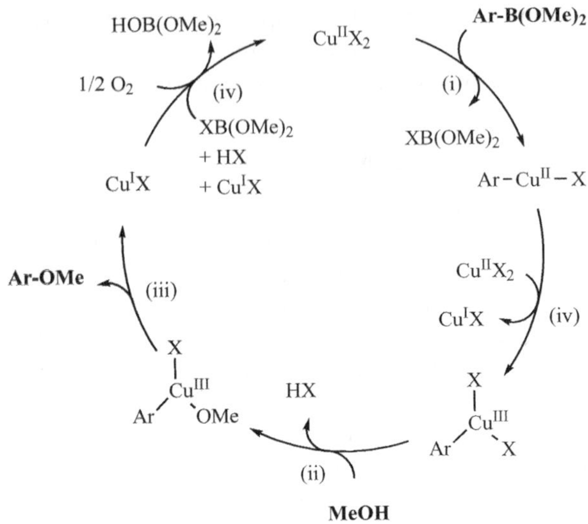

Figure 4.3 Mechanism of O-arylation of methanol with boronic methyl esters. (i) Transmetallation, (ii) ligand exchange, (iii) reductive elimination, and (iv) oxidation.

copper(I) species and oxygen was found to be only involved in the reoxidation of copper(I) to copper(II).

Support for a related reaction pathway in the case of C–N bond formation has also been reported using isolated copper(III) complexes.[37] A general mechanism is, however, not necessarily operating for every system, and extensive details concerning the intermediacy of copper(III) complexes in copper-catalyzed cross-coupling reactions as well as a discussion of the mechanism of the Chan–Lam–Evans coupling can be found in Chapter 7.

4.2.4 Scope

The development of copper(II)-promoted N- and O-arylation with boronic acids was a major breakthrough for the formation of C(aryl)-heteroatom bonds, especially because of the mild reaction conditions required and the availability of boronic acids. The initial reports by Chan, Lam, and Evans in 1998 generated immediate interest from the synthetic community and various research groups made considerable progress in developing this copper-mediated cross-coupling whose scope has been notably expanded.[5,7] Major developments will be covered in this section, starting with oxygenated nucleophiles.

4.2.4.1 C–O Bond Formation. Evans and others reported in their original work that this reaction tolerated a number of functional groups, and an expedient synthesis of thyroxine, a natural compound containing an *ortho,ortho′*-bis

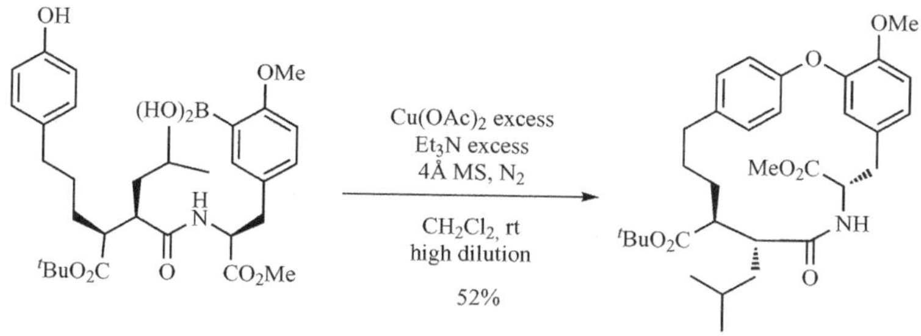

Scheme 4.19 Intramolecular O-arylation.

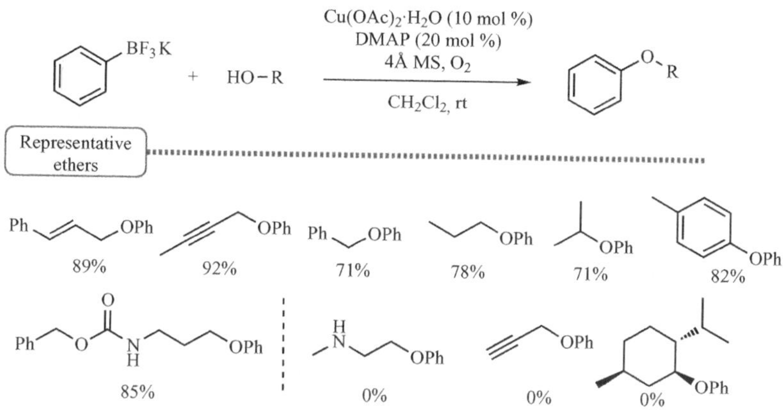

Scheme 4.20 Copper-catalyzed O-arylation of aliphatic alcohols with potassium aryltrifluoroborates.

iodinated diaryl ether, is a nice illustration of such a process.[6] As an extension to the intermolecular reaction, cyclizative O-arylation has been developed (Scheme 4.19).[38] For this intramolecular reaction, high dilution techniques were required, and the reaction was performed using an excess of copper salts in the presence of triethylamine under a nitrogen atmosphere. The same strategy was applied during the total synthesis of naturally occurring macrocyclic diarylether cycloisodityrosine.[16,39] More examples of the use of the Chan–Lam–Evans coupling and related processes for the formation of C–O bonds in the context of natural product synthesis can be found in Chapter 17.

Concerning the scope of this reaction, O-arylations were initially limited to the formation of diarylethers, and the use of stoichiometric reaction conditions was unsuccessful for aliphatic alcohols. In 2003, catalytic reaction conditions proposed by Batey overcame this major limitation, and the reaction could be smoothly performed with aliphatic alcohols under catalytic conditions (Scheme 4.20).[33] The optimal reaction conditions were established as follows: copper(II) acetate (0.1 equiv), 4-dimethylaminopyridine (DMAP, 0.2 equiv),

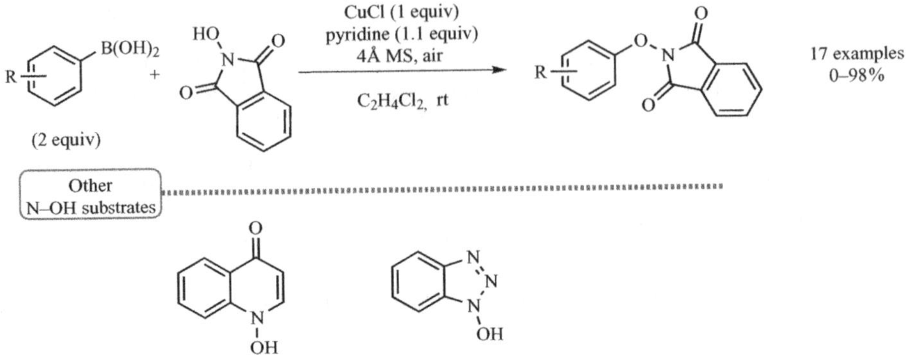

Scheme 4.21 O-Arylation of N–OH-containing heterocycles.

and 4Å molecular sieves under oxygen in dichloromethane at room temperature. Various nitrogen-containing additives were evaluated as ligands for this coupling, and it was observed that secondary or tertiary amines (*N,N*-diisopropylethylamine, tetramethylethylenediamine, triethylamine) were completely inefficient, whereas N-heterocyclic ligands (1,10-phenanthroline, pyridine, imidazole, 4-dimethylaminopyridine) did furnish the expected product. Boronic acids could be efficiently used as aryl donors, but potassium aryltrifluoroborates were the best reagents to perform this arylation. Concerning the other coupling partner, allylic, propargylic, benzylic, and aliphatic alcohols were effectively converted to their corresponding aryl ethers under the optimized conditions.

Additional O-nucleophiles have also been evaluated in the arylation reaction. For example, the arylation of *N*-hydroxyphthalimides was reported (Scheme 4.21).[40] Copper(I) chloride or copper(II) acetate performed equally well in the presence of pyridine, furnishing the corresponding aryloxyamine derivatives in up to 98% yield. Whereas most arylboronic acids could be efficiently coupled, *ortho*-fluoro and 1,4-phenylenediboronic acid did not furnish the expected products. Other N-hydroxylated precursors such as hydroxyquinolin-4(1H)-one[41] or *N*-hydroxybenzotriazoles[42] were also arylated under similar reaction conditions, therefore expanding the scope and synthetic utility of the copper-mediated arylation of oxygen nucleophiles.

Recently, Merlic and coworkers reported a single example of O-arylation of a silanol (Scheme 4.22) in a study essentially devoted to the synthesis of enol silyl ether (which is beyond the subject of the present chapter and described in Chapter 5).[43] Nevertheless, the reaction conditions are quite interesting since the reaction is best performed neat in the presence of a substoichiometric amount (0.5 equiv) of copper(II) acetate, triethylamine, and a combination of pyridine *N*-oxide and O_2 at 50°C. In order to get decent yields, the addition of 3-hexyne was mandatory. In connection with their previous work, it is believed that this additive probably acts as a ligand for copper,[44]

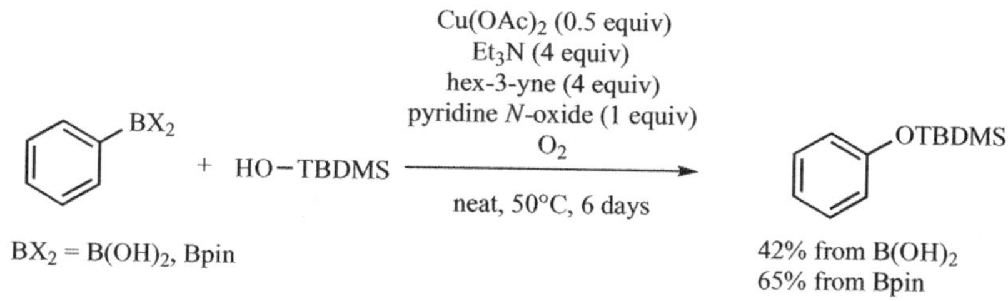

Scheme 4.22 Copper-catalyzed O-arylation of a silanol with organoboron derivatives.

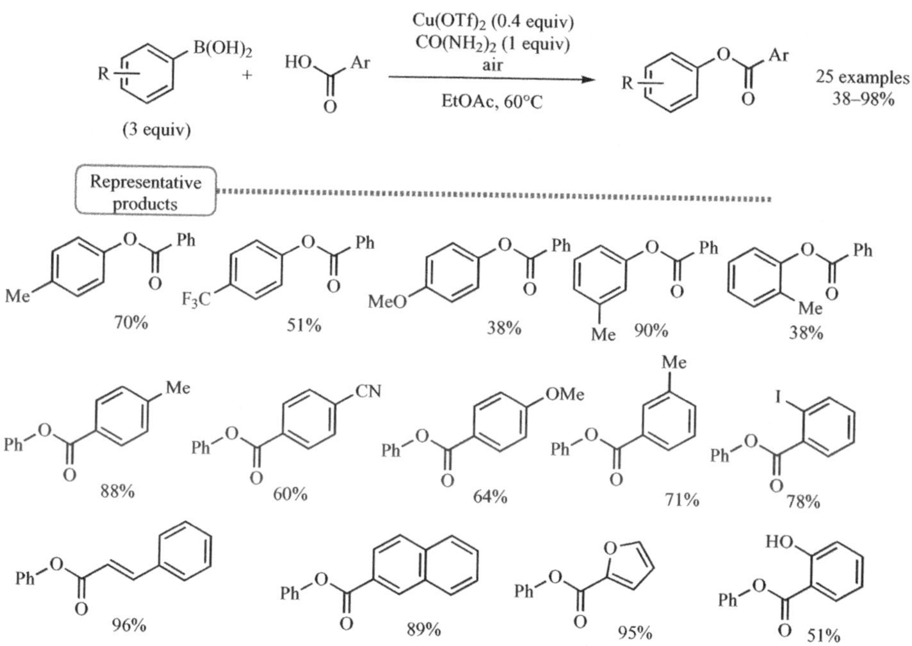

Scheme 4.23 Synthesis of phenolic esters from boronic acids.

and it is interesting to note that the phenylboronic pinacol ester furnished better yield than the corresponding boronic acid.

An interesting extension of the Chan–Lam–Evans coupling, which demonstrated that carboxylic acids could also be used as reaction partners, was reported in 2010. Phenolic esters have been synthesized from boronic acids under copper-promoted oxidative conditions (Scheme 4.23), extending the diversity of O-nucleophiles that can be arylated.[45] The reaction was performed with 0.4 equiv of copper(II) trifluoromethanesulfonate and 1 equiv of urea, which probably acts as a ligand for copper, in EtOAc at 60°C. Yields were generally good, except for 2-substituted boronic units where only

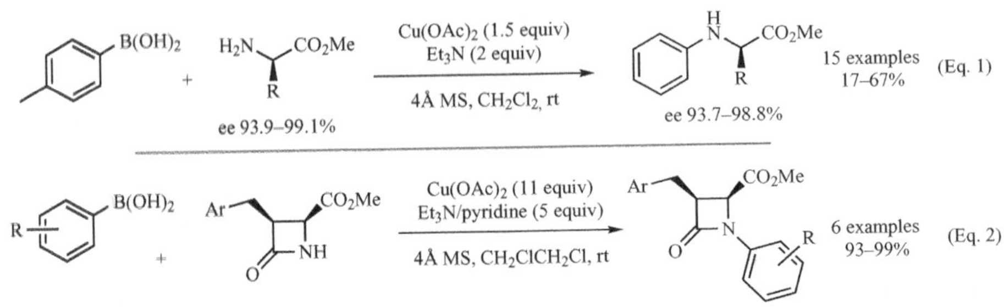

Scheme 4.24 N-Arylation of racemization-prone nitrogen-containing derivatives.

2-methyl boronic acid furnished the desired product, albeit in low yield (38%). Unfortunately, no example of the use of an aliphatic acid was reported.

4.2.4.2 C–N Bond Formation. Copper-promoted C–N bond formation from boronic acids has shown an extremely large tolerance and scope toward nitrogen-containing nucleophiles, and early studies demonstrated the possibility to arylate amines, amides, imides, ureas, carbamates, or sulfonamides. Primary amines have proven to be poor reaction partners because of competitive oxidative dealkylation. As with aliphatic alcohols, this problem was solved through the development of general catalytic reaction conditions, as reported by the Batey group.[11] It should be pointed out that even sensitive N-containing heteronucleophiles that easily epimerize, such as aminoester, were shown to be compatible with this arylation, demonstrating the mildness of the reaction (Scheme 4.24, Eq. 1).[17] Enantioenriched azetidinone ester was also arylated without racemization (Scheme 4.24, Eq. 2),[46] but related azetidinones lacking the ester were found to be unreactive under the reaction conditions, uncovering a probable internal chelation. This result is not surprising considering that secondary amides, especially acyclic ones, are usually not arylated.

A variety of π-excessive N-heterocycles such as imidazoles, triazoles, tetrazoles, benzimidazoles, and indoles, among others, were successfully arylated. The Chan–Lam–Evans reaction is one of the best methods for the synthesis of *N*-aryl-heterocycles and nicely complements their synthesis from aryl halides described in Chapter 1 (Section 1.4). Complex heterocyclic nucleobases such as cytosine, thymine, adenine, and guanine were also efficiently arylated under noncatalytic reaction conditions, but other nucleophilic functional groups of these nucleobases were protected in order to achieve a regioselective N-arylation.[47]

Discrimination between two reaction sites is not always easy to predict and is quite often challenging, but regioselective arylation has been achieved for hydantoines,[48] for 6-substituted purine (where *N*-9 arylation dominated[49]) and for 4-aminopyrimidinone where the exocyclic amine remained untouched (Fig. 4.4).[50] Selective N-arylation of aminotriazole-carboxamides was also documented to selectively occur on the amide and not on the amine.[51] In

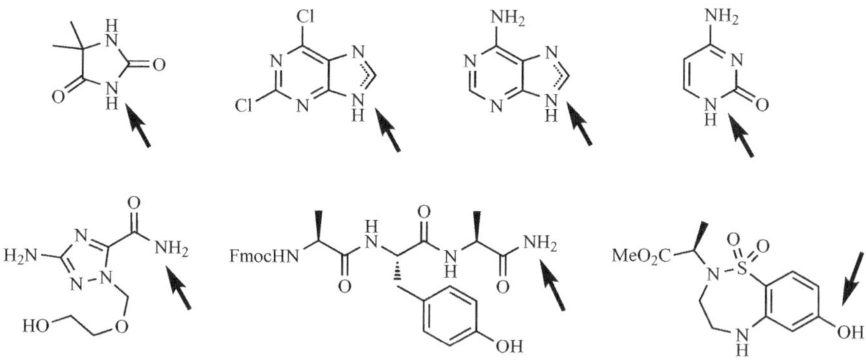

Figure 4.4 Chemoselective N- or O-arylation (arrows indicate the preferential site of arylation).

general, the N-arylation is highly favored over O-arylation, allowing the direct use of unprotected alcohols or performing the reaction using alcohols as solvents. In the case of phenol, it has been established that N-arylation of aniline was nine times faster,[7] but preferential O-arylation can predominate, since the reaction is highly sensitive to steric effects.[52]

The mildness of the Chan–Lam–Evans cross coupling has stimulated a lot of research groups, and the number of nitrogen-containing functional groups suitable for N-arylation with boronic acids or derivatives is continuously expanding (Scheme 4.25). For example, *N*-Boc-*N*-aryl-aryl hydrazines have been converted to *N*,*N*′-diaryl hydrazines under stoichiometric reaction condition using pyridine as the base (Scheme 4.25, Eq. 1),[53] and N-arylated sulfoximines could be easily prepared from the corresponding sulfoximines using room temperature, base-free, copper(II) acetate-catalyzed conditions (Scheme 4.25, Eq. 2).[54] The reaction could be performed in several solvents such as CH_2Cl_2, DMSO, or ethanol, but methanol provided the best yields. The reaction was tolerant according to the nature of the aryl unit, and substitution at the 2-position was well tolerated (with yields ranging from 62% to 96%), but the sterically crowded 2,4,6-tri-*tert*-butylphenylboronic failed to give the corresponding arylated product. Azides such as sodium azide or azidotrimethylsilane were also used as nucleophiles in catalytic arylation reaction (Scheme 4.25, Eq. 3, Eq. 4, and Eq. 5).[55] Reactions were performed in methanol at room temperature or at reflux, and were catalyzed by 0.1 equiv of copper(II) sulfate or copper(I) chloride, which furnished comparable results. Sodium azide could be used without additive, whereas trimethylsilyl azide required the addition of 1.2 equiv of tetra-*n*-butylammonium fluoride (TBAF). It is interesting to note that pinacol-derived boronic ester was shown to be a suitable reacting partner when the latter conditions were used (Scheme 4.25, Eq. 4). This azidation could be used in a one-pot fashion together with a copper-catalyzed Huisgen cycloaddition to synthesize 1-aryl-1,2,3-triazoles (Scheme 4.25, Eq. 5).

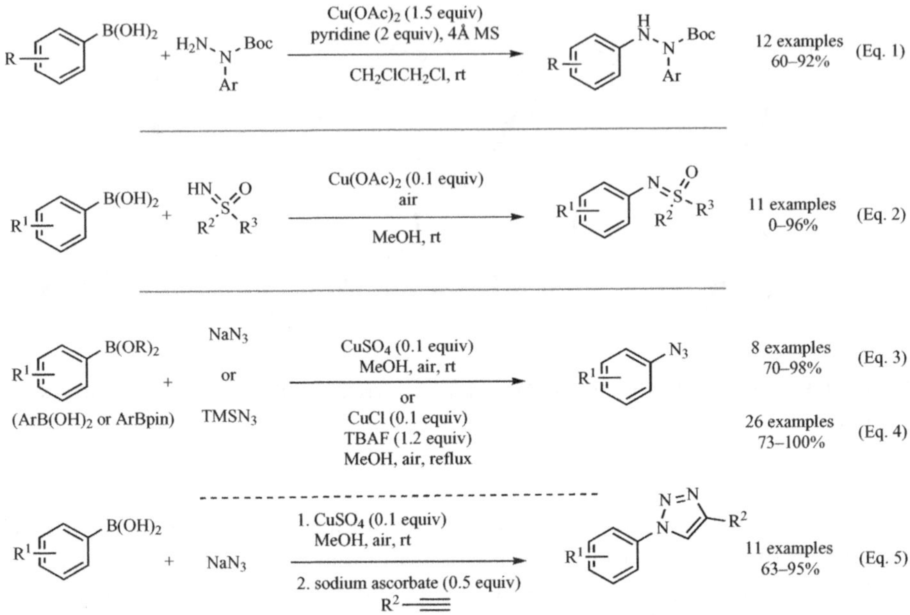

Scheme 4.25 Additional N-nucleophiles suitable for the N-arylation with arylboron reagents.

In 2009, it was reported that aniline derivatives could be simply prepared using aqueous ammonia as the nucleophile, a reaction that is an excellent alternative to the analogous copper-catalyzed arylation of aryl halides described in Chapter 1 (see Section 1.5). The reaction required only 10 mol % of copper(I) oxide without the need for a base or an additional ligand and was performed in methanol at room temperature (Scheme 4.26, Eq. 1).[56] The procedure was broad and high yielding. Methanol was again found to be the best solvent, but the reaction also occurred in ethanol, acetonitrile, DMF, and water. Based on this reaction, Fu and coworkers very recently described a general catalytic system allowing the introduction of various functional groups on aromatic rings (Scheme 4.26, Eqs. 2–6).[57] Azidonation, sulfonylation, nitration, hydroxylation, and even iodination were readily achieved based on the use of copper(I) oxide in the presence of ammonia as the catalyst, which was simply generated by premixing copper(I) oxide (0.1 equiv) with 2.5 equiv of aqueous ammonia in water under air for 30 minutes at room temperature. The water phase containing the catalytic system was recyclable after the reaction upon simple extraction. An additional base (potassium carbonate) was necessary, however, in order to neutralize the boronic acid residue released during the first batch. For azidation, sulfonylation, or hydroxylation, competitive amination was not observed when 3–5 equiv of the nucleophilic salt was used (Scheme 4.26, Eq.2, Eq. 3, and Eq.4). In the case of nitration, slight modification of the catalytic system (lowering the amount of ammonia to 1.8 equiv) significantly improved the yield (Scheme 4.26, Eq. 5). Finally, aniline could also

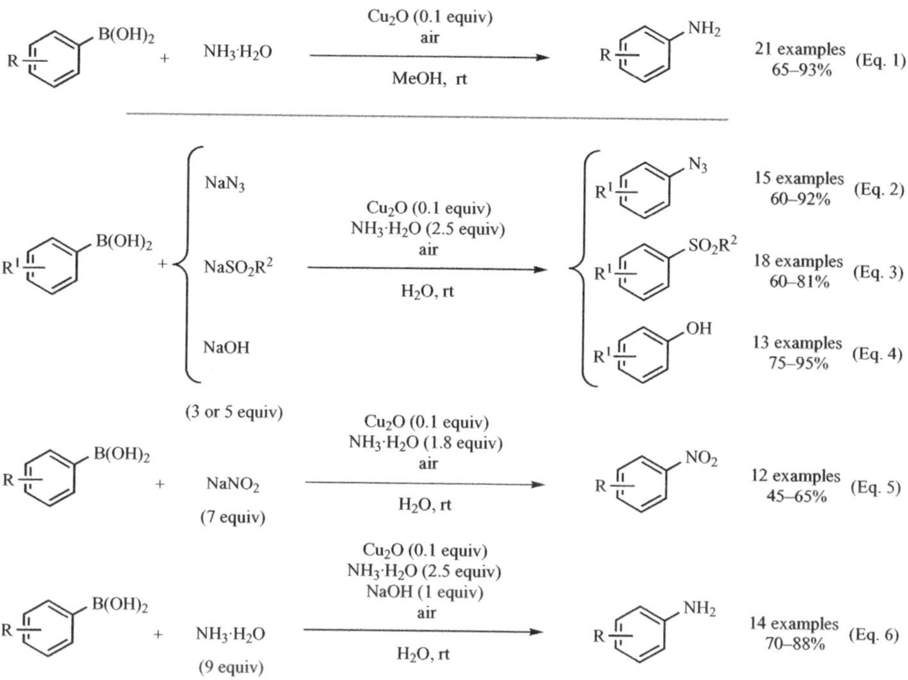

Scheme 4.26 Heteroarylation in the presence of ammonia: substrate versus ligand.

be synthesized from ammonia itself (Scheme 4.26, Eq. 6). Contrary to the related procedure in methanol, addition of a base (NaOH) was necessary, and the stoichiometry between this base and the amount of ammonia had to be adjusted in order to limit the competitive formation of phenol derivatives.

Cyanates have also recently been arylated under oxidative copper-catalyzed conditions, extending the number of salts that can be arylated.[58a] This approach allowed the direct formation of aryl carbamates from the corresponding arylboronic acids by trapping the intermediate isocyanate with an alcohol used as the solvent. Among the functionalities that have been poorly investigated or less successful, there is only a single example involving guanidines, for which an intramolecular C–N coupling involving a protected guanidine arylation has been reported.[58b] Arylation of amidines was also only recently reported and could be efficiently achieved in the presence of a catalytic amount of copper (II) acetate and a catalytic amount of sodium pivalate in DMF under aerobic conditions at 50°C.[58c]

Whereas numbers of intermolecular examples have been documented, there is only a single example of intramolecular C–N coupling involving a protected guanidine arylation.[59] Surprisingly, these procedures have not been extensively applied in the synthesis of complex natural products (see Chapter 16 for details) despite the mildness of the reaction conditions and their broad scope, which are perfectly compatible with late-stage functionalization of complex and/or sensitive substrates.

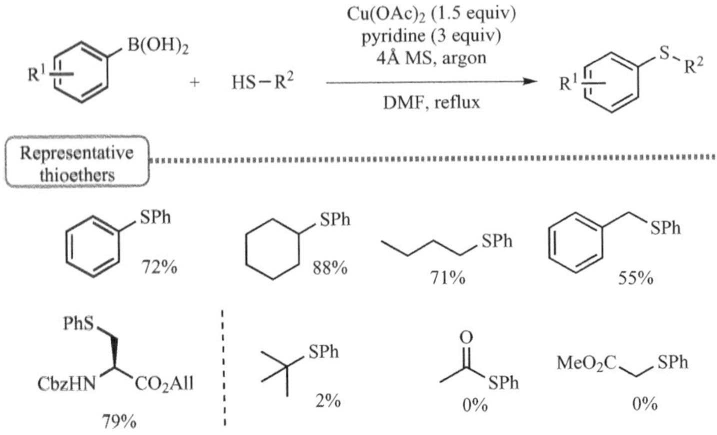

Scheme 4.27 Initial development in the copper-mediated S-arylation of thiols with boronic acids.

4.2.4.3 Other Heteronucleophiles: Sulfur Selenium, Tellurium, and Phosphorus. Besides O- and N-nucleophiles, other heteronucleophiles can also be arylated using the Chan–Lam–Evans reaction or related cross-coupling procedures. But this area is, however, still underdeveloped, especially when compared to the number of related procedures for N- or O-arylation reactions.

The first arylation of a sulfur nucleophile was reported in 2000 by Guy and coworkers (Scheme 4.27).[60] It was rapidly found that the conditions developed for O- or N-arylation were not optimal for C–S bond formation. The reaction was very sluggish, and competitive oxidation of thiols to disulfides rapidly inhibited the reaction. In the absence of an additional oxidant (air), the reaction was slow, and disulfide formation was still observed. Fortunately, it was found that in DMF and at higher temperatures the reaction was rapid enough, avoiding the competitive degradation of the thiol. Optimal conditions for the reaction involved 2 equiv of the arylboronic acid, 3 equiv of pyridine, 1.5 equiv of copper(II) acetate, and 4Å molecular sieves in DMF under argon.

Possibly inspired by Guy's work, an ingenious copper-catalyzed synthesis of diaryl thioethers from two arylboronic acids and potassium ethyl xanthogenate was recently reported. This process involved three distinct steps, two being copper-promoted S-arylations and the third one an intermediate hydrolysis (Scheme 4.28).[61a] Various symmetrical or unsymmetrical diaryl thioethers were obtained in good-to-excellent yields under relatively mild conditions. The reaction conditions were, however, drastically different from previous C–S bond formation since the reaction was performed in refluxing methanol, and only 5 mol % of copper(II) acetate or copper(I) chloride was required. According to the authors, the reaction was performed under nitrogen, leaving unanswered questions regarding the reaction mechanism. A room temperature copper-catalyzed S-arylation of thiophenols with boronic acids was also reported recently. The reaction was best performed with 5 mol % of

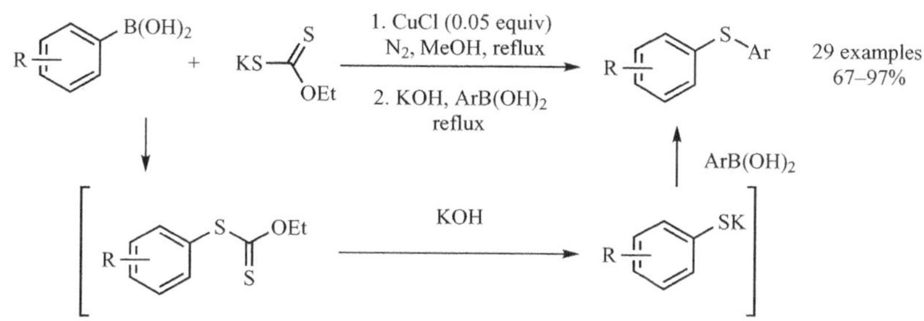

Scheme 4.28 One-pot sequential copper-catalyzed synthesis of diaryl thioethers from two distinct arylboronic acids and potassium ethyl xanthogenate.

copper(II) sulfate in combination with 5 mol % of 1,10-phenanthroline under oxygen in a 1:1 mixture of aqueous tetra-*n*-butylammonium hydroxide (40% aq) and ethanol. Although a number of boronic acids and thiophenols performed well under these conditions, the reaction could however not be extended to alkylthiols.[61b]

Cyclic thioureas or thioamides, which can be seen as masked thiols, were also effectively S-arylated using related reaction conditions (Scheme 4.29, Eq. 1).[62] 1,10-Phenanthroline was preferred over pyridine in these processes and it was shown that microwave irradiation markedly accelerated the reaction. However, contrary to the conditions developed by Guy for the arylation of thiols, air was necessary to achieve full conversion. The mechanism has been investigated and the formation of disulfides has been ascertained.[63] The following mechanism has been proposed: copper(II)-promoted oxidative formation of disulfides would occur in the first step, followed by the insertion of copper(I) into the S–S bond. The copper complex then reacted with the arylboronic acid to form the arylated thiol and a new copper(I) thiolate complex. This complex was finally oxidized to copper(II) before transmetallation and, in the presence of an oxidant, could undergo the reductive elimination. Based on this mechanism, the synthesis of unsymmetrical organochacogenides from dichalcogenides following a similar procedure, but requiring only a catalytic amount of copper/ligand, has been documented (Scheme 4.29, Eq. 2). The reaction not only allowed the formation of a C–S bond but could also be extended to the synthesis of diaryl or alkyl-aryl selenides and tellurides.[64]

Other sulfur-based nucleophiles can be used, and Evans reported in 2004 the arylation of sodium sulfinate, allowing an efficient synthesis of arylsulfones from arylboronic acids (Scheme 4.30).[65] For this stoichiometric reaction, an inorganic base, potassium carbonate, provided better yields than an organic base such as triethylamine, and the reaction was best performed in dimethylsulfoxide. Sodium alkyl or aryl sulfinates could be cross coupled to a wide range of arylboronic acids, furnishing good yields of the resulting sulfones. Heteroarylboronic acid could also be used, although with decreased efficiency. In 2007, two research groups independently published a catalytic version of this reaction (Scheme 4.30, Eq. 2 and Eq. 3).[66] Whereas a benzyl-substituted

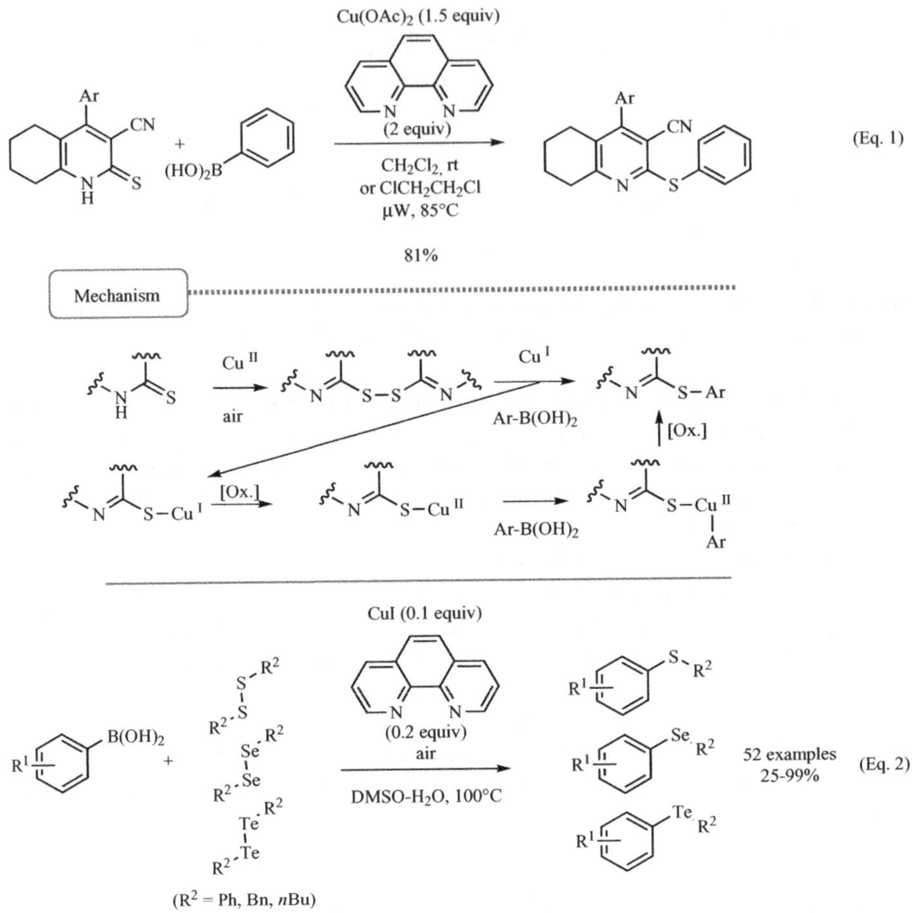

Scheme 4.29 Copper-promoted arylation of cyclic thioamides and dichalcogenides.

imidazole was used as a ligand in the reaction developed by Tse, 1,10-phenanthroline was preferred by Batey. The reaction was found to be sensitive to steric hindrance, as illustrated with the reactivity of (2,6-dimethylphenyl) boronic acid, which was efficiently coupled under Evans's condition (63%), but required gentle heating (60°C). It was found to be a poor reaction partner under the catalytic reaction conditions described by Batey (19%). The reaction can also be performed using combinations of copper(I) oxide and ammonia in water, as shown in Scheme 4.26 (Eq. 3),[57] and in ionic liquids in the absence of an additional ligand, as reported by Kantam and coworkers.[67]

The first oxidative copper-promoted arylation of a phosphorous derivative was published in 2011 and involved the arylation of dialkyl or dibenzyl phosphites with arylboronic acids (Scheme 4.31).[68] The mild catalytic reaction conditions were established as follows: copper(I) oxide (0.05 equiv), triethylamine (3 equiv), and 1,10-phenanthroline (0.1 equiv) in acetonitrile at room

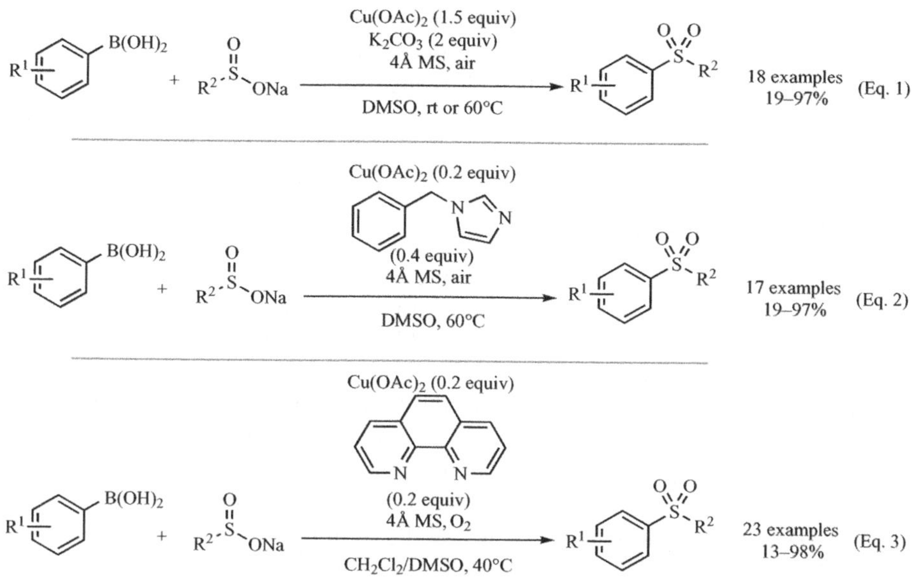

Scheme 4.30 Copper-promoted S-arylation of sodium sulfinates with boronic acids.

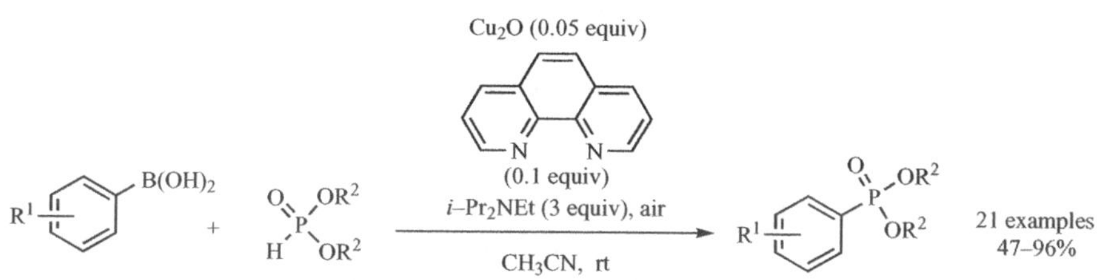

Scheme 4.31 First example of a copper-catalyzed synthesis of arylphosphonates from arylboronic acids and dialkyl phosphites.

temperature under air for 24 hours. The scope of this procedure was fairly broad but followed the general trend observed with other heteronucleophiles (*i.e*, heteroaryl). Sterically demanding arylboronic acids provided the corresponding products with reduced yields. Other procedures for the arylation of P-nucleophiles are overviewed in detail in Chapter 3.

4.3 COPPER-PROMOTED ARYL TRANSFER FROM METALLATED ARYL DERIVATIVES (NONBORON)

Arylboron reagents have been popular arylating agents because of their availability, reactivity, high functional group compatibility, and the mild reaction conditions in which the copper-mediated arylation of heteronucleophiles can be achieved using these reagents. Such processes are, however, not limited to boron-containing donors, and a number of other organo-(pseudo)-metalloids can also participate in related cross-coupling reactions. Indeed, the Chan–Lam–Evans reaction has been built on the chemistry developed with organolead and organobismuth reagents, especially following Barton's pioneering work. The use of these reagents, which can provide an efficient alternative to the use of aryl halides or arylboronic acids, will be overviewed in the next section.

4.3.1 Bismuth and Lead

This chemistry has been extensively reviewed, and since only few developments and applications appeared since the publications of these review articles, selected key points will be highlighted in the following section and have been summarized in Table 4.1.[5a,69]

Triarylbismuthanes are easily prepared by the addition of the corresponding aryllithium or aryl Grignard derivatives to bismuth trihalides and can be oxidized to the corresponding pentavalent organobismuth derivatives with a number of oxidants,[70] whereas aryllead derivatives can be obtained by a mercury(II)-catalyzed transmetallation of arylstannanes, aryltin, or arylboron derivatives with lead tetraacetate or by a direct plumbation, which is restricted to electron-rich aromatics.[69] Milder reactions, such as the copper-catalyzed cross coupling between aryl halides and metallic bismuth or the transmetallation with sodium tetraarylborate salts were also recently developed, allowing the preparation of more diversely functionalized triarylbismuthanes.[71]

Such reagents have long been known to act as arylating agents, including the arylation of heteronucleophiles, but it is only following the initial observations of Dodonov and Barton that mild and efficient copper-catalyzed reactions emerged.[72] In this remarkable work, pentavalent triarylbismuth diacetates were used in combination with a catalytic amount of copper salts for the N- or O-arylation of heteronucleophiles (Scheme 4.32, Eq. 1). A stoichiometric amount of copper diacetate was required to perform the analogous reaction using trivalent arylbismuthanes (Scheme 4.32, Eq. 2).[73] In contrast, the catalytic arylation involving aryllead reagents has been limited to N-arylation (Scheme 4.32, Eq. 3).[72d] Even if these reactions belong to oxidative processes, they do not require the use of an additional oxidant as this role is held by either the pentavalent bismuth reagent, the organolead derivative, or the stoichiometric amount of copper salt when triarylbismuthanes are used. *In situ* generation of the pentavalent bismuth reagent has been achieved from triarylbismuthane by reaction with oxone followed by one-pot catalytic C–O bond formation.[74] Ligandless conditions using copper(II) acetate are usually the most efficient, but in some cases, copper(II) pivalate markedly accelerated the reaction, probably due to solubility issues.[75]

For aryllead triacetate derivatives, reactions have been limited to the transfer of simple electron-rich arenes such as phenyl, tolyl, or methoxyphenyl residues. But more elaborated compounds, such as a tyrosine-derived organolead triacetate, were also successfully used as N-arylating agents[76] (Fig. 4.5). A relatively broad range of aromatic substituents ranging from electron-poor

TABLE 4.1 Selected Copper-Catalyzed Heteroarylation of N-, O-, and S-Nucleophiles with Organolead or Organobismuth Derivatives

Entry	Heteronucleophiles	Catalytic Systems, Additives	Conditions	Arylating Agents	References
1	Anilines, primary and secondary alkyl amines	$Cu(OAc)_2$ (0.1 equiv)	CH_2Cl_2, rt	$ArPb(OAc)_3$	[150]
2	Pyrazoles, imidazoles, benzimidazole, 1*H*-indazoles, 1,2,4-triazoles, tetrazoles	$Cu(OAc)_2$ (0.1 equiv), NaH (0 or 1.2 equiv)	CH_2Cl_2 or CH_2Cl_2-DMF, 50–90°C	$ArPb(OAc)_3$	[79]
3	3-Imidazoles	$Cu(OAc)_2$ (0.1 equiv)	CH_2Cl_2, rt	$ArPb(OAc)_3$	[76]
4	Anilines, primary and secondary alkyl amines, *N*,*N*-diphenyl-hydrazine	$Cu(OAc)_2$ (0.5 or 1 equiv)	CH_2Cl_2, rt	$Ar_3Bi(OAc)_2$	[73]
5	Indoles	$Cu(OAc)_2$ (0.1 equiv)	CH_2Cl_2, rt	$Ar_3Bi(O_2CCF_3)_2$	[80]
6	Pyrazoles, imidazoles, benzimidazole, 1*H*-indazoles, 1,2,4-triazoles, tetrazoles	$Cu(OAc)_2$ (0.2 equiv), *N*,*N*,*N*′,*N*′-tetramethyl-guanidine (2 equiv)	THF, 50°C	$Ar_3Bi(OAc)_2$	[82]
7	Phenols	$Cu(OAc)_2$ (0.1 equiv)	CH_2Cl_2, rt	$Ar_3Bi(OAc)_2$	[72c,85]
8	Carbazates, di- and trisubstituted carbamates, protected hydrazines	$Cu(OAc)_2$ (0.05 equiv)	CH_2Cl_2, –60°C or rt or reflux	$Ar_3Bi(OAc)_2$	[152]
9	Diols, functionalized alcohols	$Cu(OAc)_2$ (0.2 equiv)	THF, 50°C	$Ar_3Bi(OAc)_2$	[72b]
10	Anilines, primary and secondary alkyl amines, *N*,*N*-diphenyl-hydrazine	$Cu(OAc)_2$ (1 equiv)	CH_2Cl_2	Ar_3Bi	[73]
11	Amides, imides, sulfonamides, carbamates	$Cu(OAc)_2$ (1 equiv), Et_3N or pyridine (2 equiv)	CH_2Cl_2, rt	Ar_3Bi	[83]
12	Carbamate, protected *N*-aryl hydrazones	$Cu(OAc)_2$ (1.5 equiv), triethylamine (1.5 equiv)	CH_2Cl_2, reflux	Ar_3Bi	[85]
13	Phenols	$Cu(OAc)_2$ (1 equiv)	CH_2Cl_2, rt	Ar_3Bi	[85]
14	Tertiary alcohols	$Cu(OAc)_2$ (0.2 equiv), dicyclohexyl-methylamine (2 equiv), air	Toluene, rt	Ph_4BiF	[90b]
15	Diaryl disulfides	$Cu(OAc)_2$ (1 equiv)	DMSO, 100°C	Ar_3Bi	[91]

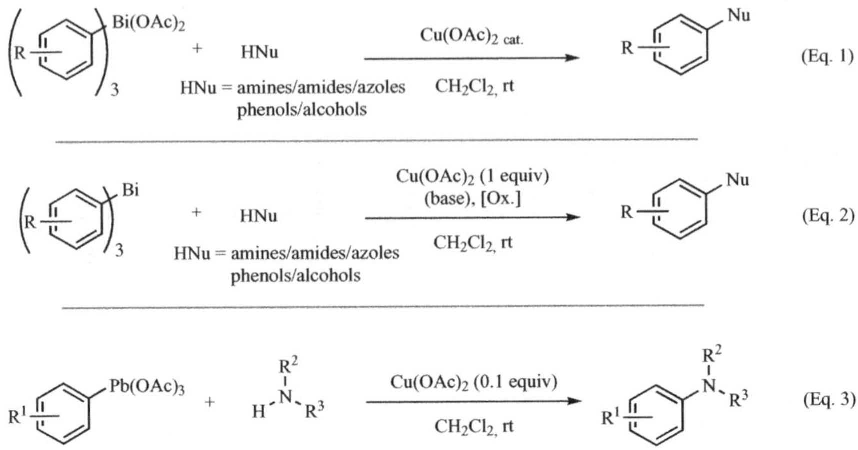

Scheme 4.32 Arylbismuth or aryllead reagents as aryl donors in copper-promoted arylation of N- and O-nucleophiles.

to electron-rich has been found to be compatible with the chemistry based on organobismuth derivatives. Following the observation made by Barton and others,[77] who reported that, in the case of unsymmetrical triarylbismuthanes, the most electron-deficient aryl group was selectively transferred, resin-bound supported triarylbismuthanes and resin-bound triarylbismuth diacetates have also been prepared and used in copper-promoted N- and O-arylation.[78] For lead- or bismuth-derived metallated arenes, *ortho*-substitution is typically well tolerated, but *ortho,ortho'*-disubstituted metallated arenes are of limited interest, usually furnishing the arylated compounds in low yields.

The N-arylation of a number of nitrogen-containing moieties, including anilines, alkyl amines, and a number of azoles, has been successfully carried out using organolead or organobismuth derivatives.[79,80] These reactions are tolerant to steric hindrance on the heteronucleophile as demonstrated, for example, during the N-arylation of mesitylamine with 2,4-dimethoxyphenyl lead triacetate[81] or with an excess (2.2 equiv) of triphenylbismuth triacetate[72c] under copper catalysis, reactions that are typically extremely difficult to perform using other techniques. The outcome of N-arylation reactions was found to be dependent on the basicity of the nitrogen function and, for less nucleophilic substrates, the addition of an inorganic base (sodium or potassium hydride) greatly improved the yields.[75,79] *N,N,N',N'*-tetramethylguanidine, an organic base that probably also acts as a ligand for copper, was reported to improve the N-arylation of benzimidazoles, 1*H*-indazoles, 1,2,4-triazoles, or tetrazoles when the arylation was performed using triphenylbismuth diacetate in the presence of a catalytic amount of copper(II) diacetate.[82] Triethylamine or pyridine was also found to be an effective promoter for the copper-promoted N-arylation of diverse N–H containing groups such as amides, imides, ureas, carbamates, and sulfonamides with triarylbismuthanes, as described by Chan. Such reaction conditions were

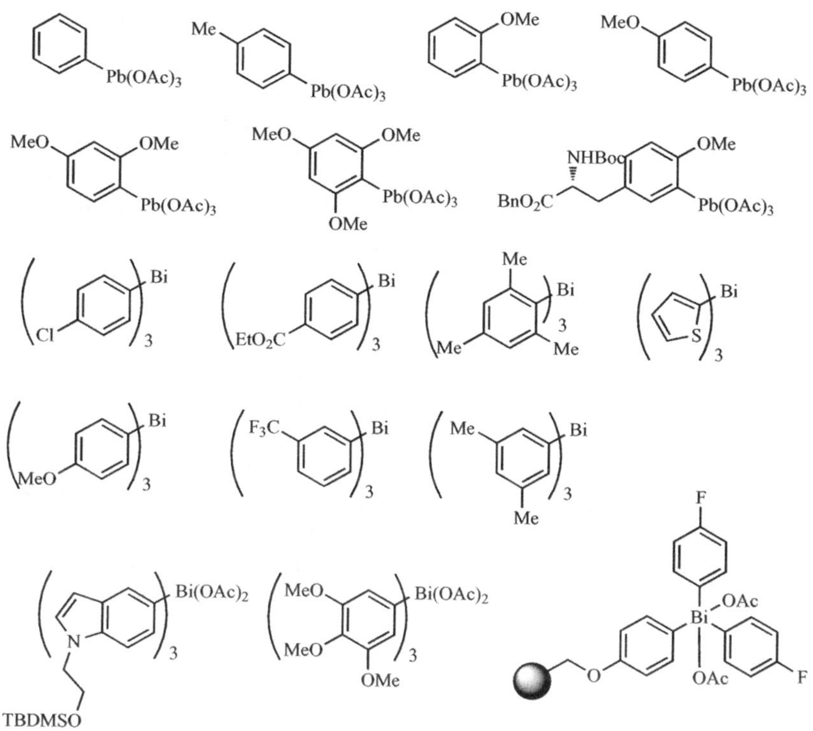

Figure 4.5 Selected and representative arylplumbane triacetates,[76,150] triarylarylbismuthanes,[84,91] and triarylarylbismuth diacetates[78,151] used in copper-promoted heteroarylation.

also effective for the arylation of *para*-nitroaniline, another challenging substrate.[83] The reduced reactivity of some N-nucleophiles was exploited to design selective N-arylation of bifunctional substrates such as aminobenzanilides[79] or aminobenzimidazoles[84] using triarylbismuthanes or aryllead triacetate in the absence of a base.

Arylbismuth derivatives have long been known to be competent reagents for the uncatalyzed O-arylation of hydroxyl groups, but a strong catalytic effect of copper salts was observed for such transformations, especially for the arylation of phenols[85] and the monoarylation of diols.[72b] Highly functionalized natural products containing a diol, such as the macrolide immunosuppressant antibiotic ascomycin, could be selectively monoarylated using triarylbismuth diacetate under copper(II) acetate-catalyzed reaction conditions, demonstrating the impressive mildness and selectivity of the reaction.[86] Even if most reactions were carried out in the presence of copper(II) acetate as the catalyst and under ligandless conditions, Brunner reported an interesting enantioselective O-arylation of meso diols (desymetrization) that was occurring in the presence of a mixed pyridine-oxazoline ligand and furnished up to 50% enantiomeric excess.[87] Simple primary and secondary alcohols furnished moderate yields in the O-arylation reaction, but the presence of a neighboring chelating group such as an amine, a hydroxyl, a carbonyl, or a sulfanyl group greatly improved the reaction efficacy.[88] As observed in the case of the related N-arylation, addition of a base was also beneficial for some O-arylation reactions, especially for poorly nucleophilic hydroxyl residues. More recently, it has been showed that the O-arylation of tertiary alcohols, even lacking any directing group, could be accomplished using a bulky aliphatic amine such as dicyclohexylmethylamine and, interestingly, it was noted that this transformation was best accomplished using tetraphenylbismuthonium reagents (Ph_4BiF[89]).[90]

A recent development in the field of heteroarylation involving triarylbismuthanes has been reported by Kurita and others in 2011, who described the first arylation of diaryl disulfides using copper(II) acetate as the catalyst, 1,10-phenanthroline as an additive, and triarylbismuthanes in dimethylsulfoxide under air at 100°C.[91] It should be highlighted that, contrary to the N- or O-arylation reactions, all three aryl groups of the starting triarylbismuthanes as well as both sulfanyl groups of the disulfide participated in the reaction, indicating that a different mechanism is probably involved in this transformation.

These bismuth or lead reagents have proven to be versatile and tolerant to many functional groups, and the reaction can be performed under mild reaction conditions. In some cases, they provided better outcomes than the parent reaction involving boronic acids. It has, for example, been established that a better regiocontrol can be achieved in the case of the arylation of substituted imidazoles with organolead donors compared to the corresponding arylboronic acids.[76] While remarkably useful in some instances, although they require the preparation of the aryl transfer agents, these methods have strong weaknesses. In most cases, the aryl donor showed limited stability, and therefore such precursors are not commercially available. In an interesting comparative study involving the arylation of hydantoin with triarylbismuthanes and arylboronic acids, it has been established that the boronic acids were better aryl donors, tolerating electron-withdrawing groups.[48] In addition, and due to their inherent oxidizing properties, pentavalent organobismuth or organolead reagents are not compatible with substrates that are prone to oxidation. And last but not least, the known toxicity of lead is a serious limitation for the use of organolead reagents. In the case of bismuth derivatives, only one out of the three aryl groups is usually transferred, which limits the use of highly functionalized aromatic units.

4.3.2 Arylstannanes and Aryl-Alkoxysilanes

Lam evaluated the use of arylstannanes and aryl-alkoxysilanes as aryl donors in copper-promoted N-arylation reactions, reagents that are less commonly used in copper-mediated arylation of heteronucleophiles. While such pseudometallated species were found to be poorly reactive, or even nonreactive under

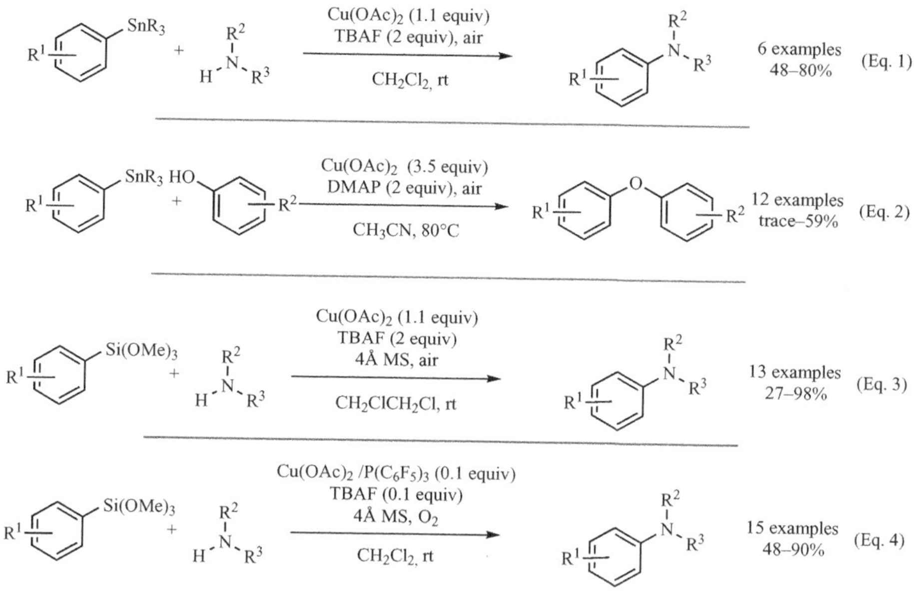

Scheme 4.33 Use of arylstannanes and arylsiloxanes for the arylation of N-nucleophiles and phenols.

standard conditions (stoichiometric copper(II)/base),[42] the addition of fluoride ions, especially tetra-*n*-butylammonium fluoride, allowed the N-arylation (Scheme 4.33) to be performed quite efficiently. Such ions are believed to form pentacoordinated stannanes or siliconate species, key intermediates in which the C–Sn or C–Si bonds are activated and susceptible to undergo the transmetallation step.[92] The reaction proceeded well with most N-nucleophiles in the case of siloxanes but proved to be more restricted for stannylated aryl donors (Scheme 4.33, Eq. 1 and Eq. 3). It was also shown that it is possible to synthesize diaryl ethers from stannylated precursors and alcohols, but for such reactions, 4-dimethylaminopyridine was preferred over tetra-*n*-butylammonium fluoride, with the reaction running at 80°C in acetonitrile (Scheme 4.33, Eq. 2).[93] In addition, conditions for catalytic N-arylation reactions involving aryl-alkoxysilanes have been recently reported. Interestingly, these conditions require only a catalytic amount of tetra-*n*-butylammonium fluoride (Scheme 4.33, Eq. 4),[94] and the addition of an electron-poor phosphine ligand, $P(C_6F_5)_3$, was essential for the process. Such conditions offer a clear improvement over the stoichiometric process, avoiding the use of an excess of the rather unfriendly tetra-*n*-butylammonium fluoride reagent.

Besides the development of new methods for the arylation of N-nucleophiles, aryl trimethoxysilanes have also been used recently in combination with less standard nucleophiles. In this context, the copper(II)-catalyzed esterification of carboxylic acids with aryl trimethoxysilanes was reported in 2010 (Scheme 4.34).[95] The reaction was best performed with a combination of 0.2 equiv of

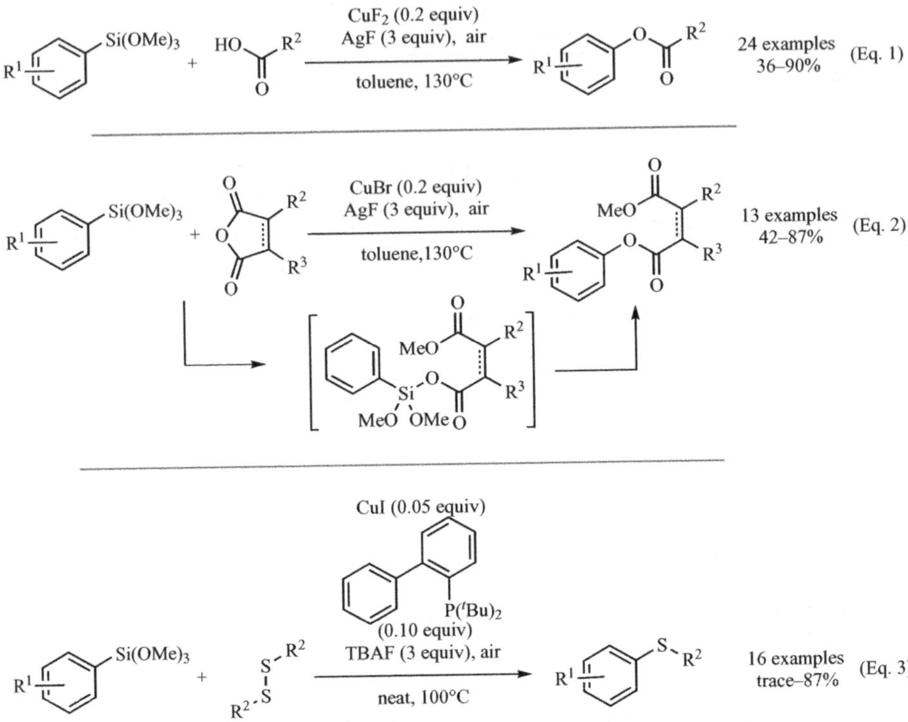

Scheme 4.34 Use of arylsiloxanes for the arylation of carboxylic acids, anhydrides, and disulfides.

copper(II) fluoride and 3 equiv of silver(I) fluoride at 130°C in toluene. This reaction was extended to the desymmetrization of cyclic anhydrides, in which a double heteroesterification takes place (Scheme 4.34, Eq. 2). It is believed that the methyl ester is initially formed through opening of the anhydride by the Lewis acidic aryl trimethoxysilane, furnishing a silyl carboxylate intermediate, which then undergoes the cross-coupling reaction.

Aryl trimethoxysilanes have also been used in oxidative C–S bond formation using disulfides as heteronucleophiles, demonstrating the parallel behavior of silicon and boron species (Scheme 4.33, Eq. 3).[96] Various phosphines or nitrogen-containing ligands were evaluated in this reaction and (2-biphenyl) di-*tert*-butylphosphine provided the best yields. It should be noted that such a procedure was performed under solvent-free conditions and at a rather high temperature.

4.3.3 Aryliodonium Salts

Diaryliodonium salts represent an important class of hypervalent reagents that have important applications in organic chemistry as versatile electrophilic arylating agents.[97] These air and moisture stable compounds are usually easily prepared from the corresponding aryl iodides through a two-step sequence involving an initial oxidation followed by ligand exchange with an arene or a metallated arene.

The copper-promoted arylation of heteronucleophiles with aryliodonium salts has a long history that can be traced back to the 1950s, though it suffered from relatively harsh reaction conditions.[98] A much milder procedure for the N-arylation of N-nucleophiles was reported only recently. As demonstrated by Kang in 2000, the room temperature N-arylation of secondary amines, anilines, and linear or cyclic amides was efficiently promoted by 0.1 equiv of copper(I) iodide and 2 equiv of sodium carbonate in dichloromethane (Scheme 3.35), showing the exceptional potential of this process that compares well to the corresponding arylations with aryl halides or boronic acids (even if the arylating agents have to be prepared, and if only one aryl group is transferred during the reaction). Various azoles such as imidazole, benzimidazole, benzotriazole or pyrazole, and lactams were best arylated under slightly different reaction conditions involving copper(II) acetylacetonate as the copper source, potassium carbonate as the optimal base, and toluene at 50°C. This reaction was extended to the synthesis of 9-arylpurines (Scheme 4.35, Eq. 2), which were efficiently produced from the corresponding diaryliodonium salts using copper(I) bromide and potassium carbonate in refluxing dichloromethane.[99] Various purines were shown to perform well in this cross-coupling reaction, but unprotected 6-amino 9*H*-purine was not suitable for such arylation. Substituted diaryliodonium salts underwent the coupling in good yields except in the case of $(3\text{-}NO_2C_6H_4)_2I^+Br^-$ where no reaction was observed because of solubility issues. Indoles could not be arylated at room temperature, but such arylation was efficiently performed at higher temperatures (140–150°C) in DMF.[100]

Base-free conditions were reported by Davydov for the N-arylation of premetallated *N*-tributyltin-5-substituted tetrazoles with diaryliodonium chlorides, but, contrary to the aforementioned conditions, this procedure requires a stoichiometric amount of copper(II) acetate (Scheme 4.36, Eq. 1).[101] Here again, no reaction was observed with electron-deficient aryliodoniums salts. Very recently, the scope of nitrogen-containing functional groups that can successfully be N-arylated with diaryliodonium salts under copper catalysis has been extended to sulfoximines, and similar reactions were achieved under base-free conditions using a catalytic amount of copper(I) bromide in aqueous polyethylene glycol 400 and accelerated by ultrasounds (Scheme 4.35, Eq. 2).[102a] Quaternarisation of nitrogen nucleophiles is also possible, as demonstrated by Gao and coworkers who reported in 2013 the synthesis of imidazolium salts from N-substituted imidazoles and diaryliodonium salts using catalytic amount of copper(II) acetate in DMF at 100°C.[102b]

Little work dealing with the arylation of alcohols or phenols using aryliodonium salts have been reported, most probably because such reactions have been known to proceed in the absence of a copper catalyst.[103] Very recently, a copper-catalyzed monoarylation of vicinal diols with diaryliodonium triflates was reported (Scheme 4.37).[104] The nature of the counter ion of the iodonium salt strongly impacted the reaction outcome, and triflate-derived iodonium

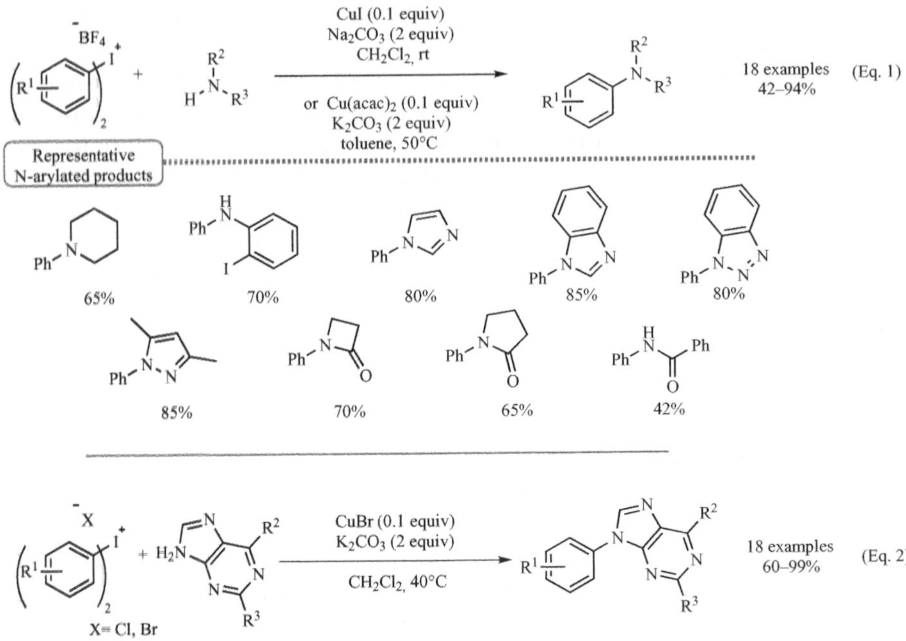

Scheme 4.35 Copper-catalyzed N-arylation involving diaryliodonium salts.

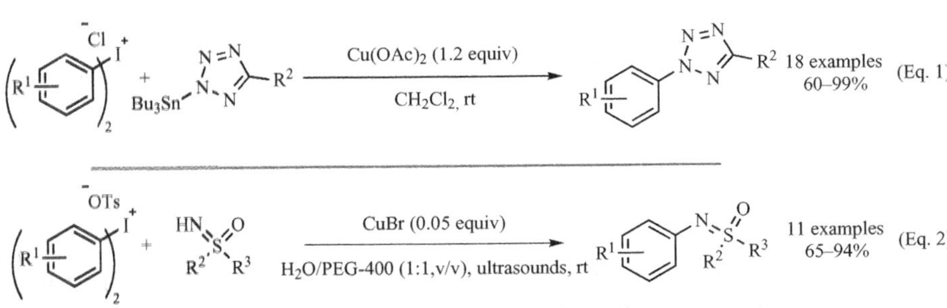

Scheme 4.36 Base-free, copper-promoted N-arylation of *N*-tributyltin-tetrazoles and sulfoximines involving diaryliodonium salts.

salts showed the highest activity when compared to non-nucleophilic anions such as BF_4^-, PF_6^-, ClO_4^-, or halides. The reaction was best performed with 0.1 equiv of copper(II) triflate in the presence of 3 equiv of sodium phosphate (Na_3PO_4) in toluene at 100°C. Selective monoarylation was achieved for 1,2-, 1,3-, 1,4-, or 1,5-diols, whereas simple alcohols were poorly arylated. For non-symmetrical diols, a mixture of products was obtained, usually favoring the arylation of the least hindered alcohol. Various aryl groups could be coupled efficiently, including heteroaryls. One drawback of the chemistry of iodonium salts relies on the fact that only one out of two aryl groups is transferred, the second one being released as an aryl iodide. However, unsymmetrical

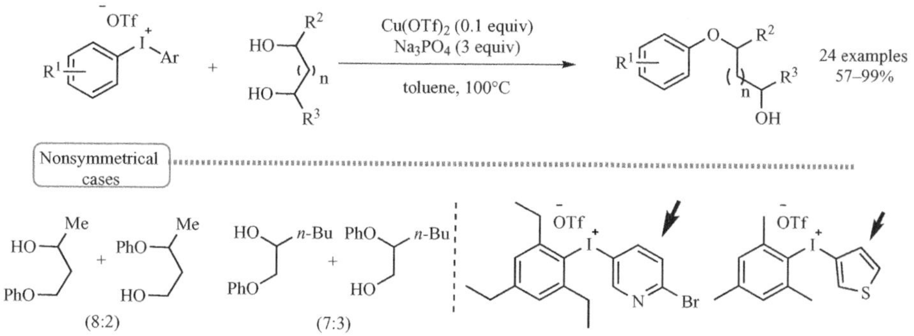

Scheme 4.37 Copper-catalyzed mono O-arylation of diols with aryliodonium triflates (arrows indicate the aryl unit selectively transferred during the arylation).

diaryliodonium salts bearing a nontransferable group such as a 2,6-disubstituted aryl group can be used as reagents, allowing an exclusive transfer of the least hindered arene unit.

As evidenced by all results overviewed in this section, the arylation of heteronucleophiles is certainly not restricted to the use of aryl halides (see Chapters 1–3 for details involving N-, O-, and P-nucleophiles, respectively) or arylboronic acids. While the former are quite often preferred because they are readily available, the latter offer milder reaction conditions in most cases. Still, they are not universal reagents, and the use of bismuth, lead, tin, silicon, or hypervalent iodide derivatives represents useful alternatives that have been demonstrated in some cases to be much more efficient than aryl halides or boronic acids, especially in the case of hypervalent iodonium salts.

Other reagents have been used for the arylation of heteronucleophiles. They include benzoic acids which, in a remarkable work by researchers at Novartis, have been recently used for the copper-catalyzed decarboxylative and oxidative arylation of N-nucleophiles under rather harsh conditions (170°C in anisole) and with restricted scope, since an *ortho*-directing group is required,[105] and with simple arenes that can be used as coupling partners in copper-mediated C–H bond functionalization with heteronucleophiles. The use of the copper-catalyzed direct amination of C–H bonds will be reviewed in Section 4.5, while the next section will be devoted to the use of masked nucleophiles, reagents that can be useful alternatives to the corresponding standard nucleophiles in some cases.

4.4 COPPER-CATALYZED ARYLATION REACTIONS INVOLVING MASKED S- AND N-NUCLEOPHILES

In most cases, the nucleophilic nature of the heteroatom is exploited for the construction of the heteroatom–aryl bond. However, methods based on an umpolung of the heteroatom recently allowed for the development of copper-promoted arylation reactions using electrophilic heteroatoms. These methods are complementary to the copper(II)-mediated oxidative aryl transfer and they avoid the use of an external oxidant, even if they are currently limited to N- and S-arylation. They will be covered in the next section, starting with masked S-nucleophiles.

4.4.1 C–S Bond Formation

A major achievement in this field was published by the Liebeskind group, who reported a mild thioetherification of arylboronic acids under nonbasic, copper(I)-catalyzed conditions (Scheme 4.38).[106] Based on Guy and coworkers' work,[60] they hypothesized that the S-arylation with boronic acids was not following the usual mechanism accepted for the Chan–Lam–Evans reaction involving N- or O-nucleophiles, but that this reaction rather involved the intermediate formation of disulfides. They indeed demonstrated that disulfides could be efficiently arylated by copper(I) species in the absence of an oxidant. This reaction required a full equivalent of copper(I)-3-methylsalicylate (CuMeSal), and it was thought that half of the disulfide was converted into a catalytically inactive copper(I) thiolate complex. In order to avoid the formation of this inactive complex, they replaced the disulfides with readily available *N*-thioimide derivatives,[107] compounds that are easily prepared from *N*-chlorosuccimide and the corresponding thiols. With such reagents, a catalytic amount of copper(I)-carboxylates could effectively promote the reaction without the need for a base, which actually inhibited the reaction. Thus, the treatment of an arylboronic acid (1.5–2.0 equiv) with *N*-(arylthio)succinimide (1.0 equiv) in the presence of 0.3 equiv CuMeSal in THF at 45–50°C afforded the corresponding arylthioethers in moderate-to-good yields (51–79%).

4.4.2 C–N Bond Formation

Besides the above-mentioned work of the Liebeskind group, the copper-promoted arylation of heteroatoms involving the umpolung of the heteroatom has mainly been developed for the formation of C–N bonds. These methods will be covered in the next section.

4.4.2.1 Synthesis of Anilines. Organometallic reagents have long been known to be readily aminated by electrophilic amination reagents.[108] The reaction of preformed organocuprates with *N*-alkyl,*O*-trimethylsilyl-hydroxylamines has been documented by Ricci in the 1990s, and has allowed the synthesis of anilines and heteroarylamine derivatives.[109] These remarkable early contributions provided a strong basis for the later use of these species in catalytic reactions.

In 2004, Johnson disclosed a new copper-catalyzed electrophilic amination reaction of diarylzinc reagents using *O*-benzoyl hydroxylamines as an electrophilic source of nitrogen and copper(II) trifluoromethanesulfonate as the catalyst at room temperature (Scheme 4.39, Eq. 1).[110] It should be noted that copper(I) chloride performed almost as efficiently in the reaction. Whereas various *O*-acyl hydroxylamines could also be arylated, related chloramines

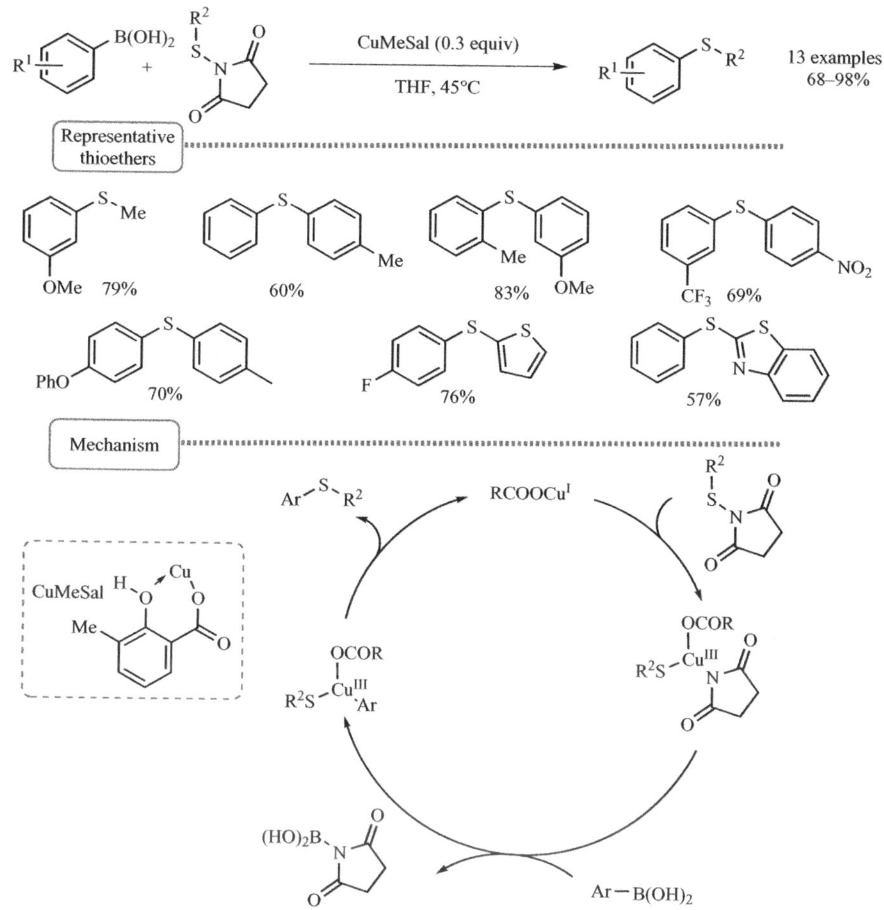

Scheme 4.38 Copper-catalyzed cross coupling of arylboronic acids with *N*-thioimides.

were not efficient reaction partners. It was also reported that an analogous reaction could not be performed from arylstannanes or arylboronic acids. The scope was extensively studied and secondary *N*-hydroxyalkyl or benzyl amine derivatives, including sterically hindered *O*-benzoyl,*N*,*N*-diisopropyl-hydroxylamine, as well as primary *O*-benzoyl,*N*-alkyl-amines were efficiently arylated. The tolerance toward the functionalization of the aromatic was broad since *ortho*-substitution and numbers of functional groups, ranging from electron-deficient nitro to electron-rich methoxy, were equally tolerated.

This reaction could be further extended to the use of Grignard reagents (slow addition) instead of diarylzinc derivatives (Scheme 4.39, Eq. 2),[111a] but *O*-benzoyl primary amine derivatives were not suitable reaction partners under such reaction conditions. From a mechanistic perspective, it has been concluded that the reaction proceed with a S_N2 mechanism of an intermediate arylcopper on the electrophilic nitrogen rather than through an oxidative

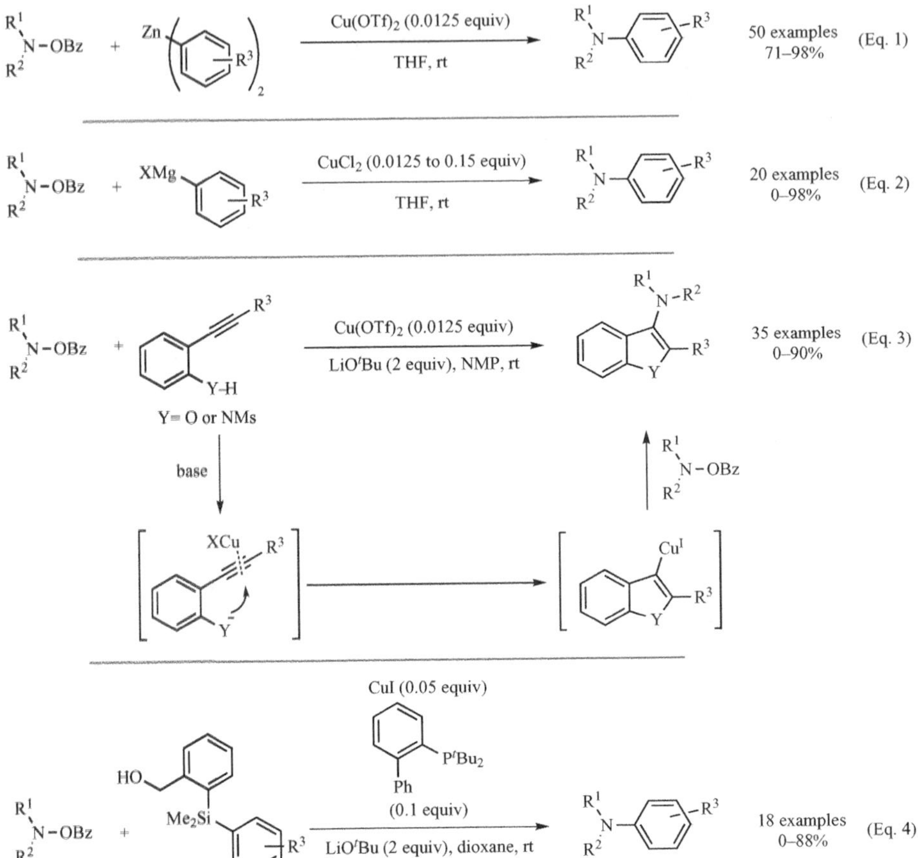

Scheme 4.39 Copper-catalyzed electrophilic amination of arylmagnesium, arylzinc and arylcopper reagents, as well as arylsilanes.

process, which might also account for the absence of reactivity observed with oxidized nitrogen heterocycles.

Based on this reaction, an annulative electrophilic amination approach to 3-aminobenzofurans and 3-aminoindoles has recently appeared in the literature (Scheme 4.39, Eq. 3).[111b,c] A strong base was required to induce the annulative cupration, which was followed by a subsequent amination. An electron-withdrawing group at the terminal position of the alkyne was not tolerated because it prevented the annulation and, in addition, a strong steric repulsion between the alkyne substituent and the incoming amine drastically reduced the yields when bulky amines were used. A convenient alternative to the use of aryl Grignard or diarylzinc reagents was reported in 2013 by Hirano, Miura, and coworkers who designed a remarkably efficient copper-catalyzed electrophilic amination of arylsilanes with hydroxylamine derivatives (Scheme 4.39, Eq. 4).[112] The success of this transformation, which proceeds at room temperature, is based on the use of aryl[(2-hydroxymethyl)phenyl]dimethylsilanes, reagents whose reactivity overcomes the limitations met with trimethylphenylsilane or trimethoxyphenylsilane during the optimization of this reaction.

An important improvement in the field of copper-catalyzed electrophilic amination has recently been reported independently by two groups, which demonstrated that such reaction was not restricted to the use of highly reactive organometallic reagents such as aryl Grignard or arylzinc reagents, but could be most efficiently performed with arylboronates (Scheme 4.40).[113]

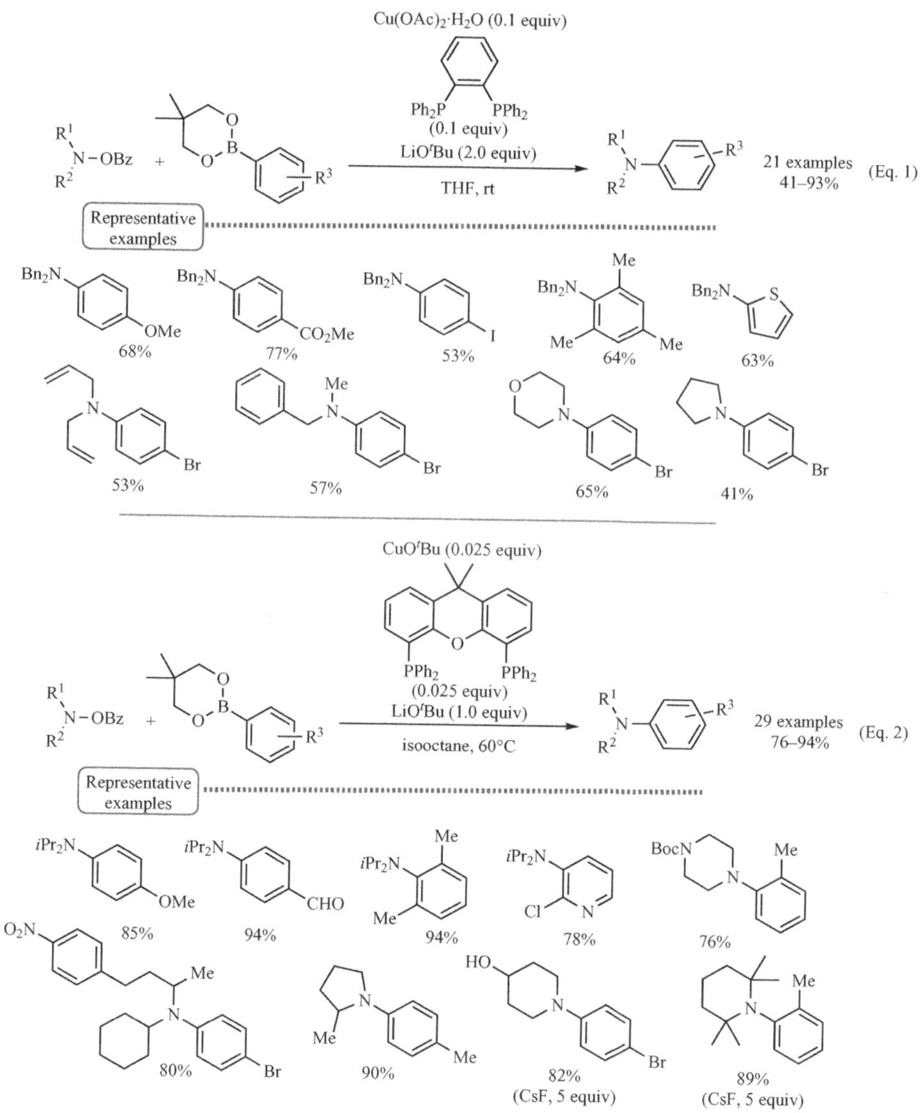

Scheme 4.40 Copper-catalyzed electrophilic amination of aryl boronates.

Both procedures used neopentylglycol boronic esters as aryl donors, lithium *tert*-butoxide as a base, and *O*-benzoyl-*N*,*N*-dialkylhydroxylamines as electrophilic aminating reagents. However, the catalytic system was different, as a combination of copper(II) acetate and 1,2-bis(diphenylphosphino)benzene in THF was used by Hirano and coworkers[113a] (Scheme 4.40, Eq. 1), whereas Lalic and coworkers[113b] used highly sensitive copper(I) *tert*-butoxide and 4,5-bis(diphenylphosphino)-9,9-dimethylxanthene in isooctane to perform the arylation (Scheme 4.40, Eq. 2). The first reaction was performed at room temperature, while the second one required gentle heating at 60°C.

These reaction conditions were shown to be tolerant to a wide range of functionalities such as iodo, carbomethoxy, formyl, or nitro groups that were previously not tolerated with more reactive organometallic reagents, or to the presence of *ortho*,*ortho′*-substituents in the aryl unit and proved to be excellent for sterically demanding secondary amines that could be arylated in high yields. In addition, by replacing lithium *tert*-butoxide with cesium fluoride, the reaction could even be extended to substrates containing an unprotected alcohol.

4.4.2.2 Synthesis of Diarylamines. Anilines were not evaluated in the procedures reported by Johnson,[110,111] most probably because of the poor stability of *O*-acyl-*N*-phenylhydroxylamines. Liebeskind proposed an interesting alternative to access diarylamines under nonoxidative conditions using nitrosoarenes as aniline surrogates, compounds that were shown to be efficient partners for the reductive amination of arylboronic acids under mild conditions (Scheme 4.41).[114] The reaction was performed in the presence of a stoichiometric amount of copper(I) salt in DMF under gentle heating. Copper(I) chloride was optimal, but other copper(I) sources were also useful. The reaction was not influenced by the electronic or steric nature of the boronic acid, and it should be noted than only 1.1 equiv of the latter was required.

The mechanism suggested for such a transformation is the following: A copper(I) species would first be involved in an oxidative addition into the N=O bond, yielding an intermediate copper(III) complex which, upon transmetallation and reductive elimination, would afford a *N*,*N*-diphenylhydroxylamine complexed to copper. This complex would be ultimately cleaved by a reducing agent such as copper(I) chloride, which accounts for the need of a full equivalent of the copper salt.

It was also possible to render the reaction catalytic in copper by adding a second equivalent of the boronic acid, which ultimately was transformed into phenol (by an oxidative copper-promoted O-arylation). Alternatively, the addition of various terminal reducing agents such as ascorbic acid or hydroquinone provided optimal results. Whereas ascorbic acid had to be added portion-wise to ensure the formation of the desired products in good yields, hydroquinone could be added at once from the beginning, though it generally slowed down the reaction and gave reduced yields.

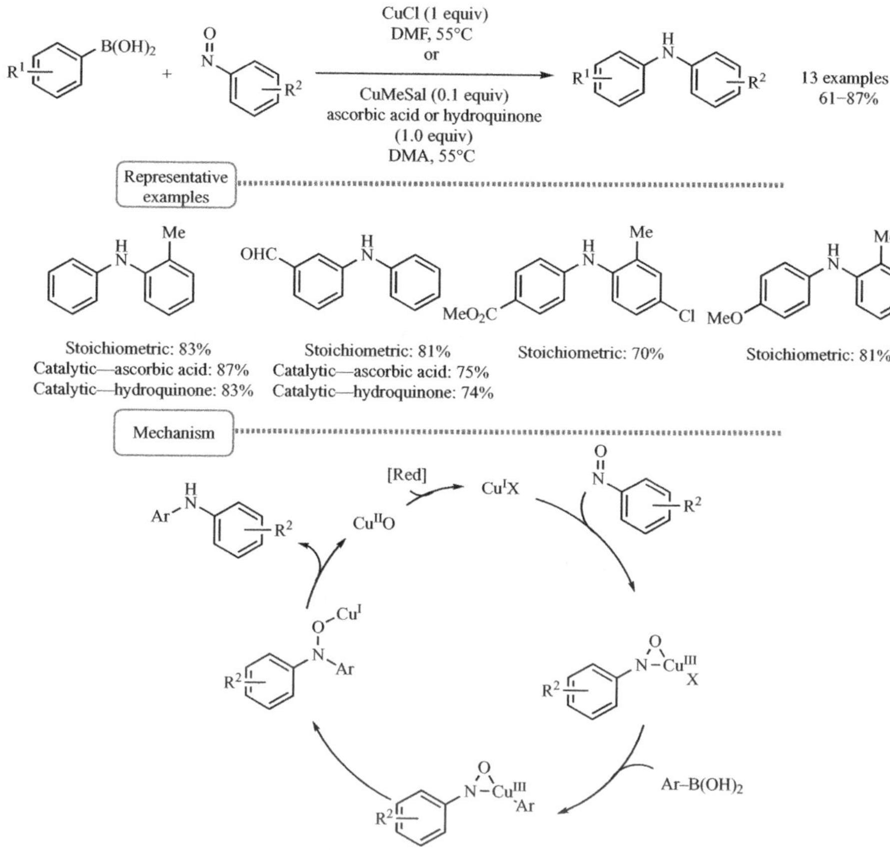

Scheme 4.41 Copper-catalyzed reductive coupling of nitrosoarenes with boronic acid.

4.4.2.3 Synthesis of* N*-aryl-imines. *N*-Aryl-imines are useful building blocks as well as protected forms of anilines. Based on early work from the late 1990s by the Narasaka group (who reported the first copper-catalyzed reaction of benzophenone derived *O*-sulfonyloximes with alkyl Grignard reagents[115] and on related arylation reactions that could be performed in the absence of copper complexes), Liebeskind and coworkers reported in 2007 an interesting method for the synthesis of *N*-aryl-imines based on a nonbasic, copper-catalyzed cross coupling involving oxime derivatives (Scheme 4.42).[116] Arylboronic acids or arylstannanes were successfully used in this transformation, which represented a net improvement over the previously reported procedures involving aryl Grignard reagents. The reaction was catalyzed by various copper(I) complexes with similar efficiencies. Equal results were obtained under argon atmosphere or under air when boronic acids were employed, while the reaction involving arylstannanes could not be performed in the presence of air. Whereas *O*-acetyl-oximes showed good reactivity, a marked

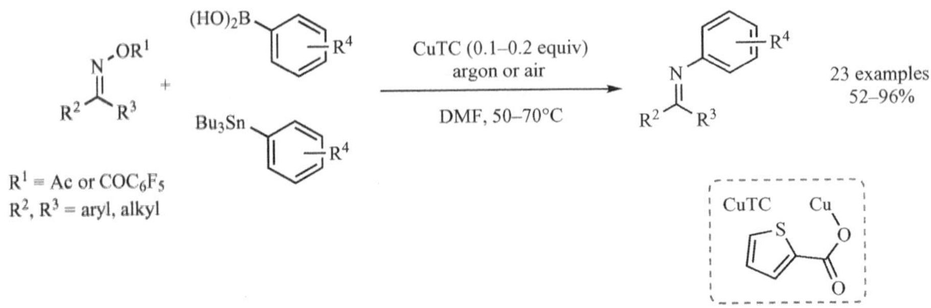

Scheme 4.42 Copper-catalyzed arylation of *O*-acyl oximes.

improvement of the reactivity was observed with the parent pentafluorobenzoyl derivatives when decreasing the amount of copper to 0.1 equiv and performing the reaction at 50°C instead of 70°C. Sensitive groups such as phenol, formyl, thiophene, or furan were well tolerated on the metallated unit and, more importantly, the reaction worked not only with benzophenone-derived oximes but also with oximes *O*-carboxylates substituted with two alkyl chains. This procedure was later on extended to a domino process for the synthesis of substituted pyridines involving a vinylation of acyloximes followed by a 6π-electrocyclization and a final oxidation.[117]

4.4.2.4 Synthesis of* N*-aryl-amides. A similar approach can be used for the preparation of *N*-aryl-carboxamides. In this context, O-acetylated hydroxamic acid derivatives were also showed to participate in copper-promoted arylation reactions with arylstannanes and boronic acids (Scheme 4.43, Eq. 1).[118] Contrary to the reaction from acyloximes, a stoichiometric amount of Cu(I) was required in this process, which was attributed to a dimerization of the amido copper complex yielding to a formation of species from which the transmetallation step would be really slow. Boronic acids appeared to be limited to electron-rich aryl substituents. Another limitation was found in the competing homocoupling, which was shown to be a serious side reaction in some cases but could be avoided by performing the reaction in a 1/1 mixture of THF/hexane where the effective concentration of the copper catalyst was lowered. Stannanes were more versatile aryl donors when associated with copper(I) diphenylphosphinate (CuDPP) as the catalyst.

While remarkable from a conceptual point of view, these reactions can hardly compete with the more classical Goldberg-type coupling involving aryl halides (see Chapter 1, Section 1.3), especially since N-substituted *O*-acetyl hydroxamic acid derivatives did not participate in the cross coupling under the conditions described by Liebeskind. A partial solution to this limitation was proposed soon after by Lei and coworkers who reported the coupling of *N*-chloro-*N*-aryl-acetamide derivatives with arylboronic acids (Scheme 4.43, Eq. 2).[119] The reaction, which took place at room temperature, required the

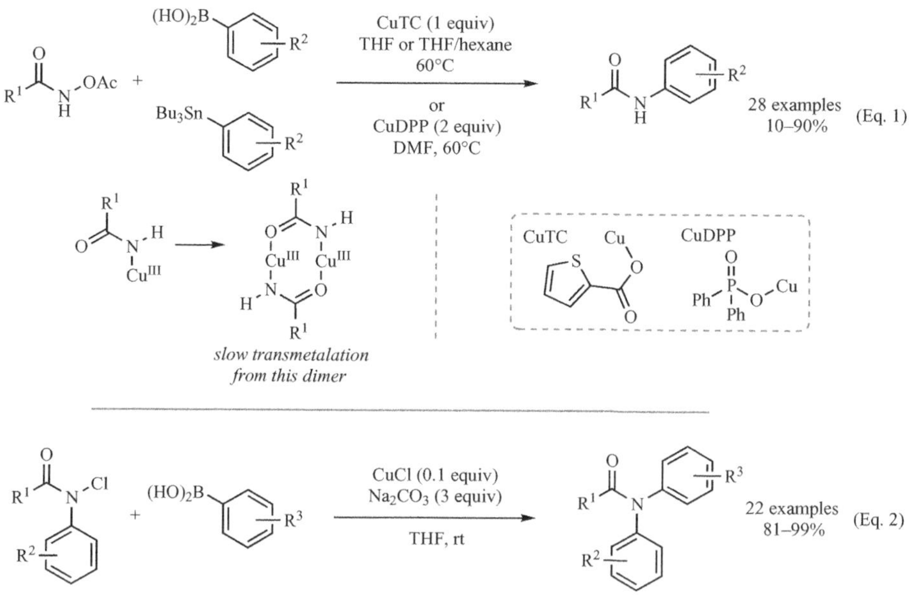

Scheme 4.43 Copper-catalyzed arylation of acyl oximes or *N*-chloro-*N*-arylacetamides.

presence of a mild base (sodium carbonate) and was truly catalytic in copper. This process therefore offers an interesting entry to *N,N*-diarylamides, compounds that can hardly be obtained using a copper-catalyzed arylation with aryl halides, which requires much harsher reaction conditions.

4.4.2.5 From Azo Derivatives. It was also demonstrated that copper(II) acetate could catalyze the addition of arylboronic acids to azo compounds under very mild, nonbasic reaction conditions, yielding protected *N*-arylhydrazines (Scheme 4.44).[120] The reaction tolerated boronic acids with various electronic natures but was very sensitive to steric parameters. THF was found to be a good solvent for the room temperature arylation of azodicarboxylates, but the reaction could also be performed in refluxing methanol, as described later on for other azo compounds.[121] Unsymetrical *N*-Boc-protected *N*-arylazo compounds were also suitable substrates for this arylation and, in this case, a regioselective arylation furnishing the corresponding 1,1-diarylhydrazines was observed. 1,2-Diphenyldiazene was, however, poorly reactive under the reaction conditions. The proposed mechanism for such transformation does not involve any change in the oxidation state of the copper, but rather an addition across the double bond followed by transmetallation with the aryl donor.

A wide range of reagents, including aryl halides and (as shown throughout this chapter) metallated or pseudometallated aromatics, have proven over the years to be versatile and valuable reagents for the arylation of

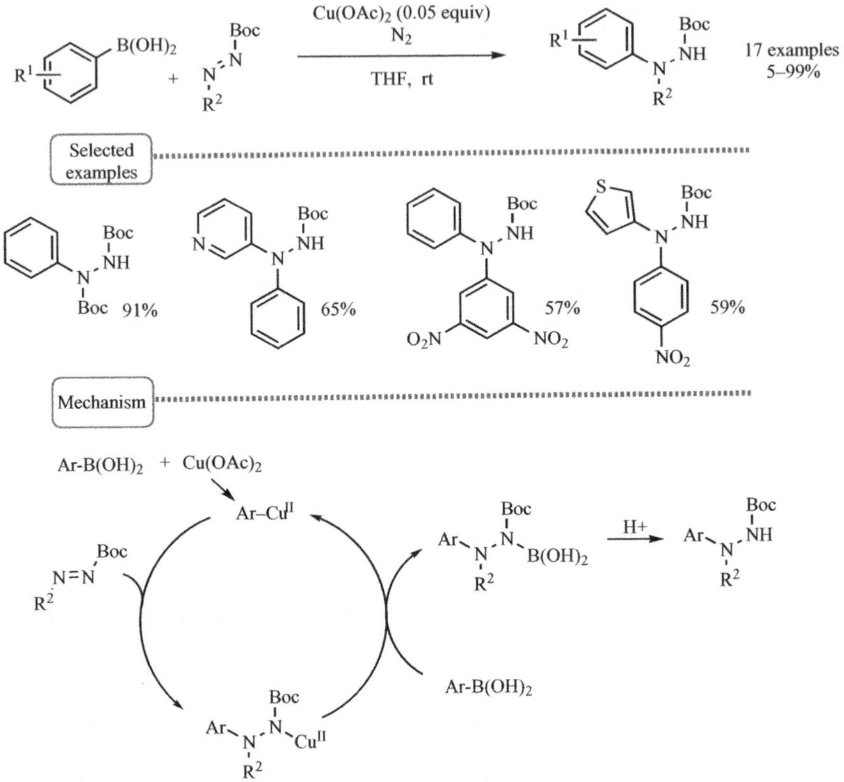

Scheme 4.44 Copper-catalyzed addition of arylboronic acids to azo compounds.

heteronucleophiles or synthetic equivalents, which clearly had a deep impact in organic synthesis as shown notably with examples of natural product synthesis or industrial processes based on these transformations and commented in details in Chapters 16, 17, and 19, respectively. However, the use of such strategies involves pre-functionalization of the aromatic ring to be transfered; and the direct formation of a C-Heteroatom bond from a C–H bond would represent the best and most efficient alternative, provided that high levels of regioselectivity could be achieved. The development of such strategies, which is still in its infancy, will be covered in the rest of the chapter.

4.5 COPPER-CATALYZED DIRECT HETEROFUNCTIONALIZATION OF AROMATIC C–H BONDS

Methods for the direct functionalization of C–H bonds are valuable transformations that do not require prefunctionalization of the substrate and fall into the "atom economical" reactions. A number of metals such as rhodium, palladium, and ruthenium have been found to promote such transformations.[122]

Investigations in this area are currently exploding and transformations involving copper have also emerged, which turned out to be quite efficient for the formation of C-heteroatom bonds. Along with rhodium and other metals, copper has been playing a central role in the development of the nitrenoid insertion into the Csp^3–H bond.[123] However, the direct copper-mediated heterofunctionalization of arenes is still in its infancy, and three main strategies have appeared in this field: the functionalization of acidic C–H bonds, the use of directing groups, and cyclizative heterofunctionalization. These strategies will be reviewed in this section.

4.5.1 Functionalization of Acidic Aromatic C–H Bonds

Mori and Schreiber independently opened the field of copper-catalyzed direct C–H amination of azoles containing an acidic site in 2009. Indeed, Mori reported the first coupling of C–H azoles with amines under aerobic copper-catalyzed reaction conditions, regioselectively affording the corresponding N-arylated amines (Scheme 4.45).[124a] Copper(II) acetate/triphenylphosphine was the optimal catalytic system, in the presence of sodium acetate as the base under an oxygen atmosphere in refluxing xylene. Heterocyclic azoles that could be used in this reaction include benzoxazoles, benzothiazoles, 4,5-disubstituted thiazoles, or N-substituted benzimidazoles and afforded the corresponding 2-aminoazoles in moderate-to-good yields. Regiospecific C-2 arylation was obtained over the possible C-4 in the case of thiazole. Only

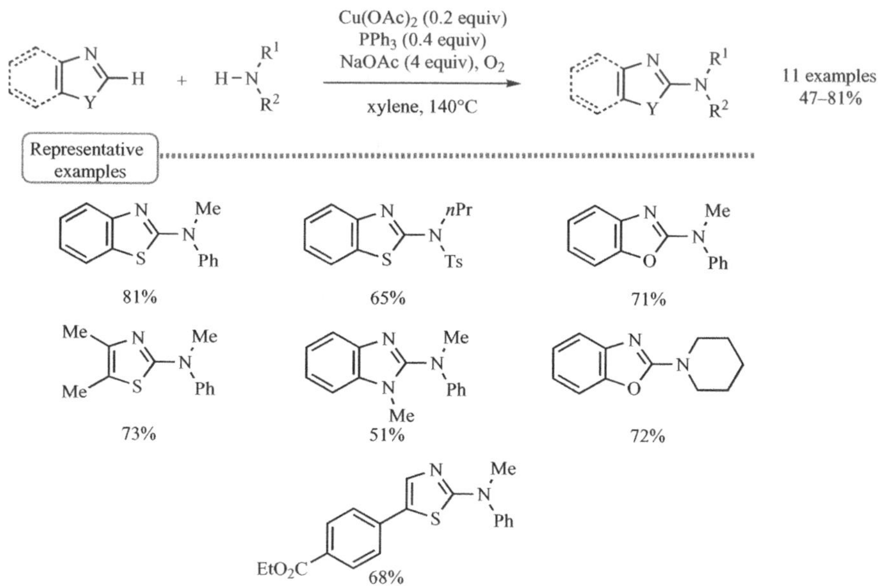

Scheme 4.45 Copper-catalyzed intermolecular amination of aryl C–H bonds in azoles with amines.

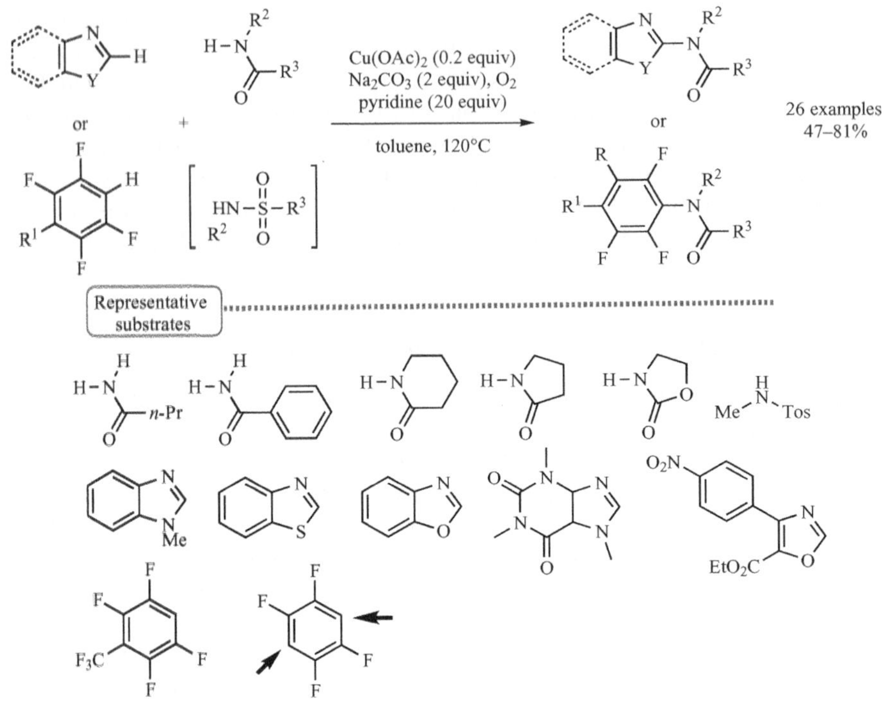

Scheme 4.46 Copper-catalyzed amidation of acidic heteroaromatic and fluoroaromatic C–H bonds.

secondary amines were evaluated in the reaction, but the scope was rather large as dialkyl, cyclic or not, aryl alkyl, or diarylamines were suitable reaction partners. Interestingly, less basic *N*-tosyl propylamine was also smoothly arylated. Alternative conditions reported in 2012, based on the use of copper(II) acetylacetonate, allowed this transformation to be perfomed under base-free conditions.[124b]

Amidation was the subject of Wang and Schreiber's report (Scheme 4.46)[125] who developed conditions involving a catalytic amount of copper(II) acetate in the presence of sodium carbonate under an oxygen atmosphere in refluxing toluene for the C–N bond formation, but, contrary to Mori's conditions, the reaction had to be performed in the presence of a large excess of pyridine (20 equiv). This additive proved to be important to avoid homocoupling of the starting heteroaromatic, which was a serious side reaction without this additive. In addition to various azaheterocycles, perfluoroarenes were also found to be suitable reaction partners for the N-arylation. Linear primary or cyclic amides, oxazolidinones, ureas, and sulfonamides could be efficiently arylated under such reaction conditions.

Extension of this procedure to the use of sulfoximines was reported by Miura, Bolm, and coworkers in 2011 (Scheme 4.47, Eq. 1).[126a] Indeed, the catalytic direct sulfoximination of azoles and polyfluoroarenes proceeds very

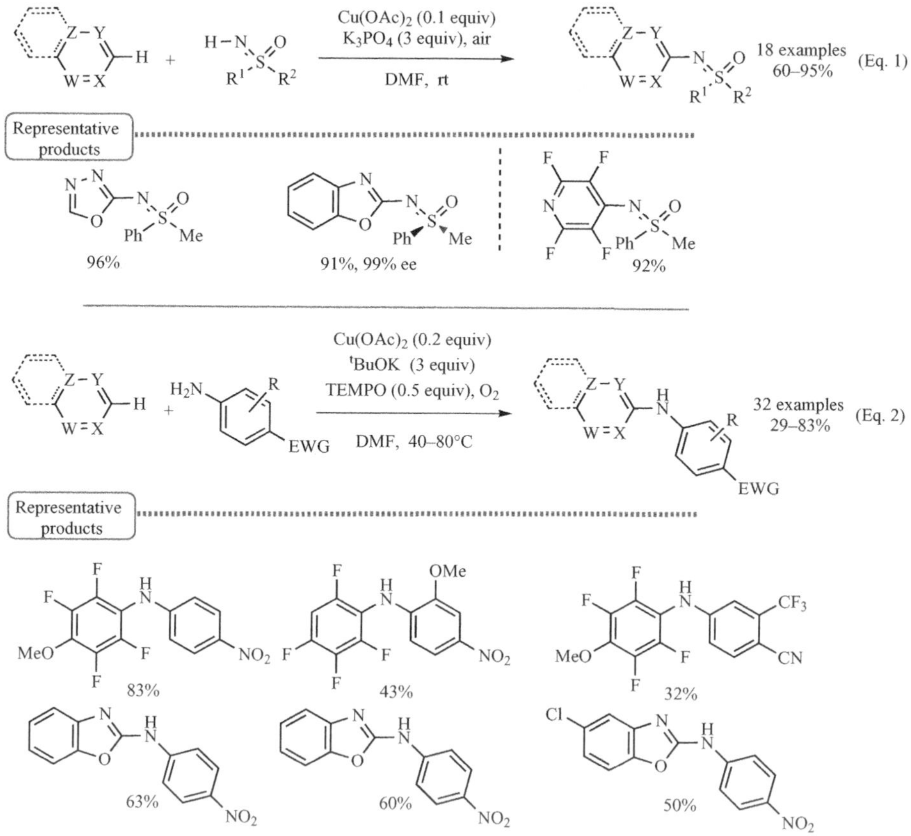

Scheme 4.47 Copper-catalyzed intermolecular sulfoximination and anilination of acidic aromatic C–H bonds.

smoothly with copper(II) acetate as the catalyst and potassium phosphate as the base, under air and at room temperature. Interestingly, no erosion of the enantiomeric excess was observed when enantioenriched sulfoximines were used, demonstrating the mildness of the conditions. Dehydrogenative coupling of C–H azoles with primary anilines was also possible (Scheme 4.47, Eq. 2).[126b] The reactions proceed at relatively low temperature for such a transformation (40–80°C) using copper(II) acetate as the catalyst and molecular oxygen and 2,2,6,6-tetramethyl-1-piperidinyl-*N*-oxide (TEMPO) as the oxidants in the presence of potassium *tert*-butoxide as the base. The process was, however, limited to anilines bearing a strong electron-withdrawing group at the *para*-position (essentially NO_2) and yields were rather moderate. Finally, it ought to be mentioned that the first examples of direct copper-catalyzed alkoxylations of C–H bonds in azoles were reported in 2013, although the intramolecular alkoxylation was found to be a lot more efficient than the intermolecular one.[127]

Hirano and Miura attempted to develop a direct coupling of azoles with N–H indoles but failed to find suitable conditions. In order to access these

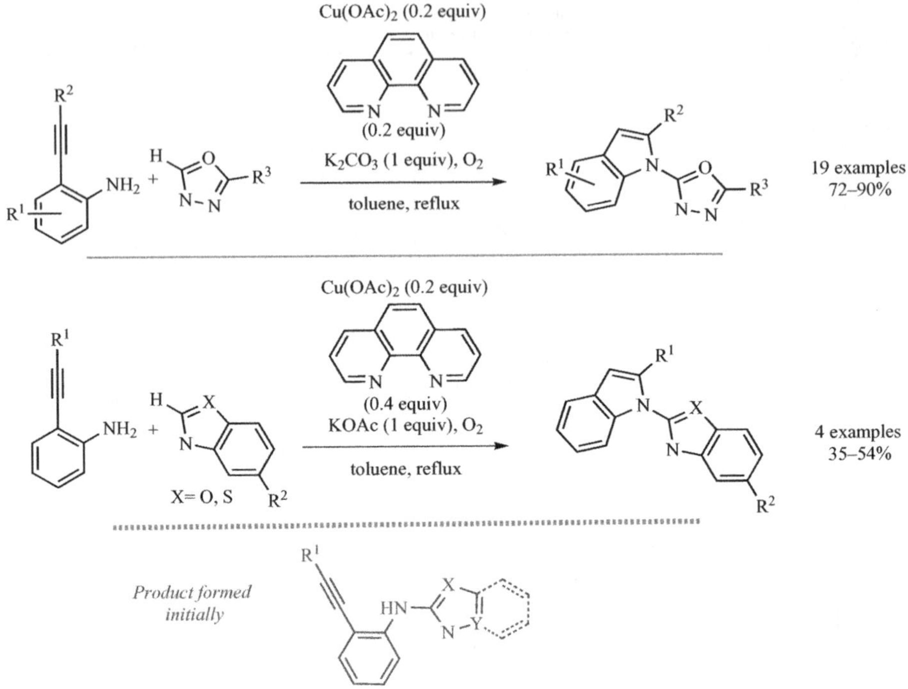

Scheme 4.48 Formation of *N*-azolyl-indoles by a domino copper-catalyzed intermolecular amination of acidic aromatic C–H bonds/intramolecular cyclization.

functionalized heterocycles, they devised an ingenious tandem copper-catalyzed reaction leading to the formation of *N*-azolyl-indoles from *ortho*-alkylnylanilines and azoles (Scheme 4.48).[128] The domino reaction, which could be effectively achieved using copper(II) acetate/1,10-phenanthroline as the catalytic system and potassium carbonate or acetate as the base in refluxing toluene under oxygen, involved first an intermolecular heteroarylation of the aniline followed by a benzannulation forming the indole core. In contrast to the work reported by Hong and coworkers[127] anilines lacking an electron-withdrawing group at the *para*-position were effectively transformed to *N*-azolyl-indoles, but this reaction could not be extended to the use of perfluoroarenes.

Most methods described for the direct amination/amidation of activated arenes were found to proceed under basic conditions, but related reactions involving tertiary alkyl amines or tertiary amides as aminating agents have been performed under acidic conditions (Scheme 4.49).[129] Generation of the secondary amines is believed to result from an *in situ* oxidative cleavage (through the formation of an imine) or from a decarbonylation under the acidic copper-catalyzed conditions, when tertiary amines or amides were used, respectively. The scope of this reaction, with respect to azoles, was limited to benzoxazoles and 1,3,4-oxadiazoles, as benzothiazole or *N*-ethyl-benzimidazole, were found to be poorly aminated, probably because of their lower acidity.

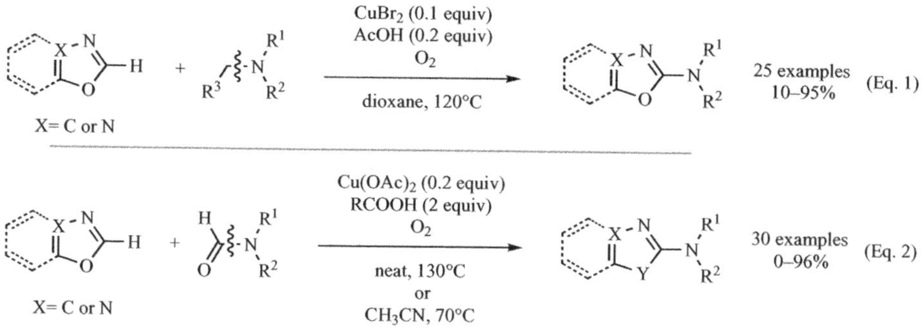

Scheme 4.49 Acid-promoted, copper-catalyzed direct amination of acidic C–H bonds in azoles.

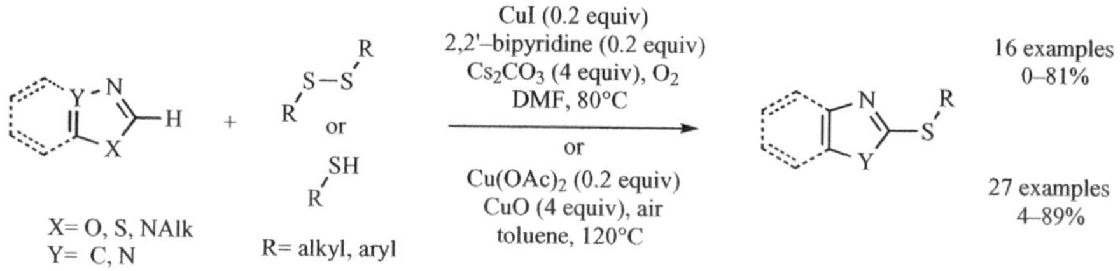

Scheme 4.50 Copper-catalyzed direct thiolation of acidic C–H bonds in azoles.

Direct functionalization of acidic C–H bonds is not limited to amination as it was recently extended to thiolation (Scheme 4.50).[130] Thiols or disulfides were equally well arylated, but electron-deficient thiophenols or bulky *tert*-butylthiol were not tolerated. Such reaction could be performed using copper(I) iodide/2,2′-bipyridine as the catalytic system and cesium carbonate as the base in DMF according to Fukuzawa.[130a] Alternative conditions involving copper(II) acetate as the catalyst and copper(I) oxide as the base (and probably as co-oxidant) in toluene were also recently disclosed.[130b] In addition to the classical range of heterocycles compatible with the aforementioned thiolation reaction conditions, which include oxazole, thiazole, imidazole, and oxadiazole, Wei reported recently a procedure allowing the sulfenylation of both N-protected and N-unprotected indoles with unsymmetrical benzothiazolyl-containing disulfides using mild conditions [CuI (10 mol %), DMSO, rt].[131]

The mechanism that has been proposed for such processes is close to the one generally accepted for the oxidative cross coupling involving metallated aryl donors with heteronucleophiles (see Section 4.2.3). In this mechanism, the classical transmetallation step would be replaced by a base-assisted cupration of the heteroaromatic. In addition, recent DFT calculations performed on a reaction involving a C–S bond formation suggested that oxygen might be involved in the hydrogen abstraction.[132]

The copper-catalyzed amination of azoles bearing an acidic C–H bond has also been achieved under nonoxidative conditions. Following an umpolung

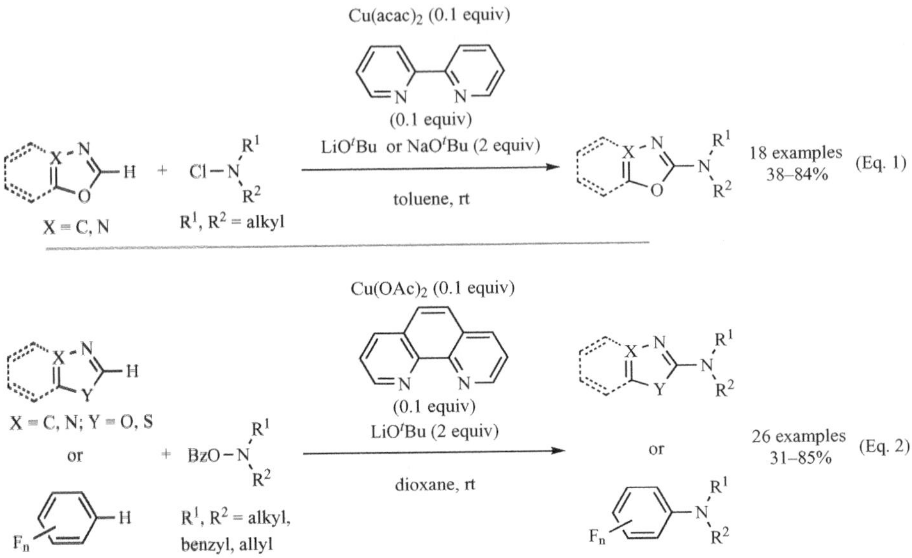

Scheme 4.51 Copper-catalyzed electrophilic amination through direct arylation of acidic C–H containing azoles and arenes with chloramines and *O*-acyl-hydroxylamines.

strategy (see Section 4.4), Miura and coworkers used chloramines as aminating agents (Scheme 4.51, Eq. 1).[133a] A combination of copper(II) acetylacetonate and 2,2′-bipyridine furnished an efficient catalytic system for the coupling of cyclic as well as acyclic chloramines with 2-aryl-1,3,4-oxadiazoles at room temperature. A relatively strong base such as lithium *tert*-butoxide was, however, required, but its nature had to be adjusted for the amination of 5-substituted benzoxazoles where sodium *tert*-butoxide substantially improved the yields. However, this process could not be extended to other azoles or perfluorobenzene derivatives.

Alternative reaction conditions with broader scope appeared soon after. Copper(II) acetate/1,10-phenanthroline proved to be a good catalytic system for the coupling of *O*-benzoyl-hydroxylamine derivatives with a range of azoles and fluorinated arenes in the presence of lithium *tert*-butoxide in dioxane at room temperature (Scheme 4.51, Eq. 2).[133b] Linear or cyclic dialkyl, benzyl-alkyl, diallyl, and primary alkyl hydroxylamines were tolerated in the reaction. Interestingly, during the optimization process, it was found that triphenylphosphine could replace 1,10-phenanthroline with close efficiency. As a note, a direct sulfonylation of quinoline *N*-oxides has been also reported using sulfonyl chlorides as the sulfonylation agents in combination with catalytic amounts of copper(I) iodide and potassium carbonate in 1,2-dichloroethane at 100°C.[133c]

4.5.2 Directed Aromatic C–H Bonds Functionalization

Copper catalysts being in general highly sensitive to the presence of chelating groups that can dramatically improve the efficiency of copper-catalyzed

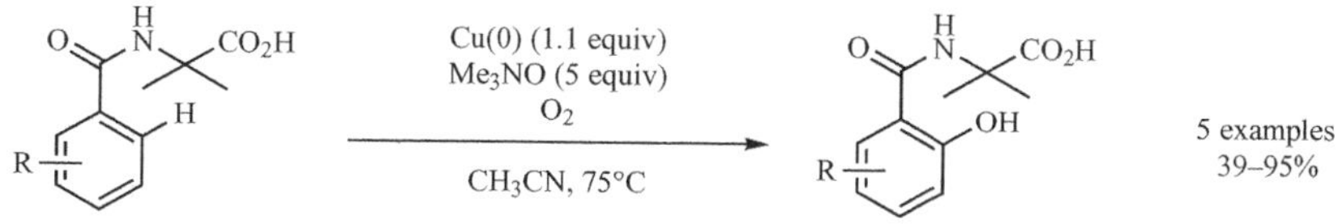

Scheme 4.52 Copper-mediated directed *ortho*-selective hydroxylation of benzamides.

processes, the *ortho*-effect in classical Ullmann–Goldberg cross-coupling reactions certainly being the most striking example of such an influence, this kind of groups can also be used for directed copper-mediated functionalization of aromatic C–H bonds. Indeed, the stoichiometric copper-promoted functionalization of aromatic rings directed by a functional group has been known for a while, the thermal decomposition of $PhCO_2Cu(II)OH$, allowing an *ortho*-selective hydroxylation, being already reported in 1962.[134] The harsh reaction conditions, the poor yield of the product, and the limited availability of the starting materials, however, seriously limited its development.

One of the first synthetically useful procedures involving copper-promoted *ortho*-functionalization of an aromatic C–H bond was reported in 1989 by Reinaud and coworkers (Scheme 4.52).[135] In this publication, a series of aromatic carboxamides possessing a pendant chelating group derived from 2-methylalanine were regioselectively hydroxylated in the presence of stoichiometric metallic copper and trimethylamine *N*-oxide under oxygen in refluxing acetonitrile. The directing group was essential for the reaction, as neither a tertiary amide analogue nor benzoic acid was hydroxylated under the reaction conditions.

Based on the ability of the pyridyl substituent to act as one of the most efficient directing groups in palladium-catalyzed processes, Yu evaluated a copper-promoted *ortho*-functionalization of 2-phenylpyridines (Scheme 4.53).[136] Initial investigation established that selective hydroxylation could be achieved by performing the reaction in the presence of a stoichiometric amount of copper(II) acetate and 1 equiv of water in acetonitrile under an atmosphere of oxygen at 130°C (Scheme 4.53, Eq. 1). Labeling experiments established that the hydroxyl group came from copper(II) acetate and it was advanced that an acetoxylated product was initially formed, but was readily hydrolyzed under the reaction conditions. The selectivity toward the monohydroxylation was explained because of the binding of the hydroxylated product to the copper salt, forming a complex that was believed to be catalytically inactive. Based on these statements, catalytic *ortho*-acetoxylation could be achieved by performing the reaction in a mixture of acetic acid and acetic anhydride that prevents the formation of phenol derivatives (Scheme 4.53, Eq. 2). Under such reaction conditions, a substantial amount of bis-acetoxylated product was however also formed.

The reaction was not limited to hydroxylation or acetoxylation. Indeed in the presence of other heteronucleophiles such as phenols, tosylamines, thiophenols, or disulfides, mono *ortho*-selective sulfamination, etherification, or thioetherification was documented (Scheme 4.53, Eq. 3). Anilines were also

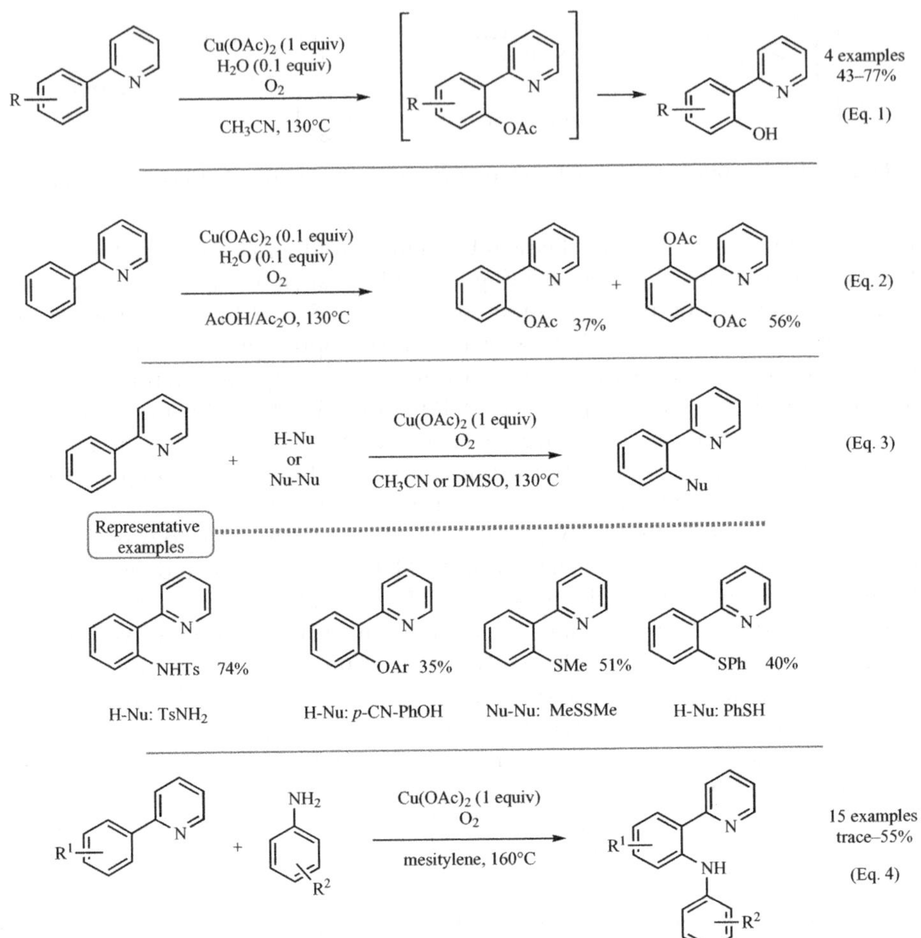

Scheme 4.53 Copper-mediated *ortho*-selective heterofunctionalization of 2-arylpyridines.

suitable nucleophiles, as reported simultaneously by Chatani and coworkers (Scheme 4.53, Eq. 4).[137]

These initial reports by the Yu and Chatani groups opened the way to the development of more efficient and diverse processes for the heterofunctionalization of 2-arylpyridines, and copper-catalyzed acyloxylation[138] reactions using acetic anhydrides, acyl chlorides, or aldehydes have been recently reported (Table 4.2).

Besides working from different precursors, these methods, complementary to the ones developed by the Yu group,[136] offered the possibility to work under neutral or basic conditions and to decrease the amount of acyloxylating reagents, whether aromatic or aliphatic, to 2 or 3 equiv. Diacyloxylation was

TABLE 4.2 Copper-Catalyzed Directed *ortho*-Acyloxylation of 2-Arylpyridines

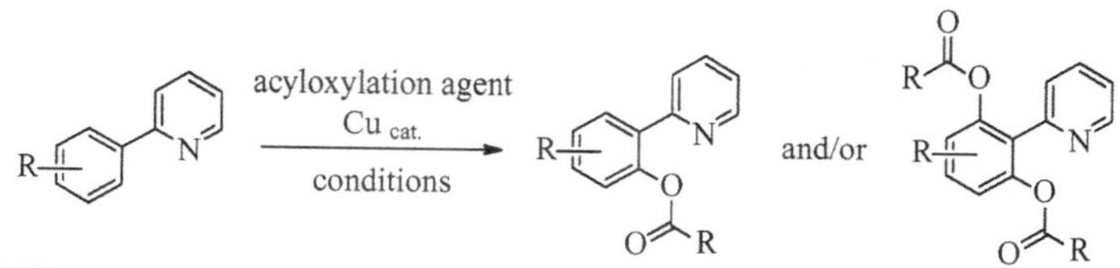

Entry	Acyloxylating Agent	Catalytic System	Conditions	Number of Examples	Yields	References
1	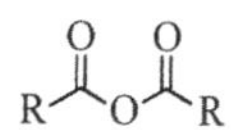 R = alkyl or aryl	$Cu(OAc)_2$ (0.1 equiv)	O_2, toluene, 145°C	19	40–85%	[138a]
2	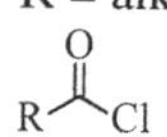 R = aryl	$Cu(OAc)_2$ (0.1 equiv)	O_2, toluene, ^{t}BuOK (2 equiv), 145°C	12	52–90%	[138b]
3	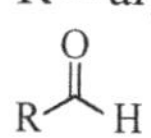 R = aryl, alkyl	$Cu(OAc)_2$ (0.2 equiv)	O_2, ^{t}BuOOH (2 equiv), PhCl, 135°C	14	28–54%	[138c]
4	$R-CH_3$ R = aryl	$Cu(OAc)_2$ (0.2 equiv)	O_2, ^{t}BuOOH (4 equiv), neat, 135°C	8	28–42%	[138d]

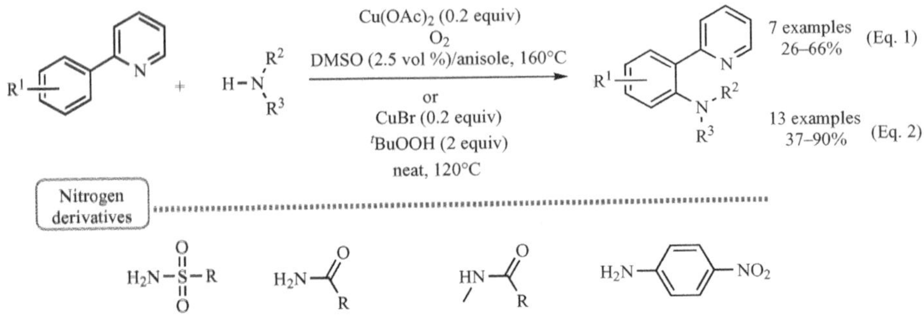

Scheme 4.54 Copper-catalyzed *ortho*-amination of 2-arylpyridines.

predominant for the reactions performed from anhydrides or acyl chlorides (Table 4.2, entries 1 and 2), but the selectivity of monoacyloxylation versus diacyloxylation was shown to be strongly influenced by the steric hindrance on the aryl group of the 2-arylpyridine. For the reaction conditions described by Huang and coworkers[138c] in which the acid is generated through an *in situ* oxidation of an aldehyde or a tolyl residue (Table 4.2, entries 3 and 4), exclusive monoacyloxylation was observed when working with aromatic aldehydes, whereas traces of the diacyloxylated product were isolated in the case of aliphatic aldehydes.

Recently, conditions for catalytic reactions involving oxygen or *tert*-butyl hydroperoxide as a terminal reoxidant in the presence of copper(II) acetate or copper(I) bromide, respectively, were also reported for the amination of 2-arylpyridines (Scheme 4.54).[139] Interestingly, reactions performed in pure dimethylsulfoxide were not efficient, but it was observed that the use of dimethylsulfoxide as an additive was crucial to achieve high yields.

Alternative directing groups could also be used in such transformations, and an interesting copper-mediated oxidative intermolecular *ortho* C–H functionalization using tetrahydropyrimidine as the directing group has been described by Ohno and coworkers in 2009 (Scheme 4.55, Eq. 1 and Eq. 2).[140] Selective *ortho*-hydroxylation or amination were possible, and intermediate 2-(tetrahydropyrimidinyl) phenols or aniline were isolated as the corresponding 3,4-dihydrobenzopyrimido-oxazinone or quinazolinone derivatives upon further reaction with triphosgene in the first case, and spontaneous cyclization in the second one, where *tert*-butyl carbamate was used as the nucleophile. Amidine analogues such as a related 2-phenyl-dihydroimidazole or 2-phenyl-1*H*-imidazole were not suitable for the hydroxylation reaction.

Liu and coworkers recently described a related copper-catalyzed directed *ortho*-nitration of aromatic rings (Scheme 4.55, Eq. 3).[141a] The reaction was carried out in the presence of copper(II) acetate as the catalyst and silver nitrate as the effective nitrating agent in 1,2,3-trichloropropane at 130°C under an atmosphere of oxygen. Besides the classical pyridine, a range of alternative directing groups has been evaluated and it has been shown that no reaction

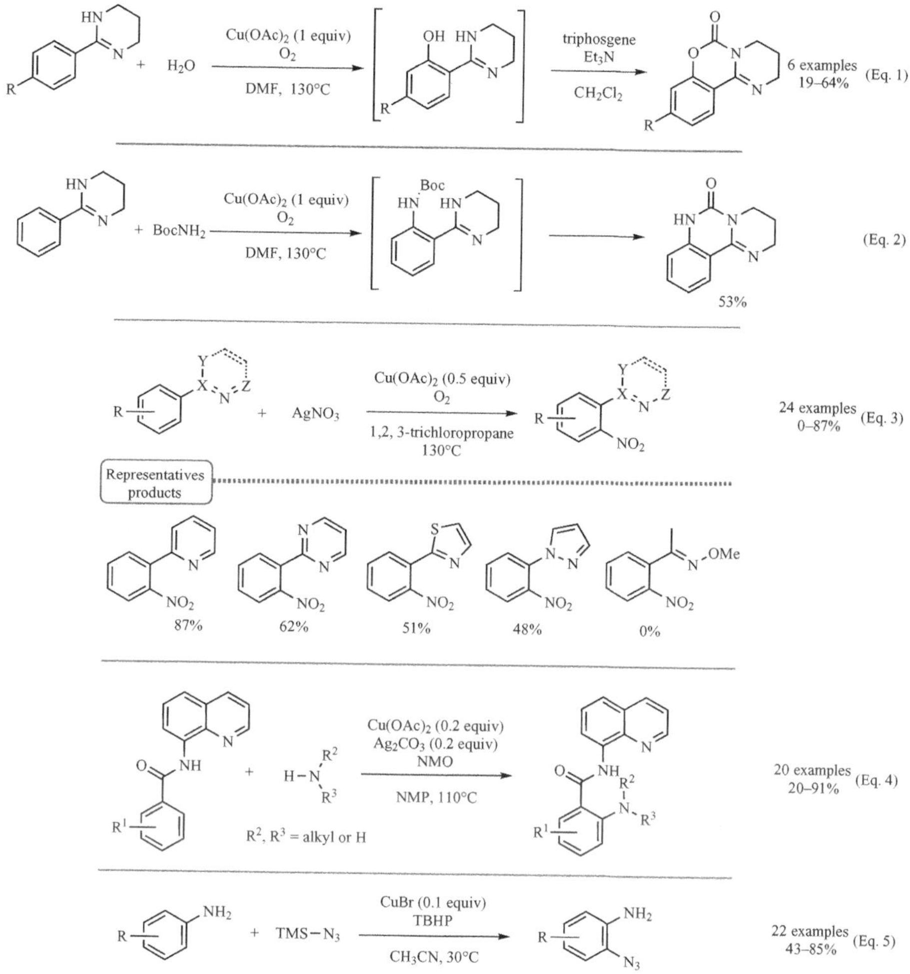

Scheme 4.55 Copper-promoted direct functionalization of aromatic C–H bonds involving various *ortho*-directing groups.

was observed in the absence of a chelating group. Pyrimidine, thiazole, and pyrazole could be used with modest-to-good efficacy, but *O*-methyl-oximes failed. Very recently, Daugulis and coworkers reported that benzamides bearing a removable 8-amidoquinoline directing group could be *ortho*-aminated with alkylamines under the action of a catalytic amount of copper(II) acetate and silver carbonate (Scheme 4.55, Eq. 4). *N*-Methylmorpholine oxide was a more effective co-oxidant than oxygen in this reaction, and secondary alkyl amines provided a better outcome than primary alkylamines.[141b] The use of this 8-amidoquinoline directing group also enabled the direct sulfenylation of C–H bonds in arenes with various disulfides—including bis(trifluoromethyl) disulfide—in the presence of substoichiometric amounts of copper(II) acetate in DMSO at 90–110°C.[141c]

Primary amines have also been recently added to the ever-growing collection of directing group, as demonstrated in 2012 by Tang and Jiao who developed a remarkably efficient direct *ortho*-azidation of anilines by reaction with trimethylsilyl azide in the presence of catalytic amounts of copper(I) bromide and *tert*-butylhydroperoxide (TBHP) in acetonitrile at room temperature (Scheme 4.55, Eq. 5).[141d] The mechanism proposed by the authors involves coordination of copper to the amine and azidation via a single electron transfer, although there is not direct proof of this mechanism.

4.5.3 Intramolecular Copper-Catalyzed Heterofunctionalization of Aromatic C–H Bonds

Heterofunctionalization of C–H bonds, whether taking advantage of the acidity of C–H bonds in various arenes or using a directing group approach, has not been limited to intermolecular reactions. Indeed, various heterocycles have been built following a benzannulation strategy featuring an intramolecular C–H heterofunctionalization, which offers new and efficient strategies in heterocyclic chemistry based on copper-mediated processes. These processes nicely complement other approaches that will be reviewed in Chapter 15 (other examples of intramolecular copper-catalyzed heterofunctionalization of aromatic C–H bonds are also described in this chapter).

In this area, the copper-catalyzed cyclization of *N*-aryl-amidines to benzimidazoles was described by Buchwald in 2008 and paved the way for further developments in the field of intramolecular heterofunctionalization.[142] The reaction proceeded using copper(II) acetate as the catalyst and acetic acid at 100°C in dimethylsulfoxide under an atmosphere of oxygen (Scheme 4.56, Eq. 1). *N*-Phenylbenzamidines were efficiently cyclized and tolerated electron-donating and electron-withdrawing substituents on both aromatic units. The benzamidine aromatic unit had to be *ortho*-substituted in order to achieve

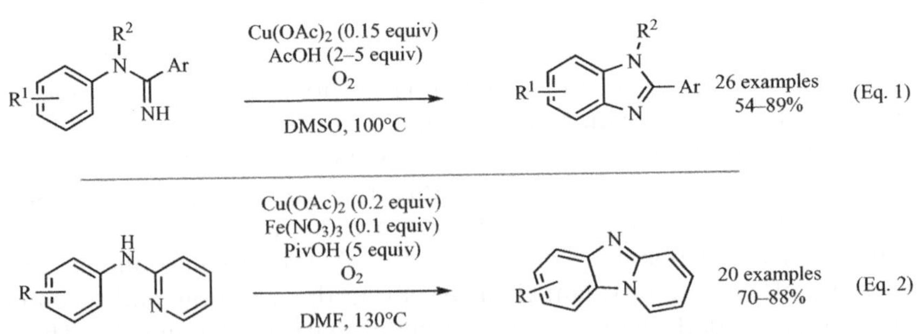

Scheme 4.56 Benzimidazole synthesis through copper-catalyzed oxidative annulation involving C–H bond functionalization.

decent conversion, which might be related to the conformation of the substrate. Concerning the substitution pattern of the starting aniline residue, it was mentioned that low conversions were observed for cyclizations leading to 4-substituted or 4,6-disubstituted products. Interestingly, the method could be extended to the preparation of *N*-methylated-2-phenylbenzimidazoles. The method also allowed the synthesis of 2-alkylbenzimidazoles but was limited to a *tert*-butyl group, most certainly to avoid problems associated with the oxidation of the benzylic position.

A direct extension of this work was provided by Zhu and coworkers who reported the synthesis of pyrido[1,2-a]benzimidazoles from *N*-aryl-2-aminopyridines (Scheme 4.56, Eq. 2).[143a] For this reaction, the addition of 10 mol % of iron(III) nitrate as a cocatalyst significantly improved the yield, and the reaction was best performed by replacing acetic acid with pivalic acid. A very similar reaction was reported almost simultaneously by the Maes group who described a more general system based on the use of a catalytic amount of 3,4,5-trifluorobenzoic acid in combination with copper(II) acetate. In contrast to the sytem reported by Zhu, this one allows the cyclization of electron-deficient and sterically hindered substrates.[143b]

A closely related process, involving an intramolecular oxidative cyclization of benzanilides to build benzoxazoles through the formation of a C–O bond, was reported by Nagasawa using copper(II) trifluoromethanesulfonate as the catalyst in *ortho*-xylene under an oxygen atmosphere or, as demonstrated later, under air when the reaction was conducted in *ortho*-dichlorobenzene at 160°C (Scheme 4.57).[144] Investigation of the reaction scope revealed that

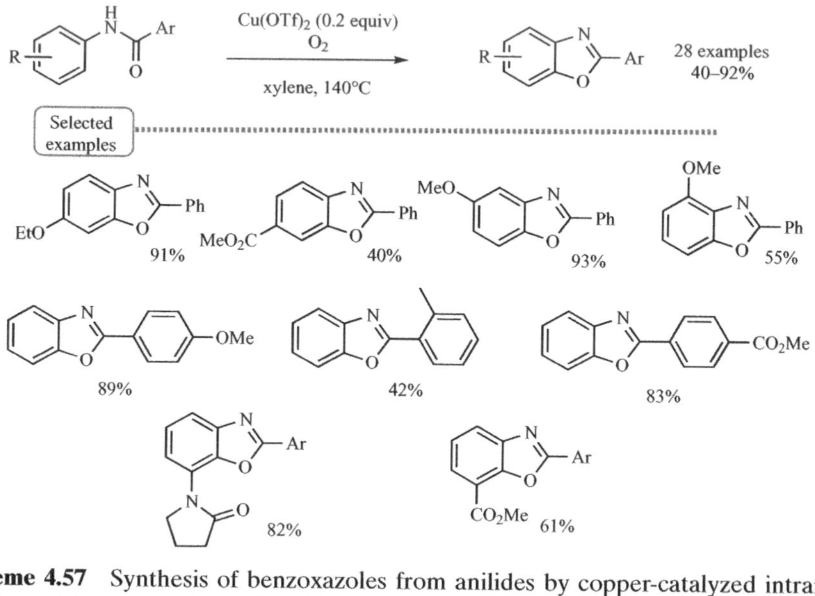

Scheme 4.57 Synthesis of benzoxazoles from anilides by copper-catalyzed intramolecular oxidative annulation involving C–H bond functionalization.

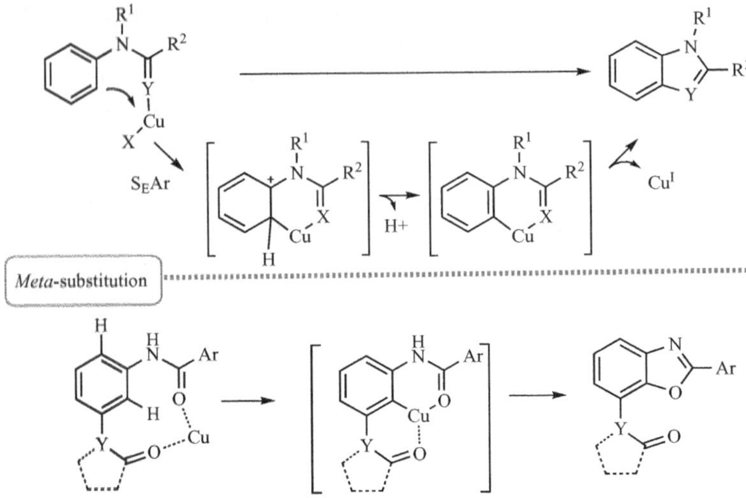

Scheme 4.58 Proposed mechanism for the oxidative heteroannulation yielding to benzimidazoles and benzoxazoles.

benzamides derived from electron-poor anilines furnished lower yields than the corresponding electron-neutral or rich ones and, contrary to Buchwald's case,[142] *ortho*-substitution on the aromatic benzamide residue generally lowered the yield. *Meta*-substituted aniline derivatives provided a divergent outcome depending on the substituent since the use of a nonchelating derivative exclusively forced the reaction to occur at the less sterically hindered position, while the formation of more sterically hindered 2,7-disubstituted benzoxazoles was achieved with chelating groups such as ester, amides, carbamates, ketones, and various heterocycles.

The mechanism involved in the formation of benzimidazoles and benzoxazoles shown in Scheme 4.56 and Scheme 4.57 has not been established, but it is accepted that coordination to the pendant heteronucleophile would initiate the reaction (Scheme 4.58). Based on the observation that electron-rich substrates tend to react faster, and that no kinetic isotopic effect was observed with deuterated substrates, it is suggested that an electrophilic aromatic substitution (S_EAr) mechanism is likely operating to afford an intermediate that can aromatize and undergo a final reductive elimination step. In the case of *meta*-substituted arenes, which present two distinct C–H bonds, the presence of a chelating group will direct the copper catalyst to the most hindered position. Not surprisingly, no information on the oxidation state of the intermediate copper complexes (+II or +III) is currently available. Chang and coworkers reported a new synthesis of carbazoles starting from readily available N-substituted amidobiaryls under either copper-catalyzed or metal-free conditions using an hypervalent iodine(III) reagent as the oxidant (Scheme 4.59, Eq. 1). This cyclization was found to be efficiently promoted when a combination of

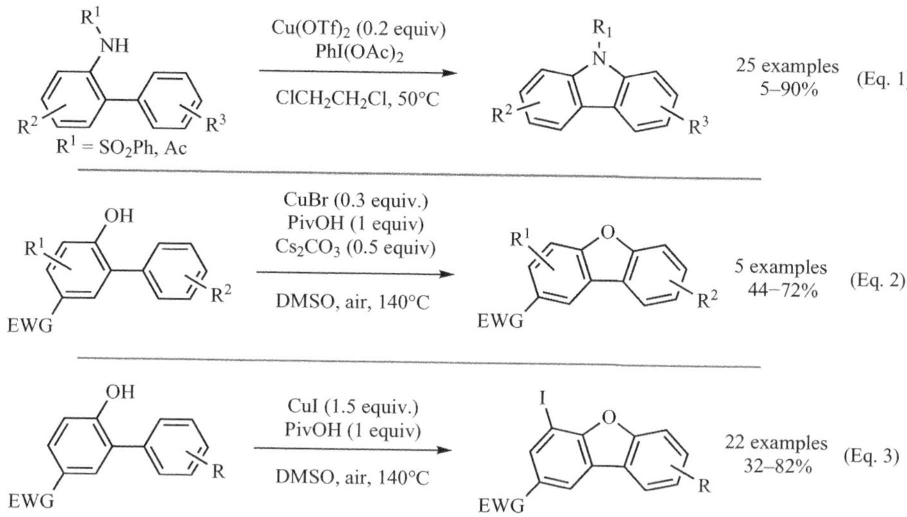

Scheme 4.59 Copper-mediated intramolecular oxidative C–N and C–O bond formation for the synthesis of carbazoles and dibenzofurans.

copper(II) triflate and phenyliodonium diacetate was used. Using these conditions, the reaction occurred rather rapidly in 1,2-dichloroethane at 50°C. The reaction was influenced by the electronic nature of the substituents of both aromatic rings and best performed when the aniline ring was substituted with electron-withdrawing groups and the other aromatic ring with electron-donating ones.[145a] A related synthesis of dibenzofurans could be accomplished using air as the oxidant and a catalytic amount of copper(I) bromide in combination with cesium carbonate and pivalic acid in DMSO at 140°C (Scheme 4.59, Eq. 2).[145b] The reaction was, however, limited to phenols bearing an electron-withdrawing group at the para-position to prevent direct oxidation of the substrate. Interestingly, iodinated dibenzofurans were obtained when the reaction was performed in the presence of 1.5 equiv of copper(I) iodide (Scheme 4.59, Eq. 3).[145c]

Tandem (or one-pot) reactions involving oxidative annulation strategies have also been reported. Building on Buchwald's cyclization of *N*-arylamidines to benzimidazoles, a copper-catalyzed N-arylation of amidines with arylboronic acids (Chan–Lam–Evans N-arylation) followed by a one-pot transformation to benzimidazoles by a C–H activation/C–N bond forming process was recently reported. Interestingly, the conditions developed for this domino transformation did not require the presence of excess acid.[58b] An elegant alternative preparation of benzoxazoles and benzimidazoles was proposed by Punniyamurthy based on the rearrangement of bisarylhydrazones or bisaryloxime ethers (Scheme 4.60).[146] Stoichiometric copper(II) trifluoromethanesulfonate was necessary to promote the reaction for the synthesis of benzimidazoles in refluxing toluene, whereas a catalytic amount of copper(II)

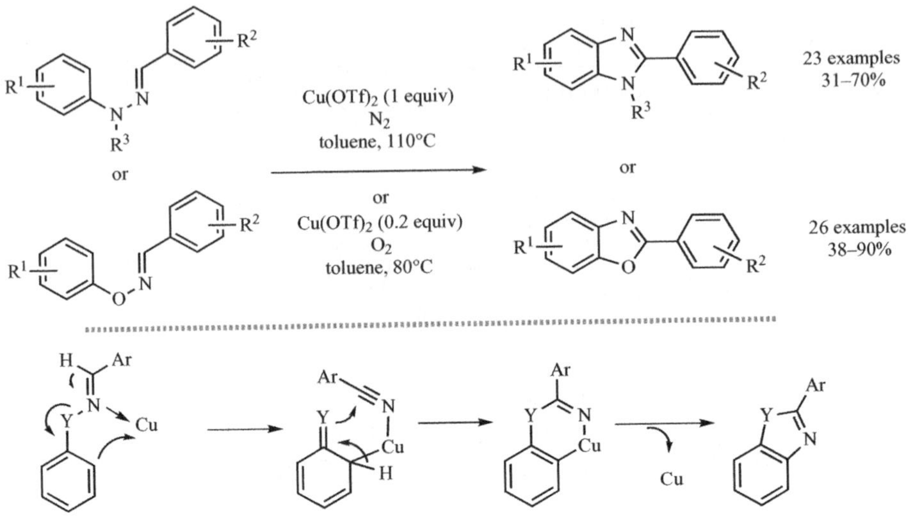

Scheme 4.60 Bisarylhydrazones and bisaryloxime ethers as precursor for the synthesis of benzoxazoles and benzimidazoles.

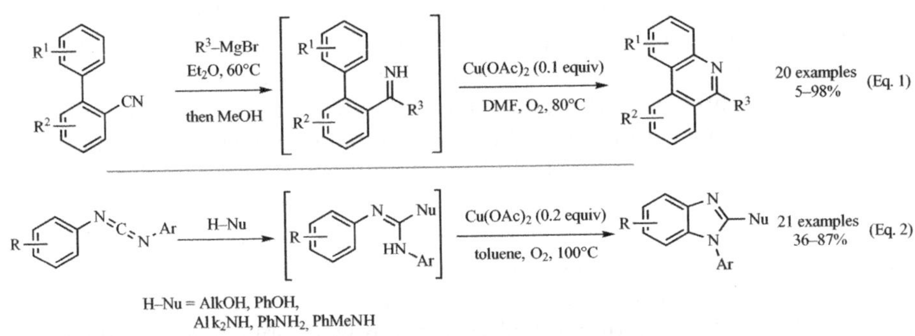

Scheme 4.61 One-pot/tandem processes involving oxidative annulation strategies and C–H bond heterofunctionalization.

trifluoromethanesulfonate under oxygen allowed for the synthesis of benzoxazoles. The suggested mechanism would involve a N–O (or N–N) bond cleavage followed by a rearrangement to form a copper-containing metallacycle, which finally undergoes a reductive elimination.

Phenanthridine derivatives can also be formed through a one-pot strategy according to Chiba's work (Scheme 4.61, Eq. 1).[147] The process relies on the addition of a Grignard reagent to biaryl nitrile substrates, which upon methanolysis, furnished an imine. Upon treatment with copper(II) acetate in DMF under oxygen, intramolecular C–H amination occurred to give the corresponding annulated products in high yields. An addition/C–H amination cascade has

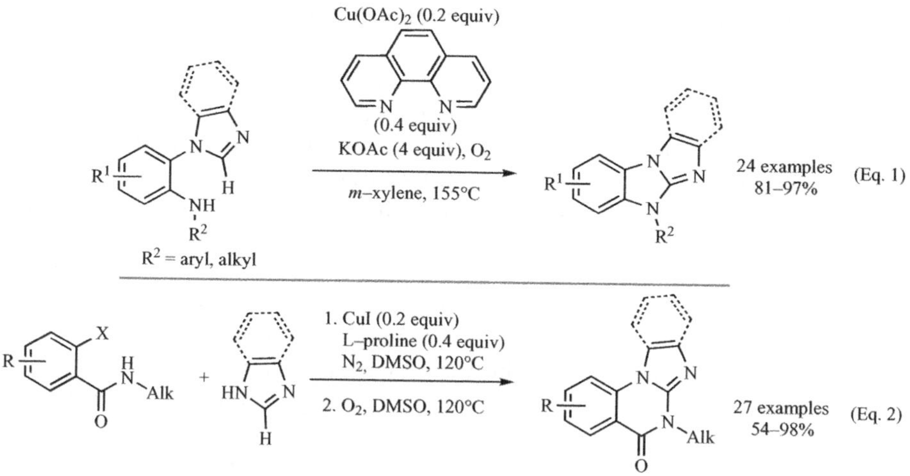

Scheme 4.62 Oxidative annulation strategies involving an intramolecular copper-promoted C–H amination/amidation of imidazole derivatives.

been described by Bao for the construction of benzimidazoles from diaryl carbodiimides with different oxygenated or aminated nucleophiles, using a copper(II) acetate/oxygen system (Scheme 4.61, Eq. 2).[148]

The synthesis of imidazobenzimidazoles and benzoimidazoquinazoline derivatives has been recently reported by Fu and coworkers following a copper-catalyzed aerobic oxidative intramolecular amination/amidation strategy, expanding the scope of the direct intramolecular functionalization of aromatic C–H bonds to azoles containing acidic bonds such as imidazoles or benzimidazoles (Scheme 4.62).[149] For the construction of imidazobenzimidazoles (Scheme 4.62, Eq. 1), intramolecular heteroarylation of the aniline was best accomplished using copper(II) acetate/1,10-phenanthroline as the catalytic system with potassium acetate as the base in *m*-xylene at 155°C under oxygen. It should be highlighted that such reaction conditions were also found to be suitable in a related intermolecular reaction (see Section 4.5.1, Scheme 4.48).[128] An interesting one-pot sequential Ullmann-type coupling followed by an intramolecular C–H amidation allowed for the development of a practical synthesis of benzoimidazoquinazolines. Initial intermolecular N-arylation was accomplished using a combination of copper(I) iodide and L-proline in the presence of cesium carbonate under a nitrogen atmosphere in DMSO at 120°C, and was followed by a copper-catalyzed intramolecular C–H amidation, which was initiated by simply submitting the crude reaction mixture to an oxygen atmosphere. Interestingly, it was noted that 2 equiv of imidazole/benzimidazole were required for this two-step process, as these heterocycles were found to be important additives for the second step of the sequence, probably acting as a ligand to generate the active catalyst.

4.6 CONCLUSION AND FUTURE PROSPECTS

Given the importance of arylated heteronucleophiles, which are used as key building blocks in the core structure of an impressive array of complex natural products and pharmaceuticals, one should not be surprised that a lot of work has been accomplished to devise efficient strategies to build such structures. Whereas the use of the classical Ullmann–Goldberg reactions or palladium-catalyzed processes has dominated the area and had a marked influence in organic synthesis (as illustrated by the impressive use of these strategies in industrial processes or for the synthesis of natural products with increasing complexity), alternative approaches have emerged. Copper complexes had a central role in these emerging methods, and we have attempted to cover them in this chapter. Whereas the development of such methods has been driven toward milder reaction conditions (as it is the case for the Chan–Lam–Evans cross coupling) or the use of precursors that would be more available and would avoid their prefunctionalization (as it is the case for the direct functionalization involving C–H bonds), no methods are universal, and a fine tuning is often needed for a specific substrate. There is no doubt, however, that these methods will play a crucial role for the formation of C(aryl)-heteroatom bonds for the years to come, and that the development of more general, milder processes will contribute to popularize these processes. The knowledge associated with the mechanisms involved in these processes is still poor, preventing the rational identification of catalytic systems with broader scope, and efforts in this direction will be crucial for the development of procedures involving alternative reagents for the arylation of heteronucleophiles with increased efficiency.

REFERENCES

[1] (a) Ullmann, F. *Ber. Dtsch. Chem. Ges.* **1903**, *36*, 2389–2391. (b) Ullmann, F. *Ber. Dtsch. Chem. Ges.* **1904**, *37*, 853–857. (c) Goldberg, I. *Ber. Dtsch. Chem. Ges.* **1906**, *39*, 1691–1696.

[2] Lindley, J. *Tetrahedron* **1984**, *40*, 1433–1456.

[3] For selected reviews involving palladium catalysis, see: (a) Hartwig, J. F. *Handbook of Organopalladium Chemistry for Organic Synthesis*, Vol. 1, Negishi, E.-I.; de Meijere, A. (eds.). Wiley-Interscience: New York, **2002**, pp. 1051–1096. (b) Muci, A. R.; Buchwald, S. L. *Topics in Current Chemistry*, Vol. 219, Miyaura, N. (ed.). Springer-Verlag: Berlin, **2002**, pp. 133–205.

[4] For selected recent reviews on copper-mediated cross-coupling reactions, see: (a) Surry, D. S.; Buchwald, S. L. *Chem. Sci.* **2011**, *1*, 13–31. (b) Monnier, F.; Taillefer, M. *Angew. Chem. Int. Ed.* **2009**, *48*, 6954–6971. (c) Evano, G.; Blanchard, N.; Toumi, M. *Chem. Rev.* **2008**, *8*, 3054–3131.

[5] (a) Finet, J.-P.; Federov, A. Y.; Combes, S.; Boyer, G. *Curr. Org. Chem.* **2002**, *6*, 597–626. (b) Ley, S. V.; Thomas, A. W. *Angew. Chem. Int. Ed.* **2003**, *42*, 5400–5449.

[6] (a) Chan, D. M. T.; Monaco, K. L.; Wang, R.-P.; Winter, M. P. *Tetrahedron Lett.* **1998**, *39*, 2933–2936. (b) Evans, D. A.; Katz, J. L.; West, T. R. *Tetrahedron Lett.* **1998**, *39*, 2937–2940. (c) Lam, P. Y. S.; Clark, C. G.; Saubern, S.; Adams, J.; Winters, M. P.; Chan, D. M. T.; Combs, A. *Tetrahedron Lett.* **1998**, *39*, 2941–2944.

[7] (a) Qiao, J. X.; Lam, P. Y. S. *Synthesis* **2011**, 829–856. (b) Thomas, A. W.; Ley, S. V. *Modern Arylation Methods*, Akermann, L. (ed.). Wiley-VCH: Weinheim, **2009**, pp. 121–153. (c) Chan, D. M. T.; Lam, P. Y. S. *Boronic Acids*, Hall, D. (ed.). Wiley-VCH: Weinheim, **2005**, pp. 205–240.

[8] Collman, J. P.; Zhong, M. *Org. Lett.* **2000**, *2*, 1233–1236.

[9] Collman, J. P.; Zhong, M.; Zhang, C.; Costranzo, S. *J. Org. Chem.* **2001**, *66*, 7892–7897.

[10] Antilla, J. C.; Buchwald, S. L. *Org. Lett.* **2001**, *3*, 2077–2079.

[11] Quach, T. D.; Batey, R. A. *Org. Lett.* **2003**, *5*, 4397–4400.

[12] (a) Lan, J.-B.; Yu, X.-Q.; You, J.-S.; Chen, L.; Yan, M.; Xie, R.-G. *Chem. Commun.* **2004**, 188–189. (b) Lan, J.-B.; Chen, X.-Q.; You, J.-S.; Xie, R.-G. *Synlett* **2004**, 1095–1097.

[13] Wentzel, M. T.; Hewgley, J. B.; Kamble, R. M.; Wall, P. D.; Kozlowski, M. C. *Adv. Synth. Catal.* **2009**, *351*, 931–937.

[14] Combs, A. P.; Saubern, S.; Lam, P. Y. S., unpublished work reported in Reference 7a.

[15] Kantam, M. L.; Neelima, B.; Reddy, C. H. V.; Neeraja, V. *J. Mol. Catal. A* **2006**, *249*, 201–206.

[16] Deng, H.; Jung, J.-K.; Liu, T.; Kuntz, K. W.; Snapper, M. L.; Hoveyda, A. H. *J. Am. Chem. Soc.* **2003**, *125*, 9032–9033.

[17] Lam, P. Y. S.; Bonne, D.; Vincent, G.; Clark, C. G.; Combs, A. P. *Tetrahedron Lett.* **2003**, *44*, 1691–1694.

[18] Santucci, L.; Triboulet, C. *J. Chem. Soc., Chem. Commun.* **1969**, 392–396.

[19] (a) Collman, J. P.; Zhong, M.; Zeng, L.; Costanzo, S. *J. Org. Chem.* **2001**, *66*, 1528–1531. (b) Joubert, N.; Baslé, E.; Vaultier, M.; Pucheault, M. *Tetrahedron Lett.* **2010**, *51*, 2994–2997.

[20] Nishiura, K.; Urawa, Y.; Soda, S. *Adv. Synth. Catal.* **2004**, *346*, 1679–1684.

[21] Sun, H.; Harms, K.; Sundermeyer, J. *J. Am. Chem. Soc.* **2004**, *126*, 9550–9551 and references cited therein.

[22] (a) Mahapatra, S.; Halfen, J. A.; Wilkinson, E. C.; Pan, G.; Wang, X.; Young, V. G.; Cramer, C. J.; Que, L.; Tolman, W. B. *J. Am. Chem. Soc.* **1996**, *118*, 11555–11574. (b) Mahapatra, S.; Halfen, J. A.; Tolman, W. B. *J. Am. Chem. Soc.* **1996**, *118*, 11575–11586.

[23] Lam, P. Y. S.; Vincent, G.; Clark, C. G.; Deudon, C. G.; Jadhav, P. K. *Tetrahedron Lett.* **2001**, *42*, 3415–3418.

[24] Simon, J.; Salzbrunn, S.; Prakash, G. K. S.; Petasis, N. A.; Olah, G. A. *J. Org. Chem.* **2001**, *66*, 633–634.

[25] McKinley, N. F.; O'Shea, D. F. *J. Org. Chem.* **2004**, *69*, 5087–5092.

[26] Singh, B. K.; Appukkuttan, P.; Claerhout, S.; Parmar, V. S.; van der Eycken, E. *Org. Lett.* **2006**, *8*, 1863–1966.

[27] Bekolo, H. *Can. J. Chem.* **2007**, *85*, 42–46.

[28] Abe, T.; Takeda, H.; Yamada, K.; Ishikura, M. *Heterocycles* **2008**, *76*, 133–136.

[29] (a) Bakkestuen, A. K.; Gundersen, L.-L. *Tetrahedron Lett.* **2003**, *44*, 3359–3362. (b) Joshi, R. A.; Patil, P. S.; Mthukrishnan, M.; Ramma, C. V.; Gurjar, M. K. *Tetrahedron Lett.* **2004**, *45*, 195–197.

[30] Chan, D. M. T.; Monaco, K. L.; Li, R.; Bonne, D.; Clark, C. G.; Lam, P. Y. S. *Tetrahedron Lett.* **2003**, *44*, 3863–3865.

[31] Voisin, A. S.; Bouillon, A.; Lancelot, J.-C.; Lesnard, A.; Rault, S. *Tetrahedron* **2006**, *62*, 6000–6005.

[32] Molander, G. A.; Ellis, N. *Acc. Chem. Res.* **2007**, *40*, 275–286.

[33] Quach, T. D.; Batey, R. A. *Org. Lett.* **2003**, *5*, 1381–1384.

[34] Yamamoto, Y.; Takizawa, M.; Yu, X.-Q.; Miyaura, N. *Angew. Chem. Int. Ed.* **2008**, *47*, 928–931.

[35] Yu, X.-Q.; Yamamoto, Y.; Miyaura, N. *Chem.-Asian J.* **2008**, *3*, 1517–1522.

[36] King, A. E.; Brunold, T. C.; Stahl, S. S. *J. Am. Chem. Soc.* **2009**, *131*, 5044–5045.

[37] (a) Huffman, L. M.; Stahl, S. S. *J. Am. Chem. Soc.* **2008**, *130*, 9196–9197. (b) King, A. E.; Huffman, L. M.; Casitas, A.; Costas, M.; Ribas, X.; Stahl, S. S. *J. Am. Chem. Soc.* **2011**, *132*, 12068–12073.

[38] Decicco, C. P.; Song, Y.; Evans, D. A. *Org. Lett.* **2001**, *3*, 1029–1032.

[39] Hitotsuyanagi, Y.; Ishikawa, H.; Naito, S.; Takeya, K. *Tetrahedron Lett.* **2003**, *44*, 5901–5903.

[40] Petrassi, H. M.; Sharpless, K. B.; Kelly, J. W. *Org. Lett.* **2001**, *3*, 139–142.

[41] Wang, Z.; Zhang, J. *Tetrahedron Lett.* **2005**, *46*, 4997–4999.

[42] Lam, P. Y. S.; Clark, C. G.; Saubern, S.; Adams, J.; Averill, K. M.; Chan, D. M. T.; Combs, A. *Synlett* **2000**, 674–677.

[43] Chan, D. G.; Winternheimer, D. J.; Merlic, C. A. *Org. Lett.* **2011**, *13*, 2778–2781.

[44] Winternheimer, D. J.; Merlic, C. A. *Org. Lett.* **2010**, *12*, 2508–2510.

[45] Zhang, L.; Zhang, G.; Zhang, M.; Cheng, J. *J. Org. Chem.* **2010**, *75*, 7472–7474.

[46] Wang, W.; Devasthale, P.; Farrelly, D.; Gu, L.; Harrity, T.; Cap, M.; Chu, C.; Kunselman, L.; Morgan, N.; Ponticiello, R.; Zebo, R.; Zhang, L.; Locke, K.; Lippy, J.; O'Malley, K.; Hosagrahara, V.; Zhang, L.; Kadiyala, P.; Chang, C.; Muckelbauer, J.; Doweyko, A. M.; Zahler, R.; Ryono, D.; Hariharan, N.; Cheng, P. T. W. *Bioorg. Med. Chem. Lett.* **2008**, *18*, 1939–1944.

[47] Jacobsen, M. F.; Knudsen, M. M.; Gothelf, K. V. *J. Org. Chem.* **2006**, *71*, 9183–9190.

[48] Hügel, H. M.; Rix, C. J.; Fleck, K. *Synlett* **2006**, 2290–2292.

[49] (a) Ding, S.; Gray, N. S.; Ding, Q.; Schultz, P. G. *Tetrahedron Lett.* **2001**, *42*, 8751–8755. (b) Bakkestuen, A. K.; Gundersen, L.-L. *Tetrahedron Lett.* **2003**, *44*, 3359–3362.

[50] Yue, Y.; Zheng, Z.-G.; Wu, B.; Xia, C.-Q.; Yu, X.-Q. *Eur. J. Org. Chem.* **2005**, 5154–5157.

[51] Li, W.; Fan, Y.; Xia, Y.; Rocchi, P.; Zhu, R.; Qu, F.; Neyts, J.; Juan, L.; Iovanna, J. L.; Peng, L. *Helv. Chim. Acta* **2009**, *92*, 1503–1513.

[52] Cherney, R. J.; Duan, J. J. W.; Voss, M. E.; Chen, L.; Wang, L.; Meyer, D. T.; Wasserman, Z. R.; Hardman, K. D.; Liu, R.-Q.; Covington, M. B.; Qian, M.; Mandlekar,

S.; Christ, D. D.; Trzaskos, J. M.; Newton, R. C.; Magolda, R. L.; Wexler, R. R.; Decicco, C. P. *J. Med. Chem.* **2003**, *46*, 1811–1823.

[53] Wang, Y. F.; Zhou, Y.; Wang, J. R.; Liu, L.; Guo, Q. X. *Chin. Chem. Lett.* **2007**, *18*, 499–501.

[54] Moessner, C.; Bolm, C. *Org. Lett.* **2005**, *7*, 2667–2669.

[55] Tao, C.-Z.; Cui, X.; Li, J.; Liu, A.-X.; Liu, L.; Guo, Q.-X. *Tetrahedron Lett.* **2007**, *48*, 3525–3529.

[56] Rao, H.; Fu, H.; Jiang, Y.; Zhao, Y. *Angew. Chem. Int. Ed.* **2009**, *48*, 1114–1116.

[57] Yang, H.; Li, Y.; Jiang, M.; Wang, J.; Fu, H. *Chem. Eur. J.* **2011**, *17*, 5652–5660.

[58] (a) Kianmehr, E.; Baghersad, M. H. *Adv. Synth. Catal.* **2011**, *353*, 2599–2603. (b) Nielsen, F. E.; Korno, H. T.; Rasmussen, K. G., inventors. Novel Process. *US patent 2004/0010142 A1, 2004 January 15.* (c) Li, J.; Benard, S.; Neuville, L. Zhu, J. *Org. Lett.* **2012**, *14*, 5980–5983.

[59] Nielsen, F. E.; Korno, H. T.; Rasmussen, K. G.; Novo Nordisk, A. S.; Nielsen, F. E.; Korno, H. T.; Rasmussen, K. G., inventors; Novo Nordisk A. S., Nielsen, F. E.; Korno, H. T.; Rasmussen, K. G., assignees. A new process for preparing fused 1,2,4-thiadiazine derivatives. PCT Int. Appl. WO 200250085 (A1), 2002 June 27.

[60] Herradura, P. S.; Pendola, K. A.; Guy, R. K. *Org. Lett.* **2000**, *2*, 2019–2022.

[61] (a) Wang, L.; Zhou, W.-Y.; Chen, S.-C.; He, M.-Y.; Chen, Q. *Synlett* **2011**, *20*, 3041–3045. (b) Xu, H.-J.; Zhao, Y.-Q.; Feng, T.; Feng, Y.-S. *J. Org. Chem.* **2012**, *77*, 2878–2884.

[62] (a) Lengar, A.; Kappe, C. O. *Org. Lett.* **2004**, *6*, 771–774. (b) Prokopcova, H.; Kappe, C. O. *Adv. Synth. Catal.* **2007**, *349*, 448–452.

[63] Prokopcova, H.; Kappe, C. O. *J. Org. Chem.* **2007**, *72*, 4440–4448.

[64] Tanigushi, N. *J. Org. Chem.* **2007**, *72*, 1241–1245.

[65] Beaulieu, C.; Guay, D.; Wang, Z.; Evans, D. A. *Tetrahedron Lett.* **2004**, *45*, 3233–3236.

[66] (a) Kar, A.; Sayyed Iliyas, A.; Lo Wei, F.; Kaiser Hanns, M.; Beller, M.; Tse, M. K. *Org. Lett.* **2007**, *9*, 3405–3408. (b) Huang, F.; Batey, R. A. *Tetrahedron* **2007**, *63*, 7667–7672.

[67] Kantam, M. L.; Neelima, B.; Sreedhar, B.; Chakravarti, R. *Synlett* **2008**, 1455–1458. Correction: *Synlett* **2008**, 1918.

[68] Zhuang, R.; Cai, J. X.; Tang, G.; Fang, M.; Zhao, Y. *Org. Lett.* **2011**, *13*, 2110–2113.

[69] (a) Abramovitch, R. A.; Barton, D. H. R.; Finet, J.-P. *Tetrahedron* **1987**, *44*, 3039–3071. (b) Finet, J.-P. *Chem. Rev.* **1989**, *89*, 1487–1501. (c) Elliot, I. G.; Konopelski, J. P. *Tetrahedron* **2001**, *57*, 5683–5705.

[70] Recent exemples: (a) Combes, S.; Finet, J.-P. *Synth. Commun.* **1996**, *26*, 4569–4575. (b) Combes, S.; Finet, J.-P. *Tetrahedron* **1998**, *54*, 4313–4318.

[71] (a) Urano, M.; Wada, S.; Suzuki, H. *Chem. Commun.* **2003**, 1202–1203. (b) Stavila, V.; Thurston, J. H.; Prieto-Centurión, D.; Whitmire, K. H. *Organometallics* **2007**, *26*, 6864–6866.

[72] For bismuth: (a) Dodonov, V. A.; Gushchin, A. V.; Brilkina, T. G. *Zhur. Obshch. Khim* **1984**, *54*, 2154–2158. (b) Barton, D. H. R.; Finet, J. P.; Pichon, C. *J. Chem. Soc. Chem. Commun.* **1986**, 65–66. (c) Barton, D. H. R.; Finet, J.-P.; Khamsi, J.

Tetrahedron Lett. **1986**, *27*, 3615–3618. (d) For lead: Barton, D. H. R.; Yadav-Bhatnagar, N.; Finet, J.-P.; Khamsi, J. *Tetrahedron Lett.* **1987**, *28*, 3111–3114.

[73] Barton, D. H. R.; Finet, J.-P.; Khamsi, J. *Tetrahedron Lett.* **1987**, *28*, 887–890.

[74] Sheppard, G. S. *Synlett* **1999**, 1207–1210.

[75] Arnauld, T.; Barton, D. H. R.; Doris, E. *Tetrahedron* **1997**, *53*, 4137–4144.

[76] Elliot, G. I.; Konopelski, J. P. *Org. Lett.* **2000**, *2*, 3055–30057.

[77] Barton, D. H. R.; Bhatnagar, N. Y.; Finet, J.-P.; Motherwell, W. B. *Tetrahedron* **1986**, *42*, 311–3122.

[78] Rasmussen, L. K.; Begtrup, M.; Ruhland, T. *J. Org. Chem.* **2004**, *69*, 6890–6893.

[79] Lopez-Alvarado, P.; Avendaiio, C.; Menkndez, J. C. *J. Org. Chem.* **1995**, *60*, 5678–5682.

[80] Barton, D. H. R.; Finet, J.-P.; Khamsi, J. *Tetrahedron Lett.* **1988**, *29*, 1115–1118.

[81] Barton, D. H. R.; Donnelly, D. M. X.; Finet, J.-P.; Guiry, P. J. *J. Chem. Soc. Perkin Trans. 1* **1991**, 2095–2102.

[82] Fedorov, A. Y.; Finet, J.-P. *Tetrahedron Lett.* **1999**, *40*, 2747–2748.

[83] Chan, D. M. T. *Tetrahedron Lett.* **1996**, *37*, 9013–9016.

[84] Sorenson, R. J. *J. Org. Chem.* **2000**, *65*, 7747–7749.

[85] Barton, D. H. R.; Finet, J.-P.; Khamsi, J.; Pichon, C. *Tetrahedron Lett.* **1986**, *27*, 3619–3622.

[86] Brands, K. M. J.; Dolling, U.-H.; Jobson, R. B.; Marchesini, G.; Reamer, R. A.; Williams, J. M. *J. Org. Chem.* **1998**, *63*, 6721–6726.

[87] Brunner, H.; Obermann, U.; Wimmer, P. *Organometallics* **1989**, *8*, 821–826.

[88] Barton, D. H. R.; Finet, J.-P.; Pichion, C. *J. Chem. Soc., Chem. Commun.* **1986**, 65–66.

[89] Ooi, T.; Goto, R.; Maruoka, K. *J. Am. Chem. Soc.* **2003**, *125*, 10494–10495.

[90] (a) Ikegai, K.; Fukumoto, K.; Mukaiyama, T. *Chem. Lett.* **2006**, *35*, 612–613. (b) Sakurai, N.; Ikegai, K.; Mukaiyama, T. *ARKIVOC* **2007**, 254–264.

[91] Yasuike, S.; Nishioka, M.; Kakusawa, N.; Kurita, J. *Tetrahedron Lett.* **2011**, *52*, 6403–6406.

[92] (a) Lam, P. Y. S.; Deudon, S.; Averill, K. M.; Li, R.; He, M. Y.; DeShong, P.; Clark, C. G. *J. Am. Chem. Soc.* **2000**, *122*, 7600–7601. (b) Lam, P. Y. S.; Vincent, G.; Bonne, D.; Clark, C. G. *Tetrahedron Lett.* **2002**, *43*, 3091–3094.

[93] Vakalopoulos, A.; Kavazoudi, X.; Schoof, J. *Tetrahedron Lett.* **2006**, *47*, 8607–8610.

[94] Lin, B.; Liu, M.; Ye, Z.; Ding, J.; Wu, H.; Cheng, J. *Org. Biomol. Chem.* **2009**, *7*, 869–873.

[95] (a) Luo, F.; Pan, C.; Qian, P.; Cheng, J. *Synthesis* **2010**, 2005–2010. (b) Luo, F.; Pan, C.; Qian, P.; Cheng, J. *J. Org. Chem.* **2010**, *75*, 5379–5381.

[96] Luo, P.-S.; Yu, R.-Y.; Zhong, P.; Li, J.-H. *Tetrahedron Lett.* **2009**, *50*, 1066–1070.

[97] For selected reviews dealing with hypervalent iodine compounds, see: (a) Stang, P. J.; Zhdankin, V. V. *Chem. Rev.* **1996**, *96*, 1123–1178. (b) Zhdankin, V. V.; Stang, P. J. *Chem. Rev.* **2002**, *102*, 5223–2584. (c) Zhdankin, V. V.; Stang, P. J. *Chem. Rev.* **2008**, *108*, 5299–5358. For a review on diaryliodonium salts, see: Merritt, E. A.; Olofsson, B. *Angew. Chem. Int Ed.* **2009**, *48*, 9052–9070.

[98] Beringer, M.; Brierley, A.; Drexler, M.; Gindler, M. E.; Lumpki, C. C. *J. Am. Chem. Soc.* **1953**, *75*, 2708–2712.

[99] Niu, H.-Y.; Xia, C.; Qu, G.-R.; Zhang, Q.; Jiang, Y.; Li, D.-Y.; Guo, H.-M. *Org. Biomol. Chem.* **2011**, *9*, 5039–5042.

[100] Zhou, T.; Chen, Z.-C. *Synth. Commun.* **2002**, *32*, 903–907.

[101] Davydov, D. V.; Beletskaya, I. P.; Semenov, B. B.; Smushkevich, Y. I. *Tetrahedron Lett.* **2002**, *43*, 6217–6219.

[102] (a) Vaddula, B.; Leazer, J.; Varna, R. S. *Adv. Synth. Catal.* **2012**, *354*, 986–990. (b) Lv, T.; Wang, Z.; You, J.; Lan, J.; Gao, G. *J. Org. Chem.* **2013**, *78*, 5723–5730.

[103] (a) Lubinkowski, J. J.; Knapczyk, J. W.; Calderon, J. L.; Petit, L. R.; McEven, W. E. *J. Org. Chem.* **1975**, *40*, 3010–3015. (b) For a recent publication, see: Jalalian, N.; Ishikawa, E. E.; Silva, L. F., Jr.; Olofsson, B. *Org. Lett.* **2011**, *13*, 1552–1555.

[104] Kuriyama, M.; Hamaguchi, N.; Onomura, O. *Chem. Eur. J.* **2012**, *18*, 1591–1594.

[105] Zhang, Y.; Patel, S.; Mainolfi, N. *Chem. Sci.* **2012**, *3*, 3196–3199.

[106] Savarin, C.; Srogl, J.; Liebeskind, L. S. *Org. Lett.* **2002**, *4*, 4309–4312.

[107] Abe, Y.; Nakabayashi, T.; Tsurugi, J. *Bull. Chem. Soc. Jpn* **1973**, *46*, 1898–1899.

[108] Erdik, E.; Ay, M. *Chem. Rev.* **1989**, *89*, 1947–1980.

[109] (a) Casarini, A.; Dembech, P.; Lazzari, D.; Marini, E.; Reginato, G.; Ricci, A.; Seconi, G. *J. Org. Chem.* **1993**, *58*, 5620–5623. (b) Bernardi, P.; Dembech, P.; Fabbri, G.; Ricci, A.; Seconi, G. *J. Org. Chem.* **1999**, *64*, 641–643.

[110] Berman, A. M.; Johnson, J. S. *J. Am. Chem. Soc.* **2004**, *126*, 5680–5681.

[111] (a) Campbell, M. J.; Johnson, J. S. *Org. Lett.* **2007**, *9*, 1521–1524. (b) Hirano, H.; Satoh, T.; Miura, M. *Org. Lett.* **2011**, *13*, 2395–2397. (c) Matsuda, N.; Hirano, K.; Satoh, T.; Miura, M. *J. Org. Chem.* **2012**, *77*, 617–625.

[112] Miki, Y.; Hirano, K.; Satoh, T.; Miura, M. *Org. Lett.* **2013**, *15*, 172–175.

[113] (a) Matsuda, N.; Hirano, K.; Satoh, T.; Miura, M. *Angew. Chem. Int. Ed.* **2012**, *51*, 3642–3645. (b) Rucker, R. P.; Whittaker, A. M.; Dang, H.; Lalic, G. *Angew. Chem. Int. Ed.* **2012**, *51*, 3953–3956.

[114] Yu, Y.; Srogl, J.; Liebeskind, L. S. *Org. Lett.* **2004**, *6*, 2631–2634.

[115] Tsutsui, H.; Hayashi, Y.; Narasaka, K. *Chem. Lett.* **1997**, 317–318.

[116] Liu, S.; Yu, Y.; Liebeskind, L. S. *Org. Lett.* **2007**, *9*, 1947–1950.

[117] Liu, S.; Liebeskind, L. S. *J. Am. Chem. Soc.* **2008**, *130*, 6918–6919.

[118] Zhang, Z.; Yu, Y.; Liebeskind, L. S. *Org. Lett.* **2008**, *10*, 3005–3008.

[119] He, C.; Chen, C.; Cheng, J.; Liu, C.; Liu, W.; Li, Q.; Lei, A. *Angew. Chem. Int. Ed.* **2008**, *47*, 6414–6417.

[120] Uemura, T.; Chatani, N. *J. Org. Chem.* **2005**, *70*, 8631–8634.

[121] Kisseljova, K.; Tsubrik, O.; Sillard, R.; Mäeorg, S.; Mäeorg, U. *Org. Lett.* **2006**, *8*, 43–45.

[122] For a recent monograph, see: (a) Akermann, L. *Modern Arylation Methods*, Wiley-VCH: Weinheim, **2009**. (b) For recent reviews on C–H functionalization using transition metals, see: Lyons, T. W.; Sanford, M. S. *Chem. Rev.* **2010**, *110*, 1147–1170. (c) Chen, X.; Engle, K. M.; Wang, D.-H.; Yu, J.-Q. *Angew. Chem. Int. Ed.* **2009**, *48*, 5094–5115. (d) Colby, D. A.; Bergman, R. G.; Ellman, J. A. *Chem. Rev.* **2010**, *110*, 624–655.

[123] For recent reviews, see: (a) Davies, H. M. L.; Mannig, J. R. *Nature* **2008**, *451*, 417–424. (b) Collet, F.; Dodd, R. H.; Dauban, P. *Chem. Commun.* **2009**, 5061–5074. (c) Doyle, M. P.; Duffy, R.; Ratnikov, M.; Zhou, L. *Chem. Rev.* **2010**, *110*, 704–724.

[124] (a) Monguchi, D.; Fujiwara, T.; Furukawa, H.; Mori, A. *Org. Lett.* **2009**, *11*, 1607–1610. (b) Wagh, Y. S.; Bhanage, B. M. *Tetrahedron Lett.* **2012**, *53*, 6500–6503.

[125] Wang, Q.; Schreiber, S. L. *Org. Lett.* **2009**, *11*, 5178–5180.

[126] (a) Miyasaka, M.; Hirano, K.; Satoh, T.; Kowalczyk, R.; Bolm, C.; Miura, M. *Org. Lett.* **2011**, *13*, 359–361. (b) Zhao, H.; Wang, M.; Su, W.; Hong, M. *Adv. Synth. Catal.* **2010**, *352*, 1301–1306.

[127] Takemura, N.; Kuninobu, Y.; Kanai, M. *Org. Lett.* **2013**, *15*, 844–847.

[128] Oda, Y.; Hirano, K.; Satoh, T.; Miura, M. *Org. Lett.* **2012**, *14*, 664–667.

[129] (a) Guo, S.; Qian, B.; Xie, Y.; Xia, C.; Huang, H. *Org. Lett.* **2011**, *13*, 522–525. (b) Li, Y.; Xie, Y.; Zhang, R.; Jin, K.; Wang, X.; Duan, C. *J. Org. Chem.* **2011**, *76*, 5444–5449.

[130] (a) Fukuzawa, S.-I.; Shimizu, E.; Atsuumi, Y.; Haga, M.; Ogata, K. *Tetrahedron Lett.* **2009**, *50*, 2374–2376. (b) Zhou, A.-X.; Liu, X.-Y.; Yang, K.; Zhao, S.-C.; Liang, Y.-M. *Org. Biomol. Chem.* **2011**, *9*, 5456–5462.

[131] Ge, W.; Wei, Y. *Synthesis* **2012**, *44*, 934–940.

[132] Ranjit, S.; Lee, R.; Heryadi, D.; Shen, C.; Wu, J.; Zhang, P.; Huang, K.-W.; Liu, X. *J. Org. Chem.* **2011**, *76*, 8999–9007.

[133] (a) Kawano, T.; Hirano, K.; Satoh, T.; Miura, M. *J. Am. Chem. Soc.* **2010**, *132*, 6900–6901. (b) Matsuda, N.; Hirano, K.; Satoh, T.; Miura, M. *Org. Lett.* **2011**, *13*, 2860–2863. (c) Wu, Z.; Song, H.; Cui, X.; Pi, C.; Du, W.; Wu, Y. *Org. Lett.* **2013**, *15*, 1270–1273.

[134] Kaeding, W. W.; Shulgin, A. T. *J. Org. Chem.* **1962**, *27*, 3551–3554.

[135] Reinaud, O.; Capdevielle, P.; Maumy, M. *Synthesis* **1989**, 612–614.

[136] Chen, X.; Hao, X.-S.; Goodhue, C. E.; Yu, J.-Q. *J. Am. Chem. Soc.* **2006**, *128*, 6790–6791.

[137] Uemura, T.; Imoto, S.; Chatani, N. *Chem. Lett.* **2006**, *35*, 842–843.

[138] (a) Wang, W.; Luo, F.; Zhang, S.; Cheng, J. *J. Org. Chem.* **2010**, *75*, 2415–2418. (b) Wang, W.; Pan, C.; Chen, F.; Cheng, J. *Chem. Commun.* **2011**, *47*, 3978–3980. (c) Bian, Y.-J.; Xiang, C.-B.; Chen, Z.-M.; Huang, Z.-Z. *Synlett* **2011**, 2407–2709.

[139] (a) Shuai, Q.; Deng, G.; Chua, Z.; Bohle, D. S.; Li, C.-J. *Adv. Synth. Catal.* **2010**, *352*, 632–636. (b) John, A.; Nicholas, K. M. *J. Org. Chem.* **2011**, *76*, 4158–4162.

[140] Mizuhara, T.; Inuki, S.; Oishi, S.; Fujii, N.; Ohno, H. *Chem. Commun.* **2009**, 3413–3415.

[141] (a) Zhang, L.; Liu, Z.; Li, H.; Fang, G.; Barry, B.-D.; Belay, T. A.; Bi, X.; Liu, Q. *Org. Lett.* **2011**, *13*, 6536–6539. (b) Tran, L. D.; Roane, J.; Daugulis, O. *Angew. Chem., Int. Ed.* **2013**, *52*, 6043–6046. (c) Tran, L. D.; Popov, I.; Daugulis, O. *J. Am. Chem. Soc.* **2012**, *134*, 18237–18240. (d) Tang, C.; Jiao, N. *J. Am. Chem. Soc.* **2012**, *134*, 18924–18927.

[142] Brasche, G.; Buchwald, S. L. *Angew. Chem. Int. Ed.* **2008**, *47*, 1932–1934.

[143] (a) Wang, H.; Wang, Y.; Peng, C.; Zhang, J.; Zhu, Q. *J. Am. Chem. Soc.* **2010**, *132*, 13217–13219. (b) Masters, K.-S.; Rauws, T. R. M.; Yadav, A. K.; Herrebout, W. A.; Ven der Veken, B.; Maes, B. U. W. *Chem. Eur. J.* **2011**, *17*, 6315–6320.

[144] (a) Ueda, S.; Nagasawa, H. *Angew. Chem. Int. Ed.* **2008**, *47*, 6411–6413. (b) Ueda, S.; Nagasawa, H. *J. Org. Chem.* **2009**, *74*, 4272–4277.

[145] (a) Cho, S. H.; Yoon, J.; Chang, S. *J. Am. Chem. Soc.* **2011**, *133*, 5996–6005. (b) Zhao, J.; Wang, Y.; He, Y.; Liu, L.; Zhu, Q. *Org. Lett.* **2012**, *14*, 1078–1081. (c) Zhao, J.; Zhang, Q.; Liu, L.; He, Y.; Li, J.; Zhu, Q. *Org. Lett.* **2012**, *14*, 5362–5365.

[146] (a) Guru, M. M.; Ali, M. A.; Punniyamurthy, T. *Org. Lett.* **2011**, *13*, 1194–1197. (b) Guru, M. M.; Ali, M. A.; Punniyamurthy, T. *J. Org. Chem.* **2011**, *76*, 5295–5308.

[147] Zhang, L.; Ang, G. Y.; Chiba, S. *Org. Lett.* **2010**, *12*, 3682–3685.

[148] He, H.-F.; Wang, Z.-J.; Bao, W. *Adv. Synth. Catal.* **2010**, *352*, 2905–2912.

[149] Wang, X.; Jin, Y.; Zhao, Y.; Zhu, L.; Fu, H. *Org. Lett.* **2012**, *14*, 452–455.

[150] Barton, D. H. R.; Donnely, D. M. X.; Finet, J.-P.; Guiry, P. J. *Tetrahedron Lett.* **1989**, *30*, 1377–1380.

[151] Finet, J.-P.; Fedorov, A. Y. *J. Organomet. Chem.* **2006**, *691*, 2386–2393.

[152] Tsubrick, O.; Mäeorg, U.; Sillard, R.; Ragnarsson, U. *Tetrahedron* **2004**, *60*, 8363–8373.

5

BEYOND ULLMANN–GOLDBERG CHEMISTRY: VINYLATION, ALKYNYLATION, AND ALLENYLATION OF HETERONUCLEOPHILES

KÉVIN JOUVIN AND GWILHERM EVANO

5.1 INTRODUCTION

If the introduction of chelating ligand over the past 10 years greatly extended the scope of the original Ullmann and Goldberg cross-coupling reactions that can now be conducted under much milder conditions together with dramatically enhanced yields,[1] it also allowed for the use of new reaction partners for the introduction of vinyl, alkyne, and allene functional groups. These include not only alkenyl, alkynyl, and allenyl halides but also vinyl dibromides, alkenyl boronic acids and boronates, potassium alkenyl and alkynyltrifluoroborates, terminal alkynes, propiolic acids, alkynyl(triaryl)bismuthonium, and alkynyl(aryl) iodonium salts. The scope of heteronucleophiles susceptible to participate in these cross-coupling reactions has also been dramatically extended: They now include nitrogen-, oxygen-, phosphorus-, and sulfur-based nucleophiles. An overview of the possibilities is shown in Figure 5.1: Great advances have been made in the synthesis of heteroatom-substituted unsaturated molecules using copper catalysis.

The synthetic potential of these transformations is now quite obvious. They have recently evolved as powerful tools, probably because they are the most efficient ones, for the synthesis of a wide array of heteroatom-substituted

Copper-Mediated Cross-Coupling Reactions, First Edition. Edited by Gwilherm Evano and Nicolas Blanchard.

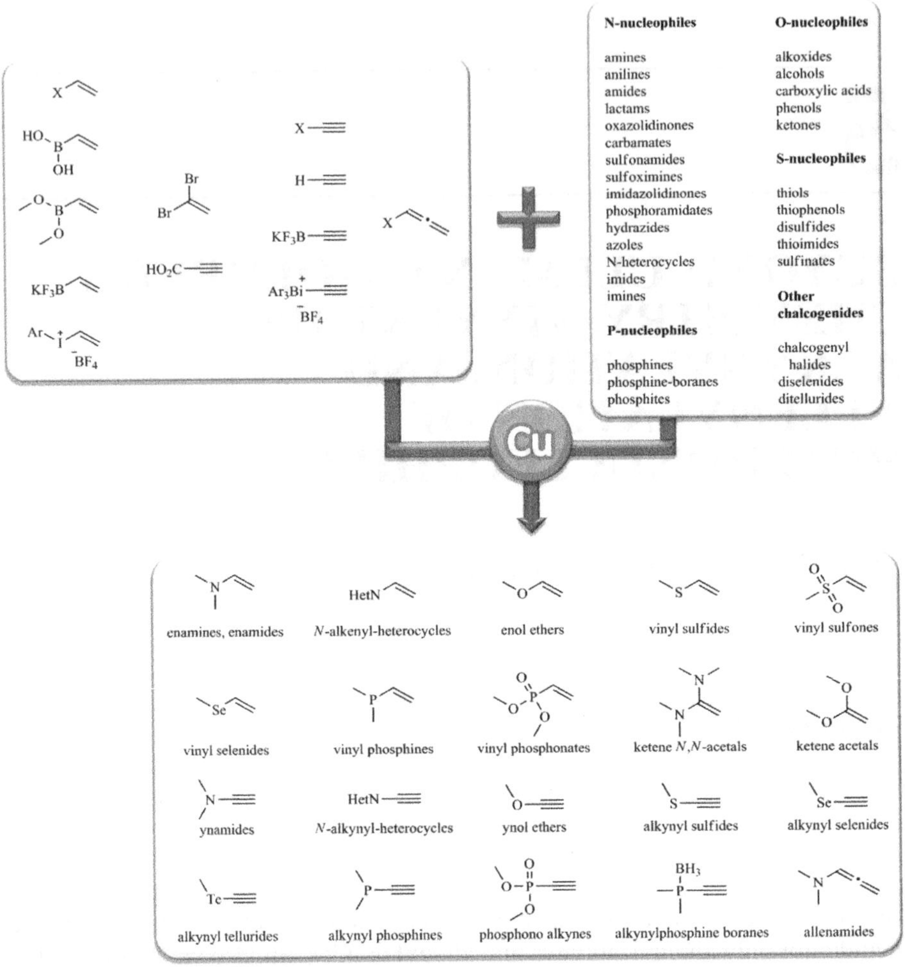

Figure 5.1 Beyond Ullmann–Goldberg chemistry: alkenylation, alkynylation, and allenylation of heteronucleophiles.

alkenes, alkynes, and allenes—building blocks of crucial importance in organic synthesis and medicinal chemistry as well. The application of these new copper-catalyzed transformations in natural product synthesis has also flourished recently. An array of copper-mediated procedures has been successfully employed to assemble many complex targets with new and efficient bond disconnections, unambiguously demonstrating the efficiency and reliability of these reactions.

Examples in this chapter have been classified according to the level of unsaturation in the product (alkene, alkyne, and allene) and are further subclassified by the nature of the bond (C–N, C–O, C–S/C–Se/C–Te, and C–P) formed during the copper-mediated process. Additionally, an overview of the state-of-the-art procedures available for each bond formation has also been included together with a brief historical perspective. This chapter is not intended to be fully exhaustive and will focus on the most general and efficient procedures.

5.2 COPPER-MEDIATED ALKENYLATION OF HETERONUCLEOPHILES: AMONG THE BEST ROUTES TO HETEROATOM-SUBSTITUTED ALKENES

Heteroatom-substituted olefins most probably represent the most useful and versatile class of alkenes. In addition to their use for the inclusion of various functionalities into organic molecules and their usefulness in medicinal chemistry and materials science, the presence of the heteroatom strongly polarizes the double bond and typically allows for highly regio- and stereoselective transformations. Enamines, enamides, enol ethers, alkenyl phosphines, or vinylphosphonates are common building blocks in organic synthesis, and the development of new transformations from these molecules clearly relies on robust procedures for their preparation. Copper catalysis has revolutionized this area and provides reliable, practical, and general methods for the synthesis of all classes of heteroatom-substituted alkenes. The synthesis of enamides using copper-catalyzed alkenylation of nitrogen nucleophiles will be reviewed in the following section.

5.2.1 Synthesis of Enamides, Enamines, and N-Alkenyl-Heterocycles

Enamides are key structural motifs in various classes of natural products as well as in especially valuable synthetic intermediates. In addition to conventional approaches for their preparation that include condensation of amides and aldehydes, hydroamination of alkynes, acylation of imines, Curtius rearrangement of α,β-unsaturated acyl azides, amide Peterson olefination, and Wittig and Horner–Wadsworth–Emmons reactions, several transition-metal-catalyzed methods for the synthesis of enamides have recently emerged. Although these protocols provide access to enamides, they often suffer from low yield, low substrate scope, or lack of stereocontrol.

For these reasons, the copper-catalyzed coupling of amides with vinyl halides, a modern variant of the Goldberg reaction recently developed, clearly appears as the most general and widely applicable reaction for the preparation of enamides and is now widely used for the synthesis of an impressive number of molecules, including especially complex ones.

From a historical perspective, Ogawa and coworkers first reported the copper iodide-promoted coupling reaction between vinyl bromides and potassium amides (which could also be generated *in situ* by deprotonation with potassium hydride) using rather harsh Ullmann–Goldberg-like conditions, that is, stoichiometric copper(I) in hexamethylphosphoramide (HMPA) at

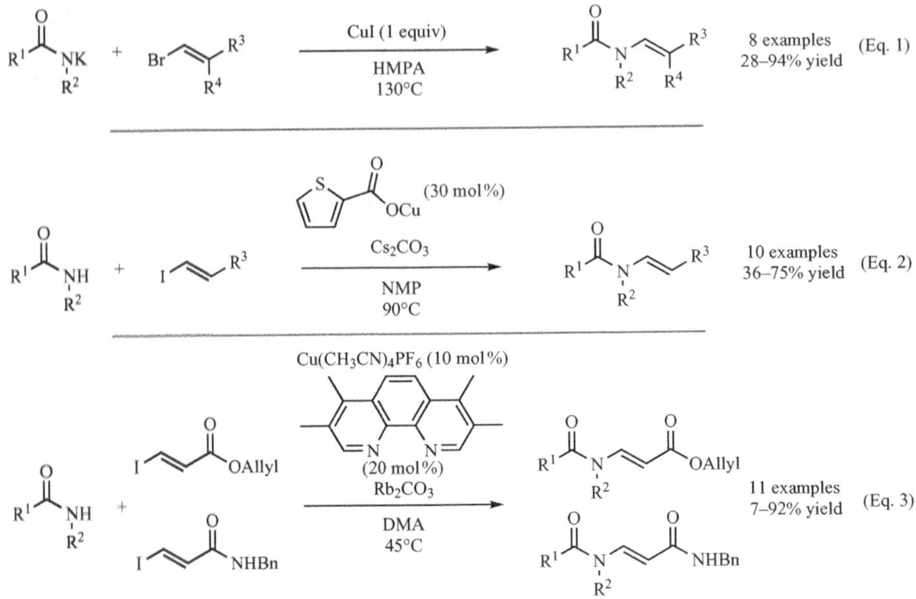

Scheme 5.1 First developments in the copper-catalyzed synthesis of enamides by alkenylation of nitrogen nucleophiles.

130°C, to afford enamides in low-to-good yields (Scheme 5.1, Eq. 1).[2] On the basis of this precedent, Porco and coworkers developed an efficient approach to the assembly of enamides using Liebeskind catalyst, copper(I) thiophene carboxylate (CuTC), cesium carbonate as the base, and terminal (*E*)-vinyl iodides in *N*-methylpyrrolidinone (NMP) at 90°C. This protocol was used for the preparation of a series of (*E*)-enamides using mild reaction conditions (Scheme 5.1, Eq. 2)[3] and could be elegantly used for the synthesis of the macrocyclic benzolactone natural product lobatamide C.[4] As an extension of this methodology and during studies aiming at the synthesis of the antibiotic CJ-15,801, another catalytic system based on the Kubas complex, $Cu(CH_3CN)_4PF_6$, 3,4,7,8-tetramethyl-1,10-phenanthroline, and rubidium carbonate in DMA at 45°C was developed for the preparation of vinylogous carbamic acids and ureas starting from amides and β-iodoacrylates or β-iodoacrylamides (Scheme 5.1, Eq. 3).[5]

Based on these early results, a series of efficient catalytic systems were reported in the early 2000s. A major breakthrough for the development of a general procedure for the synthesis of enamides by alkenylation of nitrogen nucleophiles was reported in 2003 by the Buchwald group, who extended the diamine-based catalytic system[6] first developed for the arylation of amides to the synthesis of enamides.[7] Using a combination of catalytic amounts of copper iodide and *N,N′*-dimethylethylenediamine (DMEDA) together with cesium carbonate in tetrahydrofuran (THF) at temperatures ranging from 50 to 70°C, vinyl iodides were found to be readily coupled to primary amides, lactams, and oxazolidinones to give the corresponding enamides with remarkable efficiency. Vinyl bromides were also found to be excellent reaction partners for this copper-catalyzed cross coupling, but they required the use of potassium carbonate in toluene at 110°C. An interesting feature of these conditions is that di- or trisubstituted vinyl bromides as well as (*Z*)-vinyl iodides perform well under the reaction conditions. This system is most certainly one of the most efficient to date, and has been later extended to the vinylation of amino acid amides,[8] sulfoximines,[9] bis-Boc-hydrazide,[10] *N*-alkyl-anilines, and diarylamines (sodium *tert*-butoxide in toluene at 140°C was used in this case).[11] It was featured as a key step for the synthesis of a wide range of natural products, therefore demonstrating its incredible efficiency and reliability (see Chapter 16 for details). Representative examples of the copper iodide/diamine alkenylation of nitrogen nucleophiles are shown in Scheme 5.2.

The stereoselective iodovinylation of amides by coupling with 1,2-diodoethene is also possible using this catalytic system,[12] and the use of *gem*-dibromoolefins provides, depending on the reaction conditions, a straightforward entry to α-bromoenamides or ketene *N*,*N*-acetals resulting from site-selective or double amidation reactions, respectively.[13]

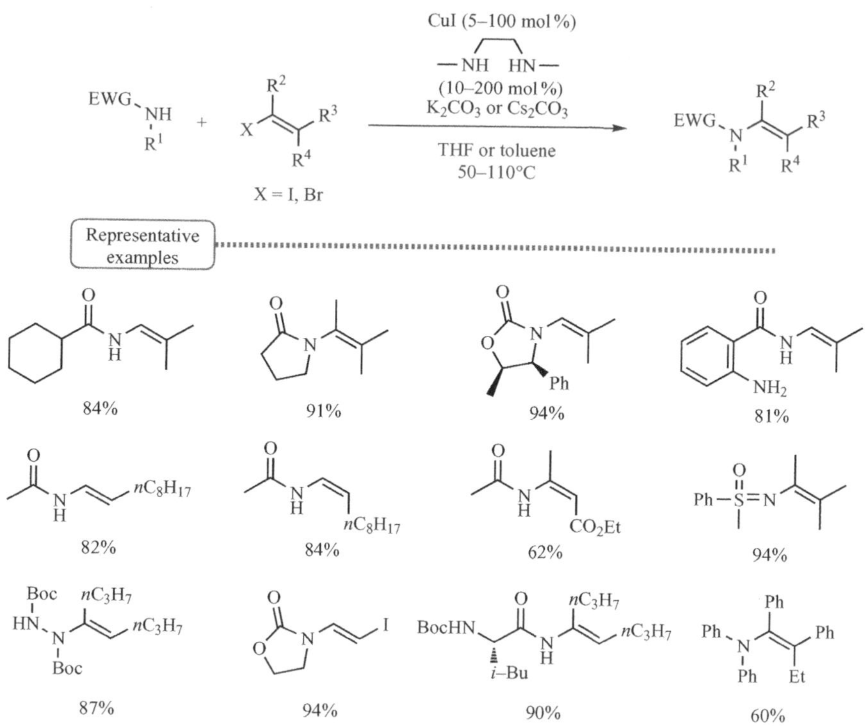

Scheme 5.2 Copper(I) iodide/diamine-catalyzed synthesis of enamides.

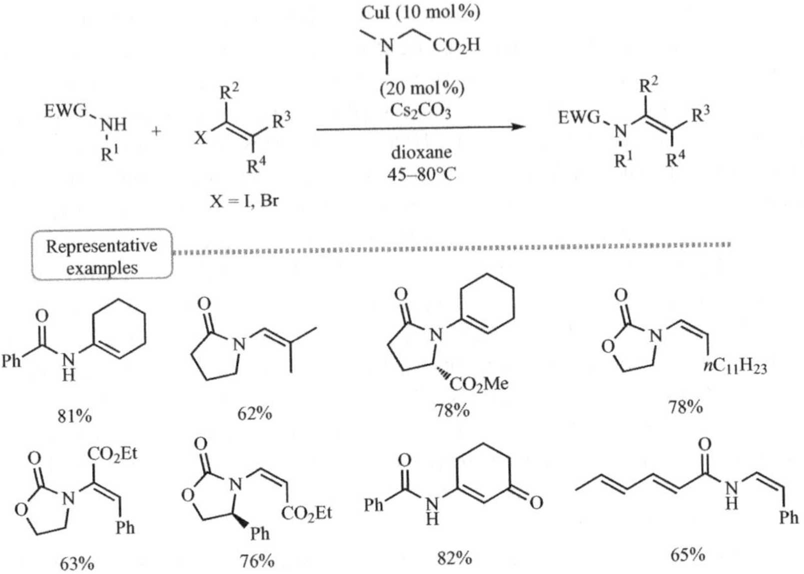

Scheme 5.3 Copper(I) iodide/amino acid-catalyzed synthesis of enamides.

An alternative to diamine ligands for copper(I) was found in the use of *N,N*-dimethylglycine. Ma and coworkers, in the continuation of their studies on the use of amino acids as promoters for copper-catalyzed Ullmann-type condensations, eventually found that vinyl iodides and bromides could be smoothly coupled with amides or oxazolidinones, using copper(I) iodide in combination with this amino acid and cesium carbonate in dioxane at temperatures ranging from 45 to 80°C (Scheme 5.3).[14]

The development of general procedures for the copper-catalyzed cross coupling between alkenyl halides and nitrogen nucleophiles had a deep impact on heterocyclic chemistry, and the development of efficient intramolecular processes enabled the synthesis of a wide range of heterocycles. Acyclic substrates possessing both a vinyl halide and a nitrogen nucleophile suitably placed could be readily cyclized, such as with the cyclization of **1** or **3** to the corresponding 2-alkylidene nitrogen heterocycles **2**[15] and **4**,[16] respectively (Scheme 5.4, Eq. 1 and Eq. 2). The intramolecular N-vinylation via *endo*-mode from **5** was found to be equally efficient and allowed for the preparation of a wide range of unsaturated heterocycles **6** (Scheme 5.4, Eq. 3).[16] Intramolecular cross coupling from related acyclic amides substituted with a vinyl iodide or bromide was also investigated, and smoothly gave the corresponding unsaturated lactams with high efficiency (Scheme 5.4, Eq. 4, Eq. 5, and Eq. 6),[17] even for the preparation of strained ring systems such as 2-alkylidene-azetidinones **8**,[18] or with functionalized substrates (since carbapenem derivatives could be easily obtained using this strategy[19]). Combined together, results from these studies indicate that the relative rate of cyclization follows the

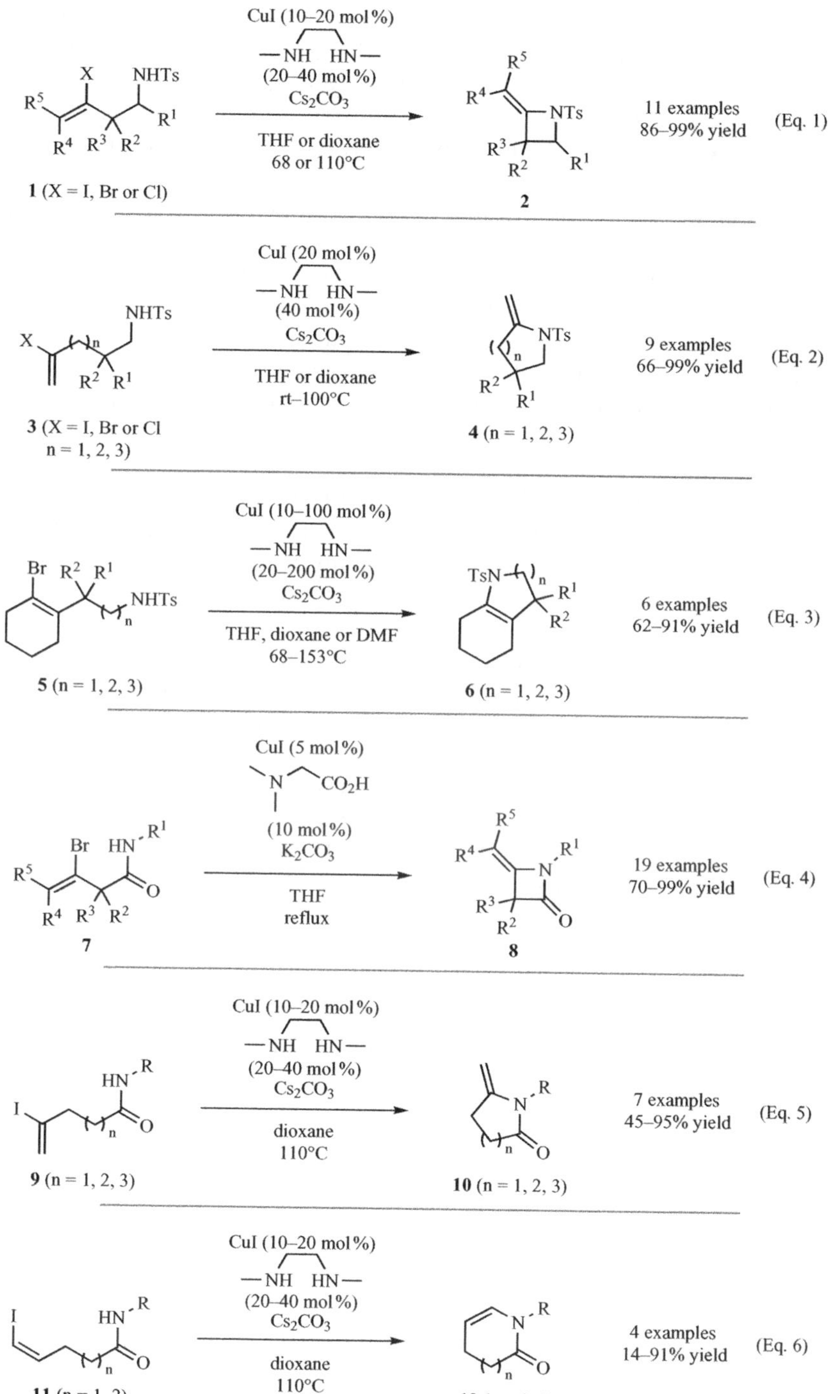

Scheme 5.4 Synthesis of unsaturated nitrogen heterocycles by intramolecular N-vinylation.

order 4-*exo* > 5-*exo* ≈ 5-*endo* > 6-*endo* > 7-*endo* > 8-*endo* (the smaller the ring size, the easier the cyclization). In addition, these reaction rates were found to be different enough so that chemoselective cyclization from substrates possessing two vinyl bromides could be efficiently used for the formation of the smaller ring system, as exemplified by the preferred formation of a β-lactam over a γ-lactam.[18]

In addition to these intramolecular cross-coupling reactions from substrates possessing both a nitrogen nucleophile and an alkenyl halide, the use of the alkenylation of amines, amides, carbamates, or hydrazines in sequential or domino processes led to the development of new strategies for the synthesis of nitrogen heterocycles that nicely complement the classical synthetic approaches. Nitrogen-containing heterocycles being privileged constituents of many bioactive natural products and compounds of interest in medicinal chemistry as well, the development of new processes for their synthesis is therefore of crucial importance. Among the advantages of reactions based on copper catalysis are the following: (i) Harsh conditions are frequently avoided; (ii) readily available starting materials can be utilized; (iii) they give access to heterocyclic cores with a high diversity using processes that are highly modulable; and (iv) copper-based catalytic systems are inexpensive and display low toxicity.

One of the simplest examples was reported in 2007 for the synthesis of 2,3-dihydropyrroles, compounds that exhibit a variety of biological and medicinal properties, and whose preparation based on classical procedures typically requires lengthy sequences or harsh reaction conditions. They can now be obtained using a one-pot N-alkylation/N-alkenylation starting from readily available 1,4-diodobut-1-ene **13** derivatives and amides or carbamates **14**. The reaction is applicable to a variety of nitrogen nucleophiles and diiodides, and furnishes the corresponding dihydropyrroles **15** in good yields using racemic *trans-N,N′*-dimethylcyclohexane-1,2-diamine as the ligand for copper iodide (Scheme 5.5).[20]

Sequential and domino processes featuring a copper-catalyzed cross coupling between a nitrogen nucleophile and a suitably functionalized vinyl halide have also been applied to the synthesis of the parent fully unsaturated pyrrole derivatives. The amination of vinyl bromides **16** possessing a ketone in the

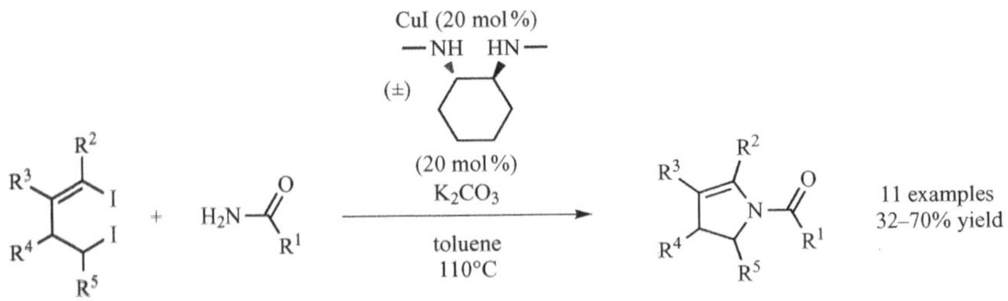

13 14 15

Scheme 5.5 Synthesis of 2,3-dihydropyrroles by sequential N-alkylation/N-alkenylation.

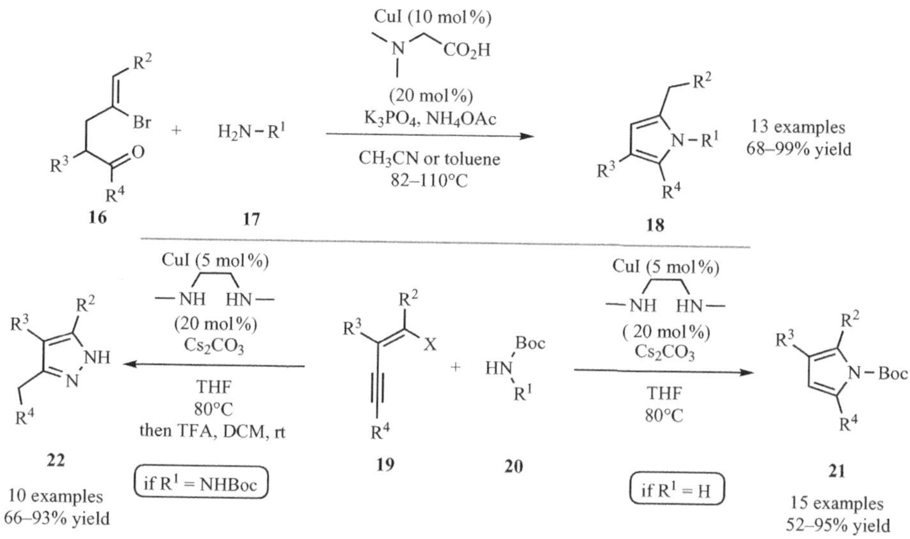

Scheme 5.6 Synthesis of pyrroles and pyrazoles by sequential N-alkenylation/condensation or hydroamidation.

γ-position results in the formation of polysubstituted pyrroles **18**, provided that ammonium acetate is used as an additive to lower the basicity of the solution (Scheme 5.6).[21] Similar products **21** can be obtained when bromo- and iodoenynes **19** are used as the reaction partners through a copper-catalyzed domino amidation/5-*endo*-dig-hydroamidation. Polycyclic pyrroles could also be prepared in high yield using this strategy by simply introducing a heterocyclic moiety either on the alkene or alkyne counterpart. The replacement of the carbamate **20** (R_1 = H) with a bis-Boc-hydrazide **20** (R_1 = NHBoc) allows for a different outcome since a 5-*exo*-dig-hydroamination is now operating after the copper-catalyzed cross coupling, yielding polysubstituted pyrazoles **22** after treatment with trifluoroacetic acid.[22]

As an alternative, and based on the bis-ene hydrazide **23** variant of the Piloty–Robinson pyrrole synthesis, a modular route to highly substituted pyrroles consists of two sequential copper-catalyzed vinylations of bis-Boc-hydrazide **20** followed by a thermal rearrangement/cyclization. A wide variety of densely functionalized pyrroles **24** can be prepared in a simple manner using this modular sequence starting from readily available starting materials (Scheme 5.7).[10]

In addition to these sequential reactions and domino processes relying on the use of a copper-catalyzed amidation of vinyl halides as the key step, other strategies based on tandem intramolecular/intermolecular cross-coupling reactions starting from bis-halides have been recently devised for the preparation of nitrogen heterocycles. Here again, pyrroles received a great deal of attention, and the double alkenylation strategy turned out to be especially

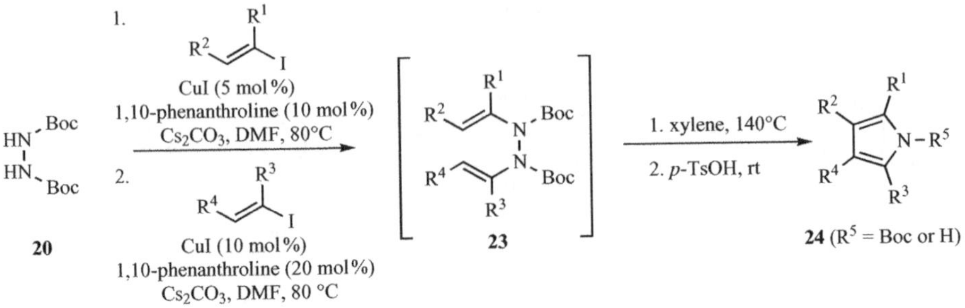

Scheme 5.7 Synthesis of pyrroles by sequential double N-alkenylation/rearrangement/cyclization: a modern variant of the Piloty–Robinson pyrrole synthesis.

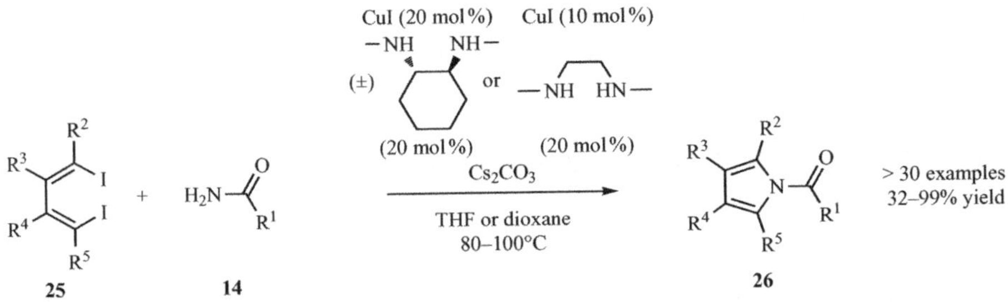

Scheme 5.8 Synthesis of pyrroles by double N-alkenylation of amides and carbamates.

efficient, providing an alternative approach to the classical Knorr, Hantzsch, Paal–Knorr syntheses. Starting from 1,4-dihalo-1,3-dienes **25**, conveniently prepared in high yields through iodination of zircona- or titanacyclopentadienes, a double amidation with carbamates or amides **14** allows the synthesis of pyrroles **26** and heteroarylpyrroles with a wide variety of functional groups and substitution patterns (Scheme 5.8).[20,23] As a note, it was shown later that anilines also perform well, yielding the corresponding *N*-aryl-pyrroles, provided that sodium *tert*-butoxide was used as the base.[24]

A double cross coupling could also be performed from 2-(2-haloalkenyl) aryl halides **27** and primary amides **17**, affording N-functionalized indoles **28** through tandem N-arylation/N-alkenylation (Scheme 5.9). Anilines, amides, and carbamates **17** are all effective coupling partners under the reaction conditions, which complement the one that can be used with palladium catalysis.[25] While the nature of the halide on the double bond was found to have little influence on the outcome of the reaction since vinyl bromides and chlorides were coupled with almost equal efficiency, the replacement of the aromatic bromide by an aromatic chloride was found to have a dramatic effect on the cyclization, and no indole would be isolated in this case. Tandem cross-coupling strategies have also been applied to the synthesis of polyheterocyclic compounds such as imidazoindolones **30** (a key structural motif in various natural

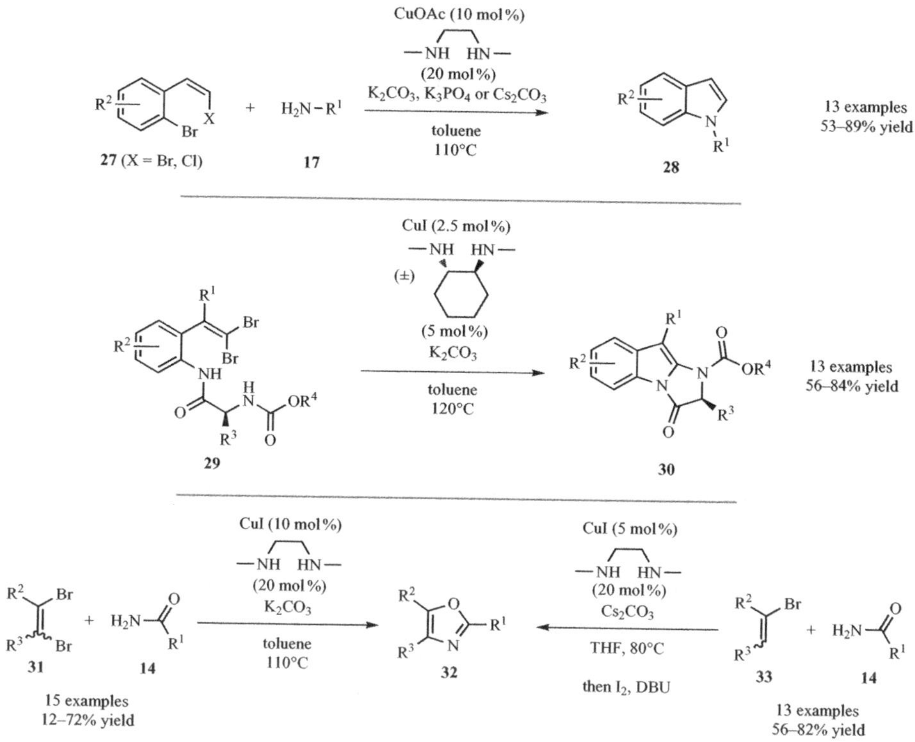

Scheme 5.9 N-Alkenylation in domino copper-catalyzed heterocyclic syntheses.

products) by double copper-catalyzed cyclization reactions,[26] as well as to the synthesis of functionalized oxazoles **32** which can be prepared in a single step and in good yields using a domino copper-catalyzed C–N and C–O cross coupling from primary amides **14** and 1,2-dihaloalkenes **31**.[27] While efficient, this oxazole synthesis is, however, not fully regioselective and affords mixtures of regioisomeric oxazoles. An interesting alternative therefore lies in the amidation of primary amides **14** with vinyl bromides **33** followed by an iodine-promoted cyclization in the presence of 1,8-diazabicyclo[5.4.0]undec-7-ene (DBU) (Scheme 5.9).[28]

Besides the use of amides, carbamates, and aromatic amines in the copper-catalyzed cross coupling with alkenyl halides, nitrogen heterocycles have also been recently included as reaction partners. This provides the most efficient entry to *N*-vinyl azoles, important building blocks in organic synthesis, polymer science, and key structural motifs for the design of new drugs. Classical syntheses of such heterocycles rely mostly on the alkylation of azoles with 1,2-dibromoethane and related compounds, followed by dehydrobromination, or on the reaction with allyl bromides followed by isomerization. The major drawbacks of such methods are the lack of generality and substrate scope together with harsh conditions and lengthy sequences. The renaissance of

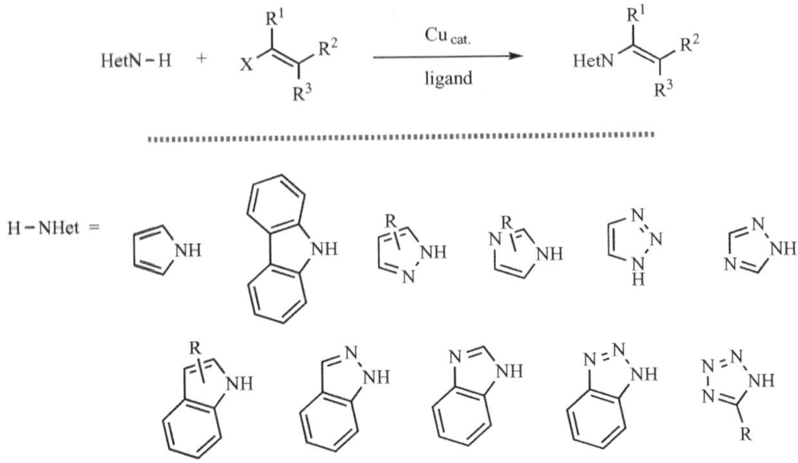

Figure 5.2 Overview of the N-alkenylation of azoles.

copper catalysis enabled the direct N-alkenylation of most π-excessive nitrogen heterocycles (general structures shown in Fig. 5.2), therefore providing a direct entry to vinyl-azoles from readily available starting materials and with high levels of efficiency and reliability. An overview of selected general transformations is given in Table 5.1.

Among all the procedures available, the combination of copper(I) iodide and 2-(pyridin-2-yl)-1*H*-benzoimidazole could be successfully applied to the vinylation of an impressive number of heterocycles including pyrazoles, imidazole, indazole, indole, benzotriazole, 1,2,4-triazole, 5-thiomethyltetrazole with styryl bromides and iodides (including electron-rich ones), and β-iodoacrylates (Table 5.1, entry 1).[29] Ethylenediamine also proved to be an efficient ligand that, in combination with copper iodide, enabled the arylation of imidazoles, indazole, pyrrole, indoles, and carbazoles with a remarkably broad scope with respect to the alkenyl halide, since even *Z*-styryl bromides, which perform only poorly with other systems and often give rise to by-products resulting from isomerization or elimination, and trisubstituted vinyl bromides, were smoothly coupled to give the corresponding vinyl-azoles (Table 5.1, entry 2).[30] Taillefer's efficient catalytic system that uses structurally simple chelating ligands also provides the vinylated products under mild and general reaction conditions with styryl bromide (Table 5.1, entry 3), while the use of more hindered substrates such as 1-bromo-2-methylpropene requires another bidentate ligand, *N*-(pyridin-2-ylmethylene)aniline.[31] It should be noted that for simple cross coupling involving *E*-styryl bromides and pyrazole, imidazole, benzimidazole, pyrrole, or indoles, the use of ligandless conditions can provide a most simple alternative when the reaction is performed in *N*,*N*-dimethylformamide (DMF) at 120°C (Table 5.1, entry 4).[32] *Z*-Bromostyrene could be coupled with retention of configuration using these conditions, although with less efficiency,

TABLE 5.1 Copper-Catalyzed Alkenylation of Nitrogen Heterocycles

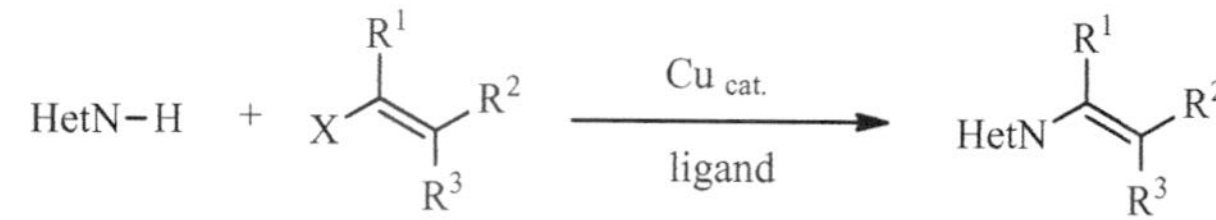

Entry	HetN-H	Alkenyl Halides	Catalytic Systems	Conditions	Number of Examples	Yields	Ref.
1	Pyrazoles, imidazole, indazole, indole, benzotriazole, 1,2,4-triazole, 5-thiomethyltetrazole	*E*-Styryl bromide, *E*-styryl iodides, *Z*-β-iodoacrylates	CuI (5 mol %), 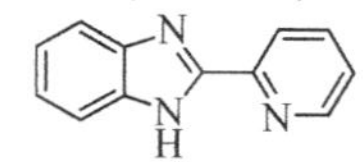 (5 mol %)	Cs_2CO_3, DMF, 40–80°C	22	81–96%	[29]
2	Imidazoles, indazole, pyrrole, indoles, carbazole	Substituted styryl bromides, substituted bromoalkenes	CuI (10 mol %), 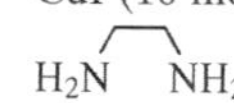 (20 mol %)	K_3PO_4, dioxane, 110°C	17	21–90%	[30]
3	Pyrazoles, imidazole, indazole, pyrrole, indole, benzotriazole, 1,2,3-triazole, 1,2,4-triazole	*E*-Styryl bromide	CuI (10 mol %), 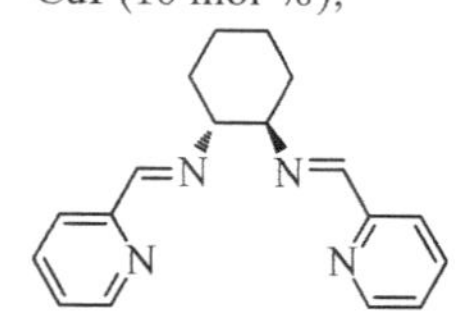 (5 mol %)	Cs_2CO_3, CH_3CN, 35–80°C or DMF, 110°C	11	38–100%	[31]
4	Pyrazole, imidazole, benzimidazole, pyrrole, indoles	Substituted styryl bromides	CuI (10 mol %)	Cs_2CO_3, DMF, 120°C	19	48–95%	[32]
5	Imidazole, benzimidazole	*E*-Styryl bromides	CuI (15 mol %), 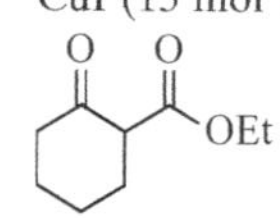 (30 mol %)	Cs_2CO_3, DMSO, 60–80°C	9	82–90%	[33]
6	Imidazoles, benzimidazole	*E*-Styryl bromides	CuI (10 mol %), 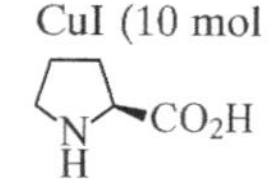 (20 mol %)	K_2CO_3, [Bmim]BF_4, 110°C	9	60–93%	[34]

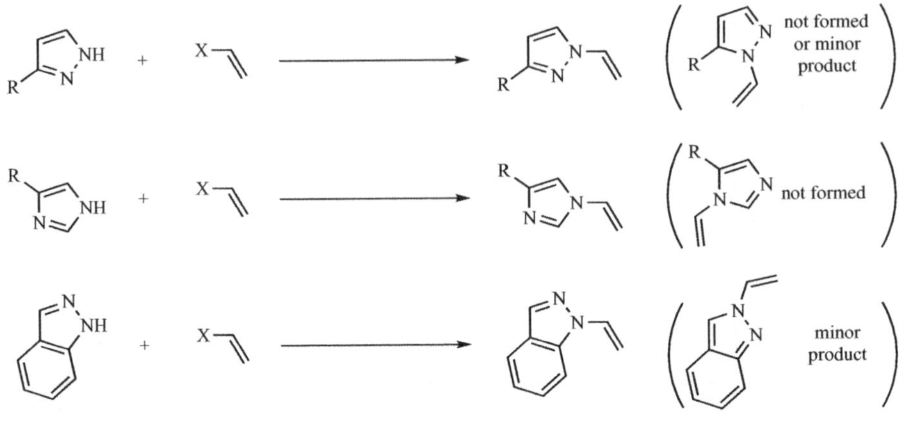

Scheme 5.10 Regioselectivity trends for the N-alkenylation of unsymmetrical N-heterocycles.

and a ligand-assisted coupling using ethylenediamine might therefore be preferable in this case. Other available catalytic systems include a ketoester as ligand (Table 5.1, entry 5)[33] or the use of copper iodide in ionic liquid (Table 5.1, entry 6),[34] and might be considered for the optimization of a given cross coupling, although their scope was studied with less details.

Several interesting trends for the N-vinylation of π-excessive nitrogen heterocycles are apparent from all these studies and could be useful tips for the preparation of *N*-vinyl-azoles. First, the reaction with "unsymmetrical" heterocycles proceeds with synthetically useful levels of regioselectivity: Pyrazoles and imidazoles predominantly give products in which the less hindered nitrogen is selectively vinylated, while indazoles are selectively vinylated at N-1 (Scheme 5.10) (as a note, the selectivity can usually be improved by switching from alkenyl iodides to alkenyl bromides, provided that they are still reactive enough). Second, the reactivity of azoles appears to be the result of a complex balance between several parameters including nucleophilicity, catalyst complexing ability, and acidity. The following order for the reactivity of N-heterocycles is emerging: pyrazole ~ imidazole ~ indazole > indole ~ pyrrole > triazoles ~ benzotriazole >> tetrazole,[31] the latter being a very weak nucleophile and a really poor coupling partner in the Ullmann-type vinylation reactions.

Lam and coworkers have reported an alternative to the use of vinyl halides and used *E*-hexenylboronic acid as room-temperature vinylating agent. Unfortunately, only five examples were reported which do not clearly allow delineation of the scope of the reaction,[35] and no extensive studies have been reported to date other than the use of the Chan–Lam–Evans procedure for the vinylation of the five DNA/RNA nucleobases[36] and the cross coupling between styrylboronic acid and azoles,[37] even though this study might provide an efficient and general procedure for the alkenylation of nitrogen nucleophiles. An excellent and remarkable alternative was found in the use of potassium alkenyltrifluoroborates. The tetracoordinate salts possess increased stability and can be used in copper-catalyzed cross-coupling reactions with amides and oxazolidinones under base-free conditions at 40°C (Scheme 5.11, Eq. 1), where *E*-enamides are obtained in excellent yields.[38] Optimization studies established two sets of conditions best suited to either high pK_a or low pK_a nitrogen nucleophiles. Lower pK_a amide substrates worked best in dichloromethane in the presence of 4Å molecular sieves, catalytic amounts of copper(II) acetate and *N*-methylimidazole as the ligand, while nucleophiles possessing higher pK_a worked best in the absence of this ligand and in a 1:1 mixture of dichloromethane and DMSO. While the cross coupling occurred with retention of alkene configuration, its major limitation was found in the use of *Z*-alkenyltrifluoroborates, which turned out to be only poorly reactive. An interesting alternative was recently reported by Chen, Xi, and coworkers who demonstrated that the combination of alkenylzirconocenes (compounds that are readily prepared by reacting an internal alkyne with Cp_2ZrEt_2 followed by alcoholysis) and *O*-benzoylhydroxylamines in the presence of catalytic amounts of copper(I) chloride in THF at room temperature turned out to be an especially efficient method for the synthesis of trisubstituted enamines (Scheme 5.11, Eq. 2).[39]

5.2.2 Synthesis of Enol Ethers

In addition to the great advances brought by the introduction of chelating ligands for copper in the Goldberg condensation reaction, which could be nicely extended to the use of alkenyl halides as coupling partners for the vinylation of various nitrogen nucleophiles, the Ullmann condensation reaction, that is, the copper-catalyzed synthesis of biarylethers, was also dramatically improved by recent developments in copper catalysis. It can now be used for the synthesis of a wide range of enol ethers (and esters) with remarkable efficiency, predictability, and reliability. With vinyl ethers being important intermediates and building blocks in organic synthesis and medicinal chemistry as well as raw materials for polymers, the development of general and cost-efficient methods for their preparation, more effective than the conventional approaches based on the addition of alcohols to acetylene under high pressure and temperature, the Michael-type addition-elimination processes, the transition metal-catalyzed vinyl transfer or the isomerization of allyl ethers, is of great importance. This section will be devoted to an overview of the state-of-the-art copper-based catalytic systems for the vinylation of O-nucleophiles, with emphasis on experimentally attractive procedures.

From a historical point of view, the first catalytic formal vinylic substitution of unactivated vinyl halides using a copper-based system, based on the pioneering work of Whitesides using stoichiometric amounts of copper alkoxides, was reported in 1992 by Keegstra, who described the coupling of sodium methoxide and butoxide with a variety of vinyl halides.[40] The success of the

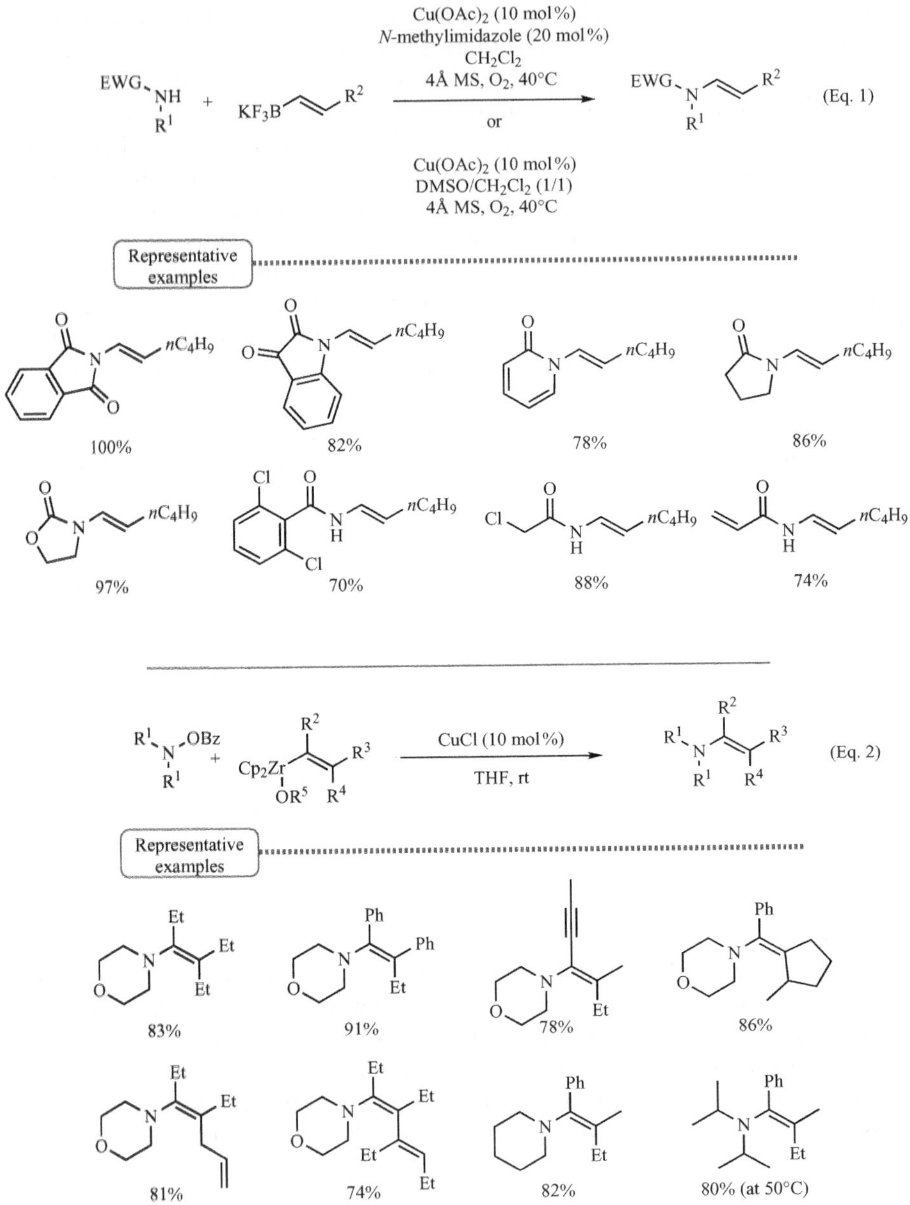

Scheme 5.11 Synthesis of enamides and enamines from potassium alkenyltrifluoroborates and alkenylzirconocenes.

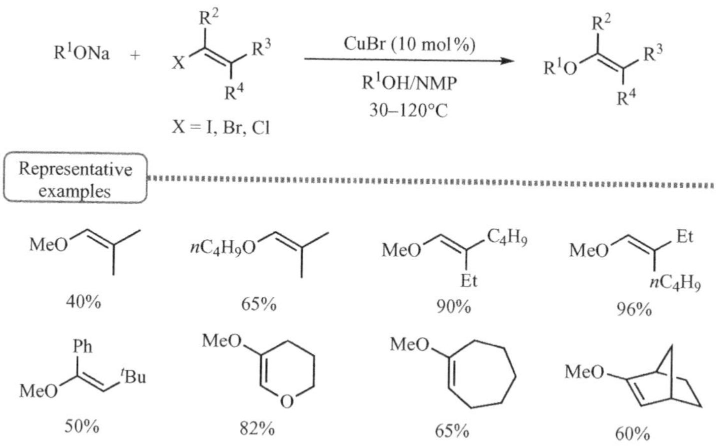

Scheme 5.12 Copper(I) bromide-catalyzed coupling between sodium alkoxides and vinylic halides.

cross coupling, which was found to be highly stereospecific and minimized the amount of by-products arising from α- and β-eliminations typically observed with the noncatalyzed reactions, strongly relied on the use of both highly concentrated solutions (ca 5 $mol.L^{-1}$) and a polar nonprotic solvent, namely, NMP (Scheme 5.12).

While remarkably efficient, this cross coupling was still limited to the use of alkoxides, which is a rather strong limitation in terms of substrate scope. The introduction of ligands allowed a dramatic softening of these conditions and phenanthroline derivatives, initially introduced for the copper-catalyzed arylation of aliphatic alcohols with aryl halides,[41,42] were found to be especially successful for the synthesis of enol ethers (Scheme 5.13). The use of cesium carbonate is a key to the success of the reaction and vinyl iodides are readily coupled to give the corresponding ethers in 67–76% yield.[43] A vinyl bromide could also be successfully subjected to the reaction conditions, albeit in a slightly lower yield. The reaction times for the different types of alcohols vary significantly, generally in the order aliphatic > allylic ≈ propargylic > benzylic. While the reaction was found to be highly stereospecific, the major limitation is that (*Z*)-vinyl iodides decompose under these reaction conditions, presumably via β-elimination due to the presence of a hydrogen atom in the *trans*-position at the double bond of the vinyl halide. This process also represents an experimentally attractive means for the preparation of allyl vinyl ethers from allylic alcohols, which enabled an especially elegant Ullmann–Claisen domino process yielding to highly functionalized ketones and aldehydes when the reaction was performed in *o*-xylene at 120°C.

Besides these examples, the use of aliphatic alcohols has not been extensively studied and most systems have been employed with phenols, compounds

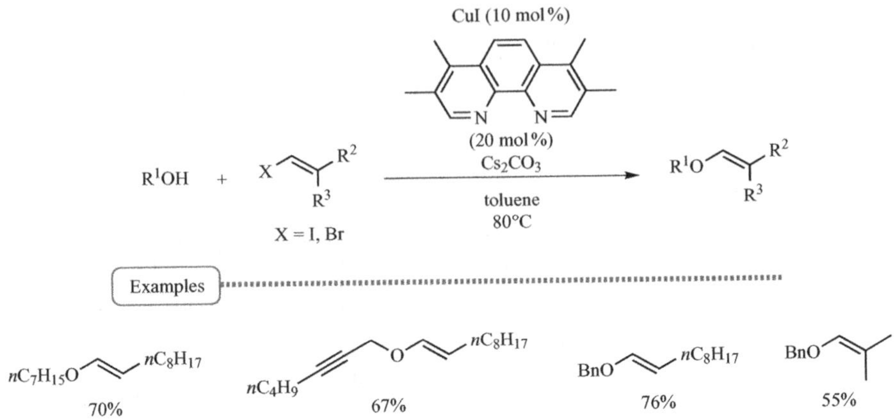

Scheme 5.13 Copper-catalyzed coupling between alcohols and alkenyl halides.

that are now readily coupled with a wide range of alkenyl halides. Most efficient and general catalytic systems are shown in Table 5.2.

These systems include the combination of copper(I) iodide and 2-(pyridin-2-yl)-1*H*-benzoimidazole, which could be successfully applied to the vinylation of all kind of phenols and naphtols with styryl bromides and iodides (including electron-rich ones) as well as β-iodoacrylates, which were readily coupled at room temperature (Table 5.2, entry 1). The nature of the substitutent on both coupling partners had little effect on the outcome of the reaction, and the stereochemistry in all ethers was retained regardless of the (*E*)- or (*Z*)-configuration.[29] Researchers at Eli Lilly reported on the use of an amino alcohol as ligand for the synthesis of aryl vinyl ethers in good-to-excellent yield and with full retention of configuration by direct coupling of vinyl halides and phenols (Table 5.2, entry 2).[44] The copper(I) iodide/*N*,*N*-dimethylglycine system, which has been successfully used for many copper-mediated processes, was also applied to the cross coupling of vinyl halides with phenols. Aryl vinyl ethers were obtained in good yields, but isomerization was observed at high temperatures or when vinyl bromides were used as reaction partners (Table 5.2, entry 3).[45] The use of a β-ketoester as ligand was also efficient with (*E*)-styryl bromides (Table 5.2, entry 4), affording the corresponding ethers in good yields and with high functional group tolerance, but failed when the *Z*-isomers were subjected to the reaction conditions. Finally, Taillefer's bis-iminopyridine ligand was also found to promote the coupling of phenols with styryl bromide in good yields, except in the case of *p*-nitro-phenol whose poor reactivity in Ullmann-type coupling reactions is well known (Table 5.2, entry 5).[31,46] As a note, butadienylphosphines[47] and phosphorus dendrimers[48] were also reported as ligands for copper(I), but a single example was only reported in each case.

Besides the acceleration of the copper-catalyzed cross coupling by chelating ligands, a clear common feature to all these procedures is the use of cesium

TABLE 5.2 Copper-Catalyzed Alkenylation of Phenols

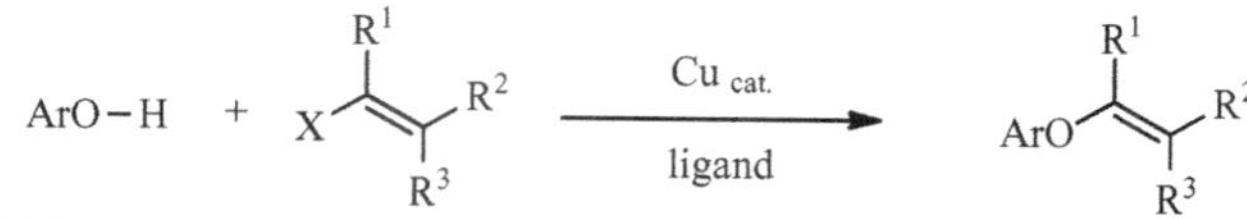

Entry	Alkenyl Halides	Catalytic Systems	Conditions	Number of Examples	Yields	Ref.
1	*E*-Styryl bromide, *E*-styryl iodides, *Z*-β-iodoacrylates	CuI (5 mol %), 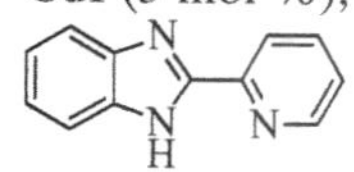 (5 mol %)	Cs_2CO_3, DMF, rt –80°C	22	81–98%	[29]
2	Polysubstituted styryl bromides, substituted bromo- and iodoalkenes	CuCl (25 mol %), 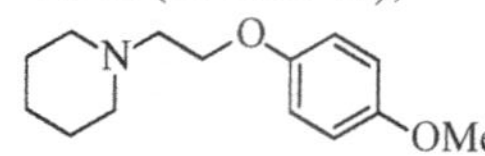 (25 mol %)	Cs_2CO_3, toluene, 110°C	10	41–94%	[44]
3	Styryl bromides, substituted bromo- and iodoalkenes	CuI (10 mol %), 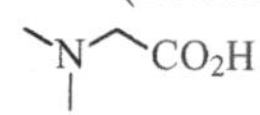 (30 mol %)	Cs_2CO_3, dioxane, 60–90°C	22	48–90%	[45]
4	*E*-Styryl bromides	CuI (15 mol %), 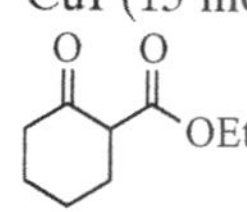 (30 mol %)	Cs_2CO_3, NMP, 60°C	8	79–95%	[33]
5	*E*-Styryl bromide	CuI (10 mol %), 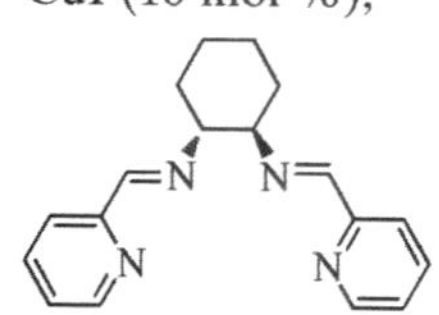 (5 mol %)	Cs_2CO_3, CH_3CN, 50°C	5	74–90%	[31,46]

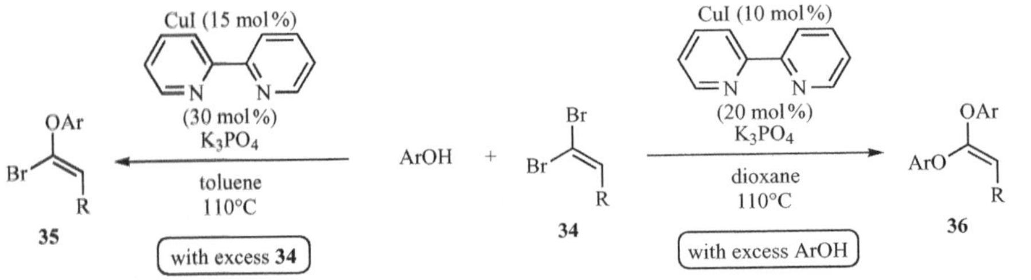

24 examples
41–81% yield

11 examples
38–88% yield

Scheme 5.14 Copper-catalyzed coupling between phenols and vinyl dibromides: synthesis of α-bromo enol ethers and ketene acetals.

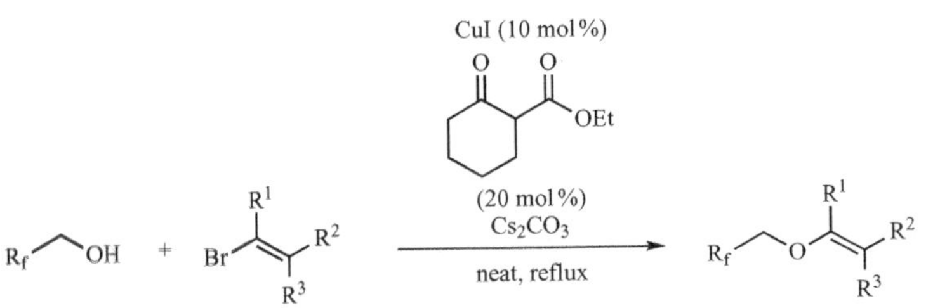

6 examples
70–95% yield

$R_f = CF_3, C_2F_5, C_7F_{15}$

Scheme 5.15 Synthesis of fluorinated vinyl ethers from alkenyl bromides.

carbonate, which is most certainly the most critical parameter for the success of the alkenylation of alcohols.

Vinyl dibromides **34** were also recently found to be remarkably efficient reaction partners: Depending on the stoichiometry of the reaction and on the solvent used, they could conveniently undergo either a site-selective or a double cross-coupling reaction, respectively, yielding α-bromo-enol ethers **35** with high levels of stereoselectivity, or ketene acetals **36** (Scheme 5.14),[49] compounds that can hardly be obtained using other processes. As a note, trifluoroethanol was also found to smoothly participate in either the site-selective or the double cross-coupling reactions.

Besides aromatic alcohols, trifluoroethanol and related higher homologues were also demonstrated to be readily coupled with alkenyl bromides, which is understandable because of their higher acidity compared with aliphatic alcohols. Fluorinated vinyl ethers can therefore be conveniently obtained, even with highly substituted systems, using a ketoester as ligand and the fluorinated alcohol as the solvent (Scheme 5.15).[50]

The intramolecular vinylation of alcohols has been studied in detail by Li and coworkers, who demonstrated the feasibility of the process for the formation of four- **38**, five- **40**, and six- **42** membered cyclic enol ethers, respectively, in excellent, good, and fair yields. An unusual preference for a 4-*exo*-ring closure in this cyclization was also observed with dibromides such as **43**

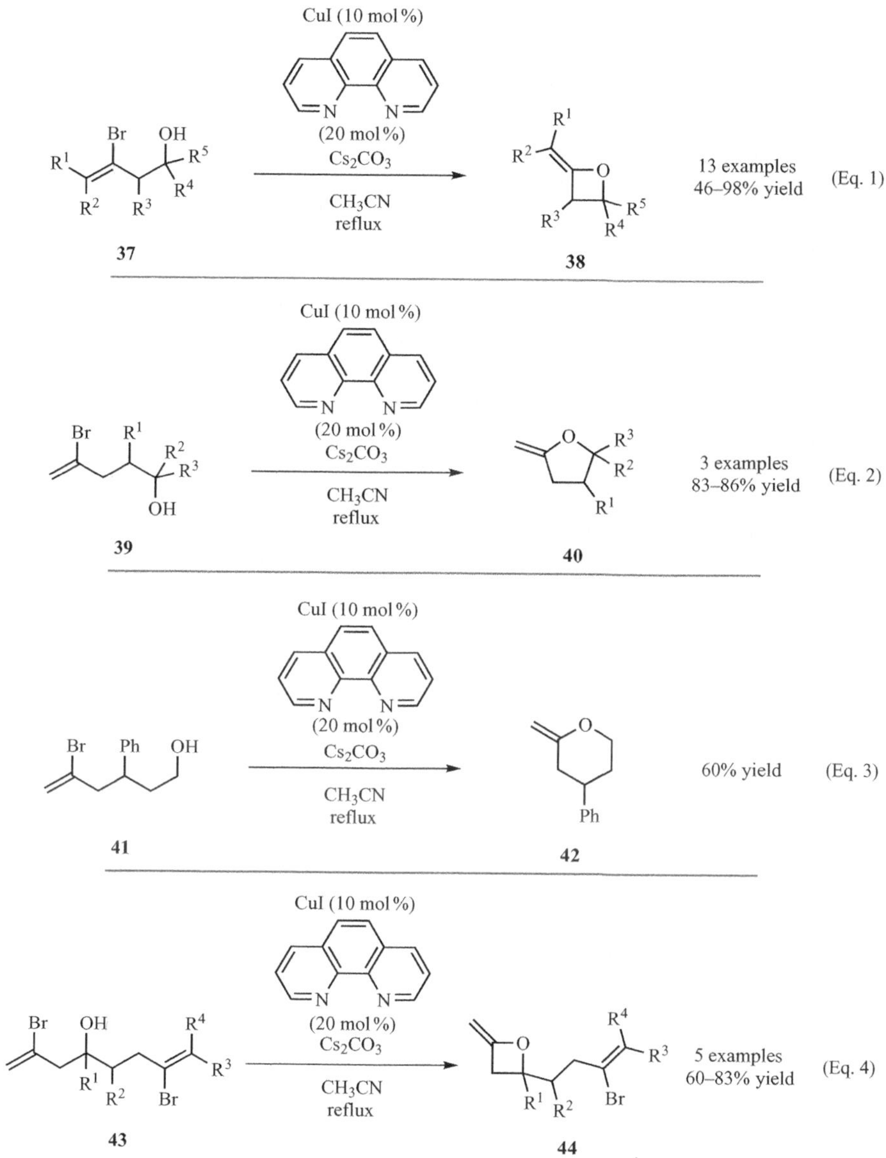

Scheme 5.16 Synthesis of cyclic enol ethers by intramolecular vinylation of alcohols.

(Scheme 5.16), which probably comes from a preferred formation of an intermediate five-membered ring metallacycle compared to a six-membered one. In this case, the differences between palladium(0) and copper(I) catalysis are truly remarkable since cyclization to a four-membered ring is not a favored process with palladium: A reversal of selectivity is observed when switching metals.[51]

β-Ketoesters can also be excellent reaction partners in O-vinylation reactions provided that they are performed in an intramolecular fashion. In place of the Hurtley-type reaction typically observed at the acidic carbon center with intermolecular couplings, O-vinylation leading to heterocyclization can be a preferred pathway depending on the substitution pattern of the starting material. The first copper-catalyzed intramolecular O-vinylation of carbonyl compounds was reported in 2005 by Fang and Li, who demonstrated that when reacted with copper iodide and DMEDA in refluxing THF, β-ketoesters such as **45** possessing a 2-bromo/chloro allyl substituent furnished five-membered cyclic enol ethers **46** without any competing C-vinylation of the activated methylene (Scheme 5.17, Eq. 1). With cyclic enol ethers being useful building blocks and constituents of biologically active natural products, this reaction has great potential. It was also applied to the synthesis of six- and seven-membered ring systems **48** by *endo* cyclization processes (Scheme 5.17, Eq. 2).[52] With the intramolecular O-vinylation being especially favored, it was later found that the need for an especially acidic hydrogen such as with ketoesters was not a strict requirement since ketones such as **49** could also be cyclized to polysubstituted furans **50** in excellent yields (Scheme 5.17, Eq. 3).[53] Vinyl chlorides could also participate in this intramolecular reaction but required twice the amount of catalyst. Finally, the intramolecular cyclization via a 6-*endo*-like mode from **51** was also studied, and proceeded smoothly to give the corresponding pyran **52** in 94% yield (Scheme 5.17, Eq. 4).

An interesting alternative to organic halides lies in the use of the corresponding boronic acids, as nicely demonstrated by Chan, Evans, and Lam and extensively described in Chapter 4. However, while the use of aryl boronic acids is well documented, detailed studies dealing with the synthesis of enol ethers by cross coupling with alkenyl boronic acids have been reported only recently. While the original publications by Lam and coworkers demonstrated the feasibility of the copper-mediated arylation of phenols with alkenyl boronic acids, only two examples were reported in 52% and 67% yields using stoichiometric amounts of copper(II) acetate.[35] While these pioneering reports spurred numerous subsequent investigations into couplings of vinylboronic acids with heteronucleophiles, these protocols do not include simple alcohols. An especially useful solution based on the direct cross coupling of alkenyl boronates was reported in 2010. Besides avoiding the generation of the corresponding boronic acids, the use of alkenyl boronates is especially appealing since they can be easily obtained by hydroboration of the corresponding terminal alkynes. While initial studies demonstrated that the oxidative coupling could be smoothly promoted by 2 equiv of copper(II) acetate, and using the alcohol as the solvent at room temperature,[54] it was later found that hex-3-yne was an especially efficient π-ligand for this system that considerably enhanced the yields, which allowed the use of only 3 equiv of the alcohol in dichloromethane (Scheme 5.18).[55] A substoichiometric system using 50 mol % of cupric acetate could also be applied to the synthesis of enol silyl ethers, provided that 1 equiv of pyridine *N*-oxide was used together with molecular oxygen.[56]

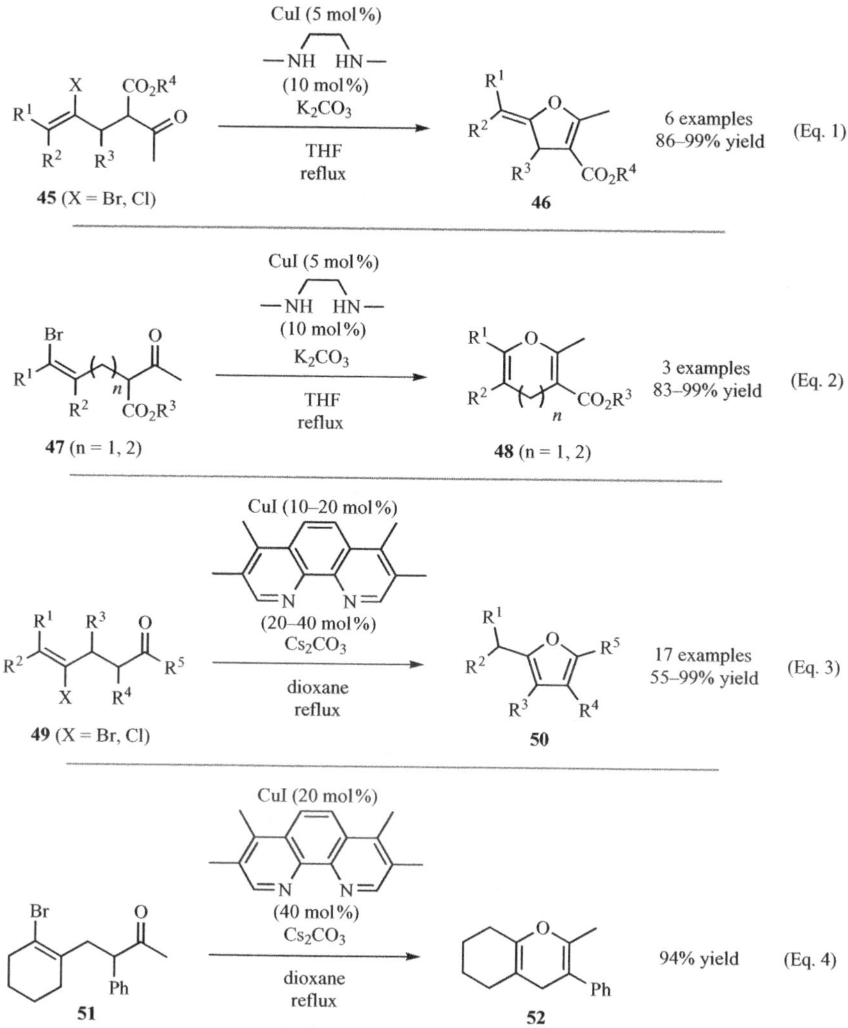

Scheme 5.17 Ketones as internal reaction partners in intramolecular O-vinylation reactions.

Potassium alkenyltrifluoroborates **54** have also been shown to be useful reagents for the vinylation of alcohols, though the scope of the reaction was not fully studied (Scheme 5.19, Eq. 1),[57] and 2,4,6-trivinylcyclotriboroxane **57** seems to be an especially useful reagent for the vinylation of phenols (Scheme 5.19, Eq. 2).[58] An especially astonishing use of alkenyl boronic acids **60** was reported in early 2011 for the direct vinylation of carboxylic acids **59**, heteronucleophiles that are quite reluctant to participate in copper-catalyzed cross-coupling reactions, and that yield rare vinyl esters **61** (Scheme 5.19, Eq. 3),[59a] compounds that can also be obtained using a remarkably efficient

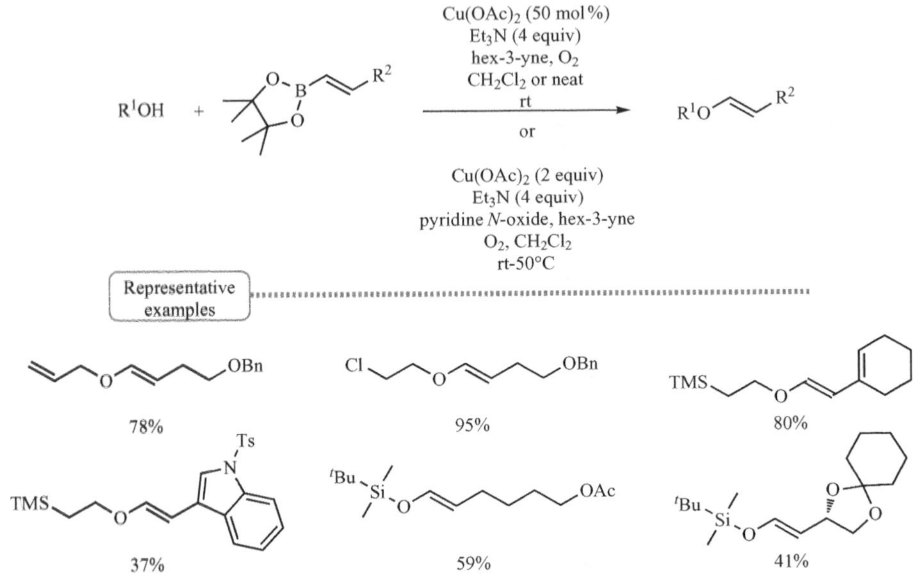

Scheme 5.18 Vinylation of alcohols with alkenyl boronates.

cross coupling between potassium alkenyltrifluoroborates and carboxylic acids, or their potassium salts, according to the procedure reported by Batey and coworkers in 2013, which is definitely one of the most efficient methods to prepare these useful building blocks (Scheme 5.19, Eq. 4).[59b] *N*-Hydroxyphthalimide has recently been shown to be smoothly O-vinylated using either stoichiometric or catalytic amounts of cupric acetate, pyridine, and sodium sulfate at room temperature in 1,2-dichloroethane in air to give the corresponding *O*-vinyl-hydroxyphthalimide derivatives which rearrange to α-hydroxyketones upon heating at 80–90°C.[60]

5.2.3 Synthesis of Vinyl Sulfides and Other Sulfur-Substituted Alkenes

Even though they are less studied than enol ethers, vinyl sulfides also play an important role as synthetic intermediates in organic chemistry. They are indeed used as enolate ions equivalents and are important intermediates for the preparation of a variety of building blocks including oxetanes, cyclopentanones, and cylopentanes. They are also found in various natural products and compounds of biological import.

Due to their importance, a number of methods for their synthesis have been reported, each with their respective limitations. The resurgence of copper catalysis has also had a deep impact on the synthesis of vinyl sulfides and related sulfur-substituted alkenes, overcoming the preconceived idea that copper-based catalysts and thiols would definitely not be a good match due to catalyst poisoning. Robust methods for the alkenylation of sulfur nucleophiles

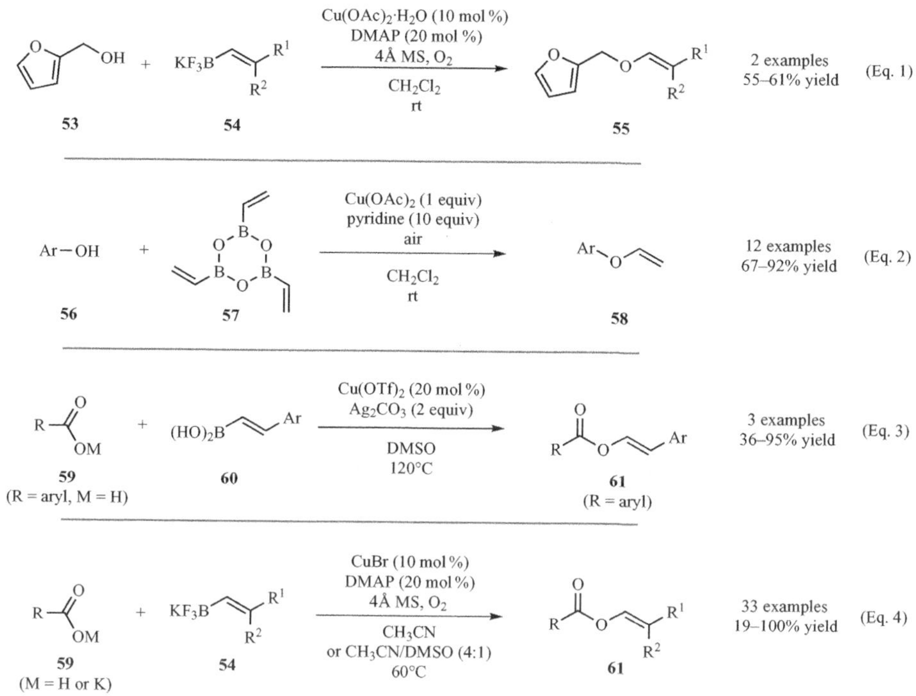

Scheme 5.19 Further uses of boron derivatives in the synthesis of vinyl ethers and esters.

have emerged since 2004 (shown in Table 5.3) and will be reviewed in this section. Methods for the preparation of vinyl selenides, whose synthesis by copper-mediated alkenylation protocols has been by far less studied, will also be covered in this section.

One of the most important and reliable catalytic systems for the preparation of vinyl sulfides by copper-catalyzed cross coupling between thiols and alkenyl halides was reported in 2004 by Venkataraman and coworkers using commercially available $[Cu(phen)(PPh_3)_2]NO_3$ (Table 5.3, entry 1).[61] The reason for the particular efficiency of this complex is its excellent solubility in organic solvents, which is in sharp contrast with most copper-based systems for which only a small amount is actually available in solution. Using 5 mol % of this preformed complex, a variety of thiols, including sterically hindered alkylthiols, thiophenols, and heteroaromatic thiols, could be successfully coupled with various representative iodoalkenes, including some *Z*-iodoolefins. While this catalytic system is still one of the most efficient ones to date, one of its only drawbacks is that its handling does require a glove box, which renders this method a bit less user-friendly.

An attractive alternative based on a most simple catalytic system was reported recently: *cis*-1,2-cyclohexane-diol, an efficient bidentate O-donor

TABLE 5.3 Copper-Catalyzed Alkenylation of Thiols

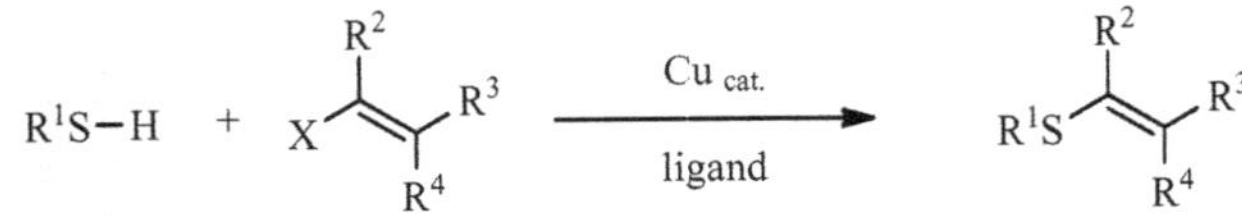

Entry	Thiols	Alkenyl Halides	Catalytic Systems	Conditions	Number of Examples	Yields	Ref.
1	Primary, secondary and tertiary thiols, thiophenols, heteroaromatic thiols	*E*-Styryl iodide, *E*-and *Z*-β-iodoacrylates, *E*-and *Z*-iodoalkenes	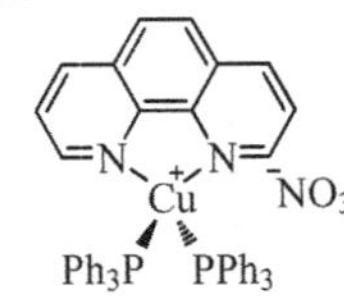 (5 mol %)	K_3PO_4, toluene, 110°C	31	80–99%	[61]
2	Primary and secondary thiols, thioalcohols, thiophenols, heteroaromatic thiols	*E*-Styryl bromide, *E*-styryl iodides, *Z*-β-iodoacrylates, *E*-cyclohexylvinyl iodide	CuI (10 mol %), 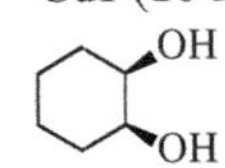 (20 mol %)	K_3PO_4, DMF, 30–50°C	33	84–98%	[62]
3	Primary and secondary thiols, thiophenols	*E*-styryl bromides, *E*-styryl iodides, *Z*-β-iodoacrylates	CuO nanoparticles (1.5 mol %)	KOH, DMSO, 80°C	19	49–97%	[63]
4	Benzylthiol, thiophenols	*E*- and Z-styryl bromides	CuI (15 mol %), 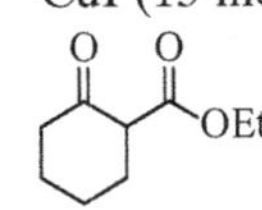 (30 mol %)	Cs_2CO_3, NMP, 60°C	7	75–92%	[33]
5	Benzylthiol, 2-mercaptoethanol, thiophenols	*E*-Styryl bromides	CuBr (10 mol %), 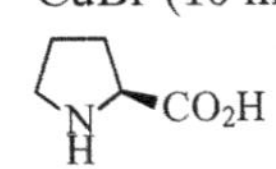 (20 mol %)	K_2CO_3, [Bmim]BF_4, 110–120°C	16	75–95%	[64]

ligand, facilitates the copper-catalyzed cross coupling of alkyl, aryl, and heteroaryl thiols with either alkyl, aryl, and heteroaryl-substituted vinyl iodides (Table 5.3, entry 2). This catalytic system is extremely versatile and provides an efficient entry to a wide range of sulfides, including electron-rich substituted ones. Vinyl bromides seem to perform well under the reaction conditions, while alkenyl chlorides, tosylates, triflates, and trifluoroborates do not participate in the cross coupling.[62] A ketoester (Table 5.3, entry 4) was also found to promote the reaction,[33] which could even be performed under ligandless conditions with CuO nanoparticles at the expense of harsher conditions, since this reusable system requires the use of potassium hydroxide in DMSO (Table 5.3, entry 3).[63] Finally, other reusable systems are based on a combination of an imidazolium-based ionic liquid with proline, which can be either generated *in situ*[64] or preformed[65] (Table 5.3, entry 5), and give moderate-to-good yields for the alkenylation of thiophenols with (*E*)- and (*Z*)-alkenyl bromides. As a note, vinyl selenides could also be obtained in the last case by cross coupling with diphenyl diselenide in the presence of zinc powder. The scope of these last three reactions, however, was not extensively studied, and the first two catalytic systems might be the first ones to try when attempting to optimize a certain thiol alkenylation reaction. It should finally be mentioned that the synthesis of vinyl sulfides in 95% ethanol was reported using L-proline as the ligand, but the only iodoalkene used in this study was 3-iodo-cyclohexenone. A competing pathway leading to the vinyl sulfide product and relying on an uncatalyzed Michael addition/elimination sequence might therefore also be operating in this case.[66]

Two completely different approaches to styryl sulfides were reported in 2010 and 2012 starting from arylpropiolic acids and arylacetylenes, respectively.[67] Decarboxylative cross coupling in the presence of 4 mol % copper(I) iodide and cesium carbonate in NMP at 90°C allowed for a clean alkenylation of alkylthiols and thiophenols. The (*Z*)-isomer was found to be predominantly formed together with typically 10% of the (*E*)-isomer in the case of thiophenols, while the use of alkylthiols resulted in a dramatic decrease in the stereoselectivity (Scheme 5.20, Eq. 1).[67] More readily available arylacetylenes were

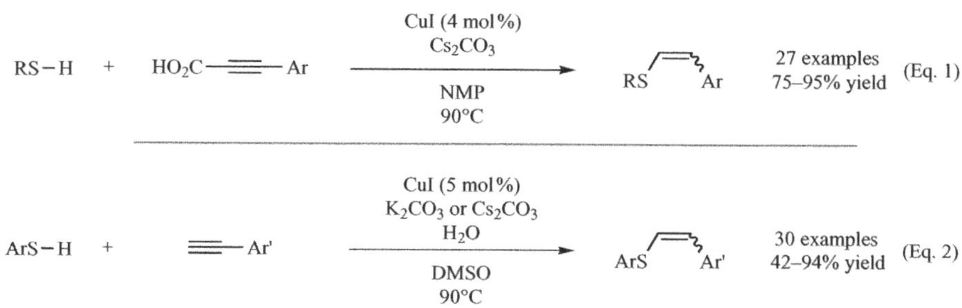

Scheme 5.20 Synthesis of alkenyl sulfides by decarboxylative cross coupling with arylpropiolic acids and reductive cross coupling with arylacetylenes.

also shown to be remarkably efficient starting materials for the preparation of (*Z*)-styryl sulfides, which could be readily obtained with reasonable selectivities (*E*/*Z* ratio in the 22/78 range) and synthetically useful yields by reaction of the corresponding arylacetylenes and thiophenols in the presence of copper(I) iodide, potassium carbonate, and water in DMSO at 90°C (Scheme 5.20, Eq. 2).[67] An interesting reversal of stereoselectivity was observed when *N*,*N*,*N′*,*N′*-tetramethylenediamine and carbon dioxide were added to the reaction mixture, although a single example was reported.

Altogether, these copper-mediated syntheses of alkenyl sulfides have found various applications in organic synthesis, as exemplified with the synthesis of thietanes **63**, tetra- and dihydrothiophenes **65**, and thiopyran **67** by intramolecular S-vinylation from precursors such as **62**, **64**, or **66** (Scheme 5.21, Eq. 1, Eq. 2, and Eq. 3),[68] or with the de novo synthesis of carbapenem derivatives

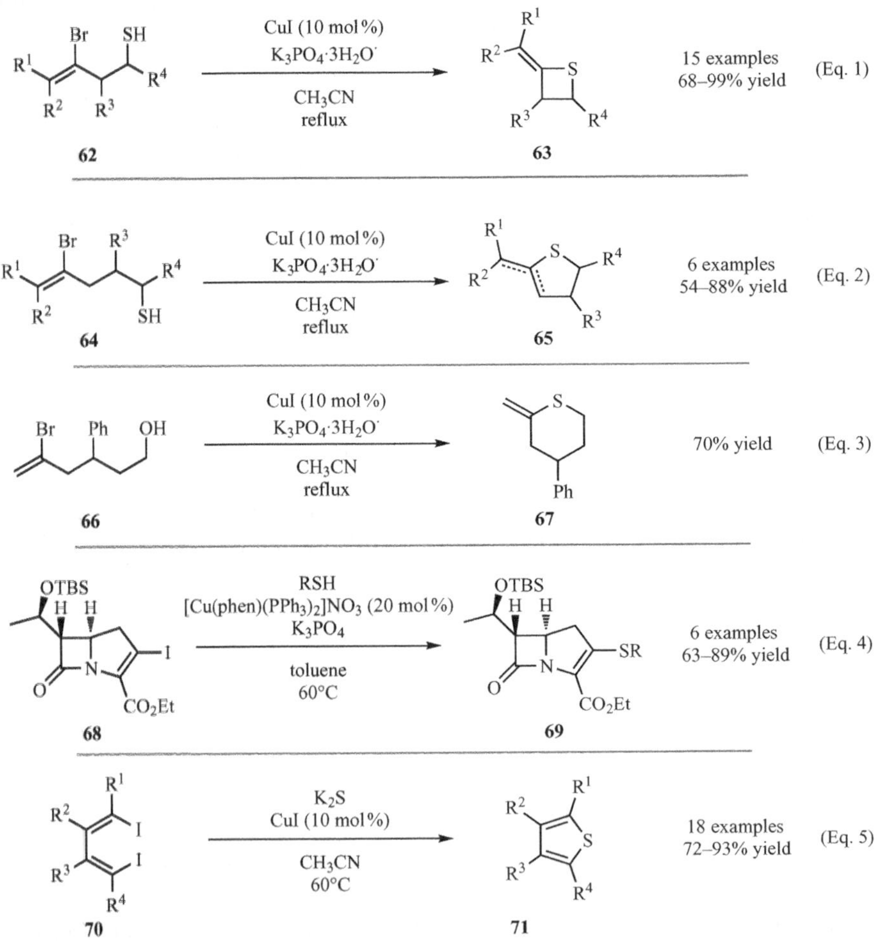

Scheme 5.21 Representative synthetic applications of the intramolecular copper-catalyzed S-vinylation.

69 by copper-mediated sulfurization of the corresponding iodoalkene **68** (Scheme 5.21, Eq. 4).[19] In the last case, it ought to be mentioned that the cyclic enamide was also installed using a copper-catalyzed reaction and that a one-pot cyclization by C–N bond formation/functionalization by C–S bond formation was also possible, starting from the corresponding vinyl bis-iodide. Finally, a double alkenylation of potassium sulfide with 1,4-diiodo-1,3-dienes **70** has been reported in 2010 (Scheme 5.21, Eq. 5).[69] This method provides a facile route to di-, tri-, and tetrasubstituted thiophenes **71** in excellent yields, and this transformation is distinguished by its simplicity and functional group tolerance.

While the use of Chan–Lam–Evans type coupling with vinylboronic acids or equivalents has resulted in the development of general and reliable methods for the synthesis of enamides and enol ethers, its use with sulfur nucleophiles has been less studied, with only isolated examples found in the literature. This is actually quite logical and is mostly due to the easy oxidation of thiols to disulfides under the oxidative conditions required in Chan–Lam–Evans coupling reactions. Thiols are indeed easily oxidized by copper(II) salts, yielding disulfides and copper(I). Based on this simple observation, several catalytic systems have been developed for the coupling of boronic acids with disulfides and have been successfully applied to the synthesis of vinyl sulfides, although the scope of these reactions has not been extensively studied. In this context, diphenyldisulfide **72** has been shown to be readily cleaved and vinylated to **74** upon treatment with the corresponding alkenyl boronic acid **73** and catalytic amounts of copper iodide and bis-pyridine in aqueous DMSO and in the presence of air (Scheme 5.22, Eq. 1).[70]

An especially interesting alternative to the use of disulfides was found with aryl-*N*-thioimides **75** in the presence of catalytic quantities of copper methylsalicylate.[71] The reaction with alkenyl boronic acid **73** takes place in the absence of a base and under mild conditions and yields the corresponding vinyl thioether **76** (Scheme 5.22, Eq. 2). The stereochemistry of the starting boronic acid is a crucial parameter since the (*Z*)-isomer of **73** did not furnish the corresponding vinyl sulfide.

Finally, sodium sulfinate salts were shown in 2007 to be excellent partners for cross-coupling reactions with alkenyl boronic acids and yielded the corresponding vinylsulfones **78** (compounds of great importance in medicinal chemistry and organic synthesis) with remarkable efficiency (Scheme 5.22, Eq. 3).[72] 1,10-Phenanthroline[72] and 1-benzylimidazole[73] display similar activities as ligands for copper(II) in this transformation.

5.2.4 Synthesis of Vinylphosphonates and Phosphine Oxides

While the formation of carbon–nitrogen, carbon–oxygen, and carbon–sulfur bonds by copper-catalyzed cross-coupling reactions from alkenyl halides and related systems has been the subject of extensive investigations in recent years, limited protocols for direct coupling with phosphorus nucleophiles have been published, and examples of copper-catalyzed vinylation of phosphines and

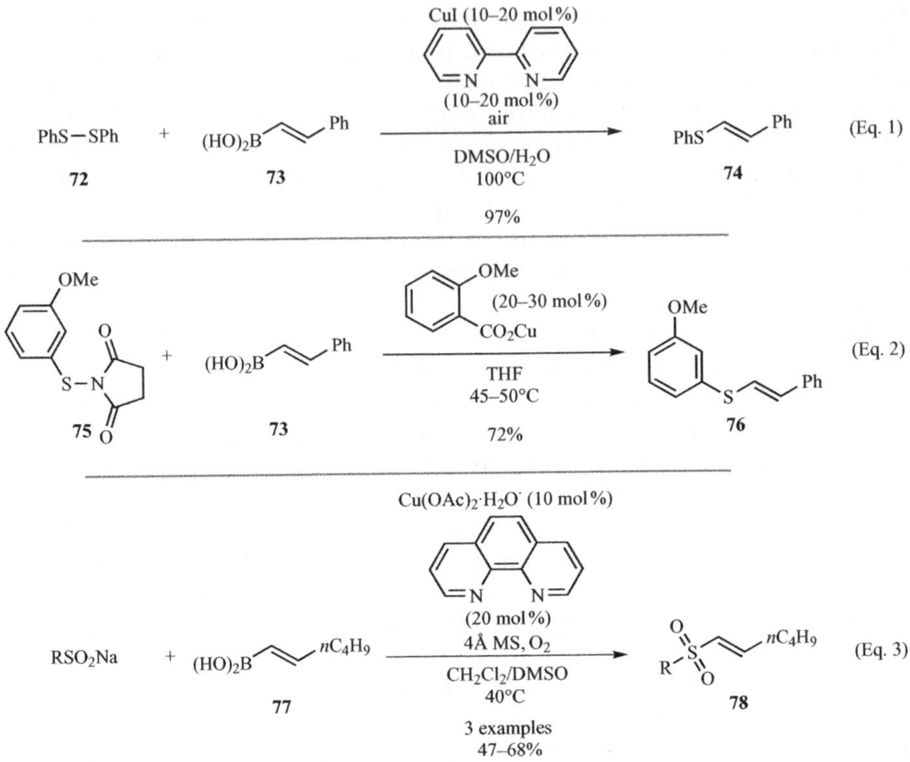

Scheme 5.22 Synthesis of vinyl sulfides and sulfones from alkenyl boronic acids.

related nucleophiles are still rare. Most examples have in fact been reported using dialkylphosphites as nucleophiles, which might simply be attributed to their greater stability compared with dialkylphosphines.

These studies have, however, resulted in the development of efficient syntheses of alkenyl phosphonates, compounds that are routinely used as building blocks for the introduction of remote phosphonate functional groups or to prepare biologically active molecules. Buchwald's copper iodide/DMEDA system was found to be especially efficient to promote the cross coupling between a wide range of alkenyl iodides and bromides and dibutyl phosphite (Scheme 5.23).[74] A variety of vinylphosphonates could be obtained in excellent yields under these conditions, including highly substituted ones and 1-alkyl-vinylphosphonates. The authors demonstrated that the reaction proceeds stereoselectively and with retention of the original stereochemistry. For example, only a negligible amount of the *E* stereoisomer was detected when *Z* vinyl iodides were used under the reaction conditions, which is of great value compared to standard methods for the synthesis of alkenyl phosphonates. The reactivity of diphenylphosphine was also evaluated: The corresponding vinylphosphine was isolated in good yield, and this method might be

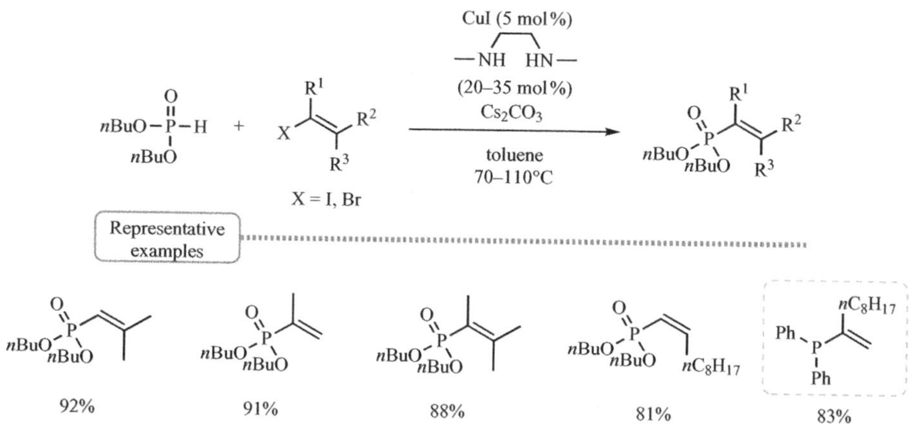

Scheme 5.23 Copper-catalyzed synthesis of vinylphosphonates from dialkylphosphites and alkenyl halides.

applicable to more sensitive dialkylphosphines since they were shown to be smoothly arylated under similar conditions.

For the synthesis of (*E*)-alkenyl phosphonates, an interesting alternative to the cross coupling with vinyl bromides was found in the use of *gem*-dibromoalkenes. This reaction has been shown to be general and provides a straightforward entry to 1-alkenylphosphonates that is complementary to previously reported synthetic routes. This method, which allows for a formal substitution of both bromine atoms, has the advantage of starting from readily available 1,1-dibromo-1-alkenes which act as attractive synthetic equivalents of (*E*)-alkenylbromides, and might be an attractive option for the synthesis of highly substituted and/or complex vinylphosphonates (Scheme 5.24).[75] As a note, the exclusive formation of vinylphosphonates from dibromoalkenes is in sharp contrast with results obtained with palladium catalysis since alkynylphosphonates are predominantly formed upon reaction of alkenyl-dibromides and dialkylphosphites.[76] The present method therefore offers a complementary catalytic system that allows the selective formation of (*E*)-alkenylphosphonates.

While the Chan–Lam coupling has not been extended to date to the synthesis of phosphorus-substituted alkenes, which can, however, be done using palladium catalysis and alkenyl boronate esters,[77] a room-temperature, base-free synthesis of vinylphosphonate by reaction of dialkylphosphites with vinyl-iodonium tetrafluoroborates has been reported (Scheme 5.25). This procedure is especially efficient for the preparation of complex and sensitive alkenyl phosphonates, and its only limitation seems to be that only aryl-substituted vinyliodonium salts can be used in this cross coupling, other ones being a lot more difficult to prepare.[78]

Cinnamic acids have also been recently introduced as partners in cross-coupling reactions involving P-nucleophiles, as demonstrated in 2011 by Liang,

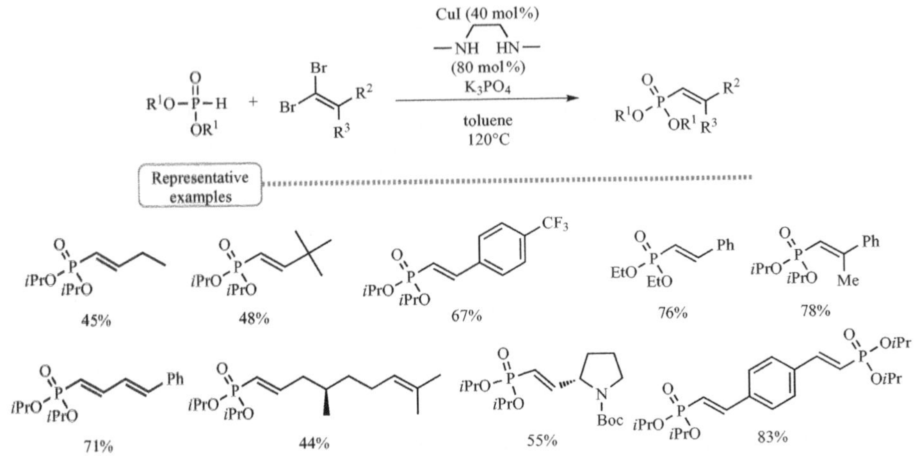

Scheme 5.24 Copper-catalyzed synthesis of vinylphosphonates from dialkylphosphites and vinyl dibromides.

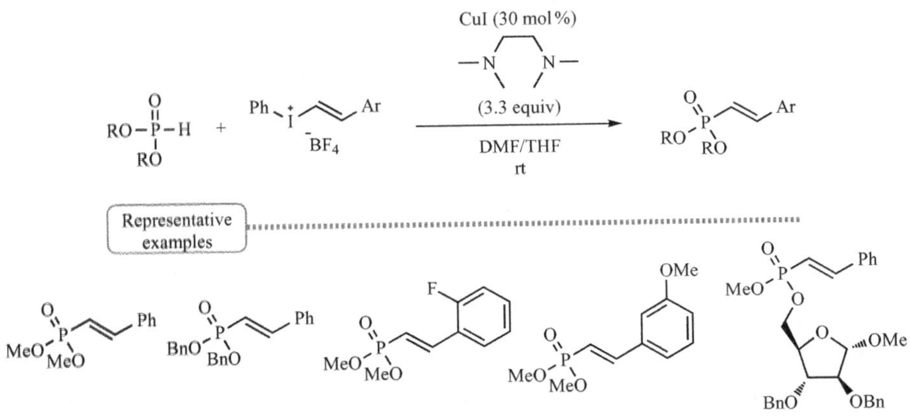

Scheme 5.25 Copper-catalyzed alkenylation of dialkylphosphites with vinyliodonium tetrafluoroborates.

Yang, and coworkers, who demonstrated the feasibility of the copper-catalyzed, silver-assisted decarboxylative cross coupling of cinnamic acid derivatives with diphenylphosphine oxide, diisopropylphosphite, and ethyl phenylphosphinate (Scheme 5.26).[79] While this provides an interesting entry to vinylphosphine derivatives, the scope of the reaction is, however, somehow limited since other alkenoic acids failed to undergo the cross coupling, which is therefore limited to the synthesis of styryl derivatives.

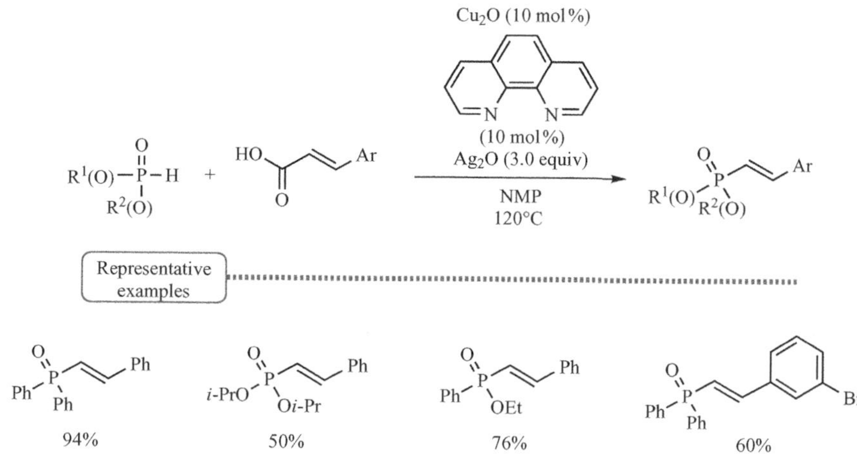

Scheme 5.26 Copper-catalyzed decarboxylative alkenylation of various P-nucleophiles with cinnamic acids.

5.3 ALKYNYLATION OF HETERONUCLEOPHILES: THE EMERGENCE OF GENERAL METHODS FOR THE SYNTHESIS OF HETEROATOM-SUBSTITUTED ALKYNES

Heteroatom-substituted acetylenes most certainly represent the most useful and versatile class of alkynes. The polarization of the triple bond generated by the presence of the heteroatom allows for highly regio- and stereoselective transformations, and these building blocks have recently been used for the development of remarkably efficient chemical transformations. They are also emerging as useful pharmacophores for the design of original compounds of pharmaceutical import.

While these compounds clearly hold a great potential in organic synthesis, their painful, stepwise, and low-yielding syntheses have clearly dramatically hampered their development, which is fairly logical: What would be the point in developing an especially elegant and efficient process from these compounds if their synthesis cannot be done on a gram scale, or is way too painful? Copper catalysis has revitalized this area, which is obvious with the chemistry of ynamides, and copper-mediated processes are now the standard (if not the classical) procedures for the synthesis of a wide range of heteroatom-substituted alkynes. This section will cover efforts in this direction, starting with ynamines/ynamides.

5.3.1 Bringing Ynamides into the New Millennium: Copper-Mediated Alkynylation of Nitrogen Nucleophiles

Because of their superior thermal stability compared to ynamines, ynamides are clearly emerging as important and versatile synthetic intermediates, and have received considerable attention over the past decade.[80] They have been shown only quite recently to be excellent substrates for ring-closing metathesis, palladium-mediated coupling reactions or rearrangements, cycloisomerizations and [2 + 2] cycloadditions, radical transformations, Pauson–Khand cyclization, diastereoselective sigmatropic rearrangements, yne-carbonyl metathesis, carbometalations, and many other useful transformations. In addition, they have been used as efficient synthons for the preparation of a wide array of molecules such as keto-imides, amino-indoles, oxazolones, carbazoles, triazoles, and benzofurans (an exhaustive list would necessitate an entire page). Up to the beginning of the twenty-first century, the research in this area was still limited, mostly because of the limited availability of these compounds whose synthesis relied either on stepwise, low-scope syntheses or on the isomerization of propargyl amides. Recent breakthroughs in the preparation of these substrates using copper-mediated processes have revitalized the chemistry of nitrogen-substituted alkynes. The beginning of the twenty-first century has witnessed an ever-increasing number of publications reporting the development of new reactions or synthetic sequences starting from ynamides.

A major breakthrough in the synthesis of ynamides was reported in 2003 by the Hsung group. Inspired by the renaissance of copper catalysis and by the work of Buchwald on the arylation of amides, Hsung and coworkers first developed a copper-catalyzed coupling between alkynyl bromides and amides using DMEDA as ligand, which clearly provided an improved synthetic access to ynamides over existing protocols (Table 5.4, entry 1).[81] However, severe limitations remained, such as the use of high temperatures and low substrate scope. Although oxazolidinones were useful in the coupling, amides were mostly poor substitutes, and sulfonamides were not suitable at all. In addressing this limitation, Danheiser and coworkers reported a useful solution using stoichiometric amounts of copper iodide along with KHMDS: This protocol allows reactions to proceed at room temperature with carbamates and sulfonamides but still requires the use of a strong base (Table 5.4, entry 2).[82] A general and mild procedure was finally published in 2004 by Hsung, who reexamined his coupling protocol by screening a variety of copper sources and ligands. The use of copper sulfate pentahydrate in combination with 1,10-phenanthroline proved to be especially successful, allowing the reaction to occur at temperatures ranging from 65 to 100°C using potassium phosphate as the base (Table 5.4, entry 3).[83] A wide range of nitrogen nucleophiles are readily alkynylated under these conditions, including reluctant ones such as acyclic secondary amides and bulky carbamates,[84] and they have been recently applied to the synthesis of *N*-phosphoryl ynamides,[85a] a new class of ynamides. Secondary amines can also be used as reaction partners for the cross coupling with bromoalkynes, as reported by Jiang and coworkers, but the intermediate ynamines that would be formed cannot be isolated and dimerize to naphthalene-1,3-diamines, therefore considerably reducing the usefulness of the process.[85b] From a practical point of view, it should be noted that it was later shown that the quality of potassium phosphate is crucial for the success of the reaction.[86]

TABLE 5.4 Copper-Mediated Synthesis of Ynamides from Haloalkynes

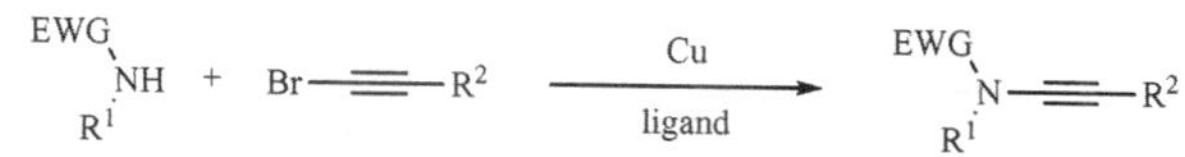

Entry	Nitrogen Nucleophiles	Catalytic Systems	Conditions	Number of Examples	Yields	Ref.
1	Oxazolidinones, cyclic secondary amides, imidazolidinones, carbamates	CuCN (5 mol %),  (10 mol %)	K_3PO_4, toluene, 110°C	23	10–85%	[81]
2	Oxazolidinones, imidazolidinones, sulfonamides, carbamates	CuI (1 equiv)	KHMDS, pyridine, rt	13	40–76%	[82]
3	Oxazolidinones, cyclic and acyclic secondary amides, imidazolidinones, sulfonamides, carbamates	$CuSO_4 \cdot 5H_2O$ (5–20 mol %), 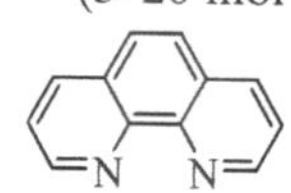 (10–40 mol %)	K_3PO_4, toluene, 65–100°C	44	37–98%	[83]
	Phosphoramidates, phosphordiamidates			17	28–95%	[85a]

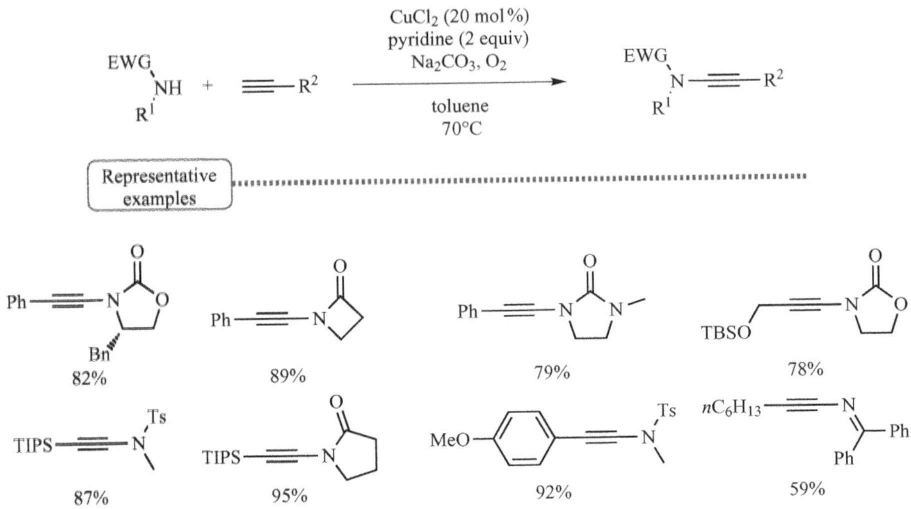

Scheme 5.27 Copper-catalyzed oxidative amidation of terminal alkynes.

These procedures are clearly among the most efficient to date. They require, however, the preparation of the starting bromoalkyne, which can be difficult in some cases, even if the vast majority of these compounds are typically formed in excellent yields by bromination of the corresponding terminal alkynes.

Following an early isolated report from the Dow Chemical Company, which reported the cupric acetate-catalyzed oxidative cross coupling between phenylacetylene and dimethylamine,[87] an especially elegant procedure that overcomes this limitation was reported by the Stahl group in 2008 and relies on an oxidative alkynylation of amides with terminal alkynes (Scheme 5.27).[88] Using this protocol, ynamides could be obtained in excellent yields, even on a large scale, by simply using oxygen as the terminal oxidant. Limitations of this procedure, which has been recently extended to the alkynylation of imines, yielding to ynimines,[89a] and sulfoximines,[89b] can be found in the use of 5 equiv of the nitrogen nucleophile, which is required to reduce the amount of Glaser–Hay dimerized products, and in the low reactivity of carbamates, acyclic amides, and ureas. Following this remarkable report, alternative catalytic systems have been reported for this oxidative alkynylation.[90] They include the use of a combination of copper(II) hydroxide and potassium bicarbonate, which allows the reduction of the equivalents of the nitrogen nucleophiles used to two or three together without the need for a slow addition of the alkyne,[90a] the copper(II) oxide/potassium chloride/4-phenylpyridine system in toluene at 80°C, which supposedly favors the formation of ynamides (although only arylacetylenes have been used in this study),[90b] and a magnetically separable copper carboxylate metal-organic framework, which could be reused five times with similar efficiencies.[90c] As a note, an efficient alternative practical

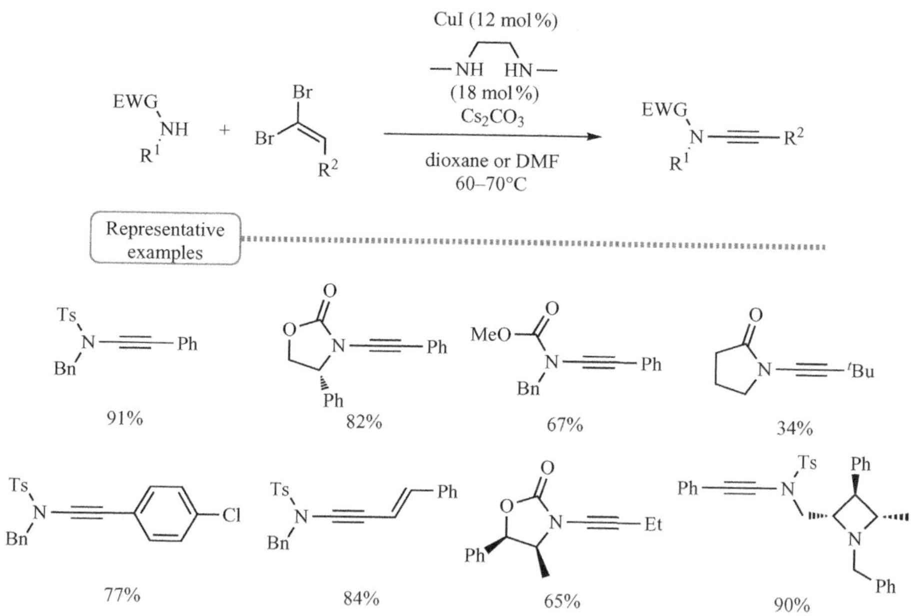

Scheme 5.28 Copper-catalyzed alkynylation with vinyl dibromides.

procedure involving the room-temperature, base-free oxidative alkynylation with preformed copper acetylides was reported in 2012.[91]

An efficient alternative to these procedures is based on the reaction of nitrogen nucleophiles with vinyl dibromides, which act as synthetic equivalents of bromoalkynes and readily available alkynylating agents (Scheme 5.28). This procedure, which has been applied to the preparation of more than a hundred ynamides to date, was found to be rather general, even on a large scale and with complex/sensitive substrates.[92] The choice of the base turned out to be crucial to avoid further reaction between the N-nucleophile and the formed ynamide yielding to the formation of ketene *N,N*-acetals.[13] The major limitation lies in the nucleophiles susceptible to undergo the cross coupling. While sulfonamides, oxazolidinones, and pyrrolidinone were found to be excellent substrates, acyclic secondary amides or ureas do not give the corresponding ynamides. Extension of this procedure to the copper-catalyzed cross coupling of less available 1,2-dibromostyrenes with sulfonamides was reported later.[93]

Besides alkynyl bromides, terminal alkynes, or *gem*-dibromoolefins, other reagents have been reported in 2010 and 2011 for the alkynylation of various nitrogen nucleophiles. To provide basic guidelines, the choice of the proper alkynylating agent depends on the following: the availability of the starting materials, the stoichiometry of the reaction (i.e., nitrogen nucleophile or alkynyl transfer agents used in excess), the ease of dimerization of the alkynylating agent (the homocoupling leading to symmetrical diynes can be a serious side reaction in some cases), the propensitivity of the formed ynamide to

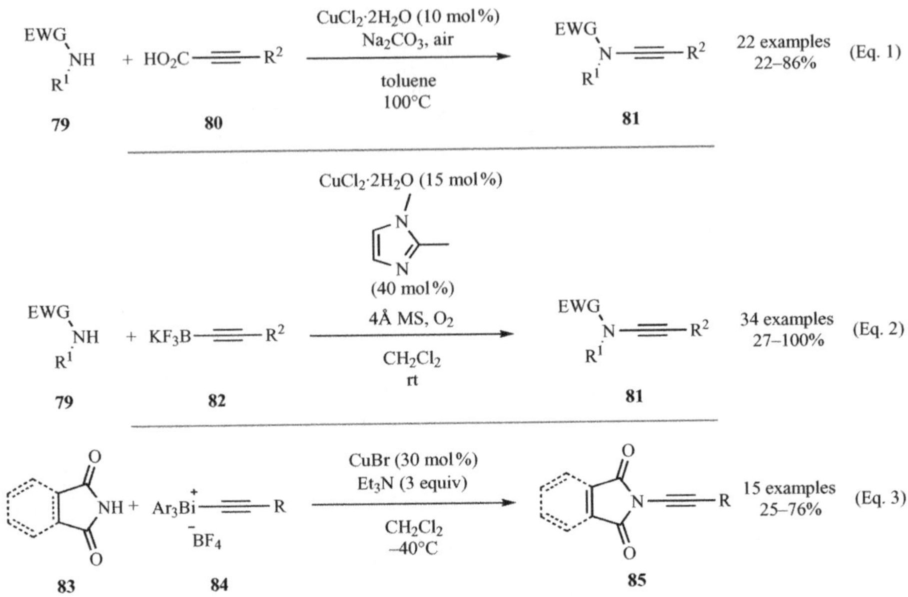

Scheme 5.29 Propiolic acids, potassium alkynyltrifluoroborates, and alkynyl(triaryl) bismuthonium salts as alternative reagent for the synthesis of ynamides.

further react with HBr leading to the formation of bromoenamides (which can be observed in rare cases when bromoalkynes or vinyl dibromides are used), and the reaction conditions, compatible or not with sensitive functional groups (thermal activation, need for a base, polar or nonpolar solvent, etc.). In this context, an interesting oxidative amidation of propiolic acids **80** under air via decarboxylative cross coupling has been reported in 2010. Upon heating the corresponding alkyl-, alkenyl-, or aryl-substituted propiolic acid **80** and the proper nitrogen nucleophile **79** in the presence of catalytic amounts of copper(II) chloride and sodium carbonate in toluene at 100°C in an open flask and in the presence of additional pyridine, in some cases, the corresponding ynamides **81** are obtained in moderate-to-good yields (Scheme 5.29, Eq. 1).[94] Potassium alkynyltrifluoroborates **82**, perfectly stable and easy to handle synthetic equivalents of the unstable alkynylboronic acids, were also shown to be excellent reaction partners that allow for a room-temperature, base-free synthesis of ynamides with a combination of copper(II) chloride and 1,2-dimethylimidazole (Scheme 5.29, Eq. 2).[95] The choice of a solvent with limited solubility for the trifluoroborate salt **82**, dichloromethane, is critical to the success of the reaction for minimizing its dimerization. A practical advantage of this procedure is that it does not require a slow addition of the alkynyl transfer agent and is compatible with various functional groups such as free alcohols or alkylchlorides. Finally, an especially efficient synthesis of ynimides **85**, synthetic equivalents of unstable and highly labile primary alk-1-ynylamines, was reported in 2011 from alkynyl(triaryl)bismuthonium tetrafluoroborates **84**, which act as remarkably efficient reagents for the synthesis of ynamides from phthalimide, succinimide, or maleimide **83**, since the reaction can be performed at –40°C (Scheme 5.29, Eq. 3). The major drawback in this case is that significant amounts of the product resulting from the competitive transfer of the aryl group of **84** can be observed in some cases, with this side reaction being minimized with *p*-tolyl of *p*-methoxy-phenyl groups.[96]

The development of copper-catalyzed reactions for the preparation of ynamides clearly provided efficient, straightforward, and reliable accesses to these useful building blocks. These procedures are without a doubt more convenient than the stepwise sequences and provided benchmarks for the development of the chemistry of ynamides. They have also been used for the preparation of various *N*-alkynyl heterocycles, which will be the subject of the following paragraphs.

While the arylation or vinylation of all types of π-excessive nitrogen heterocycles now allows for N-substitution with aromatic groups or double bonds with surgical efficiency as respectively described in Chapter 1, Section 1.4, and Chapter 5, Section 5.2.1, the alkynylation of these heterocycles has been less studied. Such a reaction provides a straightforward entry to *N*-alkynylheteroarenes, molecules that are an interesting variation on ynamines and that share with ynamides the increased stability engendered by delocalization of the nitrogen lone pair, and are emergent scaffolds in medicinal chemistry and in nonlinear optics. In addition to classical syntheses, which rely on elimination from halovinylheteroarenes, enol triflates, and trichlorovinyl derivatives, or coupling with alkynyl iodonium salts (methods that all suffer either limitations in scope or require stepwise installation of the triple bond), copper-catalyzed alkynylation of heteroarenes is emerging as an efficient and reliable tool for the preparation of *N*-alkynylheteroarenes. Most procedures described in the preceding paragraphs have been extended to the alkynylation of aromatic heterocycles, and representative procedures have been collected in Table 5.5.

Bromoalkynes have been shown to be efficient reaction partners for the arylation of pyrroles and indoles when catalytic amounts of copper(II) sulfate pentahydrate and 1,10-phenanthroline are used in combination with potassium phosphate in toluene at temperatures ranging from 70 to 85°C, yielding the corresponding *N*-alkynyl-pyrroles and indoles in excellent yield (Table 5.5, entry 1).[83] As a note, attempts at using this procedure with 2-hydroxy-pyridine resulted in the exclusive formation of the corresponding α-bromoenamide, resulting from a hydrobromination of the intermediate ynamide,[83] and the selective alkynylation of tryptamine derivatives was found to be strongly dependent on the protecting group of the tryptamine nitrogen atom. While alkynylation mostly occurred at the protected amine site when a sulfonamide was used as the protecting group, the use of a carbamate allowed for a clean and selective alkynylation of the indole.[83] The alkynylation of other azoles such as imidazoles, benzimidazoles, pyrazoles, and indazoles failed using these conditions and requires the use of copper iodide in combination with 2-acetylcyclohexanone (Table 5.5, entry 2),[97] with poly(ethylene glycol) 400 (PEG400; Table 5.5, entry 3)[98] as the ligand or ligandless conditions using a

TABLE 5.5 Copper-Catalyzed Alkynylation of Nitrogen Heterocycles

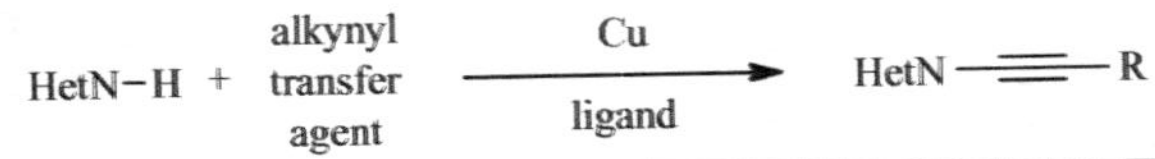

Entry	HetN-H	Alkynyl Transfer Agent	Catalytic Systems	Conditions	Number of Examples	Yields	Ref.
1	Pyrroles, indoles	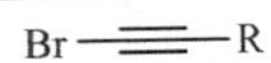	$CuSO_4{\cdot}5H_2O$ (5–20 mol %), 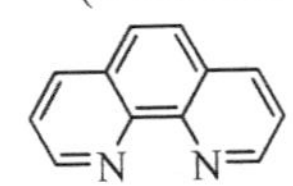 (10–40 mol %)	K_3PO_4, toluene, 70–85°C	9	50–93%	[83]
2	Imidazoles, benzimidazole		CuI (5 mol %), 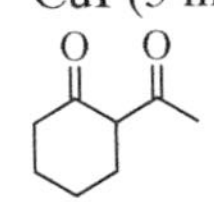 (20 mol %)	Cs_2CO_3, 4Å MS, dioxane, 50°C	15	15–90%	[97]
3	Imidazoles, benzimidazoles, pyrazoles, indazoles		CuI (5 mol %), PEG400	Cs_2CO_3, dioxane, 80°C (μW)	24	9–88%	[98]
4	Imidazoles, benzimidazole, pyrazoles		CuO (2 mol %)	KOH, dioxane, 80°C	16	63–85%	[99]

5	Benzimidazoles, indazoles	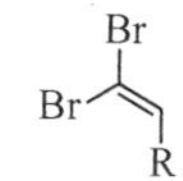	CuI (12 mol %),  (24 mol %)	Cs_2CO_3, dioxane, 80°C	19	21–97%	[100]
6	Benzimidazoles, indazoles	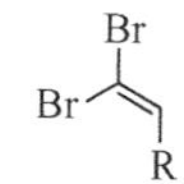	CuI (5 mol %), 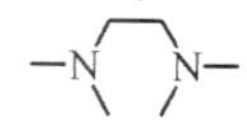 (10 mol %)	Cs_2CO_3, dioxane, 80°C	32	39–84%	[101]
7	Imidazoles, benzimidazole, pyrazoles	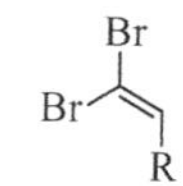	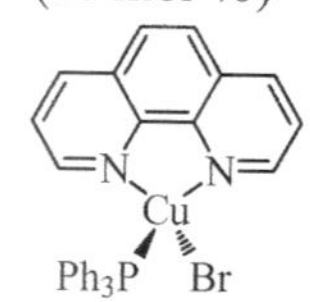 (24 mol %)	Cs_2CO_3, DMSO, 80°C	15	66–85%	[102]
8	Indoles	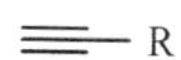	$CuCl_2$ (20 mol %)	Na_2CO_3, pyridine, O_2, toluene, 70°C	10	51–90%	[88]
9	Indoles	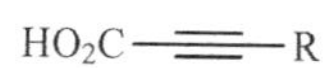	$CuCl_2{\cdot}2H_2O$ (10 mol %)	Na_2CO_3, pyridine, air, toluene, 100°C	4	39–68%	[96]

combination of copper(II) oxide and potassium hydroxide in 1,4-dioxane at 80°C (Table 5.5, entry 4).[99] Care should be taken during the alkynylation of these heterocycles with bromoalkynes since a competing direct nucleophilic addition of the nitrogen nucleophile to the alkynyl bromide yielding to heterocyclic β-bromo-enamines frequently occurs. The formation of this by-product can be avoided by simply using vinyl dibromides as alkynyl transfer agents, which were recently shown to be readily coupled with a set of benzimidazoles and indazoles to give the corresponding *N*-alkynylheteroarenes in moderate-to-good yields (Table 5.5, entry 5).[100] In the case of substituted azoles, the major product comes from a preferred alkynylation at the less hindered site, although minor amounts of the minor regioisomer are isolated in most cases. This procedure was later on extended to other azoles such as imidazoles using *N*,*N*,*N*′,*N*′-tetramethylethylenediamine (TMEDA) as the ligand (Table 5.5, entry 6),[101] which seems to be superior to DMEDA in some cases, and to the alkynylation of imidazoles and pyrazoles using a preformed soluble copper catalyst, [Cu(Phen)(PPh_3)Br] (Table 5.5, entry 7).[102]

Finally, it should be mentioned that terminal alkynes (Table 5.5, entry 8)[88] and propiolic acids (Table 5.5, entry 9)[96] were also successfully used for the alkynylation of indoles, but their use with other heterocycles, which is more challenging, has not been studied.

5.3.2 Copper-Mediated Synthesis of Ynol Ethers: The First Development

While the copper-mediated alkynylation of nitrogen nucleophiles has received considerable attention over the past decade and has clearly contributed to the development of the chemistry of ynamides, the use of oxygen nucleophiles has been left aside despite the importance of ynol ethers in organic synthesis and their limited availability. This is, however, quite understandable since most reagents that could be used for the development of a copper-mediated synthesis of ynol ethers, such as terminal alkynes or alkynyl halides, readily dimerize in the presence of copper salts in a polar cosolvent such as an alcohol.

Besides the copper-catalyzed preparation of *O*-alkynyl-sulfonates from alkynyliodonium sulfonates reported by Stang,[103] a general solution was reported by Evano and coworkers in 2012, who developed an efficient copper-catalyzed cross coupling between phenols and vinyl dibromides. In the presence of catalytic amounts of copper(I) iodide and 2,2-bipyridine, a clean and regioselective cross coupling is operative, yielding to intermediate α-bromoenol ethers, which immediately undergo a β-elimination yielding to the desired ynol ethers upon an *in situ* reaction with potassium *tert*-butoxide (Scheme 5.30).[13b,49] This reaction was found to be rather general and reliable, even on a large scale, and might provide an interesting alternative to the more classical synthesis based on a reaction of phenoxides with trichloroethene followed by elimination induced by treatment with a strong base.

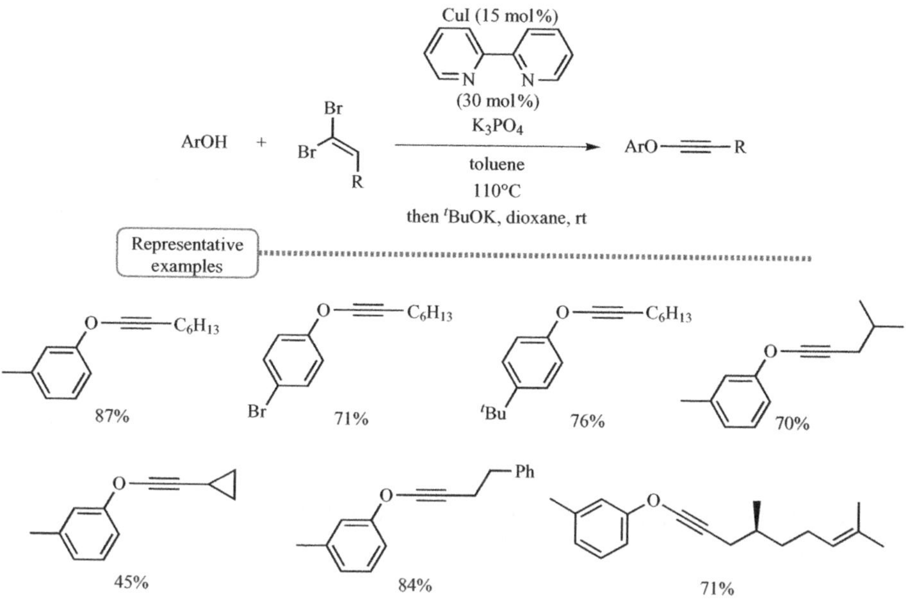

Scheme 5.30 Copper-catalyzed synthesis of ynol ethers.

5.3.3 Copper-Mediated Preparation of Alkynyl Chalcogenides

Organic chalcogens are at the core of a variety of molecules of biological, pharmaceutical interest and have found many applications in materials sciences. Among them, alkynyl chalcogenides are an interesting, understudied subclass used in a variety of organic processes such as regioselective hydrozirconation, hydroboration, hydrohalogenation, or cycloadditions. The method commonly used for their preparation relies on the stoichiometric generation of an alkynylmetal (most frequently a Grignard reagent) and its further reaction with a sulfanyl chloride, a reaction that requires strictly controlled conditions and suffers from disadvantages in terms of yields and generality.

Copper-mediated cross-coupling reactions with dichalcogenides or chalcogenyl halides allow for alternative syntheses of sulfur-, selenium-, and tellurium-substituted alkynes, even if most protocols still rely on the use of classical Ullmann-like conditions (stoichiometric amounts of copper salts and/or use of highly polar solvents), and ligand-assisted procedures are still rare in this area. Terminal alkynes **87** have been shown to participate in such reactions in 1993 when reacted with chalcogenyl bromides and chlorides **86** in the presence of 2 equiv of copper(I) iodide in DMF at room temperature and under an atmosphere of nitrogen. Under these classical conditions, the corresponding alkynyl sulfides, selenides, and tellurides **88** could be obtained in moderate-to-good yields, demonstrating that classical Ullmann-type, ligandless procedures can be quite efficient for the preparation of these useful building blocks

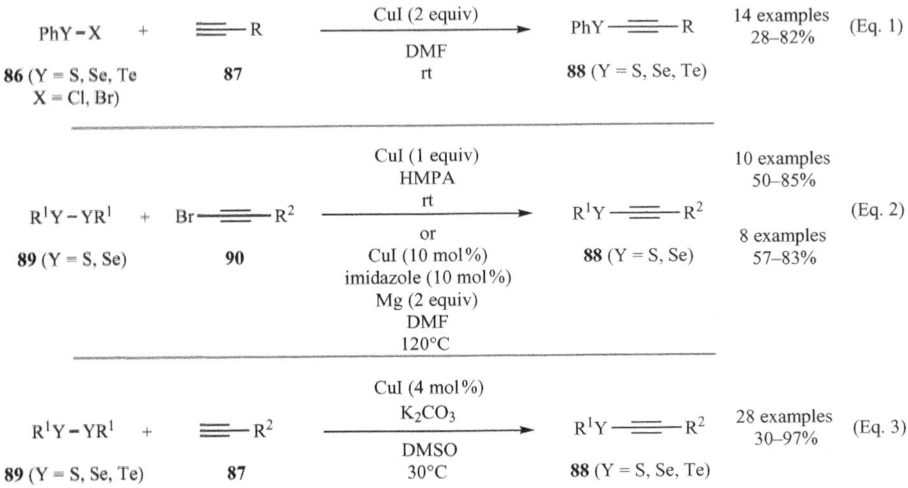

Scheme 5.31 Copper-mediated synthesis of alkynyl chalcogenides: classical but efficient Ullmann-type procedures.

(Scheme 5.31, Eq. 1).[104] Bromoalkynes **90** can also be used as cross-coupling partners in combination with dichalcogenides **89** using either stoichiometric amounts of copper iodide in HMPA at room temperature[105] or catalytic amounts of this catalyst if imidazole is used as a ligand (Scheme 5.31, Eq. 2).[106] In the second case, the reaction needs to be heated at 120°C, and excess magnesium metal, which would serve to reduce copper(I) iodide to copper(0), are needed. Finally, a smooth coupling could be observed between terminal alkynes **87** and dichalcogenides **89** by simply stirring mixtures of these reagents in the presence of 4% copper iodide and potassium carbonate (in the case of disulfides and ditellurides; this base is not required for diselenides) in DMSO at 30°C (Scheme 5.31, Eq. 3).[107] Finally, thioynol ethers can also be directly obtained from alkynyltrimethylsilanes upon treatment with stoichiometric copper(I) triflate and diphenyldisulfide, which can avoid the desilylation of the alkyne prior to its sulfurization and is useful for the synthesis of intermediates en route to thienamycin.[108]

An especially mild procedure for the neutral arylthiation of terminal acetylenes has been reported in 2010. By reaction of a terminal alkyne and a benzothiazolone with catalytic amounts of copper 3-methylsalicylate, highly substituted aromatic mercaptoacetylenes can be isolated in good yields under unprecedented neutral conditions (Scheme 5.32).[109]

5.3.4 Synthesis of Phosphorus-Substituted Alkynes: The Development of Mild Protocols Based on Copper Catalysis

Phosphorus-substituted alkynes are also attractive building blocks in organic synthesis and interesting molecules for the development of new ligands for

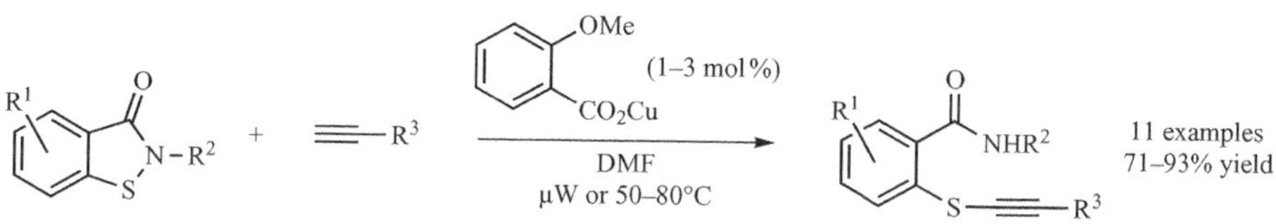

Scheme 5.32 Synthesis of mercaptoacetylenes by copper-catalyzed arylthiation of terminal acetylenes.

transition metals. The few known methods for the preparation of alkynyl phosphines, phosphine oxides, and phosphonates involve the nucleophilic substitution at the phosphorus atom of a halophosphorus derivative by a metal acetylide at low temperature. A catalytic variant, reported in 2003, consists of the copper-catalyzed cross-coupling reaction of terminal alkynes **87** with chlorophosphines **91** and provides the corresponding alkynyl-, dialkynyl-, and even trialkynylphosphines **92** with remarkable efficiency (Scheme 5.33, Eq. 1).[110] Chlorophosphines can be replaced by dialkylphosphites **93** using oxidative cross-coupling conditions and the use of copper(I) or (II) salts, which provides an especially efficient entry to alkynylphosphonates **94** (Scheme 5.33, Eq. 2).[111] This last procedure cannot, however, be extended to the use of phosphines due to their facile oxidation. An interesting alternative was therefore reported in 2011 based on the use of phosphine boranes **95**, which could be used for the first time in copper-catalyzed cross-coupling processes and seem to be excellent reaction partners, and alkynyl bromides **90** using the copper(I) iodide/1,10-phenanthroline couple (Scheme 5.33, Eq. 3).[112] Importantly, the reaction can be easily scaled up without affecting the yields, and the resulting alkynylphosphine boranes **96** were found to be perfectly stable and could even be purified by chromatography on silica gel. Propiolic acids can also be used for the synthesis of alkynylphosphine derivatives via a decarboxylative cross coupling using copper/palladium cocatalysis and assistance with silver salts.[79]

The conclusion of this section devoted to the synthesis of heteroatom-substituted alkynes is fairly obvious: Copper catalysis clearly showed the way and allowed for the development of reliable, efficient, general, and cost-effective procedures. While effort will still have to be exerted with O-, S-, and P-nucleophiles, the example of the synthesis of ynamides is striking: Only a couple of research groups in the world were involved with the chemistry of these building blocks before 2005, and only 5–10 publications per year were devoted to this research area. Copper catalysis has completely revolutionized the chemistry of ynamides, and any chemist willing to look at the chemistry of these building blocks now has an impressive choice of copper-based processes for the preparation of a wide range of ynamides. This probably accounts for the ever increasing number of publications related to ynamides, which might still be considered as lab curiosities without copper catalysis.

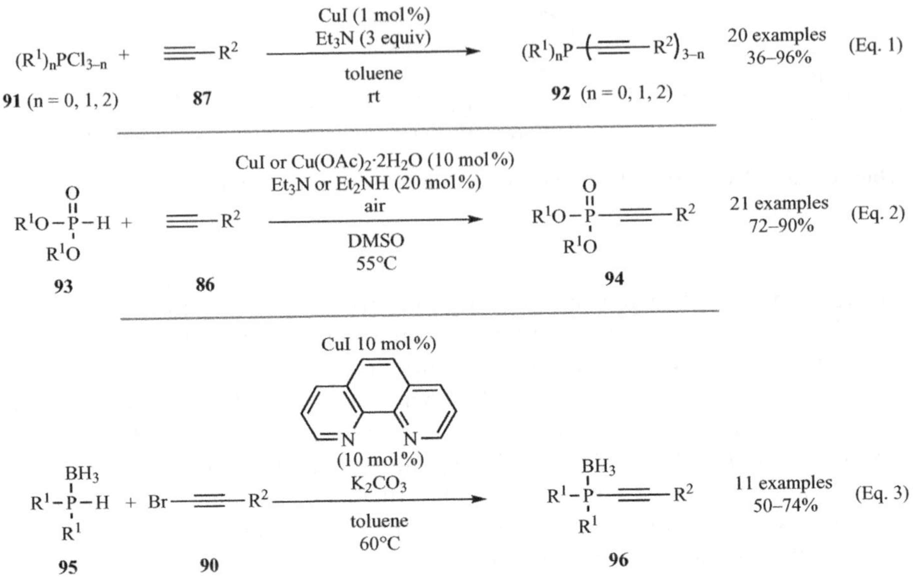

Scheme 5.33 The development of mild and general procedures for the synthesis of phosphorus-substituted alkynes with copper catalysis.

It also had a strong influence on the synthesis of heteroatom-substituted allenes, even if still restricted to allenamides whose synthesis by copper-mediated processes will be reviewed in the last section of this chapter.

5.4 ALLENYLATION OF HETERONUCLEOPHILES: NEW TOOLS FOR THE SYNTHESIS OF ALLENAMIDES

As just mentioned, the use of the copper-catalyzed processes for the synthesis of heteroatom-substituted allenes is still restricted to the cross coupling involving nitrogen nucleophiles, yielding to the formation of allenamides with remarkable efficiency.

The N-allenylation of amides, carbamates, and ureas was independently reported by the groups of Trost[113] and Hsung[114] using CuTC or CuCN, respectively, as copper sources and 1,2-diamines as ligands (Scheme 5.34). Both methods allow for efficient preparations of allenamides and are complementary. While the first catalytic system seems to be a little bit more efficient, since allenyl bromides can be used as reaction partners and since it was recently extended to the use of sulfonamides,[115] the second system was shown to be stereospecific and provides an excellent entry to optically enriched allenamides. In both cases, the reaction seems to be more sluggish with amides than with carbamates. Globally, these procedures provide an efficient tool for

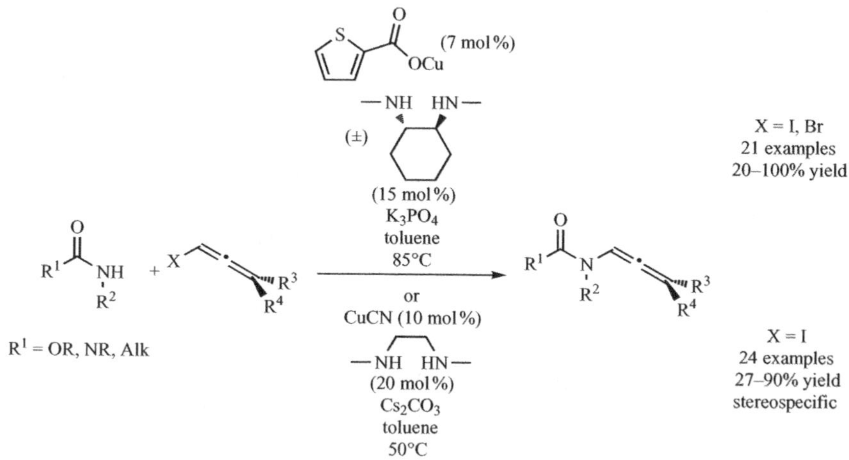

Scheme 5.34 Copper-mediated synthesis of allenamides.

the synthesis of nitrogen-substituted allenes and have also contributed to revitalizing this area. Most publications dealing with the chemistry of these building blocks now use these copper-catalyzed syntheses for their preparation.[116]

5.5 CONCLUSION AND FUTURE PROSPECTS

Considering the increasing number of applications for heteroatom-substituted alkenes, alkynes, and allenes in organic synthesis for the preparation of molecules of biological/pharmaceutical import and for the synthesis of complex natural products in materials science, efficient methods for their preparation are highly desirable. The methods we have described in this chapter based on copper catalysis provide easy and reliable access to these important building blocks. They are not, of course, completely universal, and they do indeed fail in some cases. But when they work, they are in most cases the methods of choice, allowing for the synthesis of enamides, enamines, *N*-alkenyl heterocycles, enol ethers, ketene acetals, vinyl sulfides and sulfones, phosphonoalkenes, ynamides, ynol ethers, alkynyl phosphines, phosphonoalkynes, and allenamides with remarkable efficiency and ease of operation from readily available starting materials. Future developments in this field will certainly be focused on the identification of catalytic systems with broader scope and even more practical conditions. For some obscure reasons, certain possibilities seem to have not been considered, such as the allenylation of O-, S-, and P-nucleophiles. They will also most certainly be part of the future developments in the synthesis of heteroatom-substituted unsaturated compounds.

REFERENCES

[1] For reviews, see: (a) Evano, G.; Blanchard, N.; Toumi, M. *Chem. Rev.* **2008**, *108*, 3054–3131. (b) Monnier, F.; Taillefer, M. *Angew. Chem. Int. Ed.* **2009**, *48*, 6954–6971.

[2] Ogawa, T.; Kiji, T.; Hayami, K.; Suzuki, H. *Chem. Lett.* **1991**, 1443–1446.

[3] Shen, R.; Porco, J. A., Jr. *Org. Lett.* **2000**, *2*, 1333–1336.

[4] Shen, R.; Lin, C. T.; Bowman, E. J.; Bowman, B. J.; Porco, J. A., Jr. *J. Am. Chem. Soc.* **2003**, *125*, 7889–7901.

[5] Han, C.; Shen, R.; Su, S.; Porco, J. A., Jr. *Org. Lett.* **2004**, *6*, 27–30.

[6] Surry, D. S.; Buchwald, S. L. *Chem. Sci.* **2010**, *1*, 13–31.

[7] Jiang, L.; Job, G. E.; Klapars, A.; Buchwald, S. L. *Org. Lett.* **2003**, *5*, 3667–3669.

[8] Cesati, R. R., III; Dwyer, G.; Jones, R. C.; Hayes, M. P.; Yalamanchili, P.; Casebier, D. S. *Org. Lett.* **2007**, *9*, 5617–5620.

[9] Dehli, J. R.; Bolm, C. *Adv. Synth. Catal.* **2005**, *347*, 239–242.

[10] Rivero, M. R.; Buchwald, S. L. *Org. Lett.* **2007**, *9*, 973–976.

[11] Wang, Y.; Liao, Q.; Xi, C. *Org. Lett.* **2010**, *12*, 2951–2953.

[12] Sanapo, G. F.; Daoust, B. *Tetrahedron Lett.* **2008**, *49*, 4196–4199.

[13] (a) Coste, A.; Couty, F.; Evano, G. *Org. Lett.* **2009**, *11*, 4454–4457. (b) Jouvin, K.; Coste, A.; Bayle, A.; Legrand, F.; Karthikeyan, G.; Tadiparthi, K.; Evano, G. *Organometallics* **2012**, *31*, 7933–7947.

[14] Pan, X.; Cai, Q.; Ma, D. *Org. Lett.* **2004**, *6*, 1809–1812.

[15] Lu, H.; Li, C. *Org. Lett.* **2006**, *8*, 5365–5367.

[16] Lu, H.; Yuan, X.; Zhu, S.; Sun, C.; Li, C. *J. Org. Chem.* **2008**, *73*, 8665–8668.

[17] Hu, T.; Li, C. *Org. Lett.* **2005**, *7*, 2035–2038.

[18] Zhao, Q.; Li, C. *Org. Lett.* **2008**, *10*, 4037–4040.

[19] Jiang, B.; Tian, H.; Huang, Z.-G.; Xu, M. *Org. Lett.* **2008**, *10*, 2737–2740.

[20] (a) Zhou, X.; Zhang, H.; Yuan, J.; Mai, L.; Li, Y. *Tetrahedron Lett.* **2007**, *48*, 7236–7239. (b) Li, E.; Xu, X.; Li, H.; Zhang, H.; Xu, X.; Yuan, X.; Li, Y. *Tetrahedron* **2009**, *65*, 8961–8968.

[21] Pan, Y.; Lu, H.; Fang, X.; Chen, L.; Qian, J.; Wang, J.; Li, C. *Synthesis* **2007**, 1242–1246.

[22] Martin, R.; Rivero, M. R.; Buchwald, S. L. *Angew. Chem. Int. Ed.* **2006**, *42*, 7079–7082.

[23] (a) Yuan, X.; Xu, X.; Zhou, X.; Yuan, J.; Mai, L.; Li, Y. *J. Org. Chem.* **2007**, *72*, 1510–1513. (b) Martin, R.; Larsen, C. H.; Cuenca, A.; Buchwald, S. L. *Org. Lett.* **2007**, *9*, 3379–3382.

[24] Liao, Q.; Zhang, L.; Wang, F.; Li, S.; Xi, C. *Eur. J. Org. Chem.* **2010**, 5426–5431.

[25] Hodgkinson, R. C.; Schulz, J.; Willis, M. C. *Org. Biomol. Chem.* **2009**, 432–434.

[26] Yuen, J.; Fang, Y.-Q.; Lautens, M. *Org. Lett.* **2006**, *8*, 653–656.

[27] Schuh (née Muller), K.; Glorius, F. *Synthesis* **2007**, 2297–2306.

[28] Martin, R.; Cuanca, A.; Buchwald, S. L. *Org. Lett.* **2007**, *9*, 5521–5524.

[29] Kabir, M. S.; Lorenz, M.; Namjoshi, O. A.; Cook, J. M. *Org. Lett.* **2010**, *12*, 464–467.

[30] Liao, Q.; Wang, Y.; Zhang, L.; Xi, C. *J. Org. Chem.* **2009**, *74*, 6371–6373.

[31] Taillefer, M.; Ouali, A.; Renard, B.; Spindler, J.-F. *Chem. Eur. J.* **2006**, *12*, 5301–5313.

[32] Mao, J.; Hua, Q.; Guo, J.; Shi, D.; Ji, S. *Synlett* **2008**, 2011–2016.

[33] Bao, W.; Liu, Y.; Lv, X. *Synlett* **2008**, 1911–1917.

[34] Wang, Z.; Bao, W.; Jiang, Y. *Chem. Commun.* **2005**, 2849–2851.

[35] (a) Lam, P. Y. S.; Vincent, G.; Clark, C. G.; Deudon, S.; Jadhav, P. K. *Tetrahedron Lett.* **2001**, *42*, 3415–3418. (b) Lam, P. Y. S.; Vincent, G.; Bonne, D.; Clark, C. G. *Tetrahedron Lett.* **2003**, *44*, 4927–4931.

[36] Jacobsen, M. F.; Knudsen, M. M.; Gothelf, K. V. *J. Org. Chem.* **2006**, *71*, 9183–9190.

[37] Kantam, M. L.; Venkanna, G. T.; Kumar, K. B. S.; Subrahmanyam, V. B. *Helv. Chim. Acta* **2010**, *93*, 974–979.

[38] (a) Bolshan, Y.; Batey, R. A. *Angew. Chem. Int. Ed.* **2008**, *47*, 2109–2112. (b) Bolshan, Y.; Batey, R. A. *Tetrahedron* **2010**, *66*, 5283–5294.

[39] Yan, X.; Chen, C.; Zhou, Y.; Xi, C. *Org. Lett.* **2012**, *14*, 4750–4753.

[40] Keegstra, M. A. *Tetrahedron* **1992**, *48*, 2681–2690.

[41] Wolter, M.; Nordmann, G.; Job, G. E.; Buchwald, S. L. *Org. Lett.* **2002**, *4*, 973–976.

[42] Altman, R. A.; Shafir, A.; Choi, A.; Lichtor, P. A.; Buchwald, S. L. *J. Org. Chem.* **2008**, *73*, 284–286.

[43] Nordmann, G.; Buchwald, S. L. *J. Am. Chem. Soc.* **2003**, *125*, 4978–4979.

[44] Wan, Z.; Jones, C. D.; Koenig, T. M.; Pu, Y. J.; Mitchell, D. *Tetrahedron Lett.* **2003**, *44*, 8257–8259.

[45] Ma, D.; Cai, Q.; Xie, X. *Synlett* **2005**, 1767–1770.

[46] Ouali, A.; Spindler, J.-F.; Jutand, A.; Taillefer, M. *Adv. Synth. Catal.* **2007**, *349*, 1906–1916.

[47] Kaddouri, H.; Vicente, V.; Ouali, A.; Ouazzani, F.; Taillefer, M. *Angew. Chem. Int. Ed.* **2009**, *48*, 333–336.

[48] Ouali, A.; Laurent, R.; Caminade, A.-M.; Majoral, J.-P.; Taillefer, M. *J. Am. Chem. Soc.* **2006**, *128*, 15990–15991.

[49] Jouvin, K.; Bayle, A.; Legrand, F.; Evano, G. *Org. Lett.* **2012**, *14*, 1652–1655.

[50] Vuluga, D.; Legros, J.; Crousse, B.; Bonnet-Delpon, D. *Eur. J. Org. Chem.* **2009**, 3513–3518.

[51] Fang, Y.; Li, C. *J. Am. Chem. Soc.* **2007**, *129*, 8092–8093.

[52] Fang, Y.; Li, C. *Chem. Commun.* **2005**, 3574–3576.

[53] Chen, L.; Fang, Y.; Zhao, Q.; Shi, M.; Li, C. *Tetrahedron Lett.* **2010**, *51*, 3678–3681.

[54] Shade, R. E.; Hyde, A. M.; Olsen, J.-C.; Merlic, C. A. *J. Am. Chem. Soc.* **2010**, *132*, 1202–1203.

[55] Winternheimer, D. J.; Merlic, C. A. *Org. Lett.* **2010**, *12*, 2508–2510.

[56] Chan, D. G.; Winternheimer, D. J.; Merlic, C. A. *Org. Lett.* **2011**, *13*, 2778–2781.

[57] Quach, T. D.; Batey, R. A. *Org. Lett.* **2003**, *5*, 1381–1384.

[58] McKinley, N. F.; O'Shea, D. F. *J. Org. Chem.* **2004**, *69*, 5087–5092.

[59] (a) Dai, J.-J.; Liu, J.-H.; Luo, D.-F.; Liu, L. *Chem. Commun.* **2011**, 677–679. (b) Huang, F.; Quach, T. D.; Batey, R. A. *Org. Lett.* **2013**, *15*, 3150–3153.

[60] Patil, A. S.; Mo, D.-L.; Wang, H.-Y.; Mueller, D. S.; Anderson, L. L. *Angew. Chem. Int. Ed.* **2012**, *51*, 7799–7803.

[61] Bates, C. G.; Saejueng, P.; Doherty, M. Q.; Venkataraman, D. *Org. Lett.* **2004**, *6*, 5005–5008.

[62] (a) Kabir, M. S.; Van Linn, M. L.; Monte, A.; Cook, J. M. *Org. Lett.* **2008**, *10*, 3363–3366. (b) Kabir, M. S.; Lorenz, M.; Van Linn, M. L.; Namjoshi, O. A.; Ara, S.; Cook, J. M. *J. Org. Chem.* **2010**, *75*, 3626–3643.

[63] Reddy, V. P.; Swapna, K.; Kumar, A. V.; Rao, K. R. *Synlett* **2009**, 2783–2788.

[64] Zheng, Y.; Du, X.; Bao, W. *Tetrahedron Lett.* **2006**, *47*, 1217–1220.

[65] Wang, Z.; Mo, H.; Bao, W. *Synlett* **2007**, 91–94.

[66] Guo, S.-R.; Yuan, Y.-Q. *Synth. Commun.* **2008**, *38*, 2722–2730.

[67] (a) Ranjit, S.; Duan, Z.; Zhang, P.; Liu, X. *Org. Lett.* **2010**, *12*, 4134–4136. (b) Riduan, S. N.; Ying, J. Y.; Zhang, Y. *Org. Lett.* **2012**, *14*, 1780–1783.

[68] Zhao, Q.; Li, L.; Fang, Y.; Sun, D.; Li, C. *J. Org. Chem.* **2009**, *74*, 459–462.

[69] You, W.; Yan, X.; Liao, Q.; Xi, C. *Org. Lett.* **2010**, *12*, 3930–3933.

[70] (a) Taniguchi, N. *Synlett* **2006**, 1351–1354. (b) Taniguchi, N. *J. Org. Chem.* **2007**, *72*, 1241–1245.

[71] Savarin, C.; Srogl, J.; Liebeskind, L. S. *Org. Lett.* **2002**, *4*, 4309–4312.

[72] Huang, F.; Batey, R. A. *Tetrahedron* **2007**, *63*, 7667–7672.

[73] Kar, A.; Sayyed, I. A.; Lo, W. F.; Kaiser, H. M.; Beller, M.; Tse, M. K. *Org. Lett.* **2007**, *9*, 3405–3408.

[74] Gelman, D.; Jiang, J.; Buchwald, S. L. *Org. Lett.* **2003**, *5*, 2315–2318.

[75] Evano, G.; Tadiparthi, K.; Couty, F. *Chem. Commun.* **2011**, 179–181.

[76] Lera, M.; Hayes, C. J. *Org. Lett.* **2000**, *2*, 3873–3875.

[77] Kabalka, G. W.; Guchhait, S. K. *Org. Lett.* **2003**, *5*, 729–731.

[78] Thielges, S.; Bisseret, P.; Eustache, J. *Org. Lett.* **2005**, *7*, 681–684.

[79] Hu, J.; Zhao, N.; Yang, B.; Wang, G.; Guo, L.-N.; Liang, Y.-M.; Yang, S.-D. *Chem. Eur. J.* **2011**, *17*, 5516–5521.

[80] For reviews, see: (a) Evano, G.; Coste, A.; Jouvin, K. *Angew. Chem. Int. Ed.* **2010**, *49*, 2840–2859. (b) DeKorver, K. A.; Li, H.; Lohse, A. G.; Hayashi, R.; Lu, Z.; Zhang, Y.; Hsung, R. P. *Chem. Rev.* **2010**, *110*, 5064–5106. (c) Evano, G.; Jouvin, K.; Coste, A. *Synthesis* **2013**, *45*, 17–26.

[81] Frederick, M. O.; Mulder, J. A.; Tracey, M. R.; Hsung, R. P.; Huang, J.; Kurtz, K. C. M.; Shen, L.; Douglas, C. J. *J. Am. Chem. Soc.* **2003**, *125*, 2368–2369.

[82] (a) Dunetz, J. R.; Danheiser, R. L. *Org. Lett.* **2003**, *5*, 4011–4014. (b) Kohnen, A. L.; Dunetz, J. R.; Danheiser, R. L. *Org. Synth.* **2007**, *84*, 88–101.

[83] (a) Zhang, Y.; Hsung, R. P.; Tracey, M. R.; Kurtz, K. C. M.; Vera, E. L. *Org. Lett.* **2004**, *6*, 1151–1154. (b) Zhang, X.; Zhang, Y.; Huang, J.; Hsung, R. P.; Kurtz, K. C. M.; Oppenheimer, J.; Petersen, M. E.; Sagamanova, I. K.; Shen, L.; Tracey, M. R.

J. Org. Chem. **2006**, *71*, 4170–4177. (c) Sagamanova, I. K.; Kurtz, K. C. M.; Hsung, R. P. *Org. Synth.* **2007**, *84*, 359–367.

[84] Istrate, F. M.; Buzas, A. K.; Jurberg, I. D.; Odabachian, Y.; Gagosz, F. *Org. Lett.* **2008**, *10*, 925–928.

[85] (a) DeKorver, K. A.; Walton, M. C.; North, T. D.; Hsung, R. P. *Org. Lett.* **2011**, *13*, 4862–4865. (b) Chen, Z.; Zeng, W.; Jiang, H.; Liu, L. *Org. Lett.* **2012**, *14*, 5385–5387.

[86] Dooleweerdt, K.; Birkedal, H.; Ruhland, T.; Skrydstrup, T. *J. Org. Chem.* **2008**, *73*, 9447–9450.

[87] Peterson, L. I. *Tetrahedron Lett.* **1968**, *9*, 5357–5360.

[88] Hamada, T.; Ye, X.; Stahl, S. S. *J. Am. Chem. Soc.* **2008**, *130*, 833–835.

[89] (a) Laouiti, A.; Rammah, M. M.; Rammah, M. B.; Couty, F.; Evano, G. *Org. Lett.* **2012**, *14*, 6–9. (b) Wang, L.; Huang, H.; Priebbenow, D. L.; Pan, F.-F.; Bolm, C. *Angew. Chem. Int Ed.* **2013**, *52*, 3478–3480.

[90] (a) Jin, X.; Yamaguchi, K.; Mizuno, N. *Chem. Commun.* **2012**, *48*, 4974–4976. (b) Tong, X.; Ni, G.; Deng, X.; Xia, C. *Synlett* **2012**, *23*, 2497–2500. (c) Arai, T.; Kawasaki, N.; Kanoh, H. *Synlett* **2012**, *23*, 1549–1553.

[91] (a) Jouvin, K.; Heimburger, J.; Evano, G. *Chem. Sci.* **2012**, *3*, 756–760. (b) Laouiti, A.; Jouvin, K.; Bourdreux, F.; Rammah, M. M.; Rammah, M. B.; Evano, G. *Synthesis* **2012**, *44*, 1491–1500.

[92] (a) Coste, A.; Karthikeyan, G.; Couty, F.; Evano, G. *Angew. Chem. Int. Ed.* **2009**, *48*, 4381. (b) Coste, A.; Couty, F.; Evano, G. *Org. Synth.* **2010**, *87*, 231–244.

[93] Yang, Y.; Zhang, X.; Liang, Y. *Tetrahedron Lett.* **2012**, *53*, 6557–6560.

[94] Jia, W.; Jiao, N. *Org. Lett.* **2010**, *12*, 2000–2003.

[95] Jouvin, K.; Couty, F.; Evano, G. *Org. Lett.* **2010**, *12*, 3272–3275.

[96] Sueda, T.; Oshima, A.; Teno, N. *Org. Lett.* **2011**, *13*, 3996–3999.

[97] Laroche, C.; Li, J.; Freyer, M. W.; Kerwin, S. M. *J. Org. Chem.* **2008**, *73*, 6462–6465.

[98] Burley, G. A.; Davies, D. L.; Griffith, G. A.; Lee, M.; Singh, K. *J. Org. Chem.* **2010**, *75*, 980–983.

[99] Das, B.; Reddy, G. C.; Balasubramanyam, P.; Salvanna, N. *Synthesis* **2011**, 816–820.

[100] Jouvin, K.; Evano, G. *Chim. Oggi.* **2011**, *29*, 31–33.

[101] Wang, M.-G.; Wu, J.; Shang, Z.-C. *Synlett* **2012**, *23*, 589–594.

[102] Das, B.; Salvanna, N.; Reddy, G. C.; Balasubramanyam, P. *Tetrahedron Lett.* **2011**, *52*, 6497–6500.

[103] (a) Stang, P. J.; Surber, B. W. *J. Am. Chem. Soc.* **1985**, *107*, 1452–1453. (b) Stang, P. J.; Surber, B. W.; Chen, Z.-C.; Roberts, K. A.; Anderson, A. G. *J. Am. Chem. Soc.* **1987**, *109*, 228–235.

[104] Braga, A. L.; Silveira, C. C.; Reckziegel, A.; Menezes, P. H. *Tetrahedron Lett.* **1993**, *34*, 8041–8042.

[105] Braga, A. L.; Reckziegel, A.; Menezes, P. H.; Stefani, H. A. *Tetrahedron Lett.* **1993**, *34*, 393–394.

[106] Sharma, A.; Schwab, R. S.; Braga, A. L.; Barcellos, T.; Paixão, M. W. *Tetrahedron Lett.* **2008**, *49*, 5172–5174.

[107] Bieber, L. W.; da Silva, M. F.; Mezenes, P. H. *Tetrahedron Lett.* **2004**, *45*, 2735–2737.

[108] Miyachi, N.; Shibasaki, M. *J. Org. Chem.* **1990**, *55*, 1975–1976.

[109] Henke, A.; Srogl, J. *Chem. Commun.* **2010**, 6819–6821.

[110] Afanasiev, V. V.; Beleteskaya, I. P.; Kazankova, M. A.; Efimova, I. V.; Antipin, M. M. U. *Synthesis* **2003**, 2835–2838.

[111] Gao, Y.; Wang, G.; Chen, L.; Zhao, Y.; Zhou, Y.; Han, L.-B. *J. Am. Chem. Soc.* **2009**, *131*, 7956–7957.

[112] Bernoud, E.; Alayrac, C.; Delacroix, O.; Gaumont, A.-C. *Chem. Commun.* **2011**, 3239–3241.

[113] Trost, B. M.; Stiles, D. T. *Org. Lett.* **2005**, *7*, 2117–2120.

[114] Shen, L.; Hsung, R. P.; Tracey, M. R.; Zhang, Y.; Antoline, J. E.; Zhang, X. *Org. Lett.* **2005**, *7*, 3081–3084.

[115] Persson, A. K. Å.; Johnston, E. V.; Bäckvall, J.-E. *Org. Lett.* **2009**, *11*, 3814–3817.

[116] Lu, T.; Lu, Z.; Ma, Z.-X.; Zhang, Y.; Hsung, R. P. *Chem. Rev.* **2013**, *113*, 4862–4904.

6

AROMATIC/VINYLIC FINKELSTEIN REACTION

Alicia Casitas and Xavi Ribas

6.1 INTRODUCTION

The Finkelstein reaction involves the halogen exchange in aliphatic chlorides or bromides with alkali iodide salts to produce the corresponding aliphatic iodides. This reaction, which has been known since the beginning of the twentieth century, occurs by means of an S_N2 nucleophilic substitution, and the driving force is usually the low solubility of the alkali chlorides or bromides formed during the reaction (Scheme 6.1).[1]

The analogous aromatic and vinylic halogen exchange reactions are far less known, and the use of transition-metal catalysis is generally required. The importance of this reaction resides in the extensive use of aryl halides in transition-metal-catalyzed cross-coupling reactions in organic synthesis. Examples of these are the Heck, Suzuki, Negishi, and Stille C–C cross-coupling reactions,[2,3] as well as the Buchwald–Hartwig[4,5] and Ullmann-type C-heteroatom cross couplings.[6–8] Among the different halides, the cheaper and more accessible aryl chlorides are the less-reactive ones in cross-coupling reactions, the typical order of reactivity being Cl < Br < I. There is, therefore, a high interest in developing aromatic halogen exchange methods to produce aryl bromides and iodides from the corresponding chlorides.

Few examples of aromatic halide exchanges have been reported using copper-based catalysts, as well as nickel and palladium catalysts.[9] Processes using copper catalysis, which are among the most efficient ones to date, will

Copper-Mediated Cross-Coupling Reactions, First Edition. Edited by Gwilherm Evano and Nicolas Blanchard.

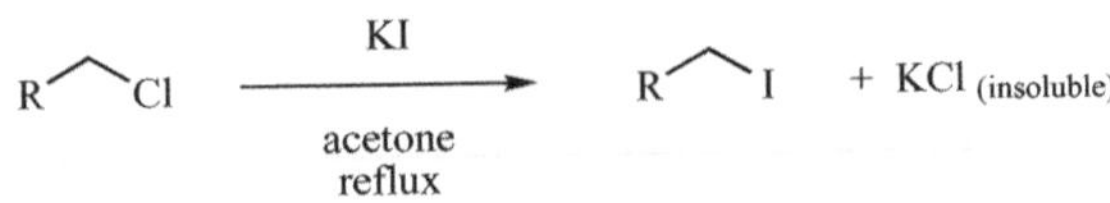

Scheme 6.1 The Finkelstein reaction.

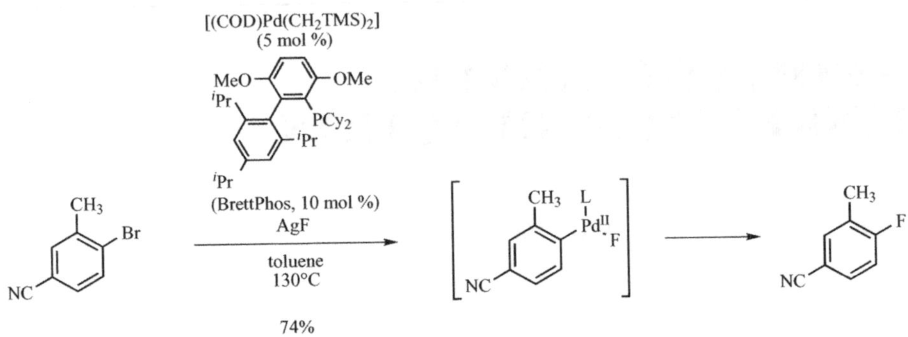

Scheme 6.2 Palladium-catalyzed fluorination of aryl bromides using fluoride sources.

be overviewed in this chapter after a brief historical perspective and a presentation of the state-of-the-art methods in this area.

Nickel-based systems have been known since the 1970s when Cramer and Coulson first published the formation of chlorobenzene from bromobenzene using $NiCl_2$ and LiCl at very high temperatures (>150°C).[10] Since then, other phosphine-Ni(II) catalysts[11] or mixtures of Ni(II) salts and Zn powder[12,13] improved the results and allowed milder experimental conditions for the synthesis of aryl iodides from aryl bromides. Mechanistic insights suggest the implication of either Ni(I) or Ni(0) as the active species for the oxidative addition, followed by halide exchange and final reductive elimination.[14]

With respect to palladium-based systems capable of performing aromatic halide exchange reactions, there is a single report where 4-chlorobenzonitrile is converted to 4-iodobenzonitrile by treatment with potassium iodide and Pd/C (3 mol %) at 130°C.[15] This procedure can only be applied to electron-deficient chlorobenzenes, moderate yields are obtained, and the role of the charcoal support remains unclear. Very recently, Lautens and coworkers have reported an aryl bromide-to-iodide exchange involved in a palladium-catalyzed domino reaction.[16] In addition, several palladium-catalyzed procedures reporting the direct halogenation of aromatic rings via C–H activation have been published recently.[9,17,18]

In a milestone work reported by Buchwald and coworkers, the authors demonstrated for the first time the feasibility of an aryl fluorination of an aryl bromide (and an aryl triflate) using a combination of a palladium catalyst and fluoride salts such as silver or cesium fluoride, as depicted in Scheme 6.2.[19] This catalytic reactivity was enabled by the use of bulky monophosphine ligands that stabilize 14-electron aryl-Pd(II)-F species, complexes that are able to form in an unprecedented manner aryl fluorides by reductive elimination at Pd(II). Following a similar methodology, Buchwald and coworkers reported the palladium-catalyzed chlorination and bromination of aryl triflates.[20]

6.2 COPPER-MEDIATED HALOGEN EXCHANGE REACTIONS IN ARYL HALIDES

6.2.1 From Early Discoveries to Most Recent Reports

Early examples of copper-mediated halogen exchanges in aryl halides were reported by Bacon and Hill in 1964.[21] Heating aryl halides in the presence of cuprous halides, Cu(I)X (X = Cl, Br, I), in various polar solvents afforded the exchanged Ar–X products. In all cases, a common reactivity trend was observed toward the exchange of heavier to lighter halogen atoms (Ar–I > Ar–Br > Ar–Cl >> Ar–F), which is explained by the stronger energy of the new Ar–X bond formed (Cl > Br > I). Twenty years later, Ogawa and coworkers reported the synthesis of aryl iodides from aryl bromides by using potassium iodide and copper(I) iodide in hexamethylphosphoramide (HMPA) at 150°C in moderate-to-good yields (Scheme 6.3).[22] Concerning the vinylic halogen exchange reactions, early examples include Axelrad and coworkers' work on the halogen exchange from vinyl bromides to vinyl chlorides in moderate yields using phosphite-ligated copper complexes.[23]

The most significant contribution to the aromatic Finkelstein reactivity was made by Buchwald and Klapars in 2002, when they managed to develop a procedure to synthesize aryl iodides from the corresponding aryl bromides in a quantitative manner.[24–26] The strategy consisted of using catalytic amounts of copper(I) iodide (5 mol %), *trans-N,N′*-dimethylcyclohexane-1,2-diamine (or other diamines in particular cases) as the ligand (10 mol %), and excess sodium iodide at 110°C in dioxane (Scheme 6.4). The presence of other

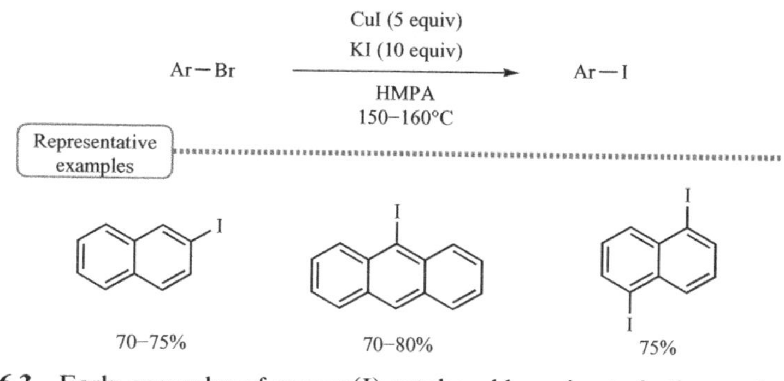

Scheme 6.3 Early examples of copper(I)-catalyzed bromine to iodine exchange reaction from aryl bromides.

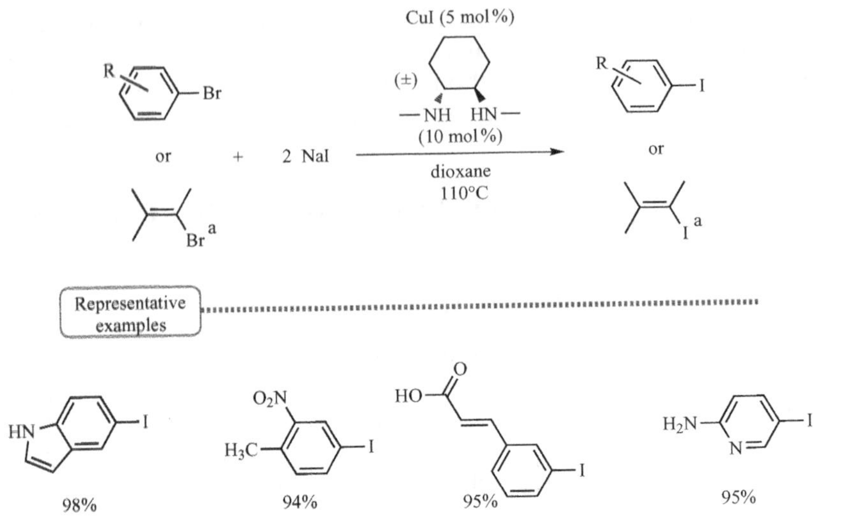

Scheme 6.4 Buchwald's breakthrough methodology for the conversion of aryl and vinyl bromides to the corresponding iodides ([a]conditions: 5 mol % CuI, 10 mol % *N,N'*-dimethylethylenediamine, 1.5 equiv NaI, *n*BuOH, 120°C).

functional groups (amides, esters, pyridines, amines, etc.) on the substrates was well tolerated. A carboxylic acid can be employed provided that hexamethyldisilazane is included in the reaction mixture to derivatize the carboxylic acid as the more soluble trimethylsilyl ester, which can be readily cleaved during workup of the reaction. Interestingly, N–H-containing substrates such as sulfonamides, amides, or indoles are also tolerated, and no competitive N-arylation is observed, which is presumably due to the absence of a strong base in the reaction mixture which prevents the N-arylation reaction. The same protocol is also operative with vinyl bromides by using *N,N'*-dimethylethylenediamine in butanol, although a single example was reported.

Buchwald's pioneer work was followed up by Willand and coworkers,[27] who developed an Ullmann–Finkelstein tandem reaction based on the previous procedure.[24] Indeed, they synthesized *N*-(iodophenyl)-amides from the corresponding secondary amides and 1-bromo-4-iodobenzene under the reaction conditions reported by the Buchwald group and with an additional base, potassium phosphate (Scheme 6.5). High-performance liquid chromatography–mass spectrometry (HPLC-MS) monitoring of the reaction revealed that cross coupling of the amide at the more reactive aryl iodide site occurred initially (Ullmann-type coupling), followed by a subsequent exchange of the aryl bromide moiety by the iodide released during the first Ullmann step (aromatic Finkelstein reaction).

Furthermore, the same group developed a more elaborated multicomponent reaction, with three consecutive copper-catalyzed transformations: chemoselective Ullmann cross coupling—aromatic Finkelstein reaction—Ullmann

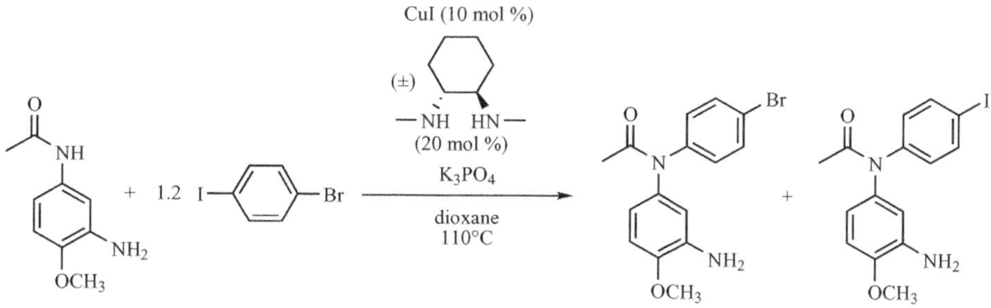

reaction time	proportions	
30 min	70%	30%
1 h	55%	45%
3 h	37%	62%
24 h	5%	95%

Scheme 6.5 Willand's methodology demonstrated the consecutive *in situ* Ullmann coupling followed by aromatic Finkelstein reaction catalyzed by copper(I) complexes.

cross coupling.[28] The methodology was designed by taking advantage of the dissymmetrical reactivity of several nucleophiles toward the cross coupling with aryl iodides (Scheme 6.6). The authors observed that, first, most acidic nucleophiles reacted faster, and second, sterically hindered nucleophiles reacted slower. With these parameters in hand, a one-pot Ullmann–Finkelstein–Ullmann reaction could be performed to obtain the double dissymmetrical cross-coupled product in high yields starting from 1-bromo-4-iodobenzene and two nitrogen nucleophiles.

A reverse reaction allowing for the synthesis of aryl chlorides from aryl bromides has been reported in 2012 by Feng and coworkers based on a combination of copper(I) oxide, proline, and tetramethylammonium chloride (2 equiv) as the catalyst, ligand, and chloride source, respectively (Scheme 6.7).[29] This system allows for an efficient conversion of a range of aryl and pyridinyl bromides to the corresponding chlorides that are isolated in satisfactory-to-good yields and provides a good alternative to less general systems based on the use of cuprous chloride or nickel dichloride.

Another example of a tandem catalysis, a hydrodehalogenation of aryl chlorides or bromides, has been reported very recently,[30] in which the first *in situ* catalytic process involves the formation of aryl iodides by an aromatic Finkelstein reaction under the experimental conditions reported by the Buchwald group.[24]

Other examples featuring an *in situ* Finkelstein reaction followed by N-[31] or S-[32] arylation have been used for the synthesis of anilides or aromatic thioethers from poorly reactive aryl bromides that can be transformed into the corresponding iodides prior to their cross coupling with heteronucleophiles.

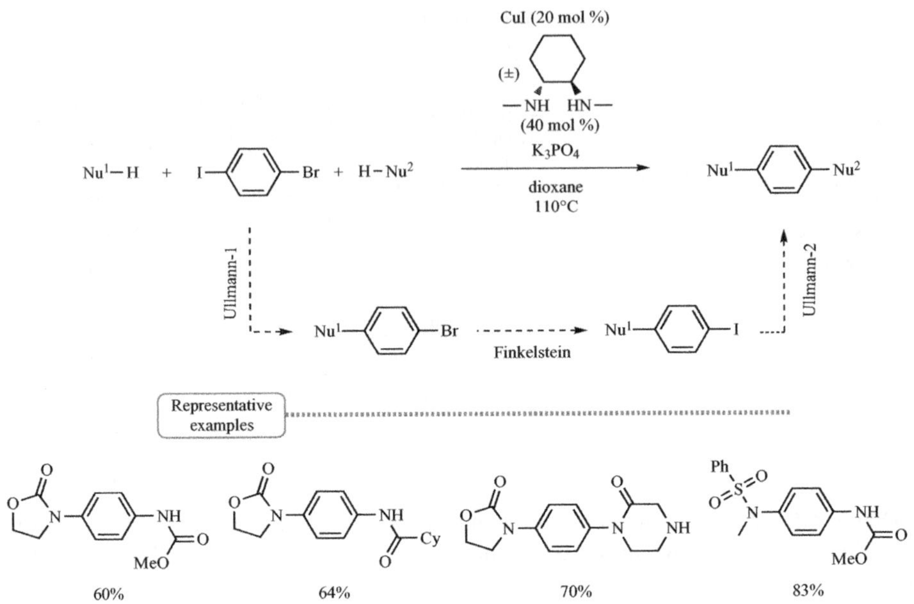

Scheme 6.6 First example of an *in situ* consecutive Ullmann–Finkelstein–Ullmann reactivity catalyzed by copper (representative examples with Nu[1] and Nu[2] on the left and right sides of the aromatic ring, respectively).

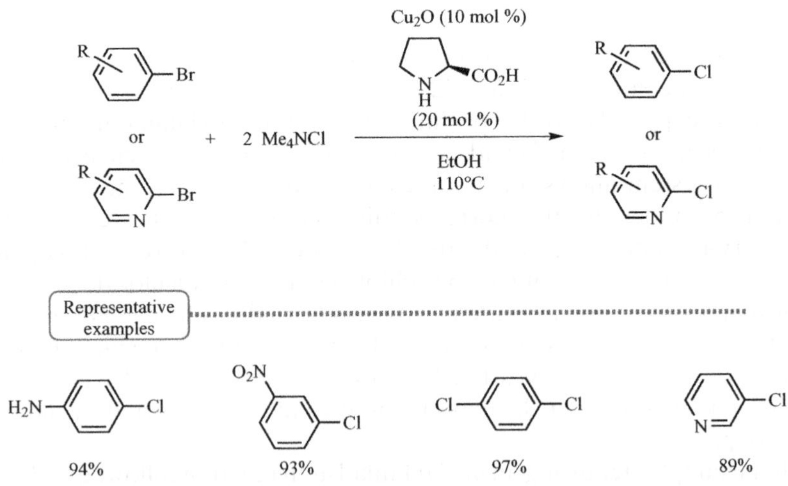

Scheme 6.7 Copper(I) oxide-catalyzed conversion of aryl and pyridinyl bromides to the corresponding chlorides using tetramethylammonium chloride.

The importance of having a synthetic tool such as the aromatic Finkelstein protocol[24] available is also illustrated by its recent use in the total synthesis of natural products as, for example, in the preparation of (±)-aspercyclide,[33] an antiinflammatory agent used in the prevention of allergic disorders, or in the synthesis of (–)-kaitocephalin,[34] a glutamate receptor antagonist studied as a potential neuroprotectant to prevent neurodegenerative disorders.

Besides the synthesis of aromatic iodides from the corresponding bromides (or chlorides), an interesting variation of the aromatic Finkelstein reaction is the copper-mediated fluorination of haloarenes, a process that is especially appealing for the preparation of a wide range of fluoroaromatics and which will be overviewed in the next section.

6.2.2 Copper-Mediated Fluorination of Haloarenes

The well-known Halex reaction is used in the industry to achieve the halogen-to-fluorine exchange of polyhalogenated aromatic substrates.[35] This transition-metal-free reaction is usually conducted with potassium fluoride in dimethylformamide or dimethylsulfoxide under harsh conditions (just below the solvent boiling point) and is limited to polyhalogenated aromatic substrates bearing electron-withdrawing groups.[36] A copper-mediated methodology for preparing fluoroarenes from haloarenes has been recently patented by Grushin.[37] In his invention, a wide range of nonactivated haloarenes can be used to exchange one or more halogen atoms by fluorine by using CuF_2, *N,N,N′,N′*-tetramethylethylenediamine (TMEDA), under high-temperature conditions (usually above 150°C) (Scheme 6.8). This is a remarkable methodology due to the difficulty of introducing fluorine atoms in aromatic rings, although the harsh conditions employed and the stoichiometric loadings of CuF_2 are clear drawbacks to the implementation of this methodology in

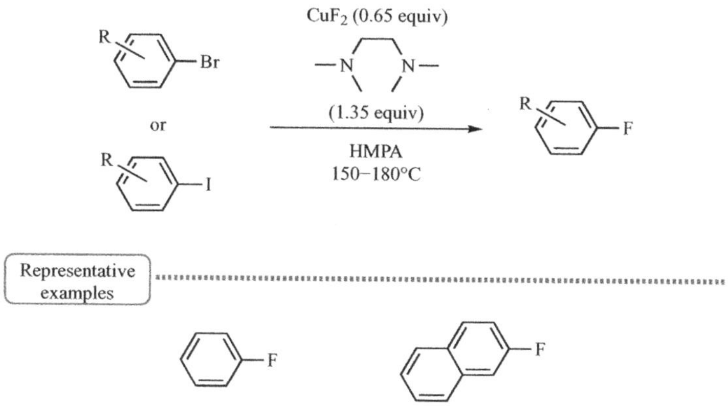

Scheme 6.8 Copper(II) fluoride-mediated aromatic fluorination of aryl bromides and iodides.

late-stage functionalization of compounds. Another CuF_2-mediated process provides the formation of aryl-F from benzene by using HF and O_2 with only H_2O as a by-product, although the yield (<30%) and the very high temperatures required (>500°C) preclude a wide implementation.[38]

6.2.3 Mechanistic Investigations in Aryl–Cu(III) Model Systems

Very recently, Ribas and coworkers have reported for the first time a copper-catalyzed halide exchange reaction under very mild reaction conditions using a family of model aryl halide triazamacrocyclic substrates.[39] All combinations of halide exchange (I, Br, Cl, F) are achieved using catalytic amounts of copper(I) salts. Initial stoichiometric studies with well-defined Ar–Cu(III)–X complexes (X = Cl, Br, I) and alkali halides MY salts (M = alkali metal, Y = F, Cl, Br, I) afforded good yields of the products resulting from the halogen exchange, Ar–Y.[40] The fundamental knowledge gained during these first studies allowed the authors to develop a catalytic version of these reactions, which could be successfully achieved in good yields when the exchange was in favor of a lighter halogen, and in moderate yields when an iodide was introduced (Table 6.1).

Strikingly, quantitative fluorination of aryl halides is also achieved catalytically at room temperature, using common fluoride sources, such as silver or

TABLE 6.1 Aromatic Finkelstein Reactions in Model Ar–X (X = Cl, Br, I) Substrates Affording Halide Exchange (Including Fluorination) under Very Mild Conditions through Ar–Cu(III)–Halide Intermediate Species

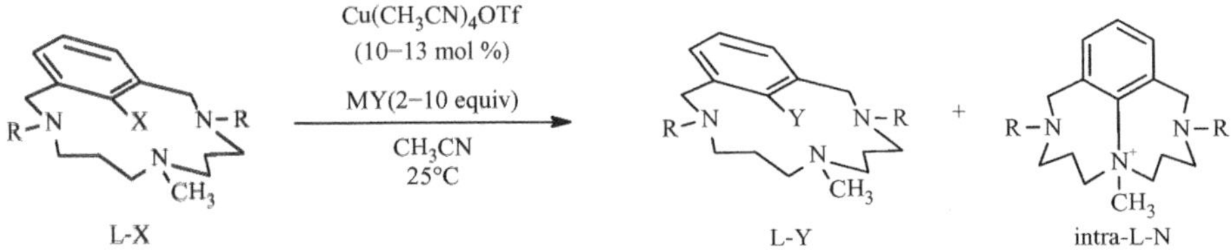

R	X	M	Y	MY (equiv)	L-Y (Yield) (%)	Intra-L-N (Yield) (%)
H	I	Bu_4N	Br	10	87	–
H	I	Bu_4N	Cl	10	99	–
H	Br	Bu_4N	Cl	10	99	–
H	Br	Na	I	10	84	–
H	Cl	Na	Br	10	37	–
H	Cl	Na	I	10	62	4
H	Cl	Ag	F	2	76	20
H	Br	Ag	F	2	71	21
CH_3[a]	Cl	Ag	F	2	98	–
CH_3[a]	Br	Ag	F	2	97	–

[a]CH_3CN:acetone (1:3) solvent mixture used.

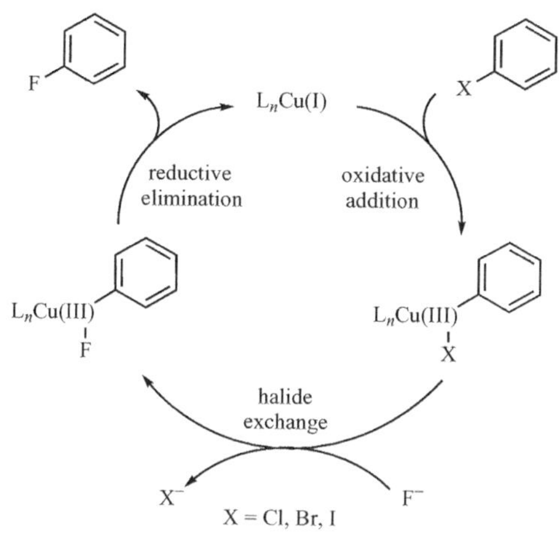

Scheme 6.9 Catalytic cycle proposed for halogen to fluorine exchange in aryl halide substrates mediated by a Cu(I)/Cu(III) system.

potassium fluoride, via the intermediacy of Ar–Cu(III)–X species.[40] Remarkably, aryl fluorides are readily formed at room temperature in CH_3CN, and isolated in good-to-excellent yields, by using silver fluoride as fluoride source (Table 6.1). Additionally, defluorination of the Ar–F model system can be also accomplished with $Cu(CH_3CN)_4OTf$ in acetone at room temperature, a reaction that is operating under a Cu(I)/Cu(III) redox pair.

Experimental and computational data support a redox Cu(I)/Cu(III) catalytic cycle involving aryl halide oxidative addition at the Cu(I) center, followed by halide exchange and reductive elimination steps (Scheme 6.9). Ideally, catalytic nucleophilic aryl fluorination should be a mild (room temperature) and rapid process to meet the requirements for late-stage aromatic fluorination as well as for the preparation of short-lived ^{18}F-radiolabeled molecular probes for positron emission tomography (PET).[41–43] This is the case for the model system reported by Ribas and coworkers, as well as for Buchwald's Pd(0)/Pd(II) system capable of fluorinating external aryl bromides substrates.[19]

6.3 MOST RECENT DEVELOPMENTS AND OVERVIEW

Aromatic halogen exchange reactions are still challenging chemical transformations since they are limited to certain substrates and experimental conditions are far from being considered mild and environmentally benign. Given the importance and high demand of aryl iodide substrates, alternative procedures have been developed very recently. Indeed, they include the Chan–Lam–Evans-type coupling with more expensive aryl boronic acids as starting materials described by Fu and coworkers,[44] a procedure that can remarkably

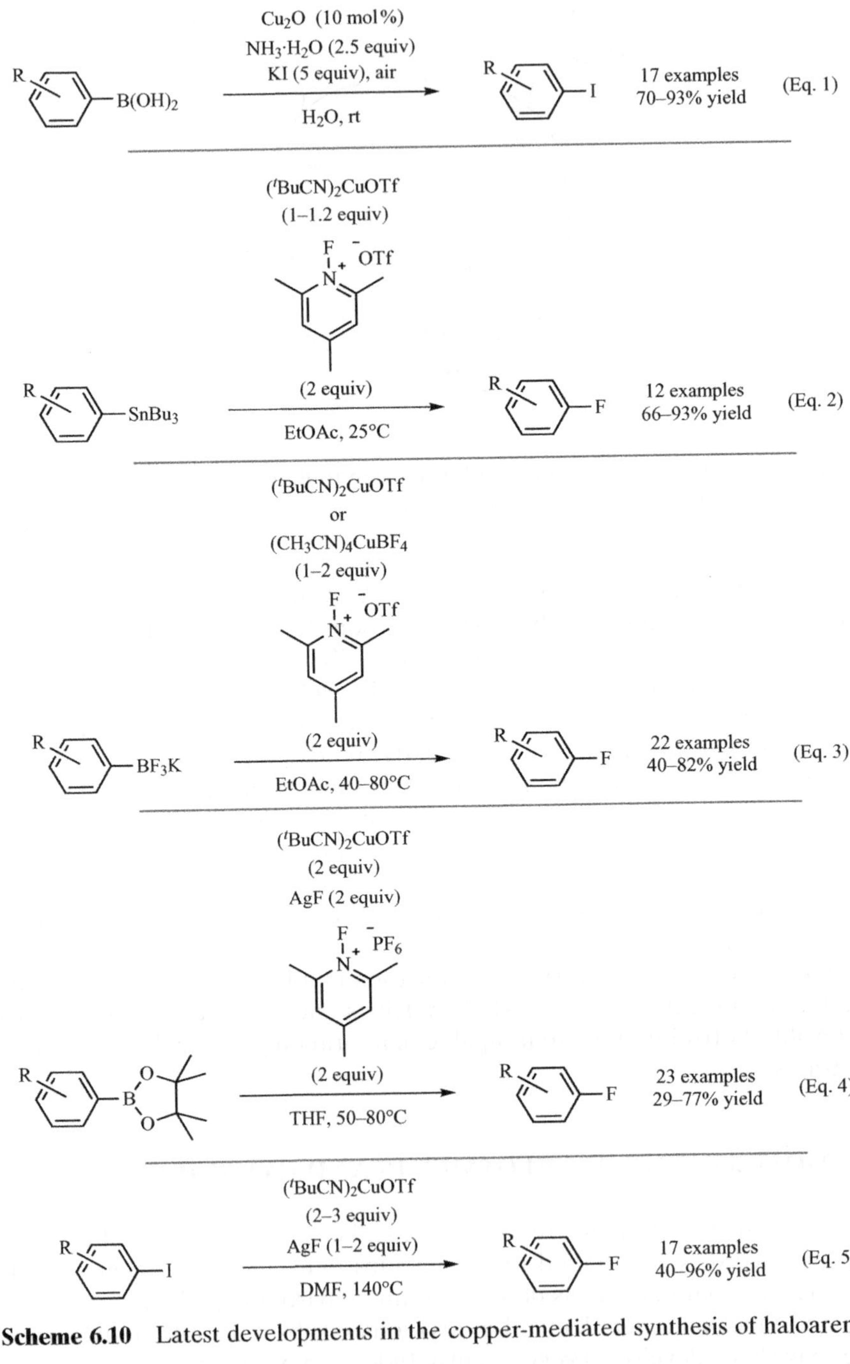

Scheme 6.10 Latest developments in the copper-mediated synthesis of haloarenes.

be conducted in water and at room temperature (Scheme 6.10, Eq. 1), the mild copper-mediated fluorination of aryl stannanes (Scheme 6.10, Eq. 2) and potassium aryl tetrafluoroborates (Scheme 6.10, Eq. 3) reported by the Sanford group,[45] and the copper-mediated fluorination or arylboronate esters reported by Hartwig and coworkers (Scheme 6.10, Eq. 4).[46] With all the reports described in this chapter, it is reasonable to point toward copper as the metal of choice for achieving this transformation, although much room for improvement remains and new systems will undoubtedly appear in the next years. Several mechanistic proposals for these aromatic halogen exchange reactions have been suggested, including oxidative addition/reductive elimination, single electron transfer, four-center metathesis-like transition state, or halide nucleophilic attack triggered by an aromatic-copper π-interaction.[9] Among them, clear mechanistic information for the oxidative addition/reductive elimination mechanism in model systems has been obtained, although more efforts are needed to clarify the mechanism in the most effective methodologies. In this line, very recently Hartwig and coworkers remarkably reported the fluorination of simple Ar-I by using stoichiometric amounts of $(^tBuCN)_2Cu^IOTf$ salt (3 equiv with respect to Ar-I) and AgF as the fluoride source (Scheme 6.10, Eq. 5).[47] The nature of the Cu^I source was crucial and the authors proposed a Cu^I/Cu^{III} redox process involving Ar-F reductive elimination from a putative aryl-Cu^{III}-F species. The latter work shows that solid steps are being achieved toward the design of practical copper-based halide exchange catalytic methodologies.

REFERENCES

[1] Finkelstein, H. *Ber. Dtsch. Chem. Ges.* **1910**, *43*, 1528–1532.

[2] Beletskaya, I. P.; Cheprakov, A. V. *Chem. Rev.* **2000**, *100*, 3009–3066.

[3] Henry, P. M. *Handbook of Organopalladium Chemistry for Organic Synthesis*, Negishi, E. (ed.). Wiley & Sons: New York, **2002**, pp. 2119–2140.

[4] Hartwig, J. F. *Acc. Chem. Res.* **1998**, *31*, 852–860.

[5] Hartwig, J. F. *Acc. Chem. Res.* **2008**, *41*, 1534–1544.

[6] Beletskaya, I. P.; Cheprakov, A. V. *Coord. Chem. Rev.* **2004**, *248*, 2337–2364.

[7] Ley, S. V.; Thomas, A. W. *Angew. Chem. Int. Ed.* **2003**, *42*, 5400–5449.

[8] Sperotto, E.; van Klink, G. P. M.; van Koten, G.; de Vries, J. G. *Dalton Trans.* **2010**, *39*, 10338–10351.

[9] Sheppard, T. D. *Org. Biomol. Chem.* **2009**, *7*, 1043–1052.

[10] Cramer, R.; Coulson, D. R. *J. Org. Chem.* **1975**, *40*, 2267–2273.

[11] Tsou, T. T.; Kochi, J. K. *J. Org. Chem.* **1980**, *45*, 1930–1937.

[12] Takagi, K.; Hayama, N.; Inokawa, S. *Bull. Chem. Soc. Jpn* **1980**, *53*, 3691–3695.

[13] Takagi, K.; Hayama, N.; Okamoto, T. *Chem. Lett.* **1978**, 191–192.

[14] Yang, S. H.; Li, C. S.; Cheng, C. H. *J. Org. Chem.* **1987**, *52*, 691–694.

[15] Thathagar, M. B.; Rothenberg, G. *Org. Biomol. Chem.* **2006**, *4*, 111–115.

[16] Newman, S. G.; Howell, J. K.; Nicolaus, N.; Lautens, M. *J. Am. Chem. Soc.* **2011**, *133*, 14916–14919.

[17] Chen, X.; Hao, X.-S.; Goodhue, C. E.; Yu, J.-Q. *J. Am. Chem. Soc.* **2006**, *128*, 6790–6791.

[18] Hull, K. L.; Anani, W. Q.; Sanford, M. S. *J. Am. Chem. Soc.* **2006**, *128*, 7134–7135.

[19] Watson, D. A.; Su, M.; Teverovskiy, G.; Zhang, Y.; García-Fortanet, J.; Kinzel, T.; Buchwald, S. L. *Science* **2009**, *325*, 1661–1664.

[20] Shen, X.; Hyde, A. M.; Buchwald, S. L. *J. Am. Chem. Soc.* **2010**, *132*, 14076–14078.

[21] Bacon, R. G. R.; Hill, H. A. O. *J. Chem. Soc.* **1964**, 1097–1107.

[22] Suzuki, H.; Kondo, A.; Inouye, M.; Ogawa, T. *Synthesis* **1986**, 121–126.

[23] Axelrad, G.; Laosooksathit, S.; Engel, R. *J. Org. Chem.* **1981**, *46*, 5200–5204.

[24] Klapars, A.; Buchwald, S. L. *J. Am. Chem. Soc.* **2002**, *124*, 14844–14845.

[25] Buchwald, S. L.; Klapars, A.; Kwong, F. Y.; Streiter, E. R.; Zanon, J., inventors; Massachusetts Institute of Technology, assignee. Copper-catalyzed formation of carbon-heteroatom and carbon–carbon bonds. U.S. patent 6,888,032, 2007 May 3.

[26] Buchwald, S. L.; Klapars, A.; Kwong, F. Y.; Streiter, E.; Zanon, J., inventors; Massachusetts Institute of Technology, assignee. Copper-catalyzed formation of carbon-heteroatom and carbon–carbon bonds. PCT Int. Appl. WO 2004013094 (A2), 2004 February 12.

[27] Toto, P.; Gesquière, J.-C.; Deprez, B.; Willand, N. *Tetrahedron Lett.* **2006**, *47*, 1181–1186.

[28] Toto, P.; Gesquière, J.-C.; Cousaert, N.; Deprez, B.; Willand, N. *Tetrahedron Lett.* **2006**, *46*, 4973–4978.

[29] Feng, X.; Qu, Y.; Han, Y.; Yu, X.; Bao, M.; Yamamoto, Y. *Chem. Commun.* **2012**, *48*, 9468–9470.

[30] Cannon, K. A.; Geuther, M. E.; Kelly, C. K.; Lin, S.; MacArthur, A. H. R. *Organometallics* **2011**, *30*, 4067–4073.

[31] Jones, C. P.; Anderson, K. W.; Buchwald, S. L. *J. Org. Chem.* **2007**, *72*, 7968–7973.

[32] Carril, M.; SanMartin, R.; Domínguez, E.; Tellitu, I. *Chem. Eur. J.* **2007**, *13*, 5100–5105.

[33] Carr, J. L.; Offermann, D. A.; Holdom, M. D.; Dusart, P.; White, A. J. P.; Beavil, A. J.; Leatherbarrow, R. J.; Lindell, S. D.; Sutton, B. J.; Spivey, A. C. *Chem. Commun.* **2010**, *46*, 1824–1826.

[34] Vaswani, R. G.; Chamberlin, A. R. *J. Org. Chem.* **2008**, *73*, 1661–1681.

[35] Langlois, B.; Gilbert, L.; Forat, G. *Ind. Chem. Lib.* **1996**, *8*, 244–292.

[36] Zhu, Y. H.; Lou, H.; Lu, P.; Liu, J. G.; Zheng, X. M. *Chin. Chem. Lett.* **2002**, *13*, 919–920.

[37] Grushin, V. V., inventor; E. I. du Pont de Nemours and Company, assignee. Processes for preparing fluoroarenes from haloarenes. U.S. patent 7,202,388. 2007 April 10.

[38] Subramanian, M. A.; Manzer, L. E. *Science* **2002**, *297*, 1665.

[39] Casitas, A.; Canta, M.; Solà, M.; Costas, M.; Ribas, X. *J. Am. Chem. Soc.* **2011**, *133*, 19386–19392.

[40] Casitas, A.; King, A. E.; Parella, T.; Costas, M.; Stahl, S. S.; Ribas, X. *Chem. Sci.* **2010**, *1*, 326–330.

[41] Miller, P. W.; Long, N. J.; Vilar, R.; Gee, A. D. *Angew. Chem. Int. Ed.* **2008**, *47*, 8998–9033.

[42] Li, L.; Hopkinson, M. N.; Yona, R. L.; Bejot, R.; Gee, A. D.; Gouverneur, V. *Chem. Sci.* **2011**, *2*, 123–131.

[43] Cai, L.; Lu, S.; Pike, V. W. *Eur. J. Org. Chem.* **2008**, 2853–2873.

[44] Yang, H.; Li, Y.; Jiang, M.; Wang, J.; Fu, H. *Chem. Eur. J.* **2011**, *17*, 5652–5660.

[45] Ye, Y.; Sanford, M. S. *J. Am. Chem. Soc.* **2013**, *135*, 4648–4651.

[46] Fier, P S.; Luo, J. Hartwig, J. F. *J. Am. Chem. Soc.* **2013**, *135*, 2552–2559.

[47] Fier, P S.; Hartwig, J. F. *J. Am. Chem. Soc.* **2012**, *134*, 10795–10798.

7

INSIGHTS INTO THE MECHANISM OF MODERN ULLMANN–GOLDBERG COUPLING REACTIONS

Alicia Casitas and Xavi Ribas

7.1 GENERAL VIEW AND KEY MECHANISTIC ASPECTS

Modern copper-catalyzed cross-coupling reactions have emerged as reliable and efficient methods for the construction of C–C and C-heteroatom bonds and are now commonly used in both industrial and academic research. The history of these reactions began more than a century ago with the early and remarkable discoveries of Fritz Ullmann[1,2] and Irma Goldberg[3] (see the Preface for details).

While remarkable from a conceptual point of view, the applicability of these copper-catalyzed formal nucleophilic aromatic substitution reactions has been limited for a long time due to the requirement of stoichiometric amounts of copper, highly polar solvents, high reaction temperatures (>200°C), and long reaction times.[4,5] However, the use of auxiliary ligands in the Ullmann condensation reaction has recently enabled the formation of carbon-heteroatom bonds under catalytic conditions and much milder reaction conditions. The enhanced solubility and stability of copper complexes due to the use of these auxiliary ligands was clearly the key to the renaissance of Ullmann coupling reactions and related copper-mediated transformations. Following important early contributions from Ma, Hauptman, and others in the late 1990s,[6–9] the

Copper-Mediated Cross-Coupling Reactions, First Edition. Edited by Gwilherm Evano and Nicolas Blanchard.

groups of Buchwald[10–12] and Taillefer[13,14] pioneered at the beginning of the twenty-first century the use of simple auxiliary ligands in combination with bases to afford cross-coupled products at lower reaction temperatures and catalyst loadings and with higher reaction rates. Since then, several auxiliary (typically bidentate) ligands such as phenanthrolines,[8] 1,2-diamines,[15] 1,3-diketones,[16,17] bipyridines,[18] and amino acids[19] have been used to promote copper-catalyzed cross-coupling reactions. The use of these auxiliary ligands has broadened the scope of these reactions and also has driven these processes up to large-scale industrial production (see Chapter 19 for details on process chemistry and copper catalysis).[20–22]

Although many mechanistic pathways have been proposed for these copper-catalyzed cross-coupling reactions,[23] experimental and computational data do not converge into a single, unified mechanism. Several key aspects are still under debate, such as the identity and oxidation state of the active copper catalyst or the activation mode of the aryl halide. These points will be extensively discussed in this chapter. As a note, this chapter focuses mainly on the mechanisms of the Ullmann-Goldberg-type cross-coupling reactions, these reactions being the ones that have been more extensively studied from a mechanistical point of view. Discussions of the mechanisms involved in other copper-catalyzed cross-coupling reactions can be found in other chapters of this book.

7.2 OXIDATION STATE OF COPPER CATALYSTS

Copper sources in three different oxidation states (0, I, and II) are found to be effective in copper-catalyzed cross-coupling reactions.[5] For instance, $CuBr_2$, $CuCl_2$, $Cu(OAc)_2$, CuI, CuBr, CuCl, or copper powder have been used in catalytic amounts;[6,24] even Cu metal has been used as coiled copper tubing for continuous flow Ullmann-type reactions.[25] Although the three oxidation states of copper are effective in Ullmann condensation reaction, there is a general agreement that Cu(I) species are the primary active catalysts. Whenever starting from Cu(II) or Cu(0) sources, active Cu(I) species are formed *in situ* by chemical reduction or oxidation processes.[26,27]

Early work from Weingarten in 1964 indicated *in situ* reduction of $CuBr_2$ to CuBr in the presence of phenoxide anions.[6] Later, Kondratov and Shein studied the interaction between Cu(II) species and amines by electron paramagnetic resonance (EPR) and showed that Cu(II) signals decayed over time.[28] It was proposed that ligand oxidation promoted the formation of Cu(I) species and the corresponding amide.

In the 1980s, Paine systematically studied the reaction of diphenylamine (Ph_2NH) with aryl halides to form triarylamines using copper catalysts with three different oxidation states.[24] He proposed that cupric salts ($CuBr_2$ or $Cu(acac)_2$) were first reduced to Cu(I) by the nucleophile before catalysis occurred. When excess of Cu(II) salts were used, the instantaneous formation of tetraphenylhydrazine (Ph_2NNPh_2) suggested the reduction of Cu(II) to

Cu(I) along with the condensation of two Ph_2N^-. Moreover, the reaction rate using copper(II) was about 40% of the rate constant using copper(I) under the same reaction conditions. When copper-metal surfaces were used, the formation of a layer of Cu_2O (360 Å) was responsible for the coupling reaction, as deduced from quantitative X-ray powder diffraction. The dissolution of the Cu_2O upon coordination with Ph_2NLi was proposed to afford the active copper(I) species in solution.

More recently, Jutand and coworkers discovered that electrogenerated [Cu(0)phen] complex is *in situ* oxidized by the aryl iodide or bromide to form $[Cu(I)phen]^+$, the active catalyst in the cross-coupling reaction, and the reduced aromatic, Ar–H, at the beginning of the reaction by an inner sphere electron transfer mechanism.[29]

7.3 IDENTITY OF THE ACTIVE COPPER(I) COMPLEX

Unraveling the precise structure of the copper(I) active catalyst in a given Ullmann-type reaction is a difficult challenge, mostly due to the variability of geometries and coordination possibilities for a d^{10} Cu(I) ion. Indeed, copper(I) can be implicated in many coordination reactions in equilibrium with the nucleophile, solvent molecules, and anions present in the reaction media. The addition of ligands of copper salts in the catalytic reactions may prevent the formation of less reactive, multiply ligated, cuprate structures.[30–33] An alternative approach is to synthesize and study well-defined, stable, and soluble copper(I) complexes, species that have been shown to have good activities in copper-catalyzed C-heteroatom bond formation reactions. This strategy will be described later on (Section 7.4.1).[34,35] Most of available experimental data favor the proposal that the nucleophile is coordinated to the Cu(I) prior to activation of the aryl halide.

In early studies, it was reported that cuprous nucleophile complexes (which contained RO-,[36–38] RS-,[39] or RCON-[40] ligands) reacted with organic halides in nucleophilic substitution reactions. For example, Kawaki and Hashimoto,[36] Whitesides and coworkers[37] and Kubota and Yamamoto[38] synthesized copper(I) alkoxide and phenoxide complexes and studied their reaction with alkyl and aryl halides to form dialkyl, alkyl aryl, and diaryl ethers. Ligand-free alkoxycopper(I) complexes were very unstable, so triphenylphospine was used as auxiliary ligand in order to enhance their stability.[38] Similar copper(I) phenoxide complexes were proposed by Weingarten in the formation of diaryl ether catalyzed by copper(I) bromide, using potassium phenoxide and bromoarenes.[6] Later on, Paine also proposed the coordination of the nucleophile to copper(I) species in the catalytic reaction of diphenylamine with aryl halides to form triarylamines;[24] it was found that the reaction was zero order in the amine nucleophile, indicating that copper(I) species in solution coordinates to the nucleophile in a fast and irreversible step to afford cuprous/nucleophile species $[Ph_2NCu(I)]$, which then react with iodobenzene in the rate-determining step. Yamamoto also isolated several Cu(I) complexes such as $[(Ph_3P)_2Cu(I)$

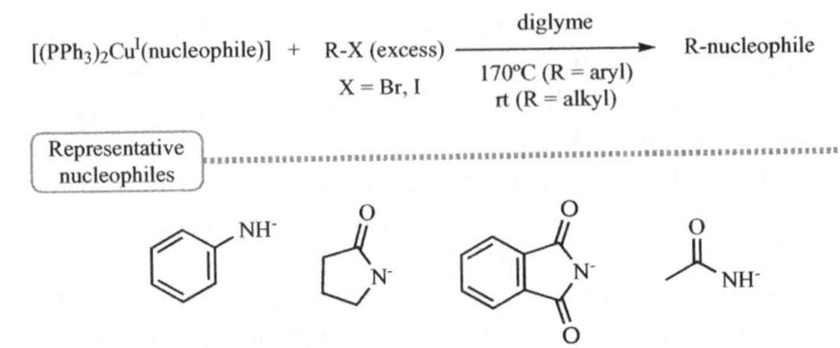

Scheme 7.1 Reactivity of preformed Cu(I)-nucleophile complexes studied by Yamamoto in C–N bond-forming reactions.

(Nucleophile)], in which the N-nucleophile was either a phthalimide, a succinimide, or an acetamide, and showed that they reacted with alkyl and aryl halides to form the corresponding coupled products (Scheme 7.1).[40]

In the last decade, the groups of Hartwig and Buchwald have synthesized and characterized series of Cu(I)amido[30,31,41,42] and Cu(I)phenoxide[43] complexes using bidentate ligands, and have shown that these complexes are chemically and kinetically competent intermediates in the corresponding N- and O-arylation coupling processes.

Hartwig and coworkers characterized series of Cu(I)imidate and Cu(I) amidate complexes (phthalimidate (phth) and pyrrolidinonate (pyrr)) containing bidentate *N,N* and *P,P* donor auxiliary ligands (L), such as phenanthroline, bipyridine, diamines, and diphosphines.[41] Related Cu(I)phenoxide complexes were synthetized using phenanthroline derivatives and 1,2-diamines as dative ligands.[43] These complexes exist in solution as dimeric ionic species $[L_2Cu][Cu(nucleophile)_2]$ in equilibrium with the neutral [LCu(nucleophile)] complexes, as determined by NMR spectroscopy and conductivity experiments (Scheme 7.2), an equilibrium that could be displaced depending on the solvent used. All complexes reacted with excess iodoarene to give the aryl–N coupled products in high yields. The three-coordinated complexes [LCu(nucleophile)] were proposed to be the active intermediates in the coupling reaction due to the lack of reactivity of $[Cu(phth)_2]^-[Bu_4N]^+$ with iodobenzene, indicating that the anionic moiety $[Cu(nucleophile)_2]^-$ is not reactive toward aryl halides.

Related studies with copper(I) phenoxide complexes and phenanthroline or cyclohexane-1,2-diamine derivatives as auxiliary ligands led to similar conclusions as in the previous C–N bond-forming reactions.[43] Again, the complexes exist in solution as dimeric ionic species $[L_2Cu]^+[Cu(OPh)_2]^-$ in equilibrium with the neutral [LCu(OPh)] species. The three-coordinated neutral species were proposed to be the active intermediates for the formation of biaryl ethers.

Buchwald and coworkers used calorimetric methods to study the kinetics of the copper(I)-catalyzed N-arylation of 2-pyrrolidinone with 3,5-dimethyliodobenzene using 1,2-diamines as auxiliary ligands (Scheme 7.3).[30,31] The

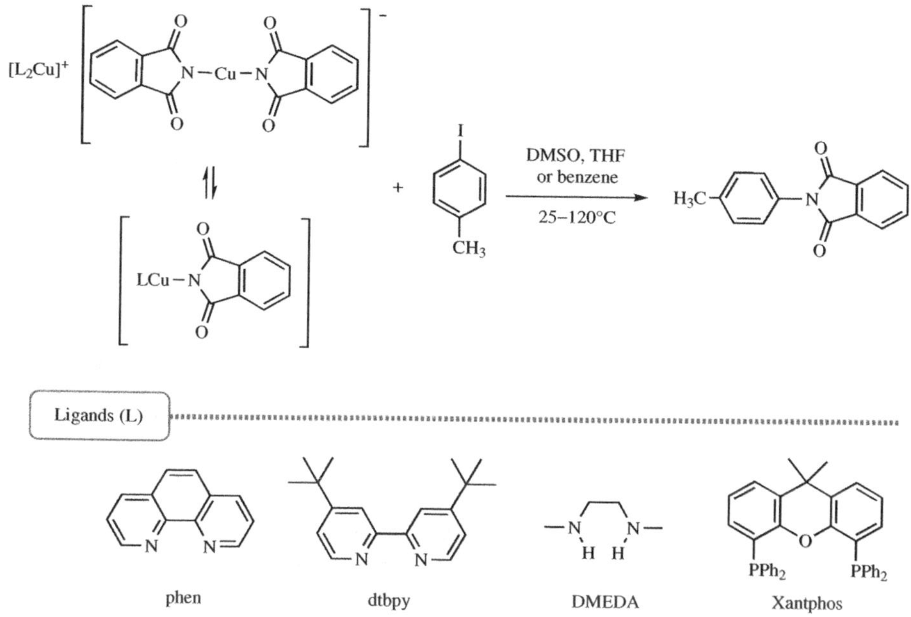

Scheme 7.2 Reaction of Cu(I)imidate complexes with aryl iodide reported by Hartwig.

dependence of the reaction rate on the amide concentration was shown to be a function of 1,2-diamine concentration: At low concentration, the reaction rate decreases upon increasing the concentration of the amide, while at high concentration, the reaction rate increases as the concentration of the amide increases. These results might be understood by the mechanism proposed in Scheme 7.3, where several copper species exist in equilibrium before the iodide activation step. At high concentration of the diamine, the formation of multiple amide ligated complex **C** is prevented, thus avoiding the sequestration of Cu(I) into this inactive copper species. The synthesis of copper(I) amidate complexes with *N,N′*-dimethylcyclohexane-1,2-diamine ligand and their studies in stoichiometric reactions with iodoarenes demonstrated that complex **B** was actually the active intermediate in the catalytic N-arylation reaction. The rate-determining step was found to be the activation of the aryl iodide by the copper(I) amidate complex **B.** Guo calculated by density functional theory (DFT) methods the concentration of multiple ligated species in solution and found that the copper(I) amidate complex [LCu(NHAc)] was predicted to be the major one in the reaction mixture, supporting the experimental data reported by the Buchwald group.[32]

Hartwig and coworkers showed that the ligand had a dramatic influence on the selectivity of the amidation of aryl iodides catalyzed by copper(I) salts and diamine ligands such as *N,N′*-dimethylethylenediamine, 1,10-phenanthroline, and 4,4′-di-*tert*-butyl-2,2′-bipyridine.[41] They performed several competitive reactions using two sterically distinct iodoarenes, *p*-tolyl iodide and *o*-tolyl

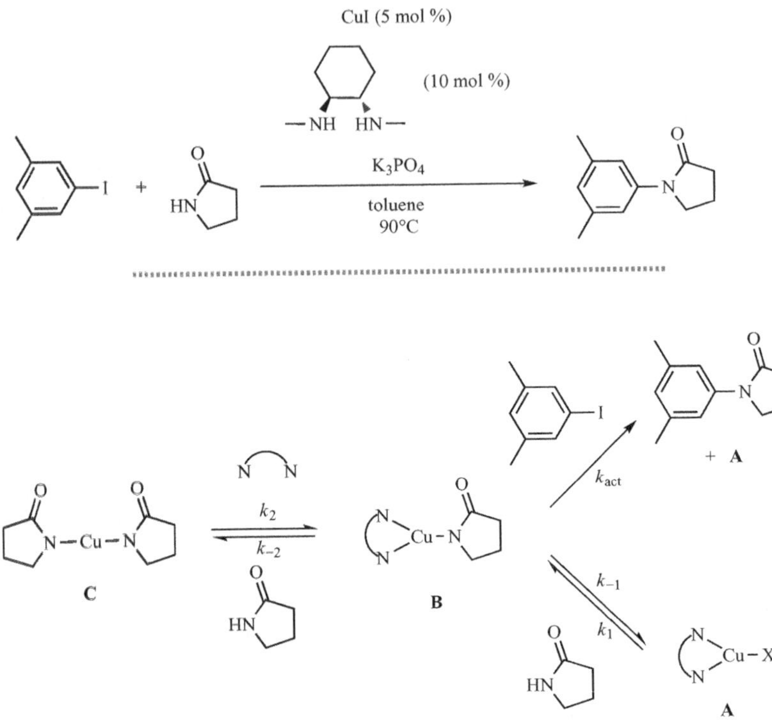

Scheme 7.3 Proposed role of the chelating diamine ligand in the copper(I)-catalyzed N-arylation of amides (Goldberg reaction) with diamine ligands (*N,N′*-dimethylcyclohexane-1,2-diamine and ethylenediamine).

iodide, and compared the product ratios from samples where only copper(I) iodide was used as the catalyst, and others where bidentate N-donor auxiliary ligands were added (Scheme 7.4). In the absence of ligand, the competitive reaction using pyrrolidinone and catalyzed by 10 mol % of copper(I) iodide at 80°C in DMSO afforded *N*-(*p*-tolyl)pyrrolidinone and *N*-(*o*-tolyl)pyrrolidinone in an approximately 1:1 ratio. In sharp contrast, when 20 mol % of *N,N′*-dimethylethylenediamine was added to the reaction mixture, the ratio of *N*-(*p*-tolyl)pyrrolidinone and *N*-(*o*-tolyl)pyrrolidinone was shifted to 83:17. These data indicate that the steric properties of the ligand coordinated to the copper center lead to a greater discrimination among substrates, favoring the less sterically hindered aryl iodide.

Buchwald also studied the ligand-directed chemoselectivities in the N- versus O-arylation of 5-amino-1-pentanol.[44] When a β-diketone ligand was used in DMF, the N-arylated product was obtained with a ratio greater than 20:1 over the product resulting from an O-arylation. In contrast, a complete reversal of chemoselectivity was observed with 3,4,7,8-tetramethylphenanthroline (Me_4-phen) since the O-arylated product was obtained in a ratio 16:1 over the N-arylated one (Scheme 7.5). The same group investigated the role

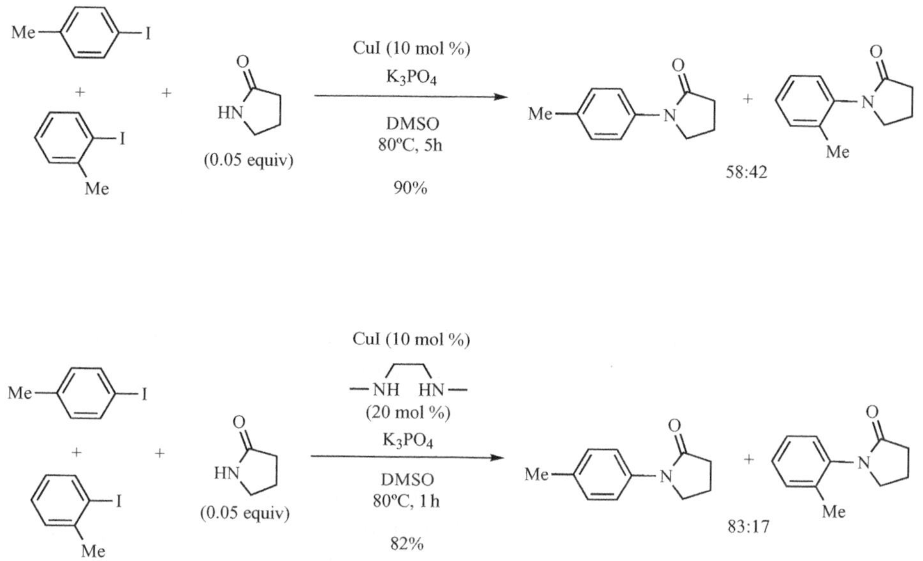

Scheme 7.4 Hartwig's competition experiments showing that auxiliary ligands can influence the selectivity of the arylation of amides.

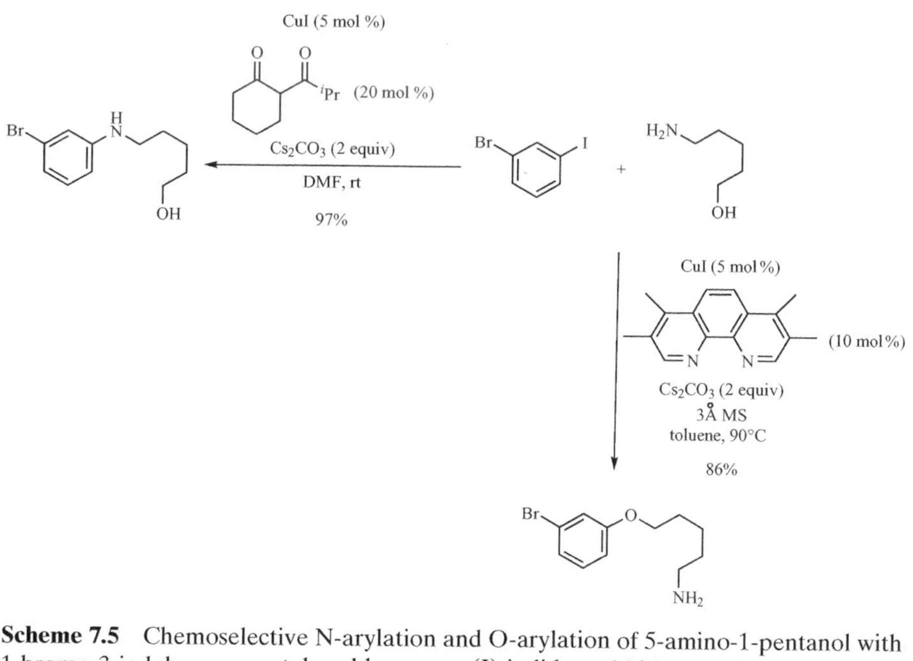

Scheme 7.5 Chemoselective N-arylation and O-arylation of 5-amino-1-pentanol with 1-bromo-3-iodobenzene catalyzed by copper(I) iodide and bidentate ligands.

of the putative Cu(I)-nucleophile initially formed by computational methods with model systems and concluded that, despite a noticeable energy difference between the LCu(I)-OMe and LCu(I)-NHMe computed model complexes, the orthogonal ligand-directed chemoselectivity occurred at the rate-limiting step, that is, the aryl halide activation step, and was not due to the nature of the LCu(I)-nucleophile complex itself.

More recently, Fu and coworkers challenged the previous theoretical results by performing another computational study using a more reliable model system.[45] They found that, indeed, the nature of the ligand already differentiates the ability of each system (either using β-diketone or phenanthroline derivatives) to follow different reaction paths: The β-diketone ligand would favor the coordination of the nucleophile by the O-residue prior to the rate-limiting aryl iodide activation, whereas the Me_4-phen ligand would allow initial aryl iodide activation followed by rate-limiting nucleophile coordination (see also Section 7.4.1).

Additional complications to the full understanding of the mechanism were described by Taillefer and coworkers about the possible existence of precatalysts or Cu(I)-reservoirs in certain systems[46] such as the one resulting from the use of tetradentate pyridine-imine ligands (Scheme 7.6). In this case, the precipitation of $[Cu(I)(L)_2]I_2$ would serve as a copper reservoir, this complex being in equilibrium with small quantities of the soluble monomeric species $[Cu(I)(L)]^+$, the active catalyst able to perform the corresponding biaryl ether formation from iodoarenes and phenols.

As can be seen from examples discussed in this section, the nature as well as the concentration of the auxiliary ligand used in Ullmann-type condensation reactions have a strong impact on the equilibriums between the different copper(I) complexes present in solution, and in the formation of the active catalyst as well. A sufficient amount of the active catalyst must be present in solution in order to obtain good yields for the coupling reaction. Furthermore, the nature of the auxiliary ligand in the LCu(I)-nucleophile complex can also

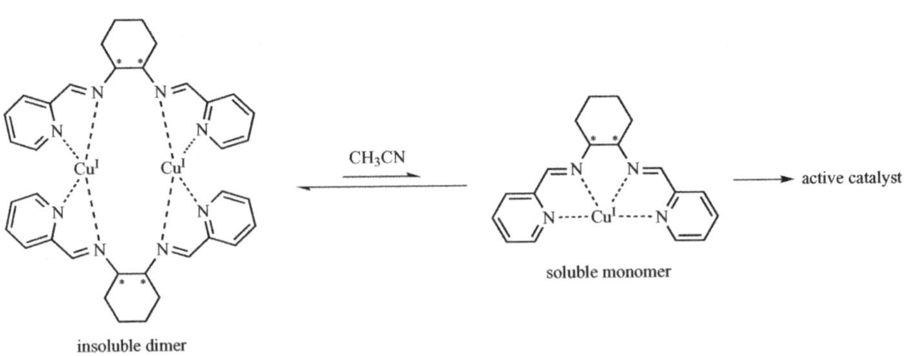

Scheme 7.6 Insoluble dimeric Cu(I) species acting as Cu(I)-reservoirs in pyridine-imine ligand-based systems. (* stands for chiral carbon center.)

determine a distinct selectivity toward sterically different aryl halide substrates. Computational studies are of great help in the comprehension of the mechanism at a molecular level, but theoretical results must be taken with caution since minor differences in the computational methods can render opposite conclusions, even for the exact same reaction.[45,47]

7.4 ACTIVATION MODE OF ARYL HALIDES BY COPPER COMPLEXES

Most reports on copper-catalyzed formal aromatic nucleophilic substitutions support the initial formation of LCu(I)-nucleophile complexes, which are the active catalysts involved in the aryl halide activation step. Since the activation of the aryl halide is usually rate limiting, the detection of intermediate species after this step has been very scarce; thus, most mechanistic proposals are drawn from kinetic and computational studies. The three most frequently invoked mechanisms for the activation of the aryl halide are (i) oxidative addition at copper(I) to form organometallic aryl–copper(III) intermediates, (ii) single-electron transfer (SET), or (iii) atom transfer (AT) resulting in the intermediacy of aryl radical species and copper(II) complexes (Scheme 7.7). Other proposals, such as concerted mechanism via σ-bond metathesis or a mechanistic pathway involving copper(I) π-complexes with the aryl halide, have also been reported, although there is less experimental evidence for these pathways.

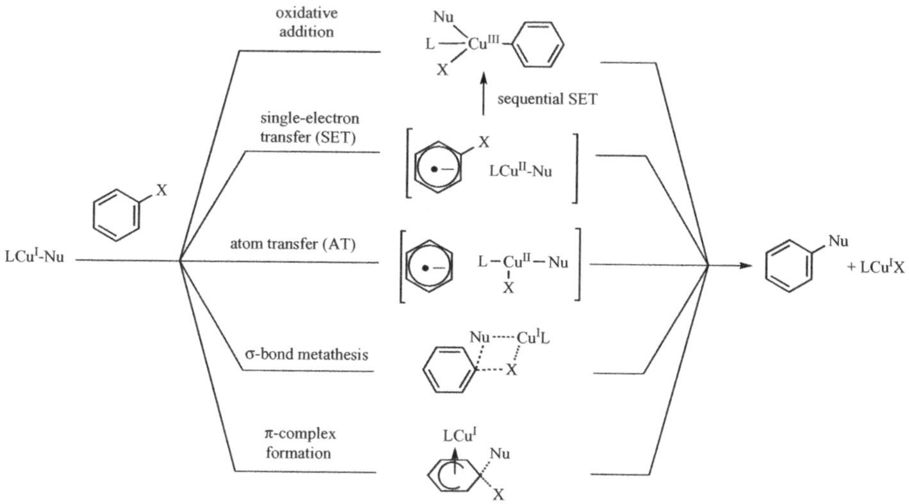

Scheme 7.7 Proposed mechanisms for Ullmann-type catalytic cross-coupling reactions.

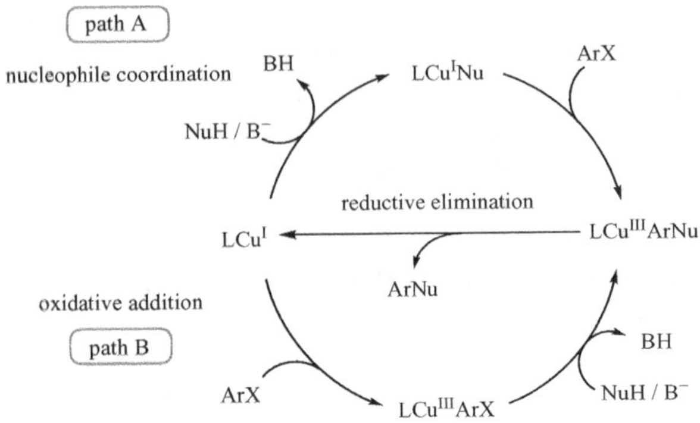

Scheme 7.8 Proposed mechanistic pathways for the Ullmann reaction involving oxidative addition/reductive elimination steps.

7.4.1 Oxidative Addition/Reductive Elimination Pathway

A Cu(I)/Cu(III) catalytic cycle based on two electron redox processes has often been proposed for Ullmann-type coupling reactions in which the activation mode of the aryl halide would consist of an oxidative addition at copper(I) to form an aryl–copper(III) intermediate. Coupling of the nucleophile and aryl moieties would give the coupled product through a reductive elimination step that would regenerate the active copper(I) complex. Two different mechanistic pathways based on the coordination of the nucleophile are proposed (Scheme 7.8). As suggested in many reports (see Section 7.3), the coordination of the base-deprotonated nucleophile to copper(I) may occur before activation of the aryl halide (path A). On the contrary, this coordination may occur after the aryl halide oxidative addition to form an aryl–copper(III) complex has occurred (path B). Experimental data, mainly based on the reactivity of isolated LCu(I)amido and LCu(I)phenoxide complexes with aryl halides (where L is a bidentate ligand), support path A as the most plausible mechanistic scenario.

However, organometallic copper(III) complexes had never been detected under the catalytic Ullmann reaction conditions until very recently (see further).[48] Copper(III) complexes are known in a limited number of compounds,[49–53] and their role in several organic transformations is under debate.[54] Very recently, organometallic copper(III) has been detected by rapid-injection NMR experiments in cuprate-based conjugate additions and nucleophilic substitution reactions.[55,56]

Despite their inherent instabilities, Cohen proposed copper(III) species as key intermediates in Ullmann chemistry for the first time in the 1970s, and since then, several groups have proposed such species with minor or none supporting mechanistic data. Cohen studied the mechanism of the copper-mediated homocoupling of *o*-bromonitrobenzene, the major product being 2,2′-dinitrobiphenyl formed together with small amounts of nitrobenzene.[57,58]

$$ArBr + Cu^{I} \underset{k_{-1}}{\overset{k_1}{\rightleftharpoons}} ArCu^{III}Br$$

$$ArCu^{III}Br + ArBr \xrightarrow{k_2} Ar_2 + Cu^{III}Br_2$$

$$ArCu^{III}Br \xrightarrow[H^+]{k_3} ArH + Cu^{III}Br$$

Scheme 7.9 Proposed mechanism for the reaction of *o*-bromonitrobenzene with copper(I).

The addition of ammonium tetrafluoroborate increased the yield of nitrobenzene and was explained by the intermediacy of organocopper complexes, which were keen to protonation. Neither the reaction rate nor the product distribution was modified by the use of radical traps, indicating that the intermediacy of radical species is unlikely. In addition, Cohen studied the Ullmann coupling reaction of vinyl halides, where the homocoupled products showed retention of the configuration, thus also arguing against the intermediacy of radicals.[58,59] With these mechanistic considerations in hand, Cohen proposed that organocopper(III) species obtained by oxidative addition from aryl or vinyl halides were key intermediate species in Ullmann reactions (Scheme 7.9).

Much more recently, Taillefer and coworkers studied the copper-catalyzed arylation of phenols using several *N*,*N*-chelating ligands in order to determine the relationship between the structure of the chelates and their catalytic activity.[18] The results were explained using a Cu(I)/Cu(III) catalytic cycle via oxidative addition/reductive elimination steps. The most efficient ligands contained one imine- and one pyridine binding site acting in a synergistic manner (Scheme 7.10). By tuning the electronic properties of the pyridine moiety of the ligand, they found that the most electron-donating ligands improved the catalytic activity, which was attributed to a more favorable oxidative addition step, as indicated by the lower oxidation potential. In contrast, the most electron-withdrawing *para*-substituents on the imine binding site favored the arylation reaction, suggesting that this binding site is related to the ligand exchange and/or reductive elimination steps, which are more favorable in electron-deficient copper(III) intermediates.

Since the putative aryl–copper(III) species are usually formed after the rate-limiting step, their detection has proved to be extremely challenging. Therefore, different research groups resorted to DFT computational methods in order to extract further mechanistic information. For example, Hartwig and coworkers have supported the intermediacy of aryl–copper(III) complexes in Ullmann reactions based on computational data and radical trapping experiments.[41–43]

In a more elaborated computational study, Guo calculated the oxidative addition/reductive elimination pathway from diamine ligated copper(I)

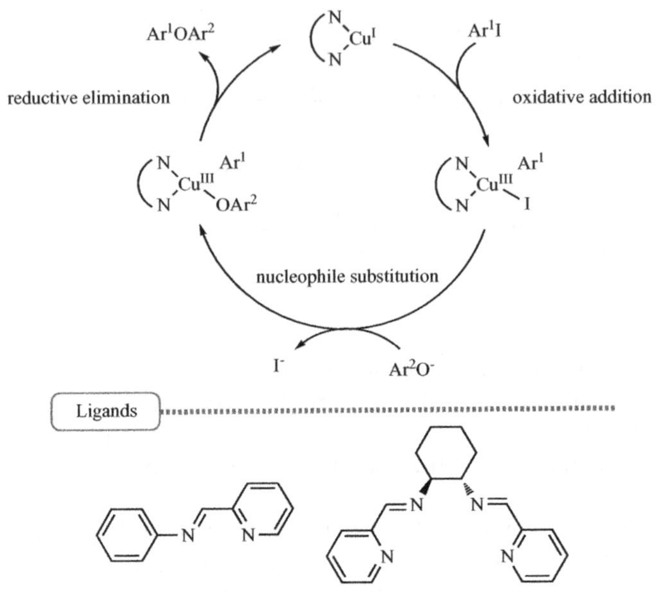

Scheme 7.10 Taillefer's mechanistic proposal for the copper-catalyzed cross-coupling reaction using *N,N*-bidentate ligands.

amidate complexes with bromobenzene (Scheme 7.11).[32] The oxidative addition of copper(I) complex into the aryl–Br bond afforded a square pyramidal pentacoordinated aryl–copper(III) intermediate, where the two coupling partners are situated in *trans*-position. After a pseudoration step, in order to obtain the phenyl and amide group in a *cis* relationship, the reductive elimination was found to have a low energetic barrier. The oxidative addition was the rate-limiting step of the overall copper(I)-catalyzed aryl amidation reaction, with overall free energy barriers among 24–35 kcal.mol^{-1} depending on the bidentate diamine ligand studied. Very recently, a related computational study for the coupling of bromobenzene with methylamine in the presence of copper/diketone ligand has been reported and is also in favor of an oxidative addition/reductive elimination pathway.[60]

Buchwald and Houk also performed a computational study of the mechanistic possibilities to explain the ligand-directed selectivities in N- versus O-arylation of 5-amino-1-pentanol (see Scheme 7.5 in Section 7.3).[44,47] The study focused on the evaluation of the energy barriers for the iodobenzene activation step through four distinct scenarios: (i) oxidative addition/reductive elimination, (ii) SET, (iii) iodine atom transfer (IAT), and (iv) σ-bond metathesis (see Scheme 7.7). The pathways showing a computed lower energy barrier were SET and IAT mechanisms and either two may occur depending on the nucleophile (Scheme 7.12; more details in Section 7.4.2). The energy barrier for oxidative addition of iodobenzene to the [LCu(I)(OMe)] model complex

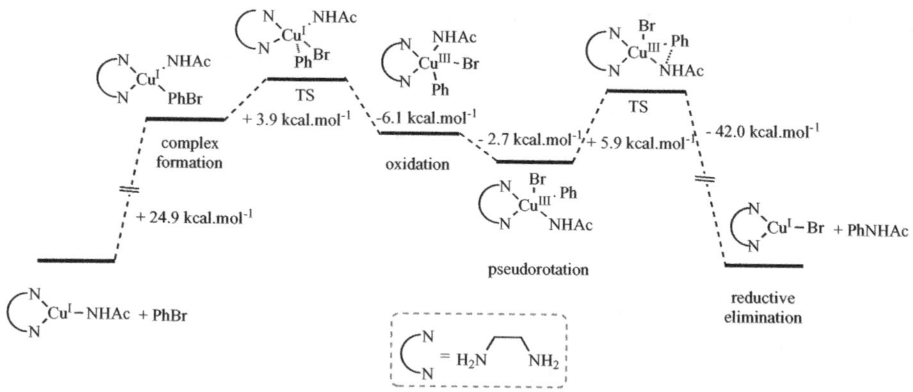

Scheme 7.11 Energetic profile calculated for the reaction of bromobenzene and acetamide catalyzed by Cu(I)/ethylenediamine.

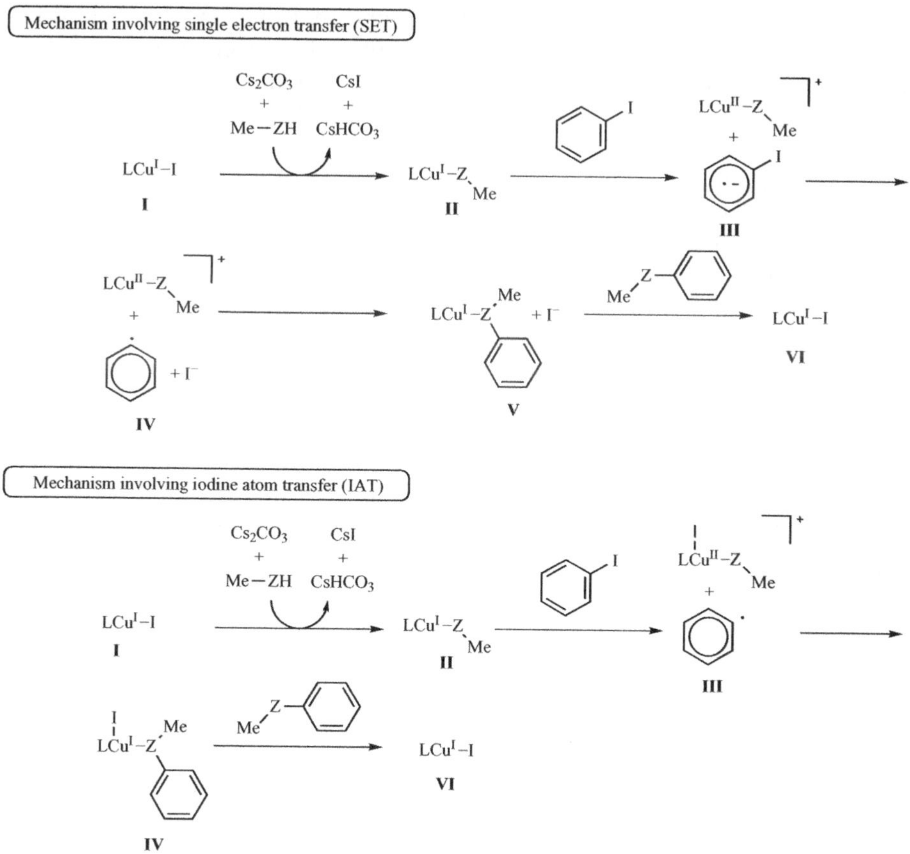

Scheme 7.12 DFT-computed mechanism for the cross coupling between iodobenzene and model MeOH and $MeNH_2$ nucleophiles, involving rate-limiting SET and/or IAT mechanistic pathways (see tabulated ΔG values for each intermediate species in Table 7.1).

TABLE 7.1 Δ*G* values (in kcal.mol^{-1}) for each intermediate species shown in Scheme 7.12

Ligand (L)	Pathway	Bound (*Z*)	Δ*G* I	Δ*G* II	Δ*G* III	Δ*G* IV	Δ*G* V	Δ*G* VI
β-Diketone	SET	O	0	2.9	27.2	20.4	–33.7	–41.3
		N	0	14.8	26.2	19.4	–50.5	–48.0
Me_4-phen	IAT	O	0	7.2	34.0	–28.9	–	–47.1
	SET	N	0	17.0	35.1	26.1	–43.9	–52.6

to form a copper(III) intermediate was found to be much higher, so this pathway was discarded (Table 7.1).

Very recently, Fu and coworkers have proposed an alternative mechanistic explanation for the ligand-dependent selectivities in copper-catalyzed N- and O-arylation reactions.[45] In their study, the model MeOH and $MeNH_2$ nucleophiles used by Buchwald were substituted by the real $NH_2(CH_2)_5OH$ nucleophile used in reaction. Surprisingly, it was found that the oxidative addition/reductive elimination pathway was the most plausible mechanism instead of SET and IAT. It was indeed found that the selectivity did not originate at the aryl halide activation step but rather from steps involving nucleophile coordination and oxidative addition. Among all possible species in equilibrium, they found by computational methods that the most favorable transition state for the oxidative addition using a β-diketone ligand did not involve the amino alcohol nucleophile but activation of the aryl halide before coordination of the nucleophile (Scheme 7.13). Oxidative addition produced a tetracoordinated Cu(III) intermediate that binds to the nucleophile only by the amine residue to form a pentacoordinated Cu(III) complex in an endergonic step (the rate-determining step). In this rate-determining step, the coordination of the amine group is favored over the coordination by the alcohol. Cesium carbonate was not a strong enough base to deprotonate the amino or alcohol groups in the free substrate so the removal of a proton from the coordinated amino alcohol substrate took place before the reductive elimination. The reductive elimination from pentacoordinated Cu(III) species to produce the final N- or O-arylation products and Cu(I) was a very favorable step, with a low energy barrier.

Calculations performed with 1,10-phenantroline (phen) ligand indicated that oxidative addition is the rate-determining step, with a transition state structure containing the copper complex coordinated to the deprotonated alcohol of the substrate. The reductive elimination to produce the C–O coupling product is again a very facile step from a kinetic point of view.

Interestingly, Jutand and coworkers performed DFT calculations on the copper(I)-catalyzed aryl–N bond formation using 1,3-diketonate ligand, and found an interesting activation of the aryl iodide by the nucleophilic cyclohexylamine ($CyNH_2$) moiety of the [(2-acetylcyclohexanone)Cu(I)(NHCy)] complex (Scheme 7.14).[61] DFT calculations favor the aryl-halide oxidative addition to form the square planar [(2-acetylcyclohexanone)Cu(III)(aryl)

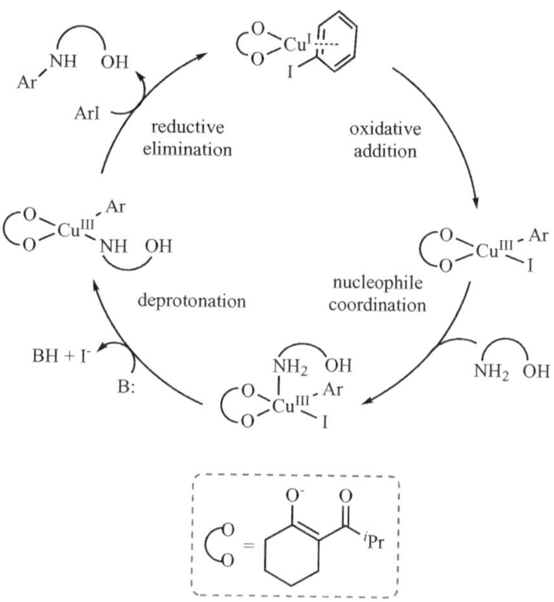

Scheme 7.13 DFT-computed mechanism for the cross coupling between aryl iodides and 5-amino-1-pentanol, involving oxidative addition and reductive elimination steps.

(NHCy)] and liberation of iodide in solution. The subsequent reductive elimination affords aryl-NHCy in moderate yields.

Very recently, Ribas and Stahl provided the first direct observation of the fundamental oxidative addition and reductive elimination steps at a copper(I) complex. They synthetized and characterized a series of pentacoordinated aryl–copper(III)–halide complexes (chloride, bromide, and iodide) based on triazamacrocyclic ligands that allow the stabilization of copper in a high oxidation state.[48] Their preparation enabled the direct observation of aryl halide reductive elimination from the aryl–copper(III)–halide complex triggered by the addition of Brønsted acids, for instance, triflic acid. The reversal step, oxidative addition of copper(I) in the aryl halide ligand to form the corresponding copper(III), is very favorable and extremely fast. *In situ* reversible interconversion between the two redox states, Cu(I)/L-X-H^+ and aryl–Cu^{III}–X, was demonstrated via sequential addition of triflic acid and proton sponge (as noncoordinating base) to an acetonitrile solution of aryl–Cu(III)–bromide complex (Scheme 7.15).

More recently, Ribas and coworkers demonstrated that a reactant-displaced equilibrium between aryl–Cu(III)–X and the corresponding aryl–X···Cu(I) species existed in solution. The addition of external ligands, such as 1,10-phenanthroline (in excess), able to trap the copper(I) displaced the equilibrium toward the aryl–X reductive elimination products in high yields.[62] Moreover, it was demonstrated that copper-catalyzed halide exchange

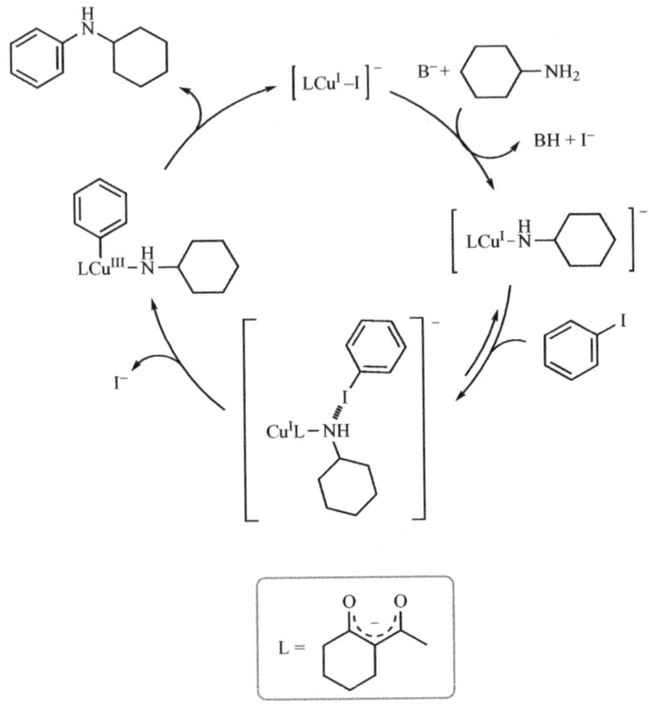

Scheme 7.14 Jutand's proposal involving 1,3-diketonate ligands and activation of the aryl iodide by the nucleophile.

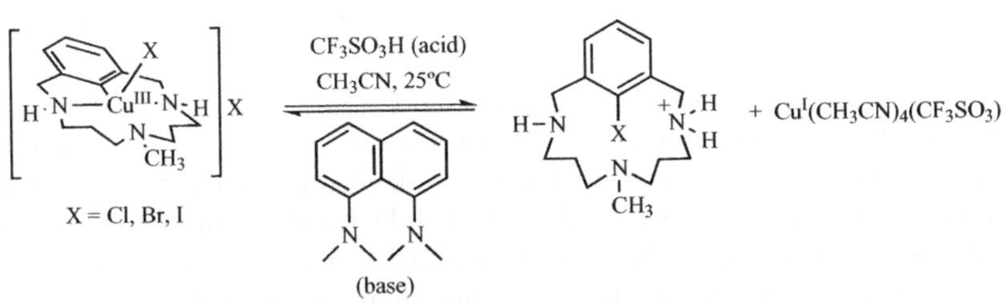

Scheme 7.15 Reductive elimination and oxidative addition in a well-defined triazamacrocyclic aryl–Cu(III) complex.

reactions within these aryl halide model systems occurred through a Cu(I)/Cu(III) catalytic cycle (see Chapter 6 for details).[62]

The groups of Stahl and Ribas also demonstrated that aryl–copper(III)–halide complexes can be plausible intermediates in copper-catalyzed C–N and C–O bond formation in the presence of many nitrogen and oxygen nucleophiles.[48,63,64] In an initial report, Huffman and Stahl studied the reactions between a triazamacrocyclic aryl–copper(III) complex (perchlorate salt) with

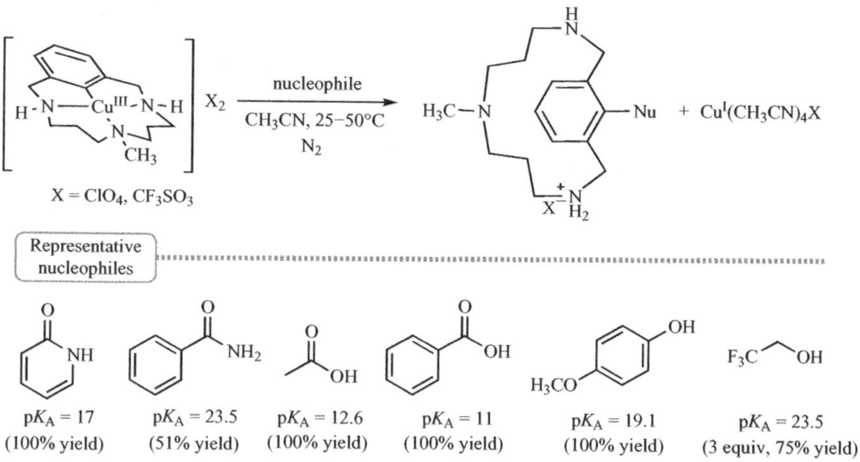

Scheme 7.16 C–N and C–O bond formation from well-defined aryl–copper(III) complexes (pK_A values referring to N–H or O–H moieties in DMSO).

several amides (N-nucleophiles) to afford the C–N reductive elimination product under mild conditions (from 25 to 50°C).[63] The reactions were faster with the more acidic nitrogen nucleophiles, implicating that proton loss took place before or during the rate-determining step. In a following study, the Stahl and Ribas groups investigated the analogous reactions with oxygen nucleophiles such as carboxylic acids, phenols, and aliphatic alcohols[64] (Scheme 7.16). On the basis of kinetic and spectroscopic data, the formation of adducts between the nucleophile and the aryl–Cu^{III} complex before the C–O bond formation step was proposed.

Wang later on reported another aryl–copper(III) complex based on a macrocyclic ligand and demonstrated its reactivity toward several anionic nucleophiles, such as carboxylates, halide salts, and cyanides, to form C-heteroatom and C–C bonds at room temperature,[65,66] demonstrating the easiness of the reductive elimination step from aryl–Cu(III) complexes reacting.

Following their studies on copper(III) complexes, Ribas and Stahl demonstrated that the cross coupling of model macrocyclic aryl-halides with N-nucleophiles could be attained catalytically at room temperature, and aryl–Cu(III)–halide intermediate species were unequivocally characterized under catalytic conditions. This observation was proof of principle for the experimental feasibility of an oxidative addition/reductive elimination Cu(I)/Cu(III) catalytic cycle at room temperature, thus supporting the a priori less favored path B in Scheme 7.8.

Additional support for the involvement of this oxidative addition/reductive elimination pathway in copper(I)-catalyzed cross-coupling reactions was provided by Hartwig and coworkers very recently, in a system where 1,10-phenanthroline ligand, copper(I) sources, and enolates were used to

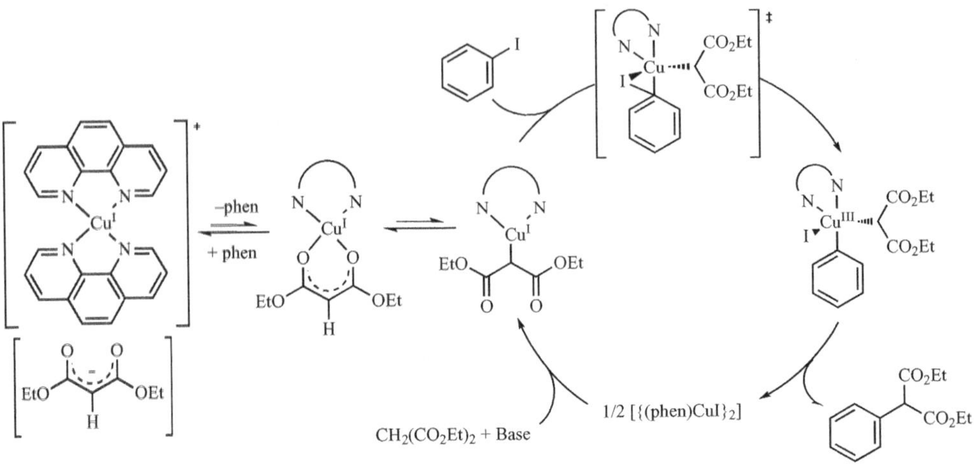

Scheme 7.17 Aryl–Cu(III) species involved in Hurtley-type C–C coupling reactions, as described by Hartwig.

achieve the α-arylation of enolates (Hurtley-type reaction; see Chapter 8 for details). Experimental and theoretical data allowed the authors to propose a mechanistic scenario involving the oxidative addition of aryl halide to [(phen)Cu(I)(enolate)], followed by reductive elimination and cross coupling to release aryl-enolate products (Scheme 7.17).[67]

7.4.2 SET Mechanism

Some authors have supported the intermediacy of radical species formed via SET in a Cu(I)/Cu(II) catalytic cycle. In this proposal, the copper(I) complex is oxidized by one-electron transfer from the aryl halide, yielding to the formation of an aryl halide radical anion. Then, the aryl radical couples with the nucleophile and the copper(II) is reduced, regenerating the catalyst and releasing the neutral coupled product (see Scheme 7.12 and Scheme 7.18). This mechanism resembles a $S_{RN}1$ radical chain pathway.

Hida reported in the 1970s the mechanistic study of the coupling reaction of 2-aminoethanol with 1-bromoanthraquinone (AQBr) catalyzed by copper(I) bromide.[68,69] EPR experiments showed the presence of two paramagnetic species in solution, copper(II) and AQBr radical anion formed by SET from copper(I) species to AQBr (AQBr + Cu(I) → $AQBr^-$ + Cu(II)). Although haloanthraquinones are very specific substrates, the presence of the reduced product anthraquinone (AQH) together with the Ar–N coupled product, as in many Ullmann condensation reactions, prompted Hida to propose the SET step as a general mechanism for the activation of aryl halides.

A SET pathway was one of Buchwald's suggested operative mechanisms for the ligand-directed selectivities in N- versus O-arylation orthogonal coupling reactions of iodobenzene with 5-amino-1-pentanol (Scheme 7.18).[47] Indeed, the SET pathway was energetically preferred over oxidative addition

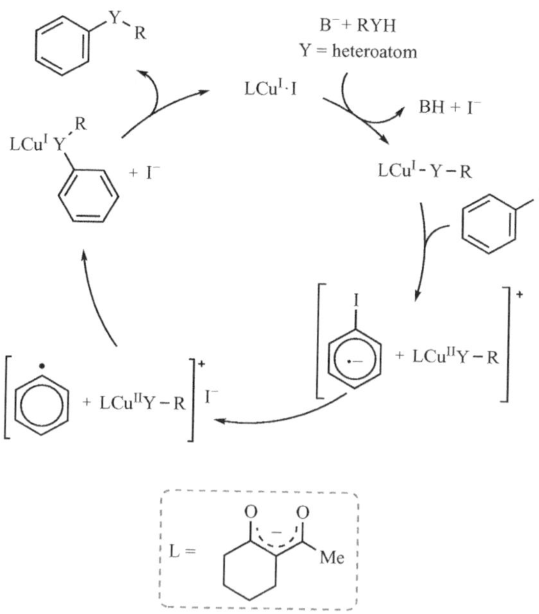

Scheme 7.18 Single-electron mechanistic pathway proposed by Buchwald.

or AT pathways (see Section 7.4.1) when β-diketone auxiliary ligand was used.

Recently, van Koten and coworkers reported the catalytic activity of a family of aminoarenethiolato-copper(I) complexes in the reaction of amine and phenolic substrates with bromobenzene derivatives.[70] Surprisingly, in these reactions, only aryl bromides were active, and almost no reactivity was found with aryl iodides or aryl chlorides. Moreover, the authors showed that the use of radical traps slowed down or even stopped the reaction. The authors also showed that certain amounts of Cu(II) were present in solution when the reactivity vanished; the addition of metallic copper to the reaction mixture restored the reactivity presumably via comproportionation with the copper(II) species in solution to regenerate the active Cu(I) species. The lack of reactivity of aryl iodides was explained by the overstabilization of Cu(I) species by the soft iodide anion, thus preventing an oxidation to Cu(II) species. These observations led the authors to propose a mechanism involving a SET from the Cu(I) center to the aryl bromide to form an aryl radical (kinetically protected by the back reaction with Cu(II)) and a Cu(II) species, subsequent reaction of the aryl radical with the amine moiety and a second SET to regenerate the Cu(I) species.

Peters, Fu, and coworkers reported in 2012 a photoluminescent copper-carbazolide complex bearing monophosphine ligands for promoting the photoinduced C–N bond forming reaction (Scheme 7.19).[71a] Upon photoexcitation of the copper-carbazolide complex, a copper-containing radical is detected by EPR spectroscopy. This radical intermediate reacts with aryl halides via SET

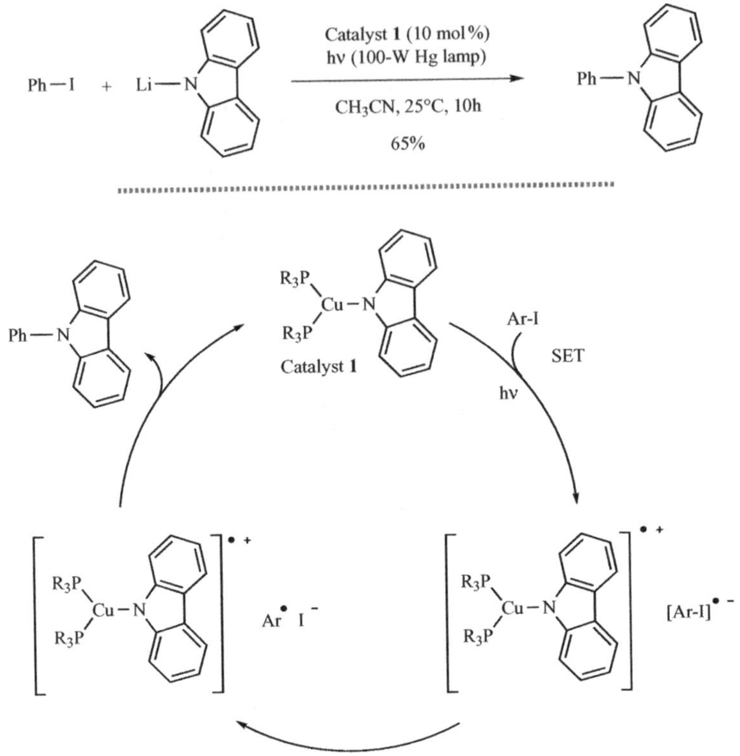

Scheme 7.19 Photoinduced Ullmann C–N coupling reactions catalyzed by a copper-carbazolide complex, along with the proposed catalytic cycle involving a photoinduced SET mechanism (*R* = *m-tolyl, phenyl*).

to afford the corresponding C–N coupling product. Interestingly, radical clock test with deuterated *o*-(allyloxy)iodobenze afforded exclusive formation of cyclized compounds as a 1:1 mixture of diastereoisomers, which is consistent with a radical pathway. This important contribution demonstrated that C–X cleavage as well as the C–N bond forming steps can proceed via SET processes. This photoinduced reaction was next extended to the development of an efficient photoinduced, copper-catalyzed cross coupling of thiols with aryl halides under remarkably mild conditions (0°C) via a SET process.[71b]

7.4.3 AT Mechanism

Kochi studied more than 50 years ago the reactivity of free aryl radicals obtained by decomposition of peroxides and iodonium salts catalyzed by copper salts.[72,73] These aryl radicals can react with copper(II) halide salts via an AT mechanism to produce the corresponding aryl halide and copper(I) species. The Kochi group proposed that the Sandmeyer and Meerwein

$$Cu^{I}X + Nu^{-} \longrightarrow Cu^{I}Nu + X^{-}$$

$$ArX + Cu^{I}Nu \longrightarrow Ar\cdot + Cu^{II}XNu$$

$$Ar\cdot + Cu^{II}XNu \longrightarrow ArNu + Cu^{I}X$$

Scheme 7.20 Kochi's proposal for ligand exchange reaction in copper(I) salts.

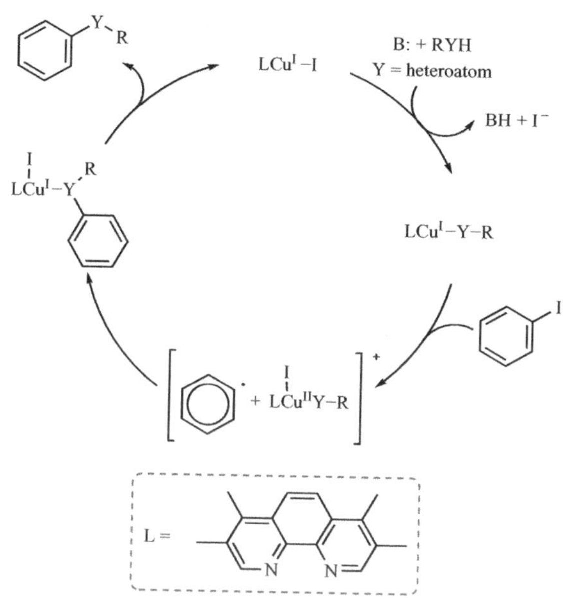

Scheme 7.21 Iodine atom transfer mechanism proposed by Buchwald on the basis of computational studies.

reactions proceed via radical chain mechanism in which copper has an important role as radical chain terminator (Scheme 7.20).

Following the comparisons of mechanisms for the ligand-directed chemoselective N- or O-arylation of 5-amino-1-pentanol with iodobenzene described by Buchwald (see previous discussion),[47] an IAT pathway was the energy favored mechanism when tetramethylphenanthroline auxiliary ligand was used (Scheme 7.21).

7.4.4 Mechanism Involving π-Complexation of Copper(I) to Aryl Halides

The first proposal of a mechanism involving nucleophilic aromatic substitution via a π-complexed organocuprate was reported in 1964 by Weingarten in the study of the copper-catalyzed coupling reaction of bromobenzene with potassium phenoxide.[6] In this mechanism, the Cu(I)-nucleophile species interacts with the π electrons of the aromatic ring in order to promote the polarization

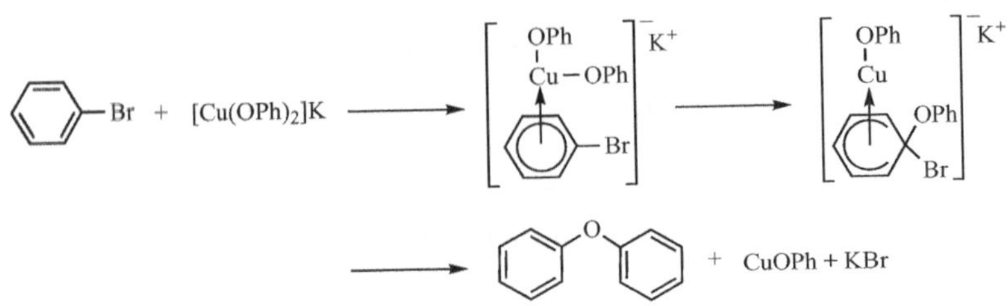

Scheme 7.22 Mechanism for the coupling reaction of bromobenzene with potassium phenoxide proposed by Weingarten.

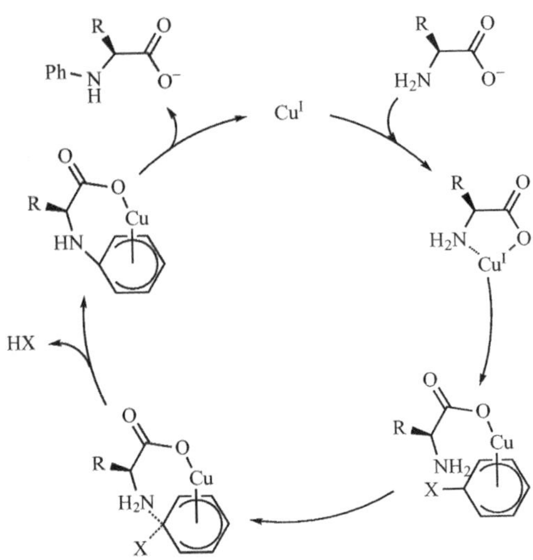

Scheme 7.23 Ma's early mechanistic proposals involving a π-complex intermediate species.

of the C–X bond. The resulting Wheland complex facilitates the substitution of the halide with the nucleophile in the ring (Scheme 7.22), while the copper species maintains its oxidation state +I.

Although experimental evidences to support this hypothesis are scarce, Weingarten proposed this mechanism based on the relative reactivity order of several aryl halides (ArI > ArBr > ArCl).[74] The relative rates found through competition experiments followed the trend of the C–X bond strength (C–I < C–Br < C–Cl) as in nucleophilic aromatic substitutions, indicating that the C–X bond cleavage is involved in the rate-determining step.

Ma and coworkers observed that α-amino acids were coupled with aryl iodides and bromides in the presence of a catalytic amount of copper(I) iodide under very mild reaction conditions.[7] The structure of the α-amino acid influenced the yield of the coupled product, and they suggested that an amino acid–copper(I) complex was formed prior to the activation of the aryl halide through a π-complex intermediate (Scheme 7.23). Recently, Ma proposed an

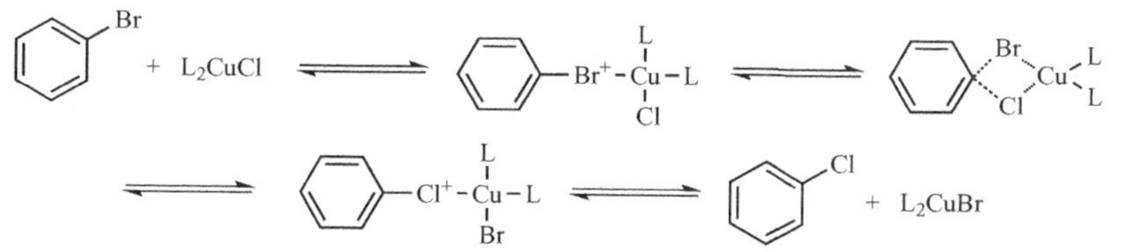

Scheme 7.24 Bacon and Hill's proposal based on a σ-bond metathesis pathway.

alternative reaction pathway based on oxidative addition/reductive elimination steps.[19]

7.4.5 Mechanism Involving σ-Bond Metathesis

In this mechanistic pathway, the copper(I) nucleophile complex activates the aryl halide via a four-centered intermediate in which the metal center forms a σ-complex with the lone electron pair of the halogen atom. The polarization of the C–X bond creates a partial positive charge on the *ipso*-carbon and facilitates the attack by the nucleophile to give the corresponding coupling product and the copper complex, which remains with an oxidation state of +I. In early studies, Bacon and Hill proposed this mechanism to explain their results in the substitution reactions between aryl halides and cuprous salts (Scheme 7.24).[75–77]

A similar mechanistic proposal was evaluated computationally by Cundari for the reaction of a well-defined NHC carbene Cu(I)-nucleophile complex, (NHC)Cu(I)(NHPh), with iodobenzene, but the higher energy barriers pointed toward an oxidative addition/reductive elimination mechanism.[78]

7.5 OVERVIEW, CONCLUSIONS, AND FUTURE PROSPECTS

Although many mechanistic proposals have been reported, the oxidative addition/reductive elimination pathway for the aryl halide activation is, in general, favored from all available experimental data. Also, a number of theoretical studies have been recently published, and some of them also favor the latter mechanism, although radical-based pathways have also been considered. Radical trapping experiments performed in some of the studies afford no differences in reactivity.[41,43,79] Thus, in general, the theoretically proposed radical-based mechanistic proposals (SET and IAT) lack experimental support in copper-catalyzed Ullmann-type coupling reactions, with some important exceptions such as the catalytic systems developed by van Koten and coworkers, where a clear inhibitory effect upon addition of radical traps was reported.[70b] Importantly, a SET mechanism has been very recently identified in a photoinduced C–N reaction using carbazolide nucleophiles.[71a] In any case, even with negative results for radical trapping experiments, the formation of short-lived intimate radicals in solvent cages cannot be completely discarded. Indeed, it is reasonable to state that the operative mechanism may differ from system to system, thus highlighting the extraordinary range of mechanistic possibilities and accessible redox states exhibited by copper.

The mechanistic comprehension of Ullmann–Goldberg and, in general, Ullmann-type coupling reactions may also help in the understanding of other copper-catalyzed reactions such as the one involving a C–H activation and further functionalization with an heteroatom nucleophile[80] or the Chan–Lam–Evans coupling reactions catalyzed by Cu(II) salts under an atmosphere of oxygen or air (see Chapter 4 for details on the Chan–Lam–Evans coupling reaction).[81] With respect to the mechanistic details for the Chan–Lam–Evans coupling, it is worth mentioning the latest mechanistic proposal for this useful transformation (Scheme 7.25): The catalytic cycle involves the rate-limiting transmetallation of the Cu(II) salt with an arylboronic acid/ester to form an aryl–Cu(II)–X complex. The latter reacts with another molecule of Cu(II)X_2 through a disproportionation reaction to afford a Cu(I)X and aryl–Cu(III)X_2, and then the latter reacts with the nucleophile to afford the reductive elimination products, namely, the coupled product Ar–Nu and Cu(I)X. Finally aerobic oxidation of the Cu(I) species generated in the catalytic cycle to Cu(II) species restores the active species and closes the cycle.[82]

Many open questions concerning Ullmann-type couplings still remain to be answered; thus, the profound understanding of the mechanistic details of copper-catalyzed coupling reactions is imperative to be able to improve the

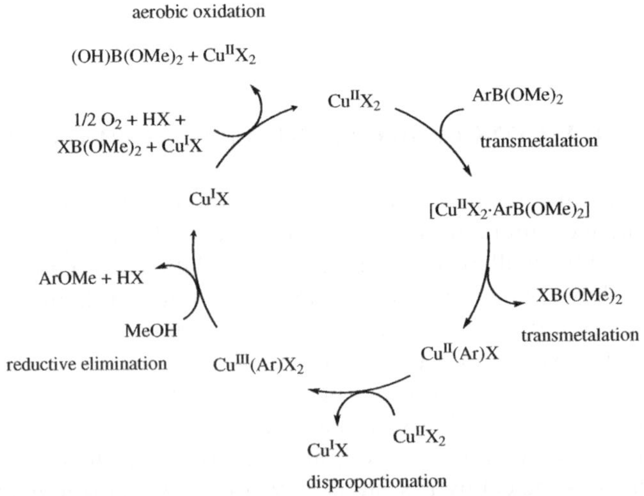

Scheme 7.25 Mechanistic proposal for Chan–Lam–Evans coupling reactions between aryl boronic acids/esters and MeOH as nucleophile, mediated by Cu(II) complexes under an atmosphere of oxygen.

catalyst design. Moreover, mechanistic comprehension of these reactions may give additional clues for other copper-catalyzed reactions, beyond Ullmann-type reactivity.

REFERENCES

[1] Ullmann, F.; Bielecki, J. *Chem. Ber.* **1901**, *34*, 2174–2178.

[2] Ullmann, F. *Ber. Dtsch. Chem. Ges.* **1903**, *36*, 2389–2391.

[3] Goldberg, I. *Ber. Dtsch. Chem. Ges.* **1906**, *39*, 1691–1696.

[4] Fanta, P. E. *Synthesis* **1974**, 9–21.

[5] Lindley, J. *Tetrahedron* **1984**, *40*, 1433–1455.

[6] Weingarten, H. *J. Am. Chem. Soc.* **1964**, *29*, 3624–3626.

[7] Ma, D.; Zhang, Y.; Yao, J.; Wu, S.; Tao, F. *J. Am. Chem. Soc.* **1998**, *120*, 12459–12467.

[8] Goodbrand, H. B.; Hu, N.-X. *J. Org. Chem.* **1999**, *64*, 670–674.

[9] Fagan, P. J.; Hauptman, E.; Shapiro, R.; Casalnuovo, A. *J. Am. Chem. Soc.* **2000**, *122*, 5043–5051.

[10] Kaplars, A.; Antilla, J. C.; Huang, X.; Buchwald, S. L. *J. Am. Chem. Soc.* **2001**, *123*, 7727–7729.

[11] Klapars, A.; Huang, X.; Buchwald, S. L. *J. Am. Chem. Soc.* **2002**, *124*, 7421–7428.

[12] Buchwald, S. L.; Klapars, A.; Antilla, J. C.; Job, G. E.; Wolter, M.; Kwong, F. Y.; Nordmann, G.; Hennessy, E. J. Patent WO 02/085838 A1, 2001.

[13] Taillefer, M.; Cristau, H.-J.; Cellier, P. P.; Spindler, J.-F.; Ouali, A. Patent Fr 840303-WO 03101966 (Pr. Nb. Fr 2002 06717), 2002.

[14] Taillefer, M.; Cristau, H.-J.; Cellier, P. P.; Spindler, J.-F. Patent Fr 2833947-WO 0353225 (Pr. Nb. Fr. 01 16547), 2001.

[15] Surry, D. S.; Buchwald, S. L. *Chem. Sci.* **2010**, *1*, 13–31.

[16] Shafir, A.; Buchwald, S. L. *J. Am. Chem. Soc.* **2006**, *128*, 8742–8743.

[17] de Lange, B.; Lambers-Verstappen, M. H.; Schmieder-van de Vondervoort, L.; Sereinig, N.; de Rijk, R.; de Vries, A. H. M.; de Vries, J. G. *Synlett* **2006**, 3105–3109.

[18] Ouali, A.; Spindler, J.-F.; Jutand, A.; Taillefer, M. *Adv. Synth. Catal.* **2007**, *349*, 1906–1916.

[19] Ma, D.; Cai, Q. *Acc. Chem. Res.* **2008**, *41*, 1450–1460.

[20] (a) Ley, S. V.; Thomas, A. W. *Angew. Chem. Int. Ed.* **2003**, *42*, 5400–5449. (b) Beletskaya, I. P.; Cheprakov, A. V. *Coord. Chem. Rev.* **2004**, *248*, 2337–2364.

[21] Evano, G.; Blanchard, N.; Toumi, M. *Chem. Rev.* **2008**, *108*, 3054–3131.

[22] Monnier, F.; Taillefer, M. *Angew. Chem. Int. Ed.* **2009**, *48*, 6954–6971.

[23] Casitas, A.; Ribas, X. *Chem. Sci.* **2013**, *4*, 2301–2318.

[24] Paine, A. J. *J. Am. Chem. Soc.* **1987**, *109*, 1496–1502.

[25] Zhang, Y.; Jamison, T. F.; Patel, S.; Mainolfi, N. *Org. Lett.* **2011**, *13*, 280–283.

[26] Sperotto, E.; van Klink, G. P. M.; van Koten, G.; de Vries, J. G. *Dalton Trans.* **2010**, *39*, 10338–10351.

[27] Franc, G.; Jutand, A. *Dalton Trans.* **2010**, *39*, 7873–7875.

[28] Kondratov, S. A.; Shein, S. M. *Zh. Org. Khim.* **1979**, *15*, 2160.

[29] Mansour, M.; Giacovazzi, R.; Ouali, A.; Taillefer, M.; Jutand, A. *Chem. Commun.* **2008**, 6051–6053.

[30] Strieter, E. R.; Blackmond, D. G.; Buchwald, S. L. *J. Am. Chem. Soc.* **2005**, *127*, 4120–4121.

[31] Strieter, E. R.; Bhayana, B.; Buchwald, S. L. *J. Am. Chem. Soc.* **2009**, *131*, 78–88.

[32] Zhang, S.-L.; Liu, L.; Fu, Y.; Guo, Q.-X. *Organometallics* **2007**, *26*, 4546–4554.

[33] Nakamura, E.; Mori, S. *Angew. Chem. Int. Ed.* **2000**, *39*, 3750–3771.

[34] Gujadhur, R. K.; Bates, C. G.; Venkataraman, D. *Org. Lett.* **2001**, *3*, 4315–4317.

[35] Haldón, E.; Alvarez, E.; Nicasio, M. C.; Pérez, P. J. *Organometallics* **2009**, *28*, 3815–3821.

[36] Kawaki, T.; Hashimoto, H. *Bull. Chem. Soc. Jpn* **1972**, *45*, 1499–1500.

[37] Whitesides, G. M.; Sadowski, J. S.; Lilburn, J. *J. Am. Chem. Soc.* **1974**, *96*, 2829–2835.

[38] Kubota, M.; Yamamoto, A. *Bull. Chem. Soc. Jpn* **1978**, *51*, 2909–2915.

[39] Adams, R.; Ferretti, A. *J. Am. Chem. Soc.* **1959**, *81*, 4927–4931.

[40] Yamamoto, T.; Ehara, Y.; Kubota, M.; Yamamoto, A. *Bull. Chem. Soc. Jpn* **1980**, *53*, 1299–1302.

[41] Tye, J. W.; Weng, Z.; Johns, A. M.; Incarvito, C. D.; Hartwig, J. F. *J. Am. Chem. Soc.* **2008**, *130*, 9971–9983.

[42] Giri, R.; Hartwig, J. F. *J. Am. Chem. Soc.* **2010**, *132*, 15860–15863.

[43] Tye, J. W.; Weng, Z.; Giri, G.; Hartwig, J. F. *Angew. Chem. Int. Ed.* **2010**, *49*, 2185–2189.

[44] Shafir, A.; Lichtor, P. A.; Buchwald, S. L. *J. Am. Chem. Soc.* **2007**, *129*, 3490–3491.

[45] Yu, H.-Z.; Jiang, Y.-Y.; Fu, Y.; Liu, L. *J. Am. Chem. Soc.* **2010**, *132*, 18078–18091.

[46] Ouali, A.; Taillefer, M.; Spindler, J.-F.; Jutand, A. *Organometallics* **2007**, *26*, 65–74.

[47] Jones, G. O.; Liu, P.; Houk, K. N.; Buchwald, S. L. *J. Am. Chem. Soc.* **2010**, *132*, 6205–6213.

[48] Casitas, A.; King, A. E.; Parella, T.; Costas, M.; Stahl, S. S.; Ribas, X. *Chem. Sci.* **2010**, *1*, 326–330.

[49] Margerum, D. W.; Chellappa, K. L.; Bossu, F. P.; Burce, G. L. *J. Am. Chem. Soc.* **1975**, *97*, 6894–6896.

[50] Hanss, J.; Beckmann, A.; Krüger, H.-J. *Eur. J. Inorg. Chem.* **1999**, 163–172.

[51] Ribas, X.; Jackson, D. A.; Donnadieu, B.; Mahía, J.; Parella, T.; Xifra, R.; Hedman, B.; Hodgson, K. O.; Llobet, A.; Stack, T. D. P. *Angew. Chem. Int. Ed.* **2002**, *41*, 2991–2994.

[52] Furuta, H.; Maeda, H.; Osuka, A. *J. Am. Chem. Soc.* **2000**, *122*, 803–807.

[53] Naumann, D.; Roy, T.; Tebbe, K.-F.; Crump, W. *Angew. Chem. Int. Ed. Engl.* **1993**, *32*, 1482–1483.

[54] Ribas, X.; Casitas, A. *Ideas in Chemistry and Molecular Science. Where Chemistry Meets Life*, Pignataro, B. (ed.). Wiley-VCH: Weinheim, **2010**, pp. 31–57.

[55] Bertz, S. H.; Cope, S.; Murphy, M.; Ogle, C. A.; Taylor, B. J. *J. Am. Chem. Soc.* **2007**, *129*, 7208–7209.

[56] Bertz, S. H.; Cope, S.; Dorton, D.; Murphy, M.; Ogle, C. A. *Angew. Chem. Int. Ed.* **2007**, *46*, 7082–7085.

[57] Cohen, T.; Cristea, I. *J. Am. Chem. Soc.* **1976**, *98*, 748–753.

[58] Cohen, T.; Cristea, I. *J. Org. Chem.* **1975**, *40*, 3649–3651.

[59] Cohen, T.; Poeth, T. *J. Am. Chem. Soc.* **1972**, *94*, 4363–4364.

[60] Zhang, Y.; Ding, Y. *Organometallics* **2011**, *30*, 633–641.

[61] Franc, G.; Cacciuttolo, Q.; Lefèvre, G.; Adamo, C.; Ciofini, I.; Jutand, A. *ChemCatChem* **2011**, *3*, 305–309.

[62] Casitas, A.; Canta, M.; Costas, M.; Solà, M.; Ribas, X. *J. Am. Chem. Soc.* **2011**, *133*, 19386–19392.

[63] Huffman, L. M.; Stahl, S. S. *J. Am. Chem. Soc.* **2008**, *130*, 9196–9197.

[64] Huffman, L. M.; Casitas, A.; Font, M.; Canta, M.; Costas, M.; Ribas, X.; Stahl, S. S. *Chem. Eur. J.* **2011**, *17*, 10643–10650.

[65] Yao, B.; Wang, D.-X.; Huang, Z.-T.; Wang, M.-X. *Chem. Commun.* **2009**, 2899–2901.

[66] Wang, Z.-L.; Zhao, L.; Wang, M.-X. *Org. Lett.* **2011**, *13*, 6560–6563.

[67] Huang, Z.; Hartwig, J. F. *Angew. Chem. Int. Ed.* **2012**, *51*, 1028–1032.

[68] Arai, S.; Hida, M.; Yamagishi, T. *Bull. Chem. Soc. Jpn* **1978**, *51*, 277–282.

[69] Arai, S.; Yamagishi, T.; Ototake, S.; Hida, M. *Bull. Chem. Soc. Jpn* **1977**, *50*, 547–548.

[70] (a) Sperotto, E.; van Klink, G. P. M.; de Vries, J. G.; van Koten, G. *Tetrahedron* **2010**, *66*, 3478–3484. (b) Sperotto, E.; van Klink, G. P. M.; de Vries, J. G.; van Koten, G. *Tetrahedron* **2010**, *66*, 9009–9020.

[71] (a) Creutz, S. E.; Lotito, K. J.; Fu, G. C.; Peters, J. C. *Science* **2012**, *338*, 647–651. (b) Uyeda, C.; Tan, Y.; Fu, G. C.; Peters, J. C. *J. Am. Chem. Soc.* **2013**, *135*, 9548–9552.

[72] Kochi, J. K. *J. Am. Chem. Soc.* **1957**, *79*, 2942–2948.

[73] Kochi, J. K.; Subramanian, R. V. *J. Am. Chem. Soc.* **1965**, *87*, 1508–1514.

[74] Weingarten, H. *J. Org. Chem.* **1964**, *29*, 977–978.

[75] Bacon, R. G. R.; Hill, H. A. O. *J. Chem. Soc.* **1964**, 1097–1107.

[76] Bacon, R. G. R.; Hill, H. A. O. *J. Chem. Soc.* **1964**, 1108–1112.

[77] Bacon, R. G. R.; Hill, H. A. O. *J. Chem. Soc.* **1964**, 1112–1119.

[78] Delp, S. A.; Goj, L. A.; Pouy, M. J.; Munro-Leighton, C.; Lee, J. P.; Gunnoe, T. B.; Cundari, T. R.; Petersen, J. L. *Organometallics* **2011**, *30*, 55–57.

[79] Bowman, W. R.; Heaney, H.; Smith, P. H. G. *Tetrahedron Lett.* **1984**, *25*, 5821–5824.

[80] Wendlandt, A. E.; Suess, A. M.; Stahl, S. S. *Angew. Chem. Int. Ed.* **2011**, *50*, 11062–11087.

[81] Qiao, J. X.; Lam, P. Y. S. *Synthesis* **2011**, 829–856.

[82] (a) King, A. E.; Brunold, T. C.; Stahl, S. S. *J. Am. Chem. Soc.* **2009**, *131*, 5044–5045. (b) King, A. E.; Ryland, B. L.; Brunold, T. C.; Stahl, S. S. *Organometallics* **2012**, *31*, 7948–7957.

PART II

FORMATION OF C–C BONDS

8

MODERN COPPER-CATALYZED HURTLEY REACTION: EFFICIENT C-ARYLATION OF CH-ACID DERIVATIVES

IRINA P. BELETSKAYA AND ALEXEY YU. FEDOROV

8.1 INTRODUCTION

Aromatic nucleophilic substitution leading to the formation of aryl–heteroatom and aryl–carbon bonds is a reaction of considerable importance in laboratory practice, as well as in industry.[1–3] These synthetic transformations have been known for more than a century, and the evolution of this field is a good illustration of the development of modern organic chemistry. Arylation reactions (a term that was popularized by Sir Derek Barton)[4,5] are closely associated now with palladium catalysis due to tremendous development of palladium-mediated, cross-coupling methodologies over the last three decades.[3,6–12] However, palladium salts are expensive; their toxicity is similar to that of the highly toxic mercury, cadmium, or osmium salts[7]; and their reactions usually require the use of various phosphines or other sophisticated ligands. Therefore, the renaissance of Ullmann's copper-catalyzed cross coupling, which has taken place during the last decade, is not surprising.[13–15]

Classical copper-mediated Ullmann reaction[16] for Ar–C bond formation and Ullmann or Ullmann–Goldberg condensations[17–19] for Ar–heteroatom bond construction have more than a 100-year history (Scheme 8.1).

Copper-Mediated Cross-Coupling Reactions, First Edition. Edited by Gwilherm Evano and Nicolas Blanchard.

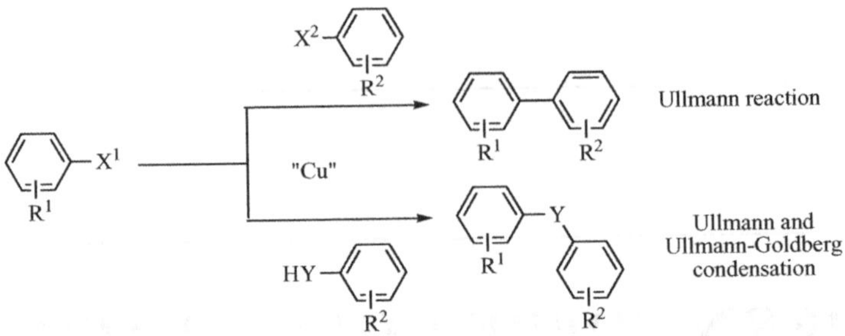

Scheme 8.1 Classical copper-catalyzed arylation reactions.

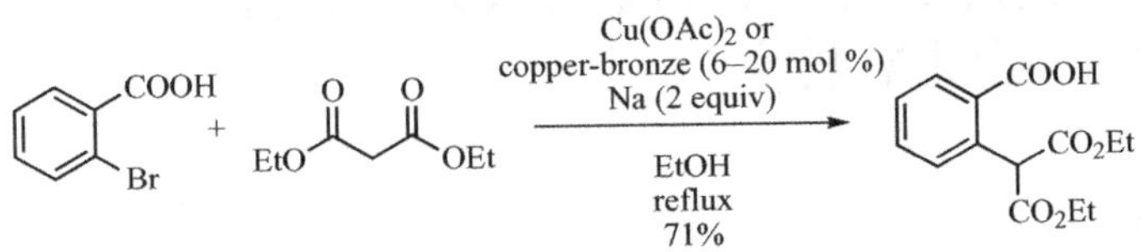

Scheme 8.2 Copper(I)-catalyzed Hurtley reaction.

The major restrictions of classical Ullmann and Goldberg protocols have always been harsh reaction conditions—in particular, high temperatures and the uses of strong bases and stoichiometric amounts of copper salts.

Almost 30 years after the pioneering discovery of Ullmann, Hurtley[20] discovered that copper bronze or copper acetate can promote the C-arylation of some families of CH-acids (Scheme 8.2) in reaction with *o*-bromobenzoic acid in the presence of sodium ethoxide as a base. Copper(I) salts are also efficient catalysts for this reaction with β-dicarbonyl derivatives, as shown in this canonical organic synthesis procedure by McKillop.[21]

Despite the very narrow scope of this reaction, limited by the application of *o*-bromobenzoic or closely related compounds, Hurtley arylation revealed as early as 1929 that copper catalysts can possess remarkable efficiencies under rather mild conditions,[20–24] which was uncommon for Ullmann-type chemistry at that time.[25] Originally, sodium acetylacetonate and *o*-bromobenzoic acid were refluxed in EtOH in the presence of catalytic amounts of copper bronze. In later works, variations of the original procedure used other systems such as copper bronze (150 mol %), acetylacetone, ^{t}BuOK as a base in refluxing ^{t}BuOH,[26] or copper(I) bromide (<10 mol %) in refluxing toluene.[27] Apparently, the ease of the Hurtley reaction is associated with a special type of substrate that can serve as a chelating ligand for copper. Such type of ligation was highly appreciated only 60 years later during the Renaissance of the Ullmann chemistry. Thus, the Hurtley reaction discussed in this review should be regarded as an early prototype for mild and efficient copper-assisted chemistry that has been developed in the last decade. Indeed, the modern version of the copper-mediated Hurtley-type arylation of carbanions is becoming very

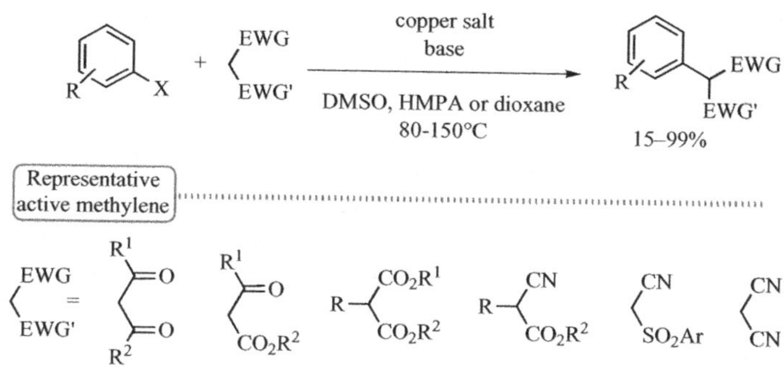

Scheme 8.3 Hurtley reaction of aryl halides with different types of CH-acids.

competitive with palladium-assisted methods of α-arylation of enolizable substrates concurrently reported in 1997 by Miura and coworkers[28] Palucki and Buchwald,[29] and Hamann and Hartwig.[30]

In this chapter, the classical Hurtley-type transformations will be exposed (Section 8.2), followed by a discussion on the importance of the different types of ligands (Section 8.3). The integration of the Hurtley reaction in cascade sequences will be presented (Section 8.4) before a critical overview of the different mechanisms that have been proposed for this reaction in the past decades (Section 8.5).

8.2 CLASSICAL HURTLEY REACTION

Early examples of copper-mediated arylation of CH-acids with nonchelating aryl halides are well known to share common deficiencies of Ullmann chemistry. Good yields are obtained only for activated aryl halides; the reactions are taking place in toxic, nonvolatile polar aprotic solvents; and preparation of the carbanion using substrate deprotonation is mandatory before the cross-coupling reaction. Finally, stoichiometric or even excess amounts of copper salts must be used in nearly all cases.

Despite these disadvantages, a rather large family of CH-acids like β-diketones,[20,21,31] β-ketoesters,[32,33] β-diesters,[20,21,34,35] β-cyanoesters,[36,37] arylsulfonylacetonitriles,[38] malononitriles,[39–41] and *N*-alkylisoquinoline-2,4-diones[42] were employed in Hurtley transformations (Scheme 8.3), sometimes with excellent yields of the desired products.

As a rule, aryl halides become more reactive with an increase in the softness of the leaving halide: ArI > ArBr > ArCl. This tendency is generally observed in the substitution of aryl halides by copper reagents.[15,43] The relative activities of copper(I) salts in these reactions are the following: CuBr > CuI > $CuBF_4(CH_3CN)_4$ > CuCl, the inefficiency of $CuBF_4(CH_3CN)_4$ being attributed to the poisoning effect of acetonitrile.[34] As noted by Kitao and coworkers[34] copper(I) bromide may be preferable because its amount can be reduced without significant decrease of the yields in contrast to copper(I) iodide.

In 1993, Miura and coworkers[44] reported an improved process for the copper-catalyzed arylation of malonitrile, ethyl cyanoacetate, and acetylacetone using catalytic amounts of copper salts and unhindered aryl iodides. Copper(I) iodide, bromide, chloride, as well as copper(I) oxide, copper(II) chloride, and copper(II) acetate were successfully used in these transformations in amounts in the 5–10 mol % range. Copper(II) oxide was, in sharp contrast, shown to be ineffective. However, some of the proposed catalytic systems require harsh conditions (DMSO, 120°C), under which malonate esters are prone to decomposition via ester hydrolysis and subsequent rapid decarboxylation.

Though this method avoids the use of additional ligands, it might be argued that any of the CH-acid substrates used under basic conditions might themselves furnish ligand assistance to form copper chelates. The addition of other ligands such as triphenylphosphine or amines inhibits the process, certainly because an excess of strong σ-donor ligands overloads the coordination sphere of copper.[44]

Only a few examples of Hurtley reaction are known in the context of total syntheses of natural products. During synthetic studies toward the originally proposed structure of diazonamide A, Konopelsky and coworkers developed a copper-catalyzed malonate arylation of *o,o′*-dibromophenol.[45a] A substoichiometric quantity of copper(I) bromide (20 mol %) was used leading to higher yields and easier purification than when a stoichiometric amount of copper salts was used.[45b]

8.3 LIGATION EFFECT IN COPPER-CATALYZED REACTIONS OF ARYL HALIDES WITH CARBANIONS

In 2002, Hennessy and Buchwald[46] observed that heating aryl iodides with malonate esters in tetrahydrofuran (THF) at 70°C in the presence of cesium carbonate and phenolic ligands under an inert atmosphere leads to the corresponding α-aryl malonates in good-to-excellent yields (Scheme 8.4). The use of phenol as a ligand resulted in the formation of a significant quantity of diaryl ether, while phenols bearing large *ortho*-substituents (^{t}Bu, *i*-Pr) slowed down the malonate arylation reaction considerably. It was found that 2-phenylphenol did not block the desired reaction and that the C–O bond formation occurred to a very small extent. Cesium carbonate could be replaced by potassium phosphate, though the latter was less efficient. The use of a twofold excess of C–H acid relative to the aryl iodide provides high yields of the desired products. Aryl bromides as well as aryl iodides bearing *ortho*-NO_2, -OH, or -NH_2 groups are not effective under these conditions.[46]

Interestingly, when the optimized reaction conditions[46] were applied to the arylation of a cyclic isopropylidene malonate, 1,3-cyclopentadienone, and

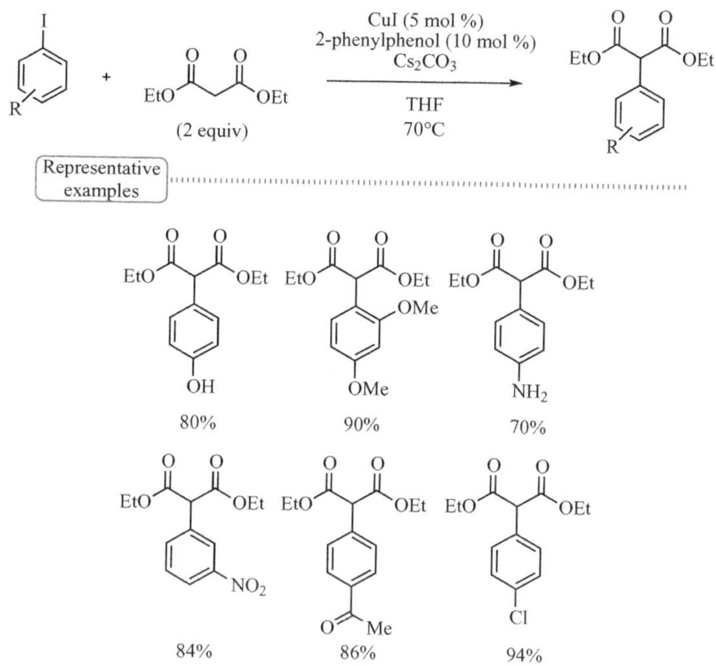

Scheme 8.4 Copper-catalyzed Hurtley arylation in the presence of phenolic ligands.

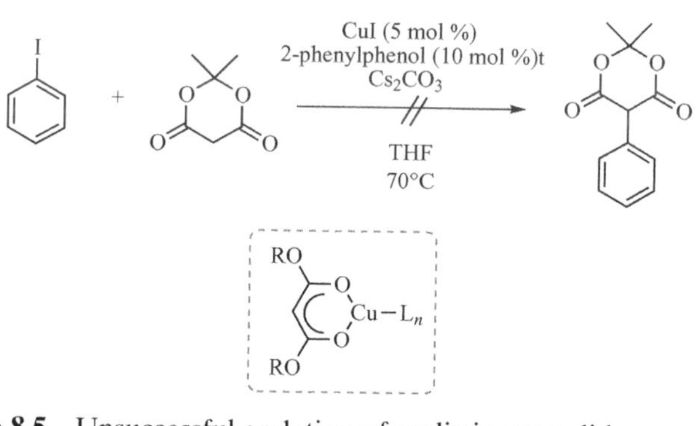

Scheme 8.5 Unsuccessful arylation of cyclic isopropylidene malonate.

1,3-cyclohexanedione, none of the desired products were observed, which was attributed by the authors to a necessary bidentate binding of the enolate to copper via the oxygen atoms, a ligation that is not possible with cyclic substrates (Scheme 8.5).

In 2007, Kwong and coworkers proposed the use of carboxylic monodentate as well as *O,O*-, *S,O*-, and *N,O*-bidentate ligands for the Hurtley arylation (Table 8.1).[47] It was shown that benzoic acid **L1** provides good catalytic activity in this transformation (entry 1) as well as *S,O*- and *N,O*-ligands **L2** and **L3**,

TABLE 8.1 Ligand Effect in Copper(I)-Catalyzed Arylation of Diethyl Malonate

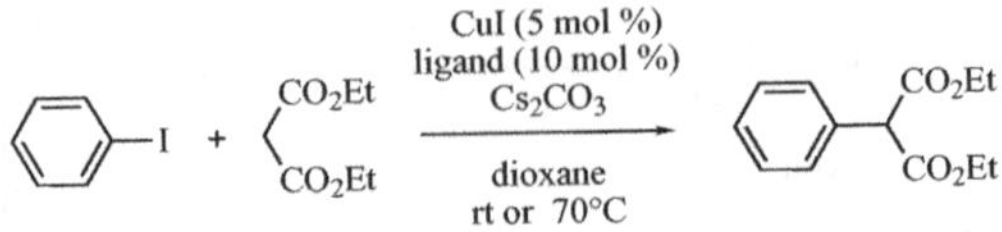

Entry	Cu:L	Ligand		Yield
1	1:2	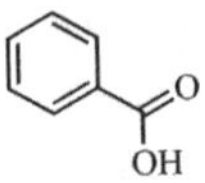	**L1**	79
2	1:2	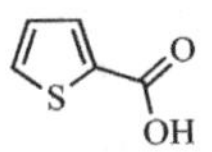	**L2**	60
3	1:2	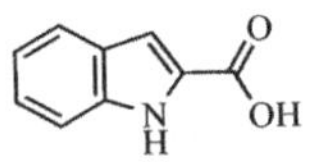	**L3**	66
4	1:2	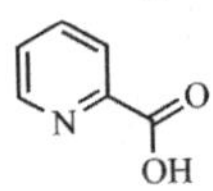	**L4**	99
5	1:2		**L4**	99[a]
6	1:10		**L4**	90[b]

[a]Reaction at rt.
[b]Cu (1 mol %).

which were also found to be rather effective (entries 2 and 3). 2-Picolinic acid **L4** showed significant rate acceleration in the arylation of malonate (entry 4)[47] and the efficiency of this ligand **L4** has been further demonstrated by its use at room temperature (entry 5). In general, the reaction proceeds smoothly to afford the desired products in good-to-high yields (73–96%) for aryl iodides bearing electron-neutral as well as electron-donor or electron-withdrawing substituents.[47,48] One should note that the catalyst loading in the case of **L4** can be reduced to 1 mol % of copper(I) iodide at 70°C, without sacrificing the product yield.[47]

Taillefer and coworkers used polydentate Chxn-Py-Al ligand (Scheme 8.6) for the synthesis of α-arylated malonic acid derivatives.[49] Its positive effect in this case is probably due to its aprotic nature and chemical inertness toward ester functionalities.[49]

Use of copper(I) iodide as a precatalyst affords good-to-excellent yields of cross-coupling products between aryl iodides and dialkylmalonates, ethyl cyanoacetate, and malononitrile in THF or acetonitrile at 70°C, using cesium carbonate as a base. Importantly, 2-alkyl-substituted diethyl malonates do not participate in the copper-catalyzed arylation process under these conditions, which was attributed to steric hindrance and decreased acidity of the substrate.[49]

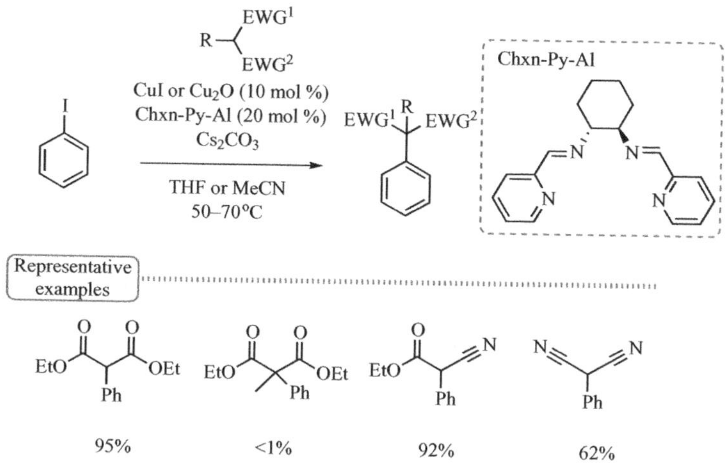

Scheme 8.6 Polydentate Chxn-Py-Al ligand in the arylation of malonic acid derivatives.

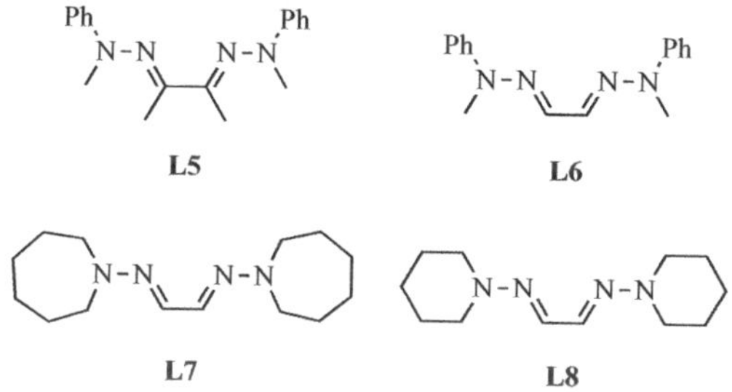

Figure 8.1 Hydrazone ligands in Hurtley arylation reaction.

Mino and coworkers proposed to use bishydrazone ligands **L5–L8** (Fig. 8.1) in the copper-mediated cross coupling of diethyl malonate with different aryl iodides.[50]

L5 was shown to be the most appropriate ligand for the Hurtley reaction in the presence of copper(I) iodide as a catalyst (5 mol %) and cesium carbonate (2 equiv) as a base in toluene at 90°C under an aerobic atmosphere (Scheme 8.7).[50] The desired products were obtained in good yields regardless of the electronic properties and the position of the substituents in the aryl coupling partner.[50]

It was found[51] that under the catalysis of copper(I) iodide (10 mol %) and L-proline (20 mol %), aryl iodides can efficiently couple with excess of β-diketones, ethyl cyanoacetate, or malononitrile in the presence of 4 equiv of potassium carbonate in DMSO at 90°C. In contrast to Buchwald observations,[46] application of L-proline as a ligand even permitted the arylation of cyclic β-diketones in good yields.[51]

Replacement of potassium carbonate by cesium carbonate allowed carrying out the Hurtley coupling at 40°C.[52] The same catalytic system is effective for

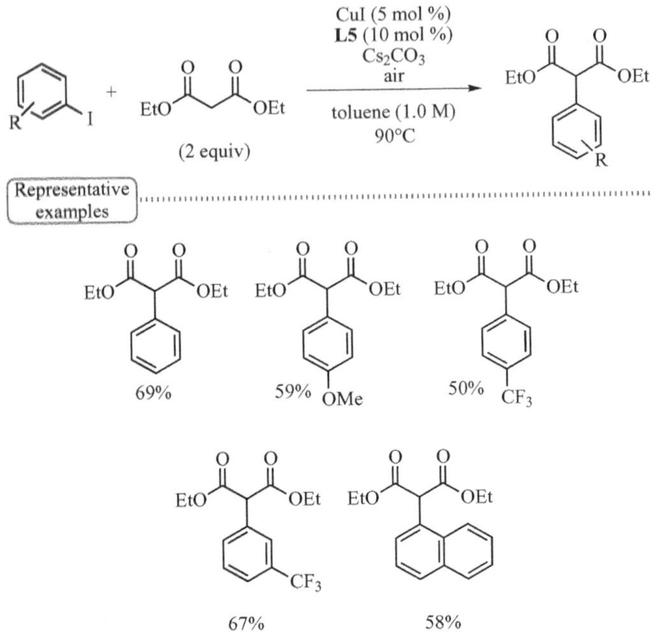

Scheme 8.7 Copper(I)-catalyzed Hurtley reaction using a bishydrazone ligand.

the coupling of CH-acids with aryl bromides at 50°C (Scheme 8.8). Indeed, aryl bromides with electron-neutral, as well as electron-rich and electron-deficient substituents reacted well under these conditions (Scheme 8.8). In these cases, good conversions were observed, although longer reaction times were required compared with aryl iodides. Attempts to facilitate the reaction by raising the temperature gave slightly lower yields. Like aryl iodides, aryl bromides with electron-withdrawing groups showed better reactivity compared with those bearing electron-donating groups.[52]

In 2006, Ma and coworkers found that the copper(I) iodide-catalyzed coupling between aryl iodides and β-ketoesters can be performed at –45°C in the presence of a proline-derived family of ligands in a dimethylformamide (DMF)/water mixture (Scheme 8.9) provided that a trifluoroacetamide substituent was present on the *ortho*-position of the aryl iodide.[53] This work represents the lowest reaction temperature for Ullmann-type reactions reported so far and the first example of highly enantioselective Hurtley arylation. The best enantioselectivity was achieved in the case of (2*S*,4*R*)-4-hydroxyproline as a ligand. The excellent enantioselectivity provided the evidence for the participation of a Cu(I)-amino acid chelate in the transition state of the reaction.

It was demonstrated that the electronic nature of substituents in aryl halides influences seriously the reaction outcome. Thus, the substrates with electron-donating groups afford the final products in high yields and enantioselectivities, while the cross coupling of aryl iodides bearing electron-withdrawing

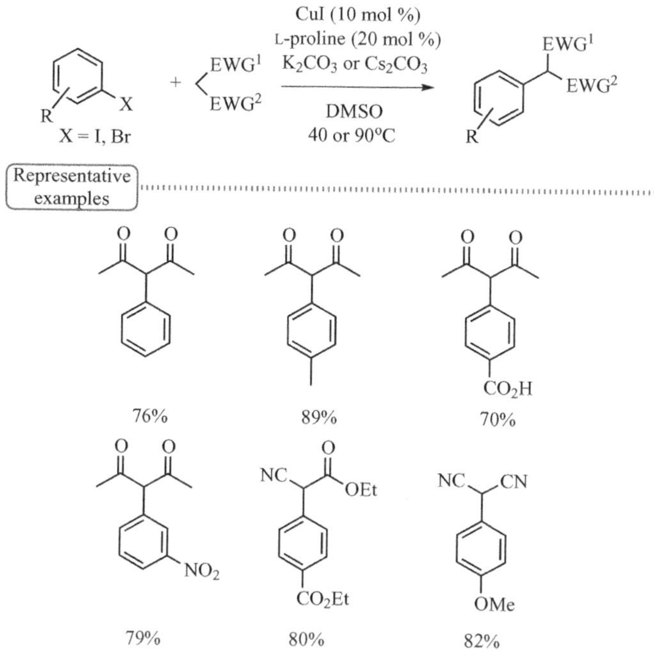

Scheme 8.8 Copper(I)-catalyzed Hurtley reaction of aryl bromides with L-proline as a ligand.

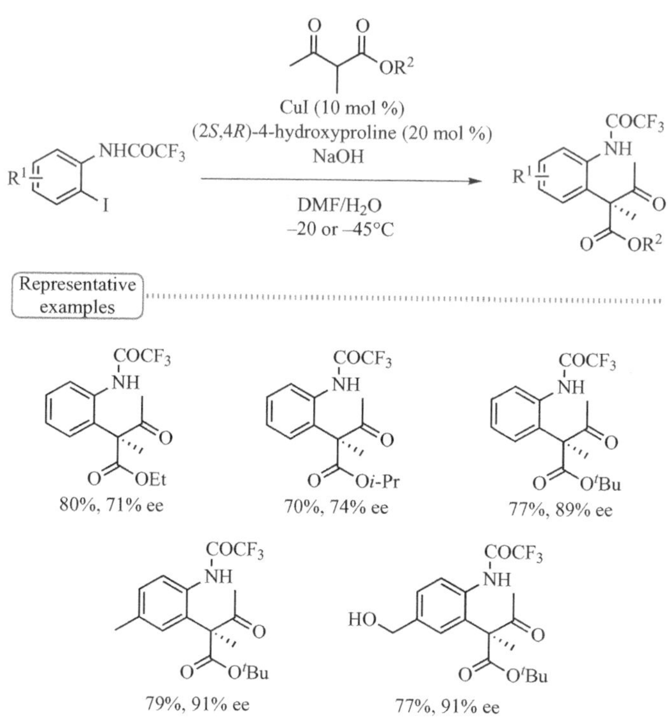

Scheme 8.9 Copper(I) iodide-catalyzed asymmetric Hurtley reaction.

substituents requires higher reaction temperatures to achieve the same degree of conversion.[53]

In addition to aryl iodides and bromides, a range of heterocyclic halides were introduced as coupling partners for the Hurtley reaction.[27,46,47] Depending on the reaction conditions, the heteroarylation products are formed in modest-to-high yields (Scheme 8.10).

A rather efficient procedure for the direct formation of a carbon–carbon bond via copper-catalyzed coupling of vinyl halides with carbanions was reported by Pei and Qian.[54] It was found that copper(I) iodide (10 mol %) and L-proline (20 mol %) as a ligand in the presence of cesium carbonate in

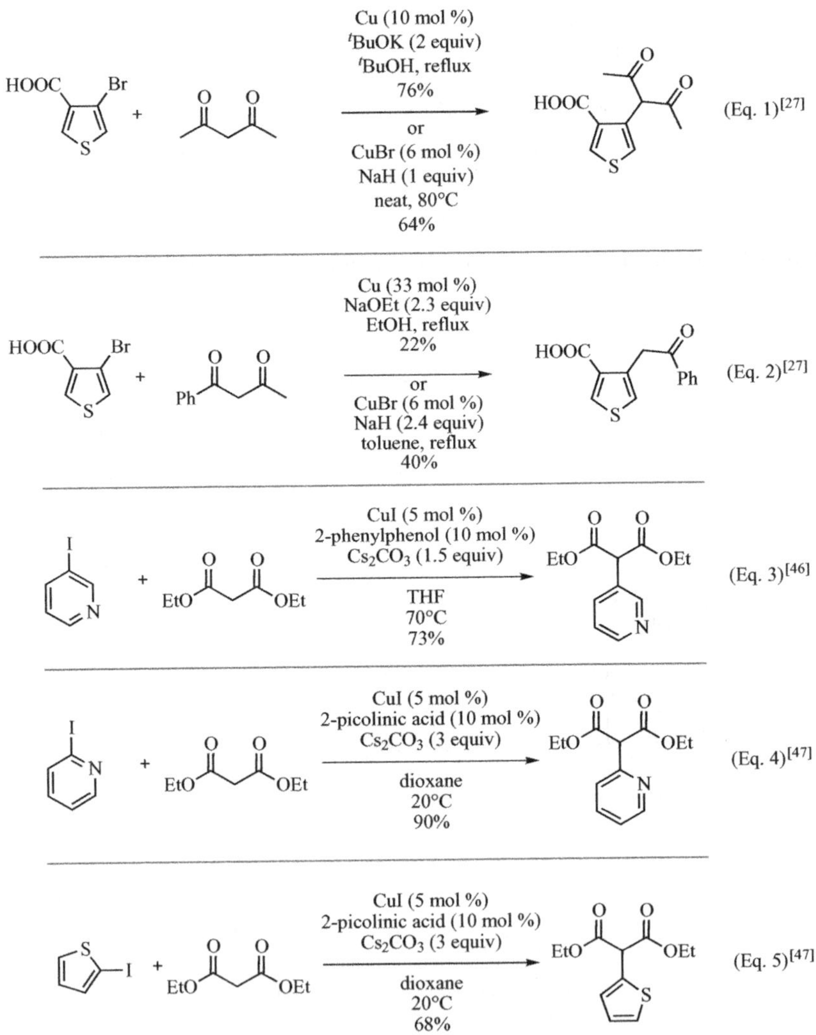

Scheme 8.10 Copper-catalyzed Hurtley heteroarylation of CH-acids.

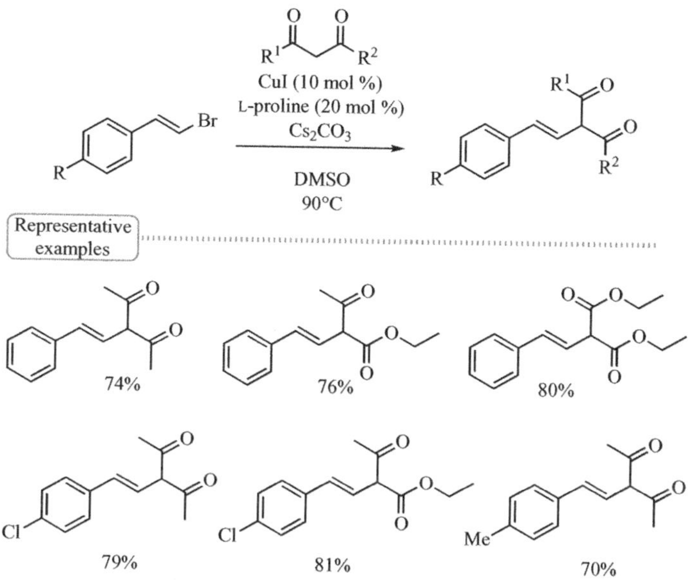

Scheme 8.11 Copper-catalyzed coupling of vinyl halides with CH-acids.

DMSO at 90°C allow the synthesis of the corresponding α-vinylated products in reasonable yields (Scheme 8.11).

(*E*)-β-Bromostyrenes substituted with electron-withdrawing groups gave the best results, while the substrates bearing electron-donor substituents (like methyl or methoxy groups) retarded the reaction and afforded lower yields of the desired products. (*Z*)-β-Bromostyrene did not react with CH-acids, probably because of the increased steric interactions.[54] It should be noted that alkyl-substituted vinyl bromides were not investigated. As a final note to this section, it should be pointed out that enolizable ketones such as deoxybenzoins are now reliable reaction partners as demonstrated by Taillefer and coworkers in late 2012.[54b]

The renaissance of the copper-mediated and copper-catalyzed transformations witnessed in the past decade clearly impacted the Hurtley reaction. As seen in this section, the introduction of various ligands largely expanded the scope of substrates. The Hurtley reaction has also been integrated into cascade sequences allowing a straightforward access to complex frameworks. This area of investigations is developed in the next section.

8.4 CASCADE REACTIONS PROCEEDING VIA A HURTLEY ARYLATION REACTION

Since the discovery of the Hurtley arylation reaction in 1929,[20] the application of this reaction to cascade transformations has been the subject of special

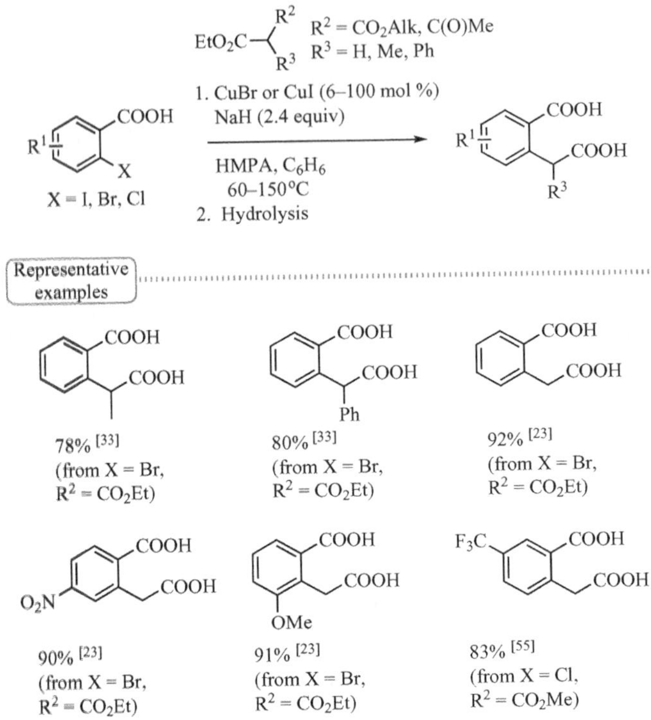

Scheme 8.12 Synthesis of α-arylalkanoic acids via a Hurtley arylation reaction.

interest. In particular, a series of papers has been devoted to the synthesis of α-arylalkanoic acids due to their pharmaceutical significance as nonsteroidal antiinflammatory drugs. The reaction pathway to arylalkanoic acids involves Hurtley arylation followed by hydrolysis (Scheme 8.12). Despite the rather harsh reaction conditions, these transformations usually afford α-arylalkanoic acids in good-to-high yields.[23,33,55]

In 2011, Fu and coworkers proposed an aerobic oxidative synthesis of aromatic carboxylic acids relying on a Hurtley arylation step.[56] According to the proposed protocol, the reaction of aryl iodide or bromide with malononitrile in the presence of copper(I) iodide (20 mol %) and L-proline (20 mol %) as a ligand led to Hurtley arylation product **3** (Scheme 8.13). *In situ* hydrolysis of the nitrile groups in derivative **3** and oxidation of tertiary C–H fragments by air in the presence of cesium carbonate afforded adduct **4**. The latter was subjected to cascade decarboxylation followed by acidification with hydrochloric acid giving aromatic carboxylic acids **7** in moderate-to-good yields.[56]

Copper-mediated Hurtley coupling of *ortho*-functionalized aryl halides with CH-acids can give an easy access to different kinds of heterocyclic compounds. In 1975, McKillop showed[24] that direct arylation of acetylacetone with 2-bromo-5-hydroxybenzoic acid **8** in the presence of copper(I) bromide as a

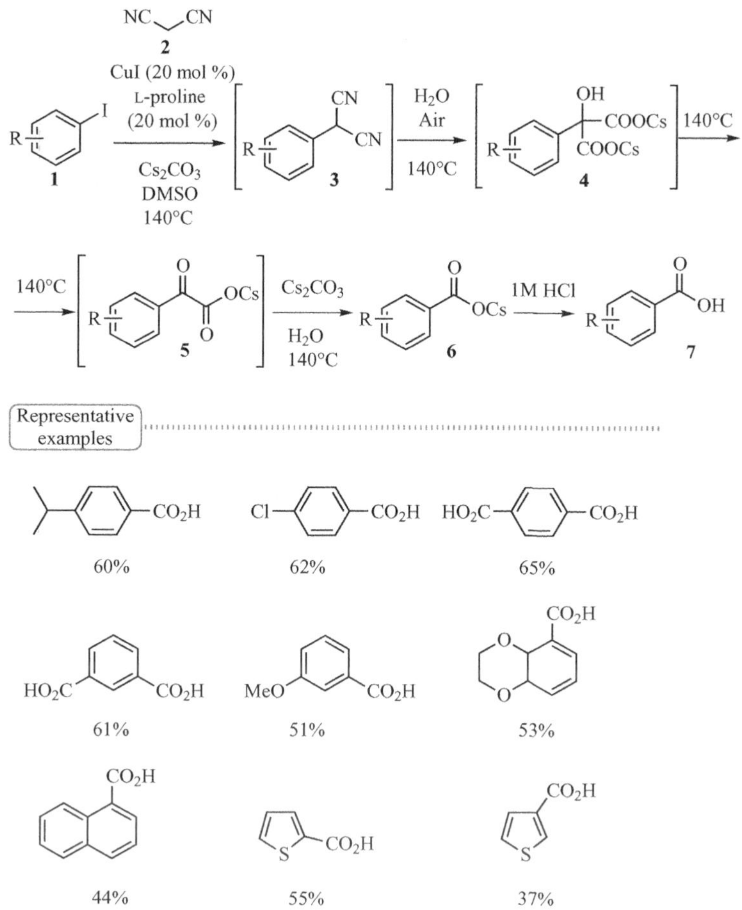

Scheme 8.13 Copper(I)-catalyzed synthesis of aromatic carboxylic acids from aryl halides and malononitrile.

catalyst and sodium hydride as a base in the absence of solvent at 80°C occurs smoothly to afford lactone **9** in good yield (Scheme 8.14, Eq. 1). Reaction of 2-bromobenzoic acid with ethyl cyanoacetate[24] under solvent-free conditions at 80°C leads to the formation of the isoquinolone derivative **10** in modest yield (Scheme 8.14, Eq. 1). In contrast, refluxing 2-bromobenzoic acid with the sodium salt of acetylacetone in DMF or dimethyl acetamide (DMA) in the presence of 7 mol % of copper(I) bromide leads to isochromanone **11** as the dominant product in 53–55% yield after the retro-Claisen deacylation-type reaction of the intermediate *o*-acetonylbenzoic acid (Scheme 8.14, Eq. 2).[32]

Copper-catalyzed reactions of 2-bromobenzoic acid with monocyclic CH-acids (e.g., 2-ethoxycarbonylcyclopentanone or 5,5-dimethylcyclohexanone-1,3-dione) in the absence of solvent or in benzene afforded tricyclic

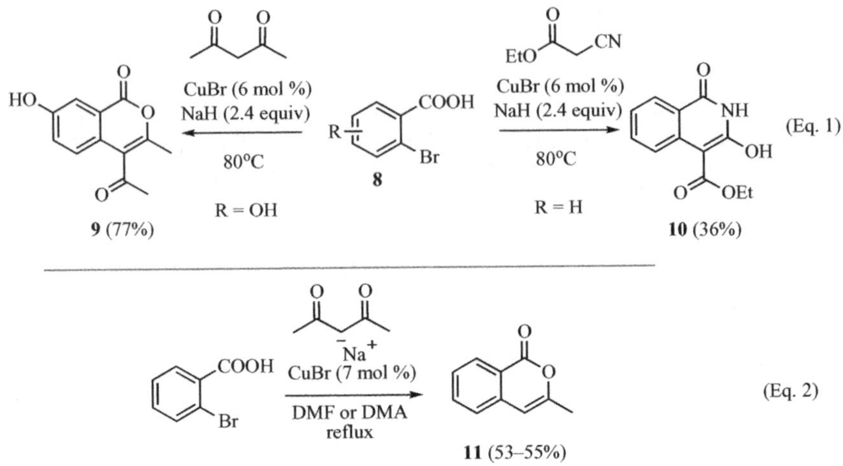

Scheme 8.14 Copper(I)-catalyzed Hurtley reaction leading to isoquinolones and isochromanones.

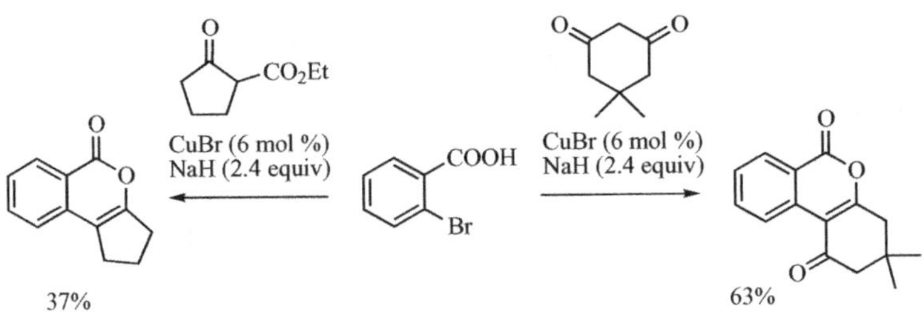

Scheme 8.15 Copper-catalyzed reactions of 2-bromobenzoic acid with monocyclic CH-acids.

lactone systems in moderate-to-good yields (Scheme 8.15).[33] These important results demonstrated that α-substituted β-dicarbonyl derivatives were reactive substrates in Hurtley arylation reaction, contrary to prior assumptions.[57]

Reaction of 2-bromobenzofuran-3-carboxylic acid with dimedone under the same conditions forms a polycyclic benzofuran in a single one-pot two-step sequence, although in moderate overall yield (Scheme 8.16).[58]

A more sustainable version of these Hurtley condensation processes has been carried out in near critical water (NCW) in a pressurized reactor under microwave heating.[59] The regioselectivity of the copper-catalyzed reaction of *ortho*-halobenzoic acid with C–H acids showed to be solvent dependent. Indeed, application of standard reflux technique in THF affords conjugated ketone **12** in 50–60% yield (Scheme 8.17), while the use of NCW under the microwave irradiation affords isochromenone **13** as a major product (55–77%) after acidification.

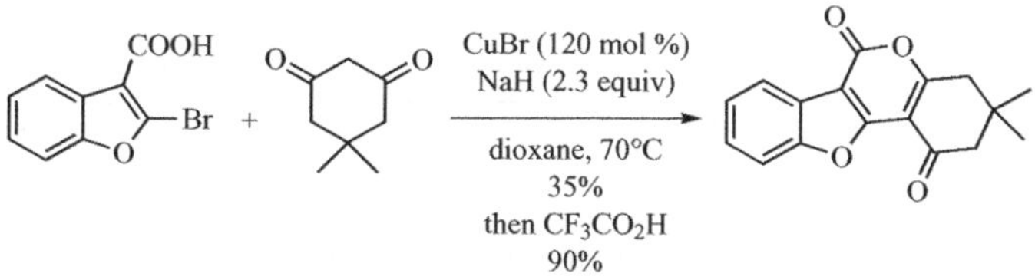

Scheme 8.16 Copper(I)-mediated synthesis of polycyclic benzofurane.

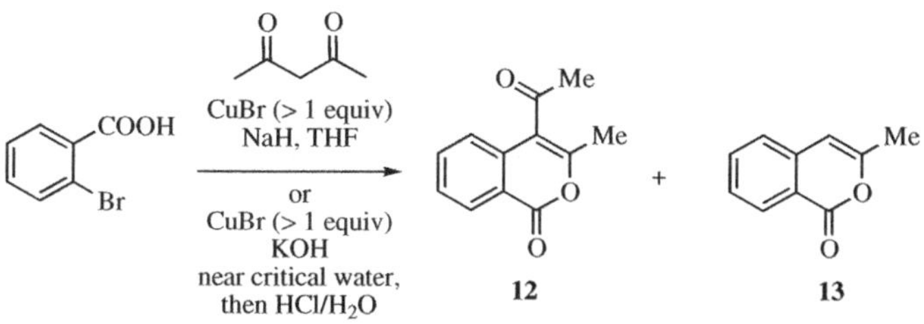

Scheme 8.17 Hutrley arylation of CH-acids in near critical water.

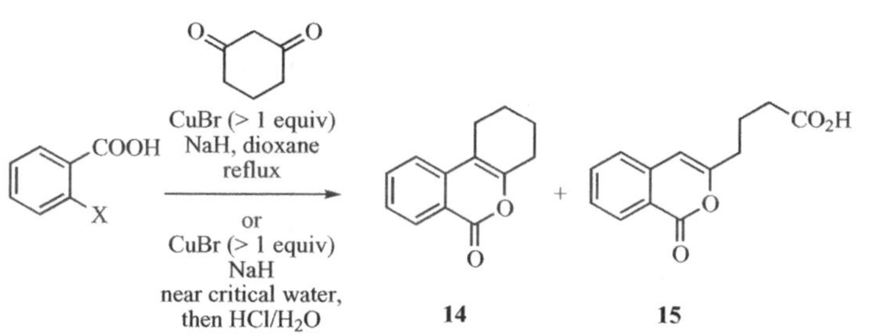

X = I or Br dioxane, reflux **14**:**15** = 60%/traces

X = I Parr pressure system, 200°C **14**:**15** = traces/55%

Scheme 8.18 Copper-mediated arylation of 1,3-cyclohexanedione in dioxane and in near critical water.

In the case of copper-mediated reaction between 2-iodobenzoic acid and 1,3-cyclohexanedione,[59] the major product formed under the classical conditions was chromenon **14** (60% yield), whereas lactone **15** was prepared almost in the same yield when hot pressurized water was used as the reaction medium (Scheme 8.18).

Interestingly, under copper catalysis, 2-bromobenzoic acid is able to react even with resorcinol[24] giving coumarin derivatives in modest yields, the formation of salicylic acid being also observed (Scheme 8.19). One should note

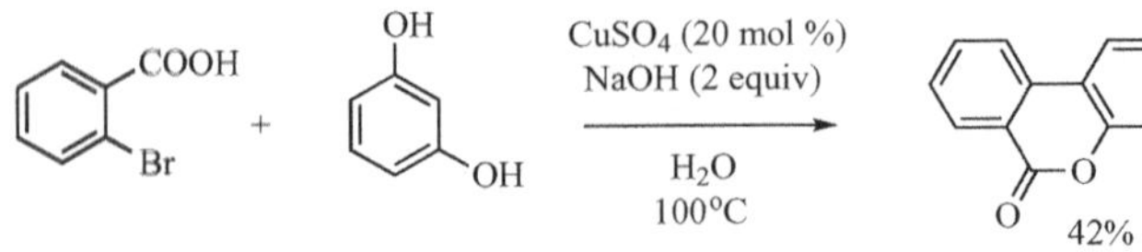

Scheme 8.19 Hurtley arylation reaction of resorcinol.

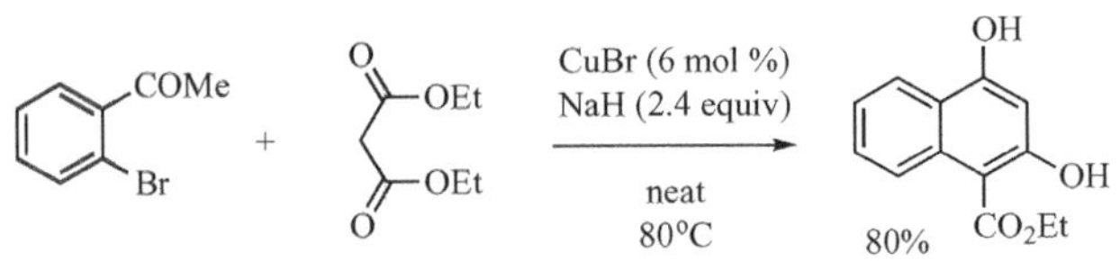

Scheme 8.20 Copper-mediated arylation of CH-acids using 2-bromoacetophenone derivatives.

that the latter reaction could be promoted also by mercury(II), cobalt(II), or manganese(II) salts.[24]

Ortho-haloarenes containing heteroatomic functionalities different from the carboxyl group are also able to undergo the copper-mediated coupling with CH-acids, affording poly- or heterocyclic compounds. Thus, 2-bromoacetophenone condenses smoothly with diethyl malonate in the presence of copper(I) bromide (6 mol %) and sodium hydride as a base in a solution of β-dicarbonyl compound to give ethyl 1,3-dihydroxy-4-naphtoate in 80% yield (Scheme 8.20).[24]

The same reaction conditions allowed the synthesis of 3-hydroxyisoquinoline *N*-oxides in moderate-to-good yields from the corresponding 2-bromoacetophenone oximes or 2-bromobenzaldoximes and malonic esters, β-keto esters, or ethyl cyanoacetate (Scheme 8.21).[60]

Setsune and coworkers have shown that *o*-bromophenol can react with sodium salts of CH-acids under copper assistance to give substituted benzofuran compounds.[45b,61] Indeed, after *o*-bromophenol was treated with the sodium salt of diethyl malonate in the presence of stoichiometric quantities of copper(I) bromide in dioxane at 70°C, 3-carbethoxy-2(3*H*)-benzofuranone **16** was isolated in 93% yield (Scheme 8.22). The use of ethyl acetoacetate and ethyl cyanoacetate instead of diethyl malonate in the above-mentioned procedure gave 3-acetyl-2-hydroxybenzofuran **17** and 3-carbethoxy-2-*o*-hydroxyanilinobenzofurane **18** in 15% and 34% yield, respectively (Scheme 8.22). The formation of the latter could be explained by an intermolecular Hurtley reaction followed by cyclization of the phenoxyde onto the nitrile. The resulting imide could then be arylated by another *o*-bromophenoxyde.

When an electrophilic reagent (e.g., β-methallyl chloride, allyl bromide, bromoacetone, or ethylbromoacetone) was added to the heated mixture of diethyl malonate, *o*-bromophenol, sodium hydride, and copper(I) bromide

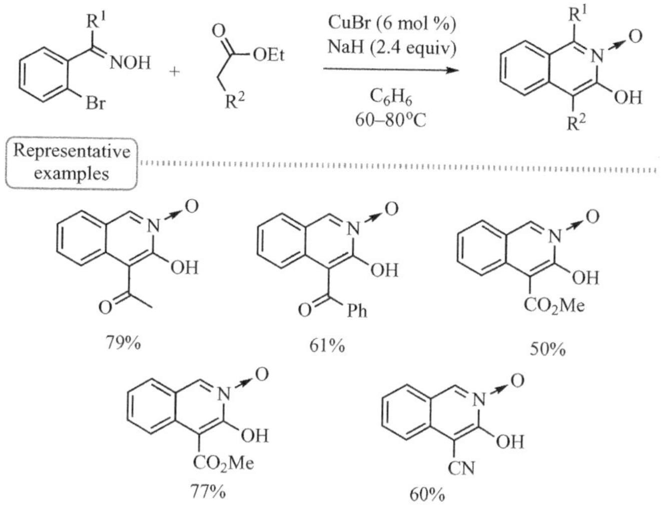

Scheme 8.21 Copper(I)-catalyzed synthesis of 3-hydroxyisoquinoline *N*-oxides.

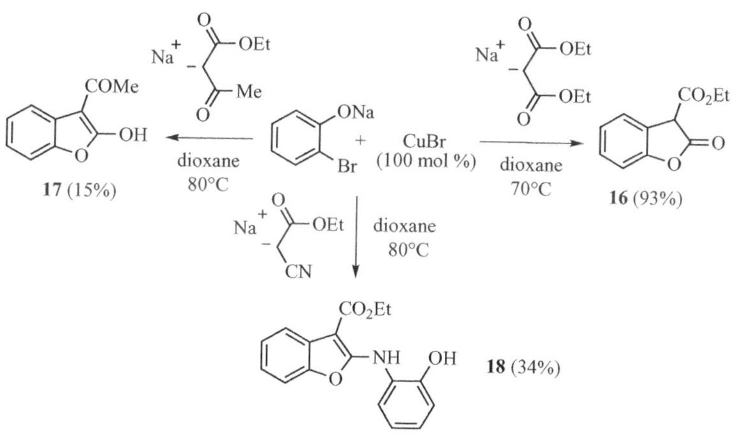

Scheme 8.22 Copper-mediated arylation of CH-acids using 2-bromophenol.

in 1,4-dioxane, the corresponding 3-substituted-3-carbethoxy-2-benzofuranones were obtained in moderate-to-good yields via a one-pot sequence (Scheme 8.23).[45]

Konopelsky and coworkers[45,62] showed that the reactions of *o*-bromophenols with malonic esters proceed readily with catalytic amounts of copper(I) (Scheme 8.24) giving improved yields of the arylation products in comparison to the application of stoichiometric catalyst quantities (compare Scheme 8.22 and Scheme 8.24). Indeed, the coupling of substituted *o*-bromophenols with malonic ester in the presence of copper(I) bromide (20 mol %) and sodium

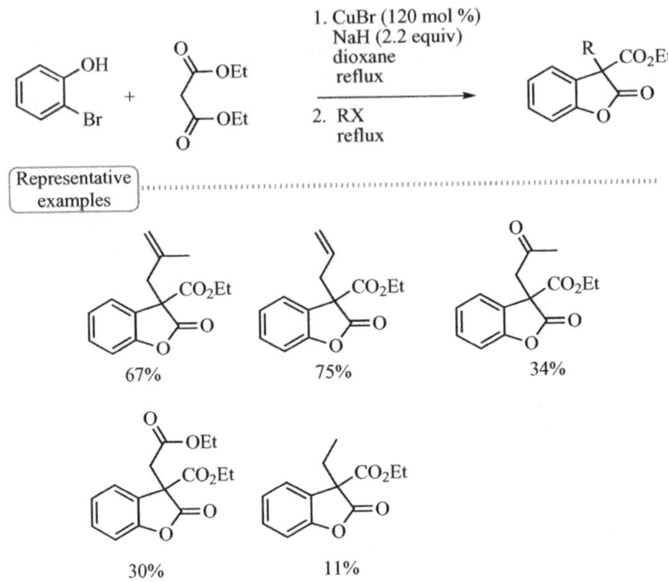

Scheme 8.23 Copper-mediated arylation of CH-acids using 2-bromophenol in the presence of electrophilic reagents.

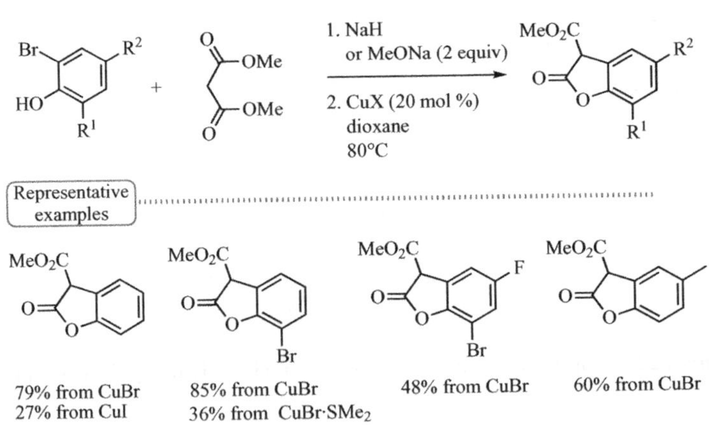

Scheme 8.24 Synthesis of 3-methoxycarbonylbenzofuran-2(3*H*)-one derivatives.

hydride or sodium methoxide as bases in dioxane at 80°C affords carbonyl-benzofuran-2(3*H*)-ones in good-to-high yields (Scheme 8.24). Attempts to use other copper(I) catalysts, including copper(I) thiophenecarboxylate, copper(I) cyanide, and CuCl(OH)-TMEDA were not successful.[45,62]

In 2011, Fu and Lu[63] proposed an elegant copper-catalyzed cascade method for the synthesis of benzimidazoisoquinoline derivatives via reactions of

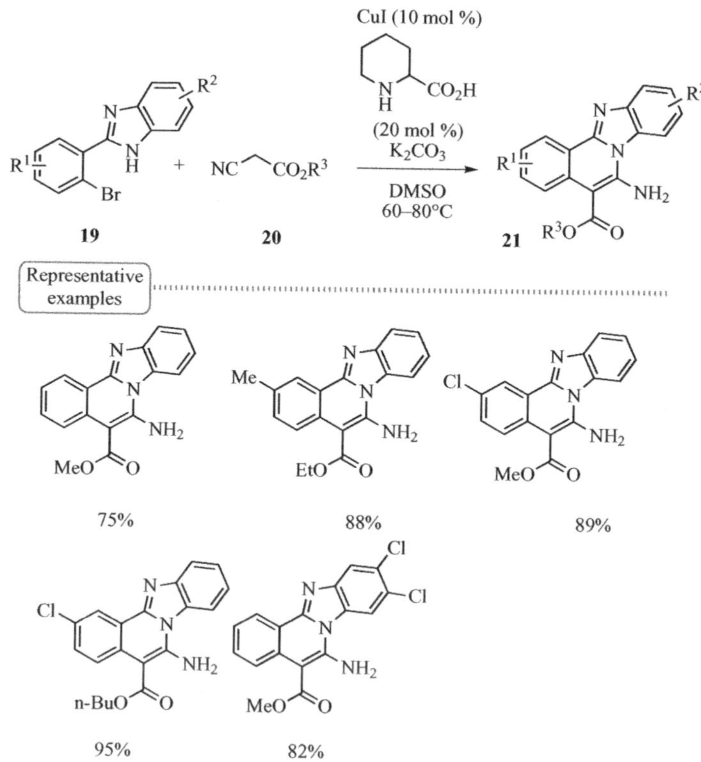

Scheme 8.25 Copper-catalyzed cascade synthesis of benzimidazoisoquinolines.

substituted 2-(2-halophenyl)benzimidazoles with alkyl cyanoacetates under mild conditions. It was shown that the catalytic system including copper(I) iodide (10 mol %), pipecolinic acid (20 mol %), potassium carbonate as a base in DMSO under a nitrogen atmosphere was the most effective (Scheme 8.25). Under these conditions, the desired alkyl-6-aminobenzimidazo[2,1-*a*] isoquinoline-5-carboxylates were obtained in good-to-excellent yields via a Hurtley reaction followed by cyclization of the benzimidazole moiety onto the nitrile.

Li and Fang reported that the reactions of δ-(2-bromophenyl)-β-keto esters in refluxing dioxane lead to the formation of 3,4-dihydronaphthalen-2(1*H*)-one derivatives via copper-mediated C-arylation.[64] The reactions readily proceed in the presence of copper(I) iodide (10 mol %), dimethylethylenediamine (DMEDA) (20 mol %), and cesium carbonate as a base to afford cyclic β-keto esters in good-to-high yields (Scheme 8.26, Eq. 1).

Interestingly, when α-methyl-substituted δ-(2-bromophenyl)-β-keto ester **22** was subjected to the same conditions as described earlier, no C-arylation product **23** was detected (Scheme 8.26, Eq. 2). Instead of the expected β-ketoester **2**, the O-arylation products **24** and **25** were isolated in an overall

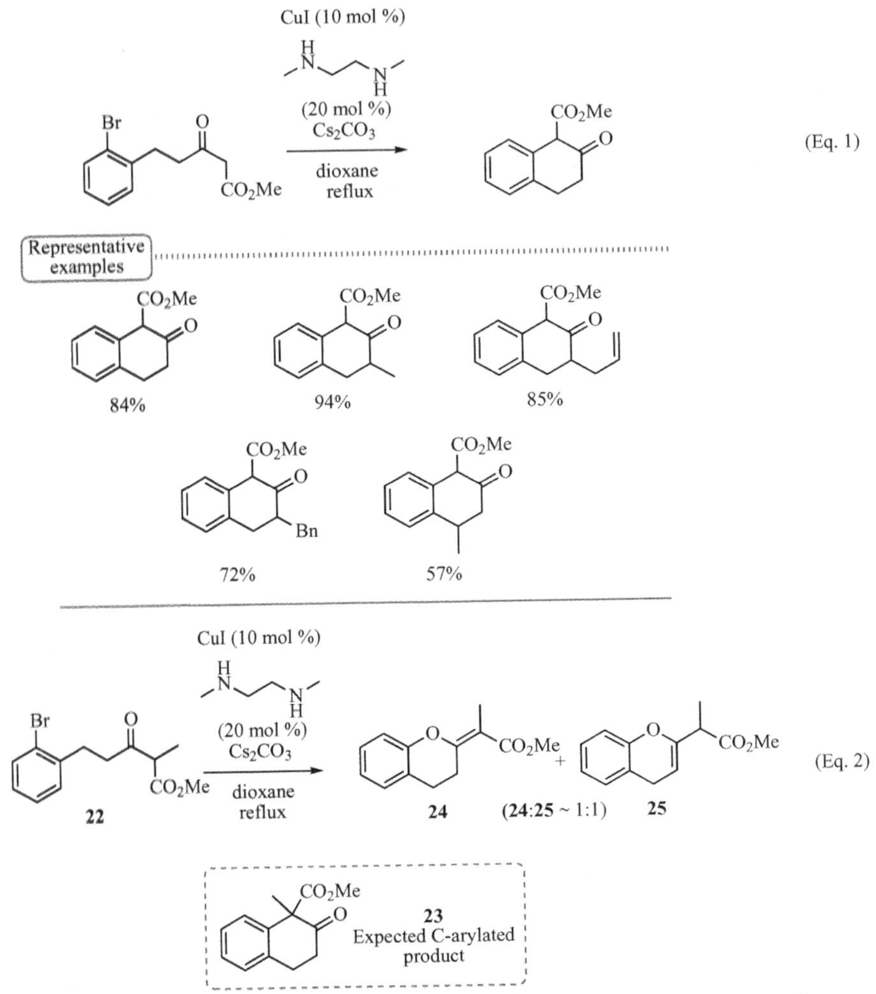

Scheme 8.26 Copper(I)-catalyzed intramolecular Hurtley reaction and competitive O-arylation.

75% yield with a ratio of ~1:1. The authors attributed the inhibition of the C-arylation reaction to the steric hindrance caused by the methyl group at the α-position.[64]

Domino sequences that integrate an inter- or intramolecular Hurtley reaction (followed by further cyclization) thus proved to be a valuable tool for the straightforward elaboration of complex polycyclic scaffolds. Besides the elegant developments of the Hurtley reaction, numerous research groups have pursued mechanistic investigations of this transformation in the past decades. An overview of the current mechanistic understanding of the Hurtley reaction is developed in the next section.

8.5 MECHANISM OF THE COPPER-CATALYZED C-ARYLATION REACTIONS

Although significant progress has been achieved in the mechanistic understanding of Ullmann reactions, (see Chapter 7, Section 7.4.1 for details), much remains to be uncovered. Indeed, the mechanisms of the copper-mediated C–H arylation may significantly differ depending on the halide or pseudohalide nature, on the ligand, nucleophile and the solvent, on the origin of the copper source and the temperature conditions. As a result, four different mechanisms have been proposed to explain the copper-catalyzed C-arylation:[15,49,65]

- copper(I)–copper(III) oxidative addition/reductive elimination pathway, which is canonical for transition metal-mediated cross-coupling reactions
- formation of aryl radical intermediates, via either single-electron transfer (SET) or halide atom transfer (AT)
- σ-bond metathesis through a four-center intermediate
- π-complexation of copper on the aryl halide.

The first two alternatives belong to reactions where a copper species changes its oxidation state during the catalytic cycle, whereas the last two pathways belong to the processes in which a copper species maintains the same oxidation state through the whole cycle.

The involvement of the oxidative addition/reductive elimination mechanism in Ullmann condensations was first proposed by Cohen in 1974.[66] This mechanism was further substantiated by several literature reports,[67–69] indicating the presence of Cu(I) and Cu(III) intermediates (see Chapter 7 for a more detailed discussion). The catalytic cycle presented in Scheme 8.27 is not a mechanism but rather an approximate presentation of what may happen based

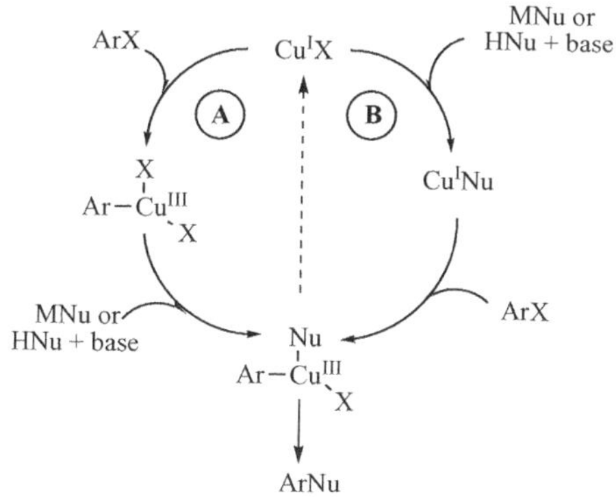

Scheme 8.27 Oxidative addition/reductive elimination mechanism for Hurtley-arylation reactions.

on analogies and general considerations and not on actual mechanistic studies. At the same time, this catalytic cycle is in a good agreement with the following experimental observations: (i) the reactivity sequence in such processes (ArI > ArBr >> ArCl) correlates with the leaving group ability of halide ions; (ii) coupling is slightly favored by electron-withdrawing groups on aryl fragment and slightly disfavored by electron-donor ones; (iii) steric hindrance on either coupling partner is rate-depressing; and (iv) aryl halides do not undergo Hurtley arylation in the absence of the copper catalyst.

The proposed mechanism involves three elementary steps: oxidative addition of the aryl halide to copper(I) generating a copper(III) intermediate,[69d] nucleophilic substitution or transmetallation of a copper bound halide by a nucleophile, and finally, reductive elimination at the last stage affords the coupling product and provides regeneration of the active catalyst.[49,70,71] Unlike palladium-promoted cross coupling in which the oxidative addition step is believed to precede the transmetallation, the ordering of the oxidative addition and transmetallation or substitution steps in the copper cycle is unknown, so either of the two possibilities can take place (pathway A or B).[15]

However, a recent and insightful report by Ribas and coworkers experimentally demonstrated that these reactions proceeded via a Cu(I)/Cu(III) cycle using a UV/Vis monitoring of the reaction (Scheme 8.28).[69b,c] In a first step, the copper(III) species is formed by direct oxidative addition of the aryl halide at copper(I) thus leading to complex **27** bearing an apical bromine atom. In a subsequent rate-limiting step, the bromine is removed and the

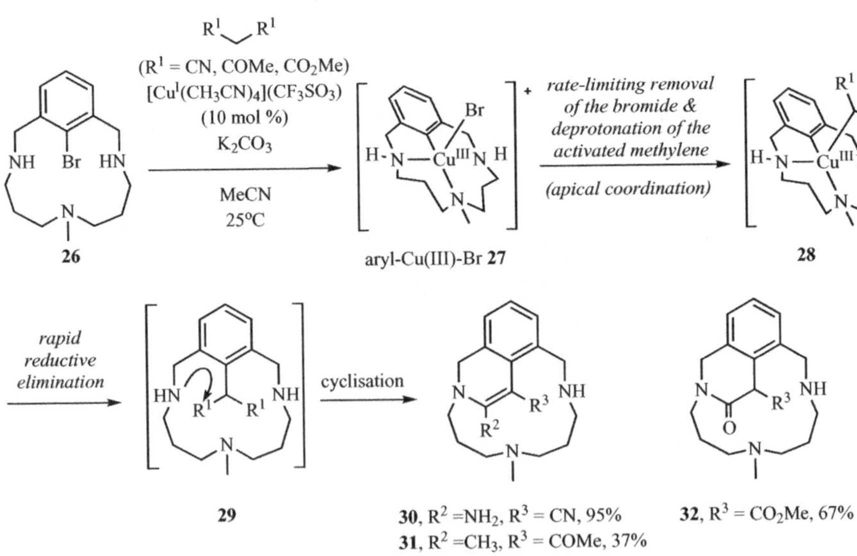

Scheme 8.28 Mechanistic investigations of the copper-catalyzed arylation reaction by Ribas and coworkers.

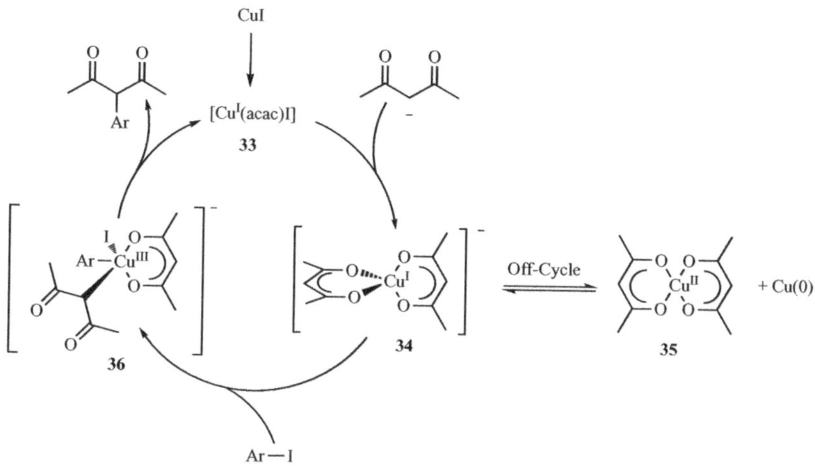

Scheme 8.29 A labile copper(I) catalyst that could engage in a Cu(I)/Cu(III) cycle or disproportionate to Cu(II) and Cu(0).

deprotonation of the active methylene derivative occurs simultaneously, delivering the corresponding copper(III) complex **28**. A rapid reductive elimination finally afforded the Hurtley product **29** that could undergo subsequent cyclization to **30–32**.

The role of the ligand in the Hurtley reaction has also attracted considerable attention and, besides the studies by Huang and Hartwig that are discussed in Chapter 7 (section 7.4.1),[70] a recent study by Lei is worth mentioning in the context of oxidative addition/reductive elimination pathway in copper(I)-catalyzed cross-coupling reactions.[69c] Using a combination of IR, in situ X-ray absorption near-edge structure (XANES) and extended X-ray absorption fine structure (EXAFS), the complex **34** between copper(I) iodide and a β-diketone nucleophile was observed (Scheme 8.29). Further kinetic studies and computational calculations demonstrated that this labile complex might be the active catalyst of the Hurtley reaction but could also undergo a rapid disproportionation to catalytically inactive copper(0) and copper(II) acetylacetonate (off-cycle). The catalytic cycle then involves an oxidative addition of the aryl halide at copper(I) thus leading to the copper(III) complex **36** that delivered the copper(I) catalyst **33** and the arylated product after reductive elimination. The dual role of the β-diketone, as a ligand and as a nucleophile, is worth noting.

Since the end of the 1930s, numerous mechanisms involving the electron transfer step were proposed to explain the copper-mediated Ullmann reactions.[72] These mechanistic proposals can be generalized in accordance with the copper-catalyzed $S_{RN}1$-type mechanism (Scheme 8.30).[65,73,74]

One should note that in the case of the Hurtley reaction, the relevance of free radical intermediates is unlikely, since in the asymmetric arylation proposed by Ma (Scheme 8.9), the final products were obtained with high

$$ArX + Cu(I)X \longrightarrow Ar^{\bullet} + Cu(II)X_2$$

$$Ar^{\bullet} + Nu^{-} \longrightarrow (ArNu)^{\bullet -}$$

$$(ArNu)^{\bullet -} + Cu(II)X_2 \longrightarrow ArNu + Cu(I)X + X^{-}$$

Scheme 8.30 Mechanisms for Hurtley reactions involving an electron transfer step.

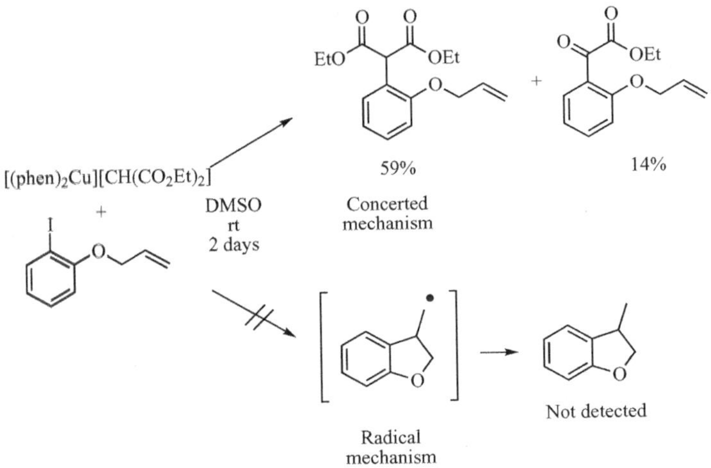

Scheme 8.31 Reaction of *o*-(allyloxy)iodobenzene with ([phen]$_2$Cu)(CH[CO$_2$Et]$_2$).

stereoselectivity.[53,54] In addition, the application of a *o*-(allyloxy)iodobenzene radical clock, known to undergo radical cyclization with a rate constant of 9.6×10^9 s^{-1} in DMSO, did not reveal the formation of any radicals (Scheme 8.31).[70]

It could, however, still be argued that under the present sets of reaction conditions, the arylation reaction can be faster than the radical trapping.[75] In this case, the intervention of radicals into a solvent cage or in the form of an undissociated complex with copper is also possible.[49]

According to the σ-bond metathesis hypothesis, the copper catalyst is assumed not to change its oxidation state during the catalytic cycle, remaining in its +1 state. In this case, one can propose the activation of the aryl halide via a four-centered intermediate between the aryl halide and copper catalyst in agreement with Scheme 8.32.[65]

And finally, in the mechanistic pathways involving π-complexation, the copper(I) species are forming a sort of electron-deficient π-arene intermediate, facilitating the substitution of the halogen atom by the nucleophilic reagent (Scheme 8.33).[76,77]

This mechanism relates to nucleophilic aromatic substitution or nucleophilic addition reactions, which are readily taking place with corresponding

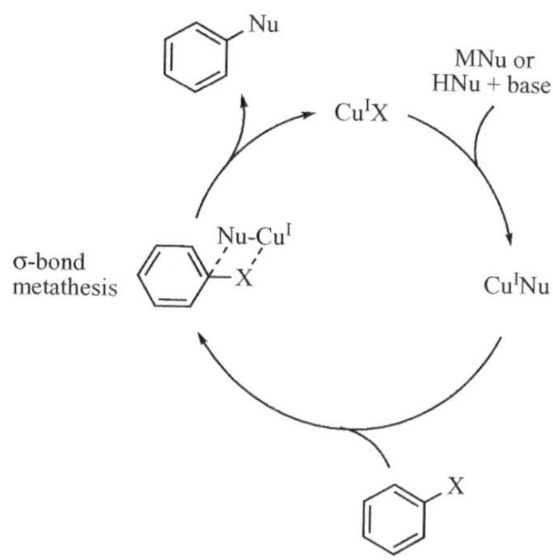

Scheme 8.32 σ-Bond metathesis hypothesis for the copper-mediated arylation reactions.

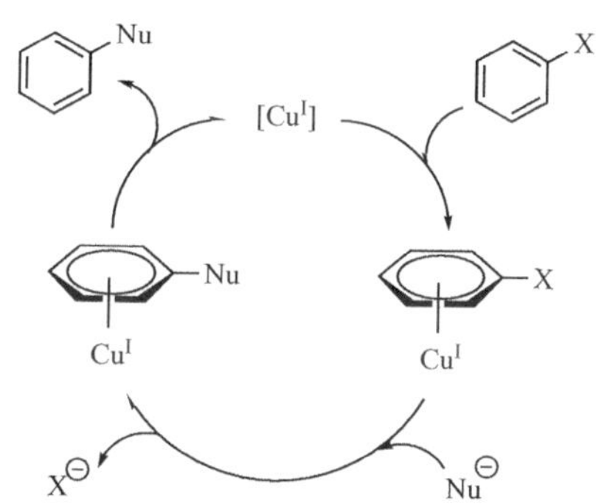

Scheme 8.33 π-Complexation mechanistic pathway for the copper-catalyzed arylation reactions.

arene–chromium tricarbonyl complexes.[78] According to Scheme 8.33, the copper species maintains its +1 oxidation state through the whole cycle.

Since the discovery of the Hurtley arylation in 1929 due to the special coordination of copper precatalyst by *o*-benzoic acid and/or by CH-acid, this reaction should be regarded as a first prototype of modern homogeneous catalysis. Indeed there are several reports demonstrating formation of the copper complexes with components of the Hurtley reaction, as well as with ancillary ligands.[46,63,69b,69c,70,74–77] The authors suppose that such ligation is playing an important role in the dissolution and stabilization of the catalytic species, as well as in the facilitation of the oxidative addition step or in the assistance for the SET process.[46,63,79–81] Despite the significant differences in the nature of copper–ligand interactions only a few studies have addressed the mechanistic importance of the ligands.[69b,69c]

8.6 CONCLUDING REMARKS

In conclusion, while the Hurtley arylation reaction stayed in the shadow for decades after its discovery in 1929, it can be regarded now as an important tool of modern organic synthesis. Indeed, the high yields of arylated products in inter- and intramolecular arylation versions, excellent regioselectivity, mild reaction conditions, the possibility of performing stereoselective arylation, use of low-cost catalysts with ability to work at low catalyst loading, and, finally, experimental simplicity made the Hurtley reaction particularly well suited for direct arylation of the different types of C–H acids.

One should note that in the past few years, the reactions of copper-mediated sp^2 C–H bond arylation in electron-deficient aromatic and heteroaromatic compounds, which are related to Hurtley arylation, have undergone explosive growth.[82–89] It was postulated that these reactions proceed by initial deprotonation of a relatively acidic sp^2 C–H bonds by more or less strong bases (^{t}BuOM with M = Li, K, Na; K_3PO_4) followed by alkali metal-copper transmetallation and coupling with an aryl or vinyl halide, and can be considered as Hurtley's arylation descendants.

Despite important contributions and significant progress in this area during the last decade, many challenges remain. To meet them, many more mechanistic studies have to be performed to identify the copper species involved in these reactions, including identifying the substrate/ligand/catalyst intermediates and relationships in this key triad. The better mechanistic understanding of the copper-mediated C–H arylation processes will be a major contribution to the further development of this field, especially for industrial-scale synthesis, for which the above-mentioned requirements for clean, mild, fast, selective, and cost-effective processes are of great concern.

REFERENCES

[1] (a) Paradisi, C. *Comprehensive Organic Synthesis*, Vol. 4, Trost, B. M.; Fleming, I. (eds.). Pergamon: Oxford, **1991**, pp. 423–450. (b) Semmelhack, M. F. *Comprehensive Organic Synthesis*, Vol. 4, Trost, B. M.; Fleming, I. (eds.). Pergamon: Oxford, **1991**, pp. 517–549.

[2] Theil, F. *Angew. Chem. Int. Ed. Engl.* **1999**, *38*, 2345–2347.

[3] Hassan, J.; Sevignon, M.; Gozzi, C.; Schulz, E.; Lemaire, M. *Chem. Rev.* **2002**, *102*, 1359–1469.

[4] Barton, D. H. R.; Finet, J.-P. *Pure Appl. Chem.* **1987**, *59*, 937–946.

[5] Abramovitch, R. A.; Barton, D. H. R.; Finet, J.-P. *Tetrahedron* **1988**, *44*, 3039–3071.

[6] Negishi, E.-I.; de Meijere, A. (eds.), *Handbook of Organopalladium Chemistry for Organic Synthesis*. Wiley: New York, **2002**.

[7] Prim, D.; Campagne, J.-M.; Joseph, D.; Andrioletti, B. *Tetrahedron* **2002**, *58*, 2041–2075.

[8] Chen, X.; Engle, K. M.; Wang, D.-H.; Yu, J.-Q. *Angew. Chem. Int. Ed. Engl.* **2009**, *48*, 5094–5115.

[9] Lyons, T. W.; Sanford, M. S. *Chem. Rev.* **2010**, *110*, 1147–1169.

[10] Sehnal, P.; Taylor, R. J. K.; Fairlamb, I. J. S. *Chem. Rev.* **2010**, *110*, 824–889.

[11] Li, B.-J.; Yu, D.-G.; Sun, C.-L.; Shi, Z.-J. *Chem. Eur. J.* **2011**, *17*, 1728–1759.

[12] Felpin, F.-X.; Nassar-Hardy, L.; Le Callonnec, F.; Fouquet, E. *Tetrahedron* **2001**, *67*, 2815–2831.

[13] Ley, S. V.; Thomas, A. *Angew. Chem. Int. Ed. Engl.* **2003**, *43*, 5400–5449.

[14] Kunz, K.; Scholz, U.; Ganzer, D. *Synlett* **2003**, 2428–2439.

[15] Beletskaya, I. P.; Cheprakov, A. V. *Coord. Chem. Rev.* **2004**, *248*, 2337–2364.

[16] Ullmann, F.; Bielecki, J. *Chem. Ber.* **1901**, *34*, 2174–2178.

[17] Ullmann, F. *Ber. Dtsch. Chem. Ges.* **1903**, *36*, 2389–2391.

[18] Ullmann, F. *Ber. Dtsch. Chem. Ges.* **1904**, *37*, 853–857.

[19] Goldberg, I. *Ber. Dtsch. Chem. Ges.* **1906**, *39*, 1691–1696.

[20] Hurtley, W. R. H. *J. Chem. Soc.* **1929**, 1870–1873.

[21] Bruggink, A.; Ray, S. J.; McKillop, A. *Org. Synth.* **1978**, *58*, 52–54.

[22] Cirigottis, K. A.; Ritchie, E.; Taylor, W. C. *Aust. J. Chem.* **1974**, *27*, 2209–2228.

[23] Bruggink, A.; McKillop, A. *Angew. Chem. Int. Ed. Engl.* **1974**, *13*, 340–341.

[24] Bruggink, A.; McKillop, A. *Tetrahedron* **1975**, *31*, 2607–2619.

[25] Lindley, J. *Tetrahedron* **1984**, *40*, 1433–1456.

[26] Ames, D. E.; Ribeiro, O. *J. Chem. Soc. Perkin Trans. 1* **1975**, 1390–1395.

[27] Schinkwin, A. E.; Whish, W. J. D.; Threadgill, M. D. *Bioorg. Med. Chem.* **1999**, *7*, 297–308.

[28] Satoh, T.; Inoch, J.; Kawamura, Y.; Miura, M.; Nomura, M. *Angew. Chem. Int. Ed. Engl.* **1997**, *36*, 1740–1742.

[29] Palucki, M.; Buchwald, S. L. *J. Am. Chem. Soc.* **1997**, *119*, 11108–11109.

[30] Hamann, B. C.; Hartwig, J. F. *J. Am. Chem. Soc.* **1997**, *119*, 12382–12383.

[31] Gao, Y.; Burke, T. R. *Synlett* **2000**, 134–136.

[32] Bacon, R. G. R.; Murray, J. C. F. *J. Chem. Soc. Perkin Trans. 1* **1975**, 1267–1272.

[33] McKillop, A.; Rao, D. P. *Synthesis* **1977**, 759–760.

[34] Setsune, J.-I.; Matsukawa, K.; Wakemoto, H.; Kitao, T. *Chem. Lett.* **1981**, *10*, 367–370.

[35] Pivsa-Art, S.; Fukui, Y.; Miura, M.; Nomura, M. *Bull. Chem. Soc. Jpn* **1996**, *69*, 2039–2042.

[36] Suzuki, H.; Kobayashi, T.; Yoshida, Y.; Osuka, A. *Chem. Lett.* **1983**, *12*, 193–194.

[37] Osuka, A.; Kobayashi, T.; Suzuki, H. *Synthesis* **1983**, 67–68.

[38] Suzuki, H.; Yi, Q.; Inoue, J.; Kusume, K.; Ogawa, T. *Chem. Lett.* **1987**, *16*, 887–890.

[39] Suzuki, H.; Kobayashi, T.; Osuka, A. *Chem. Lett.* **1983**, *12*, 589–590.

[40] Rosenau, B.; Krieger, C.; Staab, H. A. *Tetrahedron Lett.* **1985**, *26*, 2081–2084.
[41] Suzuki, H.; Koide, H.; Ogawa, T. *Bull. Chem. Soc. Jpn* **1988**, *61*, 501–504.
[42] Lowe, J. A.; Newman, M. *Synth. Commun.* **1987**, *17*, 803–807.
[43] Castro, C. E.; Gaughan, E. J.; Owsley, D. C. *J. Org. Chem.* **1966**, *31*, 4071–4078.
[44] Okuro, K.; Furuune, M.; Miura, M.; Nomura, M. *J. Org. Chem.* **1993**, *58*, 7606–7607.
[45] (a) Konopelsky, J. P.; Hottenroth, J. M.; Oltra, H. M.; Veliz, E. A.; Yang, Z.-C. *Synlett* **1996**, 609–611. (b) Setsune, J.-I.; Ueda, T.; Shikata, K.; Matsukawa, K.; Iida, T.; Kitao, T. *Tetrahedron* **1986**, *42*, 2647–2656.
[46] Hennessy, E. J.; Buchwald, S. L. *Org. Lett.* **2002**, *4*, 269–272.
[47] Yip, S. F.; Cheung, H. Y.; Zhou, Z.; Kwong, F. Y. *Org. Lett.* **2007**, *9*, 3469–3472.
[48] Johansson, C. C. C.; Colacot, T. J. *Angew. Chem. Int. Ed. Engl.* **2010**, *49*, 676–707.
[49] Cristau, H.-J.; Cellier, P. P.; Spinddler, J.-F.; Taillefer, M. *Chem. Eur. J.* **2004**, *10*, 5607–5622.
[50] Mino, T.; Yagishita, F.; Shibuya, M.; Kajiwara, K.; Shindo, H.; Sakamoto, M.; Fujita, T. *Synlett* **2009**, 2457–2460.
[51] Jiang, Y.; Wu, N.; Wu, H.; He, M. *Synlett* **2005**, 2731–2734.
[52] Xie, X.; Cai, G.; Ma, D. *Org. Lett.* **2005**, *7*, 4693–4695.
[53] (a) Xie, X.; Chen, Y.; Ma, D. *J. Am. Chem. Soc.* **2006**, *128*, 16050–16051. (b) Cai, Q.; Zhou, F. *Synlett* **2013**, 408–411.
[54] (a) Pei, L.; Qian, W. *Synlett* **2006**, 1719–1723. (b) Danoun, G.; Tlili, A.; Monnier, F.; Taillefer, M. *Angew. Chem. Int. Ed.* **2012**, *51*, 12815–12819.
[55] Quallich, G. J.; Makowski, T. W.; Sanders, A. F.; Urban, F. J.; Vazquez, E. *J. Org. Chem.* **1998**, *63*, 4116–4119.
[56] Yang, D.; Yang, H.; Fu, H. *Chem. Commun.* **2011**, *47*, 2348–2350.
[57] Ames, D. E.; Dodds, W. D. *J. Chem. Soc. Perkin Trans. 1* **1972**, 705–710.
[58] Werner, A.; Friedrichsen, W. *Synth. Commun.* **1993**, *23*, 2973–2975.
[59] Bryson, T. A.; Stewart, J. J.; Gibson, J. M.; Thomas, P. S.; Berch, J. K. *Green Chem.* **2003**, *5*, 174–176.
[60] McKillop, A.; Rao, D. P. *Synthesis* **1977**, 760–761.
[61] Setsune, J.-I.; Matsukawa, K.; Kitao, T. *Tetrahedron Lett.* **1982**, *23*, 663–666.
[62] Hang, H. C.; Drotleff, E.; Elliot, G. I.; Ritsema, T. A.; Konopelsky, J. P. *Synthesis* **1999**, 398–400.
[63] Lu, J.; Fu, H. *J. Org. Chem.* **2011**, *76*, 4600–4605.
[64] Fang, Y.; Li, C. *J. Org. Chem.* **2006**, *71*, 6427–6431.
[65] Sperotto, E.; van Klink, G. P. M.; van Koten, G.; de Vries, J. G. *Dalton Trans.* **2010**, *39*, 10338–10351.
[66] Cohen, T.; Wood, J.; Dietz, A. G. *Tetrahedron Lett.* **1974**, *15*, 3555–3558.
[67] Cohen, T.; Cristea, I. *J. Am. Chem. Soc.* **1976**, *98*, 748–753.
[68] Bowman, W. R.; Heaney, H.; Smith, P. H. G. *Tetrahedron Lett.* **1984**, *25*, 5821–5824.
[69] (a) Allred, G. D.; Liebeskind, L. S. *J. Am. Chem. Soc.* **1996**, *118*, 2748–2749. (b) Rovira, M.; Font, M.; Ribas, X. *ChemCatChem* **2013**, *5*, 687–691. (c) He, C.;

Zhang, G.; Ke, J.; Zhang, H.; Miller, J. T.; Kropf, A. J.; Lei, A. *J.Am. Chem. Soc.* **2012**, *135*, 488–493. (d) Casitas, A.; Ribas, X. *Chem.Sci.* **2013**, *4*, 2301–2318.

[70] Huang, Z.; Hartwig, J. F. *Angew. Chem. Int. Ed. Engl.* **2012**, *51*, 1028–1032.

[71] Beletskaya, I. P.; Sigeev, A. S.; Peregudov, A. S.; Petrovskii, P. V.; Khrustalev, V. N. *Chem. Lett.* **2010**, *39*, 720–722.

[72] Hey, D. H.; Waters, W. A. *Chem. Rev.* **1937**, *21*, 169.

[73] Jenkins, C. L.; Kochi, J. K. *J. Am. Chem. Soc.* **1972**, *94*, 856.

[74] Kochi, J. K. *Organometallic Mechanisms and Catalysis*, Academic Press: New York, **1978**.

[75] Fedorov, A. Y.; Finet, J.-P. *Eur. J. Org. Chem.* **2004**, 2040–2045.

[76] Ma, D.; Zhang, Y.; Yao, J.; Wu, S.; Tao, F. *J. Am. Chem. Soc.* **1998**, *120*, 12459.

[77] Ma, D.; Cai, Q. *Acc. Chem. Res.* **2008**, *41*, 1450.

[78] Pape, A. R.; Kaliappan, K. P.; Kündig, E. P. *Chem. Rev.* **2000**, *100*, 2917–2940.

[79] Aalten, H. L.; Van Koten, G.; Goubitz, K.; Stam, C. H. *Organometallics* **1989**, *8*, 2293–2299.

[80] Aalten, H. L.; Van Koten, G.; Riethorst, E. D.; Stam, C. H. *Inorg. Chem.* **1989**, *28*, 4140–4146.

[81] Aalten, H. L.; Van Koten, G.; Vrieze, K.; Van der Kerkvanhoof, A. *Rec. Trav. Chim. Pays-Bas* **1990**, *109*(2), 46–54.

[82] Do, H.-Q.; Daugulis, O. *J. Am. Chem. Soc.* **2007**, *129*, 12404–12405.

[83] Do, H.-Q.; Khan, R. M. K.; Daugulis, O. *J. Am. Chem. Soc.* **2008**, *130*, 15185–15192.

[84] Do, H.-Q.; Daugulis, O. *J. Am. Chem. Soc.* **2008**, *130*, 1128–1129.

[85] Besselievre, F.; Piguel, S.; Mahuteau-Betzer, F.; Grierson, D. S. *Org. Lett.* **2008**, *10*, 4029–4032.

[86] Ackermann, L.; Potukuchi, H. K.; Landsberg, D.; Vicente, R. *Org. Lett.* **2008**, *10*, 3081–3084.

[87] Zhao, D.; Wang, W.; Yang, F.; Lan, J.; Yang, L.; Gao, G.; You, J. *Angew. Chem. Int. Ed. Engl.* **2009**, *48*, 3296–3300.

[88] Yotphan, S.; Bergman, R. G.; Ellman, J. A. *Org. Lett.* **2009**, *11*, 1511–1514.

[89] Ackermann, L.; Vincente, R.; Kapdi, A. R. *Angew. Chem. Int. Ed. Engl.* **2009**, *48*, 9792–9826.

9

COPPER-CATALYZED CYANATIONS OF ARYL HALIDES AND RELATED COMPOUNDS

Thomas Schareina and Matthias Beller

9.1 INTRODUCTION

Benzonitriles are of general interest for organic synthesis as integral parts of dyes, herbicides, agrochemicals, pharmaceuticals, and natural products.[1] In addition, the nitrile group serves as an important intermediate for a multitude of possible transformations into other functional groups, such as benzoic acid derivatives, benzylamines, benzaldehydes, and heterocycles (Scheme 9.1).

Typically, the introduction of a cyanide group is the most direct and versatile route to the preparation of functionalized benzonitriles. For more than a century, the synthesis of benzonitriles by stoichiometric methods prevailed in laboratory and industry (Scheme 9.2). Among these methods, the Rosenmund–von Braun reaction of aryl halides (see the Preface for a historical perspective)[2,3] and the diazotization of anilines with the subsequent Sandmeyer reaction were the most popular (Scheme 9.2, Eqs. 1 and 2).[4] However, due to (over)stoichiometric amounts of metal waste being generated, such processes do not meet today's criteria of sustainable synthesis. For the ton-scale production of benzonitriles, the method of choice in industry is ammoxidation,[5] whereby the corresponding toluene derivatives are reacted with oxygen and ammonia at high temperatures (300–550°C) in the presence of heterogeneous fixed-bed catalysts (Scheme 9.2, Eq. 3). However, lack of functional group

Copper-Mediated Cross-Coupling Reactions, First Edition. Edited by Gwilherm Evano and Nicolas Blanchard.

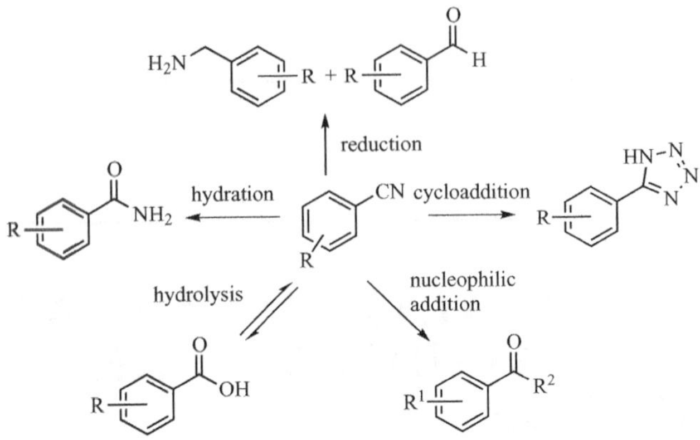

Scheme 9.1 Representative synthetic applications of benzonitriles.

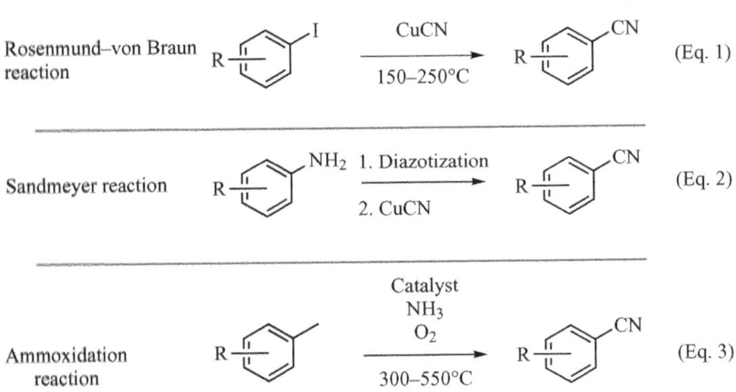

Scheme 9.2 Traditional methods for the synthesis of benzonitriles.

tolerance and harsh reaction conditions make this method less suitable for the synthesis of functionalized benzonitriles.

In the early 1970s, the introduction and development of transition-metal-catalyzed carbon–carbon coupling reactions have dramatically changed the method of functionalizations of arenes. In this context, the first palladium-catalyzed cyanation of aryl halides was introduced in 1973 by Takagi and coworkers by reacting aryl bromides and iodides with potassium cyanide as the cyanating agent.[6] Since then, various palladium, nickel, and, more recently, copper catalysts were developed for the coupling of aryl halides with cyanides. Clearly, palladium complexes have dominated as catalysts in cyanation reactions so far because they tolerate a wider variety of functional groups and are less sensitive to air and humidity than nickel catalysts. They are also more active than copper catalysts. On the other hand, copper catalysts and the nitrogen ligands used mostly for their activation are much cheaper than the palladium catalysts and the phosphine ligands necessary to activate them, in many cases also outweighing the higher cost for the more reactive substrates that are usually needed for copper. A slight downside of copper complexes as catalysts is their biological activity. While higher animals (vertebrates) are relatively inert against copper uptake, even very low concentrations can be biocidal for microorganisms, including those used in waste water treatment.*,[7,8] Thus, apart from cost arguments, the application of a low amount of the catalyst is desirable.

A general problem of metal-catalyzed cyanations is the high affinity of cyanide toward typical palladium-, nickel-, and copper-based catalysts. Often, a fast deactivation of the catalytic system is observed due to the formation of stable metal cyanide complexes; and catalysis proceeds in general with low efficiency. Notably, all mechanistic investigations in this direction have been performed with palladium-based systems.[9] One solution for overcoming this problem relies on the use of solvents in which standard cyanide sources such as sodium cyanide, potassium cyanide, and zinc(II) cyanide[10] have very low solubility.[3b] Beller and others have shown that organic (e.g., tetramethylethylenediamine [TMEDA])[11] and inorganic (e.g., zinc and zinc salts)[12] additives are beneficial for the regeneration of the catalytically active palladium metal center. Another elegant approach has been the slow dosage of the corresponding cyanide source[3g] (e.g., acetone cyanohydrin[13] or trimethylsilyl cyanide),[14] which keeps the cyanide concentration low and leads to a higher catalyst productivity. Other recent developments include microwave activation[15] and the application of novel or modified catalyst systems.[16] However, most of these developments have drawbacks such as toxicity of the cyanide source and comparably high catalyst costs (low catalyst productivity).

In the following sections, the classical copper-mediated Rosenmund–von Braun and Sandmeyer reactions are described followed by procedures using only catalytic amounts of copper salts in conjunction with mild sources of cyanide such as acetone cyanhydrin or potassium hexacyanoferrate(II). The final section is devoted to a blooming area of the cyanation reactions, the copper-catalyzed oxidative cyanation.

9.2 MODIFICATIONS AND UPDATES OF CLASSICAL CYANATION REACTIONS (ROSENMUND–VON BRAUN, SANDMEYER)

Since the first report of a copper-catalyzed cyanation of bromo-anthraquinone by the Farbwerke Hoechst Company in the patent literature in 1913,[2c] considerable progresses have been reported in the Rosenmund–von Braun reaction. Cai and coworkers were able to decrease the reaction time of this venerable transformation to 20–40 minutes by microwave heating in *N*-methylpyrrolidinone at ambient pressure (Scheme 9.3).[17] Simple standard aryl bromides were converted to the corresponding benzonitriles by reaction

* For humans, the daily uptake of 0.5 $mg.kg^{-1}$ is considered nonhazardous.

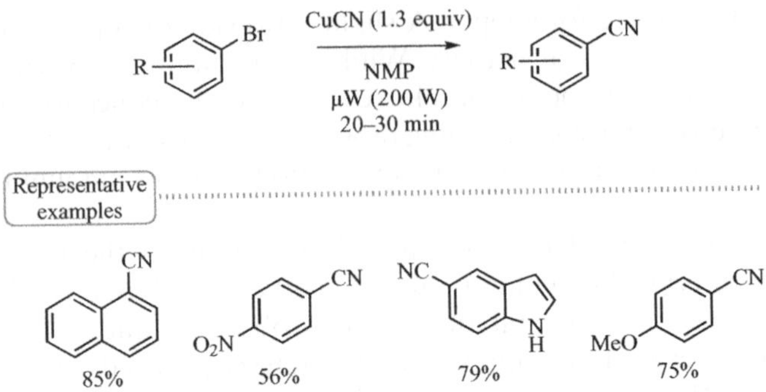

Scheme 9.3 Copper(I)-mediated cyanation reaction of aryl bromides using microwave irradiation.

with copper(I) cyanide in 68–90% yield. While the method is fast and easy to apply, still, an overstoichiometric amount of copper(I) cyanide has to be used.

Wang and coworkers identified L-proline as an effective additive to promote the Rosenmund–von Braun reaction at lower temperature (80–120°C) in dimethylformamide (DMF).[18] This modified Rosenmund–von Braun cyanation of aryl bromides exhibits good functional-group compatibility since nitro, amino, amide, hydroxyl, and ester substituents were tolerated. Once again, an overstoichiometric amount of copper(I) cyanide has to be used, which limits the usefulness of the method.

Beletskaya and coworkers introduced another interesting improvement of a classical cyanation reaction. Catalytic Sandmeyer reactions using copper(I) cyanide (10 mol %) in the presence of 1,10-phenanthroline (phen) as a ligand, copper(II) tetrafluoroborate (10 mol %, to suppress radical side reactions), and potassium cyanide as cyanating agent in acetonitrile could be performed at room temperature starting from reactive aryl diazonium tetrafluoroborate salts (Scheme 9.4).[19] However, electron-rich substrates gave lower yields of the desired aryl cyanides.

A logical evolution of the cyanation reaction of aryl halides has focused on the development of an efficient copper-catalyzed version. The efforts toward this important goal, both from an academic and an industrial point of view, are discussed in the next section.

9.3 COPPER-CATALYZED CYANATIONS OF ARYL HALIDES

After the first example of a copper-catalyzed (5 mol %) cyanation of an aryl bromide in the open literature appeared,*,[20] a general protocol for catalytic

* One example of copper-catalyzed (5 mol %) cyanation of an aryl bromide (2-bromonaphthalene) has been reported to proceed with 55% conversion when the reaction is carried out in an ionic liquid.

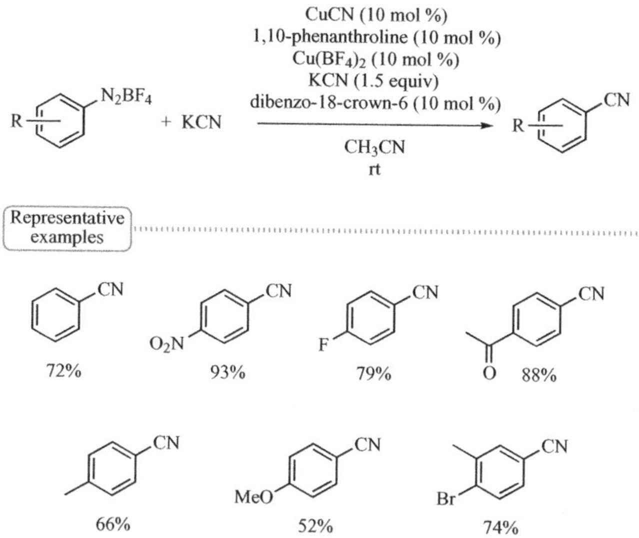

Scheme 9.4 Copper-catalyzed Sandmeyer reaction.

Rosenmund–von Braun reaction was reported by Buchwald and coworkers.[21] The latter used copper(I) iodide (10 mol %), potassium iodide (20 mol %), and *N,N′*-dimethylethylenediamine (DMEDA, 100 mol %), the cyanide source being sodium cyanide (1.2 equiv). Subsequently, Cristau, Taillefer, and coworkers[22] demonstrated that this reaction also proceeds in the presence of 1,10-phenanthroline (20 mol %) as ligand. Both methods relied upon cheap but highly toxic cyanide sources like sodium cyanide or potassium cyanide (Scheme 9.5). Both groups also stated that a catalytic amount of iodide is necessary as additive to convert aryl bromides into the corresponding aryl iodide, which then undergoes the substitution reaction with cyanide (see Chapter 6 for details on the aromatic Finkelstein reaction). It should be noted that in both cases, the corresponding patents have been issued as early as 2001.[23,24]

Additionally, Cristau, Taillefer, and coworkers reported, for the first time in the same publication, the use of acetone cyanohydrin as a cyanide source for copper-catalyzed cyanation reactions.[22] Acetone cyanohydrin decomposes quantitatively to acetone and cyanide anion in the presence of a base (Scheme 9.6). Thus, a controlled amount of cyanide can be introduced into the reaction vessel by slow addition, for example, with the help of a syringe pump, to prevent the above-mentioned deactivation of the metal catalyst. Beller and coworkers developed a similar protocol in which the use of acetone cyanohydrin allowed the reduction of the amount of catalyst by fine-tuning the addition rate of the cyanation reagent, thus avoiding catalyst poisoning.[13]

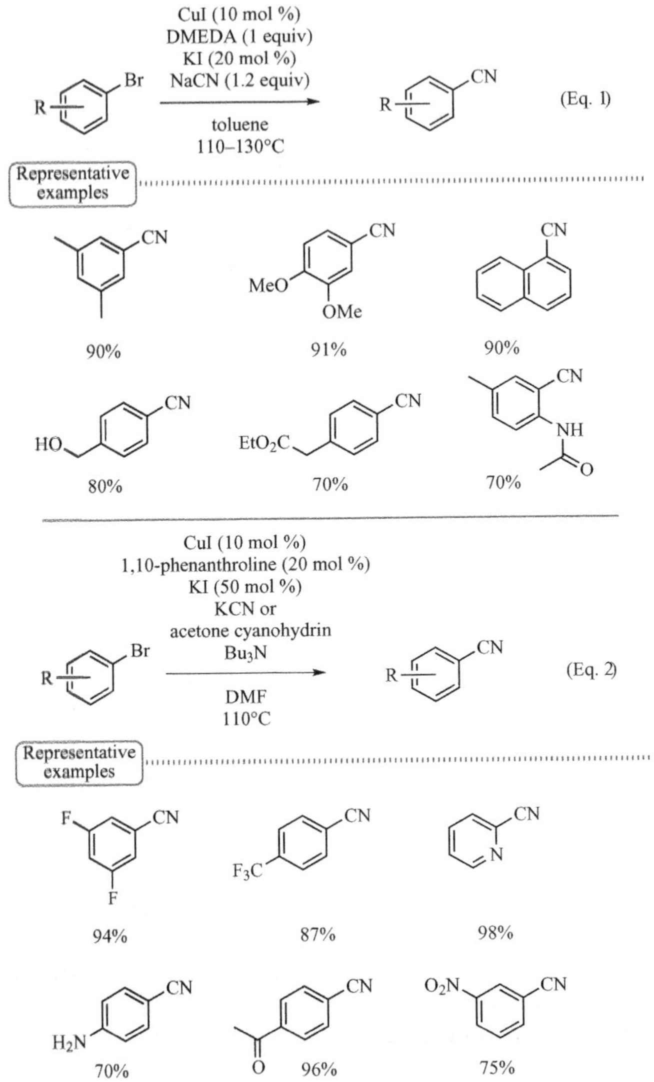

Scheme 9.5 Copper(I)-catalyzed domino halide exchange–cyanation sequence.

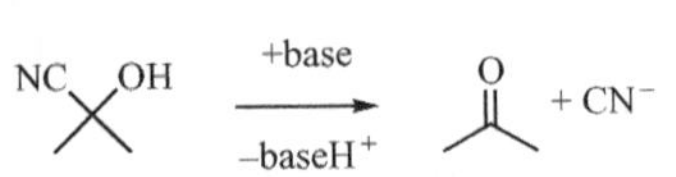

Scheme 9.6 Decomposition of acetone cyanohydrin under basic conditions.

Historically, the latest cyanation reagent for the preparation of benzonitriles is potassium hexacyanoferrate(II), $K_4[Fe(CN)_6]$.*[26] After establishing its value for the palladium-catalyzed cyanation of aryl halides,[27] Beller was also able to use it for the copper-catalyzed cyanation reaction.[28] Potassium hexacyanoferrate(II) has the advantage of being essentially the least toxic cyanide source conceivable: While all known other cyanation sources are highly poisonous (potassium cyanide has a lethal dose low of 2.86 $mg.kg^{-1}$ orally in humans), $K_4[Fe(CN)_6]$ is nontoxic (the lethal dose 50 of $K_4[Fe(CN)_6]$ is lower than the one of sodium chloride!) and is even used in food industry for metal precipitation.† Importantly, for practical applications, potassium hexacyanoferrate(II) is commercially available on ton-scale and its price per mole of cyanide is comparable to potassium cyanide.‡ Notably, all six cyanide ligands can be used for the catalytic cyanation reactions.

In the first report on the use of potassium hexacyanoferrate(II) for copper-catalyzed cyanation of aryl halides, copper(II) tetrafluoroborate and DMEDA[21] as ligand were used as the catalytic system.[28] In later reports, a modified procedure using copper(I) iodide (10 mol %) and 1-butylimidazole (200 mol %) as an additive or 1-methylimidazole as a solvent was developed (Scheme 9.7).[29] Despite comparably high reaction temperatures (up to 160°C in the majority of cases), excellent selectivities were achieved. Additionally, the scope of the cyanation reaction was increased in comparison with palladium-based methods, including the derivatization of heterocyclic and sterically hindered aromatic bromides.

More recently, other groups used 1,10-phenanthroline in combination with copper(I) iodide (20 mol %) and the same cyanide source (Scheme 9.8).[30] Aryl iodides gave marginally higher yields of benzonitrile products than aryl bromides and aryl chlorides did not react at all.

An even more simple and inexpensive ligand, ethylenediamine, was introduced by Wang and coworkers (Scheme 9.9).[31] However, the copper(II) acetate loading was relatively high (30 mol %). The method worked well for aryl bromides and highly activated aryl chlorides, and potassium iodide had to be used as an additive in this reaction.

Another option for the production of aryl nitriles from low-toxicity starting materials is the method called "combined cyanide source" by Cheng and coworkers (Scheme 9.10).[3g,32] Cyanide is produced *in situ* from ammonium salts and the solvent, DMF, and the copper mediator seems to play an

* The first use of potassium ferrocyanide as cyanating reagent in a noncatalytic reaction.

† The risk and safety statements and the Material Safety Data Sheets (MSDS) of fine chemical providers differ considerably on the topic of the toxicity of potassium hexacyanoferrate(II) and other ferrocyanides. In any case, contact with acids should be avoided due to the potential danger of HCN evolution, for the same reasons ingestion of substantial amounts should be definitely avoided. The use in the food industry is limited to low concentrations.

‡ The price per mole of cyanide is €14 for $K_4[Fe(CN)_6]$ and €11 for potassium cyanide (ACS quality both). *Source*: *Aldrich Chemicals Catalogue* 2012.

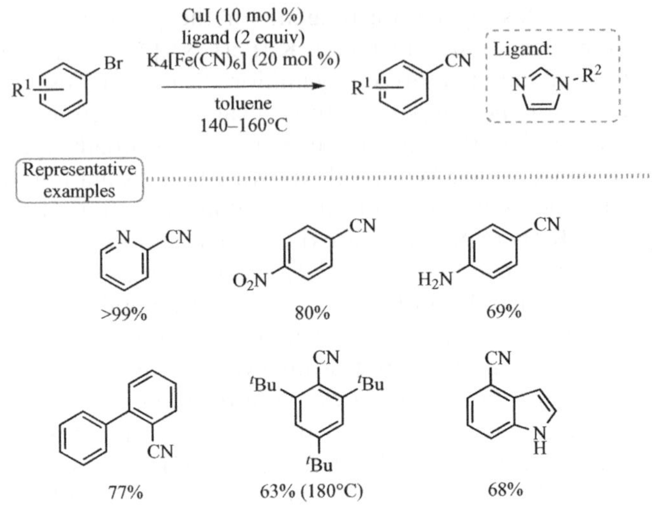

Scheme 9.7 Copper(I)-catalyzed cyanation of aryl bromides with potassium ferrocyanide and alkylimidazoles as ligands.

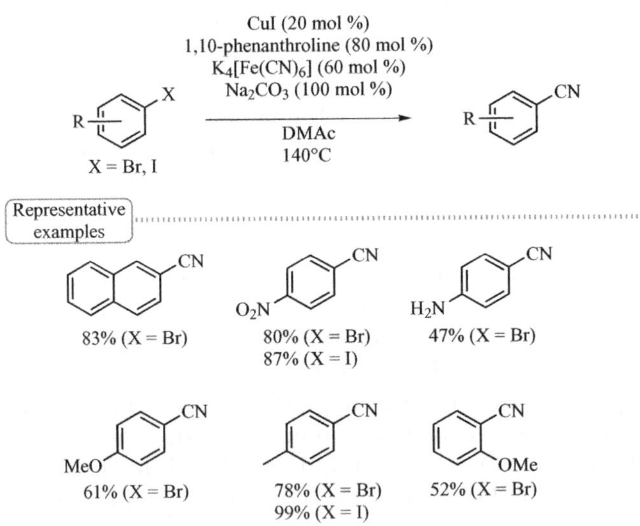

Scheme 9.8 Copper(I)-catalyzed cyanation of aryl iodides and bromides with potassium ferrocyanide and 1,10-phenanthroline as ligand.

important role in the formation of cyanide. Mostly aryl iodides were tested in this reaction, aryl bromides giving inferior results. Alkyl, alkenyl, and alkynyl iodides did not work under the standard reaction conditions.

Water as a reaction medium is a popular way to render procedures more ecologically and economically friendly. However, simple inorganic cyanides

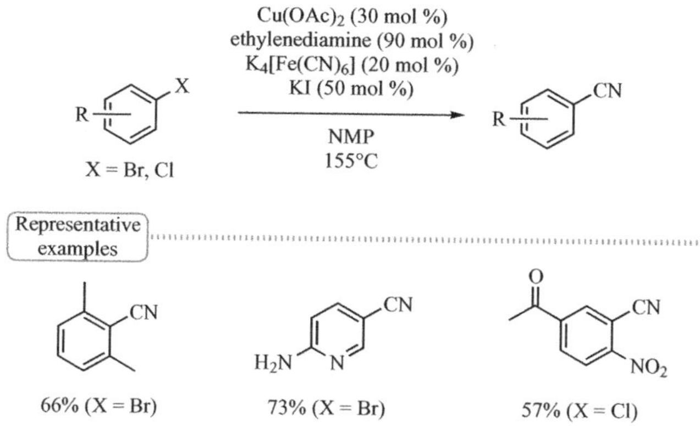

Scheme 9.9 Copper(II)-catalyzed cyanation of aryl bromides and activated aryl chlorides with ethylenediamine as ligand.

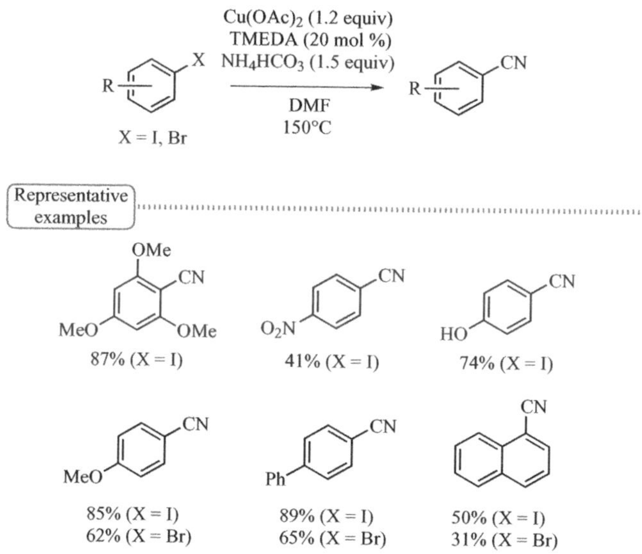

Scheme 9.10 Copper(II)-mediated cyanation of aryl halides with *in situ* generated cyanide from ammonium hydrogencarbonate and dimethylformamide.

such as sodium- and potassium cyanide decompose in water, resulting in the evolution of highly toxic gaseous hydrogen cyanide, and are therefore hardly relevant in aqueous solutions. Due to the stability of potassium hexacyanoferrate(II) in water, the cyanation reaction can also be performed in aqueous media with this reagent, as shown by the group of Wang (Scheme 9.11).[33] While all types of aryl iodides react reasonably well, only aryl bromides with strongly electron-withdrawing substituents gave the desired

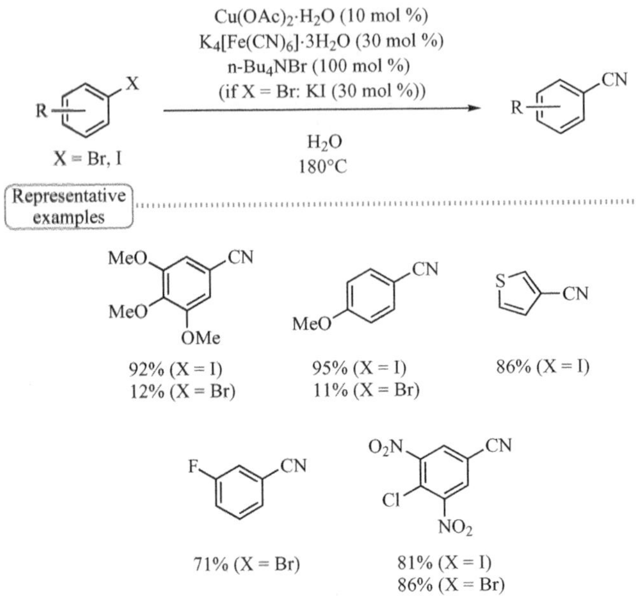

Scheme 9.11 Ligand-free copper(II)-catalyzed cyanation in aqueous media.

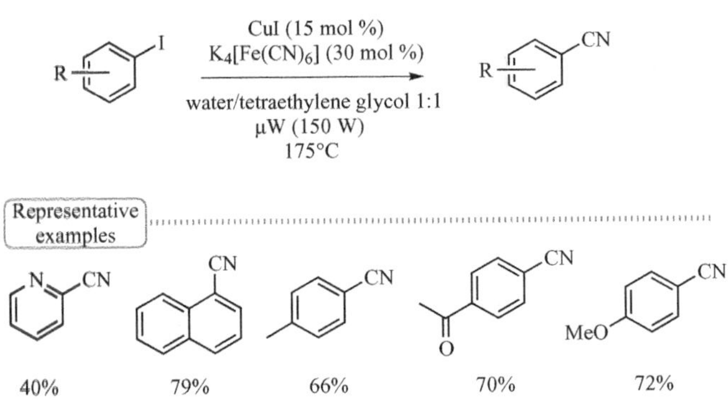

Scheme 9.12 Microwave-heated copper(I)-catalyzed cyanation of aryl iodides in aqueous media.

products. Moreover, potassium iodide (30 mol %) has to be added to ensure significant reactivity.

Besides economical and ecological aspects, water is an ideal solvent for the application of microwave heating. DeBlase and Leadbeater showed that the reaction times of catalytic cyanations could be reduced to 30 minutes thanks to microwave irradiation at 175°C (Scheme 9.12).[34] The yield of the reaction is sensitive to the water concentration, and tetraethylene glycol seems to act as a phase transfer reagent. A very similar procedure, based on palladium catalysis, has been published earlier by the same group.[35]

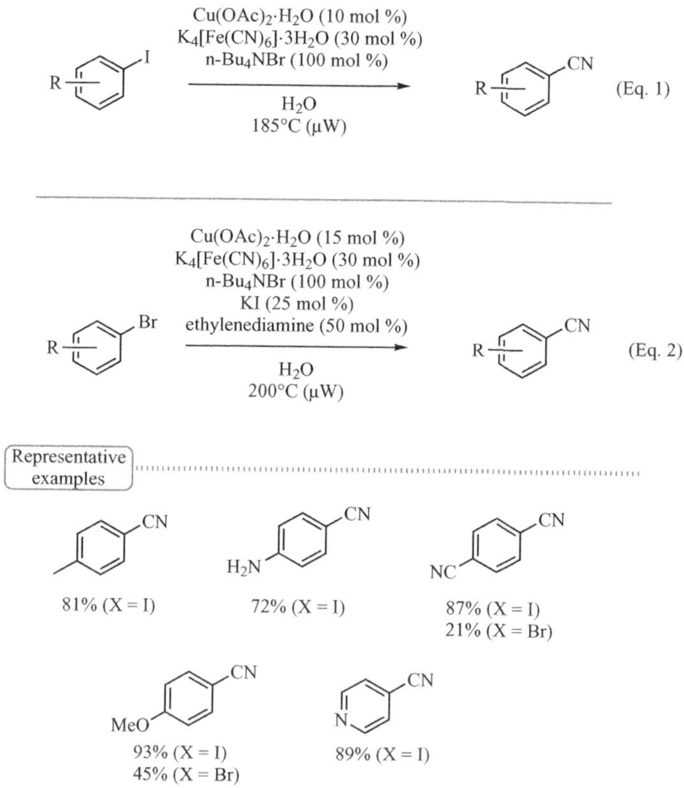

Scheme 9.13 Ligand-free copper(II)-catalyzed cyanation of aryl halides with microwave irradiation in aqueous media.

The Wang group converted mainly aryl iodides to the corresponding benzonitriles with a similar set of conditions (Scheme 9.13).[36] A corrigendum revealed some errors in the first article, resulting in slightly different conditions for the conversion of aryl bromides and aryl iodides. Aryl bromides could be cyanated in good yields with the help of additives, potassium iodide and ethylenediamine, and higher reaction temperature. After pentane extraction of the final product, the aqueous phase containing the catalytic system could be reused up to six times in the cyanation of iodobenzene with a slightly diminished yield (see Chapter 20 for an overview of reusable copper-based catalytic systems).

Another microwave-based approach by Leadbeater used $Cu_2[Fe(CN)_6]$ as both catalyst and cyanide source for the cyanation of aryl iodides (Scheme 9.14).[37] In the few model reactions of standard substrates, 0.3 equiv of $Cu_2[Fe(CN)_6]$, corresponding to 0.6 equiv catalytic copper, gave in the best case 78% yield of the nitrile product starting from 1-(4-iodophenyl)ethanone. In the case of one particularly hydrophobic substrate, 4-iodotoluene, 1 equiv of the

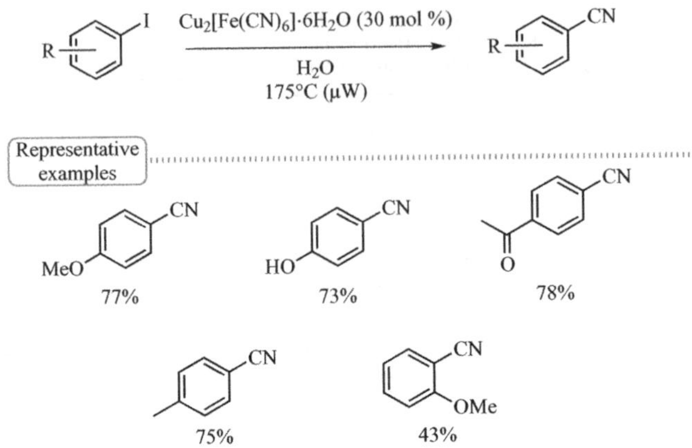

Scheme 9.14 Ligand-free copper(II)-catalyzed cyanation of aryl bromides with microwave irradiation in aqueous media.

phase transfer reagent tetrabutylammonium bromide (TBAB) had to be added to increase the yield from 35% (without TBAB) to 75% (with TBAB).

As shown by all examples described in this section, efficient copper-catalyzed cyanation reactions are now available to the synthetic chemist to prepare benzonitriles using nontoxic cyanide sources[3g] such as potassium hexacyanoferrate(II) or powerful combined copper- and cyanide sources such as $Cu_2[Fe(CN)_6]$. The last section of this chapter is devoted to copper-catalyzed oxidative cyanation reactions that are currently blossoming due to their intrinsic synthetic potential.

9.4 COPPER-MEDIATED OXIDATIVE CYANATIONS

The direct activation of aromatic C–H bonds gained increasing interest in recent years and, in this context, impressive oxidative cyanation reactions have been reported using very simple cyanide sources such as trimethylsilyl cyanide, nitromethane, benzyl cyanide, and even acetonitrile.[3g]

In a copper(II) acetate-mediated functionalization reaction with oxygen as the final oxidant, chelating 2-arylpyridines were cyanated in the *ortho*-position using trimethylsilyl cyanide (Scheme 9.15, Eq. 1) or nitromethane (Scheme 9.15, Eq. 2), although the detailed mechanism for this transformation still has to be elucidated.[38b] More recently, Wang and coworkers described a copper(I) bromide-catalyzed cyanation of 2-aryl heteroaromatics using benzyl cyanide as a mild source of cyanide (Scheme 9.15, Eq. 3).[38a] This very efficient transformation relies on the aerobic copper oxidation of the benzylic position of the latter to the corresponding benzoyl cyanide that could undergo a retro-cyanohydrination. In parallel, a copper(II) aerobic C–H functionalization that is thought to proceed via single-electron transfers allows a clean cyanation reaction. Pyridines, pyrimidines, and pyrazoles were shown to be efficient

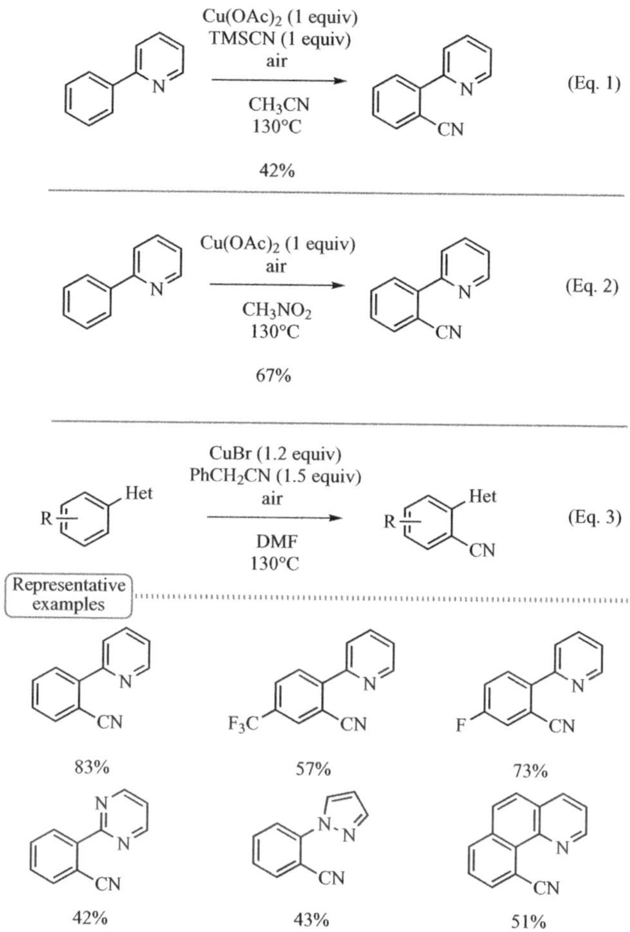

Scheme 9.15 Ligand-free copper-mediated oxidative cyanation of 2-aryl heterocycles by C–H activation using trimethylsilyl cyanide, nitromethane, and benzyl cyanide.

directing groups in this reaction. A related cyanation reaction of indoles in position 3 using benzyl cyanide and copper(I) iodide (100 mol %) in DMF under air was also reported by Kwong in 2013.[38c]

Most recently, Do and Daugulis reported a direct copper-catalyzed cyanation of heterocyclic C–H bonds with iodine as the final oxidant.[39] Heterocycles with relatively acidic hydrogens, as oxazoles, thiazoles, imidazoles, triazoles, and azole are selectively cyanated with of copper(I) cyanide (10 mol %) and sodium cyanide as stoichiometric cyanide source in the presence of the strong base lithium *tert*-butoxide (Scheme 9.16). This overall transformation into heteroaryl cyanide involves a lithiation followed by iodination, the transient heteroaryl iodide being then converted into the desired nitrile in the only copper-catalyzed step of this domino sequence.

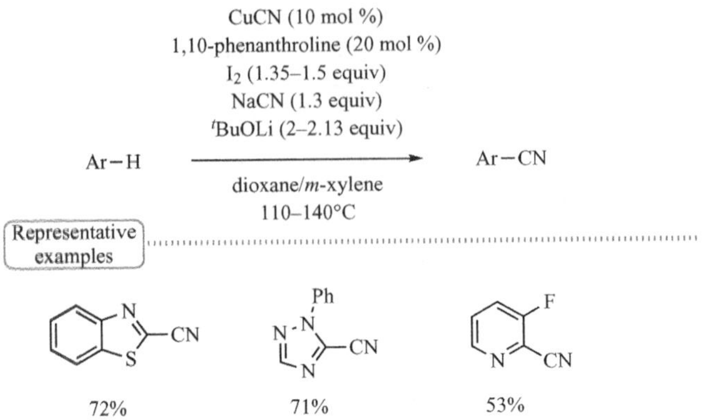

Scheme 9.16 Copper(I)-catalyzed regioselective cyanation reaction of heterocycles.

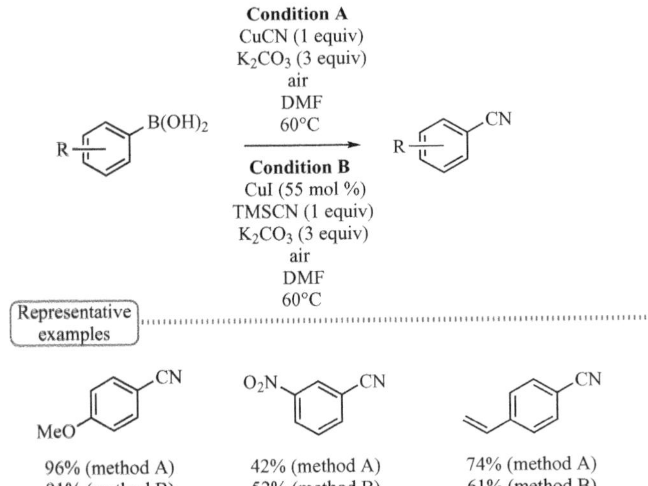

Scheme 9.17 Copper(I)-mediated oxidative cyanation reaction of boronic acids.

In addition to aryl halides, the cyanation of arylboronic acids has been achieved using copper(I) cyanide (1 equiv) and potassium carbonate in dry DMF under air at 60°C.[40] Alternatively, a combination of copper(I) cyanide (0.55 equiv), trimethylsilyl cyanide, and potassium carbonate in dry DMF can be used for this transformation. Thanks to the ease of activation of the carbon–boron bond, this method could be performed under mild reaction conditions and with an excellent functional group tolerance (Scheme 9.17).

Hartwig and coworkers developed a one-pot sequence to convert 1,3-disubstituted or 1,2,3-trisubstituted arenes and heteroarenes to aromatic and heteroaromatic nitriles by an iridium-catalyzed direct C–H borylation followed by a copper-mediated oxidative cyanation of the arene (Scheme 9.18).[41] The substrates are limited to 3,5-substituted and 3,4,5-substituted arenes, and

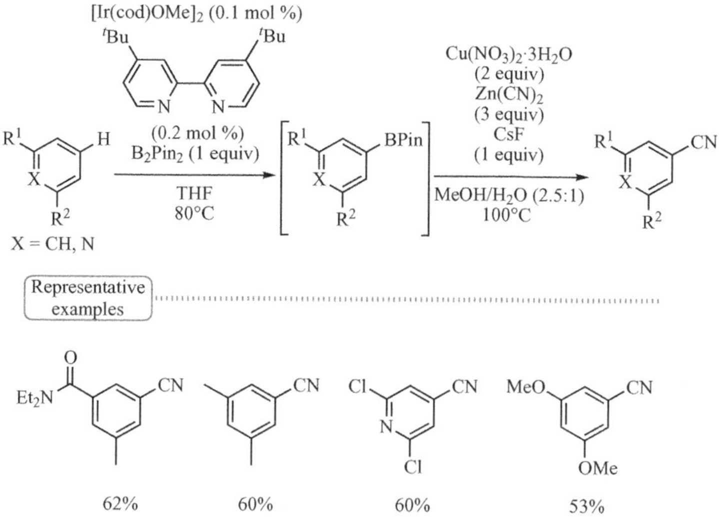

Scheme 9.18 One-pot sequential C–H borylation and cyanation of arenes.

halide, ketone, ester, amide, and protected alcohol may be present in the coupling partner.

Finally, a totally different approach developed by Lamani and Prabhu is noteworthy.[42a] In the presence of copper(I) iodide (5 mol %) and *tert*-butyl hydroperoxide in refluxing water, a wide range of primary azides are oxidized to the corresponding nitriles (Scheme 9.19, Eq. 1). The method is applicable to both aromatic and aliphatic azides. A clever extension of this concept was reported in 2012 by the same research group (Scheme 9.19, Eq. 2): Using trimethylsilyl azide as a nitrogen source in the presence of copper(II) perchlorate, benzylic and cinnamyl alcohols were smoothly converted to the corresponding nitriles in good yields via the intermediate azides that were further oxidized with dichlorodicyanoquinone.[42b]

A new development based on the combined cyanide source (see Scheme 9.10) was introduced by Chang and coworkers (Scheme 9.20).[43a] With copper(II) nitrate mediation under oxidative conditions, aryl boron compounds (boronic acids, boronate esters, and potassium trifluoroborate) as well as electron-rich aromatics were converted to the corresponding aryl nitriles. A key intermediate is ammonium iodide, which delivers the nitrogen atom of the cyanide group and the iodide as a transient leaving group for the substrate. A related transformation was reported by Chang in 2013 for the cyanation of aryl and stryrylsilanes.[43b]

Li and coworkers reported a very simple copper(I)-catalyzed oxidative cyanation of aryl halides using acetonitrile as the cyanide source in the presence of triphenylphosphine oxide as a ligand (40 mol %) and silver oxide (1 equiv) at 125°C under air (Scheme 9.21).[44] In this catalytic system, the aryl copper(II) intermediate complexes the nitrile moiety of acetonitrile, followed

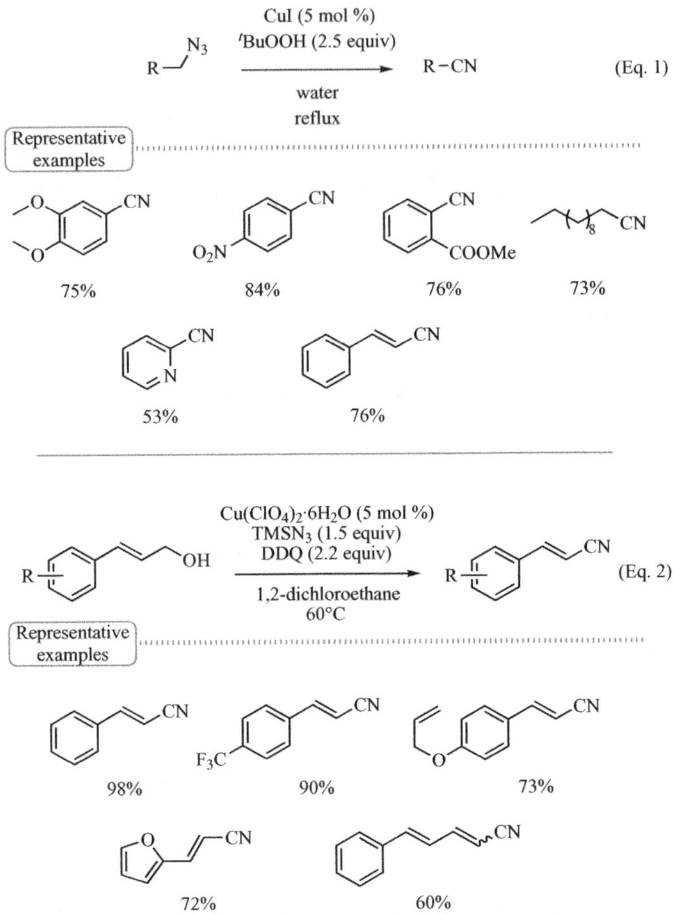

Scheme 9.19 Copper(I) and (II)-catalyzed oxidative synthesis of nitriles from azides and alcohols.

by oxidative cleavage of the carbon–carbon σ-bond of acetonitrile mediated by oxygen and silver oxide. Reductive elimination of the copper(III) intermediate finally delivers the desired nitrile and regenerates the active copper(I) catalyst. Electron-rich and electron-poor aryl iodides and bromides could be converted in moderate-to-good yields to the corresponding benzonitriles under these conditions. It is worth noting that phenolic motifs were tolerated, which is in sharp contrast with previous metal-catalyzed cyanation reactions that, upon reaction of the metal cyanide with the free phenol, liberate cyanhydric acid that immediately deactivates the catalyst. Other cyanide sources were screened such as benzyl cyanide, benzonitrile, and ethylcarbonocyanidate, all leading to the desired nitriles, although in lower yields. To close this section, a recent copper-mediated direct cyanation of aromatic C–H bond through a radical pathway ought to be mentioned (Scheme 9.22).[45] Han

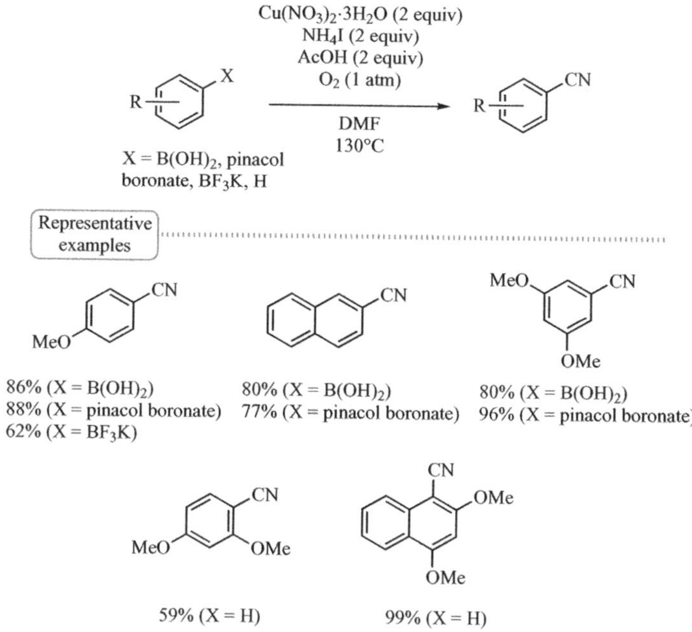

Scheme 9.20 Copper-mediated cyanation reaction of boron derivatives and electron-rich aromatics to nitriles.

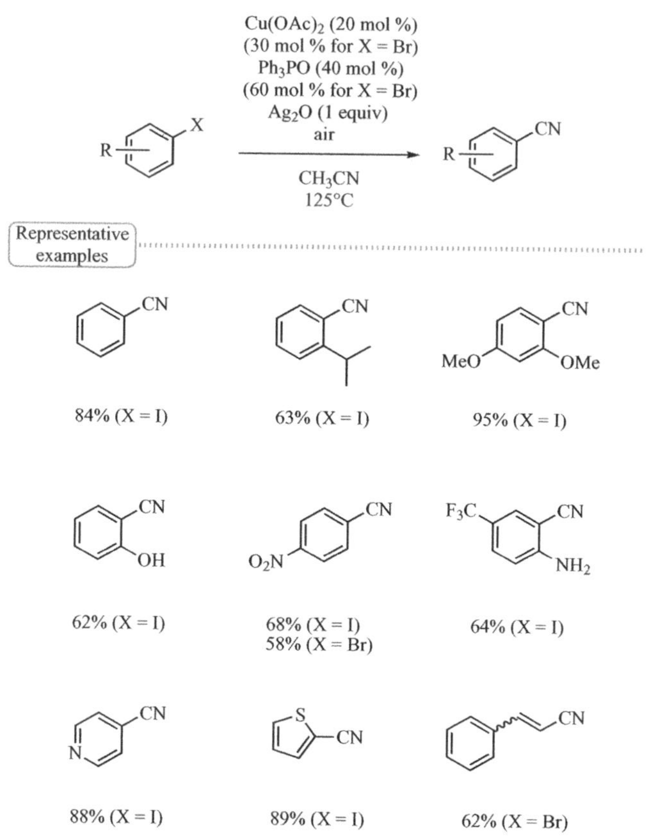

Scheme 9.21 Copper(II)-catalyzed oxidative cyanation of aryl halides using acetonitrile as the cyanide source.

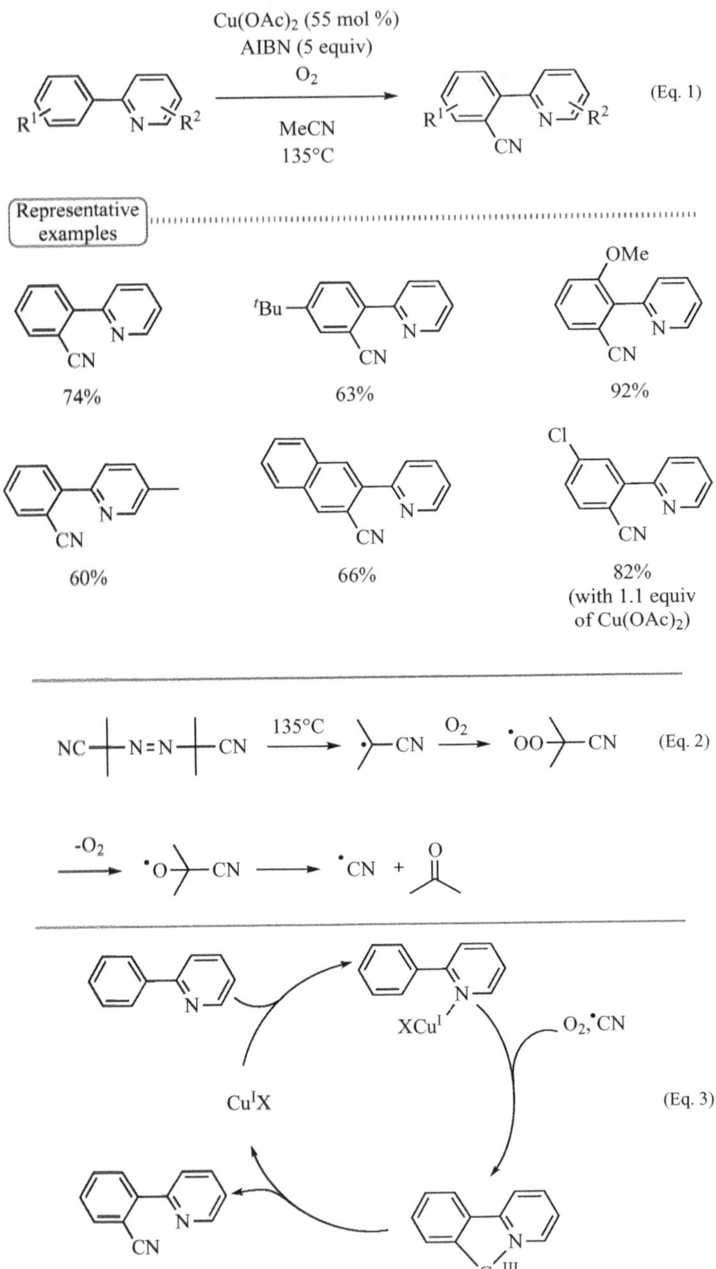

Scheme 9.22 A free-radical copper(I)-mediated direct cyanation reaction of 2-phenyl pyridines.

reported that thermolysis of 2,2′-azobisisobutyronitrile (AIBN) under an oxygen atmosphere smoothly generated a CN radical (Scheme 9.22, Eq. 2) that could combine with a copper(I) species to lead to a copper(III) species under oxidative conditions. Upon reductive elimination, the desired cyanated aromatic was obtained (Scheme 9.22, Eq. 3). This conceptually novel cyanation reaction clearly opens a new chapter between copper salts and cyanide sources.

9.5 CONCLUSION

The synthesis of nitriles continues to be an important topic in organic synthesis. In general, the synthesis of benzonitriles is still far away from being benign, although in the last two decades, significant progress has been achieved in metal-catalyzed cyanations of substituted arenes. Most notable was the introduction of less toxic cyanation agents such as potassium hexacyanoferrate(II), $K_4[Fe(CN)_6]$, or purely organic sources of cyanide such as benzyl cyanide or acetonitrile. In parallel, the development of novel catalysts with improved efficiencies still triggers a lot of interest. However, in spite of all progress, catalytic cyanations of aryl halides and related compounds are still far away from being catalytically efficient. While some of the copper-catalyzed procedures—especially using potassium hexacyanoferrate(II), $K_4[Fe(CN)_6]$, or zinc cyanide as cyanide source—have been upscaled in the pharmaceutical industry up to the kilogram scale, none of these reactions have been so far applied in industry on a larger scale. Clearly, more general and more efficient procedures are needed to achieve this. In order to be applied in the fine chemical industry, the conversion of less expensive aryl chlorides is most desirable. However, this goal is also far away from being solved in a general manner. Finally, and from an academic perspective, the mechanistic foundations of the copper-catalyzed cyanation reactions is desired. Hopefully, this knowledge will lead to improved and "greener" cyanations in the future.

REFERENCES

[1] (a) Larock, R. C. *Comprehensive Organic Transformations*. VCH: New York, **1989**. (b) Grundmann, C. *Houben–Weyl: Methoden der organischen Chemie*, Vol. E5, 4th edition, Falbe, J. (ed.). Georg Thieme Verlag: Stuttgart, **1985**, p. 1313.

[2] (a) Rosenmund, K. W.; Struck, E. *Ber. Dtsch. Chem. Ges.* **1919**, *52*, 1749–1756. (b) von Braun, J.; Manz, G. *Justus Liebigs Ann. Chem.* **1931**, *488*, 111–126. (c) Farbwerke vorm. Meister Lucius & Brüning AG. Verfahren zur Darstellung von Anthrachinon-α-nitrilen. German patents 271,790 Klasse 12o. Gruppe 10. 1913 February 18.

[3] Reviews of aryl halide cyanation methods: (a) Anbarasan, P.; Schareina, T.; Beller, M. *Chem. Soc. Rev.* **2011**, *40*, 5049–5067. (b) Sundermeier, M.; Zapf, A.; Beller, M. *Eur. J. Inorg. Chem.* **2003**, 3513–3526. (c) Ellis, G. P.; Romney-Alexander, T. M. *Chem. Rev.* **1987**, *87*, 779–794. (d) Lindley, J. *Tetrahedron* **1984**, *40*, 1433–1456.

(e) Mowry, D. T. *Chem. Rev.* **1948**, *42*, 189–283. (f) Hagedorn, F.; Gelbke, H.-P. *Ullmanns Encyklopädie der technischen Chemie*, Vol. 17, 4th edition, Bartholomé, E.; Biekert, E.; Hellmann, H.; Ley, H.; Weigert, W. M.; Weise, E. (eds.). Verlag Chemie: Weinheim, **1979**, pp. 333. (g) Kim, J.; Kim, H. J.; Chang, S. *Angew. Chem. Int. Ed.* **2012**, *51*, 11948–11959.

[4] (a) Sandmeyer, T. *Ber. Dtsch. Chem. Ges.* **1884**, *17*, 1633–1635. (b) Sandmeyer, T. *Ber. Dtsch. Chem. Ges.* **1890**, *23*, 1880–1881.

[5] For recent examples, see: (a) Martin, A.; Kalevaru, N. V.; Lücke, B.; Sans, J. *Green Chem.* **2002**, *4*, 481–485. (b) Liu, Y.; Zhong, M.; Yu, W.; Ma, Y. L. *Synth. Commun.* **2005**, *35*, 2951–2954. (c) Rombi, E.; Ferino, I.; Monaci, R.; Picciau, C.; Solinas, V.; Buzzoni, R. *Appl. Catal., A* **2004**, *266*, 73–79. (d) Lücke, B.; Narayana, K. V.; Martin, A.; Jähnisch, K. *Adv. Synth. Catal.* **2004**, *346*, 1407–1424.

[6] Takagi, K.; Okamoto, T.; Sakakibara, Y.; Oka, S. *Chem. Lett.* **1973**, *2*, 471–474.

[7] Holleman, A. F.; Wiberg, N. (eds.), *Lehrbuch der Anorganischen Chemie*, de Gruyter: Berlin, **2007**, pp. 1434.

[8] Flemming, C. A.; Trevors, J. T. *Water Air Soil Pollut.* **1989**, *44*, 143–158.

[9] (a) Dobbs, K. D.; Marshall, W. J.; Grushin, V. V. *J. Am. Chem. Soc.* **2007**, *129*, 30–31. (b) Erhardt, S.; Grushin, V. V.; Kilpatrick, A. H.; Macgregor, S. A.; Marshall, W. J.; Roe, D. C. *J. Am. Chem. Soc.* **2008**, *130*, 4828–4845.

[10] Tschaen, D. M.; Desmond, R.; King, A. O.; Fortin, M. C.; Pipik, B.; King, S.; Verhoeven, T. R. *Synth. Commun.* **1994**, *24*, 887–890.

[11] Sundermeier, M.; Zapf, A.; Beller, M.; Sans, J. *Tetrahedron Lett.* **2001**, *42*, 6707–6710.

[12] Ramnauth, J.; Bhardwaj, N.; Renton, P.; Rakhit, S.; Maddaford, S. P. *Synlett* **2003**, 2237–2239.

[13] Sundermeier, M.; Zapf, A.; Beller, M. *Angew. Chem. Int. Ed. Engl.* **2003**, *42*, 1661–1664.

[14] Sundermeier, M.; Mutyala, S.; Zapf, A.; Spannenberg, A.; Beller, M. *J. Organomet. Chem.* **2003**, *684*, 50–55.

[15] Chobanian, H. R.; Fors, B. P.; Lin, L. S. *Tetrahedron Lett.* **2006**, *47*, 3303–3305.

[16] (a) Jensen, R. S.; Gajare, A. S.; Toyota, K.; Yoshifuji, M.; Ozawa, F. *Tetrahedron Lett.* **2005**, *46*, 8645–8647. (b) Veauthier, J. M.; Carlson, C. N.; Collis, G. E.; Kiplinger, J. L.; John, K. D. *Synthesis* **2005**, 2683–2686. (c) Chidambaram, R. *Tetrahedron Lett.* **2004**, *45*, 1441–1444.

[17] Cai, L. Z.; Liu, X.; Tao, X. C.; Shen, D. *Synth. Commun.* **2004**, *34*, 1215–1221.

[18] Wang, D. P.; Kuang, L. P.; Li, Z. W.; Ding, K. *Synlett* **2008**, 69–72.

[19] For a recent catalytic variant of the Sandmeyer reaction, see: Beletskaya, I. P.; Sigeev, A. S.; Peregudov, A. S.; Petrovskii, P. V. *J. Organomet. Chem.* **2004**, *689*, 3810–3812.

[20] Wu, J. X.; Beck, B.; Ren, R. X. *Tetrahedron Lett.* **2002**, *43*, 387–389.

[21] Zanon, J.; Klapars, A.; Buchwald, S. L. *J. Am. Chem. Soc.* **2003**, *125*, 2890–2891.

[22] Cristau, H.-J.; Ouali, A.; Spindler, J.-F.; Taillefer, M. *Chem. Eur. J.* **2005**, *11*, 2483–2492.

[23] Buchwald, S. L.; Klapars, A.; Antilla, J. C.; Job, G. E.; Wolter, M.; Kwong, F. Y.; Nordmann, G.; Hennessy, E. J. inventors; Massachusetts Institute of Technology,

assignee. Copper-catalyzed formation of carbon-heteroatom and carbon-carbon bonds by arylation and vinylation of amines, amides, hydrazides, heterocycles, alcohols, enolates, and malonates, using aryl, heteroaryl, and vinyl halides and analogs. PCT Int. Appl. WO 2002085838 (A1), 2002 October 31; US 2001-286268P (P), 2001 April 24.

[24] (a) Taillefer, M.; Cristau, H.-J.; Cellier, P. P.; Spindler, J.-F. inventors; Rhodia Chimie, Fr.; Rhone Poulenc Chimie, assignee. Relatively low-temperature catalytic procedure for arylation or vinylation of nitrogen-containing nucleophilic compounds. PCT Int. Appl. WO 2003053885 (A1), 2003, July 3; FR 2001-16547 (A), 2001 December 20. (b) Taillefer, M.; Cristau, H.-J.; Cellier, P. P.; Spindler, J.-F. inventors; Rhodia Chimie, Fr.; Rhone Poulenc Chimie, assignee. Process for arylation, vinylation or alkynylation of nucleophilic compounds, in particular nitrogen-containing nucleophiles. PCT Int. Appl. WO 2003101966 (A1), 2003, December 11; FR 2002-6717 (A), 2002 May 31.

[25] Schareina, T.; Zapf, A.; Cotté, A.; Gotta, M.; Beller, M. *Adv. Synth. Catal.* **2011**, *353*, 777–780.

[26] Merz, V.; Weith, W. *Ber. Dtsch. Chem. Ges.* **1877**, *10*, 746–765.

[27] (a) Schareina, T.; Zapf, A.; Beller, M. *Chem. Commun.* **2004**, 1388–1389. (b) Schareina, T.; Zapf, A.; Beller, M. *J. Organomet. Chem.* **2004**, *689*, 4576–4583. (c) Schareina, T.; Zapf, A.; Mägerlein, W.; Müller, N.; Beller, M. *Tetrahedron Lett.* **2007**, *48*, 1087–1090. (d) Schareina, T.; Jackstell, R.; Schulz, T.; Zapf, A.; Cotté, A.; Gotta, M.; Beller, M. *Adv. Synth. Catal.* **2009**, *351*, 643–648.

[28] Schareina, T.; Zapf, A.; Beller, M. *Tetrahedron Lett.* **2005**, *46*, 2585–2588.

[29] (a) Schareina, T.; Zapf, A.; Mägerlein, W.; Müller, N.; Beller, M. *Chem. Eur. J.* **2007**, *13*, 6249–6254. (b) Schareina, T.; Zapf, A.; Mägerlein, W.; Müller, N.; Beller, M. *Synlett* **2007**, 555–558. (c) Schareina, T.; Zapf, A.; Cotté, A.; Müller, N.; Beller, M. *Synthesis* **2008**, 3351–3355.

[30] Zhu, Y. Z.; Cai, C. *J. Chem. Res. (S)* **2007**, 484–485.

[31] Ren, Y.; Liu, Z.; Zhao, S.; Tian, X.; Wang, J.; Yin, W.; He, S. *Catal. Commun.* **2009**, *10*, 768–771.

[32] Zhang, G.; Ren, X.; Chen, J.; Hu, M.; Cheng, J. *Org. Lett.* **2011**, *13*, 5004–5007.

[33] Ren, Y. L.; Zhao, S.; Tian, X. Z.; Liu, Z. F.; Wang, J. J.; Yin, W. P. *Lett. Org. Chem.* **2009**, *6*, 564–567.

[34] DeBlase, C.; Leadbeater, N. E. *Tetrahedron* **2010**, *66*, 1098–1101.

[35] Velmathi, S.; Leadbeater, N. E. *Tetrahedron Lett.* **2008**, *49*, 4693–4694.

[36] (a) Ren, Y.; Wang, W.; Zhao, S.; Tian, X.; Wang, J.; Yin, W.; Cheng, L. *Tetrahedron Lett.* **2009**, *50*, 4595–4597. (b) Ren, Y.; Wang, W.; Zhao, S.; Tian, X.; Wang, J.; Yin, W.; Cheng, L. *Tetrahedron Lett.* **2010**, *51*, 2669–2670.

[37] Mehmood, A.; Devine, W. G.; Leadbeater, N. E. *Top. Catal.* **2010**, *53*, 1073–1080.

[38] (a) Jin, J.; Wen, Q.; Lu, P.; Wang, Y. *Chem. Commun.* **2012**, *48*, 9933–9935. (b) Chen, X.; Hao, X. S.; Goodhue, C. E.; Yu, J. Q. *J. Am. Chem. Soc.* **2006**, *128*, 6790–6791. (c) Yuen, O. Y.; Choy, P. Y.; Chow, W. K.; Wong, W. T.; Kwong, F. Y. *J. Org. Chem.* **2013**, *78*, 3374–3378.

[39] Do, H.-Q.; Daugulis, O. *Org. Lett.* **2010**, *12*, 2517–2519.

[40] Zhang, G. Y.; Zhang, L. L.; Hu, M. L.; Cheng, J. A. *Adv. Synth. Catal.* **2011**, *353*, 291–294.

[41] Liskey, C. W.; Liao, X. B.; Hartwig, J. F. *J. Am. Chem. Soc.* **2010**, *132*, 11389–11391.

[42] (a) Lamani, M.; Prabhu, K. R. *Angew. Chem. Int. Ed. Engl.* **2010**, *49*, 6622–6625. (b) Rokade, B. V.; Malekar, S. K.; Prabhu, K. R. *Chem. Commun.* **2012**, *48*, 5506–5508.

[43] (a) Kim, J.; Choi, J.; Shin, K.; Chang, S. *J. Am. Chem. Soc.* **2012**, *134*, 2528–2531. (b) Wang, Z.; Chang, S. *Org. Lett.* **2013**, *15*, 1990–1993.

[44] Song, R.-J.; Wu, J.-C.; Liu, Y.; Deng, G.-B.; Wu, C.-Y.; Wei, W.-T.; Li, J.-H. *Synlett* **2012**, 2491–2496.

[45] Xu, H.; Liu, P.-T.; Li, Y.-H.; Han, F.-S. *Org. Lett.* **2013**, *15*, 3354–3357.

10

COPPER-MEDIATED ARYL–ARYL BOND FORMATION LEADING TO BIARYLS: A CENTURY AFTER THE ULLMANN BREAKTHROUGH

Yoshihiko Yamamoto

10.1 INTRODUCTION

The synthesis of functionalized biaryls has been an important research objective in organic synthesis, biaryl motifs being found in a wide range of functional molecules such as biologically active natural products,[1] optoelectronic molecules,[2] and asymmetric catalysts.[3] Naturally, transition-metal-mediated aryl–aryl bond formation reactions have been extensively investigated; and, in particular, palladium-catalyzed cross-coupling reactions have been devised and have been continuously sharpened as most efficient tools to construct functionalized biaryls.[4] The successful progress in this area of research culminated in the Nobel Prize in Chemistry 2010, which was awarded to Richard F. Heck, Ei-ichi Negishi, and Akira Suzuki for the development of palladium-catalyzed cross-coupling reactions.

In striking contrast to the tremendous success of the palladium-catalyzed cross-coupling reactions, the development of copper analogues lagged behind despite the fact that more than a century has passed since the first report of transition-metal-mediated aryl–aryl bond formation, namely, the Ullmann coupling of haloarenes with copper bronze.[5] This underdevelopment is partly associated to the difficulty in predicting and understanding the reaction

Copper-Mediated Cross-Coupling Reactions, First Edition. Edited by Gwilherm Evano and Nicolas Blanchard.

mechanisms of copper-mediated coupling, which strongly hampered the design of new catalytic systems with increased efficiencies. For instance, copper occurs in a variety of oxidation states (0–IV) and tends to be present at more than one oxidation state under certain reaction conditions and single-electron transfer processes as well as disproportionation complicates mechanistic investigations. Nevertheless, renewed interest in first-row transition metals, due to their cost-effectiveness and environmentally benign characteristics, has spurred a resurgence of Ullmann chemistry, and this has led to the recent significant progress in copper-catalyzed coupling processes.[6] Among them, C–O, C–N and C–S bond-forming reactions (described in Chapters 1 and 2) have been extensively studied,[7] while less attention has been devoted to copper-mediated aryl–aryl bond formation. Thus, a survey of the development of copper-catalyzed biaryl syntheses is expected to contribute to increased researcher interest and further advances in this area.

This chapter outlines developments in the copper-mediated aryl–aryl coupling methods. These can be classified into several major categories depending on the types of reactions and precursors involved: (i) coupling of aryl halides and diazonium salts; (ii) coupling of aryltin, arylboron, or arylsilicon reagents; (iii) coupling via arylation involving C–H or C–C bond fission; and (iv) oxidative coupling of 2-naphthols. The third category can be further divided into direct C–H arylations using prefunctionalized or nonfunctionalized arylating agents and decarboxylative coupling of benzoic acid derivatives. In addition to cross-coupling reactions, each category includes homocoupling reactions, useful tools for the construction of symmetrical biaryl frameworks from a single precursor. The conventional Ullmann reaction is still practical for large-scale production of less-functionalized biaryls as it is easily scalable due to the simple procedure and the lack of requirement for special ligands and additional promoters, although excess copper is required. The first part of this chapter is devoted to a brief survey of Ullmann coupling and related aryl–aryl bond formation reactions. The later sections survey the modern copper-mediated coupling reactions beyond the classical Ullmann biaryl synthesis.

10.2 BIARYL SYNTHESIS BY COUPLING OF ARYL HALIDES AND DIAZONIUM SALTS

10.2.1 Ullmann Couplings of Aryl Halides

The history of copper-promoted aryl–aryl bond formations can be traced back to the seminal report by Fritz Ullmann and Jean Bielecki in 1901 (see the Preface for a presentation of these early discoveries from a historical perspective).[5] They discovered that mixing *o*-iodonitrobenzene with 1.9 equiv of copper powder at 210–220°C under neat conditions formed the homocoupling product (i.e., 2,2′-dinitrobiphenyl) in 76% yield with concomitant production of CuBr (Scheme 10.1). Since then, the Ullmann coupling has been studied

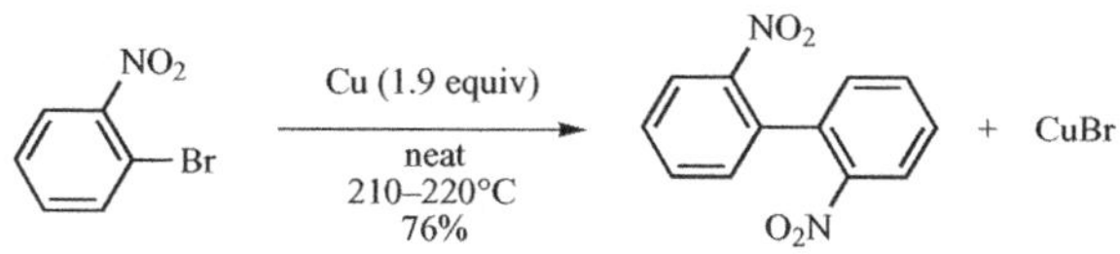

Scheme 10.1 Copper(0)-mediated homocoupling of *o*-bromonitrobenzene: the historical discovery of Fritz Ullmann (1901).

extensively, and its scope and applications have been thoroughly reviewed.[6,8] Thus, only a brief overview is given here for the classical Ullmann coupling to demonstrate the essence of this process; and then, newer Ullmann-type couplings will be overviewed.

The order of reactivity of the halogens, that is, I > Br > Cl, is common to most transition-metal-mediated cross-coupling reactions and the most challenging Ullmann coupling via the C–F bond cleavage of fluorobenzenes has not been reported to date. The activating effect of electron-withdrawing substituents such as nitro and carbomethoxy groups at the *ortho*-position was observed and elegantly used in some cases to promote the reaction under milder conditions than those initially reported by Ullmann; this effect is dramatically affected by the position of the activating group and decreases in the following order: *ortho* ≫ *para* > *meta*. Chloro, methyl, and methoxy substituents also favorably affect the reaction regardless of their positions, although no explanation has been provided. In contrast, functional groups with labile proton(s) inhibit the coupling; typical inactive substrates include phenol, aniline, and benzoic acid derivatives, which may preferentially undergo Ullmann–Goldberg condensation[6,7] or decarboxylation.[9] Polar aprotic solvents such as dimethylformamide (DMF) and *N*-methylpyrrolidinone (NMP) can be used for milder protocols, although the original synthesis was conducted without solvent, and nitroarenes as well as aromatic solvents have also been used. However, the former undergoes reductive arylation at high temperatures, and it has been reported that the Ullmann coupling of iodobenzene in nitrobenzene resulted in a triphenylamine by-product. Reductive dehalogenation also occurs as a side reaction in the presence of proton sources. For the copper source, commercially available, mechanically pulverized copper has been used with or without activation by successive washing with acetone solutions of iodine and hydrochloric acid. It has been reported that milder reaction conditions improve the yields of biaryl products. For example, the homocoupling of *o*-bromonitrobenzene over 60 hours at 60°C using a 10-fold excess of activated copper bronze improved the yield of 2,2′-dinitrobiphenyl from 76% under the original Ullmann conditions to 99.6%.[10a] The copper loading could be further reduced to 4 equiv, and the reaction rate was enhanced by presonication of the copper bronze with ultrasound in air,[10b] and copper powder that was freshly prepared by the reduction of aqueous copper sulfate with zinc powder was reported to be superior to commercial copper bronze.[11] In addition, it was also found that γ-alumina-supported copper is superior to copper powder for Ullmann coupling of *p*-chlorotoluene.[12]

In place of metallic copper(0), copper(I) ions have been employed for milder Ullmann-type coupling reactions. *o*-Iodonitrobenzene was, for example, treated with freshly prepared copper(I) triflate and 20% aqueous ammonia in acetone at ambient temperature for less than 5 minutes to produce the corresponding 2,2′-biphenyl in 92% yield.[13] A tetrakis(acetonitrile)copper(I) perchlorate catalyst was reported to be similarly effective.[13] It was also discovered that 2.5–3.0 equiv of copper(I) thiophene-2-carboxylate (CuTC) promoted inter- and intramolecular Ullmann-type couplings of iodoarenes in moderate-to-high yields, even at ambient temperature, which represents one of the most efficient coupling reactions,[14a] a typical example being 2,2′-dinitrobiphenyl, which was obtained in 92% yield from *o*-iodonitrobenzene with this procedure (Scheme 10.2). This method was also applied to the synthesis of fluorinated oligo(*para*-phenylenes) and to the enantioselective synthesis of *trans*-4,5,9,10-tetrahydroxy-9,10-dihydrophenanthrene.[14b,c]

In all examples of Ullmann coupling described up to this point, a large excess of copper was generally imperative. In contrast, when copper nanoparticles are used as the copper source, Ullman coupling of iodobenzene was executed at 150–200°C with a copper:iodobenzene ratio in the 1.6:1–1:1 range to obtain the corresponding biphenyl in good yields.[15] Moreover, a catalytic version of the Ullmann coupling was achieved using supported copper(II) diamine compound **1** (Scheme 10.3),[16] the immobilized copper(II) catalyst being easily removed and reused. Notably, various bromoarenes bearing electron-withdrawing and electron-donating substituents underwent homocoupling to give biaryls in good yields.

Besides the development of activated systems for the Ullmann biaryl synthesis, the atroposelective aryl–aryl coupling has also attracted much interest due to the wide occurrence of axially chiral biaryls in natural products and asymmetric catalysts (see Chapter 18 for an overview of copper-catalyzed aryl–aryl coupling reactions in total synthesis of natural products).[1,3,17] Accordingly, enantiomerically pure chiral biaryls, which form the axially chiral backbone of asymmetric catalysts, were synthesized by the Ullmann coupling of 2,6-disubstituted haloarenes and subsequent optical resolution of the racemic products.[18] Since optical resolution gives a maximum yield of 50% of the desired enantioenriched product and requires tedious procedures, notably for the separation of the diastereoisomers formed in the course of the reactions, the diastereoselective coupling of substrates substituted with appropriate chiral auxiliaries is more desirable and has been quite extensively studied. Toward this goal, Ullmann coupling of several haloarenes in the presence of a chiral inducer was examined; however, the observed diastereoselectivity was not sufficient and further optical resolution was required.[19] To synthesize more efficiently C_2-chiral biaryls, Meyers and coworkers introduced a chiral oxazoline as an auxiliary.[20] As shown in Scheme 10.4, valinol-derived *o*-bromooxazolinylbenzene **2** was treated with copper powder in refluxing

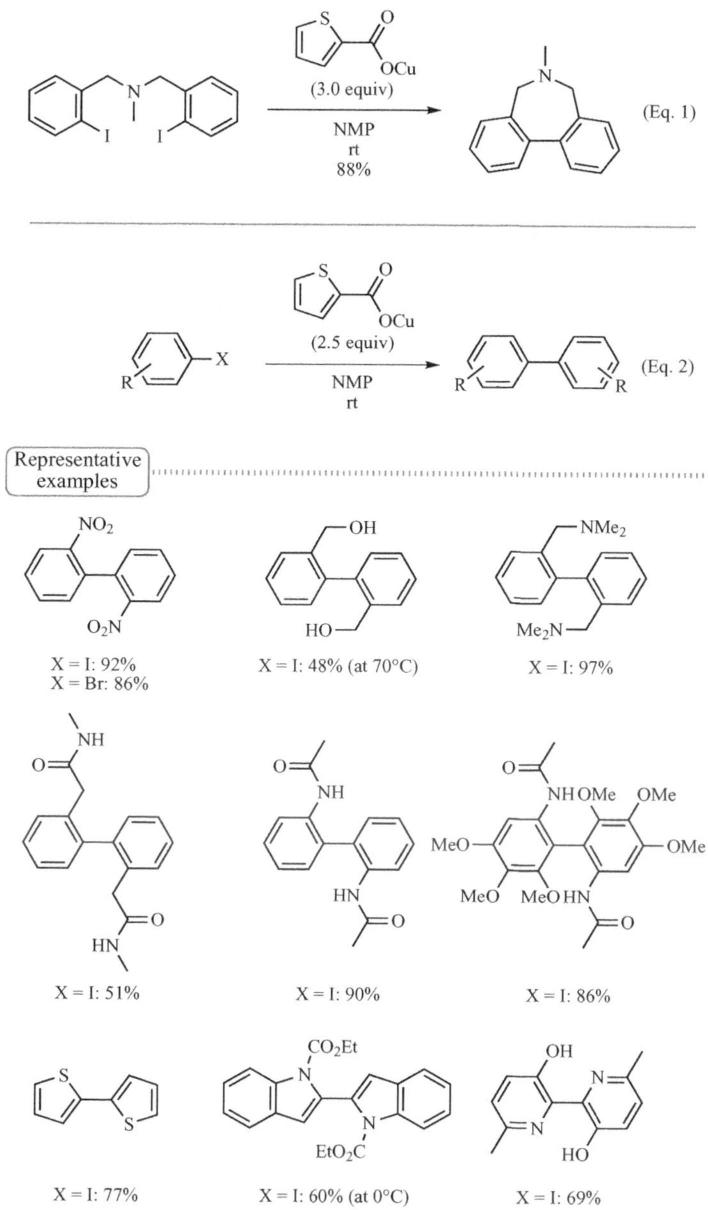

Scheme 10.2 Homocoupling of iodobenzenes mediated by copper(I) thiophene-2-carboxylate.

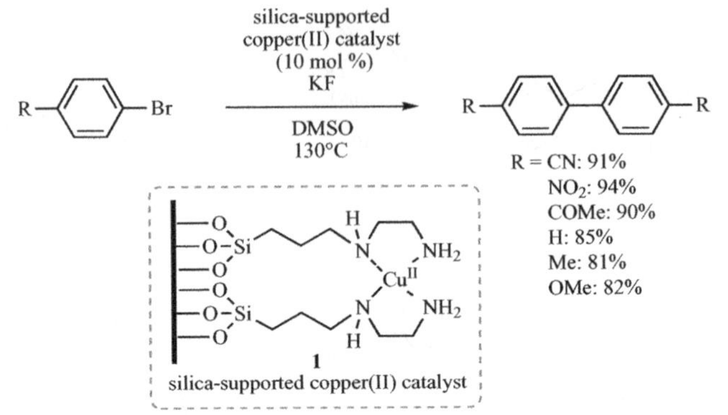

Scheme 10.3 Homocoupling of bromoarenes catalyzed by a silica-supported copper(II) diamine complex.

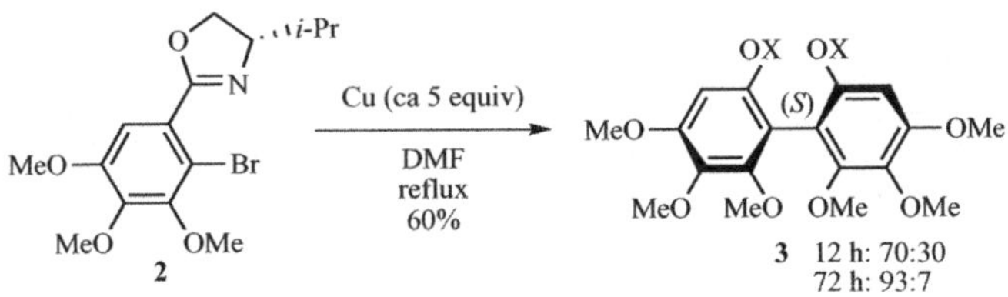

Scheme 10.4 Diastereoselective homocoupling of *o*-bromooxazolinylbenzene.

DMF to furnish the corresponding biaryl bis(oxazoline) **3**, and the diastereomeric ratio was 70:30 after 12 hours and was remarkably improved to 93:7 after 72 hours, which implies that the Ullmann coupling of oxazoline **2** first affords a copper complex of bis(oxazoline) **3** with modest diastereoselectivity followed by atropisomerization to the thermodynamically favorable (*S*)-enantiomer. The enantiopure product **3** was ultimately isolated in 60% yield and the oxazoline approach was next extended to the synthesis of binaphthyls and various natural products.[20]

An alternative promising approach is the intramolecular Ullmann coupling that involves a temporary chiral tether developed by Miyano and coworkers, who obtained (*S*)-2,2′-di(hydroxymethyl)-1,1′-binaphthyl (**6**) albeit in a moderate optical purity from the intramolecular coupling of binaphthol tethered bis(1-bromo-2-naphthoate) **4** followed by the reduction of the resulting coupling product **5** with lithium aluminum hydride (Scheme 10.5).[21] The same group also reported intramolecular cross couplings using a similar strategy.[21a] In these examples, the axial chirality of the tether was transferred to that of the newly formed binaphthyl moiety.

Tethers bearing central chirality were also extensively studied.[21c,22] For example, (*o*-iodophenyl)diphenylphosphine oxides linked by chiral diol tethers **7** underwent intramolecular coupling to furnish biaryl bis(phosphine oxides)

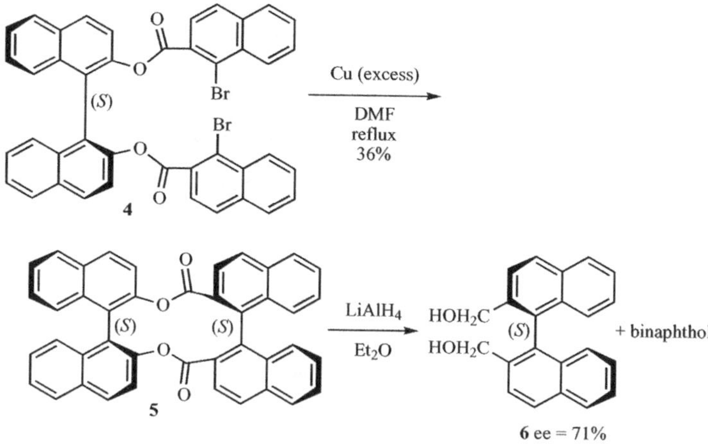

Scheme 10.5 Diastereoselective intramolecular homocoupling of BINOL-derived bis(1-bromo-2-naphthoate).

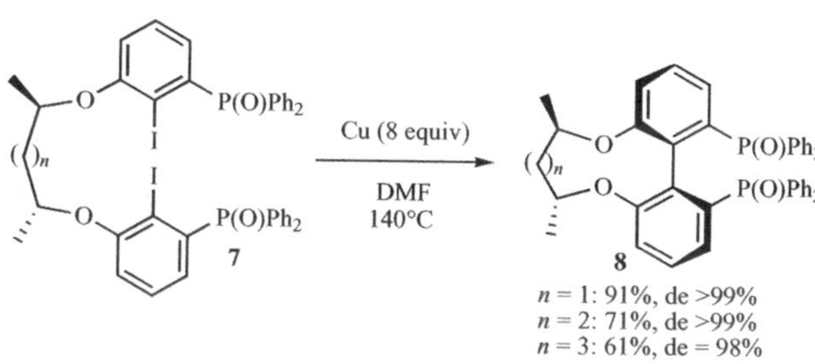

Scheme 10.6 Synthesis of chiral biaryl bis(phosphine oxides) via intramolecular diastereoselective homocoupling of (iodophenyl)diphenylphosphine oxides.

8 with an excellent diastereoselectivity of over 98% in 61–91% yields, and the smaller the tether, the higher the yield (Scheme 10.6).[22a,b] Intramolecular Ullmann coupling of 2-iodo-3,4,5-trimethoxybenzoate connected by sugar moieties was examined for the biomimetic synthesis of ellagitannins and the coupling products were generally obtained as single atropisomers, albeit in low-to-moderate yields.[23]

In terms of synthetic utility, one of the serious disadvantages of the Ullmann coupling is its little efficiency for the cross coupling of two different haloarenes to obtain unsymmetrical biaryls. Indeed, the cross Ullmann coupling is only successful if one of the coupling components is more reactive than the other, and in addition, the most reactive haloarene component must have an activating group at the position *ortho* to the halogen substituent, and an excess of the less-active component is often required. Moreover, the reaction should be carried out at temperatures below that at which the less-reactive component

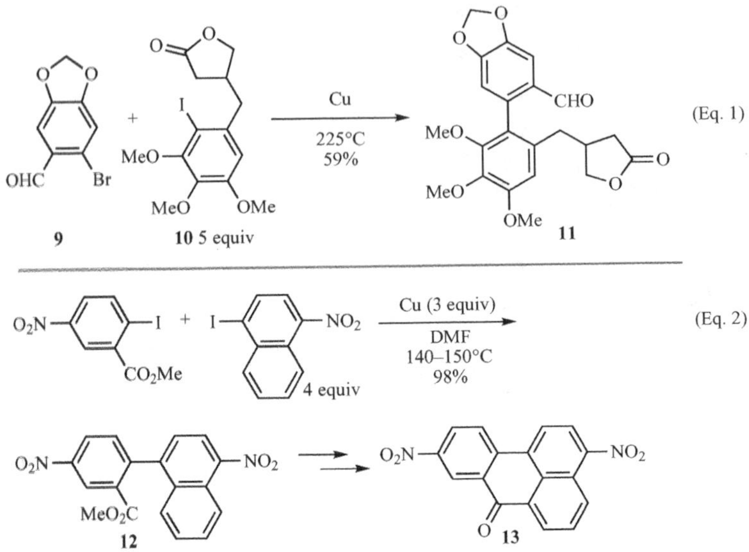

Scheme 10.7 Examples of cross Ullmann coupling of halobenzenes.

reacts with copper. One of the early examples involves the Ullmann cross coupling of 2,4-dibromonitrobenzene and iodobenzene, which gave the desired product in a moderate yield of 52% together with small amounts of 2,4-dibromonitrobenzene.[24] As another example, intermediate **11** was obtained in the total synthesis of steganone in a moderate yield from the cross coupling of 6-bromopiperonal **9** and highly congested iodoarene **10** (Scheme 10.7, Eq. 1).[25] A high-yielding procedure was later established using iodonaphthalenes as the less-reactive components.[26] This synthesis involves the dropwise addition of a DMF solution of 2-iodo-5-nitrobenzoate over 4 hours to a mixture of 1-iodo-4-nitronaphthalene and copper bronze at 140–150°C to afford the corresponding cross-coupling product **12** in a remarkable 98% yield (Scheme 10.7, Eq. 2). This product was further converted to benzanthrone derivative **13**.

10.2.2 Coupling of Organocopper Reagents Generated from Aryl Halides

An alternative approach to achieve this kind of cross coupling is the reaction of preformed arylcopper reagents with iodoarenes. Even if these reactions are outside the scope of this book, these procedures are clearly among the most efficient ones for the syntheses of biaryls to date and will undoubtedly be at the origin of developments in copper-mediated biaryl syntheses without preformed organocopper reagents for the years to come. Due to the utmost importance of these underdeveloped reactions, they should appear in this chapter and they will be overviewed in the next pages.

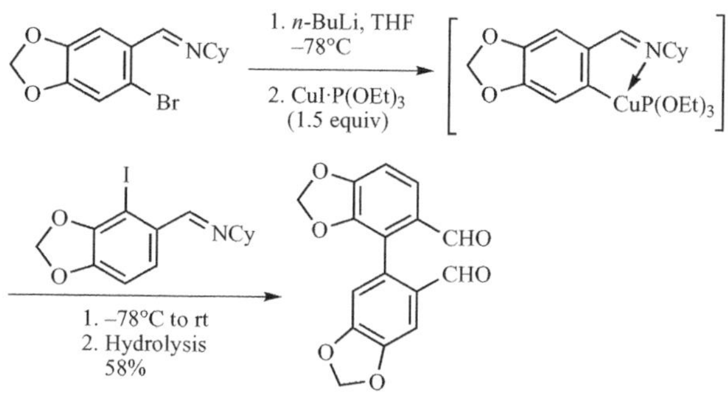

Scheme 10.8 Cross-coupling reaction using *in situ* generated organocopper(I) reagent.

Phenyl-, 2-thienyl-, 2-pyridyl-, and polyhaloarylcopper reagents successfully react with iodoarenes to furnish the corresponding cross-coupling products.[27] However, homocoupling side products were also formed from phenylcopper via a copper–halogen exchange. In response to this limitation, Ziegler and coworkers developed a selective cross-coupling protocol under mild conditions,[28] which involves the *in situ* generation of an arylcopper reagent stabilized by a phosphite ligand and an internal coordinating group such as imine or oxazoline at the *ortho*-position. The arylcopper reagent was then allowed to react at ambient temperature with an iodoarene also bearing an *ortho* imine substituent to produce the corresponding unsymmetrical biaryl in a good yield. A typical example is given in Scheme 10.8. This strategy was successfully applied to the synthesis of steganacin[28] and other natural products,[29] and in addition, the synthesis of (*S*)-(−)-*N*-acetylcolchinol was accomplished by a similar method employing an acetal as the internal coordinating group.[29d]

The oxidation of organocuprates has been known to induce C–C bond formation between the ligands on copper.[30] However, this phenomenon is in general rather useless for organic synthesis when two different ligands are involved as all combinations of the two ligands lead to three coupling products in statistical ratios except in peculiar cases. Nevertheless, Lipshutz and coworkers discovered that the aerobic oxidation of higher-order cyanocuprates (e.g., **14** in Scheme 10.9), which were prepared from sequential addition of aryllithium reagents to copper(I) cyanide in 2-methyltetrahydrofuran at −125°C, predominantly yielded cross-coupling products.[31] The success of this method is based on the formation of "kinetic" higher-order cuprates at −125°C; at higher temperatures, ligand scrambling occurs, which results in a loss of selectivity. This method is applicable to heteroaromatic systems as well as intramolecular coupling reactions, and notably, intramolecular cross-coupling reactions in tetrahydrofuran (THF) can be carried out at a higher temperature of −78°C because ligand scrambling is not as significant as it is for intermolecular couplings. Enantioselective syntheses of biaryls were also achieved by implementing a chiral tether method (Scheme 10.10).[32]

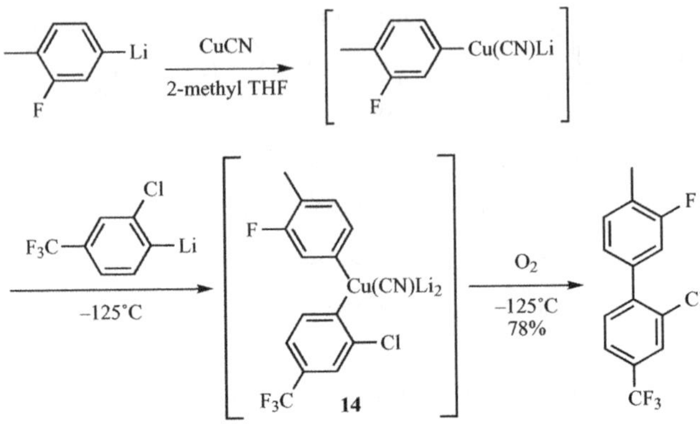

Scheme 10.9 Cross-coupling reaction of aryllithium reagents via higher-order cyanocuprate.

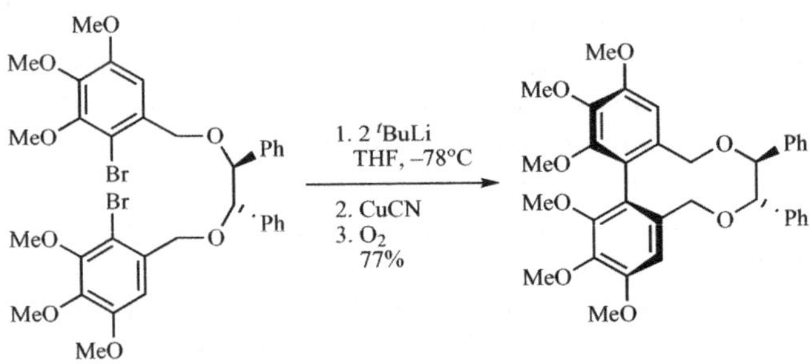

Scheme 10.10 Diastereoselective intramolecular homocoupling of arylbromide via higher-order cyanocuprate.

Recently, several significant modifications have been made to expand the scope of Lipshutz's method. Iyoda and coworkers contended with diminished yields of the homocoupling product obtained from the oxidation of (p-$ClC_6H_4)_2Cu(CN)Li_2$ with conventional oxidants such as molecular oxygen and 1,3-dinitrobenzene.[33] As a result, they screened various electron acceptors and determined that tetramethyl-p-benzoquinone was optimal. In addition, they achieved cyclocoupling via dimetallacyclic cuprates to obtain 10-membered cyclophanes. Spring and coworkers investigated the synthesis of more challenging compounds such as iodinated biaryls.[34] Toward this aim, they examined the oxidation of cuprates produced from Grignard reagents since they are obtained under mild conditions via iodide–magnesium exchange.[34a] Their optimization revealed that the reaction using $CuBr{\cdot}SMe_2$ and 3,5-dinitrobenzoic amide **15** successfully furnished the homocoupling products in moderate-to-high yields (Scheme 10.11). Notably, dihaloarenes were able to undergo homocoupling and retain one of the halogen atoms. Furthermore,

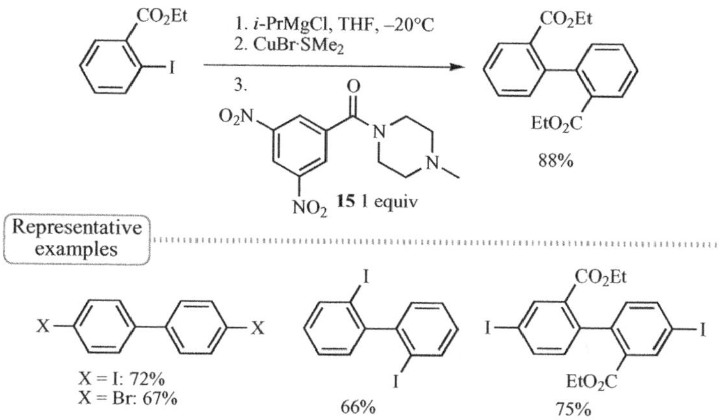

Scheme 10.11 Homocoupling of aryl Grignard reagents via cuprates.

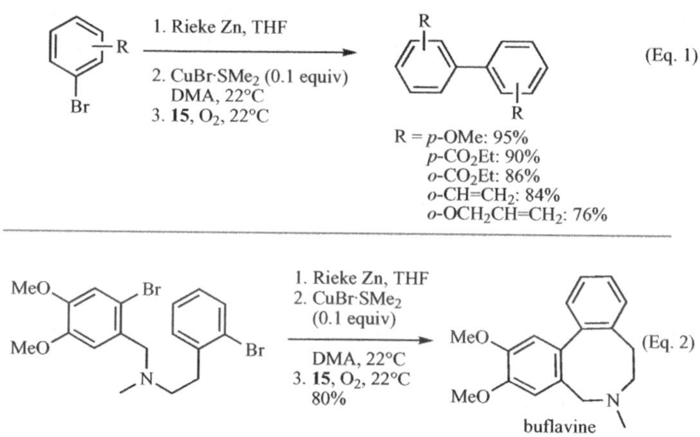

Scheme 10.12 Homocoupling and intramolecular cross coupling of arylzinc reagents via cuprates.

organozinc reagents, which are more tolerant to a wide range of functional groups than organolithium or Grignard reagents, proved to be applicable in the cuprate oxidation strategy. Spring and coworkers demonstrated that cuprates derived from the reaction of arylzinc reagents and $CuBr{\cdot}SMe_2$ were oxidized by amide **15** to give symmetrical biaryls in good yields.[35] In addition, performing the reaction under an atmosphere of oxygen enables copper loading to be reduced to 0.1 equiv, which represents the first example of the catalytic use of copper for biaryl synthesis via cuprate oxidation (Scheme 10.12, Eq. 1 and Eq. 2). Magnesium and zinc cuprate oxidation methods were also successfully applied to the synthesis of ellagitannin natural products.[36]

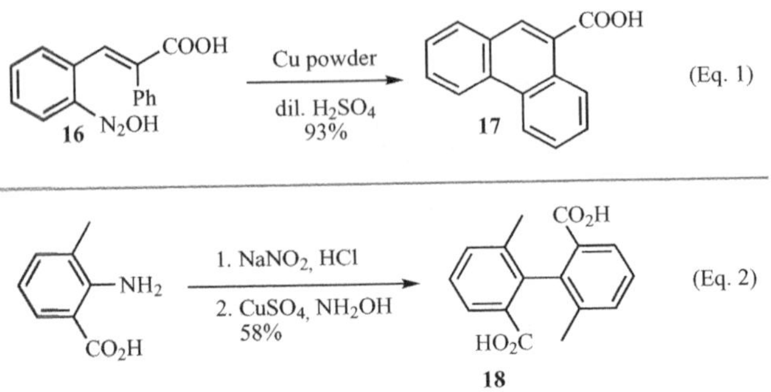

Scheme 10.13 Copper-mediated coupling of aryl diazonium compounds.

10.2.3 Coupling of Aryl Diazonium Compounds

The last part of this section briefly outlines the copper-mediated coupling of aryl diazonium compounds to obtain biaryls. In 1896, Robert Pschorr reported that the treatment of diazonium salt **16** with copper powder affords phenantherene-9-carboxylic acid **17** in 93% yield (Scheme 10.13, Eq. 1).[37] The intermolecular coupling of aryl diazonium salts with aromatic compounds is also known as the Gomberg–Bachmann reaction[38] and the intermolecular copper-mediated homocoupling of aryl diazonium salts has been applied to the synthesis of highly congested biaryls (e.g., **18** in Scheme 10.13, Eq. 2), which are difficult to access via other methods.[39]

After extensive research, it was revealed that the coupling of aryl diazonium salts proceeds using a variety of mediators such as iodide anions, iron(II) and titanium(III) ions, tetrathiafulvalene, and cobaltocene. This shows that copper is not essential for these reactions, and there are evidences that support the involvement of aryl radicals in the aryl–aryl bond formations.[38,40] However, it has also been suggested that arylcopper species, which are generated via the reaction of aryl radicals and copper ions, are involved in the aryl–aryl coupling process.[41] Cohen and coworkers observed decreased yields of biaryls and azoarenes when the copper(I)-mediated decomposition of *p*-nitrobenzenediazonium tetrafluoroborate was performed in the presence of a radical trapping agent such as iodomethane or THF.[41a] In addition, the yields of these bimolecular products rose with increases in the concentration of the copper(I) salt, and the addition of copper(II) ions increased the biaryl:azoarene ratio. From these data, biaryl formation was determined to occur via both radical and arylcopper species. The decomposition of *p*-methoxybenzenediazonium in the presence of tetrakis(acetonitrile)copper(I) in aqueous acetonitrile phosphate buffers under a variety of conditions yielded symmetrical 4,4′-dimethoxybiphenyl as the major aryl–aryl coupling product. Computational analysis of the obtained product distributions gave kinetic data corresponding to elementary processes, which indicates that the arylcopper species is involved in the aryl–aryl bond formation.[41b]

As seen in the above-mentioned examples, copper-mediated coupling reactions of aryl halides and diazonium salts to the corresponding biaryls are efficient tools for the rapid elaboration of these important motifs. The next section will focus on copper-mediated and copper-catalyzed coupling reactions of aryltin, boron, and silane derivatives that have been extensively studied in the past decade.

10.3 BIARYL SYNTHESIS BY COUPLING OF ARYLTIN, BORON, AND SILANES

10.3.1 Coupling of Organostannanes

The palladium-catalyzed cross coupling of organostannanes and organic halides, which is known as Migita–Kosugi–Stille cross coupling, has found numerous applications in organic synthesis.[4c,42] The closely related homocoupling of arylstannanes was achieved using copper(II) nitrate as a stoichiometric promoter and *p*-tolyl-, *o*-tolyl-, or *p*-anisyl-stannanes were individually treated with $Cu(NO_3)_2{\cdot}3H_2O$ (1 equiv) in THF at 23°C to obtain symmetrical biaryls in 45–67% yields.[43] In contrast, application of this method to a more hindered 2,6-dimethoxyphenyl analogue resulted in a low yield of 14%. It was also reported that benzofuran-2-ylstannane and its sulfone analogue underwent homocoupling in moderate yields,[43b] and this method was also applied to the synthesis of cyclic oligophenylenes.[44] Later, the yield and scope of copper(II)-mediated homocoupling reactions were improved using diaryldimethyltin reagents **19** as outlined in Scheme 10.14, Equation 1.[45] In the presence of 2.2 equiv of $Cu(NO_3)_2{\cdot}3H_2O$, a variety of biaryls were obtained within 30 minutes, the only exception being the reaction of an *o*-methoxyphenyl derivative, which required 4 hours. It is assumed that the coupling proceeds via monoarylcopper(II) species, that is, Ar–Cu–ONO_2, rather than diarylcopper(II), which was confirmed by a crossover experiment between two different diarylstannanes that resulted in both homo- and cross-coupling products (Scheme 10.14, Eq. 2).

Kang and coworkers established a catalytic protocol for homocoupling that employs molecular iodine as an oxidant (Scheme 10.15).[46] In the presence of 10 mol % of copper(II) chloride and half an equivalent of iodine, (hetero) arylstannanes were heated in DMF at 100°C for 4 hours to obtain symmetrical biaryl products in good yields.

Related intramolecular homocoupling of **20** is a powerful method for the synthesis of seven-membered ring compounds **21**, which are otherwise difficult to obtain (Scheme 10.16),[47] and this strategy was applied to the synthesis of tubulin-binding agents.[47] Since 2 equiv of copper(0) are produced for each carbon–carbon bond formation,[43] the researchers proposed a route comprising reversible transmetallation to form dicopper(I) intermediates **22**, which undergo disproportionation to yield copper(II) metallacycle species **23**.

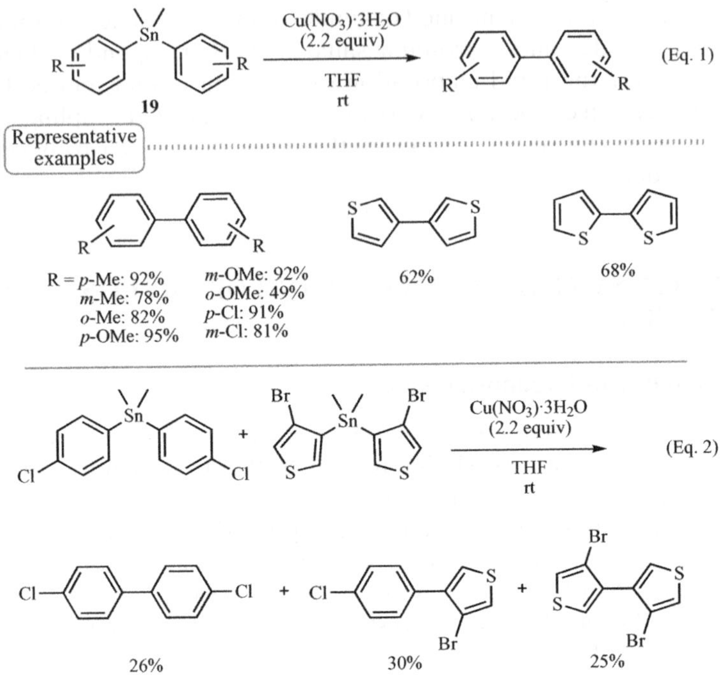

Scheme 10.14 Copper(I)-mediated biaryl synthesis using diaryltin reagents.

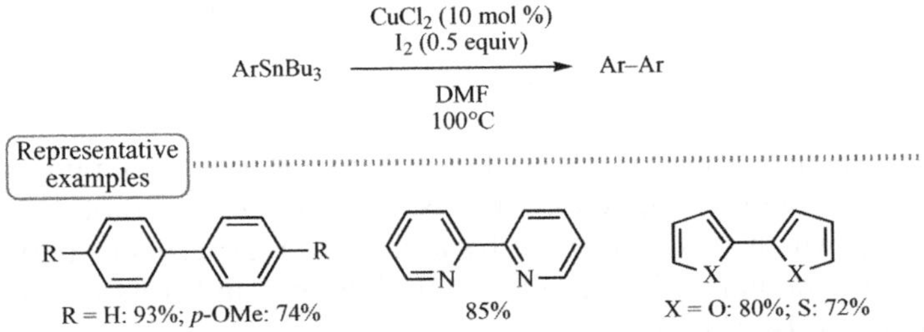

Scheme 10.15 Copper(II)-catalyzed homocoupling of arylstannanes.

Subsequent reductive elimination furnishes the desired cyclic biaryls **21**. Alternatively, the biaryls could be directly produced via binuclear reductive elimination from the dicopper(I) intermediates.[48]

Copper(I)-mediated Migita–Kosugi–Stille-type cross couplings of alkenyl- and alkyl-stannanes have been extensively developed,[49,50] and in particular, the copper(I) thiophenecarboxylate-mediated cross coupling of alkenylstannanes, which was developed by the Liebeskind group, has practical advantages including mild and simple procedure, wide substrate scope and high yields. Thus, this method was soon employed for numerous applications including the

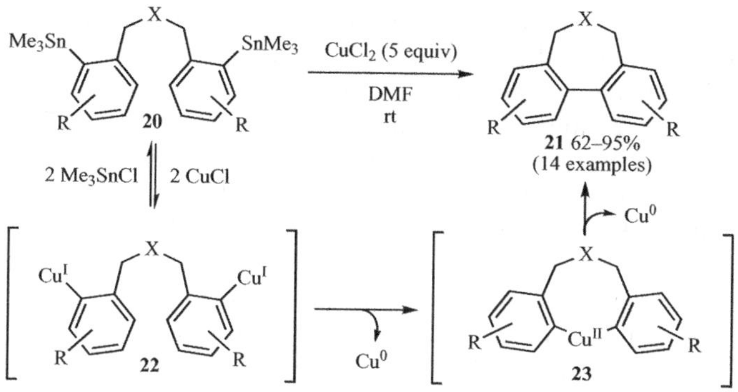

Scheme 10.16 Copper(II)-mediated intramolecular homocoupling of arylstannanes.

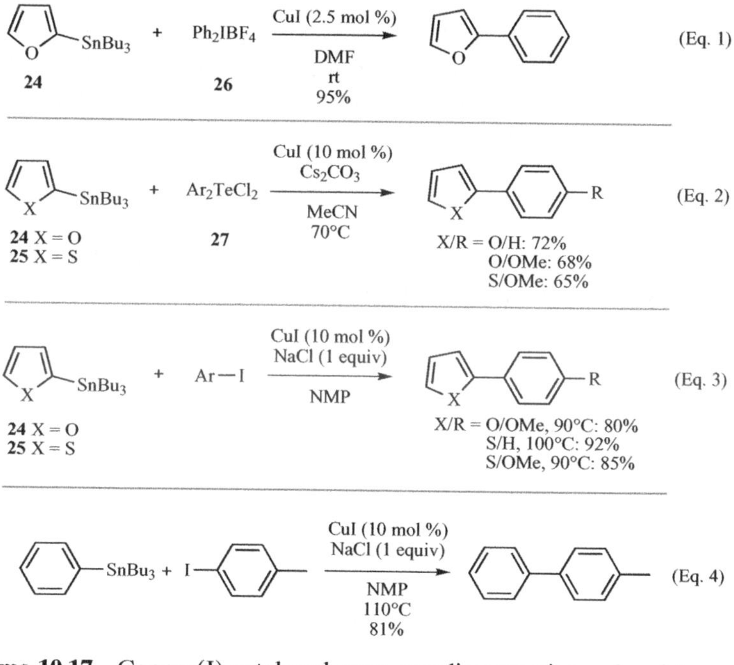

Scheme 10.17 Copper(I)-catalyzed cross-coupling reactions of arylstannanes.

total synthesis of complex natural products (see Chapter 18).[50] With respect to biaryl synthesis using this strategy, several catalytic methods have been developed,[51] and the first copper(I)-catalyzed Migita–Kosugi–Stille-type aryl–aryl coupling was achieved by Kang and coworkers using hypervalent reagents as electrophiles (Scheme 10.17, Eq. 1).[51a] The reaction involves the treatment of 2-stannylfuran **24** and diphenyliodonium tetrafluoroborate (**26**) with copper(I) iodide (2.5 mol %) in DMF at an ambient temperature for 10 minutes to obtain 2-phenylfuran in 95% yield. In this report, the scope was not expanded further, although the cross coupling of arylboronic acids was also accomplished (see next section). Similarly, the coupling of 2-furyl- or 2-thienylstannanes **24** and **25** with diaryltellurium dichlorides **27** in the presence of 2 equiv of cesium carbonate was shown to proceed at 70°C to give cross-coupling products, albeit in lower yields (Scheme 10.17, Eq. 2).[51c] The Kang group also achieved copper(I)-catalyzed Migita–Kosugi–Stille-type couplings using iodoarenes as the coupling partners;[51b] these reactions required elevated temperatures and sodium chloride as an additive to suppress the reversible transmetallation by converting tributyltin iodide to tributyltin chloride and the slow addition of the arylstannanes with a syringe pump to avoid homocoupling (Scheme 10.17, Eq. 3 and Eq. 4).

This cross coupling was later extended to the use of recyclable catalysts using copper oxide nanoparticles, tri(*o*-tolyl)phosphine, potassium fluoride, and tetrabutylammonium bromide (TBAB) (see Chapter 20 for an overview of reusable copper catalysts).[52] In this report, phenylstannane underwent cross coupling with aryl iodides and bromides that contained both electron-withdrawing and electron-donating substituents to furnish the corresponding biaryls in high yields (Scheme 10.18). Remarkably, this method also successfully converted *p*-acetyl-, *p*-nitro-, and 3,5-dimethyl-phenyl chloride to their corresponding biaryls in high yields; however, the reaction resulted in a low yield (10%) when electron-rich *p*-methoxyphenyl chloride was used. A significant advantage of the $Cu_2O/P(o\text{-tolyl})_3$/TBAB system is its recyclability; the coupling of *p*-iodoanisole and tributylphenylstannane was repeated for up to five cycles without loss of catalytic activity (average yield of 94%).

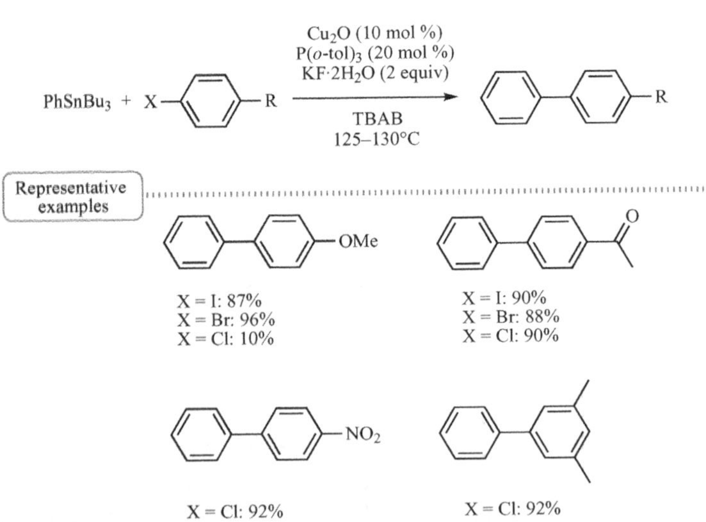

Scheme 10.18 Cross coupling of arylstannanes with aryl halides catalyzed by copper nanoparticles.

Besides these efficient copper-catalyzed Migita–Kosugi–Stille-type couplings of arylstannanes, numerous studies have been devoted to the homo- and cross-coupling reactions of arylboron derivatives. The most representative examples of this strategy are overviewed in the next section.

10.3.2 Coupling of Organoboranes

The palladium-catalyzed cross coupling of organoboronic acids and organic halides, which is known as Suzuki–Miyaura cross coupling, is one of the most powerful tools for the formation of carbon–carbon bonds, organoboron reagents being readily accessible and less harmful than the corresponding organostannanes.[53] A copper-catalyzed version of this cross coupling was realized for the first time in 1996 using hypervalent iodonium reagents. Arylboronic acids were allowed to react with diaryliodonium reagents in a mixture of dimethoxyethane (DME) and water at 35°C in the presence of copper(I) iodide (2 mol %) and sodium carbonate (1.2 equiv) to give the desired biaryls in high yields (Scheme 10.19).[51a] In each case, a selective transfer of the more

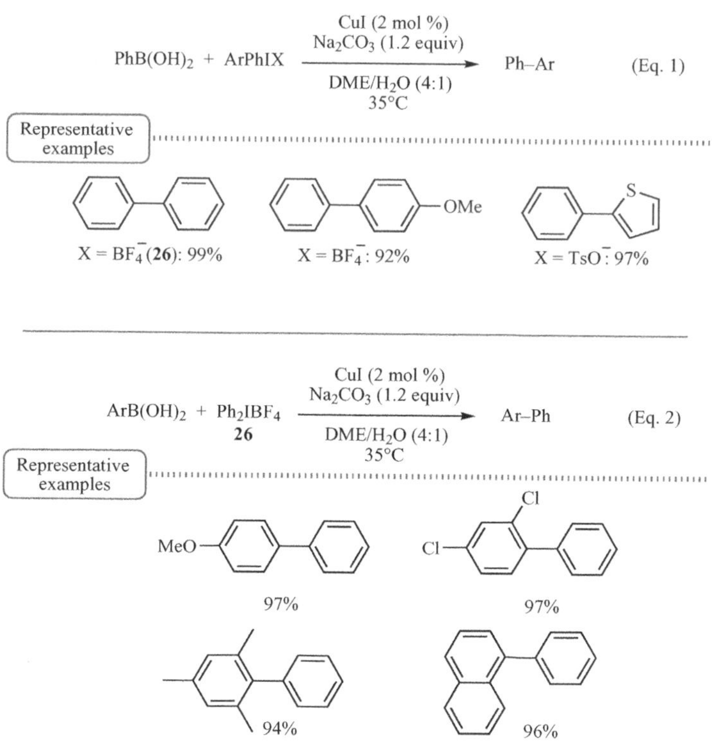

Scheme 10.19 Copper(I)-catalyzed cross coupling of arylboronic acids with diaryliodonium salts.

electron-rich (hetero)aryl group of the iodonium was observed. Interestingly, the same reaction produced diaryl ketones when executed under an atmosphere of carbon monoxide.

The Suzuki–Miyaura cross coupling of phenylboronic acid and haloarenes was also investigated using copper-based mixed nanoclusters (see Chapter 20 for more details).[54] The metal clusters, with sizes in the range of 1.6–2.1 nm, were prepared from copper, palladium, and ruthenium as well as various combinations of copper with palladium, platinum, and ruthenium. Their catalytic activities toward the Suzuki–Miyaura coupling of phenylboronic acid and iodobenzene, which was performed with a 2 mol % catalyst loading in DMF at 110°C, were evaluated. Among the nanoclusters investigated, the Cu/Pd bimetallic cluster showed a high activity that was similar to that of a palladium cluster and the copper cluster also catalyzed the cross coupling at a reasonable reaction rate, the coupling of *p*-iodotoluene being complete after 8 and 2 hours using the Cu and Cu/Pd clusters, respectively. The use of a simple copper salt for the cross coupling of haloarenes was reported by Li and coworkers, who demonstrated that the choice of an appropriate ligand was imperative for the success of the reaction.[55] In the absence of any ligand, the coupling of *p*-iodoanisole with phenylboronic acid using copper(I) iodide (10 mol %) and cesium carbonate as a base in DMF at 125–130°C for 20 hours gave the corresponding 4-methoxybiphenyl in 78% yield, while the yield improved to 98% when 1,4-diazabicyclo[2.2.2]octane (DABCO) was employed as a ligand (Scheme 10.20, Eq. 1).[55a] Triethylamine, *N,N,N′,N′*-tetramethylethylenediamine (TMEDA), or *N,N*'-dimethylethylenediamine (DMEDA) were also investigated as ligands but resulted in decreased yields of 66%, 6%, and 31%, respectively, and although phosphines are known as effective ligands for the palladium-catalyzed Suzuki–Miyaura cross coupling, the addition of neither triphenylphosphine nor tricyclohexylphosphine led to increased yields (20% and 6%, respectively). This method proved to be effective for the coupling of arylboronic acids with substituents at the *para*- and *ortho*-positions, but the use of substrates with electron-withdrawing substituents resulted in diminished yields. When bromoarenes were used as coupling partners, stoichiometric amounts of copper(I) iodide and the addition of TBAB were usually required to obtain satisfactory yields. Modified ligand-free conditions with TBAB as an additive and DMSO as the solvent enabled the catalytic cross coupling of bromoarenes and arylboronic acids albeit with limited scope (Scheme 10.20, Eq. 2).[55c] This new protocol was also effective for the coupling of heteroaromatic precursors as exemplified with the phenylation of 3-bromopyridine **28** and 5-bromopyrimidine **29** with phenylboronic acid or 2-furyl-, 2-thienyl-, or 4-pyridyl-boronic acids with iodobenzene, which were achieved in 55–83% yields (Scheme 10.20, Eq. 3 and Eq. 4).

Due to the recent focus on environmental concerns, the development of recyclable catalysts has garnered enormous interest with respect to sustainable chemistry and the immobilization of transition metals is a viable solution for the development of readily recyclable catalysts. While such methods, however,

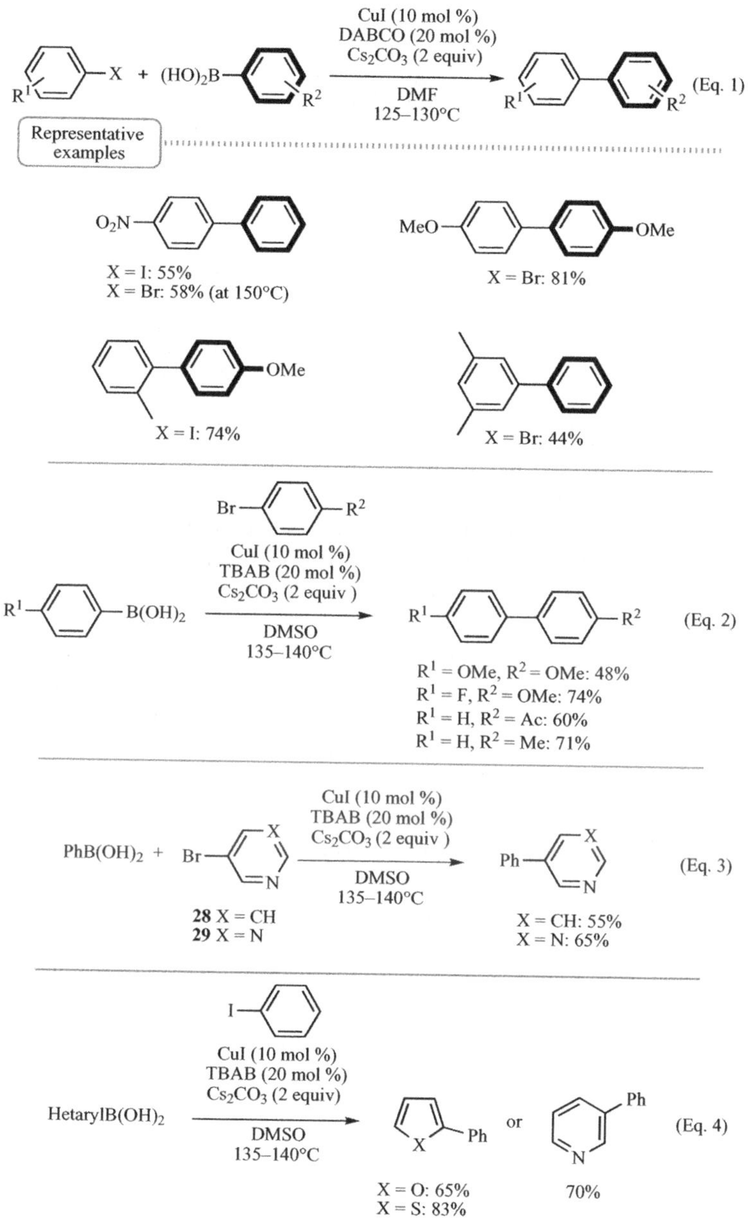

Scheme 10.20 Copper(I)-catalyzed cross coupling of arylboronic acids with aryl halides.

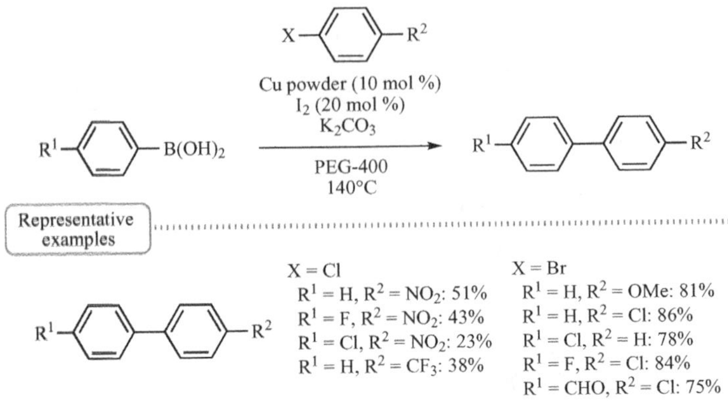

Scheme 10.21 Cross-coupling reaction of arylboronic acids with aryl halides using recyclable copper catalyst.

often decrease the catalytic activity and/or selectivity (see Chapter 20 for details), the Mao and Ji group successfully demonstrated that the use of poly(ethylene)glycol (PEG) as a solvent enables the immobilization of copper catalysts with concurrent improvement of the catalytic efficiency (Scheme 10.21).[56] The Suzuki–Miyaura cross coupling of various arylboronic acids and iodoarenes indeed occurred in good to excellent yields when the reaction was carried out in the presence of copper powder (10 mol %) and potassium carbonate (2 equiv) in PEG-400, which is considered to act as a ligand, at 110°C for 12 hours. Notably, the new catalytic system was effective for the coupling of bromoarenes and electron-deficient chloroarenes, although the yields for the latter were relatively low. For the reaction of these less-reactive haloarenes, catalytic amounts of iodine (20 mol %) were added and the reactions were performed at elevated temperatures. Recycling experiments on the coupling of phenylboronic acid with *p*-nitroiodobenzene revealed that the activity of the catalyst was maintained for at least three cycles without significant lowering of the yields (average 94%). However, the yield dropped to 66% after six cycles. Recently, a new catalytic system comprising copper oxide and 2,2′-diamino-6,6′-dimethylbiphenyl ligand was reported to be also effective for the Suzuki–Miyaura cross coupling involving aryl iodides and bromides, the advantages of this type of axially chiral ligand being its potential for future developments in asymmetric cross-coupling reactions.[57]

Besides the use of arylboronic acids in copper-catalyzed cross-coupling reactions involving various aryl halides or synthetic equivalents, they have also been used for the preparation of symmetrical biaryls by homodimerization. Indeed, the homocoupling of arylboronic acids was first accomplished with half an equivalent of copper(II) acetate in DMF at 100°C with various efficiencies (30–95%) depending on the steric bulk and orientation of the substituents, the presence of *ortho*-substituents resulting in low yields.[58a] Some arylboronic acids also underwent homocoupling when the copper(II) loading was reduced to 0.1 equiv under an atmosphere of oxygen, albeit in slightly lower yields. As

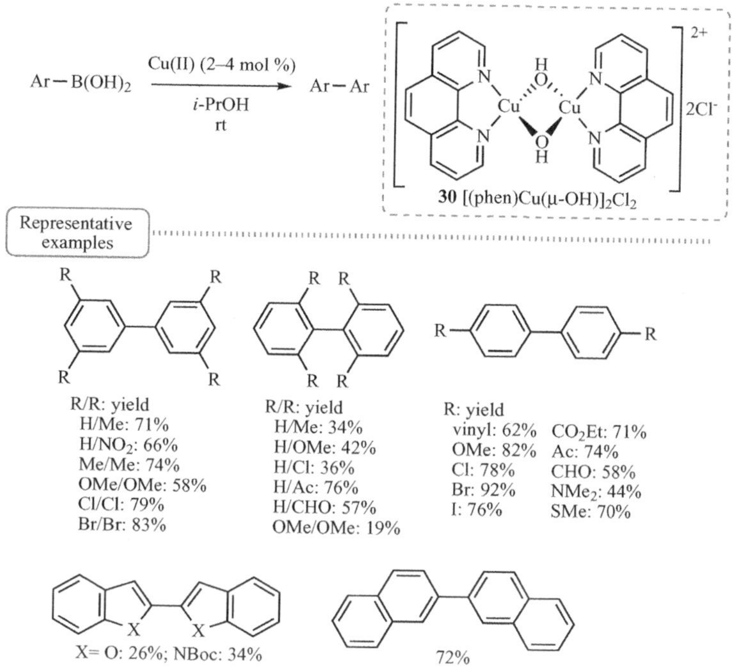

Scheme 10.22 Copper(II)-catalyzed homocoupling of arylboronic acids.

for the copper source, copper(II) sulfate can also be used for the stoichiometric homocoupling of arylboronic acids under base- and ligand-free conditions.[58b] Catalytic homocoupling with lower amounts of copper was achieved using 1,10-phenanthroline as a ligand. In the presence of copper(II) acetate (5 mol %) and 1,10-phenanthroline (6 mol %), *p*-tolylboronic acid was stirred in isopropyl alcohol at ambient temperature for 4 hours to furnish 4,4′-dimethylbiphenyl in 83% yield (Scheme 10.22).[59] Isopropyl alcohol was found to be a superior solvent to ethanol or methanol, and it was also revealed that the isolated hydroxo-bridged dinuclear Cu(II) complex [(phen)Cu(μ-OH)]$_2$Cl$_2$·3H$_2$O (**30**·3H$_2$O) showed the highest catalytic ability, the use of 2–4 mol % of this catalyst enabling the transformation of various arylboronic acids into the corresponding biaryls in moderate-to-good yields, presumably via a bimolecular reductive elimination. Nevertheless, this catalytic system was found to be less effective with *ortho*-substituted arylboronic acids and heteroaryl boronic acids as well. Later on, Cheng and Luo modified this protocol to eliminate the need for the 1,10-phenanthroline ligand and reported that the use of copper(I) chloride in methanol at room temperature was especially efficient and also improved the substrate scope.[60a] Finally, Kaboudin and coworkers next reported on the use of dinuclear (μ-hydroxo)copper(II) β-cyclodextrin complex for the homocoupling of arylboronic acids.[60b] They also proposed a binuclear reductive elimination mechanism.

10.3.3 Coupling of Organosilanes

The palladium-catalyzed cross coupling of organosilicon compounds with organic halides, which is known as Hiyama cross coupling, has received significantly less attention than those involving organotin and boron compounds (i.e., Migita–Kosugi–Stille coupling and Suzuki–Miyaura coupling), despite the low toxicity and easy handling of organosilicon compounds.[61] This is mainly due to the transmetallation from silicon to palladium that is less efficient than those from the corresponding tin or boron reagents and usually requires fluoride anion to activate the silicon–carbon bond and facilitates the transmetallation. The coupling efficiency, however, has been significantly improved by the use of silanols[62] or 2-(hydroxymethyl)phenylsilanes,[63] which enabled the application of Hiyama-type coupling to the synthesis of complex natural products.[62c]

In striking contrast to such dramatic advances in palladium-catalyzed methods, the progress in organosilicon-based coupling using copper promoters has been relatively stagnant. To date, only one example of copper-mediated Hiyama-type cross coupling has been reported by Ito and coworkers, who disclosed that copper(I) pentafluorophenoxide, prepared *in situ* from copper(I) iodide and sodium pentafluorophenoxide, is an effective promoter for this reaction (Scheme 10.23).[64] In the presence of this promoter, the reaction of trimethylsilylthiazole (**31**) with iodobenzene was performed in dimethylimidazolidinone (DMI) at 130°C for 12 hours to give the coupling product **32** in 93% yield. It should be noted that triisopropylsilyl ether was retained in coupling product **32a** since fluoride ion sources, which generally remove this

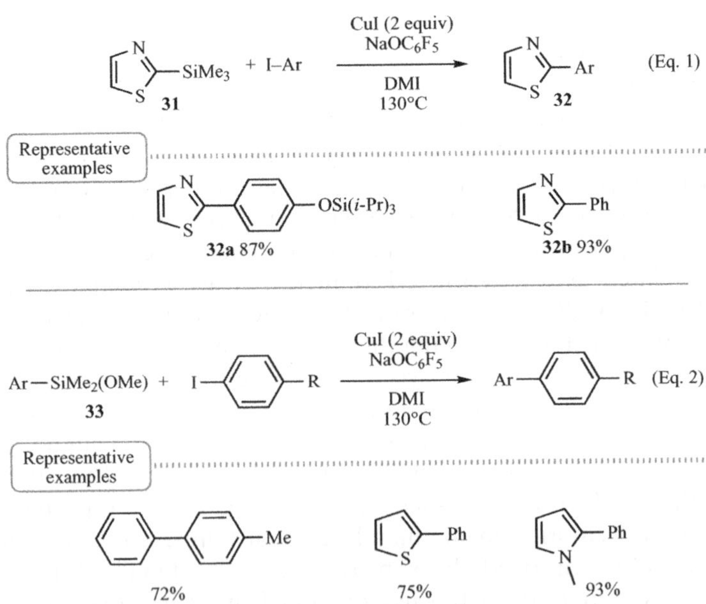

Scheme 10.23 Copper(I)-mediated cross coupling of arylsilanes with aryl iodides.

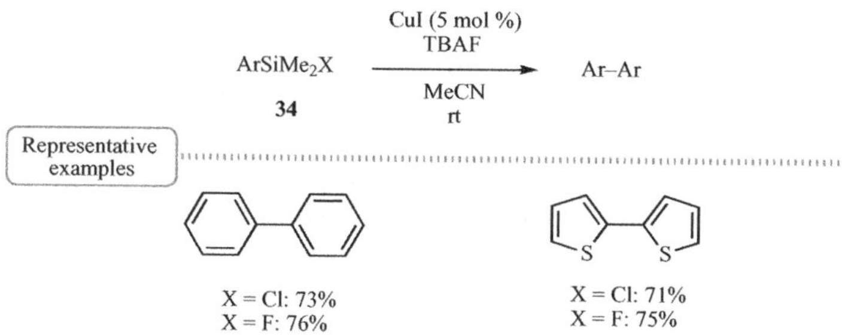

Scheme 10.24 Copper(I)-catalyzed homocoupling of arylsilanes.

protecting group, are unnecessary for the transmetallation in this case. In contrast, the coupling of phenyltrimethylsilane with *p*-iodotoluene under the optimal conditions failed and the cross coupling with these less-reactive substrates was enabled by the use of methoxydimethylsilyl analogues **33** which undergo transmetallation under the same conditions (Scheme 10.23, Eq. 2).

Meanwhile, the copper-catalyzed homocoupling of arylsilanes was being developed by Kang and coworkers, who employed aryldimethylhalosilanes **34** as substrates and tetrabutylammonium fluoride (TBAF) as an additive to promote a rapid homocoupling, even at ambient temperature (Scheme 10.24).[65] Fluoride is a slightly superior halide ligand on silicon to chloride in terms of yields, and this method was found to be also applicable to the corresponding alkenyl- and alkynyl-silanes. As a note, a stoichiometric copper(I) chloride-mediated homocoupling of arylethyldifluorosilane was also reported.[66]

As developed in the previous paragraphs, biaryls could be efficiently synthesized via copper-mediated and copper-catalyzed homocoupling and cross-coupling reactions of arylstannane, borane, and silane derivatives with different electrophiles such as aryl halides or iodonium salts. More recently, the construction of biaryls has witnessed a significant shift of paradigm with the development of direct and more practical methods relying on arene carbon–hydrogen or carbon–carbon bond fission. The most relevant copper-catalyzed strategies for biaryl elaboration are discussed in the next section.

10.4 BIARYL SYNTHESIS BY ARYLATION INVOLVING ARENE C–H OR C–C BOND FISSION

10.4.1 Direct C–H Arylation of Arenes with Prefunctionalized Arylating Reagents

Direct functionalization of C–H bonds is an ideal process that enables omission of synthetic steps such as halogenation or metallation (which can also require additional protection step) and also reduces harmful waste. The development of transition-metal-catalyzed C–H functionalization reactions entails a dramatic paradigm shift from conventional synthetic tactics based on functional-group manipulation to modern, environmentally friendly, step-economical strategies via the direct transformation of C–H bonds.[67] Toward this goal, ruthenium, rhodium, and palladium catalysts have been extensively explored for direct C–H arylation.[68] This section summarizes recent progress in copper-mediated direct C–H arylation with haloarenes or aryl organometallics, an area in which significant progress has recently been made.

An interesting hint of the feasibility of such copper-mediated process was actually published in 1960. Indeed, the Ullmann homocoupling of iodobenzene was reported to be significantly suppressed when 1,3-dinitrobenzene was used as a solvent at 195°C since the reaction furnished the expected biphenyl in a low 22% yield together with an unsymmetrical biaryl, 2,6-dinitrobiphenyl, and intact iodobenzene in 8% and 44% yields, respectively.[69a] 2,6-Dinitrobiphenyl was obviously produced by direct arylation of the C–H bond between the two nitro groups of dinitrobenzene with iodobenzene, and when 1,3,5-trinitrobenzene was used as the solvent, homocoupling was completely suppressed and the corresponding unsymmetrical biaryl was formed in 12% yield (Scheme 10.25, Eq. 1). Later, it was confirmed that the direct arylation of polynitroarenes with iodobenzene proceeded through a Meisenheimer intermediate that was formed from an intermediate phenylcopper reagent.[69b] Indeed, the reaction of 1,3,5-trinitrobenzene with 1,6-dimethoxyphenylcopper (**35**) in pyridine at 25°C resulted in a deep red solution, and acidification with acetic acid furnished the pyridinium salt of Meisenheimer anion **36** (37%), which was heated in vacuo at 110°C to afford biaryl **37** in 70% yield (Scheme 10.25, Eq. 2). Similar direct arylation of polynitroarenes was also observed when they were

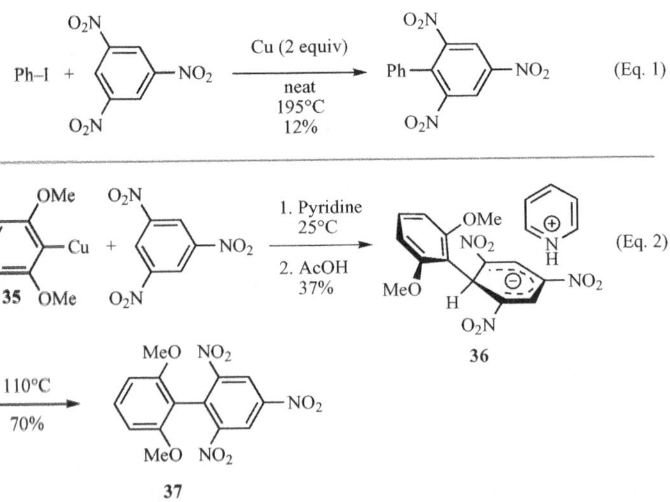

Scheme 10.25 Copper-mediated direct arylation of 1,3,5-trinitrobenzene with iodobenzene.

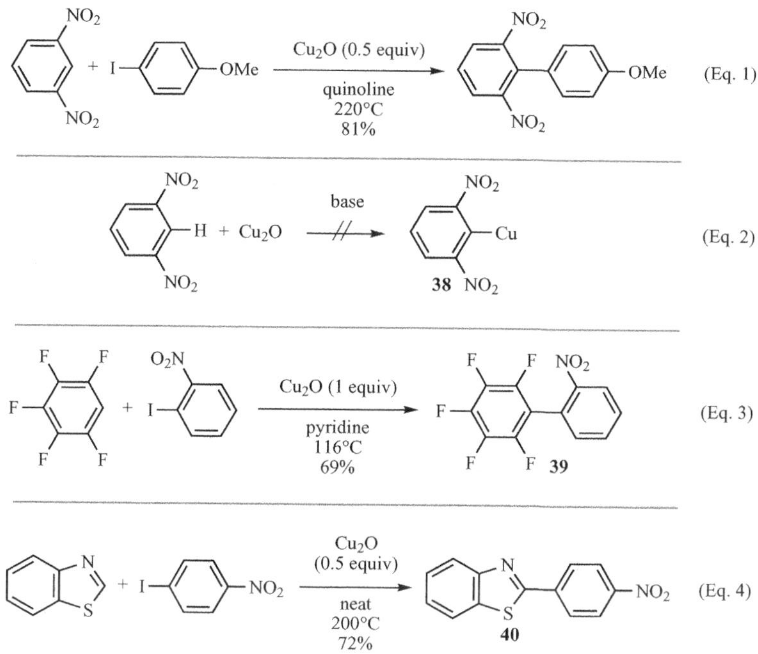

Scheme 10.26 Copper-mediated direct arylation of electron-deficient arenes with aryl iodides.

heated at 220°C with copper(I) oxide and *p*-iodoanisole in quinoline, the corresponding cross-coupling products of 1,3-dinitrobenzene or 1,3,5-trinitrobenzene being obtained in good yields (Scheme 10.26, Eq. 1).[70] While the formation of 2,6-dinitrophenylcopper (**38**) from the abstraction of the acidic 2-proton of 1,3-dinitrobenzene was proposed, attempts to prepare phenylcopper by this method failed (Scheme 10.26, Eq. 2).[70b] The cross-coupling reactions of preformed arylcopper reagents with iodoarenes were also reported,[27] and other aromatic compounds with sufficiently acidic protons were shown to undergo direct arylations via proton abstraction. Copper-mediated direct arylation of pentafluorobenzene and benzothiazole with nitrophenyl iodides was similarly performed to afford cross-coupling products **39** and **40** in 69% and 72% yields, respectively, the latter product being obtained in a higher yield (85%) using the highly sensitive copper(I) *tert*-butoxide as a promoter (Scheme 10.26, Eq. 3 and Eq. 4).[71,72]

Although these above-mentioned methods have the practical advantage of producing unsymmetrical biaryls with predictable regiochemistry via simple procedures, stoichiometric loadings of copper promoters and harsh reaction conditions are required. After these early and remarkable reports, the copper-promoted direct C–H arylation with iodoarenes was neglected for more than two decades until a study published by Miura and coworkers on

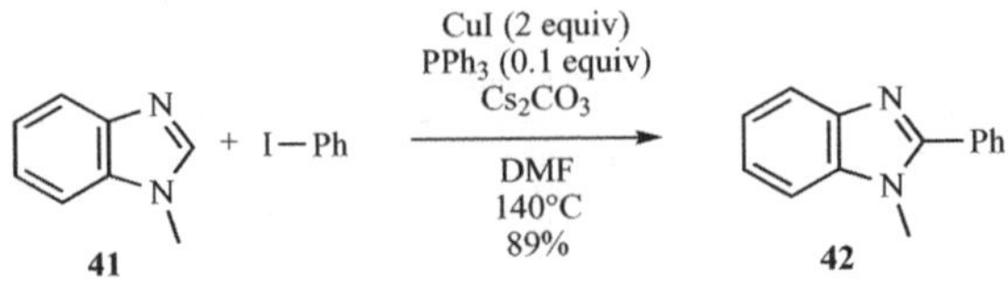

Scheme 10.27 Copper-mediated direct arylation of 1-methyl-1*H*-benzimidazole with iodobenzene.

the palladium-catalyzed direct arylation of azole with haloarenes. In this publication, it was also shown that the combination of stoichiometric loading of copper(I) iodide, triphenylphopshine, and cesium carbonate as a base promotes the coupling of azoles with iodobenzene in DMF at 120–140°C to exclusively give 2-phenylazoles, albeit in low yields, while the palladium-catalyzed reactions produce mixtures of 5-phenylazoles and 2,5-diphenylazoles.[73a] The arylation of 1-methyl-1*H*-benzimidazole **41** with iodobenzene via treatment with copper(I) iodide, triphenylphopshine, and cesium carbonate for 8 hours resulted in the 2-phenylation product **42** in 89% yield (Scheme 10.27).[73a] After extensive optimizations, the scope of direct arylation with iodoarenes using this system was extended to include benzoxazole, benzothiazole, 5-aryloxazoles, 1,3,4-oxadiazoles, and 1,2,4-triazoles.[73b–d] It should be noted that the use of extra base and phosphine ligand enables the direct arylation of azoles under milder conditions.

A further breakthrough was reported in a series of publications by Daugulis and coworkers,[74] who demonstrated that a broad spectrum of heteroarenes and polyfluorobenzenes containing C–H bonds with pK_a values below 35 underwent catalytic direct arylation with (hetero)arylbromides and iodides using the copper(I) iodide/1,10-phenanthroline catalytic system in DMF or 1,3-dimethyl-3,4,5,6-tetrahydro-2(1*H*)-pyrimidinone (DMPU) (Scheme 10.28). The choice of the base again plays a pivotal role for the success of this transformation since the use of potassium *tert*-butoxide for the coupling involving *para*-substituted haloarenes resulted in a loss of regioselectivity due to the formation of benzyne intermediates, a side reaction that could be suppressed by switching to lithium *tert*-butoxide. For less-reactive substrates, a more hindered base, lithium 3-ethyl-3-pentanoxide, was shown to be beneficial to minimize the corresponding direct aromatic nucleophilic substitution, while a weaker base such as potassium phosphate can be used for the arylation of more acidic heterocycles such as benzothiazole, pyridine *N*-oxide, and penta- and tetrafluorobenzenes. Typical examples are shown in Scheme 10.28. These cross-coupling reactions are thought to proceed via the coupling of (hetero) arylcopper(I) intermediates with haloarenes, which is supported by the formation of a phenanthroline-ligated pentafluorophenylcopper(I) complex from copper(I) iodide, pentafluorobenzene, and phenanthroline in the presence of potassium phosphate in DMF at 125°C, the structure of which is confirmed by comparing its ^{19}F nuclear magnetic resonance (NMR) spectral data with those

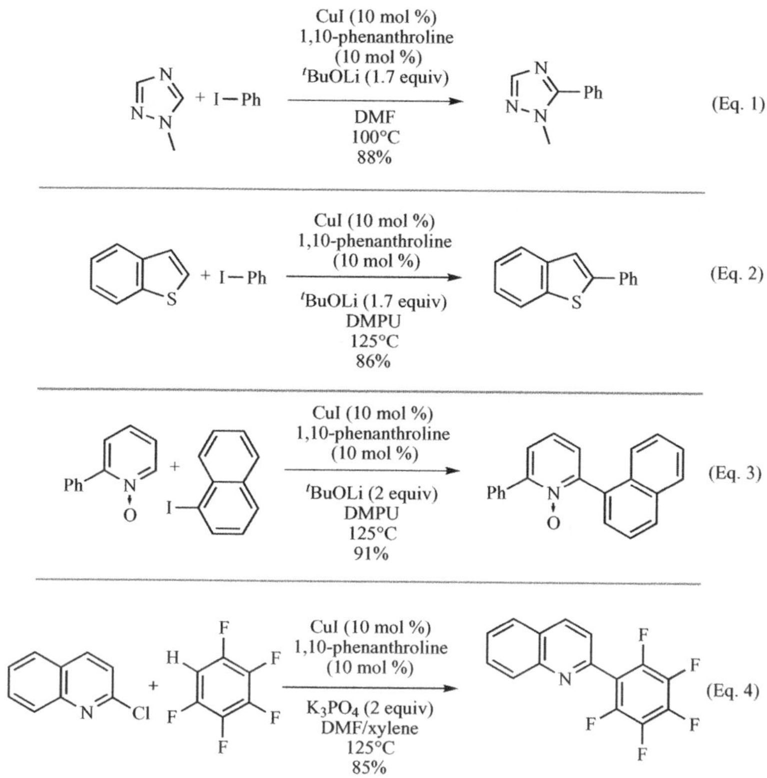

Scheme 10.28 Copper(I)-catalyzed direct arylation of heteroarenes and pentafluorobenzene.

of an authentic sample. As an extension of this method, a one-pot sequential iodination/copper-catalyzed cross coupling was also accomplished, providing a direct route to unsymmetrical biaryls from two different (hetero)aromatic precursors.[74d]

The copper-catalyzed direct C–H arylation of caffeine with bromoarenes was also investigated in detail,[75] and just as a note, one of the obtained products, that is, the adduct with *p*-bromo-*N*,*N*-dimethylaniline, exhibited strong fluorescence in chloroform solution and in the solid state and is potentially applicable as a biological imaging probe. In addition, the direct coupling of a caffeine derivative was combined with an intramolecular copper-catalyzed Ullmann-type reaction from 2-(2,2-dibromovinyl)phenol to provide a successful, high-yielding tandem route to a caffeine-derived benzofuran (Scheme 10.29).[75b]

The direct C2-arylation of 1-(pyridin-2-yl)indole-3-carboxaldehyde with iodoarenes was also achieved using subcatalytic amounts of copper(I) oxide,[76] and Ackermann and coworkers established a sequential four-component coupling route to 1,2,3-triazoles, which involves a single copper catalyst-mediated

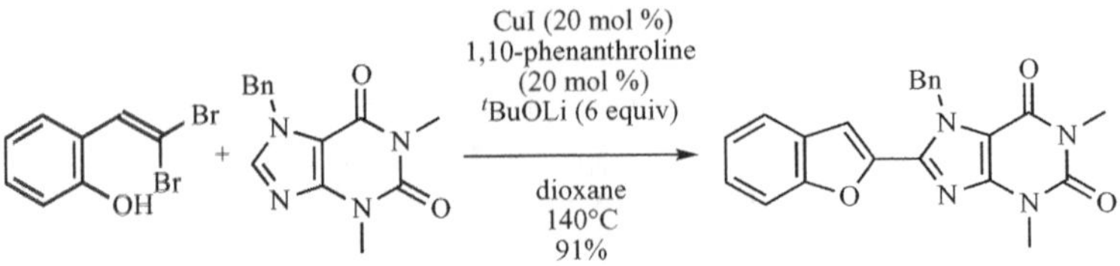

Scheme 10.29 Copper(I)-catalyzed tandem benzofuran cyclization/direct arylation.

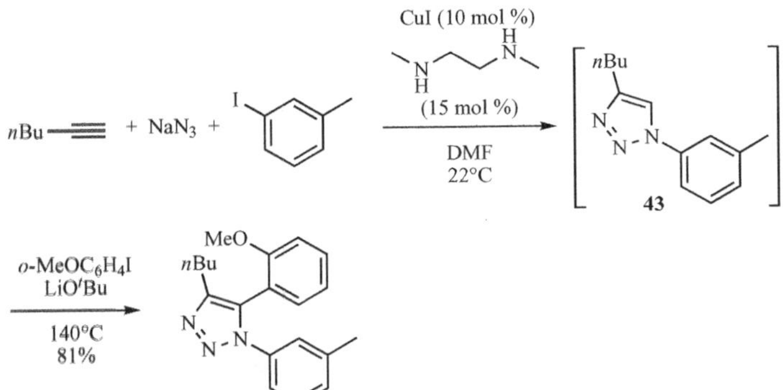

Scheme 10.30 Copper(I)-catalyzed tandem three-component azidation/dipolar cycloaddition/direct C–H arylation.

azidation of iodoarenes, [3 + 2] cycloaddition of the resulting arylazide with terminal alkynes, and subsequent direct C–H arylation of the resulting 1,2,3-triazole **43** (Scheme 10.30).[77]

As replacements for haloarenes, hypervalent compounds have been used for the copper-catalyzed direct C–H arylation of electron-rich aromatic substrates. Although these reagents are not commercially available and have to be prepared, their use typically allows for remarkably mild conditions as evidenced by results collected in this subsection.

A pioneering study published by Barton in 1988 involves the C3 arylation of indoles with triphenylbismuth bistrifluoroacetate.[78] As shown in Scheme 10.31, 2-methylindoles **44** were treated with hypervalent bismuth reagent **45**, prepared from triphenylbismuth bisacetate by ligand exchange,[78b] in the presence of copper catalysts (10 mol %) in dichloromethane at ambient temperature to afford 3-phenylindoles **46** in remarkably high yields. In the absence of the C2 methyl group, the reaction gave a *N*-phenylation product, which indicates the importance of steric shielding around the nitrogen atom. From a mechanistic perspective, indolylcopper(III) intermediates **48** were proposed to be produced from phenylcopper(III) complex **47** and indoles **44** (probably via electrophilic aromatic substitution), and this was followed by final reductive elimination to yield the C3-arylation products **46**.

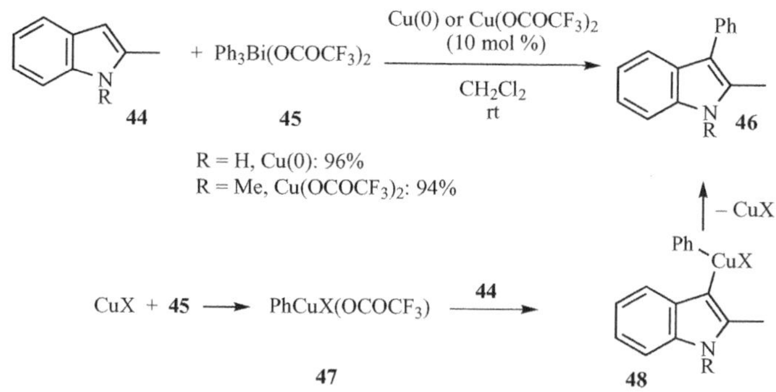

Scheme 10.31 Copper-catalyzed direct arylation of indole derivatives with triphenylbismuth reagent.

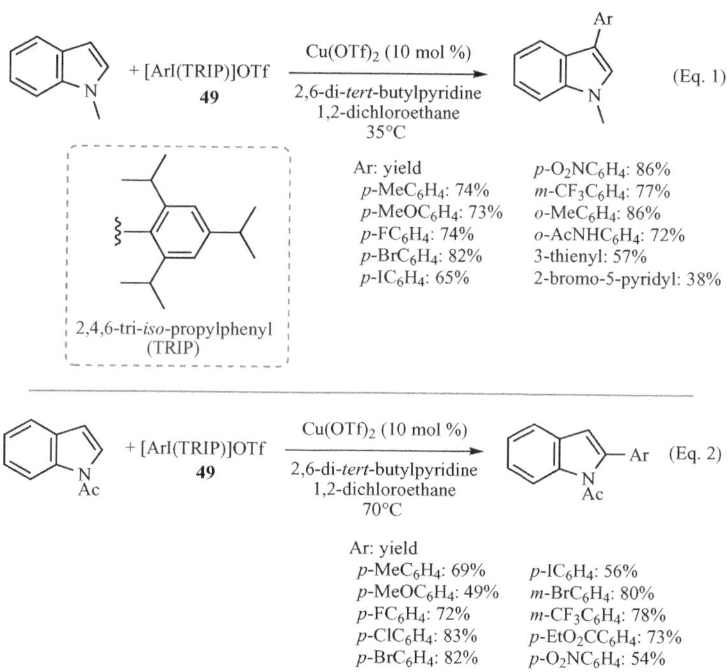

Scheme 10.32 Copper(I)-catalyzed regioselective direct arylation of indole derivatives with diaryliodonium salts.

Recently, a similar but regiocontrolled arylation of *N*-protected indoles using hypervalent iodine reagents was developed by Gaunt and coworkers (Scheme 10.32).[79a] The C3-arylation of *N*-methylindole was achieved using copper(II) triflate as the catalyst (10 mol %) and 2,6-di-*tert*-butylpyridine as the base in 1,2-dichloroethane. Nonsymmetrical diaryliodonium triflates **49**

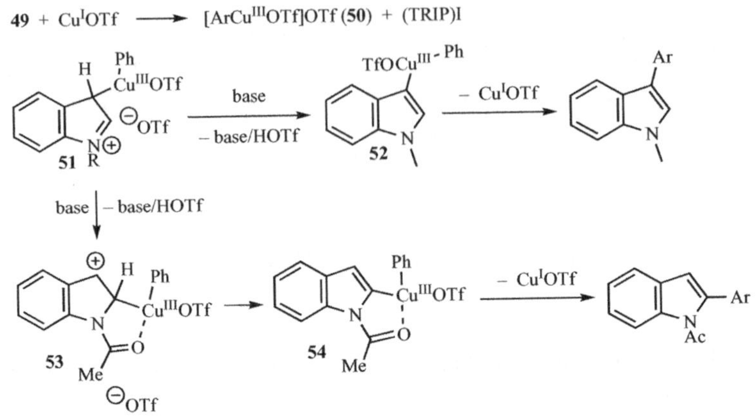

Scheme 10.33 Proposed mechanism for the copper(I)-catalyzed regioselective direct arylation of indole derivatives.

possessing a bulky 2,4,6-tri-isopropylphenyl (TRIP) as a sacrificial aryl group, were employed as arylating agents and substituted aryl and heteroaryl groups were selectively and efficiently introduced yielding the corresponding indoles in moderate-to-good yields and excellent regioselectivity. In contrast, *N*-acetylindole underwent C2-arylation under the same conditions, although the reaction had to be performed at 70°C, resulting in the corresponding products in 49–83% yields, the regioselectivity (C2:C3 = 7:1–2.6:1) being lower than those of the C3 arylations (Scheme 10.32, Eq. 2).

Scheme 10.33 outlines plausible mechanisms. Hypervalent iodine reagent **49** first reacts with copper(I) triflate as the net catalyst to produce an electrophilic arylcopper(III) species (**50**), which is trapped by the indole, reacting preferentially at C3, to yield cyclic iminium intermediate **51**. Subsequent deprotonation then gives 3-indolylcopper(III) species **52** and final reductive elimination accounts for the formation of the 3-arylindoles and regenerates copper(I) triflate. In the case of *N*-acylindole, a 1,2-migration of the phenylcopper(III) fragment occurs due to the coordination of the amide carbonyl group in **53** and the resulting 2-indolylcopper(III) species (**54**) then undergoes reductive elimination to ultimately furnish the 2-arylation product and copper(I) triflate as in the previous case.

In an outstanding study, the copper-catalyzed direct arylation with hypervalent iodine reagents has been further extended to aniline derivatives. Friedel–Crafts-type electrophilic aromatic substitutions of aniline derivatives are expected to occur at *ortho/para*-positions due to the directing effect of the electron-donating amino group and the electrophilic attack by an arylcopper(III) species is also assumed to follow this trend. However, Phipps and Gaunt reported in 2009 an unprecedented *meta*-selective direct C–H arylation of *N*-acyl anilines with diphenyliodonium triflate or tetrafluoroborate under copper catalysis (Scheme 10.34).[79b] As acyl groups on the nitrogen

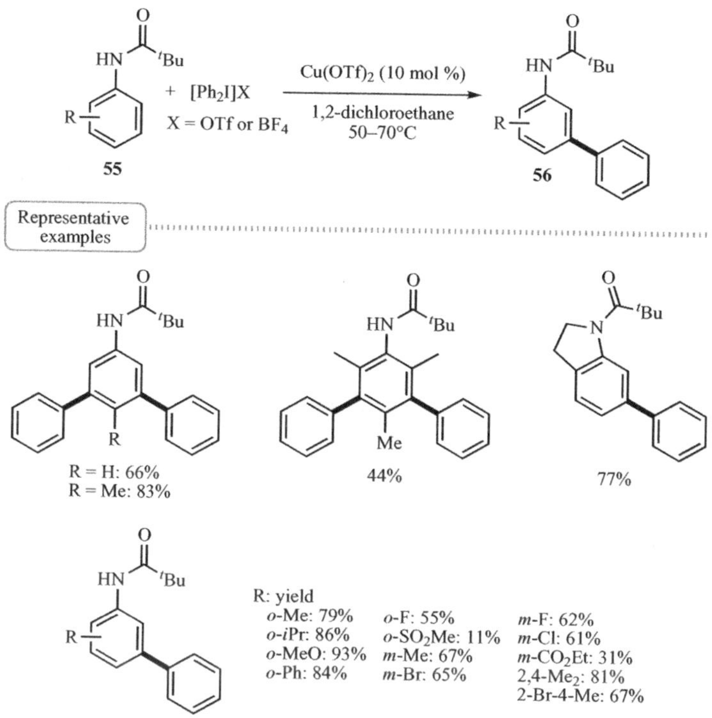

Scheme 10.34 Copper(I)-catalyzed, *meta*-selective direct arylation of aniline derivatives with diaryliodonium salts (bond formed shown in bold).

atom, benzoyl and pivaloyl groups provided the highest yields for *meta*-arylation. Pivaloyl anilides with various substituents (**55**) underwent arylation at 50–70°C to afford mono- or disubstituted products **56** in 11–93% yields. Electron-withdrawing substituents tend to diminish product yields, regardless of their position, and substrates with *ortho*-substituents were arylated at the less hindered of the two *meta*-positions. Substituted phenyl groups can also be introduced using nonsymmetrical diaryliodonium triflate possessing a bulky sacrificial mesityl group and this *meta*-selective arylation is applicable to α-arylacetamides and benzyl ketones.[79d] The *meta*-arylation of Weinreb amides **57** to obtain the *meta*-arylated products **58** is particularly useful because the amide moiety can be further transformed into aldehydes and ketones (Scheme 10.35).

This unusual *meta*-selectivity was ascribed to oxy-cupration, which involves the antiaddition of the amide carbonyl oxygen and arylcopper(III) species on the benzene ring (**59** → **60** in Scheme 10.36). Subsequent reductive elimination leads to the *meta*-arylation products and restoration of the catalytically active CuOTf species. A different mechanism involving four-membered transition state **61** was proposed on the basis of density functional theory (DFT)

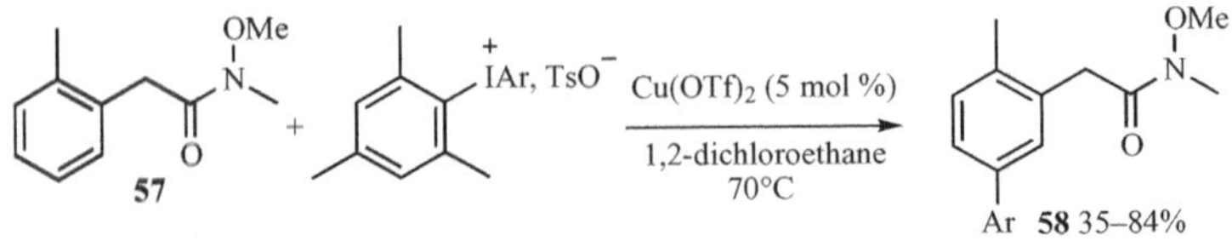

Scheme 10.35 Copper(I)-catalyzed, *meta*-selective direct arylation of Weinreb amide with diaryliodonium salts.

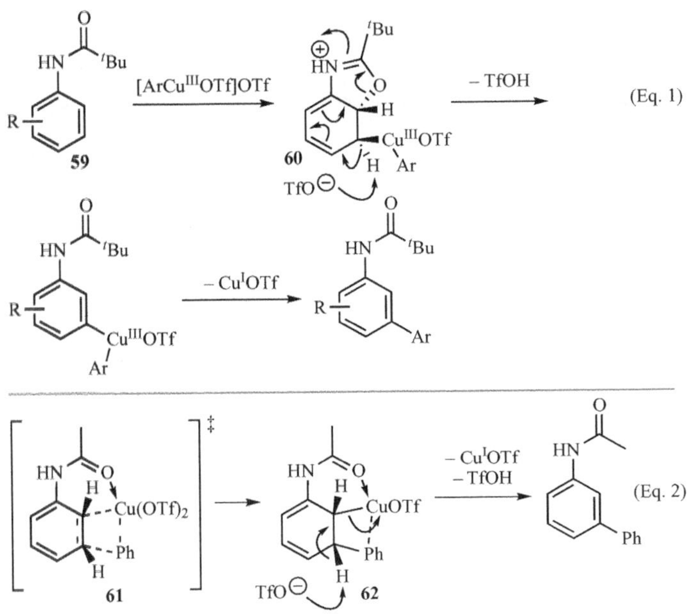

Scheme 10.36 Proposed mechanisms for the copper(I)-catalyzed *meta*-selective direct arylation of anilides.

calculations on the phenylation of acetanilide with a phenylcopper reagent.[79e] According to this proposed mechanism, the electrophilic copper(III) fragment is directed to the *ortho*-position and subsequent rearomatization of **62** via deprotonation restores copper(I) triflate as the net catalyst. It should be noted that the *meta*-arylation of Weinreb amide **57** proceeded without a copper catalyst at 80–90°C, while arylation at 70°C required the catalyst,[79d,f] which implies that the unprecedented *meta*-selectivity is attributable to a factor other than the copper catalyst. Thus, the mechanism of this remarkable transformation clearly necessitates further elucidation.

In contrast to these substrates, normal *para*-selectivity was observed for similar arylation of electron-rich phenol and anisole derivatives (Scheme 10.37, Eq. 1).[79c] When the *para*-position was blocked, arylation occurred at the *ortho*

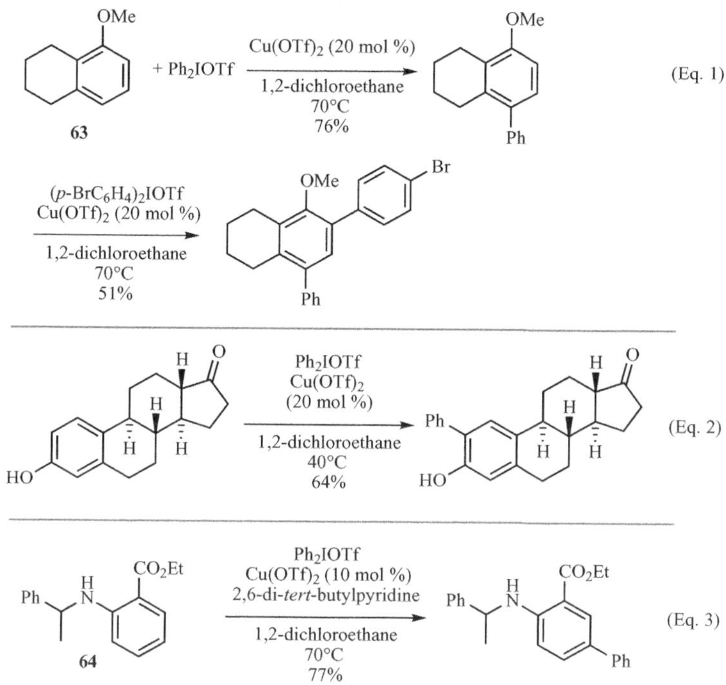

Scheme 10.37 Copper(I)-catalyzed regioselective direct arylation of electron-rich arenes.

C–H bond (Scheme 10.37, Eq. 2), and consequently, regioselective introduction of two different aryl groups was achieved by performing sequential arylations of **63** (Scheme 10.37, Eq. 1). It was also revealed that *N,N*-dibenzylaniline underwent *para*-selective arylation in the presence of a base, 2,6-di-*tert*-butylpyridine,[79c] and that for *ortho*-substituted substrates such as **64**, a single phenethyl group provided effective protection (Scheme 10.37, Eq. 3).

In the above examples, which involve hypervalent reagents, highly electrophilic copper(III) species were generated, and their electrophilic reactions with electron-rich aromatics are the origin of the regioselective C–H arylations. Related direct C–H arylations were also achieved using readily available and easy-to-handle arylboronic acids.[80a] The ability to execute the reaction under air provides a significant practical advantage, although stoichiometric amounts of copper promoter and acidic conditions are required. Using this method, 1,3,5-trimethoxybenzene was monoarylated in moderate yields, and interestingly, excess amounts of copper(II) trifluoroacetate and phenylboronic acid enabled the multiple arylations of 1-methyindole and 1-methypyrrole. Moreover, catalytic arylation of azoles with arylboronate esters was achieved under basic conditions. In this reaction, copper(II) chloride was proposed to be a net catalyst and dioxygen was utilized as a terminal oxidant to reoxidize copper(I) chloride. Arylated oxazoles and thiazoles were obtained in high

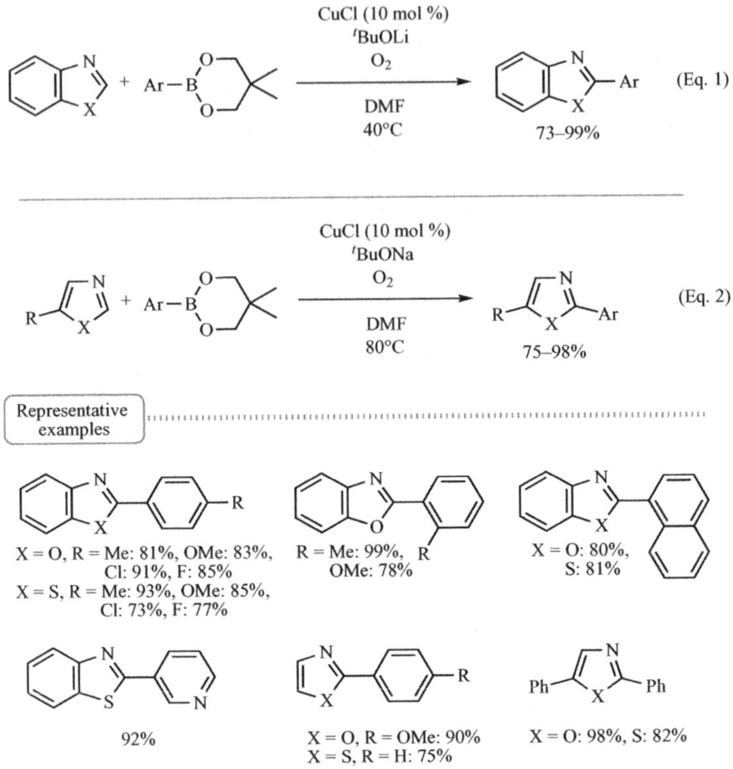

Scheme 10.38 Copper(I)-catalyzed direct arylation of azoles with arylboronates.

yields ranging from 73% to 99% and representative examples are shown in Scheme 10.38.[80b]

10.4.2 Direct C–H Arylation with Nonfunctionalized Arylating Agents

In the preceding section, the copper-mediated direct C–H arylations of arenes with haloarenes, hypervalent reagents, and arylboron reagents were shown to be reliable routes to biaryl compounds with predictable regiochemistry. They, however, require prefunctionalized aromatic compounds as arylating agents, and avoiding the use of these reagents would enable increased step economy and environmental benignancy. This section will outline the formation of biaryls that occurs via the fission of C–H bonds of both coupling components, reactions that are known as dehydrogenative or oxidative coupling reactions.

Ortho-lithiation is the conventional method for directly functionalizing aromatic C–H bonds *ortho* to the directing group,[81] and aryllithiums produced by this technique can undergo homocoupling via oxidation of cuprate

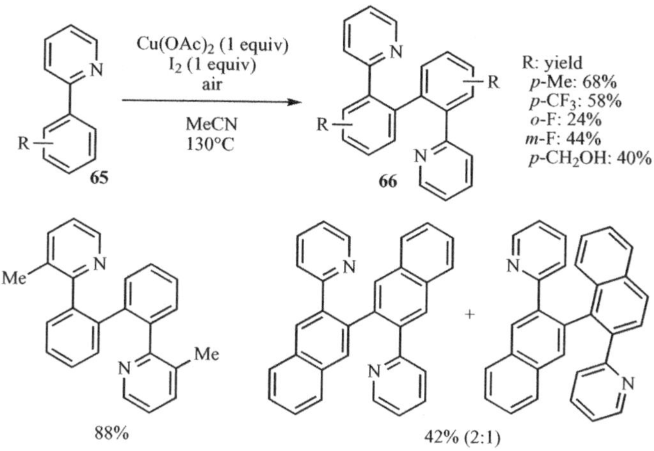

Scheme 10.39 Copper(II)-mediated dehydrogenative homocoupling of 2-arylpyridines.

intermediates as described in Section 10.2.2.[34b] This method is superior to related homocouplings via cuprate oxidation since the preparation of haloarene precursors is unnecessary; however, the use of highly reactive alkyllithium reagents is incompatible with reactive functional groups. Recently, directed homocoupling, which is presumed to proceed via *ortho*-iodination and subsequent Ullmann coupling, was discovered by Yu and coworkers.[82] They treated 2-arylpyridines **65** with 1 equiv of copper(II) acetate and 1 equiv of iodine in acetonitrile at 130°C to obtain homocoupling products **66** in varied yields (Scheme 10.39). The aryl–aryl bond formation occurred at the position *ortho* to the 2-pyridyl group; this reaction is tolerant of unprotected hydroxy groups and the highest yield of 88% was obtained for the reaction of 2-phenyl-3-methylpyridine, while a 2:1 mixture of symmetrical and unsymmetrical coupling products was formed from 2-(2-pyridyl)naphthalene.

Catalytic oxidative homocoupling without relying on a directing group was accomplished by Do and Daugulis[83a] and Yamaguchi and coworkers[83b] using molecular oxygen as a more environmentally benign oxidant. As mentioned previously, heteroarenes and polyfluorobenzenes that feature C–H bonds with pK_a values below 35 undergo catalytic direct arylation with (hetero)aryl halides in the presence of a copper(I) iodide/1,10-phenanthroline catalytic system and an appropriate base such as lithium *tert*-butoxide.[74] As an extension of this strategy, oxidative homocoupling using a catalytic amount of copper(II) chloride and bases composed of isopropylmagnesium chloride and additive(s) was developed. The reaction conditions were tolerant toward heterocycles, ester, nitrile, and nitro moieties and selected examples are shown in Scheme 10.40, (Eq. 1). Zhang and coworkers[83c] reported a similar system based on copper(I) iodide (10 mol %) or chloride (20 mol %) using silver carbonate or *tert*-butyl peroxide as oxidants (Scheme 10.40, Eq. 2 and Eq. 3).

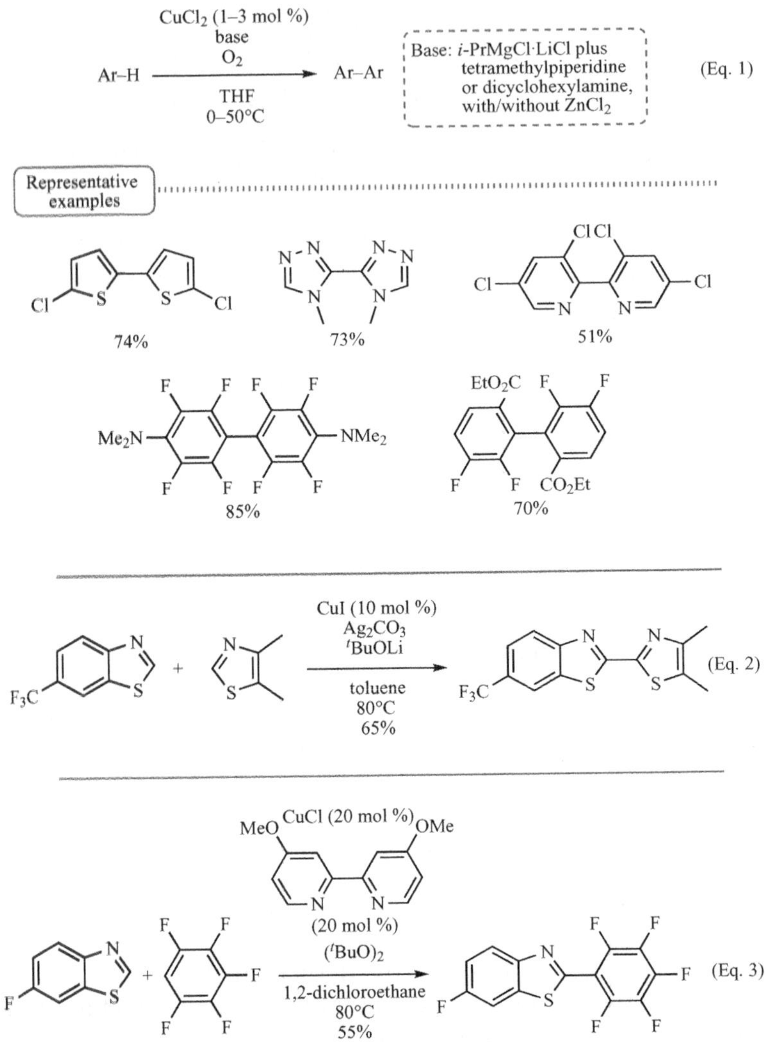

Scheme 10.40 Copper-catalyzed dehydrogenative homo- and hetero-cross coupling of heteroarenes and polyfluorobenzenes.

Subsequently, catalytic dehydrogenative cross coupling of arenes using iodine as a single oxidant was developed.[83d] The authors succeeded in suppressing undesired homocoupling by exploiting the selective iodination of one of the two coupling components followed by coupling of *in situ* formed iodoarenes with other coupling components. It is noteworthy that the single copper catalyst fulfills the dual role of promoting both the iodination of arenes and subsequent cross-coupling reactions in the presence of molecular iodine as an oxidant, and representative examples of this strategy are shown in Scheme 10.41 (Eq. 1, Eq. 2, and Eq. 3).

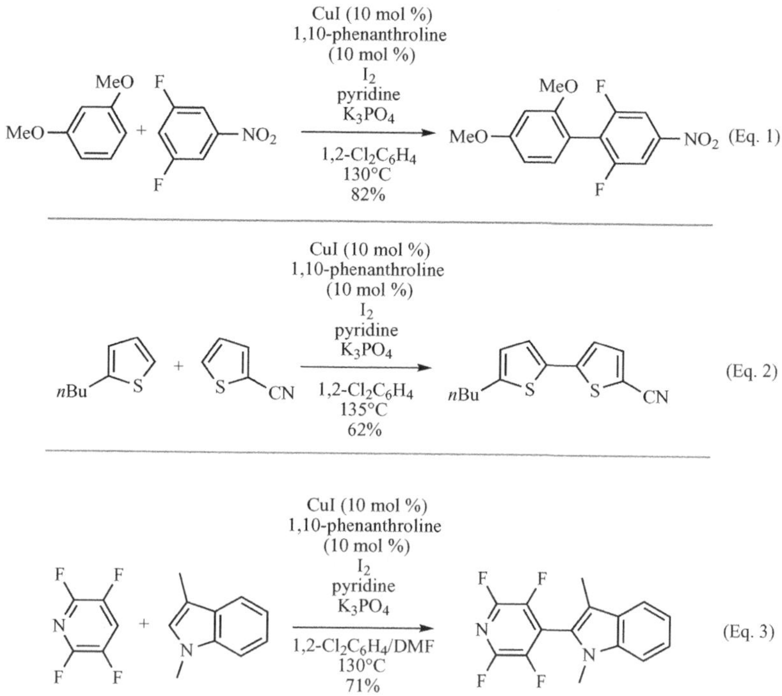

Scheme 10.41 Copper(I)-catalyzed dehydrogenative cross-coupling reactions between electron-rich (hetero)arenes and electron-deficient (hetero)arenes.

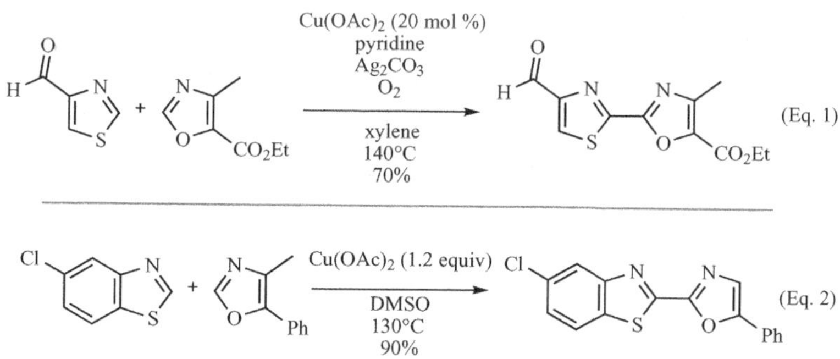

Scheme 10.42 Copper(II)-catalyzed or -mediated dehydrogenative cross-coupling reactions of heteroarenes.

It should also be noted that copper-catalyzed or copper-mediated cross coupling of heteroarenes has been reported to occur with silver carbonate as an oxidant (Scheme 10.42, Eq. 1)[84a] or in the absence of any additives provided that stoichiometric amounts of copper(II) acetate are used (Scheme 10.42, Eq. 2).[84b]

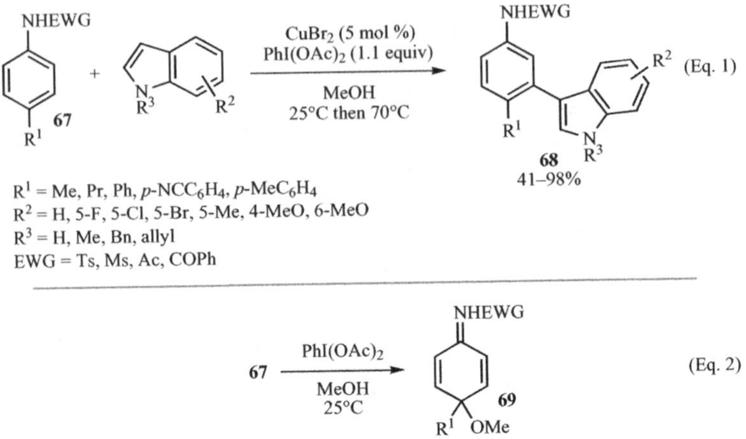

Scheme 10.43 Copper(II)-catalyzed formal dehydrogenative cross coupling of anilides and indole derivatives.

An interesting and less common dehydrogenative C3–C3 cross coupling of aniline derivatives and indoles was developed in 2010, as shown in Scheme 10.43 (Eq. 1).[85] The amino group of **67** must be protected by electron-withdrawing groups such as sulfonyl or acetyl groups and the *para*-carbon should also be substituted by alkyl or aryl groups. In this reaction, the cross coupling is preceded by oxidative methoxylation of **67** with iodobenzene diacetate to form cyclohexa-2,5-dienimines **69** (Scheme 10.43, Eq. 2), which are then trapped by indoles at C3 in the presence of copper(II) bromide (5 mol %) to deliver **68**. Therefore, the copper catalyst is most likely acting as a Lewis acid activator for **69** in this process.

The cross coupling of arenes bearing a directing group and azoles possessing an acidic proton was achieved by Miura and coworkers, although excess amounts of copper(II) acetate were required (Scheme 10.44).[86a] They discovered that the treatment of 2-phenylpyridines **65** and benzoxazole (2 equiv) with copper(II) acetate (5 equiv) and pivalic acid (PivOH, 1 equiv) in mesitylene at 170°C for 2 hours produced mono- and double cross-coupling products **70** and **71** in 45–78% combined yields with selectivities in the 7:1–20:1 range. This cross-coupling reaction could also be applied to various combinations of 2-phenylazines and azoles to obtain 1:1 adducts in moderate-to-good yields. Typical examples are summarized in Scheme 10.44. Empirical evidences including the presence of an intramolecular kinetic isotope effect on *ortho*-deuterated 2-phenylpyridine, a minimal effect of radical scavengers, and rapid H/D exchange of benzoxazole under the reaction conditions imply that the cross coupling proceeds via reversible C–H cupration of azoles, *ortho*-C–H metallation of arylazines, and reductive elimination. The same group extended this methodology to a dehydrogenative cascade reaction that provides an efficient route to heterobiaryls.[86b] Instead of the directed aromatic C–H activation of 2-phenylazines, the latter method involves the annulative

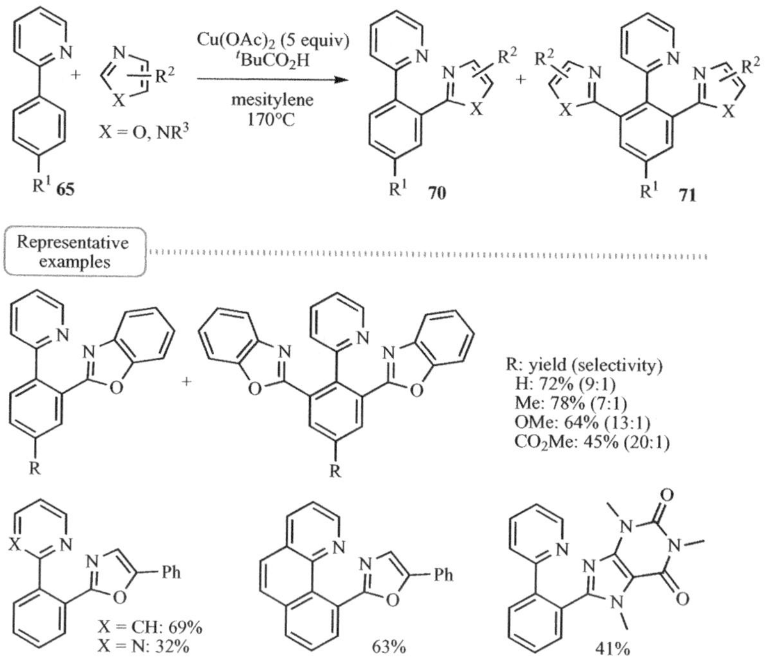

Scheme 10.44 Copper(II)-mediated dehydrogenative cross coupling of 2-phenylpyridines and azoles.

cupration of *o*-alkynylphenols to yield metallated benzofuran coupling components.

Hirano and Miura recently reported a very elegant extension of this concept to the copper-catalyzed cross-coupling reactions of substituted pyrroles and indoles with 1,3-azoles, under air. The directing group, a 2-pyrimidyl moiety, could be easily cleaved after the desired reaction by simple heating in DMSO at 100°C in the presence of sodium methoxide (Scheme 10.45).[87a] More recently, the same research group demonstrated that 8-aminoquinoline was also an excellent directing group for the copper-mediated coupling of benzoic acid derivatives with 1,3-azoles.[87b]

As seen in this section, efficient oxidative copper-catalyzed reactions are available to elaborate biaryls from nonfunctionalized arylating agents. Conceptually related is the decarboxylative coupling of benzoic acid derivatives that provides a straightforward access to biaryls from simple reagents. The recent advances in this domain are overviewed in the next subchapter.

10.4.3 Decarboxylative Coupling of Benzoic Acid Derivatives

It has long been known that copper catalyzes the decarboxylation of aromatic carboxylic acids.[9,88] Nilsson revisited in 1966 the decarboxylation of benzoic acids to investigate the mechanism of the Ullmann coupling and eventually

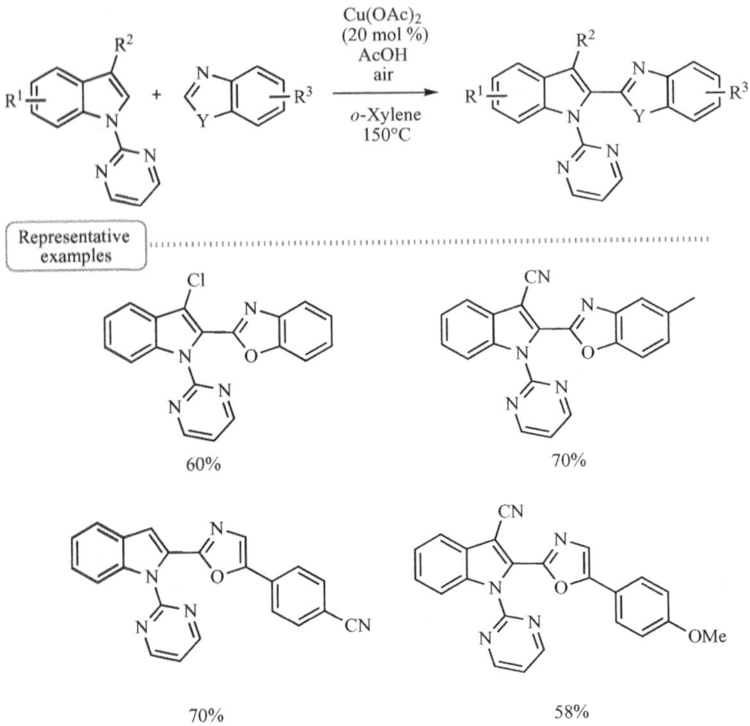

Scheme 10.45 Copper(II)-catalyzed cross coupling of indoles and 1,3-azoles.

found that *o*-nitrobenzoic acid underwent a particularly rapid decarboxylation when it was boiled with copper(I) oxide in quinoline.[89a] Since he also noted that *o*-nitrohalobenzenes are excellent substrates for Ullmann coupling, he attempted the copper-mediated decarboxylation of *o*-nitrobenzoic acid in the presence of *o*-iodoanisole in boiling quinoline and obtained the corresponding cross-coupling product **72** in 50% yield and symmetrical homocoupling products were not detected (Scheme 10.46, Eq. 1). Similar reactions with benzoic acid derivatives and iodobenzenes resulted in lower yields. Thiophene-2-carboxylic and furan-2-carboxylic acids were also subjected to the copper-promoted decarboxylation in the presence of iodobenzenes; however, the cross-coupling products were obtained in low yields (<20%) and small amounts of diarylation products were observed.[89b] A similar reaction of 2,4-dinitrobenzoic acid (**73**) with 2,6-dimethoxyiodobenzene produced cross-coupling products **74** and **75** in 45% and 10% yields, respectively (Scheme 10.46, Eq. 2).[89c] The latter product was assumed to be formed via direct C–H arylation from *m*-dinitrobenzene, which was obtained by simple decarboxylation,[70] while the former product was obtained directly from carboxylic acid **73** via a 2,4-dinitrophenylcopper(I) species. Concrete evidence for the formation of arylcopper compounds from arylcarboxylic acids was later provided independently by Sheppard and coworkers.[90]

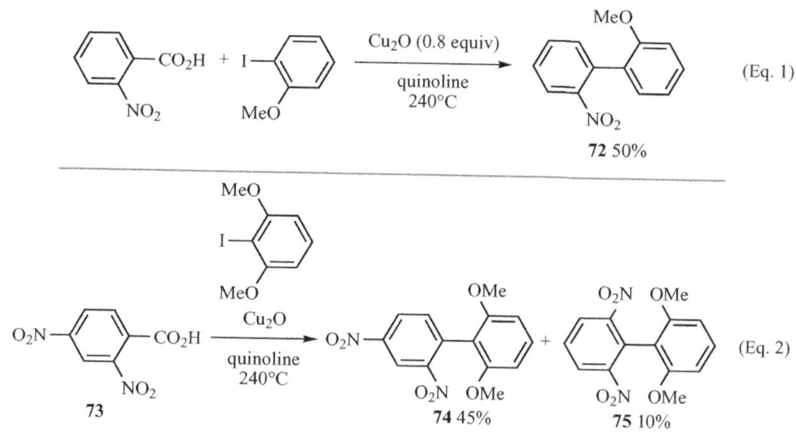

Scheme 10.46 Copper(I)-mediated decarboxylative cross-coupling reactions of benzoic acids and aryl iodides.

These early results demonstrated the feasibility of a copper-mediated decarboxylative cross coupling, and indeed, the use of readily available benzoic acids as environmentally benign surrogates for aryl organometallics or aryl halides is a fascinating tactic that is now frequently used in palladium-catalyzed (hetero)aryl–(hetero)aryl coupling.[91] Accordingly, Gooßen and coworkers developed a palladium/copper binary catalyst system for the decarboxylative cross coupling of benzoic acid derivatives with aryl halides, triflates, and tosylates, and the copper-catalyzed decarboxylation step plays a critical role in these processes.[92] Although these studies provide a powerful route to (hetero)biaryls from inexpensive benzoic acid derivatives, further improvements involving the replacement of palladium-based catalysts with inexpensive and less harmful metals should be addressed. Along this line, the decarboxylative cross coupling of potassium polyfluorobenzoates with aryl iodides and bromides was achieved using a copper catalyst (Scheme 10.47).[93] This method enabled the synthesis of various unsymmetrical (hetero)biaryls from potassium pentafluorobenzoate and iodoarenes in high yields. The cross coupling with bromoarenes was also achieved but required increased catalyst loadings and 1,10-phenanthroline as the ligand. Other fluorobenzoates were used to obtain the corresponding cross-coupling products at higher temperatures in good yields, except for monofluoro- and 2,3,4-trifluoro derivatives, which resulted in low yields. Selected examples are shown in Scheme 10.48.

The arylation reaction involving arene carbon–hydrogen or carbon–carbon bond fission detailed in this subchapter is thus a very efficient strategy to elaborate biaryls from nonfunctionnalized reagents. The impressive efficiency and selectivity of copper catalysts in this area augur well for future developments aiming at even milder conditions and functional-group tolerance. Another vibrant area of investigations is the oxidative coupling of naphthols

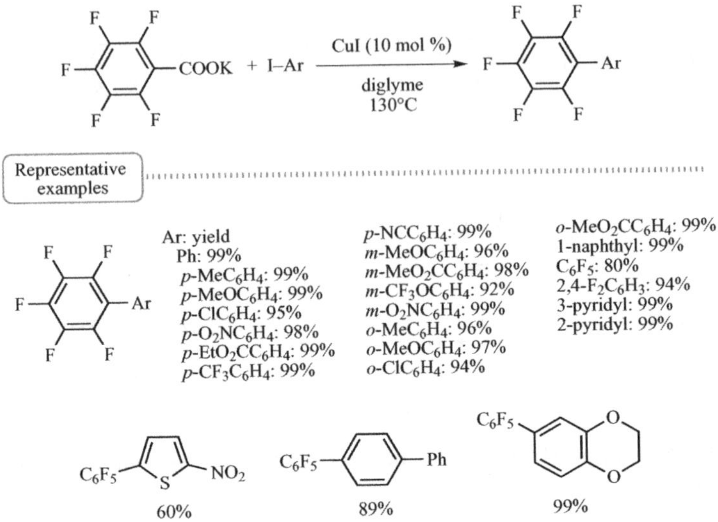

Scheme 10.47 Copper(I)-catalyzed decarboxylative cross coupling of potassium pentafluorobenzoate and aryl iodides.

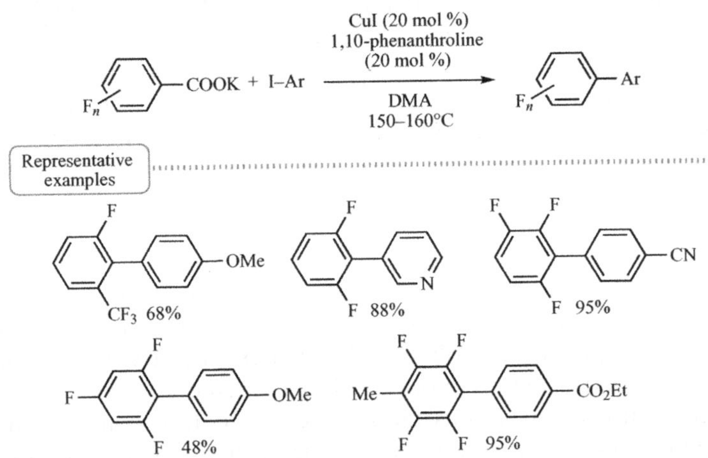

Scheme 10.48 Copper(I)-catalyzed decarboxylative cross coupling of potassium polyfluorobenzoate and aryl iodides.

that has found numerous applications in asymmetric synthesis. This exciting domain is overviewed in the next subchapter.

10.5 BIARYL SYNTHESIS BY OXIDATIVE COUPLING OF 2-NAPHTHOLS

Since they are extraordinarily versatile chiral scaffolds for the design of chiral ligands and auxiliaries in asymmetric synthesis[94] and functional materials with chiral recognition abilities,[95] an enantioselective synthetic route to 1,1′-binaphthyl-2,2′-diol (BINOL) and its derivatives has been actively sought. Among the established synthetic methods, the oxidative C1–C1′ coupling of 2-naphthols and congeners is one of the most efficient.[94b,c,96] Although various transition-metal catalysts have been developed for this method, copper-based mediators are cost-effective and highly reliable and, therefore, have led to practical applications. This section is devoted to a survey of the copper-mediated oxidative coupling of 2-naphthols with particular emphasis on enantioselective processes.

Copper-mediated oxidative transformations have been extensively studied as models for enzymatic oxidations[97] and advanced tools in modern organic synthesis.[98] The oxidation of phenols using copper complexes has been reported to produce various products including catechols, *o*/*p*-quinones, polyethers, and *cis*,*cis*-muconic acid, depending on the ligands, substrate structures, and reaction conditions. Aryl–aryl coupling products such as 4,4′-dihydroxybiphenyls and their oxidation products, diphenoquinones, and Pummerer's ketone from *p*-cresol have been reported in addition to 2,2′-biphenols. Therefore, selective coupling at the *ortho*-carbons of phenols is limited to appropriately designed reactions such as the reaction of 2,4-di(*tert*-butyl)phenol with elaborated binuclear complexes or intramolecular coupling of tethered phenols.[99] Similarly complicated reactions were also observed for the catalytic oxidations of 1- and 2-naphthols.[100] For example, in the presence of copper nitrate (2 mol %) and excess 2,4,6-collidine, 2-naphthol was oxidized in methanol under an oxygen atmosphere for 14 hours to obtain a mixture of various coupling products (Fig. 10.1) including a copper BINOL complex and collidine (yields are not shown).

The first major breakthrough that demonstrated the synthetic utility of such a process was reported by Feringa and Wynberg and involved the stoichiometric oxidation of 2-naphthol with a copper(II)/chiral amine complex.[101a] Upon treatment with 1 equiv of copper(II) nitrate and racemic α-phenethylamine ((±)-**76**, 3 equiv) in methanol at ambient temperature for 20 hours, 2-naphthol was selectively converted to BINOL in 62% yield (Scheme 10.49). This stoichiometric oxidative dimerization is quite sensitive to the substrate structure

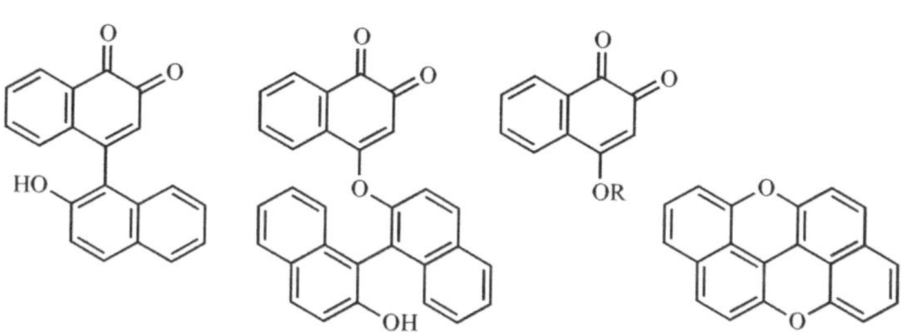

Figure 10.1 Coupling products produced by copper(I)-catalyzed aerobic oxidation of 2-naphthol.

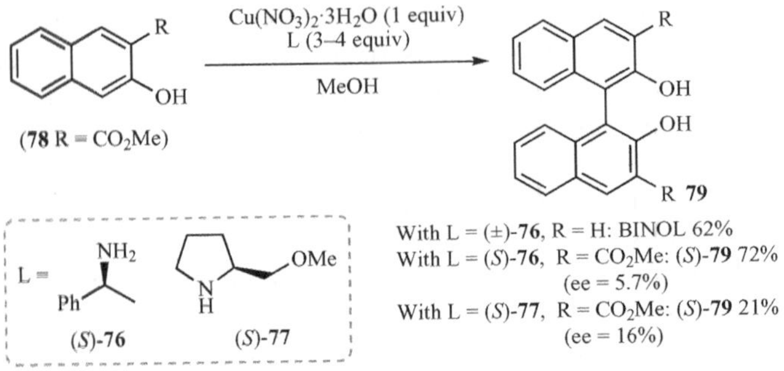

Scheme 10.49 Oxidative homocoupling of 2-naphthol derivatives mediated by copper(II) nitrate and chiral amine ligands.

as evidenced by the fact that 2,7-dihydroxynaphthalene and 3-hydroxyphenanthrene gave significantly lower yields of 20% and 25%, respectively. However, the same reaction was applied to the oxidative dimerization of 7-isopropoxy-4-methoxyphenanthren-2-ol to give the diisopropyl ether of a natural product (−)-blestriarene C in 92% yield.[102]

Naturally, it is expected that the copper(II) salt/chiral amine complex promoter could be applied to an asymmetric version of this oxidative homocoupling using enantiopure amines. Accordingly, Feringa and Wynberg attempted an asymmetric homocoupling and the use of (*S*)-**76** as an enantiopure ligand for the coupling of 2-naphthol resulted in the formation of (*S*)-BINOL with a low optical purity of 2.5% and with a moderate chemical yield of 63%.[101b] Higher, but still unsatisfactory, enantioselectivities were observed for the transformation of methyl 3-hydroxy-2-naphthoate (**78**) into **79** using (*S*)-**76** and (*S*)-2-methoxymethylpyrrolidine ((*S*)-**77**) with copper(II) chloride (Scheme 10.49). It was concluded that neither kinetic resolution nor a second-order asymmetric transformation (see below) is responsible for the observed enantioselectivity since the reaction of racemic BINOL under the same conditions resulted in 90% recovery of the racemic starting material.

The second significant improvement was the discovery of highly enantioselective oxidative homocouplings by Brussee and coworkers who employed (*S*)-(+)-amphetamine ((*S*)-**80**) in place of (*S*)-α-phenethylamine to obtain (*S*)-BINOL in 98% yield with a remarkable 96% enantiomeric excess even though these chiral amines have very similar structures (Scheme 10.50, Eq. 1).[103] Optimization of this reaction revealed that a 2-naphthol:Cu(II):amine ratio of 1:2:8 is optimal and that the enantioselectivity is dependent on the reaction temperature, the enantiomeric excess dramatically increasing from 5% to 96% within the narrow temperature range of 10–20°C. This result implies that the oxidative homocoupling step is not responsible for the high enantioselectivity, which was further confirmed by the asymmetric transformation of racemic BINOL. Indeed, the treatment of racemic BINOL with copper(II) chloride and (*S*)-**80** in methanol at 25°C for 20 hours resulted in the formation of

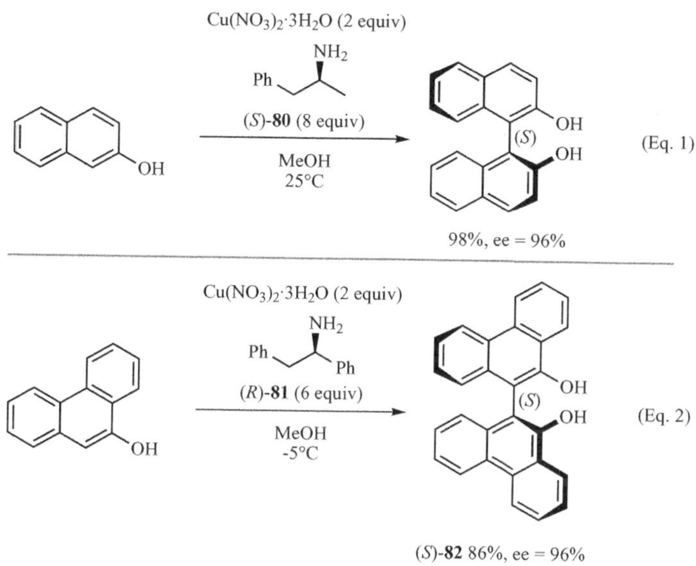

Scheme 10.50 Enantioselective oxidative homocoupling of 2-naphthol derivatives mediated by copper(II) nitrate and chiral amine ligands.

precipitates. After filtration, 87% of (−)-BINOL with 91% ee was obtained from the precipitates, while 13% of (+)-BINOL with 12% ee was recovered from the filtrate. In addition, this deracemization was not observed at 0°C. These results led to the conclusion that a second-order asymmetric transformation involving the copper(II)-mediated atropisomerization of BINOL and kinetic resolution via the selective crystallization of a copper/(−)-BINOL complex accounts for the observed high enantioselectivity. Meanwhile, a highly enantioselective homocoupling of 9-phenanthrol was achieved using a similar protocol with (*R*)-1,2-diphenylamine (**81**) as the chiral ligand (Scheme 10.50, Eq. 2).[104] In striking contrast to Brussee's report, coupling product **82** was obtained with high enantioselectivity even at −5°C. No mechanism for the observed stereoselectivity was given in this report.

Ever since highly enantioselective oxidative couplings were achieved, catalytic versions of these reactions have been sought. An initial attempt revealed that the catalytic aerobic oxidation of 2-naphthol using a copper(II) nitrate/collidine complex produced a complex product mixture.[100] Significant contributions were made by Nakajima and coworkers, who developed a new aerobic oxidation protocol using a copper chloride(hydroxyde)/TMEDA complex, which is readily prepared from copper(I) chloride and TMEDA under an oxygen atmosphere.[105] In the presence of 1 mol % catalyst, 2-naphthol underwent homocoupling in dichloromethane under an oxygen atmosphere to afford BINOL in a high yield (Scheme 10.51, Eq. 1).[105a] The same reaction can be performed under air with a prolonged reaction time, and this method

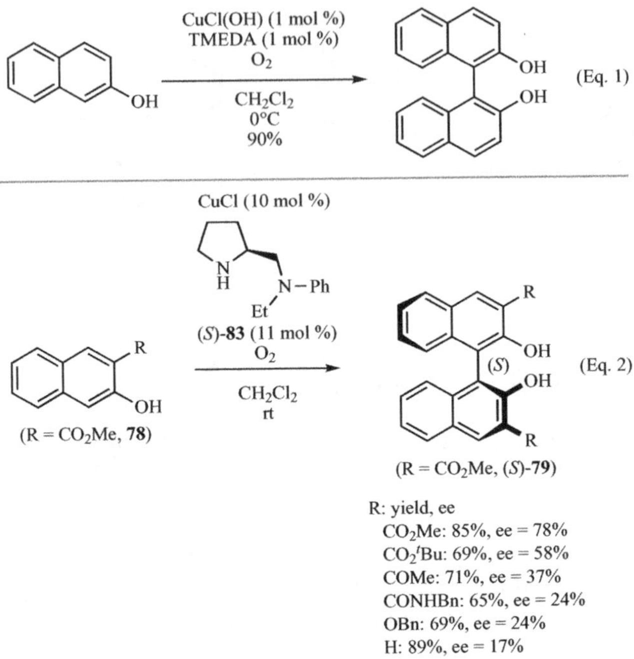

Scheme 10.51 Copper(II)-catalyzed enantioselective oxidative homocoupling of 2-naphthol derivatives using chiral diamine ligand.

proved to be readily scalable and was applied to the synthesis of 50–60 g of racemic BINOL without difficulty.[106a] Moreover, the reaction can be carried out under mild conditions, that is, without solvent at 50°C under air.[105c] In terms of the mechanism of the reaction, a gas-phase study suggested that binuclear copper complexes play an important role in the C–C bond-forming step.[106c]

Although Lipshutz and coworkers showed that 7,7′-linked bis(2-naphthol) with a chiral tether underwent a CuCl(OH)/TMEDA-catalyzed intramolecular oxidative coupling in excellent yields with reasonable diastereoselectivity,[106b] a catalytic enantioselective process was still desired and the use of chiral diamine ligands instead of TMEDA provides a viable route to this goal. Nakajima and coworkers found that (*S*)-*N*-ethyl-*N*-phenyl-2-pyrrolidinemethanamine ((*S*)-**83**) was an optimal chiral diamine ligand and successfully transformed 3-hydroxynaphthoate **78** into (*S*)-**79** in 85% yield with 78% ee (Scheme 10.51, Eq. 2).[105b,d] They also revealed that the ester substituent is indispensable for obtaining a high enantioselectivity. A bulkier *tert*-butyl ester derivative led to decreased yields and enantioselectivity, which indicates that the ester substituents act as coordinating groups. Indeed, the use of the parent 2-naphthol (R = H) as a substrate resulted in a dramatic decrease in enantioselectivity, although the chemical yield remained high (89%). As coordinating groups,

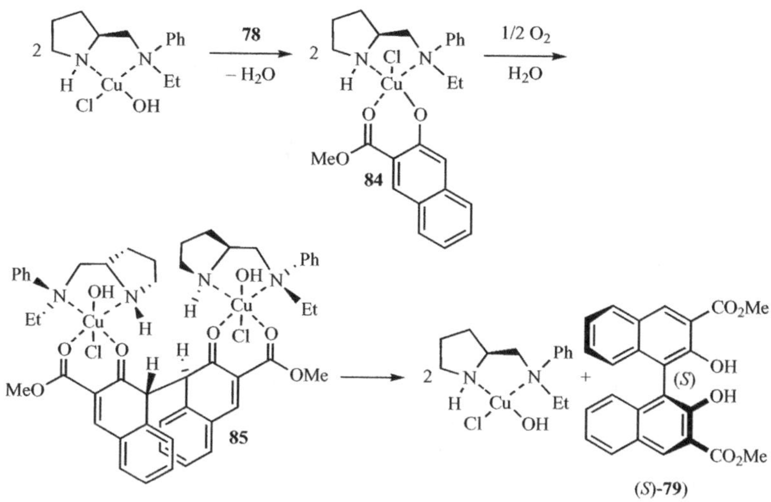

Scheme 10.52 Proposed mechanism for the copper(II)-catalyzed enantioselective oxidative homocoupling of 2-naphthol-3-carboxylate.

acetyl, *N*-benzylamide, and benzyloxy substituents were less effective than esters in terms of both yield and enantioselectivity.

According to these observations, the proposed mechanism would involve the oxidative coupling of two molecules of chiral copper complex **84**, which features a bidentate naphtholate ligand, to yield diketone intermediate **85** (Scheme 10.52). Subsequent tautomerization would convert the central chirality to axial chirality along with the dissociation of the CuCl(OH)/diamine complex to ultimately furnish (*S*)-**79**. A similar central-to-axial chirality transfer was exploited in the recently reported synthesis of axially chiral biphenols. Copper(II) chloride indeed promoted the oxidative coupling of enolates, which were derived from nonaromatizable cyclohexenones bearing a chiral tertiary carbon β to the carbonyl group.[107a] This method enables the highly enantioselective synthesis of bihydroquinones and was applied to the total synthesis of bismurrayaquinone A.[107b]

Since Nakajima's report, the copper-catalyzed asymmetric aerobic oxidative homocoupling has been significantly improved. Kozlowski and coworkers designed an efficient new chiral diamine ligand, 1,5-diaza-*cis*-decalin (**86**), for this asymmetric transformation and they identified the optimal conditions to rely on the use of 10 mol % of copper(I) iodide/(*S*,*S*)-**86** in 1,2-dichloroethane or acetonitrile at 40°C under an atmosphere of oxygen. After stirring for 2 days, naphthoate **78** was transformed into the corresponding dimer (*R*)-**79** in 85% yield with up to 93% ee (Scheme 10.53).[108] In contrast, the analogous reaction with the parent 2-naphthol afforded BINOL with only 16% ee, although the chemical yield was good (80%), which demonstrates once again the importance of the ester moiety as a coordinating group. In addition to

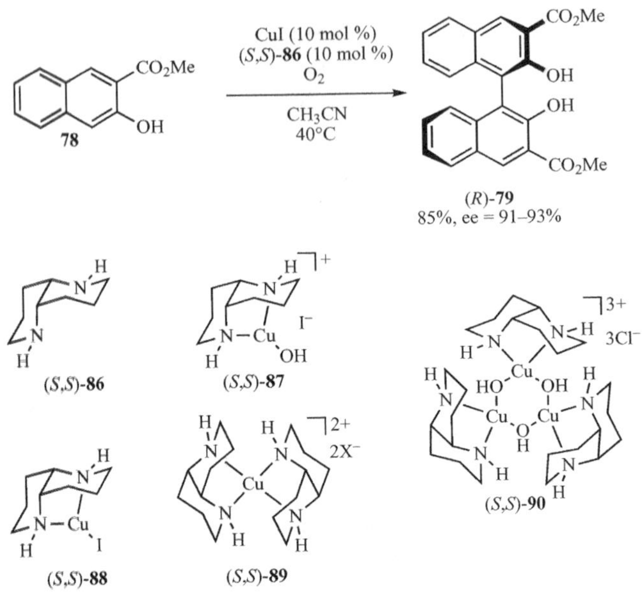

Scheme 10.53 Oxidative homocoupling of 2-naphthol-3-carboxylate using copper(I)/1,5-diaza-*cis*-decalin complexes as catalysts.

esters, secondary amides, ketones, phosphonates, and sulfones were also effective coordinating groups and gave moderate-to-high enantioselectivities. Cationic copper(II) hydroxide **87**, which was derived from neutral complex **88**, is considered to be the catalytically active species, and other complexes including monometallic **89** and trimetallic **90** were also isolated. However, $[(\mathbf{86})_2Cu]Cl_2$ proved to be catalytically inactive and the ligand:copper ratio of 1:1 is therefore critical for catalytic asymmetric oxidation. This highly enantioselective aerobic oxidative homocoupling has been applied to the total synthesis of natural products that contain axial chirality, and representative examples will be described in Chapter 18.[109] The enantioselective synthesis of binaphthyl polymers via tandem Glaser coupling/asymmetric oxidative coupling was also achieved using Kozlowski's catalyst[110] and similarly, a highly enantioselective aerobic oxidative homocoupling of naphthoate **78** leading to diester **79** in 95% yield with 94% ee was achieved using copper(II) chloride and an octahydrobinaphthyl-2,2′-diamine ligand.[111]

Schiff base complexes of copper have also been reported to be efficient catalysts for the aerobic oxidative homocoupling of 2-naphthols.[112,113] In particular, dinuclear cyclic bis(salen) complexes **91** were found to catalyze the asymmetric homocoupling of 2-naphthol to furnish (*S*)-BINOL in 79–80% yields with 69–84% ee (Scheme 10.54).[113] Closely related macrocyclic polyamine complexes **92** were superior to **91** in terms of reproducibility and stability, and gave (*S*)-BINOL with higher yields (84–92%) and enantioselectivity

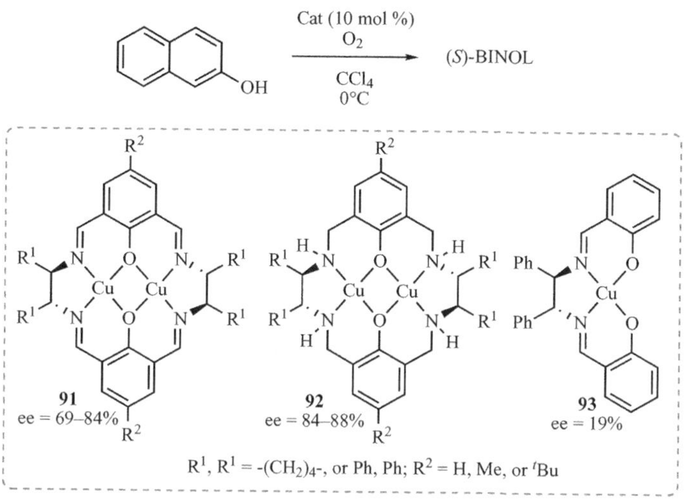

Scheme 10.54 Enantioselective oxidative homocoupling of 2-naphthol using dicopper(I)/Schiff base complexes as catalysts.

(84–88% ee). The high enantioselectivity achieved without an extra coordinating group on the 2-naphthol substrate is ascribed to the rigid dinuclear platform and accordingly, mononuclear Schiff base complex **93** gave a considerably lower enantioselectivity (19% ee).

Besides the homocoupling of naphthol derivatives, the copper-mediated oxidative cross coupling of 2-naphthols has also been investigated since the first report involving the coupling of 9-phenanthrol and 3-methoxymethoxy-2-naphthol.[114] Hovorka and coworkers generalized the cross coupling by investigating the reactivity of combinations of relatively electron-rich 2-naphthols and electron-deficient naphthoate **78** and obtained cross-coupling products with selectivities superior to 80%, while combinations of electron-rich or electron-deficient substrates led to decreased pair selectivity (around 30%) (Scheme 10.55).[115] *tert*-Butylamine was assumed to act as a base to abstract a proton from the 2-naphthols and the authors showed that copper chloride(methoxide) (4 equiv) efficiently promoted the cross-coupling reaction of 2-naphthol with **78** at 50°C in the absence of the amine, the corresponding unsymmetrical product being isolated in 86% yield. The treatment of 2-naphthol substrates with sodium methoxide followed by reaction with copper(II) chloride led to a similar result.[115c] From a mechanistic point of view, it was proposed that a radical derived from the more easily oxidized, electron-rich 2-naphthol attacks the copper naphtholate electrophile, derived from **78** (i.e., via a radical insertion mechanism). Alternatively, it was considered that cross coupling might occur in a concerted fashion via a dicopper complex with two naphtholate bridges. The authors concluded that distinction

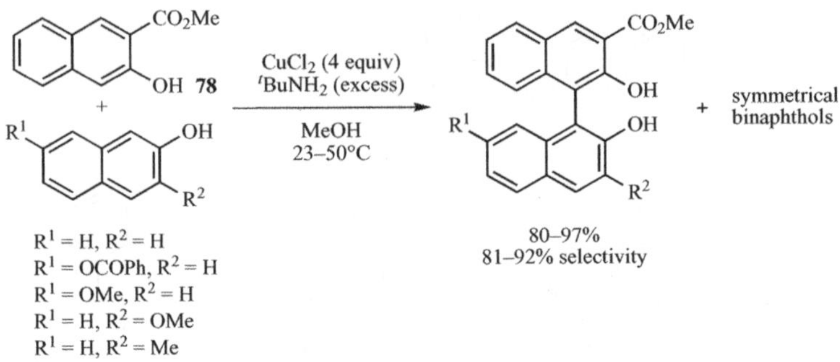

Scheme 10.55 Copper(II)-mediated oxidative cross coupling of 2-naphthol derivatives.

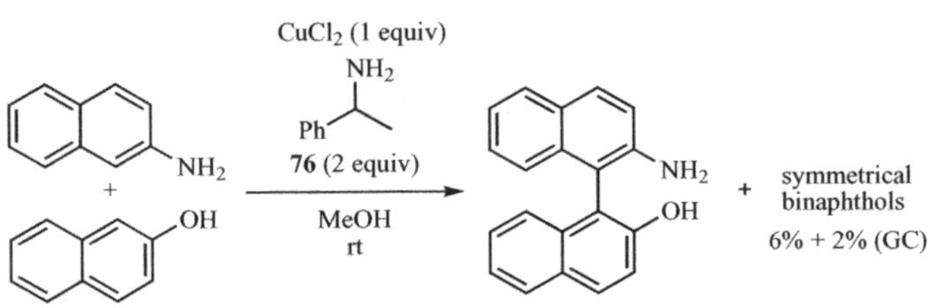

Scheme 10.56 Copper(II)-mediated oxidative cross coupling of 2-naphthol and 2-naphthylamine.

between the homolytic and heterolytic couplings or radical insertion mechanisms is futile in the latter situation.[115d]

The Kočovský group also investigated the copper-mediated couplings of 2-naphthols with 2-naphthylamine.[116] As depicted in Scheme 10.56, the highly selective cross coupling of 2-naphthylamine and 2-naphthol using 1 equiv of copper(II) chloride and α-phenethylamine (**76**) as a promoter in methanol at ambient temperature resulted in the isolation of cross-coupling product **94** in 85% yield. Interestingly, kinetic crystallization was observed albeit with moderate enantiomeric excesses, when (*R*)-(+)-**76** was used as the ligand.

Recently, N-heterocyclic carbene copper complexes were also reported to catalyze the oxidative cross coupling of 2-naphthols with 3-hydroxy-2-naphthoates using Oxone® as the terminal oxidant.[117] Habaue and coworkers introduced a chiral bisoxazoline ligand, (−)-2,2-isopropylidenebis[(4*S*)-4-phenyl-2-oxazoline] ((*S*)-PhBox, **95**), to enable catalytic asymmetric oxidative cross coupling of 2-naphthol derivatives (Scheme 10.57).[118] The aerobic oxidative cross coupling of phenyl 2-hydroxynaphthoate (**96**) and 3-benzyloxy-2-naphthol (**97**) proceeded in THF at 0°C to give (*R*)-**98** in 70% yield and 70% ee. The combination of copper(I) chloride (20 mol %), (*S*)-PhBox (20 mol %) with ytterbium(III) triflate (10 mol %) as a Lewis acid catalyst resulted in

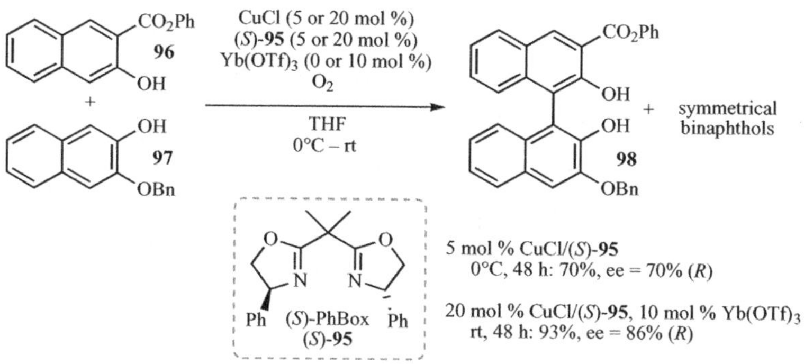

Scheme 10.57 Effect of Lewis acid catalyst on the copper(I)-catalyzed enantioselective oxidative cross coupling of 2-naphthol derivatives.

a higher yield (93%) and enantioselectivity (86% ee).[118d,e] Similarly, a selective cross coupling proceeded for the corresponding amides, which acted as electron-deficient coupling partners, although the enantioselectivity was lower.[118c,e] It was assumed that a radical species was derived from the copper catalyst and the more oxidizable naphthol **97** selectively coupled with electrophilic **96** upon activation by the Lewis acid catalyst.

These developments in the copper-mediated oxidative homo- and cross couplings of naphthol derivatives have led to many applications including the controlled synthesis of macromolecules.[110,119] Indeed, Okamoto and coworkers achieved the asymmetric oxidative coupling polymerization (AOCP) of enantiopure 3,3′-dihydroxy-2,2′-dimethoxy-1,1′-binaphthyl using copper/chiral amine catalyst systems based on the use of (*S*)-(+)-1-(2-pyrrolidinylmethyl) pyrrolidine and (−)-spartein as chiral diamine ligands.[120a] This method provides polymers with molecular weights up to 5.2×10^3 g mol^{-1}, which corresponds to approximately 24 naphthalene units. The same group extended the AOCP to various monomers including 2,3-dihydroxynaphthalene, ter-, and quarternaphthyls, a binaphthyl crown ether, and tetrahydroxybinaphthyl derivatives,[120b–h] and dinuclear copper complexes with tethered diamine ligands were examined for the polymerization of 2,3-dihydroxynaphthalene.[120f]

Bis(oxazoline) ligands proved to be effective for the oxidative coupling polymerization of 2-naphthol-derived monomers[120] and as described earlier, the copper(I) chloride/PhBox catalytic system showed high cross-coupling selectivity.[118] Accordingly, Habaue and coworkers achieved selective oxidative cross-coupling polymerizations using chiral bis(oxazoline) ligands.[121] They investigated a variety of monomers including those bearing electron-rich 2-naphthol (D: donor) and electron-deficient 3-hydroxy-2-naphthoate (A: acceptor) moieties (i.e., D–A type) and those with two electron-rich (D) or electron-deficient (A) units (i.e., D–D and A–A types). The analysis of the resultant polymers revealed that oxidative coupling proceeded in a highly selective manner in favor of cross coupling, although the enantioselectivity

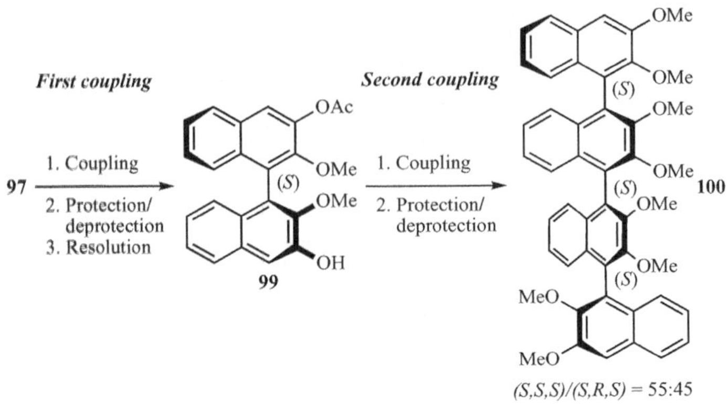

Scheme 10.58 Synthesis of quaternaphthyl derivative via repetitive oxidative homocoupling.

was estimated to be low to moderate from the model coupling reactions. Hyperbranched polymers were also obtained via the copper(I) chloride/PhBox-catalyzed oxidative coupling polymerization of monomers bearing three 2-naphthol units (i.e., D–D–D, D–D–A, and D–A–A types),[121d] and the cross-coupling selectivity was improved using a binary catalyst system composed of copper complexes and Lewis acids.[121e]

Configurationally defined oligonaphthalenes have also been synthesized using an iterative oxidative coupling method.[122] For example, the homocoupling of 3-benzyloxy-2-naphthol (**97**) afforded the homocoupling product from which optically active binaphthyl **99** was obtained as a binaphthyl substrate for the subsequent oxidation and the second homocoupling furnished a diastereomeric mixture of the corresponding quaternaphthyl derivative **100** (Schemes 10.58). Therefore, although separation of diastereomers is required at every step to obtain configurationally defined oligonaphthalenes, a highly diasteroselective homocoupling synthesis of oligonaphthalenes was achieved by exploiting asymmetric transformations provided by the copper(II) chloride/α-phenethylamine system.[122c]

Binaphthyl precursor **101a** ($n = 0$, Scheme 10.59) was subjected to homocoupling with (*S*)-α-phenethylamine ((*S*)-**76**) to afford quaternaphthyl product **102a** ($n = 0$) in 87% yield with 75% diastereomeric excess. Interestingly, the desired product (**102a**) was obtained in 58% yield with 93% diastereomeric excess from precipitates formed during the reaction, which were isolated by filtration, while the remnant filtrate yielded the same product in 15% yield with a much lower diastereomeric excess of 17%. In contrast, the use of (*R*)-**76** led to a lower yield (69%) and decreased diastereoselectivity (26% de). These results clearly imply that the observed diastereoselectivity can be attributed to atropisomerization of the newly formed axis and kinetic crystallization. The diastereoselectivity was improved to 99% for the coupling of quaternaphthyl precursor **101b** ($n = 2$), whereas the coupling of octinaphthyl precursor **101c**

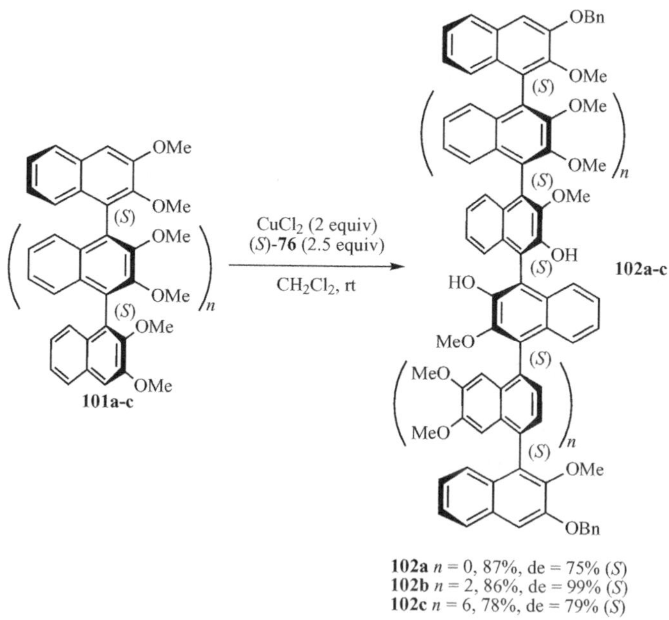

Scheme 10.59 Enantioselective synthesis of oligonaphthyls using chiral copper(II) catalyst.

($n = 6$) gave diastereomeric excess comparable to that of binaphthyl substrate **101a**, the latter coupling affording configurationally defined hexadecanaphthyl **102c** ($n = 6$) in 70% yield after purification via silica gel chromatography. After further modifications involving hydroxy protecting groups and copper catalysts, enantiopure dotriacontanaphthalene was ultimately synthesized via this bottom-up strategy, the absolute configuration of this naphthalene oligomer being unambiguously determined using the CD exciton chirality method.[122g] A wide variety of applications are expected for monodisperse, nanometer-sized molecules; hence, configurationally defined oligonaphthyls are fascinating scaffolds for functional materials.[123]

As shown earlier, the copper-mediated oxidative coupling of 2-naphthols is a highly useful tool for natural product synthesis and materials sciences. Nevertheless, residual metal contamination of the final products can be a serious problem. Accordingly, various copper catalysts supported on alumina,[124a,c] montmorillonite,[124b] mesoporous aluminosilicates,[124d,f] and polymers[124e,g] have been developed.

10.6 CONCLUSIONS AND OUTLOOK

Since the discovery of Ullmann coupling in 1901, various copper-mediated aryl–aryl coupling methods have been devised and continuously evolved into highly sophisticated tools for natural product synthesis and materials sciences. The copper-catalyzed protocols for homo- and cross coupling of organostannanes, organoboranes, or organosilanes enable the selective synthesis of functionalized biaryls under relatively mild conditions. Moreover, the development of copper-catalyzed direct C–H arylations and decarboxylative arylations result in a significant paradigm shift due to the improved accessibility to highly valuable biaryl scaffolds from inexpensive and environmentally benign starting materials and catalysts. Moreover, relatively conventional oxidative coupling of 2-naphthols has been dramatically improved to provide enantioselective access to axially chiral building blocks. Consequently, axially chiral natural products and configurationally defined naphthyl polymers and oligomers became accessible due to advances in copper-catalyzed asymmetric oxidative coupling. The exploitation of supported copper catalysts and efficient catalyst-recycling systems is becoming increasingly important in light of the focus on green chemistry. With further progress in this area, copper catalysts will ultimately play a central role in cross-coupling chemistry and may even supplant palladium catalysts in the near future.

REFERENCES

[1] (a) Loyd-Williams, P.; Giralt, E. *Chem. Soc. Rev.* **2001**, *30*, 145–157. (b) Tasler, S.; Bringmann, G. *Chem. Rec.* **2002**, *2*, 114–127. (c) Bringmann, G.; Gulder, T.; Gulder, T. A. M.; Breuning, M. *Chem. Rev.* **2011**, *111*, 563–639.

[2] (a) Martin, R. E.; Diederich, F. *Angew. Chem. Int. Ed. Engl.* **1999**, *38*, 1350–1377. (b) Pu, L. *Chem. Rev.* **2004**, *104*, 1687–1716. (c) Meier, H. *Angew. Chem. Int. Ed. Engl.* **2005**, *44*, 2482–2506. (d) Frampton, M. J.; Anderson, H. L. *Angew. Chem. Int. Ed. Engl.* **2007**, *46*, 1028–1064. (e) Grimsdale, A. C.; Chan, K. L.; Martin, R. E.; Jokisz, P. G.; Holmes, A. B. *Chem. Rev.* **2009**, *109*, 897–1091.

[3] (a) Rosini, C.; Franzini, L.; Raffaelli, A.; Salvadori, P. *Synthesis* **1992**, 503–517. (b) Che, C.-M.; Huang, J.-S. *Coord. Chem. Rev.* **2003**, *242*, 97–113. (c) Au-Yeung, T. T.-L.; Chan, S.-S.; Chan, A. S. C. *Adv. Synth. Catal.* **2003**, *345*, 537–555. (d) Kočovský, P.; Vyskočil, Š.; Smrčina, M. *Chem. Rev.* **2003**, *103*, 3213–3245. (e) Telfer, S. G.; Kuroda, R. *Coord. Chem. Rev.* **2003**, *242*, 44–46. (f) Au-Yeung, T. T.-L.; Chan, A. S. C. *Coord. Chem. Rev.* **2004**, *248*, 2151–2164. (g) Berthod, M.; Mignani, G.; Woodward, G.; Lemaire, M. *Chem. Rev.* **2005**, *105*, 1801–1836. (h) Shimizu, H.; Nagasaki, I.; Saito, T. *Tetrahedron* **2005**, *61*, 5405–5432. (i) Ding, K.; Li, X.; Ji, B.; Guo, H.; Kitamura, M. *Curr. Org. Synth.* **2005**, *2*, 499–545. (j) Ding, K.; Guo, H.; Li, X.; Yuan, Y.; Wang, Y. *Top. Catal.* **2005**, *35*, 105–116. (k) Li, Y.-M.; Kwong, F.-Y.; Yu, W.-Y.; Chan, A. S. C. *Coord. Chem. Rev.* **2007**, *251*, 2119–2144. (l) Canac, Y.; Chauvin, R. *Eur. J. Inorg. Chem.* **2010**, 2325–2335. (m) Scheker, S.; Zamfir, A.; Freund, M.; Tsogoeva, S. B. *Eur. J. Org. Chem.* **2011**, 2209–2222.

[4] (a) Stanforth, S. P. *Tetrahedron* **1998**, *54*, 263–303. (b) Hassan, J.; Sévignon, M.; Gozzi, C.; Schulz, E.; Lemaire, M. *Chem. Rev.* **2002**, *102*, 1359–1469. (c) de Meijere, A.; Diederich, F. (eds.), *Metal-Catalyzed Cross-Coupling Reactions*, 2nd edition. Wiley-VCH: Weinheim, **2004**.

[5] Ullmann, F.; Bielecki, J. *Ber. Dtsch. Chem. Ges.* **1901**, *34*, 2174–2185.

[6] (a) Beletskaya, I. P.; Cheprakov, A. V. *Coord. Chem. Rev.* **2004**, *248*, 2337–2364. (b) Evano, G.; Blanchard, N.; Toumi, M. *Chem. Rev.* **2008**, *108*, 3054–3131.

[7] (a) Finet, J.-P.; Fedorov, A. Y.; Combes, S.; Boyer, G. *Curr. Org. Chem.* **2002**, *6*, 597–626. (b) Ley, S. V.; Thomas, A. W. *Angew. Chem. Int. Ed. Engl.* **2003**, *42*, 5400–5449. (c) Kunz, K.; Scholz, U.; Ganzer, D. *Synlett* **2003**, 2428–2439. (d) Monnier, F.; Taillefer, M. *Angew. Chem. Int. Ed. Engl.* **2009**, *48*, 6954–6971. (e) Qiao, J. X.; Lam, P. Y. S. *Synthesis* **2011**, 829–856.

[8] (a) Fanta, P. E. *Chem. Rev.* **1946**, *38*, 139–196. (b) Fanta, P. E. *Chem. Rev.* **1964**, *64*, 613–632. (c) Goshaev, M.; Otroshchenko, O. S.; Sadykov, A. S. *Russ. Chem. Rev.* **1972**, *41*, 1046–1059. (d) Nelson, T.; Crouch, R. D. *Org. React.* **2004**, *63*, 265–555.

[9] (a) Gooßen, L. J.; Thiel, W. R.; Rodríguez, N.; Linder, C.; Melzer, B. *Adv. Synth. Catal.* **2007**, *349*, 2241–2246. (b) Gooßen, L. J.; Manjolinho, F.; Khan, B. A.; Rodríguez, N. *J. Org. Chem.* **2009**, *74*, 2620–2623, and cited references.

[10] (a) Rausch, M. D. *J. Org. Chem.* **1961**, *26*, 1802–1805. (b) Lindley, J.; Mason, T. J.; Lorimer, J. P. *Ultrasonics* **1987**, *25*, 45–48.

[11] (a) Gore, P. H.; Hughes, G. K. *J. Chem. Soc.* **1959**, 1615–1616. Also, see: (b) Rieke, R. D.; Rhyne, L. D. *J. Org. Chem.* **1979**, *44*, 3445–3446. (c) Ebert, G. W.; Rieke, R. D. *J. Org. Chem.* **1984**, *49*, 5280–5282.

[12] Roberge, D. M.; Hölderich, W. F. *Appl. Catal. A: Gen.* **2000**, *194–195*, 341–357.

[13] (a) Cohen, T.; Tirpak, J. G. *Tetrahedron Lett.* **1975**, *16*, 143–146. (b) Cohen, T.; Cristea, I. *J. Org. Chem.* **1975**, *40*, 3649–3651. (c) Cohen, T.; Cristea, I. *J. Am. Chem. Soc.* **1976**, *98*, 748–753.

[14] (a) Zhang, S.; Zhang, D.; Liebeskind, L. S. *J. Org. Chem.* **1997**, *62*, 2312–2313. (b) Babudri, F.; Cardone, A.; Farinola, G. M.; Naso, F. *Tetrahedron* **1998**, *54*, 14609–14616. (c) Stavrakov, G.; Keller, M.; Breit, B. *Eur. J. Org. Chem.* **2007**, 5726–5733.

[15] (a) Dhas, N. A.; Raj, C. P.; Gedanken, A. *Chem. Mater.* **1998**, *10*, 1446–1452. (b) Ponce, A. A.; Klanbunde, K. J. *J. Mol. Catal. A: Chem.* **2005**, *225*, 1–6.

[16] Wu, Q.; Wang, L. *Synthesis* **2008**, 2007–2011.

[17] (a) Bringmann, G.; Mortimer, A. J. P.; Keller, P. A.; Gresser, M. J.; Garner, J.; Breuning, M. *Angew. Chem. Int. Ed. Engl.* **2005**, *44*, 5384–5427. (b) Wallace, T. W. *Org. Biomol. Chem.* **2006**, *4*, 3197–3210. (c) Kozlowski, M. C.; Morgan, B. J.; Linton, E. C. *Chem. Soc. Rev.* **2009**, *38*, 3193–3207.

[18] (a) Schmid, R.; Foricher, J.; Cereghetti, M.; Schönholzer, P. *Helv. Chim. Acta* **1991**, *74*, 370–388. (b) Cereghetti, M.; Schmid, R.; Schönholzer, P.; Rageot, A. *Tetrahedron Lett.* **1996**, *37*, 5343–5346. (c) Gelpke, A. E. S.; Fraanje, J.; Goubitz, K.; Schenk, H.; Hiemstra, H. *Tetrahedron* **1997**, *53*, 5899–5908. (d) Pai, C.-C.; Lin, C.-W.; Lin, C.-C.; Chen, C.-C.; Chan, A. S. C. *J. Am. Chem. Soc.* **2000**, *122*, 11513–11514. (e) Pai, C.-C.; Li, Y.-M.; Zhou, Z.-Y.; Chan, A. S. C. *Tetrahedron Lett.* **2002**, *43*, 2789–2792. Also, see: (f) Wei, H.; Zhang, Y. J.; Wang, F.; Zhang, W. *Tetrahedron Asymmetry* **2008**, *19*, 482–488. (g) Wei, H.; Zhang, Y. J.; Dai, Y.; Zhang, J.; Zhang, W. *Tetrahedron Lett.* **2008**, *49*, 4106–4109. (h) Zhang, Y. J.; Wei, H.; Zhang, W. *Tetrahedron* **2009**, *65*, 1281–1286.

[19] (a) Miyano, S.; Tobita, M.; Suzuki, S.; Nishikawa, Y.; Hashimoto, H. *Chem. Lett.* **1980**, 1027–1030. (b) Qiu, L.; Qi, J.; Pai, C.-C.; Chan, S.; Zhou, Z.; Choi, M. C. K.;

Chan, A. S. C. *Org. Lett.* **2002**, *4*, 4599–4602. (c) Gorobets, E.; Wheatley, B. M. M.; Hopkins, J. M.; McDonald, R.; Keay, B. A. *Tetrahedron Lett.* **2005**, *46*, 3843–3846. (d) Rankie, D. A.; Hopkins, J. M.; Parvez, M.; Keay, B. A. *Synlett* **2009**, 2513–2517.

[20] (a) Nelson, T. D.; Meyers, A. I. *Tetrahedron Lett.* **1993**, *34*, 3061–3062. (b) Nelson, T. D.; Meyers, A. I. *J. Org. Chem.* **1994**, *59*, 2577–2580. (c) Nelson, T. D.; Meyers, A. I. *J. Org. Chem.* **1994**, *59*, 2655–2658. (d) Nelson, T. D.; Meyers, A. I. *Tetrahedron Lett.* **1994**, *35*, 3259–3262. (e) Meyers, A. I.; McKennon, M. J. *Tetrahedron Lett.* **1995**, *36*, 5869–5872. (f) Meyers, A. I.; Willemsen, J. J. *Tetrahedron Lett.* **1996**, *37*, 791–792. (g) Meyers, A. I.; Willemsen, J. J. *Chem. Commun.* **1997**, 1573–1574. (h) Meyers, A. I.; Price, A. *J. Org. Chem.* **1998**, *63*, 412–413. (i) Meyers, A. I.; Willemsen, J. J. *Tetrahedron* **1998**, *54*, 10493–10511. (j) Degnan, A. P.; Meyers, A. I. *J. Am. Chem. Soc.* **1999**, *121*, 2762–2769. (k) Meyers, A. I.; Nelson, T. D.; Moorlag, H.; Rawson, D. J.; Meier, A. *Tetrahedron* **2004**, *60*, 4459–4473. Also, see: (l) Solladié, G.; Hugelé, P.; Bartsch, R.; Skoulios, A. *Angew. Chem. Int. Ed. Engl.* **1996**, *35*, 1533–1535. (m) Solladié, G.; Hugelé, P.; Bartsch, R. *J. Org. Chem.* **1998**, *63*, 3895–3898. (n) Li, Y.; Wang, Q.; Dong, L.; Guo, X.; Wang, W.; Xie, J.; Chang, J. *Synthesis* **2009**, 3383–3390.

[21] (a) Miyano, S.; Tobita, M.; Nawa, M.; Sano, S.; Hashimoto, H. *J. Chem. Soc., Chem. Commun.* **1980**, 1233–1234. (b) Miyano, S.; Tobita, M.; Hashimoto, H. *Bull. Chem. Soc. Jpn* **1981**, *54*, 3522–3526. (c) Miyano, S.; Handa, S.; Shimizu, K.; Tagami, K.; Hashimoto, H. *Bull. Chem. Soc. Jpn* **1984**, *57*, 1943–1947. (d) Miyano, S.; Fukushima, H.; Handa, S.; Ito, H.; Hashimoto, H. *Bull. Chem. Soc. Jpn* **1988**, *61*, 3249–3254.

[22] (a) Qiu, L.; Wu, J.; Chan, S.; Au-Yeung, T. T.-L.; Ji, J. X.; Guo, R.; Pai, C.-C.; Zhou, Z.; Li, X.; Fan, Q.-H.; Chan, A. S. C. *Proc. Natl Acad. Sci. U. S. A.* **2004**, *101*, 5815–5820. (b) Qiu, L.; Kwong, F. Y.; Wu, J.; Lam, W. H.; Chan, S.; Yu, W.-Y.; Li, Y.-M.; Guo, R.; Zhou, Z.; Chan, A. S. C. *J. Am. Chem. Soc.* **2006**, *128*, 5955–5965. Also, see: (c) Gorobets, E.; McDonald, R.; Keay, B. A. *Org. Lett.* **2006**, *8*, 1483–1485. (d) Wang, C.-J.; Xu, Z.-P.; Wang, X.; Teng, H.-L. *Tetrahedron* **2010**, *66*, 3702–3706.

[23] (a) Dai, D.; Martin, O. R. *J. Org. Chem.* **1998**, *63*, 7628–7633. (b) Ikeda, Y.; Nagao, K.; Tanigakiuchi, K.; Tokumaru, G.; Tsuchiya, H.; Yamada, H. *Tetrahedron Lett.* **2004**, *45*, 487–489.

[24] Forrest, J. *J. Chem. Soc.* **1960**, 594–601.

[25] (a) Brown, E.; Robin, J.-P. *Tetrahedron Lett.* **1977**, *18*, 2015–2018. (b) Brown, E.; Dhal, R.; Robin, J.-P. *Tetrahedron Lett.* **1979**, *20*, 733–736. (c) Robin, J.-P.; Gringore, O.; Brown, E. *Tetrahedron Lett.* **1980**, *21*, 2709–2712. (d) Brown, E.; Robin, J.-P.; Dhal, R. *Tetrahedron* **1982**, *38*, 2569–2579. (e) Dhal, R.; Brown, E.; Robin, J.-P. *Tetrahedron* **1983**, *39*, 2787–2794.

[26] Suzuki, H.; Enya, T.; Hisamatsu, Y. *Synthesis* **1997**, 1273–1275.

[27] (a) Nilsson, M. *Tetrahedron Lett.* **1966**, 679–682. (b) Cairncross, A.; Sheppard, W. A. *J. Am. Chem. Soc.* **1968**, *90*, 2186–2187. (c) Nilsson, M.; Wennerström, O. *Tetrahedron Lett.* **1968**, 3307–3310. (d) DePasquale, R. J.; Tamborski, C. *J. Org. Chem.* **1969**, *34*, 1736–1740. (e) Sheppard, W. A. *J. Am. Chem. Soc.* **1970**, *92*, 5419–5422. (f) Nilsson, M.; Wennerström, O. *Acta Chem. Scand.* **1970**, *24*, 482–488. (g) Jukes, A. E.; Dua, S. S.; Gilman, H. *J. Organomet. Chem.* **1970**, *24*, 791–796.

(h) Nilsson, M.; Ullenius, C. *Acta Chem. Scand.* **1970**, *24*, 2379–2388. (i) Malmberg, H.; Nilsson, M. *Tetrahedron* **1986**, *42*, 3981–3986. Also, see: Björklund, C.; Nilsson, M.; Wennerström, O. *Acta Chem. Scand.* **1970**, *24*, 3599–3606.

[28] (a) Ziegler, F. E.; Fowler, K. W.; Kanfer, S. *J. Am. Chem. Soc.* **1976**, *98*, 8282–8283. (b) Ziegler, F. E.; Fowler, K. W.; Sinha, N. D. *Tetrahedron Lett.* **1978**, *19*, 2767–2770. (c) Ziegler, F. E.; Chliwner, I.; Fowler, K. W.; Kanfer, S. J.; Kuo, S. J.; Sinha, N. D. *J. Am. Chem. Soc.* **1980**, *102*, 790–798.

[29] (a) Kende, A. S.; Curran, D. P. *J. Am. Chem. Soc.* **1979**, *101*, 1857–1864. (b) Harrowven, D. C.; Lai, D.; Lucas, M. C. *Synthesis* **1999**, 1300–1302. (c) Stark, L. M.; Lin, X.-F.; Flippin, L. A. *J. Org. Chem.* **2000**, *65*, 3227–3230. (d) Büttner, F.; Bergmann, S.; Guénard, D.; Gust, R.; Seitz, G.; Thoret, S. *Bioorg. Med. Chem.* **2005**, *13*, 3497–3511. (e) Broady, S. D.; Golden, M. D.; Leonard, J.; Muir, J. C.; Maudet, M. *Tetrahedron Lett.* **2007**, *48*, 4627–4630.

[30] (a) Kauffmann, T. *Angew. Chem. Int. Ed. Engl.* **1974**, *13*, 291–356. (b) Surry, D. S.; Spring, D. R. *Chem. Soc. Rev.* **2006**, *35*, 218–225. (c) Iyoda, M. *Adv. Synth. Catal.* **2009**, *351*, 984–998. (d) Aves, S. J.; Spring, D. R. *The Chemistry of Organocopper Compounds*, Rappoport, Z.; Marek, I. (eds.). John Wiley & Sons: Chichester, **2009**, Chapter 12. pp. 585–602.

[31] (a) Lipshutz, B. H.; Siegmann, K.; Garcia, E. *J. Am. Chem. Soc.* **1991**, *113*, 8161–8162. (b) Lipshutz, B. H.; Siegmann, K.; Garcia, E. *Tetrahedron* **1992**, *48*, 2579–2588. (c) Lipshutz, B. H.; Siegmann, K.; Garcia, E.; Kayser, F. *J. Am. Chem. Soc.* **1993**, *115*, 9276–9282. (d) Lipshutz, B. H.; Kayser, F.; Maullin, N. *Tetrahedron Lett.* **1994**, *35*, 815–818.

[32] (a) Lipshutz, B. H.; Liu, Z.-P.; Kayser, F. *Tetrahedron Lett.* **1994**, *31*, 5567–5570. (b) Lipshutz, B. H.; Kayser, F.; Liu, Z.-P. *Angew. Chem. Int. Ed. Engl.* **1994**, *33*, 1842–1844. Also, see: (c) Lin, G.-Q.; Zhong, M. *Tetrahedron Lett.* **1997**, *38*, 1087–1090. (d) Sugimura, T.; Yamada, H.; Inoue, S.; Tai, A. *Tetrahedron Asymmetry* **1997**, *8*, 649–655. (e) Lin, G.-Q.; Zhong, M. *Tetrahedron Asymmetry* **1997**, *8*, 1369–1372. (f) Spring, D. R.; Krishnan, S.; Schreiber, S. L. *J. Am. Chem. Soc.* **2000**, *122*, 5656–5657. (g) Spring, D. R.; Krishnan, S.; Blackwell, H. E.; Schreiber, S. L. *J. Am. Chem. Soc.* **2002**, *124*, 1354–1363. (h) Michaud, G.; Bulliard, M.; Ricard, L.; Genêt, J.-P.; Marinetti, A. *Chem. Eur. J.* **2002**, *8*, 3327–3330. (i) Coleman, R. S.; Gurrala, S. R. *Org. Lett.* **2005**, *7*, 1849–1852.

[33] (a) Miyake, Y.; Wu, M.; Rhaman, M. J.; Iyoda, M. *Chem. Commun.* **2005**, 411–413. (b) Miyake, Y.; Wu, M.; Rhaman, M. J.; Kuwatani, Y.; Iyoda, M. *J. Org. Chem.* **2006**, *71*, 6110–6117. Also, see: (c) Kabir, S. M. H.; Iyoda, M. *Chem. Commun.* **2000**, 2329–2330.

[34] (a) Surry, D. S.; Su, X.; Fox, D. J.; Franckevicius, V.; Macdonald, S. J. F.; Spring, D. R. *Angew. Chem. Int. Ed. Engl.* **2005**, *44*, 1870–1873. Also, see: (b) Surry, D. S.; Fox, D. J.; Macdonald, S. J. F.; Spring, D. R. *Chem. Commun.* **2005**, 2589–2590.

[35] (a) Su, X.; Fox, D. J.; Blackwee, D. T.; Tanaka, K.; Spring, D. R. *Chem. Commun.* **2006**, 3883–3885. The Cu(II)-mediated homocoupling of arylzinc reagents was also reported; see: (b) Iyoda, M.; Kabir, S. M. H.; Vorasingha, A.; Kuwatani, Y.; Yoshida, M. *Tetrahedron Lett.* **1998**, *39*, 5393–5396. (c) Kabir, S. M. H.; Miura, M.; Sasaki, S.; Harada, G.; Kuwatani, Y.; Yoshida, M.; Iyoda, M. *Heterocycles* **2000**, *52*, 761–774. (d) Kabir, S. M. H.; Hasegawa, M.; Kuwatani, Y.; Yoshida, M.; Matsuyama, H.; Iyoda, M. *J. Chem. Soc. Perkin Trans. 1* **2001**, 159–165.

[36] (a) Su, X.; Surry, D. S.; Spandl, R. J.; Spring, D. R. *Org. Lett.* **2008**, *10*, 2593–2596. (b) Su, X.; Thomas, G. L.; Galloway, W. R. J. D.; Surry, D. S.; Spandl, R. J.; Spring, D. R. *Synthesis* **2009**, 3880–3896. (c) Zheng, S.; Laraia, L.; O'Connor, C. J.; Sorrell, D.; Tan, Y. S.; Xu, Z.; Venkitaraman, A. R.; Wu, W.; Spring, D. R. *Org. Biomol. Chem.* **2012**, *10*, 2590–2593.

[37] (a) Pschorr, R. *Ber. Dtsch. Chem. Ges.* **1896**, *29*, 496–501. (b) Duclos, R. I., Jr.; Tung, J. S.; Rapoport, H. *J. Org. Chem.* **1984**, *49*, 5243–5246. (c) Bonfand, E.; Forslund, L.; Motherwell, W. B.; Vázquez, S. *Synlett* **2000**, 475–478. For reviews, see: (d) Leake, P. H. *Chem. Rev.* **1956**, *56*, 27–48. (e) Laali, K. K.; Shokouhimehr, M. *Curr. Org. Synth.* **2009**, *6*, 193–202.

[38] (a) Gomberg, M.; Bachmann, W. E. *J. Am. Chem. Soc.* **1924**, *46*, 2339–2343. (b) Beadle, J. R.; Korzeniowski, S. H.; Rosenberg, D. E.; Garcia-Slanga, B. J.; Gokel, G. W. *J. Org. Chem.* **1984**, *49*, 1594–1603. (d) Wetzel, A.; Pratsch, G.; Kolb, R.; Heinrich, M. R. *Chem. Eur. J.* **2010**, *16*, 2547–2556.

[39] (a) Vorländer, D.; Meyer, F. *Ann.* **1902**, *320*, 122–144. (b) Atkinson, E. R.; Lawler, H. J.; Heath, J. C.; Kimball, E. H.; Read, E. R. *J. Am. Chem. Soc.* **1941**, *63*, 730–733. (c) Schmid, R.; Cereghetti, M.; Heiser, B.; Schönholzer, P.; Hansen, H.-J. *Helv. Chim. Acta* **1988**, *71*, 897–929. (d) Denmark, S. E.; Matsuhashi, H. *J. Org. Chem.* **2002**, *67*, 3479–3486. (e) Montoya-Pelaez, P. J.; Uh, Y.-S.; Lata, C.; Thompson, M. P.; Lemieux, R. P.; Crudden, C. M. *J. Org. Chem.* **2006**, *71*, 5921–5929. (f) Cepanec, I.; Litvić, M.; Udiković, J.; Pogorelić, I.; Lovrić, M. *Tetrahedron* **2007**, *63*, 5614–5621.

[40] (a) Bolton, R.; Williams, G. H. *Chem. Soc. Rev.* **1986**, *15*, 261–289. (b) Galli, C. *Chem. Rev.* **1988**, *88*, 765–792.

[41] (a) Cohen, T.; Lewarchik, R. J.; Tarino, J. Z. *J. Am. Chem. Soc.* **1974**, *96*, 7753–7760. (b) Hanson, P.; Taylor, A. B.; Walton, P. H.; Timms, A. W. *Org. Biomol. Chem.* **2007**, *5*, 679–698, and references cited.

[42] (a) Stille, J. K. *Angew. Chem. Int. Ed. Engl.* **1986**, *25*, 508–524. (b) Farina, V.; Krishnamurthy, V.; Scott, W. J. *Org. React.* **1997**, *50*, 1–652.

[43] (a) Ghosal, S.; Luke, G. P.; Kyler, K. S. *J. Org. Chem.* **1987**, *52*, 4296–4298. (b) Beddoes, R. L.; Cheeseright, T.; Wang, J.; Quayle, P. *Tetrahedron Lett.* **1995**, *36*, 283–286. Also, see: (c) Christoffers, J.; Dötz, K. H. *Chem. Ber.* **1995**, *128*, 641–643. For the homocoupling of vinylstannanes, see: (d) Piers, E.; McEachem, E. J.; Romero, M. A. *Tetrahedron Lett.* **1996**, *37*, 1173–1176. (e) Piers, E.; McEachern, E. J.; Romero, M. A.; Gladstone, P. L. *Can. J. Chem.* **1997**, *75*, 694–701.

[44] (a) Iyoda, M.; Kondo, T.; Nakao, K.; Hara, K.; Kuwatani, Y.; Yoshida, M.; Matsuyama, H. *Org. Lett.* **2000**, *2*, 2081–2083. (b) Iyoda, M.; Nakao, K.; Kondo, T.; Kuwatani, Y.; Yoshida, M.; Matsuyama, H.; Fukami, K.; Nagase, S. *Tetrahedron Lett.* **2001**, *42*, 6869–6872.

[45] Harada, G.; Yoshida, M.; Iyoda, M. *Chem. Lett.* **2000**, 160–161.

[46] Kang, S.-K.; Baik, T.-G.; Jiao, X. H.; Lee, Y.-T. *Tetrahedron Lett.* **1999**, *40*, 2383–2384.

[47] (a) Piers, E.; Yee, J. G. K.; Gladstone, P. L. *Org. Lett.* **2000**, *2*, 481–484. (b) Edwards, D. J.; Hadfield, J. A.; Wallace, T. W.; Ducki, S. *Org. Biomol. Chem.* **2011**, *9*, 219–231. For the intramolecular homocoupling of vinylstannanes, see: (c) Piers, E.; Romero, M. A. *J. Am. Chem. Soc.* **1996**, *118*, 1215–1216.

[48] (a) Cohen, H.; Meyerstein, D. *Inorg. Chem.* **1986**, *25*, 1505–1506. (b) Navon, N.; Golub, G.; Cohen, H.; Meyerstein, D. *Organometallics* **1995**, *14*, 5670–5676. (c) Goj, L. A.; Blue, E. D.; Delp, S. A.; Gunnoe, T. B.; Cundari, T. R.; Petersen, J. L. *Organometallics* **2006**, *25*, 4097–4104.

[49] (a) Piers, E.; Wong, T. *J. Org. Chem.* **1993**, *58*, 3609–3610. (b) Durr, R.; Dossu, S.; Lucchini, V.; De Lucchi, O. *Angew. Chem. Int. Ed. Engl.* **1997**, *36*, 2805–2807.

[50] (a) Allred, G. D.; Liebeskind, L. S. *J. Am. Chem. Soc.* **1996**, *118*, 2748–2749. (b) Paterson, I.; Man, J. *Tetrahedron Lett.* **1997**, *38*, 695–698. (c) Paterson, I.; Lombart, H.-G.; Allerton, C. *Org. Lett.* **1999**, *1*, 19–22. (d) Paterson, I.; Doughty, V. A.; McLeod, M. D.; Trieselmann, T. *Angew. Chem. Int. Ed. Engl.* **2000**, *39*, 1308–1312. (e) Borsato, G.; De Lucchi, O.; Fabris, F.; Groppo, L.; Lucchini, V.; Zambon, A. *J. Org. Chem.* **2002**, *67*, 7894–7897. (f) Wehlan, H.; Dauber, M.; Femaud, M.-T. M.; Schuppan, J.; Mahrwald, R.; Ziemer, B.; Juarez, M.-E.; Koert, U. *Angew. Chem. Int. Ed. Engl.* **2004**, *43*, 4597–4601. (g) Paterson, I.; Britton, R.; Delgado, O.; Meyer, A.; Poullennec, K. G. *Angew. Chem. Int. Ed. Engl.* **2004**, *43*, 4629–4633. (h) Durham, T. B.; Blanchard, N.; Savall, B. M.; Powell, N. A.; Roush, W. R. *J. Am. Chem. Soc.* **2004**, *126*, 9307–9317. (i) Falck, J. R.; Patel, P. K.; Bandyopadhyay, A. *J. Am. Chem. Soc.* **2007**, *129*, 790–793. (j) Paterson, I.; Gardner, N. M.; Poullennec, K. G.; Wright, A. E. *Bioorg. Med. Chem. Lett.* **2007**, *17*, 2443–2447. (k) König, C. M.; Gebhardt, B.; Schleth, C.; Dauber, M.; Koert, U. *Org. Lett.* **2009**, *11*, 2728–2731.

[51] (a) Kang, S.-K.; Yamaguchi, T.; Kim, T.-H.; Ho, P.-S. *J. Org. Chem.* **1996**, *61*, 9082–9083. (b) Kang, S.-K.; Kim, J.-S.; Choi, S.-C. *J. Org. Chem.* **1997**, *62*, 4208–4209. (c) Kang, S.-K.; Lee, S.-W.; Ryu, H.-C. *Chem. Commun.* **1999**, 2117–2118. For copper-catalyzed MKS-type couplings other than biaryl synthesis, see: (d) Falck, J. R.; Bhatt, R. K.; Ye, J. *J. Am. Chem. Soc.* **1995**, *117*, 5973–5982. (e) Mohapatra, S.; Bandyopadhyay, A.; Barma, D. K.; Capdevila, J. H.; Falck, J. R. *Org. Lett.* **2003**, *5*, 4759–4762. (f) Wang, Y.; Burton, D. J. *Org. Lett.* **2006**, *8*, 1109–1111.

[52] Li, J.-H.; Tang, B.-X.; Tao, L.-M.; Xie, Y.-X.; Liang, Y.; Zhang, M.-B. *J. Org. Chem.* **2006**, *71*, 7488–7490.

[53] (a) Miyaura, N.; Suzuki, A. *Chem. Rev.* **1995**, *95*, 2457–2483. (b) Chemler, S. R.; Trauner, D.; Danishefsky, S. J. *Angew. Chem. Int. Ed. Engl.* **2001**, *40*, 4544–4568. (c) Kotha, S.; Lahiri, K.; Kashinath, D. *Tetrahedron* **2002**, *58*, 9633–9695. (d) Bellina, F.; Carpita, A.; Rossi, R. *Synthesis* **2004**, 2419–2440. (e) Doucet, A. *Eur. J. Org. Chem.* **2008**, 2013–2030. (f) Alonso, F.; Beletskaya, I. P.; Yus, M. *Tetrahedron* **2008**, *64*, 3047–3101. (g) Suzuki, A. *Angew. Chem. Int. Ed. Engl.* **2011**, *50*, 6723–6737.

[54] (a) Ththagar, M. B.; Beckers, J.; Rothenberg, G. *J. Am. Chem. Soc.* **2002**, *124*, 11858–11859. (b) Ththagar, M. B.; Beckers, J.; Rothenberg, G. *Adv. Synth. Catal.* **2003**, *345*, 979–985.

[55] (a) Li, J.-H.; Wang, D.-P. *Eur. J. Org. Chem.* **2006**, 2063–2066. (b) Li, J.-H.; Li, J.-L.; Xie, Y.-X. *Synthesis* **2007**, 984–988. (c) Li, J.-H.; Li, J.-L.; Wang, D.-P.; Pi, S.-F.; Xie, Y.-X.; Zhang, M.-B.; Hu, X.-C. *J. Org. Chem.* **2007**, *72*, 2053–2057.

[56] Mao, J.; Guo, J.; Fang, F.; Ji, S.-J. *Tetrahedron* **2008**, *64*, 3905–3911.

[57] Ye, Y.-M.; Wang, B.-B.; Ma, D.; Shao, L.-X.; Lu, J.-M. *Catal. Lett.* **2010**, *139*, 141–144.

[58] (a) Demir, A. S.; Reis, Ö.; Emrullahoglu, M. *J. Org. Chem.* **2003**, *68*, 10130–10134. (b) Kaboudin, B.; Haruki, T.; Yokomatsu, T. *Synthesis* **2011**, 91–96.

[59] Kirai, N.; Yamamoto, Y. *Eur. J. Org. Chem.* **2009**, 1864–1867.

[60] (a) Cheng, G.; Luo, M. *Eur. J. Org. Chem.* **2011**, 2519–2523. (b) Kaboudin, B.; Abedi, Y.; Yokomatsu, T. *Eur. J. Org. Chem.* **2011**, 6656–6662.

[61] (a) Hatanaka, Y.; Hiyama, T. *Synlett* **1991**, 845–853. (b) Hiyama, T.; Shirakawa, E. *Top. Curr. Chem.* **2002**, *219*, 61–85. (c) Handy, C. J.; Manaso, A. S.; McElroy, W. T.; Seganish, W. M.; DeShong, P. *Tetrahedron* **2005**, *61*, 12201–1225.

[62] (a) Denmark, S. E.; Sweis, R. F. *Chem. Pharm. Bull.* **2002**, *50*, 1531–1541. (b) Denmark, S. E.; Ober, M. H. *Aldrichim. Acta* **2003**, *36*, 75–85. (c) Denmark, S. E.; Liu, J. H.-C. *Angew. Chem. Int. Ed. Engl.* **2010**, *49*, 2978–2986.

[63] (a) Nakao, Y.; Sahoo, A. K.; Imanaka, H.; Yada, A.; Hiyama, T. *Pure Appl. Chem.* **2006**, *78*, 435–440. (b) Chen, J.; Tanaka, M.; Sahoo, A. K.; Takeda, M.; Yada, A.; Nakao, Y.; Hiyama, T. *Bull. Chem. Soc. Jpn* **2010**, *83*, 554–569, and references cited.

[64] Ito, H.; Sensui, H.; Arimoto, K.; Miura, K.; Hosomi, A. *Chem. Lett.* **1997**, 639–640.

[65] Kang, S.-K.; Kim, T.-H.; Pyun, S.-J. *J. Chem. Soc. Perkin Trans. 1* **1997**, 797–798.

[66] Nishihara, Y.; Ikegashira, K.; Toriyama, F.; Mori, A.; Hiyama, T. *Bull. Chem. Soc. Jpn* **2000**, *73*, 985–990.

[67] (a) Dick, A. R.; Sanford, M. S. *Tetrahedron* **2006**, *62*, 2439–2463. (b) Seregin, I. V.; Gevorgyan, V. *Chem. Soc. Rev.* **2007**, *36*, 1173–1193. (c) Kakiuchi, F.; Kochi, T. *Synthesis* **2008**, 3013–3039. (d) Chen, X.; Engle, K. M.; Wang, D.-H.; Yu, J.-Q. *Angew. Chem. Int. Ed. Engl.* **2009**, *48*, 5094–5115. (e) Kulkarni, A. A.; Daugulis, O. *Synthesis* **2009**, 4087–4109. (f) Colby, D. A.; Bergman, R. G.; Ellman, J. A. *Chem. Rev.* **2010**, *110*, 624–655. (g) Lyons, T. W.; Sanford, M. S. *Chem. Rev.* **2010**, *110*, 1147–1169. (h) Messaoudi, S.; Brion, J.-D.; Alami, M. *Eur. J. Org. Chem.* **2010**, 6495–6516.

[68] (a) Satoh, T.; Miura, M. *Chem. Lett.* **2007**, *36*, 200–205. (b) Alberico, D.; Scott, M. E.; Lautens, M. *Chem. Rev.* **2007**, *107*, 174–238. (c) Ackermann, L.; Vicente, R.; Kapdi, A. R. *Angew. Chem. Int. Ed. Engl.* **2009**, *48*, 9792–9826. (d) McGlacken, G. P.; Bateman, L. M. *Chem. Soc. Rev.* **2009**, *38*, 2447–2464. (e) Bellina, F.; Rossi, R. *Tetrahedron* **2009**, *65*, 10269–10310.

[69] (a) Forrest, J. *J. Chem. Soc.* **1960**, 574–580. (b) Björklund, C.; Nilsson, M.; Wennerström, O. *Acta Chem. Scand.* **1970**, *24*, 3599–3606.

[70] (a) Björklund, C.; Nilsson, M. *Tetrahedron Lett.* **1966**, 675–678. (b) Björklund, C.; Nilsson, M. *Acta Chem. Scand.* **1968**, *22*, 2338–2346. (c) Björklund, C.; Nilsson, M. *Acta Chem. Scand.* **1968**, *22*, 2581–2584.

[71] Ljusberg, H.; Wahren, R. *Acta Chem. Scand.* **1973**, *27*, 2717–2721.

[72] Chodowska-Palicka, J.; Nilsson, M. *Synthesis* **1974**, 128–129.

[73] (a) Pivsa-Art, S.; Satoh, T.; Kawamura, Y.; Miura, M.; Nomura, M. *Bull. Chem. Soc. Jpn* **1998**, *71*, 467–473. (b) Yoshizumi, T.; Tsurugi, H.; Satoh, T.; Miura, M. *Tetrahedron Lett.* **2008**, *49*, 1598–1600. (c) Yoshizumi, T.; Satoh, T.; Hirano, K.; Matsuo, D.; Orita, A.; Otera, J.; Miura, M. *Tetrahedron Lett.* **2009**, *50*, 3273–3276. (d) Kawano, T.; Yoshizumi, T.; Hirano, K.; Satoh, T.; Miura, M. *Org. Lett.* **2009**, *11*, 3072–3075.

[74] (a) Do, H.-Q.; Daugulis, O. *J. Am. Chem. Soc.* **2007**, *129*, 12404–12405. (b) Do, H.-Q.; Daugulis, O. *J. Am. Chem. Soc.* **2008**, *130*, 1128–1129. (c) Do, H.-Q.; Khan, R. M. K.; Daugulis, O. *J. Am. Chem. Soc.* **2008**, *130*, 15185–15192. (d) Do, H.-Q.; Daugulis, O. *Chem. Commun.* **2009**, 6433–6435. Also see ref. [73d].

[75] (a) Zhao, D.; Wang, W.; Yang, F.; Lan, J.; Yang, L.; Gao, G.; You, J. *Angew. Chem. Int. Ed. Engl.* **2009**, *48*, 3296–3300. (b) Qin, X.; Cong, X.; Zhao, D.; You, J.; Lan, J. *Chem. Commun.* **2011**, *47*, 5611–5613.

[76] Sagnes, C.; Fournet, G.; Joseph, B. *Synlett* **2009**, 433–436.

[77] (a) Ackermann, L.; Potukuchi, H. K.; Landsberg, D.; Vicente, R. *Org. Lett.* **2008**, *10*, 3081–3084. (b) For mechanistic considerations of the azide-alkyne cycloaddition, see: Worrell, B. T.; Malik, J. A.; Fokin, V. V. *Science* **2013**, *340*, 457–460.

[78] (a) Barton, D. H. R.; Finet, J.-P.; Khamsi, J. *Tetrahedron Lett.* **1988**, *29*, 1115–1118. (b) Arnauld, T.; Barton, D. H. R.; Doris, E. *Tetrahedron* **1997**, *53*, 4137–4144.

[79] (a) Phipps, R. J.; Grimster, N. P.; Gaunt, M. J. *J. Am. Chem. Soc.* **2008**, *130*, 8172–8174. (b) Phipps, R. J.; Gaunt, M. J. *Science* **2009**, *323*, 1593–1597. (c) Ciana, C.-L.; Phipps, R. J.; Brandt, J. R.; Meyer, F.-M.; Gaunt, M. J. *Angew. Chem. Int. Ed. Engl.* **2011**, *50*, 458–462. (d) Duong, H. A.; Gilligan, R. E.; Cooke, M. L.; Phipps, R. J.; Gaunt, M. J. *Angew. Chem. Int. Ed. Engl.* **2011**, *50*, 463–466. (e) Chen, B.; Hou, X.-L.; Li, Y.-X.; Wu, Y.-D. *J. Am. Chem. Soc.* **2011**, *133*, 7668–7671. Also, see: (f) Ackermann, L.; Dell'Acqua, M.; Fenner, S.; Vicente, R.; Sandmann, R. *Org. Lett.* **2011**, *13*, 2358–2360, and cited references.

[80] (a) Ban, I.; Sudo, T.; Taniguchi, T.; Itami, K. *Org. Lett.* **2008**, *10*, 3607–3609. (b) Yang, F.; Xu, Z.; Wang, Z.; Yu, Z.; Wang, R. *Chem. Eur. J.* **2011**, *17*, 6321–6325.

[81] (a) Snieckus, V. *Chem. Rev.* **1990**, *90*, 879–933. (b) Epsztajn, J.; Jóźwiak, A.; Szcześniak, A. K. *Curr. Org. Chem.* **2006**, *10*, 1817–1848.

[82] Chen, X.; Dobereiner, G.; Hao, X.-S.; Giri, R.; Maugel, N.; Yu, J.-Q. *Tetrahedron* **2009**, *65*, 3085–3089.

[83] (a) Do, H.-Q.; Daugulis, O. *J. Am. Chem. Soc.* **2009**, *131*, 17052–17053. (b) Zhu, M.; Fujita, K.-i.; Yamaguchi, R. *Chem. Commun.* **2011**, *47*, 12876–12878. (c) Fan, S.; Chen, Z.; Zhang, X. *Org. Lett.* **2012**, *14*, 4950–4953. (d) Do, H.-Q.; Daugulis, O. *J. Am. Chem. Soc.* **2011**, *133*, 13577–13586. (e) Zou, L.-H.; Mottweiler, J.; Priebbenow, D. L.; Wang, J.; Stubenrauch, J. A.; Bolm, C. *Chem. Eur. J.* **2013**, *19*, 3302–3305.

[84] (a) Qin, X.; Feng, B.; Dong, J.; Li, X.; Xue, Y.; Lan, J.; You, J. *J. Org. Chem.* **2012**, *77*, 7677–7683. (b) Mao, Z.; Wang, Z.; Xu, Z.; Huang, F.; Yu, Z.; Wang, R. *Org. Lett.* **2012**, *14*, 3854–3857.

[85] Wang, L.; Han, Z.; Fan, R. *Adv. Synth. Catal.* **2010**, *352*, 3230–3234.

[86] (a) Kitahara, M.; Umeda, N.; Hirano, K.; Satoh, T.; Miura, M. *J. Am. Chem. Soc.* **2011**, *133*, 2160–2162. (b) Hachiya, H.; Hirano, K.; Satoh, T.; Miura, M. *Org. Lett.* **2011**, *13*, 3076–3079.

[87] (a) Nishino, M.; Hirano, K.; Satoh, T.; Miura, M. *Angew. Chem. Int. Ed. Engl.* **2012**, *51*, 6993–6997. (b) Nishino, M.; Hirano, K.; Satoh, T.; Miura, M. *Angew. Chem. Int. Ed.* **2013**, *52*, 4457–4461.

[88] (a) Shepard, A. F.; Winslow, N. R.; Johnson, J. R. *J. Am. Chem. Soc.* **1930**, *52*, 2083–2090. For recent examples, see:(b) Lisitsyn, A. S. *Appl. Catal. A: Gen.* **2007**, *332*, 166–170. (c) Gooßen, L. J.; Thiel, W. R.; Rodríguez, N.; Linder, C.; Melzer, B.

Adv. Synth. Catal. **2007**, *349*, 2241–2246. (d) Gooßen, L. J.; Manjolinho, F.; Khan, B. A.; Rodríguez, N. *J. Org. Chem.* **2009**, *74*, 2620–2623.

[89] (a) Nilsson, M. *Acta Chem. Scand.* **1966**, *20*, 423–426. (b) Nilsson, M.; Ullenius, C. *Acta Chem. Scand.* **1968**, *22*, 1998–2002. (c) Björklund, C.; Nilsson, M. *Acta Chem. Scand.* **1968**, *22*, 2585–2588.

[90] (a) Cairncross, A.; Roland, J. R.; Henderson, R. M.; Sheppard, W. A. *J. Am. Chem. Soc.* **1970**, *92*, 3187–3189. (b) Cohen, T.; Schambach, R. A. *J. Am. Chem. Soc.* **1970**, *92*, 3189–3190.

[91] Selected examples, see: (a) Forgione, P.; Brochu, M.-C.; St-Onge, M.; Thesen, K. H.; Bailey, M. D.; Bilodeau, F. *J. Am. Chem. Soc.* **2006**, *128*, 11350–11351. (b) Becht, J.-M.; Catala, C.; Le Drian, C.; Wagner, A. *Org. Lett.* **2007**, *9*, 1781–1783. (c) Voutchkova, A.; Coplin, A.; Leadbeater, N. E.; Crabtree, R. H. *Chem. Commun.* **2008**, 6312–6314. (d) Becht, J.-M.; Le Drian, C. *Org. Lett.* **2008**, *10*, 3161–3164. (e) Miyasaka, M.; Fukushima, A.; Satoh, T.; Hirano, K.; Miura, M. *Chem. Eur. J.* **2009**, *15*, 3674–3677. (f) Wang, C.; Piel, I.; Glorius, F. *J. Am. Chem. Soc.* **2009**, *131*, 4194–4196. (g) Wang, Z.; Ding, Q.; He, X.; Wu, J. *Tetrahedron* **2009**, *65*, 4635–4638. (h) Zhang, F.; Greaney, M. F. *Angew. Chem. Int. Ed. Engl.* **2010**, *49*, 2768–2771. (i) Cornella, J.; Lahlali, H.; Larrosa, I. *Chem. Commun.* **2010**, *46*, 8276–8278. (j) Gooßen, L. J.; Lange, P. P.; Rodríguez, N.; Linder, C. *Chem. Eur. J.* **2010**, *16*, 3906–3909. (k) Bilodeau, F.; Brochu, M.-C.; Guimond, N.; Thesen, K. H.; Forgione, P. *J. Org. Chem.* **2010**, *75*, 1550–1560. (l) Zhang, F.; Greaney, M. F. *Org. Lett.* **2010**, *12*, 4745–4747. (m) Dai, J.-J.; Liu, J.-H.; Luo, D.-F.; Liu, L. *Chem. Commun.* **2011**, *47*, 677–679. (n) Zhao, H.; Wei, Y.; Xu, J.; Kan, J.; Su, W.; Hong, M. *J. Org. Chem.* **2011**, *76*, 882–893. (o) Cahiez, G.; Moyeux, A.; Gager, O.; Poizat, M. *Adv. Synth. Catal.* **2013**, *355*, 790–796.

[92] (a) Gooßen, L. J.; Deng, G.; Levy, L. M. *Science* **2006**, *313*, 662–664. (b) Gooßen, L. J.; Rodríguez, N.; Melzer, B.; Linder, C.; Deng, G.; Levy, L. M. *J. Am. Chem. Soc.* **2007**, *129*, 4824–4833. (c) Gooßen, L. J.; Zimmermann, B.; Knauber, T. *Angew. Chem. Int. Ed. Engl.* **2008**, *47*, 7103–7106. (d) Gooßen, L. J.; Rodríguez, N.; Linder, C. *J. Am. Chem. Soc.* **2008**, *130*, 15248–15249. (e) Gooßen, L. J.; Linder, C.; Rodríguez, N.; Lange, P. P. *Chem. Eur. J.* **2009**, *15*, 9336–9349. (f) Gooßen, L. J.; Rodríguez, N.; Lange, P. P.; Linder, C. *Angew. Chem. Int. Ed. Engl.* **2010**, *49*, 1111–1114.

[93] Shang, R.; Fu, Y.; Wang, Y.; Xu, Q.; Yu, H.-Z.; Liu, L. *Angew. Chem. Int. Ed. Engl.* **2009**, *48*, 9350–9354.

[94] (a) Matsunaga, S.; Ohshima, T.; Shibasaki, M. *Adv. Synth. Catal.* **2002**, *344*, 3–15. (b) Chen, Y.; Yekta, S.; Yudin, A. K. *Chem. Rev.* **2003**, *103*, 3155–3211. (c) Brunel, J. M. *Chem. Rev.* **2005**, *105*, 857–897. (d) Eberhardt, L.; Armspach, D.; Harrowfield, J.; Matt, D. *Chem. Soc. Rev.* **2008**, *37*, 839–864. (e) Terada, M. *Synthesis* **2010**, 1929–1982.

[95] (a) Pu, L. *Chem. Rev.* **1998**, *98*, 2405–2494. (b) Pu, L. *Chem. Rev.* **2004**, *104*, 1687–1716.

[96] (a) Love, B. E. *Curr. Org. Synth.* **2006**, *3*, 169–185. (b) Wang, H. *Chirality* **2010**, *22*, 827–837.

[97] (a) Kitajima, N.; Moro-oka, Y. *Chem. Rev.* **1994**, *94*, 737–757. (b) Klinman, J. P. *Chem. Rev.* **1996**, *96*, 2541–2561. (c) Solomon, E. I.; Sundaram, U. M.; Machonkin, T. E. *Chem. Rev.* **1996**, *96*, 2563–2605. (d) Gamez, P.; Aubel, P. G.; Driessen, W. L.;

Reedijk, J. *Chem. Soc. Rev.* **2001**, *30*, 376–385. (e) Lewis, E. A.; Tolman, W. B. *Chem. Rev.* **2004**, *104*, 1047–1076. (f) Que, L., Jr.; Tolman, W. B. *Nature* **2008**, *455*, 333–340. (g) Himes, R. A.; Karlin, K. D. *Curr. Opin. Chem. Biol.* **2009**, *13*, 119–131.

[98] (a) Puzari, A.; Baruah, J. B. *J. Mol. Catal. A: Chem.* **2002**, *187*, 149–162. (b) Punniyamurthy, T.; Rout, L. *Coord. Chem. Rev.* **2008**, *252*, 134–154.

[99] Selected recent examples: (a) Gupta, R.; Mukherjee, R. *Tetrahedron Lett.* **2000**, *41*, 7763–7767. (b) Hosokawa, S.; Fumiyama, H.; Fukuda, H.; Fukuda, T.; Seki, M.; Tatsuta, K. *Tetrahedron Lett.* **2007**, *48*, 7305–7308. (c) Yamada, H.; Nagao, K.; Dokei, K.; Kasai, Y.; Michihata, N. *J. Am. Chem. Soc.* **2008**, *130*, 7566–7567. (d) Prokofieva, A.; Dechert, S.; Große, C.; Sheldrick, G. M.; Meyer, F. *Chem. Eur. J.* **2009**, *15*, 4994–4997. (e) Haack, P.; Limberg, C.; Ray, K.; Braun, B.; Kuhlmann, U.; Hildebrandt, P.; Herwig, C. *Inorg. Chem.* **2011**, *50*, 2133–2142.

[100] Brackman, W.; Havinga, E. *Rec. Trav. Chim. Pays-Bas* **1955**, *74*, 1021–1039.

[101] (a) Feringa, B.; Wynberg, H. *Tetrahedron Lett.* **1977**, *18*, 4447–4450. (b) Feringa, B.; Wynberg, H. *Bioorg. Chem.* **1978**, *7*, 397–408.

[102] Hattori, T.; Shimazumi, Y.; Yamabe, O.; Koshiishi, E.; Miyano, S. *Chem. Commun.* **2002**, 2234–2235.

[103] (a) Brussee, J.; Jansen, A. C. A. *Tetrahedron Lett.* **1983**, *24*, 3261–3262. (b) Brussee, J.; Groenendijk, J. L. G.; te Koppele, J. M.; Jansen, A. C. A. *Tetrahedron* **1985**, *41*, 3313–3319.

[104] (a) Yamamoto, K.; Fukushima, H.; Nakazaki, M. *J. Chem. Soc., Chem. Commun.* **1984**, 1490–1491. (b) Yamamoto, K.; Fukushima, H.; Yumioka, H.; Nakazaki, M. *Bull. Chem. Soc. Jpn* **1985**, *58*, 3633–3634.

[105] (a) Noji, M.; Nakajima, M.; Koga, K. *Tetrahedron Lett.* **1994**, *35*, 7983–7984. (b) Nakajima, M.; Kanayama, K.; Miyoshi, I.; Hashimoto, S. *Tetrahedron Lett.* **1995**, *36*, 9519–9520. (c) Nakajima, M.; Hashimoto, S.; Noji, M.; Koga, K. *Chem. Pharm. Bull.* **1994**, *46*, 1814–1815. (d) Nakajima, M.; Miyoshi, I.; Kanayama, K.; Hashimoto, S. *J. Org. Chem.* **1999**, *64*, 2264–2271.

[106] (a) Hu, Q.-S.; Vitharana, D.; Pu, L. *Tetrahedron Asymmetry* **1995**, *6*, 2123–2126. (b) Lipshutz, B. H.; James, B.; Vance, S.; Carrico, I. *Tetrahedron Lett.* **1997**, *38*, 753–756. (c) Roithová, J.; Schröder, D. *Chem. Eur. J.* **2008**, *14*, 2180–2188.

[107] (a) Guo, F.; Konkol, L. C.; Thomson, R. J. *J. Am. Chem. Soc.* **2011**, *133*, 18–20. (b) Konkol, L. C.; Guo, F.; Sarjeant, A. A.; Thomson, R. J. *Angew. Chem. Int. Ed. Engl.* **2011**, *50*, 9931–9934.

[108] (a) Li, X.; Yang, J.; Kozlowski, M. C. *Org. Lett.* **2001**, *3*, 1137–1140. (b) Kozlowski, M. C.; Li, X.; Carroll, P. J.; Xu, Z. *Organometallics* **2002**, *21*, 4513–4522. (c) Li, X.; Hewgley, J. B.; Mulrooney, C. A.; Yang, J.; Kozlowski, M. C. *J. Org. Chem.* **2003**, *68*, 5500–5511.

[109] (a) Mulrooney, C. A.; Li, X.; DiVirgilio, E. S.; Kozlowski, M. C. *J. Am. Chem. Soc.* **2003**, *125*, 6856–6857. (b) DiVirgilio, E. S.; Dugan, E. C.; Mulrooney, C. A.; Kozlowski, M. C. *Org. Lett.* **2007**, *9*, 385–388. (c) Kozlowski, M. C.; Dugan, E. C.; DiVirgilio, E. S.; Maksimenka, K.; Bringmann, G. *Adv. Synth. Catal.* **2007**, *349*, 583–594. (d) O'Brien, E. M.; Morgan, B. J.; Kozlowski, M. C. *Angew. Chem. Int. Ed. Engl.* **2008**, *47*, 6877–6880. (e) Morgan, B. J.; Dey, S.; Johnson, S. W.; Kozlowski, M. C. *J. Am. Chem. Soc.* **2009**, *131*, 9413–9425. (f) Mulrooney, C. A.; Morgan, B. J.; Li, X.; Kozlowski, M. C. *J. Org. Chem.* **2010**, *75*, 16–29. (g) Morgan, B. J.;

Mulrooney, C. A.; Kozlowski, M. C. *J. Org. Chem.* **2010**, *75*, 44–56. (h) O'Brien, E. M.; Morgan, B. J.; Mulrooney, C. A.; Carroll, P. J.; Kozlowski, M. C. *J. Org. Chem.* **2010**, *75*, 57–68.

[110] (a) Xie, X.; Phuan, P.-W.; Kozlowski, M. C. *Angew. Chem. Int. Ed. Engl.* **2003**, *42*, 2168–2170. (b) Morgan, B. J.; Xie, X.; Phuan, P.-W.; Kozlowski, M. C. *J. Org. Chem.* **2007**, *72*, 6172–6182.

[111] Kim, K. H.; Lee, D.-W.; Lee, Y.-S.; Ko, D.-H.; Ha, D.-C. *Tetrahedron* **2004**, *60*, 9037–9042.

[112] Sharma, V. B.; Jain, S. L.; Sain, B. *J. Mol. Catal. A: Chem.* **2004**, *219*, 61–64.

[113] Gao, J.; Reibenspies, J. H.; Martell, A. E. *Angew. Chem. Int. Ed. Engl.* **2003**, *42*, 6008–6012.

[114] Yamamoto, K.; Yumioka, H.; Okamoto, Y.; Chikamatsu, H. *J. Chem. Soc., Chem. Commun.* **1987**, 168–169.

[115] (a) Hovorka, M.; Günterová, J.; Závada, J. *Tetrahedron Lett.* **1990**, *31*, 413–416. (b) Hovorka, M.; Závada, J. *Org. Prep. Proced. Int.* **1991**, *23*, 200–203. (c) Hovorka, M.; Ščigel, R.; Gunterová, J.; Tichý, M.; Závada, J. *Tetrahedron* **1992**, *48*, 9503–9516. (d) Hovorka, M.; Závada, J. *Tetrahedron* **1992**, *48*, 9517–9530.

[116] (a) Smrčina, M.; Lorenc, M.; Hanuš, V.; Kočovský, P. *Synlett* **1991**, 231–232. (b) Smrčina, M.; Lorenc, M.; Hanuš, V.; Sedmera, P.; Kočovský, P. *J. Org. Chem.* **1992**, *57*, 1917–1920. (c) Smrčina, M.; Poláková, J.; Vyskočil, Š.; Kočovský, P. *J. Org. Chem.* **1993**, *58*, 4534–4538. (d) Smrčina, M.; Vyskočil, Š.; Máca, B.; Polášek, M.; Claxton, T. A.; Abbott, A. P.; Kočovský, P. *J. Org. Chem.* **1994**, *59*, 2156–2163.

[117] Grandbois, A.; Mayer, M.-È.; Bédard, M.; Collins, S. K.; Michel, T. *Chem. Eur. J.* **2009**, *15*, 9655–9659.

[118] (a) Temma, T.; Habaue, S. *Tetrahedron Lett.* **2005**, *46*, 5655–5657. (b) Temma, T.; Hatano, B.; Habaue, S. *Tetrahedron* **2006**, *62*, 8559–8563. (c) Habaue, S.; Takahashi, Y.; Temma, T. *Tetrahedron Lett.* **2007**, *48*, 7301–7304. (d) Habaue, S.; Temma, T.; Sugiyama, Y.; Yan, P. *Tetrahedron Lett.* **2007**, *48*, 8595–8598. (e) Yan, P.; Sugiyama, Y.; Takahashi, Y.; Kinemuchi, H.; Temma, T.; Habaue, S. *Tetrahedron* **2008**, *64*, 4325–4331.

[119] (a) Amou, S.; Takeuchi, K.; Asai, M.; Niizeki, K.; Okada, T.; Seino, M.; Haba, O.; Ueda, M. *J. Polym. Sci. Part A Polym. Chem.* **1999**, *37*, 3702–3709. (b) Sasada, Y.; Shibasaki, Y.; Suzuki, M.; Ueda, M. *Polymer* **2003**, *44*, 355–360.

[120] (a) Habaue, S.; Seko, T.; Okamoto, Y. *Macromolecules* **2002**, *35*, 2437–2439. (b) Habaue, S.; Seko, T.; Okamoto, Y. *Macromolecules* **2003**, *36*, 2604–2608. (c) Habaue, S.; Seko, T.; Isonaga, M.; Ajiro, H.; Okamoto, Y. *Polym. J.* **2003**, *35*, 592–597. (d) Habaue, S.; Seko, T.; Okamoto, Y. *Polymer* **2003**, *44*, 7377–7381. (e) Habaue, S.; Ajiro, H.; Yoshii, Y.; Hirasa, T. *J. Polym. Sci. Part A Polym. Chem.* **2004**, *42*, 4528–4534. (f) Habaue, S.; Muraoka, R.; Aikawa, A.; Murakami, S.; Higashimura, H. *J. Polym. Sci. Part A Polym. Chem.* **2005**, *43*, 1635–1640. (g) Habaue, S.; Ishikawa, K. *Polym. Bull.* **2005**, *55*, 243–250. (h) Habaue, S.; Ishikawa, K.; Aikawa, A.; Murakami, S.; Hatano, B. *Polym. Bull.* **2006**, *57*, 305–312.

[121] (a) Temma, T.; Habaue, S. *J. Polym. Sci. Part A Polym. Chem.* **2005**, *43*, 6287–6294. (b) Temma, T.; Hatano, B.; Habaue, S. *Polymer* **2006**, *47*, 1845–1851. (c) Temma, T.; Takahashi, Y.; Yoshii, Y.; Habaue, S. *Polym. J.* **2007**, *39*, 524–530. (d) Temma,

T.; Habaue, S. *J. Polym. Sci. Part A Polym. Chem.* **2008**, *46*, 1034–1041. (e) Yan, P.; Temma, T.; Habaue, S. *Polym. J.* **2008**, *40*, 710–715.

[122] (a) Tanaka, K.; Furuta, T.; Fuji, K.; Miwa, Y.; Taga, T. *Tetrahedron Asymmetry* **1996**, *7*, 2199–2202. (b) Fuji, K.; Furuta, T.; Tanaka, K. *Org. Lett.* **2001**, *3*, 169–171. (c) Tsubaki, K.; Miura, M.; Morikawa, H.; Tanaka, H.; Kawabata, T.; Furuta, T.; Tanaka, K.; Fuji, K. *J. Am. Chem. Soc.* **2003**, *125*, 16200–16201. (d) Furuta, T.; Tanaka, K.; Tsubaki, K.; Fuji, K. *Tetrahedron* **2004**, *60*, 4431–4441. (e) Tsubaki, K.; Tanaka, H.; Takaishi, K.; Miura, M.; Morikawa, H.; Furuta, T.; Tanaka, K.; Fuji, K.; Sasamori, T.; Tokitoh, N.; Kawabata, T. *J. Org. Chem.* **2006**, *71*, 6579–6587. (f) Tsubaki, K.; Takaishi, K.; Tanaka, H.; Miura, M.; Kawabata, T. *Org. Lett.* **2006**, *8*, 2587–2590. (g) Tsubaki, K.; Takaishi, K.; Sue, D.; Kawabata, T. *J. Org. Chem.* **2007**, *72*, 4238–4241. (h) Sue, D.; Takaishi, K.; Harada, T.; Kuroda, R.; Kawabata, T.; Tsubaki, K. *J. Org. Chem.* **2009**, *74*, 3940–3943. (i) Takaishi, K.; Sue, D.; Kuwahara, S.; Harada, N.; Kawabata, T.; Tsubaki, K. *Tetrahedron* **2009**, *65*, 6135–6140.

[123] (a) Fuji, K.; Furuta, T.; Otsubo, T.; Tanaka, K. *Tetrahedron Lett.* **1999**, *40*, 3001–3004. (b) Tsubaki, K.; Tanaka, H.; Furuta, T.; Kinoshita, T.; Fuji, K. *Tetrahedron Lett.* **2000**, *41*, 6089–6093. (c) Tsubaki, K.; Tanaka, H.; Furuta, T.; Tanaka, K.; Kinoshita, T.; Fuji, K. *Tetrahedron* **2002**, *58*, 5611–5617. (d) Tsubaki, K.; Miura, M.; Nakamura, A.; Kawabata, T. *Tetrahedron Lett.* **2006**, *47*, 1241–1244. (e) Pieraccini, S.; Ferrarini, A.; Fuji, K.; Gottarelli, G.; Lena, S.; Tsubaki, K.; Spada, G. P. *Chem. Eur. J.* **2006**, *12*, 1121–1126. (f) Tsubaki, K.; Takaishi, K.; Sue, D.; Matsuda, K.; Kanemitsu, Y.; Kawabata, T. *J. Org. Chem.* **2008**, *73*, 4279–4282.

[124] (a) Sakamoto, T.; Yonehara, H.; Pac, C. *J. Org. Chem.* **1994**, *59*, 6859–6861. (b) Kantam, M. L.; Santhi, P. L. *Synth. Commun.* **1996**, *26*, 3075–3079. (c) Sakamoto, T.; Yonehara, H.; Pac, C. *J. Org. Chem.* **1997**, *62*, 3194–3199. (d) Armengol, E.; Corma, A.; García, H.; Primo, J. *Eur. J. Org. Chem.* **1999**, 1915–1920. (e) Mastrorilli, P.; Muscio, F.; Suranna, G. P.; Nobile, C. F.; Latronico, M. *J. Mol. Catal. A: Chem.* **2001**, *165*, 81–87. (f) Prasad, M. R.; Kamalakar, G.; Kulkarni, S. J.; Raghavan, K. V. *J. Mol. Catal. A: Chem.* **2002**, *180*, 109–123. (g) Reddy, K. R.; Rajgopal, K.; Kantam, M. L. *Catal. Lett.* **2007**, *114*, 36–40.

11

COPPER-CATALYZED ALKYNYLATION, ALKENYLATION, AND ALLYLATION REACTIONS OF ARYL DERIVATIVES

REN-JIE SONG AND JIN-HENG LI

11.1 INTRODUCTION

As with other copper-mediated transformations, the copper-mediated alkynylation and alkenylation reactions of aromatic derivatives have seen tremendous progress in recent years, notably due to the introduction of a variety of simple organic ligands that allowed a significant softening of the reaction conditions. Reliable and robust procedures for the synthesis of alkynyl-, alkenyl-, and allyl-substituted aromatics have recently emerged and constitute remarkably efficient synthetic tools. This chapter will cover the most general copper-catalyzed syntheses of arylalkynes and arylalkenes by Stille-, Sonogashira-, and Suzuki–Miyaura-type cross-coupling reactions, as well as key recent developments in the copper-mediated allylation of aryl derivatives. In addition, recent advances in copper-mediated C–H bond activation and decarboxylation reactions for the synthesis of arylalkynes and arylalkenes will be presented.

Copper-Mediated Cross-Coupling Reactions, First Edition. Edited by Gwilherm Evano and Nicolas Blanchard.

11.2 COPPER-CATALYZED ALKYNYLATION OF ARYL DERIVATIVES

Alkynes are most useful building blocks in organic synthesis as well as basic functional groups that are found in many natural products and bioactive compounds.[1] The Stille reaction, the Sonogashira reaction, and the direct functionalization of C–H bonds represent most powerful and straightforward methods for the construction of arylalkynes, and they have been extensively studied. In addition, these methods have been also widely used for the synthesis of natural products, biologically active molecules, and materials. They have been recently found to be conveniently catalyzed by copper salts. Key and representative advances in this area will be overviewed in this subchapter, in which examples have been classified according to the nature of the coupling partners, starting with organotin reagents.

11.2.1 Synthesis of Aryl–Ynes by Stille-Type Cross-Coupling Reaction

The transition-metal-catalyzed cross coupling of organotin compounds with electrophiles, namely, the Stille cross-coupling reaction first reported by John Kenneth Stille and David Milstein in 1977, has emerged as an extremely powerful tool for carbon–carbon bond formation (Scheme 11.1). This palladium-catalyzed cross-coupling reaction is usually performed under an inert gas atmosphere, using dried and degassed solvent, because oxygen can cause the oxidation of the palladium catalyst and promote homocoupling of stannylated derivatives. These side reactions often lead to a decrease in the yield of the desired cross-coupling product.[2]

In 1990, Liebeskind and coworkers found that the use of a copper(I) cocatalyst could dramatically enhance the reaction rate of a palladium-catalyzed Stille cross coupling (Scheme 11.2).[3] The practical utility of the "copper effect" was immediately recognized, and the underlying mechanisms have triggered intense research efforts (see Chapter 18, Section 18.4.4 for a more detailed discussion).[2b]

In a remarkable report published in 1997, Kang and coworkers demonstrated that the Stille coupling could be catalyzed by a catalytic amount of copper(I) iodide (10 mol %) in the presence of stoichiometric sodium chloride without palladium catalysts at all (Scheme 11.3).[4]

These conditions were later on extended to the Stille-type alkynylation of a polymer-bound aryl iodide with stannylated alkyne, affording the

$$R^1\text{-}Sn(R^2)_3 \; + \; R^3\text{-}X \xrightarrow{[Pd]} R^1\text{-}R^3 + X\text{-}Sn(R^2)_3$$

X = I, Br, Cl, OTf, CF_3SO_3

Scheme 11.1 Palladium-catalyzed Stille reaction.

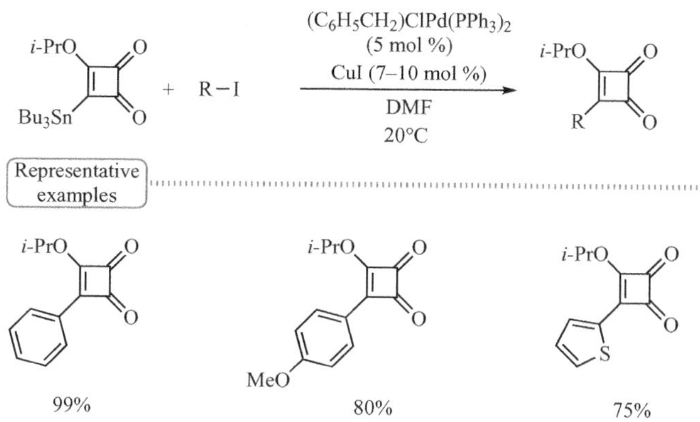

Scheme 11.2 Copper cocatalyzed Stille cross coupling of stannylcyclobutenedione with aryl iodides.

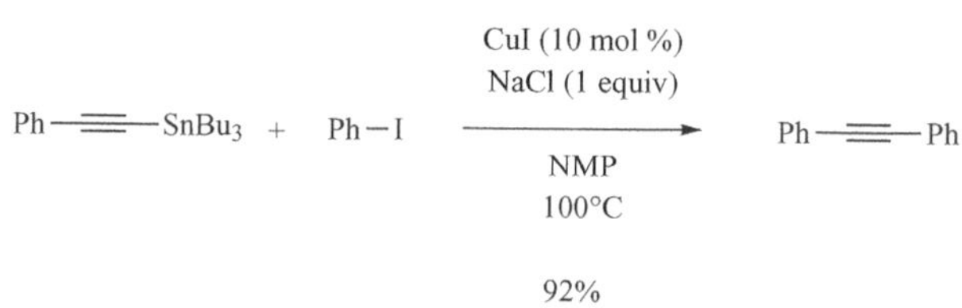

Scheme 11.3 Early example of copper-catalyzed cross coupling of alkynyltin derivatives with aryl iodides.

corresponding arylalkyne in 78% yield after cleavage from the resin by transesterification (Scheme 11.4).[5]

The scope of aryl halides that can be used in this copper-catalyzed Stille cross coupling was extended in 2006 by Li and coworkers (Scheme 11.5):[6] aryl bromides could undergo the Stille reaction using a catalytic amount of copper(I) oxide nanoparticles combined with tri-*o*-tolylphosphine as the ligand (20 mol %), although good yields were obtained only from activated aryl bromides. It is noteworthy that the $Cu_2O/P(o\text{-tol})_3/n\text{-}Bu_4NBr$ catalytic system can be recovered and reused at least three times without any loss of catalytic activity (see Chapter 20 for an overview of reusable catalytic systems).

The polarity of the reagents can be reversed and 1-iodoalkynes were shown to undergo the cross-coupling reactions with arylorganotin reagents in the presence of copper(I) iodide (10 mol %) in dimethylformamide (DMF) at room temperature for 6 hours. The desired conjugated arylated alkynes were obtained in good yields provided that the 1-iodoalkynes were slowly added to the reaction mixture containing the aryl- and heteroaryl tin derivatives. (Scheme 11.6).[7] Although the scope with respect to the alkynyl halides is quite limited, this report provided a novel and alternative route to internal aromatic alkynes.

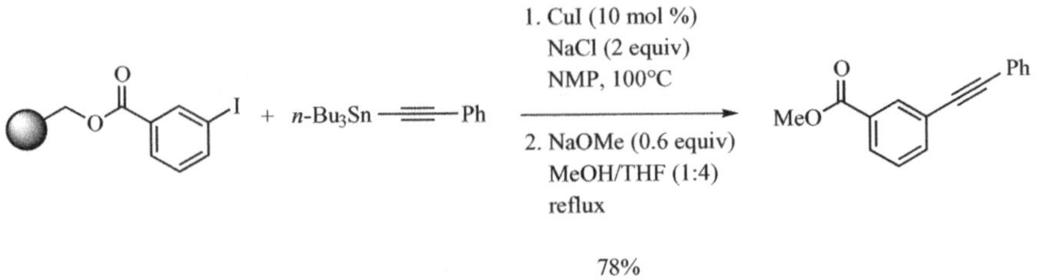

Scheme 11.4 Copper(I) iodide-catalyzed cross coupling of polymer-bounded aryl iodide with an ethynylstannane.

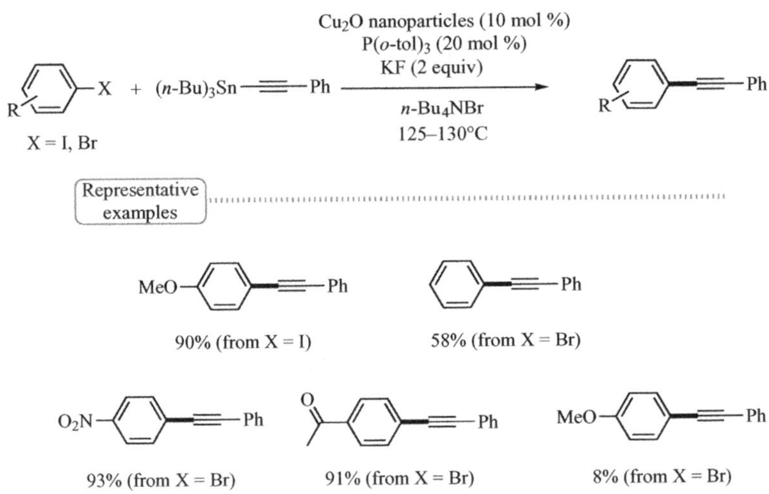

Scheme 11.5 Cu_2O nanoparticle-catalyzed Stille cross couplings of aryl halides with alkynyltin derivatives in n-Bu_4NBr.

In addition, highly active electrophiles such as alkynyliodonium tetrafluoroborates, which, however, have to be synthesized, were also found to be excellent reaction partners for the copper-catalyzed Stille cross coupling with aryl- and heteroaryltin derivatives in aqueous 1,2-dimethoxyethane (DME) under mild conditions (room temperature, 3 hours) (Scheme 11.7).[8] As a note, arylboronic acids were also found to be efficient reaction partners in this cross coupling under similar reaction conditions.

The major inherent drawback associated with the use of the Stille cross coupling and related reactions such as the ones described in this section are the use of organotin reagents and the generation of highly toxic waste. The development of other copper-mediated processes involving other reagents has been also recently developed and represents a useful alternative. Recent developments in the copper-catalyzed Sonogashira-type cross coupling involving aryl halides will be described in the following paragraphs.

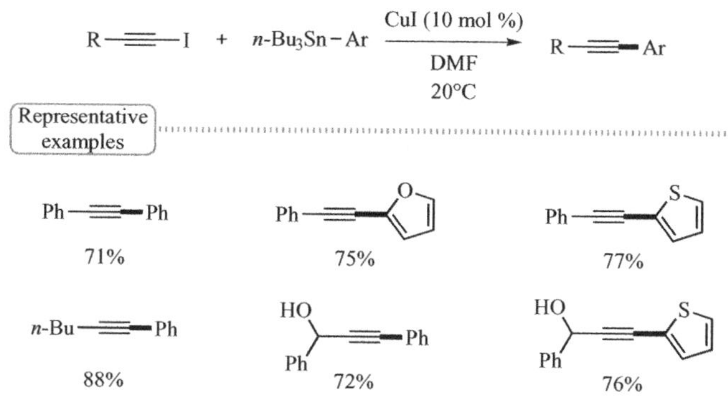

Scheme 11.6 Copper-catalyzed cross coupling between 1-iodoalkynes and aryltin reagents.

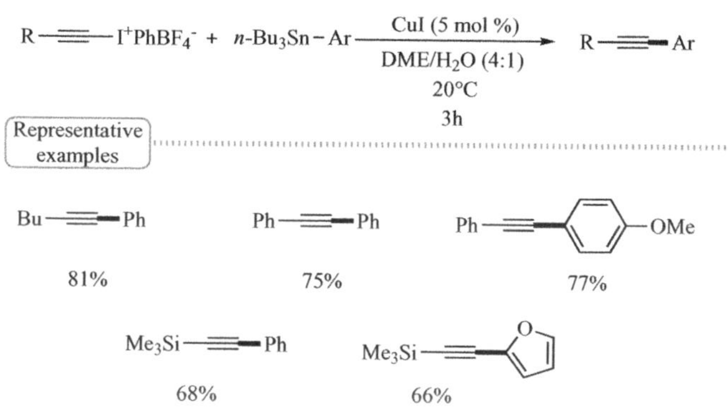

Scheme 11.7 Copper-catalyzed Stille cross coupling of alkynyliodonium tetrafluoroborates with aryltin reagents.

11.2.2 Synthesis of Aryl–Ynes by Sonogashira-Type Cross-Coupling Reaction

The Sonogashira coupling is the reaction of terminal alkynes with organohalides leading to new substituted alkynes. This reaction was first reported by Kenkichi Sonogashira and coworkers in 1975 (Scheme 11.8).[9] Since then, the Sonogashira coupling has been a central tool for the straightforward synthesis of biologically and/or synthetically important alkynes. Generally, this reaction utilizes a palladium complex (at the 0 or +2 oxidation state) and a copper(I) halide (often copper(I) iodide) as the catalytic system, which is problematic for industrial use due to the evolution of the price of palladium (see Chapter 19, Fig. 19.2). These limitations have elicited intense research efforts, including the use of copper(I) alone as the catalyst for the cross-coupling reactions of aryl halides and terminal alkynes.

In comparison with the common palladium-catalyzed coupling reactions, which are sensitive toward oxygen or moisture, the copper-mediated

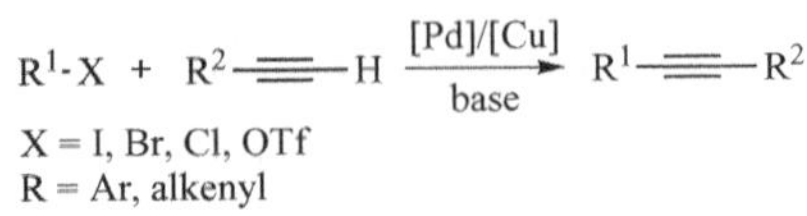

Scheme 11.8 The Sonogashira-type cross-coupling reaction.

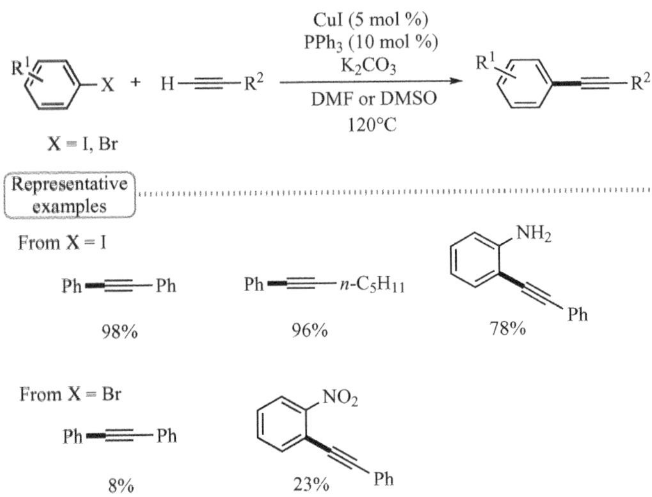

Scheme 11.9 Copper-catalyzed Sonogashira-type reaction of aryl halides with terminal alkynes.

Sonogashira-type alkynylation reaction can be conducted with satisfactory yields under aerobic conditions to produce the desired alkynes, provided that the nature and/or the stoichiometry of the ligand is appropriate.

In 1992, Miura and coworkers found that aryl iodides smoothly reacted with terminal alkynes and a catalytic amount of copper(I) iodide using potassium carbonate as the base and triphenylphosphine, giving the corresponding coupled products in good yields without the need of an additional palladium catalyst (Scheme 11.9).[10] This protocol represents the first truly copper-catalyzed Sonogashira alkynylation reaction. The scope and limitations of the copper-catalyzed Sonogashira-type reactions including the stereochemistry of the reaction were thoroughly investigated, and it was notably found that vinyl halides were also suitable substrates, while aryl bromides gave poor yields. Other copper catalysts, such as copper(I) bromide, copper(I) chloride, or copper(II) acetate were also effective, but copper(I) oxide was ineffective.[11]

The 1992 reaction mechanism of the copper-catalyzed Sonogashira alkynylation reaction proposed by Miura is shown in Scheme 11.10, where ligand **L** may be a neutral ligand (e.g., triphenylphosphine), the solvent, and/or the alkyne.[11a] In the first step, the terminal alkyne is deprotonated by potassium carbonate in the presence of copper(I) iodide, thereby generating the liganded copper acetylide **2** and then the coordinatively unsaturated copper acetylide **3**.

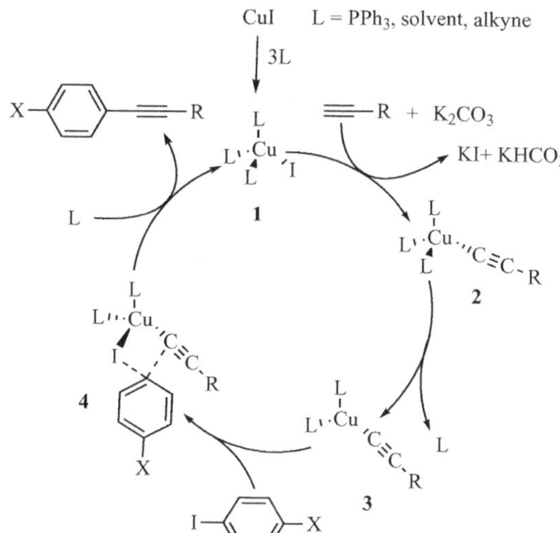

Scheme 11.10 The 1992 reaction mechanism of the copper-catalyzed Sonogashira alkynylation reaction or aryl iodides (see also Scheme 11.22).

Upon reaction of the latter with the aryl iodide partner, a four-centered transition state **4** might be operative by analogy with the Castro–Stephens reaction.[11b] A final ligand exchange would then regenerate the active copper(I) catalyst. However, recent mechanistic investigations by Bolm demonstrated that a different transition state should be considered, as discussed in Scheme 11.22.

While Miura used phosphine ligands to promote the reaction, Ma and Liu developed a palladium- and phosphine-free reaction condition for the Sonogashira-type alkynylation, in which copper(I) iodide (10 mol %) and *N*,*N*-dimethylglycine hydrochloride (30 mol %) was used as the catalytic system.[12] The 3/1 ratio between ligand and copper(I) was necessary to limit the homocoupling of alkyne. The functional group tolerance of this transformation is particularly noteworthy, and as the ligand is water-soluble, workup of this copper-catalyzed cross coupling is very simple and straightforward (Scheme 11.11). Overall, it is one of the most useful copper-catalyzed Sonogashira-type cross coupling for the synthesis of aromatic alkynes.

Pyrimidines could also be used as ligands for the copper-catalyzed Sonogashira-type alkynylation reaction. Indeed, Li and coworkers have reported that a catalytic system comprising 2-amino-4,6-dimethoxypyrimidine (20 mol %) and copper(II) acetate (10 mol %) was particularly effective in promoting the cross coupling between aryl iodides and terminal alkynes under solvent-free and aerobic conditions (Scheme 11.12).[13] Aryl bromides were less-reactive partners in the reaction, although the yields could be improved when 1 equiv of tetrabutylammonium bromide (TBAB) was added to the catalytic system.

Subsequently, Li and coworkers found that 1,4-diazabicyclo[2.2.2]octane (DABCO) was also an efficient ligand for the Sonogashira-type alkynylation

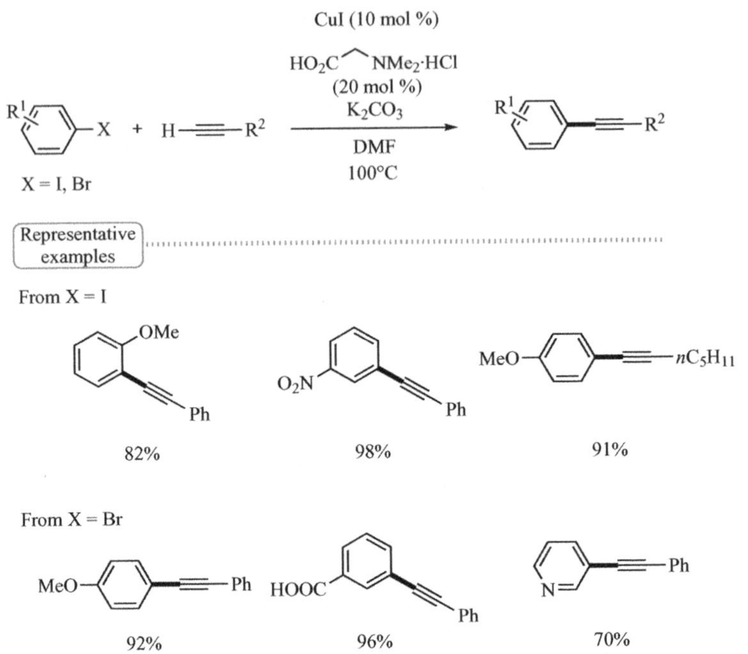

Scheme 11.11 Sonogashira-type alkynylation in the presence of copper(I) iodide and *N*,*N*-dimethylglycine hydrochloride.

reaction using copper catalysts (Scheme 11.13).[14] Interestingly, the scope was extended to less activated aryl chlorides.

Capitalizing on the well-known *ortho*-effect in copper-mediated transformations, Zhao and coworkers have developed an efficient copper-catalyzed Sonogashira alkynylation method using *o*-iodoacetanilide derivatives and alkynes at room temperature, affording the corresponding arylated alkynes in good-to-excellent yields (Scheme 11.14).[15] The high catalyst and ligand loadings (30 and 100 mol %, respectively) were required to prevent cyclization to the corresponding indole. Overall, it should be noted that this inexpensive CuI/*N*-methylpyrrolidine-2-carboxamide catalyst system shows excellent functional group compatibility.

Recently, a number of ligands were discovered to promote the copper-catalyzed Sonogashira alkynylation reaction. For example, Mao and coworkers reported a copper-catalyzed cross coupling of aryl halides and heteroaryl halides with phenylacetylene in DMF at 110–130°C that afforded the corresponding alkyne products in satisfactory-to-good yields when 8-hydroxyquinoline was used a ligand (Scheme 11.15).[16] Copper(I) iodide (10 mol %) and 8-hydroxyquinoline (20 mol %) generated *in situ* a bifunctional catalyst that is very efficient at promoting this carbon–carbon bond formation. It is worth noting that tetrabutylammomium bromide could greatly improve the yields of the reaction when using aryl chlorides. Mao and

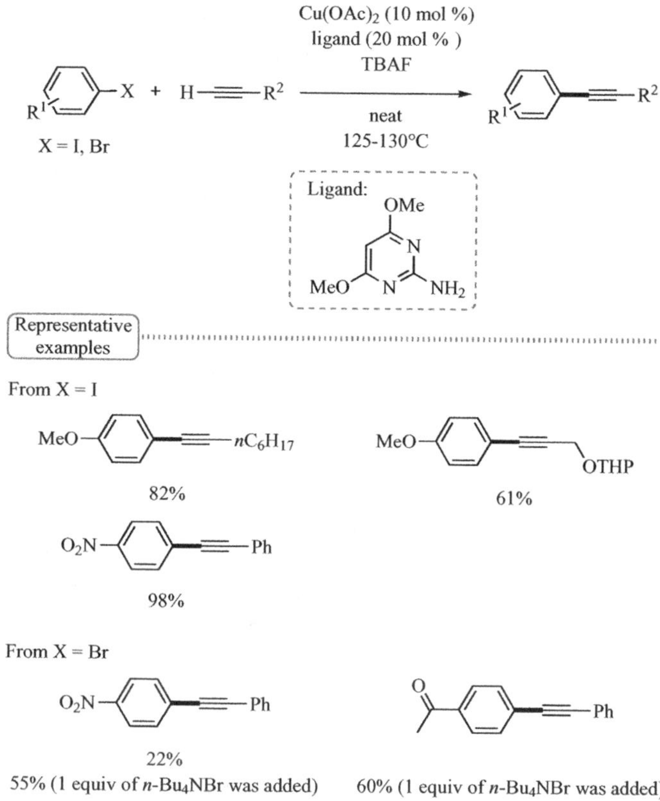

Scheme 11.12 Copper(II)-catalyzed Sonogashira cross-coupling reaction using pyrimidine as a ligand.

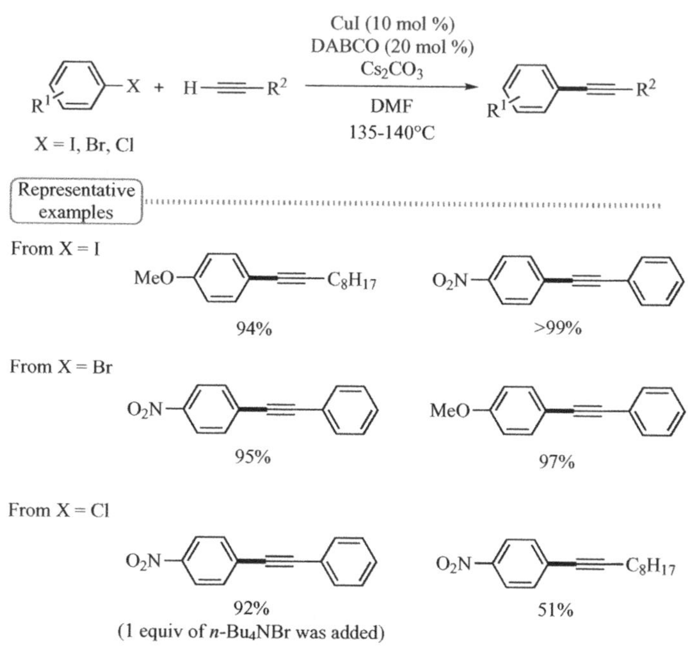

Scheme 11.13 DABCO as an efficient ligand for the copper-catalyzed Sonogashira-type alkynylation reaction.

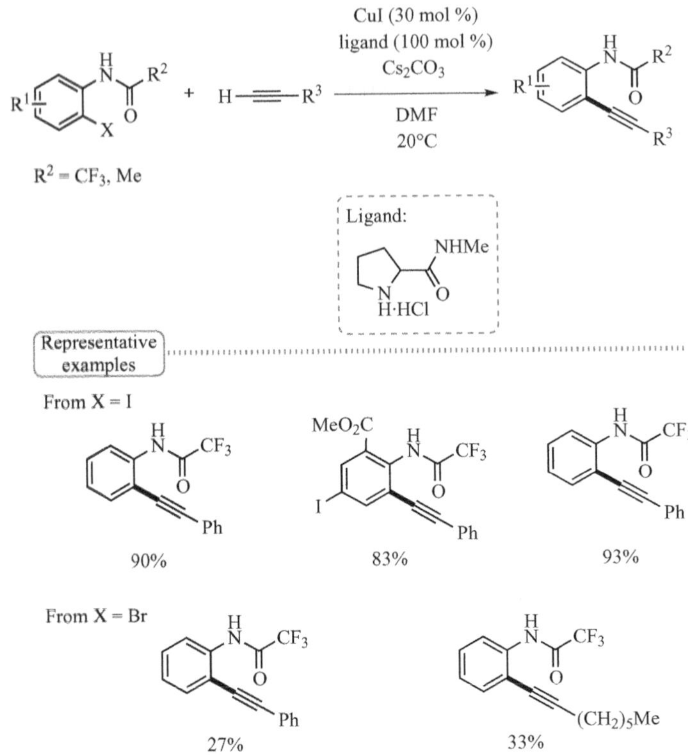

Scheme 11.14 Copper(I)-catalyzed Sonogashira alkynylation using an amino carboxamide as the ligand.

coworkers also employed racemic 1,1′-bi(2-naphthol) (BINOL) as a ligand to improve the copper(I) bromide-catalyzed Sonogashira-type alkynylation reaction.[17] However, the yields were generally lower.

Taillefer and coworkers also reported a mild and efficient catalysis of the Sonogashira-type alkynylation reactions using copper(II) acetylacetonate (10 mol %) and 1,3-diphenyl-1,3-propanedione (30 mol %) as the catalyst (Scheme 11.16).[18] This method, which is among the most general ones, was applicable to the coupling of a wide range of substituted aryl iodides with either alkyl- or aryl-substituted terminal alkynes.

Sekar and coworkers demonstrated that 1,1′-binaphtyl-2,2′-diamine (BINAM) ligands were quite effective at promoting the copper(I) catalyzed Sonogashira alkynylation reactions between aryl halides and terminal alkynes under rather harsh reaction conditions (Scheme 11.17).[19] Notably, aryl bromides could also react with terminal alkynes leading to the corresponding arylated alkynes.

The same group disclosed a second efficient catalytic system consisting of copper(I) bromide (20 mol %) and 1,8-diazabicyclo[5.4.0]undec-7-ene (DBU, 20 mol %) as a ligand, under quite harsh conditions, since the reaction was performed in DMF at 140°C (Scheme 11.18).[20]

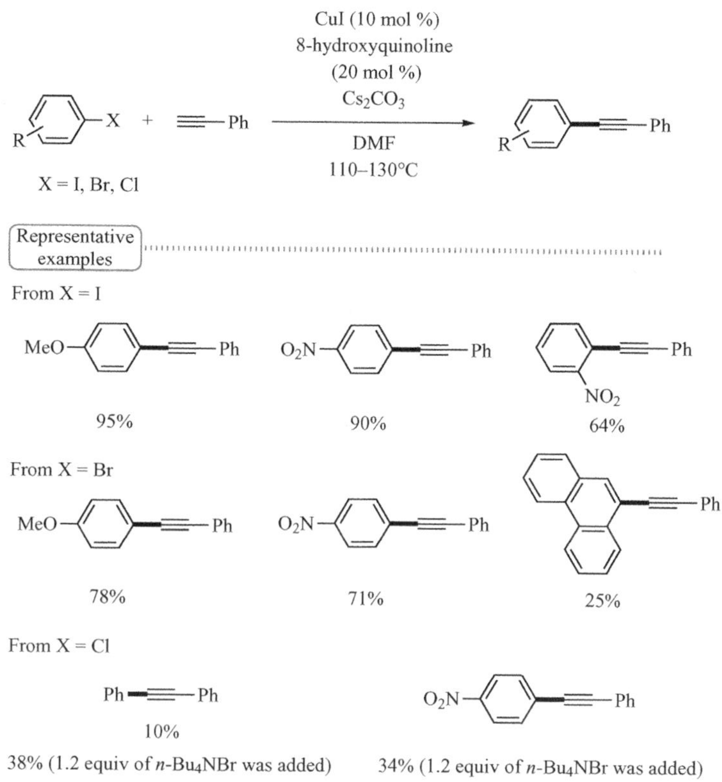

Scheme 11.15 Copper(I)-catalyzed Sonogashira alkynylation using 8-hydroxyquinoline as the ligand.

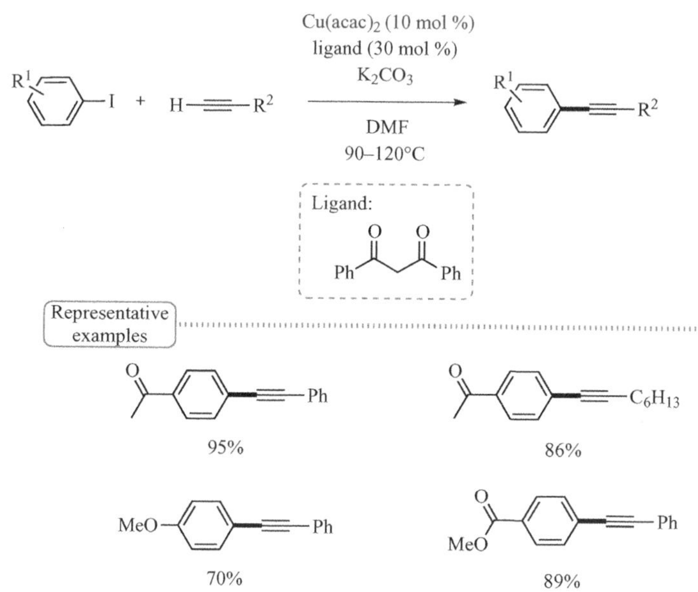

Scheme 11.16 Copper(II)-catalyzed Sonogashira alkynylation using 1,3-diphenyl-1,3-propanedione as the ligand.

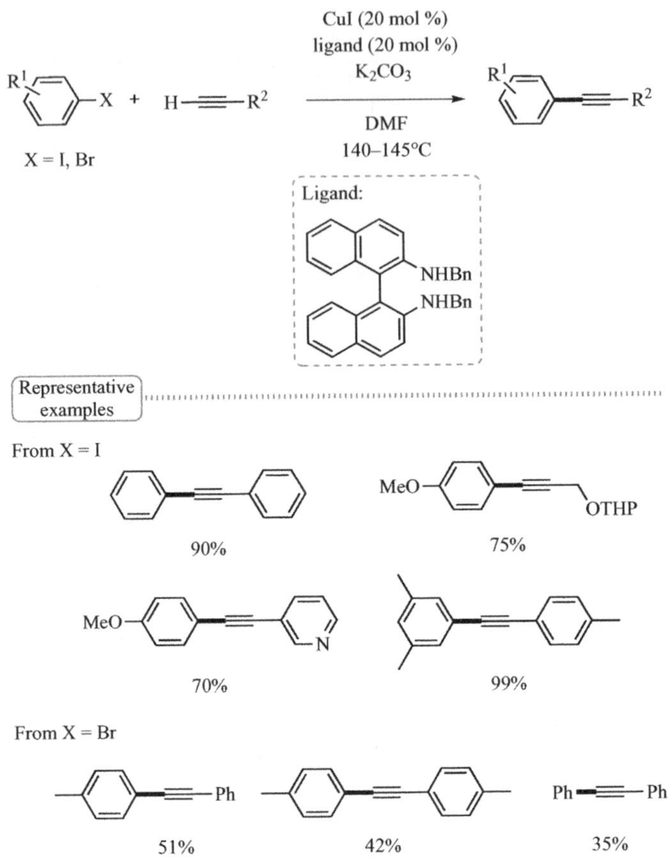

Scheme 11.17 Copper(I)-catalyzed Sonogashira alkynylation using BINAM as a ligand.

Chen and coworkers showed that salicylic acid is a powerful ligand in the copper(II) chloride Sonogashira-type alkynylation reaction of haloarenes with terminal alkynes in DMF at 130°C (Scheme 11.19).[21] The authors postulated that this bidentate *O*,*O*-donor ligand might activate copper(II) chloride during the alkynylation reaction.

To the best of our knowledge, di-*ortho*-substituted aryl halides were only rarely employed as coupling partners in the Sonogashira reaction.[16] Recently, Lee and coworkers described that the combination of copper(I) oxide (1–2.5 mol %) with the diphosphine ligand xantphos (1–2.5 mol %) served as a powerful catalytic system for the alkynylation reactions of aryl iodides with terminal alkynes (Scheme 11.20).[22] The catalytic system tolerates a broad range of functional groups, such as enolizable ketones, esters, nitro group, unprotected amines, chloride, bromide, and heterocycles. Moreover, a highly sterically demanding substrate, 2-ethyl-6-methyliodobenzene, was also suitable for this Sonogashira reaction, which renders this method highly valuable.

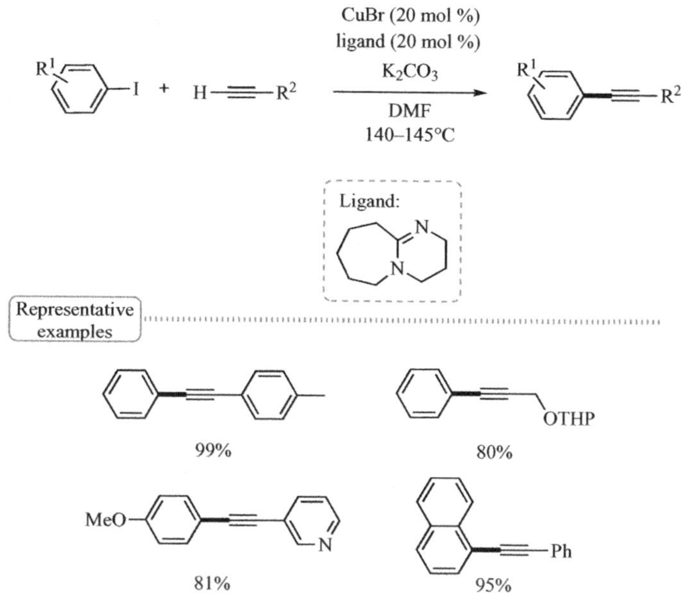

Scheme 11.18 Copper(I)-catalyzed Sonogashira alkynylation using DBU as the ligand.

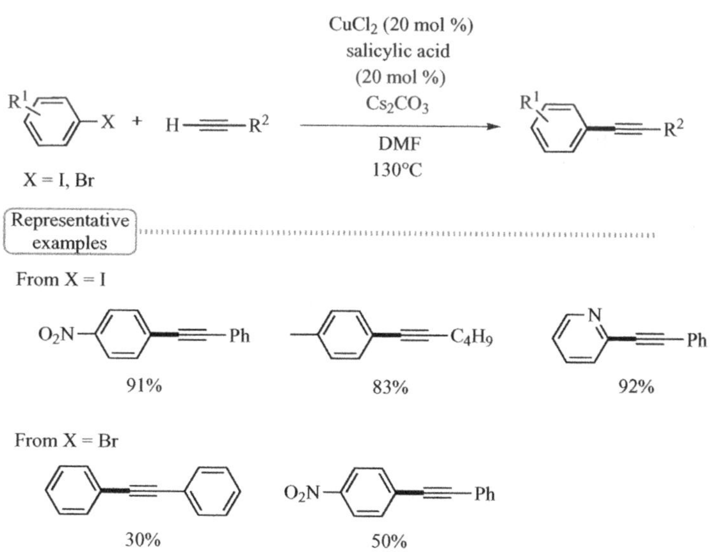

Scheme 11.19 Copper(II)-catalyzed Sonogashira alkynylation using salicylic acid as the ligand.

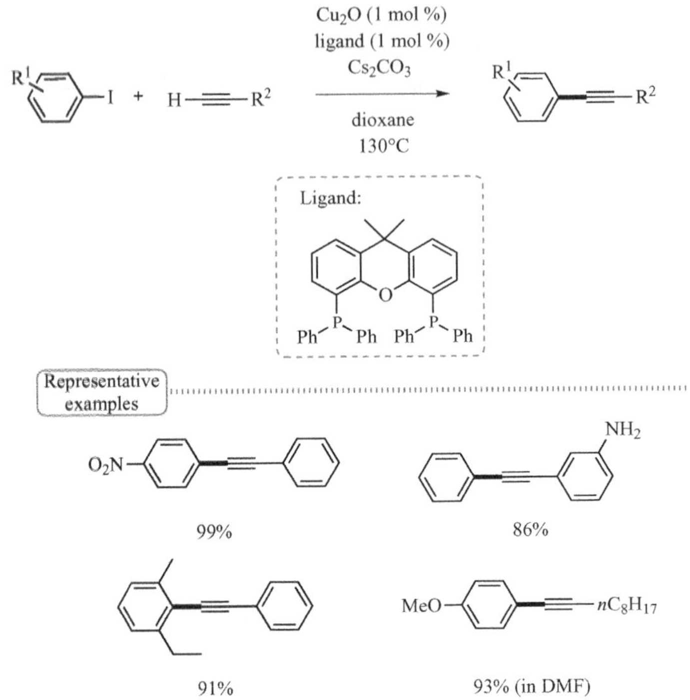

Scheme 11.20 Copper(I)-catalyzed Sonogashira alkynylation using xantphos as the ligand.

Traditionally, copper-mediated coupling reactions have some drawbacks: most copper(I) salts are insoluble in organic solvents, and hence, the coupling reactions are often heterogeneous. In 2001, Venkataraman and coworkers reported that a stable and soluble copper(I) complex could catalyze the Sonogashira reaction with the aid of a base (Scheme 11.21).[23] Both electron-rich and electron-poor aryl iodides were smoothly coupled with phenylacetylene, leading to the corresponding diaryl-substituted alkynes in good yields.

Another frequent drawback of copper-catalyzed reaction is the high catalyst loading typically used (often above 5 mol %). Trying to overcome this limitation, Bolm and coworkers have demonstrated that the Sonogashira reaction could in fact be performed using a submole percent copper loading in the presence of 30 mol % of *N,N'*-dimethylethylenediamine (DMEDA) as a ligand (Scheme 11.22, Eq. 1).[24a] The very low catalyst loading of $[Cu(DMEDA)_2]Cl_2{\cdot}H_2O$ (0.5–0.05 mol %) is especially noteworthy. Kinetic measurements and density functional theory (DFT) calculations showed later on that a concerted breaking of the aryl halide and elaboration of the carbon-carbon bond was operating in a rate-limiting step. No stable copper(III) intermediate was found after the transition state (Scheme 11.22, Eq. 2).[24b]

The third drawback often met with copper catalysis is the usually low reactivity of aryl chlorides. As already shown earlier, additives such as

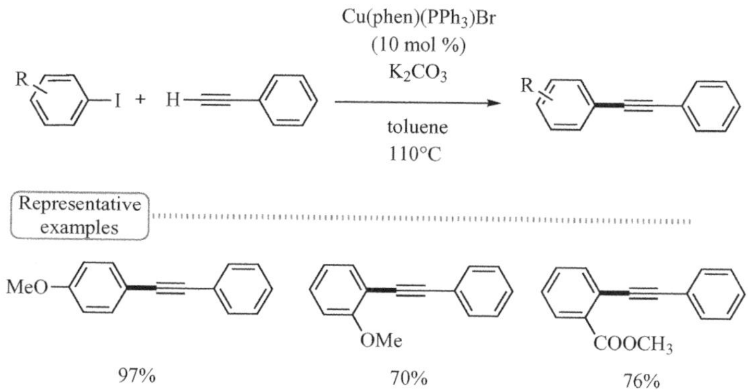

Scheme 11.21 Venkataraman's copper(I)-catalyzed Sonogashira alkynylation using a soluble, preformed copper complex.

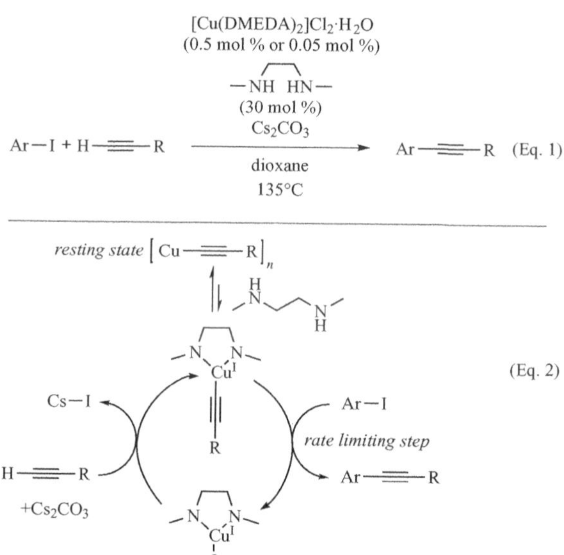

Scheme 11.22 Very low catalyst loading in the copper(II)-catalyzed Sonogashira cross coupling.

tetrabutylammonium bromide can improve their cross coupling, but more reactive catalytic systems would be attractive options. In this context, Kantam and coworkers found that a variety of haloarenes could be reactive partners, including activated aryl chlorides, in the Sonogashira reaction using the bis(μ-iodo)bis[(−)-sparteine]dicopper(I) catalyst (5 mol %) under mild conditions (Scheme 11.23).[25]

Recently, Mao and coworkers developed an efficient alternative process for the Sonogashira-type cross-coupling reaction using $[Cu(acac)_2]\cdot H_2O$

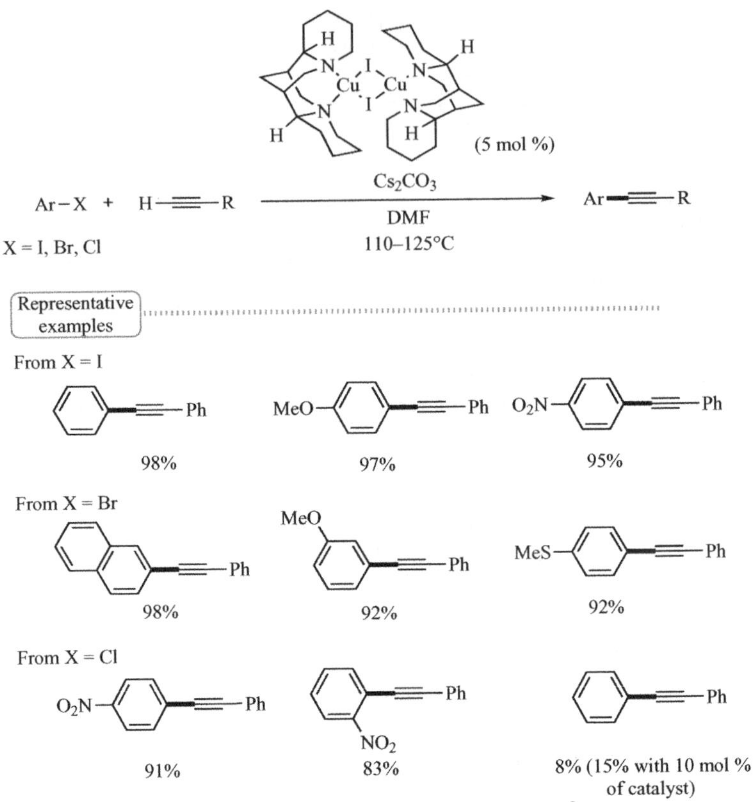

Scheme 11.23 Bis(μ-iodo)bis[(−)-sparteine]dicopper(I)-catalyzed Sonogashira alkynylation.

(10 mol %) as the catalyst, affording the corresponding alkynes in high yields (Scheme 11.24).[26] The reaction was carried out under an inert gas atmosphere and no homocoupling (Glaser type) product was detected, but the coupling had to be performed at 140°C in DMSO.

The number of ligands able to catalyze this reaction being obviously ever expanding, Shao and coworkers also reported that 2,2′-diamino-6,6′-dimethylbiphenyl was a good ligand for the copper(I) iodide-catalyzed Sonogashira coupling reaction. Both aryl iodides and bromides were reacted with terminal alkynes to furnish the corresponding products in moderate-to-excellent yields, even if high reaction temperatures in a polar solvent were required with this system (Scheme 11.25).[27]

As seen with all examples collected, there are many ligands available to date to promote the copper-catalyzed cross coupling between aryl halides and terminal alkynes, which renders the choice of the catalytic system rather tricky. For activated aryl iodides (and bromides), the simplest and commercially available ligands are usually sufficient, and there is clearly no need to spend

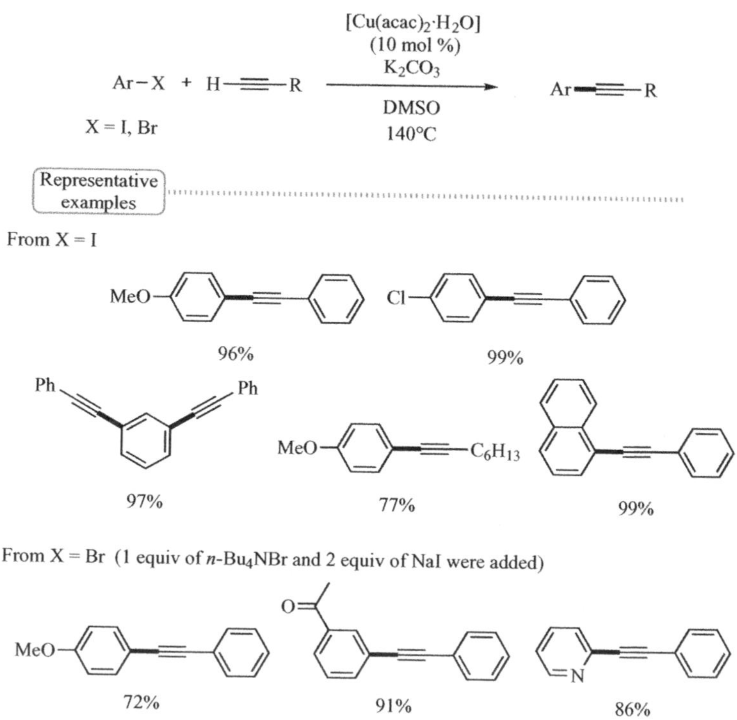

Scheme 11.24 Cu(acac)$_2$-catalyzed Sonogashira alkynylation reaction.

time preparing a ligand that would not be readily available. For more challenging systems such as aryl chlorides or sterically hindered aryl halides, more complex systems whose advantages have been outlined might be needed. They are, however, usually readily synthesized.

The early reported copper-based catalytic systems were not recovered and reused because of their solubilities in harmful organic solvents (often DMF) and their instability after a few catalytic cycles. Moreover, alkynols were found to be unsuitable substrates, and cross coupling involving heteroaryl halides remained largely unexplored. Li and coworkers developed an efficient and reusable catalytic system based on copper(I) nanoparticles with triphenylphosphine as a ligand in tetrabutylammonium bromide for the alkynylation reaction of aryl and heteroaryl halides with terminal alkynes (Scheme 11.26).[28] The results demonstrated that the octahedral Cu_2O nanoparticles were the most effective catalyst for the reaction: A variety of aryl and heteroaryl halides could be reacted with alkynes in the presence of these octahedral Cu_2O nanoparticles (10 mol %), triphenylphosphine (20 mol %), potassium carbonate (2 equiv), and tetrabutylammonium bromide (TBAB). It is noteworthy that the Cu_2O/PPh_3/TBAB system can be recovered and reused several times without loss of activity. Furthermore, the scope of the substrates was nicely

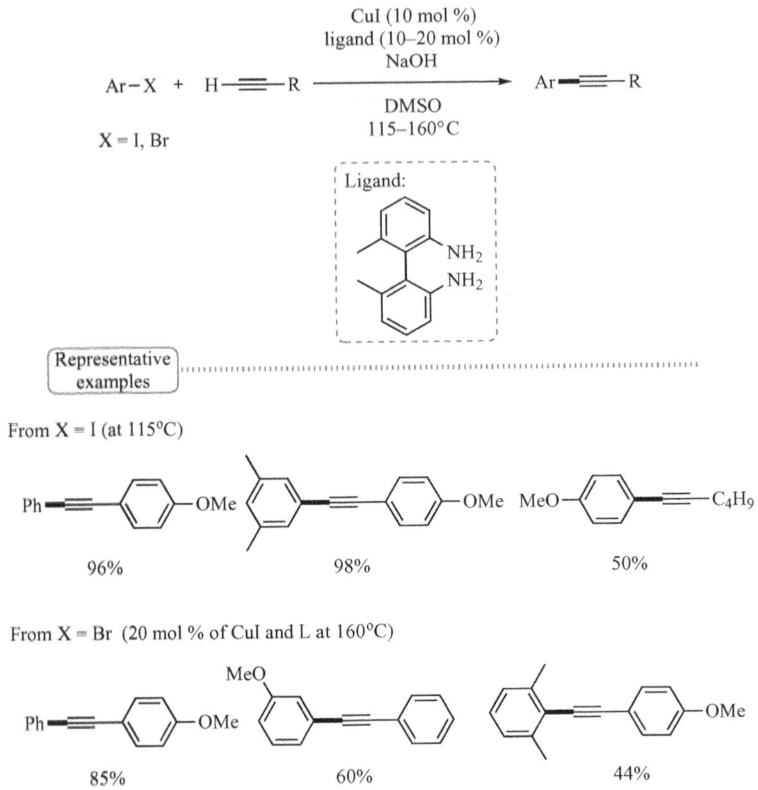

Scheme 11.25 Copper(I)-catalyzed Sonogashira reaction using 2,2′-diamino-6,6′-dimethylbiphenyl as a ligand.

expanded to alkynols (a comprehensive coverage on reusable copper catalysts in copper-mediated reactions can be found in Chapter 20).

Another recoverable system was developed by Rothenberg and coworkers, who used copper nanoclusters to catalyze the Sonogashira alkynylation reaction of aryl iodides or bromides with alkynes (Scheme 11.27).[29] The clusters were simple to prepare, stable, and displayed different catalytic properties from their homogeneous analogues. The recycling of the active catalyst was also studied and it was shown that the cross coupling of phenylacetylene and *p*-iodotrifluorotoluene could be catalyzed by the same copper cluster three times in a row without any deactivation (final turnover number [TON] of 73).

In most of these copper-mediated cross-coupling reactions, organic solvents are employed as the media, often creating safety, health, and environmental issues owing to their flammability, toxicity, and volatility. The use of water or aqueous solvents represents one of the most environmentally benign alternatives to organic solvents for the copper-catalyzed reactions.

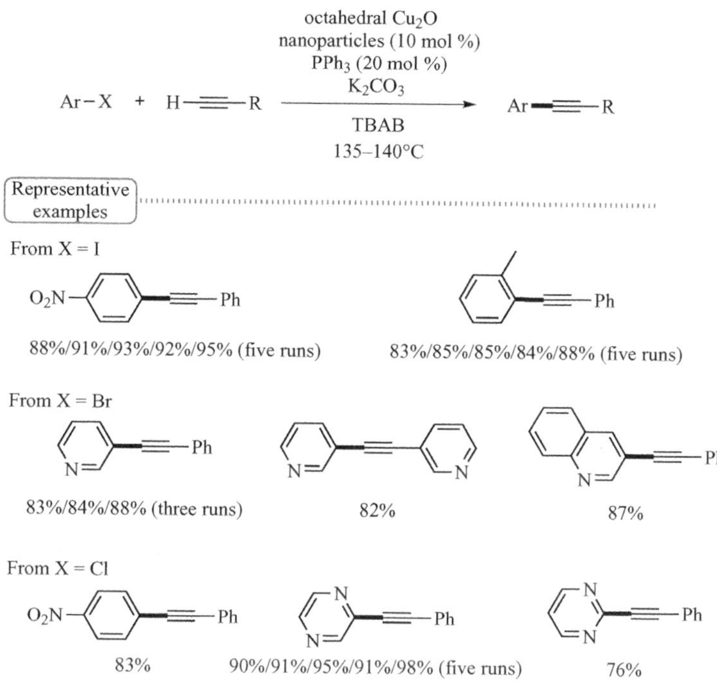

Scheme 11.26 Reusable octahedral copper(I) nanoparticle-catalyzed Sonogashira reaction.

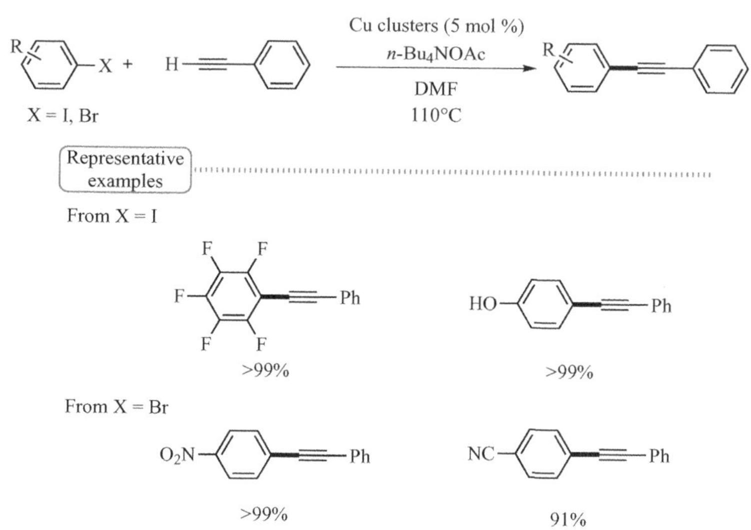

Scheme 11.27 Reusable copper nanocluster-catalyzed Sonogashira reaction.

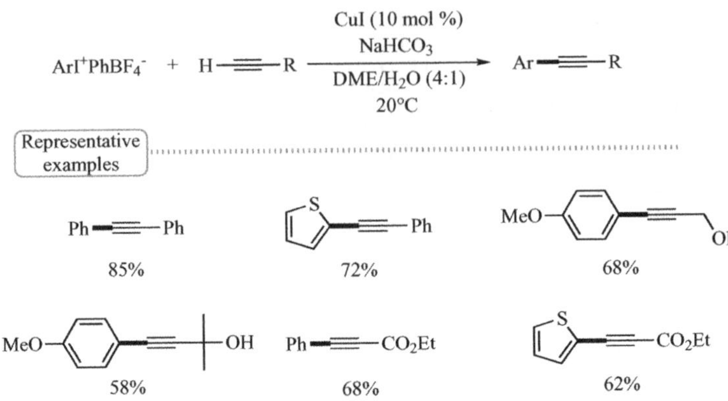

Scheme 11.28 Copper(I)-catalyzed Sonogashira reaction with hypervalent iodonium salts.

Kang and coworkers have developed a new and mild Sonogashira alkynylation protocol for the synthesis of disubstituted arylated alkynes in DME/H_2O (4:1) utilizing copper(I) iodide (10 mol %) as the catalyst and hypervalent iodonium salts as the electrophilic coupling partner (Scheme 11.28).[30] The scope of the reaction is quite interesting since aryl-, alkyl-, and ester-substituted alkynes could be employed for this purpose. In addition, 2-methyl-3-butyn-2-ol could nicely be used as surrogates for terminal alkynes under basic conditions (sodium hydroxide) with remarkable efficiencies. Hypervalent alkenyl iodonium salts could also be employed (not shown). A limitation of this procedure is that it requires the synthesis of the hypervalent iodonium salt, which is clearly less practical than starting from more available aryl- or alkenyl halides. The reaction can, however, be performed at room temperature in 30 minutes with these reagents, which is clearly a marked advantage over other methods.

While water was only used as a minor cosolvent in the previous system, Liu and coworkers subsequently described another system in which only water was used as the solvent for the alkynylation reaction in the presence of copper(I) iodide (5 mol %), triphenylphosphine (15 mol %) and potassium hydroxide as the base (Scheme 11.29).[31] Both electron-rich and electron-deficient aromatic iodides could successfully undergo the reaction with aryl- and alkyl-substituted alkynes.

Wan and coworkers reported an alternative procedure for the copper-catalyzed Sonogashira alkynylation reaction in water, under microwave irradiation, or at reflux (Scheme 11.30).[32] A stoichiometric amount of tetrabutylammonium bromide was necessary to enhance the yields.

Lamaty and coworkers found that the copper-catalyzed Sonogashira reaction could be performed in poly(ethylene)glycol (PEG) with various molecular weights in the 300–3400 range under microwave activation (Scheme

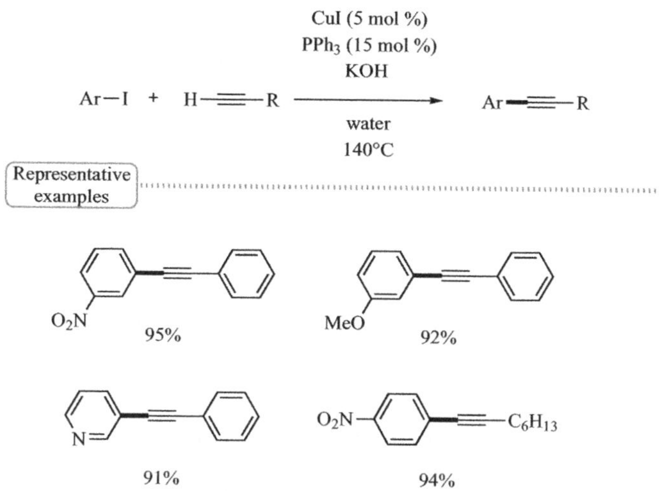

Scheme 11.29 Copper(I)-catalyzed Sonogashira reaction in water.

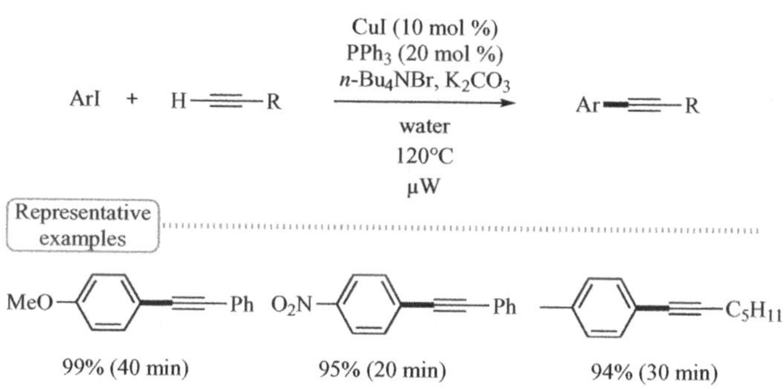

Scheme 11.30 Microwave assisted copper(I)-catalyzed Sonogashira reaction in water.

11.31).[33] A variety of substituted diarylacetylenes were synthesized in the presence of copper(I) iodide (10 mol %) and potassium carbonate, PEG most certainly acting as a chelating ligand for copper. This environmentally more friendly procedure might prove useful in the near future, especially since the recovery of the PEG-based catalytic system is based on a precipitation step.

As seen with all results described in this section, tremendous efforts have been devoted to the development of copper-based systems for the Sonogashira reaction that can be performed without the need for a palladium catalyst. In most cases, heating at elevated temperatures in a polar solvent is, however, needed, which might be problematic for sensitive or fragile substrates. Efforts will clearly have to be devoted to the design of more efficient copper-based catalysts to answer this challenging problem. Another option is to use other

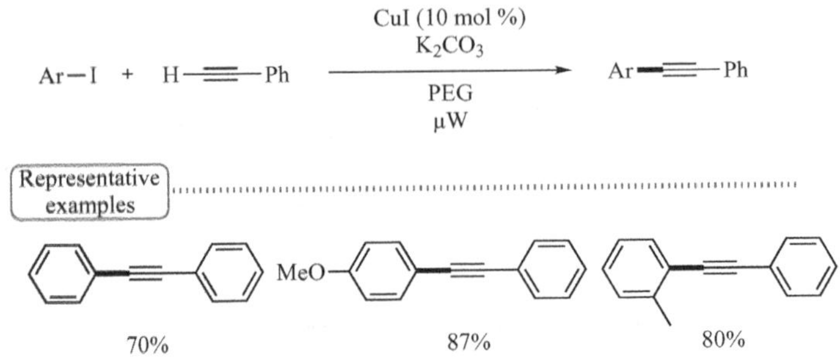

Scheme 11.31 Copper(I)-catalyzed Sonogashira reaction in poly(ethylene glycol).

coupling partners than aryl halides. Indeed, arylboronic acids have been recently shown to be readily coupled with terminal alkynes under mild oxidative conditions. This strategy will be overviewed in the next section.

11.2.3 Synthesis of Aryl–Ynes by Oxidative Cross Coupling of Alkynes and Arylboronic Acids

The synthesis of alkynyl-substituted aromatics by oxidative cross coupling between arylboronic acids and terminal alkynes is especially attractive from a conceptual point of view since it does not involve any oxidative addition step and could therefore be performed under mild conditions using copper catalysis, provided that the easy homodimerization of both starting materials can be minimized. After early developments of this reaction based on the use of palladium catalysts, this concept was actually validated in 2006 by Cheng and coworkers, who reported a remarkable copper(I)-catalyzed, ligand-free, oxidative cross coupling of arylboronic acids and terminal alkynes.[34] Using a catalytic amount of copper(I) iodide (15 mol %) and cesium carbonate in 1,2-dichloroethane at 80°C in combination with silver(I) oxide as the oxidant, a range of activated alkynes (arylacetylenes, propiolates, and conjugated acetylenic amides) was readily coupled to a series of arylboronic acids in moderate-to-good yields (Scheme 11.32). This simple alternative to the traditional Sonogashira reaction performed well regardless of the electronic properties of the starting arylboronic acid. However, steric hindrance of the arylboronic acid had a detrimental effect on the reaction and the reactivity of alkyl-substituted alkynes was not reported.

Milder conditions were reported a year later by the Fu group, who demonstrated the crucial effect of additives for this oxidative cross coupling. The protocol used inexpensive copper(I) oxide as the catalyst, air as the stoichiometric oxidant, and pyridine (and methanol with some substrates) as the additive.[35] Under these conditions, the cross coupling could be performed with a wide range of substrates at room temperature in chloroform (Scheme 11.33). The electronic effect of substituents on the aromatic boronic acids had little

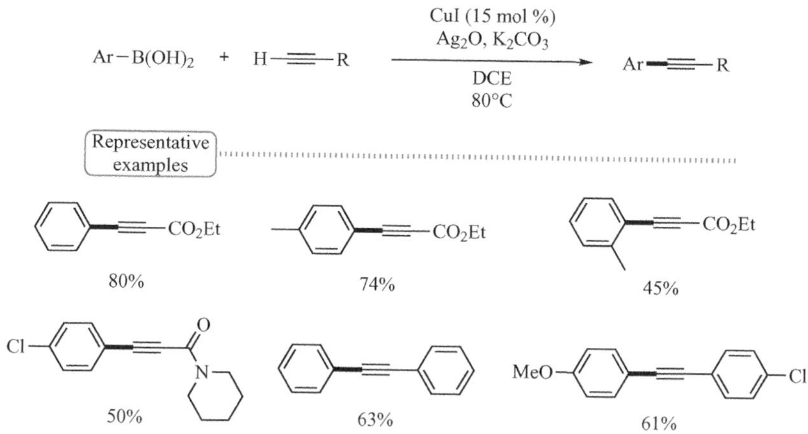

Scheme 11.32 Copper(I)-catalyzed oxidative cross coupling of arylboronic acids with terminal alkynes.

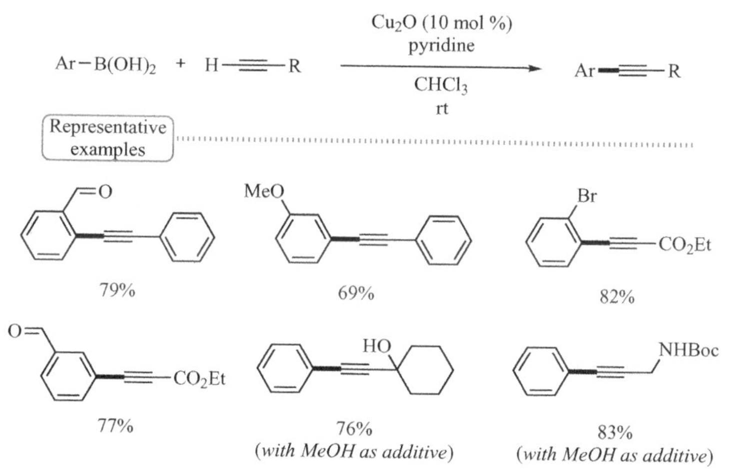

Scheme 11.33 Crucial role of additives in the room-temperature copper(I)-catalyzed oxidative cross coupling of arylboronic acids with terminal alkynes.

influence on the outcome of the reaction, and the reaction rates were more influenced by the nature of the terminal alkynes, the one containing unprotected alcohols or carbamates requiring small amount of methanol as extra additive.

Another advance in this area was reported in 2011 by Kobayashi and coworkers, who demonstrated the dramatic rate enhancement of the oxidative cross coupling with 2,6-lutidine.[36] Using this reagent as an additive or a cosolvent in methanol allowed for a smooth copper(I)-bromide-catalyzed alkynylation of a series of arylboronic acids with an impressive range of terminal alkynes with remarkable selectivities and yields (Scheme 11.34). The low

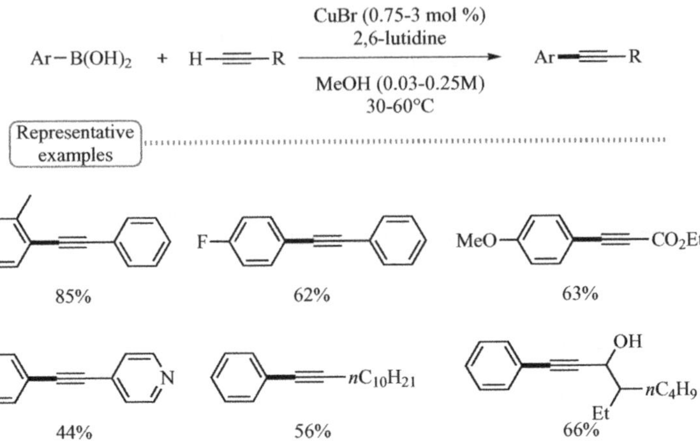

Scheme 11.34 Remarkable selectivity of the copper(I)-catalyzed oxidative cross coupling of arylboronic acids with terminal alkynes using 2,6-lutidine as an additive.

catalyst loading (0.75–3 mol %) is another astonishing feature of this system, which, in combination with the concentration of the reaction mixture, was key to suppress the homocoupling side reactions.

As evidenced with these three recent developments, the copper-catalyzed oxidative cross coupling of alkynes and arylboronic acids has clearly evolved recently into a powerful method for the synthesis of aryl-substituted alkynes with remarkable efficiency and under mild conditions. Compared with the more classical approach involving oxidative addition from alkynyl bromides (which have been recently shown to be catalyzed by catalytic amounts of copper(I) iodide in ethanol at 80°C[37a]) or 1,1-dibromoalkene in the presence of copper(I) iodide (10 mol %), 8-hydroxyquinoline (10 mol %) and potassium phosphate as a base,[37b] or metallation/oxidation from the corresponding terminal alkynes, the reaction conditions are milder, which renders this approach particularly attractive for the preparation of sensitive substrates.

Besides the use of this strategy, other methods have also been reported, such as the decarboxylative cross coupling involving propiolic acids or the direct functionalization of C–H bonds in activated arenes. The former strategy will be briefly described in the following section.

11.2.4 Synthesis of Aryl–Ynes by Decarboxylative Cross Coupling from Propiolic Acids

Decarboxylative cross-coupling reactions have been extremely studied recently, and various classes of carboxylic acids have been shown to be excellent reaction partners for various transformations, mostly associated with palladium catalysis, copper salts being often used for the decarboxylation step. Among all carboxylic acids that have been shown to participate in such

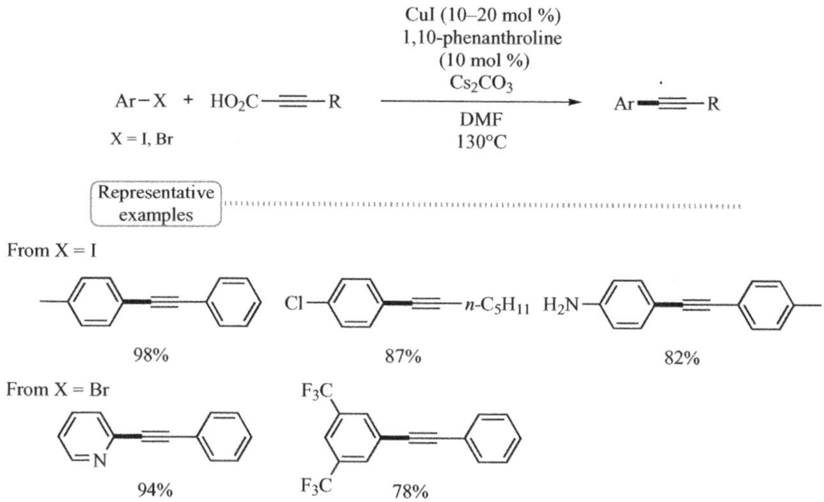

Scheme 11.35 Copper(I)-catalyzed decarboxylative cross coupling of propiolic acids with aryl halides.

coupling reactions, propiolic acids (or alkynoic acids) have been demonstrated to be rather efficient reagents for the introduction of a triple bond and therefore act as efficient alkynes surrogates.

Indeed, they have recently been shown to participate in copper-catalyzed cross coupling with aryl halides without the assistance of palladium complexes.

Initial efforts in this area were reported by You, Xue, and coworkers in 2010. In a seminal publication, they demonstrated that a wide range of aryl iodides and activated aryl bromides could be efficiently alkynylated by various propiolic acids in the presence of copper(I) iodide, 1,10-phenanthroline and cesium carbonate in DMF at 130°C (Scheme 11.35).[38] This reaction is quite attractive in terms of scope and functional group tolerance, although an elevated reaction temperature is required, which is inherent to the decarboxylation step. Worthy of note is that aliphatic propiolic acids could be effectively coupled in good-to-excellent yields and that treatment of 2-iodophenol with alkynyl carboxylic acids under these conditions yields to the corresponding benzofurans after *in situ* cyclization.

A year later, the Mao and Jiao groups independently reported alternative conditions for this decarboxylative cross coupling based on the use of copper(I) iodide (2 mol %)/triphenylphosphine (4 mol %) in DMSO at 90–110°C[39a] (Scheme 11.36, Eq. 1) or copper(I) bromide (1–5 mol %)/dibenzoylmethane (1–5 mol %) in the presence of potassium carbonate, air, and sodium nitrite in DMF at 120°C (Scheme 11.36, Eq. 2),[39b] respectively. These systems are also fairly efficient, although in the second case, aryl bromides require palladium cocatalysis. In addition, the reactivity of alkyl-substituted propiolic acids

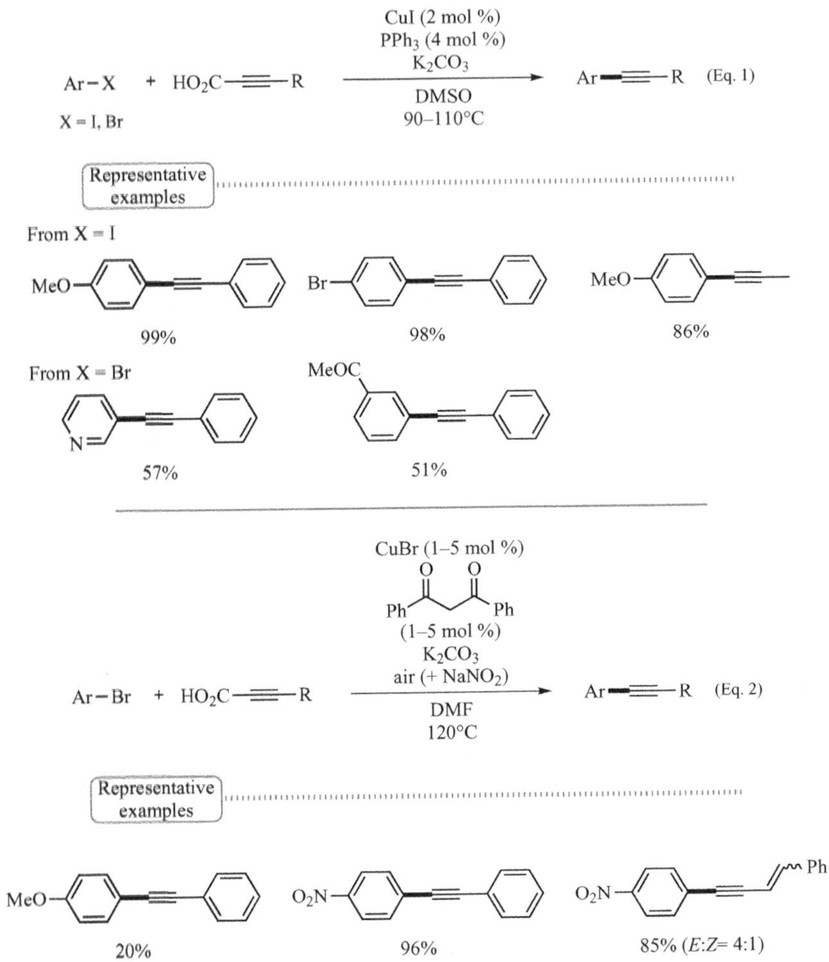

Scheme 11.36 Copper(I)-catalyzed decarboxylative cross coupling of propiolic acids and aryl halides.

has not been addressed and the role of air and sodium nitrite is rather unclear. As a note, this decarboxylative cross coupling has also been recently shown to be efficiently performed in water using a quite low catalyst loading[40a] and could also use aryl boronic acids as coupling partners.[40b]

The copper-catalyzed decarboxylative cross coupling therefore represents an interesting strategy for the synthesis of aryl–ynes provided that the propiolic acids needed are readily available and that the substrates are compatible with the high temperatures required for these cross-coupling reactions. An advantage of these methods is the higher boiling points of propiolic acids compared to the ones of the corresponding alkynes, which can be quite interesting when the latter are too volatile.

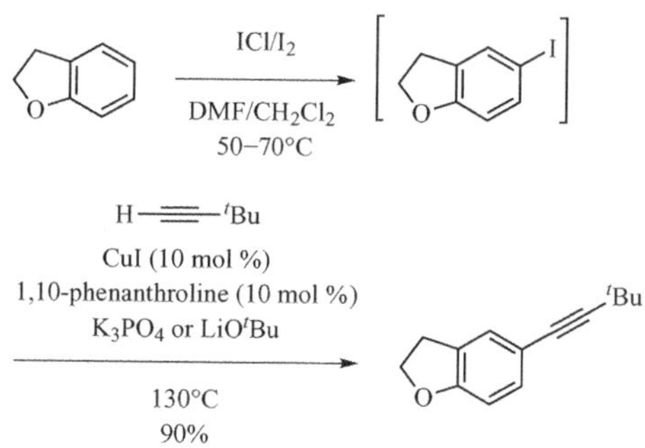

Scheme 11.37 Daugulis copper(I)-catalyzed formal C–H bond functionalization of benzofurans.

Before closing this subchapter on the synthesis of aryl–ynes, a last strategy that has also been developed over the last couple of years and that clearly holds promises will be described: the direct alkynylation of C–H bonds in activated arenes.

11.2.5 Synthesis of Aryl–Ynes by Direct C–H Functionalization of Arenes

Transition-metal-catalyzed direct C–H functionalization has gained enormous attention over the past decade because it presents many advantages over traditional cross-coupling reactions.[41] Logically, copper-mediated oxidative direct carbon–carbon bond formation has spurred a lot of interest and the most relevant examples are presented in the following paragraphs.[41d]

In 2009, Do and Daugulis reported a single example of the synthesis of an arylated alkyne by a formal copper-catalyzed C–H alkynylation reaction (Scheme 11.37).[42] The reaction was conducted by an initial electrophilic iodination of benzofuran, followed by arylation of the newly formed carbon–iodide bond with copper(I) iodide (10 mol %) and phenanthroline (10 mol %).

A key report in the synthesis of aromatic alkynes by direct copper-mediated alkynylation of acidic C–H bonds was described by Besselièvre and Piguel in 2009 when they developed an efficient system for the direct alkynylation reaction of various heterocycles (Scheme 11.38, Eq. 1).[43a] This straightforward protocol relies on CuBr·SMe_2 (15 mol %), DPEPhos (15 mol %), and lithium *tert*-butoxide in dioxane at 120°C. It is worth noting that these reaction conditions were also successfully applied to both electron-rich and electron-deficient oxazoles. Furthermore, this work illustrated once again that inexpensive copper salts are able to replace palladium catalysis in the field of C–H bond activation. A year later, the same group reported that *gem*-dibromoalkenes could also be used for the direct alkynylation of azoles under the exact same reaction conditions but using a larger amount of lithium *tert*-butoxide to promote the elimination involved in the process (Scheme 11.38, Eq. 2).[43b]

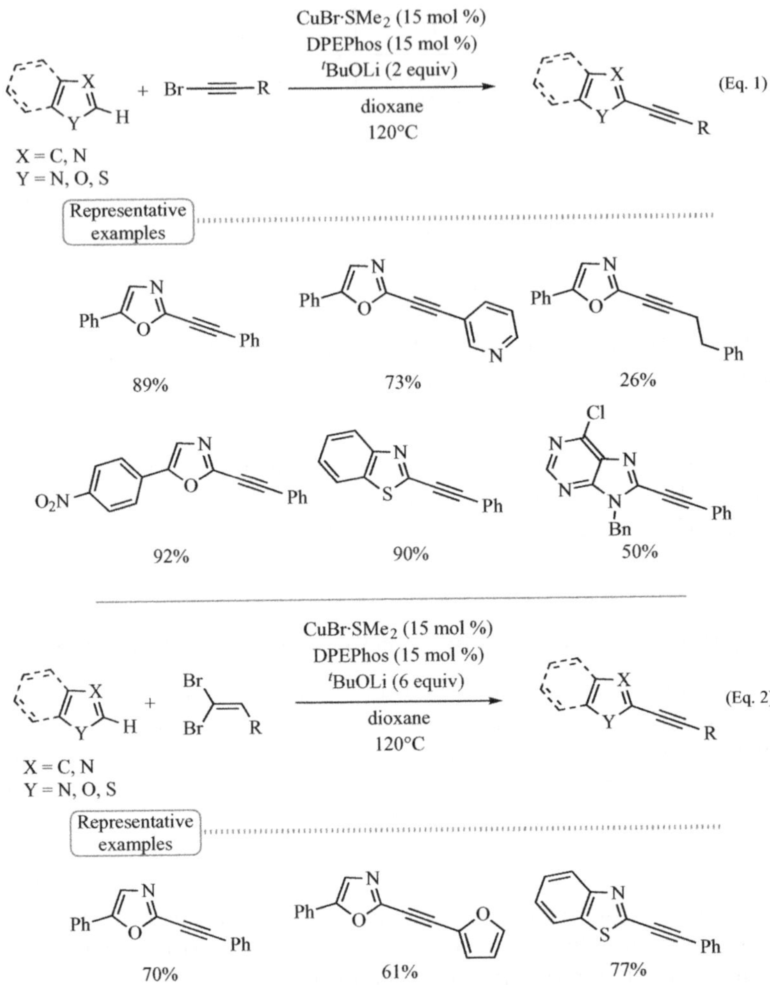

Scheme 11.38 Piguel's direct alkynylation reaction of azoles.

In a subsequent study, Miura and coworkers described a milder system for the copper-catalyzed direct alkynylation of 1,3,4-oxadiazoles with alkynyl bromides, allowing the creation of the corresponding heteroaryl-alkynyl linkage in good yields (Scheme 11.39).[44] It is important to note that the reaction can proceed at room temperature.

In continuation of their studies, the Miura group next reported a copper-mediated direct oxidative alkynylation of 1,3,4-oxadiazoles and oxazoles with terminal alkynes through the formal union of sp^2 C–H and sp C–H bonds, furnishing the corresponding alkynylazoles in good yields, under much harsher conditions than in the previous procedure, however (Scheme 11.40).[45]

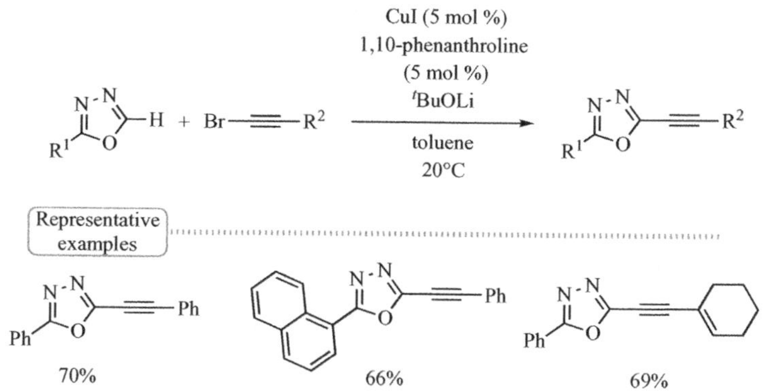

Scheme 11.39 Miura's copper(I)-catalyzed room-temperature direct alkynylation of 1,3,4-oxadiazoles with alkynyl bromides.

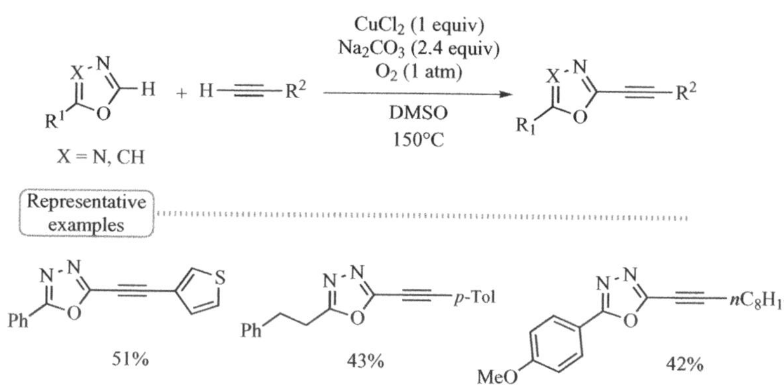

Scheme 11.40 Miura's copper(II)-mediated direct alkynylation of 1,3,4-oxadiazoles with terminal alkynes.

Control experiments suggested that the reaction could, quite logically, proceed in the presence of only a substoichiometric amount of copper(II) chloride under an oxygen atmosphere: The copper(II) chloride/*N*,*N′*-dimethylethylenediamine catalytic system was indeed found to efficiently mediate this transformation, at the expense of a rather high catalyst loading (Scheme 11.41).

A plausible reaction mechanism for this transformation is shown in Scheme 11.42. Formation of copper acetylide **8** in a first step is followed by the cupration of an oxazole or oxadiazole using sodium carbonate, thereby generating the [Cu^{II}-(alkynyl)(heteroaryl)] intermediate **9**. Oxidation of the latter by oxygen or by copper(II) should lead to a copper(III) intermediate which upon reductive elimination should give the desired products **5** and **6** together with copper(I), which is subsequently reoxidized to the catalytically active

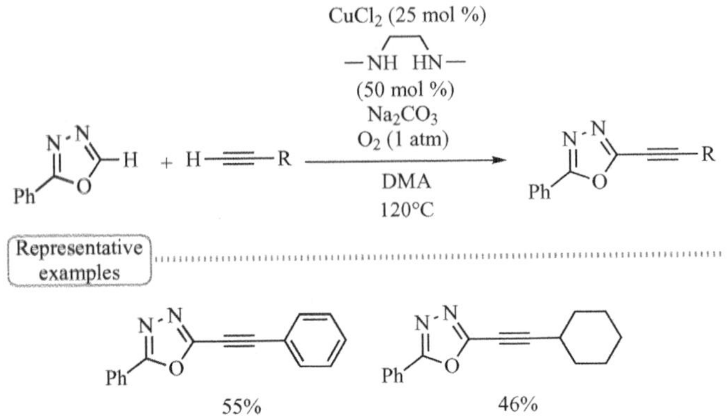

Scheme 11.41 Miura's preliminary results for the copper(II)-catalyzed direct alkynylation of 1,3,4-oxadiazoles with terminal alkynes.

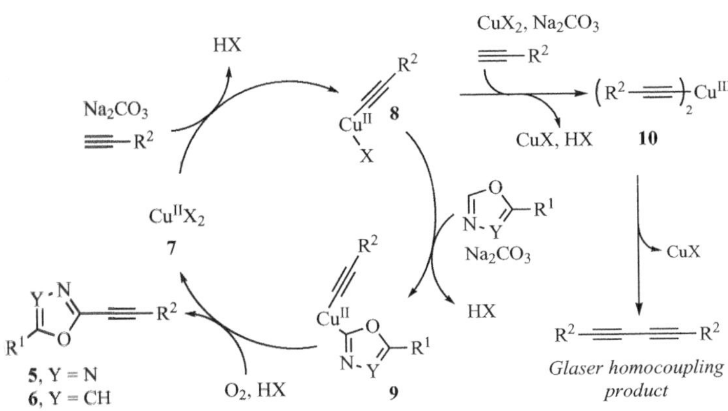

Scheme 11.42 A plausible reaction mechanism for the copper(II)-mediated direct cross coupling of 1,3,4-oxadiazoles with terminal alkynes.

copper(II) salt **7**. In order to minimize the classical Glaser homocoupling reaction of terminal alkynes, a slow addition technique of the alkyne partner is mandatory (for mechanistic insights of the Glaser reaction and of its variants, see Chapter 12, Section 12.2). The exact role of the oxygen atmosphere was not clearly understood, but the authors hypothesized that it might facilitate the reductive elimination step (by generation a copper(III) intermediate; see Chapter 7 for a detailed discussion of the mechanism of the related Chan–Lam–Evans reaction).

Su and coworkers have also described a copper(II)-catalyzed method for the construction of C(sp^2)–C(sp) bond via a direct alkynylation of highly activated aromatic C–H bond of perfluoroarenes with a broad range of terminal alkynes as the coupling partners (Scheme 11.43).[46] This protocol used molecular oxygen as the terminal oxidizing reagent and enables the direct

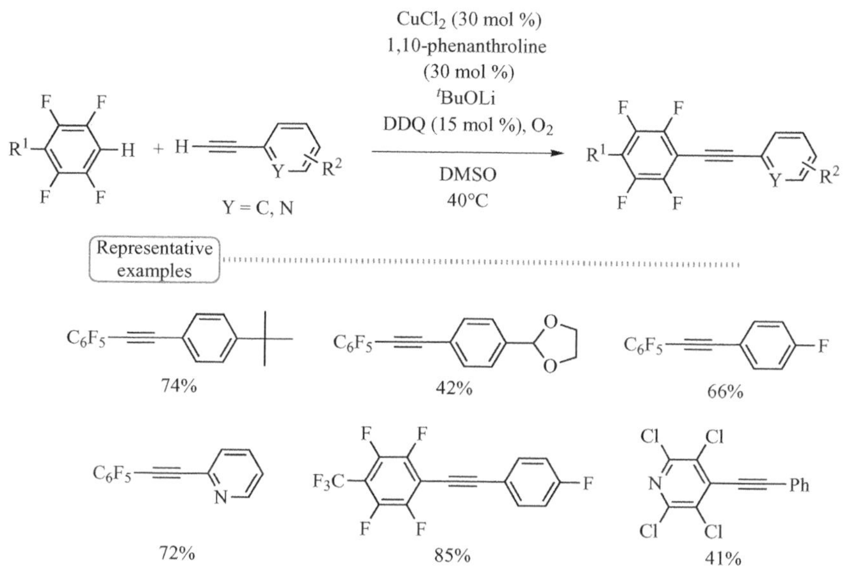

Scheme 11.43 Copper(II)-catalyzed direct alkynylation of polyfluoroaryl C–H bond.

alkynylation of electron-deficient polyfluoroarenes (and tetrachloropyridine) under mild conditions. The products, fluorinated arylalkynes, are valuable intermediates in organic synthesis and material science.

A possible mechanism for this copper-catalyzed direct alkynylation of polyfluoroarenes with terminal alkynes has been proposed by the authors and is shown in Scheme 11.44. The formation of the key Cu(alkynyl)(fluoroaryl) complex **11**, which gives the desired cross-coupling products via reductive elimination, is expected to compete with the formation of the symmetrically substituted Cu(fluoroaryl)$_2$ complex **12** and Cu(alkynyl)$_2$ complex **13**. Reductive eliminations from the latter two give rise to homocoupling products (bipolyfluoroaryls and 1,3-diynes).

As described in this subchapter, tremendous advances have been reported in a short period of time in the development of efficient and general processes for the synthesis or aryl–ynes by the means of copper catalysis. A variety of aromatic partners including aryl halides, stannanes, hypervalent iodonium salts, boronic acids, or activated C–H arenes can now be readily alkynylated by simple alkynes as well as alkynyl halides, stannanes, or hypervalent iodonium salts. Depending on the availability of the starting materials, the required reaction conditions and on the substitution pattern of the aryl–yne one wishes to prepare, at least one of the methods should definitely be most useful.

Copper catalysis also had a deep impact on the synthesis of styrene derivatives, compounds for which there are also now various copper-mediated processes available for their synthesis. The most significant strategies in this area will be the subject of the next section.

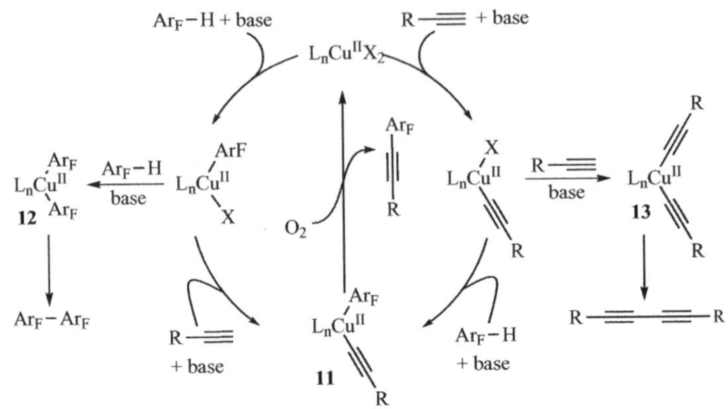

Scheme 11.44 A possible mechanism for the copper(II)-catalyzed direct alkynylation of polyfluoroarenes with terminal alkynes.

11.3 COPPER-CATALYZED ALKENYLATION OF ARYL DERIVATIVES

Copper-mediated Stille-, Heck-, and Suzuki–Miyaura-type alkenylation reactions have emerged as the most efficient tools for the introduction of an aryl substituent to an alkene via carbon–carbon bond formation, and most of these reactions can now be performed with copper as the only catalyst, without the need of additional palladium complexes.

11.3.1 Synthesis of Aryl–Enes by Stille-Type Cross Coupling

As shown in Scheme 11.45, Allred and Liebeskind demonstrated that 1.5 equiv of copper(I) thiophene-2-carboxylate (CuTC) was an efficient mediator for the alkenylation reactions of aryl-, heteroaryl-, and alkenylstannanes with alkenyl or activated aryl iodides in *N*-methyl pyrrolidinone (NMP) at 0°C or room temperature.[47a] The reaction is stereospecific and could tolerate many functional groups.

Lin and coworkers proposed a mechanism for the CuTC-mediated Stille alkenylation reaction of vinylstannane (R^1-$SnMe_3$) with vinyl iodide (R^2-I) using density functional theory calculations (Scheme 11.46).[47b] The first step is the transmetallation between the solvent-CuTC complex **14** and R^1-$SnMe_3$, giving the copper intermediate Sol–Cu-R^1 (**15**). The second step is a one-step process: oxidative addition of R^2-I to Sol–Cu-R^1 and reductive elimination providing the desired coupled product and Sol–Cu–I (**16**). Further ligand exchange with thiophenecarboxylate regenerates the catalytically active copper(I) species **14**.

Aryl–alkene bond formation could also be catalyzed by copper(I) iodide (10 mol %) without any added ligand, starting from organotin derivatives and aryl or vinyl iodides, in the presence of sodium chloride (Scheme 11.47).[4]

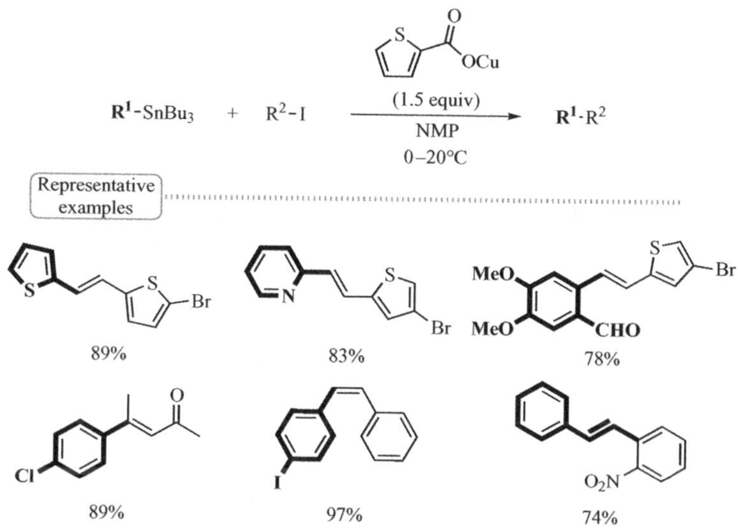

Scheme 11.45 Copper(I) 2-thiophenecarboxylate (CuTC)-mediated synthesis of styrene derivatives by Stille-type cross coupling (group originating from the organotin reagent shown in bold).

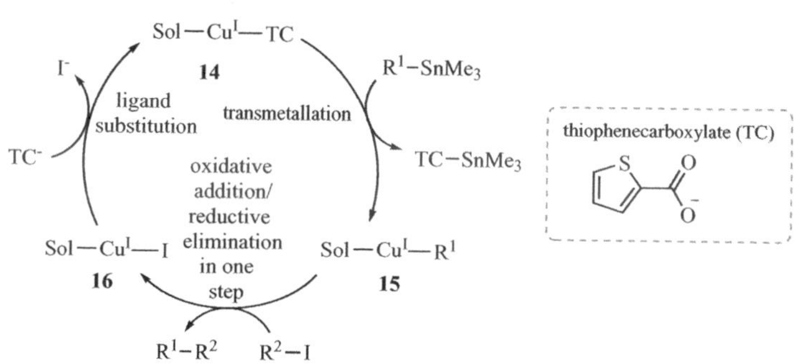

Scheme 11.46 Lin's mechanism for the CuTC-mediated the Stille alkenylation reaction.

This catalytic system was later extended to the alkenylation reaction of vinylstannanes with polymer-bounded aryl iodides (Scheme 11.48).[5] It is worth noting that organostannanes must be added slowly by a syringe pump to improve the yields, which limits the practical use of this cross coupling.

Copper nanoparticle-catalyzed Stille-type alkenylation reaction between aryl halides and vinyltin derivatives for the synthesis of alkenes could also be performed in inexpensive ionic liquids such as tetrabutylammonium bromide (Scheme 11.49).[6] In the presence of copper(I) oxide nanoparticles (10 mol %) and tri-*o*-tolylphosphine (20 mol %), 1-bromo-4-nitrobenzene underwent the Stille reaction with tributyl(vinyl)stannane smoothly in 89% yield.

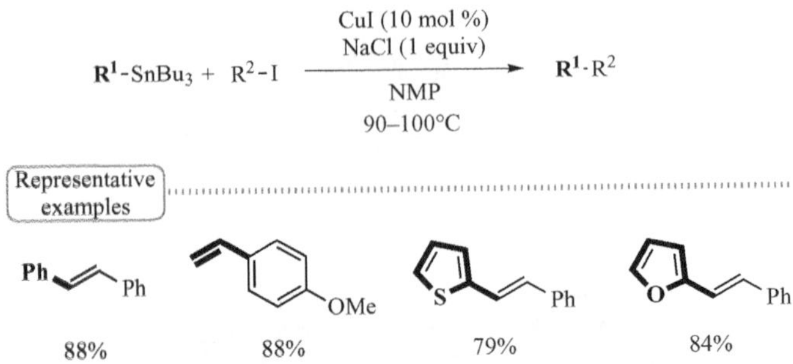

Scheme 11.47 Copper(I) iodide-catalyzed synthesis of styrene derivatives by Stille-type cross coupling (group originating from the organotin reagent shown in bold).

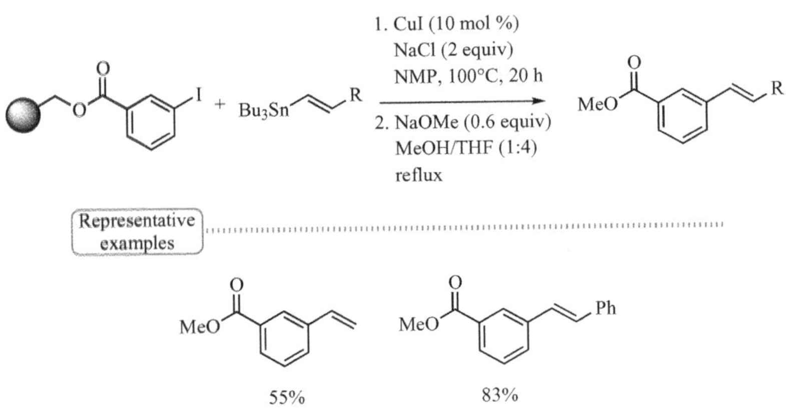

Scheme 11.48 Copper(I)-catalyzed alkenylation reaction of vinylstannanes with polymer-bounded aryl iodide.

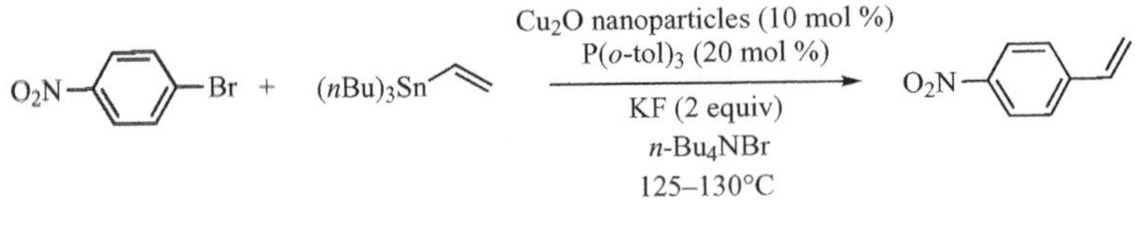

Scheme 11.49 Cu_2O nanoparticle-catalyzed Stille reaction for the synthesis of 1-nitro-4-vinylbenzene in ionic liquid.

11.3.2 Synthesis of Aryl–Enes by Suzuki–Miyaura-Type Cross Coupling

The Suzuki–Miyaura cross coupling is the reaction between an organoboronic acid and an organohalide. The Suzuki–Miyaura reaction was first published in 1979 by Akira Suzuki and Norio Miyaura,[48] the former being awarded the Nobel Prize in Chemistry (in 2010) for the discovery and development of this reaction.

The most common catalytic system used for this transformation is based on palladium combined with a ligand (often a phosphine). However, both the palladium catalysts and the ligands could be quite expensive. From the economic and environmental points of view, the development of cheaper metals in place of palladium-based catalysts is extremely attractive. Although copper has been widely used as the catalyst in many cross-coupling transformations, little attention was paid to the use of copper as the catalyst for the Suzuki–Miyaura cross-coupling reaction (see Chapter 10 for a detailed account on the synthesis of biaryls using this approach), which has been only scarcely explored for the synthesis of styrene derivatives.

Li and coworkers have reported the Suzuki–Miyaura alkenylation reaction of vinyl halides with arylboronic acids in the presence of copper(I) iodide (10 mol %), DABCO as a ligand (20 mol %), and tetrabutylammonium bromide to afford the corresponding 1,1- or 1,2-diarylethenes in moderate-to-good yields (Scheme 11.50).[14] One exception is the cross coupling of

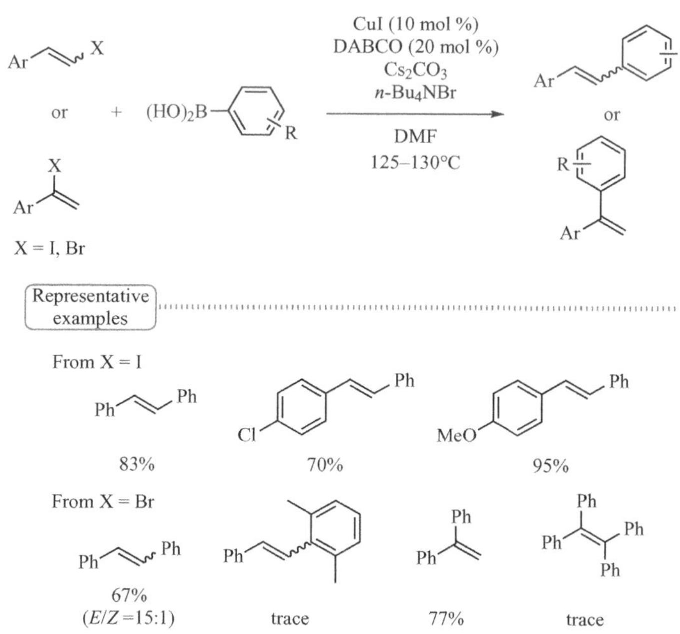

Scheme 11.50 Copper(I)-catalyzed Suzuki–Miyaura cross coupling for the synthesis of diarylethenes using DABCO as a ligand.

1-(2-bromovinyl)benzene with the highly sterically demanding 2,6-dimethylphenylboronic acid that leads only to traces of the targeted compound. It is noteworthy that 1-(1-bromovinyl)benzene can be coupled with phenylboronic acid in 77% yield under the standard conditions.

The same authors found that the reaction also proceeds without DABCO, in DMSO at 135°C. Tetrabutylammonium bromide (20 mol %) was mandatory in this copper(I)-catalyzed cross-coupling reaction (Scheme 11.51).[49]

Li and coworkers proposed a possible catalytic cycle outlined in Scheme 11.52. The reaction of the active catalyst **17** with vinyl halides would afford

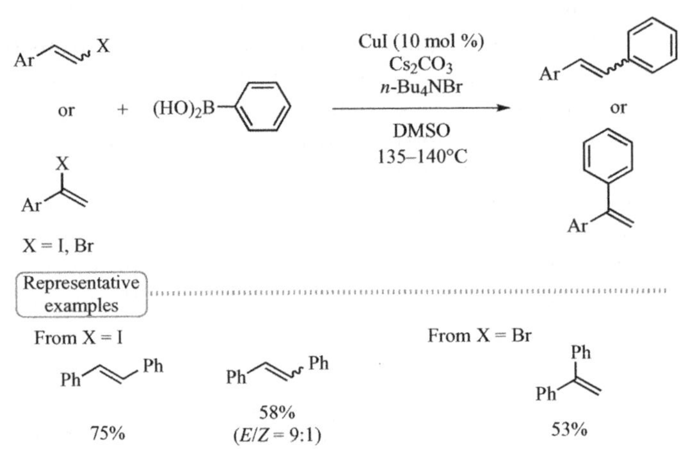

Scheme 11.51 Copper(I)-catalyzed Suzuki–Miyaura cross coupling for the synthesis of diarylethenes in the presence of TBAB.

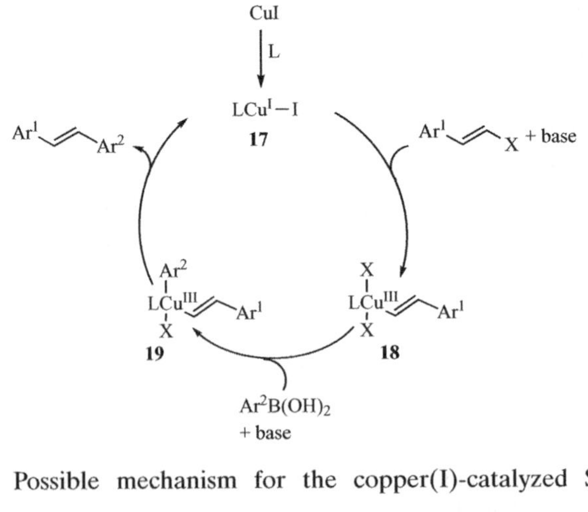

Scheme 11.52 Possible mechanism for the copper(I)-catalyzed Suzuki–Miyaura reaction.

intermediate copper(III) **18**. Transmetallation of this intermediate **18** with the arylboronic acid would then follow, yielding copper(III) complex **19**. Subsequent reductive elimination would then give the coupled product and the active copper(I) complex **17**. This mechanism is just a proposal and is not based on any experimentation. A more detailed presentation of related mechanisms can be found in Chapter 7.

While copper-catalyzed Stille or Suzuki–Miyaura cross-coupling reactions represent efficient entries to styrene derivatives, they require prefunctionalization of the starting alkene coupling partner, either as an alkenyl metal or as an alkenyl iodide. An interesting alternative, which is devoid of this constraint, is the use of the Heck cross coupling, which can also now be performed under copper catalysis. Recent and significant examples in this area will be overviewed in the next section.

11.3.3 Synthesis of Aryl–Enes by Heck-Type Cross Coupling

The Heck cross coupling is the reaction of an organohalide (or pseudohalide) with an alkene to form a newly substituted alkene (Scheme 11.53) and was named after Tsutomu Mizoroki and Richard F. Heck, the latter being awarded the 2010 Nobel Prize in Chemistry for the discovery and development of this reaction.

Many transition-metal complexes, such as palladium, nickel, cobalt, rhodium, or iridium complex can catalyze the reaction of aryl halides with alkenes to give the coupled products in high yields, and copper-based catalysts are also part of this list since the late 1990s.

Indeed, Iyer and coworkers reported the first copper-catalyzed Heck reaction of aryl and vinyl iodides with olefins via both inter- and intramolecular processes in 1997. The catalytic system is very simple: copper(I) iodide (10 mol %) or copper(I) bromide (10 mol %) and potassium carbonate in NMP at 150°C (Scheme 11.54).[50] Interestingly, phosphine and other ligands were not required for the reaction. Under the optimal conditions, aryl iodides could react with activated alkenes (styrene derivatives and acrylates) to form the corresponding aromatic alkenes in good yield. However, less-activated aryl bromides and chlorides could not undergo the reaction under the same conditions.

As a minor note, the cyclization of *N*-methyl-*N*-allyl-2-iodobenzamide afforded *N*-methyl-4-methylisoquinolone in 54% yield in the presence of copper(I) iodide (10 mol %) (Scheme 11.55).

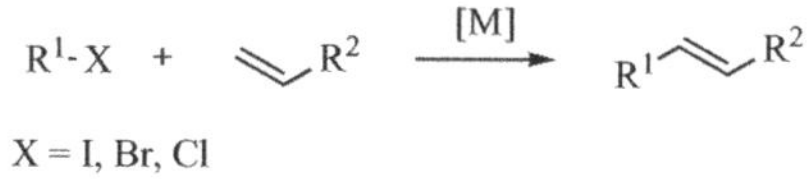

Scheme 11.53 The Heck-type cross-coupling reaction.

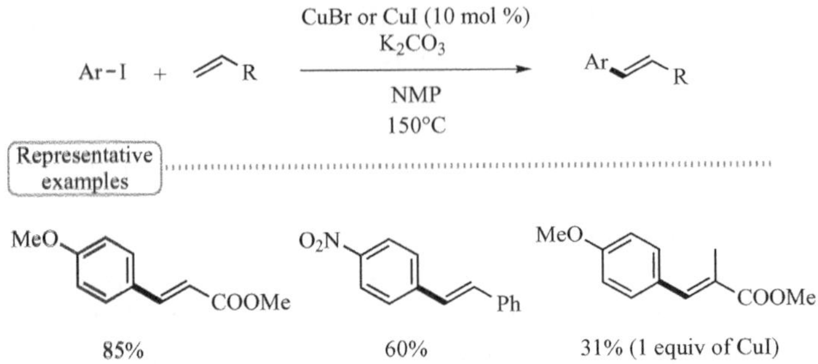

Scheme 11.54 Copper(I)-catalyzed Heck-type vinylation of aryl iodides.

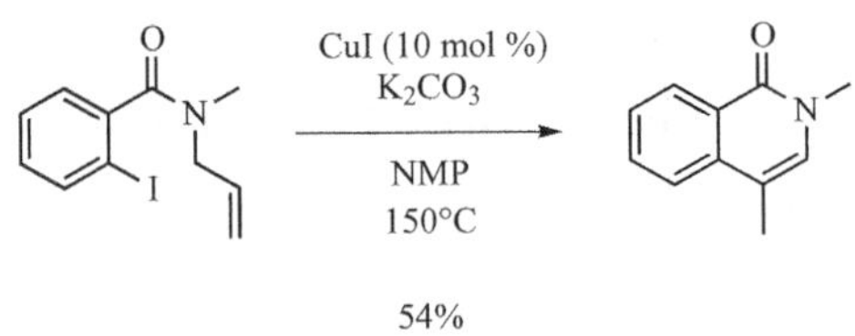

Scheme 11.55 Copper(I)-catalyzed cyclization of *N*-methyl-*N*-allyl-2-iodobenzamide.

Iyer and Thakur also demonstrated that a heterogeneous catalyst, Cu/Al_2O_3, could catalyze the reaction of aryl and vinyl halides with olefins to yield the corresponding substituted alkenes in high yields (Scheme 11.56).[51] Interestingly, the catalytic system can be recycled: Three to four cycles could be carried out, but with a rapid decrease in yield. In addition, considerable leaching from the supported catalyst was demonstrated.

Zheng and coworkers have also reported that the silica-supported poly-γ-aminopropylsilane copper(II) complex (abbreviated as "Si"-NH_2·$Cu(OAc)_2$) could catalyze the alkenylation of substituted aryl iodides with olefins in NMP at 130–150°C, yielding the corresponding *trans*-alkene in high yields (Scheme 11.57).[52] This complex exhibited a high reactivity and stereoselectivity and could be reused after washing without loss of activity. It is interesting to note that an induction period of more than 2 hours was observed in the first run for this catalyst.

Li and coworkers have developed a mild route for the alkenylation reaction using copper(I) iodide (10 mol %), DABCO (20 mol %), and potassium carbonate in ethanol (Scheme 11.58).[53] Thanks to the DABCO ligand, the Heck-type alkenylation reaction was carried out at 80°C for aryl iodides and 100°C for activated aryl bromides, affording the desired products in moderate-to-excellent yields. However, deactivated aryl bromides were not reactive enough.

As with most systems, green variants have rapidly appeared using recyclable systems or aqueous systems. In this context, Calò and coworkers found

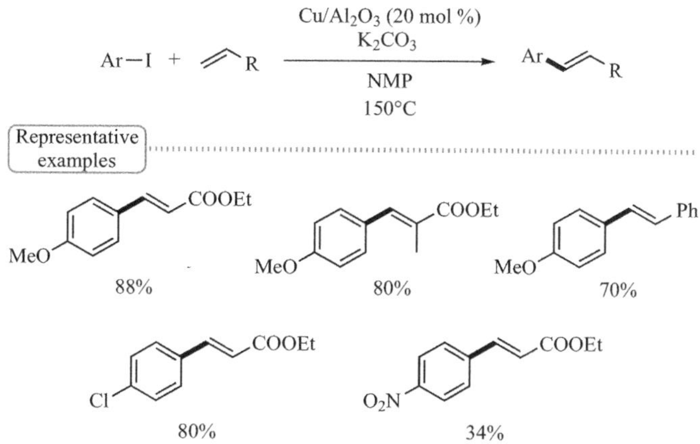

Scheme 11.56 Cu/Al_2O_3-catalyzed Heck reaction with aryl iodides and activated alkenes.

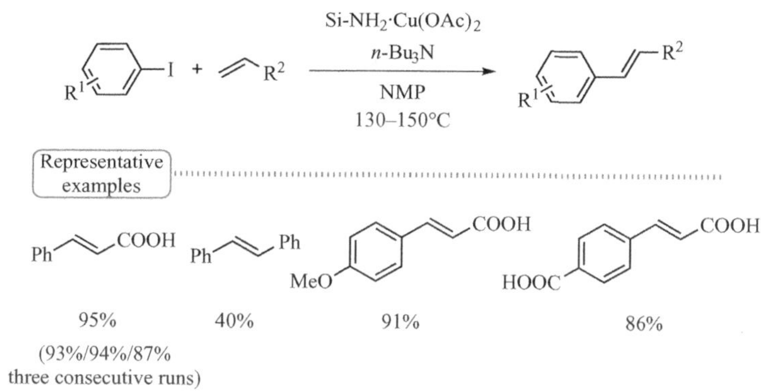

Scheme 11.57 Silica-supported copper(II) catalyst in the Heck reaction of aryl iodides with activated alkenes.

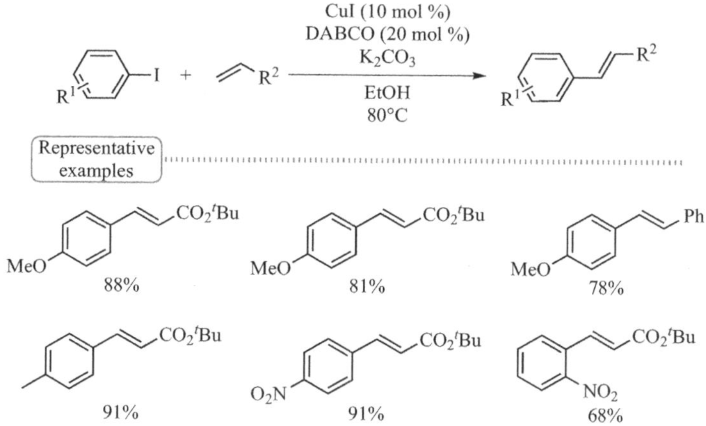

Scheme 11.58 Copper(I)-catalyzed Heck reaction for the synthesis of aryl–enes with DABCO as the ligand.

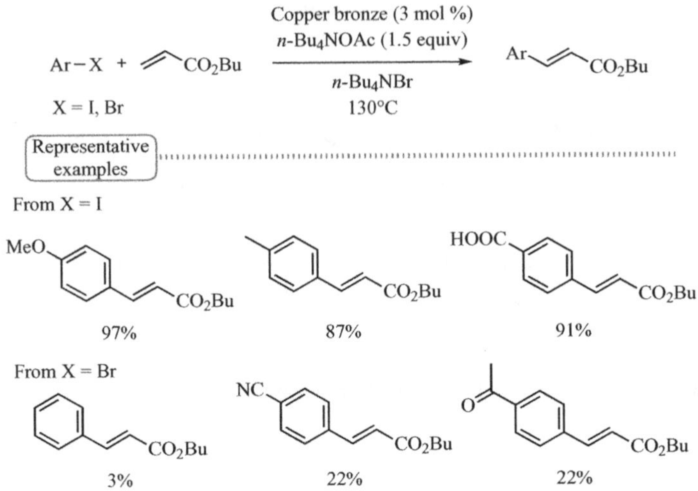

Scheme 11.59 Copper bronze-catalyzed Heck reaction in ionic liquid.

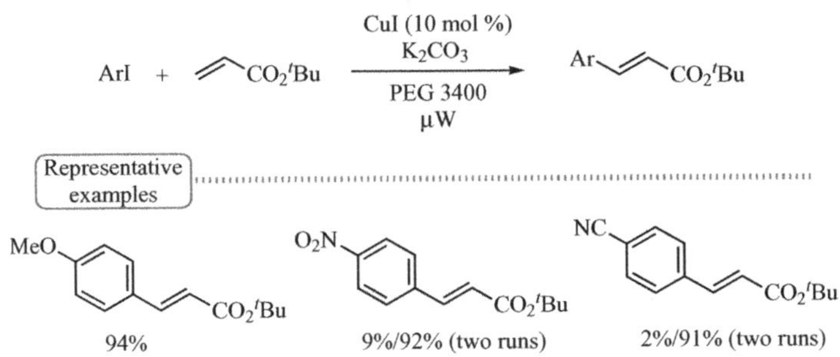

Scheme 11.60 Copper(I)-catalyzed Heck reaction of different aryl halides in PEG.

that copper nanocolloids, derived from the reaction of iodobenzene with copper bronze, could catalyze the reactions of aryl iodides or activated bromides with acrylates in the ionic liquid tetrabutylammonium bromide using tetrabutylammonium acetate as a base (Scheme 11.59).[54] These copper nanoparticles displayed an excellent catalytic activity over 20 reuse (turnover numbers of ca. 490), a stability that was attributed by the authors to the surrounding of the nanoparticles by these salts, which would impose a coulombic barrier for collision and would therefore impede the formation of clusters and aggregation.

The copper-catalyzed Heck-type alkenylation reaction usually requires high boiling point organic solvents, such as DMF or NMP. Recently, Lamaty and coworkers reported a copper(I) iodide-catalyzed Heck reaction in the presence of potassium carbonate and PEG 3400 under microwave condition. It was noted that the catalysts/solvent system could be recycled and that better results were usually obtained after recycling (Scheme 11.60).[55]

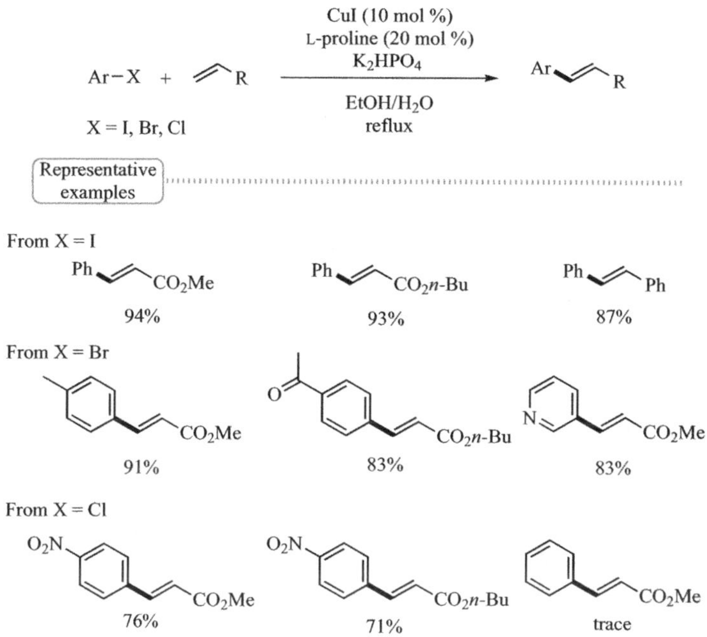

Scheme 11.61 Efficient copper(I)-catalyzed alkenylation reactions of aryl halides using L-proline as the ligand.

Guo and coworkers reported a real improvement for the copper-catalyzed Heck reaction of acrylic esters or styrene with aryl halides, which allowed the obtention of the coupling products in good-to-excellent yields in an aqueous two-phase system at only 75–85°C using L-proline as the ligand (Scheme 11.61).[56] This method provides a new, simple, and inexpensive access to the Heck-type alkenylation: A variety of aryl halides, including activated aryl chlorides, smoothly underwent the reaction with acrylic esters or styrene thanks to copper(I) iodide (10 mol %), L-proline (20 mol %), and dipotassium hydrogenphosphate in moderate-to-excellent yields.

Not surprisingly, the evolution of strategies based on copper-mediated transformations for the synthesis of alkenyl aromatics closely follows the one observed with their alkynyl counterparts. The copper-catalyzed direct alkenylation of acidic C–H bonds in arenes has been therefore also investigated recently and will be briefly discussed, with focus on the most significant results, in the coming paragraphs.

11.3.4 Synthesis of Aryl–Enes by Direct C–H Functionalization of Arenes

In 2008, the copper-catalyzed direct C–H alkenylation of 5-aryl-oxazoles and related heterocycles (oxazole, benzoxazole, and benzothiazole) with various

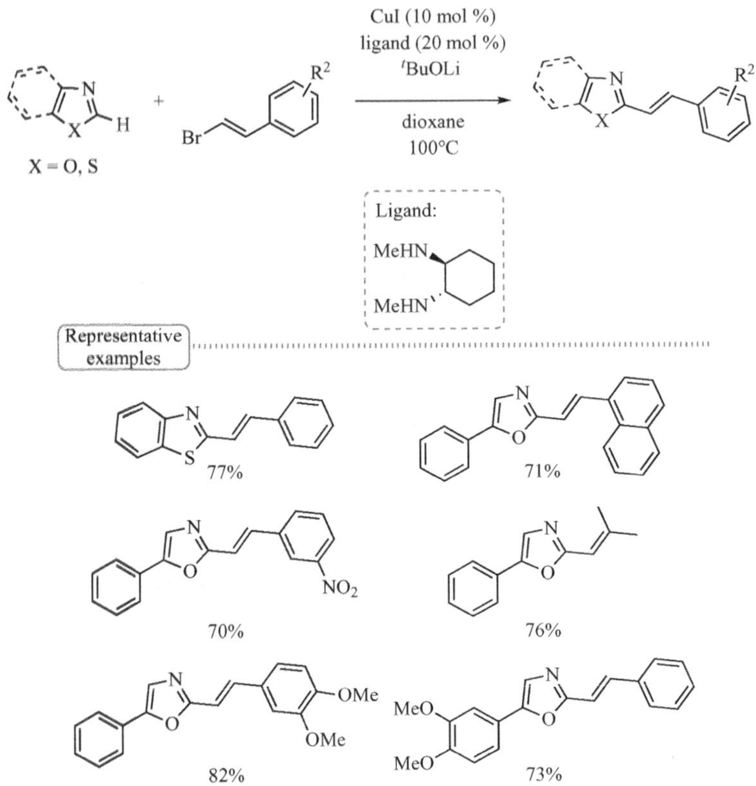

Scheme 11.62 Copper(I)-catalyzed direct alkenylation of oxazoles and related heterocycles with bromoalkenes.

styryl bromides or iodides was reported by Piguel and coworkers (Scheme 11.62).[57] The method was found to be both regio- and stereoselective and tolerated a variety of functional groups. A wide range of 2-*E*-vinyl-substituted oxazoles including the highly fluorescent alkaloid annuloline were obtained in high yields using a combination of copper(I) iodide, racemic *trans*-*N*,*N*′-dimethylcyclohexane-1,2-diamine, and lithium *tert*-butoxide in dioxane at 100°C.

Recent works demonstrated that the C–H alkenylation reactions could be applicable to activated pyridinium species. Indeed, Charette and coworkers reported the first copper(II)-catalyzed direct alkenylation of *N*-iminopyridinium ylides with inexpensive copper salts and using potassium phosphate as the base (Scheme 11.63).[58] This remarkable reaction allowed for the direct alkenylation of a wide range of substrates with high efficiency and in a most straightforward manner.

It was noted that the *N*–*N* bond could be cleaved in two elementary steps (methylation with methyl iodide followed by reductive cleavage of the N–N

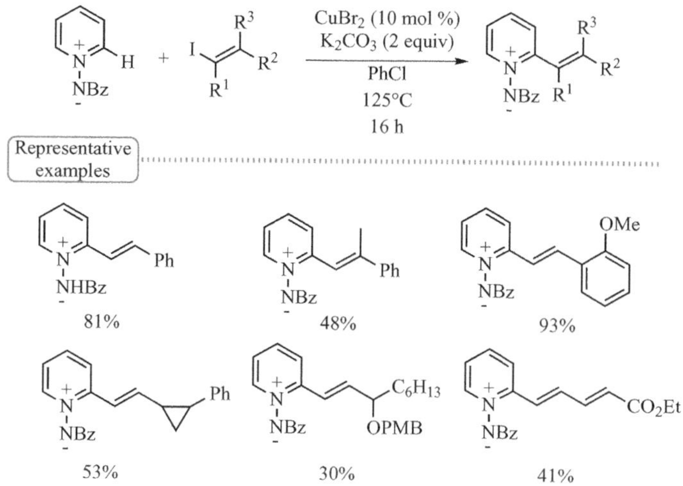

Scheme 11.63 Copper(II)-catalyzed direct alkenylation of *N*-iminopyridinium ylides.

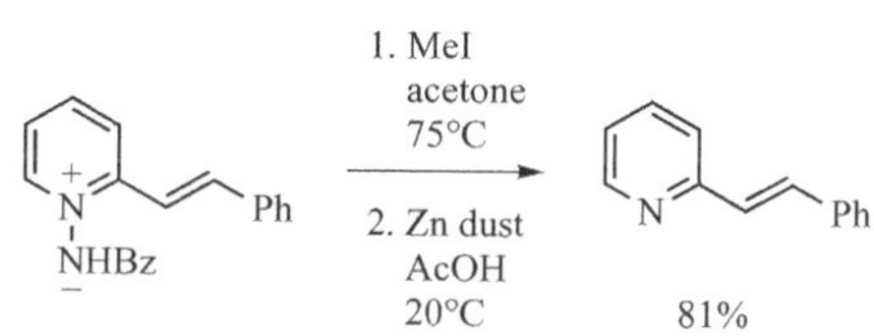

Scheme 11.64 Reductive cleavage of the N–N bond to 2-alkenyl-pyridines.

bond with zinc dust) to provide the corresponding 2-alkenylated pyridines (Scheme 11.64).

A possible mechanism suggested by Charette is illustrated in Scheme 11.65. Initially, Cu(II) would be reduced to Cu(I) with the aid of the pyridinium ylide (step **A**). Next, a carbonate–halide exchange would generate $CuCO_3K$ (step **B**), which can promote a deprotonation/metallation of the pyridinium to generate an organocupracycle stabilized by the Lewis basic iminobenzoyl moiety (step **C**). Finally, oxidative addition (step **D**) into the alkenyl iodide and subsequent reductive elimination (step **E**) would afford the corresponding 2-alkenylated derivative.

The Daugulis group has also recently described a direct copper-catalyzed alkenylation of activated aromatics. Using a combination of copper(I) iodide (10 mol %), 1,10-phenanthroline (10 mol %), and potassium phosphate in DMF/xylene at 120–140°C, C–H bonds in polyfluoroarenes could be readily activated to introduce a double bond at this position (Scheme 11.66).[59]

Before closing this subchapter, a last strategy that has been only little explored ought to be mentioned: the decarboxylative cross coupling, which is described in the next subsection.

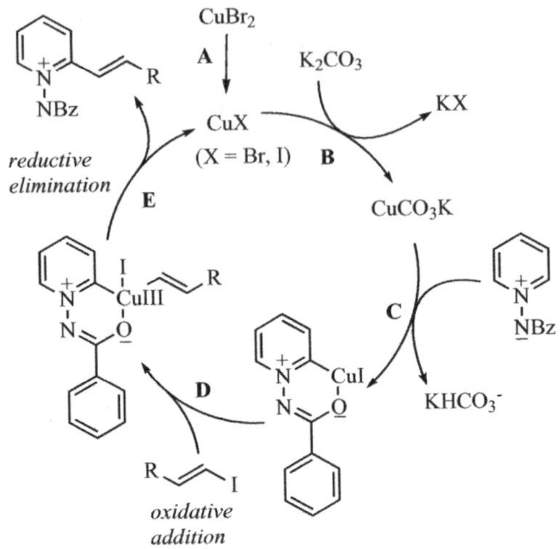

Scheme 11.65 Proposed mechanism for the copper-catalyzed direct alkenylation of *N*-iminopyridinium ylides.

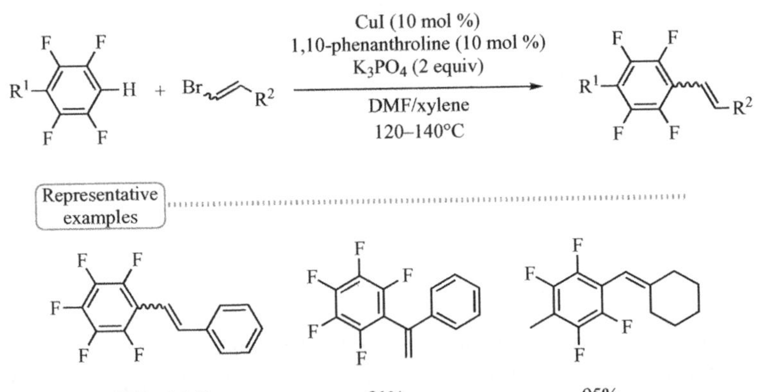

Scheme 11.66 Copper(I)-catalyzed direct alkenylation of polyfluoroarenes.

11.3.5 Synthesis of Aryl–Enes by Decarboxylative Cross Coupling

The first example of copper-catalyzed decarboxylation was reported by Shepard and coworkers in 1930.[60] Halogenated furancarboxylic acids were decarboxylated in the presence of copper or copper salt at high temperatures (about 250°C). The importance of transition-metal-catalyzed decarboxylative cross-coupling chemistry has grown rapidly in recent years. Carboxylic acids are available at low cost in great structural diversity both from natural and synthetic sources, and are easy to store and handle. Most of the decarboxylative alkenylation reactions were performed using a palladium catalyst in the presence of a phosphane ligand and alternative systems are therefore needed.

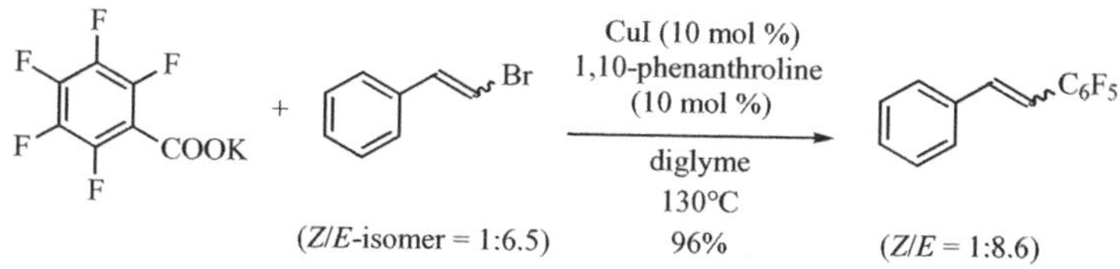

Scheme 11.67 Copper(I)-catalyzed the decarboxylative alkenylation of potassium perfluorobenzoate.

In 2009, Gooβen and coworkers reported the copper-catalyzed decarboxylation of aromatic carboxylic acids. Nonetheless, there has not been any example for the synthesis of alkenes by copper-catalyzed decarboxylative alkenylation reaction.[61] Liu and coworkers have achieved the copper-catalyzed decarboxylative alkenylation of potassium perfluorobenzoate with styryl bromide in the presence of phenanthroline (10 mol %) in diglyme at 130°C. An excellent yield was obtained, although the scope of this transformation is somehow quite restricted (Scheme 11.67).[62]

As evidenced with representative examples overviewed in this subchapter, the renaissance of copper catalysis also had a deep impact on the construction of Csp^2(aryl)-Csp^2(alkenyl) bonds. A variety of strategies, reagents, and conditions based on readily available, inexpensive copper-based systems are now available and should expedite the synthesis of styrene derivatives.

The last bond construction that will be considered in this chapter is the Csp^2(aryl)–Csp^3(allyl) one. Efficient procedures have been described for the formation of this kind of bonds using copper-based catalytic systems, and representative recent examples have been collected in the next section.

11.4 COPPER-CATALYZED STRATEGIES FOR THE FORMATION OF ALLYL–ARYL BONDS

The allylation reaction is one of the most fundamental and important transformations in organic synthesis because allyl moieties are easily manipulated to a diversity of functional groups. To date, various transition-metal catalysts such as palladium, rhodium, iridium, and copper have been employed to enable the introduction of allyl groups to several aryl metal compounds, including magnesium, aluminum, zinc, and boron reagents.

The use of preformed, nonisolable metal reagents is outside the scope of this book and will therefore not be covered. Instead, recent significant advances in the copper-mediated allylation of arylboron reagents will be highlighted in this section, which does not intend to be exhaustive, together with the direct allylation of C–H bonds of activated arenes.

The Lalic group has reported in 2010 an efficient copper-catalyzed S_N2'-selective arylation of allylic chlorides using arylboronic esters as nucleophiles

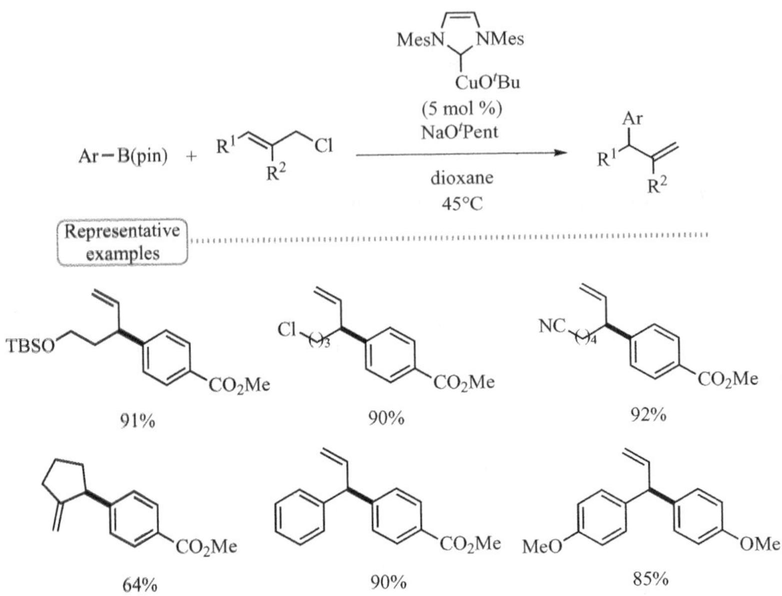

Scheme 11.68 Copper-catalyzed S_N2'-selective aryl allylation using arylboronic esters as nucleophiles.

(Scheme 11.68).[63] This approach is particularly appealing considering the availability, stability, and excellent functional group compatibility of the starting arylboronic esters. In addition, the results are the first examples of the use of arylboronic esters in copper-catalyzed allylation reactions.

Ohmiya and Sawamura have developed a copper(II)-catalyzed γ-selective and stereospecific allyl–aryl cross-coupling reaction between (*Z*)-acyclic or cyclic allylic phosphates and arylboronates to give the corresponding cross-coupled products possessing benzylic and allylic stereogenic centers (Scheme 11.69).[64] This copper-based catalytic system is complementary to the reported palladium-based one in that it tolerates (*Z*)-acyclic and cyclic allylic systems. Notably, the use of arylmetal reagents as nucleophilic coupling partners has not been well exploited previously in copper chemistry due to the poor nucleophilicity of the arylcopper species relative to that of the alkylcopper species.

Finally, Shintani and Hayashi have described a copper(I)/*N*-heterocyclic carbene complex-catalyzed asymmetric allylic substitution of allyl phosphates with arylboronic acid esters to construct both tertiary and quaternary carbon stereocenters with high regio- and enantioselectivities (Scheme 11.70).[65]

Recently, the copper-mediated C–C bond formation by C–H functionalization has grown rapidly and could be considered as a complementary process to the conventional cross-coupling methodologies. Elegant direct C–H arylations, alkenylations, alkynylations, and alkylations were achieved with great efficiency. On the other hand, the corresponding allylation reaction is still rare

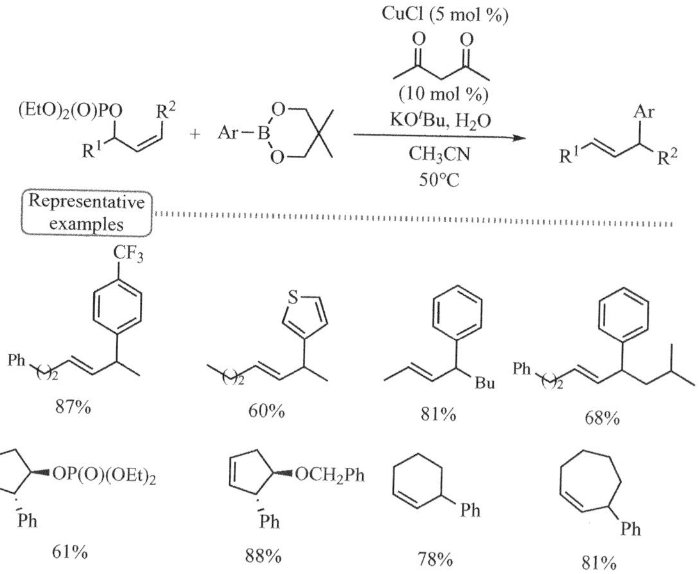

Scheme 11.69 Copper(I)-catalyzed aryl allylation with cyclic allylic phosphates.

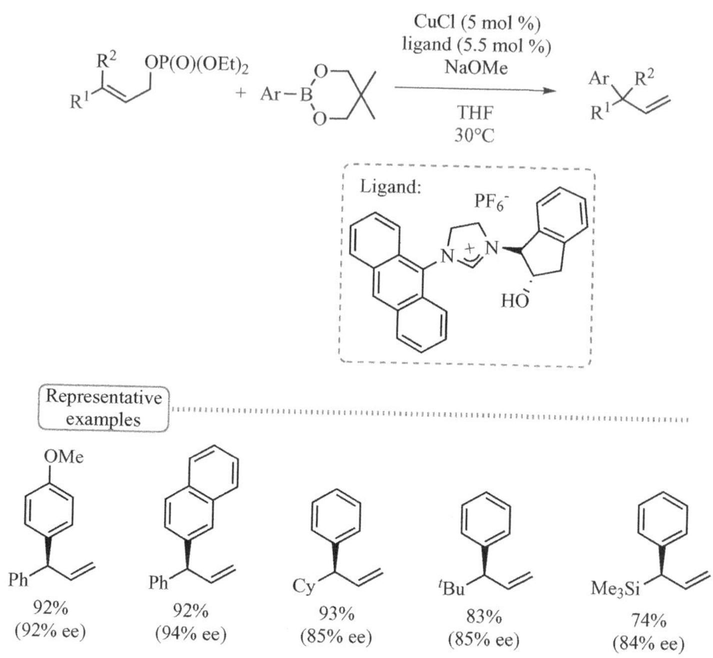

Scheme 11.70 Copper(I)-catalyzed asymmetric allylic substitution of allyl phosphates with arylboronates.

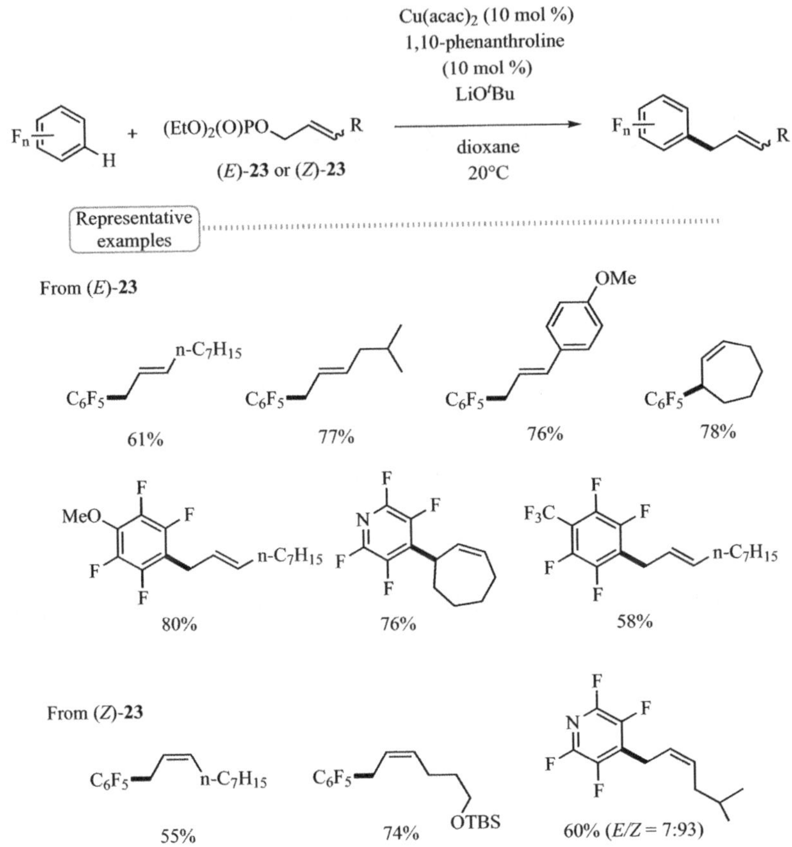

Scheme 11.71 Copper(II)-catalyzed direct C–H allylation of fluoroarenes with allylic phosphates.

despite its potential. Hirano and coworkers reported a rare example of direct C–H allylation of polyfluoroarenes with (*E*)- and (*Z*)-allyl phosphates in the presence of $Cu(acac)_2$ (10 mol %), 1,10-phenanthroline (10 mol %), and lithium *tert*-butoxide (2 equiv) (Scheme 11.71).[66] The proposed catalytic system provided a straightforward access to stereodefined substituted allyl arenes that contain fluorinated aromatic cores, central building blocks in materials and life sciences, and demonstrated the feasibility of the direct allylation of acidic aromatic C–H bonds.

The authors also investigated the regiospecificity of the reaction under the standard reaction conditions from branched allylic phosphate **25** (Scheme 11.72). The reaction proceeded smoothly to furnish the linear allylated product **26** exclusively with moderate *E*/*Z* selectivity, indicating that a π-allyl copper intermediate was probably involved in the catalytic cycle.

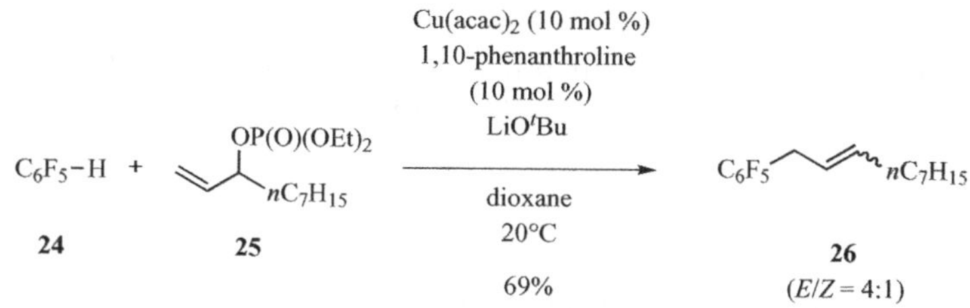

Scheme 11.72 Regiospecificity of the copper(II)-catalyzed direct C–H allylation of fluoroarenes with allylic phosphates.

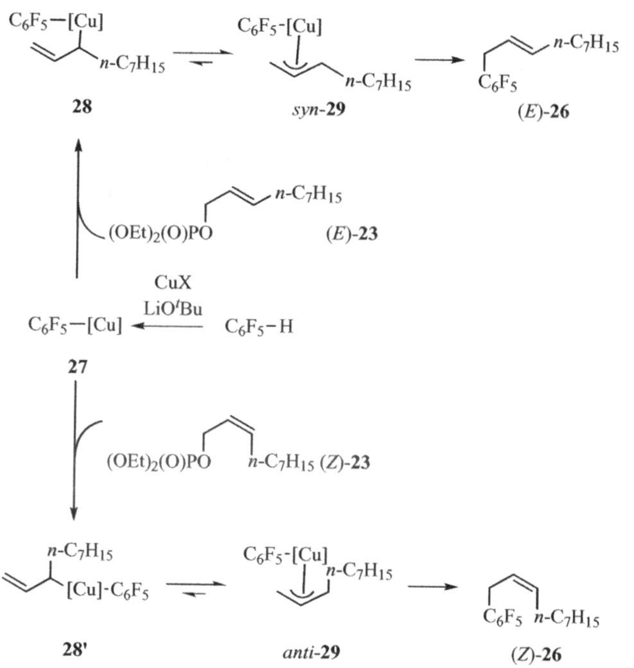

Scheme 11.73 Proposed mechanism for the direct C–H allylation of polyfluoroarenes with allylic phosphates.

A possible reaction mechanism was proposed on the basis of these results and literature precedents (Scheme 11.73). The mechanism would first involve a lithium *tert*-butoxide-assisted direct cupration of pentafluorobenzene to give the pentafluorophenylcopper intermediate **27**, in which copper is liganded to phenanthroline, acetylacetonate, or phosphate. Subsequently, oxidative addition of (*E*)- or (*Z*)-**23** would proceed in a S_N2' manner followed by rapid σ–π conversion to give the π-allyl copper species *syn*-**29** or *anti*-**29**, respectively. Finally, the corresponding allylated products (*E*)- or (*Z*)-**26** would be generated by reductive elimination. The σ–π interconversion between σ-allyl isomers **28** and **28′** needs to be much faster than bond rotation for the stereochemical information of the starting allyl phosphate **23** to be transferred into the product.

11.5 CONCLUSION AND OUTLOOK

Copper-catalyzed alkynylation, alkenylation, and allylation reactions of aryl derivatives are now essential methodologies in organic synthesis from both practical and economical points of view. Importantly, these copper-catalyzed C–C bond formation reactions have emerged as key steps in the total syntheses of various natural products, bioactive molecules, and functional materials (see Chapter 18 for a detailed discussion). Up to now, numerous alkynylation, alkenylation, and allylation reactions of aryl derivatives have been reported and copper is clearly an interesting alternative to palladium catalysis in a number of transformations. However, these processes still often require reaction conditions that are still harsh and call for high catalyst loadings, which will clearly have to be improved in the near future.

More recently, a shift in paradigm was noted with the spectacular development of copper-catalyzed direct cross-coupling reactions of aromatic C–H bonds. The generalization of ever more efficient and sustainable copper-catalyzed alkynylation, alkenylation, and allylation (including asymmetric allylation) reactions of aryl C–H bond is one of the most exciting challenges in the near future.

REFERENCES

[1] (a) Evano, G.; Blanchard, N.; Toumi, M. *Chem. Rev.* **2008**, *108*, 3054–3131. (b) Chinchilla, R.; Nájera, C. *Chem. Soc. Rev.* **2011**, *40*, 5084–5121. (c) Monnier, F.; Taillefer, M. *Angew. Chem. Int. Ed. Engl.* **2008**, *47*, 3096–3099. (d) Yeung, C. S.; Dong, V. M. *Chem. Rev.* **2011**, *111*, 1215–1292.

[2] (a) Stille, J. K. *Angew. Chem. Int. Ed. Engl.* **1986**, *25*, 508–524. (b) Espinet, P.; Echavarren, A. M. *Angew. Chem. Int. Ed. Engl.* **2004**, *43*, 4704–4734.

[3] (a) Liebeskind, L. S.; Fengl, R. W. *J. Org. Chem.* **1990**, *55*, 5359–5364. (b) Farina, V.; Kapadia, S.; Krishnan, B.; Wang, C.; Liebeskind, L. S. *J. Org. Chem.* **1994**, *59*, 5905–5911.

[4] Kang, S.-K.; Kim, J.-S.; Choi, S.-C. *J. Org. Chem.* **1997**, *62*, 4208–4209.

[5] Kang, S.-K.; Kim, J.-S.; Yoon, S.-K.; Lim, K.-H.; Yoon, S. S. *Tetrahedron Lett.* **1998**, *39*, 3011–3012.

[6] Li, J.-H.; Tang, B.-X.; Tao, L.-M.; Xie, Y.-X.; Liang, Y.; Zhang, M.-B. *J. Org. Chem.* **2006**, *71*, 7488–7490.

[7] Kang, S.-K.; Kim, W.-Y.; Jiao, X.-H. *Synthesis* **1998**, 1252–1254.

[8] Yu, S.-M.; Kweon, J.-H.; Ho, P.-S.; Kang, S.-C.; Lee, G. Y. *Synlett* **2005**, 2631–2634.

[9] Sonogashira, K.; Tohda, Y.; Hagihara, N. *Tetrahedron Lett.* **1975**, *16*, 4467–4470.

[10] Okuro, K.; Furuune, M.; Miura, M.; Nomura, M. *Tetrahedron Lett.* **1992**, *33*, 5363–5364.

[11] (a) Okuro, K.; Furuune, M.; Enna, M.; Miura, M.; Nomura, M. *J. Org. Chem.* **1993**, *58*, 4716–4721. (b) Stephens, R. D.; Castro, C. E. *J. Org. Chem.* **1963**, *28*, 3313–3315.

[12] Ma, D.; Liu, F. *Chem. Commun.* **2004**, 1934–1935.

[13] Xie, Y.-X.; Deng, C.-L.; Pi, S.-F.; Li, J.-H.; Yin, D.-L. *Chin. J. Chem.* **2006**, *24*, 1290–1294.

[14] Li, J.-H.; Li, J.-L.; Wang, D.-P.; Pi, S.-F.; Xie, Y.-X.; Zhang, M.-B.; Hu, X.-C. *J. Org. Chem.* **2007**, *72*, 2053–2057.

[15] Jiang, H.-M.; Fu, H.; Qiao, R.-Z.; Jiang, Y.-Y.; Zhao, Y.-F. *Synthesis* **2008**, 2417–2426.

[16] Wu, M.-Y.; Mao, J.-C.; Guo, J.; Ji, S. *Eur. J. Org. Chem.* **2008**, 4050–4054.

[17] Mao, J.-C.; Guo, J.; Ji, S.-J. *J. Mol. Catal. A: Chem.* **2008**, *284*, 85–88.

[18] Monnier, F.; Turtaut, F.; Duroure, L.; Taillefer, M. *Org. Lett.* **2008**, *10*, 3203–3206.

[19] Thakur, K. G.; Jaseer, E. A.; Naidu, A. B.; Sekar, G. *Tetrahedron Lett.* **2009**, *50*, 2865–2869.

[20] Thakur, K. G.; Sekar, G. *Synthesis* **2009**, 2785–2789.

[21] Chen, H.-J.; Lin, Z.-Y.; Li, M.-Y.; Lian, R.-J.; Xue, Q.-W.; Chung, J.-L.; Chen, S.-C.; Chen, Y.-J. *Tetrahedron* **2010**, *66*, 7755–7761.

[22] Lin, C.-H.; Wang, Y.-J.; Lee, C.-F. *Eur. J. Org. Chem.* **2010**, 4368–4371.

[23] Gujadhur, R. K.; Bates, C. G.; Venkataraman, D. *Org. Lett.* **2001**, *3*, 4315–4317.

[24] (a) Zuidema, E.; Bolm, C. *Chem. Eur. J.* **2010**, *16*, 4181–4185. (b) Zou, L.-H.; Johansson, A. J.; Zuidema, E.; Bolm, C. *Chem. Eur. J.* **2013**, *19*, 8144–8152.

[25] Priyadarshini, S.; Joseph, P. J. A.; Srinivas, P.; Maheswaran, H.; Kantam, M. L.; Bhargava, S. *Tetrahedron Lett.* **2011**, *52*, 1615–1618.

[26] Li, T.-Y.; Qu, X.-M.; Xie, G.-L.; Mao, J.-C. *Chem. Asian J.* **2011**, *6*, 1325–1330.

[27] Wang, B.-B.; Ye, Y.-M.; Chen, J.-J.; Zhou, X.-X.; Lu, J.-M.; Shao, L.-X. *Bull. Chem. Soc. Jpn* **2011**, *84*, 526–530.

[28] Tang, B.-X.; Wang, F.; Li, J.-H.; Xie, Y.-X.; Zhang, M.-B. *J. Org. Chem.* **2007**, *72*, 6294–6297.

[29] Thathagar, M. B.; Beckers, J.; Rothenberg, G. *Green Chem.* **2004**, *6*, 215–218.

[30] Kang, S. K.; Yoon, S. K.; Kim, Y. M. *Org. Lett.* **2001**, *3*, 2697–2699.

[31] Guan, J.-T.; Yu, G.-A.; Chen, L.; Weng, T.-Q.; Yuan, J.-J.; Liu, S.-H. *Appl. Organomet. Chem.* **2009**, *23*, 75–77.

[32] Chen, G.; Zhu, X.-H.; Cai, J.-W.; Wan, Y.-Q. *Synth. Commun.* **2007**, *37*, 1355–1361.

[33] Colacino, E.; Daïch, L.; Martinez, J.; Lamaty, F. *Synlett* **2007**, 1279–1283.

[34] Pan, C.; Luo, F.; Wang, W.; Ye, Z.; Cheng, J. *Tetrahedron Lett.* **2009**, *50*, 5044–5046.

[35] Rao, H.; Fu, H.; Jiang, Y.; Zhao, Y. *Adv. Synth. Catal.* **2010**, *352*, 458–462.

[36] Yasukawa, T.; Miyamura, H.; Kobayashi, S. *Org. Biomol. Chem.* **2011**, *9*, 6208–6210.

[37] (a) Wang, S.; Wang, M.; Wang, L.; Wang, B.; Li, P.; Yang, J. *Tetrahedron* **2011**, *67*, 4800–4806. (b) Liu, J.; Dai, F.; Yang, Z.; Wang, S.; Xie, K.; Wang, A.; Chen, X.; Tan, Z. *Tetrahedron Lett.* **2012**, *53*, 5678–5683.

[38] Zhao, D.; Gao, C.; Su, X.; He, Y.; You, J.; Xue, Y. *Chem. Commun.* **2010**, *46*, 9049–9051.

[39] (a) Qu, X.; Li, T.; Sun, P.; Zhu, Y.; Yang, H.; Mao, J. *Org. Biomol. Chem.* **2011**, *9*, 6938–6942. (b) Pan, D.; Zhang, C.; Ding, S.; Jiao, N. *Eur. J. Org. Chem.* **2011**, 4751–4755.

[40] (a) Li, T.; Sun, P.; Yang, H.; Yan, Y.; Lu, L.; Mao, J. *Tetrahedron* **2012**, *68*, 6413–6419. (b) Shi, L.; Jia, W.; Li, X.; Jiao, N. *Tetrahedron Lett.* **2013**, *54*, 1951–1955.

[41] (a) Crabtree, R. H. *Chem. Rev.* **2010**, *110*, 575–575. (b) McMurray, L.; O'Hara, F.; Gaunt, M. J. *Chem. Soc. Rev.* **2011**, *40*, 1885–1898. (c) Wendlandt, A. E.; Suess, A. M.; Stahl, S. S. *Angew. Chem. Int. Ed. Engl.* **2011**, *50*, 11062–11087. (d) Hirano, K.; Miura, M. *Chem. Commun.* **2012**, *48*, 10704–10714.

[42] Do, H.; Daugulis, O. *Chem. Commun.* **2009**, 6433–6435.

[43] (a) Besselièvre, F.; Piguel, S. *Angew. Chem. Int. Ed. Engl.* **2009**, *48*, 9553–9556. (b) Beatriz, P. B.; Besselièvre, F.; Piguel, S.; Lebrequier, S. *Org. Lett.* **2010**, *12*, 4038–4041.

[44] Kawano, T.; Matsuyama, N.; Hirano, K.; Satoh, T.; Miura, M. *J. Org. Chem.* **2010**, *75*, 1764–1766.

[45] Kitahara, M.; Hirano, K.; Tsurugi, H.; Satoh, T.; Miura, M. *Chem. Eur. J.* **2010**, *16*, 1772–1775.

[46] Wei, Y.; Zhao, H.-Q.; Kan, J.; Su, W.-P.; Hong, M.-C. *J. Am. Chem. Soc.* **2010**, *132*, 2522–2523.

[47] (a) Allred, G. D.; Liebeskind, L. S. *J. Am. Chem. Soc.* **1996**, *118*, 2748–2749. (b) Wang, M.; Lin, Z. *Organometallics* **2010**, *29*, 3077–3084.

[48] Miyaura, N.; Suzuki, A. *Chem. Commun.* **1979**, *19*, 866–867.

[49] Li, J.-H.; Li, J.-L.; Xie, Y.-X. *Synthesis* **2007**, 984–988.

[50] Iyer, S.; Ramesh, C.; Sarkar, A.; Wadgaonkar, P. P. *Tetrahedron Lett.* **1997**, *38*, 8113–8116.

[51] Iyer, S.; Thakur, V. V. *J. Mol. Catal. A: Chem.* **2000**, *157*, 275–278.

[52] Yang, Y.-F.; Zhou, R.-X.; Zhao, S.-F.; Li, Q.-L.; Zheng, X.-M. *J. Mol. Catal. A: Chem.* **2003**, *192*, 303–306.

[53] Li, J.-H.; Wang, D.-P.; Xie, Y.-X. *Tetrahedron Lett.* **2005**, *46*, 4941–4944.

[54] Calò, V.; Nacci, A.; Monopoli, A.; Leva, E.; Cilffi, N. *Org. Lett.* **2005**, *7*, 617–620.

[55] Declerck, V.; Martinez, J.; Lamaty, F. *Synlett* **2006**, 3029–3032.

[56] Guo, S.; Yung, Y.; Wang, Y.; Zheng, Y. *Chin. J. Org. Chem.* **2010**, *30*, 60–65.

[57] Besselièvre, F.; Piguel, S.; Mahuteau-betzer, F.; Grierson, D. S. *Org. Lett.* **2008**, *10*, 4029–4032.

[58] Mousseau, J.; Bull, J. A.; Charette, B. *Angew. Chem. Int. Ed. Engl.* **2010**, *49*, 1115–1118.

[59] (a) Do, H.-Q.; Daugulis, O. *J. Am. Chem. Soc.* **2008**, *130*, 1128–1129. (b) Do, H.-Q.; Khan, R. M. K.; Daugulis, O. *J. Am. Chem. Soc.* **2008**, *130*, 15185–15192.

[60] Shepard, A. F.; Winslow, N. R.; Johnson, J. R. *J. Am. Chem. Soc.* **1930**, *52*, 2083–2090.

[61] Gooβen, L. J.; Manjolinho, F.; Khan, B. A.; Rodríguez, N. *J. Org. Chem.* **2009**, *74*, 2620–2623.

[62] Shang, R.; Fu, Y.; Wang, Y.; Xu, Q.; Yu, H.-Z.; Liu, L. *Angew. Chem. Int. Ed. Engl.* **2009**, *48*, 9350–9354.

[63] Whittaker, A. M.; Rucker, R. P.; Lalic, G. *Org. Lett.* **2010**, *12*, 3216–3218.

[64] Ohmiya, H.; Yokokawa, N.; Sawamura, M. *Org. Lett.* **2010**, *12*, 2438–2440.

[65] Shintani, R.; Takatsu, K.; Takeda, M.; Hayashi, T. *Angew. Chem. Int. Ed. Engl.* **2011**, *50*, 8656–8659.

[66] Yao, T.; Hirano, K.; Satoh, T.; Miura, M. *Angew. Chem. Int. Ed. Engl.* **2011**, *50*, 2990–2994.

12

COPPER-CATALYZED ALKYNYLATION AND ALKENYLATION REACTIONS OF ALKYNYL DERIVATIVES: NEW ACCESS TO DIYNES AND ENYNES

Ruimao Hua

12.1 INTRODUCTION

Diynes and enynes represent important classes of organic compounds. In particular, conjugated 1,3-diynes and 1,3-enynes are not only important building blocks in organic synthesis and organic materials[1] but also are the structural functional units of many natural products.[2] Therefore, numerous accesses to diynes and enynes have been developed, and the alkynylation and alkenylation reactions of alkynyl derivatives catalyzed by palladium with or without the use of copper salts as cocatalysts have become standard synthetic methods.[3] In the past decade, cuprous or cupric salts, with or without external ligands, have found widespread applications in carbon–carbon or carbon–heteroatom bond formation.[2,4] The nontoxicity, ease of handling, and low cost of these versatile catalytic systems make them attractive for the elaboration of sensitive substrates such as diynes or enynes. The formation of these two classes of structural units, catalyzed by copper(I) or copper(II) salts, are reviewed in this chapter.

Copper-Mediated Cross-Coupling Reactions, First Edition. Edited by Gwilherm Evano and Nicolas Blanchard.

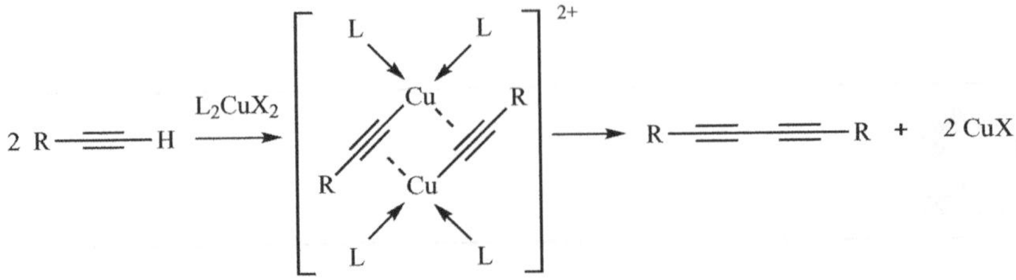

Scheme 12.1 Bohlmann's dimeric copper acetylides in the Glaser coupling reaction.

12.2 COPPER-CATALYZED SYNTHESIS OF SYMMETRICAL AND UNSYMMETRICAL 1,3-DIYNES

The homo- and heterocoupling reactions of alkynes is the most efficient and direct route to symmetrical and unsymmetrical 1,3-diynes. The oxidative homocoupling of terminal alkynes is an efficient reaction allowing the synthesis of symmetrical 1,3-diynes, and the first example of such type of coupling reaction was reported by Glaser in 1869 with the use of cuprous chloride as catalyst and aqueous ammonia as the base under an oxygen atmosphere.[5] A very stimulating historical perspective on this venerable reaction, including a discussion on the different mechanisms, can be found in a 2000 review by Diederich and coworkers.[1a] Although several mechanisms have been proposed, the most currently accepted one has been reported by Bohlmann and coworkers in 1964 and relies on dimeric copper acetylides as intermediates in the oxidative acetylenic coupling (Scheme 12.1).[6]

A similar catalytic system for the oxidative homocoupling of terminal alkynes was reported by Hay in 1960 with the use of an atmosphere of oxygen and CuCl·TMEDA (tetramethylethylenediamine [TMEDA]) as the catalyst, which is soluble in a range of organic solvents, resulting in a much more practical coupling reaction (Scheme 12.2, Eq. 1).[7] Recent density functional theory (DFT) calculations have demonstrated that a bimetallic process relying on copper(I), (II), and (III) oxidation states was operating (Scheme 12.2, Eq. 2).[6b–d] ^{13}C NMR kinetics experiments have also been recently reported supporting a Cu(I)/Cu(II)/Cu(III)/Cu(I) cycle (Eq. 3).[6e]

In the presence of copper(I), terminal alkynes also underwent the oxidative homocoupling reaction to give symmetrical 1,3-diynes (in the presence of an equivalent or excess amount of copper(II) acetate) (Eglinton coupling reaction) (Scheme 12.3).[8]

Beifuss and coworkers studied the Glaser-type reaction under different conditions and found that the yields of 1,3-diynes greatly depended on the nature of the base and ligand.[9] The optimal catalytic system was found to consist of copper(I) chloride (2 mol %), bases such as 1,8-diazabicyclo[5.4.0]undec-7-ene (DBU) or 1,4-diazabicyclo[2.2.2]octane (DABCO) (1 equiv), and TMEDA (1.5 mol %) as the ligand. Aromatic alkynes could be oxidatively homocoupled in high yields at room temperature in acetonitrile (Scheme 12.4,

(Eq. 1)

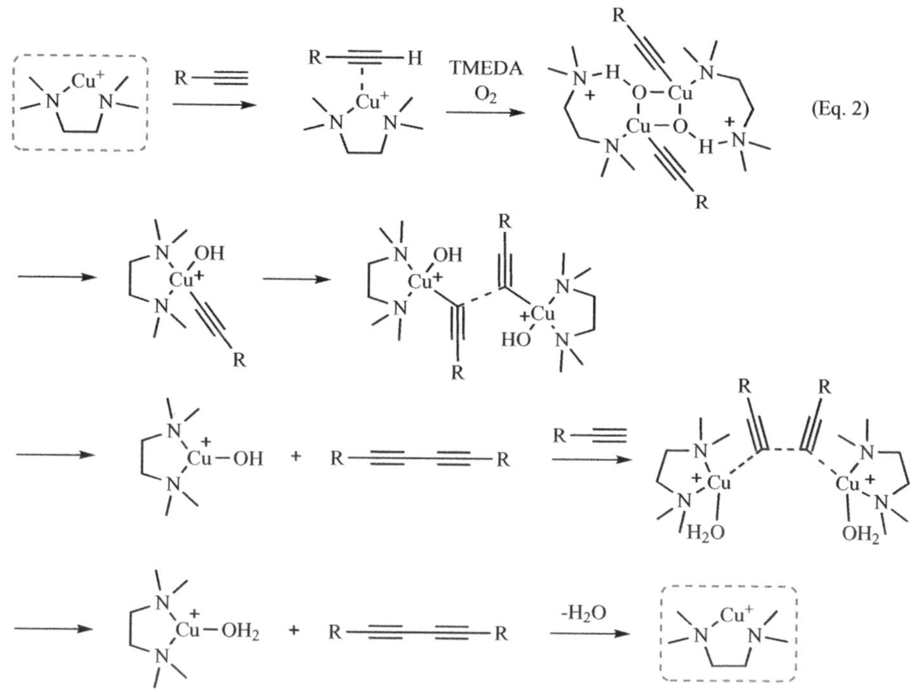

(Eq. 2)

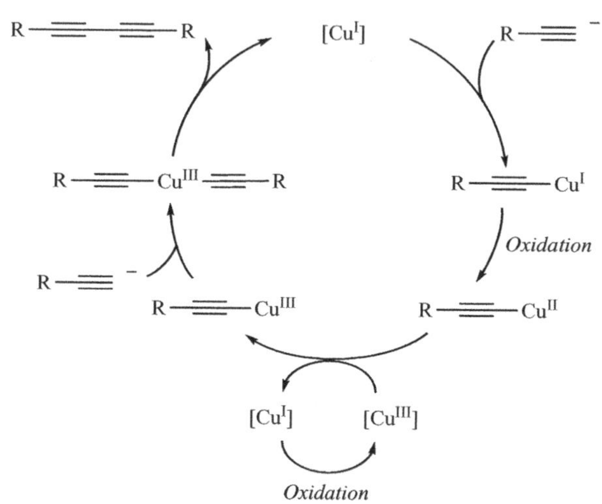

(Eq. 3)

Scheme 12.2 The Hay coupling reaction and a proposed mechanism based on DFT and spectroscopic studies.

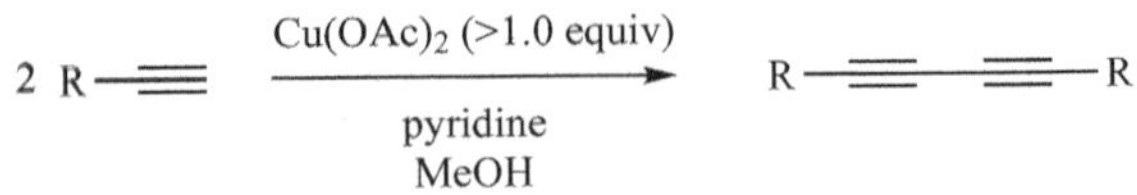

Scheme 12.3 The Eglinton coupling reaction.

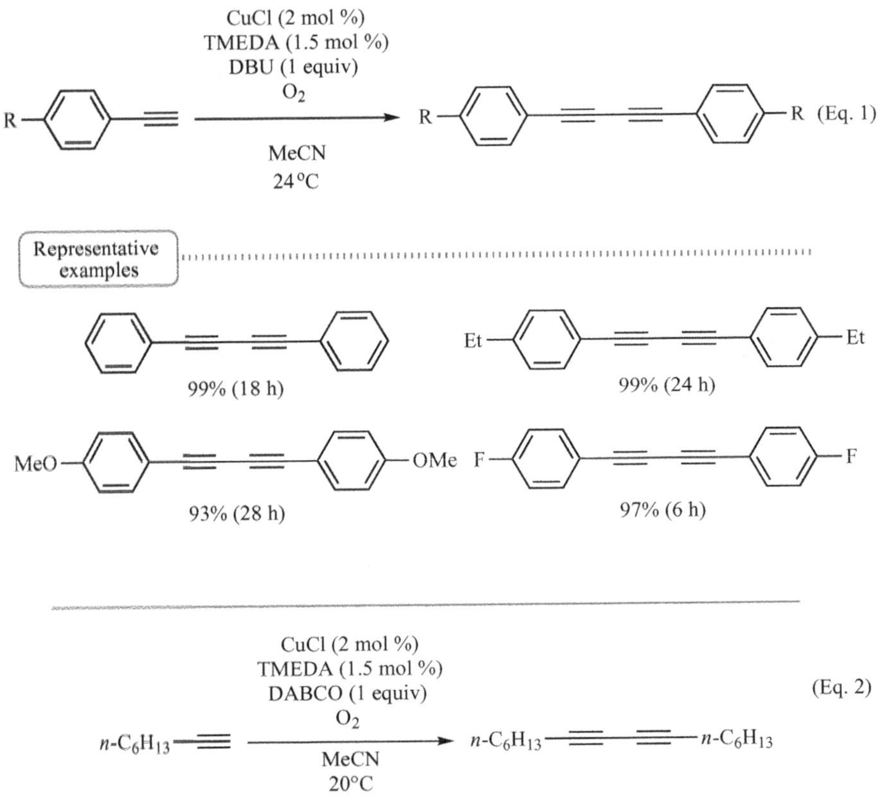

Scheme 12.4 Influence of the base in the Glaser-type homocoupling reaction.

Eq. 1). Less acidic alkynes, such as aliphatic alkynes that are traditionally less reactive, could also be used in this transformation (Scheme 12.4, Eq. 2).

Hua and coworkers reported that neither copper(I) chloride or iodide nor copper(II) chloride had catalytic activity for the homocoupling of terminal alkynes. However, in the presence of a catalytic amounts of piperidine (10 mol %), copper(I) chloride showed high catalytic activity at 60°C in toluene under an open air atmosphere affording symmetrical 1,4-disubstituted-1,3-diynes in high yields (Scheme 12.5).[10] Under the optimized reaction

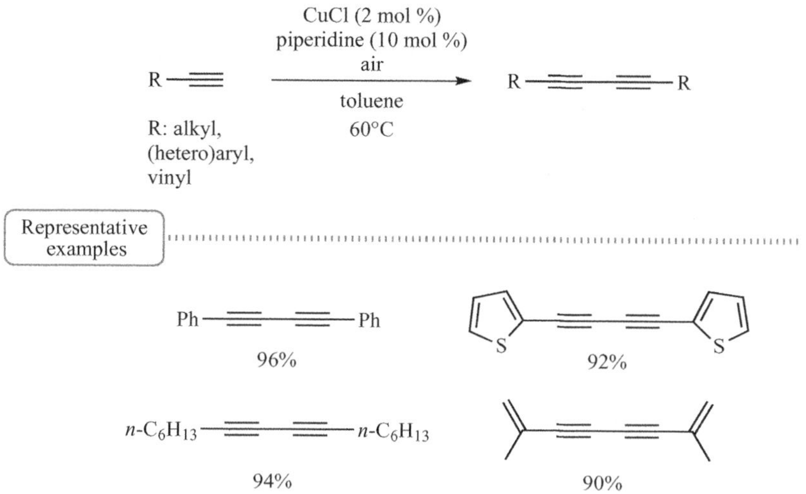

Scheme 12.5 Hua's copper(I or II)-catalyzed homocoupling of alkynes.

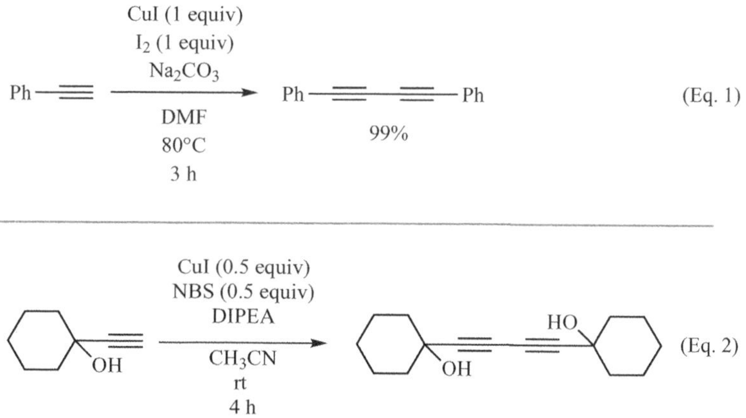

Scheme 12.6 Iodine and *N*-bromosuccinimide as oxidants in Glaser-type reactions.

conditions, copper(II) chloride/piperidine showed low catalytic activity. However, $CuCl_2$/DBU,[11] $CuCl_2/Et_3N$,[12] and $Cu(OAc)_2$/piperidine[13] could efficiently catalyze the homocoupling reactions of terminal alkynes.

Iodine[14] and *N*-bromosuccinimide (NBS)[15] could be also employed as oxidants in the copper(I) iodide-mediated oxidative homocoupling of terminal alkynes (Scheme 12.6). The great attractiveness of this catalytic system is its simplicity and efficiency due to the use of solid oxidants, which are soluble in organic solvents. In the case of NBS as an oxidant, the homocoupling reaction occurred at ambient temperature, and a variety of functionalized terminal alkynes including sugar- and amino acid-containing alkynes could react smoothly to give functionalized 1,3-diynes in good yields.

Macrocycles linked by acetylenic units are important compounds in supramolecular chemistry and organic materials science, and copper-mediated oxidative homocoupling reaction has become a general synthetic approach to such type of macrocyle. As shown in Scheme 12.7, excess copper(I) chloride

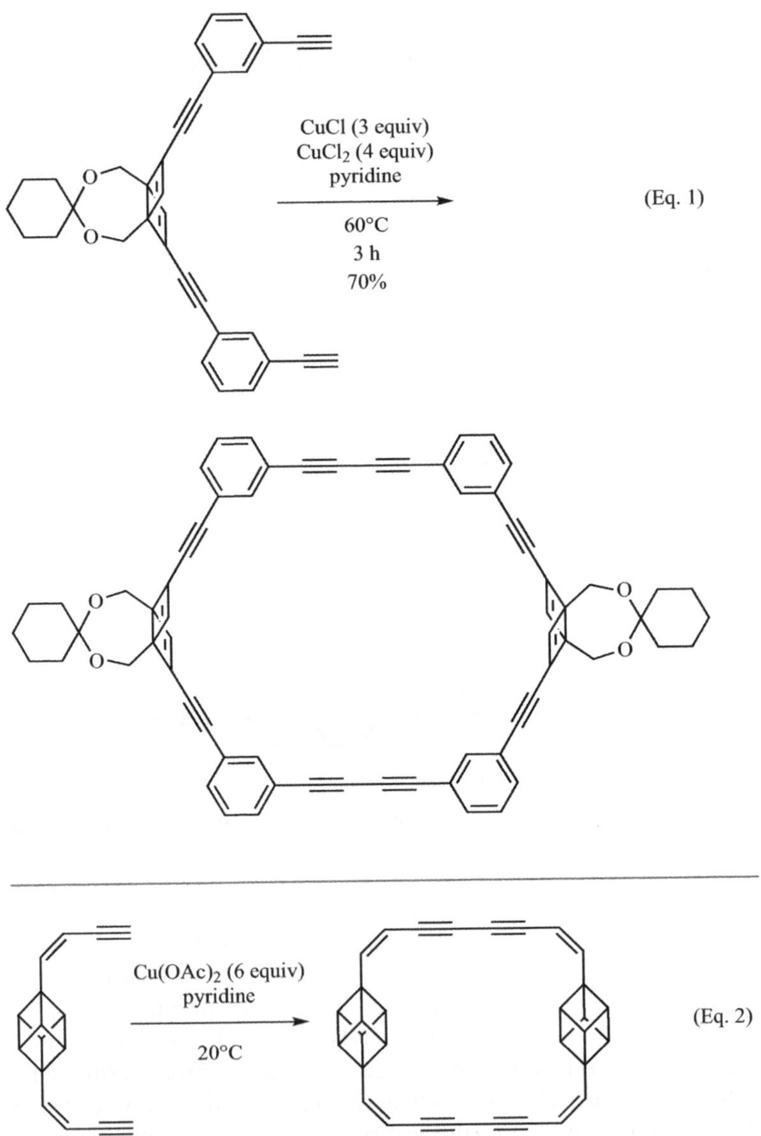

Scheme 12.7 Macrocycle synthesis via terminal alkyne dimerization reaction.

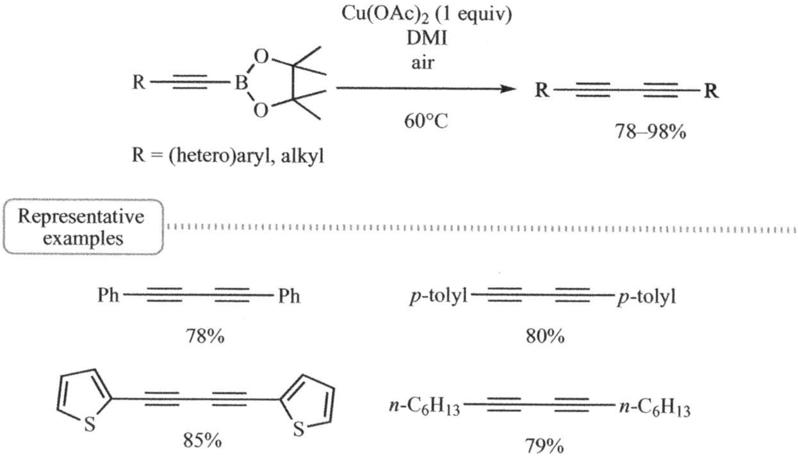

Scheme 12.8 Copper(II)-mediated homocoupling reaction of alkynylboronates.

and copper(II) chloride[16] or copper(II) acetate[17] were applied for the dimerization reaction of terminal alkynes, leading to 1,3-diyne-linked macrocycles in good yields.

Besides the oxidative homocoupling of terminal alkynes, both alkynylboronates and alkynyltrifluoroborates were found to be practical and versatile precursors for the synthesis of 1,3-diynes via the activation of the C(*sp*)–B bond in the presence of copper salts under aerobic conditions. As shown in Scheme 12.8, various symmetrical 1,3-diynes could be synthesized by copper(II) acetate-mediated homocoupling reaction of alkynylboronates in good-to-excellent yields in aprotic polar solvents such as 1,3-dimethyl-2-imidazolidinone (DMI). Copper(I) iodide and copper(I) thiophene-2-carboxylate (CuTC) also showed high catalytic activity.[18]

Paixão and coworkers studied the homocoupling reaction of alkynyltrifluoroborates in the presence of a catalytic amount of copper salts under different conditions and found that not only copper(I) and copper(II) salts showed different catalytic activity but also that the homocoupling reaction depended greatly on the solvents.[19] As shown in Scheme 12.9 (Eq. 1), both copper(II) acetate and copper(I) chloride could efficiently catalyze the homocoupling reaction of potassium phenylethynyltrifluoroborate in DMSO, affording 1,4-diphenyl-1,3-butadiyne in high yield. On the other hand, copper(I) iodide or cyanide and copper(II) chloride showed low catalytic activity. In addition, in *N*,*N*-dimethylformamide (DMF) and tetrahydrofuran (THF), the catalytic activity of copper(II) acetate decreased significantly, which be due to the lower solubility of the starting trifluoroborates in the solvents. It is interesting to note that under the optimized conditions, the homocoupling reaction occurred at C(*sp*)–B bond selectively and left the C(*sp*)–H bond untouched (Scheme 12.9, Eq. 2).

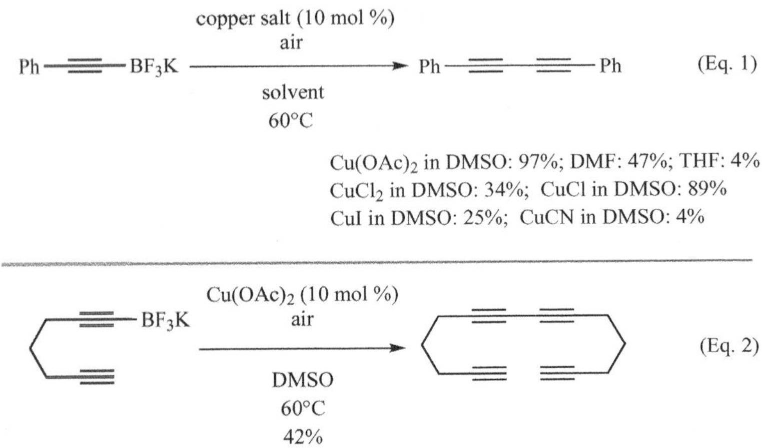

Scheme 12.9 Copper(I or II)-catalyzed homocoupling reaction of alkynyltrifluoroborates.

The coupling reaction of terminal alkynes can be also performed in environmentally benign solvents such as supercritical carbon dioxide, water, and ionic liquids. For instance, Jiang and coworkers studied the homocoupling reactions of various terminal alkynes in supercritical carbon dioxide using MeOH as a cosolvent and found that, when sodium acetate was used as a base, 1,3-butadiynes could be obtained in good-to-high yields (Scheme 12.10, Eq. 1).[20] In order to get a high yield of the desired coupling products, the addition of MeOH was crucial to increase the solubility of reactants. Copper(II) chloride was superior to other cupric salts (i.e., copper(II) sulfate, copper(II) nitrate, copper(II) acetate, and copper(II) bromide) possibly due to its higher solubility in $scCO_2$/MeOH medium. Under similar reaction conditions, copper(I) chloride also showed high catalytic activity for the Cadiot–Chodkiewicz coupling reaction (the heterocoupling process between a 1-bromoalkyne and a terminal alkyne in the presence of a copper(I) catalyst and an amine), giving unsymmetrical 1,3-diynes in high yield (Scheme 12.10, Eq. 2).[21]

Copper(II) chloride also showed high catalytic activity for the homocoupling of terminal alkynes in near-critical water (200°C) without the use of organic solvent and base, as reported by Wang and coworkers in 2004 (Scheme 12.11).[22]

In ionic liquids, cuprous halides could be used as efficient catalysts for the Glaser coupling reaction. As shown in Scheme 12.12, copper(I) chloride catalyzed the homocoupling of terminal alkynes in [bmim]PF_6, with the use of TMEDA as ligand, at room temperature to afford smoothly 1,3-butadiynes in excellent yields.[23] The considerable advantages in ionic liquid are enhanced reaction rates and the ease of recovery and reuse of the catalytic system up to six reuses without any noticeable decrease in yield.[23] Very interestingly,

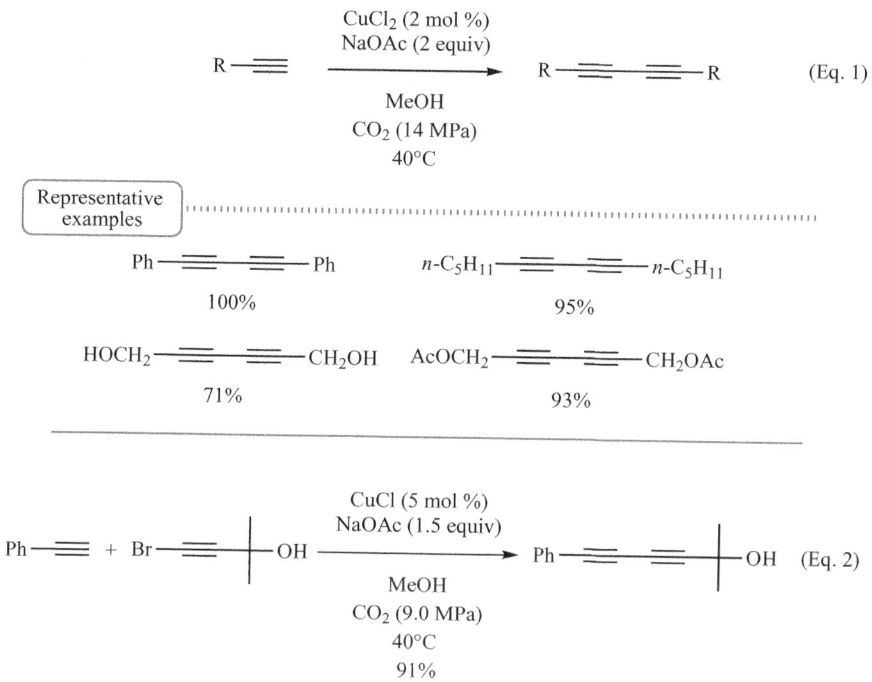

Scheme 12.10 Copper(I and II)-catalyzed synthesis of 1,3-diynes in supercritical carbon dioxide.

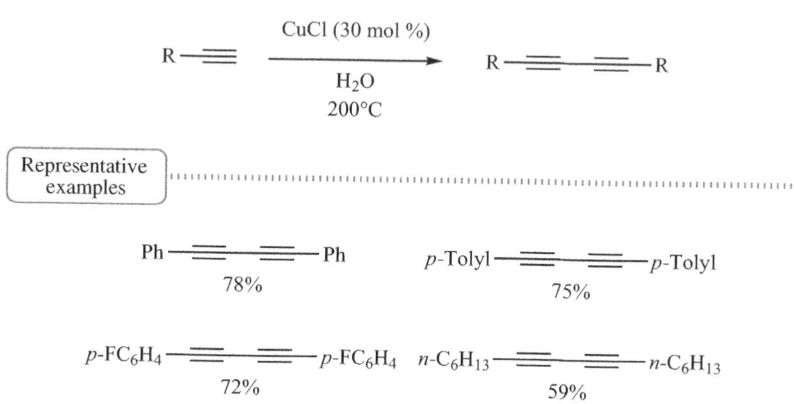

Scheme 12.11 Homocoupling reaction of terminal alkynes in near-critical water.

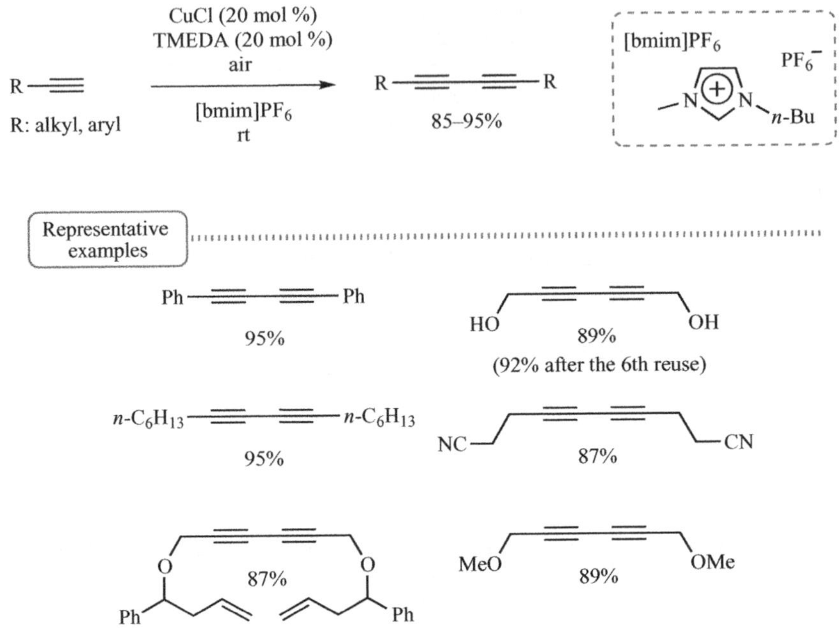

Scheme 12.12 Copper(I)-catalyzed homocoupling of terminal alkynes in ionic liquid.

copper(I) iodide also showed good catalytic activity in [bmim]OH in the absence of both ligand and base.[24]

Glaser, Hay, and Eglinton coupling reactions are efficient methods for the synthesis of symmetrical 1,3-diynes, but they are not practical for the preparation of unsymmetrical 1,3-diynes via intermolecular oxidative heterocoupling of two different terminal alkynes due to the difficulty controlling the chemoselectivity of the cross-coupling event. Therefore, the copper-catalyzed cross-coupling reaction of an alkynyl halide with a terminal alkyne is one of the most important and efficient methods in this area.

The pioneering work on the synthesis of unsymmetrical 1,3-diynes was reported by Cadiot and Chodkiewicz in the 1950s through the reaction of an alkynyl halide with a cuprous acetylide (Scheme 12.13).[25] A base, usually an amine, is also mandatory in this cross-coupling reaction. Hydroxylamine is also often used as an additive to reduce any cupric ions in the reaction media. The proposed mechanism includes the oxidative addition of the haloalkyne to copper(I) to form a copper(III) intermediate,[25b] and subsequent reductive elimination affords 1,3-diyne.

The preparation of terminal polyacetylenes is not trivial using the classical Cadiot–Chodkiewicz conditions since the acidity of the cross-coupling product is greater than the one of the starting terminal alkyne, and an appropriate protecting group is therefore required.[26] Walton demonstrated that a triethylsilyl group is particularly well suited in this regard. Copper(I) chloride could

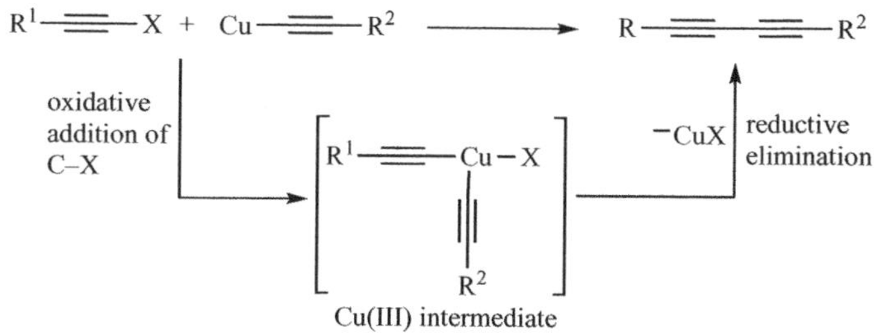

Scheme 12.13 Proposed mechanism for Cadiot–Chodkiewicz coupling reaction.

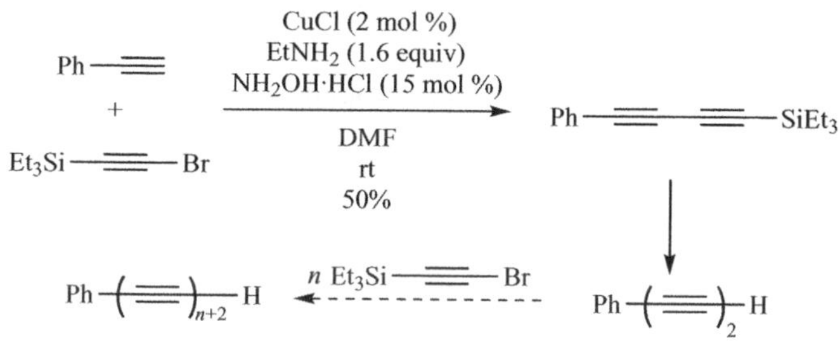

Scheme 12.14 1-Bromo-2-triethylsilyl acetylene in the Cadiot–Chodkiewicz cross-coupling reactions.

efficiently catalyze the Cadiot–Chodkiewicz cross-coupling reaction of phenylacetylene with 1-bromo-2-triethylsilyl acetylene at room temperature to afford unsymmetrical triethylsilyl-substituted 1,3-diynes in moderate yield (Scheme 12.14).[27] Since the carbon–silicium bond could be easily converted into a carbon–hydrogen bond, important application of such type of cross-coupling reactions can be envisaged, such as the synthesis of unique linear conjugated polyynes, which possess an interesting potential in electronic and optical materials.

Under essentially similar conditions, Marino and Nguyen studied the cross-coupling reaction of trialkylsilyl acetylenes with different bromoalkynes leaving the carbon–silicium bond intact. A variety of unsymmetrical 1,3-diynes was formed in good yields, providing an efficient route to these synthetically useful motifs (Scheme 12.15).[28]

Hosomi and coworkers found that the trimethylsilyl group of alkynylsilanes could be smoothly activated by copper(I) chloride to give the corresponding copper(I) acetylides in DMI or DMF.[29] Similar results were published simultaneously by Mori and coworkers[30a–c] and based on these observations, they studied the copper(I) chloride-catalyzed cross-coupling reaction of alkynylsilanes with chloroalkynes in DMF, developing an alternative route to synthesize a variety of unsymmetrical 1,4-biaryl-1,3-butadiynes in moderate-to-good yields (Scheme 12.16).[30d] It is worthy to note that this cross-coupling reaction

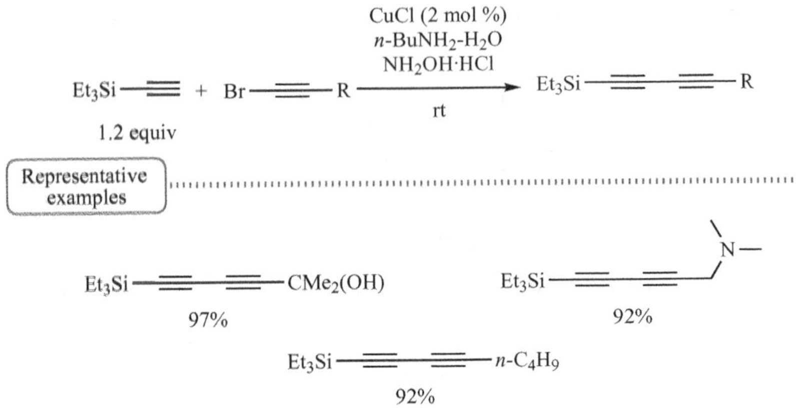

Scheme 12.15 Silylated alkynes in Cadiot–Chodkiewicz cross-coupling reaction.

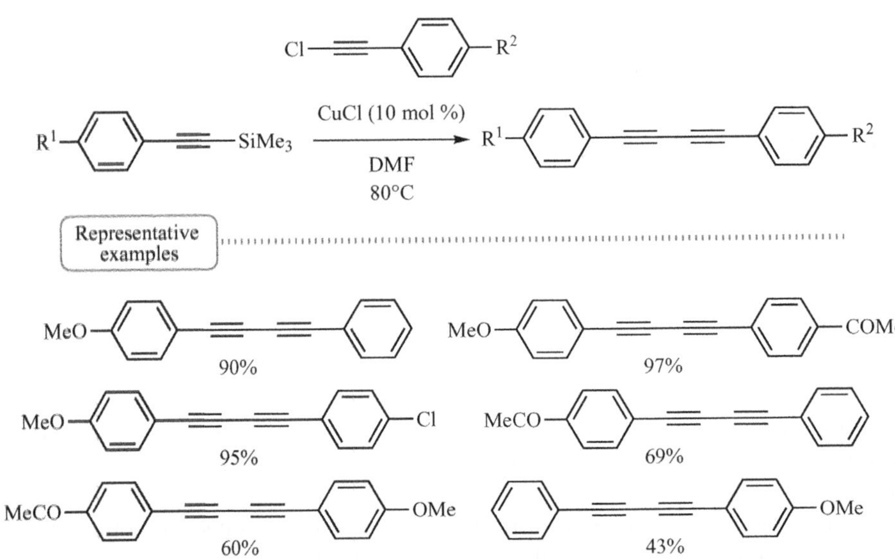

Scheme 12.16 Copper(I)-catalyzed cross coupling of silylated alkynes with chloroalkynes.

proceeds without any base, which significantly expands the scope of this transformation.

Some natural products containing the 1,3-butadiynyl moiety show interesting important biological and physiological properties.[2,31] The copper-catalyzed heterocoupling reactions of terminal alkynes have been applied as key steps in the total synthesis of such type of natural products, which is discussed in Chapter 18 in more details.

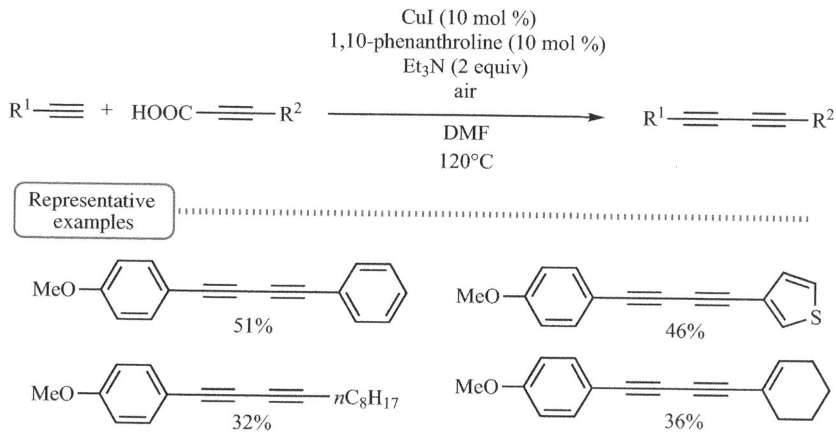

Scheme 12.17 Copper(I)-catalyzed decarboxylative cross coupling of 3-aryl propiolic acids.

The transition-metal-catalyzed decarboxylative coupling reaction is an efficient carbon–carbon bond formation reaction, which has been recently extensively studied.[32] Yu and coworkers examined the cross-coupling reaction of 3-aryl propiolic acid with terminal alkynes in the presence of copper(I) iodide (10 mol %) with the use of 1,10-phenanthroline (10 mol %) as ligand, at 120°C under air. This new route allowed the synthesis of unsymmetrical 1,4-disubstituted-1,3-butadiynes in moderate yields via decarboxylation (Scheme 12.17).[33a] As a logical extension, the use of 1,1-dibromoalkenes instead of terminal alkynes proved possible.[33b]

In the Cadiot–Chodkiewicz cross-coupling reaction, the formation of symmetrical homocoupled products sometimes is a serious side reaction that decreases the yield of the desired unsymmetrical product and seriously complicates the purification. In order to easily get rid of the dimerization product of the haloalkyne partner, Schore and coworkers developed a copper(I) chloride-catalyzed coupling reaction of polymer-bound bromo- or iodoacetylene with 1-octyne to give unsymmetrical 1,3-diynol after removal from the polymer by hydrolysis. The formation of symmetrical 1,3-diynes is suppressed compared to classical homogeneous conditions.[34a] As a final note to this section, the synthesis of unsymmetrical 1,3-diynes has been reported from *cis*-styrenyl bromides and terminal alkynes in the presence of copper(I) hydroxyapatite. The use of the *trans*-styrenyl bromides on the other hand delivered the corresponding 1,3-enyne.[34b]

In order to recover copper catalysts, a variety of heterogeneous copper catalysts such as CuAl hydrotalcite,[35] Cu(I)-zeolites,[36] $Cu(OH)_x/TIO_2$,[37] and copper nanoparticles[38] have been successfully applied in the Glaser and Cadiot–Chodkiewicz coupling reactions. An overview of reusable copper-based catalytic systems can be found in Chapter 20.

12.3 COPPER-CATALYZED SYNTHESIS OF 1,4-DIYNES

Copper-mediated cross coupling of acetylenic Grignard reagents with propargyl halides is one of the traditional methods for the formation of 1,4-diynes.[39] In 1965, Cadiot and coworkers showed that the copper-mediated cross coupling of a terminal alkyne with a propargyl halide provided a direct and efficient method for the synthesis of the corresponding 1,4-diyne (Scheme 12.18, Eq. 1).[40] However, this transformation was still very limited in scope.[39,40] More recently, Linstrumelle and coworkers expanded the scope of this reaction by showing that propargylic bromides or iodides or tosylates could react with alkynes in the presence of sodium carbonate and tetrabutylammonium chloride in DMF or acetonitrile as shown in Scheme 12.18 (Eq. 2).[41] As an example, the copper(I) iodide-mediated cross coupling of 1-heptyne with chloro-2-octyne under basic conditions at room temperature afforded 6,9-pentadecadiyne in an impressive 93% yield (Scheme 12.18, Eq. 3).[42]

As described in this short subchapter, only a handful of efficient strategies have been developed for the straightforward synthesis of 1,4-diynes using simple reaction conditions. The situation is clearly different for the copper-mediated or copper-catalyzed synthesis of 1,3-enynes. The most representative cross-coupling reactions leading to this central motif are presented in the next chapter.

12.4 SYNTHESIS OF 1,3-ENYNES BY DIRECT REACTION OF VINYL HALIDES WITH ALKYNES

Early examples on the copper-mediated direct cross-coupling reaction of vinyl halides with terminal alkynes were reported by Suzuki and coworkers in 1989.[43] It was found that in hexamethylphosphoric triamide (HMPA), copper(I) iodide and copper(I) bromide could efficiently mediate the reaction of vinyl iodides with 1-octyne, while copper(I) chloride showed only modest activity (Scheme 12.19, Eq. 1). In the case of vinyl bromides, copper(I) iodide and copper(II) acetate were efficient, but copper(I) bromide and $Cu(acac)_2$ showed low activity, and CuCl was inefficient (Scheme 12.19, Eq. 2). Under the reaction conditions, all the cross-coupling reactions proceeded with excellent stereoselectivity and with retention of the alkene's configuration.

Three years later, Miura and coworkers reported the copper-catalyzed cross coupling of vinyl halides with terminal alkynes to furnish conjugated 1,3-enynes.[44] In the presence of a catalytic amount of copper(I) iodide and triphenylphosphine as ligand and with the use of potassium carbonate as base, the coupling reactions occurred smoothly to give 1,3-enynes in good yields (Scheme 12.20).

In addition to phosphine ligand, Liu and Ma reported that *N*,*N*-dimethylglycine·HCl was also an efficient ligand in copper(I) iodide-catalyzed

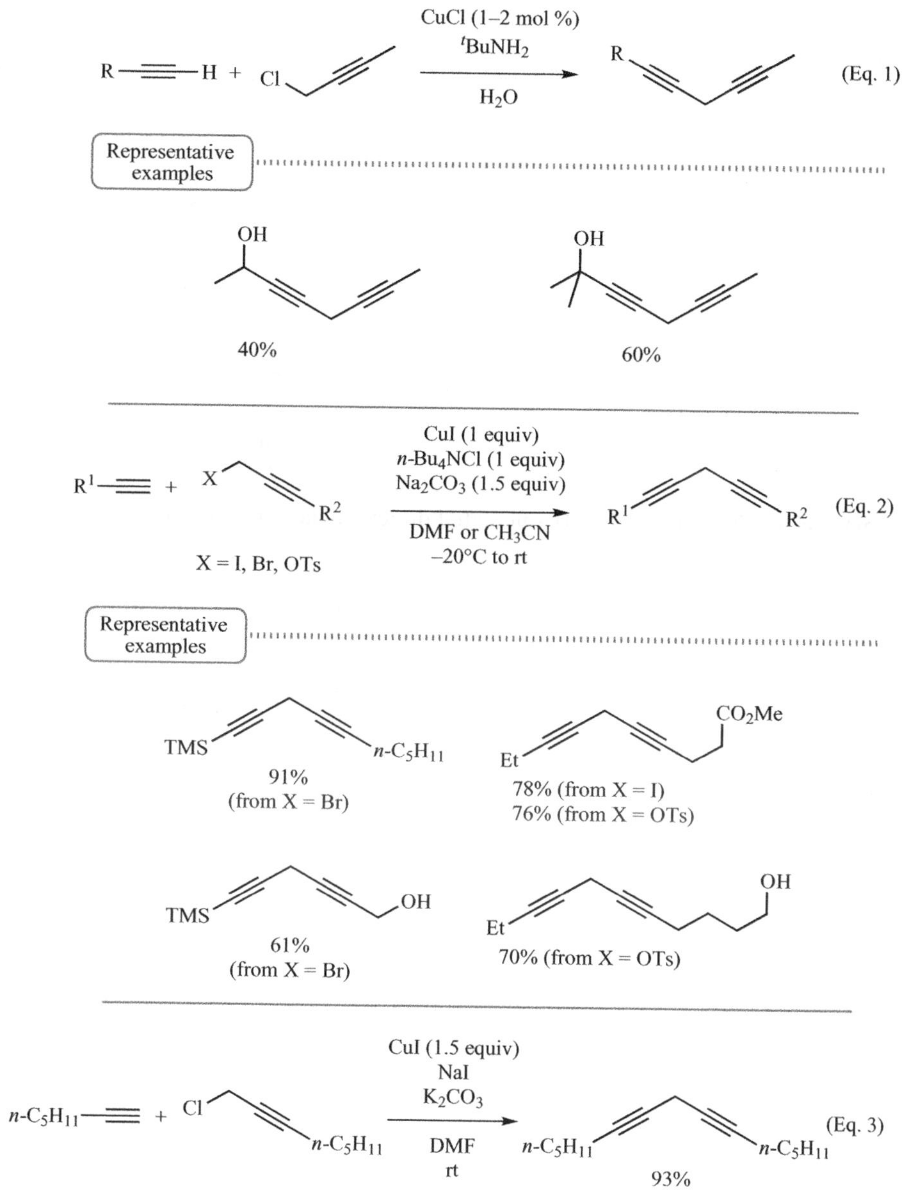

Scheme 12.18 Cadiot–Chodkiewicz and Linstrumelle syntheses of 1,4-diynes.

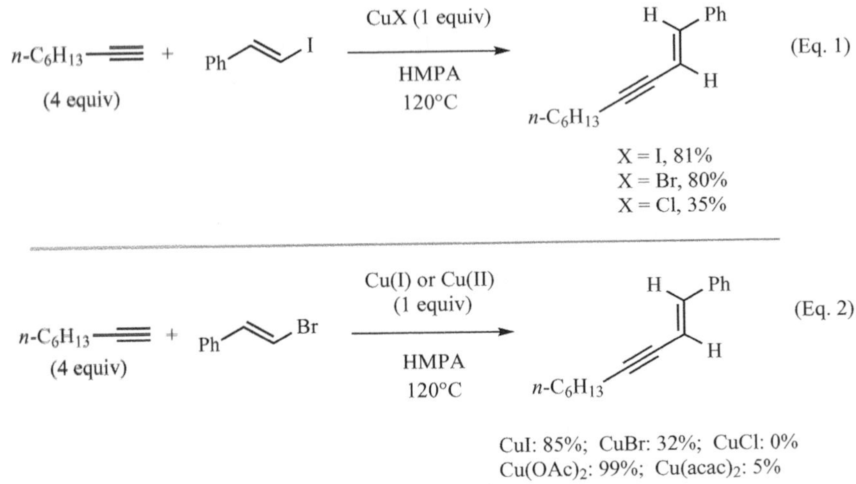

Scheme 12.19 Copper(I or II)-mediated direct vinylation of terminal alkynes.

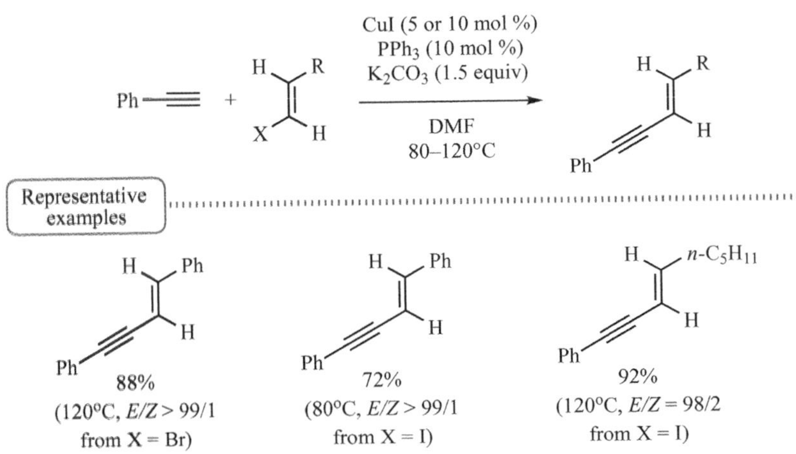

Scheme 12.20 Copper(I)-catalyzed direct vinylation of terminal alkynes.

cross coupling of terminal alkyne with electron-rich and electron-deficient vinyl iodides in dioxane. The corresponding conjugated 1,3-enynes were obtained in good-to-high yields (Scheme 12.21).[45] The geometry of the olefin was also generally preserved, leading to *Z*- and *E*-enynes exclusively. Moreover, other catalytic systems, such as copper(I) iodide/8-hydroxyquinoline[46] or copper(I) iodide/1,10-phenanthroline,[47] also showed high catalytic activity for the direct cross coupling of vinyl halides with alkynes.

Shao and Shi reported a ligand-free copper(I) iodide-catalyzed cross-coupling reactions of 2,4-halo-but-1-enes with terminal alkynes that provided

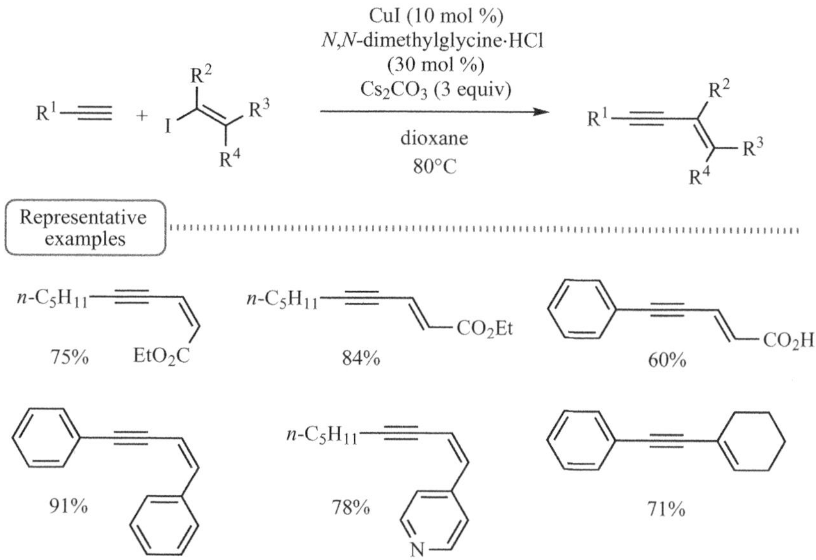

Scheme 12.21 Copper(I) iodide/amino acid-catalyzed synthesis of 1,3-enynes.

a simple and efficient procedure for the synthesis of 2-alkynyl-buta-1,3-dienes in high yields (Scheme 12.22, Eq. 1).[48] To get insights into the reaction mechanism, control experiments were carried out using 1,1-diphenyl-2-iodo-but-1-ene (with or without 1,1-diphenyl-buta-1,3-diene as a potential ligand; Scheme 12.22, Eq. 2) or 1,1-diphenyl-2-iodo-buta-1,3-diene (Scheme 12.22, Eq. 3). In the former case, only 10–12% of the desired product was obtained, demonstrating that 1,1-diphenyl-buta-1,3-diene was not a ligand. In the latter case, the yield of the expected 2-alkynyl-buta-1,3-diene reached 95%, indicating that 1,1-diphenyl-2-iodo-buta-1,3-diene was a key reaction intermediate.

Venkataraman and coworkers studied the catalytic activity of well-defined copper(I) complexes and copper(I) salts with additives in the cross-coupling reaction of aryl- or alkyl-substituted acetylenes with vinyl iodides (Scheme 12.23).[49] It was found that [Cu(bipy)PPh_3Br] (bipy = 2,2′-bipyridine) and [Cu(phen)$(PPh_3)_2$]NO_3 (phen = 1,10-phenanthroline) showed much higher catalytic activity than a mixture of CuI/bipy, CuI/phen, and CuI/phen/PPh_3. As a general rule, the (*E*)-isomers cross coupled more slowly than the corresponding (*Z*)-isomers (Scheme 12.23, Eq. 2). An enhancement in reaction rate could be obtained when [Cu(phen)$(PPh_3)_2$]NO_3 was used as a catalyst in combination with cesium carbonate as a base. The considerable advantages of this catalytic system include the retention of configuration of alkene, tolerance of a variety of functional groups, and good-to-excellent yields for various enynes.

Miura's catalytic system (copper(I) iodide/PPh_3) could also promote the intramolecular cross-coupling reaction of vinyl halides with terminal alkynes in DMF to produce macrocyclic compounds embedding a 1,3-enyne moiety in

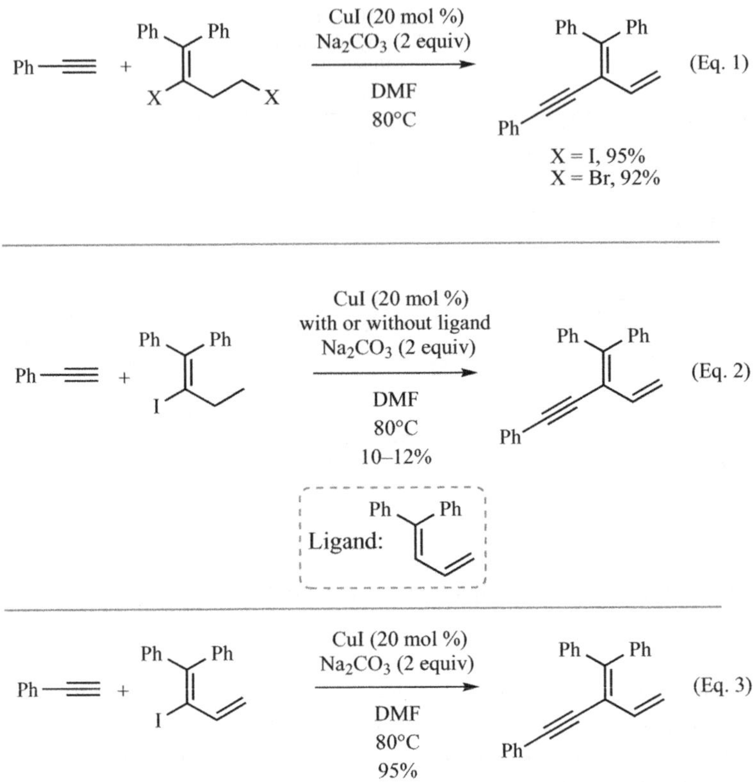

Scheme 12.22 Ligand-free copper(I) iodide-catalyzed of 2-alkynyl-buta-1,3-dienes.

fair yield (Scheme 12.24).[50] This transformation was used as a key step in model studies toward the total synthesis of oximidines I and II (see Chapter 18 for a more detailed discussion).

In addition, in the presence of an excess of copper(I) chloride, the direct cross coupling of alkynylsilane with vinyl iodides was an alternatively efficient route to 1,3-enynes. As shown in Scheme 12.25, the copper(I) chloride-mediated cross-coupling reaction of 1-trimethylsilyl acetylenes with (*E*)-iodo-alkenes in dimethylimidazolidinone (DMI) gave the corresponding (*E*)-1,3-enynes in good-to-excellent yields.[51] In order to perform the cross-coupling reaction, an excess of copper(I) chloride (2.2 equiv) was required since when 1 equiv of copper(I) chloride was used, no cross-coupling product could be detected.

Moreover, 1,3-enynes could also be obtained via a copper(I) iodide-catalyzed cross-coupling reaction of terminal alkynes with very potent electrophiles such as alkenyliodonium salts (e.g., (*E*)-PhCH = CHI$^+$PhBF$_4^-$) in a mixture of solvents (DME/H_2O, 4:1) at room temperature.[52]

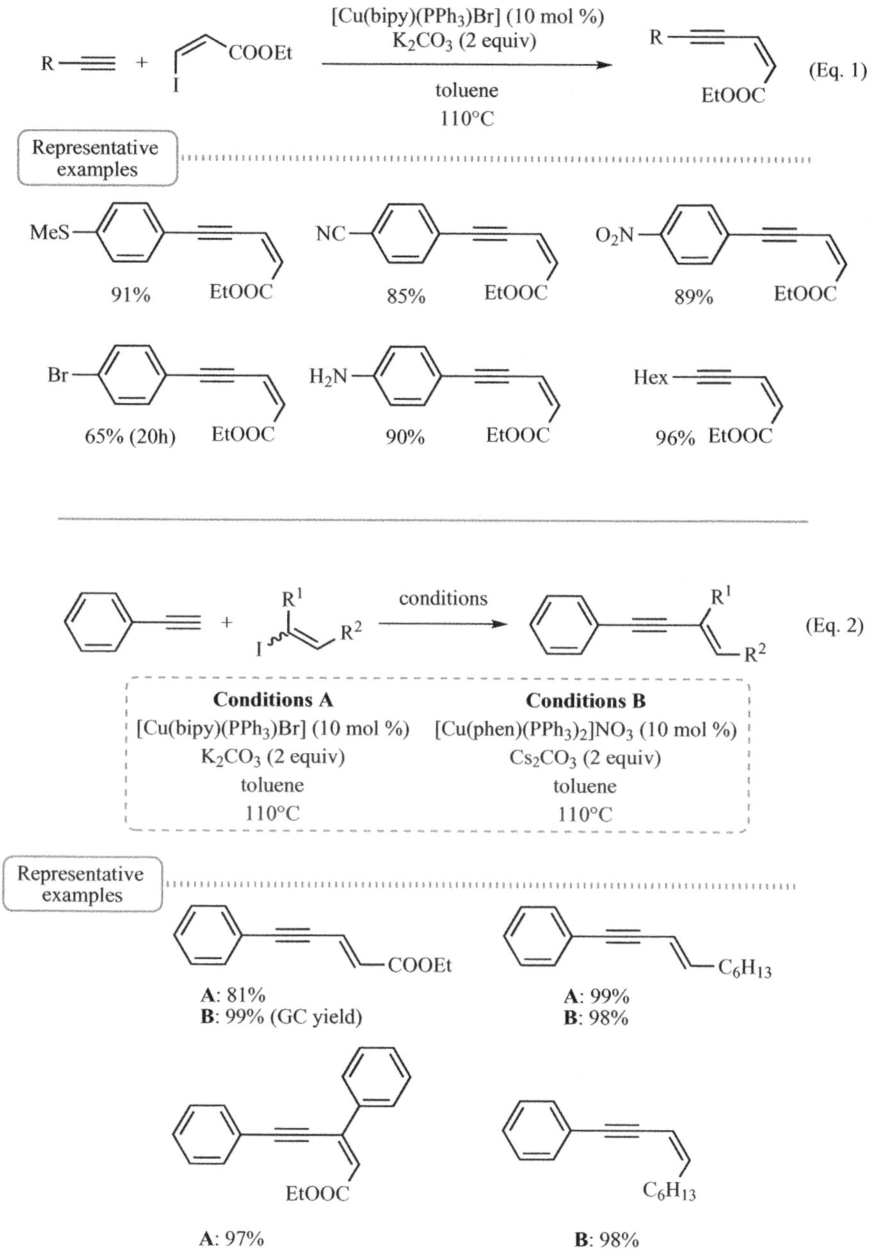

Scheme 12.23 Venkataraman's copper(I)-catalyzed synthesis of 1,3-enynes.

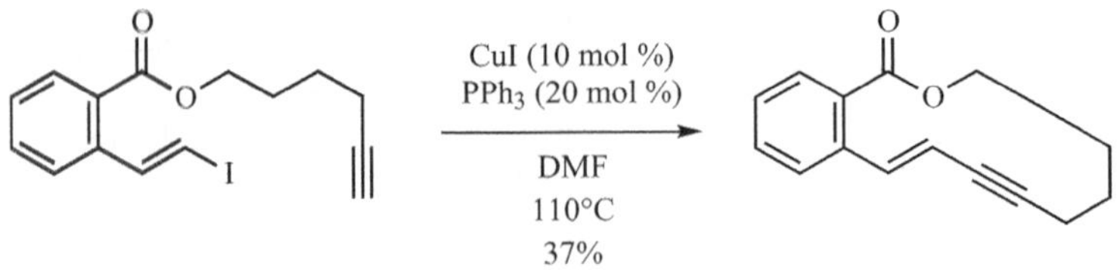

Scheme 12.24 An approach of oximidines I and II using Miura's catalytic system.

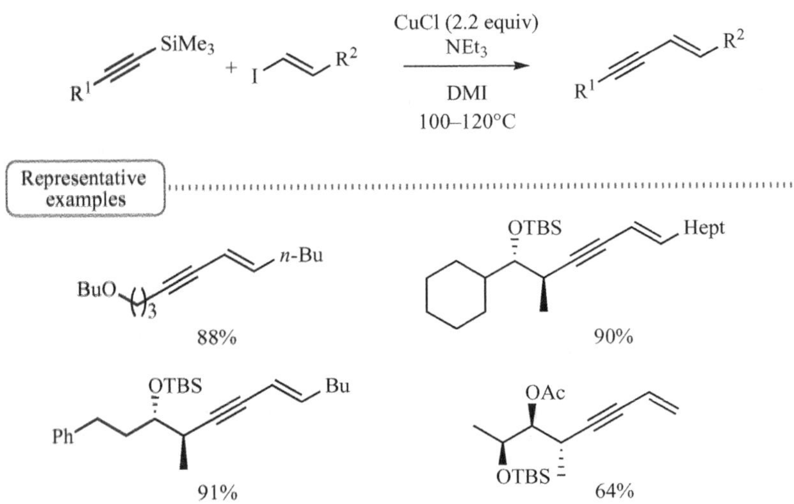

Scheme 12.25 Marshall's direct cross coupling of alkynylsilanes with vinyl iodides.

Overall, several efficient intra- or intermolecular copper-mediated or copper-catalyzed cross-coupling reactions have been developed for the synthesis of 1,3-enynes form terminal alkynes and vinyl halides, relying on simple ligands such as triphenylphosphine or *N*,*N*-dimethylglycine. Silylated alkynes are also reactive partners, thus expanding the scope of possible reactants. Similar progress has been reported in the copper-catalyzed synthesis of 1,3- and 1,4-enynes. Relevant examples will be critically discussed in the next chapters.

12.5 SYNTHESIS OF 1,3-ENYNES BY STILLE-TYPE CROSS-COUPLING REACTION

Compared to palladium- and palladium/copper-catalyzed Stille-type cross-coupling of alkynyltin derivatives, the examples of syntheses of 1,3-enynes through a similar type of copper-catalyzed cross-coupling reaction are rare. Kang and coworkers studied the copper- and manganese-catalyzed cross coupling of organostannanes with alkenyl iodides in the presence of sodium

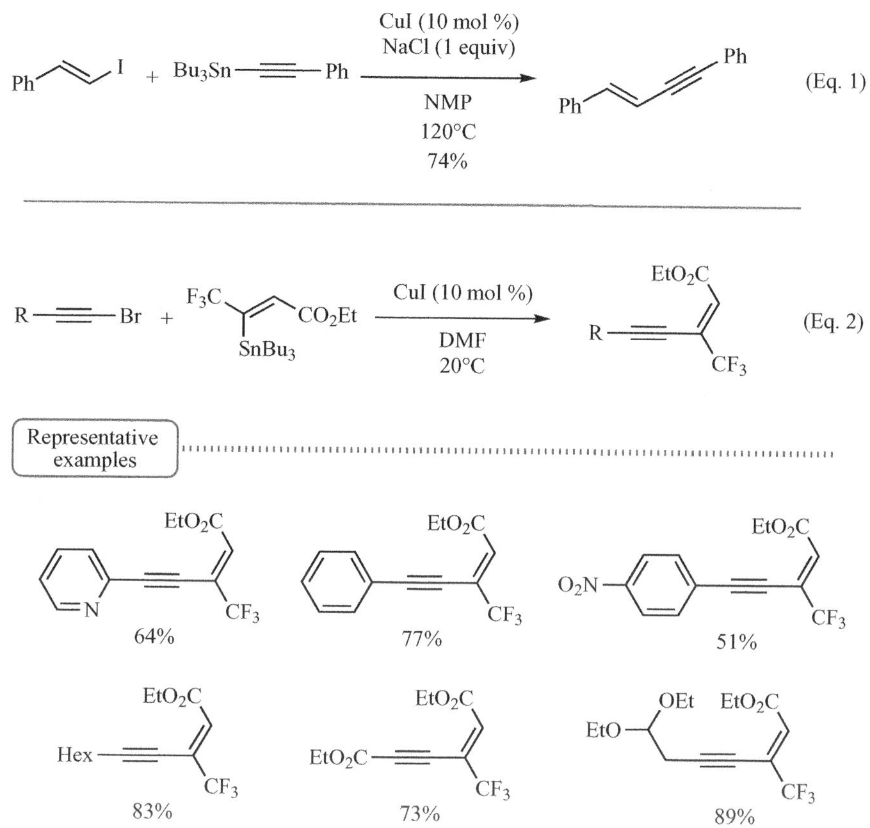

Scheme 12.26 Copper(I)-catalyzed Stille reaction for the synthesis of 1,3-enynes.

chloride and found that the catalytic system was also efficient for the cross coupling of (2-phenylethynyl)tributylstannane with (*E*)-β-styryl iodide in *N*-methyl-2-pyrrolidone (NMP) at 120°C to give the corresponding 1,4-diphenyl-1,3-enyne in 74% yield (Scheme 12.26, Eq. 1).[53a] More recently, Abarbri and coworkers reported on an easy access to (*E*)-3-trifluoromethylbut-2-en-3-ynoate via a copper(I)-catalyzed Stille coupling of alkynyl bromides and β-tributylstannyl-α,β-acrylates (Scheme 12.26, Eq. 2).[53b] The excellent stereoselectivity as well as the functional group tolerance should prove very useful for the rapid elaboration of trifluoromethyl-substituted 1,3-enynes. In addition, control experiments demonstrated the tin–copper transmetallation and the relative stability of the *cis* copper intermediate that could be presumably ascribed to a coordination of copper by the ester carbonyl.

Although only isolated examples of copper-catalyzed synthesis of 1,3-enynes by Stille cross-couplings have been reported, the simplicity of the reaction conditions and the good overall yields are noticeable features of this transformation. In this context, the Suzuki–Miyaura cross-coupling reaction has also been investigated and will be discussed in the next chapter.

12.6 SYNTHESIS OF 1,3-ENYNES BY THE SUZUKI–MIYAURA-TYPE CROSS-COUPLING REACTION

The first copper-catalyzed Suzuki–Miyaura-type[54] cross-coupling reaction to afford 1,3-enyne was reported in 1983 by Arase and coworkers.[55] They found that copper(II) acetylacetonate showed good catalytic activity for the cross-coupling reaction of alkenyldicyclohexylborane (which was pretreated with sodium hydroxyde to generate the corresponding boronate) with 1-bromoalkyne. The conjugated (*E*)-enynes were obtained in moderate-to-good yield in a highly stereoselective manner (Scheme 12.27), similar to the cases where $Pd(PPh_3)_4$ was employed as the catalyst.[56]

Recently, Hoshi and Shirakawa studied a similar cross-coupling reaction from (*E*)- and (*Z*)-1-alkenyldialkylboranes with (trimethylsilyl)ethynyl bromide in the presence of copper(II) acetylacetonate using sodium methoxide or $LiOH{\cdot}H_2O$ as the base (Scheme 12.28, Eq. 1 and Eq. 2).[57] It was found that the nature of the base dramatically influenced the nature of the product. When sodium methoxide was used (Scheme 12.28, Eq. 1), the reaction exclusively afforded both (*E*)- and (*Z*)-3-alken-1-ynes with high regio- and stereoselectivity. However, when lithine was employed as a base, the trimethylsilyl group resisted to the reaction conditions and both (*E*)- and (*Z*)-1-trimethylsilyl-3-alken-1-ynes were formed regio- and stereoselectively (Eq. 2).[58]

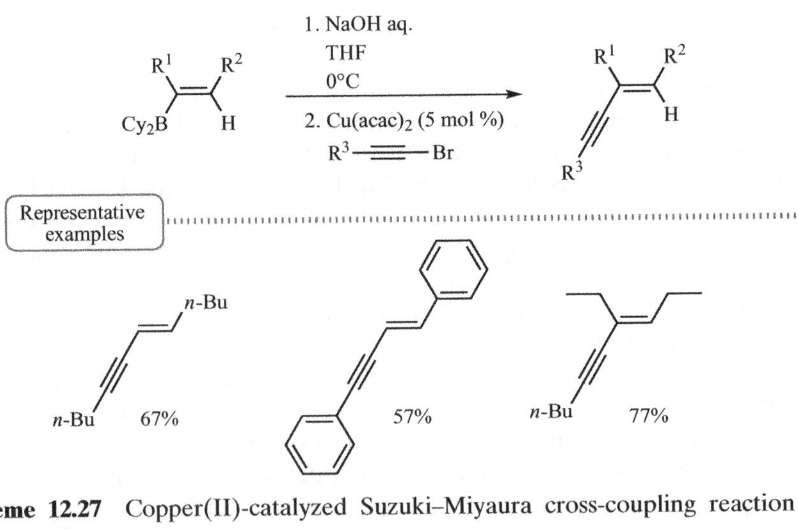

Scheme 12.27 Copper(II)-catalyzed Suzuki–Miyaura cross-coupling reaction with 1-bromoalkyne.

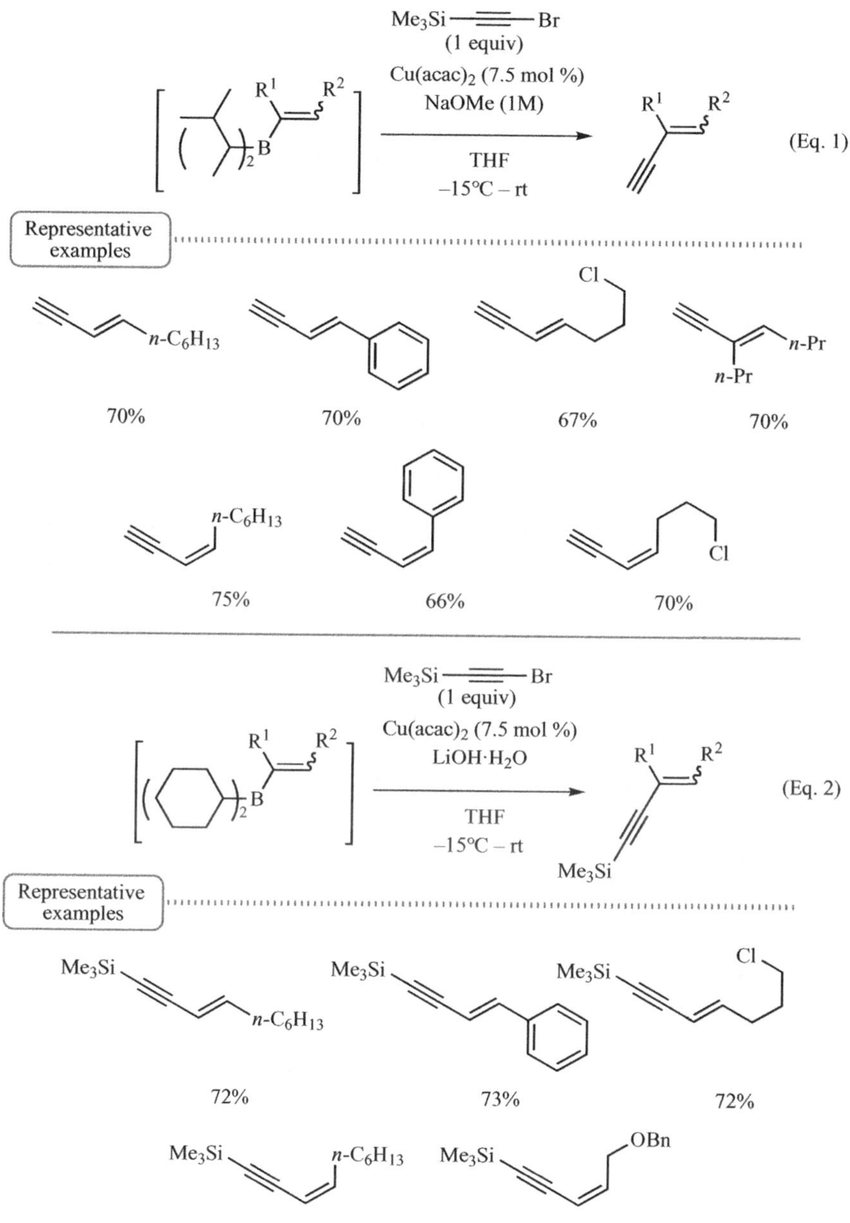

Scheme 12.28 Influence of the base in the copper-catalyzed Suzuki–Miyaura synthesis of 1,3-enynes.

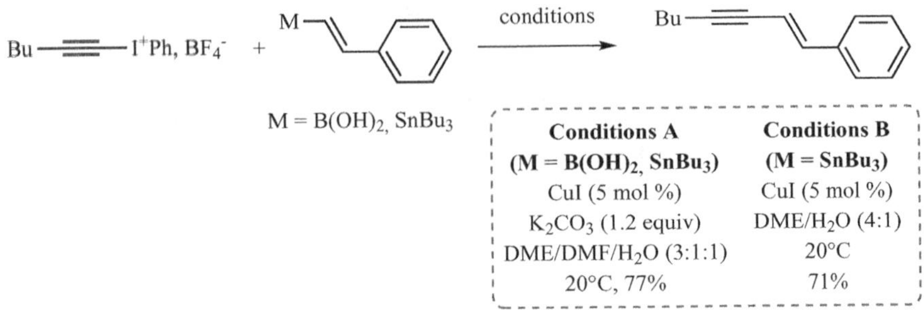

Scheme 12.29 Copper(I)-catalyzed synthesis of 1,3-dienes from alkynyliodonium salt.

Yu and coworkers studied the copper-catalyzed cross coupling of alkynyliodonium tetrafluoroborates with aryl- and vinylboronic acids and vinyltin derivatives under different reaction conditions (Scheme 12.29). It was found that under the optimized conditions, the cross coupling of (*E*)-phenylethenylboronic acid catalyzed by copper(I) iodide in a mixture of solvents (DME/DMF/H_2O = 3:1:1) at 20°C afforded the corresponding 1,3-enyne derivative in 77% (Scheme 12.29, condition A). Vinyltin derivatives could also be cross coupled with a slightly diminished yield (71%, condition B), although the scope of the reaction was not fully investigated.[59]

Thus, the copper-catalyzed Suzuki–Miyaura cross coupling proved to be a valuable tool to synthesize 1,3-enynes from 1-bromoalkynes or alkynyliodonium salts and vinylboronate derivatives. The copper-catalyzed synthesis of nonconjugated enynes, such as 1,4-enynes, has also attracted a great deal of attention. A selection of the most relevant examples of this strategy is discussed in the next subchapter.

12.7 SYNTHESIS OF 1,4-ENYNES BY ALLYLATION REACTION OF TERMINAL ALKYNES

It is well known that the reaction of alkynylcopper(I) organometallics with allylic halides is an efficient reaction for the synthesis of 1,4-enynes.[60] Therefore, the copper(I)-catalyzed allylation of terminal alkynes with allylic halides has been extensively studied and became an important process for the synthesis of 1,4-enynes. For instance, Grushin and Alper investigated the cross-coupling reaction of terminal alkynes with allylic bromides in the presence of copper(I) chloride and a phase-transfer catalyst with the use of aqueous sodium hydroxyde as a base, and found that the yield of the desired 1,4-enyne depended on the nature of alkyne, the structures of the allylic bromides, and of the phase-transfer catalyst.[61] As shown in Scheme 12.30 (Eq. 1), the reaction of aliphatic acetylenes with allyl bromide afforded the expected 1,4-enyne derivatives as the exclusive products with the use of A-336 as the

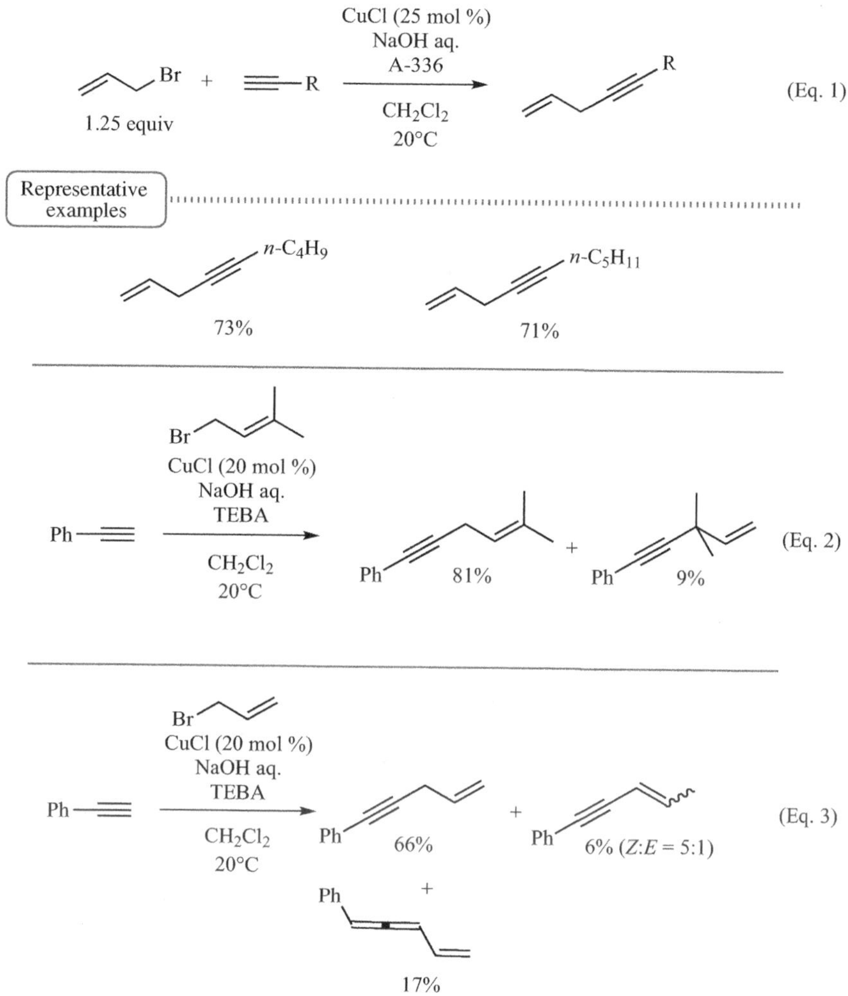

Scheme 12.30 Copper(I)-catalyzed synthesis of 1,4-enynes using phase-transfer catalysis.

phase-transfer catalyst in good yields. On the other hand, the reaction of phenylacetylene with allyl bromide or l-bromo-3-methyl-but-2-ene in the presence of triethylbenzylammonium chloride (TEBA) produced a mixture of two or three isomeric products arising from competitive S_N2/S_N2' processes or double bond isomerization (Scheme 12.30, Eq. 2 and Eq. 3). However, the use of a relatively weak base such as potassium carbonate (Scheme 12.31, Eq. 1)[62] and potassium carbonate/sodium sulfite (Scheme 12.31, Eq. 2)[63] resulted in a significant improvement of the chemoselectivity in the copper(I) iodide-catalyzed reaction of phenylacetylene with allyl bromide. The desired 1,4-enyne derivative was obtained in high or quantitative yield.

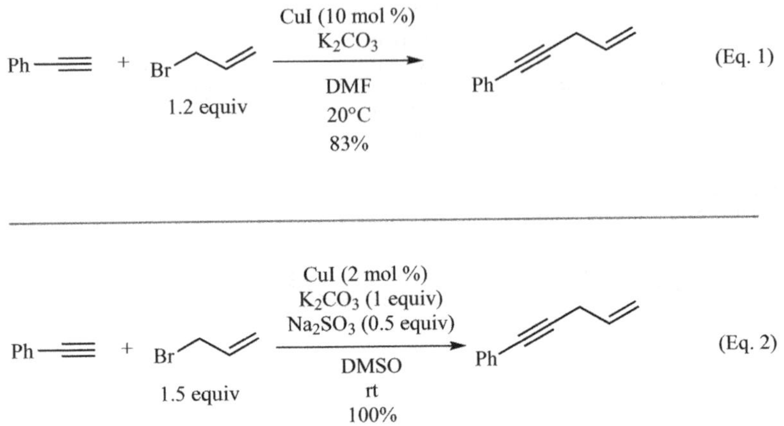

Scheme 12.31 Copper(I)-catalyzed synthesis of 1,4-enynes.

12.8 CONCLUSION

Copper salts have recently triggered intense interest as versatile catalysts in organic synthesis.[4] This chapter gives a summary of recent advances in the application of copper salts as promoters or catalysts in important coupling reactions such as Glaser/Hay/Eglinton coupling, Cadiot–Chodkiewicz coupling, Stille and Suzuki–Miyaura-type cross-coupling reactions, and allylation of terminal alkynes that provide efficient accesses to diynes and enynes. Although significant progresses have been achieved in this field, developing much simpler and much more efficient copper-based catalytic systems for the alkynylation and alkenylation of alkynyl derivatives is still an interesting and challenging research topic since diynes and enynes are important building blocks in organic synthesis and functional organic materials.

REFERENCES

[1] (a) Siemsen, P.; Livingston, R. C.; Diederich, F. *Angew. Chem. Int. Ed. Engl.* **2000**, *39*, 2632–2657. (b) Saito, S.; Yamamoto, Y. *Chem. Rev.* **2000**, *100*, 2901–2915. (c) Aubert, C.; Buisine, O.; Malacria, M. *Chem. Rev.* **2002**, *102*, 813–834. (d) Liu, J.; Lam, J. W. Y.; Tang, B. Z. *Chem. Rev.* **2009**, *109*, 5799–5867.

[2] (a) Shi Shun, A. L. K.; Tykwinski, R. R. *Angew. Chem. Int. Ed. Engl.* **2006**, *45*, 1034–1057. (b) Evano, G.; Blanchard, N.; Toumi, M. *Chem. Rev.* **2008**, *108*, 3054–3131.

[3] (a) Negishi, E.; Anastasia, L. *Chem. Rev.* **2003**, *103*, 1979–2017. (b) Doucet, H.; Hierso, J. C. *Angew. Chem. Int. Ed. Engl.* **2007**, *46*, 834–871.

[4] (a) Ley, S. V.; Thomas, A. W. *Angew. Chem. Int. Ed. Engl.* **2003**, *42*, 5400–5449. (b) Beletskaya, I. P.; Cheprakov, A. V. *Coord. Chem. Rev.* **2004**, *248*, 2337–2364.

(c) Bock, V. D.; Hiemstra, H.; van Maarseveen, J. H. *Eur. J. Org. Chem.* **2006**, 51–68. (d) Chemler, S. R.; Fuller, P. H. *Chem. Soc. Rev.* **2007**, *36*, 1153–1160. (e) Poulsen, T. B.; Jørgensen, K. A. *Chem. Rev.* **2008**, *108*, 2903–2915. (f) Monnier, F.; Taillefer, M. *Angew. Chem. Int. Ed. Engl.* **2008**, *47*, 3096–3099. (g) Carril, M.; SanMartin, R.; Domínguez, E. *Chem. Soc. Rev.* **2008**, *37*, 639–647. (h) Jerphagnon, T.; Pizzuti, M. G.; Minnaard, A. J.; Feringa, B. L. *Chem. Soc. Rev.* **2009**, *38*, 1039–1075.

[5] Glaser, C. *Ber. Dtsch. Chem. Ges.* **1869**, *2*, 422–424.

[6] (a) Bohlmann, F.; Schönowsky, H.; Inhoffen, E.; Grau, G. *Chem. Ber.* **1964**, *97*, 794–800. (b) Fomina, L.; Vazquez, B.; Tkatchouk, E.; Fomine, S. *Tetrahedron* **2002**, *58*, 6741–6747. (c) Wendlandt, A. E.; Suess, A. M.; Stahl, S. S. *Angew. Chem. Int. Ed. Engl.* **2011**, *50*, 11062–11087. (d) Schauer, P. A.; Low, P. J. *Eur. J. Inorg. Chem.* **2012**, 390–411. (e) Vilhelmsen, M. H.; Jensen, J.; Tortzen, C. G.; Nielsen, M. B. *Eur. J. Org. Chem.* **2013**, 701–711.

[7] (a) Hay, A. S. *J. Org. Chem.* **1960**, *25*, 1275–1276. (b) Hay, A. S. *J. Org. Chem.* **1962**, *27*, 3320–3321.

[8] (a) Eglinton, G.; Galbraith, A. R. *Chem. Ind.* **1956**, 737–738. (b) Behr, O. M.; Eglinton, G.; Galbraith, A. R.; Raphael, R. A. *J. Chem. Soc.* **1960**, 3614–3625.

[9] Adimurthy, S.; Malakar, C. C.; Beifuss, U. *J. Org. Chem.* **2009**, *74*, 5648–5651.

[10] Zheng, Q.; Hua, R.; Wan, Y. *Appl. Organomet. Chem.* **2010**, *24*, 314–316.

[11] Balamurugan, R.; Naveen, N.; Manojveer, S.; Nama, M. V. *Aust. J. Chem.* **2011**, *64*, 567–575.

[12] Wang, D.; Li, J.; Li, N.; Gao, T.; Hou, S.; Chen, B. *Green Chem.* **2010**, *12*, 45–48.

[13] Balaraman, K.; Kesavan, V. *Synthesis* **2010**, 3461–3466.

[14] Li, D.; Yin, K.; Li, J.; Jia, X. *Tetrahedron Lett.* **2008**, *49*, 5918–5919.

[15] Li, L.; Wang, J.; Zhang, G.; Liu, Q. *Tetrahedron Lett.* **2009**, *50*, 4033–4036.

[16] Ohkita, M.; Ando, K.; Suzuki, T.; Tsuji, T. *J. Org. Chem.* **2000**, *65*, 4385–4390.

[17] Moriarty, R. M.; Pavlovic, D. *J. Org. Chem.* **2004**, *69*, 5501–5504.

[18] Nishihara, Y.; Okamoto, M.; Inoue, Y.; Miyazaki, M.; Miyasaka, M.; Takagi, K. *Tetrahedron Lett.* **2005**, *46*, 8661–8664.

[19] Paixão, M. W.; Weber, M.; Braga, A. L.; de Azeredo, J. B.; Deobald, A. M.; Stefani, H. A. *Tetrahedron Lett.* **2008**, *49*, 2366–2370.

[20] (a) Li, J.; Jiang, H. *Chem. Commun.* **1999**, 2369–2370. (b) Jiang, H.; Tang, J.; Wang, A.; Deng, G.; Yang, S. *Synthesis* **2006**, 1155–1161.

[21] Jiang, H.; Wang, A. *Synthesis* **2007**, 1649–1654.

[22] Li, P.-H.; Yan, J.-C.; Wang, M.; Wang, L. *Chin. J. Chem.* **2004**, *22*, 219–221.

[23] Yadav, J. S.; Reddy, B. V. S.; Reddy, K. B.; Gayathri, K. U.; Prasad, A. R. *Tetrahedron Lett.* **2003**, *44*, 6493–6496.

[24] Ranu, B. C.; Banerjee, S. *Lett. Org. Chem.* **2006**, *3*, 607–609.

[25] (a) Chodkiewicz, W.; Cadiot, P. *C. R. Hebd. Seances Acad. Sci.* **1955**, *241*, 1055–1057. (b) Casitas, A.; Ribas, X. *Chem. Sci.* **2013**, *4*, 2301–2318.

[26] Jenny, N. M.; Mayor, M.; Eaton, T. R. *Eur. J. Org. Chem.* **2011**, 4965–4983.

[27] (a) Eastmond, R.; Walton, D. R. M. *Tetrahedron* **1972**, *28*, 4591–4599. (b) Johnson, T. R.; Walton, D. R. M. *Tetrahedron* **1972**, *28*, 5221–5236.

[28] Marino, J. P.; Nguyen, H. N. *J. Org. Chem.* **2002**, *67*, 6841–6844.

[29] Ito, H.; Arimoto, K.; Sensui, H.; Hosomi, A. *Tetrahedron Lett.* **1997**, *38*, 3977–3980.

[30] (a) Ikegashira, K.; Nishihara, Y.; Hirabayashi, K.; Mori, A.; Hiyama, T. *Chem. Commun.* **1997**, 1039–1040. (b) Nishihara, Y.; Ikegashira, K.; Mori, A.; Hiyama, T. *Chem. Lett.* **1997**, *26*, 1233–1234. (c) Mori, A.; Fujita, A.; Nishihara, Y.; Hiyama, T. *Chem. Commun.* **1997**, 2159–2160. (d) Nishihara, Y.; Ikegashira, K.; Mori, A.; Hiyama, T. *Tetrahedron Lett.* **1998**, *39*, 4075–4078.

[31] (a) Hansen, L.; Boll, P. M. *Phytochemistry* **1986**, *25*, 285–293. (b) Ito, A.; Cui, B.; Chavez, D.; Chai, H. B.; Shin, Y. G.; Kawanishi, K.; Kardono, L. B.; Riswan, S.; Farnsworth, N. R.; Cordell, G. A.; Pezzuto, J. M.; Kinghorn, A. D. *J. Nat. Prod.* **2001**, *64*, 246–248.

[32] (a) Tanaka, D.; Romerril, S. P.; Myers, A. G. *J. Am. Chem. Soc.* **2005**, *127*, 10323–10333. (b) Baudoin, O. *Angew. Chem. Int. Ed. Engl.* **2007**, *46*, 1373–1375. (c) Bonesi, S. M.; Fagnoni, M.; Albini, A. *Angew. Chem. Int. Ed. Engl.* **2008**, *47*, 10022–10025.

[33] (a) Yu, M.; Pan, D.; Jia, W.; Chen, W.; Jiao, N. *Tetrahedron Lett.* **2010**, *51*, 1287–1290. (b) Huang, Z.; Shang, R.; Zhang, Z.-R.; Tan, X.-D.; Xiao, X.; Fu, Y. *J. Org. Chem.* **2013**, *78*, 4551–4557.

[34] (a) Montierth, J. M.; DeMario, D. R.; Kurth, M. J.; Schore, N. E. *Tetrahedron* **1998**, *54*, 11741–11748. (b) Saha, D.; Chatterjee, T.; Mukherjee, M.; Ranu, B. C. *J. Org. Chem.* **2012**, *77*, 9379–9383.

[35] Zhu, B. C.; Jiang, X. Z. *Appl. Organomet. Chem.* **2007**, *21*, 345–349.

[36] Kuhn, P.; Alix, A.; Kumarraja, M.; Louis, B.; Pale, P.; Sommer, J. *Eur. J. Org. Chem.* **2009**, 423–429.

[37] Oishi, T.; Atayama, T.; Yamaguchi, K.; Mizuno, N. *Chem. Eur. J.* **2009**, *15*, 7539–7542.

[38] Nador, F.; Fortunato, L.; Moglie, Y.; Vitale, C.; Radivoy, G. *Synthesis* **2009**, 4027–4031.

[39] Tedeschi, C.; Saccavini, C.; Maurette, L.; Soleilhavoup, M.; Chauvin, R. *J. Organomet. Chem.* **2003**, *670*, 151–169.

[40] Sevin, M. M. A.; Chodkiewicz, W.; Cadiot, P. *Tetrahedron Lett.* **1965**, 1953–1959.

[41] (a) Jeffery, T.; Gueugnot, S.; Linstrumelle, G. *Tetrahedron Lett.* **1992**, *33*, 5757–5760. (b) Dasse, O.; Mahadevan, A.; Han, L.; Martin, B. R.; Marzo, V. D.; Razdan, R. K. *Tetrahedron* **2000**, *56*, 9195–9202. (c) Montel, F.; Beaudegnies, R.; Kessabi, J.; Martin, B.; Muller, E. *Org. Lett.* **2006**, *8*, 1905–1908.

[42] Tallman, K. A.; Roschek, B., Jr.; Porter, N. A. *J. Am. Chem. Soc.* **2004**, *126*, 9240–9241.

[43] Ogawa, T.; Kusume, K.; Tanaka, M.; Hayami, K.; Suzuki, H. *Synth. Commun.* **1989**, *19*, 2199–2207.

[44] (a) Okuro, K.; Furuune, M.; Miura, M.; Nomura, M. *Tetrahedron Lett.* **1992**, *33*, 5363–5364. (b) Okuro, K.; Furuune, M.; Enna, M.; Miura, M.; Nomura, M. *J. Org. Chem.* **1993**, *58*, 4716–4721.

[45] Liu, F.; Ma, D. *J. Org. Chem.* **2007**, *72*, 4844–4850.

[46] Wu, M.; Mao, J.; Guo, J.; Ji, S. *Eur. J. Org. Chem.* **2008**, 4050–4054.

[47] Liu, Y.; Yang, J.; Bao, W. *Eur. J. Org. Chem.* **2009**, 5317–5320.

[48] Shao, L. X.; Shi, M. *Tetrahedron* **2007**, *63*, 11938–11942.

[49] (a) Bates, C. G.; Saejueng, P.; Venkataraman, D. *Org. Lett.* **2004**, *6*, 1441–1444. (b) Saejueng, P.; Bates, C. G.; Venkataraman, D. *Synthesis* **2005**, 1706–1712.

[50] Coleman, R. S.; Garg, R. *Org. Lett.* **2001**, *3*, 3487–3490.

[51] Marshall, J. A.; Chobanian, H. R.; Yanik, M. M. *Org. Lett.* **2001**, *3*, 4107–4110.

[52] Kang, S. K.; Yoon, S. K.; Kim, Y. M. *Org. Lett.* **2001**, *3*, 2697–2699.

[53] (a) Kang, S. K.; Kim, J. S.; Choi, S. C. *J. Org. Chem.* **1997**, *62*, 4208–4209. (b) Zine, K.; Petrignet, J.; Thibonnet, J.; Abarbri, M. *Synlett* **2012**, 755–759.

[54] (a) Suzuki, A. *Pure Appl. Chem.* **1991**, *63*, 419–422. (b) Miyaura, N.; Suzuki, A. *Chem. Rev.* **1995**, *95*, 2457–2483. (c) Suzuki, A. *J. Organomet. Chem.* **1999**, *576*, 147–168.

[55] Hoshi, M.; Masuda, Y.; Arase, A. *Bull. Chem. Soc. Jpn* **1983**, *56*, 2855–2856.

[56] (a) Miyaura, N.; Yamada, K.; Suzuki, A. *Tetrahedron Lett.* **1979**, *20*, 3437–3440. (b) Miyaura, N.; Suginome, H.; Suzuki, A. *Tetrahedron Lett.* **1981**, *22*, 127–130.

[57] Hoshi, M.; Shirakawa, K. *Synlett* **2002**, 1101–1104.

[58] Hoshi, M.; Kawamura, N.; Shirakawa, K. *Synthesis* **2006**, 1961–1970.

[59] Yu, C.-M.; Kweon, J.-H.; Ho, P.-S.; Kang, S.-C.; Lee, G. Y. *Synlett* **2005**, 2631–2634.

[60] Normant, J. F. *Synthesis* **1972**, 63–80.

[61] Grushin, V. V.; Alper, H. *J. Org. Chem.* **1992**, *57*, 2188–2192.

[62] Bumagin, N. A.; Ponomarev, A. B.; Beletskaya, I. P. *Russ. Chem. Bull.* **1987**, *36*, 1445–1448.

[63] Bieber, L. W.; da Silva, M. F. *Tetrahedron Lett.* **2007**, *48*, 7088–7090.

13

COPPER-MEDIATED ALKENYLATION REACTION OF ALKENYL DERIVATIVES: A STRAIGHTFORWARD ELABORATION OF 1,3-DIENES

Hao Li, Songbai Liu, and Lanny S. Liebeskind

13.1 INTRODUCTION

1,3-Dienes are synthetically important motifs in chemistry. They are found in natural products, medicinal compounds, and novel materials; and they enable the construction of more complex systems, usually through Diels–Alder reactions or other synthetic transformations. The early work on the copper-mediated synthesis of 1,3-dienes focused on the homocoupling of alkenyl derivatives to deliver symmetrical 1,3-dienes. Recent developments on copper-mediated 1,3-diene synthesis have focused on the straightforward elaboration of functionalized unsymmetrical 1,3-dienes via stereoselective copper-mediated cross couplings and the application of these versatile coupling reactions in natural product synthesis. In this chapter, research progress in the area of copper-mediated alkenylation of alkenyl derivatives in the context of 1,3-diene synthesis is discussed.

Copper-Mediated Cross-Coupling Reactions, First Edition. Edited by Gwilherm Evano and Nicolas Blanchard.

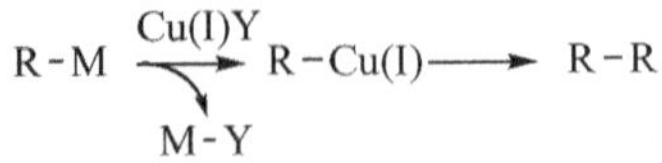

R = vinyl
M = B, Si, Sn
Y = Cl, OTf, TC, etc.

Scheme 13.1 Proposed mechanism for copper-mediated homocoupling of vinyl metal derivatives.

13.2 SYMMETRICAL 1,3-DIENES BY HOMOCOUPLING REACTION OF VINYL DERIVATIVES

A straightforward tactic to deliver symmetrical 1,3-dienes is the copper-mediated homocoupling of various vinyl derivatives, including vinylboranes, vinylsilanes, vinylstannanes, and vinyl halides. This approach tolerates many functionalities, including ester, nitro, hydroxyl, amine, imine, and amide, and often has complimentary selectivity to various palladium-catalyzed coupling protocols. Conceptual simplicity, mild reaction conditions, and high efficiency contribute to the power of this approach to symmetrical 1,3-dienes.

According to studies available so far, a plausible mechanistic pathway for copper-mediated homocoupling of vinyl metal derivatives is suggested in Scheme 13.1.[1] In general, a vinyl group is transferred from vinyl metal reagents to preformed copper(I) species, which generates the requisite organocopper(I) intermediates that are then converted to 1,3-dienes. In contrast, the reductive homocoupling of vinyl halides is assumed to involve copper(III) intermediates, which would undergo ligand exchange followed by reductive elimination to afford 1,3-dienes. Characteristic features of the homocoupling of each separate vinyl derivative are discussed in the following sections.

13.2.1 Coupling of Vinylboranes

Dialkenylchloroboranes, first introduced by Yamamoto in 1975, are readily prepared by the addition of chloroborane to acetylenes. The dialkenylchloroboranes participate in an alkenyl homocoupling reaction at 0°C upon treatment with 3 equiv of methylcopper to afford (*E*,*E*)-1,3-dienes in excellent yields, and the reaction proceeds with retention of stereochemistry (the stereochemical purity of products is higher than 99%) (Scheme 13.2).[2,3] Other vinylborane species such as alkenyldisiamylboranes or alkenyldicyclohexylboranes are also suitable reaction partners but generate only moderate yields of the corresponding diene products, which could be attributed to the steric crowding of the boron atom. In contrast to vinylboranes, trialkylboranes and dialkylchloroboranes are completely inactive in the homocoupling reaction.

High stereoselectivity, mild reaction conditions, and high yields render this reaction a valuable method for achieving synthesis of isomerically pure (*E*,*E*)-1,3-dienes starting from readily available acetylenes. The reaction conditions

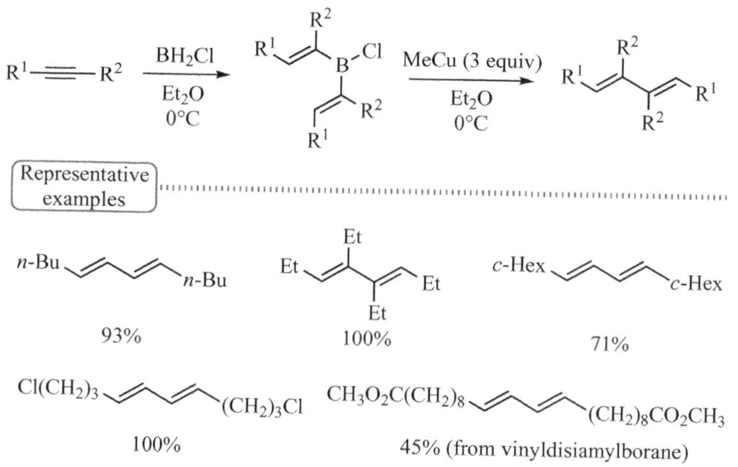

Scheme 13.2 Copper-mediated homocoupling of dialkenylchloroboranes.

suggest that a variety of functionalities could be tolerated. As a key step in the reaction, a facile and stereoselective dissociation of the alkenyl copper(I) species from the *in situ* generated copper trimethylalkenylborate ate complexes is suggested.

In 1980, Brown demonstrated that the facile transmetallation of alkenyl groups from boron to copper was facilitated by the addition of sodium methoxide to the preformed alkenylboron reagents.[4] Under Brown's conditions, alkenyldialkylboranes (which failed to react under Yamamoto's reaction conditions) readily reacted with copper(I) bromide at 0°C when activated by addition of sodium methoxide to produce symmetrical 1,3-dienes in excellent yields (Scheme 13.3). The stereospecific nature of this homocoupling reaction suggests the same alkenyl copper(I) mechanism as with Yamamoto's reaction system described earlier.

13.2.2 Coupling of Vinylsilanes

One of the first examples of a copper-mediated homocoupling of vinylsilanes was discovered by Kumada and coworkers in 1979.[5] Solid-state dipotassium alkenylpentafluorosilicates, which are derived from the platinum-catalyzed hydrosilylation of acetylenes followed by the addition of potassium fluoride, can undergo copper(I) chloride-mediated homocoupling reactions at 200–300°C under reduced pressure to yield 1,3-dienes. Later on, Mori and coworkers reported a milder procedure employing alkenylfluorosilanes, which smoothly dimerize at ambient temperature to give the corresponding 1,3-dienes when exposed to a stoichiometric quantity of copper(I) chloride in an aprotic and highly polar solvent such as *N*,*N*-dimethylformamide (DMF) or dimethyl sulfoxide (DMSO) (Scheme 13.4).[6,7] A catalytic amount of CuCl

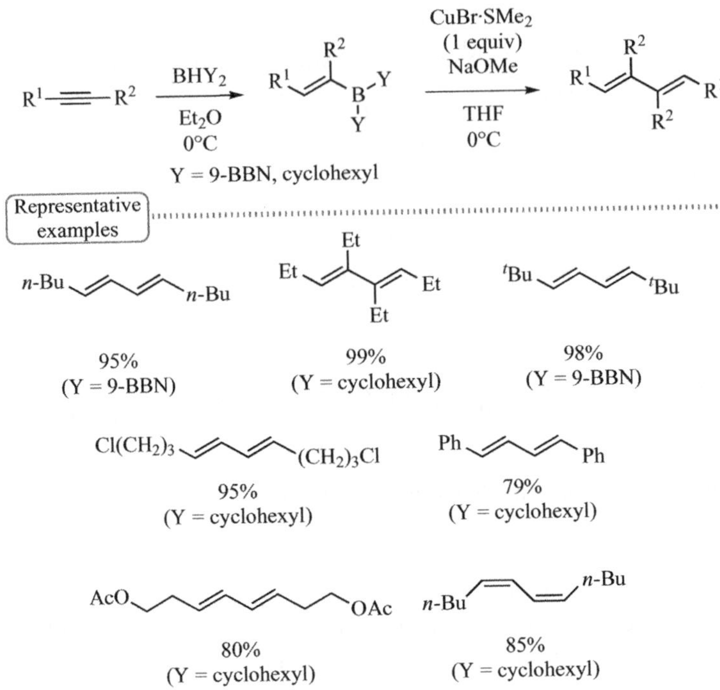

Scheme 13.3 Copper-mediated alkenyl homocoupling of alkenyldialkylboranes.

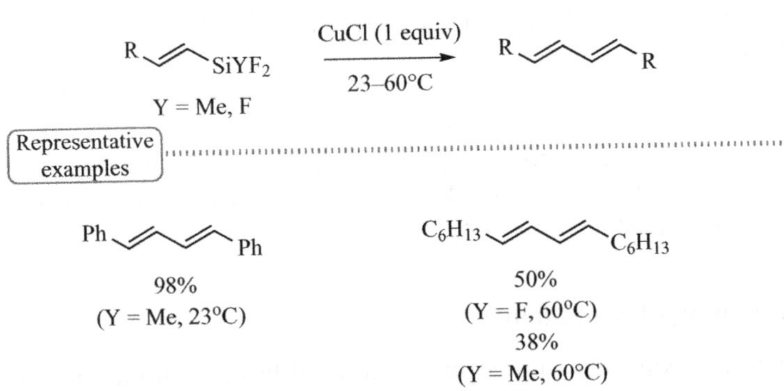

Scheme 13.4 Copper-mediated homocoupling of vinylsilanes.

(10 mol %) is also effective when prolonged reaction times are involved, whereas copper(II) salts like $CuCl_2$ are inactive in the reaction. The reactivity of alkenylfluorosilanes is highly dependent on the number of fluorine atoms on silicon, with two or three fluorine atoms required for an effective homocoupling transformation. A coordination of polar solvents to the organosilicon species to facilitate silicon to copper transmetallation was suggested by the authors.

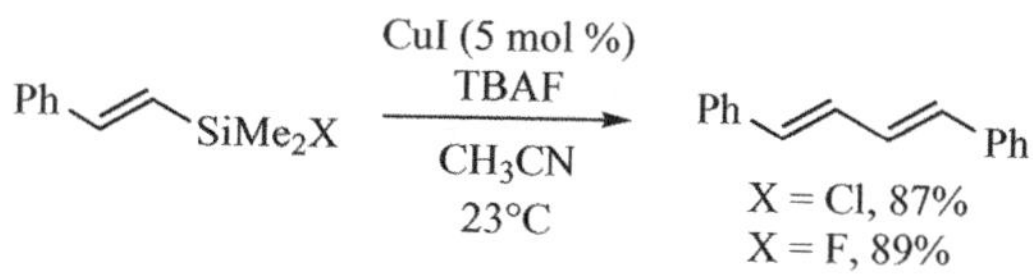

Scheme 13.5 Copper-catalyzed homocoupling of vinylsilanes.

In 1997, Kang demonstrated that those alkenyldimethylchlorosilanes or alkenyldimethylfluorosilanes that failed to dimerize under Mori's conditions could be successfully dimerized to afford 1,3-dienes upon activation with tetrabutylammonium fluoride (TBAF), at ambient temperature in the presence of a catalytic amount (5 mol %) of copper(I) iodide in acetonitrile (Scheme 13.5).[8] The speculative mechanism suggested by the authors involves oxidative addition of an activated alkenyldimethylfluorosilane (RCH=CHSiMe$_2$X) to the copper(I) salt to afford an organocopper(III) complex [RCH=CHCu(III) SiMe$_2$XI]. Subsequent transmetallation from pentacoordinated silicates, generated by TBAF and organosilanes, would then produce a diorganocopper(III) species [(RCH=CH)$_2$Cu(III)I], which would then undergo reductive elimination to afford the desired symmetrical 1,3-diene product and the regenerated catalyst copper(I) iodide. However, Mori's study mentioned earlier implies that transmetallation from an activated alkenyldimethylfluorosilane–TBAF complex to copper(I) to form an alkenyl copper(I) intermediate followed by oxidative homocoupling may be more plausible.

Recently, Itami and coworkers developed a stoichiometric copper(I) iodide- and cesium fluoride-mediated stereospecific homocoupling of alkenyl (2-pyridyl)silanes for the synthesis of structurally and electronically diverse multisubstituted 1,3-dienes in moderate to high yields with virtually complete retention of stereochemistry (Scheme 13.6).[9] The promoting effect of the 2-pyridyl directing group in the homocoupling was revealed by the fact that the corresponding phenyl analogue remained unaffected under otherwise identical conditions. Palladium-catalyzed, copper-mediated homocouplings of vinylsilanes using stoichiometric copper(II) chloride have also been reported.[10,11] The role of copper(II) chloride in these reactions is to act as an oxidant to regenerate active palladium(II) species.

13.2.3 Coupling of Vinylstannanes

In 1987, Kyler and coworkers discovered a copper(II) nitrate-mediated stereospecific homocoupling of alkenyl stannanes to produce 1,3-dienes at ambient temperature in tetrahydrofuran (THF) (Scheme 13.7).[12] A transmetallation of the alkenyl group from tin to copper to deliver an organocopper(II) species was proposed. tri-*n*-Butyltin nitrate was also isolated in high yield.

Later on, Quayle and coworkers applied this reaction to the preparation of highly functionalized 1,3-dienes.[13] A range of functional groups such as

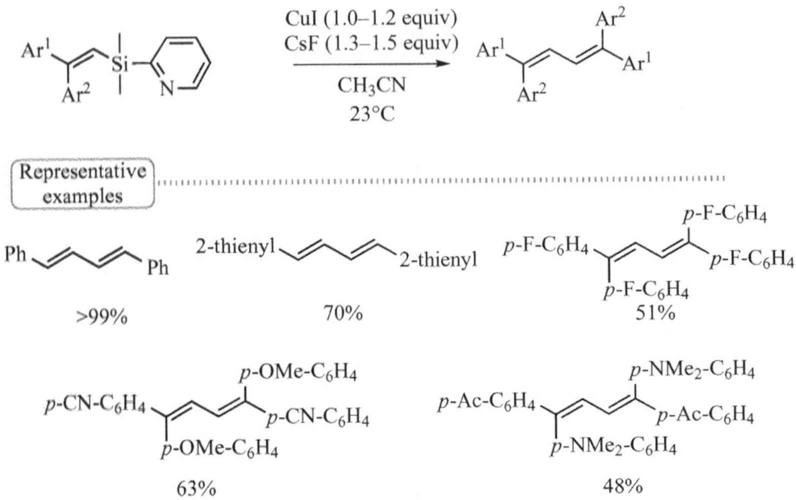

Scheme 13.6 Copper-mediated homocoupling of alkenyl(2-pyridyl)silanes.

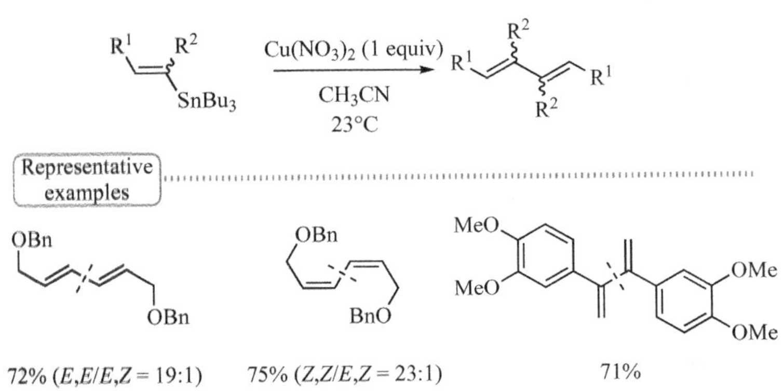

Scheme 13.7 Copper-mediated homocoupling of alkenyl stannanes.

hydroxyl, thioether, enol ether, ester, aryl bromide, aryl iodide, and trichloromethyl are tolerated in this reaction (Scheme 13.8). Both intermolecular and intramolecular coupling reactions occur at ambient temperature over short reaction times (usually 10 minutes).

In 1996, Piers and coworkers found that excess copper(I) chloride effectively mediated the dimerization of alkenyltrimethylstannanes at ambient temperature in DMF. Extensive studies demonstrated the effectiveness of this reaction in intramolecular (Scheme 13.9) and intermolecular (Scheme 13.10) manners, and its synthetic versatility was illustrated by the construction of four-, five-, six-, seven-, and eight-membered rings and conjugated

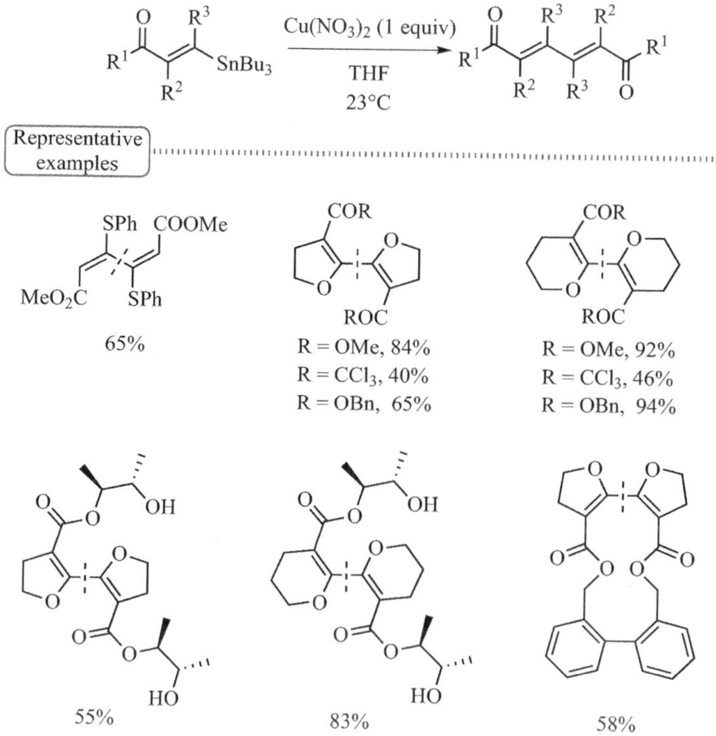

Scheme 13.8 Copper-mediated homocoupling of alkenyl stannanes with electron-withdrawing substituents.

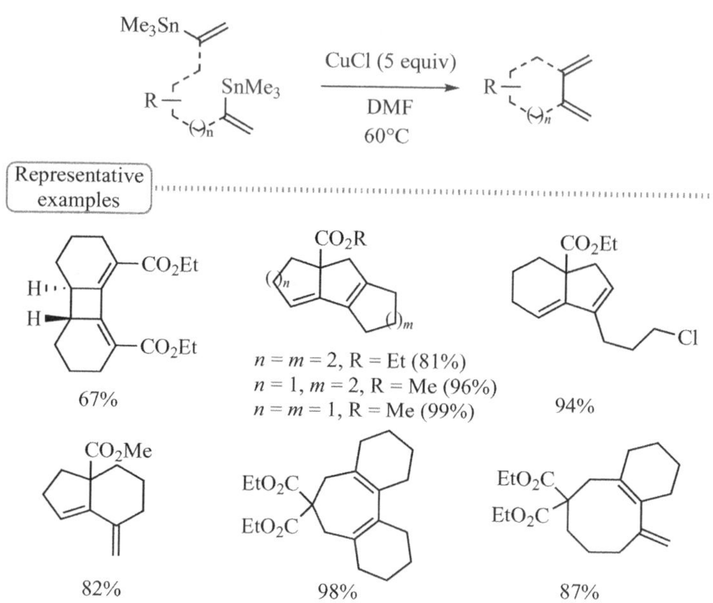

Scheme 13.9 Copper-mediated intramolecular dimerization of alkenyltrimethylstannanes.

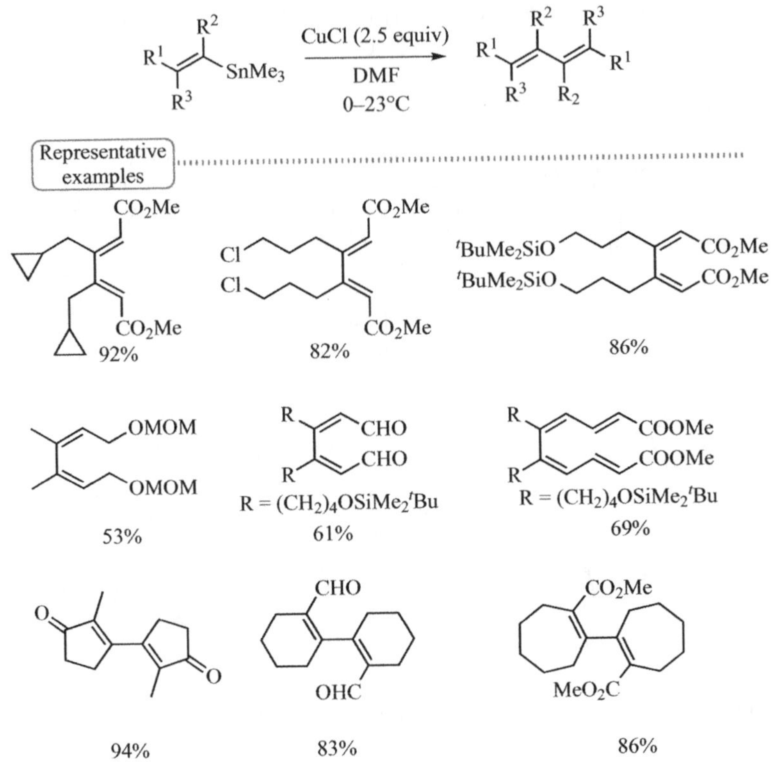

Scheme 13.10 Copper-mediated intermolecular dimerization of alkenyltrimethylstannanes.

tetraenes.[14–16] As expected, this reaction tolerates a number of functionalities including triorganosilyl, hydroxyl, ether, ester, and chloride. Although a transmetallation of the alkenyl group from tin to copper to form an alkenyl copper(I) species was suggested by the authors, the formation of copper metal was observed during the reaction and the exact mechanism was not determined. In Corey and coworkers' study of a copper(I) chloride-modified Stille reaction, one example of a homocoupling of a vinylstannane was reported in support of a tin to copper transmetallation.[17]

Later on, Blumenthal and Burton revealed the superiority of copper(II) over copper(I) when used for the homocoupling of fluorinated vinylstannanes (in contrast to nonfluorinated vinylstannanes).[18] Employing anhydrous copper(II) acetate prepared by azeotropic reflux of the hydrate in benzene, 1,2-difluorovinylstannanes are stereospecifically transformed to 1,3-dienes in good-to-excellent yields under an oxygen atmosphere at ambient temperature in DMF (Scheme 13.11). The deleterious effect of water on the reaction was assumed to be a reflection of the moisture-sensitive nature of the fluorinated vinylcopper(I) intermediates.

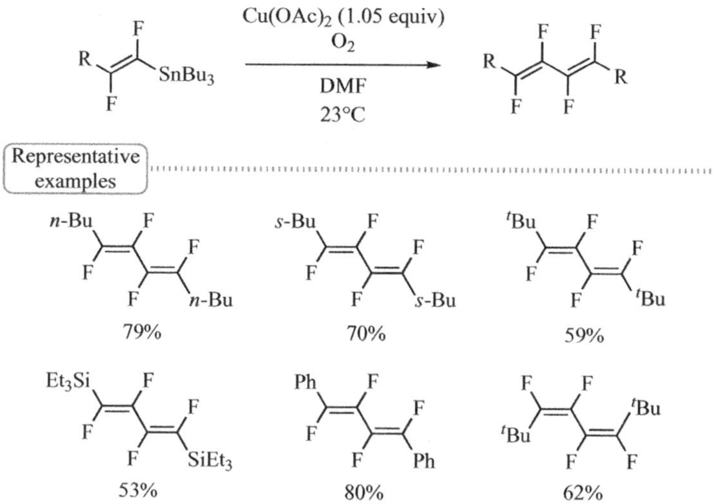

Scheme 13.11 Copper-mediated homocoupling of fluorinated vinylstannanes.

When Lucchi and coworkers attempted to construct fully substituted benzene rings using a copper-mediated cyclotrimerization of *vic*-bromo (trimethylstannyl)bicycloalkenes, they observed that typical copper salts, such as copper(I) chloride and copper(II) nitrate, which usually work very well in other couplings, failed to provide the desired products in satisfactory yields. Minimization of by-products and a beneficial influence on stereoselectivity was achieved in some cases by the nontrivial adjustment of reaction parameters including the addition of cosolvents and comediators.[19–22] Fortunately, they finally found copper(I)-thiophene-2-carboxylate (CuTC), originally developed by Allred and Liebeskind,[23] generally afforded the desired cyclotrimerization products in nearly quantitative yields (Scheme 13.12). Encouraged by the successful CuTC-mediated cyclotrimerization, they further investigated the influence of substrate structures on stereoselectivity and product reactivity in the subsequent Diels–Alder reaction.

To realize a catalytic version of the homocoupling reaction of vinylstannanes, Itoh attempted electro-oxidation to regenerate the active copper(I) or copper(II) species from copper metal produced by reductive homocoupling during reaction, although only low-to-moderate yields were achieved (Scheme 13.13, Eq. 1).[24] The yields were improved by Kang and coworkers using iodine as an oxidant (Scheme 13.13, Eq. 2).[25]

13.2.4 Coupling of Vinyl Halides

Studies of the homocoupling of vinyl halides were initiated by Cohen and Cristea in 1975. They found that treatment of diethyl iodofumarate in dry acetone saturated by gaseous ammonia with copper(I) triflate afforded the

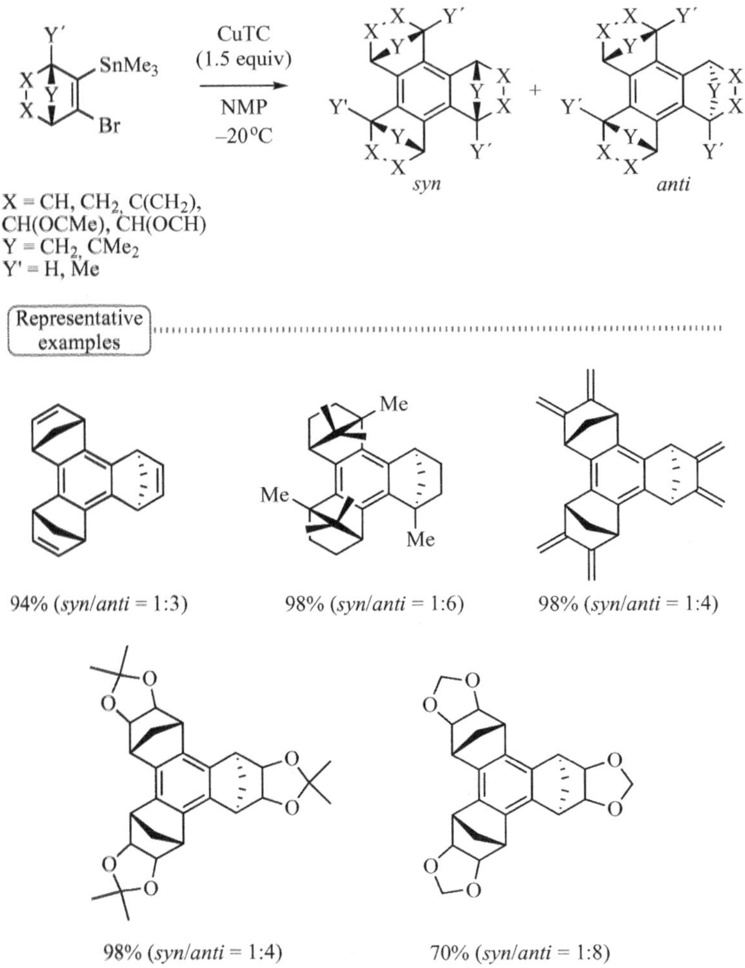

Scheme 13.12 Copper-mediated cyclotrimerization of *vic*-bromo(trimethylstannyl)-bicycloalkenes.

corresponding 1,3-dienes in excellent yield.[26] However, when the reaction was conducted in acetone with aqueous ammonia, in which homocoupling of aryl halides worked very well, the protodeiodination product, diethyl fumarate, was the predominant product formed.

In 1997, Liebeskind and coworkers significantly advanced the vinyl halide homocoupling reaction. Using a nonhalide copper(I) source, copper(I) thiophenecarboxylate, an inexpensive, easy-to-prepare, reasonably air stable, nonhydroscopic solid, the authors realized a general, efficient, and stereospecific reductive homocoupling of alkenyl iodides at ambient temperature in *N*-methylpyrrolidinone (Scheme 13.14).[27] The formation of an organocopper(III) intermediate via reversible oxidative addition of the C–I bond to

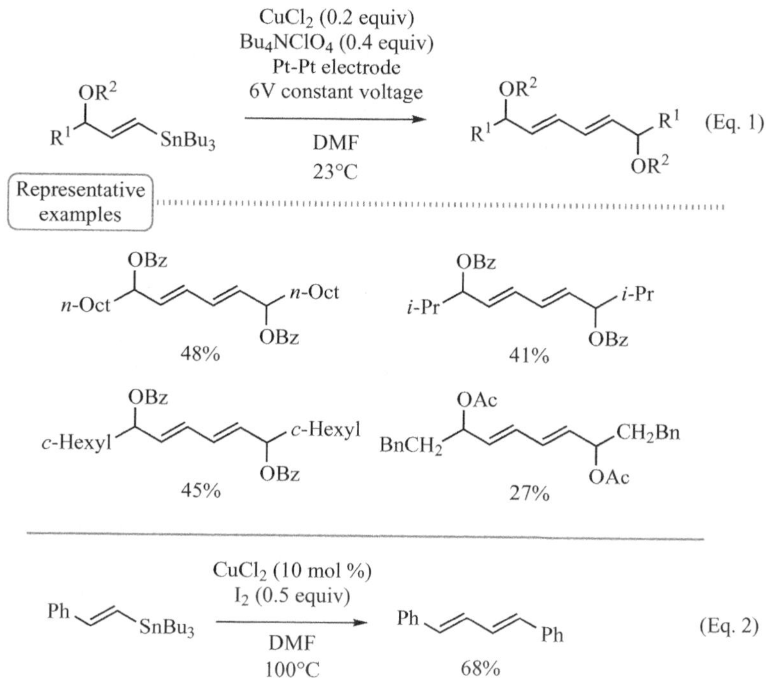

Scheme 13.13 Copper-catalyzed homocoupling of vinylstannanes using electro-oxidation and iodine as an oxidant.

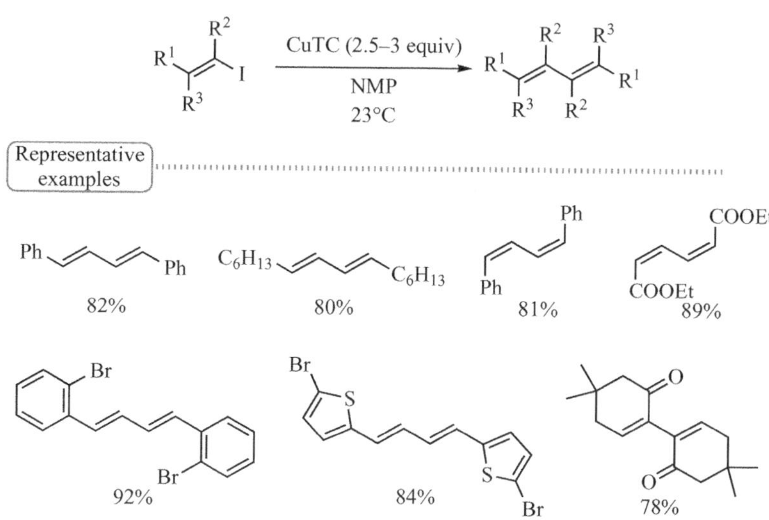

Scheme 13.14 Copper-mediated reductive homocoupling of alkenyl iodides.

the copper(I) reagent was suggested by the authors, which was supported by halide exchange between a vinyl iodide and a copper(I) halide when copper(I) bromide or copper(I) chloride was used. The mildness of this reaction renders it tolerable of many functionalities including nitro, hydroxyl, ester, amide, and carbamate, indicating broad synthetic utility in complex molecule synthesis.

13.2.5 Miscellaneous

Schwartz and coworkers disclosed the preparation of alkenyl copper(I) reagents via transmetallation from readily available alkenylzirconium complexes in 1977. Addition of alkenylzirconium reagents to copper(I) chloride at 0°C produces an alkenyl copper(I) species, which subsequently decomposes to a copper mirror and the corresponding 1,3-diene in high yield with full retention of stereochemistry.[28] The easily prepared alkenyl copper(I) species is of broader synthetic utility and can be used in conjugate additions to α,β-unsaturated ketones as illustrated in the study.

Radivoy and coworkers suggests that, regardless of the actual reductive coupling mechanism, copper nanoparticles, generated *in situ* from commercially available copper(II) chloride with an excess of lithium powder and a catalytic amount (5 mol %) of 4,4′-di-*tert*-butylbiphenyl (DTBB) as electron carrier, can promote the homocoupling of alkenyl Grignard reagents in tetrahydrofuran at ambient temperature.[29]

13.3 UNSYMMETRICAL 1,3-DIENES BY CROSS-COUPLING REACTIONS

Since the original documentation of the beneficial influence of cocatalytic copper(I) on palladium-catalyzed Stille cross-coupling reactions,[30–32] the application of copper-cocatalyzed and copper-mediated cross-coupling reactions in the stereoselective synthesis of functionalized unsymmetrical 1,3-dienes has been a fruitful field for organic chemists.[33–35] Traditional approaches for the synthesis of unsymmetrical 1,3-dienes typically involve Wittig olefination of α,β-unsaturated carbonyl compounds,[36–38] but often suffer from limited scope or poor stereoselectivity. More recently, the elaboration of unsymmetrical 1,3-dienes focuses on transition-metal-catalyzed carbon–carbon bond formation including the alkenylation of alkenyl derivatives, alkene–alkyne couplings, and enyne metathesis.[39–41] Among these varied approaches, the palladium-catalyzed alkenylation of stereodefined alkenyl derivatives using vinyl organometallic reagents has been extensively studied both in methodology development and natural product synthesis.[42] Compared with well-established palladium-catalyzed coupling reactions, copper-mediated 1,3-diene synthesis can be carried out at low reaction temperature (i.e., 0°C or ambient temperature) with rapid reaction rates, good chemo- and stereoselectivity,

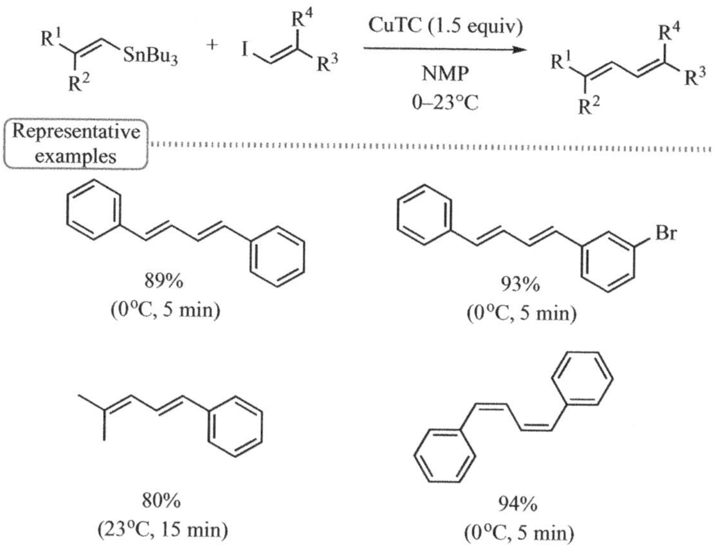

Scheme 13.15 CuTC-mediated Stille-type coupling of alkenyl iodides and vinylstannanes.

and excellent functional group compatibility.[33,35,43] Thus, the copper-mediated alkenylation of alkenyl derivatives is of particular interest to the organic chemistry community in the context of the synthesis of highly functionalized complex molecules with sensitive 1,3-diene motifs (see Chapter 18 for applications of this strategy in total synthesis of natural products).[34,44]

13.3.1 Synthesis of 1,3-Dienes by Stille-Type Coupling

13.3.1.1 Copper-Mediated Stille-Type Coupling. The Stille–Migita–Kosugi reaction, that is, the cross coupling of organohalides and organostannanes, has been widely used in organic synthesis.[45–48] One of the first examples of copper-mediated intermolecular alkenylation of alkenyl derivatives was discovered by Liebeskind using stoichiometric CuTC[23] (Scheme 13.15). This general Stille-type cross coupling of stereodefined alkenyl iodides and vinylstannanes proceeds at 0°C or room temperature to afford 1,3-dienes in high yields within 30 minutes and with complete retention of stereochemistries of both the alkenyl iodide and the organostannane. Aryl iodides and bromides are untouched in this extremely mild, efficient copper-mediated reaction displaying a unique chemoselectivity not achievable with palladium catalysis. In some specific cases, a catalytic amount (20 mol %) of a copper(I) salt was effective when the reaction was carried out in the presence of stoichiometric cesium fluoride at 60°C in *N*-methyl pyrrolidinone (NMP).[23]

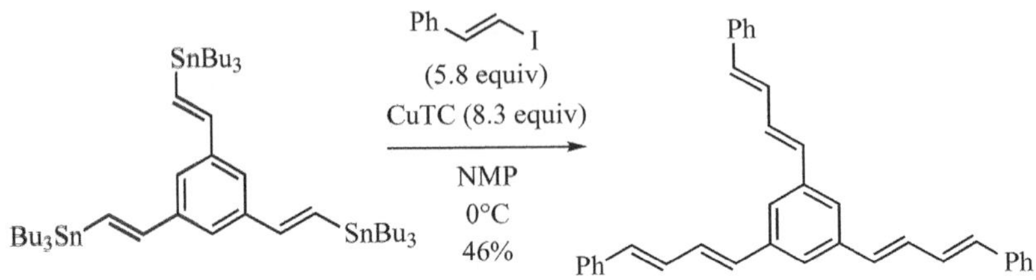

Scheme 13.16 CuTC-mediated cross coupling in the synthesis of a trifunctional conjugated diene.

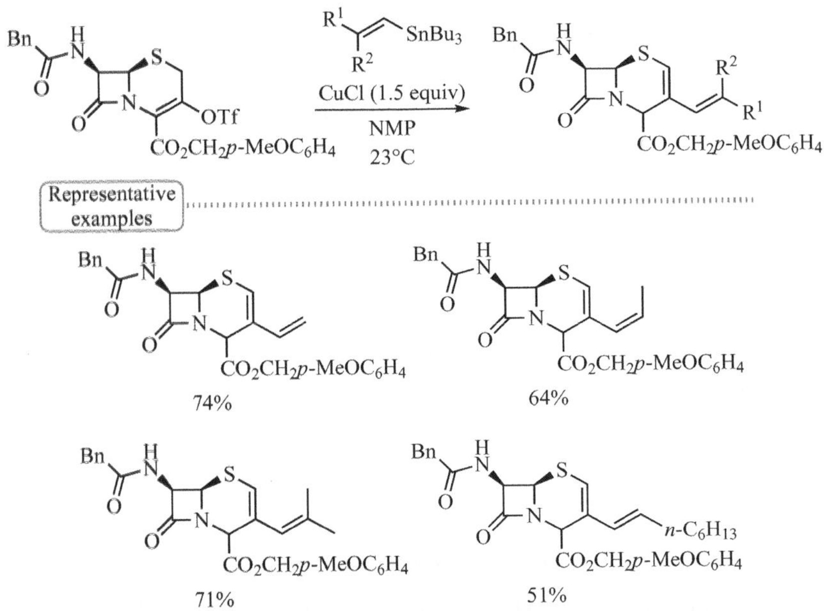

Scheme 13.17 Copper-mediated alkenylation of 3-trifluoromethanesulfonyloxy-Δ^3-cephems with alkenyl stannanes.

Ko and coworkers described an application of the CuTC-mediated cross coupling in the synthesis of a trifunctional conjugated diene monomer (Scheme 13.16).[49] Interestingly, Stille coupling conditions employing tetrakis(triphenylphopshine)palladium(0) and cesium fluoride proved to be ineffective in this system. Thus, the desired product was obtained in moderate yield only when stoichiometric CuTC was used.

A stoichiometric copper(I) chloride-promoted alkenylation of 3-trifluoromethanesulfonyloxy-Δ^3-cephems with alkenyl stannanes was reported by Torii and coworkers (Scheme 13.17).[50] Instead of vinyl iodides, which are typically used for an effective copper-mediated alkenylation, a vinyl triflate

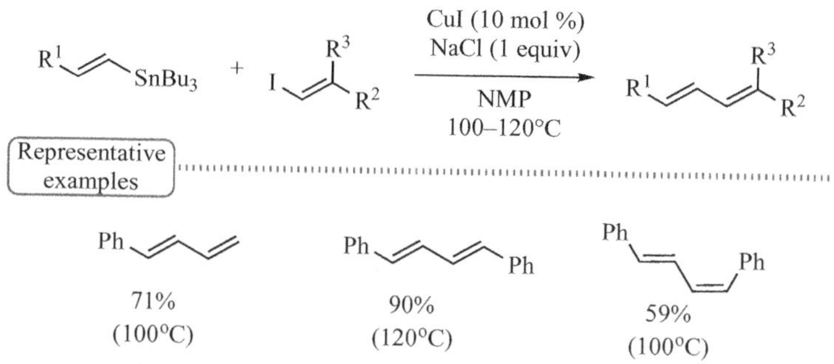

Scheme 13.18 Copper-catalyzed Stille-type coupling of alkenyl iodides and vinylstannanes.

coupled well with various alkenyl stannanes to afford 3-alkenyl-Δ^2-cephems with migration of the endocyclic double bond. The related palladium-catalyzed alkenylation often provides 3-alkenyl-Δ^3-cephems without double bond migration, suggesting a mechanistic subtlety in this copper(I) chloride-promoted alkenylation.

The copper-mediated alkenylation of alkenyl derivatives was also realized using only a catalytic amount of the copper source. A palladium-free cross coupling of vinyl iodides and vinylstannanes was carried out using only 10 mol % of copper(I) iodide as the catalyst when stoichiometric sodium chloride was used as an additive (Scheme 13.18).[51] This coupling reaction could serve as a valuable alternative to the Stille coupling, although high temperatures are required for an efficient reaction.

The synthesis of unsymmetrical 1,3-dienes via copper-mediated intramolecular Stille coupling was first reported by Piers and Wong (Scheme 13.19).[52] Employing copper(I) chloride as the copper(I) source and DMF as the solvent of choice, novel cyclic conjugated dienes contained within four-, five-, and six-membered rings were prepared efficiently in a stereospecific fashion. This coupling reaction tolerates a number of functional groups including esters, unprotected alcohols, and silyl ethers. Control experiments revealed that the copper(I) chloride-mediated intramolecular reactions are generally faster than the corresponding palladium-catalyzed couplings, and alkenyl bromides are reactive in the coupling process as well.

13.3.1.2 Palladium-Catalyzed Copper-Promoted Stille-Type Coupling. The beneficial effect of cocatalytic copper in traditional palladium-catalyzed Stille reactions was first emphasized in 1990,[32] although an earlier example can be found in the work of Marino and Long for the cross coupling of a vinylstannane and a vinyl tosylate in their study of the synthesis of dihydrocompactin.[53] Since then, various copper salts have been applied as cocatalysts and mediators in the cross coupling of organostannanes with organohalides

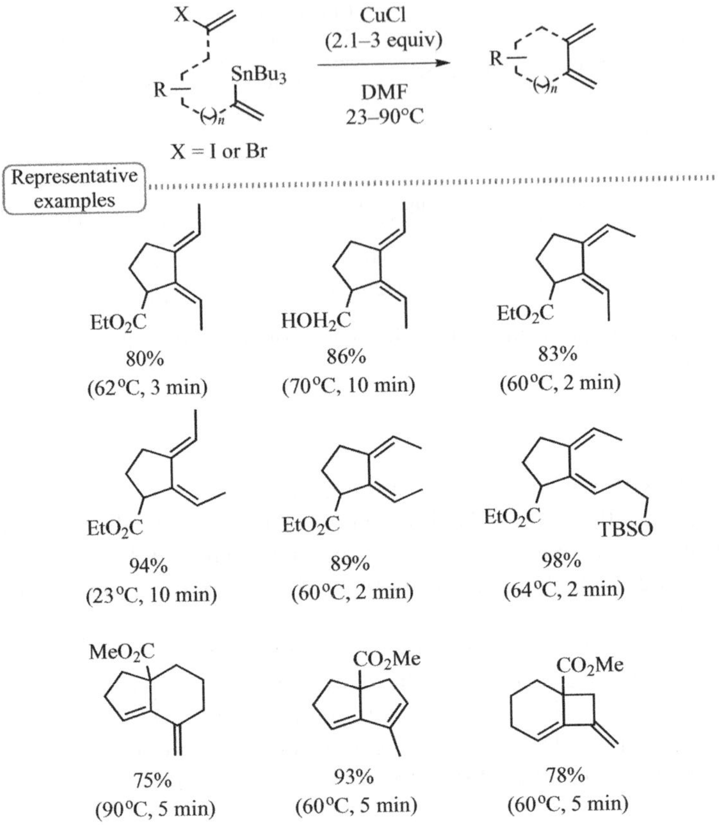

Scheme 13.19 Copper-mediated intramolecular cross coupling of vinyl halides with vinylstannanes.

and triflates to dramatically increase the rates of the reaction and allow the coupling reactions to proceed under mild reaction conditions.[35,54]

Common copper sources used in the palladium/copper-cocatalyzed Stille-type coupling include copper(I) chloride, copper(I) iodide, and CuTC. In most cases, only a catalytic amount of copper salt is required for an effective reaction, while stoichiometric copper can also be used to achieve the maximum coupling efficiency for difficult substrates that suffer from low reaction rates and side reactions, such as destannylation and homocoupling.

The beneficial influence of cocatalytic copper(I) in the Stille cross coupling was first studied by Liebeskind using 3-(propan-2-yloxy)-4-(tributylstannyl) cyclobut-3-ene-1,2-dione as substrate (Scheme 13.20).[32] When a catalytic amount of copper(I) iodide was included in the cross coupling of vinyl iodides and 3-stannyl cyclobutenediones, the reaction could be carried out at room temperature, while the corresponding palladium-catalyzed reaction required higher temperatures and longer reaction times. Highly functionalized alkenyl

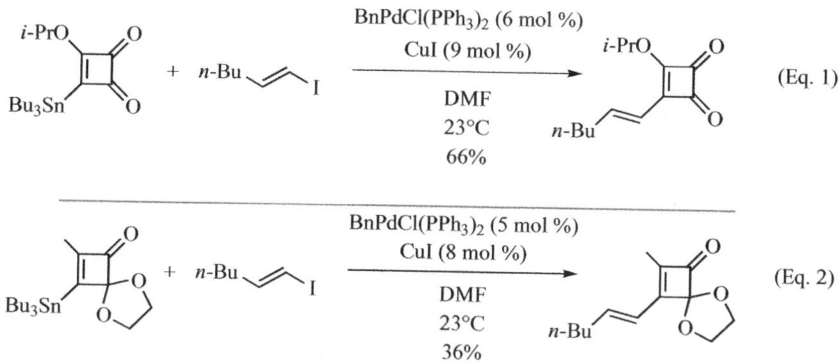

Scheme 13.20 Palladium-catalyzed copper-promoted cross coupling of vinyl iodides and 3-(propan-2-yloxy)-4-(tributylstannyl)cyclobut-3-ene-1,2-dione.

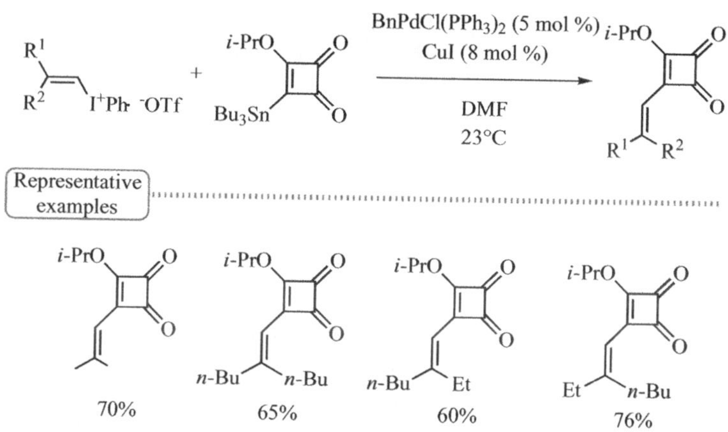

Scheme 13.21 Palladium-catalyzed copper-promoted cross coupling of vinyl iodonium triflates and 3-stannyl cyclobutenediones.

cyclobutenediones were easily prepared via this general copper-cocatalyzed Stille-type coupling.

Extension of the palladium/copper-cocatalyzed cross coupling to alkenyliodonium salts was developed by Stang and coworkers (Scheme 13.21).[55a] Vinyl iodonium triflates, compounds that could be prepared from cyano(phenyl) iodonium tosylate and the desired vinylstannane,[55b] couple well with 3-tri-*n*-butylstannyl cyclobutenedione to afford stereodefined products in good yields under exceedingly mild conditions. It is worth noting that iodonium salts show higher reactivity over iodobenzene and a simple vinyl triflate, which provides selective coupling conditions that are orthogonal to other common coupling partners.

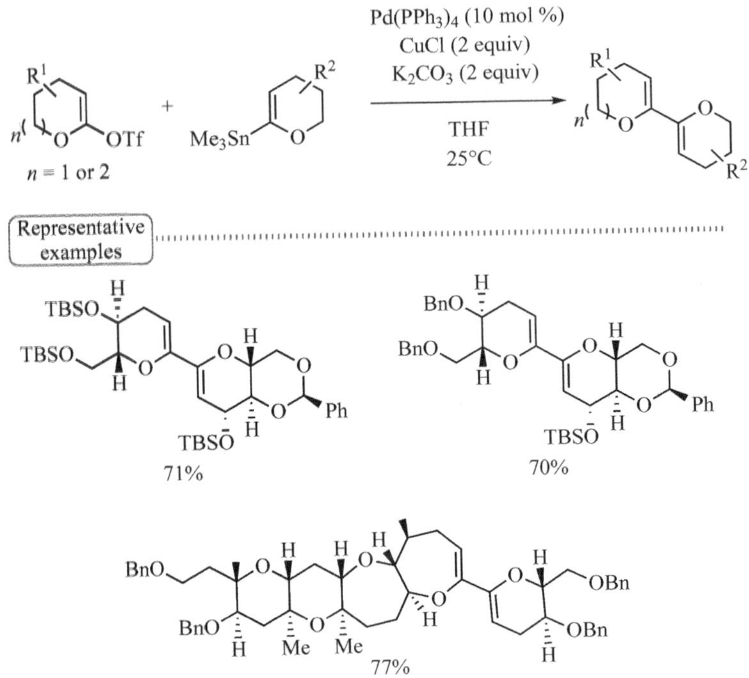

Scheme 13.22 Palladium-catalyzed copper-promoted cross coupling of stannyl enol ethers and enol triflates.

Later, Nicolaou and coworkers described a stoichiometric copper(I)-promoted Stille cross coupling of stannyl enol ethers and enol triflates in the construction of complex polyether frameworks (Scheme 13.22).[56] The cross coupling of glycal partners often suffer from low reaction rates and undesired homocoupling of the reactant. To develop a practical and efficient cross-coupling procedure, stoichiometric copper(I) chloride was added to the reaction system, and the reaction proceeded at 25°C within 1–2 hours to deliver complex bisglycals in good yields.

Recently, a stereospecific metal-selective coupling of sterically encumbered vinyl iodides and bis-metalated olefins was demonstrated by Burke and coworkers (Scheme 13.23).[57] This extremely mild cross coupling provides a facile approach to highly functionalized stereodefined unsymmetrical 1,3-dienes with bulky substituents. Notably, *N*-methyliminodiacetic acid (MIDA) boronates and alkenyl germanium are completely untouched under these modified Stille coupling conditions.

Piers and coworkers reported a palladium/copper-cocatalyzed intramolecular Stille coupling as the key step in the rapid assembly of novel, functionalized tricycles containing dienyl esters (Scheme 13.24).[58] Surprisingly, only 1 mol % of copper(I) iodide is used for an efficient coupling reaction, although higher

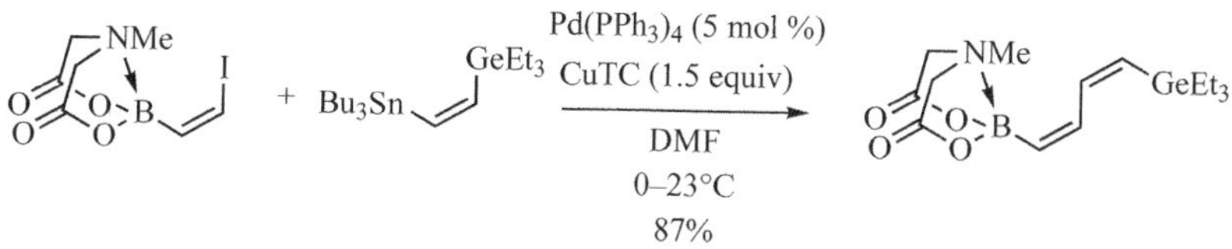

Scheme 13.23 Palladium-catalyzed, copper-promoted, metal-selective coupling of vinyl iodides and bis-metalated olefins.

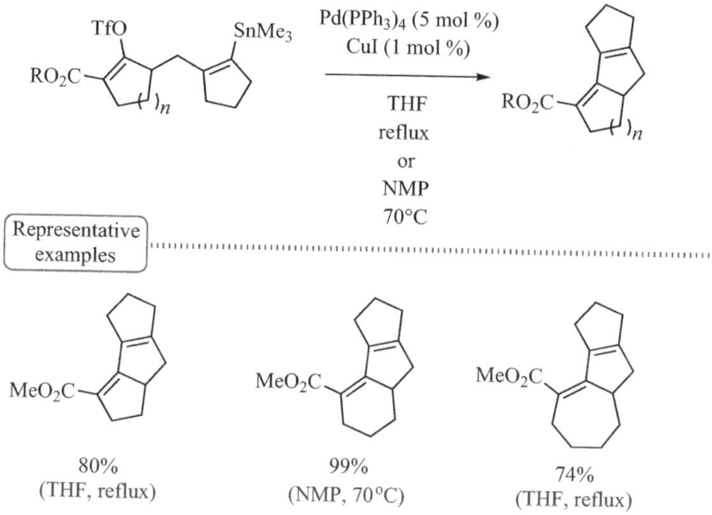

Scheme 13.24 Palladium-catalyzed copper-cocatalyzed intramolecular Stille coupling.

copper loading is usually required in intermolecular Stille coupling.[31,59–61] Structurally interesting polycyclic diene esters including the linearly fused triquinane, 6-5-5 and 7-5-5 tricyclic frameworks are rapidly assembled using this procedure, which clearly represents an attractive synthetic protocol in the area of polyquinane synthesis.

The copper-promoted Stille coupling has been extended to the synthesis of fluorinated 1,3-dienes. The stereoselective cross coupling of (*E*)-1,2-difluorovinylstannes with vinyl halides using 4 mol % tetrakis(triphenylphosphine)palladium(0) and 0.5 equiv of copper(I) iodide leads to difluorinated or tetrafluorinated dienes in good-to-excellent yields (Scheme 13.25).[62] This mild procedure provides a straightforward approach to multifluorinated dienes, which are otherwise difficult to synthesize. Similarly, a variety of monofluorinated and difluorinated unsymmetrical 1,3-dienes were prepared by the groups of Burton[63] and Chen.[64] Other types of heteroatom-substituted 1,3-dienes such as 1,3-dienyl sulfides and 1,3-dienyl sulfones can be accessed via

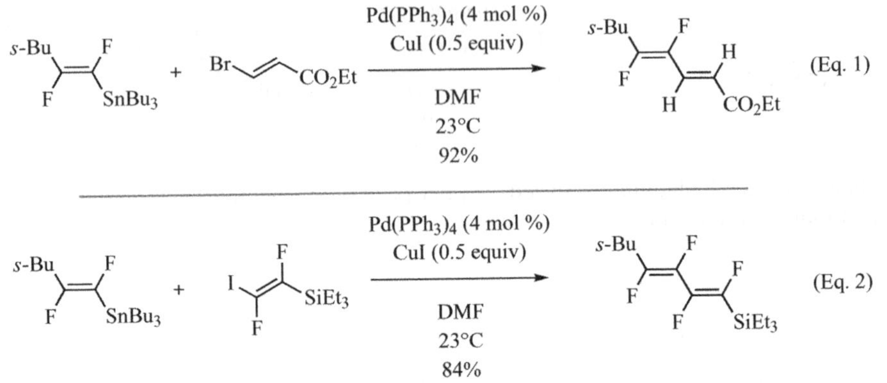

Scheme 13.25 Palladium-catalyzed copper-promoted Stille coupling in the synthesis of fluorinated 1,3-dienes.

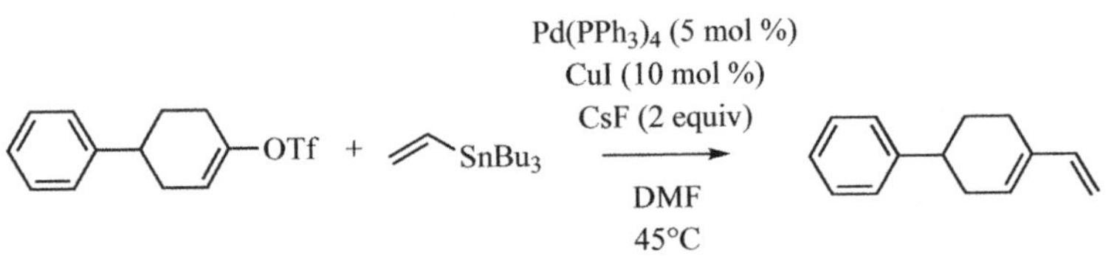

Scheme 13.26 Palladium-catalyzed, copper-cocatalyzed Stille coupling using stoichiometric cesium fluoride.

copper-promoted alkenylation of vinyl iodides with (*E*)-α-stannylvinyl sulfides[65] or (*E*)-α-stannylvinyl sulfones.[66]

13.3.1.3 Effect of Additives in Palladium-Catalyzed Copper-Promoted Stille-Type Coupling. Significant enhancement of the Stille reaction using a combination of catalytic copper(I) iodide and stoichiometric cesium fluoride was observed by Baldwin and coworkers (Scheme 13.26).[60] These improved conditions can be utilized in the synthesis of sterically hindered systems and promote electronically disfavored coupling processes. The addition of cesium fluoride likely facilitates the removal of the tributyltin iodide by-product as the insoluble tributyltin fluoride and enhances the cross coupling.

In 2008, Trauner and coworkers reported an elegant elaboration of highly functionalized trienes employing modified Liebeskind coupling conditions (Scheme 13.27).[67] Both stoichiometric copper(I) thiophenecarboxylate and stoichiometric cesium fluoride were used in this case, and the reaction proceeded stereospecifically at 0°C to afford the thermally sensitive triene products.

Taking a lead from Liebeskind's disclosure of copper(I) diphenylphosphinate (CuDPP) as an effective reagent for the cross coupling of organotin reagents that is facilitated by precipitation of $Ph_2PO_2SnBu_3$,[68–70] Fürstner and coworkers showed that the phosphinate salt $[Ph_2PO_2][NBu_4]$ can be used

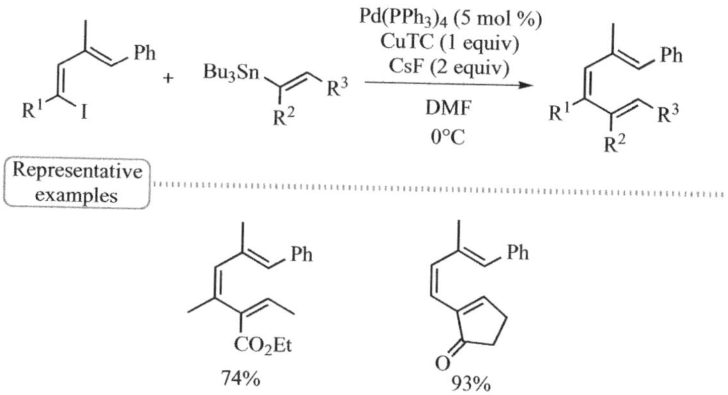

Scheme 13.27 Palladium-catalyzed, copper-promoted cross coupling in the synthesis of highly functionalized trienes.

to enhance challenging Stille cross-coupling reactions via removal of Bu_3SnI (Scheme 13.28).[71] Using this procedure, bulky alkenyl stannanes containing fragile functional groups could be coupled efficiently with substituted vinyl iodides to give the corresponding complex diene products at ambient temperature. Importantly, this fluoride-free procotol tolerates frequently used *C*- and *O*-silyl protecting groups.

In a synthetic study of steroidal analogues (Scheme 13.29),[72] stoichiometric lithium chloride was found to enhance palladium/copper-cocatalyzed cross coupling of bicyclic vinylstannanes and vinyl triflates to furnish tetracyclic 1,3-dienes in good yields. Other palladium sources such as tetrakis(triphenylphosphine)palladium(0) proved to be inefficient in this system.

Palladium-catalyzed, copper-cocatalyzed Stille cross couplings are valuable transformations for the efficient synthesis of 1,3-dienes. In challenging cases, additives such as fluorides or phosphinate salts proved essential. The Suzuki–Miyaura cross-coupling reaction is another central transformation that is developed in the next subchapter.

13.3.2 Synthesis of 1,3-Dienes by Suzuki-Type Coupling

13.3.2.1 Copper-Mediated Suzuki-Type Coupling. Although the Suzuki–Miyaura reaction is an extremely versatile method for the formation of carbon–carbon bonds,[73,74] the Suzuki-type copper-mediated alkenylation of alkenyl derivatives is rare in the literature (compared to the use of copper salts in Stille-type reactions). Kang and coworkers[75] described a copper-catalyzed cross coupling of highly reactive alkenyliodonium tetrafluoroborates and vinyl boronates or alkenyl-9-borabicyclononane under mild conditions (Scheme 13.30). This reaction was carried out in water and stoichiometric sodium carbonate was required to promote the putative boron to copper transmetallation.

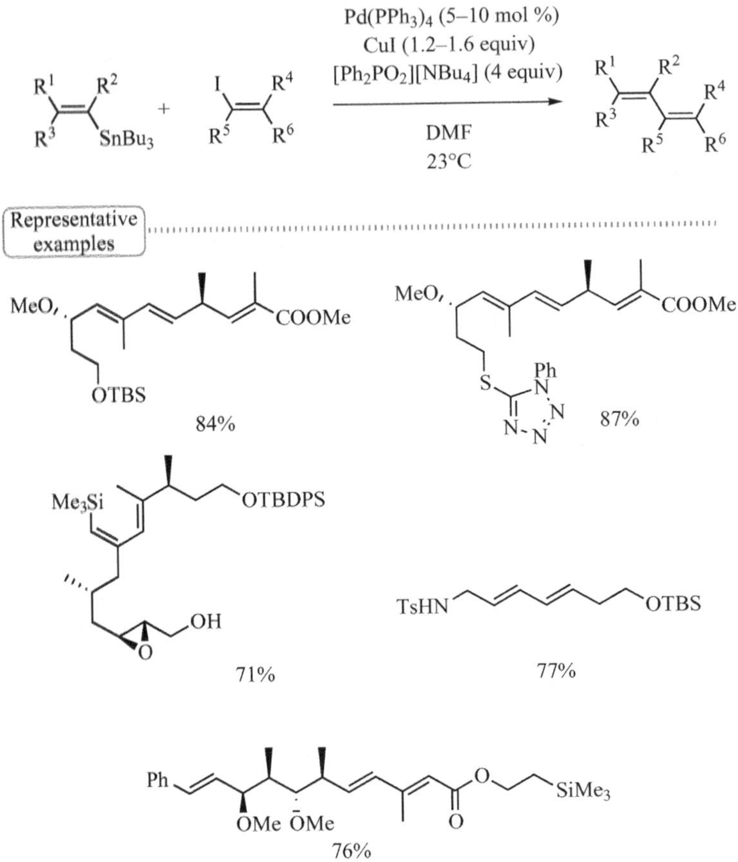

Scheme 13.28 Phosphinate salt-promoted Stille coupling.

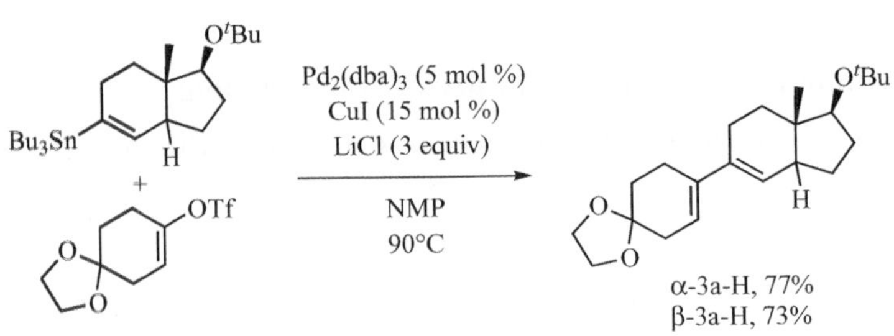

Scheme 13.29 Lithium chloride-promoted cross coupling of bicyclic vinylstannanes and vinyl triflates.

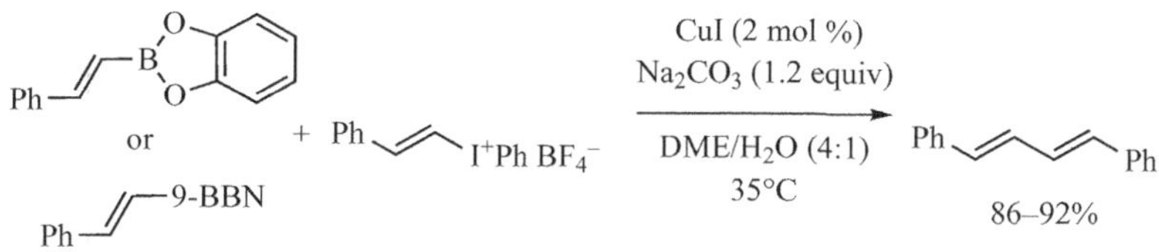

Scheme 13.30 Copper-catalyzed cross coupling of alkenyliodonium tetrafluoroborates and vinyl boronates.

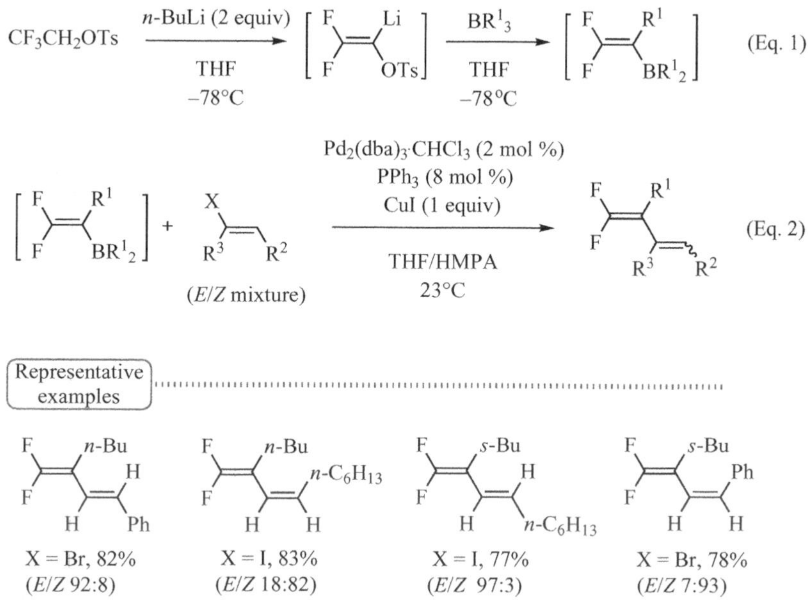

Scheme 13.31 Palladium-catalyzed copper-promoted Suzuki-type coupling in the synthesis of fluorinated 1,3-dienes.

13.3.2.2 Palladium-Catalyzed Copper-Promoted Suzuki-Type Coupling. A one-pot procedure for the synthesis of fluorinated 1,3-dienes from trifluoroethyl tosylate employing palladium/copper-cocatalyzed Suzuki-type coupling was described by Ichikawa and coworkers (Scheme 13.31).[76] The procedure involves the difluoroalkene formation via an elimination reaction followed by the formation of a boronate intermediate that transfers a R^1 group, leading to a vinylboron derivative. Palladium-catalyzed, copper-promoted alkenylation of the latter with various vinyl halides possessing aryl- and alkyl substituents delivered the targeted 1,1-difluoro-1,3-dienes in good yields. The reaction occurs at room temperature and stoichiometric copper(I) iodide is used to promote the Suzuki-type coupling. It is worth noting that the stereochemistry of the vinyl halide is completely preserved in the final product.

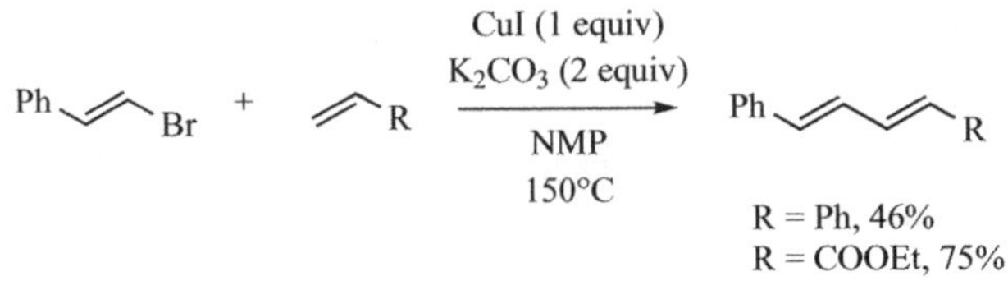

Scheme 13.32 Copper-promoted Heck reaction of (*E*)-β-bromostyrene and activated olefins.

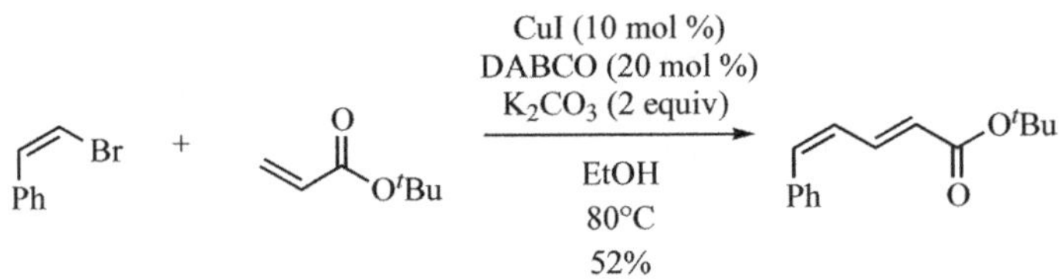

Scheme 13.33 Copper-catalyzed Heck-type cross coupling of (*Z*)-β-bromostyrene and acrylates.

13.3.3 Synthesis of 1,3-Dienes by Heck-Type Coupling

13.3.3.1 Copper-Mediated Heck-Type Coupling. The Heck reaction is a powerful synthetic tool for organic chemists.[77,78] The first copper-promoted Heck reaction between vinyl iodides and olefins was reported by Iyer and coworkers in 1997 (Scheme 13.32).[79] Stoichiometric copper(I) iodide promoted the vinylation of (*E*)-β-bromostyrene with activated olefins at 150°C to afford substituted dienes in synthetically useful yields. However, no vinylation was observed with unactivated alkenes such as cyclooctene.

Later on, an improved Heck-type cross coupling of (*Z*)-β-bromostyrene and acrylates using 10 mol % of copper(I) iodide and 20 mol % of 1,4-diazabicyclo[2.2.2]octane (DABCO) was demonstrated by Li and coworkers (Scheme 13.33).[80] A stereodefined functionalized diene was synthesized in moderate yield via this copper-catalyzed vinylation at lower temperature in a protic solvent. The diene product yield was increased to 88% when 1 equiv of copper(I) iodide was used. However, the scope of the copper-catalyzed Heck reaction is still limited to activated olefins.

13.3.3.2 Palladium-Catalyzed Copper-Promoted Heck-Type Coupling. Extension of the copper-promoted Heck reaction to unactivated alkenes was first achieved by Loh and coworkers via C–H bond activation (Scheme 13.34).[81] This palladium-catalyzed, copper-promoted direct cross coupling of simple olefins with acrylates provides a very mild, efficient method for the synthesis of dienoates, wherein stoichiometric copper(II) acetate and oxygen are used as the terminal oxidants to regenerate the active palladium(II) catalyst. The steric properties of substituents on both the alkene and the acrylate were shown to have a strong influence on the reaction yields, and only 2-substituted olefins could be used in this cross coupling.

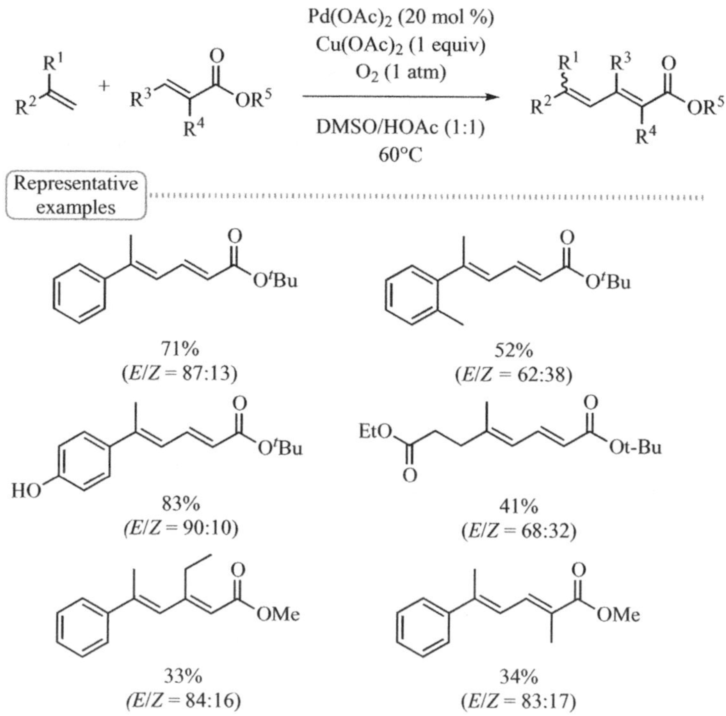

Scheme 13.34 Palladium-catalyzed, copper-promoted cross coupling of unactivated alkenes and acrylates.

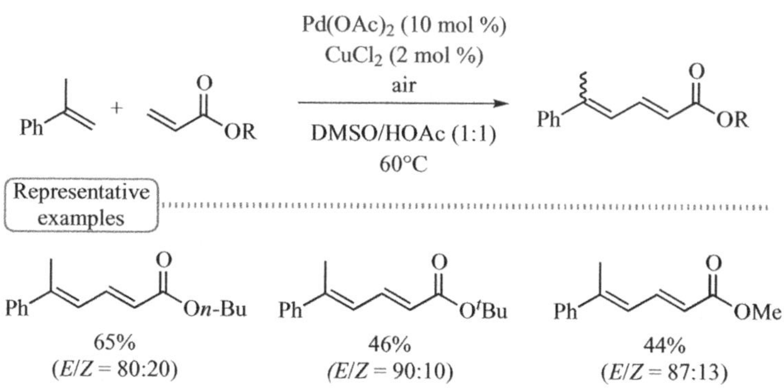

Scheme 13.35 Palladium-catalyzed, copper-promoted cross coupling of alkenes and acrylates using air as the terminal oxidant.

Djakovitch and coworkers reported a related Heck reaction employing only a catalytic amount of copper(II) chloride and air as the terminal oxidant (Scheme 13.35).[82] The scope and efficiency of the reaction are similar to the above-mentioned procedure,[81] although a longer reaction time is required.

A copper-promoted Heck reaction is also involved in the synthesis of chloro-substituted 1,3-dienes developed very recently by Huang and coworkers

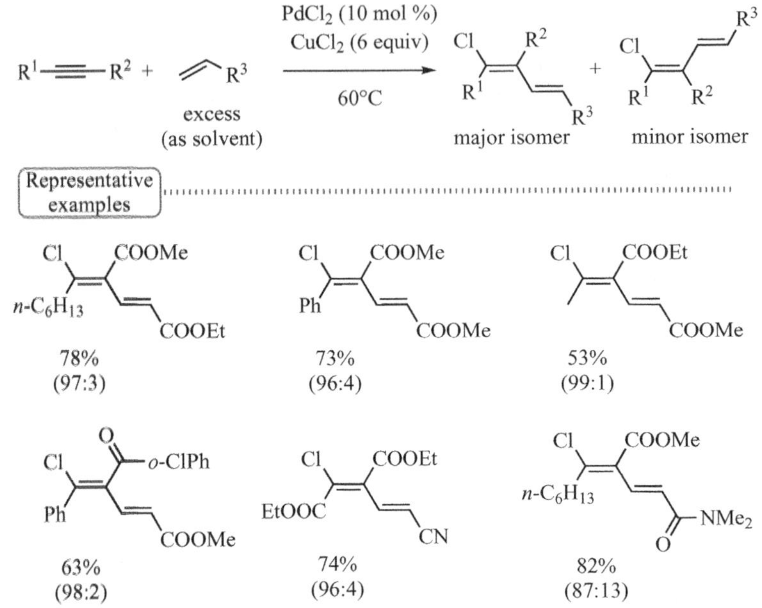

Scheme 13.36 Palladium-catalyzed, copper-promoted Heck-type coupling in the synthesis of chloro-substituted 1,3-dienes.

(Scheme 13.36).[83] This palladium-catalyzed highly regio- and stereoselective cascade reaction of alkynes, copper(II) chloride, and alkenes proceeds under mild conditions via a chloropalladation/Heck reaction sequence. Excess copper(II) chloride is required for this three-component coupling reaction due to its dual role as both the terminal oxidant to regenerate active palladium(II) catalyst and the chloride source for the *trans*-chloropalladation of alkynes.

13.4 CONCLUSIONS

Since the discovery of copper-mediated homocoupling reactions of vinyl derivatives in the 1970s, the copper-mediated alkenylation of alkenyl derivatives has emerged as an important tool in the synthesis of highly functionalized symmetrical and unsymmetrical 1,3-dienes. Copper-mediated 1,3-diene synthesis can often be carried out under very mild reaction conditions with rapid reaction rates, good chemo- and stereoselectivity, excellent functional group compatibility, and high efficiency. One of the major advancements in this field is the introduction of CuTC as an effective catalyst/cocatalyst/mediator in the synthesis of 1,3-dienes by Stille-type cross couplings as nicely exemplified with elegant applications of these facile coupling reactions in the synthesis of highly functionalized complex molecules and natural products with sensitive 1,3-diene motifs. Opportunity remains for the development of new copper catalysts with higher efficiencies and with expanded reaction scope.

REFERENCES

[1] Stefani, H. A.; Guarezemini, A. S.; Cella, R. *Tetrahedron* **2010**, *66*, 7871–7918 and references cited therein.

[2] Yamamoto, Y.; Yatagai, H.; Maruyama, K.; Sonoda, A.; Murahashi, S.-I. *J. Am. Chem. Soc.* **1977**, *99*, 5652–5656.

[3] Yamamoto, Y.; Yatagai, H.; Moritani, I. *J. Am. Chem. Soc.* **1975**, *97*, 5606–5607.

[4] Campbell, J. B.; Brown, H. C. *J. Org. Chem.* **1980**, *45*, 549–550.

[5] Yoshida, J.-I.; Tamao, K.; Kakui, T.; Kumada, M. *Tetrahedron Lett.* **1979**, *20*, 1141–1144.

[6] Ikegashira, K.; Nishihara, Y.; Hirabayashi, K.; Mori, A.; Hiyama, T. *Chem. Commun.* **1997**, 1039–1040.

[7] Nishihara, Y.; Ikegashira, K.; Toriyama, F.; Mori, A.; Hiyama, T. *Bull. Chem. Soc. Jpn.* **2000**, *73*, 985–990.

[8] Kang, S.-K.; Kim, T.-H.; Pyun, S.-J. *J. Chem. Soc. Perkin Trans. 1* **1997**, 797–798.

[9] Itami, K.; Ushiogi, Y.; Nokami, T.; Ohashi, Y.; Yoshida, J.-I. *Org. Lett.* **2004**, *6*, 3695–3698.

[10] Cicco, S. R.; Farinola, G. M.; Martinelli, C.; Naso, F.; Tiecco, M. *Eur. J. Org. Chem.* **2010**, 2275–2279.

[11] Babudri, F.; Cicciomessere, A. R.; Farinola, G. M.; Fiandanese, V.; Marchese, G.; Musio, R.; Naso, F.; Sciacovelli, O. *J. Org. Chem.* **1997**, *62*, 3291–3298.

[12] Ghosal, S.; Luke, G. P.; Kyler, K. S. *J. Org. Chem.* **1987**, *52*, 4296–4298.

[13] Beddoes, R. L.; Cheeseright, T.; Wang, J.; Quayle, P. *Tetrahedron Lett.* **1995**, *36*, 283–286.

[14] Piers, E.; McEachern, E. J.; Romero, M. A. *Tetrahedron Lett.* **1996**, *37*, 1173–1176.

[15] Piers, E.; McEachern, E. J.; Romero, M. A.; Gladstone, P. L. *Can. J. Chem.* **1997**, *75*, 694–701.

[16] Piers, E.; Gladstone, P. L.; Yee, J. G. K.; McEachern, E. J. *Tetrahedron* **1998**, *54*, 10609–10626.

[17] Han, X.; Stoltz, B. M.; Corey, E. J. *J. Am. Chem. Soc.* **1999**, *121*, 7600–7605.

[18] Blumenthal, E. J.; Burton, D. J. *Isr. J. Chem.* **1999**, *39*, 109–115.

[19] Cossu, S.; Cimenti, C.; Paola, P.; Paulon, A.; Lucchi, O. D. *Angew. Chem. Int. Ed. Engl.* **2001**, *40*, 4086–4088.

[20] Borsato, G.; Lucchi, O. D.; Fabris, F.; Groppo, L.; Lucchini, V.; Zambona, A. *J. Org. Chem.* **2002**, *67*, 7894–7897.

[21] Borsato, G.; Lucchi, O. D.; Fabris, F.; Lucchini, V.; Pasqualottia, M.; Zambona, A. *Tetrahedron Lett.* **2003**, *44*, 561–563.

[22] Fabris, F.; Zambrini, L.; Rosso, E.; Lucchi, O. D. *Eur. J. Org. Chem.* **2004**, 3313–3322.

[23] Allred, G. D.; Liebeskind, L. S. *J. Am. Chem. Soc.* **1996**, *118*, 2748–2749.

[24] Itoh, T.; Emoto, S.; Kondo, M.; Ohara, H.; Tanaka, H.; Torii, S. *Electrochim. Acta* **1997**, *42*, 2133–2142.

[25] Kang, S.-K.; Baik, T.-G.; Jiao, X. H.; Lee, Y.-T. *Tetrahedron Lett.* **1999**, *40*, 2383–2384.

[26] Cohen, T.; Cristea, I. *J. Org. Chem.* **1975**, *40*, 3649–3651.

[27] Zhang, S.; Zhang, D.; Liebeskind, L. S. *J. Org. Chem.* **1997**, *62*, 2312–2313.

[28] Yoshifuji, M.; Loots, M. J.; Schwartz, J. *Tetrahedron Lett.* **1977**, *28*, 1303–1306.

[29] Moglie, Y.; Mascaró, E.; Nador, F.; Vitale, C.; Radivoy, G. *Synth. Commun.* **2008**, *38*, 3861–3874.

[30] Roth, G. P.; Farina, V. *Tetrahedron Lett.* **1995**, *36*, 2191–2194.

[31] Farina, V.; Kapadia, S.; Krishnan, B.; Wang, C.; Liebeskind, L. S. *J. Org. Chem.* **1994**, *59*, 5905–5911.

[32] Liebeskind, L. S.; Fengl, R. W. *J. Org. Chem.* **1990**, *55*, 5359–5364.

[33] Beletskaya, I. P.; Cheprakov, A. V. *Coord. Chem. Rev.* **2004**, *248*, 2337–2364.

[34] Evano, G.; Blanchard, N.; Toumi, M. *Chem. Rev.* **2008**, *108*, 3054–3131.

[35] Peng, Y.; Li, W.-D. Z. *Eur. J. Org. Chem.* **2010**, 6703–6718.

[36] Maryanoff, B. E.; Reitz, A. B. *Chem. Rev.* **1989**, *89*, 863–927.

[37] Kelly, S. E. *Comprehensive Organic Synthesis*, Vol. 1, Trost, B. M.; Fleming, I. (eds.). Pergamon: Oxford, **1991**, pp. 729–817.

[38] Edmonds, M.; Abell, A. *Modern Carbonyl Olefination*, Takeda, T. (ed.). Wiley-VCH: Weinheim, Germany, **2004**, pp. 1–17.

[39] Trost, B. M.; Toste, F. D.; Pinkerton, A. B. *Chem. Rev.* **2001**, *101*, 2067–2096.

[40] Aubert, C.; Buisine, O.; Malacria, M. *Chem. Rev.* **2002**, *102*, 813–834.

[41] de Meijere, A.; Diederich, F. (eds.), *Metal-Catalyzed Cross-Coupling Reactions*. Wiley-VCH: Weinheim, Germany, **2004**.

[42] Negishi, E.-I. (ed.), *Handbook of Organopalladium Chemistry for Organic Synthesis*. Wiley-Interscience: New York, **2002**.

[43] Knochel, P.; Betzemeier, B. *Modern Organocopper Chemistry*, Krause, N. (ed.). Wiley-VCH Verlag: Weinheim, **2002**, pp. 45–78.

[44] Wipf, P. *Synthesis* **1993**, 537–557.

[45] Kosugi, M.; Shimizu, Y.; Migita, T. *Chem. Lett.* **1977**, *6*, 1423–1424.

[46] Kosugi, M.; Sasazawa, K.; Shimizu, Y.; Migita, T. *Chem. Lett.* **1977**, *6*, 301–302.

[47] Stille, J. K. *Angew. Chem. Int. Ed. Engl.* **1986**, *25*, 508–524.

[48] Mitchell, T. N. *Synthesis* **1992**, 803–815.

[49] Ko, J.; Kang, S. O.; Jung, I.; Lee, T. *Synthesis* **2005**, 986–992.

[50] Tanaka, H.; Sumida, S.-I.; Torii, S. *Tetrahedron Lett.* **1996**, *37*, 5967–5970.

[51] Kang, S.-K.; Kim, J.-S.; Choi, S.-C. *J. Org. Chem.* **1997**, *62*, 4208–4209.

[52] Piers, E.; Wong, T. *J. Org. Chem.* **1993**, *58*, 3609–3610.

[53] Marino, J. P.; Long, J. K. *J. Am. Chem. Soc.* **1988**, *110*, 7916–7917.

[54] Casado, A. L.; Espinet, P. *Organometallics* **2003**, *22*, 1305–1309.

[55] (a) Hinkle, R. J.; Poulter, G. T.; Stang, P. J. *J. Am. Chem. Soc.* **1993**, *115*, 11626–11627. (b) Hinkle, R. J.; Stang, P. J. *Synthesis* **1994**, 313–316.

[56] Nicolaou, K. C.; Sato, M.; Miller, N. D.; Gunzner, J. L.; Renaud, J.; Uiitersteller, E. *Angew. Chem. Int. Ed. Engl.* **1996**, *35*, 889–891.

[57] Lee, S. J.; Anderson, T. M.; Burke, M. D. *Angew. Chem. Int. Ed. Engl.* **2010**, *49*, 8860–8863.

[58] Piers, E.; Romero, M. A.; Walker, S. D. *Synlett* **1999**, 1082–1084.

[59] Farina, V. *Pure Appl. Chem.* **1996**, *68*, 73–78.

[60] Mee, S. P.; Lee, V.; Baldwin, J. E. *Angew. Chem. Int. Ed. Engl.* **2004**, *43*, 1132–1136.

[61] Mee, S. P.; Lee, V.; Baldwin, J. E. *Chem. Eur. J.* **2005**, *11*, 3294–3308.

[62] Lu, L.; Burton, D. J. *Tetrahedron Lett.* **1997**, *38*, 7673–7676.

[63] Zhang, X.; Lu, L.; Burton, D. J. *Collect. Czech. Chem. Commun.* **2002**, *67*, 1247–1261.

[64] Chen, C.; Wilcoxen, K.; Zhu, Y.-F.; Kim, K.; McCarthy, J. R. *J. Org. Chem.* **1999**, *64*, 3476–3482.

[65] Cai, M.-Z.; Wang, D.; Wang, P.-P. *J. Organomet. Chem.* **2006**, *691*, 737–740.

[66] Cai, M.-Z.; Chen, G.-Q.; Hao, W.-Y.; Wang, D. *J. Organomet. Chem.* **2007**, *692*, 1125–1130.

[67] Bishop, L. M.; Barbarow, J. E.; Bergman, R. G.; Trauner, D. *Angew. Chem. Int. Ed. Engl.* **2008**, *47*, 8100–8103.

[68] Srogl, J.; Allred, G. D.; Liebeskind, L. S. *J. Am. Chem. Soc.* **1997**, *119*, 12376–12377.

[69] Wittenberg, R.; Srogl, J.; Egi, M.; Liebeskind, L. S. *Org. Lett.* **2003**, *5*, 3033–3035.

[70] Li, H.; Yang, H.; Liebeskind, L. S. *Org. Lett.* **2008**, *10*, 4375–4378.

[71] Fürstner, A.; Funel, J. A.; Tremblay, M.; Bouchez, L. C.; Nevado, C.; Waser, M.; Ackerstaff, J.; Stimson, C. C. *Chem. Commun.* **2008**, 2873–2875.

[72] Sünnemann, H. W.; Banwell, M. G.; de Meijere, A. *Chem. Eur. J.* **2008**, *14*, 7236–7249.

[73] Suzuki, A. *J. Organomet. Chem.* **1999**, *576*, 147–168.

[74] Miyaura, N.; Suzuki, A. *Chem. Rev.* **1995**, *95*, 2457–2483.

[75] Kang, S.-K.; Yamaguchi, T.; Kim, T.-H.; Ho, P.-S. *J. Org. Chem.* **1996**, *61*, 9082–9083.

[76] Ichikawa, J.; Ikeura, C.; Minami, T. *Synlett* **1992**, 739–740.

[77] Beletskaya, I. P.; Cheprakov, A. V. *Chem. Rev.* **2000**, *100*, 3009–3066.

[78] Bedford, R. B.; Cazin, C. S. J.; Holder, D. *Coord. Chem. Rev.* **2004**, *248*, 2283–2321.

[79] Iyer, S.; Ramesh, C.; Sarkar, A.; Wadgaonkar, P. P. *Tetrahedron Lett.* **1997**, *38*, 8113–8116.

[80] Li, J.-H.; Wang, D.-P.; Xie, Y.-X. *Tetrahedron Lett.* **2005**, *46*, 4941–4944.

[81] Xu, Y.-H.; Lu, J.; Loh, T.-P. *J. Am. Chem. Soc.* **2009**, *131*, 1372–1373.

[82] Al-Maksoud, W.; Djakovitch, L.; Jahjah, M.; Pinel, C. *Sci. China Chem.* **2010**, *53*, 1927–1931.

[83] Huang, J. M.; Dong, Y.; Wang, X. X.; Luo, H. C. *Chem. Commun.* **2010**, *46*, 1035–1037.

14

EMERGING AREAS IN COPPER-MEDIATED TRIFLUOROMETHYLATIONS OF ARYL DERIVATIVES: CATALYTIC AND OXIDATIVE CROSS-COUPLING PROCESSES

Kévin Jouvin, Céline Guissart, Cédric Theunissen, and Gwilherm Evano

14.1 INTRODUCTION

The incorporation of trifluoromethyl groups into organic molecules has a profound effect on their chemical and physical properties and can serve to dramatically modify their behavior, including their bioavailability, lipophilicity, receptor binding selectivity, and metabolic stability. It is therefore not surprising that trifluoromethyl groups can be found in key pharmaceuticals; and notable examples include efavirenz (Sustiva®), an HIV reverse transcriptase inhibitor; fluoxetine (Prozac®), an antidepressant; celecoxib (Celebrex®), a nonsteroidal antiinflammatory; or lansoprazole (Prevacid®), a proton-pump inhibitor used in the treatment of ulcers and acid reflux disease (Fig. 14.1). Trifluoromethylation has also been used in the design of agrochemicals with improved properties such as the fungicide trifloxystrobin (Flint® or Delaro®) and the bleaching herbicide diflufenican (Alister® or Azur®) or for the development of liquid crystals, a famous example being the ZLI-2857 mixture, a component of liquid crystals used in LCD screens.

Copper-Mediated Cross-Coupling Reactions, First Edition. Edited by Gwilherm Evano and Nicolas Blanchard.

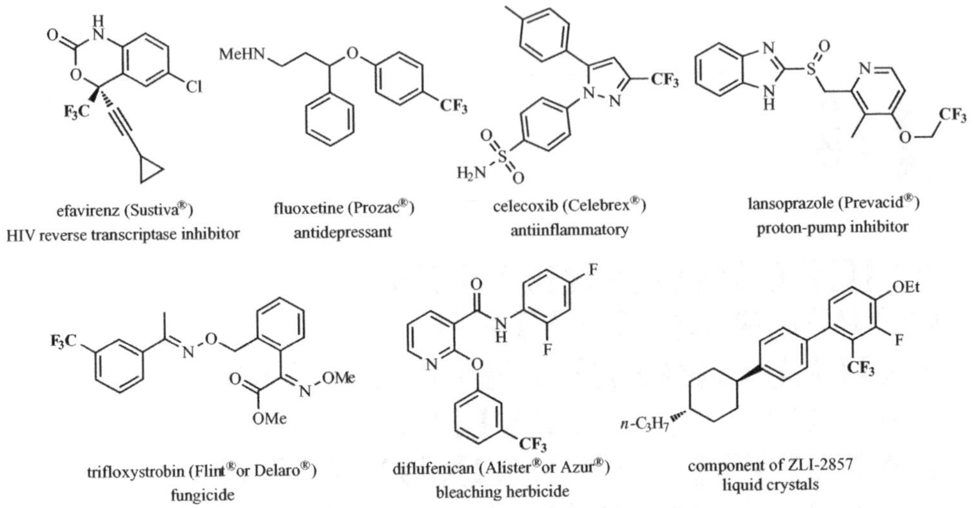

Figure 14.1 Representative trifluoromethylated drugs, agrochemicals, and molecules for materials.

Despite the widespread occurrence of trifluoromethylaryl subunits, the development of general, practical, mild, and/or catalytic methods for the selective trifluoromethylation of aromatics has been a long-lasting task. Historical developments include the Swarts reaction, which allows for the transformation of trichlorotoluene derivatives to their trifluoro-congeners upon treatment with antimony pentafluoride and anhydrous hydrogen fluoride—not the mildest conditions—and the development of reagents aimed at the selective trifluoromethylation of aromatics using radical, electrophilic, and nucleophilic reagents.[1] Unfortunately, none of these methods is generally applicable and their scope is limited, even if recent advances in trifluoromethylation with trifluoromethyl sulfonium salts, hypervalent iodine-based compounds, Langlois reagent, or with photoredox catalysis seem to be quite promising.

One of the most efficient methods for the regioselective trifluoromethylation of aryl halides relies on the generation of trifluoromethylcopper, $CuCF_3$, or related species. These reactive intermediates are generally produced *in situ* from stoichiometric or even excess metallic copper or copper salts and triflurornethylating agents, the most widely applicable procedure being based on the use of the Ruppert–Prakash reagent, CF_3SiMe_3, in combination with a base and/or a source of fluoride ions. The chemistry of these fluorinated organocopper reagents has been extensively reviewed recently[1,2] and will, therefore, not be covered in this chapter. Despite the remarkable synthetic utility of procedures based on this approach, the major drawbacks are the need for stoichiometric amounts of copper, harsh reaction conditions, and/or limited substrate scope. Besides remarkable and recent development in the preparation of simple, stable, easily accessible, and/or well-defined copper(I)-CF_3 complexes[3] or precursors,[4] or in the use of copper-mediated trifluoromethylation of aryl halides and arylboronic acids with trifluoromethylsulfonium salts using stoichiometric amounts of copper salts,[5] crucial advances in this area have been recently reported, based either on copper-catalyzed trifluoromethylation processes or on oxidative trifluoromethylation reactions.[6] These two emerging and rapidly expanding areas will be overviewed in this chapter. In some cases, representative and key examples dealing with the introduction of other fluorinated chains such as difluoromethyl or perfluoroalkyls or with cross coupling yielding to trifluoromethylated alkenes or alkynes will also be commented upon briefly. Other reactions such as the trifluoromethylation of alkenes (with loss or migration of the double bond) or benzylic, allylic, propargylic, and other activated positions are outside the scope of this chapter (and this book) and will not be covered.

14.2 COPPER-CATALYZED TRIFLUOROMETHYLATION OF ARYL HALIDES: A LONG-LASTING QUEST FINALLY REACHED

Whereas recent achievements demonstrated the feasibility of a catalytic trifluoromethylation of aryl halides using nickel, palladium, or silver catalysis,[7] the development of a related copper-catalyzed process appeared to be a lot more difficult. Following pioneering studies from Chen and Wu, who reported in 1989 the first copper-catalyzed trifluoromethylation of aryl iodides with 2,2-difluoro-2-(fluorosulfonyl)acetate,[8] the first general procedure was only reported in 2009. Indeed, Amii and coworkers demonstrated that the reaction rate for the copper-mediated process can be dramatically enhanced, as with most other copper-based systems, by the addition of 1,10-phenanthroline: The increased electron density at copper, resulting in higher nucleophilicity at the CF_3 ligand, combined with the stabilization and solubilizing effect of the ligand, allows for the cross coupling to proceed before significant decomposition of CF_3SiEt_3. This observation led to the discovery of the first copper-catalyzed trifluoromethylation based on Ruppert–Prakash-like reagent: In the presence of 10 mol % of copper(I) iodide, 10 mol % of 1,10-phenanthroline and 2 equiv of potassium fluoride in a 1:1 mixture of *N*-methyl-pyrrolidinone (NMP) and *N*,*N*-dimethylformamide (DMF) at 60°C, the trifluoromethylation of aromatic iodides with CF_3SiEt_3 occurred smoothly, although good conversions were only observed with electron-poor aryl iodides and 2-iodo-heteroaromatics (Scheme 14.1).[9] α-Triethylsilyl-difluoroacetate could also be used for the introduction of a carboethoxy difluoromethylene group, an excellent precursor of the corresponding difluoromethylene, although the stoichiometric version of the reaction definitely seems to be more efficient.[10]

While this work demonstrated the feasibility of a copper-catalyzed trifluoromethylation reaction and was indeed a remarkable breakthrough, the use of the more economical Ruppert–Prakash reagent, CF_3SiMe_3, resulted in much lower yields, presumably due to its higher propensity to undergo self-decomposition in the presence of a fluoride source under the reaction

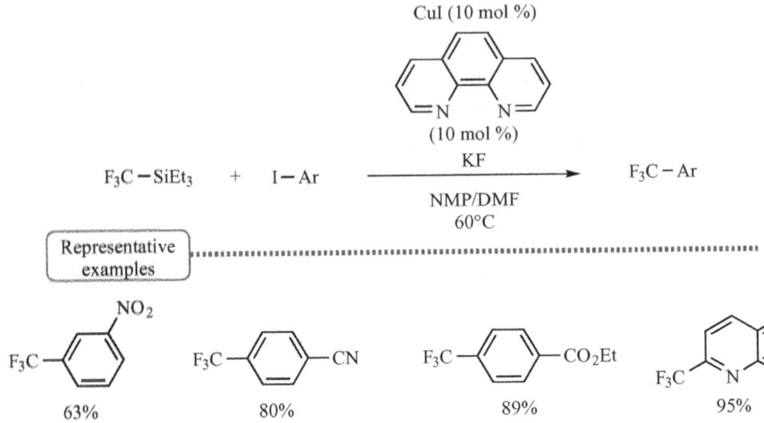

Scheme 14.1 CuI/1,10-phenanthroline-catalyzed trifluoromethylation of aryl iodides with CF_3SiEt_3.

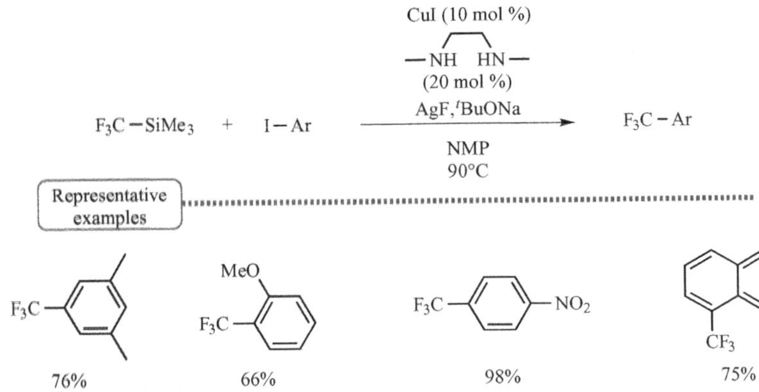

Scheme 14.2 CuI/*N,N′*-dimethylethylenediamine-catalyzed trifluoromethylation of aryl iodides with Ruppert–Prakash reagent CF_3SiMe_3.

conditions. A solution to this problem was reported in 2011 and was found in the use of a cooperative effect of silver additives: By using stoichiometric amounts of silver fluoride, an intermediate trifluoromethylsilver complex would be generated before the transmetallation to copper, therefore avoiding the presence of the free trifluoromethyl anion, which would easily decompose before the cross coupling (Scheme 14.2). The scope of this second generation procedure is fairly broad and trifluoromethyl aromatics can be obtained in moderate to good yields whatever the substitution pattern of the starting aryl iodide.[11] This procedure has not been extended to date to the catalytic trifluoromethylation of nonactivated alkenyl halides, which still require the use of classical, stoichiometric, Ullmann-like conditions.[12]

To avoid problems associated with the decomposition of trifluoromethyltrialkylsilanes and/or to develop trifluoromethylation reagents that would be cheaper and/or more practical, several alternative approaches have been reported in 2011.

In this context, Gooßen reported the use of potassium (trifluoromethyl) trimethoxyborate, $CF_3B(OMe)_3K$, a crystalline, stable, and easy to handle reagent readily obtained in nearly quantitative yields by simply mixing trimethylborate, Ruppert–Prakash reagent, and potassium fluoride. This trifluoromethylation reagent allows the conversion of various aryl iodides into the corresponding benzotrifluorides in high yields under relatively mild, base-free conditions in the presence of catalytic amounts of copper(I) iodide and 1,10-phenanthroline. The reaction is applicable to both electron-rich and electron-deficient arenes, as well as heteroarenes, and tolerates a broad range of functional groups (Scheme 14.3, Eq. 1).[13] Trifluoromethyltrialkylsilanes can also be conveniently replaced by stable, fluoral-derived aminals, compounds that are easily prepared from commercially available fluoral hemiacetal $CF_3CH(OH)(OEt)$ and morpholine, and allow for a slow release of the trifluoromethyl anion during the course of the reaction (Scheme 14.3, Eq. 2).[14] Trifluoroacetates probably being the most convenient, inexpensive, and readily

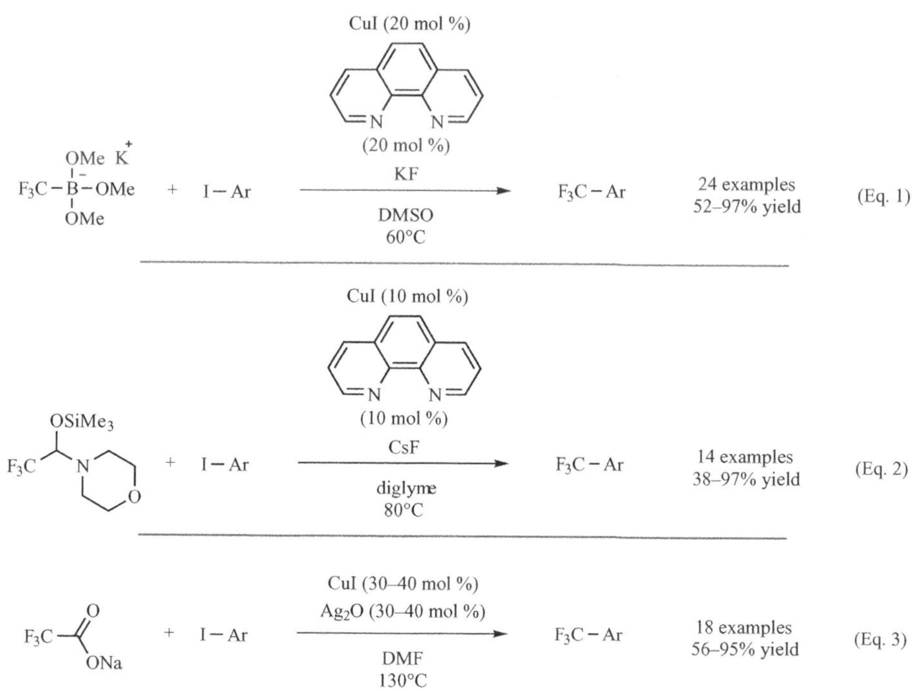

Scheme 14.3 Copper-catalyzed trifluoromethylation of aryl iodides with a trifluoromethylborate salt, a fluoral aminal, and sodium trifluoroacetate.

available source of trifluoromethyl, their use as reaction partners in copper-catalyzed trifluoromethylation reactions is especially appealing. They can indeed participate in copper-mediated decarboxylative trifluoromethylations, provided that silver(I) oxide, which presumably serves both for an intermediate transmetallation step and to accelerate the reductive elimination by precipitation of silver(I) iodide, is used as a co-catalyst (Scheme 14.3, Eq. 3).[15]

A complementary approach was developed based on a formal umpolung of the cross coupling. It can be achieved by switching from aryl iodides to the corresponding arylboronic acids, which are now used in combination with electrophilic trifluoromethylation reagents such as Umemoto's trifluoromethyl sulfonium (Scheme 14.4, Eq. 1)[16] or Togni's hypervalent iodine-based reagent (Scheme 14.4, Eq. 2).[17] In both cases, the reaction conditions are remarkably mild (from 0 to 35°C) and are efficiently catalyzed by copper(I) complexes combined with electron-rich aromatic nitrogen ligands such as 2,4,6-trimethylpyridine or 1,10-phenanthroline. The scope of these cross-coupling reactions is relatively broad and they can, in addition, nicely be extended to the trifluoromethylation of alkenylboronic acids and terminal alkynes as well,[18] providing useful entries to trifluoromethylated alkenes and alkynes, compounds that can also be obtained by copper-catalyzed decarboxylative cross coupling between alkenoic acids and Togni's reagent[19] or copper-catalyzed oxidative trifluoromethylation of alkynes with Ruppert–Prakash reagent (see next section), respectively. As a note, the last reaction has also been used in a one-pot two-step sequential iridium-catalyzed borylation of arenes/copper-catalyzed trifluoromethylation of the resulting boronate esters, an interesting process that allows for the formal trifluoromethylation of aromatic C–H bonds and that was nicely used for the late-stage trifluoromethylation of complex molecules such as vitamin B_3, vitamin E, or estrone,[20] and for the room temperature trifluoromethylation of potassium aryl trifluoroborates (Scheme 14.4, Eq. 3)[21] and potassium alkynyl trifluoroborates (Scheme 14.4, Eq. 4).[22]

All these methods represent significant progress and provide useful techniques for the preparation of trifluoromethyl aromatics, especially when compared to the traditional Swarts reaction. Despite these advances, most of these processes still suffer from limitations associated with the use of expensive or noncommercially available reagents, the need for heating and/or a base, the presence of a source of fluoride, or limited substrate scope. Trying to address these limitations, the Sanford group developed an especially interesting method for the trifluoromethylation of arylboronic acids with commercially available and relatively inexpensive iodotrifluoromethane. This reagent could be efficiently activated by photoredox catalysis using catalytic amounts of tris(2,2′-bipyridine)dichlororuthenium(II) hexahydrate with irradiation from 26-W household lightbulbs, allowing for the reduction of CF_3I to the corresponding trifluoromethyl radical. The merger of this photocatalytic step with copper catalysis to activate the arylboronic acid provides a mild and general method for their trifluoromethylation. The method has been applied to the

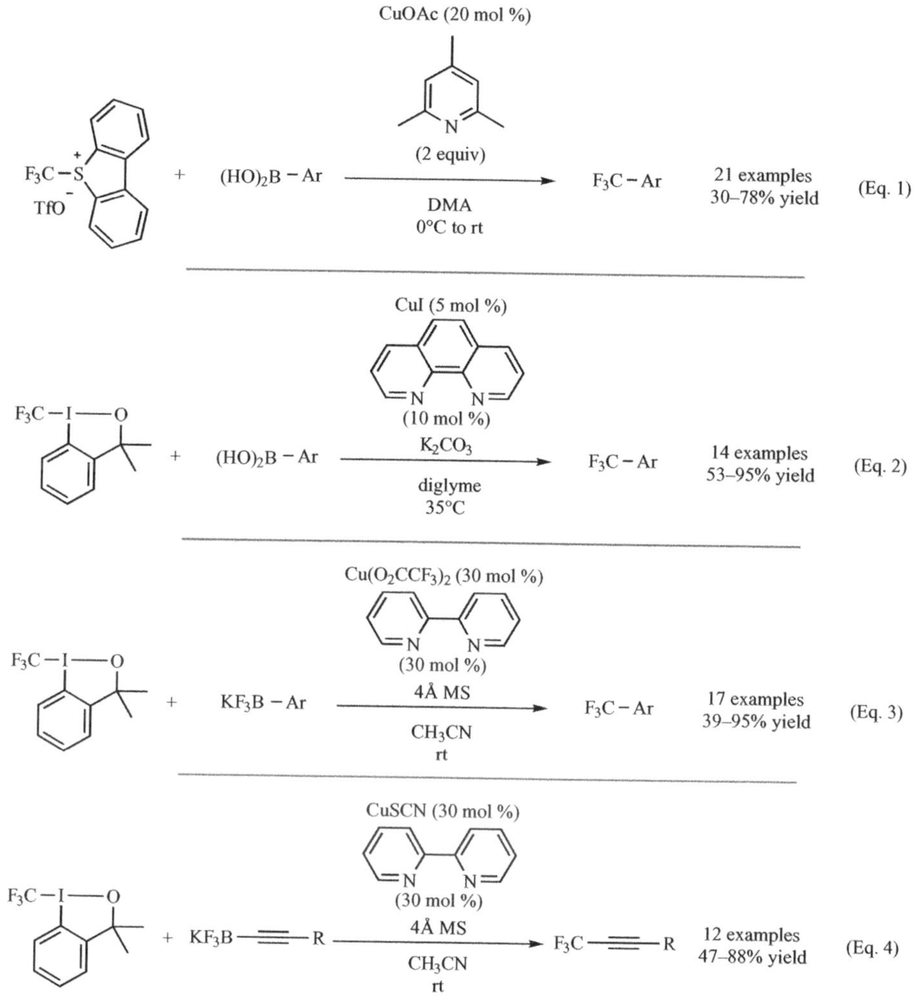

Scheme 14.4 Copper-catalyzed trifluoromethylation of arylboronic acids and potassium aryl/alkynyl trifluoroborates with electrophilic reagents.

trifluoromethylation of a wide range of electronically diverse aromatic and heteroaromatic substrates and could be successfully extended to the introduction of perfluoroalkyl chains (Scheme 14.5).[23]

An especially elegant and efficient copper-catalyzed process was reported in 2011 for the perfluoroalkylation of aryl iodides with readily available 1*H*-perfluoroalkanes. Upon treatment with TMP_2Zn, an intermediate bis(perfluoroalkyl)zinc species would be generated whose subsequent transmetallation with copper(I) chloride in the presence of 1,10-phenanthroline and in *N*,*N*′-dimethylpropylene urea (DMPU) would form a transient anionic

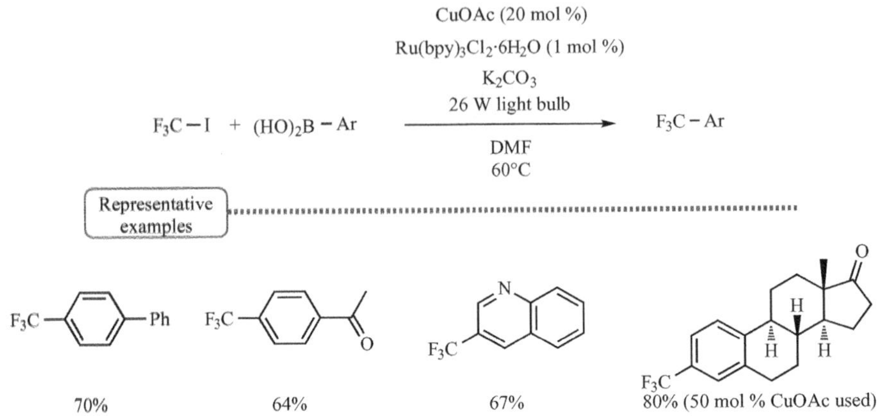

Scheme 14.5 Merging visible-light photocatalysis and copper catalysis for the trifluoromethylation of arylboronic acids.

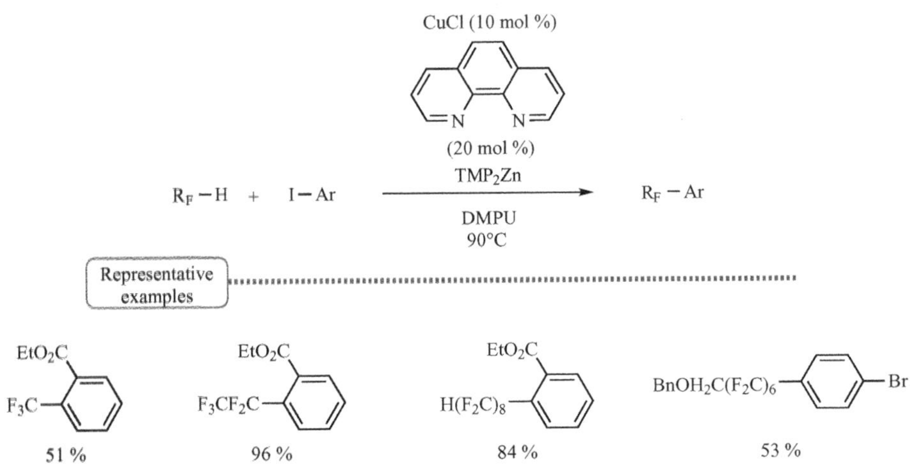

Scheme 14.6 CuCl/1,10-phenanthroline-catalyzed perfluoroalkylation of aryl iodides with 1*H*-perfluoroalkanes.

perfluoroalkylcopper chloride that would finally react with the aryl iodide, either directly or via a neutral perfluoroalkyl complex, to give the corresponding perfluoroalkylated arenes in excellent yields, except in the case of 2,6-disubstituted electron-rich aryl iodides. Notably, this reaction is also efficient with trifluoromethane and displays high selectivity for aryl iodides since unactivated bromoarenes are unaffected by the reaction conditions (Scheme 14.6).[24]

Indoles can also be efficiently and selectively trifluoromethylated at the C2 position using Togni's hypervalent iodine reagent in the presence of catalytic amounts of copper(I) acetate in methanol at room temperature. The reaction

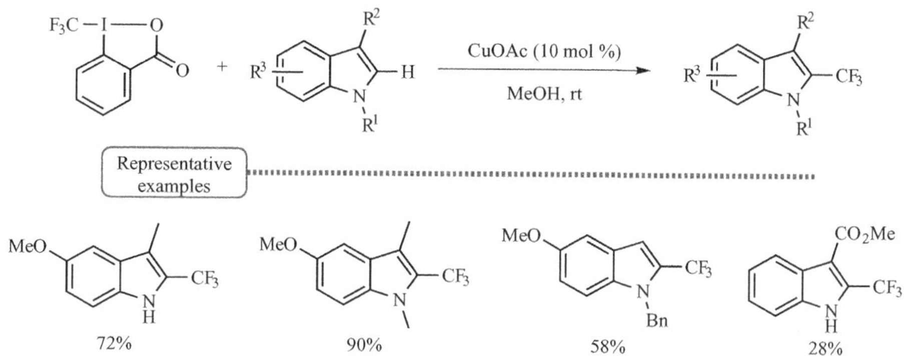

Scheme 14.7 CuOAc-catalyzed regioselective trifluoromethylation of indoles with Togni's reagent.

uses only a slight excess of Togni's reagent under mild conditions and tolerates various functional groups, as exemplified by the efficient trifluoromethylation of tryptamine, melatonin, or tryptophan derivatives. The presence of an electron-withdrawing group at C3, however, diminished the reaction rate and yields. Among all metals screened, iron, cobalt, nickel, and zinc were also found to promote the reaction, as well as copper(II) acetate, although with lower efficiency (Scheme 14.7).[25]

While extensive progress has been made, the development of alternative copper-mediated trifluoromethylations remains an attractive goal, especially for the design of systems that allow the reactions to proceed under especially mild conditions. Recent reports in this area tend to show that copper-based oxidative trifluoromethylation reactions seem to meet this requirement: They will be described in the next section of this chapter.

14.3 COPPER-MEDIATED OXIDATIVE TRIFLUOROMETHYLATION REACTIONS

Besides the development of catalytic processes for the trifluoromethylation/perfluoroalkylation of arenes, the extension of the Chan–Lam–Evans coupling to the use of a synthetic equivalent of the trifluoromethyl carbanion CF_3^- is another challenging task, mostly due to serious side reactions involving the boronic acid such as protodeborylation, halogenation, or dimerization. It is, however, an especially appealing option due to the mildness of the reaction conditions typically required for the copper-mediated cross-coupling processes involving boronic acids. In this context, two close copper-based systems have been reported in 2010 and 2011 for the reaction between arylboronic acids and Ruppert–Prakash reagent, CF_3SiMe_3. Indeed, the reaction can be promoted by copper(I) triflate in combination with 1,10-phenanthroline, silver carbonate, which acts as an oxidant for copper species, and potassium fluoride,

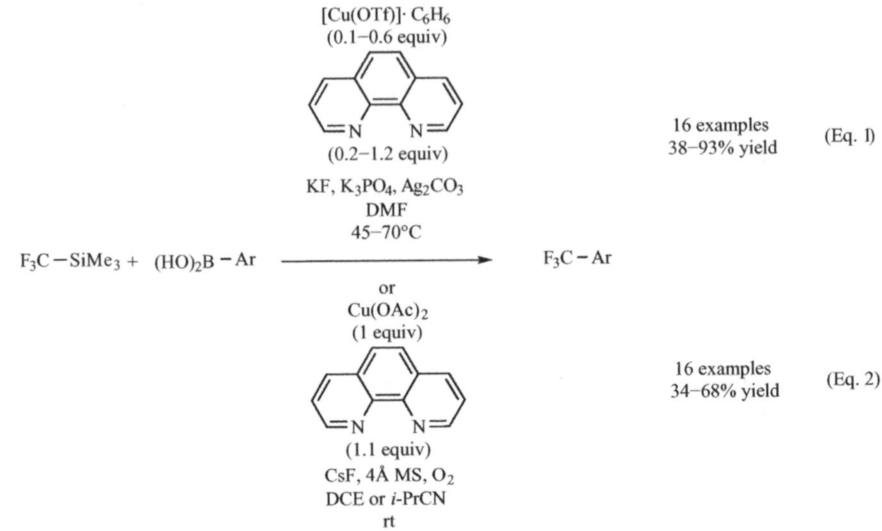

Scheme 14.8 Copper-mediated oxidative trifluoromethylation of arylboronic acids: a most useful extension of the Chan–Lam–Evans cross coupling.

which allows for the generation of CF_3^- from CF_3SiMe_3, in the presence of additional potassium phosphate in DMF at 45°C (Scheme 14.8, Eq. 1).[26a] Modified conditions were reported 2 years later, and the same authors demonstrated that the catalyst loading could be reduced to 10 mol % and that the slow addition of 3 equiv of Ruppert–Prakash reagent could be replaced by the addition of 2 equiv in one portion with only a slight decrease in the reaction yields (typically 5–15%).[26b] A milder and probably more practical protocol was reported in 2011 based on the use of copper(II) acetate—which precludes the preliminary oxidation of copper(I) species to copper(II) complexes required before the transmetallation step with arylboronic acids—1,10-phenanthroline, cesium fluoride, and simply using molecular oxygen as the terminal oxidant (Scheme 14.8, Eq. 2).[27] In contrast to most methods for the synthesis of benzotrifluorides, the mild conditions of this method provide a high level of functional group tolerance, an especially remarkable example being the chemoselective trifluoromethylation of a boronic acid bearing a methyl ketone without competing trifluoromethylation of the ketone. As a note, this strategy has been recently extended to the trifluoromethylation of primary and secondary alkylboronic acids.[28]

Another shelf-stable trifluoromethylating agent, potassium (trifluoromethyl)trimethoxyborate, that was elegantly used by Gooßen for the trifluoromethylation of aryl iodides described in the previous section (see Scheme 14.3, Eq. 1), can also be used for the oxidative trifluoromethylation of arylboronates. In combination with 1 equiv of cuprous acetate in DMSO at 60°C under an atmosphere of oxygen, this reagent readily transfers its trifluoromethyl group to a variety of aromatic, pinacol-derived boronate esters

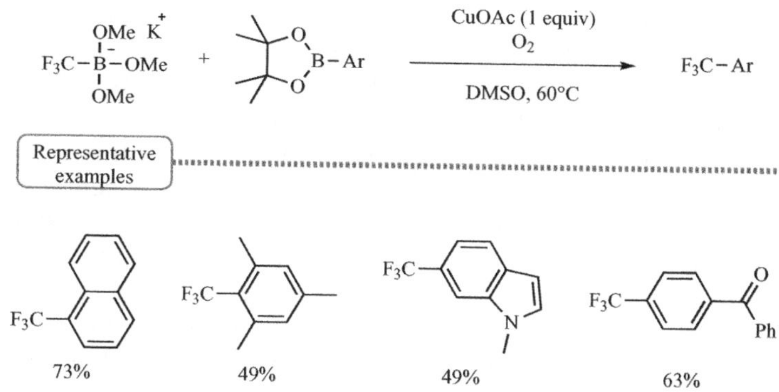

Scheme 14.9 Copper-mediated oxidative trifluoromethylation of arylboronates with potassium (trifluoromethyl)trimethoxyborate.

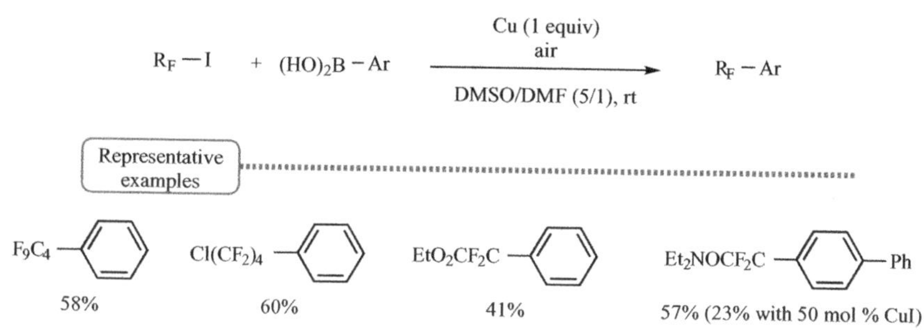

Scheme 14.10 Copper-mediated aerobic room temperature perfluoroalkylation of arylboronic acids with perfluoroalkyl iodides.

(Scheme 14.9).[29] This user-friendly protocol compares well with other techniques and, interestingly, no protodeborylation products could be detected. However, substitution of the boronate by a methoxy group originating from the trifluoromethylating agent was observed in some cases.

In early 2012, the oxidative copper-mediated functionalization of arylboronic acids was extended to the introduction of perfluoroalkyl chains using perfluoroalkyl iodides together with stoichiometric metallic copper under ligandless conditions and in the presence of air at room temperature (Scheme 14.10). The reaction is fairly efficient and tolerates a wide range of functional groups, and, most interestingly, mechanistic studies suggest that the perfluoroalkylcopper active species arise from intermediate arylcopper reagents by either an oxidative addition/reductive elimination mechanism or a nucleophilic substitution via an iodate intermediate.[30] The use of substoichiometric amounts of copper(I) iodide (0.5 equiv) was also investigated but resulted in lower yields.

In addition to boronic acids that were recently demonstrated to be remarkably efficient reaction partners in various oxidative copper-mediated trifluoromethylation processes, as shown by all results discussed in the previous paragraphs, heteroarenes possessing an acidic C–H bond were also recently shown to be readily transformed to their trifluoromethylated derivatives. The main advantage of methods relying on this strategy clearly is the absence of prefunctionalization, even if the substrates susceptible to be used in these C–H trifluoromethylation are still rather specific and limited. In this context, heteroarenes and highly electron-deficient arenes have been recently shown to be efficiently trifluoromethylated through direct C–H activation using a combination of a copper(II) catalyst and Ruppert–Prakash reagent. The oxidative trifluoromethylation of 1,3,4-oxadiazoles could be achieved using catalytic amounts of copper(II) acetate and 1,10-phenanthroline in the presence of sodium *tert*-butoxide and sodium acetate as bases and air or di-*tert*-butyl peroxide as the oxidants (Scheme 14.11, Eq. 1). Other azoles such as benzoxazoles,

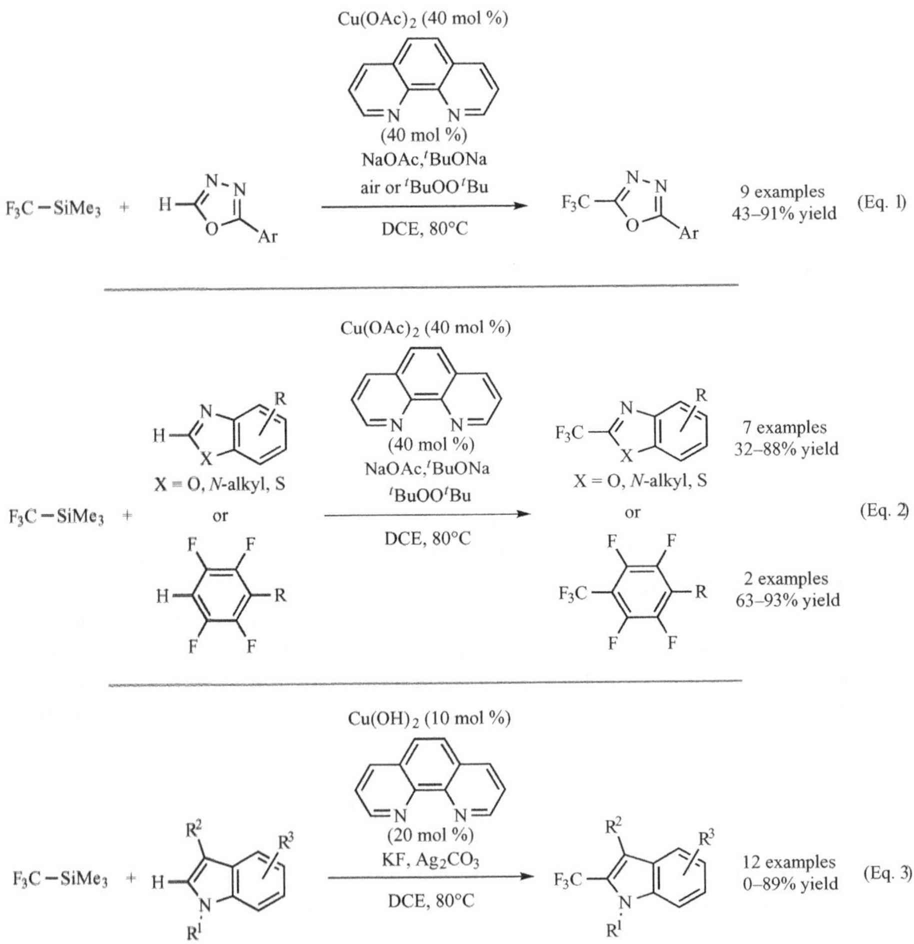

Scheme 14.11 Copper-catalyzed oxidative trifluoromethylation of activated aromatic C–H bonds with Ruppert–Prakash reagent.

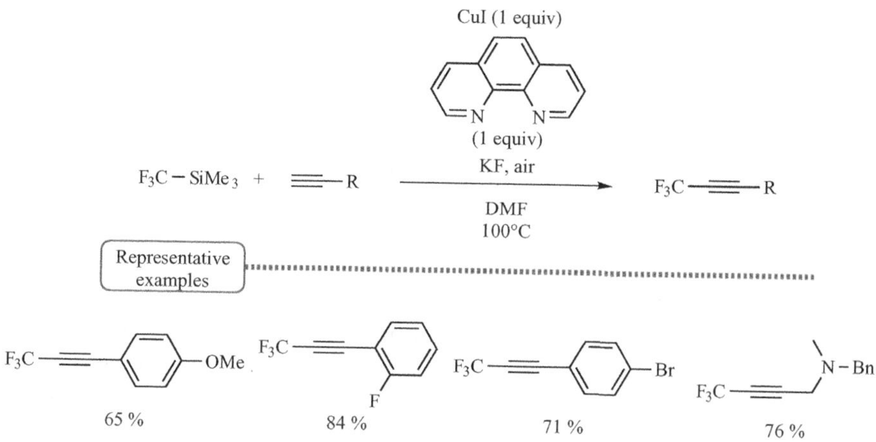

Scheme 14.12 Copper-mediated oxidative trifluoromethylation of terminal alkynes with CF_3SiMe_3.

benzimidazoles, or benzothiazole as well as polyfluoroarenes could also be trifluoromethylated provided that di-*tert*-butyl peroxide was used as the oxidant (Scheme 14.11, Eq. 2), while indoles required the use of copper(II) hydroxide as the catalyst and silver carbonate as the oxidant in the presence of potassium fluoride (Scheme 14.11, Eq. 3).[31]

Finally, copper-mediated cross-coupling processes under oxidative conditions have been recently extended to the trifluoromethylation of terminal alkynes upon reaction with stoichiometric amounts of copper iodide, 1,10-phenanthroline, excess Ruppert–Prakash reagent, and potassium fluoride in DMF at 100°C under air (Scheme 14.12).[32] This reaction provides a general, straightforward, and practically useful method to prepare trifluoromethylated alkynes, although the excess of the expensive nucleophilic trifluoromethylating agent, the requirement for its slow addition by syringe pump, and the high reaction temperature left some room for further improvements. These limitations were later on addressed, and it was first shown that catalytic amounts of both copper(I) iodide and 1,10-phenanthroline could be used and that 2 equiv of the Ruppert–Prakash reagent could be added in one portion at the beginning of the reaction, at the expense of the yields that were significantly lower in most cases.[26b] A third generation procedure was finally published at the beginning of 2012, and it was shown that the oxidative trifluoromethylation of terminal alkynes could be efficiently performed at room temperature by first mixing copper(I) chloride, potassium *tert*-butoxide, 1,10-phenanthroline, and silver carbonate (or oxygen provided that 4Å molecular sieves was added) in DMF at room temperature followed by consecutive addition of 2 equiv of TMS-CF_3 and, after 1 hour at room temperature, the terminal alkynes. Using this improved procedure, the desired trifluoromethylated aryl-substituted alkynes could be obtained in reasonable to good yields.[33] As a note, this oxidative cross coupling with terminal alkynes was also extended to the preparation of α,α-difluoropropargylphosphonates by replacing TMS-CF_3 with α-trifluoromethyl-difluoromethylphosphonates.[34]

14.4 CONCLUSION AND FUTURE PROSPECTS

Copper-catalyzed trifluoromethylation reactions and copper-mediated oxidative processes have greatly expanded the toolbox of chemists looking for efficient, robust, and reliable reactions for the selective introduction of a trifluoromethyl group or related functionalities onto an sp^2 or even sp carbon atom. The stage is now set for continued optimization and more efficient processes to increase the scope of these transformations. Future efforts will also be focused on the use of other halides or pseudohalides since aryl bromides, chlorides, and tosylates still do not participate in these trifluoromethylations. Catalyst development and loading as well as the use of other CF_3 sources will also most certainly be part of future prospects in this area as well as the extension of these procedures to the introduction of other fluorinated chains such as difluoromethyl[35] or trifluoromethylthiol[36] groups. Finally, as with other copper-catalyzed cross-coupling reactions, the use of strategies based on radical processes or single electron transfers might also prove useful for the development of copper-catalyzed trifluoromethylation under mild conditions. A recent report in this direction, published in 2013 by the Liu group who demonstrated that trifluoromethylated alkenes could be efficiently obtained by a copper-catalyzed decarboxylative cross coupling from alkenoic acids and Langlois' reagent, already demonstrated the viability of this approach.[37]

REFERENCES

[1] For reviews on the synthesis of trifluoromethyl aromatics, see: (a) Schlosser, M. *Angew. Chem. Int. Ed. Engl.* **2006**, *45*, 5432–5446. (b) Ma, J.-A.; Cahard, D. *J. Fluor. Chem.* **2007**, *128*, 975–996. (c) Lundgren, R. J.; Stradiotto, M. *Angew. Chem. Int. Ed. Engl.* **2010**, *49*, 9322–9324. (d) Roy, S.; Gregg, B. T.; Gribble, G. W.; Le, V.-D.; Roy, S. *Tetrahedron* **2011**, *44*, 2161–2195. (e) Furuya, T.; Kamlet, A. S.; Ritter, T. *Nature* **2011**, *473*, 470–477. (f) Tomashenko, O. A.; Grushin, V. V. *Chem. Rev.* **2011**, *111*, 4475–4521.

[2] For general references on fluorinated organocopper reagents, see: (a) Burton, D. J.; Yang, Z.-Y. *Tetrahedron* **1992**, *48*, 189–275. (b) McClinton, M. A.; McClinton, D. A. *Tetrahedron* **1992**, *48*, 6555–6666. (c) Davis, C. R.; Burton, D. J. *Patai Series: The Chemistry of Functional Groups. The Chemistry of Organocopper Compounds*, Vol. 24, Rappoport, Z.; Marek, I. (eds.). John Wiley & Sons Ltd: Chichester, **2009**, pp. 991–1032.

[3] (a) Dubinina, G. G.; Furutachi, H.; Vicic, D. A. *J. Am. Chem. Soc.* **2008**, *130*, 8600–8601. (b) Dubinina, G .G.; Ogikubo, J.; Vicic, D. A. *Organometallics* **2008**, *27*, 6233–6235. (c) Morimoto, H.; Tsubogo, T.; Litvinas, N. D.; Hartwig, J. F. *Angew.*

Chem. Int. Ed. Engl. **2011**, *50*, 3793–3798. (d) Tomashenko, O. A.; Escudero-Adán, E. C.; Belmonte, M. M.; Grushin, V. V. *Angew. Chem. Int. Ed. Engl.* **2011**, *50*, 7655–7659. (e) Zanardi, A.; Nokinov, M. A.; Martin, E.; Benet-Buchholz, J.; Grushin, V. V. *J. Am. Chem. Soc.* **2012**, *133*, 20901–20913. (f) Litvinas, N. D.; Fier, P. S.; Hartwig, J. F. *Angew. Chem. Int. Ed. Engl.* **2012**, *51*, 536–539. (g) Novák, P.; Lishchynskyi, A.; Grushin, V. V. *Angew. Chem. Int. Ed. Engl.* **2012**, *51*, 7767–7770.

[4] For recent examples, see: (a) Kremlev, M. M.; Tyrra, W.; Mushta, A. I.; Naumann, D.; Yagupolskii, Y. L. *J. Fluor. Chem.* **2010**, *131*, 212–216. (b) Mc Reynolds, K. A.; Lewis, R. S.; Ackerman, L. K. G.; Dubinina, G. G.; Brennessel, W. W.; Vicic, D. A. *J. Fluor. Chem.* **2010**, *131*, 1108–1112. (c) Kremlev, M. M.; Mushta, A. I.; Tyrra, W.; Yagupolskii, Y. L.; Naumann, D.; Möller, A. *J. Fluor. Chem.* **2012**, *133*, 67–71.

[5] (a) Zhang, C.-P.; Wang, Z.-L.; Chen, Q.-Y.; Zhang, C.-T.; Gu, Y.-C.; Xiao, J.-C. *Angew. Chem. Int. Ed. Engl.* **2011**, *50*, 1896–1900. (b) Zhang, C.-P.; Cai, J.; Zhou, C.-B.; Wang, X.-P.; Zheng, X.; Gu, Y.-C.; Xiao, J.-C. *Chem. Commun.* **2011**, *47*, 9516–9518. (c) Ye, Y.; Künzi, S. A.; Sanford, M. S. *Org. Lett.* **2012**, *14*, 4979–4981.

[6] For a review on copper-mediated formation of trifluoromethylated arenes that appeared after this chapter was written, see: Liu, T.; Shen, Q. *Eur. J. Org. Chem.* **2012**, 6679–6687.

[7] For examples, see: (a) Grushin, V. V.; Marshall, W. J. *J. Am. Chem. Soc.* **2006**, *128*, 12644–12645. (b) Grushin, V. V. *Acc. Chem. Res.* **2010**, *43*, 160–171. (c) Dubinina, G. G.; Brennessel, W. W.; Miller, J. L.; Vicic, D. A. *Organometallics* **2008**, *27*, 3933–3938. (d) Ball, N. D.; Kampf, J. W.; Sanford, M. S. *J. Am. Chem. Soc.* **2010**, *132*, 2878–2879. (e) Wang, X.; Truesdale, L.; Yu, J.-Q. *J. Am. Chem. Soc.* **2010**, *132*, 3648–3649. (f) Cho, E. J.; Senecal, T. D.; Kinzel, T.; Zhang, Y.; Watson, D. A.; Buchwald, S. L. *Science* **2010**, *328*, 1679–1681. (g) Mu, X.; Chen, S.; Zhen, X.; Liu, G. *Chem. Eur. J.* **2011**, *17*, 6039–6042. (h) Loy, R. N.; Sanford, M. S. *Org. Lett.* **2011**, *13*, 2548–2551. (i) Ye, Y.; Lee, S. H.; Sanford, M. S. *Org. Lett.* **2011**, *13*, 5464–5467.

[8] Chen, Q.-Y.; Wu, S.-W. *J. Chem. Soc., Chem. Commun.* **1989**, 705–706.

[9] Oishi, M.; Kondo, H.; Amii, H. *Chem. Commun.* **2009**, 1909–1911.

[10] Fujikawan, K.; Fujioka, Y.; Kobayashi, A.; Amii, H. *Org. Lett.* **2011**, *13*, 5560–5563.

[11] Weng, Z.; Lee, R.; Jia, W.; Yuan, Y.; Wang, W.; Feng, X.; Huang, K.-W. *Organometallics* **2011**, *30*, 3229–3232.

[12] Hafner, A.; Bräse, S. *Adv. Synth. Catal.* **2011**, *353*, 3044–3048.

[13] Knauber, T.; Arikan, F.; Röschenthaler, G.-V.; Gooßen, L. J. *Chem. Eur. J.* **2011**, *17*, 2689–2697.

[14] Kondo, H.; Oishi, M.; Fujikawa, K.; Amii, H. *Adv. Synth. Catal.* **2011**, *353*, 1247–1252.

[15] Li, Y.; Chen, T.; Wang, H.; Zhang, R.; Jin, K.; Wang, X.; Duan, C. *Synlett* **2011**, 1713–1716.

[16] Xu, J.; Luo, D.-F.; Liu, Z.-J.; Gong, T.-J.; Fu, Y.; Liu, L. *Chem. Commun.* **2011**, *47*, 4300–4302.

[17] Liu, T.; Shen, Q. *Org. Lett.* **2011**, *13*, 2342–2345.

[18] Weng, Z.; Li, H.; He, W.; Yao, L.-F.; Tan, J.; Chen, J.; Yuan, Y.; Huang, K.-W. *Tetrahedron* **2012**, *68*, 2527–2531.

[19] He, Z.; Luo, T.; Hu, M.; Cao, Y.; Hu, J. *Angew. Chem. Int. Ed. Engl.* **2012**, *51*, 3944–3947.

[20] Liu, T.; Shao, X.; Wu, Y.; Shen, Q. *Angew. Chem. Int. Ed. Engl.* **2012**, *51*, 540–543.

[21] Huang, Y.; Fang, X.; Lin, X.; Li, H.; He, W.; Huang, K.-W.; Yuan, Y.; Weng, Z. *Tetrahedron* **2012**, *68*, 9949–9953.

[22] Zheng, H.; Huang, Y.; Wang, Z.; Li, H.; Huang, K.-W.; Yuan, Y.; Wenig, Z. *Tetrahedron Lett.* **2012**, *53*, 6646–6649.

[23] Ye, T.; Sanford, M. *J. Am. Chem. Soc.* **2012**, *134*, 9034–9037.

[24] Popov, I.; Lindeman, S.; Daugulis, O. *J. Am. Chem. Soc.* **2011**, *133*, 9286–9289.

[25] Shimizu, R.; Egami, H.; Nagi, T.; Chae, J.; Hamashima, Y.; Sodeoka, M. *Tetrahedron Lett.* **2010**, *51*, 5947–5949.

[26] (a) Chu, L.; Qing, F.-L. *Org. Lett.* **2010**, *12*, 5060–5063. (b) Jiang, X.; Chu, L.; Qing, F.-L. *J. Org. Chem.* **2012**, *77*, 1251–1257.

[27] Senecal, T. D.; Parsons, A. T.; Buchwald, S. L. *J. Org. Chem.* **2011**, *76*, 1174–1176.

[28] Xu, J.; Xiao, B.; Xie, C.-Q.; Luo, D.-F.; Liu, L.; Fu, Y. *Angew. Chem. Int. Ed.* **2012**, *51*, 12551–12554.

[29] Khan, B. A.; Buba, A. E.; Gooßen, L. J. *Chem. Eur. J.* **2012**, *18*, 1577–1581.

[30] Qi, Q.; Shen, Q.; Lu, L. *J. Am. Chem. Soc.* **2012**, *134*, 6548–6551.

[31] Chu, L.; Qing, F.-L. *J. Am. Chem. Soc.* **2012**, *134*, 1298–1304.

[32] Chu, L.; Qing, F.-L. *J. Am. Chem. Soc.* **2010**, *132*, 7262–7263.

[33] Zhang, K.; Qiu, X.-L.; Huang, Y.; Qing, F.-L. *Eur. J. Org. Chem.* **2012**, 58–61.

[34] Jiang, X.; Chu, L.; Qing, F.-L. *Org. Lett.* **2012**, *14*, 2870–2873.

[35] For a recent and key development in the copper-mediated (stoichiometric) difluoromethylation of aryl and vinyl iodides, see: Fier, P. S.; Hartwig, J. F. *J. Am. Chem. Soc.* **2012**, *134*, 5524–5527.

[36] For recent and key developments in the copper-catalyzed oxidative trifluoromethylthiolation of arylboronic acids, see: (a) Chen, C.; Xie, Y.; Chu, L.; Wang, R.-W.; Zhang, X.; Qing, F.-L. *Angew. Chem. Int. Ed. Engl.* **2012**, *51*, 2492–2495. (b) Zhang, C.-P.; Vici, D. A. *Chem. Asian J.* **2012**, *7*, 1756–1758.

[37] Li, Z.; Cui, Z.; Liu, Z.-Q. *Org. Lett.* **2013**, *15*, 406–409.

PART III

APPLICATIONS OF COPPER-CATALYZED CROSS-COUPLING REACTIONS: HETEROCYCLES, NATURAL PRODUCTS, PROCESS, AND SUSTAINABLE CHEMISTRY

15

COPPER-MEDIATED CYCLIZATION REACTIONS: NEW ENTRIES TO HETEROCYCLES

Daoshan Yang and Hua Fu

15.1 INTRODUCTION

Heterocycles are ubiquitous in natural products and synthetic molecules. They display a high degree of structural diversity and possess numerous biological and pharmacological activities, making them attractive candidates for screening in drug discovery programs. Therefore, the development of novel, efficient, and practical methods for their synthesis has still stimulated an impressive number of research groups. Nowadays, a number of well-established methods are available to elaborate a wide range of heterocycles, and among those methods, the ones that rely on the use of transition-metal catalysis are certainly the most attractive since they allow direct and straightforward formation of various heterocycles from readily available starting materials under mild conditions.[1] In this field, copper-mediated reactions are especially attractive due to the low cost of copper-based catalytic systems, their availability, and their remarkable efficiency for the formation of C–heteroatom bonds, as shown in Chapters 1–5. The introduction of chelating ligands for copper salts greatly improved the reaction conditions compared with the original publications by Ullmann and Goldberg, which is perfectly illustrated by the numerous reports published during the last decade.[2–4] In this chapter, the development of new and efficient entries to heterocycles based on copper-catalyzed cyclization

Copper-Mediated Cross-Coupling Reactions, First Edition. Edited by Gwilherm Evano and Nicolas Blanchard.

reactions will be detailed. These reactions have been classified according to the nature of the bond formed during the process (C–N, C–O, and C–C). The use of copper-catalyzed reactions in tandem sequences, or in association with a second copper-mediated cross-coupling step, leading to complex heterocyclic structures will be also presented. The role of these procedures in the emergence of new processes in heterocyclic chemistry will be discussed throughout the entire chapter. Since this topic is especially wide and extensively studied, this chapter does not intend to be completely exhaustive. Most representative examples illustrative of new strategies in heterocyclic chemistry based on the use of copper catalysis have been compiled in this chapter and will be overviewed in the following pages. As a note, additional copper-mediated synthesis of heterocycles based on an intramolecular aromatic C–H bond heterofunctionalization can be found in Chapter 4, Section 4.5.3.

15.2 CYCLIZATION BY C–N BOND FORMATION

N-Heterocycles are at the core structure of a variety of natural products and biologically active molecules. Most drugs actually incorporate at least one nitrogen heterocycle within their structures, and they are known to dramatically increase their solubility and to allow the formation of salts, two characteristics that have a major influence on their absorption and bioavailability.[5] Consequently, N-heterocycles are often considered as privileged targets by pharmaceutical industries and medicinal chemists. Various strategies for their synthesis have been recently designed using a copper-catalyzed intramolecular C–N bond formation as the key step. The main advantage with these intramolecular amination reactions, compared with their intermolecular counterparts, is that they are usually highly favored, except in certain cases where steric hindrance or ring strain can be problematic. For example, a five-membered ring formation via intramolecular amidation of an aryl bromide in the presence of copper(I) iodide and *N,N'*-dimethylethylenediamine could be performed at room temperature (Scheme 15.1, Eq. 1) in quantitative yield, and even an aryl chloride, which is a poor partner in intermolecular processes, can provide an 88% yield of the corresponding cyclized product, although heating at 100°C for 23 hours was necessary (Scheme 15.1, Eq. 2).[6] This reaction was also shown to proceed smoothly with unprotected primary amines using diethylsalicylamide,[7] proline,[8] or 2-isobutyrylcyclohexanone[9] as ligands for copper iodide. These results clearly indicate that intramolecular amidation reactions are, not surprisingly, more facile than the corresponding intermolecular versions. These intramolecular arylation reactions of amines, amides, or carbamates and other N-nucleophiles have found many synthetic applications that will be discussed in this section. As an important note, ligandless conditions for the formation of indolines have also been developed by Fukuyama and coworkers, who reported that a unique combination of copper iodide and cesium acetate in DMSO could efficiently promote the intramolecular amination of aryl halides under relatively mild conditions (Scheme 15.1, Eq. 3). The reaction proceeds at room temperature

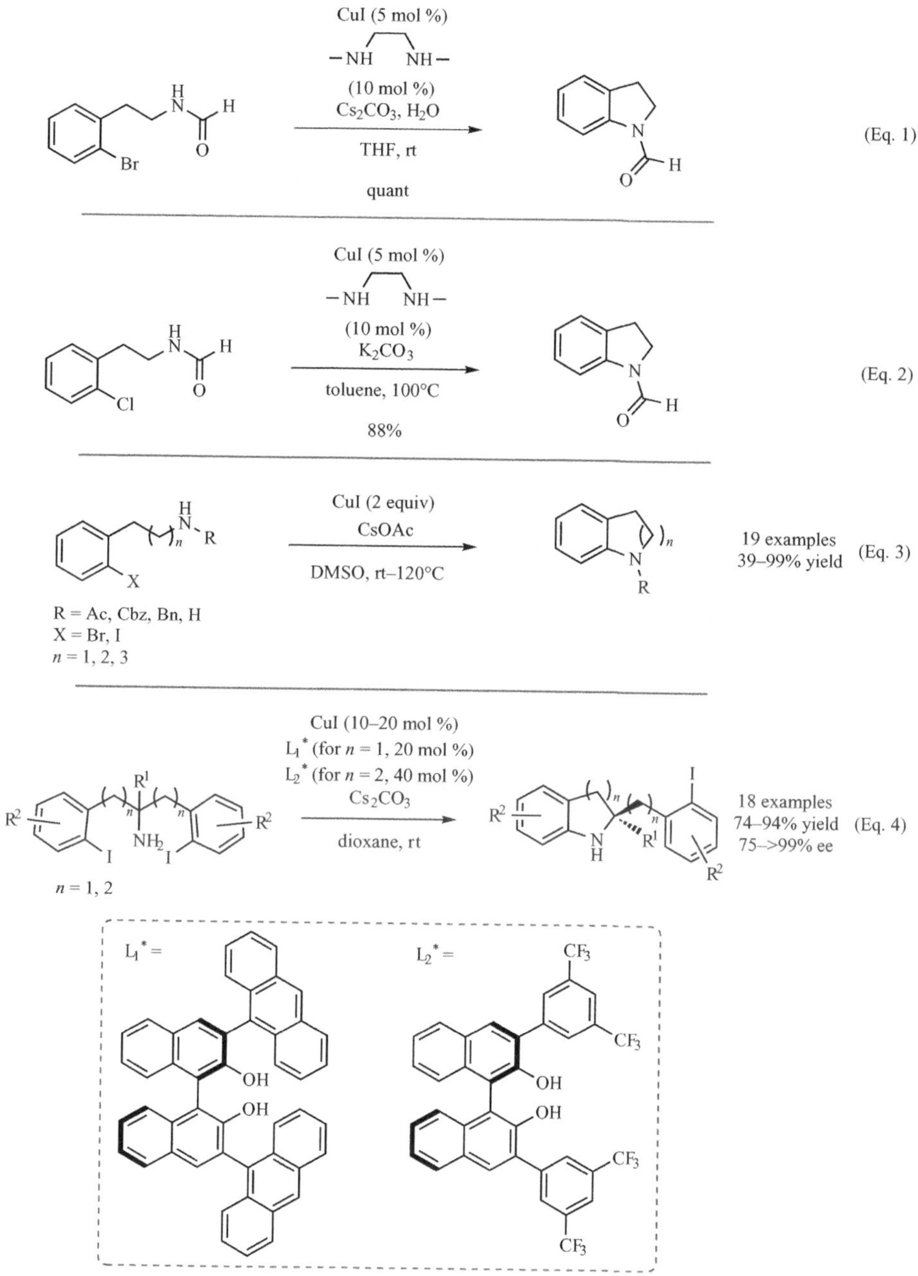

Scheme 15.1 Intramolecular aryl amination/amidation for the synthesis of indolines and higher related ring systems.

with primary or *N*-benzyl amines but requires high temperatures with other amine derivatives, and has been applied to the formation of five-, six-, and seven-membered rings.[10] Remarkably, halogens at the *meta*-position were retained, providing a definitive advantage over palladium-catalyzed systems.

Even more remarkably, this kind of cyclization can now also be performed in an enantioselective way as demonstrated in 2012 by Yu and coworkers, who devised an especially elegant, highly enantioselective, copper-catalyzed intramolecular Ullmann C–N coupling reaction. The asymmetric desymmetrization of 1,3-bis(2-iodoaryl)propan-2-amines catalyzed by a combination of copper(I) iodide and (*R*)-BINOL-derived ligands led to the enantioselective formation of indolines and tetrahydroquinolines in high yields and excellent enantiomeric excesses (Scheme 15.1, Eq. 4).[11]

15.2.1 Copper-Catalyzed Synthesis of Pyrroloindole Derivatives

The pyrroloindole skeleton is a key structural motif that appears in the core structure of an impressive number of biologically active alkaloids such as the potent vasodilator amauromine, the insecticide okaramine C, the anticancer agent phakellistatin, and other related polycyclic natural products. It also represents an especially interesting platform for drug discovery and possesses various sites for introducing molecular diversity. While the classical approach toward this polycyclic ring system mainly relies on Witkop's oxidative cycloaromatization of tryptophan derivatives, this method usually performs well with simple systems, but its use with complex starting materials can be problematic. Aromatization, protecting group migration, and oxidation can dramatically lower yields. Trying to address these limitations and capitalizing on intramolecular C–N bond formation reactions shown in Scheme 15.1, Evano and coworkers reported a straightforward entry to these heterocycles using a copper-mediated cyclization from readily available iodo-tryptophan (or tryptamine) derivatives (Scheme 15.2).[12] A combination of copper(I) iodide and *N*,*N*′-dimethylethylenediamine or (±)-*trans*-cyclohexane-1,2-diamine together with potassium phosphate or potassium carbonate in toluene or *N*,*N*-dimethylformamide (DMF) was found to best promote the cyclization, which could not be catalyzed by other metals such as palladium or iron. Interestingly, the optical purity of the starting material is fully retained during the process, and the use of 2-iodo-tryptophan-derived diketopiperazines provided an efficient entry to tetra-, penta-, and even heptacyclic nitrogen heterocycles. This strategy was later on extended to the preparation of pyrido-indoles and was used as a key step for the synthesis of chaetominine (see Chapter 16, Scheme 16.53, for details).

15.2.2 Copper-Catalyzed Synthesis of Benzimidazole Derivatives

Benzimidazoles and their derivatives play an important role in biological and medicinal chemistry. Selected representative molecules featuring this skeleton

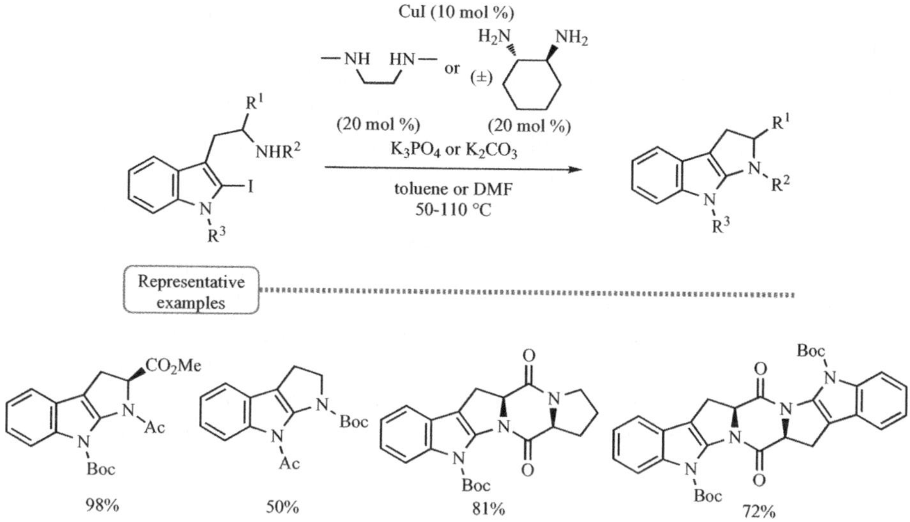

Scheme 15.2 Copper-catalyzed synthesis of pyrroloindoles.

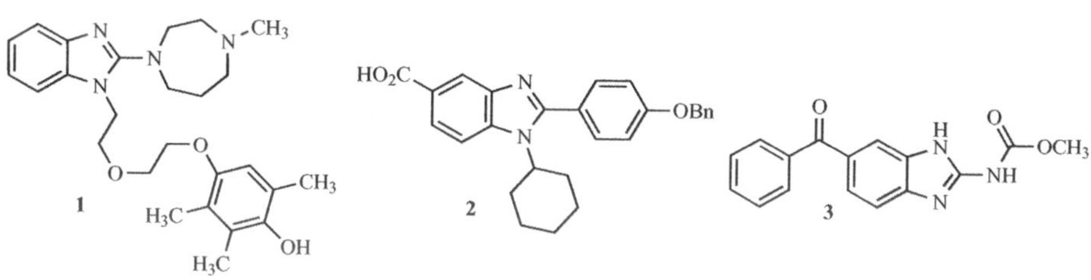

Figure 15.1 Structures of biologically and pharmacologically significant representative benzimidazole derivatives.

are shown in Figure 15.1 and give an idea of the potential of this class of compounds. For example, the 5-lipoxygenase inhibitor **1** is an antiallergic agent; the hepatitis C virus (HCV) NS5B polymerase inhibitor **2** is used as a lead compound; and mebendazole **3** is a potent antibiotic. The construction of the benzimidazole motif has received considerable attention and has been recently achieved using various strategies based on the use of copper-mediated transformations.

One of the classical methods used to synthesize benzimidazole derivatives relies on the condensation of 1,2-diaminoarenes with carboxylic acids or aldehydes,[13] and substituted *o*-nitroanilines may also be used as starting materials in place of 1,2-diaminoarenes under reductive conditions.[14] However, these methods often suffer from limited scope, harsh conditions, complex chemical processes, and/or low yields. Major improvements in this field were made using transition-metal catalysis, which allows the synthesis of benzimidazole derivatives under milder conditions.[15] Considering the availability and low toxicity of copper catalysts and their ligands,[2–4,16] Ma and coworkers reported a highly efficient and practical approach to 1,2-disubstituted benzimidazoles using a copper-catalyzed tandem aryl amination/condensation sequence (Scheme 15.3).[17] First, the coupling of 2-haloacetanilides with amines in the presence of copper(I) iodide (10 mol %), L-proline (20 mol %), and potassium carbonate gave the corresponding aryl amination product. Then, an increase of both the reaction time and temperature or the addition of acetic acid to the reaction mixture at 50°C promoted an intramolecular condensation between the newly formed amine and the adjacent amide, providing the corresponding benzimidazole (Scheme 15.3). Importantly, the first step proceeds under especially mild conditions, demonstrating the dramatic *ortho*-effect of the acetamide on the copper-catalyzed amination. The scope

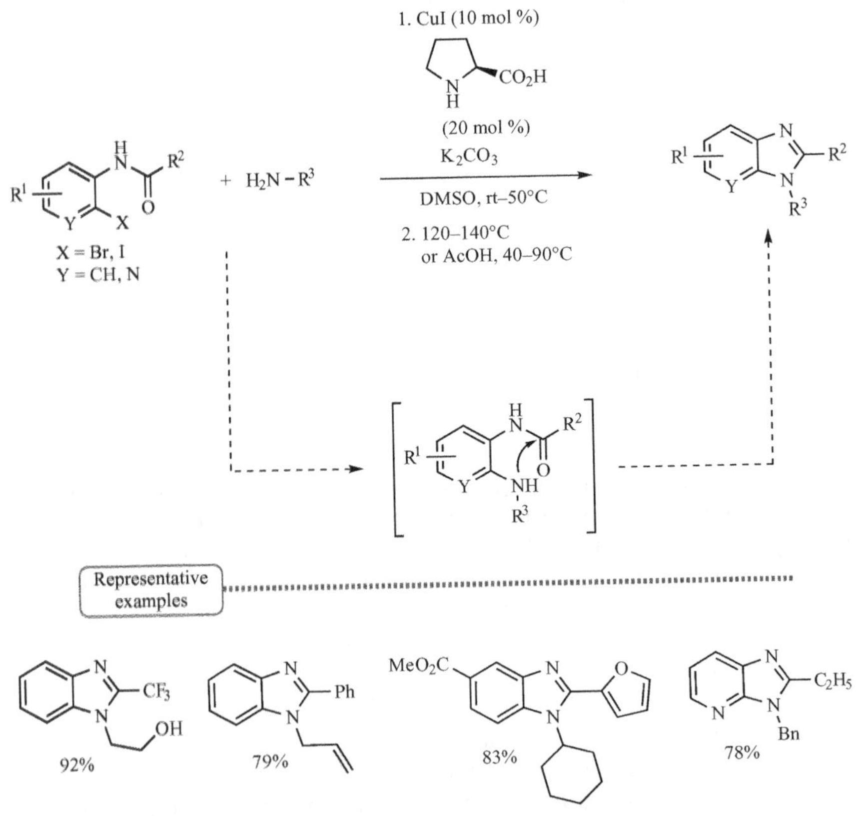

Scheme 15.3 Copper-catalyzed synthesis of 1,2-disubstituted benzimidazoles.

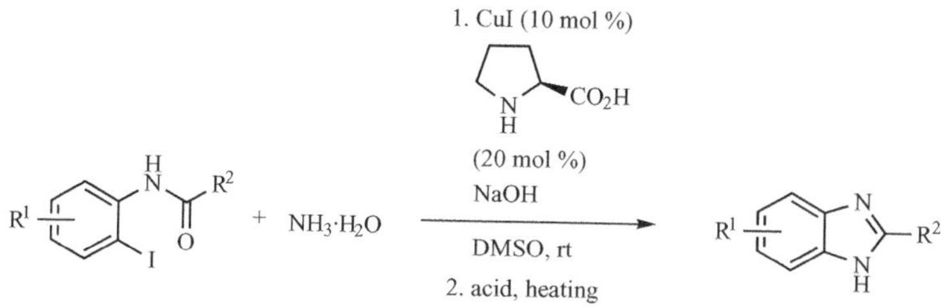

Scheme 15.4 Copper-catalyzed synthesis of benzimidazoles possessing a free N–H.

of the reaction is fairly broad (the corresponding benzimidazoles being obtained in good-to-excellent yields) and pyridine derivatives can also be used as starting materials.

While this method was originally restricted to the synthesis of *N*-substituted benzimidazoles, in 2009, Ma and coworkers extended their method to the preparation of benzimidazoles possessing a free N–H, compounds that could be easily obtained with similar reaction conditions but using ammonia as the nucleophile for the copper-catalyzed amination step (Scheme 15.4).[18]

In 2008, Fu and coworkers described an efficient alternative to this strategy, allowing for the preparation of 2-substituted benzimidazoles via copper-catalyzed cascade reactions starting from *o*-haloacetanilide derivatives and amidine hydrochlorides (Scheme 15.5).[19] Inexpensive copper(I) bromide was used as the catalyst in the presence of cesium carbonate, and no ligand was required when the reaction was performed in DMSO. The proposed mechanism would first involve an Ullmann-type coupling between the *o*-haloacetanilide and the amidine leading to the N-arylated product. Further hydrolysis of the amide group under basic conditions followed by ring closure and elimination of a molecule of ammonia would provide the desired 2-substituted-benzimidazole. The yields are usually fairly good, and a notable feature of this reaction is that selective activation of the halide *ortho* to the acetamide moiety in the presence of an aromatic bromide can be achieved, yielding the corresponding brominated benzimidazole, and therefore allowing for further functionalization.

The synthesis of benzimidazoles can also be achieved in a fully intramolecular way, starting from *N*-(2-bromoaryl)-amidines, as reported in 2009 by Punniyamurthy and coworkers. The catalytic system was rather unusual and was based on copper(II) oxide nanoparticles. In combination with potassium hydroxide in DMSO at 110°C, the starting *N*-(2-bromoaryl)-amidines were readily cyclized to the corresponding substituted benzimidazoles (Scheme 15.6).[20] *N*-(2-Bromoaryl)-guanidines were also found to be suitable reaction partners, allowing for the formation of 2-aminobenzimidazoles with similar efficiency. It is noteworthy that no ligand was required for this reaction, and that the catalyst can be recovered and recycled without loss of activity.

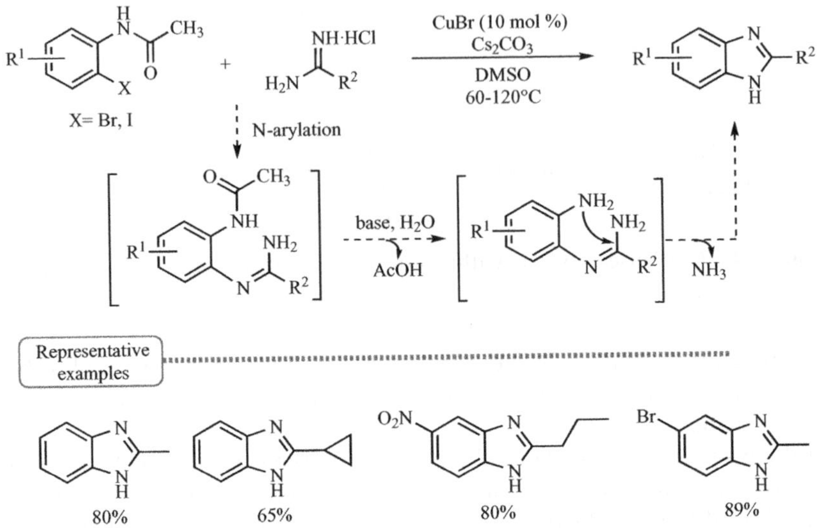

Scheme 15.5 Copper-catalyzed synthesis of 2-substituted-benzimidazoles via a cascade from *o*-halo-acetanilides and amidine hydrochlorides.

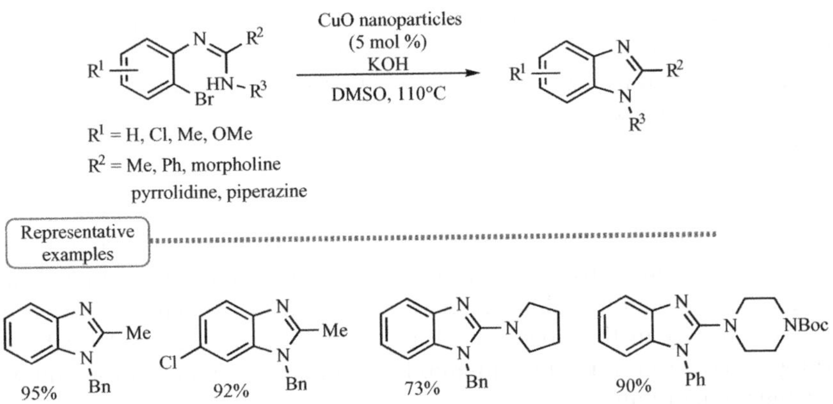

Scheme 15.6 Synthesis of substituted benzimidazoles by intramolecular copper-mediated C–N bond formation with CuO nanoparticles.

A more classical system that is equally efficient was reported early on by the Batey group for the synthesis of 2-aminobenzimidazoles from similar substrates, using a combination of copper(I) iodide, 1,10-phenanthroline, and cesium carbonate in 1,2-dimethoxyethane (DME). Using these optimized conditions, the cyclization between the aryl halide and guanidine moieties proceeded smoothly, affording the bicyclic products in good yields (Scheme 15.7).[21] Interestingly, the copper-based catalytic system was found to be

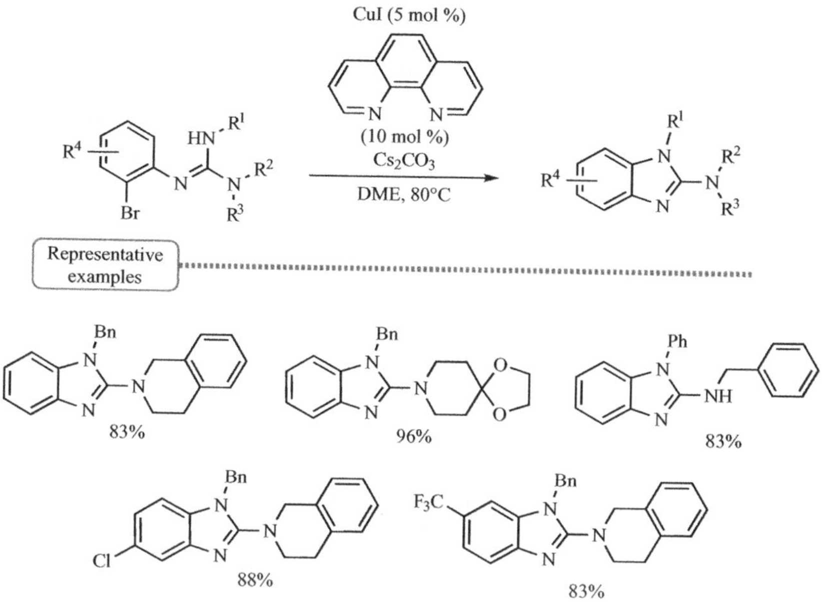

Scheme 15.7 Copper-catalyzed intramolecular aryl guanidinylation to 2-aminobenzimidazoles.

generally superior to the use of palladium catalysts, especially when less substituted starting materials were used ($R^3 = H$), where regioselective cyclizations could be achieved in high yields under the CuI/1,10-phenanthroline-catalyzed conditions, whereas the use of palladium tetrakis(triphenylphosphine) resulted in the formation of regioisomeric products. An additional aryl bromide on the starting material was not tolerated and was partially reduced during the course of the cyclization.

In 2009, the Bao group described a close process where the *N*-(2-bromoaryl)-guanidines were generated *in situ* by the addition of an amine to a carbodiimide. Proline or 1,10-phenanthroline could be used as ligands together with copper(I) iodide, and this efficient tandem process allowed for the synthesis of various 2-aminobenzimidazole derivatives with excellent efficiency (Scheme 15.8).[22]

Ila and coworkers reported an extension of this work for the synthesis of pyrazolo[1,5-*a*]benzimidazoles from the corresponding 5-(2-bromoanilino)-pyrazole precursors (prepared by refluxing the appropriate *N*,*S*-acetals with hydrazine hydrate in *tert*-butanol in the presence of a catalytic amount of acetic acid) via a copper-catalyzed intramolecular C–N bond formation (Scheme 15.9).[23] Copper(I) iodide and L-proline were used as the copper source and the ligand, respectively, but the reaction required the use of a stronger base (sodium hydride), presumably because of the lower nucleophilicity of the pyrazole, in DMF at 90°C.

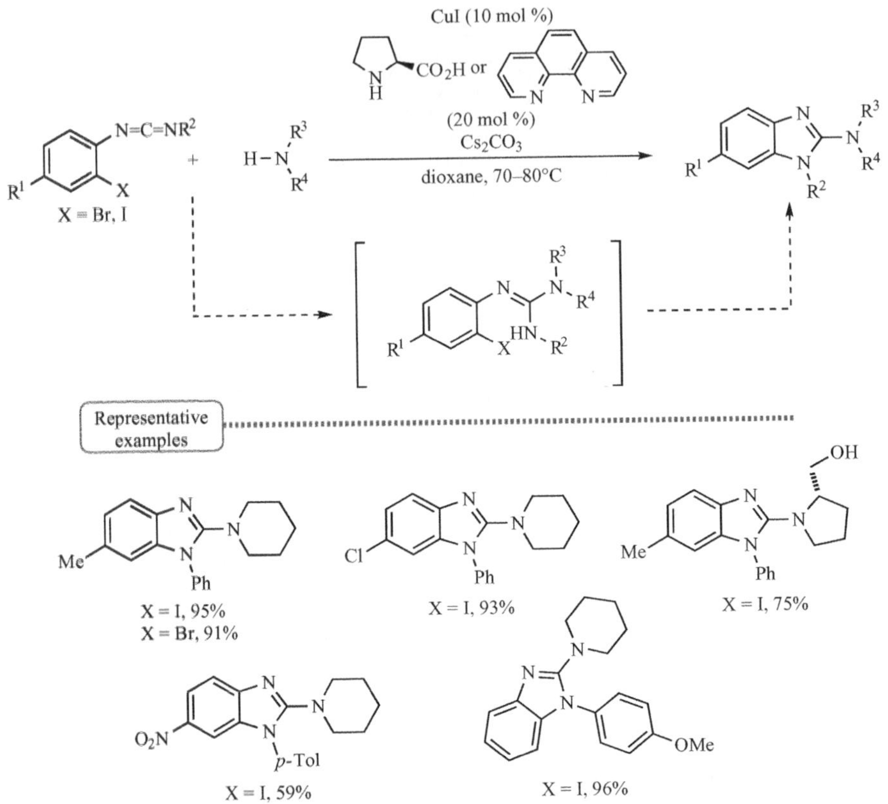

Scheme 15.8 Copper-catalyzed tandem synthesis of 2-aminobenzimidazoles from *N*-(2-haloaryl)-carbodiimides and amines.

In 2008, Brasche and Buchwald reported a major breakthrough that considerably simplifies the synthesis of benzimidazoles since it allows starting from simple, readily available starting materials (*N*-aryl-amidines). They were indeed able to optimize an efficient copper-catalyzed C–H functionalization/C–N bond-forming approach to 2-arylbenzimidazoles via an intramolecular cyclization from *N*-aryl-benzamidines using copper(II) acetate as the catalyst and acetic acid as an additive under oxygen, the latter being required for the reductive elimination step (Scheme 15.10).[24a] The reaction was usually high yielding in DMSO at 100°C and a variety of functional groups were tolerated, but only aryl and *tert*-butyl groups can be incorporated at the 2-position of the final benzimidazole. An extension of this procedure was reported in 2012 by Neuville, Zhu, and coworkers, who demonstrated that benzimidazoles could be obtained from amidines and arylboronic acids upon treatment with copper(II) acetate, sodium pivalate, and air in DMF at 50°C, yielding an intermediate *N*-aryl-amidine that could undergo the intramolecular C–H amination when the temperature of the reaction mixture was raised to 120°C.[24b] Following

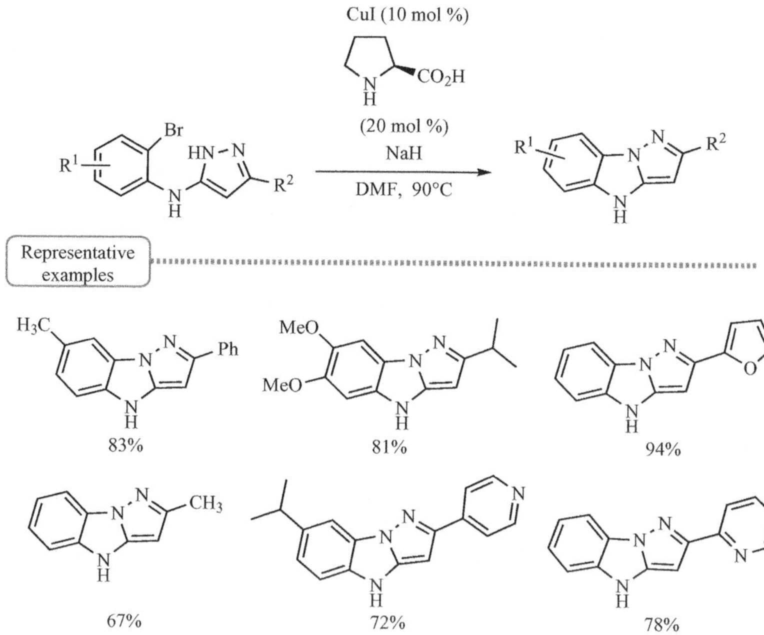

Scheme 15.9 Copper-catalyzed synthesis of pyrazolo[1,5-*a*]benzimidazoles.

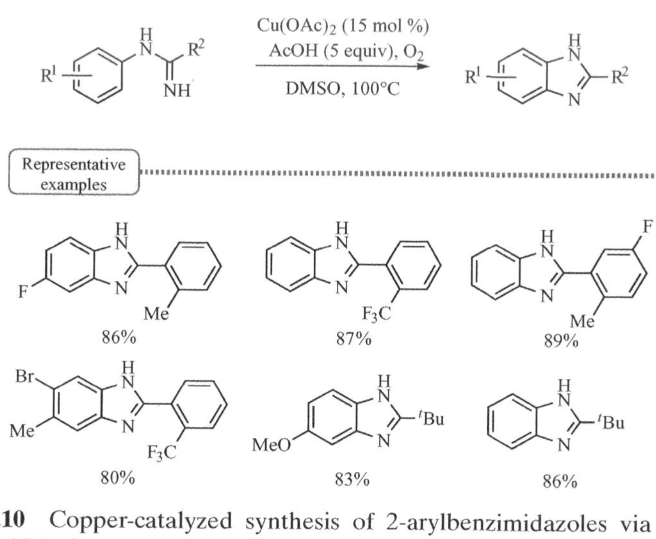

Scheme 15.10 Copper-catalyzed synthesis of 2-arylbenzimidazoles via C–H functionalization/C–N bond formation.

Buchwald's remarkable report, related strategies for the synthesis of nitrogen-containing heterocycles based on an intramolecular aromatic C–H bond amination have appeared: They include, for example, the preparation of pyrazoles and indazoles from the corresponding vinylic or aromatic hydrazones,[24c] and the synthesis of acridones from 2-aminobenzophenones.[24d]

As evidenced by these examples, a variety of routes is now available for the synthesis of an impressive number of benzimidazoles, and copper-catalyzed processes are clearly among the most efficient ones for the elaboration of this useful scaffold. Other processes based on double copper-catalyzed cross coupling have been reported and will be discussed in Section 15.5.2.

15.2.3 Copper-Catalyzed Synthesis of 1,3-Dihydrobenzimidazol-2-One Derivatives

1,3-Dihydrobenzimidazol-2-one derivatives constituted another class, rather close to benzimidazoles, of useful N-heterocycles that are found in many biologically active compounds. For example, they have been used as HIV-1 reverse transcriptase (RT) non-nucleoside inhibitors, calcitonin gene-related peptide (CGRP) receptor antagonists, vasopressin 1α receptor antagonists, or p38 mitogen-activated protein (MAP) kinase inhibitors. The development of general methods for the preparation of these compounds has thus drawn much attention, especially in the areas of drug discovery and medicinal chemistry. As a gross simplification, the main classical method used to synthesize *N*-substituted 1,3-dihydrobenzimidazol-2-ones involves nucleophilic aromatic substitution from 2-fluoronitrobenzenes with amines followed by reduction and cyclization with carbonyldiimidazole.[25] However, this method is hampered by the limited number of commercially available 2-fluoronitrobenzenes and the need for a stepwise sequence. Moreover, further elaboration involving alkylation, or arylation, of the nitrogen atoms involves protection/deprotection steps that render the sequence even lengthier.[26] During the last decade, convenient copper-catalyzed procedures have been developed to synthesize *N*-substituted 1,3-dihydrobenzimidazole-2-ones in a much more convergent and straightforward manner.

In continuation of their work on the synthesis of benzimidazoles described earlier, Ma and coworkers described in 2007 an efficient copper(I) iodide/amino acid-catalyzed assembly of *N*-substituted 1,3-dihydrobenzimidazol-2-ones from the cross coupling of methyl *N*-(2-haloaryl)carbamates with amines and subsequent condensation (Scheme 15.11).[27] The *trans*-4-hydroxy-L-proline ligand afforded the best results with methyl *N*-(2-bromoaryl)carbamates when the N-arylation reaction was performed at 70°C (Scheme 15.11, Eq. 1) and comparison with the related reaction performed from *N*-(2-haloaryl)carboxamides shown in Scheme 15.3, indicates that the carbamate has a weaker *ortho*-effect than the carboxamide.[17,28] Slight modifications of the reaction conditions allow one to perform the reaction from methyl *N*-(2-iodoaryl)

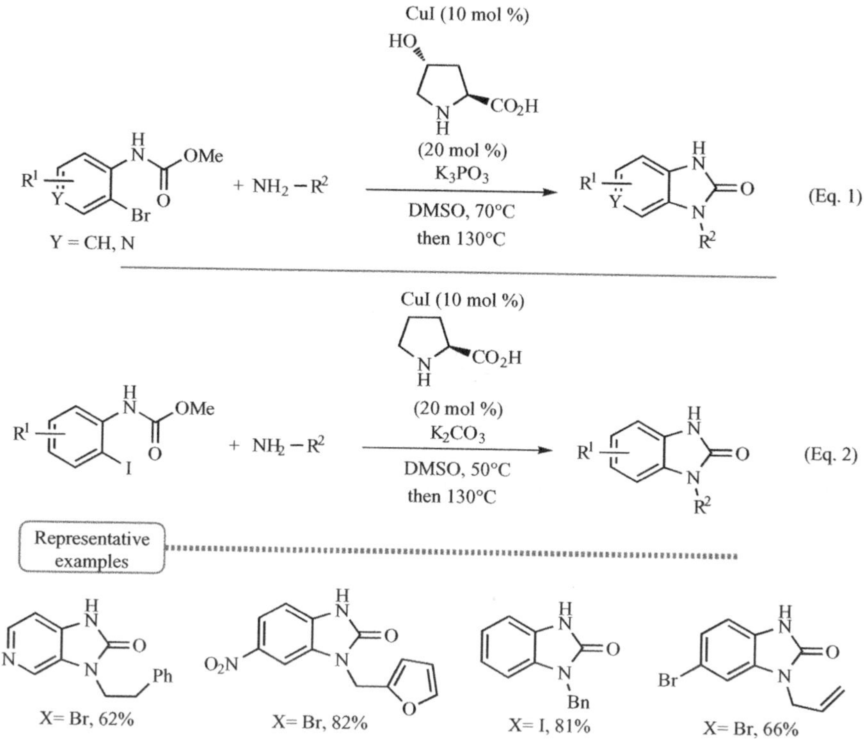

Scheme 15.11 Copper-catalyzed synthesis of *N*-substituted-1,3-dihydrobenzimidazol-2-ones.

carbamates by using L-proline as the ligand and performing the reaction at 50°C (Scheme 15.11, Eq. 2). Interestingly, various functional groups including vinyl, nitro, carboxylate, amide, ester, ketone, and silyl ethers were tolerated, and the benzene ring can be replaced by a pyridine, yielding to the formation of imidazo[4,5-*c*]pyridin-2-ones with similar efficiency.

As with benzimidazoles, other strategies based on copper-mediated cyclization reactions appeared later, and in 2008, Liu and coworkers reported an intramolecular strategy. They indeed designed an efficient method for synthesizing *N*-substituted-1,3-dihydrobenzimidazol-2-ones starting from *N*′-substituted-*N*-(2-haloaryl)ureas, which can be easily prepared by the reaction of 2-haloanilines either with triphosgene and amines or with isocyanates. From these precursors, the formation of the dihydrobenzimidazol-2-one core was promoted by copper(I) iodide in the presence of 1,8-diazabicycloundec-7-ene (DBU), in DMSO and under microwave irradiation (Scheme 15.12).[29] A variety of functional groups was well tolerated under these conditions, including aryl, heteroaryl, and alkyl groups on the urea.

A really close approach to *N*-substituted-1,3-dihydrobenzimidazol-2-ones was reported almost simultaneously by San Martin and Domínguez and is

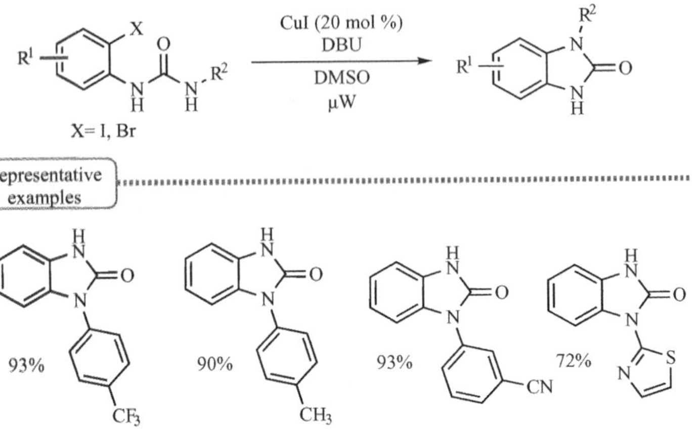

Scheme 15.12 Copper-catalyzed intramolecular cyclization to *N*-substituted-1,3-dihydrobenzimidazol-2-ones under microwave heating.

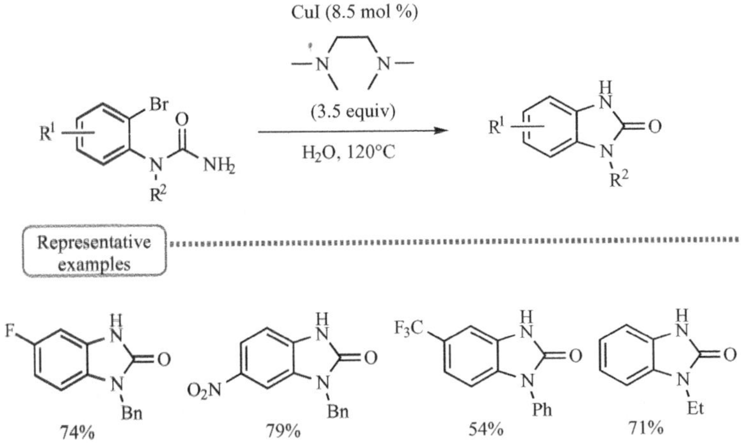

Scheme 15.13 Copper-catalyzed intramolecular cyclization to *N*-substituted-1,3-dihydrobenzimidazol-2-ones in water.

shown in Scheme 15.13.[30] The reaction of *N*-(2-bromoaryl)ureas with copper(I) iodide in the presence of excess *N,N,N′,N′*-tetramethylethylenediamine (TMEDA) was carried out in water to afford the desired benzimidazolones in good-to-excellent yields. As a note, this method showed similar efficiency with a range of copper salts such as CuCl, CuBr, $Cu(OAc)_2{\cdot}H_2O$, $Cu(OTf)_2$, and $Cu(ClO_4)_2{\cdot}6H_2O$, which is rather unusual for copper-based catalytic systems. Besides the reaction conditions, the major difference between this approach and the previous one is the substitution pattern of the starting material, since the *N*-substituent can be either on the nitrogen atom involved in the copper-catalyzed cyclization or on the other one, which completely changes the starting

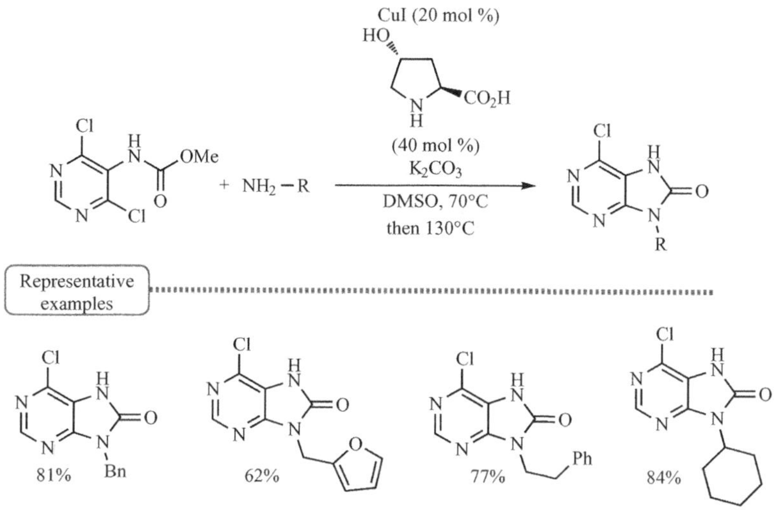

Scheme 15.14 Copper-catalyzed synthesis of 6,9-disubstituted purin-8-ones.

materials from which these precursors can be prepared. Based on their availability, one route can be preferable to the other.

6,9-Disubstituted-purin-8-ones, which are structurally related to 1,3-dihydrobenzimidazol-2-ones, are also used as therapeutic agents, and their synthesis has been the topic of several studies.[31,32] However, the methods available for their preparation are rather unattractive because of long reaction times and low yields. In 2010, Sun and Zhong reported an interesting alternative based on a copper-catalyzed approach to 6,9-disubstituted-purin-8-ones, a process that is related to the synthesis of *N*-substituted-1,3-dihydrobenzimidazol-2-ones reported by the Ma group and discussed previously (Scheme 15.11). The catalytic system consisted of a combination of copper(I) iodide and *trans*-4-hydroxy-L-proline and efficiently promoted the coupling/cyclization sequence between *N*-(dichloropyrimidin-5-yl)carbamates and amines to afford the cyclized 9-substituted-6-chloropurin-8-ones (Scheme 15.14).[33] A microwave-assisted amination procedure was then carried out to give 6,9-disubstituted purin-8-ones in moderate-to-good yields.

15.2.4 Copper-Catalyzed Synthesis of Quinazolines

Quinazolines have attracted much attention because of their therapeutic activities as antitubercular, antiviral, or anticancer agents, as well as ALK5 inhibitors. They are also used as ligands for benzodiazepine and gamma-aminobutyric acid (GABA) receptors in the central nervous system (CNS) and as DNA binders. Various approaches to quinazolines based on classical transformations have been developed,[34] and they sometimes suffer from drawbacks such as harsh conditions, complex and/or stepwise procedures, the lack of available

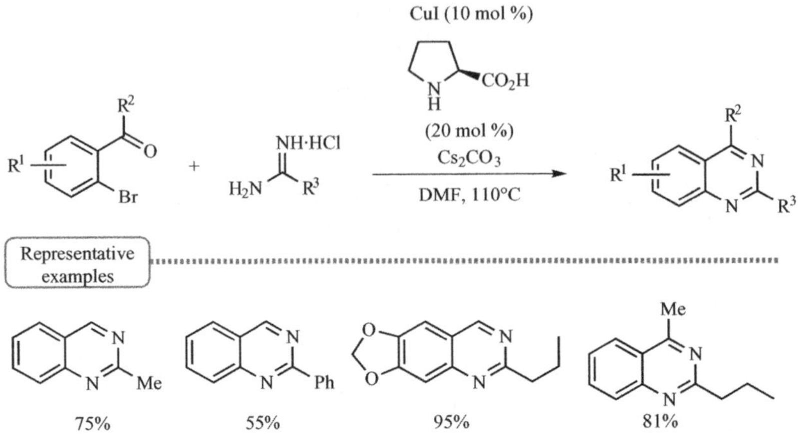

Scheme 15.15 Copper-catalyzed synthesis of quinazoline derivatives.

starting materials, and the use of expensive catalysts, which somewhat limit their applications. Recently, copper-catalyzed Ullmann-type reactions have been applied to the synthesis of quinazolines, and they will be reviewed in this section.

In 2008, Fu and coworkers reported an efficient method to synthesize quinazoline derivatives via a cascade reaction from readily available starting materials, amidine hydrochlorides and substituted 2-bromo-benzaldehydes or (2-bromoaryl)ketones, involving a copper-catalyzed intermolecular arylation of the amidine followed by condensation (Scheme 15.15).[35] Upon studying the reaction scope, it was observed that (i) the 2-bromobenzaldehydes were more reactive than the 2-bromophenylketones; (ii) 6-bromo-1,3-benzodioxole-5-carboxaldehyde was more efficient than 2-bromobenzaldehyde; (iii) 2′-bromoacetophenone gave higher yields than 2-bromobenzophenone; and (iv) aliphatic amidines were better substrates than their aromatic counterparts. In 2010, Truong and coworkers demonstrated that substituted 2-iodobenzaldehydes could be condensed with amidine hydrochlorides following a similar approach under ligandless conditions, and the resulting quinazolines were obtained in good-to-excellent yields.[36]

In continuation of these studies, Fu's group extended their procedure to the use of guanidines and developed a convenient copper-catalyzed method for synthesizing 2-aminoquinazoline derivatives (Scheme 15.16).[37] Using readily available substituted 2-bromo-benzaldehydes, (2-bromoaryl)ketones, and guanidines as starting materials and an inexpensive catalytic system based on copper(I) iodide and L-proline, various 2-amino-quinazoline derivatives were prepared in moderate-to-good yields.

To extend the starting materials that could be used for the preparation of quinazolines, they later reported another copper-catalyzed approach starting from substituted (2-bromophenyl)-methylamines and amides under air (Scheme 15.17).[38] The sequence involved a copper-catalyzed sequential Ullmann-type

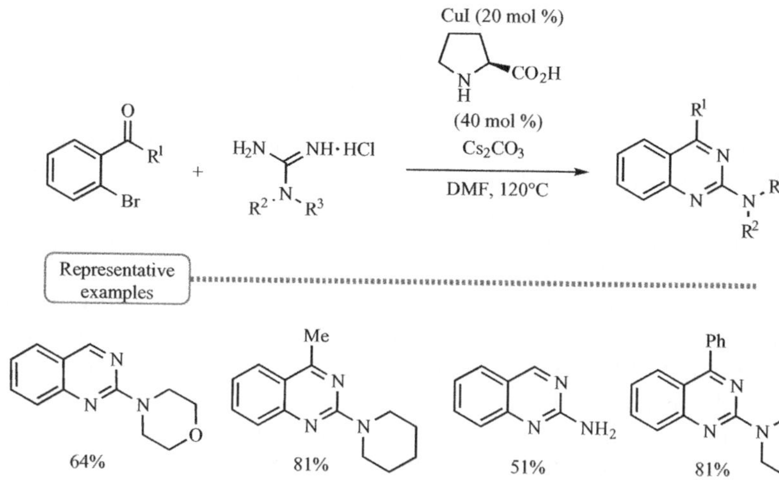

Scheme 15.16 Copper-catalyzed cascade synthesis of 2-aminoquinazoline derivatives.

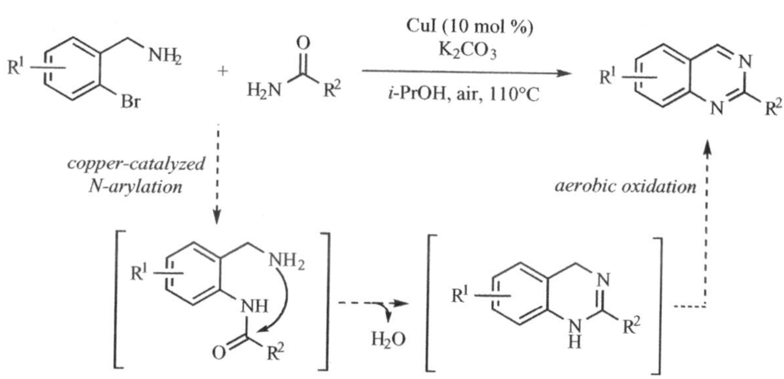

Scheme 15.17 Copper-catalyzed cascade synthesis of quinazoline derivatives under oxidative conditions.

cross coupling under ligand-free and additive-free conditions, followed by condensation of the pendant aminomethyl group onto the carbonyl group, followed by aromatization under aerobic conditions. The examined substrates provided the corresponding quinazolines in moderate-to-good yields at 110°C, and the starting (2-bromophenyl)methylamines containing electron-donating groups showed slightly lower reactivity. Concerning the amides that could be used in this transformation, the ones containing electron-withdrawing groups gave higher yields than the ones containing electron-donating groups. The cascade reactions could tolerate various functional groups including aromatic fluorides, nitro, and ethers, as well as the presence of other heterocycles.

Such approaches are not restricted to simple quinazolines or their 2-amino derivatives but can also be used for the preparation of quinazolines fused to other ring systems, such as in benzimidazoquinazolines, compounds that represent a useful platform in drug discovery and display various biological activities. They indeed exhibit a wide range of therapeutic activities, such as anticancer, antiviral, antimicrobial, antiinflammatory, and anticonvulsant properties. However, only a few methods are available to prepare this class of molecules. In 2011, Fu and coworkers used a combination of copper(I) iodide (10 mol %) and potassium phosphate under ligandless conditions in dichloromethane or DMSO to promote a cascade involving intermolecular arylation of the guanidine followed by intramolecular condensation starting from substituted 2-(2-halophenyl)benzoimidazoles (formed from the reaction of substituted benzene-1,2-diamines with 2-halobenzoic acids under acid conditions[39]), amidines, and guanidines. The process worked very well at room temperature with substituted 2-(2-bromophenyl)benzoimidazoles and amidines, while 2-(2-chlorophenyl)benzoimidazole and guanidine proved to be good substrates but required higher temperatures to react (80–100°C). Benzimidazoquinazoline and benzimidazo[1,2-*c*]quinazolin-5-amine derivatives were obtained in good-to-excellent yields (Scheme 15.18).[40a] By replacing the benzimidazole in the starting material with a pyrazole, pyrazoloquinazolines can be obtained upon treatment with copper(I) iodide in DMF at 100°C.[40b]

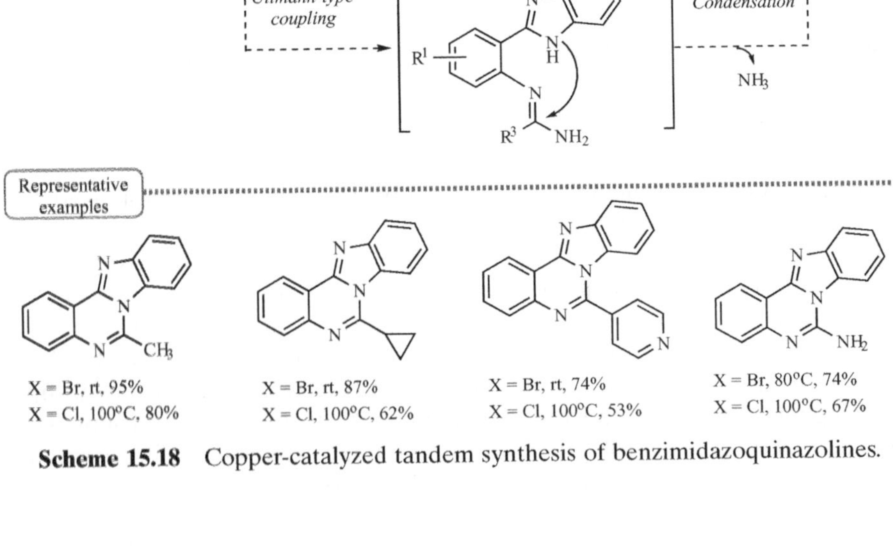

Scheme 15.18 Copper-catalyzed tandem synthesis of benzimidazoquinazolines.

15.2.5 Copper-Catalyzed Synthesis of Quinazolinone Derivatives

Quinazolinone derivatives are also widely found in natural products and display important biological and medicinal activities. For example, they have been shown to be efficient antiinflammatory, antifungal, hypnotic, sedative, analgesic, anticonvulsant, antibacterial, or anticancer agents as well as α-amino-3-hydroxy-5-methyl-4-isoxazole propionic acid (AMPA) receptor antagonists. Although there are efficient methods for the preparation of quinazolinone derivatives, mostly based on the reaction of *ortho*-amino-arylcarboxamides with orthoformates, the precursors required for this approach, that is, *ortho*-amino- or *ortho*-nitrobenzoic acid derivatives, are sometimes difficult to prepare. Alternative methods that provide efficient routes to quinazolinones from other starting materials are highly valuable. In recent years, the use of copper-mediated intramolecular arylation of N-nucleophiles allowed for the development of such procedures, and various efficient and convenient copper-catalyzed approaches to quinazolinones have been developed.

In this context, the Fu group reported a highly efficient method to synthesize quinazolinone derivatives in 2009. The protocol used copper(I) iodide as the catalyst under ligandless conditions in DMF and readily available 2-halobenzoic acid derivatives and amidines as starting materials.[41] This quinazolinone synthesis, which proceeds via an intermolecular amidine arylation/condensation, readily proceeds at room temperature with bromo- and iodobenzoic acid derivatives and allows for the isolation of the corresponding quinazolinones in high yields. Less-reactive substrates such as 2-chlorobenzoic acid or guanidines could also be used in this process, provided that the reaction temperature was raised to 80°C (Scheme 15.19). The reaction conditions are notably mild in all cases, which can be attributed to the accelerating effect of

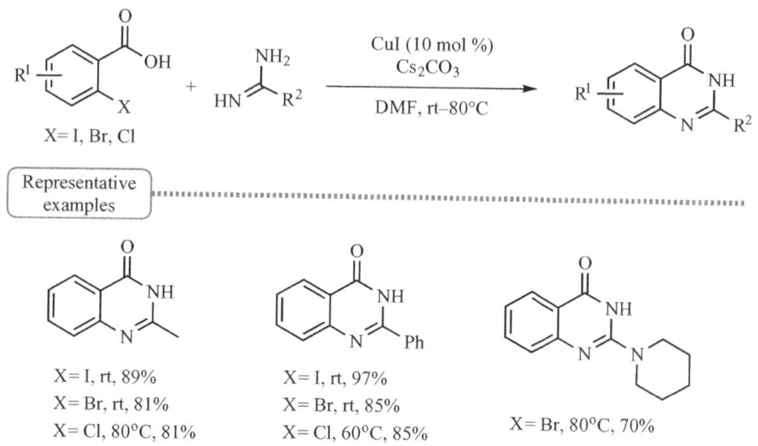

Scheme 15.19 Copper-catalyzed cascade synthesis of quinazolinone derivatives.

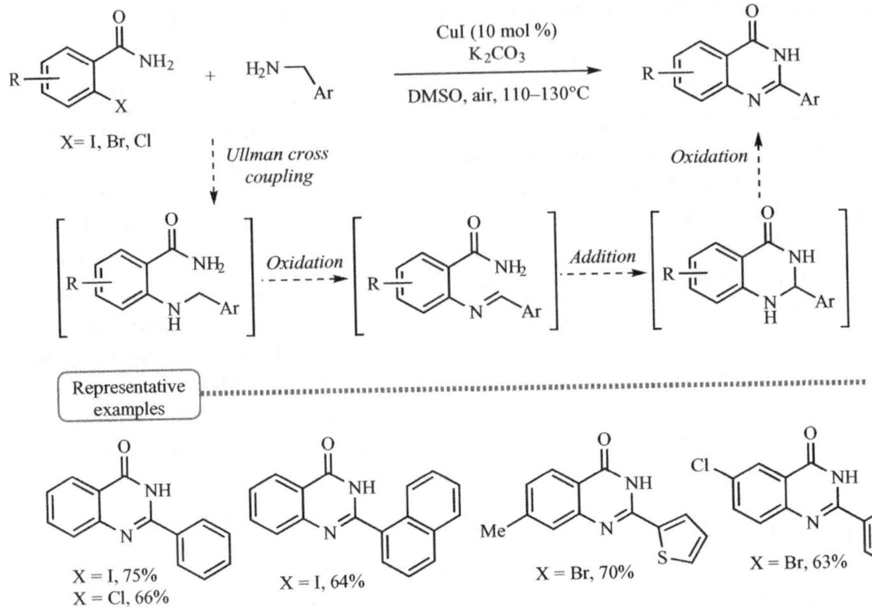

Scheme 15.20 Copper-catalyzed domino synthesis of quinazolinone derivatives via Ullmann-type coupling and aerobic oxidative formal C–H activation.

the carboxylic acid, a group that is well known for its strong *ortho*-effect in copper-catalyzed cross-coupling reactions.

Capitalizing on their work on quinazoline synthesis under oxidative conditions (see Scheme 15.17), Fu and coworkers described 2 years later an alternative entry to 2-aryl-quinazolinones starting from different precursors: 2-halobenzamides and benzylic amines. This efficient route is readily catalyzed by copper(I) iodide alone in the presence of potassium carbonate and air and proceeds through a copper-catalyzed arylation of the benzylic amine/oxidation to the corresponding imine/addition/oxidation sequence (Scheme 15.20).[42]

15.2.6 Copper-Catalyzed Synthesis of 1,2,4-Benzothiadiazine 1,1-Dioxides

In 2009, Fu and coworkers developed a simple and practical method for the synthesis of 1,2,4-benzothiadizaine 1,1-dioxide derivatives via a copper-catalyzed cascade reaction of substituted 2-haloarylsulfonamides with amidines (Scheme 15.21).[43] A wide range of cyclized products can be easily and readily obtained from simple starting materials and inexpensive reagents in good yields. Several reaction pathways can account for the formation of 1,2,4-benzothiadizaine 1,1-dioxide derivatives, and the reaction of ^{15}N-labeled 2-bromo-4-trifluoromethyl-benzenesulfonamide with butyramidine hydrochloride led to a fully ^{15}N-labeled 6-trifluoromethyl-3-propyl-2*H*-1,2,4-benzothiadiazine-1,1-dioxide. This result allows postulating the following

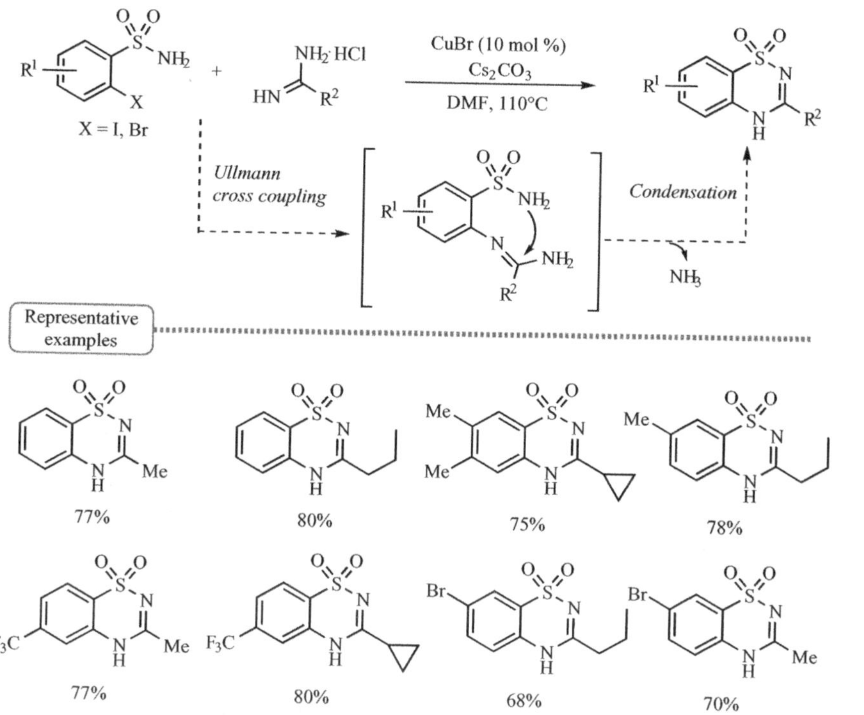

Scheme 15.21 Copper-catalyzed synthesis of benzothiadiazine-1,1-dioxide derivatives.

mechanism: An Ullmann cross coupling of the substituted 2-halobenzene-sulfonamide with amidine would first produce an N-arylated amidine intermediate, and subsequent intramolecular nucleophilic addition of the amino group of sulfonamide to the amidine group in the presence of cesium carbonate would afford the cyclized product and release 1 equiv of ammonia.

15.2.7 Copper-Catalyzed Synthesis of 1-Aryl-2-Thioxo-2,3-Dihydroquinazolin-4(1H)-Ones

Xi and coworkers reported a similar route for the one-pot synthesis of 2-thioxo-2,3-dihydroquinazolin-4(1*H*)-ones using readily available 2-bromobenzamides and isothiocyanates as the starting materials, copper(I) iodide/*N,N′*-dimethylethylenediamine as the catalytic system, cesium carbonate as the base, and toluene as the solvent at 120°C.[44] The reaction is likely to proceed via a sequential intermolecular addition of the *o*-bromobenzamides to isothiocyanates followed by a copper-catalyzed intramolecular C–N bond formation, and yields the title compounds in fairly good yields (Scheme 15.22).

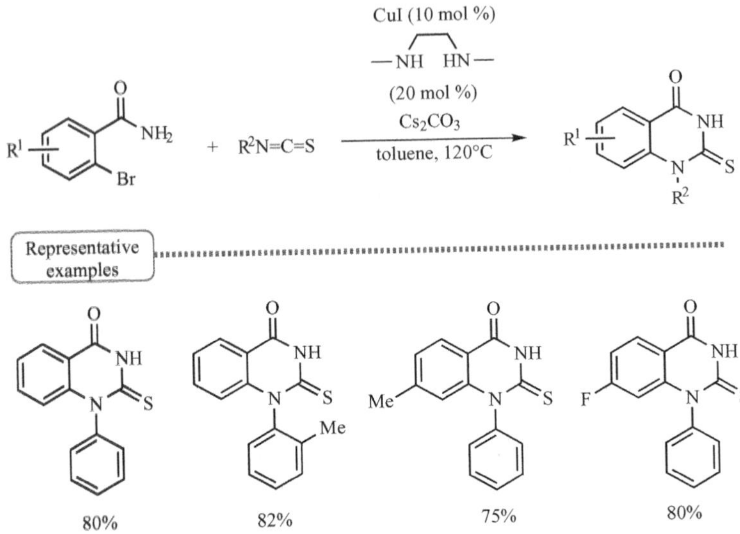

Scheme 15.22 Copper-catalyzed synthesis of 1-aryl-2-thioxo-2,3-dihydroquinazolin-4(1*H*)-ones.

15.2.8 Copper-Catalyzed Synthesis of Indole Derivatives

Indole derivatives display outstanding pharmacological properties, such as analgesic, antioxidant, and psychotropic effects. In addition, they are commonly found in a wide range of natural products, and their structural diversity is especially attractive.[45] Traditional methods to synthesize indoles include, among others, the classical Fisher indole synthesis,[46] sequential reduction/cyclization,[47] and metal-catalyzed heteroannulation processes.[48] Nevertheless, the preparation of substituted indoles is still not a trivial task, and the use of copper catalysis, and more specifically copper-mediated cyclization reactions involving intramolecular C–N bond formation, provided useful, original, and valuable alternatives as well as access to indoles possessing substitution patterns that can hardly be obtained using other routes.

An example was published in 2008 by Karchava and coworkers, who prepared 1-amino-indole-3-carboxylates using a copper-catalyzed intramolecular N-arylation reaction starting from an enehydrazino ester substituted with a bromoaryl group (Scheme 15.23). The reaction readily proceeded under ligandless conditions using 5 mol % of copper(I) iodide and potassium phosphate in DMF at 85°C.[49a] A related copper-mediated cyclization yielding to 1-sulfonyl-3-arylindoles from the corresponding acyclic N-sulfonyl-enamines was reported recently, therefore expanding the scope of this cyclization and the range of substrates that can be obtained through this strategy.[49b]

Indole-2-carboxylates can also be prepared through a ligand-free copper-catalyzed cascade reaction starting from (2-haloaryl)-aldehydes or ketones with

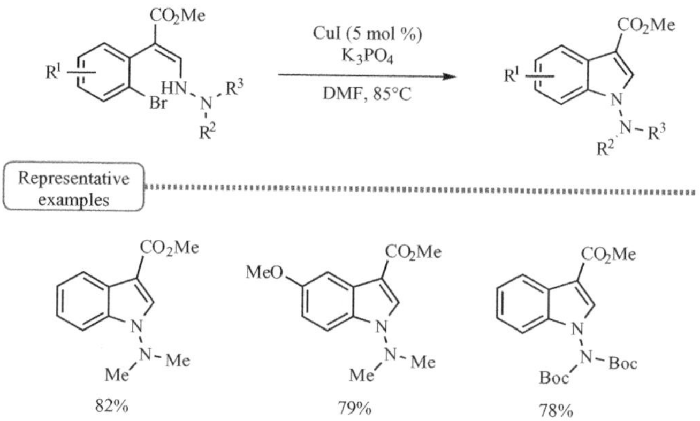

Scheme 15.23 Copper-catalyzed intramolecular N-arylation leading to 1-amino-indole-3-carboxylates.

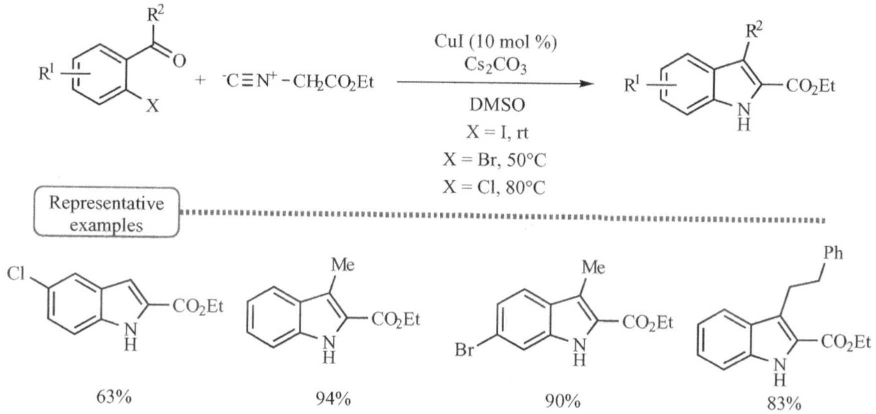

Scheme 15.24 Copper-catalyzed synthesis of indole-2-carboxylic acid esters.

ethyl isocyanoacetate, as reported by Cai, Ding, and coworkers in 2009 (Scheme 15.24).[50] The reactions were carried out at room temperature and 50°C for iodo- and bromo-substituted substrates, respectively. The use of chlorinated substrates required higher temperature but still gave the desired products in reasonable yields. The method, which involves a copper-catalyzed condensation/coupling/deformylation cascade, tolerated various functional groups and therefore provides a synthetically useful entry to indole-2-carboxylates.

Koenig and coworkers reported in 2010 a related approach that has the advantage of starting from *N*-acetylglycine ethyl ester instead of ethyl isocyanoacetate. Using a ligand-free, copper-catalyzed cascade sequence, (2-haloaryl)-aldehydes and *N*-acetylglycine ethyl ester were smoothly coupled to afford indole-2-carboxylic esters (Scheme 15.25).[51] The reaction proceeds via an

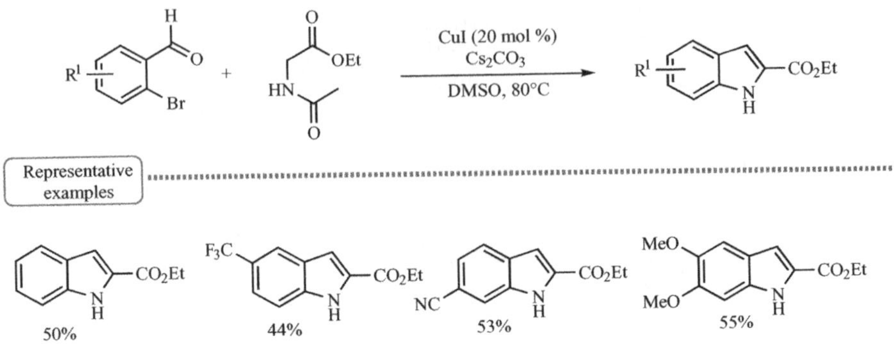

Scheme 15.25 Copper-catalyzed synthesis of indole-2-carboxylic esters from (2-haloaryl)-aldehydes and glycine amidoesters.

intermolecular Knoevenagel condensation followed by subsequent copper-catalyzed N-arylation and hydrolysis.

Besides these three efficient approaches to indole derivatives that allow for the preparation of a wide range of indoles with various substitution patterns, other routes, complementary to the one described in this section, have also been developed based on a key copper-mediated step, but this time with the formation of a C–C bond. They will be reviewed in Section 15.4.1.

15.2.9 Copper-Catalyzed Formation of Cyclic Enamides: Efficient Routes to Unsaturated Lactams, Alkylideneazetidines, and Dihydropyrroles

As already mentioned in the introduction of this section, intramolecular copper-catalyzed C–N bond formation is especially suitable for the formation of heterocycles such as indolines or tetrahydroquinolines (Scheme 15.1), compounds that are typically obtained in excellent yields (even starting from aryl chlorides) and under rather mild conditions. This approach has also been used for the synthesis of various lactam derivatives and is in fact a method of choice for the formation of cyclic enamides and related compounds that can be readily obtained from the corresponding acyclic compounds possessing both an N-nucleophile and a vinyl halide.

This approach has been explored in 2005 by the Hu and Li group, who reported an elegant copper-catalyzed intramolecular vinylation of acyclic iodovinyl-amides to lactams (Scheme 15.26).[52] Lactams ranging in size from five- to seven-member rings were prepared in moderate to excellent yields using Buchwald's conditions (copper(I) iodide [20 mol %], *N*,*N*′-dimethylethanediamine [40 mol %], and cesium carbonate in dioxane), and depending on the substitution pattern of the starting material, the double bond could be either *endo* or *exo*.

This approach was next extended to the development of a general and highly efficient approach to alkylideneazetidines via a related copper-catalyzed

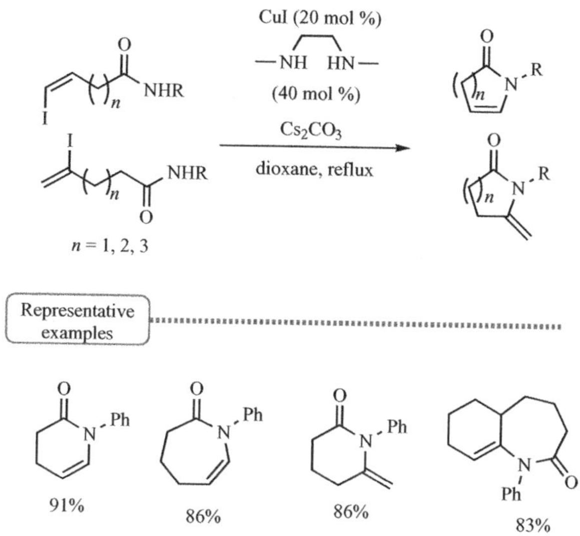

Scheme 15.26 Copper-catalyzed synthesis of unsaturated lactams by intramolecular enamidation.

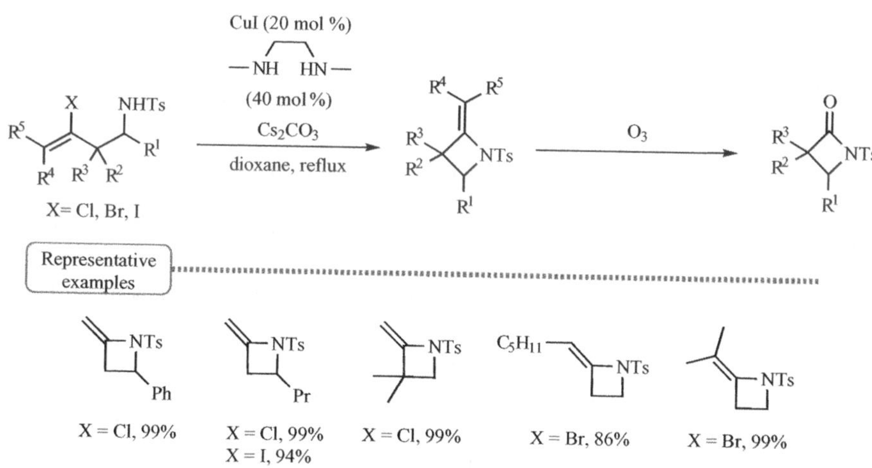

Scheme 15.27 Copper-catalyzed synthesis of 2-alkylideneazetidines.

intramolecular N-vinylation (Scheme 15.27).[53] The starting *N*-tosyl-3-halo-buten-3-ylamines underwent a remarkably efficient copper-catalyzed Ullmann-Goldberg-type coupling to afford the corresponding 2-alkylideneazetidines, which could be readily converted to the corresponding β-lactams by ozonolysis. In addition to providing an interesting entry to 2-alkylideneazetidines, this work also demonstrates that copper-mediated cyclization reactions are certainly not restricted to the formation of five-membered or higher ring systems

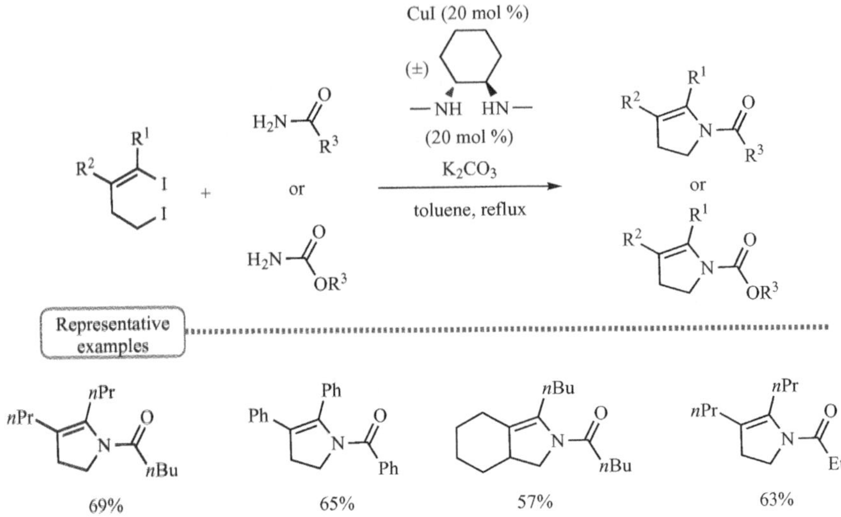

Scheme 15.28 Alkylation/copper-catalyzed N-alkenylation to dihydropyrroles.

and that the reductive elimination from the intermediate five-membered metallacycle seems to be highly favorable. This point will also be discussed in Section 15.3.2.

A related approach was reported in 2010 for the synthesis of dihydropyrrole derivatives using a sequential alkylation/copper-catalyzed N-vinylation from 1,4-diiodobut-1-enes and carbamates. This approach furnishes the corresponding protected dihydropyrroles with moderate-to-good yields but is strongly dependent on the availability of the starting materials that can be tricky to prepare depending on their substitution pattern (Scheme 15.28).[54]

Before closing this section, it ought to be mentioned that copper-catalyzed C–N bond formation can also be used for the preparation of macrocyclic ynamides, as reported in 2006 by Hsung and coworkers, who investigated the copper-catalyzed intramolecular amidation of bromoalkynes. Cyclization under relatively mild conditions gave the desired macrocyclic ynamides with a ring size ranging from 9 to 19 (Scheme 15.29). Such macrocycles were used as precursor for the formation of macrocyclic (*Z*)-enamides via hydrogenation of the ynamide in the presence of Lindlar catalyst.[55]

15.2.10 Copper-Catalyzed Synthesis of 1,4-Benzoxazines

1,4-Benzoxazine derivatives show various biological activities. For example, they have been used as potential intracellular calcium antagonists, vasodilator agents, and potassium channel activators. In 2009, Sekar and coworkers developed a practical copper-catalyzed domino approach to 3,4-dihydro-2*H*-1,4-benzoxazines.[56] A variety of *trans*-1,4-benzoxazines were readily constructed

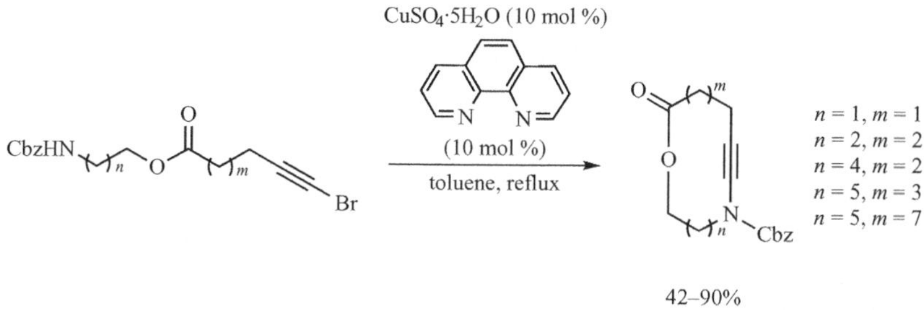

Scheme 15.29 Cyclization by copper-catalyzed intramolecular bromoalkyne amination.

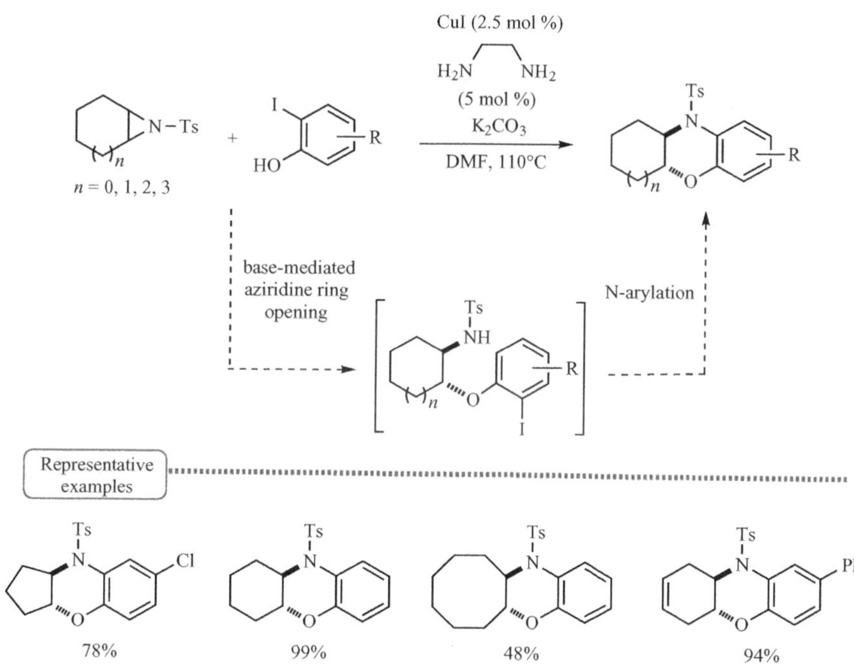

Scheme 15.30 Copper-catalyzed domino ring opening-intramolecular arylation yielding to *trans*-1,4-benzoxazines.

in good-to-excellent yields starting from readily available precursors: aziridines and *o*-iodophenols (Scheme 15.30). It was logically assumed that the reaction proceeded via a base-mediated nucleophilic ring opening of the aziridine by the phenoxide, followed by an intramolecular arylation of the resulting metallated sulfonamide to yield the cyclized product.

As evidence with representative syntheses of N-heterocycles described in the first part of this chapter, the use of copper-mediated intramolecular C–N bond formation allowed the design of remarkably efficient syntheses of an impressive range of nitrogen heterocycles. While some of these procedures nicely complement more classical approaches, some of them also provide straightforward entries to heterocyclic compounds that could hardly be obtained using other methods. This area is expanding at a fast rate, and there is no doubt that copper-mediated synthesis of nitrogen heterocycles will find applications in various areas of research, especially for the preparation of compounds of pharmaceutical or agrochemical import.

The development of efficient procedures for the copper-catalyzed arylation or vinylation of alcohols also had a strong impact on heterocyclic synthesis, and various efficient syntheses of oxygenated heterocycles have been recently devised using intramolecular versions of these reactions. These processes will be reviewed in the following section.

15.3 CYCLIZATION BY C–O BOND FORMATION

Copper-catalyzed Ullmann-type C–O bond formation has recently drawn much attention, leading to the development of efficient and practical methods for the formation of C–O bonds from various precursors and O-nucleophiles (see Chapters 2 and 5, Section 5.2.2 for details).[57] These methods also provide facile routes to synthesize oxygen-containing heterocycles, which will be reviewed in this section.

15.3.1 Copper-Catalyzed Synthesis of Benzoxazole Derivatives

Benzoxazole derivatives are important heterocycles found in a variety of natural products and used as fluorescent probes as well as antibacterial, antimicrobial, and anticancer agents. Classical approaches to benzoxazoles typically rely on the coupling of 2-aminophenol derivatives with carboxylic acids[58] or aldehydes under oxidative conditions.[59] Although these methods are useful, they are limited by two main drawbacks: the limited availability of the required substituted *o*-aminophenols and the harsh reaction conditions.[60]

During the last decade, several approaches to benzoxazoles via various strategies based on a copper-catalyzed intramolecular O-arylation of *o*-halobenzanilides have been developed.[20,61,62] In 2006, Evindar and Batey showed that the use of a catalytic system made of copper(I) iodide and 1,10-phenanthroline in combination with cesium carbonate enabled the formation of benzoxaoles from *ortho*-haloanilide precursors with remarkable efficiency (Scheme 15.31, Eq. 1). The cyclization is usually high yielding, the approach is readily applied to the synthesis of substituted benzoxazoles through incorporation of appropriately positioned substituents on the starting *ortho*-haloanilide precursors, and this cyclization can be extended to the preparation of benzothiazoles. Interestingly, the 1,10-phenanthroline ligand can be replaced by *N,N′*-dimethylethylenediamine, but, in general, the former showed

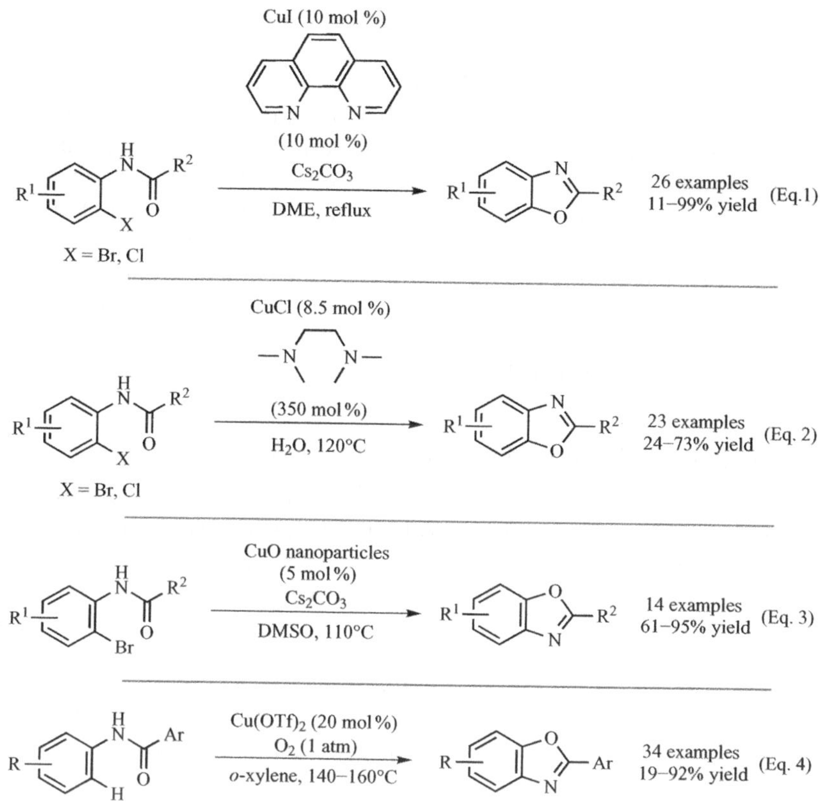

Scheme 15.31 Copper-catalyzed synthesis of benzoxazoles via intramolecular cyclization of *o*-halobenzanilides and benzanilides.

greater substrate tolerance. Most examples utilized the cyclization of *ortho*-bromoanilides, but *ortho*-iodoanilides and *ortho*-chloroanilides also undergo the cyclization under these conditions, and the reaction rate follows the standard order I > Br > Cl. As a note, the cyclization cannot be efficiently promoted by palladium complexes, which differs from related cyclization for the preparation of other heterocycles such as benzimidazoles or benzothiazoles.[61] In 2007, San Martin and coworkers reported an alternative system to perform this cyclization in water using copper(I) chloride as the precatalyst (Scheme 15.31, Eq. 2).[62] A notable feature of these studies is the use of the intramolecular copper-catalyzed arylation using aryl chlorides, compounds that have been far less studied than their iodo- and bromo-analogues, despite their lower cost and greater availability, and that usually perform well in intramolecular processes.

In 2009, Punniyamurthy and coworkers reported the use of catalytic amounts of copper(II) oxide nanoparticles to perform the same transformation (Scheme 15.31, Eq. 3).[20] With this system, the reactions were carried out under air and ligand-free conditions, and the CuO nanoparticles could be recovered and recycled without loss of activity.

The evolution of strategies for the copper-catalyzed synthesis of benzoxazoles is similar to the one observed with benzimidazoles (see Section 15.2.2). Processes based on transition-metal-catalyzed C–H functionalization/C–heteroatom bond formation, which have been used for the formation of indazoles,[63] benzothiazoles,[64] indolines,[65] benzothiophenes,[66] and benzimidazoles,[24] have been elegantly applied to the preparation of benzoxazoles. Indeed, in 2009, Ueda and Nagasawa designed an efficient copper(II)-catalyzed regioselective protocol to access benzoxazoles from readily available benzanilides, therefore avoiding the need for their pre-functionalization with a halogen (Scheme 15.31, Eq. 4).[67]

15.3.2 Copper-Catalyzed Synthesis of Cyclic Enol and Aryl Ethers

The intramolecular vinylation of alcohols has been studied in detail by Li and coworkers, who demonstrated the feasibility of the process for the formation of four- (Scheme 15.32, Eq. 1), five- (Scheme 15.32, Eq. 2), and six- (Scheme 15.32, Eq. 3) membered cyclic enol ethers in excellent, good, and fair yields, respectively. An unusual preference for 4-*exo*-ring closure in this cyclization was also observed with alcohols possessing vinylic bromides in the β and γ-positions (Scheme 15.32, Eq. 4). In this last case, the differences between palladium(0) and copper(I) catalysis are truly remarkable since cyclization to a four-membered ring is not a favored process with palladium: A reversal of selectivity is observed when switching metals.[68]

Ketones and β-ketoesters can also be excellent reaction partners in O-vinylation reactions provided that they are performed in an intramolecular fashion. In 2005, Fang and Li reported the use of the copper(I) iodide/*N*,*N*′-dimethylethylenediamine catalytic system to carry out an intramolecular O-vinylation of ketones with vinyl bromides, which provides an efficient entry to multisubstituted furans (Scheme 15.33, Eq. 1).[69] Good yields were obtained under these mild conditions, and the strategy was extended to the formation of cyclic alkenyl ethers embedded in higher ring systems (six- and seven-membered rings) by changing the position of the halogen on the double bond (Scheme 15.33, Eq. 2). The intramolecular O-vinylation being especially favored, it was later found that the need for an especially acidic hydrogen such as that present in ketoesters was not a strict requirement since ketones could also be cyclized to polysubstituted furans in excellent yields using vinyl chlorides, provided that the amount of the catalyst was increased (see Chapter 5, Scheme 5.17, and corresponding text for details).[70]

This strategy is also a viable one with aryl halides and provides an interesting entry to 4*H*-1-benzopyrans, an important constituent of many biologically active products for which there is clearly a lack of general synthetic methods. In 2006, and in continuation of their studies, the Li group reported a

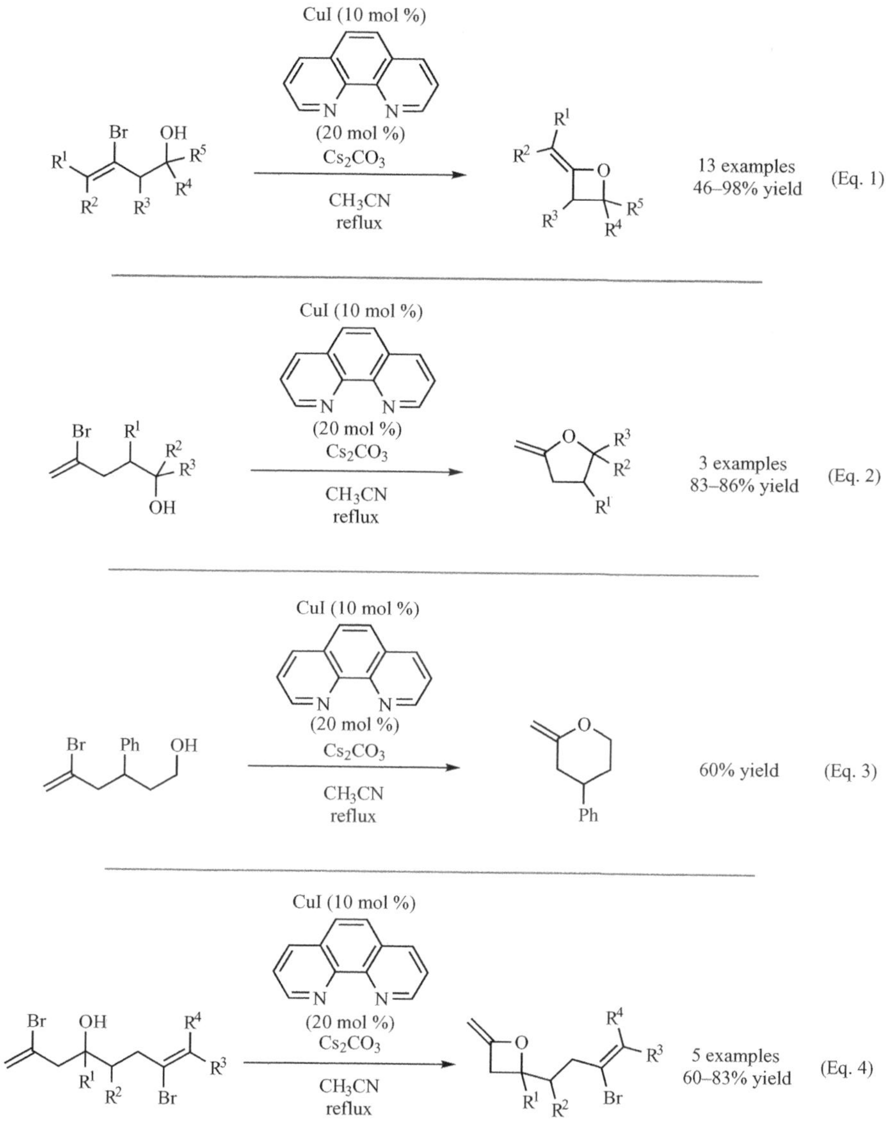

Scheme 15.32 Synthesis of cyclic enol ethers by intramolecular vinylation of alcohols.

copper(I)-catalyzed intramolecular O-arylation of aryl bromides with 1,3-dicarbonyls, a reaction that readily proceeds in the presence of catalytic amounts of copper(I) iodide and *N*,*N*′-dimethylethylenediamine to afford various substituted benzopyrans in good yields (Scheme 15.34).[71a] This strategy was later on extended to the synthesis of 2,3,4,9-tetrahydro-1*H*-xanthen-1-ones

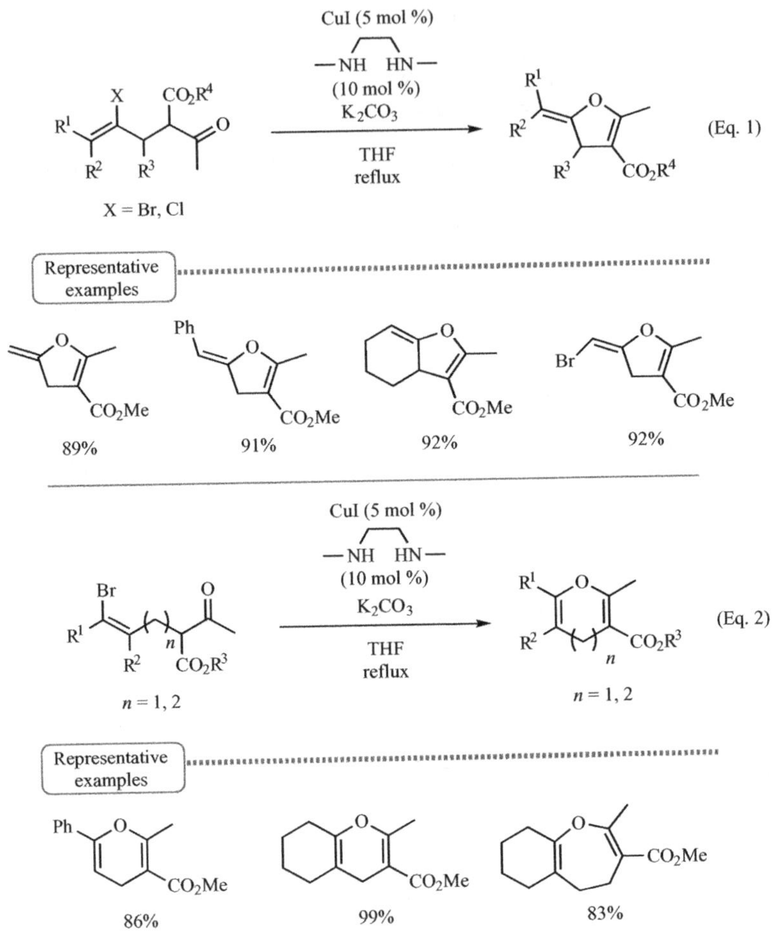

Scheme 15.33 Copper-catalyzed intramolecular O-vinylation of ketoesters with vinyl halides.

by intramolecular O-arylation of readily available 2-(2-bromobenzyl)cyclohexane-1,3-diones upon treatment with 0.5 mol % of copper(I) chloride in the presence of cesium carbonate and pivalic acid in DMF at 130°C.[71b]

A related strategy had in fact been reported earlier by Chen and Dormer for the synthesis of benzo[*b*]furans via a copper-catalyzed intramolecular cyclization from *o*-halo benzylic ketones (Scheme 15.35).[72a] This process proved to be exceptionally efficient with a wide variety of benzylic ketones and could be extended to benzylic aldehydes and heteroaromatics. The method also tolerated various functional groups and afforded the corresponding benzofurans in 71–99% yields. By preforming an imine from the exact same *o*-halo benzylic ketones and an amine before the copper-mediated cyclization, indoles can also be obtained.[72b]

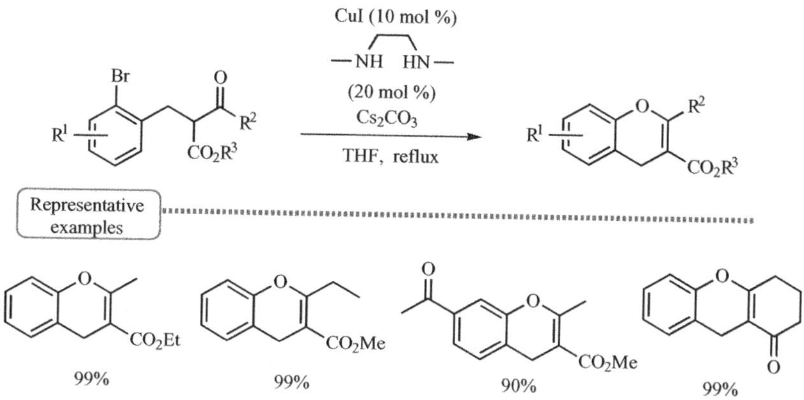

Scheme 15.34 Copper-catalyzed intramolecular O-arylation leading to 4*H*-1-benzopyrans.

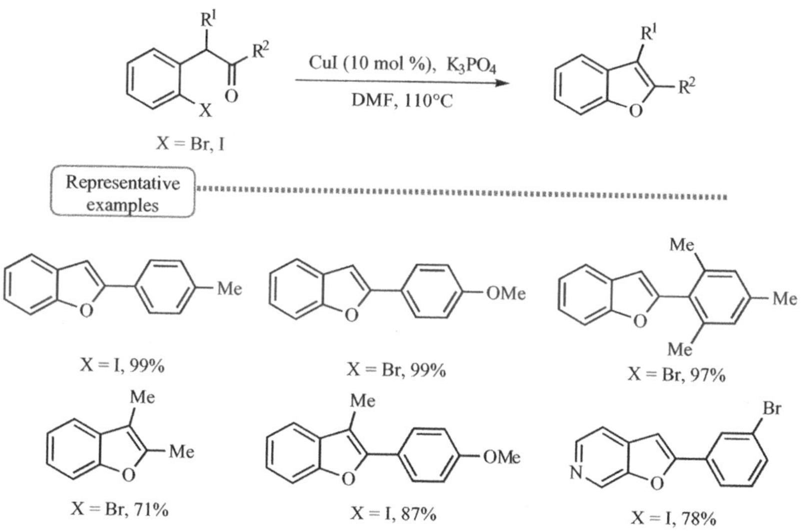

Scheme 15.35 Copper-catalyzed synthesis of benzo[*b*]furans.

15.3.3 Copper-Catalyzed Synthesis of Enol Lactones and Benzopyranones

Enol lactones are found in many biologically active natural products and are useful intermediates. Traditionally, their preparation involves a sequential halolactonization of alkynoic or alkenoic acids followed by dehydrohalogenation. An alternative entry to these building blocks has been recently developed based on a copper-catalyzed intramolecular cross coupling using carboxylic acids as the nucleophiles. While these nucleophiles are poor partners in intermolecular copper-catalyzed cross-coupling processes, their use in copper-catalyzed cyclization reactions is much more efficient. Capitalizing on this

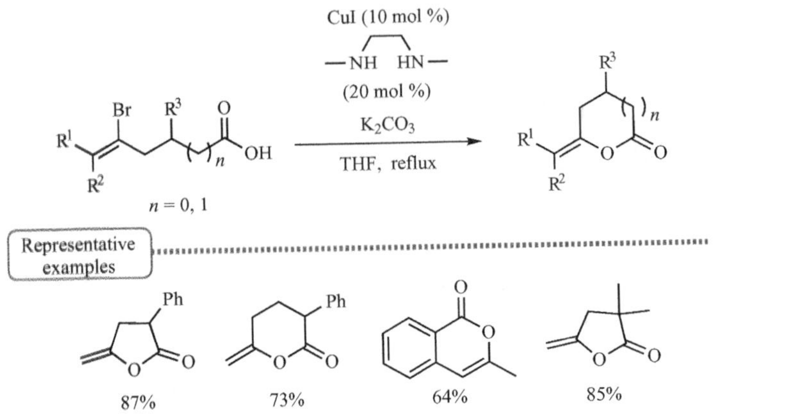

Scheme 15.36 Copper-catalyzed intramolecular O-vinylation of carboxylic acids leading to enol lactones.

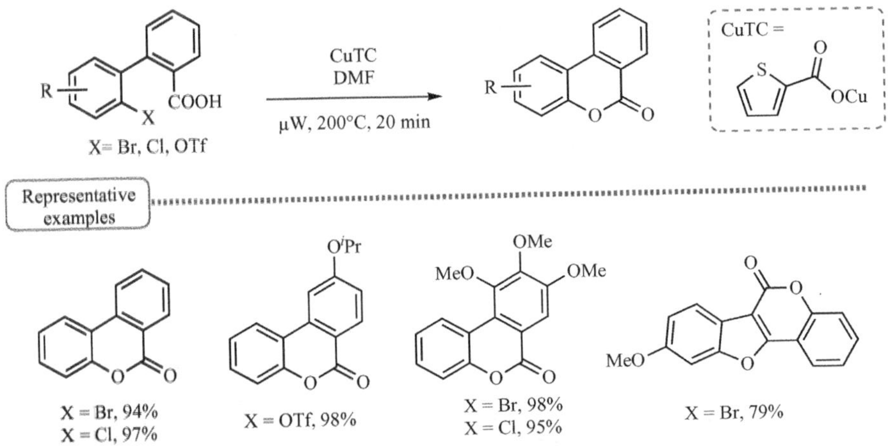

Scheme 15.37 Copper-catalyzed synthesis of benzopyranone derivatives by intramolecular arylation of carboxylic acids.

observation, Li and coworkers demonstrated in 2009 that five- and six-membered enol lactones could be smoothly prepared, in a mild and efficient manner, using a copper(I)-catalyzed intramolecular O-vinylation of carboxylic acids (Scheme 15.36).[73]

In a similar way, in 2007, Ruchirawat and coworkers developed a copper-mediated intramolecular lactonization procedure leading to the formation of benzopyranones by intramolecular arylation of a carboxylate (Scheme 15.37).[74] The use of a stoichiometric amount of copper(I)-thiophene-2-carboxylate (CuTC) under ligand- and base-free conditions at 200°C under microwave irradiation proved to be the optimum conditions. Other copper sources (CuI, CuBr, CuCl) were tested in both stoichiometric and catalytic amounts as well as in the presence or absence of a base and a ligand, but the desired benzopyranones were generally obtained in poor yields.

As with nitrogen nucleophiles, the use of intramolecular copper-mediated arylation and vinylation of O-nucleophiles offers new opportunities in heterocyclic synthesis that should find many applications in various domains. The use of intramolecular processes is especially interesting since coupling partners such as aryl chlorides or carboxylic acids, which typically react poorly in intermolecular processes, can be efficiently used, enabling the preparation of heterocycles from readily available starting materials.

Great progress has also been made recently in copper-mediated C–C bond formation, and efficient catalytic systems are now available to perform the Hurtley and Rosenmund–von Braun reactions under mild conditions and to form aryl–aryl, aryl–alkyne, aryl–alkene, and many other C–C bonds as well. The development of these processes has been extensively covered in Chapters 8–13 and has been extensively applied in heterocyclic synthesis. Representative examples will be reviewed in the next section.

15.4 CYCLIZATION BY C–C BOND FORMATION

Among all copper-mediated C–C bond-forming processes, copper-catalyzed coupling between aryl halides and activated methylene compounds has drawn a lot of attention recently. Robust systems, which have been overviewed in Chapter 8, have been developed to perform these reactions under mild conditions.[75] They also provide, in combination with a step where a carbon-heteroatom bond is formed, convenient access to a range of heterocycles. Selected examples are reviewed in the following paragraphs.

15.4.1 Copper-Catalyzed Synthesis of Indole Derivatives

Copper-catalyzed arylation of β-ketoesters to construct indoles has recently emerged as an especially efficient tactic and has stimulated various research groups.[28,76–79] In 2007, Kirihata and coworkers developed a relatively simple procedure to synthesize 2,3-disubstituted indoles from readily available 2-iodoaniline and a range of β-ketoesters using copper(I) iodide/1,1′-binaphthyl-2,2′-diol (BINOL) as the catalytic system (Scheme 15.38).[76] The cyclization/condensation sequence proceeds in good-to-high yields but, unfortunately, no example using substituted 2-iodoanilines was reported in this study.

That same year, Ma and coworkers described a mild process to prepare 2,3-disubstituted indoles that nicely complements the work of Kirihata and allows the synthesis to be performed at room temperature.[77] In this sequential procedure, the copper(I) iodide/L-proline-catalyzed cross coupling between 2-iodotrifluoroacetanilides and β-ketoesters (and amides) is followed by an acidic hydrolysis to deliver the disubstituted indoles after condensation (Scheme 15.39). A wide range of functional groups was tolerated under these coupling conditions such as ketones, esters, nitro, iodo, chloro, olefin, or benzyl moieties.

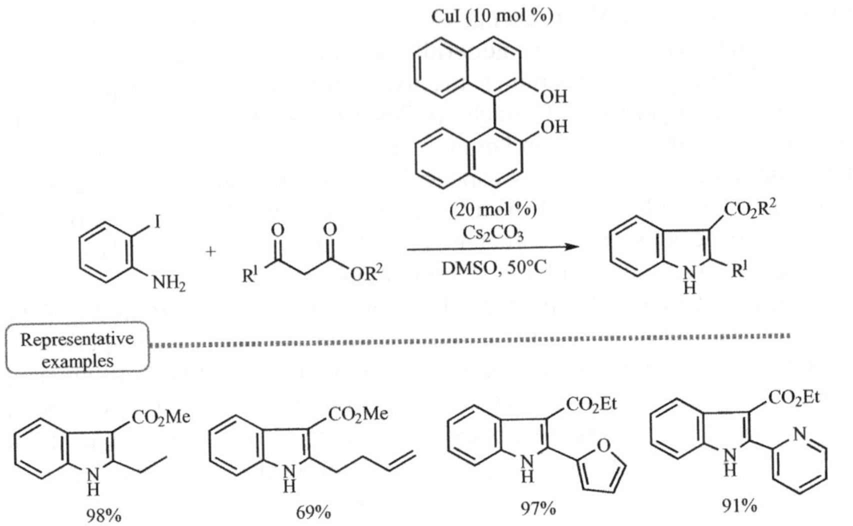

Scheme 15.38 Copper-catalyzed synthesis of 2,3-disubstituted indoles from 2-iodoaniline and β-ketoesters.

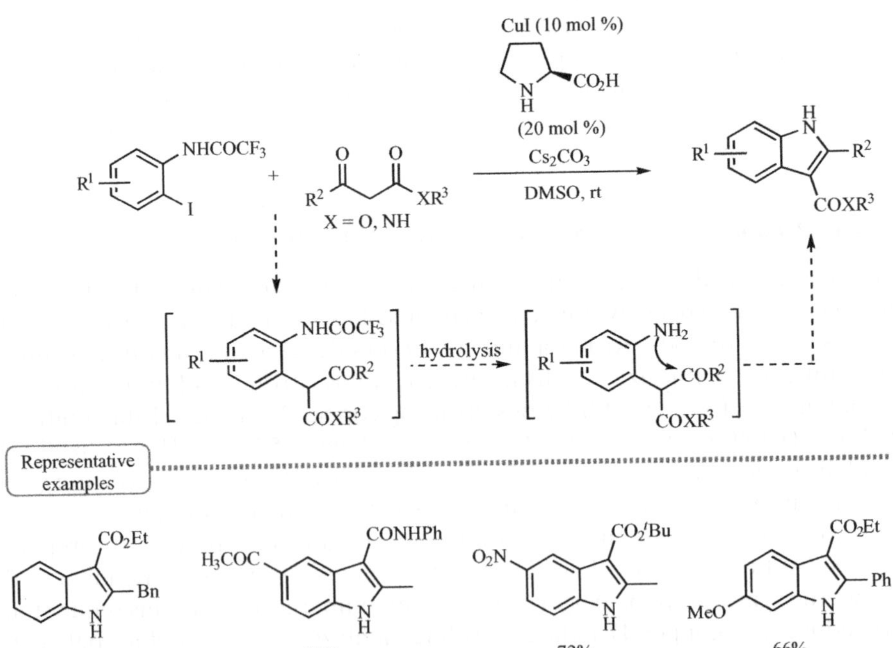

Scheme 15.39 Copper-catalyzed cascade synthesis of indoles from 2-iodotrifluoroacetanilides and β-ketoesters.

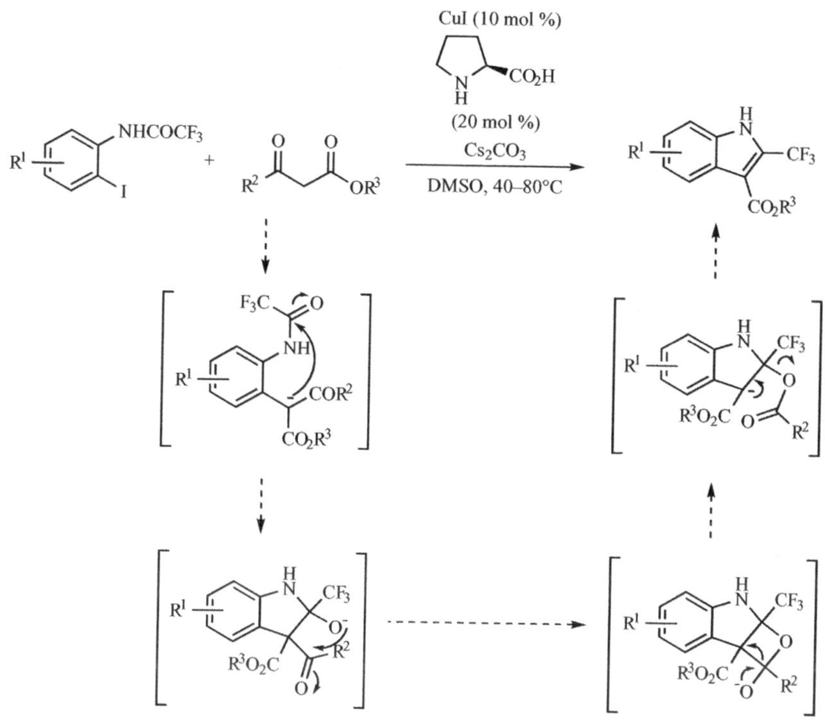

Scheme 15.40 Copper-catalyzed cascade synthesis of 2-(trifluoromethyl)indoles from 2-iodotrifluoroacetanilides and β-ketoesters.

An interesting switch in the reaction pathway was noted by the authors when they carried out this cross coupling from the same reaction partners in anhydrous DMSO at 40–80°C. Under these conditions, substituted 2-(trifluoromethyl)indoles were obtained in good-to-excellent yields. The reactions are likely to proceed via a sequential cross-coupling/condensation/deacylation mechanism (Scheme 15.40).[78]

In 2010, Fu and coworkers extended this procedure and developed a simple and efficient approach to 2-amino-1*H*-indole-3-carboxylates and 2-amino-1*H*-indole-3-carbonitriles (Scheme 15.41).[79] The cascade reactions of substituted *N*-(2-halophenyl)-2,2,2-trifluoroacetamides with alkyl 2-cyanoacetates or malononitrile in the presence of catalytic amounts of copper(I) iodide and L-proline as the catalytic system, and potassium carbonate as the base in DMSO, were performed under mild conditions. The corresponding indole derivatives containing amino, carboxylate, and nitrile groups were obtained in good yields.

15.4.2 Copper-Catalyzed Synthesis of 3-Acyloxindoles

3-Acyloxindoles represent a widespread structural motif that can be found in a number of pharmaceutically promising compounds such as GSK3 kinase

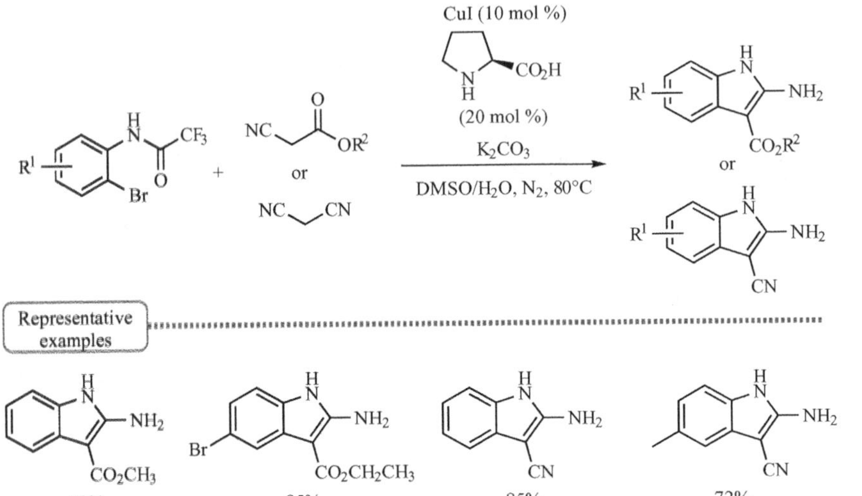

Scheme 15.41 Copper-catalyzed synthesis of 2-amino-1H-indole-3-carboxylates and 2-amino-1H-indole-3-carbonitriles.

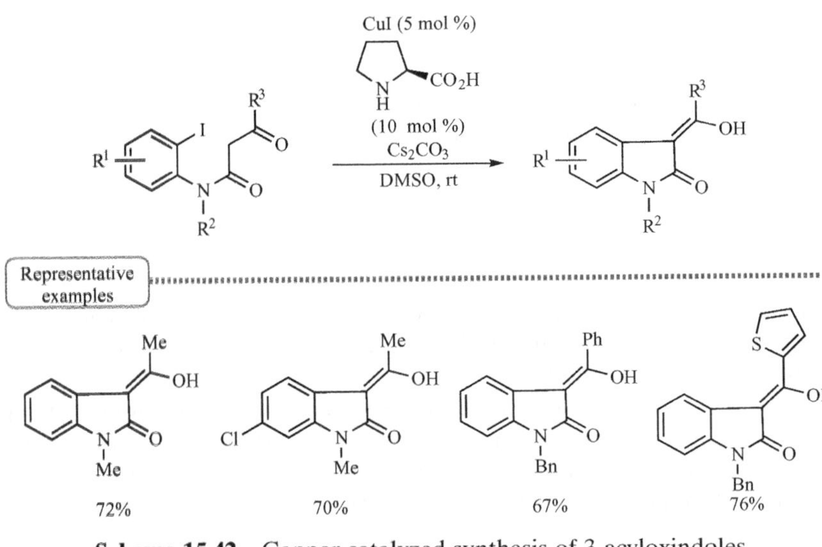

Scheme 15.42 Copper-catalyzed synthesis of 3-acyloxindoles.

inhibitors, influenza endonuclease inhibitors, and tenidap. In 2006, an elegant method for the preparation of 3-acyloxindoles was reported by Lu and Ma based on an intramolecular arylation from N-(2-iodoaryl)-β-ketocarboxamides catalyzed at room temperature by the copper(I) iodide/L-proline couple (Scheme 15.42).[28] The results showed that both electron-donating and electron-withdrawing substituents were equally well tolerated in the reaction and that substituted oxindoles (at the 1-, 3-, 4-, 5-, and 6-positions) could be formed by using the appropriate amide precursors.

15.4.3 Copper-Catalyzed Synthesis of Isoquinoline and Isoquinolinone Derivatives

The isoquinoline nucleus is found in various alkaloids and various compounds of pharmaceutical imports such as anesthetics, antihypertension agents, antifungal agents, disinfectants, and vasodilators. Here, again, various routes to this heterocyclic system relying on classical transformations are available, but they suffer in general from harsh reaction conditions, and alternatives to these processes are therefore highly desirable. In continuation of their work on the development of new copper-catalyzed processes, Ma and coworkers established an efficient procedure to prepare this type of skeleton from readily available starting materials. Starting from 2-halobenzylamines and β-ketoesters or 1,3-diketones, the catalyzed cross-coupling reaction took place in the presence of potassium carbonate in isopropyl alcohol at 90°C without the need for an additional ligand (this role being certainly assumed by the diketone or ketoester) and was followed by a mild aerobic dehydrogenation. Moderate-to-good yields were obtained under these conditions, and various degrees of substitution were tolerated on both partners (Scheme 15.43).[80]

In 2009, Fu and coworkers developed a simple and efficient method for synthesizing 3,4-disubstituted isoquinolin-1(2*H*)-one derivatives (Scheme 15.44).[81] The copper-catalyzed cross coupling between substituted 2-halobenzamides and β-ketoesters was performed under mild conditions without any additional ligand or additive, with the starting β-ketoester most certainly acting as a ligand for copper in this case. This method provides a practical strategy for the preparation of isoquinolin-1(2*H*)-one derivatives, useful building blocks in organic and medicinal chemistry.

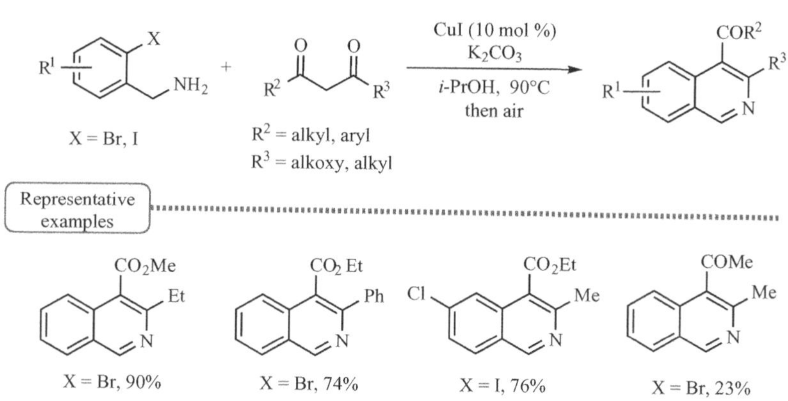

Scheme 15.43 Copper-catalyzed synthesis of isoquinolines from 2-halobenzylamines and β-ketoesters or 1,3-diketones.

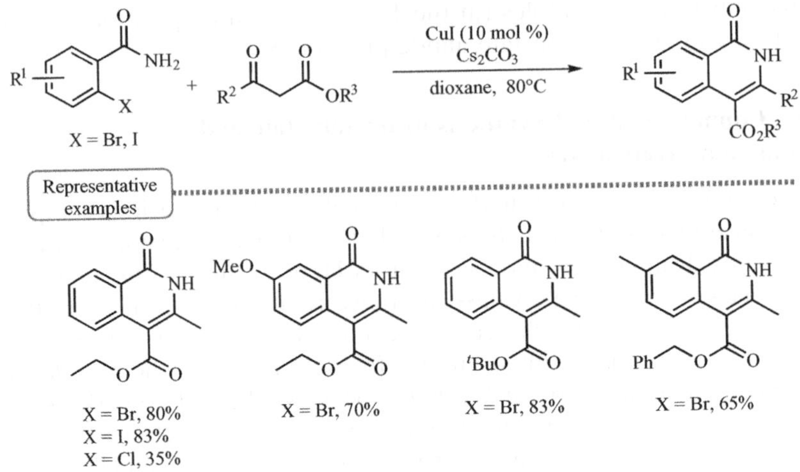

Scheme 15.44 Copper-catalyzed synthesis of isoquinolin-1(2*H*)-one derivatives from 2-halobenzamides and β-ketoesters.

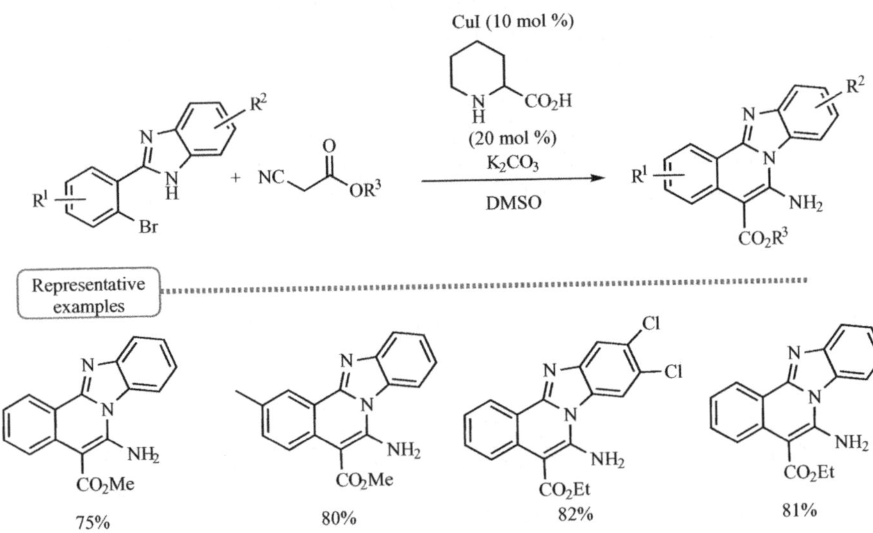

Scheme 15.45 Copper-catalyzed cascade synthesis of 6-aminobenzimidazo[2,1-*a*] isoquinoline-5-carboxylates.

This procedure can be extended to substrates possessing N-nucleophiles other than the carboxamide. The extension of this procedure to benzimidazoles was indeed reported by the same group, who developed a convenient and efficient copper-catalyzed cascade method for the synthesis of benzimidazoisoquinoline derivatives from substituted 2-(2-halophenyl)benzoimidazoles with alkyl cyanoacetates under mild conditions (Scheme 15.45).[82]

Using copper(I) iodide as the precatalyst, pipecolinic acid as the ligand, potassium carbonate as the base, and DMSO as the solvent, the targeted alkyl 6-aminobenzimidazo[2,1-*a*]isoquinoline-5-carboxylates resulting from an intermolecular Hurtley reaction followed by addition of the imidazole to the nitrile were obtained in good-to-excellent yields.

In 2011, a related reaction was reported by Fu and coworkers, who developed a copper-catalyzed synthesis of imidazoisoquinoline-2,5-diones via a tandem sequence starting from amino acid-derived 2-bromobenzamides and substituted phenylacetonitriles or malononitrile (Scheme 15.46).[83] No ligand or additive was required to induce the copper-catalyzed C-arylation due to the *ortho*-effect of the amide group. Functional groups such as amide, chloride, ether, hydroxy, imine, and nitrile were well tolerated under the reaction conditions, and the mechanism is likely to proceed via a C-arylation reaction of the activated methylene, followed by intramolecular addition of the amide moiety

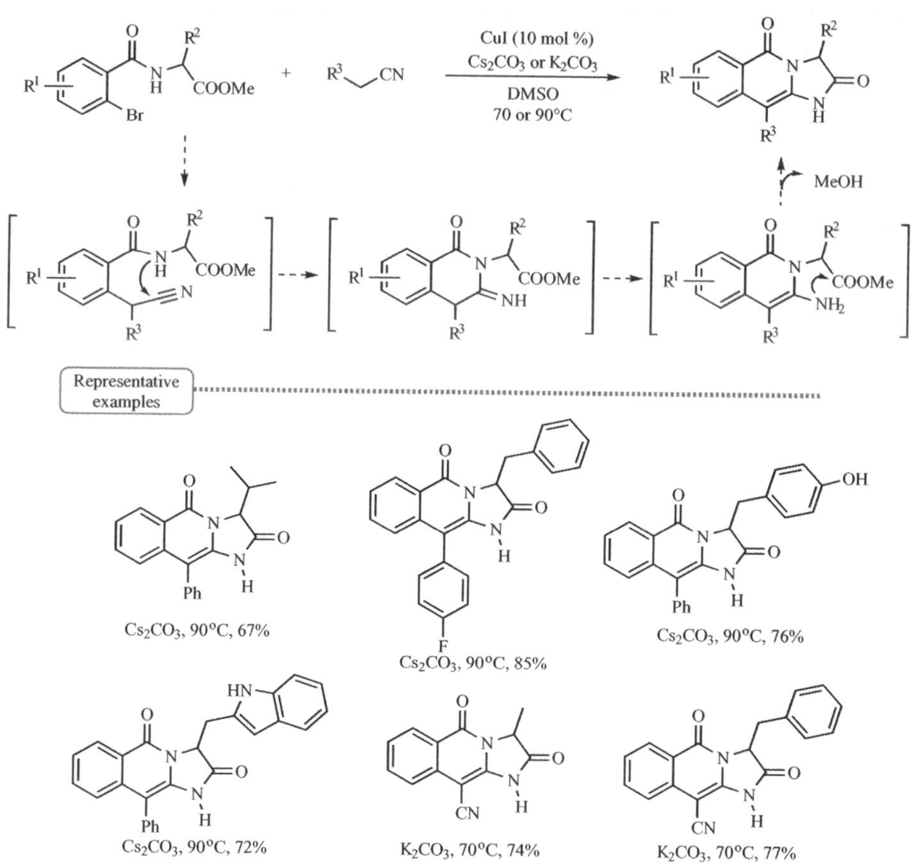

Scheme 15.46 Copper-catalyzed tandem synthesis of amino acid-derived imidazoisoquinoline-2,5-diones.

onto the nitrile group to give the corresponding imine. Further isomerization of this intermediate imine and intramolecular nucleophilic attack of the released amine onto the ester would afford the polycyclic compounds. Interestingly, substrates derived from tyrosine or tryptophan also gave the corresponding products in good yields.

As shown by the results collected in this section, the development of efficient systems that allow performing the Hurtley reaction under mild conditions had a deep impact in heterocyclic chemistry, and efficient routes to various scaffolds, some of them being more exotic than others, have recently emerged based on this transformation.

The copper-catalyzed Sonogashira-type cross coupling (see Chapter 11, Section 11.1, for details) has also been used to develop new routes to various heterocyclic skeletons in combination with a step where a C–heteroatom bond is formed. The next section is devoted to examples illustrating this strategy.

15.4.4 Heterocyclic Synthesis Based on a Copper-Catalyzed Sonogashira-Type Cross Coupling

In 2009, the Ma group reported a convenient approach to 3-methyleneisoindolin-1-ones via a copper-catalyzed Sonogashira-like[84] cross coupling between 2-bromobenzamides and terminal alkynes (Scheme 15.47).[85] This coupling/condensation domino process was found to be efficiently catalyzed by a copper(I) iodide/L-proline system (the copper salt also playing a key role in the activation of the triple bond after the cross coupling) and in a polar solvent, which might contribute to the partial dissolution of the catalyst (*i*-PrOH, DMF,

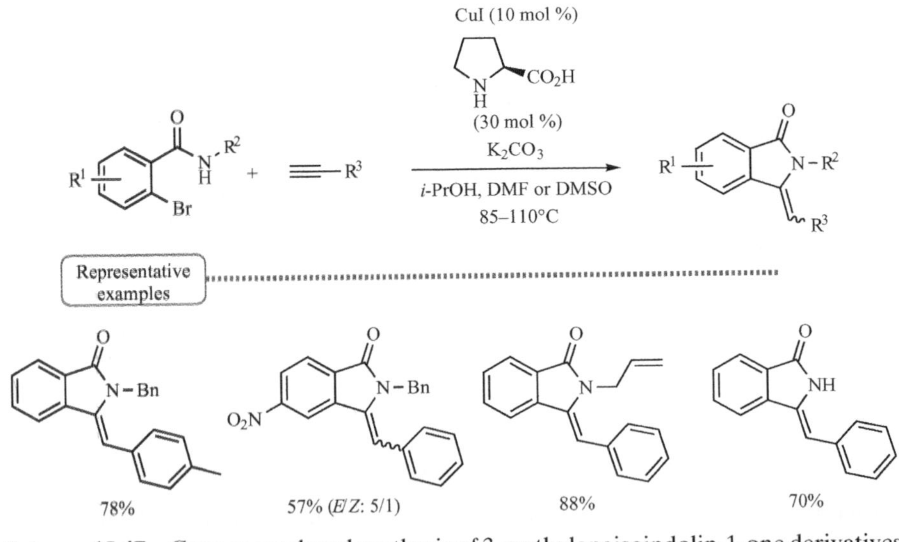

Scheme 15.47 Copper-catalyzed synthesis of 3-methyleneisoindolin-1-one derivatives.

and DMSO) at 85–110°C. The presence of subtituents on both partners gave substituted 3-methyleneisoindolin-1-ones, the *Z*-isomer being formed exclusively in most cases, in moderate-to-good yields. This procedure was then nicely implemented in an elegant synthesis of lennoxamine.

In a similar vein, Abarbri, Parrain, and coworkers described in 2009 a remarkably straightforward entry to γ-alkylidenebutenolides and isocoumarins by simply utilizing a (*Z*)-3-iodopropenoic acid with a terminal alkyne in the presence of catalytic amounts of copper(I) iodide in DMF at 55°C. A series of (*Z*)-5-alkylidene- or arylidene-5*H*-furan-2-ones resulting from a domino coupling–heterocyclization reaction could be obtained with good yields (Scheme 15.48, Eq. 1), and the method could be extended to the coupling of *o*-iodobenzoic acid with terminal alkynes (Scheme 15.48, Eq. 2).[86] In this case, however, mixtures of five- and six-membered ring products resulting from 6-*endo-dig* and 5-*exo-dig* competitive cyclization were obtained in most cases.

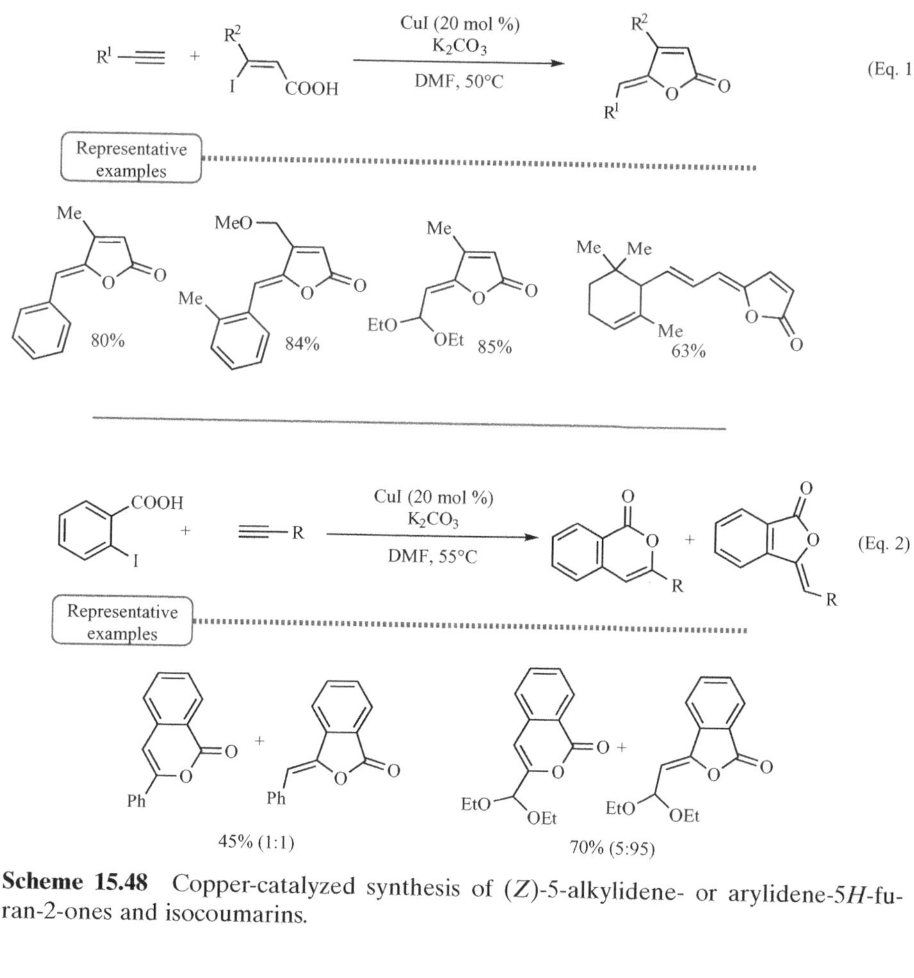

Scheme 15.48 Copper-catalyzed synthesis of (*Z*)-5-alkylidene- or arylidene-5*H*-furan-2-ones and isocoumarins.

15.5 COPPER-CATALYZED DOUBLE CROSS-COUPLING REACTIONS FOR THE ASSEMBLY OF HETEROCYCLES

Copper-catalyzed C–N, C–O, or C–C bond formation has been extensively used to develop new routes to various heterocyclic systems. As described in the previous sections, they have also been used in tandem sequences, which allow for the rapid elaboration of nitrogen or oxygen heterocycles from readily available starting materials and with a limited number of steps. In addition to these strategies, methods relying on a double copper-catalyzed cross-coupling reaction have recently emerged, and, in addition to their elegance, they also provide practical routes for the assembly of various heterocycles, such as benzimidazoles, benzoxazoles, benzofurans, pyrroles, or thiophene derivatives, provided that the starting dihalides can be easily prepared. Representative methods using this strategy will be discussed in this section starting with benzoxazoles, compounds for which the double copper-catalyzed cross-coupling strategy was used for the first time.

15.5.1 Synthesis of Benzoxazoles and Oxazoles by Double Copper-Catalyzed Cross Coupling

In 2004, Altenhoff and Glorius described a domino C–N, C–O copper-catalyzed process leading to benzoxazoles from *o*-dihaloarobenzenes and primary amides (Scheme 15.49).[87] Even if some primary amides did not react under

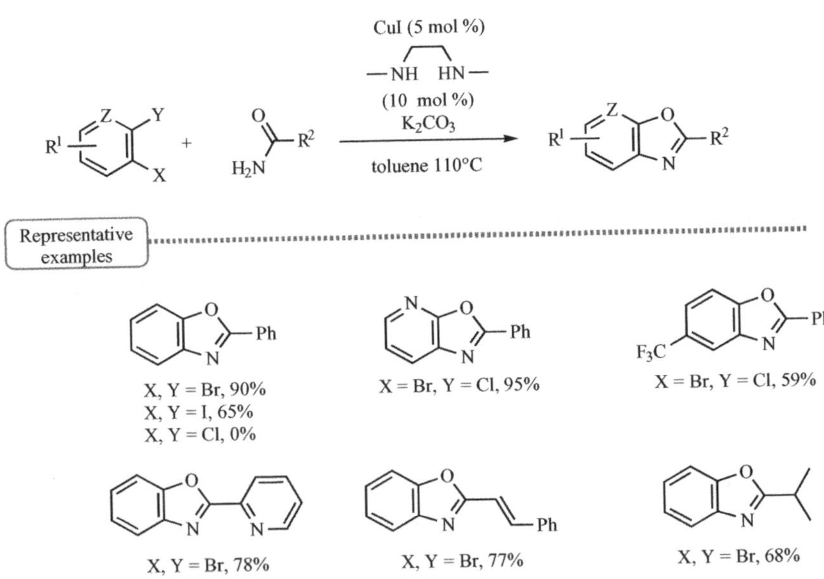

Scheme 15.49 Copper-catalyzed synthesis of benzoxazoles through domino formation of C–N and C–O bonds.

the reaction conditions, good-to-excellent yields were obtained with a broad scope of substrates, making this method a valuable tool to prepare benzoxazoles. While symmetrical aromatic dibromides or diiodides are efficiently coupled with primary amides to give the corresponding benzoxazoles, aryl dichlorides are not reactive under the reaction conditions. Starting from substituted *o*-bromochlorobenzenes, this domino reaction allows the formation of non-symmetrical benzoxazoles with excellent yields and complete regioselectivity, with the first reaction of the domino process involving the N-arylation of the carboxamide with the aryl bromide. As a note, despite many palladium-based catalysts being known to catalyze C–N and C–O bond formations, all palladium complexes that were tested failed to produce the desired products.

This strategy was implemented 3 years later to the preparation of oxazoles by replacing the aryl dihalides by the corresponding alkenyl dihalides (Scheme 15.50). While the preparation of 1,2-dihaloalkenes possessing two different halogens requires several steps, which would diminish the overall attractiveness of the oxazole synthesis starting from these building blocks (even if their use would certainly ensure high levels of regioselectivity), the corresponding dibrominated alkenes were easily prepared in high yields from the corresponding alkynes by treatment with bromine. The double cross coupling using Buchwald's conditions proceeded smoothly from these alkenyl dibromides, and the corresponding 2,5-disubstituted oxazoles were obtained in good yields. Although efficient, this oxazole synthesis is, however, not fully regioselective and affords mixtures of regioisomeric oxazoles in favor of the 2,5-disubstituted (ratio of 2,5-/2,4-oxazoles varies from 5/1 to 14/1).[88] Intriguingly, *E*-, *Z*-, or mixtures of *E*- and *Z*-configured dihalogenated olefins all reacted to form the desired oxazoles. In some cases, an isomerization of *E*- and Z-alkenes was

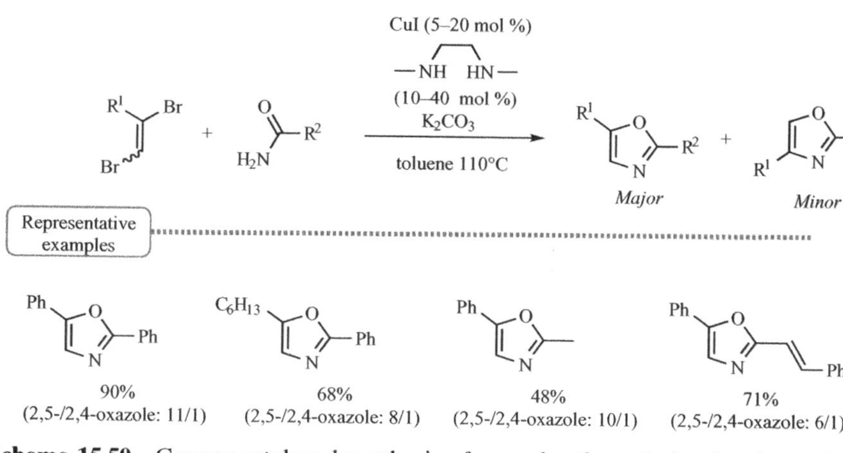

Scheme 15.50 Copper-catalyzed synthesis of oxazoles through domino formation of C–N and C–O bonds.

observed under the reaction conditions, but not systematically, indicating that isomerization prior to enamide formation does not seem to be a prerequisite for oxazole formation. Many functional groups such as halides, amino-, methoxy-, and silyl groups were tolerated, and, in addition, the use of a trimethylsilylated olefin allowed the selective preparation of 2,4-disubstituted oxazoles after acid-mediated protodesilylation. Substituents such as vinyl, chloro, fluoro, nitro, carboxylate, ketone, and silyl ether groups were suitable for this reaction and offered a wide range of substituted oxazoles in moderate-to-good yields. Attempts at expanding this double cross coupling to the use of thioamides, urea, pyrrolidinone, or carbamates were not successful. An interesting alternative that overcomes the regioselectivity issues lies in the vinylation of primary amides with vinyl bromides followed by either an iodine-promoted cyclization in the presence of DBU[89] or a copper(II) bromide/ethyl nicotinate-catalyzed oxidative cyclization.[90]

15.5.2 Synthesis of Benzimidazoles by Double Copper-Catalyzed Cross Coupling

In 2009, Deng and coworkers extended this strategy to the preparation of benzimidazole derivatives starting from 1,2-dihaloarenes, and amidines or guanidines (Scheme 15.51, Eq. 1).[91] Using copper(I) iodide as the catalyst, *N,N′*-dimethylethylenediamine as the ligand, cesium carbonate as the base, and *N,N*-dimethylacetamide (DMA) or *N*-methylpyrrolidin-2-one (NMP) as the solvent required high temperatures and gave the corresponding benzimidazoles in moderate yields only starting from 1,2-dihaloarenes, where the two halogen atoms were not differentiated. A year later, the Deng group described an improved regioselective copper-catalyzed tandem amination reaction

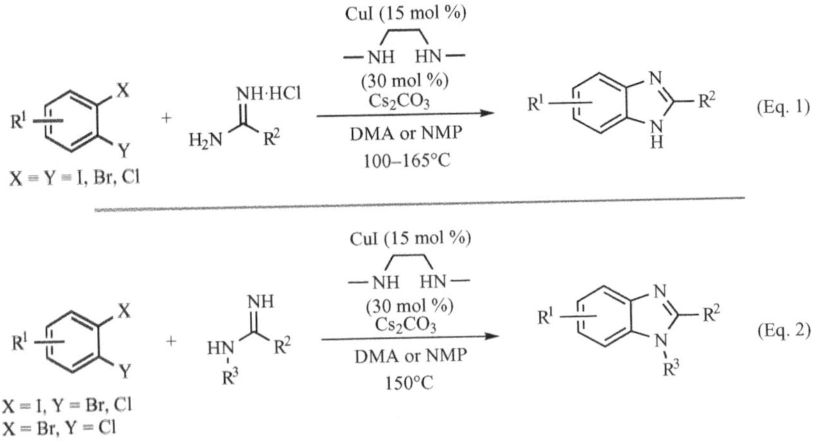

Scheme 15.51 Copper-catalyzed synthesis of benzimidazole derivatives through sequential formation of two C–N bonds.

between 1,2-dihaloarenes possessing two different halides and *N*-substituted amidines. The yields were good, although high reaction temperatures were still needed, and the regiochemical outcome was achieved thanks to a combination of N^1/N^2 chemoselectivity on the amidine and reactivity-controlled X/Y chemoselectivity on the 1,2-dihaloarenes (Scheme 15.51, Eq. 2).[92]

15.5.3 Synthesis of Benzofurans by Double Copper-Catalyzed Cross Coupling

In 2007, the Ma group reported an efficient copper-catalyzed cascade process to form 2,3-disubstituted benzofurans involving sequential intermolecular formation of a C–C bond followed by intramolecular formation of a C–O one (Scheme 15.52).[93] This copper-catalyzed double cross coupling between 1-bromo-2-iodobenzenes and β-ketoesters was performed without additional ligand or additive under mild conditions, with the β-ketoester certainly acting as the ligand for the copper salt. Substituents such as vinyl, chloro, fluoro, nitro, carboxylate, ketone, and silyl ether groups were suitable for this reaction and offered a wide range of substituted benzofurans in moderate-to-good yields. As in the previous cases, the regioselectivity is governed by the nature of the two substrates, with the arylation of the activated methylene with the most reactive halogen being the first step of the process. Subsequent intramolecular O-arylation of the ketone with the remaining halide then allows for the formation of the five-membered oxygenated heterocycle.

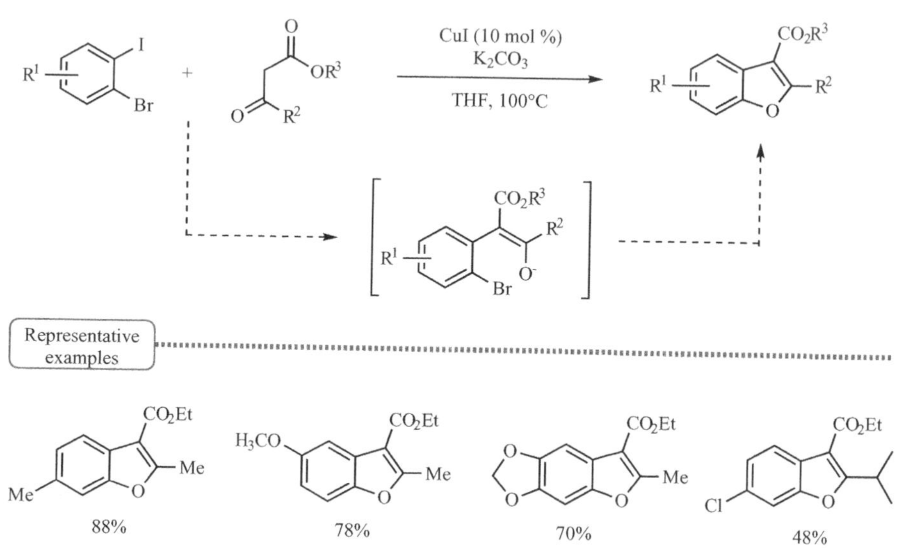

Scheme 15.52 Synthesis of substituted benzofurans through double copper-catalyzed cross coupling involving C–C and C–O bonds formation.

15.5.4 Synthesis of Benzimidazo[1,2-*b*]isoquinolin-11-one Derivatives by Double Copper-Catalyzed Cross Coupling

Benzimidazoisoquinolinone derivatives are valuable substrates and display interesting biological activities: They have, for example, been used as potent anti-*Trypanosoma cruzi* agents. However, methods for their preparation remain scarce and are not general enough to fully exploit this scaffold. In 2010, Fu and coworkers described an efficient and practical copper-catalyzed tandem synthesis of benzimidazo[1,2-*b*]isoquinolin-11-one derivatives (Scheme 15.53).[94] The reactions of substituted 2-halo-*N*-(2-halophenyl)benzamides with alkyl 2-cyanoacetates or malononitrile were performed under mild conditions without addition of any ligand or additive. A possible mechanism for

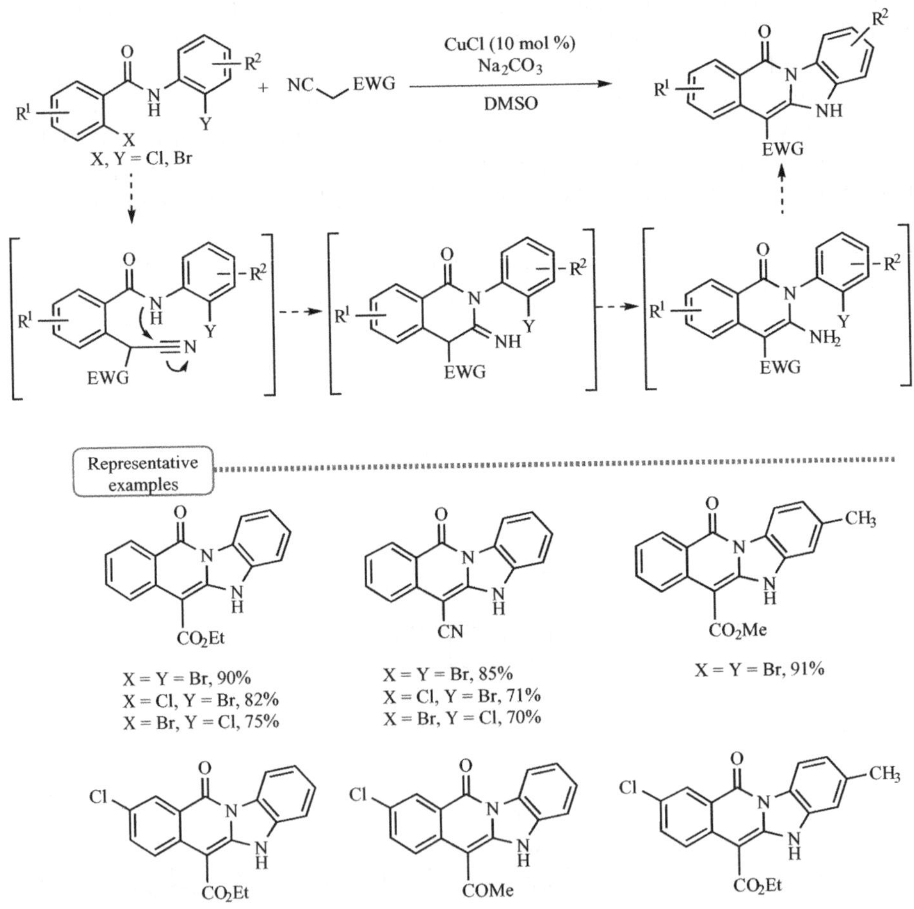

Scheme 15.53 Synthesis of benzimidazo[1,2-*b*]isoquinolin-11-one derivatives through double copper-catalyzed cross coupling involving C–C and C–N bonds formation.

the formation of benzimidazo[1,2-*b*]isoquinolin-11-one derivatives was proposed and is shown in Scheme 15.53. First, a copper-catalyzed Ullmann-type coupling between the starting 2-halo-*N*-(2-halophenyl)benzamide and alkyl 2-cyanoacetate (C-arylation of alkyl 2-cyanoacetate) would provide the C-arylated product. Then, a base-promoted nucleophilic attack of the amide to the nitrile followed by isomerization would generate a conjugated amino group whose subsequent intramolecular N-arylation would generate the tetracyclic product.

15.5.5 Synthesis of Pyrrole and Thiophene Derivatives by Double Copper-Catalyzed Cross Coupling

Besides dihaloarenes and alkenyl 1,2-dihalides, 1,4-dihalo-1,3-dienes are also suitable substrates for the synthesis of five-membered ring heterocycles using double copper-catalyzed cross-coupling reactions. Depending on their substitution patterns, they are more or less readily accessible, with the method of choice for their synthesis relying on iodination of zircona- or titanacyclopentadienes, metallacycles that are available through zirconium-mediated cross coupling of alkynes[95] or treatment of (η^2-propene)Ti(OiPr)$_2$ with internal and terminal alkynes,[96] respectively.

In 2007, Buchwald and coworkers reported a highly efficient synthesis of pyrroles and heteroarylpyrroles through double copper-catalyzed cascade C–N bond formation from such substrates (Scheme 15.54).[97] The protocol used 1,4-dihalo-1,3-dienes and *tert*-butylcarbamate as the starting materials and Buchwald's classical catalytic system based on copper(I) iodide and

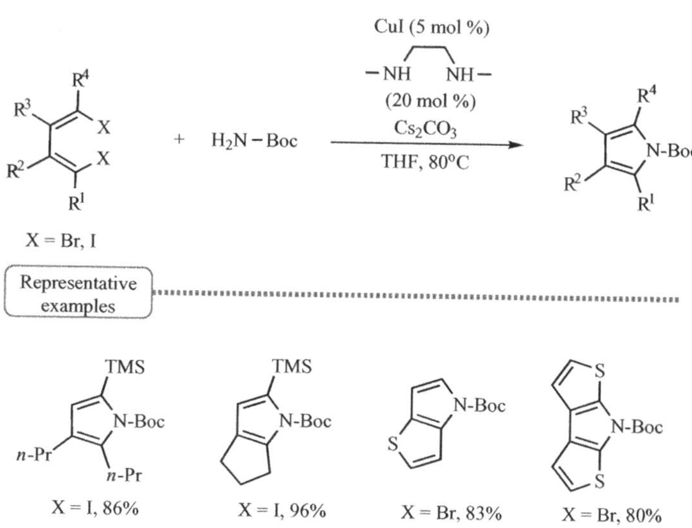

Scheme 15.54 Synthesis of *N*-Boc-pyrroles through double copper-catalyzed C–N bond formation.

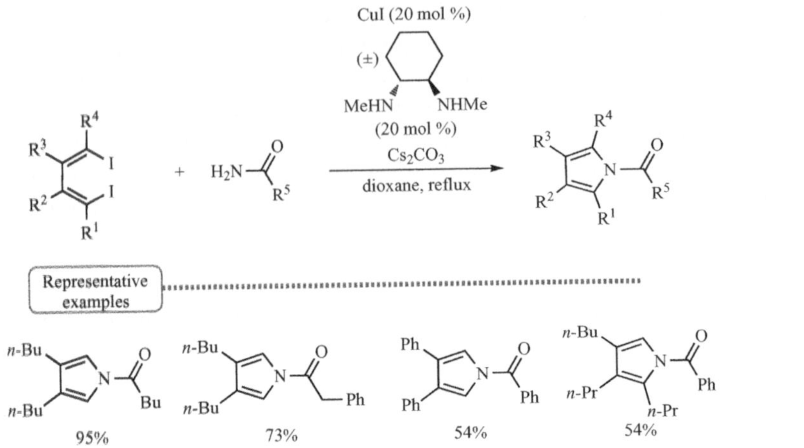

Scheme 15.55 Synthesis of *N*-acyl-pyrroles through double copper-catalyzed C–N bond formation.

N,N′-dimethylethylenediamine. The transformation is distinguished by its mild conditions, allowing the tolerance of a wide variety of functional groups in a range of substitution patterns, and is also efficient for the preparation of heteroaryl- and sterically encumbered pyrroles.

That same year, Li and coworkers described a related copper-catalyzed synthesis of *N*-acylpyrroles by cross coupling between (1*Z*,3*Z*)-1,4-diiodo-1,3-dienes and amides (Scheme 15.55).[98] The protocol used copper(I) iodide as the catalyst, *trans-N,N′*-dimethylcyclohexane-1,2-diamine as the ligand, and cesium carbonate as the base and afforded di- or trisubstituted *N*-acylpyrroles in good-to-excellent yields. Stoichiometric amounts of both the copper salt and the ligand were, however, required when the reaction was performed from ethyl carbamate.

This reaction was later on extended to the preparation of *N*-arylpyrroles by Xi and coworkers by double copper-catalyzed alkenylation of anilines with dienyl diiodides (Scheme 15.56).[99] A copper(I) iodide/*N,N′*-dimethylethylenediamine catalytic system was used, in the presence of sodium *tert*-butoxide, to perform the sequential intermolecular/intramolecular process. Other copper sources were also tested, such as copper(I) bromide or copper(I) chloride, but proved to be less effective precatalysts. The substitution patterns of the aromatic amines had virtually no effect on the yields.

This methodology was finally applied to the synthesis of various thiophenes under similar conditions. The double copper-catalyzed C–S bond formation proved equally efficient in the absence of ligand and additive under rather harsh conditions, however, since the reaction had to be performed at 140°C (Scheme 15.57).[100]

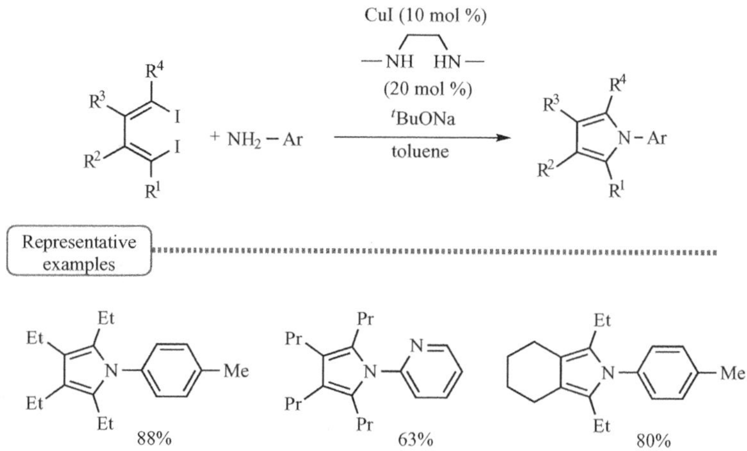

Scheme 15.56 Synthesis of *N*-aryl-pyrroles through double copper-catalyzed C–N bond formation.

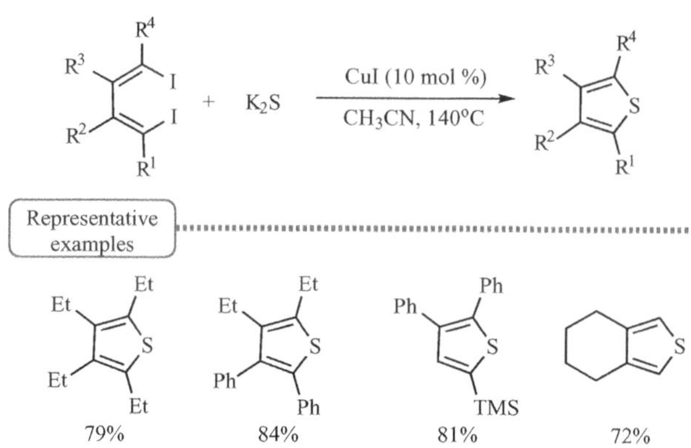

Scheme 15.57 Synthesis of thiophenes through double copper-catalyzed C–S bond formation.

15.6 CONCLUSION AND FUTURE PROSPECTS

Heterocyclic synthesis involving transition-metal complexes has become commonly used in the past decade because transition-metal-catalyzed reactions can directly build complicated molecules from readily accessible starting materials under mild conditions. The formation of heterocycles by using various transition metals such as palladium, nickel, ruthenium, and rhodium has been extensively investigated and documented in the literature and provides new efficient entries to various heterocyclic systems. One should keep in mind, however, that another challenge for the development of new routes in heterocyclic synthesis is to develop methods that are cost-efficient. Heterocyclic compounds are mostly used in the pharmaceutical, agrochemical, and polymer industries, where the cost of starting materials has a strong impact on many processes. Copper catalysis is therefore an excellent and attractive alternative to the use of other metals that are also, in most cases, more toxic than copper.

As discussed in this chapter, copper-mediated processes for the synthesis of heterocycles are original, straightforward, and remarkably efficient for the preparation of a wide range of heterocycles, and the development of straightforward cyclization procedures allows for new directions in heterocyclic synthesis. Altogether, these copper-mediated syntheses of heterocycles present various advantages compared to other traditional pathways, such as the use of inexpensive catalysts and readily available starting materials under relatively mild conditions. Usually, a good tolerance toward a wide range of functional groups is observed. The synthesis of heterocycles based on copper-catalyzed reactions remains a very attractive and promising field as highlighted by the ever-increasing number of publications in this area. We believed that many other methods should soon complement the ones discussed in this chapter, especially those based on copper-mediated C–H functionalization. Lowering the catalyst loading, which is still rather high in all processes, will also represent an important challenge for the years to come.

REFERENCES

[1] Patil, N. T.; Yamamoto, Y. *Chem. Rev.* **2008**, *108*, 3395–3442.

[2] Ley, S. V.; Thomas, A. W. *Angew. Chem., Int. Ed.* **2003**, *42*, 5400–5449.

[3] Beletskaya, I. P.; Cheprakov, A. V. *Coord. Chem. Rev.* **2004**, *248*, 2337–2364.

[4] Evano, G.; Blanchard, N.; Toumi, M. *Chem. Rev.* **2008**, *108*, 3054–3131.

[5] Leeson, P. D.; Springthorpe, B. *Nat. Rev. Drug Discovery* **2007**, *6*, 881–890.

[6] Klapars, A.; Huang, X.; Buchwald, S. L. *J. Am. Chem. Soc.* **2002**, *124*, 7421–7428.

[7] Kwong, F. Y.; Buchwald, S. L. *Org. Lett.* **2003**, *5*, 793–796.

[8] Zhang, H.; Cai, Q.; Ma, D. *J. Org. Chem.* **2005**, *70*, 5164–5173.

[9] Shafir, A.; Buchwald, S. L. *J. Am. Chem. Soc.* **2006**, *128*, 8742–8743.

[10] Yamada, K.; Kubo, T.; Tokuyama, H.; Fukuyama, T. *Synlett* **2002**, 231–234.

[11] Zhou, F.; Guo, J.; Liu, J.; Ding, K.; Yu, S.; Cai, C. *J. Am. Chem. Soc.* **2012**, *134*, 14326–14329.

[12] Coste, A.; Toumi, M.; Wright, K.; Razafimahaléo, V.; Couty, F.; Marrot, J.; Evano, G. *Org. Lett.* **2008**, *10*, 3841–3844.

[13] (a) Meng, L.; Fettinger, J. C.; Kurth, M. J. *Org. Lett.* **2007**, *9*, 5055–5058; (b) Liao, Y.-L.; Lin, C.-Y.; Wong, K.-T.; Hou, T.-H.; Hung, W.-Y. *Org. Lett.* **2007**, *9*, 4511–4514.

[14] Yang, D.; Fokas, D.; Li, J.; Yu, L.; Baldino, C. M. *Synthesis* **2005**, 47–56 and references cited therein.

[15] Zheng, N.; Anderson, K. W.; Huang, X.; Nguyen, H. N.; Buchwald, S. L. *Angew. Chem. Int. Ed.* **2007**, *46*, 7509–7512.

[16] For additional recent reviews on the use of copper-catalyzed Ullmann cross-coupling reactions, see: (a) Ma, D.; Cai, Q. *Acc. Chem. Res.* **2008**, *41*, 1450–1460; (b) Kunz, K.; Scholz, U.; Ganzer, D. *Synlett* **2003**, 2428–2439; (c) Rao, H.; Fu, H. *Synlett* **2011**, 745–769.

[17] Zou, B.; Yuan, Q.; Ma, D. *Angew. Chem. Int. Ed.* **2007**, *46*, 2598–2601.

[18] Diao, X.; Wang, Y.; Jiang, Y.; Ma, D. *J. Org. Chem.* **2009**, *74*, 7974–7977.

[19] Yang, D.; Fu, H.; Hu, L.; Jiang, Y.; Zhao, Y. *J. Org. Chem.* **2008**, *73*, 7841–7844.

[20] Saha, P.; Ramana, T.; Purkait, N.; Ali, M. A.; Paul, R.; Punniyamurthy, T. *J. Org. Chem.* **2009**, *74*, 8719–8725.

[21] Evindar, G.; Batey, R. A. *Org. Lett.* **2003**, *5*, 133–136.

[22] Lv, X.; Bao, W. *J. Org. Chem.* **2009**, *74*, 5618–5621.

[23] Kumar, S.; Ila, H.; Junjappa, H. *J. Org. Chem.* **2009**, *74*, 7046–7051.

[24] (a) Brasche, G.; Buchwald, S. L. *Angew. Chem. Int. Ed.* **2008**, *47*, 1932–1934. (b) Li, J.; Bénard, S.; Neuville, L.; Zhu, J. *Org. Lett.* **2012**, *14*, 5980–5983. (c) Li, X.; He, L.; Chen, H.; Wu, W.; Jiang, H. *J. Org. Chem.* **2013**, *78*, 3636–3646. (d) Huang, P.-C.; Parthasarathy, K.; Cheng, C.-H. *Chem. Eur. J.* **2013**, *19*, 460–464.

[25] For a representative example, see: Kawamoto, H.; Nakashima, H.; Kato, T.; Arai, S.; Kamata, K.; Iwasawa, Y. *Tetrahedron* **2001**, *57*, 981–986.

[26] For a representative example, see: Li, Q.; Li, T.; Woods, K. W.; Gu, W.-Z.; Cohen, J.; Stoll, V. S.; Galicia, T.; Hutchins, C.; Frost, D.; Rosenberg, S. H.; Sham, H. L. *Bioorg. Med. Chem. Lett.* **2005**, *15*, 2918–2922.

[27] Zou, B.; Yuan, Q.; Ma, D. *Org. Lett.* **2007**, *9*, 4291–4294.

[28] Lu, B.; Ma, D. *Org. Lett.* **2006**, *8*, 6115–6118.

[29] Li, Z.; Sun, H.; Jiang, H.; Liu, H. *Org. Lett.* **2008**, *10*, 3263–3266.

[30] Barbero, N.; Carril, M.; SanMartin, R.; Domínguez, E. *Tetrahedron* **2008**, *64*, 7283–7288.

[31] Kutsuma, T.; Sakai, Y.; Ouchi, H.; Shiina, N.; Yamagishi, T.; Shibuya, S.; Yokomatsu, T. *Heterocycles* **2005**, *65*, 1967–1973.

[32] Isobe, Y.; Kurimoto, A.; Tobe, M.; Hashimoto, K.; Nakamura, T.; Norimura, K.; Ogita, H.; Takaku, H. *J. Med. Chem.* **2006**, *49*, 2088–2095.

[33] Zhong, Q.; Sun, L. *Tetrahedron* **2010**, *66*, 5107–5111.

[34] (a) Lewis, J. C.; Wiedemann, S. H.; Bergman, R. G.; Ellman, J. A. *Org. Lett.* **2004**, *6*, 35–38; (b) Shreder, K. R.; Wong, M. S.; Nomanbhoy, T.; Leventhal, P. S.; Fuller, S. R. *Org. Lett.* **2004**, *6*, 3715–3718; (c) Wiedemann, S. H.; Ellman, J. A.; Bergman, R. G. *J. Org. Chem.* **2006**, *71*, 1969–1976; (d) Mhaske, S. B.; Argade, N. P. *Tetrahedron* **2006**, *62*, 9787–9826 and references cited therein.

[35] Huang, C.; Fu, Y.; Fu, H.; Jiang, Y.; Zhao, Y. *Chem. Commun.* **2008**, 6333–6335.

[36] Truong, V. L.; Morrow, M. *Tetrahedron Lett.* **2010**, *51*, 758–760.

[37] Huang, X.; Yang, H.; Fu, H.; Qiao, R.; Zhao, Y. *Synthesis* **2009**, 2679–2688.

[38] Wang, C.; Li, S.; Liu, H.; Jiang, Y.; Fu, H. *J. Org. Chem.* **2010**, *75*, 7936–7938.

[39] Reddy, K. R.; Krishna, G. G. *Tetrahedron Lett.* **2005**, *46*, 661–663.

[40] (a) Xu, S.; Lu, J.; Fu, H. *Chem. Commun.* **2011**, *47*, 5596–5598. (b) Guo, S.; Wang, J.; Fan, X.; Zhang, X.; Guo, D. *J. Org. Chem.* **2013**, *78*, 3262–3270.

[41] Liu, X.; Fu, H.; Jiang, Y.; Zhao, Y. *Angew. Chem. Int. Ed.* **2009**, *48*, 348–351.

[42] Xu, W.; Jin, Y.; Liu, H.; Jiang, Y.; Fu, H. *Org. Lett.* **2011**, *13*, 1274–1277.

[43] Yang, D.; Liu, H.; Yang, H.; Fu, H.; Hu, L.; Jiang, Y.; Zhao, Y. *Adv. Synth. Catal.* **2009**, *351*, 1999–2004.

[44] Wang, F.; Zhao, P.; Xi, C. *Tetrahedron Lett.* **2011**, *52*, 231–235.

[45] For a review article, see: Humphrey, G. R.; Kuethe, J. T. *Chem. Rev.* **2006**, *106*, 2875–2911.

[46] Phillips, R. R. *Org. React.* **1959**, *10*, 143–178.

[47] Tsoungas, P. G.; Diplas, A. I. *Curr. Org. Chem.* **2004**, *8*, 1579–1606.

[48] Zeni, G.; Larock, R. C. *Chem. Rev.* **2006**, *106*, 4644–4680.

[49] (a) Melkonyan, F.; Topolyan, A.; Yurovskaya, M.; Karchava, A. *Eur. J. Org. Chem.* **2008**, 5952–5956. (b) Selander, N.; Worrell, B. T.; Chuprakov, S.; Velaparthi, S.; Fokin, V. V. *J. Am. Chem. Soc.* **2012**, *134*, 14670–14673.

[50] Cai, Q.; Li, Z.; Wei, J.; Ha, C.; Pei, D.; Ding, K. *Chem. Commun.* **2009**, 7581–7583.

[51] Koenig, S. G.; Dankwardt, J. W.; Liu, Y.; Zhao, H.; Singh, S. P. *Tetrahedron Lett.* **2010**, *51*, 6549–6551.

[52] Hu, T.; Li, C. *Org. Lett.* **2005**, *7*, 2035–2038.

[53] Lu, H.; Li, C. *Org. Lett.* **2006**, *8*, 5365–5367.

[54] Li, E.; Xu, X.; Li, H.; Zhang, H.; Xu, X.; Yuan, X.; Li, Y. *Tetrahedron* **2009**, *65*, 8961–8968.

[55] Zhang, X.; Zhang, Y.; Huang, J.; Hsung, R. P.; Kurtz, K. C. M.; Oppenheimer, J.; Petersen, M. E.; Sagamanova, I. K.; Shen, L.; Traceyn, M. R. *J. Org. Chem.* **2006**, *71*, 4170–4177.

[56] Rao, R. K.; Naidu, A. B.; Sekar, G. *Org. Lett.* **2009**, *11*, 1923–1926.

[57] (a) Lu, X.; Bao, W. *J. Org. Chem.* **2007**, *72*, 3863–3867; (b) Lipshutz, B. H.; Unger, J. B.; Taft, B. R. *Org. Lett.* **2007**, *9*, 1089–1092; (c) Gujadhur, R. K.; Bates, C. G.; Venkataraman, D. *Org. Lett.* **2001**, *3*, 4315–4317; (d) Ma, D.; Cai, Q. *Org. Lett.* **2003**, *5*, 3799–3802; (e) Fagan, P. J.; Hauptman, E.; Shapiro, R.; Casalnuovo, A. *J. Am. Chem. Soc.* **2000**, *122*, 5043–5051; (f) Marcoux, J.-F.; Doye, S.; Buchwald, S. L. *J. Am. Chem. Soc.* **1997**, *119*, 10539–10540; (g) Cai, Q.; Zou, B.; Ma, D. *Angew. Chem. Int. Ed.* **2006**, *45*, 1276–1279; (h) Gujadhur, R.; Venkataraman, D. *Synth. Commun.* **2001**, *31*, 2865–2879.

[58] (a) Pottorf, R. S.; Chadha, N. K.; Katkevics, M.; Ozola, V.; Suna, E.; Ghane, H.; Regberg, T.; Player, M. R. *Tetrahedron Lett.* **2003**, *44*, 175–178; (b) Bougrin, K.; Loupy, A.; Soufiaoui, M. *Tetrahedron* **1998**, *54*, 8055–8064.

[59] (a) Tale, R. H. *Org. Lett.* **2002**, *4*, 1641–1642; (b) Kawashita, Y.; Nakamichi, N.; Kawabata, H.; Hayashi, M. *Org. Lett.* **2003**, *5*, 3713–3715; (c) Varma, R. S.; Kumar, D. *J. Heterocyclic Chem.* **1998**, *35*, 1539–1540.

[60] Hein, D. W.; Alheim, R. J.; Leavitt, J. J. *J. Am. Chem. Soc.* **1957**, *79*, 427–429.

[61] Evindar, G.; Batey, R. A. *J. Org. Chem.* **2006**, *71*, 1802–1808.

[62] Barbero, N.; Carril, M.; SanMartin, R.; Dominguez, E. *Tetrahedron* **2007**, *63*, 10425–10432.

[63] Inamoto, K.; Saito, T.; Katsuno, M.; Sakamoto, T.; Hiroya, K. *Org. Lett.* **2007**, *9*, 2931–2934.

[64] Inamoto, K.; Hasegawa, C.; Hiroya, K.; Doi, T. *Org. Lett.* **2008**, *10*, 5147–5150.

[65] Li, J.-J.; Mei, T.-S.; Yu, J.-Q. *Angew. Chem. Int. Ed.* **2008**, *47*, 6452–6455.

[66] Inamoto, K.; Arai, Y.; Hiroya, K.; Doi, T. *Chem. Commun.* **2008**, 5529–5531.

[67] Ueda, S.; Nagasawa, H. *J. Org. Chem.* **2009**, *74*, 4272–4277.

[68] Fang, Y.; Li, C. *J. Am. Chem. Soc.* **2007**, *129*, 8092–8093.

[69] Fang, Y.-W.; Li, C.-Z. *Chem. Commun.* **2005**, 3574–3576.

[70] Chen, L.; Fang, Y.; Zhao, Q.; Shi, M.; Li, C. *Tetrahedron Lett.* **2010**, *51*, 3678–3681.

[71] (a) Fang, Y.-W.; Li, C.-Z. *J. Org. Chem.* **2006**, *71*, 6427–6431. (b) Sudheendran, K.; Malakar, C. C.; Conrad, J.; Beifuss, U. *J. Org. Chem.* **2012**, *77*, 10194–10210.

[72] (a) Chen, C.-Y.; Dormer, P. G. *J. Org. Chem.* **2005**, *70*, 6964–6967. (b) Besandre, R.; Jaimes, M.; May, J. A. *Org. Lett.* **2013**, *15*, 1666–1669.

[73] Sun, C.-H.; Fang, Y.-W.; Li, S.; Zhang, Y.; Zhao, Q.-W.; Zhu, S.-N.; Li, C.-Z. *Org. Lett.* **2009**, *11*, 4084–4087.

[74] Thasana, N.; Worayuthakarn, R.; Kradanrat, P.; Hohn, E.; Young, L.; Ruchirawat, S. *J. Org. Chem.* **2007**, *72*, 9379–9382.

[75] For representative examples, see: (a) Xie, X.; Cai, G.; Ma, D. *Org. Lett.* **2005**, *7*, 4693–4695; (b) Ma, D.; Cai, Q. *Synlett* **2004**, 128–130; (c) Zhu, W.; Ma, D. *Chem. Commun.* **2004**, 888–889; (d) Jiang, Y.; Wu, N.; Wu, H.; He, M. *Synlett* **2005**, 2703–2734; (e) Hennessy, E. J.; Buchwald, S. L. *Org. Lett.* **2002**, *4*, 269–272. (f) Fang, Y.; Li, C. *J. Org. Chem.* **2006**, *71*, 6427–6431; (g) Cristau, H.-J.; Cellier, P. P.; Spindler, J.-F.; Taillefer, M. *Chem. Eur. J.* **2004**, *10*, 5607–5622.

[76] Tanimori, S.; Ura, H.; Kirihata, M. *Eur. J. Org. Chem.* **2007**, *72*, 3977–3980.

[77] Chen, Y.; Xie, X.; Ma, D. *J. Org. Chem.* **2007**, *72*, 9329–9234.

[78] Chen, Y.; Wang, Y.; Sun, Z.; Ma, D. *Org. Lett.* **2008**, *10*, 625–628.

[79] Yang, X.; Fu, H.; Qiao, R.; Jiang, Y.; Zhao, Y. *Adv. Synth. Catal.* **2010**, *352*, 1033–1038.

[80] Wang, B.; Lu, B.; Jiang, Y.; Zhang, Y.; Ma, D. *Org. Lett.* **2008**, *10*, 2761–2763.

[81] Wang, F.; Liu, H.; Jiang, Y.; Zhao, Y. *Org. Lett.* **2009**, *11*, 2469–2472.

[82] Lu, J.; Fu, H. *J. Org. Chem.* **2011**, *76*, 4600–4605.

[83] Liu, T.; Wang, R.; Yang, H.; Fu, H. *Chem. Eur. J.* **2011**, *17*, 6765–6771.

[84] For examples of palladium-free Sonogashira-type cross coupling, see: (a) Ma, D.; Liu, F. *Chem. Commun.* **2004**, 1934–1935; (b) Li, J.; Li, J.; Wang, D.; Pi, S.; Xie, Y.; Zhang, M.; Hu, X. *J. Org. Chem.* **2007**, *72*, 2053–2057; (c) Chen, G.; Zhu, X.; Cai, J.; Wan, Y. *Synth. Commun.* **2007**, *37*, 1355–1361; (d) Guan, J. T.; Yu, G.-A.; Chen, L.; Weng, T. Q.; Yuan, J. J.; Liu, S. H. *Appl. Organomet. Chem.* **2009**, *23*, 75–77.

[85] Li, L.; Wang, M.; Zhang, X.; Jiang, Y.; Ma, D. *Org. Lett.* **2009**, *11*, 1309–1311.

[86] Inack-Ngi, S.; Rahmani, R.; Commeiras, L.; Chouraqui, G.; Thibonnet, J.; Duchêne, A.; Abarbri, M.; Parrain, J.-L. *Adv. Synth. Catal.* **2009**, *351*, 779–788.

[87] Altenhoff, G.; Glorius, F. *Adv. Synth. Catal.* **2004**, *346*, 1661–1164.

[88] Schuh, K. M.; Glorius, F. *Synthesis* **2007**, 2297–2306.

[89] Martin, R.; Cuanca, A.; Buchwald, S. L. *Org. Lett.* **2007**, *9*, 5521–5524.

[90] Cheung, C. W.; Buchwald, S. L. *J. Org. Chem.* **2012**, *77*, 7526–7537.

[91] Deng, X.; McAllister, H.; Mani, N. S. *J. Org. Chem.* **2009**, *74*, 5472–5475.

[92] Deng, X.; Mani, N. S. *Eur. J. Org. Chem.* **2010**, *680*, 33, 680–686.

[93] Lu, B.; Wang, B.; Zhang, Y.; Ma, D. *J. Org. Chem.* **2007**, *72*, 5337–5341.

[94] Lu, J.; Gong, X.; Yang, H.; Fu, H. *Chem. Commun.* **2010**, *46*, 4172–4174.

[95] Buchwald, S. L.; Nielsen, R. B. *J. Am. Chem. Soc.* **1989**, *111*, 2870–2874.

[96] Hamada, T.; Suzuki, D.; Urabe, H.; Sato, F. *J. Am. Chem. Soc.* **1999**, *121*, 7342–7344.

[97] Martin, R.; Larsen, C. H.; Cuenca, A.; Buchwald, S. L. *Org. Lett.* **2007**, *9*, 3379–3382.

[98] Yuan, X.; Xu, X.; Zhou, X.; Yuan, J.; Mai, L.; Li, Y. *J. Org. Chem.* **2007**, *72*, 1510–1513.

[99] Liao, Q.; Zhang, L.; Wang, F.; Li, S.; Xi, C. *Eur. J. Org. Chem.* **2010**, 5426–5431.

[100] You, W.; Yan, X.; Liao, Q.; Xi, C. *Org. Lett.* **2010**, *12*, 3930–3933.

16

APPLICATION OF COPPER-MEDIATED C–N BOND FORMATION IN COMPLEX MOLECULES SYNTHESIS

JIHOON LEE AND JAMES S. PANEK

16.1 INTRODUCTION

The development of copper-mediated cross-coupling reactions for the formation of C–N bonds has led to numerous reports of its application in the synthesis of natural products and nitrogen-containing bioactive molecules. This strategy, which has been used in the synthesis of an impressive array of complex alkaloids, has several advantages. One of them is that synthetic equivalents (cf. $-NO_2$, $-N_3$) of the amine group can be unmasked prior to C–N bond formation, thus allowing it to be introduced at a late stage in the synthesis. As a result, redundant protection–deprotection processes of an amine (or its synthetic equivalent) can be circumvented, and as a result, the unnecessary loss of materials is avoided. Moreover, relatively unreactive amides can successfully undergo N-arylation or N-vinylation to convert aryl or vinyl halides into anilides and enamides.

The successful application of an intramolecular reaction that allows for a mild and efficient pathway to access nitrogen-containing heterocycles has also been demonstrated by the synthesis of a wide range of biologically active natural products. In fact, the development and execution of a copper-mediated C–N bond formation can be used as an efficient strategy for macrocyclization in the assembly of medium- to large-sized lactams. Several examples are now

Copper-Mediated Cross-Coupling Reactions, First Edition. Edited by Gwilherm Evano and Nicolas Blanchard.

available that support this notion and are illustrative of the advantages of employing a copper-mediated C–N bond-forming reaction at any stage of a complex molecule synthesis. The development of this methodology has created a new retrosynthetic analytical disconnection and has been shown to significantly simplify syntheses of complex alkaloids and macrocyclic systems. In this chapter, we intend to survey recent and selected syntheses of natural products and bioactive molecules that highlight the copper-promoted C–N bond construction methodology.

This chapter is organized into six subsections, beginning with the preparation of anilines by aryl amination. A section on the preparation of anilides through aryl amidation is then followed by a section overviewing the use of copper-mediated arylation of nitrogen heterocycles in total synthesis. The fourth section is focused on the synthesis of enamides, a difficult functional group to assemble that is greatly simplified by the use of copper-mediated cross-coupling procedures. A section on ynamidation follows and nicely illustrates another effective procedure for the synthesis of ynamides. Chapter 16 concludes with intramolecular amination and amidation reactions to assemble small- and medium-sized rings, and with macrocyclization strategies for the formation of macrocyclic lactams.

16.2 ARYL AMINATION IN COMPLEX MOLECULE SYNTHESIS

Ma and coworkers reported an early and especially elegant example of a copper-mediated amination and its application in complex molecule synthesis in 1998.[1] More specifically, they developed an efficient copper-mediated N-arylation of optically pure α-amino acids with aryl halides under modified Ullmann conditions. This methodology is advantageous in that reaction conditions are relatively simple, mild, and economical compared to established amination methodologies involving organotin or organoborane reagents or palladium-catalyzed amination, and enantiomeric purities of configurationally sensitive amino acids are completely transferred to the products. It is postulated that formation of a stable five-membered cyclic complex between the amino acid and copper is involved in the reaction and facilitates the copper-mediated N-arylation. This concept was later used to develop an efficient amino acid-promoted, copper-catalyzed arylation of N-nucleophiles (see Chapter 1 for details).

Benzolactam-V8 **6**, a core structure of a new class of protein kinase C activators, was synthesized starting with the amination of 2-iodobenzyl alcohol **2** with L-valine **1** as the key step of the synthesis, a reaction that efficiently produced the key intermediate **3** in 86% yield (Scheme 16.1).[1] Then, a simple manipulation to introduce two additional carbon atoms as well as the amine moiety was achieved using an aldol-type reaction with methyl 2-nitroacetate, and subsequent diastereoselective reduction of **4** produced the advanced intermediate **5** whose reduction finally furnished benzolactam V8 **6**.

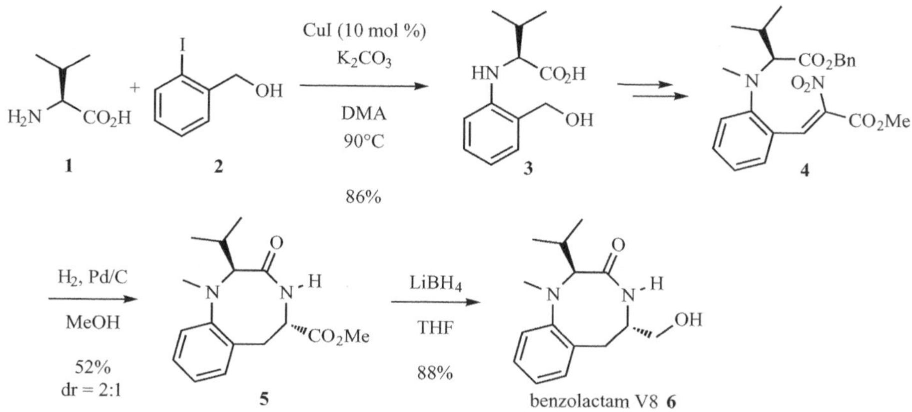

Scheme 16.1 Synthesis of benzolactam V8.

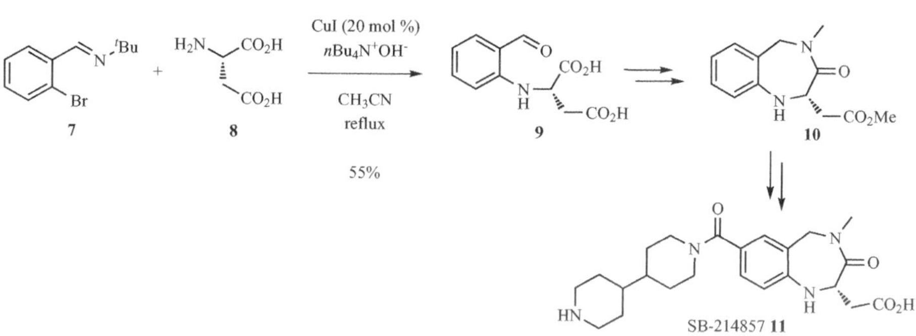

Scheme 16.2 Synthesis of SB-214857.

Following this remarkable report, two syntheses of SB-214857, a potent glycoprotein IIb/IIIa receptor antagonist, were reported independently by the groups of Hayes[2] and Ma.[3] In Hayes' synthesis, protected 2-bromobenzaldehyde **7** was coupled with aspartic acid tetrabutylammonium salt, generated *in situ* by treating the amino acid **8** with tetrabutylammonium hydroxide to overcome the poor solubility of the amino acid reaction partner, in the presence of substoichiometric amounts of copper iodide (Scheme 16.2). Although slight racemization was observed (ee > 90%) and longer reaction time increased the amount of racemized product under the described reaction conditions, this synthesis is an illustration of the effectiveness of copper-promoted aryl amination and exemplifies an efficient strategy to access enantioenriched SB-214857 **11**, especially when compared to an earlier synthesis employing kinetic resolution.[4] In a related approach, Ma and Xia completed the synthesis of **11** employing an intramolecular variant of the aryl amination as a key step (see Scheme 16.43 for details).[3]

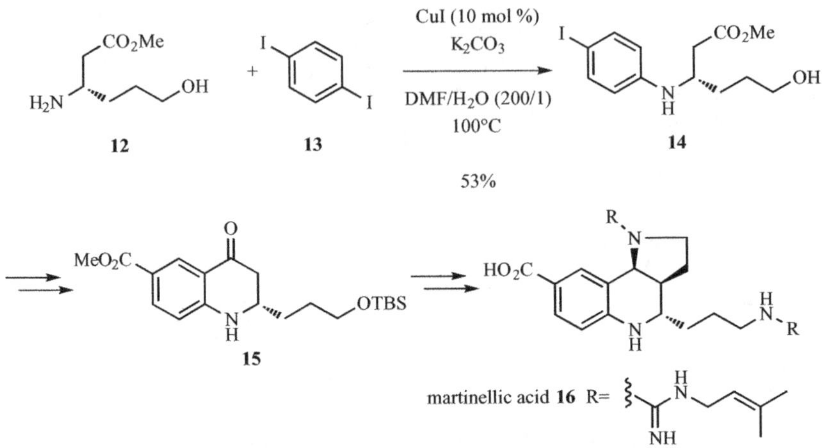

Scheme 16.3 Synthesis of martinellic acid.

β-Amino acid derivatives have also been employed as substrates for copper-mediated amination with aromatic halides. For instance, martinellic acid **16**, a naturally occurring bradykinin receptor antagonist, was synthesized by a copper-mediated aryl amination of 1,4-diiodobenzene **13** with β-amino ester **12** (Scheme 16.3).[5] The coupling reaction proceeded successfully to give the iodoaniline derivative **14** under milder conditions than those of conventional Ullmann-type couplings. Another notable feature of this amination is its selectivity since the secondary amine in **12** chemoselectively underwent the arylation in the presence of the free hydroxyl group. The quinoline system of the natural product **16** was next constructed by Friedel–Crafts-type cyclization after hydrolysis of methyl ester.

As extensively described in Chapters 1 and 5, the development of ligand-assisted, copper-mediated amination procedures represents a significant breakthrough in the field of organic syntheses by establishing the amination as an especially powerful and reliable strategy. This amination is generally carried out under relatively mild reaction conditions and with a high level of functional group tolerance. Over the last 20 years, these coupling methodologies have been frequently applied to the syntheses of natural products and bioactive nonnatural molecules (or potential pharmaceuticals) and have often considerably shortened the preparation of these compounds.

As an example, a cyclopropyl derivative of iprodione **21** was prepared utilizing a copper-catalyzed, ligand-assisted cross-coupling methodology in a key step (Scheme 16.4).[6] Amination of dichloroiodobenzene **17** with cyclopropylamine derivative **18** in the presence of ethylene glycol as the ligand provided the aniline **19** in moderate yield, contaminated, however, with an inseparable O-arylation product (6%) resulting from competitive arylation of the ligand, a product that could be removed in the next step and whose formation could be avoided using other conditions published after this synthesis and extensively described in Chapter 1.

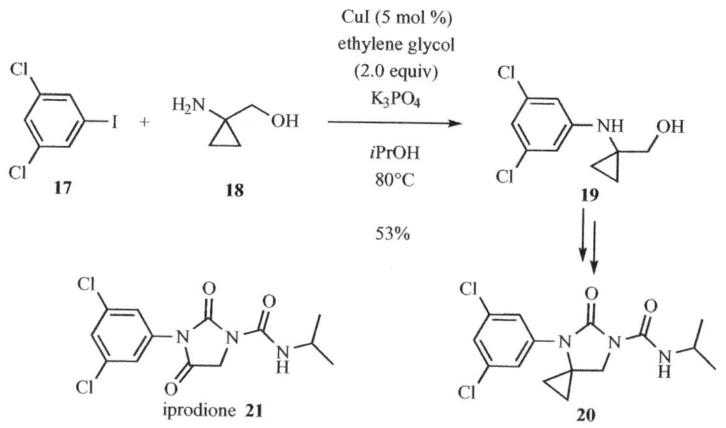

Scheme 16.4 Synthesis of a cyclopropyl derivative of iprodione.

A related example of a copper-mediated, ligand-assisted aryl amination was featured in the syntheses of less lipophilic analogues **25** of tariquidar **22a**. These analogues were synthesized from the bromo derivative **22b**, which was readily prepared through a series of amide couplings starting from an amino benzoic acid, by using copper-catalyzed N-arylation reactions in the presence of various ligands (Scheme 16.5).[7] After screening several copper sources and ligands, cyclic secondary amine **23** and primary amine **24** were found to undergo arylation to give the desired products **25** in moderate-to-high yields, respectively, using combinations of copper(I) bromide dimethylsulfide complex/proline **L1** and copper(I) iodide/2-isobutyryl cyclohexanone **L2**, while the corresponding palladium-catalyzed reactions afforded only disappointingly small amounts of the desired products. In the case of acyclic secondary amines, however, suitable reaction conditions that provide significant amounts of products could not be found, which is still one of the major limitations in copper-catalyzed C–N bond-forming reactions. This synthesis provides support and offers opportunities for copper-mediated coupling to be used as a reliable strategy for the production of bioactive compounds bearing functionalized *N*-aryl ring systems.

In addition to ligand-assisted procedures that are still not as general for the arylation of amines as they are for the cross coupling involving other N-nucleophiles such as amides or carbamates, ligandless Ullmann coupling conditions can be efficiently used in the preparation of substituted anilines. For instance, in the total synthesis of the natural product megistophylline I **30**,[8a] which exhibits antibacterial activity against gram-positive bacteria, the ligandless Ullmann protocol was successfully applied to give a highly substituted diarylamine (Scheme 16.6). The amination of 2-iodobenzoic acid **27** with aniline **26** in the presence of copper(II) acetate and subsequent cyclization of the resulting diarylamine furnished the acridanone **28** in 73% yield. In the final stage of this synthesis, the prenyl side chain was introduced through a thermally promoted aromatic Claisen rearrangement of the corresponding

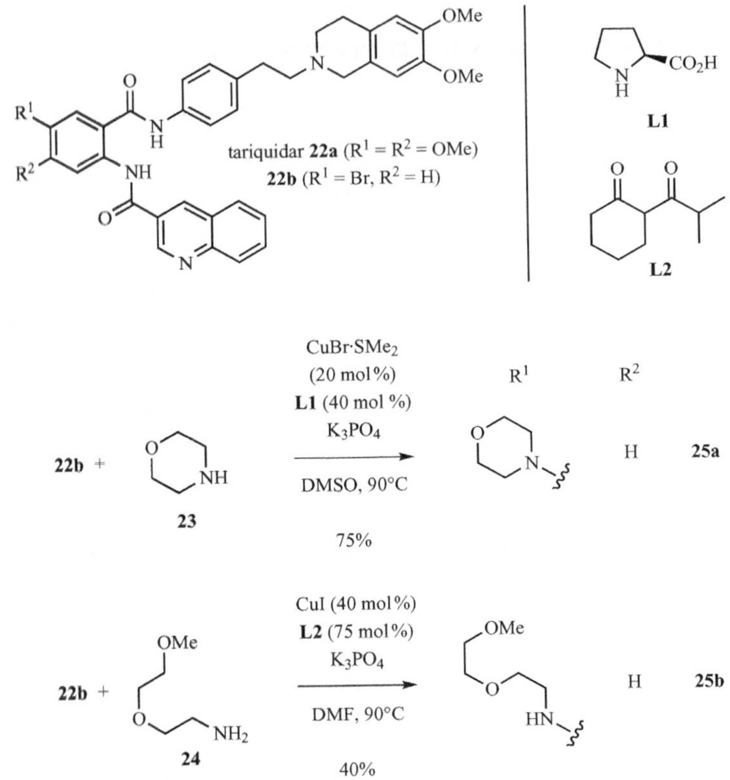

Scheme 16.5 Syntheses of analogues of tariquidar.

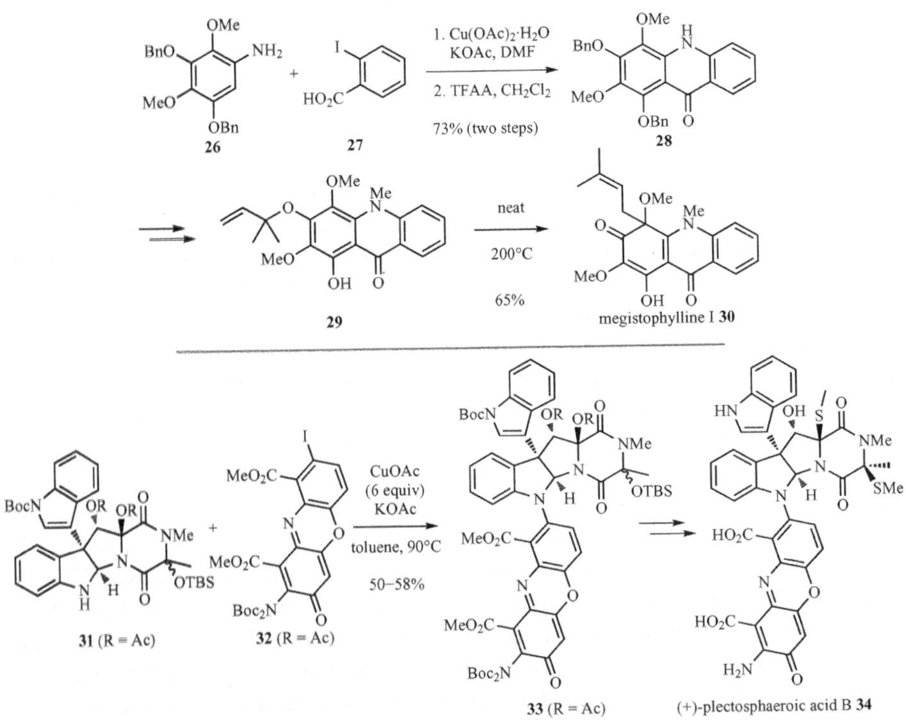

Scheme 16.6 Synthesis of megistophylline I and (+)-plectosphaeroic acid B.

prenyl ether to give the desired natural product **30**. It should be noted that the high efficiency of the amination step might also be due to the *ortho*-effect provided by the carboxylic acid in the starting aryl iodide **27**.

Another remarkably efficient example of a ligandless arylation of an amine was reported in 2013 by the Overman group during their study of the total synthesis of (+)-plectosphaeroic acid B **34**, a secondary metabolite possessing a tryptohan-derived epitrithiodioxopiperazine core and a 2-aminophenoxazin-3-one subunit isolated from the fungus *Plectosphaerella cucumerina.* A key step of the synthesis they devised relied on the union of highly functionalized fragments **31** and **32**. While model studies with a simpler indoline fragment demonstrated the possibility of such an approach using an arylation with **32** using 3 equiv of Liebeskind catalyst copper(I) thiophene-2-carboxylate CuTC and potassium carbonate in toluene at 90°C, the use of this system for the arylation of **31** with **32** resulted in inefficient conversion to the coupled product **33** and formation of byproducts resulting from the addition of thiopene-2-carboxylate to the angular *N,O*-acetal. The use of copper(I) acetate in combination with potassium acetate in toluene at 90°C was found to be more efficient for this cross coupling and gave the desired advanced intermediate **33** in 50–58% yield.[8b]

From the examples illustrated in this section, the copper-mediated cross coupling of amines with aryl halides exhibits a unique reactivity allowing for mild and selective aryl amination starting from aromatics possessing various halides, as nicely illustrated in Scheme 16.3 and Scheme 16.4. The procedures developed over the past 10 years or so have found elegant applications in the synthesis of various natural products, even if the amidation of aryl halides, which will be discussed in the next section, has been more commonly used in the context of total synthesis, which is probably due to the more general catalytic systems available to perform this reaction.

16.3 ARYL AMIDATION IN COMPLEX MOLECULE SYNTHESIS

Buchwald's key contribution to the development of practical copper-mediated amidation and the introduction of simple, readily available diamine ligands for copper allowed for the exploitation of new strategies toward the synthesis of various alkaloids.[9] While classical syntheses of anilides mainly rely on the coupling reactions of acids with anilines, copper-mediated processes typically occur between amides and aryl halides, mostly in the presence of diamine ligands. This reaction is conducted under relatively mild conditions compared to those of classical Goldberg couplings, and exhibits a broad range of reaction scope and functional group tolerance. Thus, amide functionality can be introduced at the late stage of the synthesis in the presence of other functional groups, which strengthens the importance of this methodology in diversity-oriented syntheses as well as in the syntheses of complex natural products. With the copper-catalyzed amidation conditions, additionally, aryl halides are

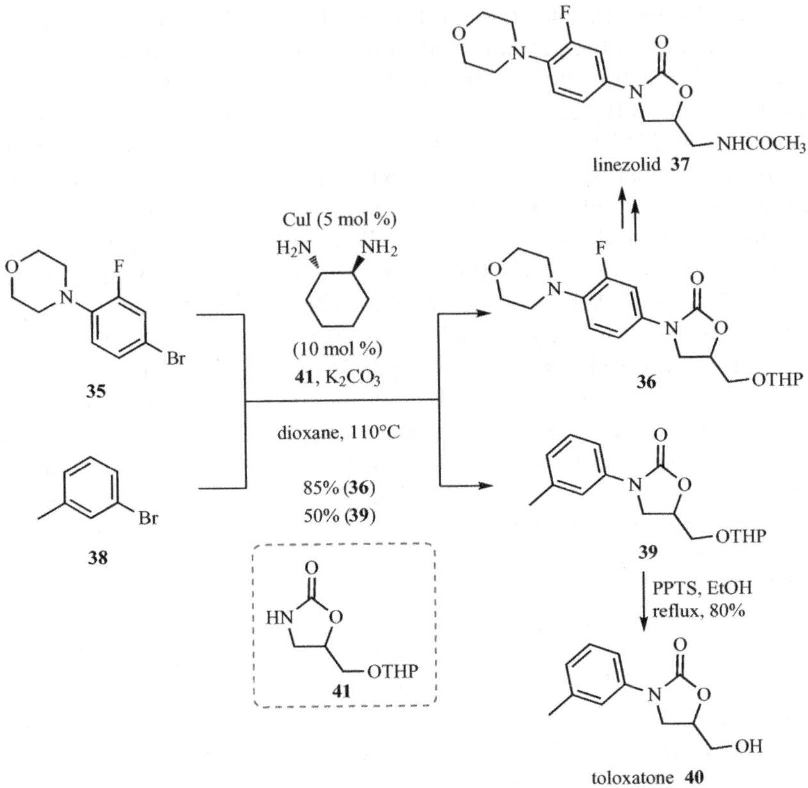

Scheme 16.7 Syntheses of oxazolidinone antibiotics linezolid and toloxatone.

successfully coupled with a wide range of other N-nucleophiles such as carbamates, oxazolidinones, and sulfonamides, which have been demonstrated to be good alternatives to prepare arylcarbamates and related protected aniline derivatives.

Advantages of ligand-assisted copper-mediated amidation have been cleverly applied in the syntheses of various oxazolidinone antibiotics, such as linezolid **37** and toloxatone **40** (Scheme 16.7).[10] While significant effort has been devoted to the construction of the oxazolidinone after amide bond formation in previous syntheses,[11] two fragments (aryl bromides **35** and **38**) were simply coupled with oxazolidinone **41** in the presence of 5 mol % of copper iodide and 10 mol % of *trans*-cyclohexane-1,2-diamine, which efficiently afforded the desired *N*-aryl oxazolidinones **36** and **39**.

The first reported application of a copper-mediated amidation in the synthesis of a complex natural product appeared from the Panek group in 2004.[12] In that regard, the synthesis of the macrocyclic core **46** of cytotrienins **47**, which were isolated from the fermentation broth of *Streptomyces* sp. RK95-74, has been reported utilizing a copper-mediated aryl amidation and ring-closing metathesis (RCM) as key steps (Scheme 16.8). While previous anilide

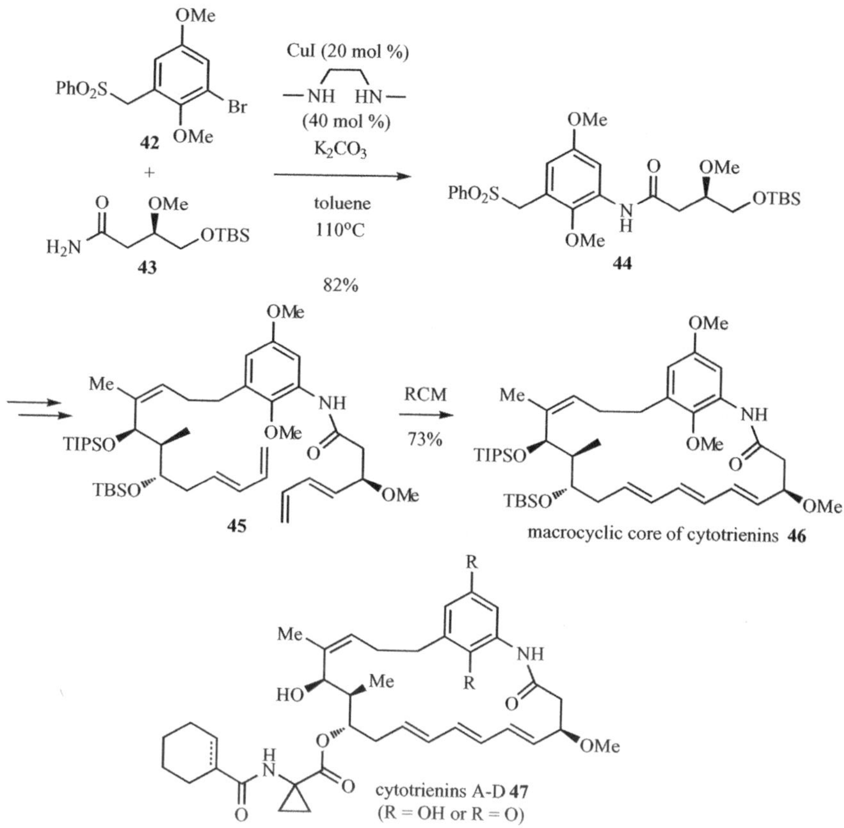

Scheme 16.8 Synthesis of the macrocyclic core of cytotrienins.

syntheses in complex molecule synthesis have mostly been achieved by acylation of anilines, anilide construction, in this synthesis, was carried out by the ligand-assisted copper-mediated amidation of aryl bromide **42** with amide **43** in the presence of *N*,*N*′-dimethylethylenediamine (DMEDA), which produced the corresponding anilide **44** in high yield. The reaction was scaled to multi-gram batches, using a 1:1 ratio of reaction partners. This synthesis is illustrative of a new strategy in which an aryl–N bond in a complex molecule is divided into an aryl halide and an amide, which is now commonly used in organic syntheses.

Buchwald and Chae reported an enantioselective synthesis of indolodioxane U86192A **52**,[13] a potent antihypertensive agent, through application of their N-aryl coupling methodology and subsequent preparation of the indole nucleus (Scheme 16.9). Thus, the indole core in **52** was constructed through a copper-catalyzed chemoselective amidation of Boc-protected hydrazine with symmetrical dibromoaryliodide **48**, yielding the *N*-aryl-hydrazine **49** which then underwent a Fisher indolization followed by a selective debromination in the presence of a palladium catalyst to give **51**.

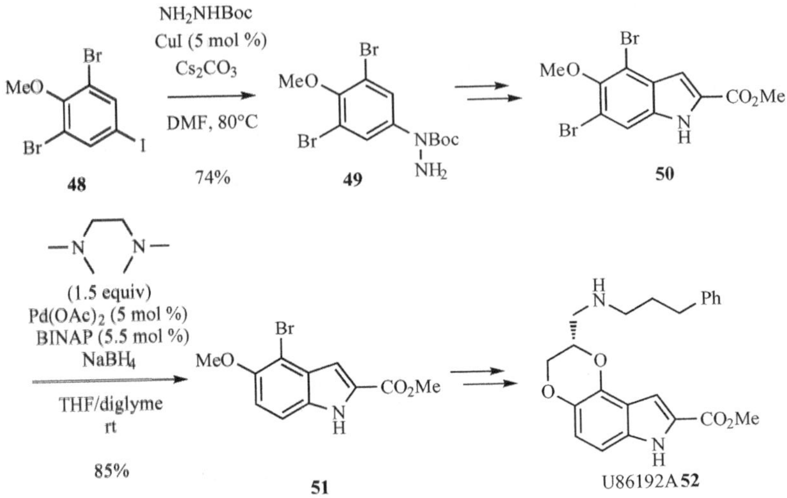

Scheme 16.9 Synthesis of U86192A.

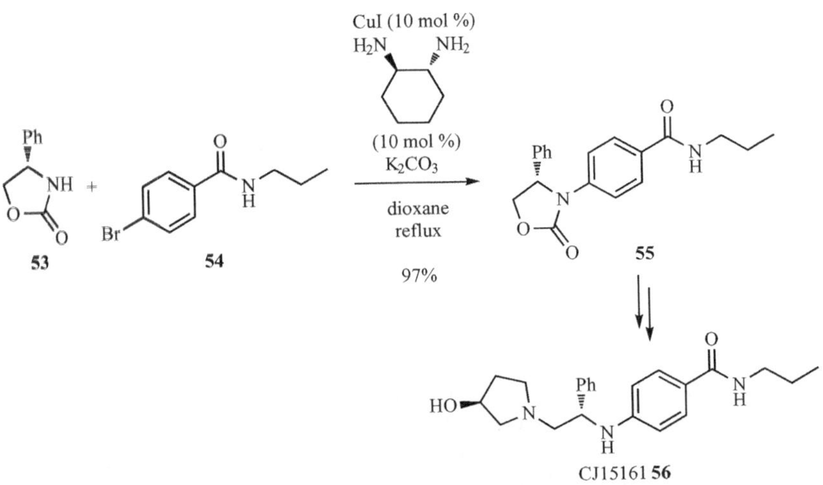

Scheme 16.10 Synthesis of CJ15161.

Ghosh's synthesis of CJ15161 **56**,[14] a κ-opioid receptor, is another illustrative example of the ligand-assisted, copper-catalyzed amidation of aryl halides, and also highlights the potential of this amidation as a shortcut for classical aniline synthesis. In this synthesis, oxazolidinone **53** was coupled with 4-bromoaryl amide **54** under Buchwald's conditions to construct an enantioenriched aniline scaffold (Scheme 16.10). Using this approach, the limitations of the original route to **56**, including low yield and regioselectivity issues during an epoxide ring-opening by an amine to form a chiral amino side chain,[15] were simply detoured. By way of comparison, an analogous route

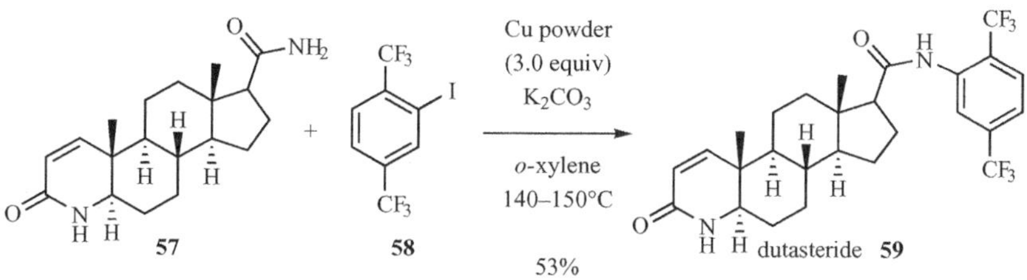

Scheme 16.11 Synthesis of dutasteride.

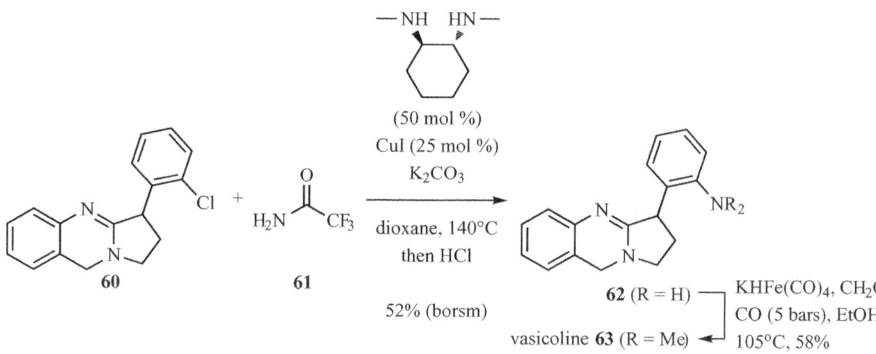

Scheme 16.12 Synthesis of vasicoline.

using a palladium-catalyzed coupling afforded the aryl oxazolidinone product **55** in moderate yield only.

The azasteroid dutasteride **59**, a marketed drug for the treatment of prostate diseases sold under the brand name Avodart™, has been efficiently synthesized from a commercially available starting material in five steps (Scheme 16.11).[16] Right-handed bis-trifluoromethylanilide was successfully installed at the last step using a copper-mediated amidation. After screening several conditions, it was found that the use of copper powder and potassium carbonate in *o*-xylene at a temperature of 140–150°C afforded dutasteride **59** in 63% crude yield, and then a series of recrystallizations yielded **59** in high purity. Interestingly, the terminal amide in **57** was selectively coupled with aryl iodide **58** in the presence of the lactam functionality, resulting from a reactivity difference between the primary and secondary amide.

In addition, the copper(I)-mediated amidation starting from labile amides can be used as a method to introduce a synthetic equivalent of an amine. An example of this reaction was found in the total synthesis of vasicoline **63** reported by Bergman and Ellman where a rhodium-catalyzed direct C–H addition of quinazolines to alkenes was employed as a key step.[17] The amino group in the aromatic side chain was introduced through copper catalysis using a weakly nucleophilic trifluoroacetamide **61** as the amine source, which was followed by hydrolysis of the amide under acidic conditions (Scheme 16.12). The same amidation utilizing substoichiometric amounts of a palladium

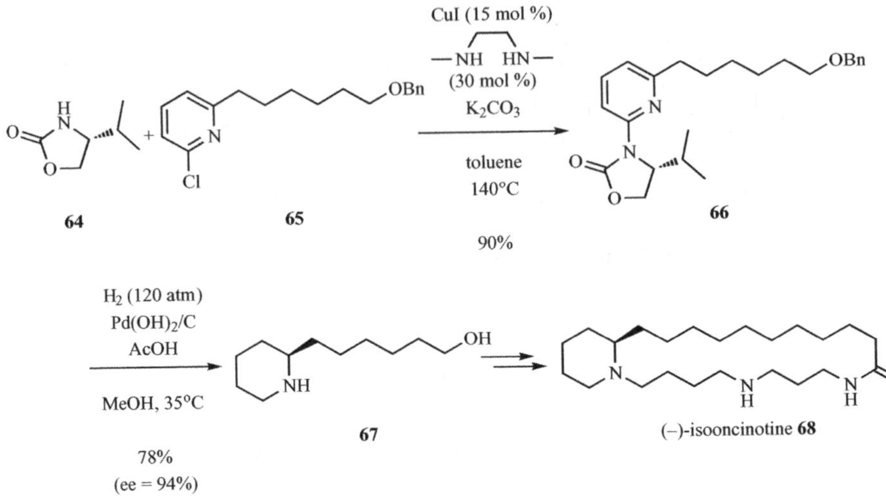

Scheme 16.13 Synthesis of (–)-isooncinotine.

catalyst was found to be inefficient starting from aryl chloride **60**. Direct amination techniques with ammonia reported later and described in Chapter 1, Section 1.5, might now allow for a direct amination without the need for an additional deprotection step.

The use of an oxazolidinone represents another example where a weak basic nitrogen nucleophile is used as the amine source, and has been illustrated by Fürstner and coworkers in the total synthesis of isooncinotine **68** (Scheme 16.13).[18] A chiral auxiliary, Evans' oxazolidinone **64**, was conveniently introduced to 2-chloropyridine **65** under Buchwald's conditions and was used to control the facial selectivity in the hydrogenation process of the resulting 2,5-disubstituted pyridine **66**. Concomitant removal of the auxiliary during hydrogenation under acidic conditions generated the 2-alkyl piperidine **67**, a direct precursor of (–)-isooncinotine **68**, with an excellent optical purity (94% ee).

The acidic hydrolysis product of amicetin, cytimidine **74**, which retains its antibacterial activity against *Mycobacterium tuberculosis H37Rv*, has been synthesized through a one-pot copper-mediated amidation cascade by the Looper's laboratory (Scheme 16.14).[19] Initially, they envisioned a chemoselective one-step amidation to afford the cytimidine precursor, diamide **73**, from bromopyrimidine **69**, considering steric differences between amides **70** and **72**. However, their initial efforts failed to afford diamide **73**, as each amidation required different reaction conditions. Therefore, a one-pot sequential amidation was investigated, upon which amidation of bromopyrimidine **69** with 4-iodobenzamide **70** in the presence of 1,10-phenanthroline, and subsequent addition of amide **72** and DMEDA to the reaction mixture successfully provided the cytimidine precursor **73**.

The C–N cross-coupling reaction between aromatic halides and amides definitely represents a powerful transformation that has been nicely and

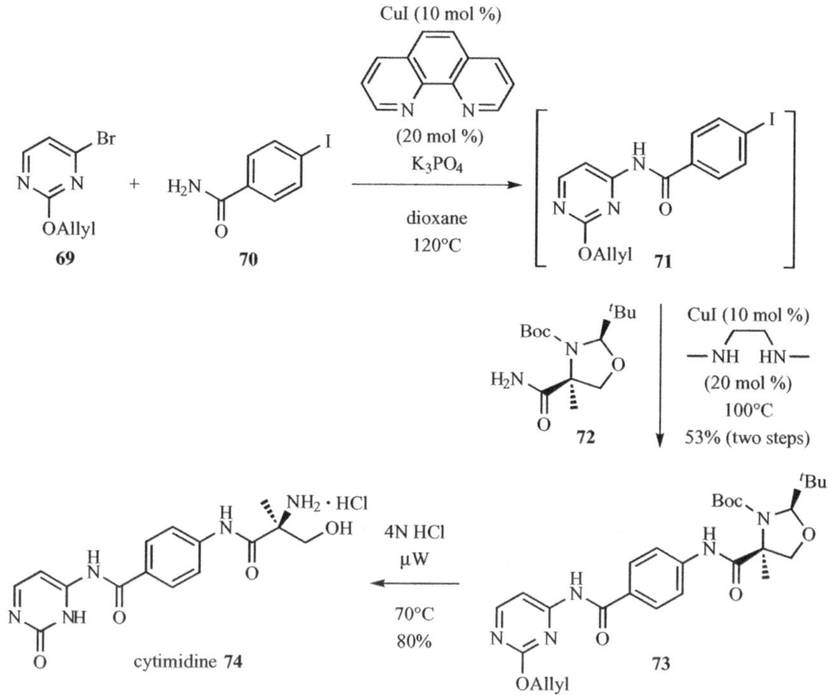

Scheme 16.14 Synthesis of cytimidine.

efficiently applied to the synthesis of various complex molecules and natural products. This strategy is now commonly used, generally with high efficiency, and provides an especially interesting alternative to the classical anilide syntheses. The major limitation of this reaction, which is the limited reactivity of acyclic secondary amides, needs to be taken into account when one is planning to use such a step at a late stage of a synthesis. The arylation of N-heterocycles also had a deep impact in strategies and on the choice of key disconnections in natural product synthesis, and will be covered in the next section.

16.4 ARYLATION OF N-HETEROCYCLES IN COMPLEX MOLECULE SYNTHESIS

The arylation of N-heterocycles has also been used recently in the synthesis of complex molecules and was efficiently featured in various total syntheses. One remarkable example of this category appeared from the Abbott Laboratories. Naphthalenoid H_3 antagonist **78**, a potent agent for cognition and attention deficit/hyperactivity disorder (ADHD), was synthesized using a copper-mediated coupling of naphthalenyl bromide **75** with pyridazinone **76** in multikilogram quantities (Scheme 16.15).[20] Through considerable efforts to find optimal amidation conditions, it was found that the combination of copper(I) chloride as the copper source, 8-hydroxyquinoline as the ligand, and

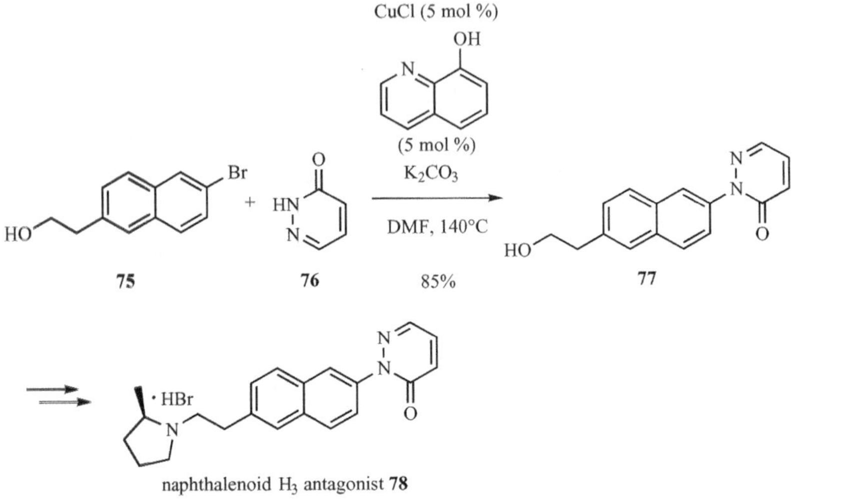

Scheme 16.15 Synthesis of naphthalenoid H_3 antagonist.

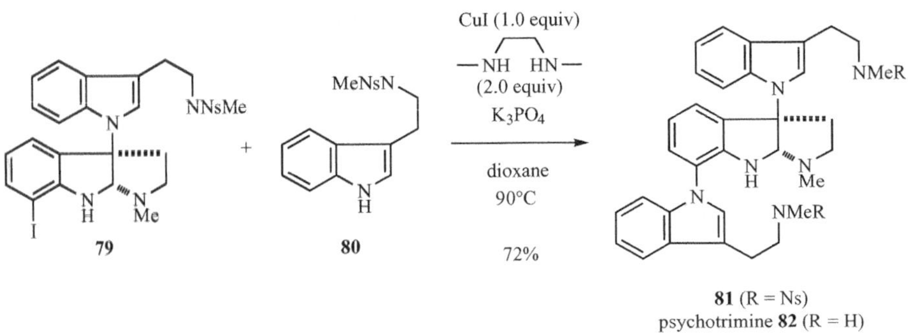

Scheme 16.16 Synthesis of (±)-psychotrimine.

potassium carbonate in DMF gives the best yield without any detectable O-arylation by-product. Additionally, another batch of screening with various palladium catalysts and ligands revealed that palladium catalysis is totally inefficient for this N-arylation.

To the best of our knowledge, the total synthesis of (±)-psychotrimine **82**, reported in 2008 by Takayama and coworkers, is the first example of the use of an indole as the nucleophile for a copper-mediated arylation in natural product synthesis (Scheme 16.16).[21] Copper(I)-mediated intra- and intermolecular aminations (the intramolecular one is described in Section 16.7, Scheme 16.47) were employed as key steps to install the pyrrolidinoindoline core as well as the tryptamine side chain. In this synthesis, iodoindoline **79** was successfully coupled with tryptamine **80** under Buchwald's conditions to give an advanced intermediate **81**, which required a simple deprotection step (removal of the nosyl group) to afford the natural product **82**.

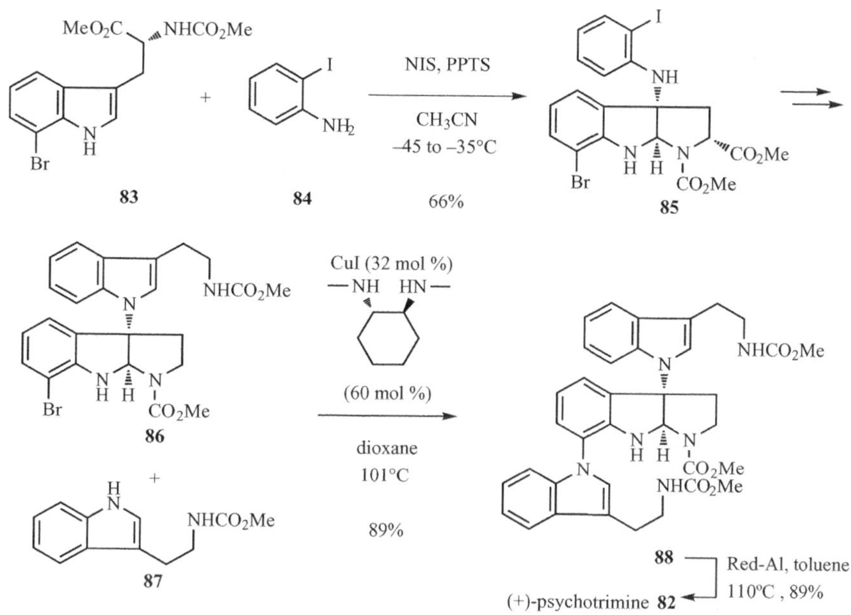

Scheme 16.17 Synthesis of enantioenriched psychotrimine.

A similar disconnection strategy, which envisioned a late-stage copper-mediated amination for the installation of the tryptamine side chain, was also employed in the asymmetric total synthesis of (+)-psychotrimine reported by Baran and coworkers, utilizing an indole–aniline coupling and late-stage copper-mediated amidation (Scheme 16.17).[22] In this synthesis, the indole moiety of 7-bromo-D-tryptophan **83** was coupled with *o*-iodoaniline **84** in the presence of *N*-iodosuccinimide to construct indoline **85**, which was subjected to Larock annulation and subsequent Barton decarboxylation to afford the amination precursor **86**. Selective amination of the indole nitrogen in the presence of the carbamate and indoline nitrogen atoms under conventional copper-mediated amination conditions successfully furnished the desired coupling product **88** with remarkable efficiency.

In another useful contribution, Sarli and coworker introduced the use of "microwave assistance" for the copper-mediated aryl amination reactions, which was extended to the total synthesis of (±)-marinopyrrole A **92**. This work demonstrated its utility in the synthesis of complex heterocycles and was elegantly used to assemble the bis-pyrrole framework of marinopyrrole A through microwave-assisted N-arylation of a pyrrole (Scheme 16.18).[23] Their model revealed that conventional amidations developed by Buchwald and Fukuyama were less effective in the case of N-arylation of a pyrrole, but the use of ligandless amidation conditions under microwave irradiation yielded the desired coupling product **91**. This intermediate bis-pyrrole **91**, therefore, was prepared in 43% yield by a copper(II)-mediated amination between

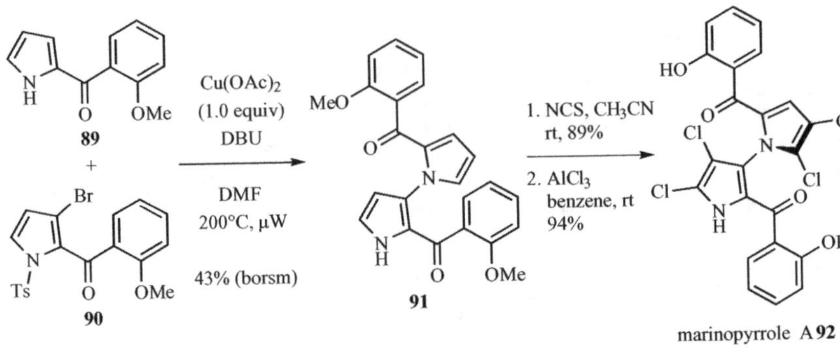

Scheme 16.18 Synthesis of (±)-marinopyrrole A.

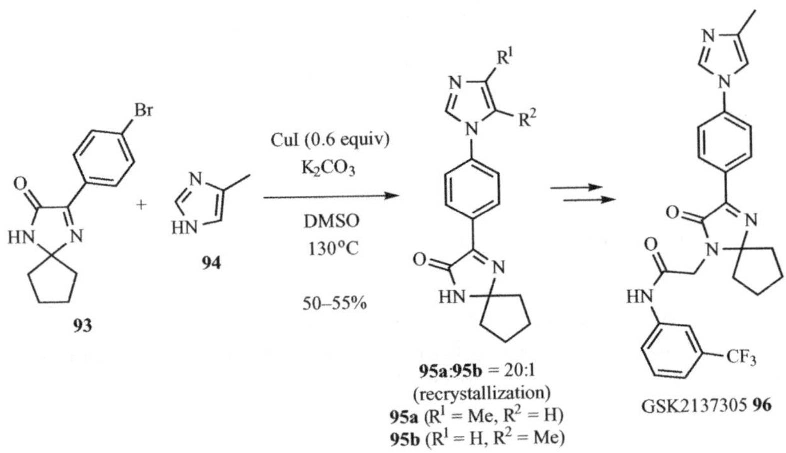

Scheme 16.19 Synthesis of GSK2137305.

pyrroles **89** and **90** under microwave heating, and further chlorination followed by demethylation provided the natural product, marinopyrrole A **92**.

Graham and coworkers at GSK developed a practical route to GSK2137305 **96**, a spirocyclic glycine transporter type-1 inhibitor, employing a ligandless amination as a key transformation to assemble the biaryl intermediate (Scheme 16.19).[24] Although a histidine-assisted protocol was developed in their initial study to couple 4-methyl imidazole **94** with bromophenylimidazolidinone **93** on a multikilogram scale, the tedious and inefficient isolation process involving removal of the ligand and copper catalyst prompted them to develop ligandless amidation conditions. The ligandless copper-mediated amidation proceeded well in DMSO at 130°C using substoichiometric amounts of copper(I) iodide, which provided intermediate **95a** in 50–55% yield with a 20:1 regioselectivity after recrystallization of the crude reaction mixture.

In addition to these types of N-arylations between aryl halides and N-heterocycles, aryl amination/amidation involving aryl metal reagents, which has been extensively described in Chapter 4, is also a useful alternative because these amidations usually proceed under milder conditions. A nice example of

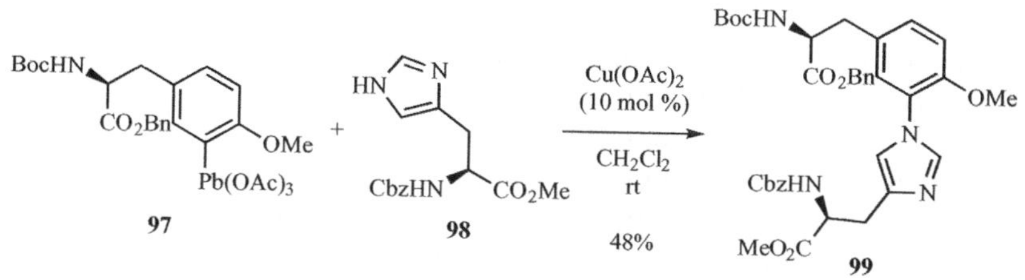

Scheme 16.20 Synthesis of a histidine-tyrosine dipeptide found in the active site of cytochrome *c* oxidase.

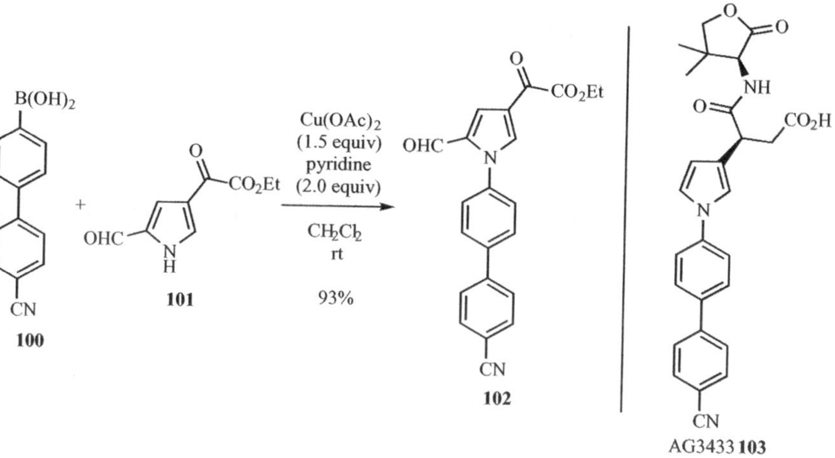

Scheme 16.21 Synthesis of a key intermediate of AG3433.

this strategy was reported for the synthesis of the histidine-tyrosine dipeptide side chain **99**, found in the active site of cytochrome *c* oxidase, which was synthesized through a copper(II)-mediated amination of tyrosine-derived lead reagent **97** with N-protected histidine **98** at room temperature (Scheme 16.20).[25] Excellent regioselectivity for the N1 nitrogen atom and absence of racemization of the amino acid reaction partners were observed, as determined by ^{1}H and ^{13}C NMR analysis.

Another example using an aryl metal reagent was described in the synthesis of a key intermediate of AG3433 **103**, a potent inhibitor of matrix metalloproteases (MMPs) currently under preclinical evaluation (Scheme 16.21).[26] Aryl boronic acid **100**, which was prepared from a commercially available source in two steps, was coupled with the electron-deficient pyrrole **101** to give the N-arylation product **102** in high yield. This organometallic reagent-based methodology is also advantageous in that the reaction proceeds at room temperature, usually has a broad range of functional group tolerance, and is effective for a wide range of heterocycles, although it requires stoichiometric amounts of copper in most cases.

The synthetic utility of the copper-mediated arylation of N-heterocycles stems primarily from the shortcomings of older existing methods, such as aromatic nucleophilic substitution, for the synthesis of C(aryl)–N bonds, existing protocols suffering from limited substrate scope and poor functional group compatibility. The development of copper-mediated processes facilitated the synthesis of complex *N*-aryl heterocycles, replacing to an extent less selective methods such as the classical Golberg reaction and nuclophilic aromatic substitution, while significantly expanding the scope of reaction partners that can be used in copper-catalyzed C–N bond-forming reactions.

While a wide range of N-nucleophiles can be successfully coupled to various aryl halides with excellent efficiencies, as demonstrated by the use of these procedures in the synthesis of complex natural products, the use of modified conditions for the Ullmann and Goldberg reactions also allows the use of other coupling partners such as vinyl halides, leading to the development of especially efficient strategies for the formation of enamides. These strategies have also been extensively used for the total synthesis of various natural products that will be discussed in the next section.

16.5 VINYL AMIDATION IN COMPLEX MOLECULE SYNTHESIS

Since enamide functionality is frequently found in a structurally diverse array of naturally occurring products, the development of simple and efficient synthetic routes to enamides has been of considerable interest to synthetic chemists. Efficient syntheses of these functional groups present challenges as these groups are substantially labile to the broad range of reaction conditions including acidic and basic conditions. Moreover, classical enamide formation, comprising condensations of aldehydes with corresponding amides, organophosphonium-based olefinations, and Curtius-type rearrangements, are usually low yielding and suffer from issues associated with the control of the double bond geometry. Accordingly, providing a usable solution employing classical synthetic tools is a difficult task, and the development of copper-mediated cross-coupling technologies has had a deep impact in this area, as covered extensively in Chapter 5.

In that regard, Shen and Porco in 2000 made a significant improvement in a copper-catalyzed vinylic substitution reaction of vinyl iodides with amides.[27] In their initial study, the use of copper(I)-thiophene-2-carboxylate (CuTC) and cesium carbonate in NMP successfully allowed the formation of enamides under mild reaction conditions, after which the utility of this methodology was demonstrated by the subsequent total syntheses of a series of salicylate enamides.

For instance, salicylihalamides, marine-derived natural products that exhibit potent cytotoxicity against a range of tumor cell lines, was synthesized by Fürstner and coworkers using a late-stage copper-mediated formation of the enamide side chain as one of the key steps of the synthesis (Scheme 16.22).[28] The 12-membered macrocycle in this natural product was constructed by the means of RCM with only moderate *E*/*Z* selectivity (2:1). A copper-mediated amidation of an inseparable mixture of vinyl iodides **105** (E:Z = 9:1) with the

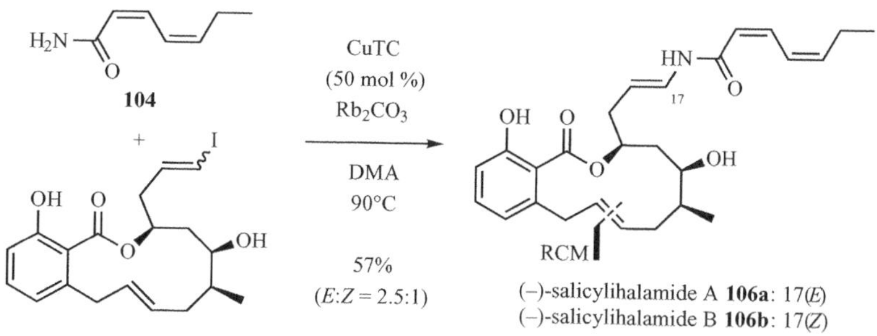

105 (E:Z = 9:1)

Scheme 16.22 Synthesis of salicylihalamides.

conjugated amide **104** under modified Porco's conditions, in which rubidium carbonate was used as a base rather than cesium carbonate, afforded a mixture of salicylihalamide A **106a** and B **106b** in 57% yield and with a 2.5:1 ratio. The reaction using cesium carbonate failed to provide even a trace of the desired product, necessitating further development. Although there was degradation of the integrity of olefin geometry during the course of the reaction, it is noteworthy that this synthesis is the first example applying the copper-catalyzed amidation of alkenyl iodides to natural product synthesis.

After establishing a viable method to construct enamides, Porco and coworkers pursued several total syntheses of various members of the salicylate class of natural products. Among their objectives, the first accomplishment was the total synthesis of the antitumor macrolide lobatamide C **110**, which is characterized by a 15-membered macrocyclic lactone and an enamide-bearing side chain terminated with an oxime (Scheme 16.23).[29] The enamide fragment **109** was prepared through the CuTC-mediated cross coupling of vinyl iodide **108** with (E)-O-methyloxime amide **107**, in the presence of both 1,10-phenanthroline and dibenzylideneacetone (dba) with 4.5 to 1 of E/Z selectivity. Although the labile enamide functionality was introduced in the middle of the synthesis, it endured the multistep synthesis afterward, where acidic and basic conditions were used, without signs of decomposition or loss of geometry integrity.

Two additional total syntheses of oximidine II **112**[30] and III **114**,[31] natural products exhibiting selective inhibition of mammalian vacuolar-type proton ATPase, appeared from the Porco laboratory as a continuing study of copper-mediated enamide synthesis (Scheme 16.24). In both cases, a similar strategy was used in that the 12-membered ring core was constructed by RCM and the labile enamide side chain was introduced by a copper-mediated vinyl amidation at the late stage. In the synthesis of oximidine II **112**, it was learned that coupling of pure (Z)-vinyl iodide **111** with oxime amide (Z) **107** under modified reaction conditions, in which excess quantities of potassium carbonate and DMEDA were employed, proceeded to give the desired product with retention of olefin geometry. In the case of oximidine III **114**, however, excess amounts of potassium carbonate and extended reaction time led to the

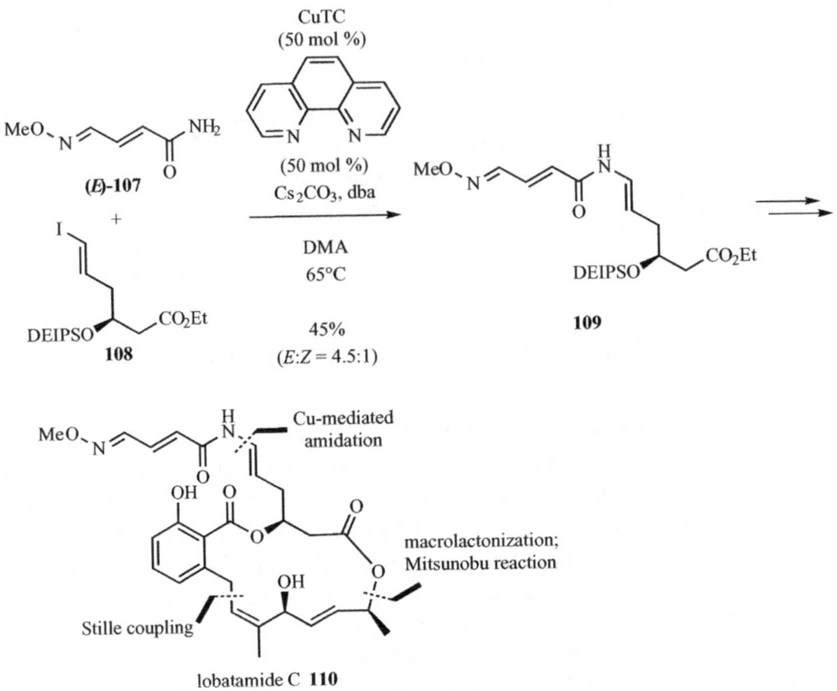

Scheme 16.23 Synthesis of lobatamide C.

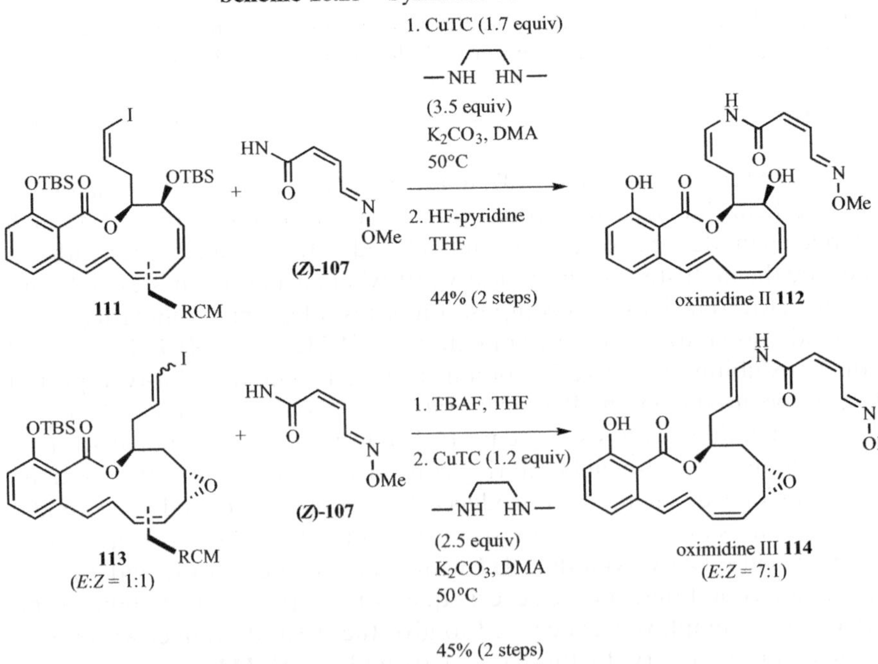

Scheme 16.24 Synthesis of oximidines II and III.

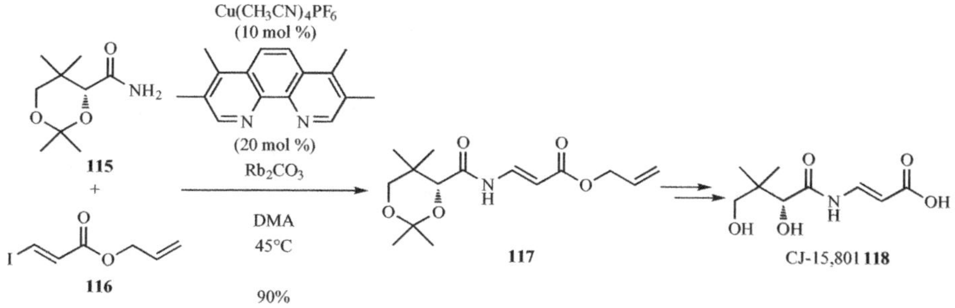

Scheme 16.25 Synthesis of CJ-15,801.

decomposition of the sensitive epoxide in **113**, and the amidation of 1:1 mixture of (*E*)- and (*Z*)-vinyl iodides **113** afforded the desired enamide product in 45% yield with an *E*:*Z* ratio of 7:1.

This strategy was next extended by the Porco group to include the formation of *N*-acyl vinylogous carbamic acids and ureas (Scheme 16.25).[32] Their efforts to find optimal reaction conditions revealed that a combination of $Cu(CH_3CN)_4PF_6$ (Kubas salt), rubidium carbonate, and 3,4,7,8-tetramethyl-1,10-phenanthroline in DMA at 45°C gives the best results. This method was applied to the synthesis of the antibiotic natural product CJ-15,801 **118**, where a copper-catalyzed coupling of (*E*)-allyl-β-iodoacrylate **116** with amide **115** provided *N*-acyl vinylogous carbamic acid **117** in 90% yield, which then underwent a series of deprotection steps to eventually give the final product, CJ-15,801 **118**.

Apicularen A is another representative example of a synthesis that employed a copper-mediated amidation as a key step in total synthesis. Three independent total syntheses have been reported from the groups of Nicolaou,[33] Panek,[34] and Uenishi.[35] Each of the syntheses featuring a copper(I)-catalyzed formation of the enamide moiety starting from a vinyl iodide and the same unsaturated amide **104**, although the unsaturated enamide functionality was introduced at different stages of the synthesis, their comparison offers a nice overview of different copper-based systems for the synthesis of enamides with complex substrates.

The first total synthesis of apicularen A **121** was reported in 2002 by Nicolaou and coworkers utilizing Brown's asymmetric allylboration methodology to install the oxygen-bearing stereocenters of the pyran ring system (Scheme 16.26).[33] In their initial approach, the enamide side chain was to be assembled by condensation of an aldehyde with an amide. However, this reaction only gave a ketene *N*,*N*-acetal resulting from a double addition of the amide to the starting aldehyde in 75% yield, and further efforts to convert this product to the desired enamide were unsuccessful, resulting in the revision of the authors' synthetic plans. In a second generation synthesis, Porco's copper-mediated cross coupling, starting from vinyl iodide **119** and amide **104**, was employed to construct the enamide side chain in the middle of the stage, which

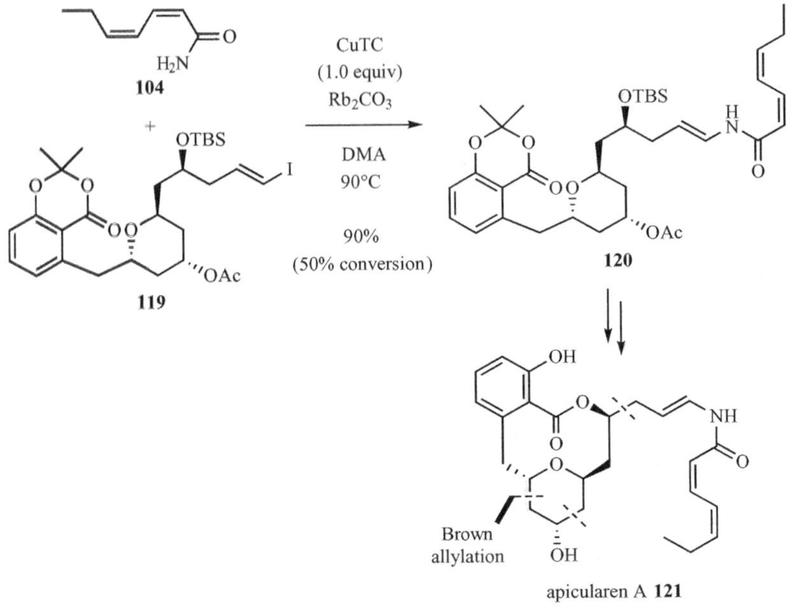

Scheme 16.26 Nicolaou's synthesis of apicularen A.

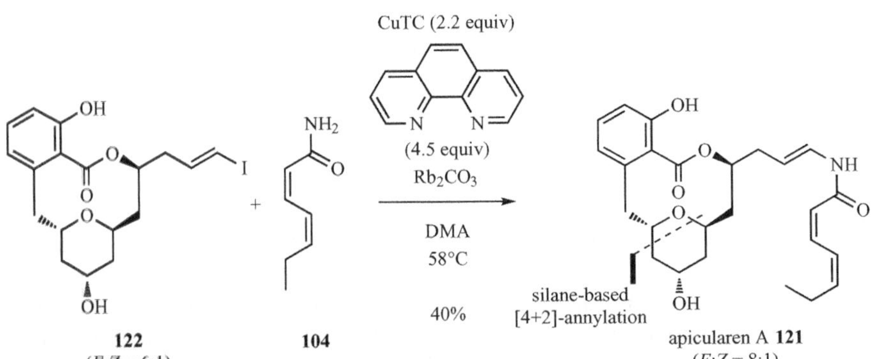

Scheme 16.27 Panek's synthesis of apicularen A.

successfully afforded the desired vinyl amidation product **120** in 90% yield (E:Z = 10:1), at 50% conversion, however.

In another elegant approach to apicularen A, Panek and Su reported a total synthesis of apicularen A highlighting the author's silane-based [4+2]-annulation methodology as a key step to construct the tetrahydropyran core (Scheme 16.27).[34] A CuTC-catalyzed vinyl amidation of vinyl iodide **122** with the unsaturated amide **104** in the presence of 1,10-phenanthroline as the ligand was employed in the final stage of the apicularen A synthesis and afforded a 40% yield of the coupling product with an E/Z ratio of 8:1. Interestingly, the same coupling reactions in the absence of the ligand promoted only the decomposition of the starting vinyl iodide **122**.

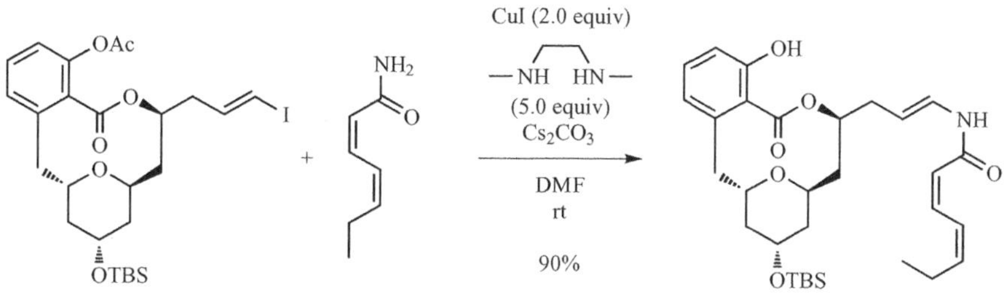

123 104 124

Scheme 16.28 Uenishi's synthesis of apicularen A.

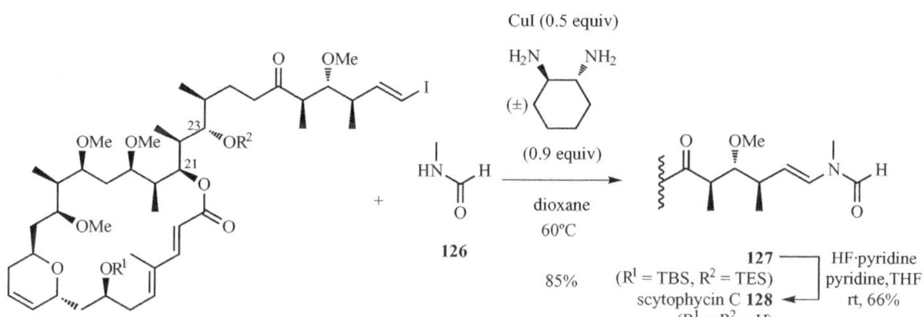

125
(R^1 = TBS, R^2 = TES)

Scheme 16.29 Installation of the enamide functionality of scytophycin C.

A recent report from the Uenishi laboratory provided an improved protocol for the introduction of the enamide side chain at the late stages of the synthesis (Scheme 16.28).[35] In this case, the use of a modification of Buchwald's catalytic system relying on the use of excess quantities of copper(I) iodide (2.0 equiv) and DMEDA (5.0 equiv) at room temperature, which is quite remarkable, successfully gave an advanced intermediate **124** with excellent yield and remarkable selectivity, which was only one step away from the natural product. In this reaction, the use of substoichiometric amounts of copper iodide significantly decreased the formation of the desired enamide product.

As nicely demonstrated in Uenishi's synthesis of apicularen A, the Buchwald-type diamine ligand-assisted copper(I) iodide-catalyzed coupling can also be used as an efficient methodology to access vinyl amides and is probably the most efficient procedure to date. In fact, the use of Buchwald's protocol to access such enamides had been applied in earlier total syntheses, and examples that will be described in the next pages clearly demonstrate its synthetic utility and efficiency.

In 2003, Miyashita and coworkers published the total synthesis of scytophycin C **128** employing a copper(I)-mediated, diamine-assisted vinyl amidation at the end stage of the synthesis (Scheme 16.29).[36] In this work, the 22-membered macrolactone core was constructed through Yamaguchi macrolactonization from the corresponding acyclic intermediate possessing

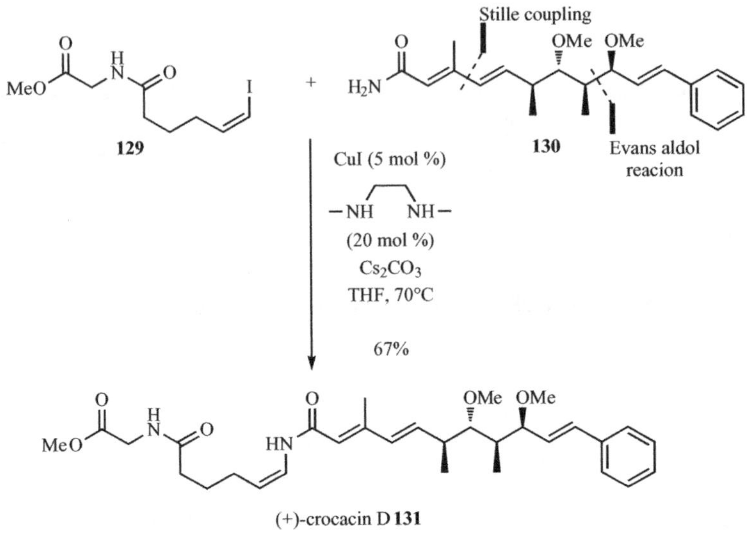

Scheme 16.30 Synthesis of (+)-crocacin D.

unprotected hydroxyl group at both C21 and C23 to give a mixture of 22- and 24-membered macrolactones. Subsequent equilibration to afford the more-stable 22-membered macrocycle under acidic conditions led the desired macrocycle in 75% yield. Once again, the use of Buchwald's amidation of vinyl iodide **125** with *N*-methylformamide **126** provided the desired enamide **127** in 85% yield, demonstrating the remarkable efficiency of the copper-mediated enamide formation, even if a quite high catalyst loading was required with a challenging substrate such as **125**.

The synthesis of crocacin D offers an excellent example of the application of a copper-mediated amidation from the union of the two main fragments **129** and **130** (Scheme 16.30).[37] The initial attempts to join fragments by a dehydrative condensation between an aldehyde with an amide were unsuccessful. In the revised plan, unsaturated amide **130**, which was prepared by Stille coupling, was coupled with (*Z*)-vinyl iodide **129** in the presence of 5 mol % of copper iodide and 20 mol % of DMEDA, which successfully afforded the natural product **131**, with retention of the geometry of the double bond, in 67% yield.

Ma and coworkers have reported the total syntheses of ziziphine N **135**,[38] abyssenine B **139a**, and mucronine E **139b**,[39] which are members of macrocyclic peptide alkaloids, using a ligand-assisted, copper-catalyzed vinyl amidation as a key step for the assembly of acyclic precursors **134** and **138** (Scheme 16.31). In this approach, vinyl iodides **132** and **136**, which were generated from the corresponding aldehydes by conventional Wittig olefination, were coupled with α-amino amides **133** and **137** under the standard conditions developed by the Ma group (copper iodide/*N*,*N*-dimethylglycine, cesium carbonate, dioxane, 80°C) with high *E*/*Z* selectivity since no equilibration of the double

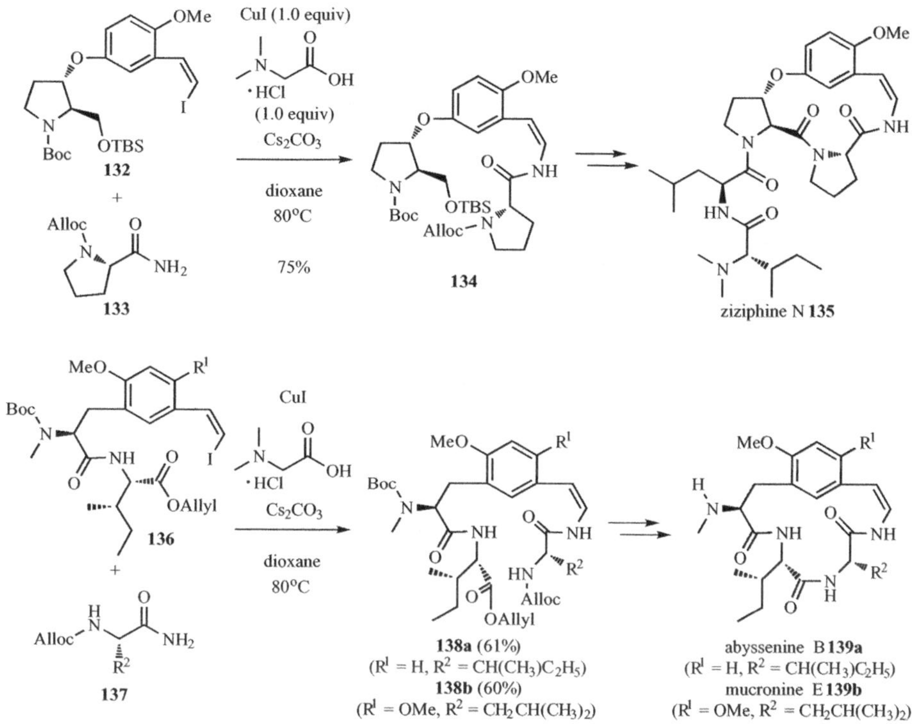

Scheme 16.31 Syntheses of ziziphine N, abyssenine B, and mucronine E.

bond occurred during the cross coupling. An interesting approach to the related cyclopeptide alkaloid paliurine F **259** was reported by the Evano group employing an intramolecular variant of the vinyl amidation reaction, where only catalytic amounts of copper and ligand were needed (see Scheme 16.58 for details). Compared with previous syntheses of related natural products where the enamide was painfully installed at the end of the synthesis starting from the corresponding β-amino alcohol and using a stepwise and low-yielding sequence including activation of the alcohol followed by transformation to the corresponding selenide, oxidation, and thermal elimination, it is rather clear that the copper-mediated formation of the enamide is highly preferable in considerably shortening the synthesis.

Weinreb and coworkers have described the total synthesis of ageladine A **145**, a zinc matrix metalloproteinase inhibitor, using a biomimetically inspired 6π-2-azatriene electrocyclization as the key step to furnish the imidazolopyridine core (Scheme 16.32).[40] In this synthesis, amidation of vinyl iodide **140** with dibromopyrrole carboxamide **141** under Buchwald's conditions gave the corresponding enamide **142** in 92% with excellent stereo- and chemoselectivity. A crucial 6π-cyclization in mesitylene at 145°C afforded the imidazolopyridine **144** in 44% yield (based on recovered starting material) with a small amount of olefin isomerized product (*E*)-**143**.

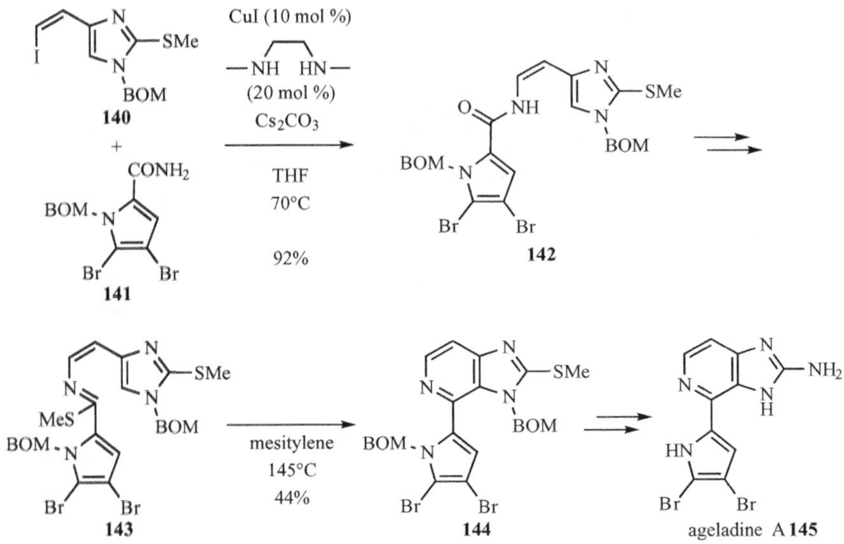

Scheme 16.32 Synthesis of ageladine A.

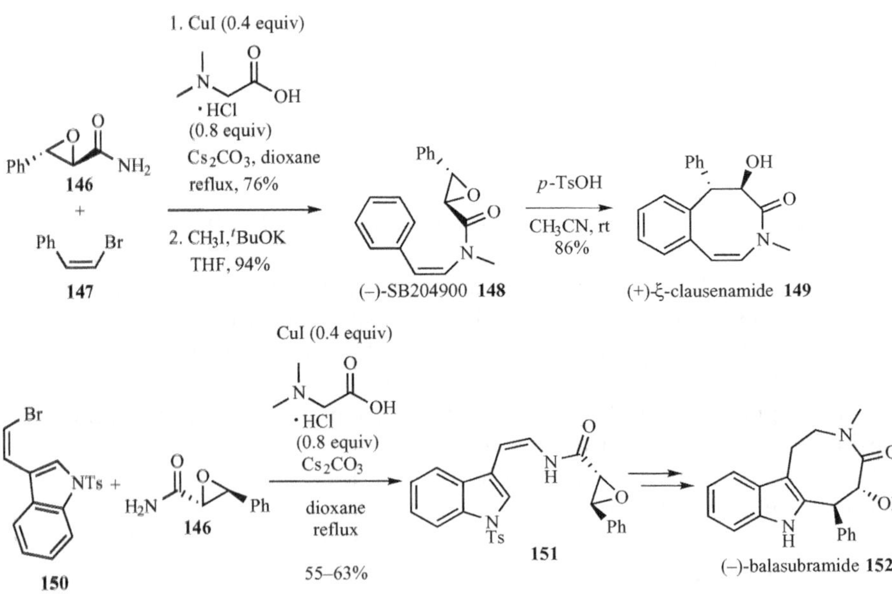

Scheme 16.33 Syntheses of (–)-SB204900, (+)-ξ-clausenamide, and (–)-balasubramide.

The *Clausena* alkaloids (–)-SB204900 **148**, (+)-ξ-clausenamide **149**, and (–)-balasubramide **152** were synthesized employing Ma's vinyl amidation conditions as a key step for the preparation of the acyclic enamides **148** and **151** (Scheme 16.33).[41] In that regard, initial experiments to define optimal conditions revealed that *N,N*-dimethylglycine was the only effective ligand for the formation of the enamide, while diamine ligands introduced by Buchwald and coworkers gave only trace amounts of the products. Having this preliminary result in hand, vinylation of enantioenriched amide **146**, which was prepared by enzymatic hydrolysis of the corresponding racemic epoxy nitrile with bromostyrene **147**, led to the resulting enamide in 76%. Further N-methylation afforded (–)-SB-204900 **148**, which was converted to (+)-ξ-clausenamide **149** via an acid-promoted unusual 8-*endo*-aryl-epoxide cyclization. Also, (–)-balasubramide **152** was synthesized through the same protocol by amidation of vinyl bromide **150** with **146** followed by 8-*endo*-aryl-epoxide cyclization. It should be noted that the methylation of the amide after the cross coupling in the synthesis of **148** is a smart approach, the *N*-methyl analogue of **146** certainly being much more difficult to couple with **147** because of the lowest reactivity of acyclic secondary amides in copper-catalyzed cross-coupling reactions.

The total synthesis of palmerolide A **155**, a polyketide secondary metabolite exhibiting various activities, is another excellent illustration of the copper(I)-mediated vinyl amidation at the final stage of the synthesis (Scheme 16.34).[42] In the initial retrosynthesis, palmerolide A **155** was separated into four main fragments, and the macrocyclic core was assembled through Stille coupling, Yamaguchi esterification, and RCM. The use of a vinyl amidation reaction of the fully deprotected advanced intermediate **154** with amide **153** using Buchwald's conditions provided the natural product in 50% yield with 38% recovered starting material and 10% of the by-product that lost the carbamate functionality. The double-bond geometry was fully retained, and the presence of free alcohols and a carbamate does not seem to have a detrimental influence on the outcome of the reaction which proceeded at room temperature but,

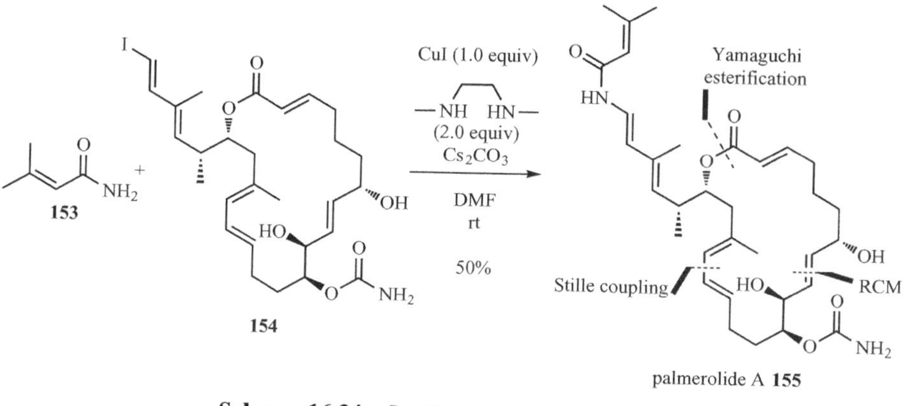

Scheme 16.34 Synthesis of palmerolide A.

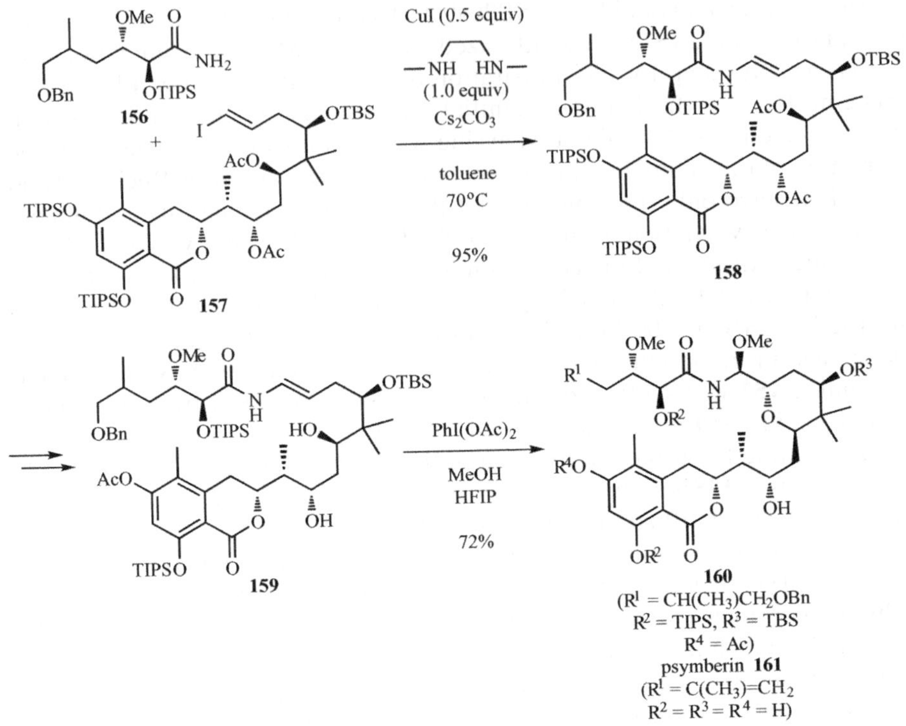

Scheme 16.35 Synthesis of psymberin.

however, required stoichiometric amounts of both the copper source and ligand. As an extension of this study, a small library composed of different enamide analogues of palmerolide A has been prepared using the late-stage copper(I)-mediated amidation of vinyl iodide **154** with several aryl and aliphatic amides.[43]

Another example of a late-stage copper-mediated vinyl amidation on complex substrates has been reported by the Huang group in the course of the total synthesis of psymberin **161** utilizing their $PhI(OAc)_2$-mediated oxidative cyclization as a key cyclization step (Scheme 16.35).[44] The required enamide functionality in this cyclization was introduced by a remarkably efficient vinyl amidation between **156** and **157** yielding the corresponding enamide **158** in 95% yield using substoichiometric amounts of copper(I) iodide, a precursor that next underwent the $PhI(OAc)_2$-mediated cyclization to afford an intermediate tetrahydropyran **160** in 72% yield.

Enamide analogues **165** of cruentaren A **166**, a natural product exhibiting powerful cytotoxicity in a cellular assay with the L929 cell line with an IC_{50} value of 1.2 ng.mL^{-1}, were prepared via copper-mediated vinyl amidation of vinyl iodide **162** with amide **163** (Scheme 16.36).[45] Once again, a late-stage introduction of the corresponding amide to (*E*)-vinyl iodide under Buchwald's

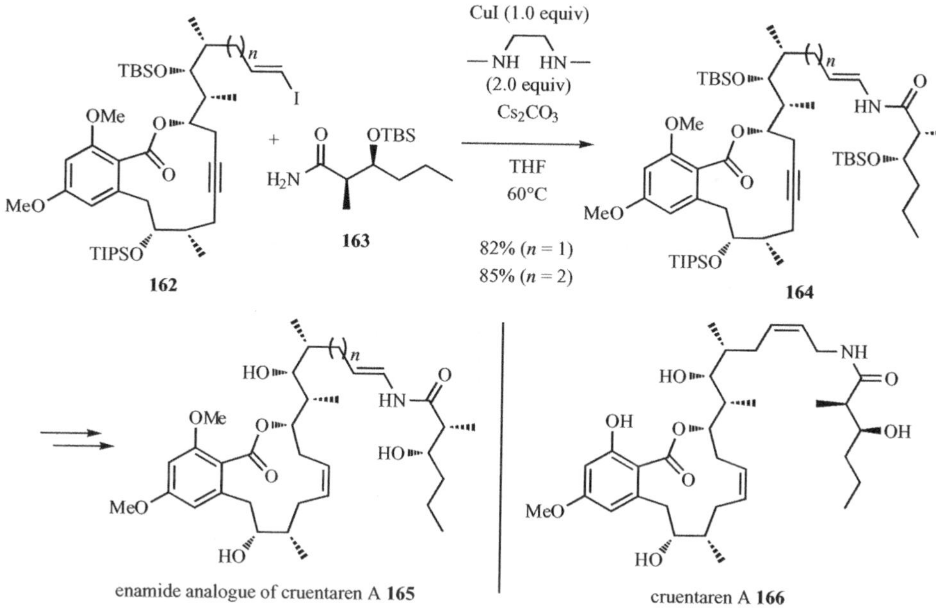

Scheme 16.36 Synthesis of enamide analogues of cruentaren A.

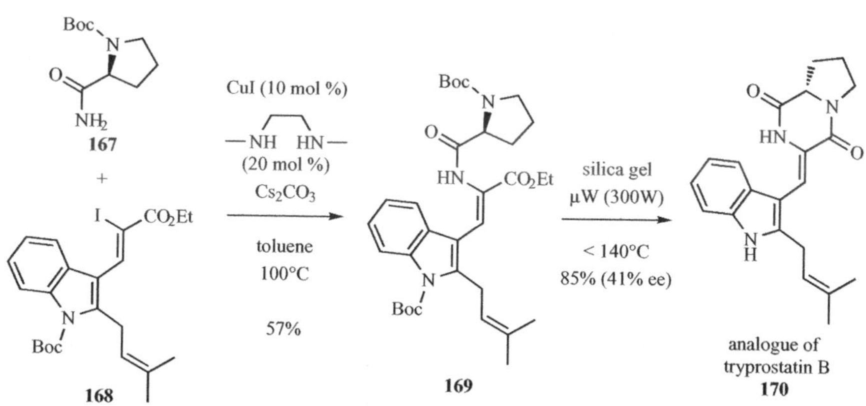

Scheme 16.37 Synthesis of an analogue of tryprostatin B.

conditions resulted in the assembly of the enamide side chain in high yield despite the presence of sensitive functional groups, but still required the use of stoichiometric amounts of both the copper salt and the ligand.

Additional substituents on the vinyl iodide are well tolerated for the cross coupling, as shown in the synthesis of the unsaturated tryprostatin analogue precursor, Pro-ΔTrp dipeptide **169**, which was prepared through amidation of disubstituted vinyl iodide **168** with *N*-Boc prolinamide **167**. Microwave-assisted formation of the diketopiperazine provided the novel analogue of tryprostatin B **170** (Scheme 16.37).[46]

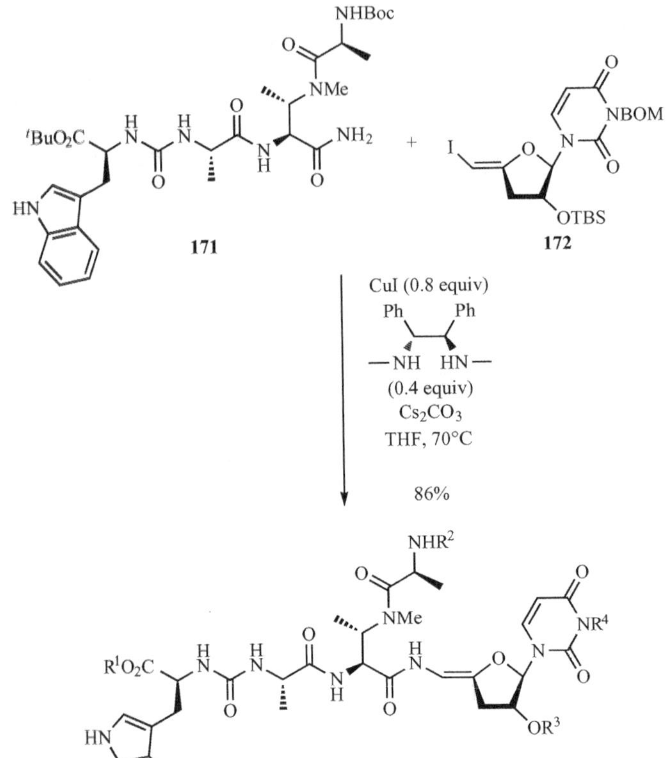

171 **172**

173
(R^1 = ^{t}Bu, R^2 = Boc, R^3 = TBS, R^4 = BOM)
pacidamycin D **174**
(R^1 = R^2 = R^3 = R^4 = H)

Scheme 16.38 Synthesis of pacidamycin D.

An amino acid-derived α-amino amide could be coupled with even more complex and sensitive vinyl halides such as β-alkoxyvinyliodide **172**. The first total synthesis of pacidamycin D **174**, a uridylpeptide antibiotic exhibiting activity against certain strains of *Pseudomonas* with an MIC of 1.5–12.5 $\mu g.mL^{-1}$, has been accomplished featuring an especially interesting copper-mediated amidation of complex vinyl iodide **172** with amide **171** to unite the right- and left-hand fragments of **174** (Scheme 16.38).[47] Initial studies revealed that the use of DMEDA as the ligand gave only a trace amount of the expected product, the major by-product resulting from cyclization of the primary amide on the carbamate, presumably through formation of a carboxamide-copper(I) complex on the terminal amide of **171** and subsequent intramolecular transamidation to the *tert*-butyl ester functionality. However, utilization of an unusual bulky ligand to suppress the transamidation afforded **173** in 86% yield.

Vinyl amidation can be used not only to construct stereochemically well-defined enamide functional groups but also to introduce nitrogen-containing

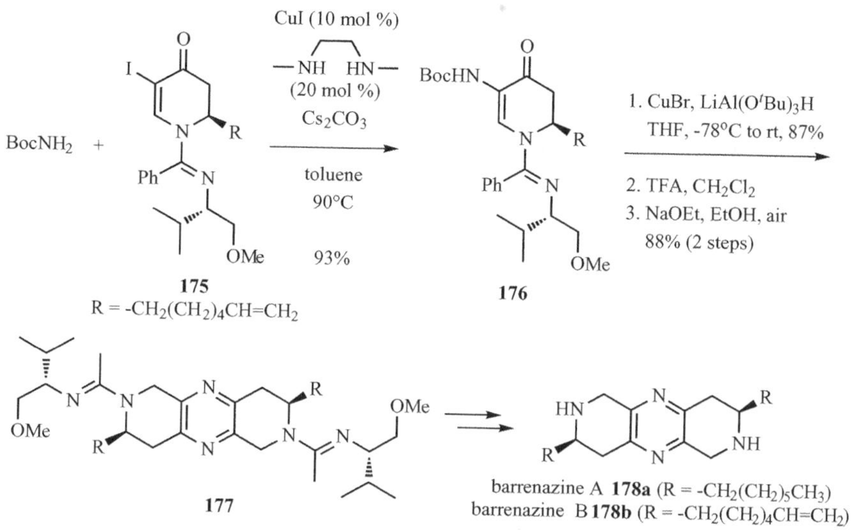

Scheme 16.39 Synthesis of barrenazines A and B.

groups into a molecule or other functional group equivalents. In that regard, Charette and coworker employed this concept in their total syntheses of (−)-barrenazines **178** (Scheme 16.39).[48] Specifically, the vinyl carbamoylation of the iodopyridone **175** with *tert*-butyl carbamate in the presence of a copper(I)-catalyst afforded the *N*-Boc derivative **176**, which was subjected to selective 1,4-reduction and subsequent oxidative dimerization to construct the pyrazine core **177** of barrenazines **178**. It is noteworthy that vinyl carbamoylation is used as a mild and selective method for insertion of nitrogen atoms.

Another interesting example is found in the total synthesis of the galbulimima alkaloid 13 **183**, published by the Movassaghi's laboratory (Scheme 16.40).[49] In this synthesis, a copper-catalyzed cross coupling was used for the introduction of a masked carbonyl. Thus, vinyl amidation of bromodiene **179** with oxazolidinone **180** afforded the desired triene **181**, which underwent an intramolecular Diels–Alder reaction and a series of intramolecular cyclizations to give a pentacyclic compound **182**. Oxidative cleavage of the masked carbonyl (enamide) into an α,β-unsaturated ketone followed by removal of the Cbz-protecting group provided galbulimima alkaloid 13 **183**.

Very recently, the Movassaghi group employed this strategy once again in their total synthesis of (−)-himandrine **188**, which is structurally related to the galbulimima alkaloid 13 **183**.[50] Similarly, vinyl amidation of vinyl bromide **184** with 2-azetidinone **185** was used as a method to introduce a masked ketone, which was subjected to the intramolecular Diels–Alder reaction and a series of aldol cyclization steps to give a pentacyclic intermediate **187**. Finally, the unmasking of enamide to ketone and subsequent oxidative spirocyclization using *N*-chlorosuccinimide (NCS) in CH_3CN accomplished the total synthesis of (−)-himandrine **188**.

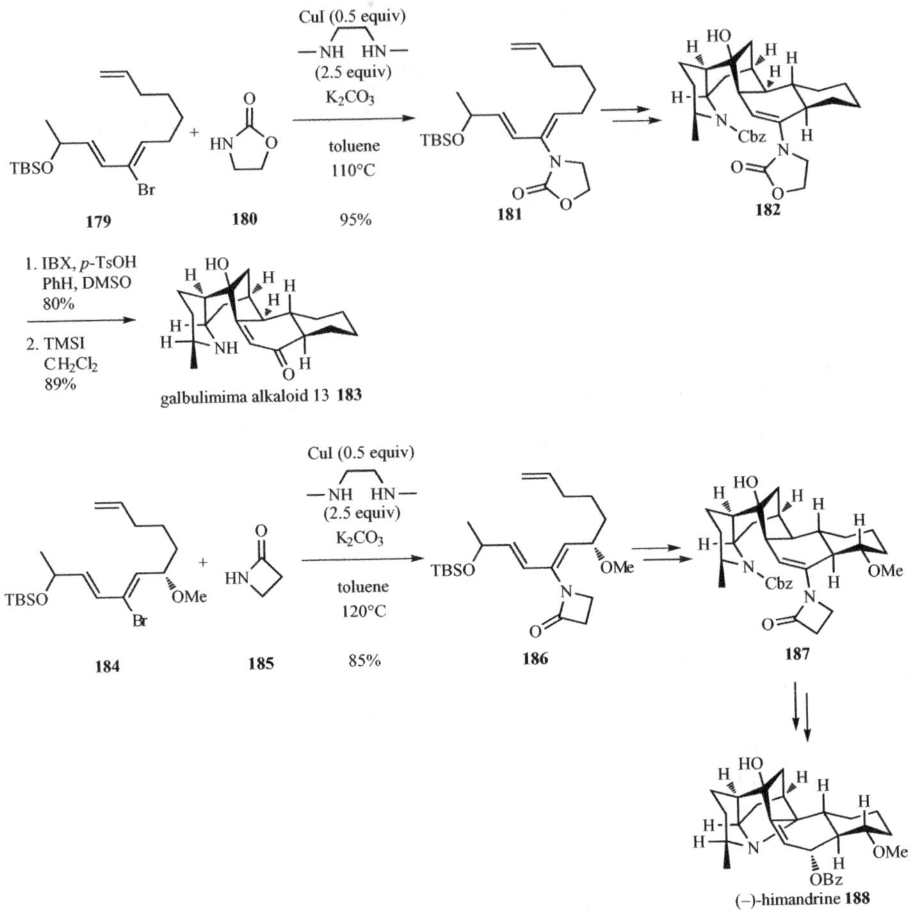

Scheme 16.40 Synthesis of galbulimima alkaloid 13 and (–)-himandrine.

16.6 ALKYNE AMIDATION IN COMPLEX MOLECULE SYNTHESIS

Recently, ynamides have been emerging as useful reaction partners to construct alkaloids, notably because of their stability compared to ynamines and their use as precursors of keteniminium ions. Hsung's contribution to the development of a practical and easy access to versatile ynamides using copper catalysis provides a shortcut to overcome the synthetic difficulties of the classical preparations of ynamides involving isomerization of propagylic amides, elimination from bromoenamides, and alkynylation with alkynyl iodonium salts, and has created new opportunities for complex alkaloid syntheses.[51]

Despite the recently established utility of copper-promoted cross-coupling reactions for the synthesis of ynamides, which have been extensively described in Chapter 5, their use in complex molecule synthesis remains rather

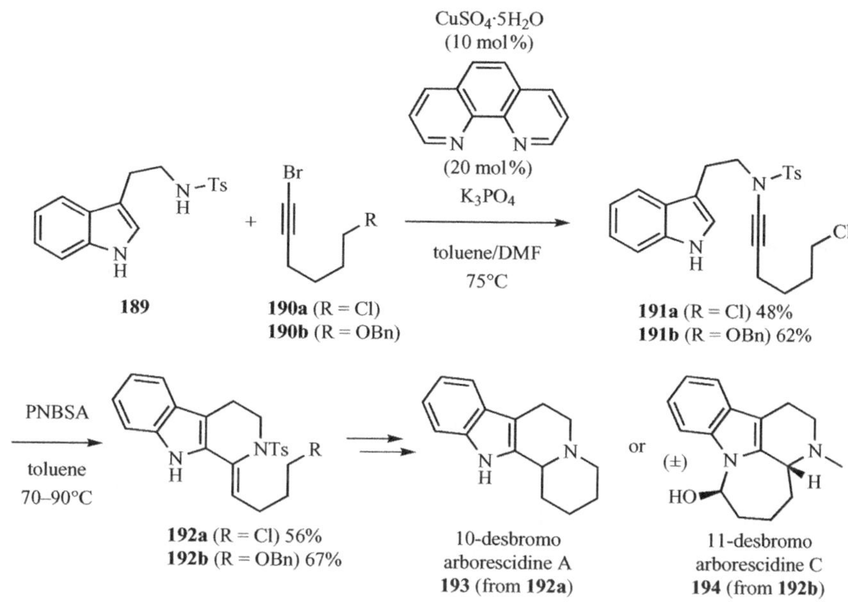

Scheme 16.41 Synthesis of 10-desbromoarborescidine A and 11-desbromoarborescidine C.

underdeveloped. One of the first examples in this category is the total syntheses of desbromoarborescidines A **193** and C **194** reported in 2005 (Scheme 16.41).[52] In these syntheses, carbon-tethered arene-ynamides **191** were prepared in moderate to reasonable yields by Hsung's second-generation protocol ($CuSO_4$, K_3PO_4, and 1,10-phenanthroline), and were subjected to a keteniminium Pictet–Spengler cyclization in the presence of 15 mol % of *p*-nitrobenzenesulfonic acid (PNBSA) to construct the tricyclic heterocycles **192**.

Another application of the use of the copper-mediated ynamide synthesis in natural product synthesis is nicely illustrated in Danheiser's formal synthesis of (+)-FR66979 **201** and (+)-FR900482 **202**,[53] highly potent antitumor agents isolated by researchers at Fujisawa Pharmaceutical Co. Intermediate ynamide **197** was constructed by Hsung's $CuSO_4$-catalyzed coupling of allylcarbamate **196** with bromoalkyne **195** (Scheme 16.42). A key tandem benzannulation is next initiated with electrocyclic cleavage of **198** and [2 + 2]-cycloaddition of the resulting vinylketene with ynamide **197** to furnish an intermediate cyclobutene, after which a series of electrocyclic cleavage of the cyclobutene and subsequent 6-electron electrocyclic closure and tautomerization to conjugation provided the highly substituted derivative benzene **199**. Then, RCM of side chains afforded the end-game intermediate **200**, which was previously transformed to the natural product FR66979 **201** and FR900482 **202**.

The development of efficient, robust, and reliable copper-based systems for the formation of C–N bonds using a wide range of N-nucleophiles that can be smoothly coupled with aryl, vinyl, or alkynyl halides clearly had a deep impact

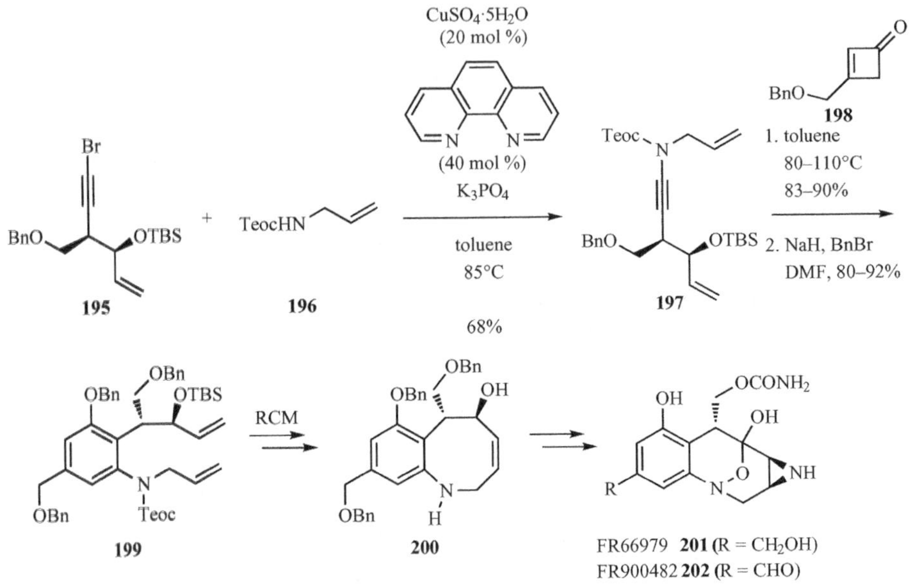

Scheme 16.42 Formal synthesis of FR66979 and FR900482.

on natural product synthesis, and these methods are now commonly integrated in a synthetic plan even at late stages of the synthesis, which clearly demonstrates both their efficiency and reliability. The methods have also been used intramolecularly, where they provide efficient entries to an impressive range of nitrogen heterocycles with excellent efficiencies. The intramolecular copper-mediated formation of a C–N bond for the synthesis of complex natural products will be discussed in the following section.

16.7 INTRAMOLECULAR C–N BOND FORMATION IN NATURAL PRODUCT SYNTHESIS

16.7.1 Copper-Mediated Synthesis of Small- to Medium-Sized N-Heterocycles in Total Synthesis

The successful development of efficient and atom-economical methodologies for copper-mediated C–N bond construction led to the extension to the intramolecular version, the corresponding N-arylated cyclic cores being frequently found in various natural products as well as bioactive molecules. Recently, efficient ligandless or ligand-assisted copper-mediated intramolecular amidations have been extensively studied, and an increasing number of total syntheses have been accomplished using these powerful coupling methods for the formation of various N-heterocyclic systems in a key step.

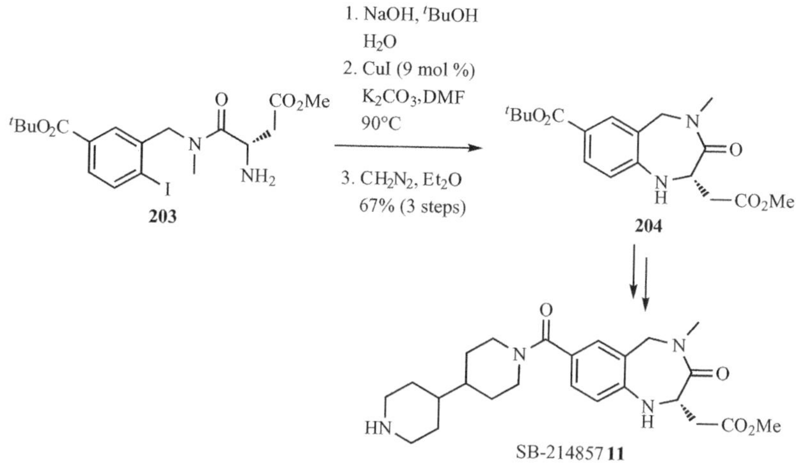

Scheme 16.43 Synthesis of SB-214857 by intramolecular copper-mediated amination.

In 2001, Ma and Xia reported an improved synthetic protocol for the synthesis of SB-214857 **11** by introducing an intramolecular aryl amination to access the bicyclic framework of the target (Scheme 16.43).[3] The intramolecular coupling reaction occurred under the amino acid-assisted amination conditions they developed, which have been discussed in Chapter 1. Thus, methyl ester in **203** was saponified to give the corresponding amino acid, which was next exposed to catalytic amounts of copper(I) iodide under relatively mild conditions and was subsequently treated with diazomethane to give the desired bicyclic intermediate **204** in 67% yield without racemization. Not surprisingly, the intramolecular amidation overwhelms the intermolecular reaction used in Hayes's synthesis (see Scheme 16.2) in terms of yield and preservation of chirality.

Fukuyama and coworkers have also developed remarkably efficient conditions for ligandless copper-mediated intramolecular aryl amination using copper iodide and cesium acetate in DMSO.[54] In this case, the cyclization is straightforward to access 5- and 6-membered N-heterocycles, even though limited success was met in the case of a seven-membered ring formation. Subsequently, this development led to several elegant total syntheses from the Fukuyama group; the total syntheses of (+)-duocarmycin A **213** and SA **216**[55] and (+)-yatakemycin **219**,[56] where chiral indoline subunits in these natural products were synthesized through a series of key intramolecular aminations with remarkable efficiencies, are excellent and striking examples of applications of this methodology.

In that regard, (+)-duocarmycin A **213**, a potent antitumor antibiotic isolated from a *Streptomyces* species, was synthesized using the ligandless copper-mediated intramolecular amination (Scheme 16.44).[55] In this synthesis, the chiral indoline **206** was first assembled by a chemoselective

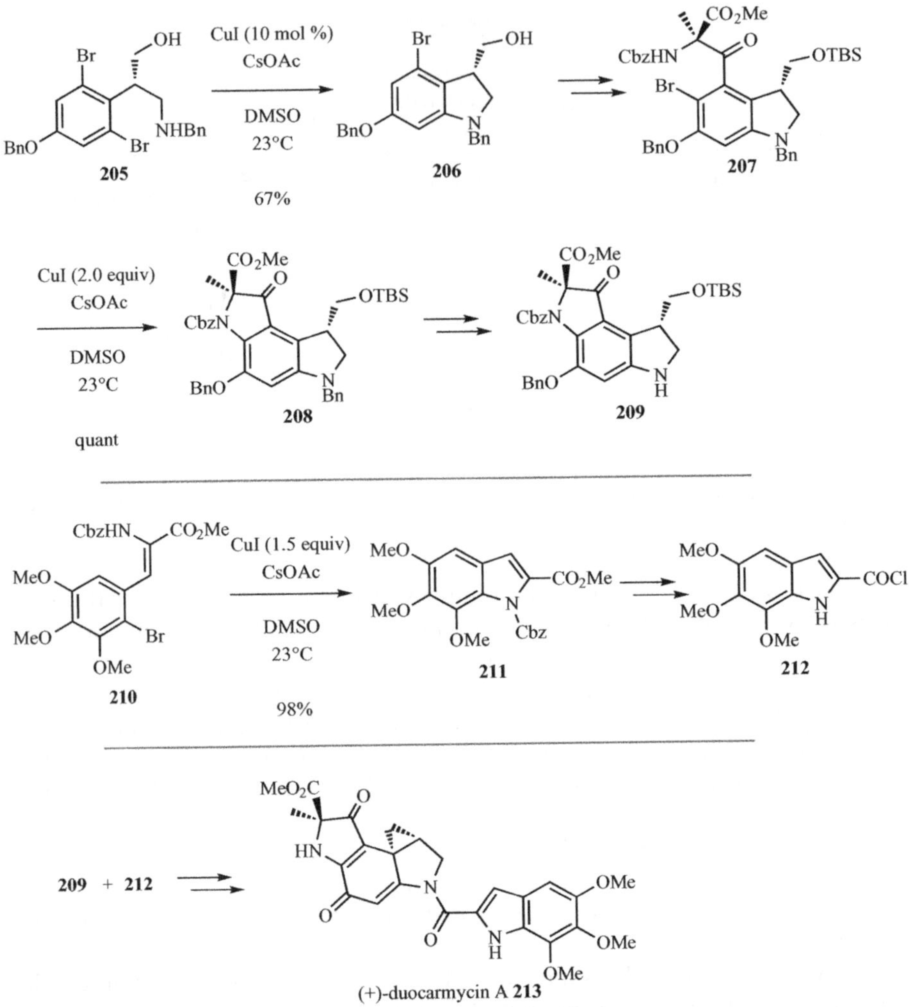

Scheme 16.44 Synthesis of (+)-duocarmycin A.

intramolecular amination from aminoethyl dibromobenzene **205** under the catalysis of 10 mol % of copper(I) iodide in the presence of cesium acetate in DMSO at room temperature. While the use of palladium catalysis gave the desired indoline product **206** with poor efficiency, mostly due to a second oxidative addition of the palladium catalyst to remaining bromide, the coupling reaction catalyzed by copper(I) iodide turned out to be the best method, and successfully gave the cyclized product **206** in 67% yield at room temperature. After several transformations, the second amination, which involves a carbamate instead of an amine, required the use of 2 equiv of copper(I) iodide but provided the cyclized pyrrolidinoindolinone **208** in quantitative

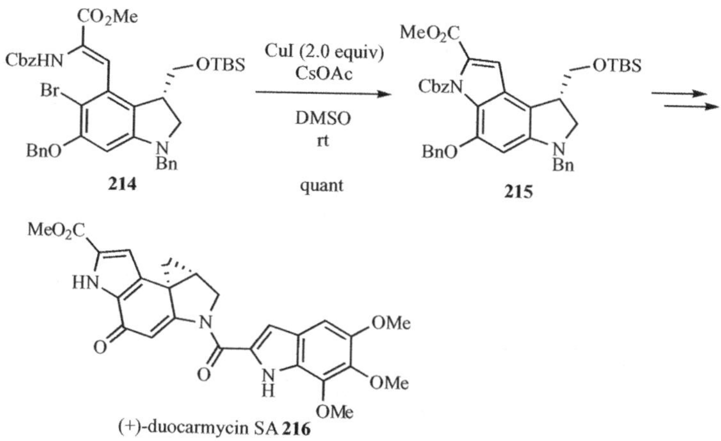

Scheme 16.45 Synthesis of (+)-duocarmycin SA.

yield. The left-hand fragment **212** was prepared by a third intramolecular amination starting from carbamate-protected enaminoester **210** with a similar efficiency. Finally, duocarmycin A **213** was achieved through a simple coupling between the amine in **209** and acyl chloride **212** and subsequent S_N2-type cyclopropanation.

Structurally related (+)-duocarmycin SA **216** was synthesized using a similar intramolecular amination strategy from substrate **214** in the presence of copper iodide (1.0 equiv) at room temperature,[55] which was followed by fragment coupling and cyclopropanation to provide duocarmycin SA **216** (Scheme 16.45).

The impressive total synthesis of (+)-yatakemycin **219** reported by Fukuyama and coworkers is probably one of the best syntheses highlighting the efficiency of intramolecular copper-mediated amination reactions, since all five nitrogen heterocycles have been installed with impressive efficiency using their ligandless protocol (Scheme 16.46).[56] While the syntheses of duocarmycins depended on the formation of various five-membered ring indole or indoline derivatives, in this synthesis, the cyclization into a six-membered ring was conducted under their ligandless copper-mediated coupling conditions. The synthesis started with the intramolecular amidation of substrate **217**, which proceeded smoothly to afford the corresponding tetrahydroquinoline **218**, the formation of the six-membered ring system requiring gentle heating at 60°C. This material subsequently underwent S_N2-type ring contraction to form a spirocyclopropane at the end stage of the synthesis. The bond disconnections shown in Scheme 16.46 indicate the formation of indole and indoline moieties under copper-mediated amidation. In all cases, intramolecular amination of Ns-protected amines was especially successful and systematically afforded the cyclized desired products in high yields, which resulted in a subgram-scale preparation of **219** in 13% overall yield over 20 steps.

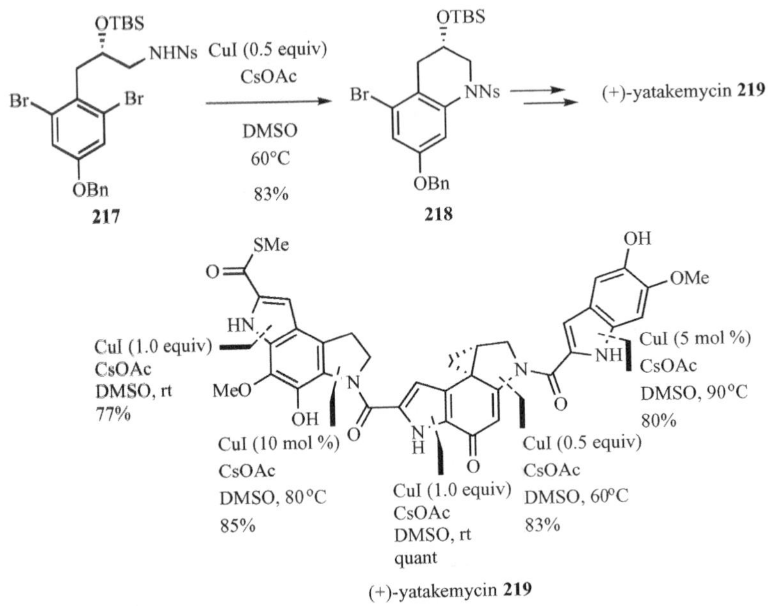

Scheme 16.46 Synthesis of (+)-yatakemycin.

The ligand-assisted and ligandless amination procedures can be complementary to each other. One interesting example is found in Takayama's total synthesis of psychotrimine **82**, which features intramolecular (already described in Scheme 16.16) and intermolecular amination reactions as key steps to install two out of the six Csp^2–N bonds of the natural product, the other ones being already present in the starting materials used (Scheme 16.47).[21] In this synthesis, pyrrolidinoindoline **221** was synthesized by intramolecular amination starting from the corresponding amidine **220** under ligandless conditions close to the ones established by Fukyama. While a diamine ligand-assisted amination gave the desired cyclization product **221** in disappointingly low yields (10% in the best case), the combination of copper iodide and potassium phosphate in the absence of a diamine ligand provided pyrrolidinoindoline **221** in 91% yield, clearly a substantial improvement in reaction efficiency and one that is especially marked for intramolecular reactions. Note that for all the "ligandless" procedures, DMSO is used as the solvent: Its coordination to copper might play a crucial role in these reactions.

Similar conditions were used by Fürstner and coworkers in the total syntheses of the aporphine alkaloids *O*-methyldehydroisopiline **224** and its dimer 7,7′-bisdehydro-*O*-methylisopiline **225** employing an indium-mediated cycloisomerization/intramolecular copper-mediated amination sequence (Scheme 16.48).[57] A six-*endo* cyclization of bromoalkyne **222** with 1 equiv of indium trichloride followed by deprotection of the amine afforded bromophenanthrene **223** in 87% yield, which smoothly underwent intramolecular amination under conditions previously described by Fukuyama, yielding the desired

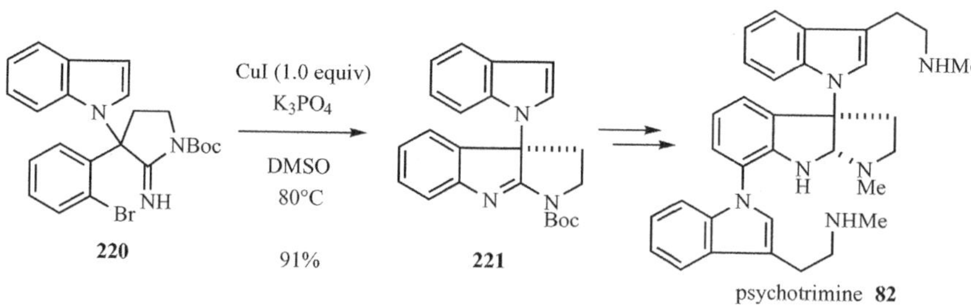

Scheme 16.47 Synthesis of psychotrimine.

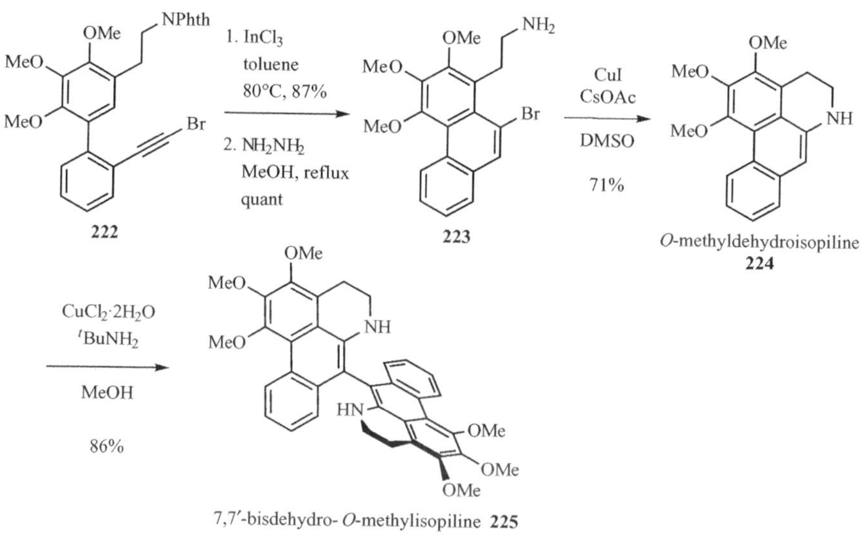

Scheme 16.48 Synthesis of *O*-methyldehydroisopiline and 7,7′-bisdehydro-*O*-methylisopiline.

O-methyldehydroisopiline **224**. Finally, oxidative dimerization of **224** in the presence of copper(II) chloride afforded the dimeric natural product **225** in 86% yield.

An elegant application of Fukuyama's amidation also appeared in the synthesis of cholesteryl ester transfer protein (CETP) inhibitor **228** where the key intramolecular amination under the conventional conditions (CuI, Cs_2CO_3 in DMSO) reported by Fukuyama were used to construct the quinoline **227** in an enantioenriched form and in nearly quantitative yield, once again highlighting the efficiency of these intramolecular cross-coupling reactions (Scheme 16.49).[58]

Another application of the ligandless intramolecular aryl amination was reported by Tokuyama and coworkers who elegantly synthesized PDE-II **233** using an elimination/double aryl amidation strategy (Scheme 16.50).[59] The intramolecular amination of **229** under Fukuyama's conditions surprisingly

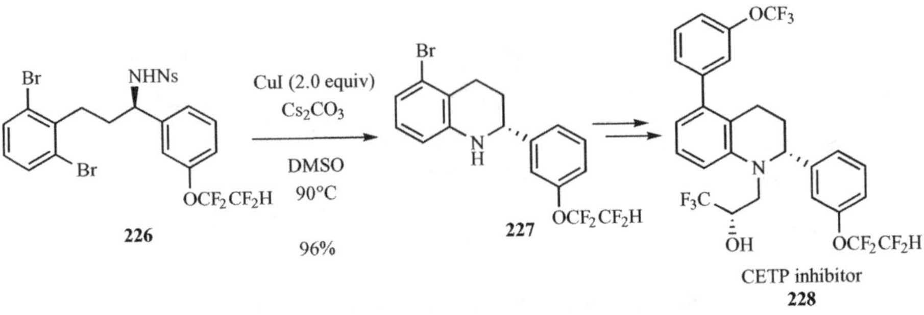

Scheme 16.49 Synthesis of a CETP inhibitor.

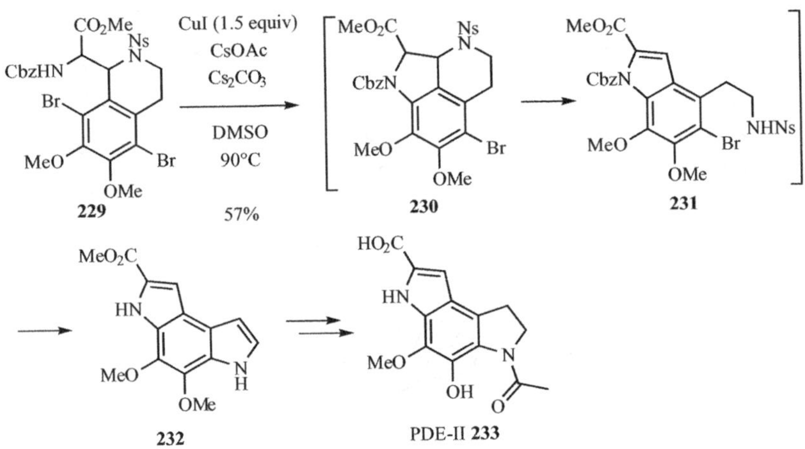

Scheme 16.50 Synthesis of PDE-II.

furnished a pyrroloindole **232** through a five-step, one-pot sequence, including double aryl amination, β-elimination, deprotection of a Cbz group, and removal of an Ns group followed by rearomatization. Presumably, this reaction is initiated by the arylation of the Cbz carbamate, and subsequent β-elimination of the resulting tricyclic intermediate **230** affords the unsaturated ester **231**. The tandem amination/deprotection of Cbz, and subsequent cleavage of the nosyl group with concomitant aromatization, furnishes pyrroloindole **232**. It is noteworthy that this is the first example implementing a long-sequence cascade reaction under copper-mediated coupling conditions.

The ligand-assisted amidation conditions, which have extensively been used in intermolecular processes as described in previous sections, were also proven to be useful in their intramolecular versions. Derivatives of the imidazoindolone core of the potent cholecystokinin antagonist asperlicin **236** and antifungal agent fumiquinazoline **237** were rapidly prepared by tandem intramolecular copper-mediated amidation (Scheme 16.51).[60] Although significant amounts

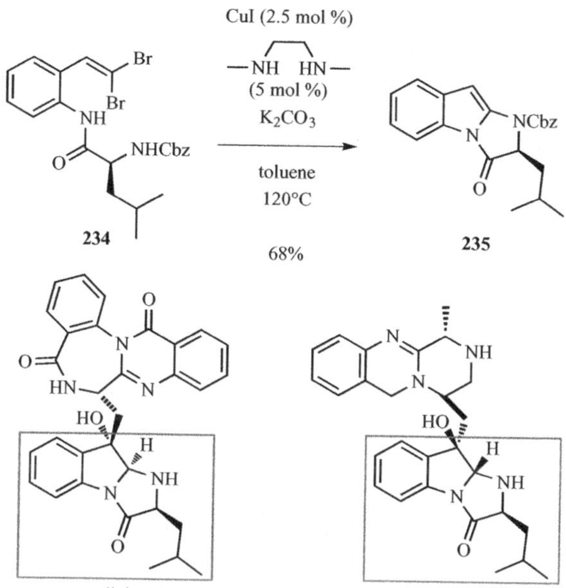

Scheme 16.51 Synthesis of the imidazoindolone core of asperlicin and fumiquinazoline.

of racemization were observed in the reaction of substrates bearing electron-withdrawing groups in either the aromatic ring or the side chain, the cyclization of *gem*-dibromovinyl compounds with a simple aliphatic side chain, such as in the leucine derivative **234**, proceeded efficiently under conditions described by Buchwald to afford imidazoindolone **235** with good enantiomeric excess. Tandem C–N bond formation under palladium-catalyzed amidation conditions was not successful and only afforded a monocyclized 2-bromoindole in moderate yield. Importantly, all parameters were found to be important for the reaction to proceed in good yields and without racemization since studies performed with a substrate similar to **234** but possessing a methyl group in place of the isobutyl side chain showed that racemic *trans*-1,2-cyclohexyldiamine was found to be equally efficient, while *N*-(*n*-butyl)-ethylenediamine and *N*,*N*,*N*′-trimethylethylenediamine resulted in the poorest yields. Screening a range of bases (K_2CO_3, K_3PO_4, Cs_2CO_3, and DABCO) showed that the first one was the most efficient. The greatest effect on yield and enantiomeric excess of the product was observed by simply changing the ratio and quantity of ligand, copper(I) iodide, and base, and notably showed that an excess of the diamine ligand is necessary to prevent multiple ligation of the amide to copper.

The synthesis of analogues of the antimitotic agent ceratamine A **240**, a microtubule-stabilizing alkaloid isolated from the marine sponge *Pseudoceratina* sp., was accomplished employing a copper-mediated, ligand-assisted intramolecular N-vinylation as a key step (Scheme 16.52).[61] The intramolecular

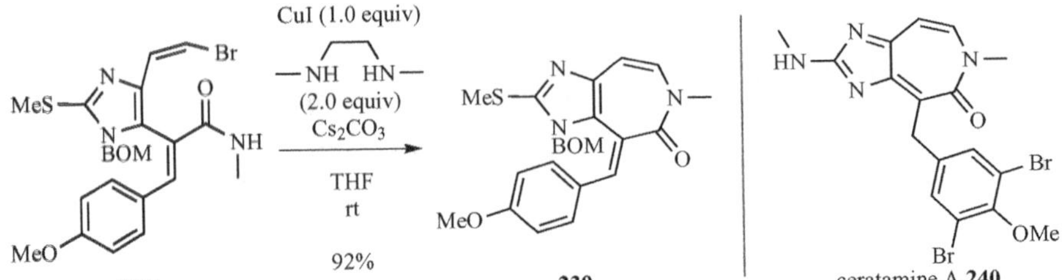

Scheme 16.52 Synthesis of an analogue of ceratamine A.

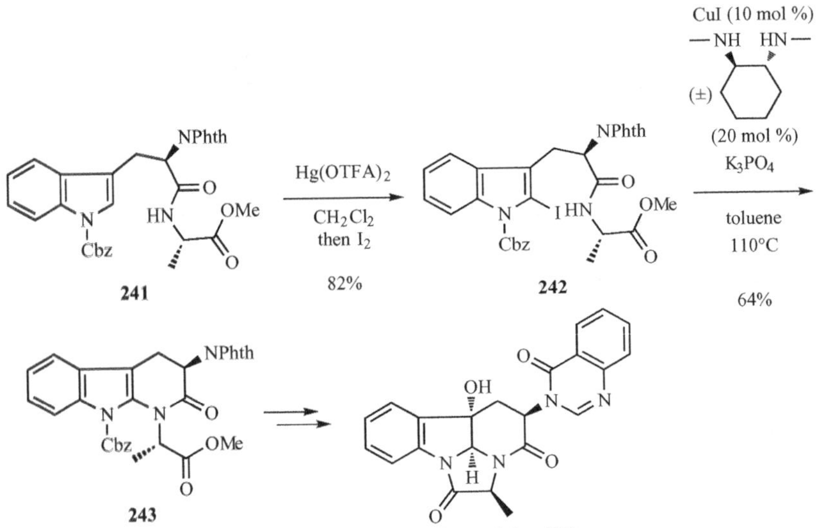

Scheme 16.53 Synthesis of chaetominine.

amidation of the (*Z*)-vinylbromide with the *N*-methyl amide functionality in **238** produced a seven-membered enamide **239** in a remarkable 92% yield, despite the use of an acyclic secondary amide as reaction partner, probably due to the intramolecular reaction overcoming the low reactivity of such substrates and also accounting for the especially low temperature at which this reaction was performed.

In a recent publication, Evano and coworkers have achieved a total synthesis of the cytotoxic alkaloid chaetominine **244** using a copper-catalyzed intramolecular coupling in the presence of *trans*-*N*,*N′*-dimethylcyclohexane-1,2-diamine (Scheme 16.53).[62a] The fully protected dipeptide **241** was prepared by standard coupling between D-tryptophan and L-alanine methyl ester, and underwent a regioselective iodination at the C2 position upon treatment with $Hg(OTFA)_2$ and I_2 to afford the corresponding 2-iodoindole **242**. Through a series of reaction screenings to find the optimal cyclization condition, it was

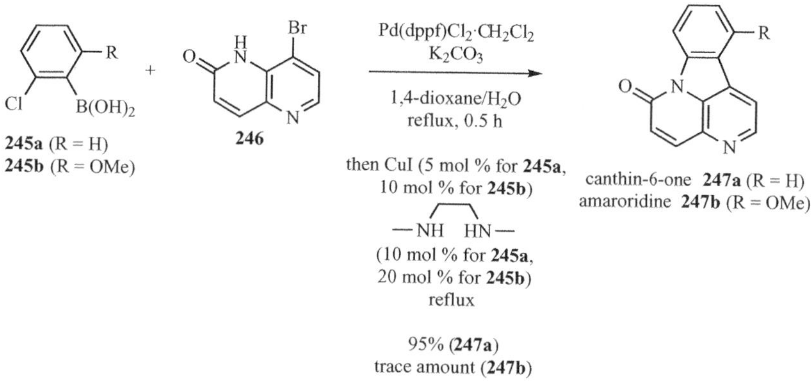

Scheme 16.54 Synthesis of canthin-6-one and amaroridine.

found that the combination of copper(I) iodide, *trans-N,N′*-dimethylcyclohexane-1,2-diamine, and potassium phosphate at 110°C in toluene gave the desired tricyclic product **243** in good yield with small amounts of reduced product **241**, which is able to be recycled, without epimerization at the two sensitive stereocenters. Other coupling reactions with palladium or iron-based catalysts were found to be totally inefficient. It was also found that protection of the indole nucleus was key to the success of the cyclization, which did not proceed otherwise, probably because of competitive chelation of the copper catalyst at the indole N–H, and that, compared to the formation of five-membered ring systems from similar substrates,[62b] the reaction was slower and specifically required the use of *trans-N,N′*-dimethylcyclohexane-1,2-diamine.

The total syntheses of canthin-6-one **247a** and amaroridine **247b** are interesting examples implementing a one-pot sequential Suzuki–Miyaura coupling and a copper-mediated intramolecular amidation (Scheme 16.54).[63] The sequence was initiated by the Suzuki–Miyaura coupling of 2-chlorophenylboronic acid **245a** with bromonaphthyridone **246** to furnish an intermediate arylnaphthyridone, which was subsequently treated with a mixture of copper iodide and DMEDA to afford the desired natural product, canthin-6-one **247a**, in 95% yield. Interestingly, no amidation product was formed during the course of the Suzuki coupling. However, identical reactions utilizing sterically demanding 2,6-disubstituted phenylboronic acids such as **245b** led to a significant decrease in yield. Thus, amaroridine **247b** was synthesized in higher yield through a stepwise Suzuki coupling and intramolecular amidation (73%, three steps).

Buchwald's amidation protocol was also applied in the total synthesis of the alkaloid (+)-isatisine A **250** by the Panek group (Scheme 16.55).[64] In that regard, the intramolecular coupling between the cyclic secondary amide and the aryl bromide functionalities in **248** with copper(I) iodide (0.5 equiv) and *N,N′*-dimethylethylenediamine (1.0 equiv) proceeded smoothly to generate the tetracyclic advanced intermediate **249** in 90% yield. Further concomitant

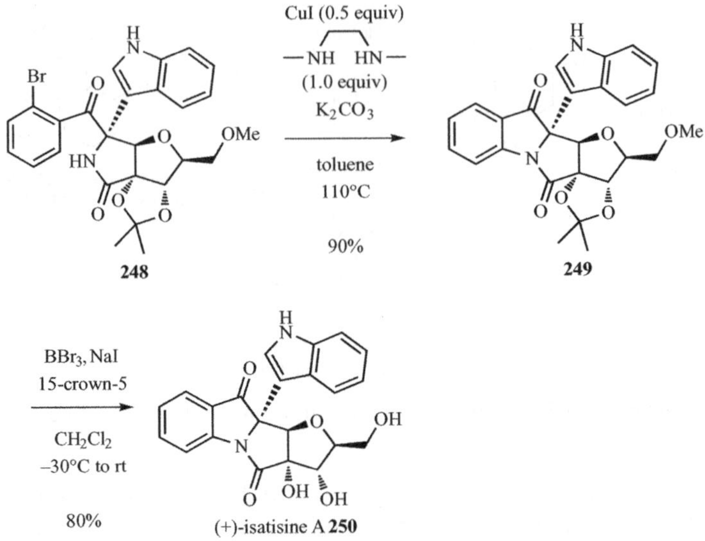

Scheme 16.55 Synthesis of (+)-isatisine A.

cleavage of the acetonide and methyl ether using BBr_3 in the presence of a crown ether accomplished the total synthesis of (+)-isatisine A **250**.

Finally, the synthesis of circumdatins H **252a** and J **252b** were accomplished through a proline-assisted copper-mediated intramolecular amidation in the final step (Scheme 16.56).[65] One particularly interesting feature of this amidation is the use of sodium hydride, which was determined to be the optimal base after extensive screening. Remarkably, exposure of quinazolinones **251**, which possess a potentially epimerizable stereocenter, to the optimal conditions (CuI, proline, NaH in DMF at 120°C) furnished circumdatins H **252a** and J **252b** with almost complete preservation of the optical purity and good yields. The diminished acidity responsible for the absence of epimerization was attributed to the π-conjugation of the methoxy group with the imine moiety.

16.7.2 Copper-Mediated Macrocyclization by C–N Bond Formation in Natural Product Synthesis

Although several useful methods have been developed over the past several decades for the formation of large heterocycles, the development of new, efficient, and selective macrocyclization methodologies remains an important task for the syntheses of natural products and bioactive molecules. In this regard, the successful development of copper-mediated C–N bond formation processes for macrocyclization provides especially attractive routes toward the construction of macrolactams, as well as new and reliable bond disconnections for retrosynthetic analysis.

The first application of copper-mediated amination/amidation for macrocyclization appeared in the synthesis of reblastatin by Panek and coworkers in

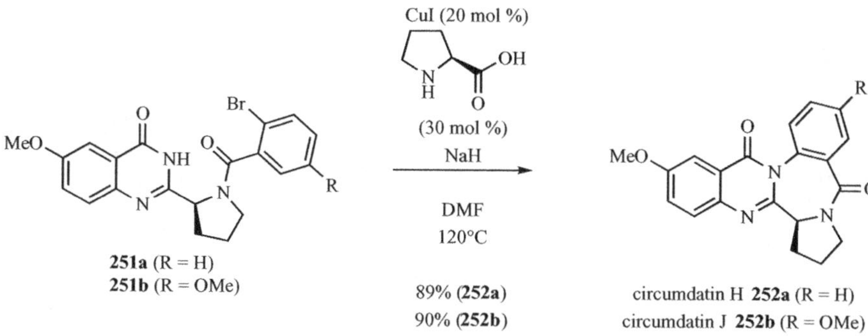

Scheme 16.56 Synthesis of circumdatins H and J.

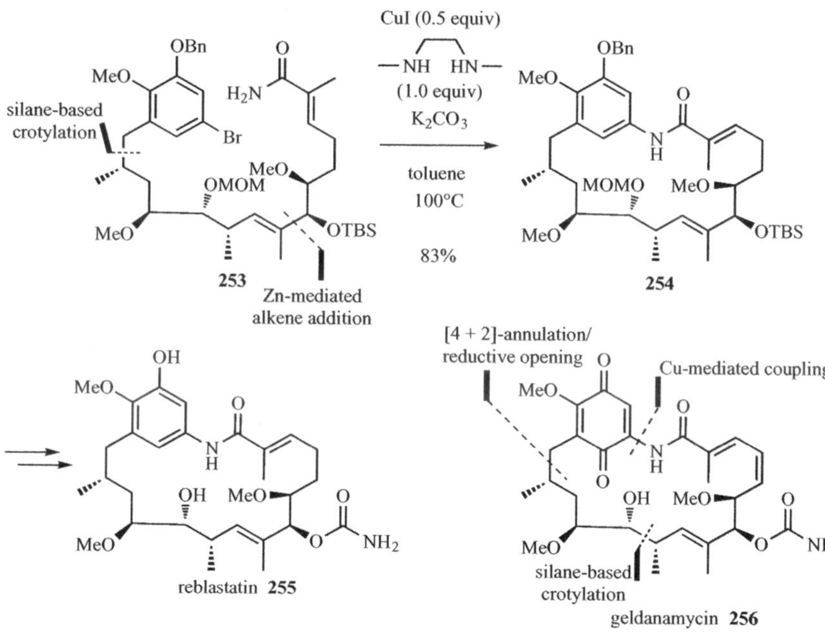

Scheme 16.57 Synthesis of reblastatin and geldanamycin.

2005 (Scheme 16.57).[66] Reblastatin **255**, a benzenoid ansamycin-like cell cycle inhibitor macrolactam-19 membered ring, was constructed through an intramolecular copper(I)-mediated N-arylation strategy for macrolactamization. Acyclic precursor **253**, which was prepared through their silane-based crotylation and subsequent chelation-controlled Zn-mediated alkene addition to an aldehyde, was subjected to modified Buchwald's amidation conditions (50 mol % CuI, 1.0 equiv of DMEDA, 3.0 equiv of K_2CO_3 in toluene) to forge the desired macrolactam **254** in 83% yield.

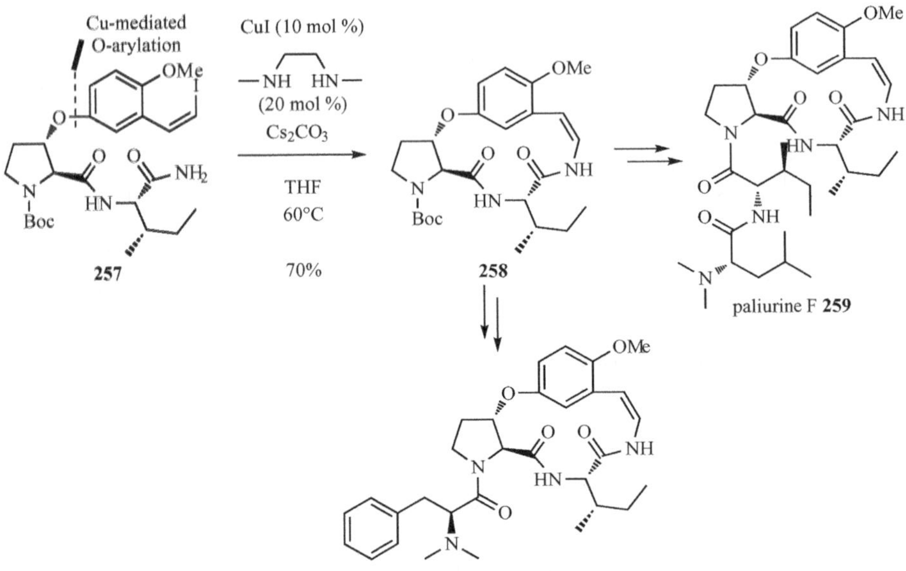

Scheme 16.58 Synthesis of paliurines E and F.

The utility of a copper-mediated macrolactamization was demonstrated once again in the total synthesis of geldamycin **256**,[67] a natural product exhibiting Hsp 90 inhibition effect. Similarly, a macrolactam was obtained by intramolecular copper-mediated amidation under the same conditions described in the synthesis of reblastatin **255**. Panek's successful trials provide a new strategy for the synthesis of macrolactams, in which an aryl C–N bond serves as a key retrosynthetic disconnection.

A clever application of a copper-mediated macrolactamization was employed in the total synthesis of paliurine F **259**,[68] a member of the cyclopeptide alkaloid family of natural products, which exhibits useful biological activity (Scheme 16.58). In that context, Evano envisioned that the enamide functionality in **259** could be introduced by intramolecular amidation to furnish the macrolactam core. Efforts to find optimal reaction conditions revealed that exposure of iodoamide **257**, which is prepared by another copper-mediated O-arylation (see Chapter 17, Scheme 17.21, and corresponding text for details), under Buchwald's conditions provided a 13-membered macrolactam **258** in 70% yield along with 20% of recovered starting material. Interestingly, the application of Ma's conditions based on the use of *N,N*-dimethylglycine or proline as the ligands, which gave a vinyl amidation product in the intermolecular version of the total synthesis of ziziphine N **135** (see Scheme 16.31), failed to afford any macroenamidation product in this case. Other strategies for macrocyclization such as RCM, oxidative alkene amidation, or even simple cyclodehydration from an α-amido-ω-carboxamide were found to be less efficient.[69]

The successful development of this macroenamidation protocol led to consecutive results in the total syntheses of other cyclopeptide alkaloids. The 13-membered cyclopeptides paliurine F **259** and E **260** were synthesized from the same intermediate macrolactam **258**. In addition, ziziphine N **135** and Q **262** were also synthesized from a similar macrolactam **261**, establishing a divergent approach to this class of natural products with increased efficiency compared with previous syntheses of cyclopeptide alkaloids, where a lengthy stepwise installation of the enamide after formation of the macrocycle through standard macrolactamization techniques was used (Scheme 16.59).[69]

Abyssenine A **264**[70] and mucronine E **139b**,[71] which differ only in the nature of substituents on the aryl group and the size of macrocycle, were also synthesized utilizing an intramolecular copper-mediated vinyl amidation (Scheme 16.60). Intramolecular enamidation of the corresponding acyclic precursors **263** proceeded smoothly to yield 15-membered macrocycles in high yields, and with no detectable epimerization despite the presence of three sensitive stereocenters, which are then converted to the corresponding natural products **264** and **139b** by cleavage of the Boc protecting group.

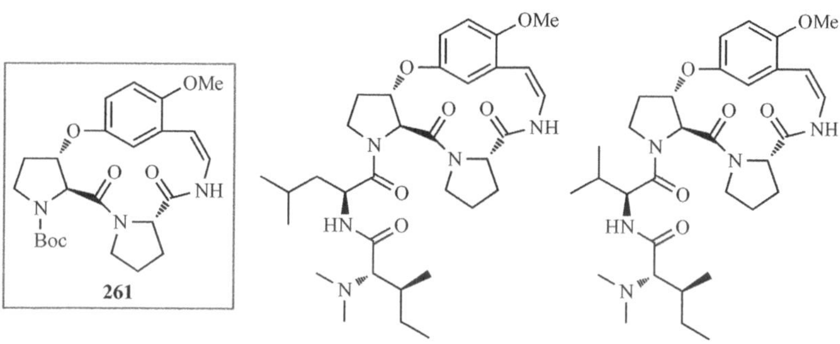

Scheme 16.59 Macrolactam core of ziziphine N and Q.

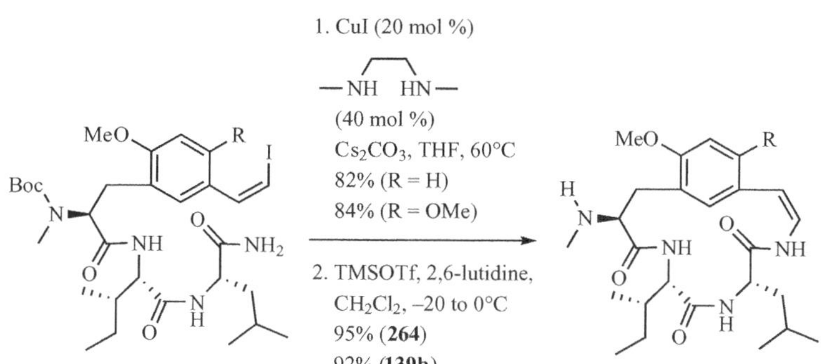

Scheme 16.60 Synthesis of abyssenine A and mucronine E.

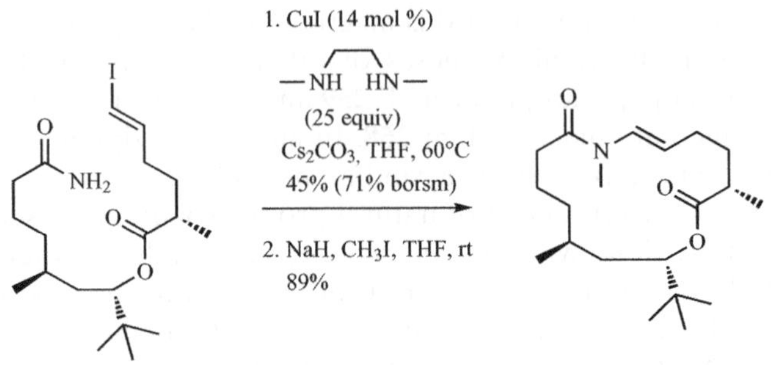

265 **266**

Scheme 16.61 Synthesis of palmyrolide A.

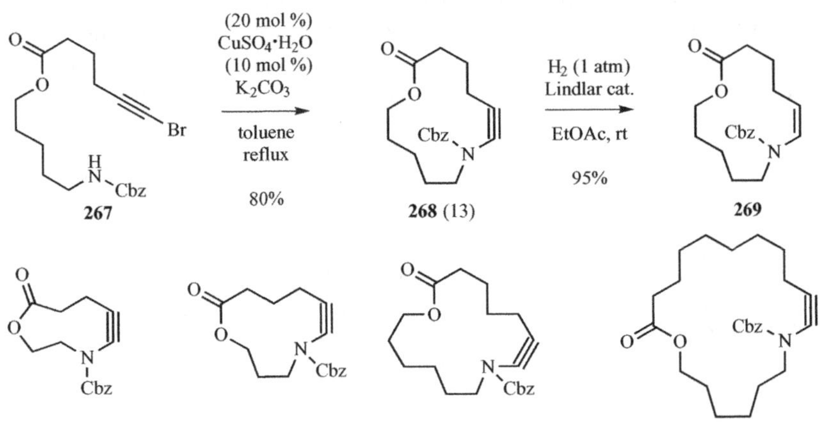

270 (9); 76% **271** (11); 90% **272** (15); 45% **273** (19); 42%

Scheme 16.62 Syntheses of various sizes of macrocyclic ynamides using an intramolecular copper(I)-mediated ynamidation (ring size indicated between brackets).

Another example of the use of this macroenamidation procedure in natural product synthesis was found in the total synthesis of palmyrolide A **266**, reported by the Maio laboratory in 2012 (Scheme 16.61).[72] Late-stage intramolecular vinyl amidation of the acyclic precursor **265** produced the desired 15-membered macrocycle in a moderate yield with retention of (*E*)-geometry of the vinyl iodide. Simple N-methylation of the resulting enamide successfully achieved the first total synthesis of (+)-palmyrolide A **266**. Significantly, it is the first example that demonstrates the potential of copper-mediated enamidation to construct a *trans*-enamide macrocycle. Interestingly, the use of palladium catalysis failed to promote the macrocyclization, and other strategies such as dehydrative cyclization and intramolecular alkyne amidation, respectively, starting from the corresponding acyclic α-amido-ω-aldehyde or α-amido-ω-alkyne, were also inefficient.[73]

To date, the use of an intramolecular copper-mediated ynamidation as a method for macrocyclization has not been reported in the context of natural product synthesis. However, Hsung and coworkers disclosed successful model studies for macroynamidations, yielding macrocyclic ynamides that, upon reduction, nicely produce the corresponding (*Z*)-enamides (Scheme 16.62).[74] Nine- to 19-membered cycles (**268**, **270-273**) were prepared from the corresponding acyclic Cbz-protected primary amines using their reaction conditions, which provided 9-, 11-, and 13-membered macrocycles in excellent yields. In the cases of 15- and 19-membered rings, however, the desired macrocycles were constructed with moderate efficiency.

16.8 SUMMARY AND OUTLOOK

As proven by the number of examples collected in this chapter, the development of efficient copper-mediated cross couplings for the formation of C–N bonds has provided straightforward and efficient routes to complex natural products. In many cases, it allowed for the development of new pathways to overcome existing synthetic shortcomings, previously addressed through classical synthetic methods. For instance, enamide functionality, which was previously synthesized by a condensation or elimination method with poor efficiency, can be simply constructed by copper-mediated coupling with a high level of *E*/*Z* selectivity. One of the latest total syntheses using this strategy, reported during the edition of this chapter, is a remarkable example of the input of copper catalysis in natural product synthesis. Indeed, in their total synthesis of Yaku'amide A, a complex polypeptide bearing stereodefined dehydroaminoacids, Inoue and coworkers showcased the efficiency of the copper-catalyzed enamide synthesis since all four dehydroaminoacid subunits of this natural product, which are usually installed through inefficient and/or lengthy sequences, were efficiently prepared using amidation of vinyl iodides (Scheme 16.63).[75] Also, excellent coupling efficiency and a broad range of functional group tolerance in intramolecular reactions have led to an increasing number of applications in complex molecule synthesis. Although intramolecular copper-mediated ynamide formation has not been well developed, the existing evidence suggests that this efficient and economic approach to the formation of C–N bonds will be further exploited in the context of complex molecule synthesis. Further developments in copper catalysis that would allow for the formation of C–N bonds with even higher efficiency, lower catalyst loading, milder reaction conditions, and greater substrate scope should definitely be most useful for the design of efficient and straightforward routes to complex natural products. There is no doubt that copper catalysis will continue playing a major role in the synthesis of alkaloids and other natural products for the years to come and will also be used more frequently in industry, notably for the preparation of bioactive molecules.

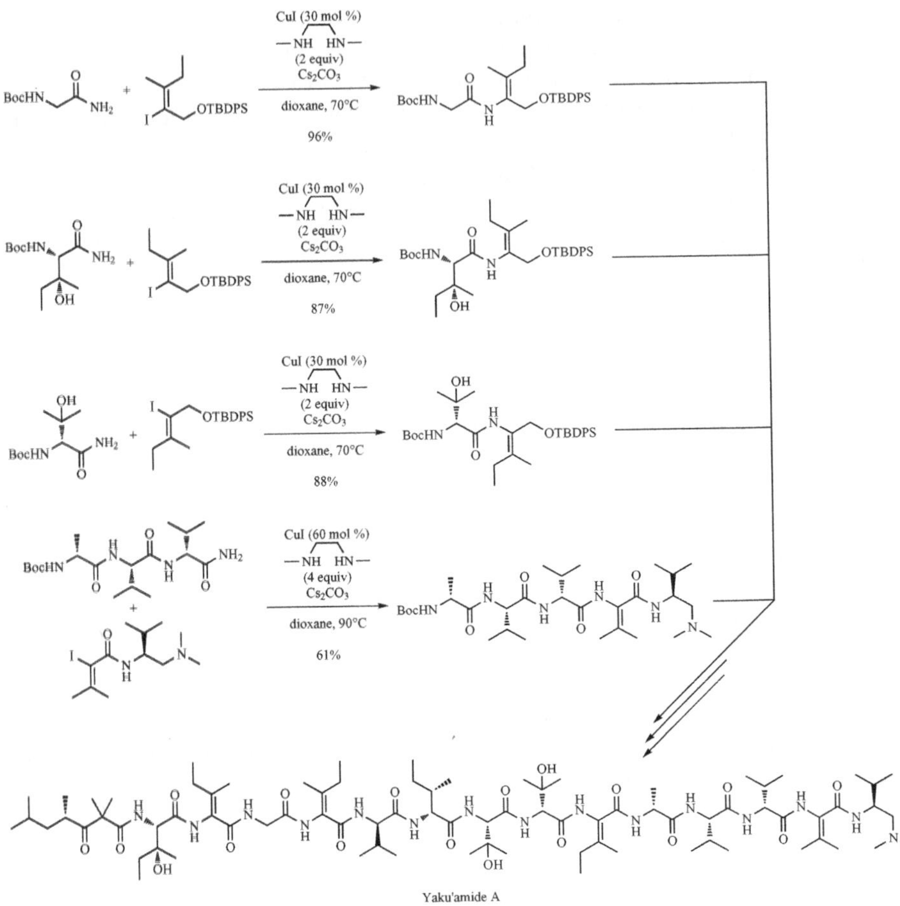

Scheme 16.63 One last example showcasing the efficiency of copper-catalyzed cross-coupling reactions in total synthesis: the total synthesis of Yaku'amide A.

REFERENCES

[1] Ma, D.; Zhang, Y.; Yao, J.; Wu, S.; Tao, F. *J. Am. Chem. Soc.* **1998**, *120*, 12459–12467.

[2] Clement, J.-B.; Hayes, J. F.; Sheldrake, H. M.; Sheldrake, P. W.; Wells, A. S. *Synlett* **2001**, 1423–1427.

[3] Ma, D.; Xia, C. *Org. Lett.* **2001**, *3*, 2583–2586.

[4] (a) Andrews, I. P.; Atkins, R. J.; Badham, N. F.; Bellingham, R. K.; Breen, G. F.; Carey, J. S.; Etridge, S. K.; Hayes, J. F.; Hussain, N.; Morgan, D. O.; Share, A. C.; Smith, S. A. C.; Walsgrove, T. C.; Wells, A. S. *Tetrahedron Lett.* **2001**, *42*, 4915–4917; (b) Hayes, J. F. *Synlett* **1999**, *S1*, 865–866.

[5] (a) Ma, D.; Xia, C.; Jiang, J.; Zhang, J.; Tang, W. *J. Org. Chem.* **2003**, *68*, 442–451; (b) Ma, D.; Xia, C.; Jiang, J.; Zhang, J. *Org. Lett.* **2001**, *3*, 2189–2191.

[6] Brackmann, F.; Es-Sayed, M.; de Meijere, A. *Eur. J. Org. Chem.* **2005**, 2250–2258.

[7] Egger, M.; Li, X.; Müller, C.; Bernhardt, G.; Buschauer, A.; König, B. *Eur. J. Org. Chem.* **2007**, 2643–2649.

[8] (a) Nishihama, Y.; Ishikawa, Y.; Nishiyama, S. *Tetrahedron Lett.* **2009**, *50*, 2801–2804. (b) Jabri, S. Y.; Overman, L. E. *J. Am. Chem. Soc.* **2013**, *135*, 4231–4234.

[9] (a) Strieter, E. R.; Blackmond, D. G.; Buchwald, S. L. *J. Am. Chem. Soc.* **2005**, *127*, 4120–4121; (b) Klapars, A.; Huang, X.; Buchwald, S. L. *J. Am. Chem. Soc.* **2002**, *124*, 7421–7428; (c) Klapars, A.; Antilla, J. C.; Huang, X.; Buchwald, S. L. *J. Am. Chem. Soc.* **2001**, *123*, 7727–7729; (d) Surry, D. S.; Buchwald, S. L. *Chem. Sci.* **2010**, *1*, 13–31.

[10] Mallesham, B.; Rajesh, B. M.; Reddy, P. R.; Srinivas, D.; Trehan, S. *Org. Lett.* **2002**, *5*, 963–965.

[11] (a) Selvakumar, N.; Srinivas, D.; Khera, M. K.; Kumar, M. S.; Mamidi, N. V. S. R.; Sarnaik, H.; Chandrasekar, C.; Rao, B. S.; Raheem, M. A.; Das, J.; Iqbal, J.; Rajagopalan, R. *J. Med. Chem.* **2002**, *45*, 3953–3962; (b) Jegham, S.; Nedelec, A.; Burnier, P. H.; Guminski, Y.; Puech, F.; Koenig, J. J.; George, P. *Tetrahedron Lett.* **1998**, *39*, 4453–4454; (c) Schaus, S. E.; Jacobsen, E. N. *Tetrahedron Lett.* **1996**, *37*, 7937–7940; (d) Park, C.-H.; Brittelli, D. R.; Wang, C. L.-J.; Marsh, F. D.; Gregory, W. A.; Wuonola, M. A.; McRipley, R. J.; Eberly, V. S.; Slee, A. M.; Forbes, M. *J. Med. Chem.* **1992**, *35*, 1156–1165.

[12] Evano, G.; Schaus, J. V.; Panek, J. S. *Org. Lett.* **2004**, *6*, 525–528.

[13] Chae, J.; Buchwald, S. L. *J. Org. Chem.* **2004**, *69*, 3336–3339.

[14] Ghosh, A.; Sieser, J. E.; Caron, S.; Couturier, M.; Dupont-Gaudet, K.; Girardin, M. *J. Org. Chem.* **2006**, *71*, 1258–1261.

[15] Couturier, M.; Tucker, J. L.; Andresen, B. M.; DeVries, K. M.; Vanderplasa, B. C.; Ito, F. *Tetrahedron: Asymmetry* **2003**, *14*, 3517–3523.

[16] Satyanarayana, K.; Srinivas, K.; Himabindu, V.; Reddy, G. M. *Org. Process Res. Dev.* **2007**, *11*, 842–845.

[17] Wiedemann, S. H.; Ellman, J. A.; Bergman, R. G. *J. Org. Lett.* **2006**, *71*, 1969–1976.

[18] Scheiper, B.; Glorius, F.; Leitner, A.; Fürstner, A. *Proc. Natl. Acad. Sci. U.S.A.* **2004**, *101*, 11960–11965.

[19] Serrano, C. M.; Looper, R. E. *Org. Lett.* **2011**, *13*, 5000–5003.

[20] Pu, Y.-M.; Ku, Y.-Y.; Grieme, T.; Black, L. A.; Bhatia, A. V.; Cowart, M. *Org. Process Res. Dev.* **2007**, *11*, 1004–1009.

[21] Matsuda, Y.; Kitajima, M.; Takayama, H. *Org. Lett.* **2008**, *10*, 125–128.

[22] (a) Foo, K.; Newhouse, T.; Mori, I.; Takayama, H.; Baran, P. S. *Angew. Chem. Int. Ed. Engl.* **2011**, *50*, 2716–2719; (b) Newhouse, T.; Lewis, C. A.; Eastman, K. J.; Baran, P. S. *J. Am. Chem. Soc.* **2010**, *132*, 7119–7137; (c) Newhouse, T.; Baran, P. S. *J. Am. Chem. Soc.* **2008**, *130*, 10886–10887.

[23] Kanakis, A. A.; Sarli, V. *Org. Lett.* **2010**, *12*, 4872–4875.

[24] Graham, J. P.; Langlade, N.; Northall, J. M.; Roberts, A. J.; Whitehead, A. J. *Org. Process Res. Dev.* **2011**, *15*, 44–48.

[25] Elliott, G. I.; Konopelski, J. P. *Org. Lett.* **2000**, *2*, 3055–3057.

[26] Yu, S.; Saenz, J.; Srirangam, J. K. *J. Org. Lett.* **2002**, *67*, 1699–1702.

[27] Shen, R.; Porco, J. A., Jr. *Org. Lett.* **2000**, *2*, 1333–1336.
[28] Fürstner, A.; Dierkes, T.; Thiel, O. R.; Blanda, G. *Chem. Eur. J.* **2001**, *7*, 5286–5298.
[29] Shen, R.; Lin, C. T.; Porco, J. A., Jr. *J. Am. Chem. Soc.* **2002**, *124*, 5650–5651.
[30] Wang, X.; Porco, J. A., Jr. *J. Am. Chem. Soc.* **2003**, *125*, 6040–6041.
[31] Wang, X.; Bowman, E. J.; Bowman, B. J.; Porco, J. A., Jr. *Angew. Chem. Int. Ed. Engl.* **2004**, *43*, 3601–3605.
[32] Han, C.; Shen, R.; Su, S.; Porco, J. A., Jr. *Org. Lett.* **2004**, *6*, 27–30.
[33] Nicolaou, K. C.; Kim, D. W.; Baati, R. *Angew. Chem. Int. Ed. Engl.* **2002**, *41*, 3701–3704.
[34] Su, Q.; Panek, J. S. *J. Am. Chem. Soc.* **2004**, *126*, 2425–2430.
[35] Palimkar, S. S.; Uenishi, J. *Org. Lett.* **2010**, *12*, 4160–4163.
[36] Nakamura, R.; Tanino, K.; Miyashita, M. *Org. Lett.* **2003**, *5*, 3583–3586.
[37] Dias, L. C.; de Oliveira, L. G.; Vilcachagua, J. D.; Nigsch, F. *J. Org. Chem.* **2005**, *70*, 2225–2234.
[38] He, G.; Wang, J.; Ma, D. *Org. Lett.* **2007**, *9*, 1367–1369.
[39] Wang, J.; Schaeffler, L.; He, G.; Ma, D. *Tetrahedron Lett.* **2007**, *48*, 6717–6721.
[40] Meketa, M. L.; Weinreb, S. M. *Org. Lett.* **2007**, *9*, 853–855.
[41] Yang, L.; Deng, G.; Wang, D.-X.; Huang, Z.-T.; Zhu, J.-P.; Wang, M.-X. *Org. Lett.* **2007**, *9*, 1387–1390.
[42] (a) Nicolaou, K. C.; Sun, Y.-P.; Guduru, R.; Banerji, B.; Chen, D. Y.-K. *J. Am. Chem. Soc.* **2008**, *130*, 3633–3644; (b) Nicolaou, K. C.; Guduru, R.; Sun, Y.-P.; Banerji, B.; Chen, D. Y.-K. *Angew. Chem. Int. Ed. Engl.* **2007**, *46*, 5896–5900.
[43] Nicolaou, K. C.; Leung, G. Y. C.; Dethe, D. H.; Guduru, R.; Sun, Y.-P.; Lim, C. S.; Chen, D. Y.-K. *J. Am. Chem. Soc.* **2008**, *130*, 10019–10023.
[44] Huang, X.; Shao, N.; Palani, A.; Aslanian, R.; Buevich, A. *Org. Lett.* **2007**, *9*, 2597–2600.
[45] Vintonyak, V. V.; Cal, M.; Lay, F.; Kunze, B.; Sasse, F.; Maier, M. E. *Chem. Eur. J.* **2008**, *14*, 3709–3720.
[46] Wagger, J.; Svete, J.; Stanovnik, B. *Synthesis* **2008**, 1436–1442.
[47] (a) Okamoto, K.; Sakagami, M.; Feng, F.; Togame, H.; Takemoto, H.; Ichikawa, S.; Matsuda, A. *J. Org. Chem.* **2012**, *77*, 1367–1377; (b) Okamoto, K.; Sakagami, M.; Feng, F.; Togame, H.; Takemoto, H.; Ichikawa, S.; Matsuda, A. *Org. Lett.* **2011**, *13*, 5240–5243.
[48] Focken, T.; Charette, A. B. *Org. Lett.* **2006**, *8*, 2985–2988.
[49] Movassaghi, M.; Hunt, D. K.; Tjandra, M. *J. Am. Chem. Soc.* **2006**, *128*, 8126–8127.
[50] Movassaghi, M.; Tjandra, M.; Qi, J. *J. Am. Chem. Soc.* **2009**, *131*, 9648–9650.
[51] (a) Evano, G.; Coste, A.; Jouvin, K. *Angew. Chem. Int. Ed. Engl.* **2010**, *49*, 2840–2859; (b) DeKorver, K. A.; Li, H.; Lohse, A. G.; Hayashi, R.; Lu, Z.; Zhang, Y.; Hsung, R. P. *Chem. Rev.* **2010**, *110*, 5064–5106.
[52] Zhang, Y.; Hsung, R. P.; Zhang, X.; Huang, J.; Slafer, B. W.; Davis, A. *Org. Lett.* **2005**, 7, 1047–1050.

[53] Mak, X. Y.; Crombie, A. L.; Danheiser, R. L. *J. Org. Chem.* **2011**, *76*, 1852–1873.

[54] Yamada, K.; Kubo, T.; Tokuyama, H.; Fukuyama, T. *Synlett* **2002**, 231–234.

[55] Yamada, K.; Kurokawa, T.; Tokuyama, H.; Fukuyama, T. *J. Am. Chem. Soc.* **2003**, *125*, 6630–6631.

[56] Okano, K.; Tokuyama, H.; Fukuyama, T. *J. Am. Chem. Soc.* **2006**, *128*, 7136–7137.

[57] Fürstner, A.; Mamane, V. *Chem. Commun.* **2003**, *39*, 2112–2113.

[58] Rano, T. A.; Kuo, G.-H. *Org. Lett.* **2009**, *11*, 2812–2815.

[59] Okano, K.; Mitsuhashi, N.; Tokuyama, H. *Chem. Commun.* **2010**, *46*, 2641–2643.

[60] Yuen, J.; Fang, Y.-Q.; Lautens, M. *Org. Lett.* **2006**, *8*, 653–656.

[61] (a) Nodwell, M.; Pereira, A.; Riffell, J. L.; Zimmerman, C.; Patrick, B. O.; Roberge, M.; Andersen, R. J. *J. Org. Lett.* **2009**, *74*, 995–1006; (b) Nodwell, M.; Riffell, J. L.; Roberge, M.; Andersen, R. J. *Org. Lett.* **2008**, *10*, 1051–1054.

[62] (a) Toumi, M.; Couty, F.; Marrot, J.; Evano, G. *Org. Lett.* **2008**, *10*, 5027–5030. (b) Coste, A.; Toumi, M.; Wright, K.; Razafimahaléo, V.; Couty, F.; Marrot, J.; Evano, G. *Org. Lett.* **2008**, *10*, 3841–3844.

[63] Gollner, A.; Koutentis, P. A. *Org. Lett.* **2010**, *12*, 1352–1355.

[64] Lee, J.; Panek, J. S. *Org. Lett.* **2011**, *13*, 502–505.

[65] Kshirsagar, U. A.; Argade, N. P. *Org. Lett.* **2010**, *12*, 3716–3719.

[66] Wrona, I. E.; Gabarda, A. E.; Evano, G.; Panek, J. S. *J. Am. Chem. Soc.* **2005**, *127*, 15026–15027.

[67] Qin, H.-L.; Panek, J. S. *Org. Lett.* **2008**, *10*, 2477–2479.

[68] Toumi, M.; Couty, F.; Evano, G. *Angew. Chem. Int. Ed. Engl.* **2007**, *46*, 572–575.

[69] Toumi, M.; Rincheval, V.; Young, A.; Gergeres, D.; Turos, E.; Couty, F.; Mignotte, B.; Evano, G. *Eur. J. Org. Chem.* **2009**, 3368–3386.

[70] Toumi, M.; Couty, F.; Evano, G. *J. Org. Chem.* **2007**, *72*, 9003–9009.

[71] Toumi, M.; Couty, F.; Evano, G. *Synlett* **2008**, 29–32.

[72] Tello-Aburto, R.; Johnson, E. M.; Valdez, C. K.; Maio, W. A. *Org. Lett.* **2012**, *14*, 2150–5153.

[73] Tello-Aburto, R.; Newar, T. D.; Maio, W. A. *J. Org. Chem.* **2012**, *77*, 6271–6289.

[74] Zhang, X.; Zhang, Y.; Huang, J.; Hsung, R. P.; Kurtz, K. C. M.; Oppenheimer, J.; Petersen, M. E.; Sagamanova, I. K.; Shen, L.; Tracey, M. R. *J. Org. Lett.* **2006**, *71*, 4170–4177.

[75] Kuranaga, T.; Sesoko, Y.; Sakata, K.; Maeda, N.; Hayata, A.; Inoue, M. *J. Am. Chem. Soc.* **2013**, *135*, 5467–5474.

17

NATURAL PRODUCTS AND C–O/C–S BOND-FORMING REACTIONS: COPPER SHOWED THE WAY

Doron Pappo

17.1 INTRODUCTION

The art of natural product synthesis plays an essential role in the tremendous development that the fields of organic and organometallic chemistry have made.[1] On one hand, total synthesis brings to light new paradigms whose solutions hasten the invention of new tools and technologies. On the other hand, total synthesis serves as an arena for examining the true values of novel methods in a "real-life" situation where high selectivity and efficiency are essential. Moreover, the value of any method is measured by its expression in the synthesis of complex molecules. In that context, the science of copper-mediated C–O bond formation has been widely used in natural product synthesis, rendering the Ullmann condensation a major synthetic tool allowing for new, straightforward, and efficient bond disconnections.

The widespread use of the Ullmann C–O bond-forming reactions in the total synthesis of natural products can be traced back to the beginning of the previous century. As discussed in the Preface of this book, Fritz Ullmann and Irma Goldberg reported between 1903 and 1906 that copper could catalyze the reactions between aryl halides and anilines or benzamides to form *N,N*-diarylamines, as well as those of phenol salts to form diaryl ethers. In 1920, Rosenmund and Harms obtained salicylic acid by boiling an aqueous solution of sodium *o*-bromobenzoate in the presence of sodium β-naphthalene sulfonate for 4 hours with "Naturkupfer C" (an oil-dispersed metallic copper) as

Copper-Mediated Cross-Coupling Reactions, First Edition. Edited by Gwilherm Evano and Nicolas Blanchard.

the catalyst.[2] Nine years later, Hurtley[3] reported in his breakthrough publication that full conversion could be reached within 30 minutes provided that the reaction was performed in an aqueous solution of sodium acetate using copper acetate as the catalyst. From the synthesis of salicylic acid to the synthesis of vancomycin and other complex natural products by the Nicolaou group, the Ullmann ether bond formation reaction led the way in many natural product synthesis campaigns.

As extensively discussed in the first part of this book, the Ullmann condensation reaction has enjoyed a renaissance over the last two decades. The demand for new techniques and methods for the synthesis of aryl–O–aryl and aryl–O–alkyl bonds (see Chapter 2), which fulfill the demands of modern chemistry where high regio- and chemoselectivity are required, was satisfied by pioneer researchers who developed the missing synthetic tools to further their total synthesis projects. The synthesis of compounds of the glycopeptide class of antibiotics, including vancomycin—the most famous member of this family—is an example of a long-standing synthetic problem that was solved only with the advent of modified copper-mediated reactions.

Today, many Ullmann condensation protocols are available for the synthetic chemist who wishes to perform such a coupling reaction. This chapter highlights some of the most important applications of the copper-mediated C–O bond formation in the context of natural products and their total synthesis. Quite often, the original general procedures that were described by the inventors must be modified and tailored for a single specific step in the synthetic scheme, and much effort is expended to improve the coupling efficiency. New insights into the reaction mechanism are usually found during this process, which often results in the development of improved and modified methods. In this chapter, we will highlight these disclosures as well as the use of copper-mediated C–S bond formation in natural product synthesis, a strategy that is clearly still underrated.

Applications of copper-mediated C–O bond-forming reactions in natural product synthesis are classified in this chapter by reaction types according to the following order:

- synthesis of naturally occurring diaryl ethers by arylation of phenols;
- intramolecular diaryl ether bond-forming reactions;
- arylation of alcohols;
- vinylation of alcohols;
- copper-mediated C–S bond-forming reactions in natural product synthesis.

17.2 TOTAL SYNTHESIS OF NATURALLY OCCURRING DIARYL ETHERS BY ARYLATION OF PHENOLS

The diaryl ether is a common motif found in many natural products, as well as in human-designed materials of pharmaceutical,[4] herbicidal,[5] and technological interest.[6] Over the years, many successful syntheses of bioactive natural products possessing diaryl ether functionalities were achieved through copper-mediated reactions, palladium-catalyzed reactions, nucleophilic aromatic substitution (S_NAr) reactions, or oxidative phenol couplings.[7] In many cases, several approaches based on different disconnections and conditions were successfully applied for the synthesis of a specific natural product, and in some examples, the use of other synthetic methods were found to be even more effective than the copper-mediated protocols. However, in the context of this book, which deals with Ullmann coupling, we will only touch upon those syntheses in which copper chemistry was utilized.

The bisbenzylisoquinoline alkaloids, a large family of biologically active natural bases found in plants, are constructed from two dimeric tetrahydroisoquinoline subunits joined via diaryl ether bonds. Several members of the family were synthesized using Ullmann protocols, such as (±)-cepharanthine[8] and (±)-cycleanine.[9] The synthesis of phaeanthine (**7**) and isotetrandrine (**8**), two bisbenzylisoquinoline alkaloids with a diastereomeric relationship (Scheme 17.1), was achieved by Inubushi in 1968 via two sequential intermolecular Ullmann condensations.[10] First, *R*-(−)-**1** was reacted

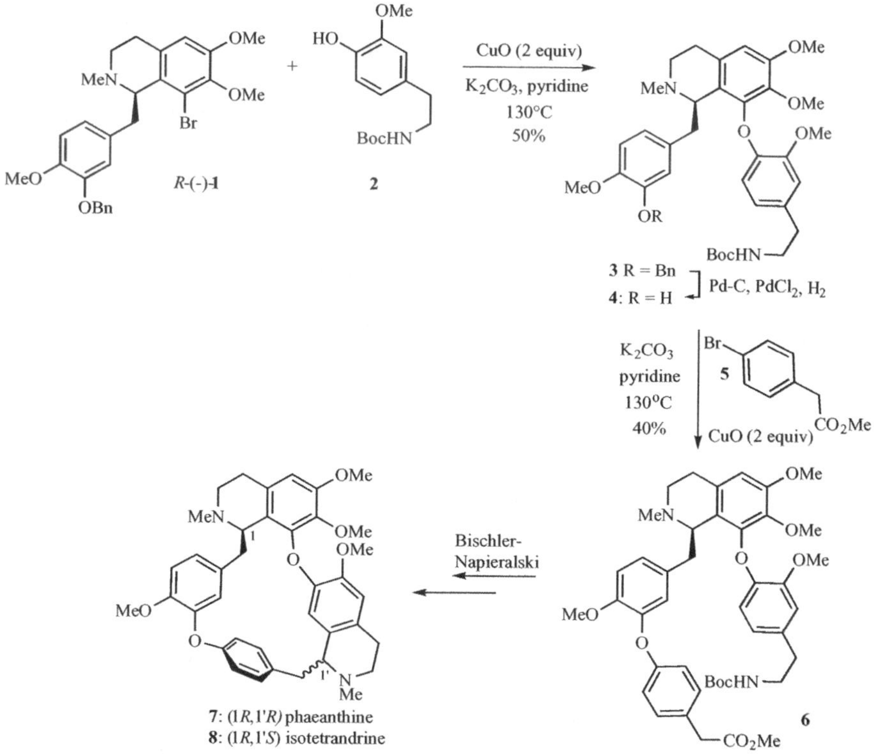

Scheme 17.1 Inubushi's synthesis of phaeanthine and isotetrandrine.

with dopamine derivative **2**, producing intermediate **3** (CuO, K_2CO_3, pyridine, 130°C, 50% yield), which was reacted again after a single debenzylation step with aryl bromide **5** under similar conditions to afford pentacyclic precursor **6** in 40% yield, which was further converted to the target dimeric alkaloids.

Since then, many successful copper-mediated transformations based on classical Ullmann conditions (CuO, K_2CO_3, pyridine) have been applied in natural product synthesis. Even today, these conditions are widely used, as exemplified by the total synthesis of perrottetin E,[11] marchantin I,[12] aspercyclide C,[13–15] and combretastatin D1 (Fig. 17.1) and D2.[16]

The success of a classical Ullmann condensation reaction is dependent on the reactivity of the coupling partners. In some cases, substitution around the reaction center can have a large influence on the outcome of the cross coupling, as demonstrated in Wipf and Jung's campaign toward the total synthesis of diepoxin σ (**17**), a highly oxygenated antifungal and anticancer natural product (Scheme 17.2).[17] An early attempt to couple tetralone **9** with 1-iodo-8-methoxynaphthalene (**10**) failed, possibly due to the lack of nucleophilicity of the deactivated tetralone hydroxyl group under the standard reaction conditions and/or chelation of the copper with the proximal ketone. However, once tetralone **9** was reduced to the corresponding diol **11**, the phenol group was no longer deactivated by the carbonyl group and the reaction with **10** (Cu_2O, pyridine, reflux, 63% yield) proceeded smoothly to give intermediate **12**, which was then oxidized with PCC (pyridinium chlorochromate) to afford naphthyl ether **13**. Unfortunately, this route failed to produce the natural product, and an alternative scheme was examined. In this approach, naphthyl ether **15** was prepared from the coupling of diol **14** and 1-iodo-8-methoxynaphthalene (**10**) under classic Ullmann conditions (Cu_2O, pyridine, reflux) in 70% yield. After several synthetic steps, naphthyl ether **15** provided diepoxide **16**, which was successfully converted to diepoxin σ (**17**) via retro-Diels–Alder reaction and removal of the *tert*-butyldimethylsilyl (TBS) group.

In 1997, Buchwald and coworkers introduced a crucial development that had a deep impact on the efficiency of copper-mediated C–O cross-coupling reactions. By using cesium carbonate as the base, they demonstrated that the reaction efficiency improved dramatically.[18] Since this discovery, cesium carbonate has become the most widely used base for copper-mediated reactions performed with alcohols, and many improved procedures were developed. Although this new finding clearly extended the reaction scope by enabling the coupling of electron-deficient aryl halides, which were less active toward copper chemistry, the reaction, however, remains sensitive to small electronic changes in the substrates as well as to the nature of the neighboring groups around the reaction center.

This was highlighted in the total synthesis of verbenachalcone (**26**, Scheme 17.3),[19] a natural product reported to enhance nerve growth factors' ability to stimulate neurite outgrowths in PC12D cells. In order to synthesize this naturally occurring biaryl ether, the Cuny group examined two model

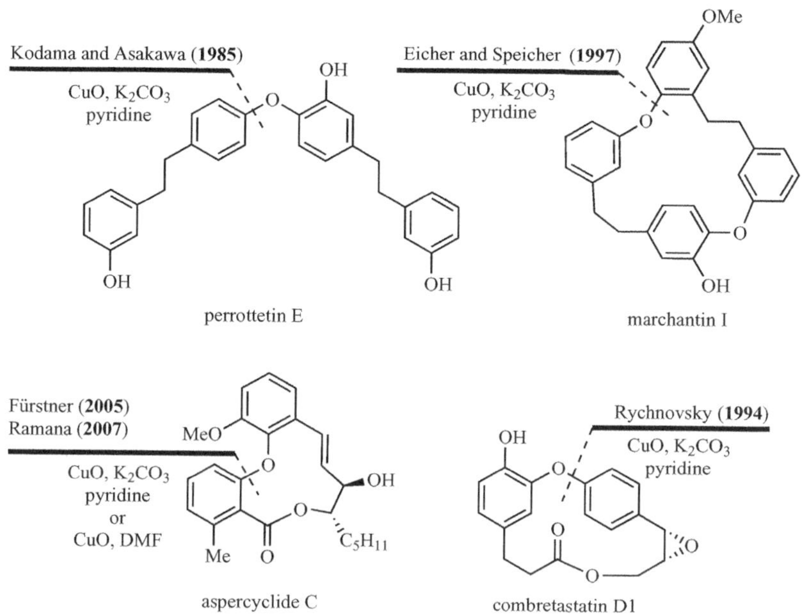

Figure 17.1 Classical Ullmann-based natural product syntheses.

systems for the construction of 2-benzyloxy-2′-methoxydiphenylether **22**. While 2-bromoanisole derivative **18** failed to react with 2-benzyloxyphenol derivative **19** ($Cu(OTf)_2 \cdot PhMe$, Cs_2CO_3, pyridine, 110°C), 2-benzyloxybromobenzene derivative **20** did condense with 2-methoxyphenol derivative **21** under identical reaction conditions, affording the coupling product **22** in moderate yield (51%). In this case, both the location of the methyl substitution, either at the *para-* or *meta*-position relative to the reacting C and O centers, and the choice of protecting group for the neighboring phenol had a crucial influence on the outcome of the reaction. Based on these exploratory studies, verbenachalcone (**26**) was prepared starting from the condensation of ethyl 3-(4-hydroxy-3-methoxyphenyl)propionate (**23**) with aryl bromide **24** ($Cu(OTf)_2 \cdot PhMe$, Cs_2CO_3, pyridine, 110°C, 24 hours), providing intermediate **25** in 78% yield.

As part of the studies toward the total synthesis of obovatol (**30**, Scheme 17.4), a natural product with manifold biological activities, the Jung group[20] examined the coupling between aryl bromide **27** and 4-allylphenol **28**. In the course of their study, they found that an undesired double bond migration occurred, affording a mixture of side products **29b** and **29c**. It was found that various isomers could be isolated under different reaction conditions (entries 2, 3, 4, and 5) and that fortunately, when Ma's conditions were utilized (entry 1: CuI, *N,N*-dimethylglycine hydrochloride, Cs_2CO_3, dioxane, 90°C),[21,22] the desired coupling product **29a** could be obtained in 75% yield.

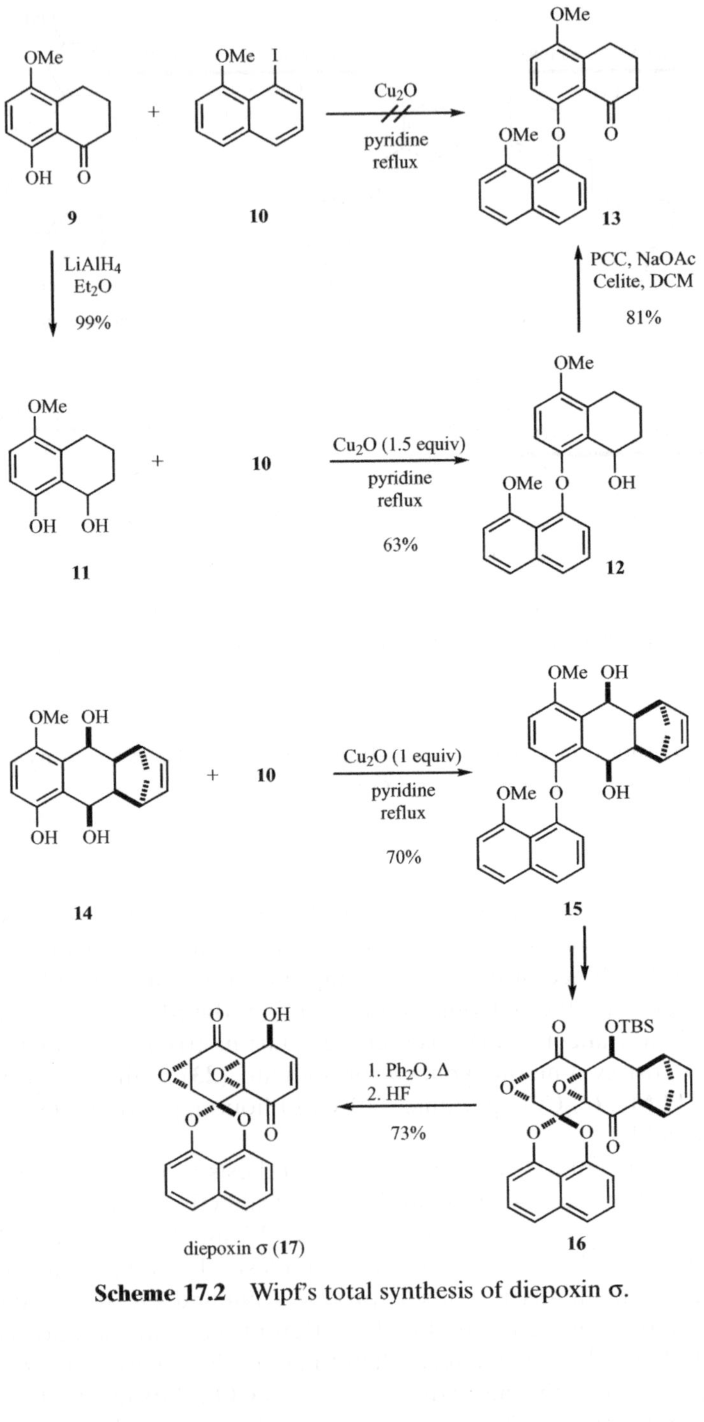

Scheme 17.2 Wipf's total synthesis of diepoxin σ.

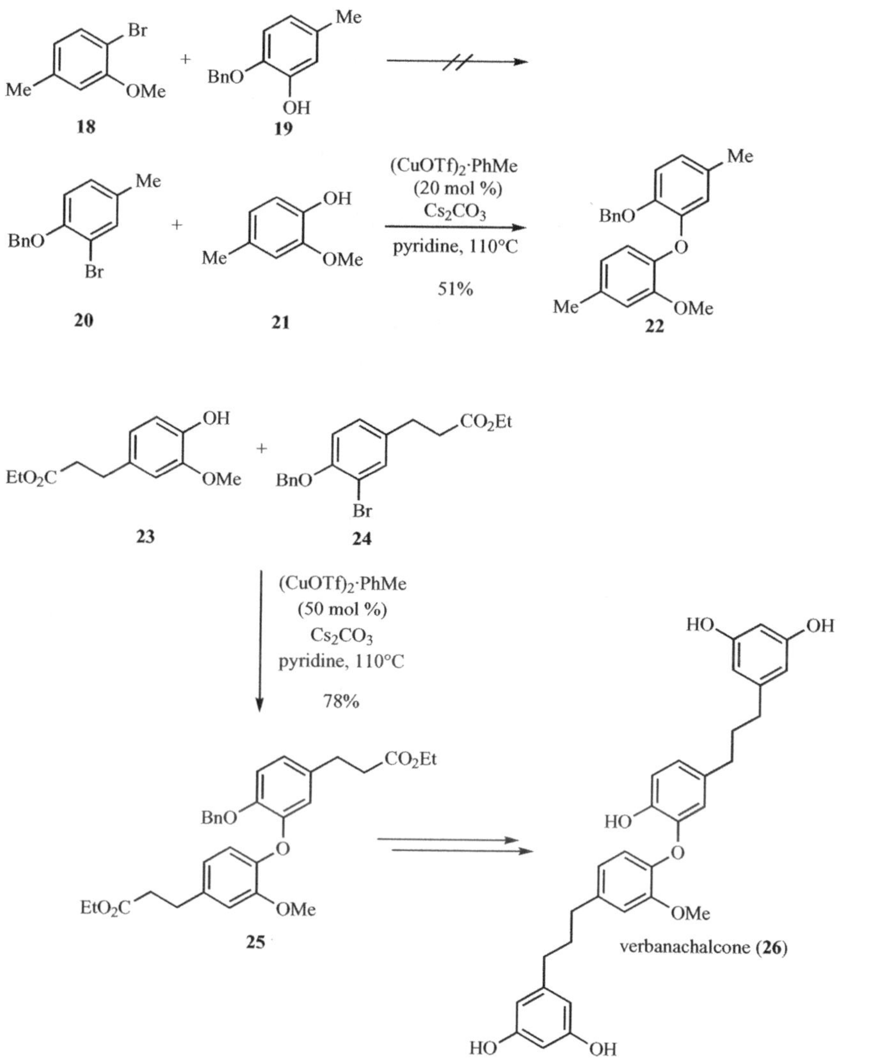

Scheme 17.3 Cuny's total synthesis of verbanachalcone.

The influence of strongly chelating solvents has been known since the discovery of copper-mediated reactions, and the use of pyridine[23] was found to be essential in Spivey and coworkers' total synthesis of aspercyclide A (**34**, Scheme 17.5).[24] In their attempts to condense bulky aryl bromide **31** with bromophenol **32**, the authors were unable to identify any palladium-mediated conditions for the formation of the key biaryl ether, and attempts to use most of the copper-mediated Ullmann reaction protocols resulted in

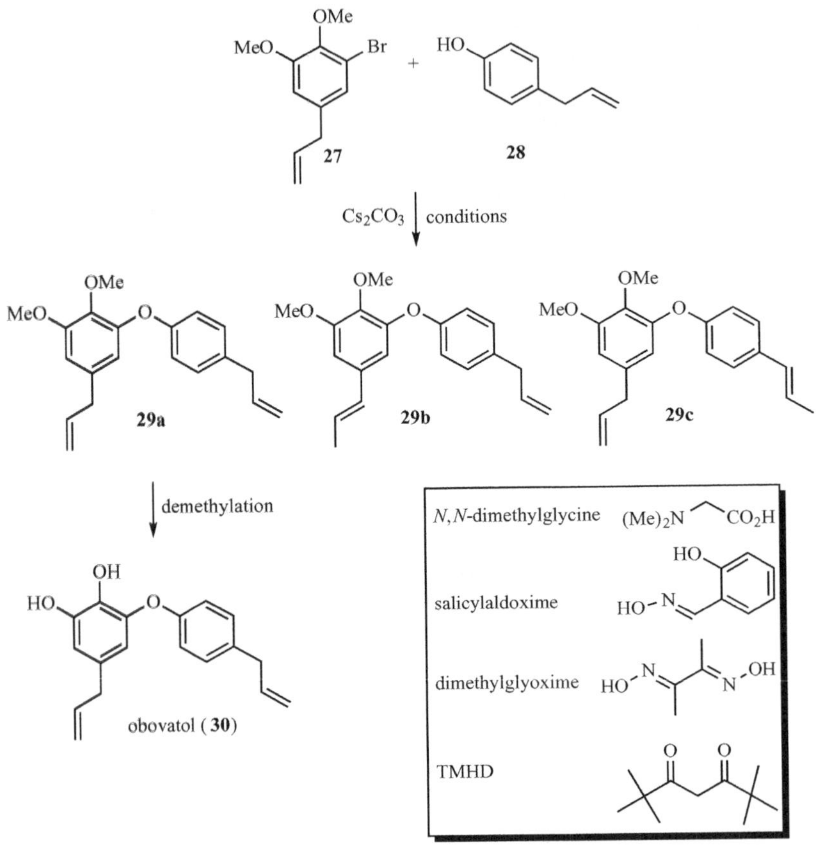

Coupling conditions

Entry	[Cu]	Ligands	Solvent	Temp	Product (Yield, %)
1	CuI	*N,N*-dimethylglycine	dioxane	90°C	**29a** (75)
2	CuI	*N,N*-dimethylglycine	CH_3CN	110°C	**29b** (60)
3	CuBr	salicylaldoxime	CH_3CN	110°C	**29b** (55)
4	CuBr	dimethylglyoxime	CH_3CN	110°C	**29a**:**29b** = 1:1 (50)
5	CuCl	TMHD	NMP	130°C	**29c** (65)

Scheme 17.4 Jung's total synthesis of obovatol.

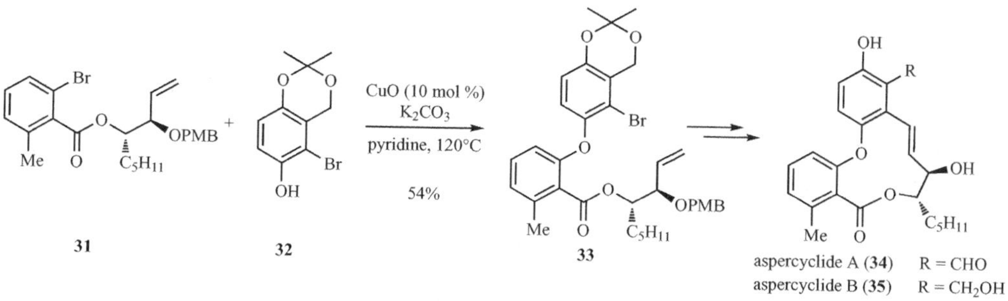

Scheme 17.5 Spivey's total synthesis of aspercyclide A.

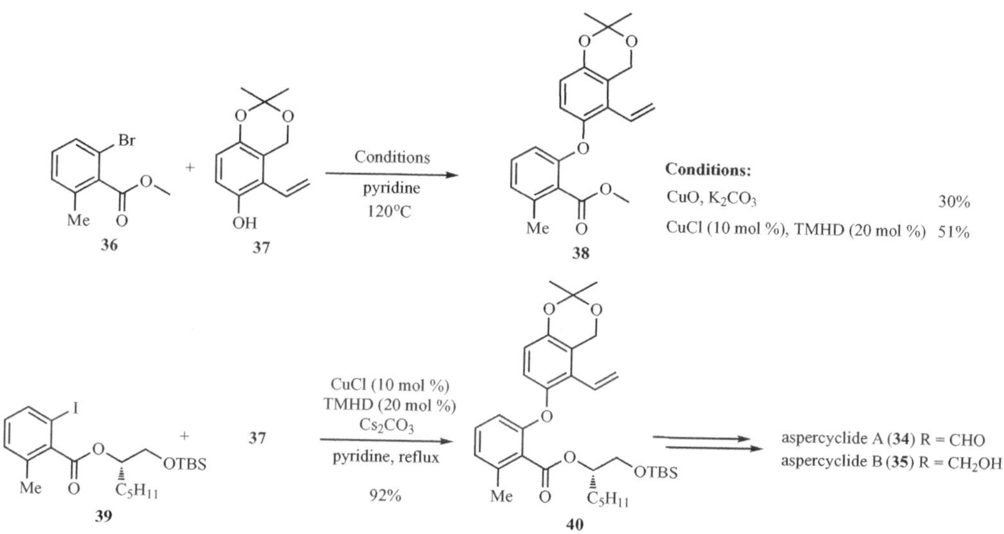

Scheme 17.6 Fürstner's total synthesis of aspercyclides A and B.

reductive dehalogenation. However, when pyridine was employed as the solvent (CuO, K_2CO_3, pyridine, 120°C), biaryl ether **33**, a key intermediate toward the aspercyclides, was obtained in 54% yield.

In sharp contrast, the same reaction conditions gave only poor results in Fürstner's approach to aspercyclides A (**34**) and B (**35**) (Scheme 17.6).[14] In model studies, Fürstner and coworkers attempted to couple methyl-(2-bromo-6-methyl)benzoate **36** and phenol **37** using the CuO/K_2CO_3/pyridine system, but these conditions resulted in low yield only (<30%). Better results were obtained when Song's modified conditions[25] were applied (CuCl, 2,2,6,6-tetramethylheptane-3,5-dione [TMHD], Cs_2CO_3, pyridine, reflux), improving the yield of this transformation to 51%. In order to increase the efficiency of his transformation, Fürstner replaced the bromine atom, which gave limited results in his model study, with a more reactive iodine atom. Fortunately, this tactic paid off and the coupling precursor **40** was obtained in excellent yield (92%) from aryl iodide **39** and phenol **37** (CuCl, TMHD, Cs_2CO_3, pyridine, reflux).

Recently, Jung and Bräse demonstrated the feasibility of the synthesis of diaryl ethers on solid phase for the preparation of a library of natural products and their precursors.[26] The limited number of commercially available arylboronic acids (which were required for their synthesis), together with the necessity of repeating the coupling procedure several times, convinced them to concentrate their efforts on performing the reaction using aryl halides as coupling partners. The successful application of Song's coupling conditions[25] to the solid-phase synthesis procedure enabled the synthesis of hernandial (**43**), spiraformin A (**48**), and several members of the aristogin family – aristogin A (**45**) and aristogin C – which were previously prepared using the Ullmann reaction in water.[27] Condensation between the immobilized cinnamic acid derivative **41** and 2-bromo-4,5-dimethoxybenzaldehyde or methyl 4-iodobenzoate (CuCl, Cs_2CO_3, TMHD, DMF/MeCN (2:1), 120°C) afforded resin-bound hernandial **42** and aristogin A **44** precursors. Ozonolytic cleavage, followed by a purification step, provided hernandial **43** and aristogin **45** in 23% and 59% yields, respectively, based on Merrifield resin. Inverse coupling modes are also possible, as demonstrated in the synthesis of spiraformin A (**48**). In this case, resin-bound 3-(4-bromophenyl)propanoic acid **46** was coupled with isovanillin under similar conditions, affording coupling precursor **47**, which in turn afforded the natural product in 21% overall yield based on the Merrifield resin (Scheme 17.7).

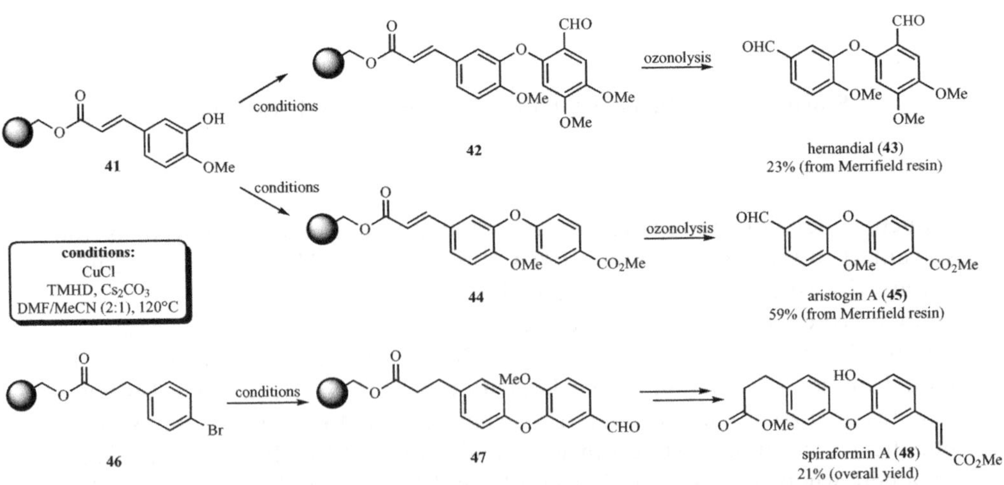

Scheme 17.7 Jung and Bräse's synthesis of naturally occurring biaryl ethers on solid support.

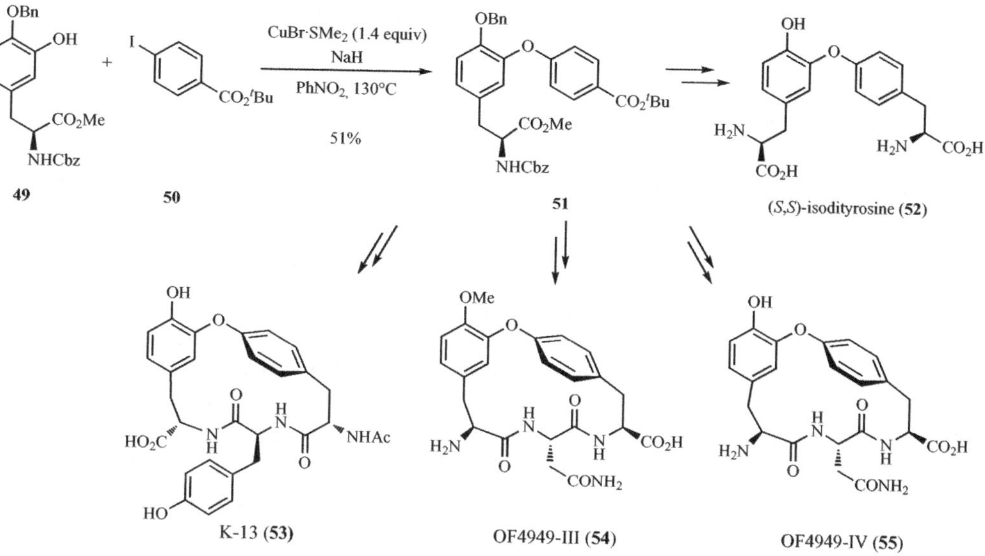

Scheme 17.8 Boger's total synthesis of isodityrosine, K-13, OF4949-III, and OF4949-IV.

During the last decades, the Ullmann reaction has undergone nothing short of an evolution. A nice illustration is the evolution of the synthesis of naturally occurring macrocycles bearing both a diaryl ether and sensitive amino acids, such as (*S*,*S*)-isodityrosine (**52**), K-13 (**53**), OF4949-III (**54**), and OF4949-IV (**55**) (Scheme 17.8).[28] In the early campaigns, arduous multistep sequences were needed to prepare the isodityrosine motif. Due to the harsh reaction conditions, the sensitive amino acid stereocenters were racemized prior to the diaryl ether bond formation.[28] Introduction of new technologies in the field of copper-mediated cross-coupling reactions enabled the formation of the diaryl ether bond in the presence of sensitive amino acids, allowing for a direct coupling of two amino acid derivatives. Thus, the syntheses of these important natural products and other members of the glycopeptide class of antibiotics, including vancomycin, were greatly simplified.

As described in Chapter 2, the traditional Ullmann condensation reaction involves heating aryl halides and phenols in the presence of copper and a strong base at high temperatures and long reaction times. These harsh conditions are not compatible with many substrates, especially those that contain sensitive functional groups and epimerizable stereocenters such as amino acids. (*S*,*S*)-Isodityrosine (**52**) is a naturally occurring amino acid and a key element in a large family of natural products that contain endocyclic diaryl ethers and exhibit various biological activities. Early attempts to synthesize isodityrosine through a direct Ullmann condensation reaction between protected phenylalanine-derived aryl halides and tyrosine derivatives were, as expected, unsuccessful. To circumvent this, long multistep routes, based on copper-mediated transformations, were developed by the groups of Schmidt,[29] Evans,[30] Rao,[31] and Boger[32] to make progress toward the syntheses of (*S*,*S*)-isodityrosine (**52**) and related compounds, such as K-13 (**53**), OF4949-III (**54**), and OF4949-IV (**55**).

In their 1990 synthesis, the Boger group coupled L-Dopa derivative **49** (Scheme 17.8) with electron-deficient *p*-iodobenzoate **50** using copper(I) bromide dimethylsulfide complex in nitrobenzene at 130°C, to afford precursor **51** in 51% yield without racemization. Further synthetic manipulations were needed in order to complete the assembly of the second amino acid and thus complete the total synthesis of (*S*,*S*)-isodityrosine (**52**) and three other cycloisodityrosine-derivative agents, K-13 (**53**), OF4949-III (**54**), and OF4949-IV (**55**).[32]

In 2006, Ma and coworkers[33] once again confronted the challenge of coupling L-tyrosine- and L-phenylalanine-derived iodides, a task that, in its initial attempts by Enomoto (CuO, K_2CO_3, pyridine, 130°C), succeeded in only 1.5% yield.[34] Utilizing an improved Ullmann reaction protocol, the group first studied the parameters influencing the racemization of tyrosine derivatives **57a–d** (Scheme 17.9) during their coupling with *p*-iodoacetophenone (CuI, *N*,*N*-dimethylglycine, Cs_2CO_3, dioxane, 90°C). Under these relatively mild

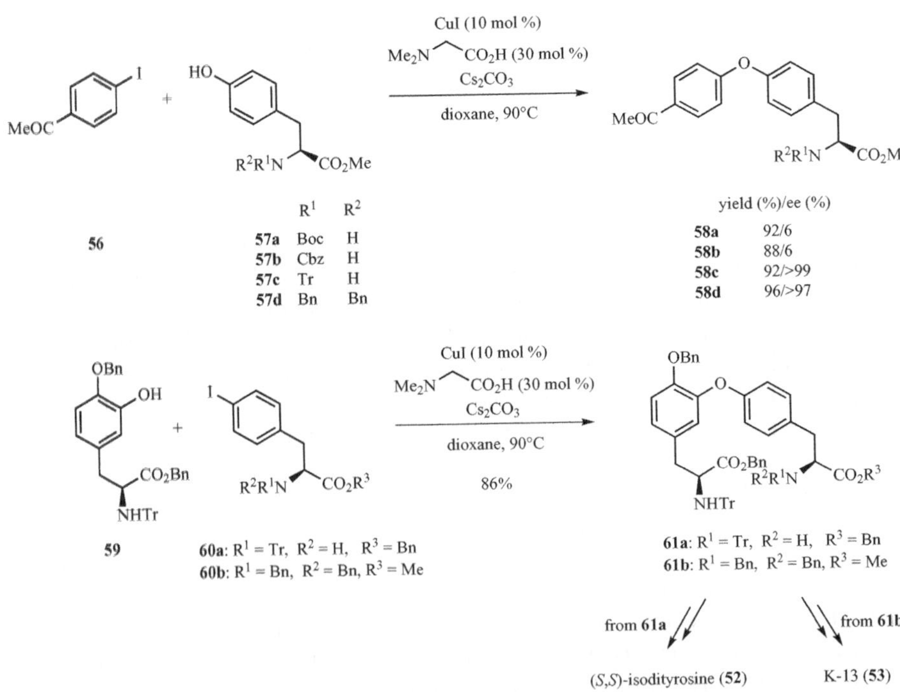

Scheme 17.9 Ma's total synthesis of isodityrosine and K-13.

conditions, *N*-Boc and *N*-CBz tyrosine derivatives (**58a** and **58b**) were not stable and were almost completely racemized (6% ee). On the other hand, the *N*–Tr and *N,N*-dibenzyl tyrosine derivatives (**57c** and **57d**) were found to be configurationally stable and the coupling products (**58c** and **58d**) were obtained in high yield and optical purity (92% and 96% yield, respectively, >97% ee). Based on these findings, L-Dopa derivative **59** was reacted with suitably protected phenylalanine-derived aryliodides **60a** and **60b**, affording coupling product **61a** (86%), a precursor of (*S,S*)-isodityrosine (**52**), and **61b**, which was used for the total synthesis of K-13 (**53**).[33,35]

One of the most efficient syntheses of (*S,S*)-isodityrosine (**52**) to date was reported by Jung and Lazarova[36] and is based on the cupric acetate-promoted condensation of arylboronic acids with phenols, a reaction simultaneously developed by Chan, Lam, and Evans (see Chapter 4 for details, and the next pages for the application of the Chan–Lam–Evans coupling in natural product synthesis). In the key step of the synthesis, phenylalanine-derived boronate **64**, readily prepared by palladium-catalyzed Miyaura borylation of **62** with bis(pinacolato)diboron, was successfully condensed with L-Dopa derivative **63** ($Cu(OAc)_2$, pyridine, 4Å molecular sieves, dichloromethane, room temperature) to afford precursor **65** in 71% yield, which was readily transformed to the natural product. Under these mild conditions, the danger of epimerization decreases and sensitive amino acids that tend to epimerize under the previously described methods are perfectly suitable reaction partners (Scheme 17.10).

As detailed in Chapter 4, the introduction in 1998 of arylboronic acids as surrogates for aryl halides in the Ullmann condensation reaction by Chan,[37] Evans,[38] and Lam[39] was a significant breakthrough and marked the beginning of a new era in the copper-mediated arylation of phenols. In the general

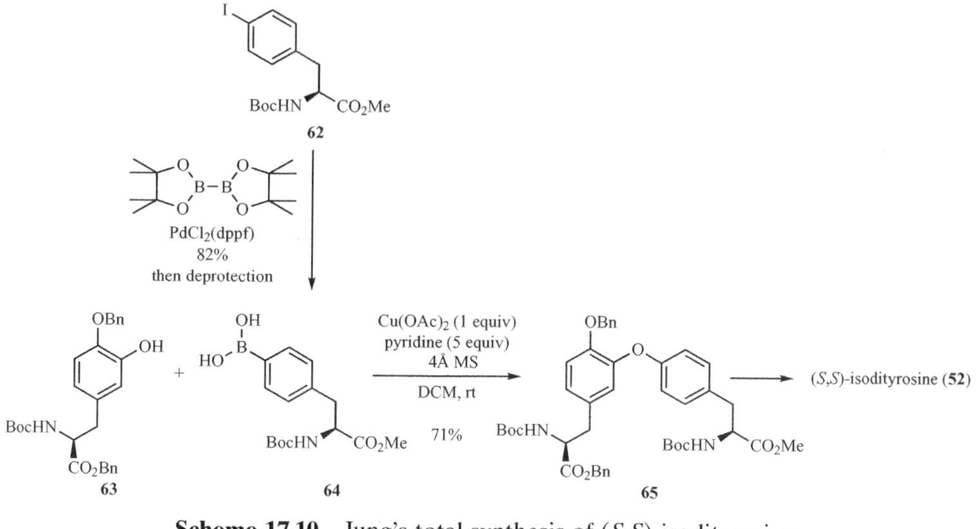

Scheme 17.10 Jung's total synthesis of (*S,S*)-isodityrosine.

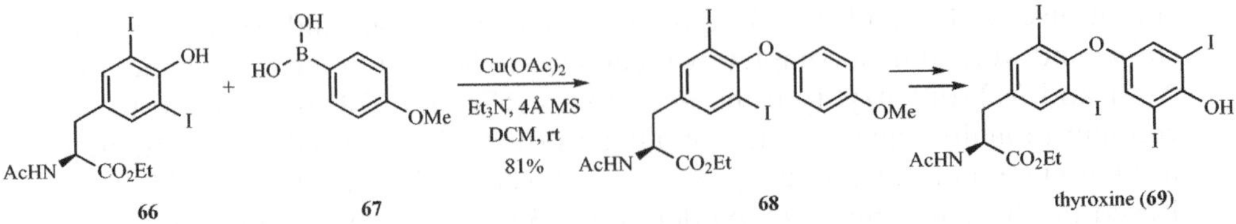

Scheme 17.11 Evans's formal synthesis of thyroxine.

reaction procedure, an arylboronic acid (typically used in excess) is coupled with a phenol using a stoichiometric amount of $Cu(OAc)_2$ in the presence of triethylamine, pyridine, or 4-dimethylaminopyridine (DMAP) as chelating bases and 4Å molecular sieves in dichloromethane under an atmosphere of air or molecular oxygen at room temperature. Under these conditions, many base-sensitive functional groups are perfectly stable and the reaction is less sensitive to steric and electronic effects, which are serious concerns when aryl halides are used as coupling partners (see further). Since their back-to-back publications in 1998, many natural products have been prepared using this powerful method.

In their proof of concept manuscript, the Evans group demonstrated the scope of the technique by performing a formal total synthesis of L-thyroxine (**69**, Scheme 17.11).[38] The successful coupling helped to evaluate the impact of steric hindrance around the reaction center and the tolerance of sensitive amino acid groups to the reaction conditions. In their synthesis, tyrosine derivative **68**, an intermediate in Hem's synthesis,[40] was prepared via the condensation of diiodotyrosine derivative **66** and 4-methoxyphenylboronic acid (**67**) in 81% yield.

In 2001, the Evans group utilized their method as a key step in the total synthesis of a much more challenging substrate, teicoplanin aglycon (**73**, Scheme 17.12).[41] Teicoplanin is a member of the large family of glycopeptide antibiotics and, together with vancomycin, is one of the only two representatives of this family that are clinically used for the treatment of methicillin-resistant *Staphylococcus aureus* infections. These two compounds are considered to be the antibiotics of last resort against this pathogen and their synthesis has attracted a great deal of attention. While the use of nucleophilic aromatic substitution has been the method of choice for the construction of the diaryl ether bonds of rings 6, 4, and 2, it was recognized that the Chan–Lam–Evans reaction might be an efficient tool to join rings 1 and 3. The condensation of arylglycine boronic acid **70** with phenol **71** under the conditions developed by the group ($Cu(OAc)_2$, pyridine, 4Å molecular sieves, dichloromethane, O_2, room temperature) proceeded smoothly to provide **72** in 80% yield. The fact that no epimerization of the highly sensitive arylglycine stereocenter was detected is astonishing, emphasizing this method's superiority over other diaryl ether bond-forming methods.

The development of the Chan–Lam–Evans reaction had a strong impact on the synthesis of complex diaryl ethers and has been successfully integrated

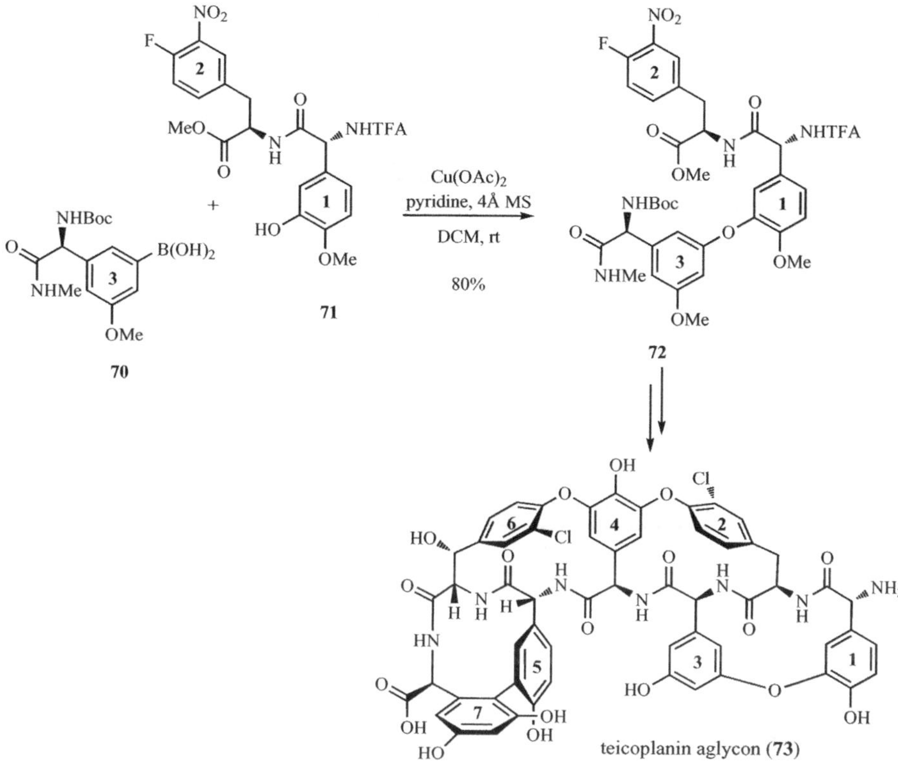

Scheme 17.12 Evans's synthesis of teicoplanin aglycon.

into several natural product syntheses, as shown in Figure 17.2. They include the proposed structure of puetuberosanol,[42] rodgersinol,[43] combretastatin D2,[44] and tejedine.[45]

Although the Chan–Lam–Evans reaction was found to be efficient and robust, it is not free of complications. It was indeed observed that when the arylboronic acid possesses a heteroatom in the *ortho* position, the yield of the reaction drops due to the competing formation of protodeborylated and oxidized products.[38]

Pulcherosine (**82**, Scheme 17.13), an oxidatively coupled trimer of tyrosine, as well as other related natural products, were prepared by Hutton and coworkers.[46] In their early attempts to couple tyrosine 3-boronic acid **74** with protected tyrosine **75** ($Cu(OAc)_2$, DMAP, and molecular sieves), isodityrosine derivative **76** was produced in only 33% yield together with BnO-protected tyrosine **77** (15%), generated by protodeborylation of **74**, and trace amounts of L-Dopa derivative **78**, formed by oxidation of **74**. Fortunately, by switching the phenol and the boronic acid coupling partners to L-Dopa derivative **79** and phenylalanine 4-boronic acid **80**, a more efficient condensation reaction occurred, and protected isodityrosine **81** was obtained in 80% yield.

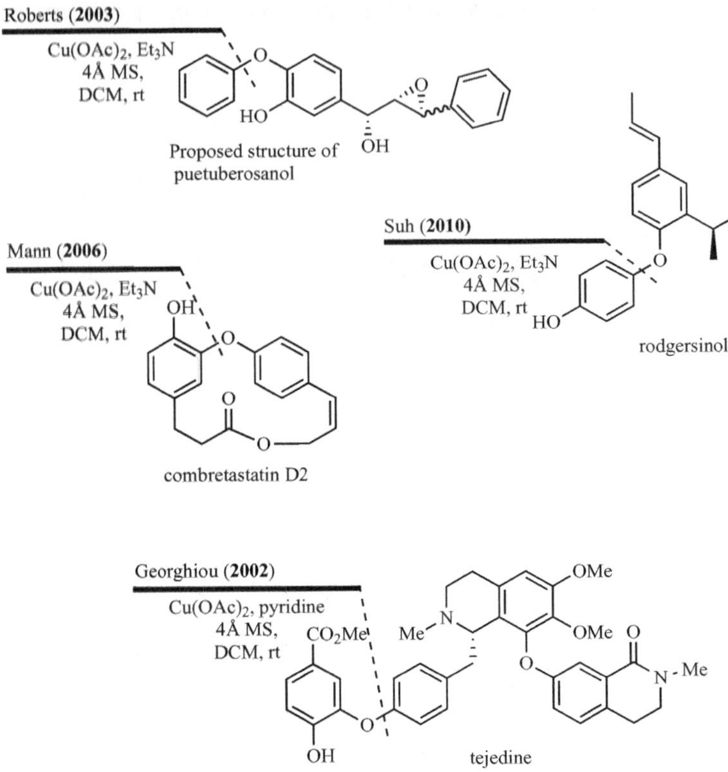

Figure 17.2 Natural product syntheses based on the Chan–Lam–Evans reaction.

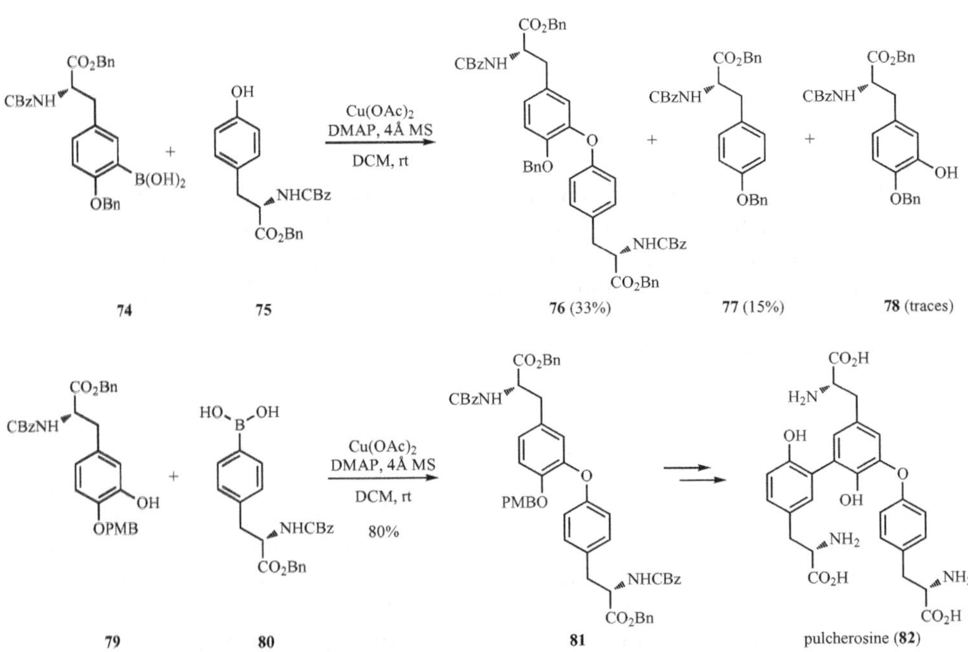

Scheme 17.13 Hutton's total synthesis of pulcherosine.

17.3 INTRAMOLECULAR DIARYL ETHER BOND-FORMING REACTIONS

17.3.1 Macrocyclization

The development of new synthetic methods and protocols for the copper-mediated formation of diaryl ether moieties was readily integrated by synthetic chemists and encouraged the use of new bond disconnections, notably for key macrocyclization steps in various natural product syntheses.

In this context, the Boger group accomplished the synthesis of bouvardin (**85**, Scheme 17.14),[47] a bicyclic hexapeptide with potent antibiotic and antitumor activity, by using an intramolecular Ullmann condensation for the key macrocyclization step. Early attempts to prepare dipeptide **84** through a variety of methods, including conventional macrolactamization techniques, Ullmann macrocyclizations, and intramolecular oxidative phenol coupling, all failed to provide the 14-membered cycloisodityrosine subunit. Attempts to promote the cyclization in DMF (NaH, CuBr·SMe_2, 180°C), under conditions developed by the same group for the total synthesis of piperazinomycin (Fig. 17.3),[48] were also unsuccessful, as well as was the use of methyl copper to stoichiometrically generate the cuprous phenoxide salt of **83**, a protocol successfully applied by Boger in the total synthesis of combretastatin D2 (Fig. 17.3).[49] Eventually, the Boger group found that the cuprous phenoxide salt of **83** had to be generated *in situ* (1.1 equiv of NaH, 10 equiv of CuBr·SMe_2, 130°C) in a diluted solution (0.004 M) of rigorously dried 2,6-lutidine in order to get a successful macrocyclization (37% yield). The utility of 2,6-lutidine in the synthesis of **85**, or of collidine in the synthesis of deoxybouvardin (**86**) and related compounds,[50,51] both unconventional solvents for Ullmann condensation reactions, serves a dual purpose since they act both as bases and as chelating solvents for the copper complexes.

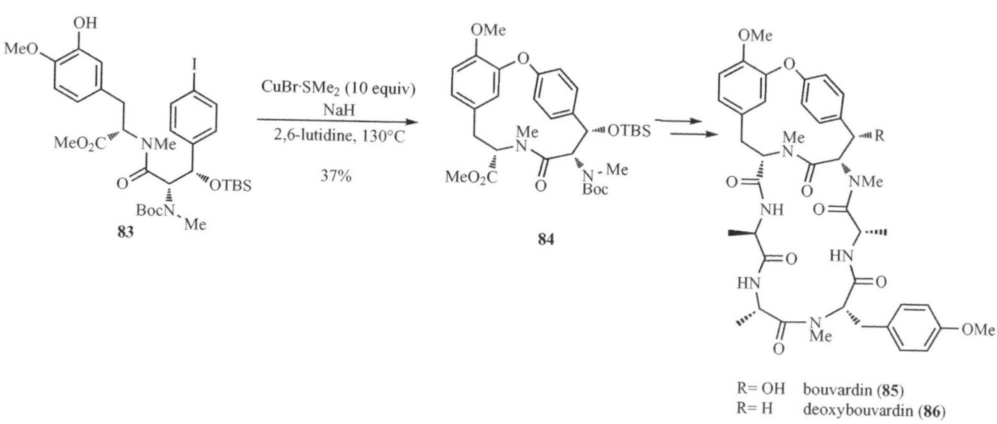

Scheme 17.14 Boger's total synthesis of bouvardin.

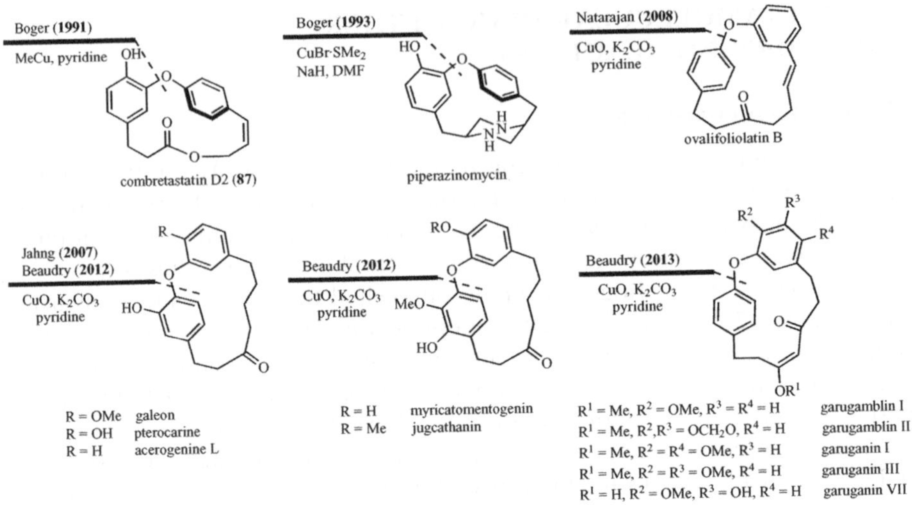

Figure 17.3 Ullmann macrocyclization-based syntheses.

Several other cyclic natural products have been prepared according to related strategies, and in all cases, the key to their syntheses turned out to be an implementation of an intramolecular Ullmann macrocyclization reaction (Fig. 17.3). The 15-membered ring of combrestastatin D2 (**87**, Fig. 17.3), an antineoplastic agent, was formed by the Boger group[49] using 1.5 equiv of methyl copper in a diluted solution of pyridine, under reflux conditions (37% yield), a step that was recently reinvestigated by the Pettit group.[52] Other applications of Ullmann macrocyclization reactions can be found in the total synthesis of piperazinomycin by the Boger group,[48] ovalifoliolatin B by Kumar and Natarajan[53] and Galleon, and pterocarine and acerogenine by Jahng and coworkers (Fig. 17.3).[54] Recently, Shen reported a total synthesis of the proposed structure of engelhardione, which also led to the structural revision of this macrocyclic natural product. Based on their synthetic work, it was found that the correct structure of the previously reported engelhardione should be that of pterocarine (Fig. 17.3).[55a] It also ought to be mentioned that recent additions to this collection of natural products prepared by intramolecular Ullmann macrocyclization reactions include garuganin and garugamblin diaryl ether heptanoids (and related natural products)[55b–d] and that, in some particular cases, the macrocyclization can be performed at high concentrations.[55e]

Chloropeptin I (**90**, Scheme 17.15), a natural product with notable activity against HIV, was synthesized through the joint efforts of Snapper and Hoveyda.[56] One of the key steps in their synthesis was the macrocyclization of boronic acid **88** using the Chan–Lam–Evans reaction. In their attempt to apply the conditions previously reported for the copper-mediated formation

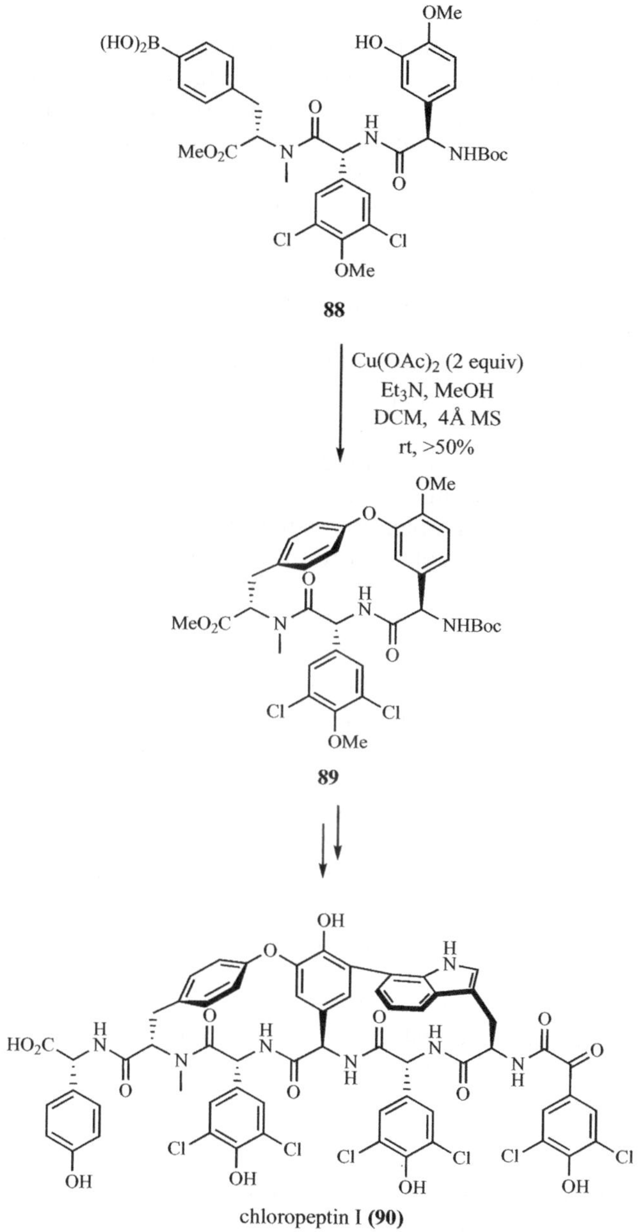

Scheme 17.15 Snapper and Hoveyda's synthesis of chloropeptin I.

of biaryl ethers ($Cu(OAc)_2$ [2 equiv], Et_3N [10 equiv], 4Å MS, dichloromethane [0.01 M], 22°C, 5 hours), the reaction, when carried out on a multigram scale, gave only 15–20% yield of cyclic peptide **89**. Extensive experimentation led them to determine that the presence of 10 equiv of methanol (relative to the substrate) was crucial for efficient biaryl ether formation in terms of reaction time (3–6 hours with methanol vs. over 24 hours without) and yield (50%). The authors suggested that the positive influence of methanol may be due to the *in situ* formation of the dimethyl boronate or to the increased solubility of the copper salt.

Since the pioneering report by Hurtley in 1929 that aryl halides possessing halogens in the *ortho*-position to a carboxylic acid are much more reactive than common aryl halides,[3] considerable efforts have been devoted to identify suitable *ortho*-substituents that can promote the Ullmann coupling reaction.[33,57,58]

The Ma group found that *ortho*-amide groups, and mainly trifluoroacetanilide, have a strong accelerating effect on copper-mediated diaryl ether bond-forming reactions. For example, the coupling between 2-haloacetanilide derivatives and phenols takes place at room temperature, while in the absence of this *ortho*-substituent, much higher temperatures (at least 80°C) are required.[22,35] Based on these findings, synthetic intermediates for isodityrosine (**52**) and K-13 (**53**, Scheme 17.8) were readily prepared through the coupling of suitably protected 3-bromo-4-(trifluoroacetamide)phenylalanine and tyrosine derivatives. Similarly, protected isodityrosine **91** (Scheme 17.16) was easily obtained in 80% yield under especially mild reaction conditions (CuI [0.3 equiv], *N,N*-dimethylglycine [1 equiv], Cs_2CO_3 [3 equiv], dioxane, room temperature, 14 hours). While attempts to apply these conditions to the macrocyclization of dipeptide **92** failed, a solution was found in the use of a different solvent system (MeCN/pyridine: 4/1) and a higher reaction temperature (45°C), which allowed for the formation of the desired 14-membered cyclic peptide **93** (which could serve as an intermediate for the synthesis of RP-66453[59]) in 42% yield. This *ortho*-effect of the trifluoroacetamide group was later on implemented in a total synthesis of K-13 in which the macrocyclization step from tripeptide **94**, via the postulated intermediate **A**, provided the 17-membered cyclic peptide **95** in 50% yield at room temperature (Scheme 17.16). Five more steps were required in order to convert the amide to a phenol group and for the completion of the total synthesis of K-13.

In the spectacular total synthesis of vancomycin (**100**, Scheme 17.17), the most important and well-known member of the glycopeptide class of antibiotics, the Nicolaou group developed a mild copper-mediated arylation method based on the ingenious incorporation of a triazene unit in the *ortho*-position to the bromine in the central D-ring.[60,61] The triazene group assisted the intramolecular coupling, which occurred under mild conditions, without the requirement for a strong base, and within short reaction times, providing the cyclized diaryl ether in good yields, especially when considering the type of reaction and the complexity of the substrate. In addition to promoting the copper-mediated cyclization, the triazene group is easy to introduce, stable

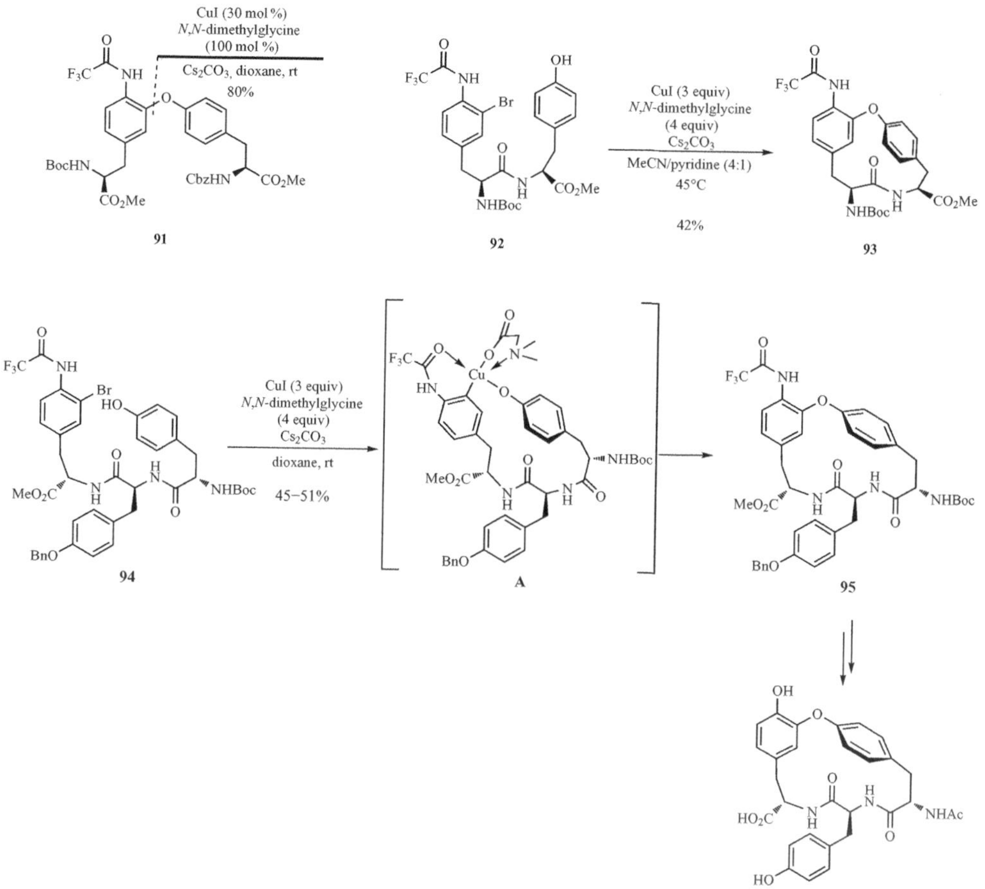

Scheme 17.16 Ma's application of the trifluoroacetamide *ortho*-effect in the synthesis of K-13.

under a wide range of conditions, and is easily transformed to variety of functional groups such as halides, amines, and phenols. In this milestone synthesis, during the first triazene-driven ring closing step between rings C and D (**96**, CuBr·DMS, K_2CO_3, pyridine, CH_3CN, 80°C, 15 minutes), a 1:1 mixture of the two atropoisomers **97a** and **97b** was obtained. After several more steps, the connectivity between rings D and E was established under the same reaction conditions, obtaining derivatives **99a** and **b** in good atroposelectivity (1:3 ratio, respectively) from precursor **98**.

17.3.2 Small- to Medium-Sized Oxygenated Heterocycles

The intramolecular arylation of phenols and carboxylic acids has also been used in total synthesis and in the pharmacological industry. Although most examples deal with macrocylization, there are cases in which high degrees of

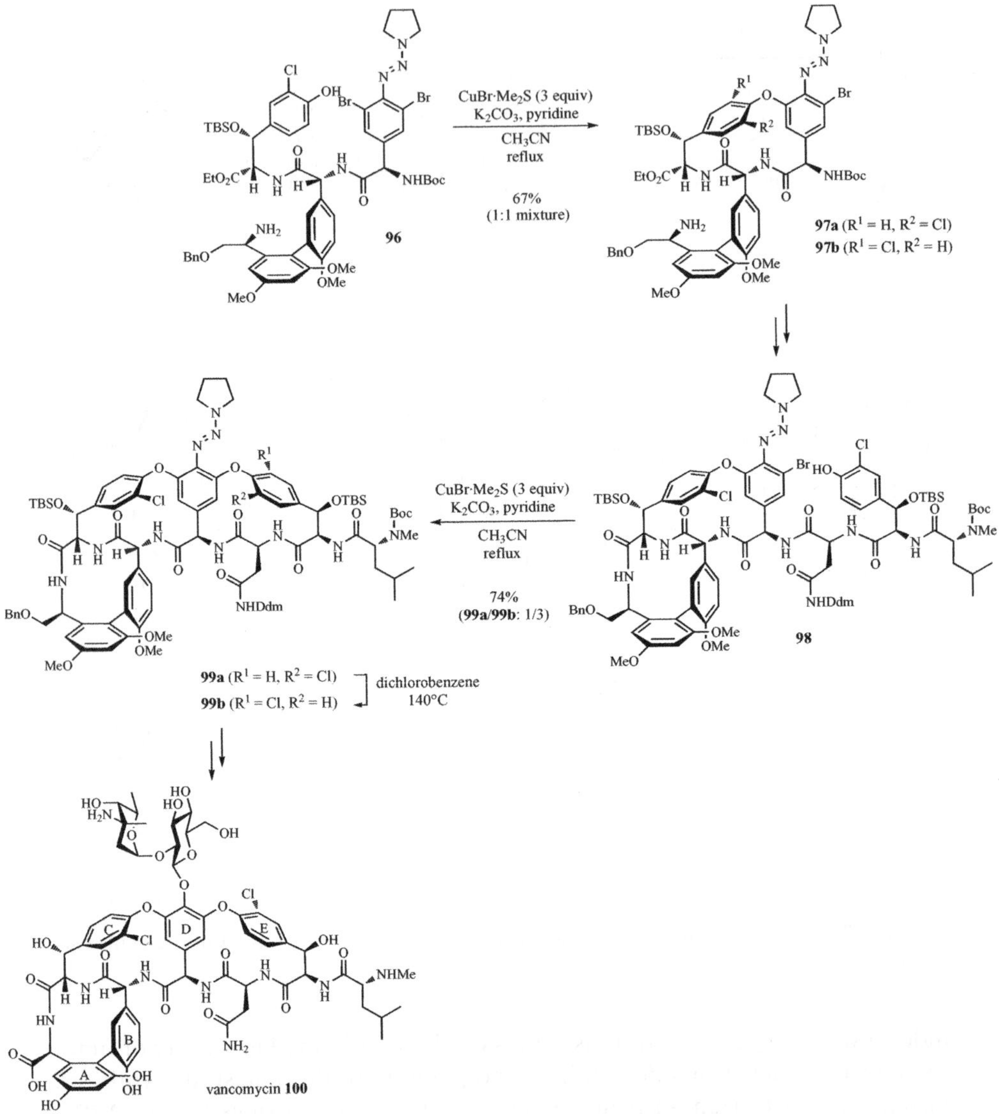

Scheme 17.17 Nicolaou's total synthesis of Vancomycin.

efficiency and chemoselectivity were obtained when copper-mediated reactions were used in the cyclization to small- and medium-sized ring systems.

In this context, the Thasana group reported a rare example of a copper-mediated lactonization by intramolecular coupling between a carboxylic acid and an aryl bromide (Scheme 17.18).[62] They successfully prepared benzopyranones and isolamellarines **102a,b** using a combination of Cu(I)-mediated and microwave-assisted reactions. In their synthesis, **101a** or **101b** were mixed with 2 equiv of Liebeskind catalyst CuTC in DMF at 200°C under microwave

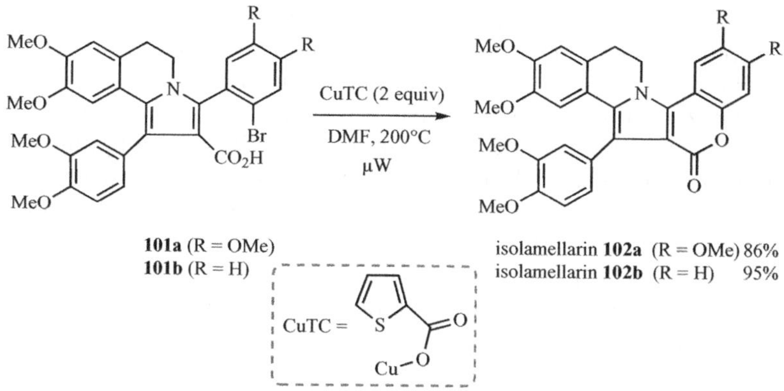

Scheme 17.18 Thasana's synthesis of isolamellarines.

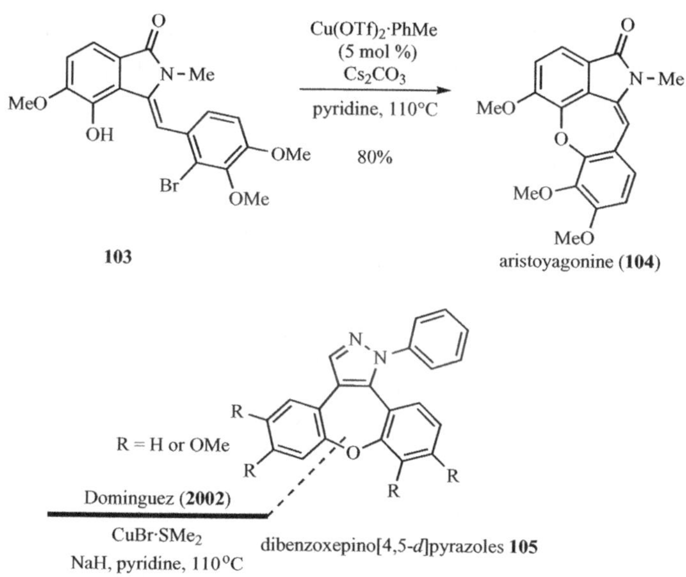

Scheme 17.19 Seven-membered ring cyclization: Couture's synthesis of aristoyagonine and formation of dibenzoxepines.

irradiation, without the need for a base or an additional ligand, which smoothly afforded **102a** and **102b** in 86% and 95% yields, respectively.

The intramolecular arylation of phenols has also been used in the context of the total synthesis of various natural products such as aristoyagonine (**104**, Scheme 17.19), a cularine alkaloid that incorporates a dibenzoxepine ring system and displays biological activity against various cell lines. Couture and coworkers[63] developed a total synthesis of this alkaloid through a 7-membered ring closure of isoindolinone **103**. Although the configuration of the central double bond in **103** was assigned *Z*, stereochemical considerations were not crucial for the creation of the diaryl ether linkage, and the construction of the dibenzoxepine ring system was ensured using a combination of $Cu(OTf)_2$·PhMe, Cs_2CO_3, and pyridine at 110°C, yielding the cyclized product in 80% yield.

Dominguez conducted an intensive investigation on the formation of a library of dibenzoxapines **105**[64] based on the Ullmann coupling reaction (Scheme 17.19). It was found that Boger's conditions (NaH, CuBr·SMe_2, pyridine, 110°C) gave the best results and suitably promoted the cyclization to the seven-membered ring systems, which were formed with high efficiency. Notably, an alternative approach to the target dibenzoxepine derivatives **105** based on a palladium-catalyzed C–O bond formation was found to be limited in terms of yields and steric demands.

17.4 ARYLATION OF ALCOHOLS

The synthesis of aryl alkyl ethers via a copper-mediated reaction between aryl halides and aliphatic alcohols is still not an easy task to perform. The low reactivity of aliphatic alcohols demands the use of strong bases, such as alkoxides or sodium hydride typically used in large quantities, high temperature, highly polar aprotic solvents, and stoichiometric amounts of copper salts.[65] Although improved procedures based on efficient ligand-assisted, copper-mediated arylation reactions and the use of cesium carbonate were developed (see Chapter 2 for details),[66] this type of reaction still cannot compete with other traditional reaction schemes that afford the same products. S_NAr reactions between alkoxides and activated aryl halides, Mitsunobu reactions between phenols and alcohols, or S_N2 displacements between phenols and alkyl halides are all considered safer routes when the synthesis of a complex natural product is designed, even if they sometimes require longer reaction sequences or additional steps. However, in some cases, there is justification for choosing the Ullmann condensation reaction over other methods: (i) when the reaction is performed on a complex substrate obtained from a natural source, which has an alcohol group already installed on it, and a direct coupling is needed in order to preserve one stereocenter, or (ii) when the synthetic path must go through an aryl halide intermediate in order to simplify the entire synthetic scheme. In these cases, copper has recently proved to be the metal of choice for the arylation and vinylation of alcohols.

The utility of the classical Ullmann coupling for the formation of aryl alkyl ether bonds in the synthesis of complex natural products remains limited. Sharpless and coworkers applied standard copper-mediated conditions (CuI, NaH, pyridine, DMSO, 113°C, 70 hours) for the arylation of dihydroquinine (DHQ) **106** (Scheme 17.20) and dihydroquinidine (DHQD) with 9-iodophenanthrene (**107**). The reaction was performed on a 50-g scale and required equimolar amounts of copper(I) iodide and 2 equiv of pyridine, affording DHQ and DHQD phenanthryl ethers **108** as their hydrochlorides in 70–73% yields.[67] In the syntheses of elacomine (**111**, Scheme 17.20) and

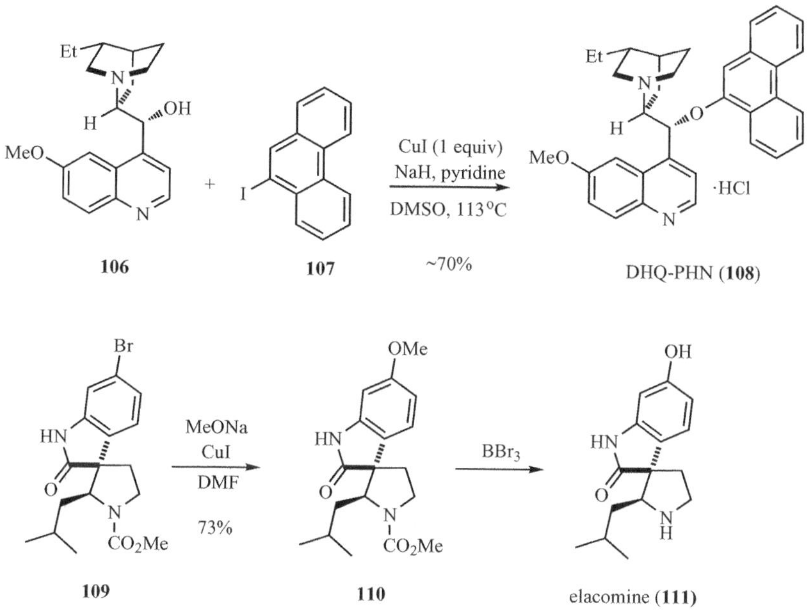

Scheme 17.20 Arylation of dihydroquinine by Sharpless and Horne total synthesis of elacomine.

isoelacomine by the Horne group,[68] aryl bromide **109** was prepared in only three steps from tryptamine hydrobromide and then was subjected to a copper-catalyzed methoxylation reaction (MeONa, CuI, DMF) to afford 6-methoxyoxindole derivative **110** in good yield (73%), whose subsequent deprotection afforded elacomine (**111**). Recent advances in the direct hydroxylation of aryl halides with hydroxide ions might allow for a direct introduction of the phenol group (see Chapter 2 for details) without the need for a subsequent, late-stage demethylation.

During the 1990s, triaryl organobismuth reagents were introduced as coupling partners and efficient aryl-transfer agents in copper-mediated intermolecular reactions[69,70] for the preparation of aryl ethers from secondary alcohols, even with complex natural products and pharmaceuticals (see Chapter 4 for details on the arylation with triaryl organobismuth reagents).

In this area, and in order to reduce undesired side effects associated with the clinical use of the powerful immunosuppressant macrolide FK-506, researchers at Merck applied this technology for the preparation of aryl ether derivatives of this natural product. The pentavalent triaryl bismuth diacetate reagent was first prepared *in situ* by reacting a Grignard or organolithium reagent with $BiCl_3$ (Scheme 17.21), followed by direct oxidation with peracetic acid. FK-506 (**112**) was next added to this reagent, which, in the presence of $Cu(OAc)_2$, reacted with the 32-hydroxy group to afford mainly the product possessing a 32-aryloxy group **115** (10–65% yields), leaving the alcohol at C-24 untouched.[71,72] Subsequently, an optimized process for the preparation of

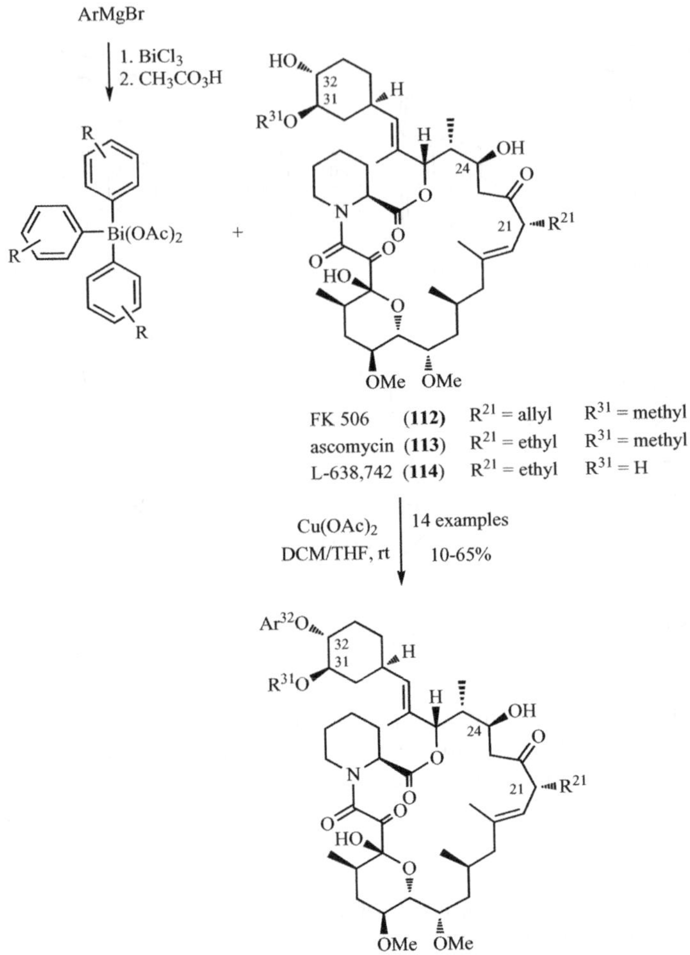

115

L-732,531 (**116**) R^{21} = Et
R^{31} = Me
Ar^{32} = 1-(2-hydroxyethyl-1*H*-indol-5-yl)

Scheme 17.21 Merck's synthesis of FK 506 analogues.

L-732,531 **116**,[73] a 1-(2-hydroxyethyl-1*H*-indol-5-yl)oxy derivative of ascomycin, was reported by the same group. In this work, the novel macrolide **116** was prepared by reacting ascomycin with an indolyl-derived bismuth reagent, which gave the desired product **116** in 55–60% yields.

In a similar work by Pietri and coworkers, a three-step synthesis of the 7-*O*-(4-methylphenyl) derivative of ginkgolide C **120** (Scheme 17.22) was reported, including O-arylation of the 7-hydroxyl group of **117** with tristolylbismuth diacetate **118** in the presence of catalytic amounts of copper

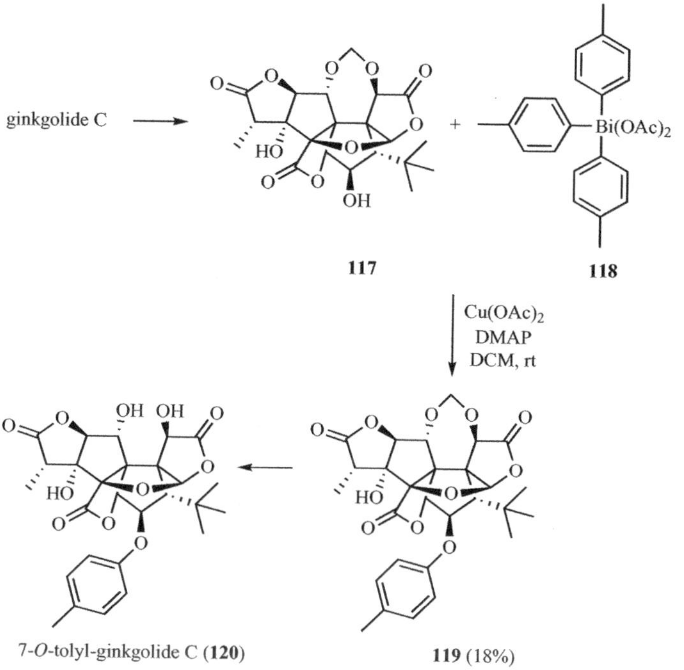

Scheme 17.22 Pietri's synthesis of 7-*O*-(4-methylphenyl) ginkgolide C.

diacetate (0.1 equiv) and 4-dimethylaminopyridine (0.1 equiv) to produce **119** in 18.3% yield.[74] In this case, the aryl group was introduced on the concave face-oriented and sterically hindered hydroxy group. Despite the poor results obtained in some cases, the complexity of the substrates and the high chemo- and regioselectivity obtained in these examples demonstrate the importance of these aryl bismuth reagents as coupling partners in the Ullmann condensation.

In 2007, Evano applied the ligand-assisted arylation of alcohols introduced earlier on by Buchwald and coworkers[66] (see Chapter 2 for details) to the syntheses of paliurine F (**128**, Scheme 17.23),[75] abyssenine A (**134**, Scheme 17.24),[76] and other related cyclopeptide alkaloids.[77]

To assemble the aryl alkyl ether bond in paliurine F, a model copper-mediated reaction between iodobenzene and *trans*-3-hydroxyproline derivative **121** was first examined. Unfortunately, this transformation failed to produce the desired coupling product since it immediately underwent β-elimination to yield the endocyclic enamide **122** and phenol. Therefore, an alternative approach was studied and protected hydroxy-pyrrolidine fragments were examined.[77] After extensive research, it was found that the best results were obtained when protected pyrrolidine **123** was reacted with aryl iodide **126** in the presence of a catalytic amount of CuI (10 mol %) and 1,10-phenanthroline (20 mol %), with Cs_2CO_3 as the base in toluene at 125°C. These conditions afforded the phenyl pyrrolidine ether **127** in 75% yield. The

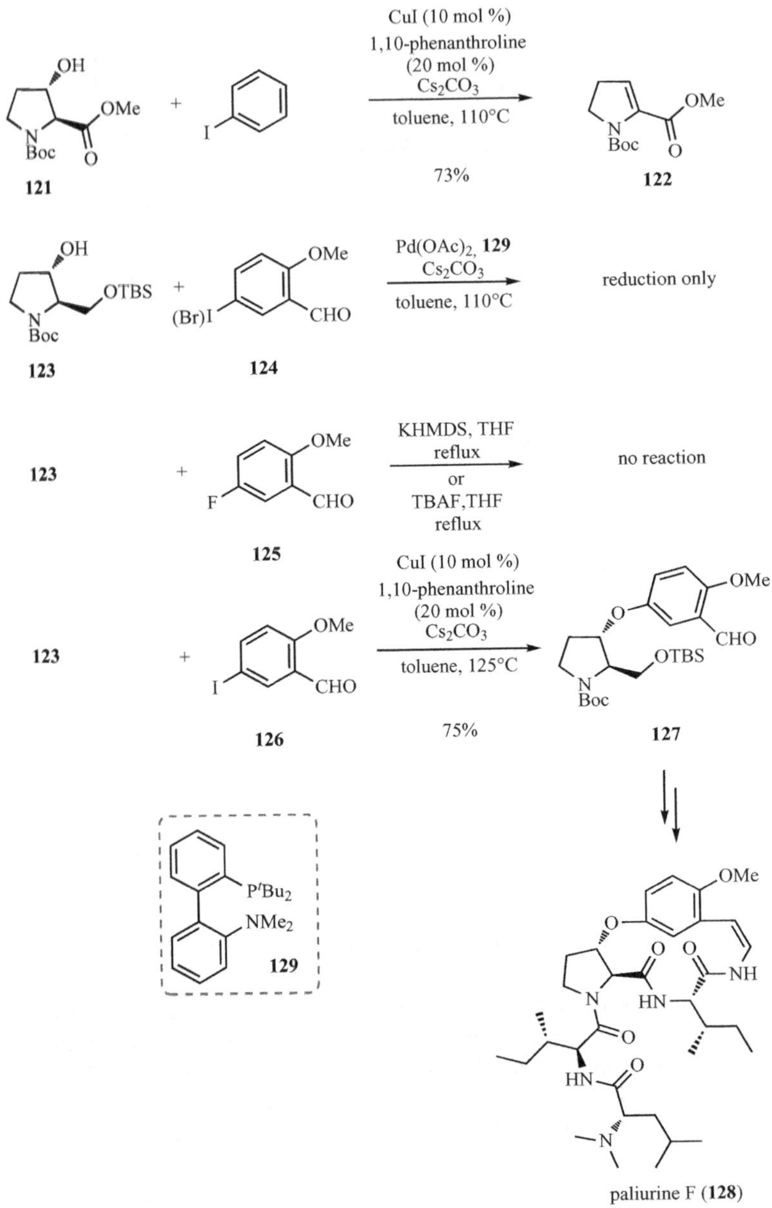

Scheme 17.23 Evano's total synthesis of paliurine F.

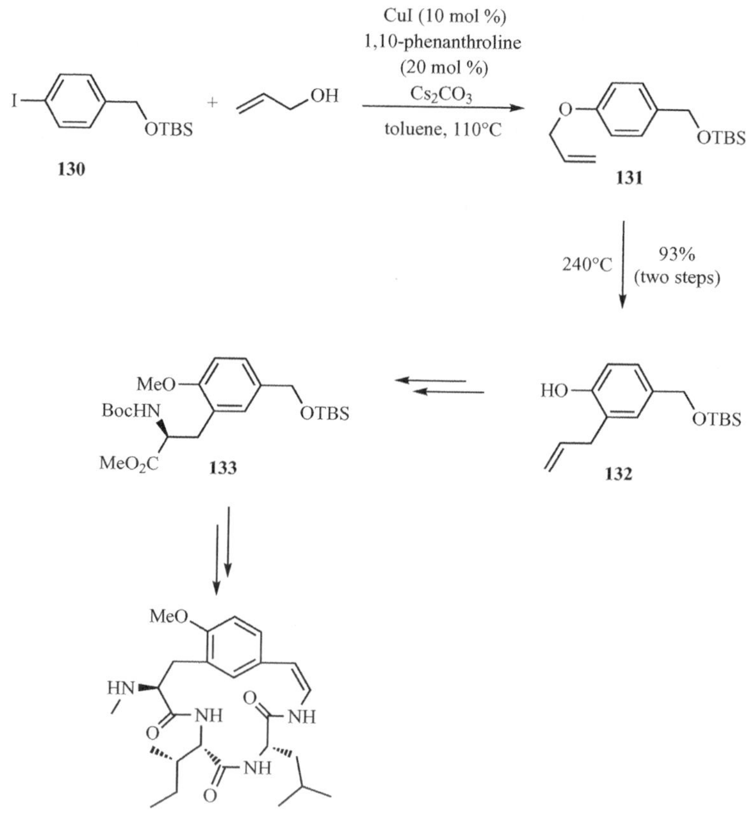

Scheme 17.24 Evano's total synthesis of abyssenine A.

efficiency of this copper-catalyzed arylation procedure was compared with the one of alternative palladium-mediated protocols or nucleophilic aromatic substitution reactions. When pyrrolidine **123** was treated with aromatic iodide (or bromide) **124** in the presence of palladium acetate, cesium carbonate, and phosphine ligands (**129** and others), the reaction mostly led to a reductive dehalogenation of the aryl halide. In addition, no reaction was observed when **123** reacted with aromatic fluoride **125** in the presence of potassium hexamethyldisilazane (KHMDS) or tetrabutylammonium fluoride (TBAF), rendering this copper-mediated reaction the reaction of choice for this type of condensation. Impressively, based on this methodology, the Evano group achieved the total syntheses of paliurines E and F, ziziphines N and Q, and mucronine E, in average overall yields of about 10%.[77]

The ligand-assisted arylation reaction strategy was extended by Evano to the synthesis of abyssenine A as well (Scheme 17.24).[76,77] In order to prepare the trisubstituted phenol **132**, that is, the aromatic core of the target molecule, an approach involving the condensation of aryl iodide **130** with allyl alcohol (CuI, 1,10-phenanthroline, Cs_2CO_3, toluene, 110°C), followed by a Claisen rearrangement (240°C) of coupling intermediate **131**, was successfully applied. This Ullmann–Claisen rearrangement sequence efficiently afforded phenol **132** in 93% yield (two steps), enabling access to abyssenine A via 15 steps (longest linear sequence) and in excellent overall yield (35%).

Besides aliphatic alcohols, other oxygenated nucleophiles have been used as coupling partners for copper-mediated arylation reactions in the context of natural product synthesis. A two-step approach to *O*-arylhydroxylamines via copper-mediated condensation of arylboronic acids and *N*-hydroxyphthalimide was developed by Kelly.[78] In this route, an arylboronic acid (3 equiv) is coupled with a *N*-hydroxylamine (1 equiv) using CuCl (1 equiv) and molecular sieves in dichloroethane under an atmosphere of air. This modified Chan–Lam–Evans protocol was efficiently utilized for the large-scale production of starting materials in Naito's syntheses of stemofuran A and eupomatenoid.[79]

The intermolecular version of this reaction has also been used in the synthesis of various products such as corsifuran A (**136**, Scheme 17.25), a natural product containing a 4′,5-dioxygenated-2-arylbenzofuran skeleton. As part of their asymmetric synthetic strategy, Jones and coworkers[80] prepared enantiopure alcohol **135** and examined several metal-based cycloetherification methods. Screening various palladium catalysts in conjunction with popular ligands and bases revealed that the choice of both the ligand and the base is extremely important, in terms of conversion and side product formation. For example, when 2-di-*tert*-butylphosphino-2′-(*N*,*N*-dimethylamino)-biphenyl (**129**, Scheme 17.23) was used as the ligand (Pd_2dba_3, NaO^tBu, toluene, reflux), a complete conversion to the desired product **136** was observed (100% yield). However, extensive racemization of the sensitive benzylic alcohol had occurred, resulting in a cyclized product with 45% ee only. As an alternative approach aimed to circumvent the racemization process, an intramolecular Ullmann

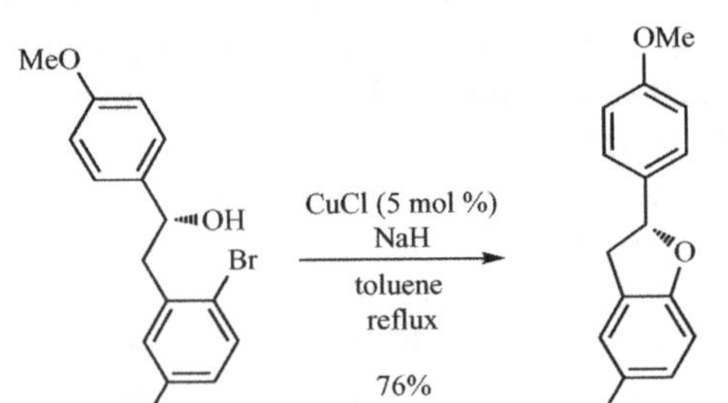

135 corsifuran A (**136**) (100% ee)

Scheme 17.25 Jones's total synthesis of corsifuran A.

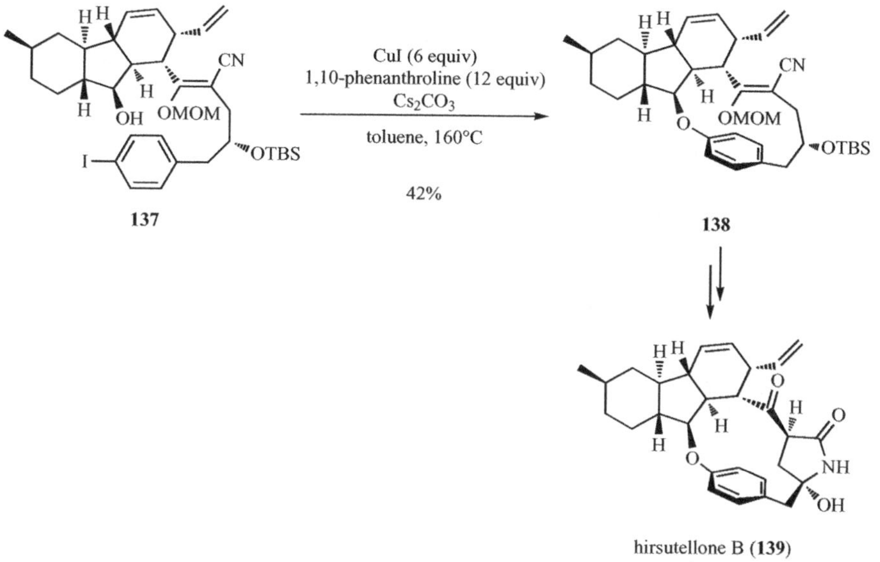

Scheme 17.26 Uchiro's total synthesis of hirsutellone B.

etherification was applied (CuCl, NaH, toluene, reflux), affording the natural product in 76% yield, and with full conservation of the optical purity (ee 100%).

Recently, the Uchiro group successfully completed an impressive synthesis of hirsutellone B (**139**, Scheme 17.26) through a brave 13-membered ring macrocyclization.[81] Impressively, when aryl iodide **137** was subjected to modified Buchwald coupling conditions (CuI [6 equiv], 1,10-phenanthroline [12 equiv], and Cs_2CO_3 in toluene at 160°C), the highly strained 13-membered macrocycle was formed in 42% yield, affording intermediate **138**, which gave the desired natural product **139** after several additional synthetic steps.

17.5 VINYLATION OF ALCOHOLS

While vinyl ethers are not frequently found in natural products or can be installed when necessary through classical methods, the copper-catalyzed vinylation of alcohols has found elegant applications in natural product synthesis in combination with sigmatropic rearrangements. The Claisen rearrangement is among the most powerful tools for the construction of carbon–carbon bonds. However, the lack of general approaches for the stereoselective synthesis of simple allyl vinyl ethers has continually frustrated synthetic chemists. The report by Buchwald in 2003 that allyl vinyl ether can be obtained by Ullmann condensation of allyl halides and allylic alcohols was a significant breakthrough. More exciting was the fact that these ethers undergo a [3,3]-sigmatropic rearrangement (Claisen rearrangement) with high stereoselectivity, producing compounds with vicinal quaternary stereocenters upon

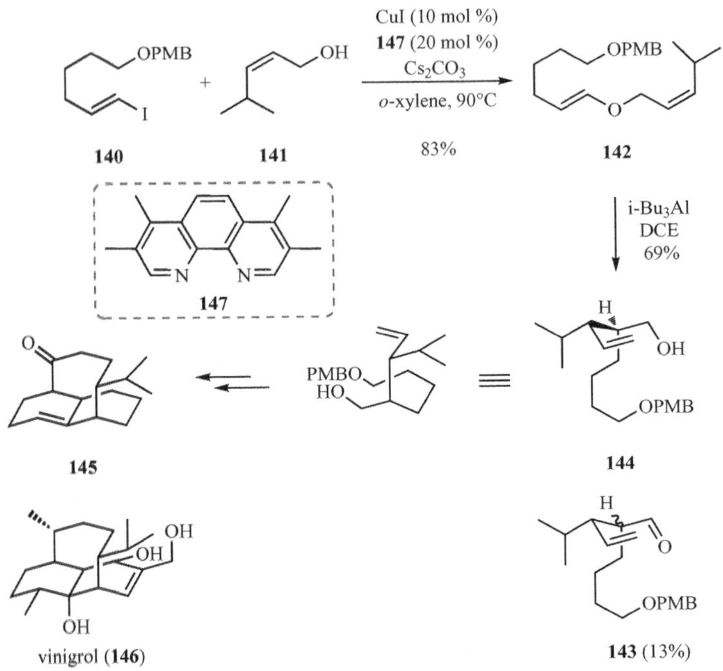

Scheme 17.27 Barriault's synthesis of the tricyclic core of vinigrol.

heating the reaction mixture.[82] This domino process (Ullmann–Claisen rearrangement) was quickly adopted by synthetic chemists and applied to the syntheses of various natural products.

The efficiency of this approach was demonstrated by the installation of two out of the eight contiguous stereocenters of vinigrol (**146**, Scheme 17.27) that were installed by Barriault in his synthesis of the tricyclic core of **146** using the Ullmann–Claisen rearrangement method.[83] The synthesis began with the coupling between vinyl iodide **140** and allyl alcohol **141**, using the conditions developed by Buchwald (CuI, Cs_2CO_3, tetramethylphenanthroline (**147**), *o*-xylene, 90°C), affording enol ether **142** (83% yield) along with aldehyde **143** (13%) as a mixture of epimers (details concerning the synthesis of enol ethers can be found in Chapter 5, Section 5.2.2). It was evidenced that the reaction is particularly sensitive to the temperature, and elevating the temperature above 90°C caused an increase in the Claisen rearrangement product **143**. The sensitivity of aldehyde **143** toward epimerization focused the attention of the authors on the use of Lewis acids, which can catalyze the Claisen rearrangement while at the same time reducing the formed aldehyde. To this end, treatment of enol ether **142** with triisobutylaluminum in dichloromethane afforded alcohol **144** as the sole detectable diastereoisomer.

Another brilliant application of the Ullmann coupling/Claisen rearrangement domino process was featured in Kobayashi's synthesis of pyrrolidinoindoline natural products.[84] His approach to these important alkaloids was

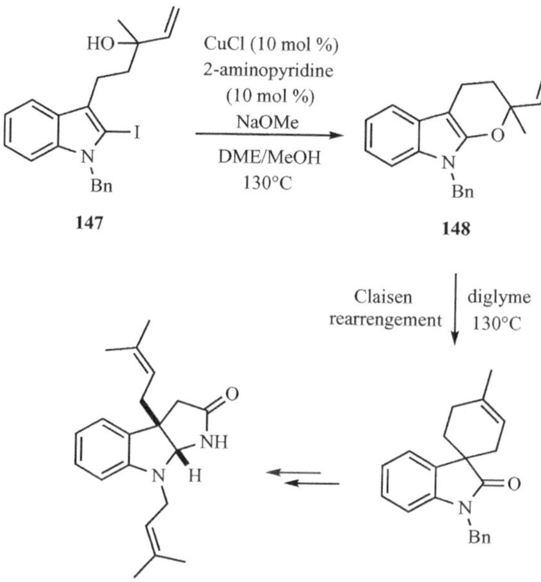

147 148

debromoflustramide B (**150**) **149** (77% from **147**)

Scheme 17.28 Kobayashi's total synthesis of debromoflustramide B.

based on the recognition that spirocyclic oxindoles such as **149** (Scheme 17.28) are key intermediates in their synthesis and that these intermediates might be obtained using a Ullmann coupling/Claisen rearrangement from readily available starting materials **147**. He indeed demonstrated that when iodoindole **147** was subjected to modified Hauptman coupling conditions (CuCl, 2-aminopyridine and NaOMe),[85] an intramolecular Ullmann condensation reaction takes place, affording intermediate vinylpyranoindole **148**, which immediately rearranges to 3-spiro-2-oxindole **149** upon heating at 130°C for 1 hour. Several pyrrolidinoindoline natural products such as debromoflustramide B (**150**), debromoflustramine E,[86] and (−)-flustramine B[87] were prepared based on this especially elegant one-pot operation.

17.6 COPPER-MEDIATED C–S BOND FORMATION IN NATURAL PRODUCT SYNTHESIS

While the formation of C–O bonds using copper-catalyzed processes has been extensively used in natural product synthesis, either as a key step or for the preparation of basic starting materials, copper catalysis has found up to now only a few applications for the formation of C–S bonds in total synthesis. This might be due to both the limited numbers of natural products possessing a C(sp_2)–S moiety and the preconceived idea that copper and sulfur are not a good match, although efficient systems are now available for the copper-catalyzed arylation/vinylation of various S-nucleophiles (see Chapters 1 and 5, respectively).

In addition to the use of copper-catalyzed or copper-mediated arylation of thiols and related S-nucleophiles for the preparation of biologically relevant compounds such as *S*-phenyl-cysteine,[88] HSP90 inhibitors,[89] the GlyT1 inhibitor Lu AA20465,[90] or the p38 mitogen-activated protein kinase inhibitor VX-745,[91] there are to date only two representative examples of the use of copper-mediated formation of C–S bonds in natural product synthesis, both reported in 2012.

In a remarkably elegant total synthesis of (±)-mevashuntin **155**, a structurally unique naturally occurring pyrano-napthoquinone-thaizolone, based on a late-stage Diels–Alder reaction to form the tetracyclic core of the natural product between a highly functionalized diene and bromoquinone **154**, an efficient copper-mediated protocol for the preparation of the latter was devised by Nawrat and Moody (Scheme 17.29).[92] Based on previous results from the Patel group,[93] the brominated benzothiazolone quinone **153** could be efficiently obtained by a copper-catalyzed cyclization starting from an *ortho*-bromoarylthiocarbamate generated *in situ* from the corresponding isothiocyanate **151**. This cyclization, which was smoothly catalyzed by a combination of copper(I) iodide and 1,10-phenanthroline, furnished an intermediate 2-ethoxy-benzothiazole **152** that was converted to the corresponding benzothiazolone **153** by *in situ* hydrolysis using concentrated hydrochloric acid. Using these conditions, the desired dienophile, which could be isolated by simple filtration and did not require additional purification, could be obtained in 96% yield.

Koert's total synthesis of lodopyrione **158**, a polycyclic cytotoxic alkaloid isolated in 2009 from the marine bacterial strain CNQ490, is another illustration of the efficiency of copper-mediated C–S bond-forming reactions. To

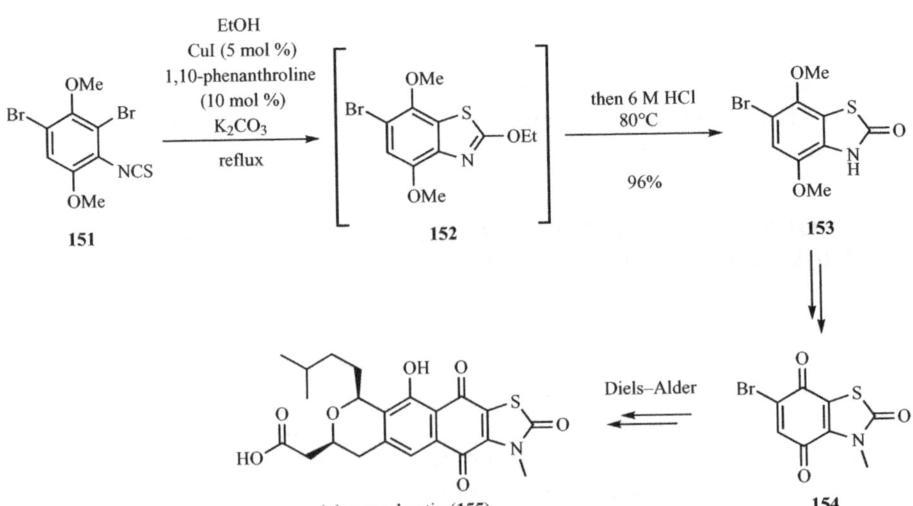

Scheme 17.29 Moody's total synthesis of (±)-mevashuntin.

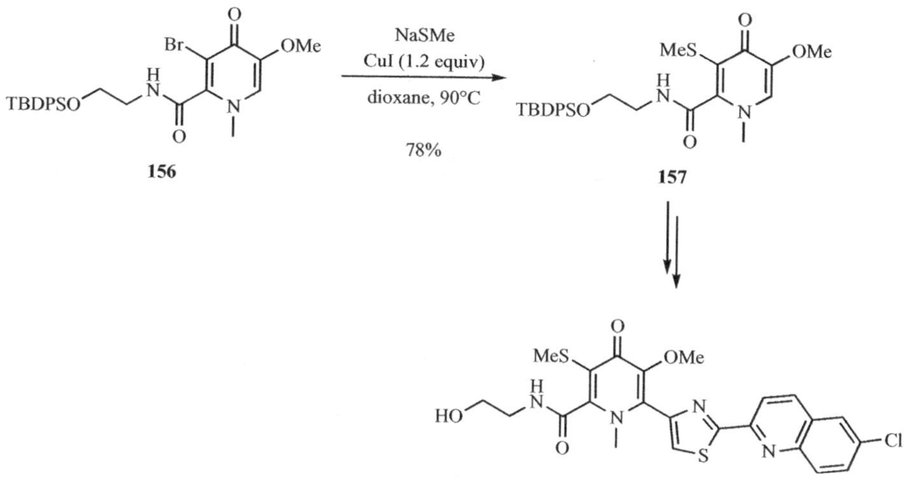

Iodopyridone (158)

Scheme 17.30 Koert's total synthesis of Iodopyridone.

fully elaborate the unprecedented pentasubstituted 4-pyridone core **157** of the natural product, the methylthio substitutent was introduced in 78% yield from the corresponding brominated 4-pyridone **156** using sodium thiomethoxide and a stoichiometric amount of copper(I) iodide in dioxane at 90°C (Scheme 17.30).[94] Interestingly, the use of NaSMe alone gave very low yields, even at elevated temperature, and resulted in the cleavage of both the methyl ester and silyl ether.

17.7 CONCLUSION AND FUTURE PROSPECTS

Over the last two decades, we have witnessed an impressive renaissance in copper chemistry that can be evaluated by the ever-increasing number of new methods and protocols for the formation of aryl–O–aryl, aryl–O–alkyl and alkene–O–allyl (enol ethers) bonds as well as for their sulfur analogues. These new techniques have been examined in natural product synthesis, and in many cases, they turned out to be the missing piece in solving long-standing synthetic problems. The Chan–Lam–Evans reaction, the introduction of Cs_2CO_3 and the ligand-assisted reaction by Buchwald, and the development of new *ortho*-substituent groups with accelerating effect are all milestones in the development of the Ullmann coupling up to its mature state. They have all been successfully incorporated into syntheses of complex natural compounds that were described in this chapter. However, we still need to confront several synthetic challenges in order to allow more reliable bond disconnections based on the Ullmann reaction.[95] Hopefully, we will continue to see an increase in the number of applications in natural product syntheses in which copper leads the way.

REFERENCES

[1] (a) *Classics in Total Synthesis: Targets, Strategies, Methods*, Nicolaou, K. C.; Sorensen, E. J. (eds.). VCH: New York, **1996**, pp. 1–19. (b) *Classics in Total Synthesis II: More Targets, Strategies, Methods*, Nicolaou, K. C.; Snyder, S. A. (eds.). Wiley-VCH: Weinheim, **2003**, pp. 1–13. (c) *Classics in Total Synthesis III: Further Targets, Strategies, Methods*, Nicolaou, K. C.; Chen, J. S. (eds.). Wiley-VCH: Weinheim, **2011**, pp. 1–10.

[2] Rosenmund, K. W.; Harms, H. *Ber. Dtsch. Chem. Ges. B* **1920**, *53B*, 2226–2240.

[3] Hurtley, W. R. H. *J. Chem. Soc.* **1929**, 1870–1873.

[4] (a) Hodous, B. L.; Geuns-Meyer, S. D.; Hughes, P. E.; Albrecht, B. K.; Bellon, S.; Bready, J.; Caenepeel, S.; Cee, V. J.; Chaffee, S. C.; Coxon, A.; Emery, M.; Fretland, J.; Gallant, P.; Gu, Y.; Hoffman, D.; Johnson, R. E.; Kendall, R.; Kim, J. L.; Long, A. M.; Morrison, M.; Olivieri, P. R.; Patel, V. F.; Polverino, A.; Rose, P.; Tempest, P.; Wang, L.; Whittington, D. A.; Zhao, H. *J. Med. Chem.* **2007**, *50*, 611–626. (b) Caron, S.; Do, N. M.; Sieser, J. E.; Whritenour, D. C.; Hill, P. D. *Org. Process Res. Dev.* **2009**, *13*, 324–330. (c) Liu, L.; Siegmund, A.; Xi, N.; Kaplan-Lefko, P.; Rex, K.; Chen, A.; Lin, J.; Moriguchi, J.; Berry, L.; Huang, L.; Teffera, Y.; Yang, Y.; Zhang, Y.; Bellon, S. F.; Lee, M.; Shimanovich, R.; Bak, A.; Dominguez, C.; Norman, M. H.; Harmange, J.-C.; Dussault, I.; Kim, T.-S. *J. Med. Chem.* **2008**, *51*, 3688–3691. (d) Wada, C. K.; Holms, J. H.; Curtin, M. L.; Dai, Y.; Florjancic, A. S.; Garland, R. B.; Guo, Y.; Heyman, H. R.; Stacey, J. R.; Steinman, D. H.; Albert, D. H.; Bouska, J. J.; Elmore, I. N.; Goodfellow, C. L.; Marcotte, P. A.; Tapang, P.; Morgan, D. W.; Michaelides, M. R.; Davidsen, S. K. *J. Med. Chem.* **2002**, *45*, 219–232.

[5] (a) Draper, W. M.; Casida, J. E. *J. Agric. Food Chem.* **1983**, *31*, 227–231. (b) Draper, W. M.; Casida, J. E. *J. Agric. Food Chem.* **1983**, *31*, 1201–1207. (c) Scrano, L.; Bufo, S. A.; D'Auria, M.; Meallier, P.; Behechti, A.; Shramm, K. W. *J. Environ. Qual.* **2002**, *31*, 268–274.

[6] (a) Laskoski, M.; Dominguez, D. D.; Keller, T. M. *J. Polym. Sci. Part A Polym. Chem.* **2006**, *44*, 4559–4565. (b) Labadie, J. W.; Hedrick, J. L.; Ueda, M. In *ACS Symp. Ser.: Step-Growth Polymers for High-Performance Materials, New Synthetic Methods*, Vol. 624, Hedrick, J. L.; Labadie, J. W. (eds.). ACS: Washington, **1996**, pp. 210–225.

[7] Pitsinos, E. N.; Vidali, V. P.; Couladouros, E. A. *Eur. J. Org. Chem.* **2011**, *11*, 1207–1222.

[8] (a) Tomita, M.; Fujitani, K.; Aoyagi, Y. *Tetrahedron Lett.* **1967**, *8*, 1201–1206. (b) Tomita, M.; Fujitani, K.; Aoyagi, Y.; Kajita, Y. *Chem. Pharm. Bull.* **1968**, *16*, 217–226.

[9] (a) Tomita, M.; Fujitani, K.; Aoyagi, Y. *Tetrahedron Lett.* **1966**, *7*, 4243–4248. (b) Tomita, M.; Fujitani, K.; Aoyagi, Y. *Chem. Pharm. Bull.* **1968**, *16*, 62–69.

[10] (a) Inubushi, Y.; Masaki, Y.; Matsumoto, S.; Takami, F. *Tetrahedron Lett.* **1968**, *9*, 3399–3402. (b) Inubushi, Y.; Masaki, Y.; Matsumoto, S.; Takami, F. *J. Chem. Soc. C* **1969**, 1547–1556.

[11] Toyota, M.; Tori, M.; Takikawa, K.; Shiobara, Y.; Kodama, M.; Asakawa, Y. *Tetrahedron Lett.* **1985**, *26*, 6097–6100.

[12] Eicher, T.; Fey, S.; Puhl, W.; Buechel, E.; Speicher, A. *Eur. J. Org. Chem.* **1998**, 877–888.

[13] Fuerstner, A.; Mueller, C. *Chem. Commun.* **2005**, 5583–5585.

[14] Pospíšil, J.; Müller, C.; Fürstner, A. *Chem. Eur. J.* **2009**, *15*, 5956–5968.

[15] Ramana, C. V.; Mondal, M. A.; Puranik, V. G.; Gurjar, M. K. *Tetrahedron Lett.* **2007**, *48*, 7524–7527.

[16] (a) Rychnovsky, S. D.; Hwang, K. *Tetrahedron Lett.* **1994**, *35*, 8927–8930. (b) Rychnovsky, S. D.; Hwang, K. *J. Org. Chem.* **1994**, *59*, 5414–5418.

[17] Wipf, P.; Jung, J. K. *J. Org. Chem.* **2000**, *65*, 6319–6339.

[18] Marcoux, J. F.; Doye, S.; Buchwald, S. L. *J. Am. Chem. Soc.* **1997**, *119*, 10539–10540.

[19] Xing, X.; Padmanaban, D.; Yeh, L.-A.; Cuny, G. D. *Tetrahedron* **2002**, *58*, 7903–7910.

[20] Kwak, J.-H.; In, J.-K.; Lee, M.-S.; Choi, E.-H.; Lee, H.; Hong, J. T.; Yun, Y.-P.; Lee, S. J.; Seo, S.-Y.; Suh, Y.-G.; Jung, J.-K. *Arch. Pharmacal Res.* **2008**, *31*, 1559–1563.

[21] Ma, D.; Cai, Q. *Org. Lett.* **2003**, *5*, 3799–3802.

[22] Cai, Q.; Zou, B.; Ma, D. *Angew. Chem., Int. Ed.* **2006**, *45*, 1276–1279.

[23] Williams, A. L.; Kinney, R. E.; Bridger, R. F. *J. Org. Chem.* **1967**, *32*, 2501–2505.

[24] Carr, J. L.; Offermann, D. A.; Holdom, M. D.; Dusart, P.; White, A. J. P.; Beavil, A. J.; Leatherbarrow, R. J.; Lindell, S. D.; Sutton, B. J.; Spivey, A. C. *Chem. Commun.* **2010**, *46*, 1824–1826.

[25] Buck, E.; Song, Z. J.; Tschaen, D.; Dormer, P. G.; Volante, R. P.; Reider, P. J. *Org. Lett.* **2002**, *4*, 1623–1626.

[26] Jung, N.; Bräse, S. *Eur. J. Org. Chem.* **2009**, 4494–4502.

[27] Gao, H.-Y.; Ha, C.-Y. *Synth. Commun.* **2006**, *36*, 3283–3286.

[28] Rao, A. V. R.; Gurjar, M. K.; Reddy, K. L.; Rao, A. S. *Chem. Rev.* **1995**, *95*, 2135–2167.

[29] Schmidt, U.; Weller, D.; Holder, A.; Lieberknecht, A. *Tetrahedron Lett.* **1988**, *29*, 3227–3230.

[30] Evans, D. A.; Ellman, J. A. *J. Am. Chem. Soc.* **1989**, *111*, 1063–1072.

[31] Rao, A. V. R.; Chakraborty, T. K.; Reddy, K. L.; Rao, A. S. *Tetrahedron Lett.* **1992**, *33*, 4799–4802.

[32] (a) Boger, D. L.; Yohannes, D. *J. Org. Chem.* **1990**, *55*, 6000–6017. (b) Boger, D. L.; Yohannes, D. *J. Org. Chem.* **1989**, *54*, 2498–2502.

[33] Cai, Q.; He, G.; Ma, D. *J. Org. Chem.* **2006**, *71*, 5268–5273.

[34] Sano, S.; Ikai, K.; Katayama, K.; Takesako, K.; Nakamura, T.; Obayashi, A.; Ezure, Y.; Enomoto, H. *J. Antibiot.* **1986**, *39*, 1685–1696.

[35] Ma, D.; Cai, Q. *Acc. Chem. Res.* **2008**, *41*, 1450–1460.

[36] Jung, M. E.; Lazarova, T. I. *J. Org. Chem.* **1999**, *64*, 2976–2977.

[37] Chan, D. M. T.; Monaco, K. L.; Wang, R.-P.; Winters, M. P. *Tetrahedron Lett.* **1998**, *39*, 2933–2936.

[38] Evans, D. A.; Katz, J. L.; West, T. R. *Tetrahedron Lett.* **1998**, *39*, 2937–2940.

[39] Lam, P. Y. S.; Clark, C. G.; Saubern, S.; Adams, J.; Winters, M. P.; Chan, D. M. T.; Combs, A. *Tetrahedron Lett.* **1998**, *39*, 2941–2944.

[40] Chalmers, J. R.; Dickson, G. T.; Elks, J.; Hems, B. A. *J. Chem. Soc.* **1949**, 3424–3433.

[41] Evans, D. A.; Katz, J. L.; Peterson, G. S.; Hintermann, T. *J. Am. Chem. Soc.* **2001**, *123*, 12411–12413.

[42] Gillmore, A.; Lauret, C.; Roberts, S. M. *Tetrahedron* **2003**, *59*, 4363–4375.

[43] Jung, J.-W.; Jang, J.; Seo, S.-Y.; Jung, J.-K.; Suh, Y.-G. *Tetrahedron* **2010**, *66*, 6826–6831.

[44] Cousin, D.; Mann, J.; Nieuwenhuyzen, M.; van der Berg, H. *Org. Biomol. Chem.* **2006**, *4*, 54–62.

[45] Wang, Y.-C.; Georghiou, P. E. *Org. Lett.* **2002**, *4*, 2675–2678.

[46] Skaff, O.; Jolliffe, K. A.; Hutton, C. A. *J. Org. Chem.* **2005**, *70*, 7353–7363.

[47] Boger, D. L.; Patane, M. A.; Zhou, J. *J. Am. Chem. Soc.* **1994**, *116*, 8544–8556.

[48] Boger, D. L.; Zhou, J. *J. Am. Chem. Soc.* **1993**, *115*, 11426–11433.

[49] Boger, D. L.; Sakya, S. M.; Yohannes, D. *J. Org. Chem.* **1991**, *56*, 4204–4207.

[50] Boger, D. L.; Yohannes, D.; Zhou, J.; Patane, M. A. *J. Am. Chem. Soc.* **1993**, *115*, 3420–3430.

[51] Boger, D. L.; Zhou, J. *J. Am. Chem. Soc.* **1995**, *117*, 7364–7378.

[52] Pettit, G. R.; Quistorf, P. D.; Fry, J. A.; Herald, D. L.; Hamel, E.; Chapuis, J.-C. *J. Nat. Prod.* **2009**, *72*, 876–883.

[53] Kumar, G. D. K.; Natarajan, A. *Tetrahedron Lett.* **2008**, *49*, 2103–2105.

[54] Jeong, B.-S.; Wang, Q.; Son, J. K.; Jahng, Y. *Eur. J. Org. Chem.* **2007**, 1338–1344.

[55] (a) Shen, L.; Sun, D. *Tetrahedron Lett.* **2011**, *52*, 4570–4574. (b) Salih, M. Q.; Beaudry, C. M. *Org. Lett.* **2012**, *14*, 4026–4029. (c) Zhu, Z.-Q.; Salih, M. Q.; Fynn, E.; Bain, A. D.; Beaudry, C. M. *J. Org. Chem.* **2013**, *78*, 2881–2896. (d) Zhu, Z.-Q.; Beaudry, C. M. *J. Org. Chem.* **2013**, *78*, 3336–3341. (e) Collins, J. C.; Farley, K. A.; Limberakis, C.; Liras, S.; Price, D.; James, K. *J. Org. Chem.* **2012**, *77*, 11079–11090.

[56] Deng, H.; Jung, J.-K.; Liu, T.; Kuntz, K. W.; Snapper, M. L.; Hoveyda, A. H. *J. Am. Chem. Soc.* **2003**, *125*, 9032–9034.

[57] Nicolaou, K. C.; Boddy, C. N. C.; Natarajan, S.; Yue, T. Y.; Li, H.; Bräse, S.; Ramanjulu, J. M. *J. Am. Chem. Soc.* **1997**, *119*, 3421–3422.

[58] Kalinin, A. V.; Bower, J. F.; Riebel, P.; Snieckus, V. *J. Org. Chem.* **1999**, *64*, 2986–2987.

[59] (a) Helynck, G.; Dubertret, C.; Frechet, D.; Leboul, J. *J. Antibiot.* **1998**, *51*, 512–514. (b) Boisnard, S.; Zhu, J. *Tetrahedron Lett.* **2002**, *43*, 2577–2580. (c) Bois-Choussy, M.; Cristau, P.; Zhu, J. *Angew. Chem., Int. Ed.* **2003**, *42*, 4238–4241.

[60] Nicolaou, K. C.; Mitchell, H. J.; Jain, N. F.; Winssinger, N.; Hughes, R.; Bando, T. *Angew. Chem., Int. Ed.* **1999**, *38*, 240–244.

[61] (a) Nicolaou, K. C.; Li, H.; Boddy, C. N. C.; Ramanjulu, J. M.; Yue, T.-Y.; Natarajan, S.; Chu, X.-J.; Braese, S.; Rubsam, F. *Chem. Eur. J.* **1999**, *5*, 2584–2601. (b) Nicolaou, K. C.; Boddy, C. N. C.; Li, H.; Koumbis, A. E.; Hughes, R.; Natarajan, S.; Jain, N. F.; Ramanjulu, J. M.; Braese, S.; Solomon, M. E. *Chem. Eur. J.* **1999**, *5*, 2602–2621. (c) Nicolaou, K. C.; Koumbis, A. E.; Takayanagi, M.; Natarajan, S.; Jain, N. F.; Bando, T.; Li, H.; Hughues, R. *Chem. Eur. J.* **1999**, *5*, 2622–2647. (d) Nicolaou, K. C.; Mitchell, H. J.; Jain, N. F.; Bando, T.; Hughes, R.; Winssinger, N.; Natarajan, S.; Koumbis, A. E. *Chem. Eur. J.* **1999**, *5*, 2648–2667.

[62] Thasana, N.; Worayuthakarn, R.; Kradanrat, P.; Hohn, E.; Young, L.; Ruchirawat, S. *J. Org. Chem.* **2007**, *72*, 9379–9382.

[63] Moreau, A.; Couture, A.; Deniau, E.; Grandclaudon, P. *J. Org. Chem.* **2004**, *69*, 4527–4530.

[64] Olivera, R.; SanMartin, R.; Churruca, F.; Domínguez, E. *J. Org. Chem.* **2002**, *67*, 7215–7225.

[65] Lindley, J. *Tetrahedron* **1984**, *40*, 1433–1456.

[66] Wolter, M.; Nordmann, G.; Job, G. E.; Buchwald, S. L. *Org. Lett.* **2002**, *4*, 973–976.

[67] Amberg, W.; Bennani, Y. L.; Chadha, R. K.; Crispino, G. A.; Davis, W. D.; Hartung, J.; Jeong, K. S.; Ogino, Y.; Shibata, T.; Sharpless, K. B. *J. Org. Chem.* **1993**, *58*, 844–849.

[68] Miyake, F. Y.; Yakushijin, K.; Horne, D. A. *Org. Lett.* **2004**, *6*, 711–713.

[69] Barton, D. H. R.; Finet, J.-P.; Pichon, C. *J. Chem. Soc., Chem. Commun.* **1986**, 65–66.

[70] Dodonov, V. A.; Gushchin, A. V.; Brilkina, T. G. *Zh. Obshch. Khim.* **1985**, *55*, 2514–2519.

[71] Sinclair, P. J.; Wong, F.; Wyvratt, M.; Staruch, M. J.; Dumont, F. *Bioorg. Med. Chem. Lett.* **1995**, *5*, 1035–1038.

[72] Sinclair, P. J.; Wong, F.; Staruch, M. J.; Wiederrecht, G.; Parsons, W. H.; Dumont, F.; Wyvratt, M. *Bioorg. Med. Chem. Lett.* **1996**, *6*, 2193–2196.

[73] Brands, K. M. J.; Dolling, U.-H.; Jobson, R. B.; Marchesini, G.; Reamer, R. A.; Williams, J. M. *J. Org. Chem.* **1998**, *63*, 6721–6726.

[74] Pietri, S.; Liebgott, T.; Finet, J. P.; Culcasi, M.; Billottet, L.; Bernard-Henriet, C. *Drug Dev. Res.* **2001**, *54*, 191–201.

[75] Toumi, M.; Couty, F.; Evano, G. *Angew. Chem., Int. Ed.* **2007**, *46*, 572–575.

[76] Toumi, M.; Couty, F.; Evano, G. *J. Org. Chem.* **2007**, *72*, 9003–9009.

[77] Toumi, M.; Rincheval, V.; Young, A.; Gergeres, D.; Turos, E.; Couty, F.; Mignotte, B.; Evano, G. *Eur. J. Org. Chem.* **2009**, *2009*, 3368–3386.

[78] Petrassi, H. M.; Sharpless, K. B.; Kelly, J. W. *Org. Lett.* **2001**, *3*, 139–142.

[79] Miyata, O.; Takeda, N.; Naito, T. *Org. Lett.* **2004**, *6*, 1761–1763.

[80] Adams, H.; Gilmore, N. J.; Jones, S.; Muldowney, M. P.; von Reuss, S. H.; Vemula, R. *Org. Lett.* **2008**, *10*, 1457–1460.

[81] Rittle, J.; Green, M. T. *Science* **2010**, *330*, 933–937.

[82] Nordmann, G.; Buchwald, S. L. *J. Am. Chem. Soc.* **2003**, *125*, 4978–4979.

[83] Grisé, C.; Tessier, G.; Barriault, L. *Org. Lett.* **2007**, *9*, 1545–1548.

[84] Miyamoto, H.; Okawa, Y.; Nakazaki, A.; Kobayashi, S. *Angew. Chem., Int. Ed.* **2006**, *45*, 2274–2277.

[85] Fagan, P. J.; Hauptman, E.; Shapiro, R.; Casalnuovo, A. *J. Am. Chem. Soc.* **2000**, *122*, 5043–5051.

[86] Miyamoto, H.; Okawa, Y.; Nakazaki, A.; Kobayashi, S. *Tetrahedron Lett.* **2007**, *48*, 1805–1808.

[87] Hirano, T.; Iwakiri, K.; Miyamoto, H.; Nakazaki, A.; Kobayashi, S. *Heterocycles* **2009**, *79*, 805–820.

[88] Herradura, P. S.; Pendola, K. A.; Guy, R. K. *Org. Lett.* **2000**, *2*, 2019–2022.
[89] Llauger, L.; He, H.; Kim, J.; Aguirre, J.; Rosen, N.; Peters, U.; Davies, P.; Chiosis, G. *J. Med. Chem.* **2005**, *48*, 2892–2905.
[90] Krafft, E. A.; Pinard, E.; Thomas, A. W. *Tetrahedron Lett.* **2006**, *47*, 5355–5357.
[91] Bagley, M. C.; Davis, T.; Dix, M. C.; Rokicki, M. J.; Kipling, D. *Bioorg. Med. Chem. Lett.* **2007**, *17*, 5107–5110.
[92] Nawrat, C. C.; Moody, C. J. *Org. Lett.* **2012**, *14*, 1484–1487.
[93] Murru, S.; Mondal, P.; Yella, R.; Patel, B. K. *Eur. J. Org. Chem.* **2009**, 5406–5413.
[94] Burckhardt, T.; Harms, K.; Koert, U. *Org. Lett.* **2012**, *14*, 4674–4677.
[95] Wu, B.; He, S.; Wu, X.-D.; Wu, D.-K.; Pan, Y.-J. *Helv. Chim. Acta* **2007**, *90*, 1586–1592.

18

COPPER-CATALYZED C–C BOND FORMATION IN NATURAL PRODUCT SYNTHESIS: ELEGANT AND EFFICIENT SOLUTIONS TO A KEY BOND DISCONNECTION

MORGAN DONNARD AND NICOLAS BLANCHARD

18.1 INTRODUCTION

Copper-mediated C–C bond-forming reactions are among the oldest metal-mediated coupling transformations, Ullmann's biaryl coupling being more than a century old. However, the last decade has seen an exponential increase of new and clever copper-based methodologies with improved efficiencies. Conditions are often really mild and tolerant to many functional groups, and the catalytic systems used are undoubtedly less expensive than many other metal complexes used in related processes. Thus, if palladium-catalyzed cross couplings have firmly occupied the first rank in the area of natural product total synthesis, copper-catalyzed or copper-mediated coupling reactions have recently become classical and essential alternatives in the chemist's synthetic toolbox.[1,2]

As in Chapters 16 and 17, which are dedicated to the copper-mediated formation of carbon–nitrogen and carbon–oxygen bonds in total synthesis, respectively, the aim of this chapter is to present and discuss the usual and recent copper-mediated carbon–carbon bond formation processes and representative total syntheses of biologically relevant natural products. Whenever

Copper-Mediated Cross-Coupling Reactions, First Edition. Edited by Gwilherm Evano and Nicolas Blanchard.

possible, a comparison with palladium-catalyzed transformations will be presented. This concise overview is dedicated to the formation of biaryls, enyne, polyenes, and polyynes, which will be described in the following order:

- synthesis of naturally occurring biaryls by copper-mediated cross coupling
- copper-catalyzed enyne formation in natural product synthesis
- synthesis of naturally occurring polyenes: copper-mediated strategies and copper(I) salts as efficient cocatalysts for the Stille cross-coupling reaction
- synthesis of naturally occurring polyynes by copper-mediated cross coupling.

18.2 NATURAL BIARYLS BY COPPER-CATALYZED CROSS COUPLING

Among natural products, compounds bearing a biaryl unit undeniably represent one of the most significant classes. More than its innumerable members, its particular structures as well as its specific biological activities make this unique group of molecules essential. Two representative examples are shown in Figure 18.1. As simple proof, the ever-increasing interest for these compounds has resulted in more than 1000 publications over the last decade.[3] From a chemist's point of view, these targets represent a real synthetic

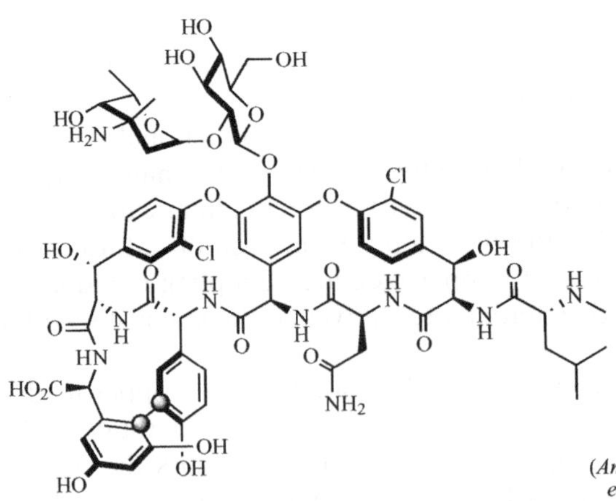

Vancomycin (1)

(Powerful antibiotic)

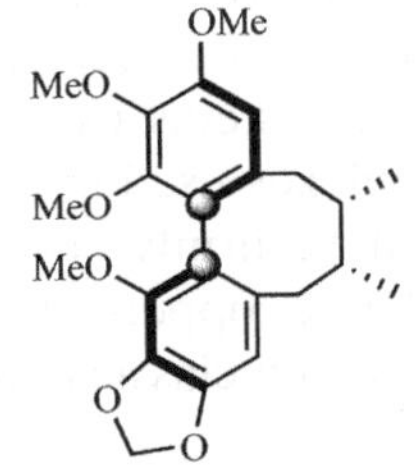

Schizandrin B (2)

(Antitumor agent against liver cancer, enhancement of hepatic function)

Figure 18.1 Representative naturally occurring diaryl compounds and their biological activities.

challenge. In most cases, in addition to asymmetric construction of classical stereocenters, the axial chirality induced by the aryl–aryl C–C bond has to be controlled to form a single atropodiastereoisomer. It is interesting to note that this specific point can be strongly complicated by the occurrence of a bridge between both aryl groups. As an answer to this synthetic challenge, several methodologies have been developed over the last few years; and, even if palladium-catalyzed reactions still represent the main tool for aryl–aryl couplings, very interesting copper-mediated asymmetric cross-coupling reactions have recently emerged and represent an efficient alternative solution for the formation of these specific C–C bonds in total synthesis of natural products.

18.2.1 Classical Ullmann Cross-Coupling Reaction

In the field of total synthesis of biaryl-based natural products, the old[4] but outstanding Ullmann coupling has been extensively used. This reaction involving aryl halides and copper metal has been well documented and appreciably tamed in order to make it a fantastic modernized tool for the synthesis of C(aryl)–C(aryl) bonds as overviewed in Chapter 10.[2,5] Variants of this coupling involving the preformation of organometallic species are beyond the scope of this chapter and will not be treated (e.g., Ziegler–Ullmann reaction). It is essential to note at this point that the use of copper(0) for the Ullmann coupling is limiting the possibilities of chirality control of the biaryl C–C bond to substrate-induced diastereoselectivity or to postcoupling resolution.

The first method for the preparation of stereoenriched coupling products is based on the use of a temporary chiral auxiliary. In 1997, Meyers and coworker reported the asymmetric total synthesis of (*S*)-gossypol (Scheme 18.1),[6] the main coloring pigment of cottonseeds using a sequence including a highly diastereoselective Ullmann reaction as a key step. In order to control the configuration of the biaryl bond, a *tert*-butyl oxazoline has been used as chiral auxiliary and the copper-mediated coupling procured the desired (a*S*-*S*)-**4** in 30 minutes with a diastereomeric ratio of 11:1 and 81% yield. The oxazoline has then been easily removed and (*S*)-gossypol **5** has been obtained efficiently in a few more steps (Scheme 18.1).

The second method to control the configuration of the major atropoisomer formed is making use of the stereocenters present in the substrate. These atroposelective Ullmann couplings are essentially realized in order to synthesize bridged biaryl compounds, and this specific structural element can, in some cases, seriously increase the general degree of synthesis difficulty. An interesting example that nicely illustrates the subtlety of this biaryl formation is the challenging synthesis of corilagin (Scheme 18.2). The latter features a central D-glucopyranose moiety with a notable all-axial substituent conformation. A direct coupling would clearly be difficult as it would request a highly energetic ring-flip into a skew boat conformation in order to fit the natural product stereostructure. Yamada and coworkers proposed as a solution the use of a temporarily ring-opened equivalent of the pyranose core in order to realize

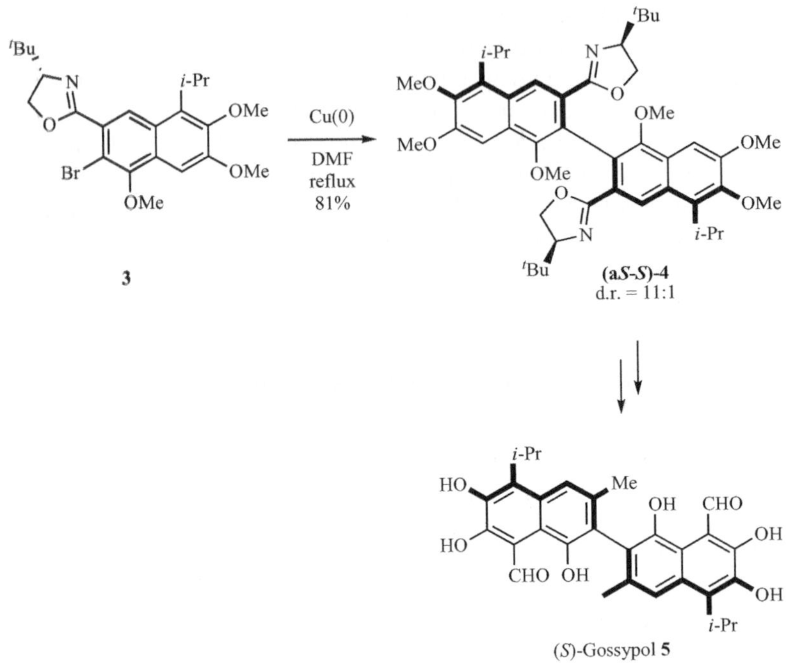

Scheme 18.1 Meyers' total synthesis of (*S*)-gossypol via diastereoselective Ullmann homocoupling using temporary chiral auxiliary.

the coupling step.[7] Transformation of **6** under classical Ullmann conditions led to the atroposelective formation of (a*R*)-7 in 48% yield, a compound that was then easily converted into nonamethylcorilagin **8**.

The last method requires an additional step of kinetic resolution in order to obtain optically pure biaryl compounds. In 2003, Molander and coworkers reported the total synthesis of (+)-isoschizandrin (Scheme 18.3), a molecule isolated from fruits of *Schizandra chinesis* and presenting an antiulcer activity in rats.[8] Interestingly, the synthetic sequence involved a racemic classical Ullmann coupling followed by a Cannizzaro reaction that led to the formation of the lactone-bridged biaryl product **11**. The well-established atroposelective Bringmann's kinetic resolution[9] involving a (*R*)-2-methyl-CBS-oxazaborolidine reduction has then been applied in order to obtain the desired optically pure seven-membered lactone (+)-**13** in 61% yield. (+)-Isoschizandrin **14** was then secured in seven more steps. The major limitation of this strategy lies in the fact that it often requires laborious additional steps to recycle the undesired stereoisomer in order to obtain synthetically useful yields (in this specific case, the transformation of the undesired enantioenriched diol (−)-**12** into racemic lactone **11** requires four steps).

If Ullmann reaction represents the classical efficient methodology in order to synthesize biaryl compounds, it finds its limits when atropoenriched

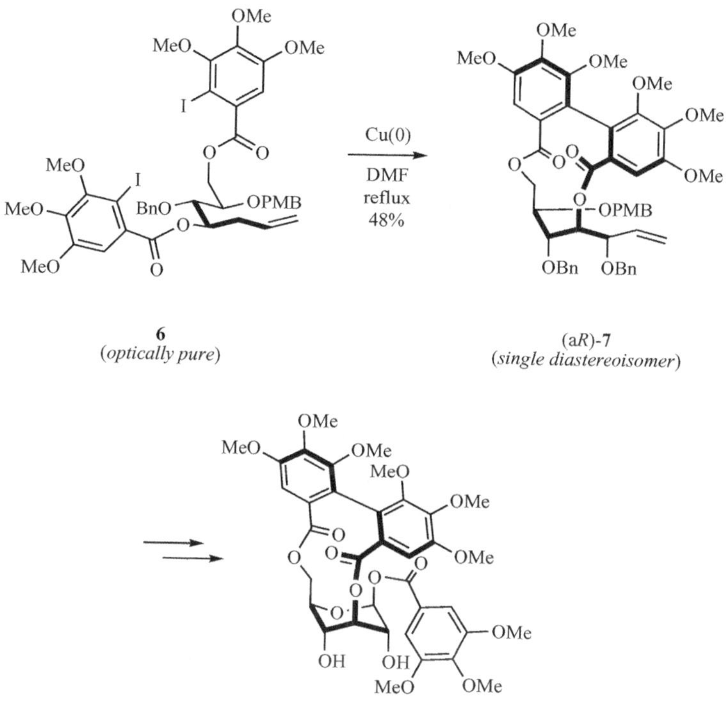

Scheme 18.2 Yamada's synthesis of nonamethylcorilagin via atroposelective Ullmann cross coupling.

molecules are targeted. The development of alternative approaches, mainly inspired by nature such as the oxidative couplings of phenols, opened the way to a major breakthrough: the copper-catalyzed induced enantioselective biaryl coupling.

18.2.2 Oxidative Biaryl Coupling of Phenols and Naphthols

The similarity of oxidative couplings of phenols and naphthols with a significant number of biosynthetic routes has inspired several elegant natural product total syntheses even if these reactions can be quite challenging, especially in the case of strongly deactivated aromatic rings. Usually performed under mild conditions, these transformations tolerate many functional groups and do not rely on tedious preactivation of the starting materials such as with a halide or a boronic acid. In contrast, this absence of prefunctionalized carbon centers can have a detrimental impact on regioselectivity when both steric and electronic factors are not strong enough. However, it is important to keep in mind that if *para*- and *ortho*-positions are reactive coupling sites,

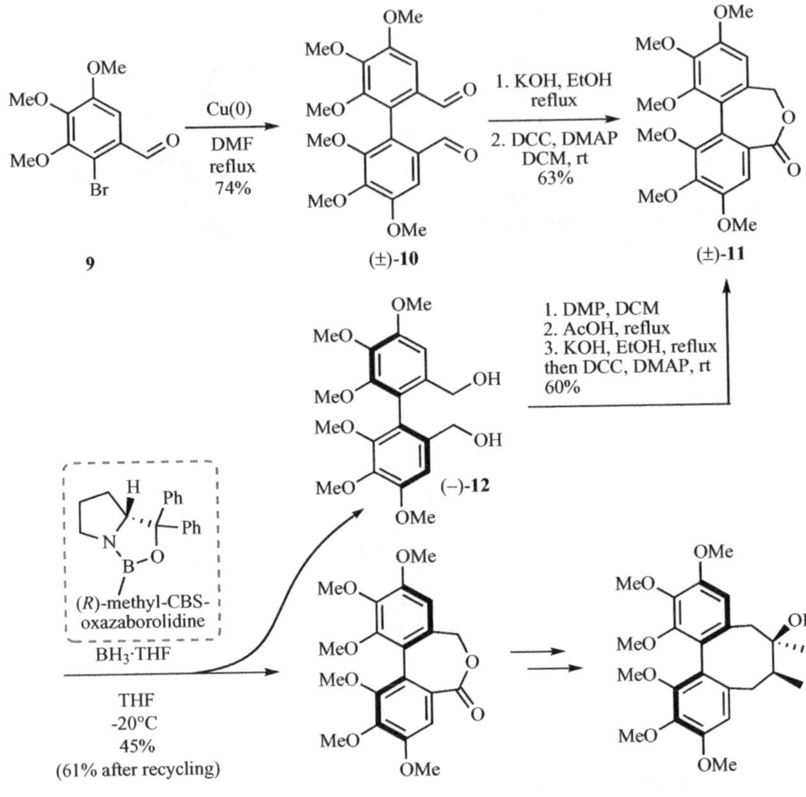

Scheme 18.3 Molander's synthesis of (+)-isoschizandrin via Ullmann homocoupling/atroposelective kinetic resolution.

meta ones are not, which is a pertinent starting point to elaborate discriminating strategies.[10]

These reactions can be performed under two close types of activation as copper species can be used as mediating agent or catalysts. If enantioselective versions of these copper-mediated oxidative couplings have been developed, discriminations were not good enough to make these methodologies a viable strategic tool in the total synthesis of natural homochiral compounds. In contrast, these couplings can be realized in a really high diastereoselective manner. For example, Yamada and coworkers reported in 2012 the total synthesis of (+)-davidiin[11] (Scheme 18.4), a compound that inhibits the ligand binding to a μ-opioid receptor, via an elegant oxidative copper-mediated coupling of two phenol moieties bridged by a skew-boat (axial-rich) glucose. The sequence started with conformationally locked sugar **15** that was converted in four steps to the desired substrate bearing two phenolic rings **16**. Then, the 12-membered

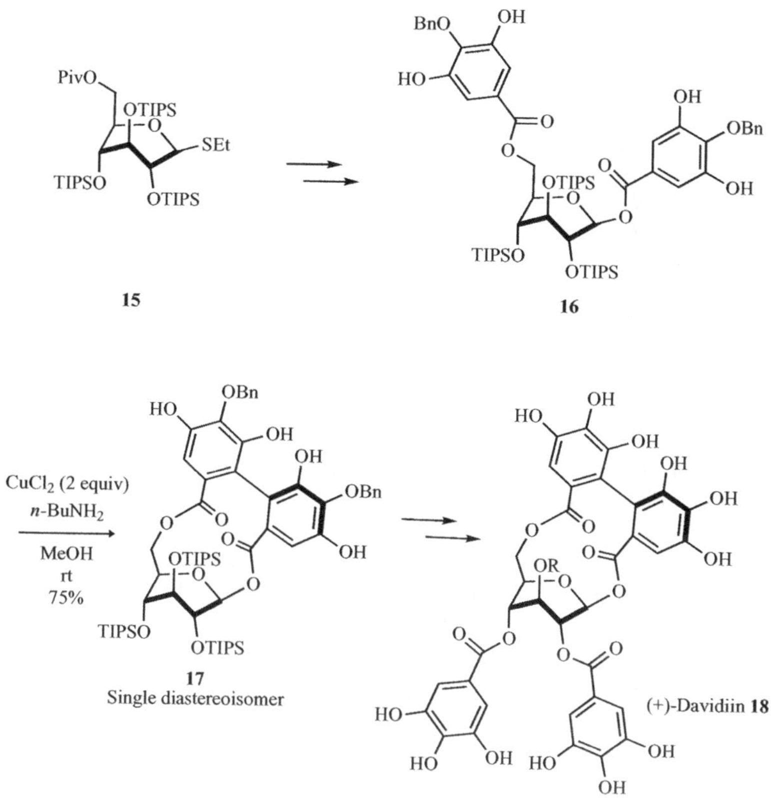

Scheme 18.4 Yamada's synthesis of (+)-davidiin.

bislactone **17** was obtained using copper(II) chloride (2 equiv) and butylamine in methanol in 75% yield as a single diastereoisomer, indicating that the sp^3 central chirality of the pyranose was totally transferred to the biaryl axial chirality (*S* according to circular dichroism prediction). This efficient key transformation allowed a concise total synthesis of (+)-davidiin **18** in nine steps and 19% overall yield. This approach offers an interesting alternative to the Ullmann coupling used for the synthesis of structurally similar compounds like corilagin and its derivatives (see Scheme 18.2).

In terms of asymmetric copper-catalyzed oxidative bi-naphthol couplings applied to total synthesis of natural products, the major input in the field is due to the Kozlowski group. After having computationally identified 1,5-diaza-*cis*-decalin as an optimal ligand to achieve catalytically these couplings, they have developed, rationalized, and applied in total synthesis the powerful catalyst **21** (Scheme 18.5).[12] During their investigations, substitution with an electron-withdrawing group at the C3 position appeared to be crucial in order to obtain efficient coupling as well as excellent enantioselectivity as copper seems to form a chelate with C2- and C3-substituents. Even so, many

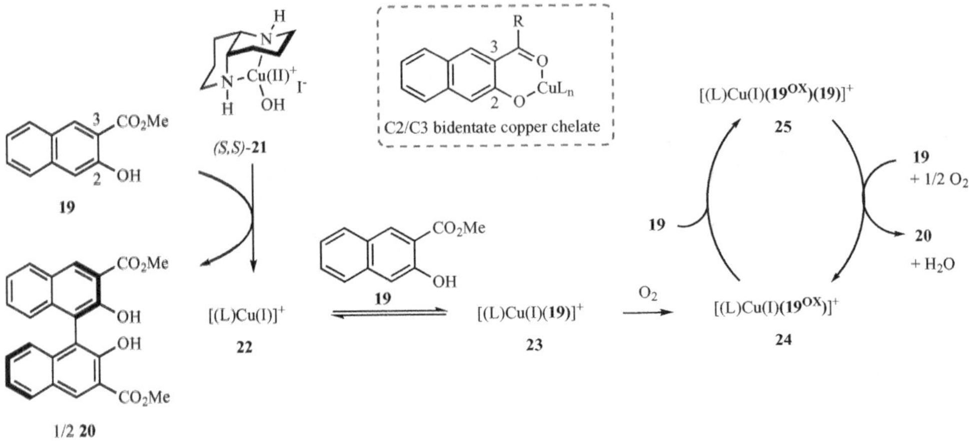

Scheme 18.5 Mechanism for asymmetric copper-catalyzed biaryl coupling developed by Kozlowski and coworkers.

functional groups are permitted such as ester, ketone, amide, phosphonyl, or sulfonyl.[13] In addition, kinetic studies have demonstrated that the mechanism is more complex than just a simple two-step process based on coupling of partners followed by reoxidation of the catalyst (Scheme 18.5).[14] Indeed, if a first step is the coupling of two naphthol moieties **19** to form biaryl **20**, in a second step, the generated reduced copper complex **22** reacts again with naphthol **19** in order to produce species **23** that is subsequently oxidized by oxygen to an activated catalyst **24** that is then performing the catalytic cycle until full consumption of the substrate.

Kozlowski and coworkers have then successfully applied this methodology as a key step to the total synthesis of challenging targets like (*R*)- and (*S*)-nigerone[15] (Scheme 18.6), a bis-naphthopyrone derivative, or natural perylenequinones[16] (e.g., (+)-calphostin D[17] [Scheme 18.6]), a family of potent kinase C inhibitors. These studies on total synthesis of natural products significantly participated in determining the scope and limits of these oxidative couplings, especially in terms of substituents required on substrates. Furthermore, if atroposelectivities were not always complete, an additional trituration step usually furnishes the desired compound as a single isomer. More specifically, during nigerone **28** synthesis effort (Scheme 18.6), a rigorous study on coupling partners demonstrated that flavasperone **26** possesses an ideal strained geometry that is required in order to perform an efficient atroposelective oxidative coupling. In addition, total synthesis of both isomers allowed Kozlowski's group to demonstrate, thanks to CD comparison, that nigerone **28** naturally occurring atropoisomer is *M* and not *P* as previously reported. In the case of perylenequinones, the key oxidative coupling step allowed the authors to synthesize the whole family starting from the common biaryl intermediate **30** selectively obtained as *M* or *P* enantiomer according to the configuration of the copper(II) catalyst.

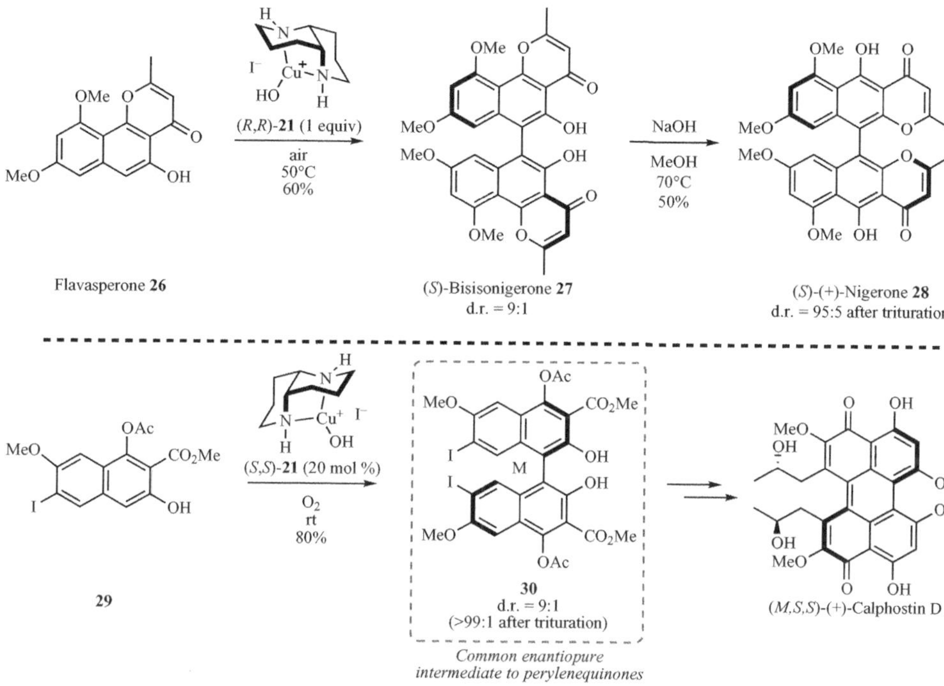

Scheme 18.6 Kozlowski's (*S*)-(+)-nigerone and (*M*,*S*,*S*)-(+)-calphostin D syntheses. Examples of asymmetric copper-catalyzed couplings of naphthols.

More recently, interests have been focused on the first asymmetric synthesis of bisanthraquinones. In 2012, Podlesny and Kozlowski reported the first enantioselective total synthesis of (*S*)-bisoranjidiol[18] (Scheme 18.7), a compound that presents remarkable photosensitizing properties with potential antimicrobial and anticancer activities. The strategy was based on the formation of the bisnaphthol intermediate **33**, a compound that is much simpler to obtain via an oxidative coupling than bisanthraquinone as anthraquinones are particularly resistant to oxidation. Subsequently, an oxidation of the two naphthols in **33** to 1,4-naphtoquinones **34** followed by a Diels–Alder/aromatization sequence transformed efficiently compounds **33** into bisanthraquinone **36**. (*S*)-Bisoranjidiol **37** has been finally synthesized in 12 steps and 4% overall yield.

18.3 COPPER-CATALYZED 1,3-ENYNE FORMATION

If a significant number of copper-mediated ene–yne coupling methods have been developed over the last decade, examples of natural product total syntheses involving these reactions as the key step are more than rare. This has mainly to do with rivalry with well-established organometallic couplings such

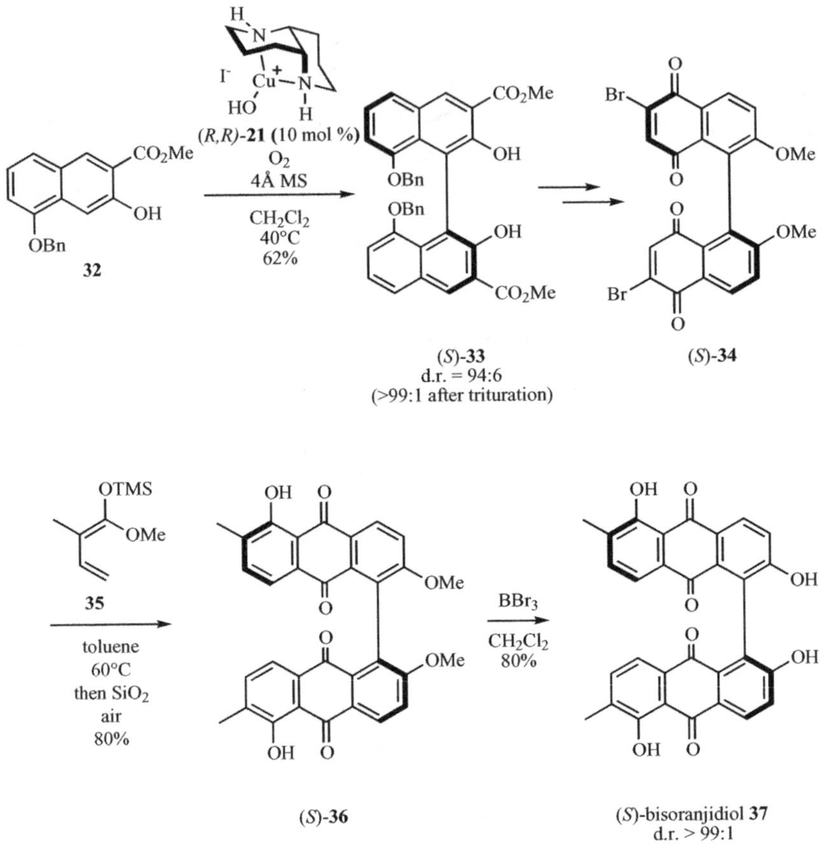

Scheme 18.7 Kozlowski's asymmetric total synthesis of natural bisanthraquinone (*S*)-bisoranjidiol.

as palladium- or rhodium-catalyzed reactions despite the efficiency of the related copper-mediated processes.[19]

One of these scarce examples is the total synthesis reported in 2006 by Ma and coworkers of one of the unnatural isomers of FR252921 (Scheme 18.8), a powerful immunosuppressive agent isolated from *Pseudomonas fluorescens* (n° 408813).[20] An interesting sequence involved a copper(I)-catalyzed coupling step[21] between alkyne **38** and allyl β-iodoacrylate **39** in order to obtain enyne **40** in 86% yield. The latter was subsequently isomerized by the action of triphenylphosphine and phenol at 50°C, leading to triene **41** in 95% yield. Finally, FR252921 **42** was nicely obtained in seven more steps.

Another nice example is the oximidine II total synthesis reported in 2011 by Georg and coworkers[22] in which the ene–yne coupling step is a nicely improved reductive Castro–Stephens cyclization (Scheme 18.9) that had been first investigated on a model substrate by Coleman in 2001.[23] At this time, the

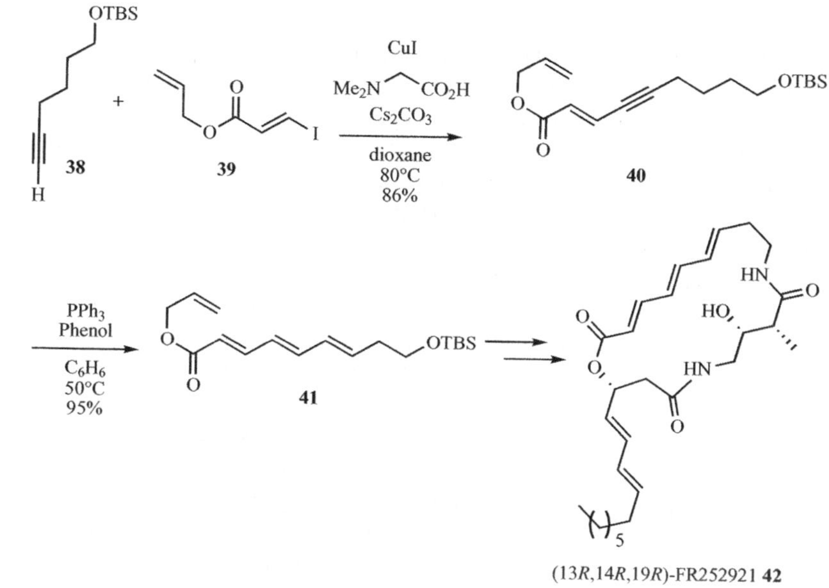

Scheme 18.8 Ma's synthesis of FR252921.

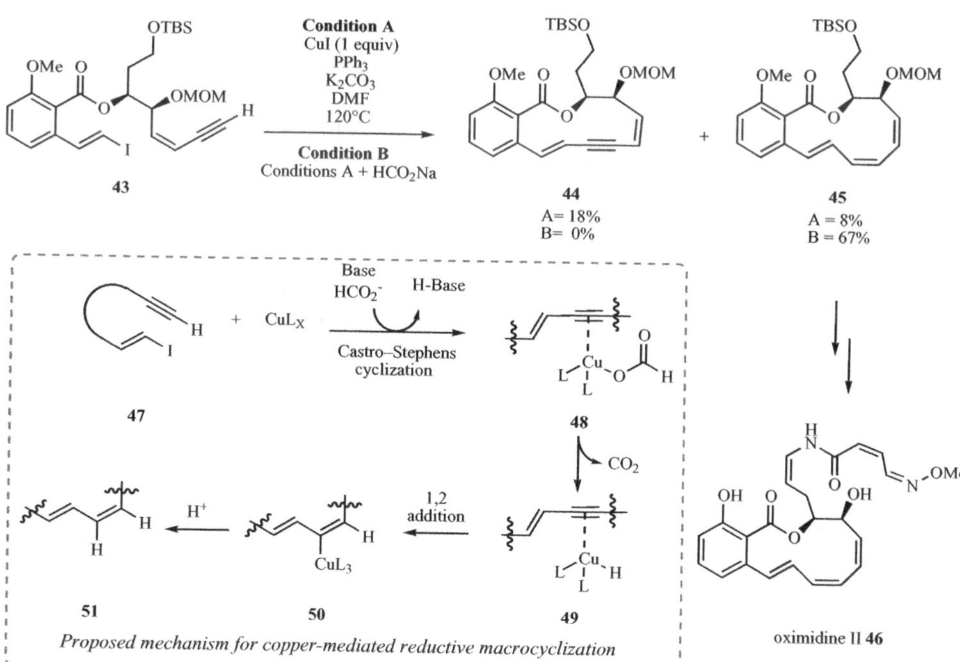

Scheme 18.9 Georg's synthesis of oximidine II via a copper(I)-mediated reductive Castro–Stephens macrocyclization.

desired ene–yne macrolactone had been only obtained in modest 35% yield on a model system (not shown), but this copper-mediated approach appeared to be much more efficient than standard macrolactonization strategies. The Georg group increased the yield of this cyclization by using sodium formate as an additive that directly induced a reduction of the C–C triple bond in order to obtain the required trienic macrocycle **45** in 67% yield. Mechanism of this new transformation has been elucidated thanks to studies essentially based on deuterium incorporation. First, a Castro–Stephens reaction of **47** forms the strained diene–yne macrocycle **48** with the copper remaining coordinated to the alkyne. Then, expulsion of carbon dioxide from formate induces the generation of a copper(I) hydride species **49**, which then undergoes a 1,2-addition on the C–C triple bond to give the vinyl-copper derivative **50**. Finally, a last protonation step yields the targeted triene **51**. Using this elegant strategy, oximidine II **46** has been nicely synthesized in 14 steps, including the original reductive copper-mediated ene–yne macrocyclization.

If ene–yne copper-mediated couplings in total synthesis are presently like gold dust, recent elegant syntheses such as Georg's preparation of oximidine II should change these transformations into common and reliable tools in natural product syntheses.

The synthesis of biaryl compounds was already an important field of research for organic chemists a century ago and it is fascinating to observe that more than a hundred years after the discovery of Ullmann, this area of investigation is still very exciting for the scientific community. More interestingly, the renaissance of copper-catalyzed or copper-mediated transformations drove the development of new atroposelective biaryl coupling reactions that allowed synthetic chemists to design elegant and efficient total syntheses of challenging natural products. We can assume that in a century, copper, probably in another way, will still be used as a key tool in the synthesis of biaryl-containing natural compounds.

The next section is dedicated to copper-catalyzed stereoselective syntheses of central motifs of natural products: 1,3-dienes, 1,3,5-trienes, and extended polyenes with selected relevant examples. The synthetic strategies featuring a copper-catalyzed cross-coupling reaction often show complimentary reactivity to the palladium-based transformations. Whenever possible, the parallel between copper- and palladium-based cross couplings will be discussed.

18.4 COPPER-MEDIATED SYNTHESIS OF DIENES, TRIENES, AND EXTENDED POLYENES

18.4.1 Copper-Mediated Synthesis of 1,3-Dienes

Marine macrolides are a unique source of biologically active natural products that have triggered intense synthetic efforts for the past decades. The recently discovered actin-binding (−)-reidispongiolide A **55** from the reidispongiolide

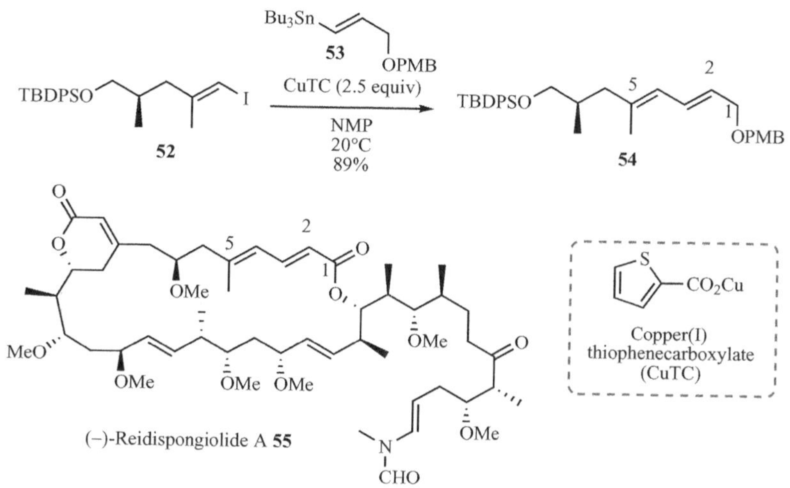

Scheme 18.10 Paterson's total synthesis of (−)-reidispongiolide A.

family has been targeted by the Paterson group with the aim of establishing its relative and absolute configurations as well as designing a suitable modular synthetic approach to simplified analogues (Scheme 18.10).[24] A key disconnection was a Stille cross coupling between (*E*)-vinyliodide **52** and (*E*)-vinylstannane **53** in order to establish the desired (2*E*,4*E*)-dienoate corresponding to the C2-C5 region of (−)-reidispongiolide A. While palladium-catalyzed cross-coupling reactions proved unsatisfactory, copper(I) thiophenecarboxylate (Liebeskind promoter)[25] mediated coupling led to 89% of the targeted 1,3-dienic system **54** at room temperature. The classical (undesired) homocoupling of the vinylstannane partner **53** was minimized thanks to its slow addition to the reaction mixture.

Monorhizopodin **59** and 16-*epi*-monorhizopodin are monomeric units of rhizopodin, a polyketide natural product isolated from the myxobacterium *Myxococcus stipitatus*. To get insights into the contribution of the macrocycle to the binding affinity to actin, Nicolaou devised an elegant approach to monorhizopodin featuring a high-yielding copper(I) thiophenecarboxylate-mediated Stille coupling to forge the C6-C9 dienic system of **59** (Scheme 18.11).[26]

In the presence of 10 equiv of copper(I) thiophenecarboxylate in *N*-methyl-2-pyrrolidinone (NMP) at 0°C, the cross coupling of vinylstannane **56** and vinyliodide **57** delivered the desired 1,3-diene **58** in 75% yield. The functional group tolerance is particularly noteworthy since an unprotected alcohol, two ethers (methyl- and silyl-), as well as a methyl ester are present. Further elaboration into monorhizopodin and 16-*epi*-monorhizopodin was then pursued and preliminary biological studies were thus possible.

The same disconnection was used in Chakraborty 2012 synthesis of the monomeric unit of rhizopodin on more advanced synthetic intermediates **60** and **61** (Scheme 18.12).[27] Standard Stille couplings with a diversity

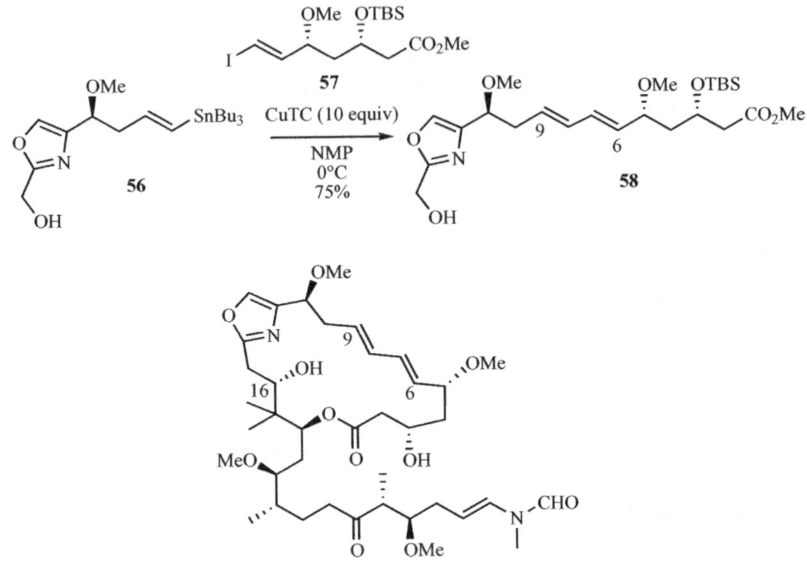

Scheme 18.11 Nicolaou's total synthesis of monorhizopodin.

of palladium(0) or palladium(II) precatalysts led to low yields of the desired 1,3-triene **62** (30%). Extensive survey of palladium catalysts, copper(I) sources and solvents led to the identification of $Pd(PPh_3)_4$ (5 mol %) and copper(I) thiophenecarboxylate (1.2 equiv) as an efficient catalytic system in tetrahydrofuran-dimethylformamide (THF-DMF) at room temperature. An excellent yield of **62** was obtained in only 30 minutes. Copper(I) thiophenecarboxylate was used this time as a cocatalyst of a palladium-catalyzed Stille coupling, a strategy that will be developed in detail in Section 18.4.4.

Copper-mediated elaboration of (*E*,*E*)-1,3-dienic units was also investigated by Brückner while designing an efficient route to γ-lactones **66** embedding a quaternary stereocenter (Scheme 18.13).[28] A sterically congested vinylstannane **64** underwent a smooth cross coupling with *trans*-1-iodobut-2-ene **65** in the presence of a slight excess of copper(I) thiophenecarboxylate. The stereochemical integrity of the vinyl iodide was preserved with Liebeskind's promoter, contrary to standard palladium-catalyzed Stille methods.

The plecomacrolide family encompasses a diversity of biologically active natural products such as formamicin **70** or the concanamycins. A unified strategy for the elaboration of the dienic units of these macrolides has emerged relying on the copper(I)-mediated Stille cross-coupling reaction. In the first total synthesis of the vacuolar H^+-ATPase inhibitor formamicin **70**, Roush reported a straightforward union of a sterically hindered trisubstituted vinylstannane **67** with (*E*)-β-iodoacrylate **68** in the presence of copper(I)

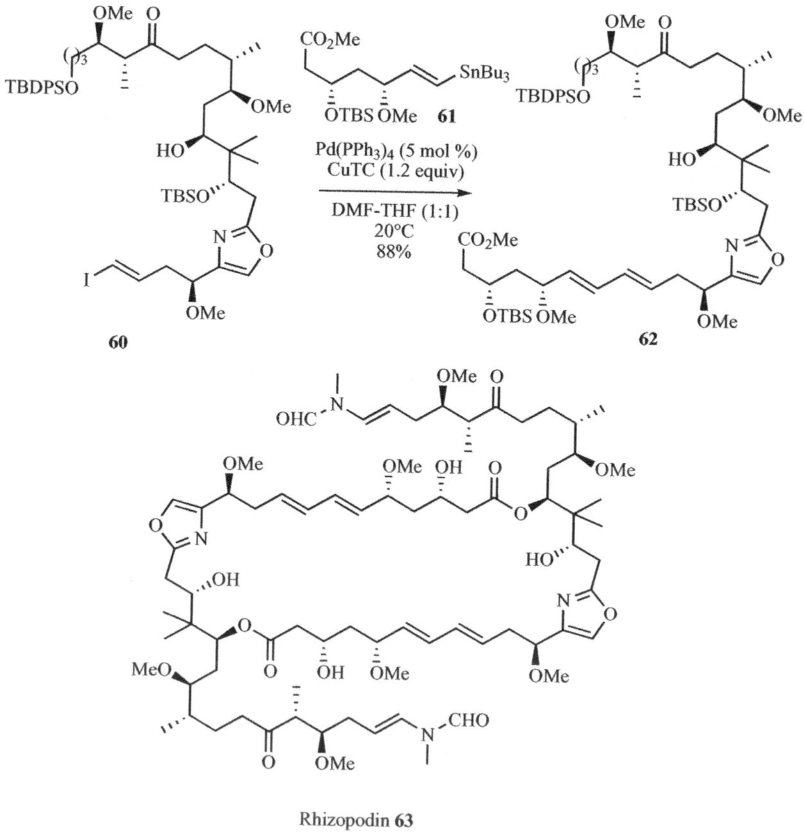

Scheme 18.12 Chakraborty's synthesis of the monomeric unit of rhizopodin.

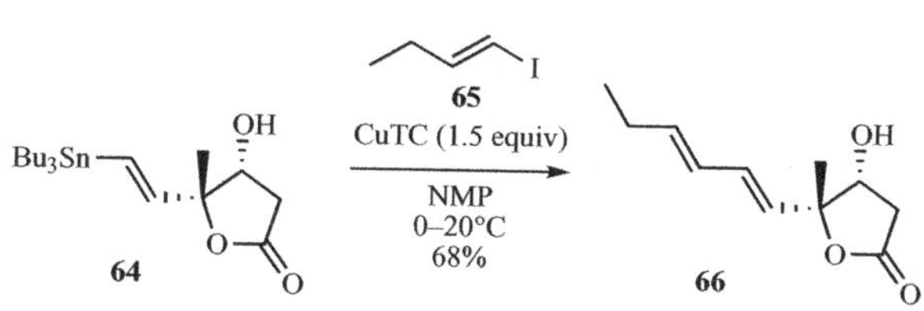

Scheme 18.13 Brückner's synthesis of γ-lactones bearing a quaternary stereocenter.

thiophenecarboxylate (Scheme 18.14).[29] A synthetically useful yield of the formamicin C1-C11 fragment **69** was obtained in the presence of a stoichiometric quantity of the tin scavenger tetrabutylammonium diphenylphosphinate.[25] As noted previously, the functional group tolerance is exceptional since the sensitive trisubstituted vinylsilane and the seven-membered ketal remained untouched.

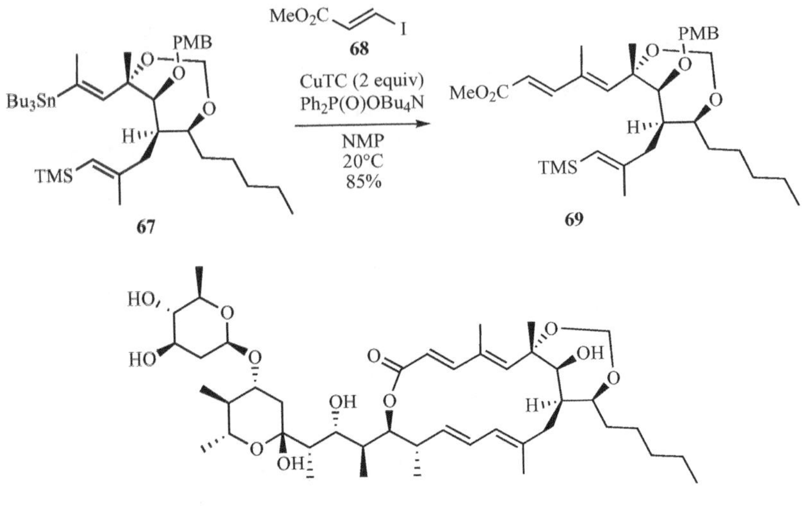

Scheme 18.14 Roush's total synthesis of the plecomacrolide formamicin.

Another member of the plecomacrolide family, (+)-concanamycin F **76** has been studied by Paterson since 2000.[30] The synthetic blueprint featured a copper(I)-mediated Stille coupling for the elaboration of the C12-C15 dienic unit (Scheme 18.15). Intra- and intermolecular strategies were evaluated, highlighting the crucial role of protecting groups in the challenging formation of the macrocyclic ring.

Actually, intramolecular copper(I)-mediated Stille coupling has often been a dead end in the total synthesis of complex natural products. An example is provided in Scheme 18.15, where attempts to build the C13-C14 σ-bond from precursors **71a,b** led to starting material and/or protodestannylation. On the other hand, it is worth noting that classical palladium-catalyzed Stille coupling was effective with precursor **71b** leading to 64% of macrolactone **72b**.

In sharp contrast, the same set of protecting group is perfectly compatible with an *intermolecular* version of this copper(I) thiophenecarboxylate-mediated coupling. A threefold excess of vinyl iodide **73a,b** was required in this reaction, the unreacted material being recovered in the end. Palladium-catalyzed cross coupling of **73a** was much less efficient.

The stereochemical integrity of the coupling partners being conserved in copper(I) thiophenecarboxylate-mediated coupling reactions, numerous applications to the synthesis of (*E*,*Z*)- or (*Z*,*Z*)-1,3-dienes were reported in the context of total synthesis. Paterson reported notable examples during the elegant total synthesis of a series of double hybrids (e.g., **80**) of discodermolide **81** and dictyostatin **82**, two natural products that possess similar microtubule-stabilizing anticancer activities (Scheme 18.16).[31] In the presence of an excess

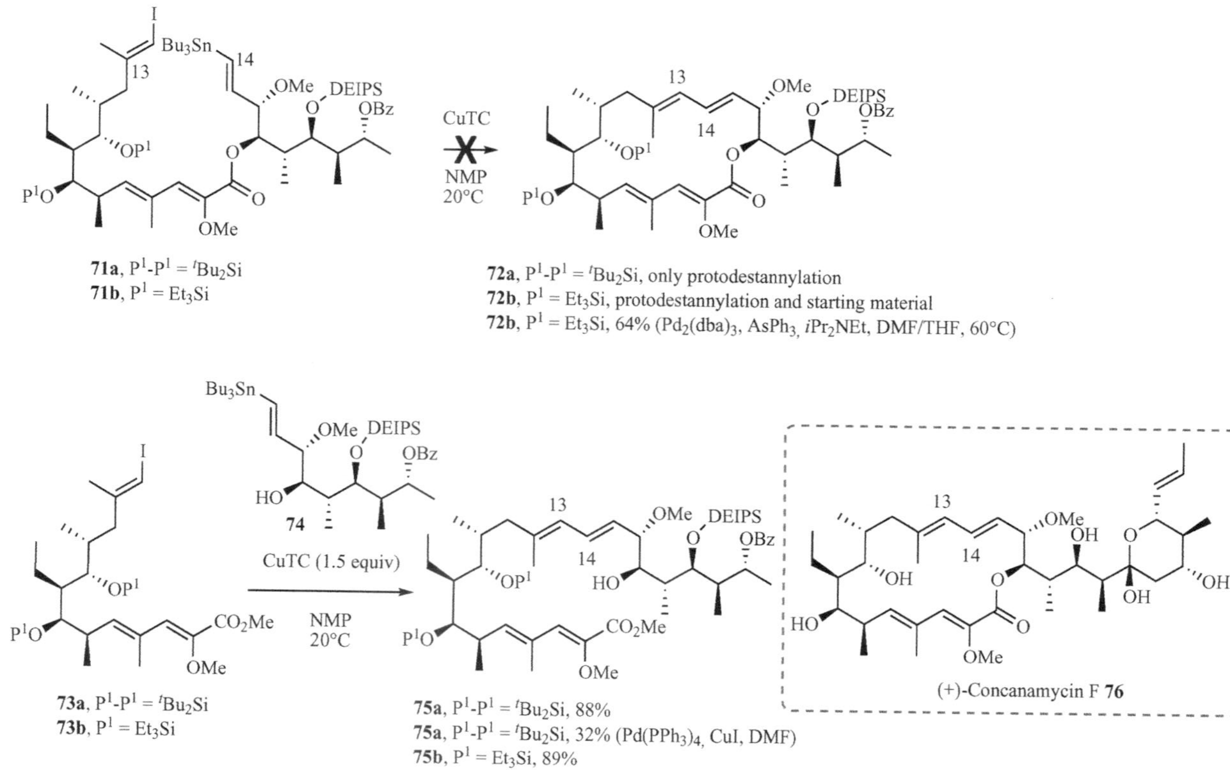

Scheme 18.15 Paterson's total synthesis of the plecomacrolide (+)-concanamycin F.

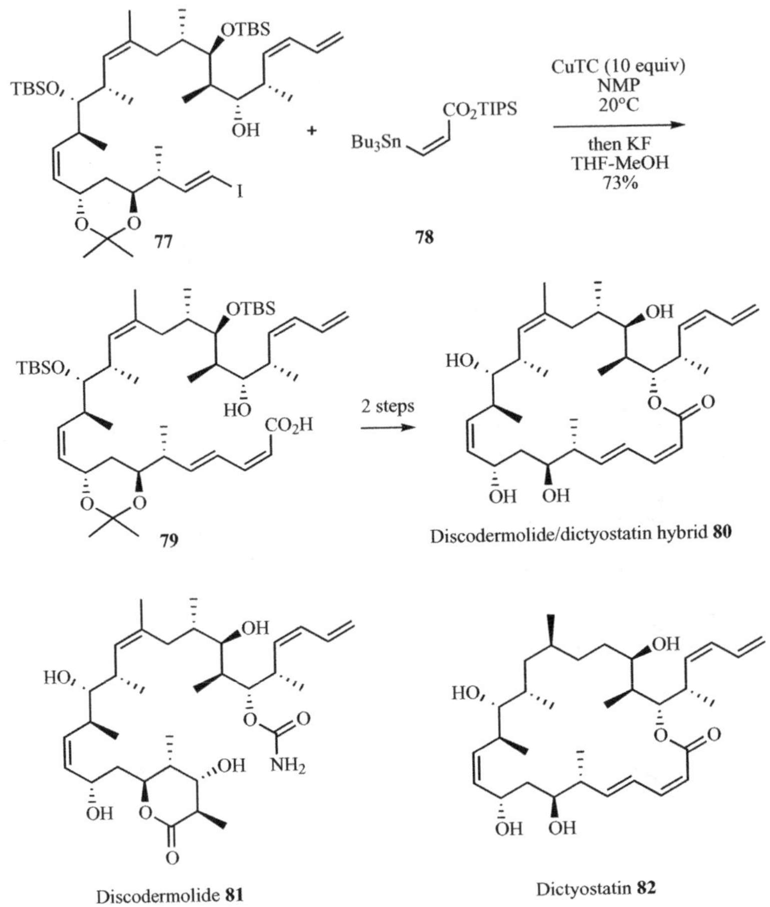

Scheme 18.16 Paterson's synthesis of discodermolide/dictyostatin hybrids.

of the more accessible vinylstannane **78** (4 equiv) and copper(I) thiophenecarboxylate (10 equiv), the intermolecular cross coupling with **77** delivered the desired (*Z*,*E*)-1,3-diene **79** in 73% yield for two steps with complete preservation of the stereochemistry. Silylethers, 1,3-acetonide, and an unprotected alcohol were tolerated.

Such an excess of copper(I) thiophenecarboxylate is not a prerequisite for successful elaboration of polyenic motifs. In the very elegant total synthesis of neurymenolide A **86**, an unusual polyunsaturated α-pyrone isolated from the red alga *Neurymenia fraxinifolia*, Fürstner used only a slight excess of Liebeskind's promoter in NMP at 0°C for the synthesis of (*E*,*Z*)-1,3-diene intermediate **85** (Scheme 18.17).[32] After acetylation of the allylic alcohol, an excellent yield of the targeted diene was obtained.

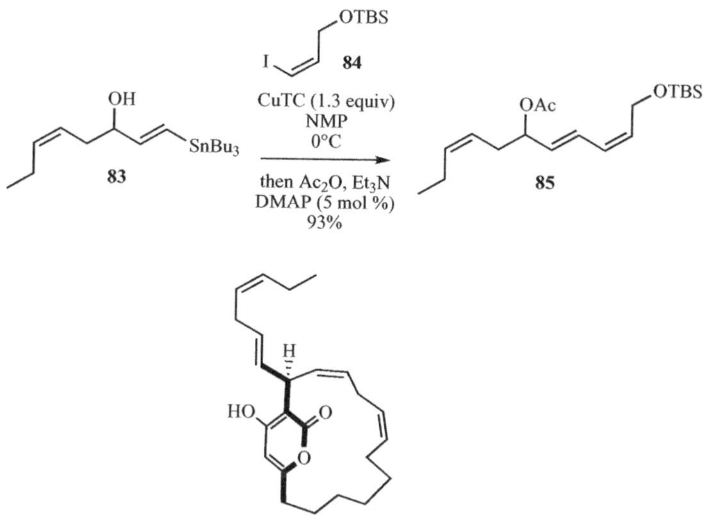

Neurymenolide A **86**

Scheme 18.17 Fürstner's total synthesis of neurymenolide A.

Thermodynamically less stable (*Z*,*Z*)-1,3-dienes are structural motifs found in several natural products such as phoslactomycin A**89**, isolated from *Streptomyces nigrescens*. The structural proximity of the phoslactomycins A–F with related biologically active compounds such as leustroducsin, fostriecin, cytostatin, and sultriecin that have already succumbed to total synthesis allowed a comparison of the different synthetic blueprints. This elegant and insightful analysis by Koert ended up in a highly convergent elaboration of phoslactomycin A**89** relying on a late-stage assembly of the C12-C15 1,3-dienic unit via a copper(I) thiophenecarboxylate-mediated Stille coupling of vinyliodide **87** and vinylstannane **88** (Scheme 18.18).[33]

The functional group compatibility of Liebeskind's promoter is especially noteworthy in this context. The C9-protected phosphate of **87** is absolutely stable under the reaction conditions and a single palladium-catalyzed deprotection step completed the synthesis of this challenging natural compound. On the opposite, classical palladium-catalyzed Stille coupling reported in the (+)-phoslactomycin B **93**[34] and leustroducsin B **94**[35] total synthesis could not handle a protected phosphate at C9, thus requiring a multiple step end-game process (Scheme 18.19).

18.4.2 Copper-Mediated Synthesis of 1,3,5-Trienes

Trienic motifs with stereodefined configurations are efficiently prepared via copper(I)-mediated Stille coupling as recently demonstrated by the groups of

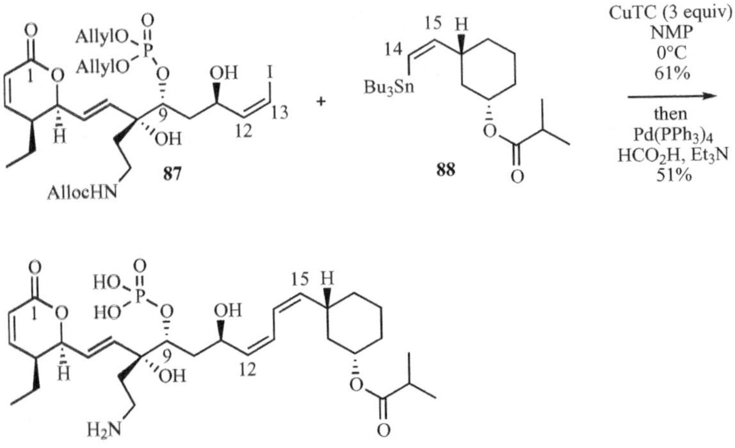

Phoslactomycin A **89**

Scheme 18.18 Koert's total synthesis of phoslactomycin A.

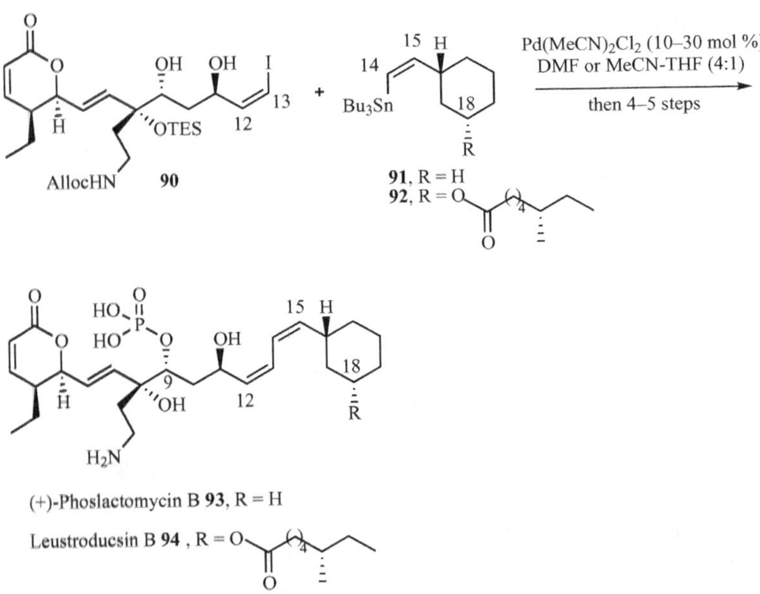

(+)-Phoslactomycin B **93**, R = H

Leustroducsin B **94**, R = O

Scheme 18.19 Palladium-catalyzed Stille cross couplings in the total syntheses of (+)-phoslactomycin B and leustroducsin B.

Nicolaou[36] and Scheidt[37] in their total syntheses of hirsutellone B **98** and (–)-exiguolide **101**, respectively.

Hirsutellone B **98** is a promising lead compound for antituberculosis drugs isolated from the fungus *Hirsutella nivea* BCC 2594, whose tricyclic core could be assembled through an epoxide opening/[4+2] cycloaddition tandem

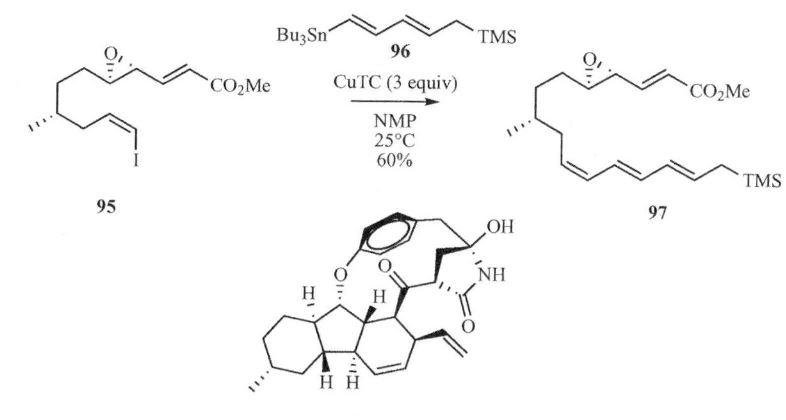

Hirsutellone B **98**

Scheme 18.20 Nicolaou's total synthesis of hirsutellone B.

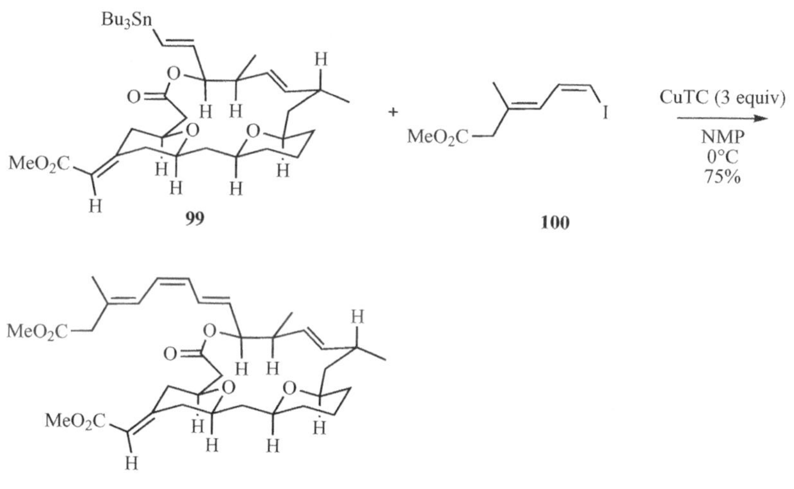

(−)-Exiguolide **101**

Scheme 18.21 Scheidt's total synthesis of (−)-exiguolide.

sequence of an appropriate epoxytriene. The cyclization precursor was efficiently prepared on a 2 mmol scale by a stereoselective Stille coupling of vinyliodide **95** (*Z*/*E* = 5:1) with dienylstannane **96** using an excess of copper(I) thiophenecarboxylate (3 equiv) in NMP at room temperature (Scheme 18.20). A *Z*/*E* isomer ratio of around 5:1 was observed for the triene **97**. Interestingly, the sensitive γ,δ-epoxyacrylate motif of **97** was left untouched.

Scheidt used a copper(I)-mediated Stille cross coupling as the ultimate step in his total synthesis of (−)-exiguolide **101**, a potential antitumor agent isolated from the marine sponge *Geodia exigua* Thiele (Scheme 18.21). In the presence of 3 equiv of copper(I) thiophenecarboxylate in NMP (40 mM) at 0°C, a good yield of the targeted natural product featuring the desired (*E*,*Z*,*E*)-triene was obtained via the coupling of vinylstannane **99** and dienyliodide **100**. On the other hand, palladium-catalyzed cross coupling led to complex mixtures. The stereochemical integrity of the triene was not compromised using Liebeskind's promoter, which proved essential for the observed growth inhibition of a number of cancer cell lines by (−)-exiguolide, notably A549 lung cancer cells (26 ± 6% at 1 μM).

18.4.3 Copper-Mediated Synthesis of Extended Polyenes

Extended polyenic systems could also be accessed by copper(I)-mediated cross-coupling reactions. Mycolactones A/B **106** are virulence factors secreted by *Mycobacterium ulcerans*, the causative microorganism of the necrotic skin disease Buruli ulcer. This unique family of macrolides has been the topic of intense synthetic efforts since 2002 and the groups of Kishi,[38] Altmann,[39] and Negishi[40] have reported total syntheses of mycolactones A/B.[41] In 2011, a function-oriented synthesis of mycolactone analogues has been developed by Blanchard and coworkers[42] allowing the first description of the structure–activity relationships of analogs of these seducing macrolidic polyketides. One of the key steps of the synthesis of the C1′-C16′ southern fragment **105** was a copper(I) thiophenecarboxylate-mediated cross coupling between iodotrienoate **104** and dienylstannane **103** (Scheme 18.22).

The simplicity of the experimental protocol combined with the good overall yield for the two-step process (**102** → **105**, 48%) compares favorably with the Negishi palladium-catalyzed union reported by Negishi (**107** → **110**, Scheme 18.23).[40] Although the stereochemical integrity of the sensitive pentaenic motif is completely preserved in the latter strategy (**110**, *E*,*E*, *E*,*E*,*E* > 98%), further equilibration to the naturally occurring and thermodynamically more stable Z-$\Delta^{4',5'}$/E-$\Delta^{4',5'}$ = 60:40 ratio could not be avoided during the final deprotection steps of Negishi's total synthesis of mycolactones A/B. It is worth noting that in Blanchard's example of pentaenic construction exclusively mediated by copper salts (Scheme 18.22), Liebeskind's promoter allowed a redox economy approach since the ester function at C1′ was tolerated.

18.4.4 Copper Salts as Essential Cocatalysts of Palladium-Catalyzed Stille Cross-Coupling Reactions

Copper(I) thiophenecarboxylate proved over the years to be an essential promoter in Stille cross-coupling reactions, allowing the efficient elaboration of dienes, trienes, tetraenes, and even pentaenes with conservation of the stereochemical integrity. However, palladium cocatalysis is sometimes mandatory, although it is difficult to anticipate it. The beneficial effect of copper(I) salts in classical palladium-catalyzed Stille coupling is well known and has

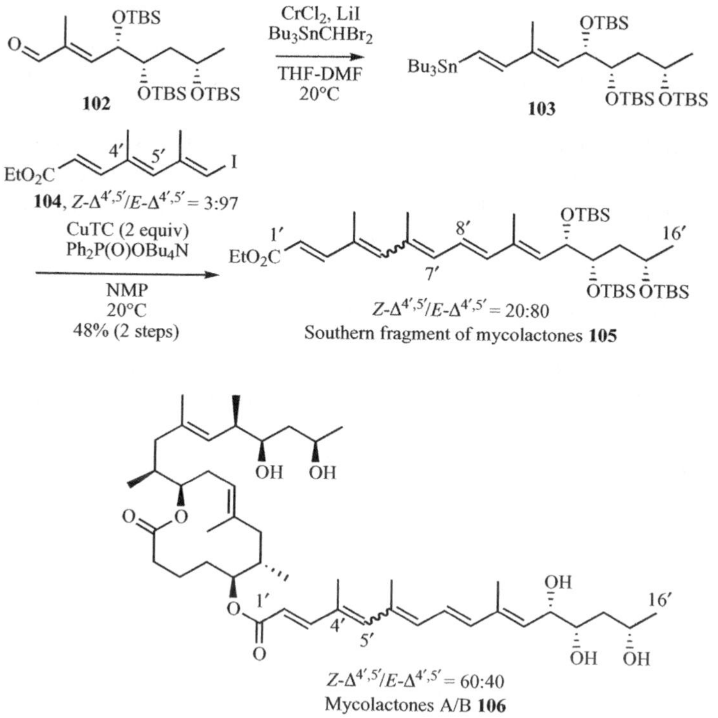

Scheme 18.22 Blanchard's synthesis of the southern fragment of mycolactones A/B.

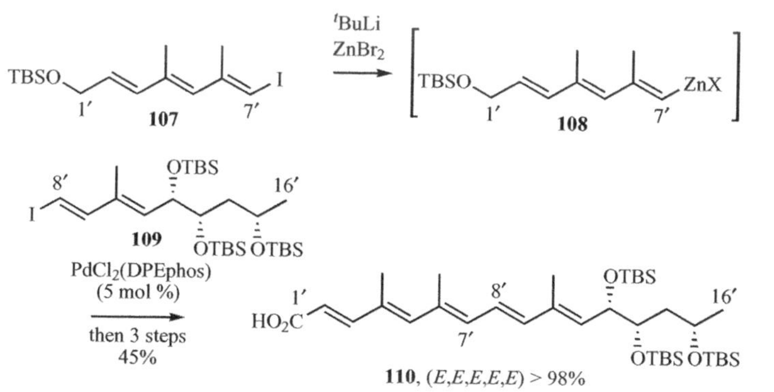

Scheme 18.23 Negishi's synthesis of the southern fragment of mycolactones A/B.

triggered intense mechanistic studies.[43] Two protocols using (i) copper(I) iodide/cesium fluoride or (ii) copper(I) thiophenecarboxylate/tetrabutylammonium diphenylphosphinate in conjunction with a palladium source have emerged as useful tools in the straightforward assembly of stereodefined polyenic motifs.[44,45]

In 2004, Baldwin showed that aryliodides, bromides and triflates were reactive coupling partners in the presence of vinyl- and aryltin derivatives provided that copper(I) iodide and cesium fluoride were used as additives to a palladium source.[44] The observed rate enhancement was attributed to (i) an enhanced nucleophilicity of an organocopper reagent arising from a preliminary transmetallation and (ii) an exaltation of the "copper effect" through the formation of insoluble tin fluoride by-products that were no longer able to engage in a (reversible) transmetallation event with a vinyl- or aryltin derivative (Scheme 18.24).

The so-called Baldwin protocol was rapidly adopted by the synthetic community. During studies directed toward 19-*nor*-clerodanes such as 5β-hydroxy-*cis*-dehydrocrotonin **115**, a group of biologically active diterpenes, Williams prepared (2*E*,4*Z*,6*E*)-trienoate **113** by Stille coupling of (*Z*,*E*)-dienylstannane **111** and (*E*)-vinyltriflate **112**. An excellent yield of the 6π-electrocyclization precursor **113** was obtained at 45°C in 1 hour and a half (Scheme 18.25).[45]

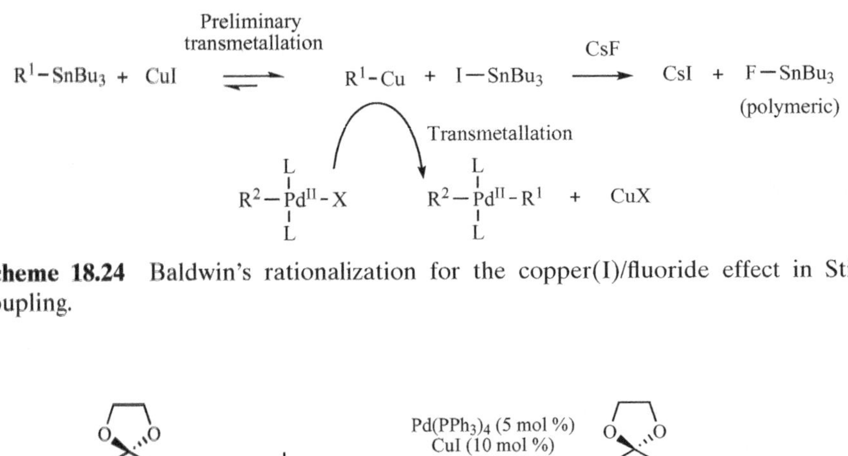

Scheme 18.24 Baldwin's rationalization for the copper(I)/fluoride effect in Stille coupling.

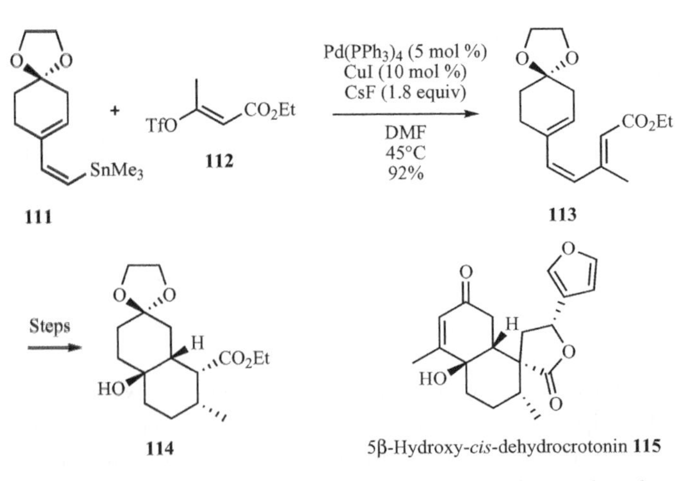

Scheme 18.25 Williams' synthetic approach of 19-*nor*-clerodanes.

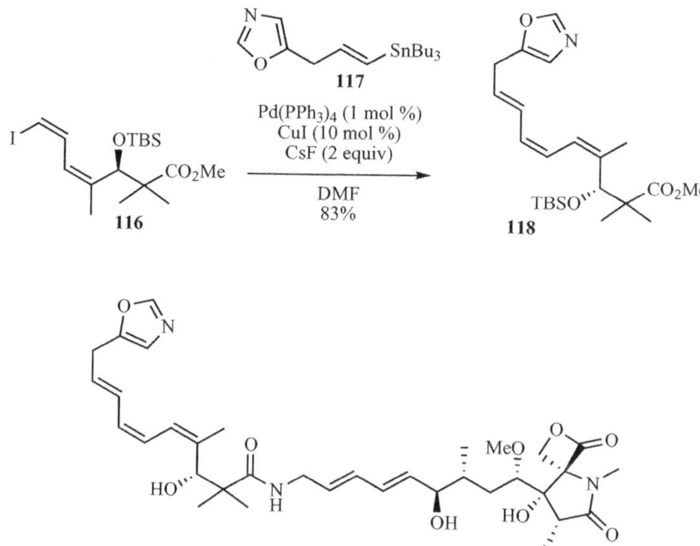

Scheme 18.26 Hatakeyama's total synthesis of oxazolomycin A.

In the first total synthesis of oxazolomycin A**119**, a structurally novel spiro-β-lactone-γ-lactam antibiotic, Hatakeyama reported a challenging Stille coupling using the Baldwin conditions between (*Z*,*E*)-dienyliodide **116** bearing an allylic silylether and vinylstannane **117** (Scheme 18.26). A good yield of the targeted (*E*,*Z*,*E*)-triene **118** was obtained at room temperature. The tolerance of the silylether in the presence of 2 equiv of cesium fluoride is worthy of note. The copper(I) effect is once again central to the success of this cross coupling since in its absence, isomerization of the trienic system of **118** was observed.[46]

Taylor, Hwang, and Ryu used a similar synthetic blueprint in their total syntheses of members of the inthomycin family. (−)-Inthomycin C, a (*E*,*E*,*E*)-trienic β-hydroxyamide isolated from *Streptomyces* sp., was prepared in 85% yield using a palladium(0) source, cesium fluoride, and copper(I) iodide in warm, degassed DMF for three hours and a half.[47] (+)-Inthomycin B featuring a (*E*,*E*,*Z*)-trienic system was prepared in quantitative yield using this strategy, without isomerization. However, the use of copper(I) thiophenecarboxylate as the only catalytic species led only to low yields of isomerized trienes.[48]

Stereodefined pentaenes are also efficiently synthesized using the Baldwin protocol of the Stille reaction. Wada demonstrated that stannyl esters **121** were reactive partners in cross-coupling reactions with vinyltriflates such as **120**, allowing the rapid elaboration of a library of 9-*cis*-retinoic acid analogues **122** (Scheme 18.27).[49]

Besides classical palladium precatalysts, Fürstner introduced in 2008 the most versatile protocol for this kind of cross coupling by combining $Pd(PPh_3)_4$ (5–70 mol %), copper(I) thiophenecarboxylate (1.2–3 equiv) and $[Ph_2PO_2]$

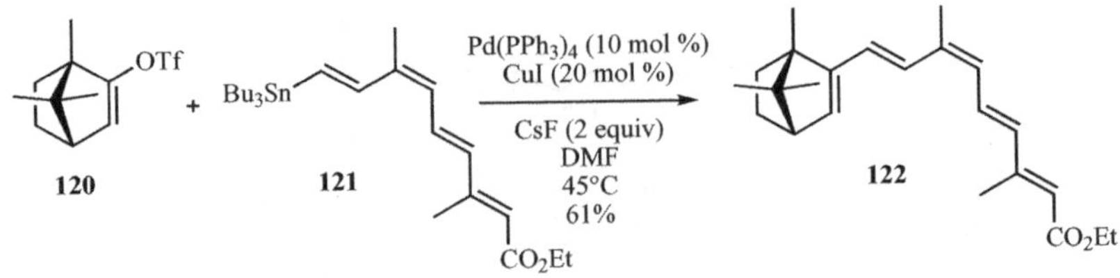

Scheme 18.27 Wada's synthesis of 9-*cis*-retinoic acid analogues.

$[NBu_4]$[25] (1.1–4 equiv) in DMF at room temperature. This report capitalizes on seminal observations by Liebeskind on the ability of a diphenylphosphinate salt to sequester tin halide by-products that accumulate in the reaction medium during the cross-coupling reaction and that are responsible for driving the pre-equilibrium to the left (see Scheme 18.24).[25] Additionally, the tin thiophenecarboxylate is less susceptible to engage in a transmetallation equilibrium.

Using this three-component catalytic system, challenging Stille cross coupling was accomplished with impressive yields and functional group compatibility.[50] The total synthesis campaign of amphidinolides B1, B4, G1, and H1 by the group of Fürstner is particularly relevant in this context since a sensitive s-*cis*-1,3-diene motif is embedded within these exceptionally potent anticancer agents from dinoflagellate strains of the genus *Amphidinium*.[51] Early studies on amphidinolide B **128** by Cid and Pattenden (Scheme 18.28)[52] showed that a late-stage copper(I) thiophenecarboxylate-mediated Stille macrocyclization was plagued by dimerization of the vinylstannane moiety of **123** or protodestannylation. Only the intermolecular approach met with success when copper(I) thiophenecarboxylate was used as the only cross-coupling mediator. A 60% yield of diene **127** was then obtained at room temperature. On the other hand, palladium-catalyzed conditions led to decomposition of **125**.

Based on this precedent, Fürstner was able to elaborate amphidinolides B1, B4, G1, and H1 through an intermolecular Stille coupling, as demonstrated in Scheme 18.29 for amphidinolide B4. The three-component catalytic system delivered advanced intermediate **131** in excellent yield, leaving the sensitive epoxyalcohol moiety untouched.

In the total synthesis and structure revision of (−)-brevenal **136**, Sasaki relied on a late-stage palladium-catalyzed copper(I) cocatalyzed Stille reaction to forge the highly substituted (*E*,*E*)-dienic moiety, starting from vinyliodide **133** and vinylstannane **134** (Scheme 18.30).[53] Interestingly, triphenyl arsine (80 mol %) was used as an additional external ligand leading to the desired cross-coupled product **135** in 60% yield.

Cocatalysis of classical Stille conditions was also used by Burke for the elaboration of a (*E*,*E*,*E*)-trienic building block in a stereocontrolled access to (−)-peridinin **140** (Scheme 18.31). Union of dienyliodide **137** and vinylstannane

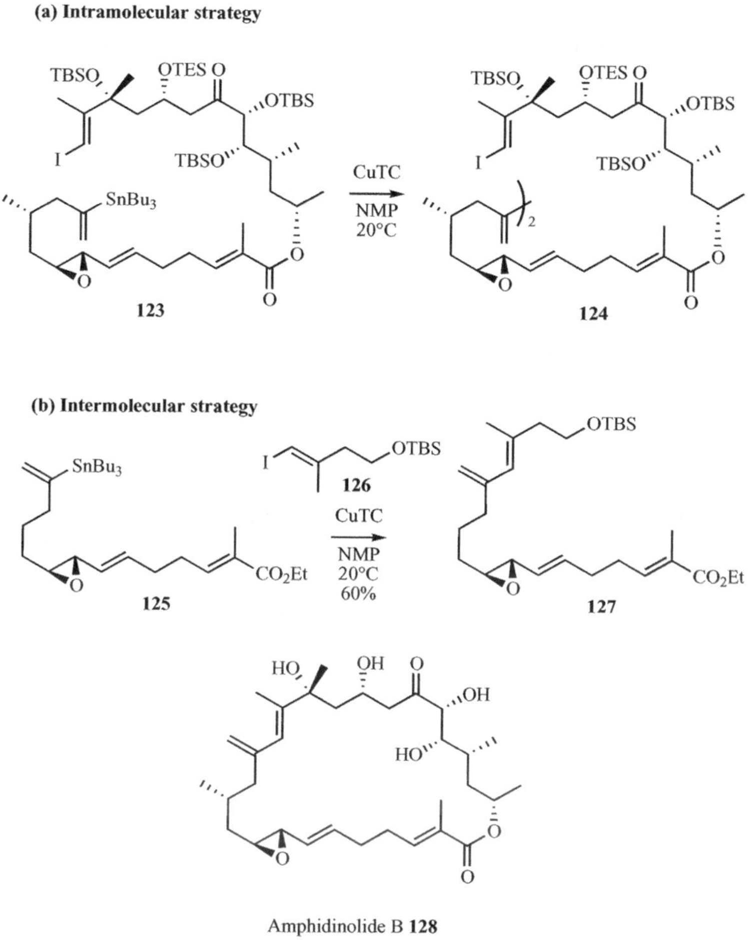

Scheme 18.28 Pattenden's synthetic approach to amphidinolide B featuring intra- and intermolecular Stille couplings.

138 led to the targeted triene **139** in 58% yield. The exceptional stability of the carbon–boron and carbon–germanium bonds is worthy of note.[54]

The efficient synthesis of the (*Z*,*E*,*Z*,*E*)-tetraenic motif of chivosazole F **144** and chivosazole A **145**, polyenic macrolides isolated from the myxobacterium *Sorangium cellulosum*, was reported by Paterson via a modified Stille coupling (Scheme 18.32).[55] Using Fürstner conditions, the (*Z*,*E*)-dienyliodide **141** cross-coupled with (*Z*,*E*)-dienyltin derivative **142** in 85% yield at 0°C leaving the (*Z*)-vinylbromide motif untouched. The rate-accelerating effect of the copper(I) cocatalyst is absolutely essential here. In its absence, complete isomerization of dienyliodide **141** occurs before the cross-coupling event, leading to the (*Z*,*E*,*E*,*E*)-tetraene in only 42% yield.

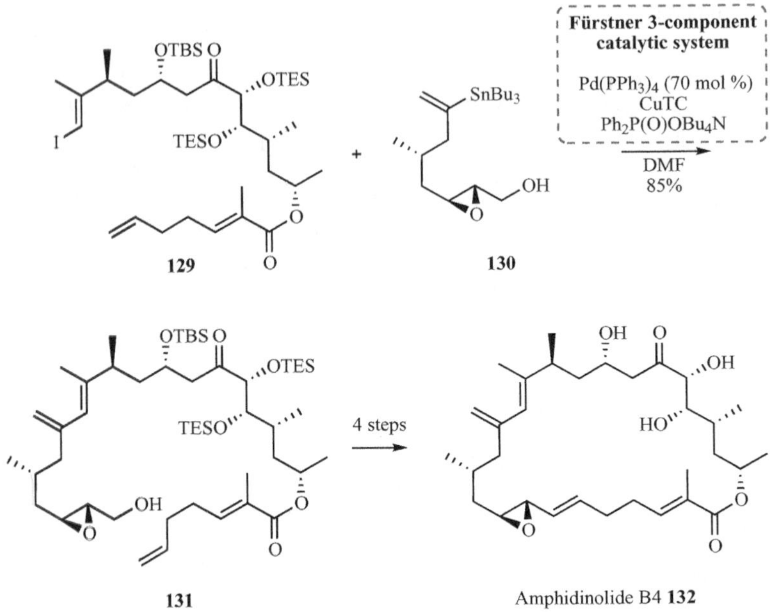

Scheme 18.29 Fürstner's total synthesis of amphidinolide B4.

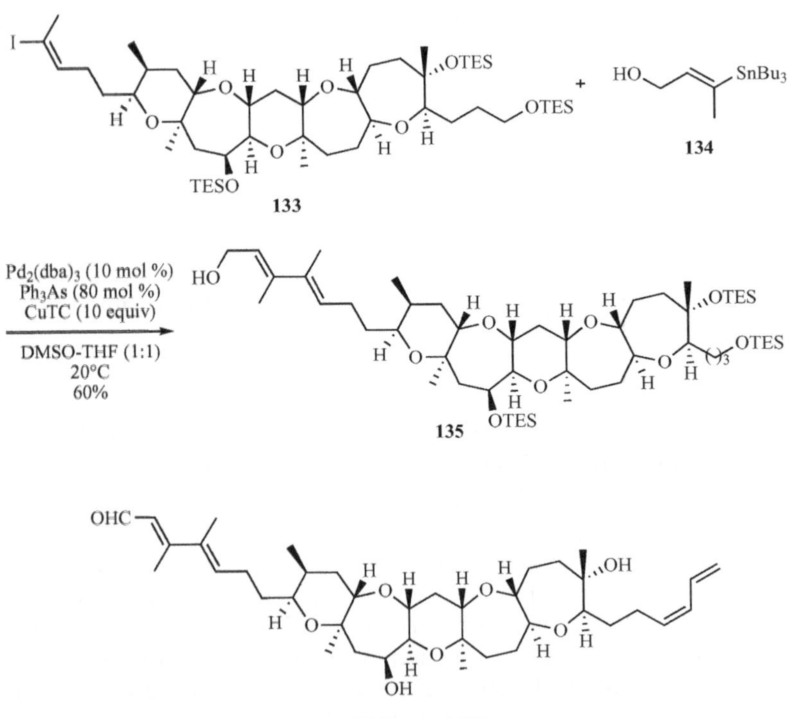

Scheme 18.30 Sasaki's total synthesis of (–)-brevenal.

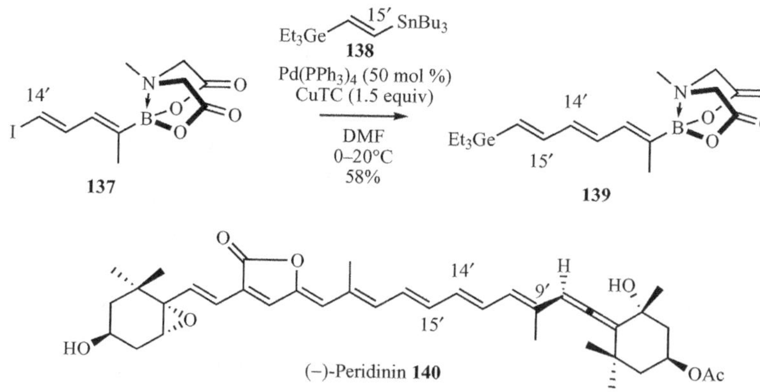

Scheme 18.31 Burke's total synthesis of (−)-peridinin.

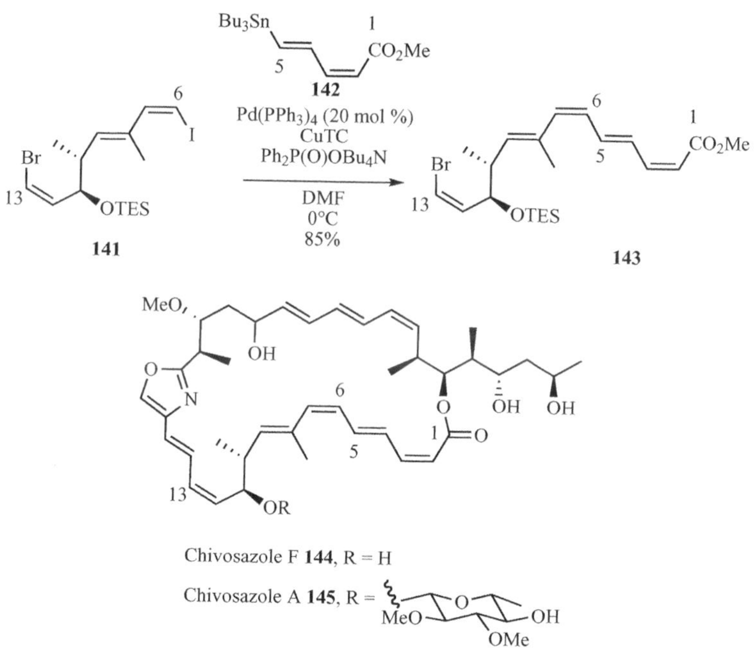

Scheme 18.32 Paterson's synthesis of the C1-C13 fragment of chivosazoles F and A.

18.5 COPPER-CATALYZED SYNTHESIS OF 1,N-POLYYNES NATURAL PRODUCTS

Polyynes natural products constitute a vast class of biologically active compounds found in plants, marine organisms, bacteria, and fungi. Since their first isolation at the beginning of the 19th century, elucidations of their structures as well as studies of their biological activities have parallel chemical synthesis efforts. Seminal contributions by Bohlmann and coworkers[56] and Jones and Thaller[57] highlighted the importance of these particularly seducing structural motifs.

From a strategic perspective, polyynes have been prepared mainly through the classical Cadiot–Chodkiewicz reaction,[58] a heterocoupling process between a 1-bromoalkyne and a terminal alkyne in the presence of a copper(I) catalyst and an amine. The Eglinton protocol is also a central methodology involving a copper(II) salt (usually acetate) in excess, pyridine to trap the acetic acid formed during the coupling reaction, and methanol to solubilize the cupric ion–pyridine complex.[59] A third class of oxidative acetylenic coupling reaction that has found widespread use in total synthesis is the Glaser–Hay reaction, which makes use of catalytic amount of copper(I) salt and *N,N,N′,N′*-tetramethylethylenediamine as a ligand.[60]

The aim of this subchapter is to give the reader an overview of the copper-catalyzed elaboration of polyyne natural products; a comprehensive coverage of the dozens of accomplished total syntheses is beyond the scope of this book. Excellent recent reviews were dedicated to this exciting field of research, notably the seminal reviews by Diederich and Tykwinski,[61] the latter being focused on naturally occurring polyynes.[61]

18.5.1 Cadiot–Chodkiewicz Strategies

In the field of naturally occurring polyynes, the Cadiot–Chodkiewicz heterocoupling reaction enjoys great popularity due to its mildness and functional group orthogonality. Among the numerous total syntheses relying on this key step, the straightforward elaboration of (–)-minquartynoic **150** acid by Gung is especially noteworthy (Scheme 18.33).[62] This potent anthelmintic tetrayne carboxylic acid has been isolated from the stem bark of *Minquartia guianensis*

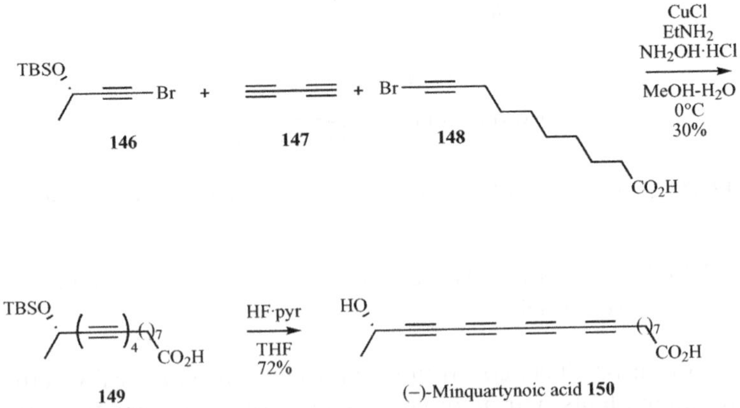

Scheme 18.33 Gung's total synthesis of (–)-minquartynoic acid.

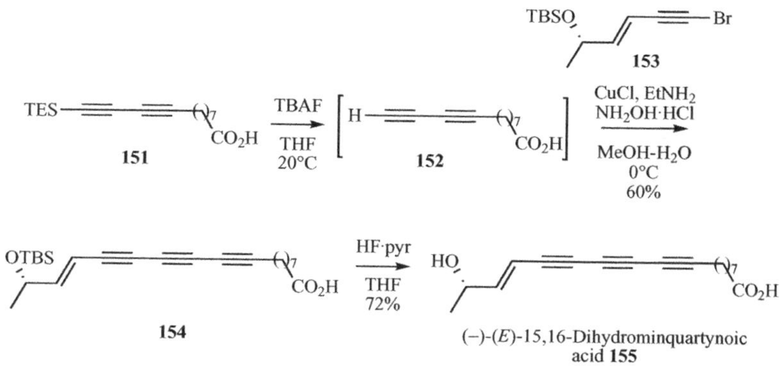

Scheme 18.34 Gung's total synthesis of (–)-(*E*)-15,16-dihydrominquartynoic acid.

and from the twigs of *Ochanostachys amentacea*. Using a one-pot three-component Cadiot–Chodkiewicz reaction, the desired tetrayne **149** was isolated in 30% yield. It was necessary to run the reaction at 0°C to prevent rapid decomposition of butadiyne **147** by copper(I) chloride. A single step was then required to access the targeted natural product.

A convergent strategy was also used by the same group in the total synthesis of (–)-(*E*)-15,16-dihydrominquartynoic acid **155** (Scheme 18.34).[63] As the terminal diyne motif of **152** required for the alkyne heterocoupling reaction was quite unstable, a one-pot desilylation–coupling procedure was developed. After removal of the triethylsilyl group of carboxylic acid **151**, copper(I) chloride (5 mol %), hydroxylamine hydrochloride (5 mol %, required to reduce copper(II) back to copper(I)), and a 1:1 mixture of methanol and diethylamine were added, followed by slow addition of the bromo-enyne partner **153** at 0°C. These optimized conditions led to the desired enetriyne **154** in 60% yield, the direct precursor of (*E*)-15,16-dihydrominquartynoic acid **155**.

The Alami modification of the Cadiot–Chodkiewicz coupling implies the use of alkynyl iodide instead of bromide and copper(I) iodide (10 mol %) in pyrrolidine or piperidine at room temperature.[64] Compared to classical Cadiot–Chodkiewicz conditions (copper(I) chloride, 5 mol %; hydroxylamine, 30 mol %; diethylamine in aqueous methanol), better yields are generally obtained, especially in the case of aliphatic alkynes that are usually less reactive in this type of coupling. López used Alami's modification of the Cadiot–Chodkiewicz reaction in her total synthesis of the callyberynes A–C, bioactive marine polyynes isolated from Japanese *Callyspongia* sp. (scheme 18.35).[65] The cross coupling between terminal enyne **156** and 1-iodoalkyne **157** in the presence of copper(I) iodide (10 mol %) in piperidine for 1 hour led to the direct precursor of callyberyne A **159** in 60% yield. By contrast, classical Cadiot–Chodkiewicz conditions were quite inefficient, leading to less than 25% of the desired polyyne.

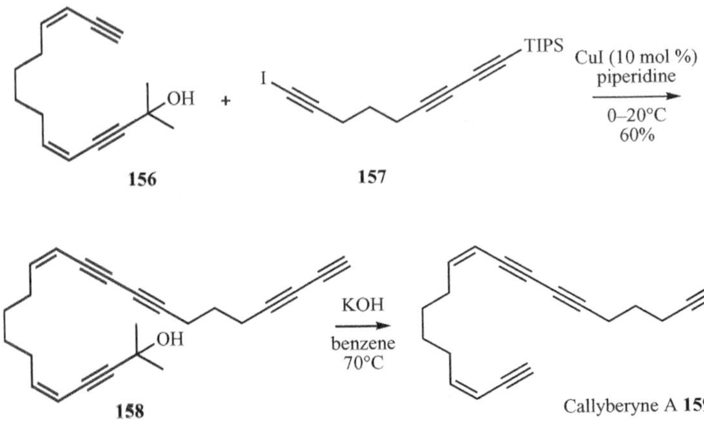

Scheme 18.35 Alami's modification of the Cadiot–Chodkiewicz in López total synthesis of callyberyne A.

18.5.2 Eglinton Strategies

The Eglinton acetylenic coupling has been a cornerstone of the elegant total synthesis of the endiandric acids A–D by Nicolaou.[1,66] Endiandric acids are racemic natural products that were isolated in the early 1980s from *Endiandra introrsa* (Lauraceae), an Australian plant. Their quite intricate architecture features two to four fused carbocyclic rings that were thought to originate from achiral polyenic precursors through consecutive 8π-electron electrocyclization, 6π-electron electrocyclization, and intramolecular Diels–Alder cycloaddition. In a "biomimetic" approach of endiandric acid A, Nicolaou synthesized cascade precursor **163** thanks to an Eglinton coupling between terminal enyne **160** and terminal alkyne **161** (in fivefold excess) (Scheme 18.36). In the presence of copper(II) acetate (2 equiv) in a pyridine/methanol 1:1 mixture at 25°C, good yield of the desired diyne **162** was obtained. This diyne was, in turn, transformed into endiandric acid A methyl esters **164** in a few steps including the key 8π-electron electrocyclization/6π-electron electrocyclization/intramolecular Diels–Alder cycloaddition sequence, thus demonstrating the validity of the 1980s hypothesis about their biosynthesis.

In 2007, Myers reported on the total synthesis of the proposed structure of kedarcidin chromophore **167**, a complex nine-membered ring enediyne possessing exceptional DNA cleaving, antibiotic and antitumor properties, that could undergo a Bergman cycloaromatization upon nucleophilic epoxide opening (Scheme 18.37).[67] Modified Eglinton conditions were key to the challenging intramolecular oxidative acetylenic coupling. In the presence of copper(II) acetate (61 equiv), copper(I) iodide (15 equiv) in pyridine/THF (1:2) at 60°C, formation of the targeted diyne **166** from the open chain precursor **165** was pleasingly observed in 59% yield in only 5 minutes. Extensive optimization of this intramolecular Glaser–Eglinton coupling demonstrated

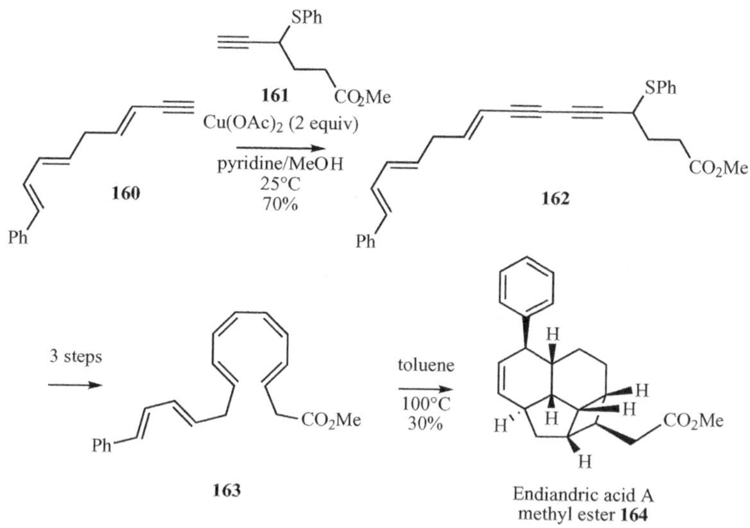

Scheme 18.36 Nicolaou's "biomimetic" total synthesis of endiandric acid A methyl ester.

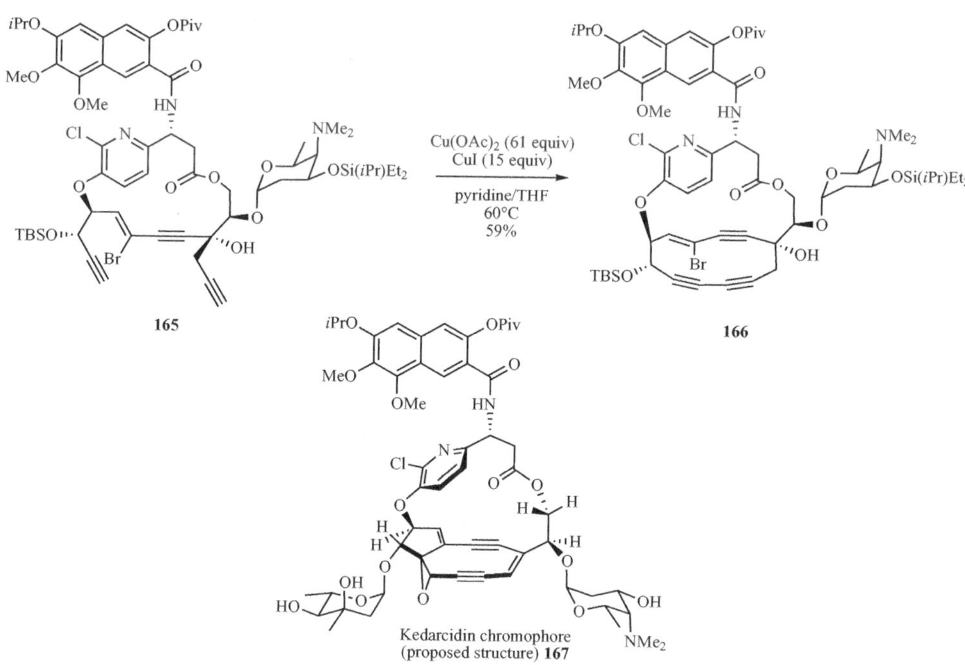

Scheme 18.37 Myers' synthesis of the proposed structure of kedarcidin chromophore.

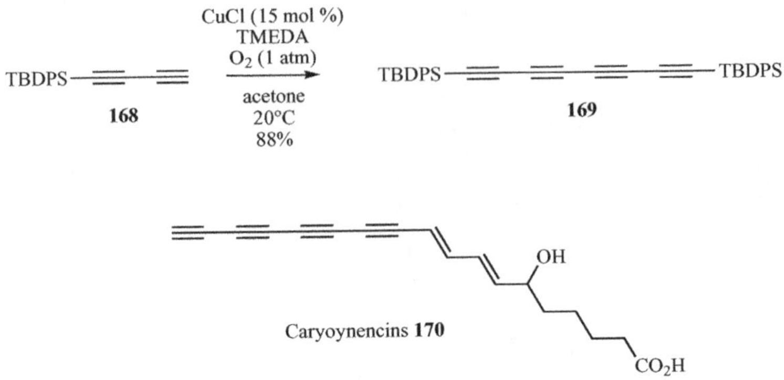

Scheme 18.38 Yamaguchi's total synthesis of the caryoynencins.

the importance of the additional copper(I) source that is not required in the historical 1959 protocol.[59] Furthermore, the order of mixing of the copper(I) and (II) salts proved crucial. The exceptional functional group tolerance is worthy of note, highlighting once again the efficiency of oxidative acetylenic couplings in the context of total synthesis.

18.5.3 Hay Strategies

Only a few examples of intermolecular Hay oxidative couplings were reported in the context of natural product total syntheses. In the total synthesis of caryoynencins **170**, polyynes antibiotics isolated from the plant pathogen *Pseudomonas caryophylli*, Yamaguchi relied on this key step to homocouple the silylated 1,3-diyne **168** (Scheme 18.38). In the presence of copper(I) chloride (15 mol %) and *N,N,N′,N′*-tetramethylethylenediamine in acetone under an atmosphere of oxygen, the desired tetrayne **169** was isolated in 88% yield. Further transformations led to caryoynencins **170**.[68]

The intramolecular Hay coupling has recently been applied to the derivation of natural products into three types of hybrids: symmetrical open, macrocyclic, and caged. This conceptually elegant approach reported by Ramírez-López and coworkers relies on Hay and Eglinton couplings.[69] A single example is provided further with the synthesis of a C_2-symmetric bis(alkynyl)-tethered homohybrid **172** derived from reserpine **171** (Scheme 18.39). This strategy should prove useful in the context of diversity-oriented synthesis based on the privileged scaffolds that are natural products.

In 2011, the intramolecular Hay oxidative coupling has witnessed an important conceptual breakthrough. Classically, the intramolecular oxidative acetylenic couplings required very high dilution to prevent undesired oligomerization reactions and slow addition of the reactants using a syringe pump was common practice. These serious limitations hampered the use of this useful synthetic disconnection on a large scale. Collins proposed an elegant alternative to this problem via a phase separation strategy allowing the macrocylization reaction

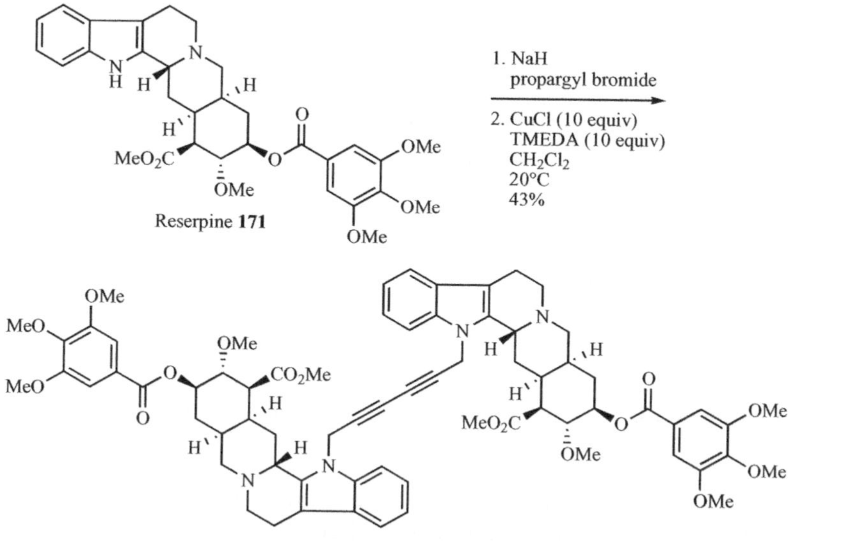

Reserpine-derived C$_2$-symmetric homohybrid **172**

Scheme 18.39 Glaser–Hay coupling in the synthesis of a reserpine-derived hybrid.

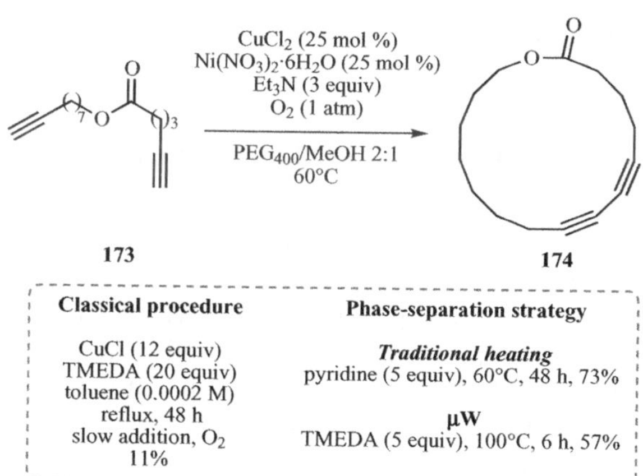

Scheme 18.40 Collins' phase-separation strategy in Glaser–Hay coupling.

to occur in a liquid/liquid microenvironment where the relative concentration of the cyclization precursor is low (Scheme 18.40).[70] A combination of PEG$_{400}$ and methanol (2:1, 0.03 M) proved very efficient to control the effective molarity during this oxidative alkyne coupling in the presence of copper(II) chloride (25 mol %), nickel(II) nitrate (25 mol %), triethylamine (3 equiv), and pyridine (5 equiv) at 60°C for 2 days under an atmosphere of oxygen. The use of

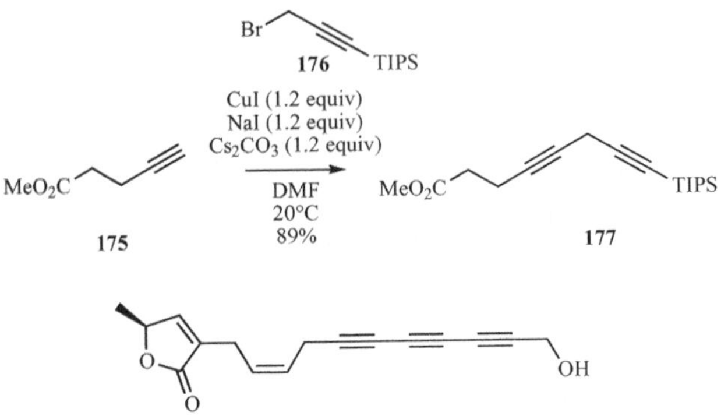

Trocheliophorolide D **178**

Scheme 18.41 Kim's synthetic approach of trocheliophorolide D.

a nickel(II) salt was central to the success of this coupling and was based on pioneering observations by Lei and coworkers[71]

Using this catalytic system, diyne **173** could be easily cyclized into the corresponding 16-membered macrolactone **174**. Compared to a classical monophasic reaction, an astonishing increase of the concentration could be obtained while the yield was greatly improved from 11% to 73%. It is worthy of note that the 2 days required for this modified Hay coupling could be reduced to only 6 hours by using microwave irradiation with a marginally lower yield.[72]

18.5.4 Copper-Catalyzed Synthesis of Skipped Dienes

Skipped (or 1,4-) diynes can be obtained via the classical Linstrumelle copper(I)-mediated displacement of propargyl halides or tosylates by terminal alkynes.[73] Among the numerous applications of this efficient transformation in the total synthesis of natural products, the work of Kim is representative (Scheme 18.41).[74] In this approach of the proposed structure of trocheliophorolide D **178**, a butenolide lipid isolated the soft coral *Sarcophyton trocheliophorum*, Kim used the Linstrumelle reaction for the elaboration of the skipped diyne **177** on a 26 mmol scale with an excellent yield of 89%, thus highlighting the reliability of this protocol for preparative transformations.

18.6 CONCLUSIONS AND FUTURE PROSPECTS

As illustrated in this chapter with selected recent examples, copper-mediated cross-coupling reactions are central tools for the straightforward elaboration of biarylic, polyenic, or polyynic motifs embedded in complex natural products. The exceptional functional group compatibility and the mild reaction conditions often reported are keys to this success.

Among the cross-coupling strategies reviewed, the central role of Liebeskind's promoter (CuTC, copper(I) thiophenecarboxylate) in Stille couplings is especially worthy of note. First reported in 1996, CuTC has been used in an impressive number of cross-coupling reactions leading to stereodefined dienic, trienic, and extended polyenic motifs. In addition, copper(I) salts such as copper(I) iodide or copper(I) thiophenecarboxylate could also be essential cocatalysts of classical palladium-catalyzed Stille coupling.

Oxidative alkyne coupling reactions have also been crucial for the synthesis of advanced intermediates or for the elaboration of complex naturally occurring polyynes. Recent breakthrough in terms of reaction efficiency reported in the Glaser–Hay reaction should prove useful in the near future for the rapid assembly of complex targets.

As demonstrated during the past 20 years, copper-mediated cross-coupling reactions have become a powerful strategy at hand for the synthetic chemists. Improved catalytic systems with exceptional functional group tolerance will certainly continue to appear, enabling the ever more rapid and practical total syntheses of biologically relevant natural products.

REFERENCES

[1] (a) In *Classics in Total Synthesis: Targets, Strategies, Methods*, Nicolaou, K. C.; Sorensen, E. J. (eds.). VCH: New York, **1996**. (b) In *Classics in Total Synthesis II: More Targets, Strategies, Methods*, Nicolaou, K. C.; Snyder, S. A. (eds.). Wiley-VCH: Weinheim, **2003**. (c) In *Classics in Total Synthesis III: Further Targets, Strategies, Methods*, Nicolaou, K. C.; Chen, J. S. (eds.). Wiley-VCH: Weinheim, **2011**.

[2] Evano, G.; Blanchard, N.; Toumi, M. *Chem. Rev.* **2008**, *108*, 3054–3131.

[3] Bringmann, G.; Gulder, T.; Gulder, T. A. M.; Breuning, M. *Chem. Rev.* **2011**, *111*, 563–639.

[4] Ullmann, F.; Jean Bielecki, J. *Chem. Ber.* **1901**, *34*, 2174–2185.

[5] For recent reviews, see: (a) Monnier, F.; Tailler, M. *Angew. Chem., Int. Ed.* **2009**, *48*, 6954–6971. (b) Hassan, J.; Sévignon, M.; Gozzi, C.; Schulz, E.; Lemaire, M. *Chem. Rev.* **2002**, *102*, 1359–1470.

[6] (a) Meyers, A. I.; Willemsen, J. J. *Chem. Commun.* **1998**, 1573–1574. (b) Meyers, A. I.; Willemsen, J. J. *Tetrahedron* **1998**, *54*, 10493–10511.

[7] (a) Ikeda, Y.; Nagao, K.; Tanigakiuchi, K.; Tokumaru, G.; Tsichiya, H.; Yamada, H. *Tetrahedron Lett.* **2004**, *45*, 487–489. (b) Yamada, H.; Nagao, K.; Dokei, K.; Kasai, Y.; Michihata, N. *J. Am. Chem. Soc.* **2008**, *130*, 7566–7567.

[8] Molander, G. A.; George, K. M.; Monovich, L. G. *J. Org. Chem.* **2003**, *68*, 9533–9540.

[9] (a) Bringmann, G.; Hinrichs, J. *Tetrahedron: Asymmetry* **1997**, *8*, 4121–4126. (b) Bringmann, G.; Hinrichs, J.; Pabst, T.; Henschel, P.; Peters, K.; Peters, E.-M. *Synthesis* **2001**, 155–167.

[10] For a recent review, see: Kozlowski, M. C.; Morgan, B. J.; Linton, E. C. *Chem. Soc. Rev.* **2009**, *38*, 3193–3207.

[11] Kasai, Y.; Michihata, N.; Nishimura, H.; Hirokane, T.; Yamada, H. *Angew. Chem. Int. Ed.* **2012**, *51*, 8026–8029.

[12] Li, X.; Yang, J.; Kozlowski, M. C. *Org. Lett.* **2001**, *3*, 1137–1140.

[13] Li, X.; Hewgley, J. B.; Mulrooney, C.; Yang, J.; Kozlowski, M. C. *J. Org. Chem.* **2003**, *68*, 5500–5511.

[14] Hewgley, J. B.; Stahl, S. S.; Kozlowski, M. C. *J. Am. Chem. Soc.* **2008**, *130*, 12232–12233.

[15] (a) DiVirgilio, E. S.; Dugan, E. C.; Mulrooney, C. A.; Kozlowski, M. C. *Org. Lett.* **2007**, *9*, 385–388. (b) Kozlowski, M. C.; Dugan, E. C.; DiVirgilio, E. S.; Maksimenka, K.; Bringmann, G. *Adv. Synth. Catal.* **2007**, *349*, 583–594.

[16] Mulrooney, C. A.; O'Brien, E. M.; Morgan, B. J.; Kozlowski, M. C. *Eur. J. Org. Chem.* **2012**, 3887–3904.

[17] (a) Morgan, B. J.; Dey, S.; Johnson, S. W.; Kozlowski, M. C. *J. Am. Chem. Soc.* **2009**, *131*, 9413–9425. (b) Mulrooney, C. A.; Morgan, B. J.; Li, X.; Kozlowski, M. C. *J. Org. Chem.* **2010**, *75*, 16–29. (c) Morgan, B. J.; Mulrooney, C. A.; O'Brien, E. M.; Kozlowski, M. C. *J. Org. Chem.* **2010**, *75*, 30–43.

[18] Podlesny, E. E.; Kozlowski, M. C. *Org. Lett.* **2012**, *14*, 1408–1411.

[19] Examples in total synthesis: (a) Kotora, M.; Negishi, E. *Tetrahedron Lett.* **1996**, *37*, 9041–9042. (b) Mulzer, J.; Berger, M. *J. Org. Chem.* **2004**, *69*, 891–898.

[20] Yu, S.; Liu, F.; Ma, D. *Tetrahedron Lett.* **2006**, *47*, 9155–9157.

[21] Ma, D.; Liu, F. *Chem. Commun.* **2004**, 1934–1935.

[22] Schneider, C. M.; Kriangsak, K.; Li, W.; Spletstoser, J. T.; Haack, T.; Georg, G. I. *Angew. Chem. Int. Ed.* **2011**, *50*, 7855–7857.

[23] Coleman, R. S.; Garg, R. *Org. Lett.* **2001**, *3*, 3487–3490.

[24] Paterson, I.; Ashton, K.; Britton, R.; Cecere, G.; Chouraqui, G.; Florence, G. J.; Knust, H.; Stafford, J. *Chem. Asian. J.* **2008**, *3*, 367–387.

[25] (a) Allred, G. D.; Liebeskind, L. S. *J. Am. Chem. Soc.* **1996**, *118*, 2748–2749. (b) Srogl, J.; Allred, G. D.; Liebeskind, L. S. *J. Am. Chem. Soc.* **1997**, *119*, 12376–12377.

[26] Nicolaou, K. C.; Jiang, X.; Lindsay-Scott, P. J.; Corbu, A.; Yamashiro, S.; Bacconi, A.; Fowler, V. M. *Angew. Chem. Int. Ed.* **2011**, *50*, 1139–1144.

[27] Pulukuri, K. K.; Chakraborty, T. K. *Org. Lett.* **2012**, *14*, 2858–2861.

[28] Burghart-Stoll, H.; Kapferer, T.; Brückner, R. *Org. Lett.* **2011**, *13*, 1016–1019.

[29] Durham, T. B.; Blanchard, N.; Savall, B. M.; Powell, N. A.; Roush, W. R. *J. Am. Chem. Soc.* **2004**, *126*, 9307–9317.

[30] Paterson, I.; Steadman (née Doughty), V. A.; McLeod, M. D.; Trieselmann, T. *Tetrahedron* **2011**, *67*, 10119–10128.

[31] (a) Florence, G. J.; Gardner, N. M.; Paterson, I. *Nat. Prod. Rep.* **2008**, *25*, 342–375. (b) Paterson, I.; Gardner, N. M.; Poullennec, K. G.; Wright, A. E. *J. Nat. Prod.* **2008**, *71*, 364–369. (c) Paterson, I.; Naylor, G. J.; Wright, A. E. *Chem. Commun.* **2008**, 4628–4630. (d) Paterson, I.; Gardner, N. M.; Guzmán, E.; Wright, A. E. *Bioorg. Med. Chem.* **2009**, *17*, 2282–2289. (e) Paterson, I.; Britton, R.; Delgado, O.; Gardner, N. M.; Meyer, A.; Naylor, G. J.; Poullennec, K. G. *Tetrahedron* **2010**, *66*, 6534–6545. (f) Paterson, I.; Naylor, G. J.; Gardner, N. M.; Guzmán, E.; Wright, A. E. *Chem. Asian. J.* **2011**, *6*, 459–473.

[32] Chaladaj, W.; Corbet, M.; Fürstner, A. *Angew. Chem. Int. Ed.* **2012**, *51*, 6929–6933.

[33] (a) König, C. M.; Gebhardt, B.; Schleth, C.; Dauber, M.; Koert, U. *Org. Lett.* **2009**, *11*, 2728–2731. (b) Gebhardt, B.; König, C. M.; Schleth, C.; Dauber, M.; Koert, U. *Chem. Eur. J.* **2010**, *16*, 5934–5941.

[34] Shibahara, S.; Fujino, M.; Tashiro, Y.; Takahashi, K.; Ishihara, J.; Hatakeyama, S. *Org. Lett.* **2008**, *10*, 2139–2142.

[35] Miyashita, K.; Tsunemi, T.; Hosokawa, T.; Ikejiri, M.; Imanishi, T. *J. Org. Chem.* **2008**, *73*, 5360–5370.

[36] Nicolaou, K. C.; Sarlah, D.; Wu, R. T.; Zhan, W. *Angew. Chem. Int. Ed.* **2009**, *48*, 6870–6874.

[37] Crane, E. A.; Zabawa, T. P.; Farmer, R. L.; Scheidt, K. A. *Angew. Chem. Int. Ed.* **2011**, *50*, 9112–9115.

[38] Kishi, Y. *Proc. Natl. Acad. Sci. USA* **2011**, *108*, 6703–6708, and references cited therein.

[39] (a) Feyen, F.; Jantsch, A.; Altmann, K. H. *Synlett* **2007**, 415–418. (b) Gersbach, P.; Jantsch, A.; Feyen, F.; Scherr, N.; Dangy, J.-P.; Pluschke, G.; Altmann, K.-H. *Chem. Eur. J.* **2011**, *17*, 13017–13031.

[40] (a) Yin, N.; Wang, G.; Qian, M. X.; Negishi, E. *Angew. Chem. Int. Ed.* **2006**, *45*, 2916–2920. (b) Wang, G. W.; Yin, N.; Negishi, E. *Chem. Eur. J.* **2011**, *17*, 4118–4130.

[41] For synthetic efforts toward mycolactones A/B, see: (a) Alexander, M. D.; Fontaine, S. D.; La Clair, J. J.; DiPasquale, A. G.; Rheingold, A. L.; Burkart, M. D. *Chem. Commun.* **2006**, 4602–4604. (b) Ko, K. S.; Alexander, M. D.; Fontaine, S. D.; Biggs-Houck, J. E.; La Clair, J. J.; Burkart, M. D. *Org. Biomol. Chem.* **2010**, *8*, 5159–5165. (c) Gurjar, M. K.; Cherian, J. *Heterocycles* **2001**, *55*, 1095–1103. (d) van Summeren, R. P.; Feringa, B. L.; Minnaard, A. J. *Org. Biomol. Chem.* **2005**, *3*, 2524–2533.

[42] Chany, A.-C.; Casarotto, V.; Schmitt, M.; Tarnus, C.; Guenin-Macé, L.; Demangel, C.; Mirguet, O.; Eustache, J.; Blanchard, N. *Chem. Eur. J.* **2011**, *17*, 14413–14419.

[43] (a) Espinet, P.; Echavarren, A. M. *Angew. Chem. Int. Ed.* **2004**, *43*, 4704–4734. (b) Pérez-Temprano, M. H.; Casares, J. A.; Espinet, P. *Chem. Eur. J.* **2012**, *18*, 1864–1884.

[44] (a) Mee, S. P. H.; Lee, V.; Baldwin, J. E. *Angew. Chem. Int. Ed.* **2004**, *43*, 1132–1136. (b) Mee, S. P. H.; Lee, V.; Baldwin, J. E. *Chem. Eur. J.* **2005**, *11*, 3294–3308.

[45] Mirzayans, P. M.; Pouwer, R. H.; Williams, C. M.; Bernhardt, P. V. *Eur. J. Org. Chem.* **2012**, 1633–1638.

[46] Eto, K.; Yoshino, M.; Takahashi, K.; Ishihara, J.; Hatakeyama, S. *Org. Lett.* **2011**, *13*, 5398–5401.

[47] Senapati, B. K.; Gao, L.; Lee, S. I.; Hwang, G.-S.; Ryu, D. H. *Org. Lett.* **2010**, *12*, 5088–5091.

[48] Webb, M. R.; Addie, M. S.; Crawforth, C. M.; Dale, J. W.; Franci, X.; Pizzonero, M.; Donald, C.; Taylor, R. J. K. *Tetrahedron* **2008**, *64*, 4778–4791.

[49] (a) Okitsu, T.; Iwatsuka, K.; Wada, A. *Chem. Commun.* **2008**, 6330–6332. (b) Okitsu, T.; Sato, K.; Iwatsuka, K.; Sawada, N.; Nakagawa, K.; Okano, T.; Yamada, S.; Kakuta, H.; Wada, A. *Bioorg. Med. Chem.* **2011**, *19*, 2939–2949.

[50] (a) Fürstner, A.; Funel, J.-A.; Tremblay, M.; Bouchez, L. C.; Nevado, C.; Waser, M.; Ackerstaff, J.; Stimson, C. C. *Chem. Commun.* **2008**, 2873–2875. For selected total syntheses using this three-component catalytic system, see: (b) Francais, A.; Leyva, A.; Etxebarria-Jardi, G.; Ley, S. V. *Org. Lett.* **2009**, *12*, 340–343. (c) Gagnepain, J.; Moulin, E.; Fürstner, A. *Chem. Eur. J.* **2011**, *17*, 6964–6972.

[51] Fürstner, A.; Bouchez, L. C.; Morency, L.; Funel, J.-A.; Liepins, V.; Porée, F.-H.; Gilmour, R.; Laurich, D.; Beaufils, F.; Tamiya, M. *Chem. Eur. J.* **2009**, *15*, 3983–4010.

[52] Cid, M. B.; Pattenden, G. *Tetrahedron Lett.* **2000**, *41*, 7373–7378.

[53] Fuwa, H.; Ebine, M.; Bourdelais, A. J.; Baden, D. G.; Sasaki, M. *J. Am. Chem. Soc.* **2006**, *128*, 16989–16999.

[54] Woerly, E. M.; Cherney, A. H.; Davis, E. K.; Burke, M. D. *J. Am. Chem. Soc.* **2010**, *132*, 6941–6943.

[55] Paterson, I.; Kan, S. B. J.; Gibson, L. J. *Org. Lett.* **2010**, *12*, 3724–3727.

[56] Bohlmann, F.; Burkhardt, H.; Zdero, C. (eds.), *Naturally Occurring Acetylenes*, Academic Press: London and New York, **1973**.

[57] Jones, E. R. H.; Thaller, V. *Natural acetylenes*. Chap. 14. In *The Chemistry of the Carbon-Carbon Triple Bond, Part 2*, Patai, S. (ed.). Wiley: New York, **1978**, pp. 621–633.

[58] Chodkiewicz, W.; Cadiot, P. *C. R. Hebd. Seances Acad. Sci.* **1955**, *241*, 1055–1057.

[59] Eglinton, G.; Galbraith, A. R. *J. Chem. Soc.* **1959**, 889–896.

[60] Hay, A. S. *J. Org. Chem.* **1962**, *27*, 3320–3321.

[61] (a) Shi Shun, A. L. K.; Tykwinski, R. R. *Angew. Chem. Int. Ed.* **2006**, *45*, 1034–1057. (b) Siemsen, P.; Livingston, R. C.; Diederich, F. *Angew. Chem. Int. Ed.* **2000**, *39*, 2632–2657. (c) Diederich, F.; Stang, P.; Tykwinski, R. R. *Acetylene Chemistry: Chemistry, Biology, and Material Science*, John Wiley & Sons: Hoboken, NJ, **2006**.

[62] (a) Gung, B. W.; Dickson, H. *Org. Lett.* **2002**, *4*, 2517–2519. (b) Gung, B. W. *C. R. Chimie* **2009**, *12*, 489–505.

[63] Gung, B. W.; Kumi, G. *J. Org. Chem.* **2004**, *69*, 3488–3492.

[64] Alami, M.; Ferri, F. *Tetrahedron Lett.* **1996**, *37*, 2763–2766.

[65] López, S.; Fernández-Trillo, F.; Midón, P.; Castedo, L.; Saá, C. *J. Org. Chem.* **2006**, *71*, 2802–2810.

[66] (a) Nicolaou, K. C.; Petasis, N. A.; Zipkin, R. E.; Uenishi, J. *J. Am. Chem. Soc.* **1982**, *104*, 5555–5557. (b) Nicolaou, K. C.; Petasis, N. A.; Uenishi, J.; Zipkin, R. E. *J. Am. Chem. Soc.* **1982**, *104*, 5557–5558. (c) Nicolaou, K. C.; Zipkin, R. E.; Petasis, N. A. *J. Am. Chem. Soc.* **1982**, *104*, 5558–5560. (d) Nicolaou, K. C.; Petasis, N. A.; Zipkin, R. E. *J. Am. Chem. Soc.* **1982**, *104*, 5560–5562.

[67] (a) Ren, F.; Hogan, P. C.; Anderson, A. J.; Myers, A. G. *J. Am. Chem. Soc.* **2007**, *129*, 5381–5383. (b) Myers, A. G.; Hogan, P. C.; Hurd, A. R.; Goldberg, S. D. *Angew. Chem. Int. Ed.* **2002**, *41*, 1062–1067. (c) Myers, A. G.; Goldberg, S. D. *Angew. Chem. Int. Ed.* **2000**, *39*, 2732–2735.

[68] Yamaguchi, M.; Park, H.-J.; Hirama, M.; Torisu, K.; Nakamura, S.; Minami, T.; Nishihara, H.; Hiraoka, T. *Bull. Chem. Soc. Jpn* **1994**, *67*, 1717–1725.

[69] (a) Ramírez-López, P.; de la Torre, M. C.; Montenegro, H. E.; Asenjo, M.; Sierra, M. A. *Org. Lett.* **2008**, *10*, 3555–3558. (b) Montenegro, H. E.; Ramírez-López, P.; de la Torre, M. C.; Asenjo, M.; Sierra, M. A. *Chem. Eur. J.* **2010**, *16*, 3798–3814.

[70] Bédard, A.-C.; Collins, S. K. *J. Am. Chem. Soc.* **2011**, *133*, 19976–19981.

[71] (a) Yin, W.; He, C.; Chen, M.; Zhang, H.; Lei, A. *Org. Lett.* **2008**, *11*, 709–712. (b) Shi, W.; Liu, C.; Lei, A. *Chem. Soc. Rev.* **2011**, *40*, 2761–2776. (c) Crowley, J. D.; Goldup, S. M.; Gowans, N. D.; Leigh, D. A.; Ronaldson, V. E.; Slawin, A. M. Z. *J. Am. Chem. Soc.* **2010**, *132*, 6243–6248.

[72] Bédard, A.-C.; Collins, S. K. *Chem. Commun.* **2012**, *48*, 6420–6422.

[73] Jeffery, T.; Gueugnot, S.; Linstrumelle, G. *Tetrahedron Lett.* **1992**, *33*, 5757–5760.

[74] Hwang, S.; Kim, J. H.; Kim, H. S.; Kim, S. *Eur. J. Org. Chem.* **2011**, 7414–7418.

19

PROCESS CHEMISTRY AND COPPER CATALYSIS

Klaus Kunz and Norbert Lui

19.1 INTRODUCTION AND SCOPE

The palladium-catalyzed coupling of (aryl)C–N and (aryl)C–O bonds has attracted tremendous attention in the area of synthetic organic chemistry over the past 15 years and was awarded with one of the last publicly accepted named reactions (*Buchwald–Hartwig reaction*).[1] In the meantime, many groups have contributed to the development of numerous robust and sustainable laboratory procedures for multiple substrate classes.

However, in the late 1990s, chemists were less aware that excellent catalytic protocols of the related copper-based Ullmann and Goldberg reactions had been explored and published 100 years ago (Scheme 19.1).[2]

In the last decade, academia initiated a real renaissance of the Ullmann and Goldberg reactions by offering protocols allowing milder reaction conditions than those used in classical Ullmann–Goldberg chemistry. This development, basically driven by the introduction of tailor-made ligands, was documented not only in a series of publications and reviews[3] but also in some basic patent applications by the groups of Buchwald,[4] Taillefer,[5] and Bayer AG.[6] (Fig. 19.1).

Establishing the ligand toolbox for a broad range of copper-catalyzed coupling reactions, academia also paved the way for a significant reconsideration in industrial laboratories and pilot plants, even though the copper-based technology has never really disappeared since its discovery in 1901.[2] To

Copper-Mediated Cross-Coupling Reactions, First Edition. Edited by Gwilherm Evano and Nicolas Blanchard.

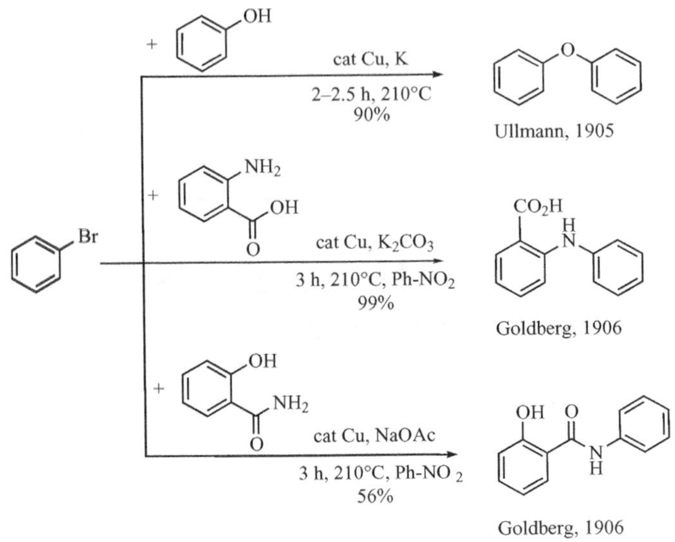

Scheme 19.1 Ullmann ether synthesis, Ullmann–Goldberg amidation, and Goldberg amination.

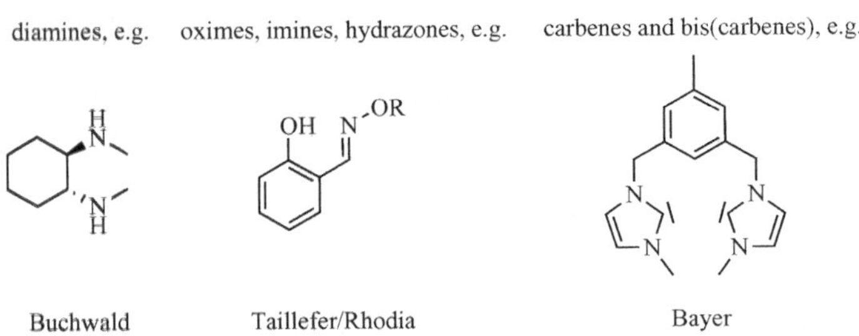

Figure 19.1 Basic ligand classes from the patent literature.

summarize this long history in a comprehensive and complete way is a challenging task for several reasons:

(a) In the case of process optimization for life science products, specific patents usually refer only to the improvement of single steps, while the entire process is not described. Therefore, the authors have also considered publications from scientific journals with process development.

(b) The quality of data as found in the patents is rather heterogeneous. Thus, a straightforward comparison of protocols in terms of applicability, scale-up potential, degree of optimization, and costs remains difficult.

(c) In most of the cases, it is not possible to investigate whether the protocols described in the patents are also practically employed in production processes. In the case of highly competitive products, companies tend to patent a range of synthetic approaches in order to prevent others' access to the respective molecules.

For this chapter, the authors have collected material from both scientific journals and patent applications from the past 15 years to illustrate the potential of the Ullmann–Goldberg reactions in modern industrial scale-up processes. Early research-focused patents about copper-catalyzed syntheses of new chemical classes and respective biological activities are not considered.

19.2 COPPER VERSUS PALLADIUM

The great challenge for the process chemist is to develop a sustainable, robust, and economically feasible synthetic pathway to the respective target molecule. A comparison of the 2006–2011 price development of copper and palladium[7] reveals the most striking advantage of copper; due to its low price, the actual metal price is basically irrelevant for the cost of goods (COGs) in copper-catalyzed reactions (Fig. 19.2). Easy handling and the option to employ heterogeneous species (e.g., oxides) may also support the choice of copper.

While both methodologies are compatible with a large range of aryl halides, amines, amides, and alcohols/phenols, palladium catalysts are more effective for the coupling of tosylates and triflates. If lower temperatures are required (e.g., to avoid the formation of side or tare products), the use of palladium may also prove to be advantageous. Issues specific to copper-based methodologies from an industrial point of view are purification and waste treatment due to the higher catalyst loadings and the readily formed stable complexes with products, intermediates, and reagents.[3h,8] Among a variety of methods for copper complex decomposition and workup of reaction mixtures, oxidation of copper(I) to copper(II) with iron salts or complexation of copper(I) and (II) with aqueous ethylenediamine or aqueous ammonia and Na_2EDTA has proven to be most useful.

19.3 APPLICATIONS

19.3.1 Materials Sciences

A series of patents has been published about the copper-catalyzed synthesis of key intermediates for high-performance materials. The low costs of the copper catalysts play a key role in the economics of these processes. In many cases, the protocols are based on previously elaborated procedures from academic laboratories, as illustrated by DuPont's approach to alkoxy-substituted

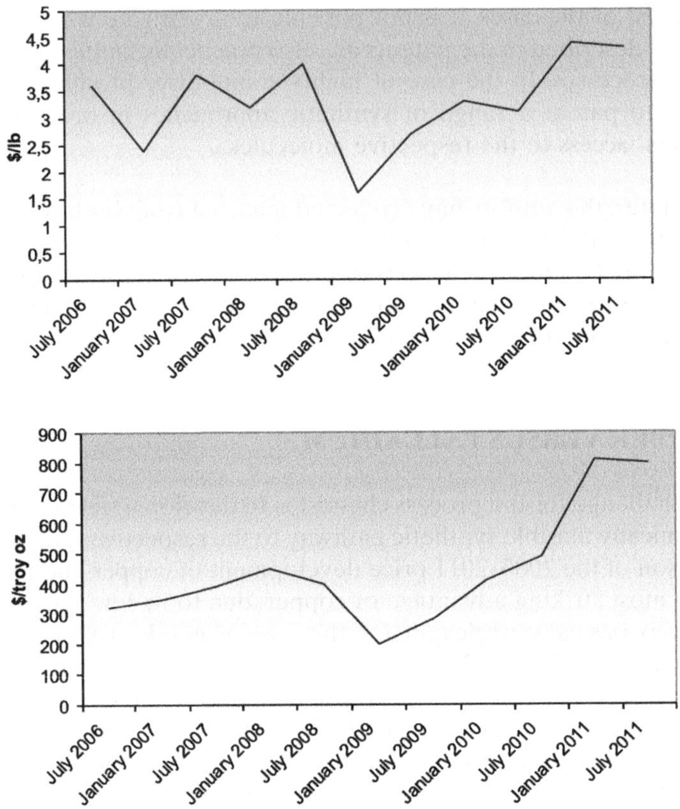

Figure 19.2 2006–2011 price development of copper (top) and palladium (bottom)/ (1 lb = 454 g; 1 troy oz = 31 g).

terephtalic acids, which are further employed in a condensation reaction to polypyridobisimidazoles (high-tech fibers, Scheme 19.2).[9] The use of 2,2′,6,6′-tetramethylheptanedione-3,5 as a bidendate ligand has been published before.[10]

In a recent modification, fluorinated alkoxy substituents (2,2,2-trifluoroethoxy or 2,2,3,3,-tetrafluoropropoxy) were successfully introduced with amino acids as ligands (valine, proline, *N*-methylanthranilic acid). The corresponding fluorinated polymers show improved and more durable oil and water persistence.[11]

DuPont also reported on an optimized process to 2,6-aminopyridine based on a copper-catalyzed amination of 2-amino-6-chloropyridine,[12a] improving older copper-based procedures[12b] via the addition of ammonium acetate as a buffer (Scheme 19.3). In a control experiment without buffer, the isolated yield dropped from 81% down to 69%. The chelating properties of the substrate superseded the use of an additional ligand.

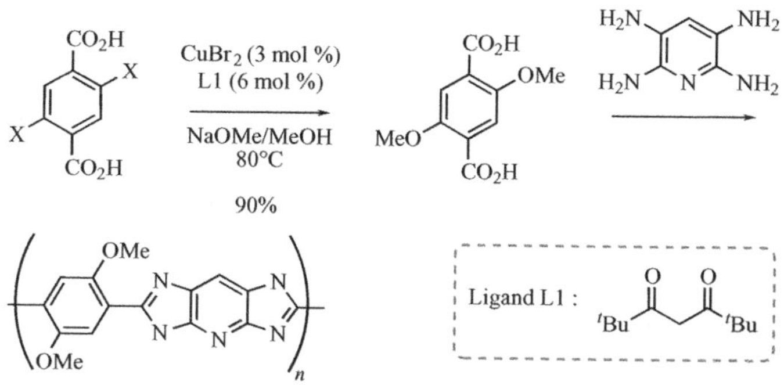

Scheme 19.2 Copper(II)-catalyzed synthesis of polypyridobisimidazoles by DuPont.

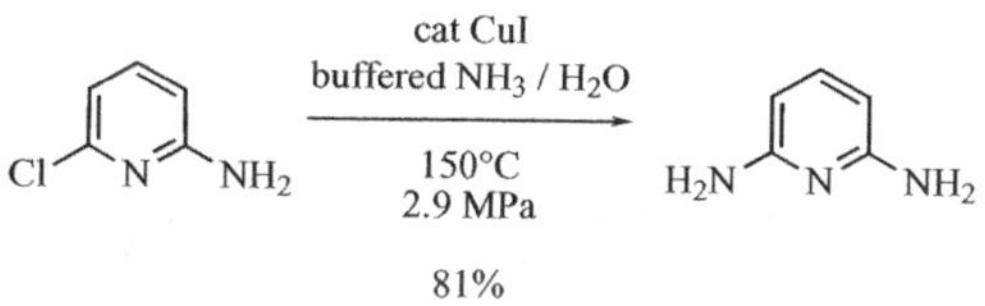

Scheme 19.3 DuPont's improved approach to diaminopyridine.

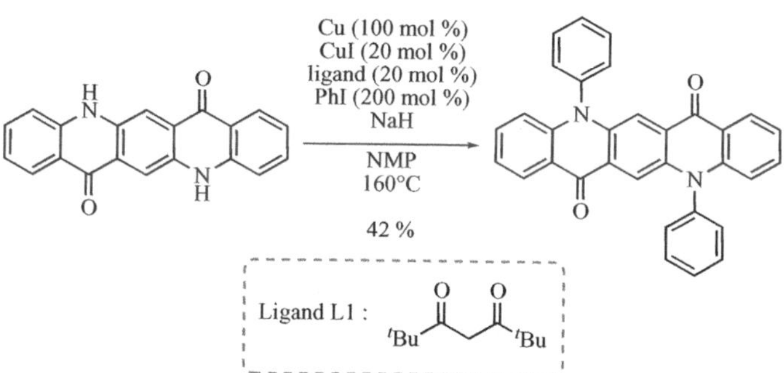

Scheme 19.4 Copper(0)/copper(I) synthesis of diarylquinacridones by Eastman Kodak.

A mixture of copper (100 mol %) and copper iodide (20 mol %) proved to be an efficient catalytic system in Kodak's protocol for the double arylation reaction of quinacridones to diarylquinacridones (Scheme 19.4), which has found applications as color pigments. Again, 2,2′,6,6′-tetramethyl-3,5-heptanedione proved to be the best choice as a ligand even though the optimized isolated yield of 42% was only moderate.[13]

4-Aminodiphenylamine (4-ADPA) is an important intermediate for the production of stabilizers for rubbers and polymers. For the synthesis of the precursor 4-nitrodiphenylamine (4-NDPA), innovative Ullmann couplings

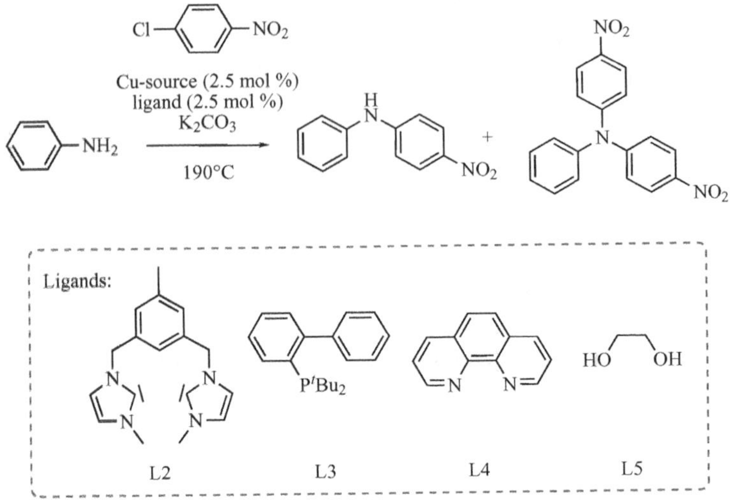

Cu source/ligand	Conversion of 4-nitrochlorobenzene (%)	Yield of 4-nitrodiphenylamine (%)	Selectivity (%)
$CuBr_2/L_2$	78	76	98
$[Cu(L_3)Br]_2$	90	89	99
CuO/L_4	68	60	89
CuO/L_5	28	27	96

Scheme 19.5 Selective approach to 4-NDPA by Bayer.

with new metal–ligand combinations were patented by Bayer. Building on the concept of copper/phosphine catalysts as reported by Venkataraman and coworkers[14] more electron-rich phosphines with a higher steric demand led to excellent yields and selectivities.[15] In a further patent application, copper/bis-carbene combinations have been reported for the first time.[16] Both catalytic systems showed a better performance than the phenanthroline or ethylene glycol-based standards (Scheme 19.5).[17]

Copper-mediated cross-coupling reactions were extensively employed in the production of polyarylamines and polyaryldiamines, which have found several applications as high-tech materials. In a recent example, researchers of Sankyo added tetrabutylphosphonium bromide to facilitate the straightforward Ullmann coupling reaction to polyaryldiamines (Scheme 19.6).[18]

Electroluminescent 4,4′-biscarbazolylarenes have been prepared by a cross-coupling reaction of 1,4-dihaloarenes and carbazoles in the presence of copper and potassium carbonate in *N*,*N*-dimethylformamide (DMF). The product is alternatively accessible if tetrahydrocarbazole is coupled with dihaloarenes before dehydrogenation.[19]

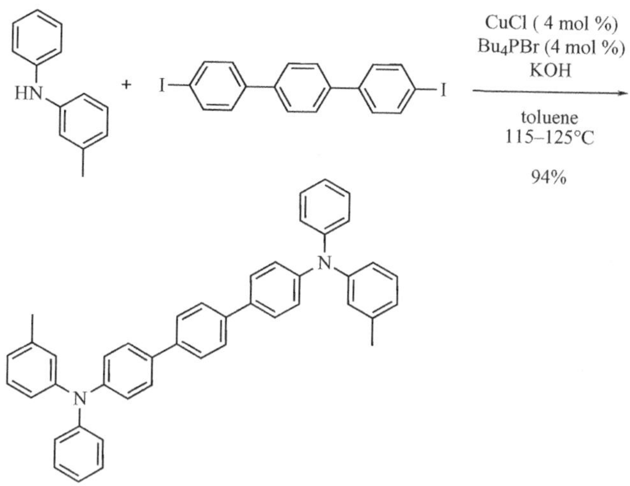

Scheme 19.6 Copper(I)-catalyzed synthesis of polyaryldiamines by Sankyo.

19.3.2 Pharmaceutical Chemistry

Copper-mediated cross-coupling reactions have elicited a great deal of interest from the pharmaceutical industry. A considerable number of publications reporting impressive applications of Ullmann couplings have been published and reviewed;[3f] some examples will be highlighted in the following paragraphs. Patented approaches to key intermediates will be described as well.

GlaxoSmithKline reported on a process development for the new corticotropin release factor (CFR) agonist GW876008 on a 100 kg scale.[20] Of particular note, the final step is a copper(I)-catalyzed C–N bond formation. One breakthrough toward an economically feasible approach was the replacement of the well-established but expensive *N,N′*-dimethylcyclohexanediamine by the cheaper *N,N*-dimethylglycine. This ligand modification allowed a reduction of 10% of the COGs. In order to avoid a double arylation reaction, the protection of the free nitrogen atom of the imidazolidinone moiety was crucial (Scheme 19.7). Finally, metal contamination was reduced to a very low level (5–45 ppm) by employing 1 equiv of the ligand as a copper scavenger during recrystallization.

An elegant example of a copper-catalyzed aryl-*N*-(heterocycle) cross-coupling reaction was reported by Abbott in the histamine H_3 receptor (H_3R) antagonist area of research.[21] Classical imidazole-based H_3R antagonists are thwarted by unfavorable profiles and recent developments aimed at replacing the imidazole motif by a various N-heterocycles, such as pyridazinones (Scheme 19.8). After extensive optimization,[22] the cross-coupling reaction of 2-(6-bromonaphtalen-2-yl)ethanol with pyridazinone occurred smoothly in the presence of copper(I) chloride (5 mol %), 8-hydroxyquinoline (5 mol %), and potassium carbonate in DMF at 140°C. The desired product was obtained

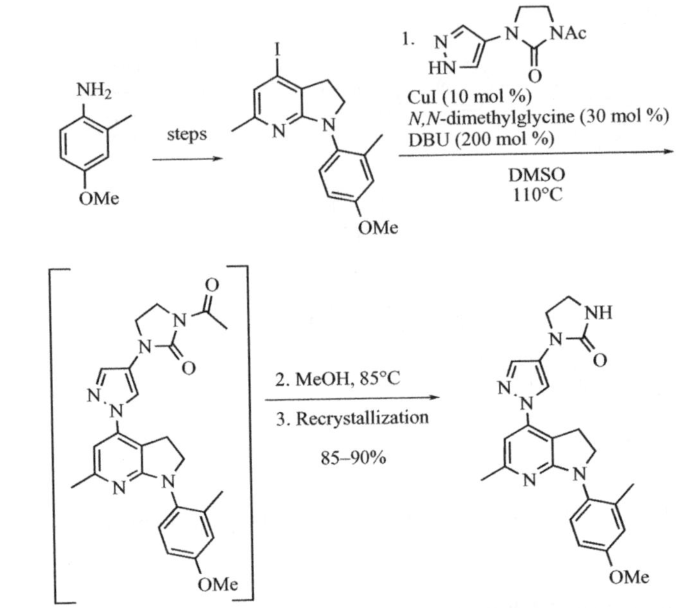

Scheme 19.7 Copper(I)-catalyzed cross coupling as the final step of GW876008 by GlaxoSmithKline.

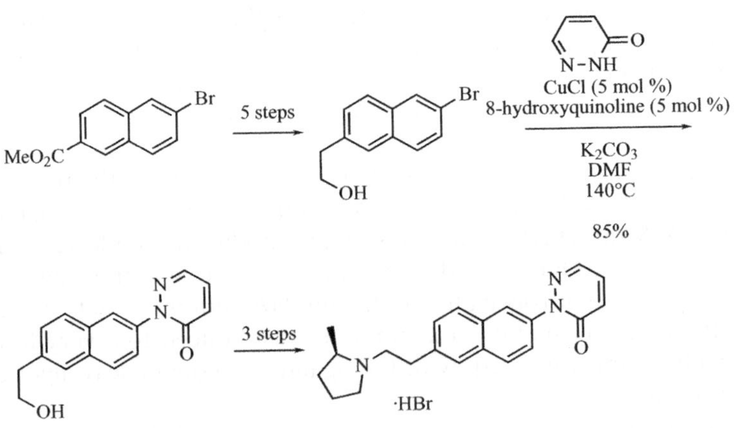

Scheme 19.8 Abbott's large scale synthesis of a new H_3R agonist.

in 85% yield on a 13kg scale. It is worthy of note that the corresponding palladium-catalyzed transformation was not successful, presumably due to the low nucleophilicity of pyridazinone.

A very efficient example of copper-catalyzed amidation reaction of aryl halides was reported by Pfizer during the development of the drug candidate

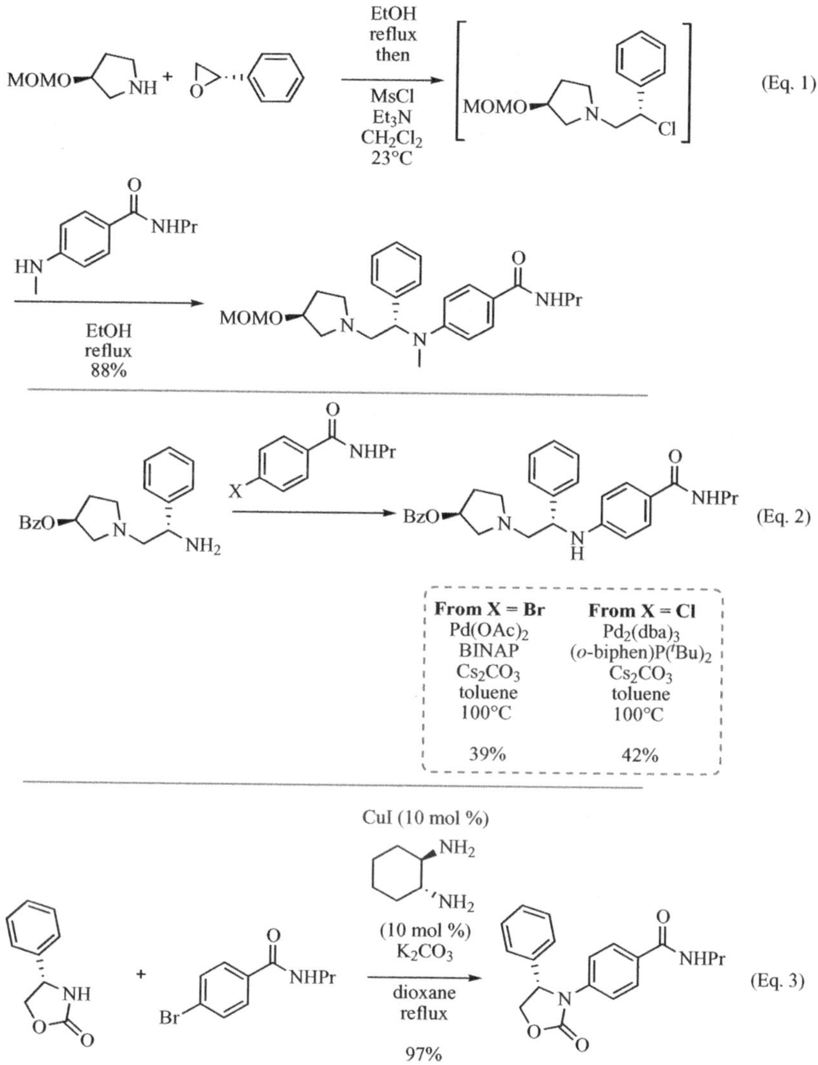

Scheme 19.9 Three generations of Pfizer' synthesis of CJ-15,161.

CJ-15,161, a κ-opioid receptor agonist (Scheme 19.9).[23] The original synthesis from the medicinal chemistry laboratories featured a nucleophilic attack of an aniline derivative onto an aziridinium intermediate with an excellent yield (Eq. 1).[23a] However, multiple chromatographic separations were required and prevented an easy scale-up of this synthetic blueprint. Alternative routes based on palladium-catalyzed amidation reactions performed poorly (Eq. 2).[23b] It was eventually found that a copper(I) iodide-catalyzed amidation reaction of an activated arylbromide in the presence of 1,2-diaminocyclohexane

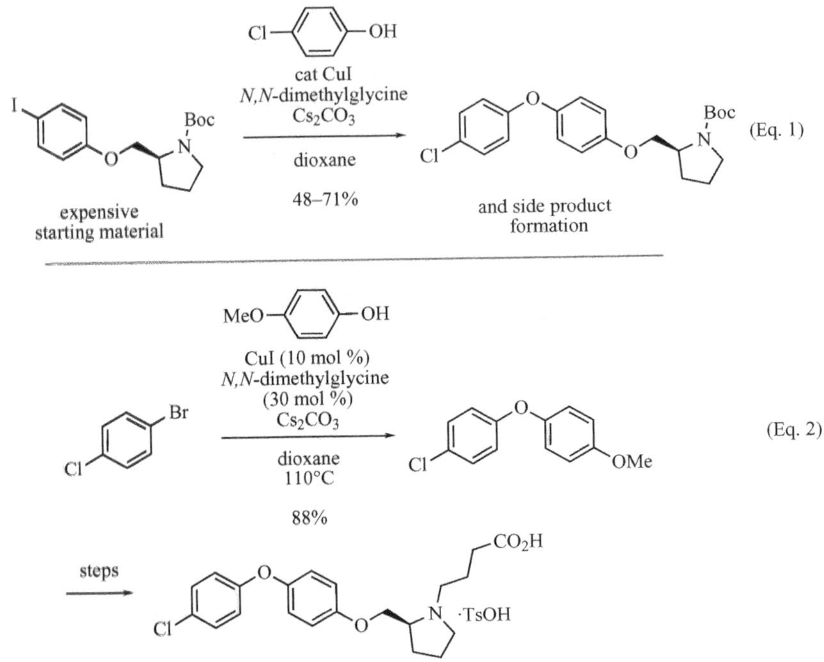

Scheme 19.10 Copper(I)-catalyzed Ullmann ether synthesis approach by deCode.

(10 mol %) and potassium carbonate in refluxing dioxane led to the desired amide in almost quantitative yield on a 300 mmol scale (Eq. 3). Several steps were then necessary to intercept known intermediates in this series.

The advantages and disadvantages of Ullmann couplings are both illustrated in deCode's development of a scalable synthetic approach to the leukotriene A4 hydrolase inhibitor DG-051B.[24] While the original laboratory synthesis employed copper(I)-catalyzed cross coupling at a later stage with significant side product formation and loss of expensive material (Scheme 19.10, Eq. 1), the preferred approach includes the C–O coupling reaction right at the beginning (Scheme 19.10, Eq. 2). The reaction was catalyzed by copper(I) iodide with yields in the 90% range. In addition, addition of a solubilizing ligand such as *N,N*-dimethylglycine greatly improved the rate and the temperature control.

An interesting O-arylation reaction was described recently by Merck, offering a simple approach to 4-chloro-2-methoxypyridine as a starting material for the synthesis of a new potent and selective renin inhibitor (treatment of hypertension, Scheme 19.10).[25] The classical S_NAr reaction of 2,4-dichloropyridine with sodium methoxide afforded the desired product, albeit in low yield. However, a catalytic system, comprising copper(I) iodide (2 mol %) and *N,N′*-dimethylethylenediamine (10 mol %) in toluene at 110°C, delivered the desired 4-chloro-2-methoxypyridine with high selectivity in 74% yield (Scheme 19.11).

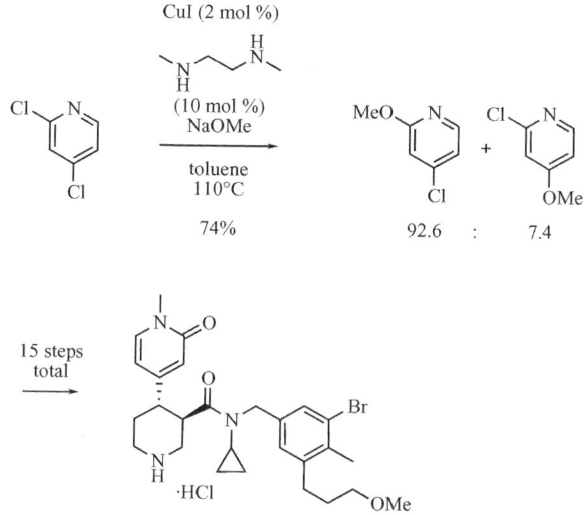

Scheme 19.11 Copper(I)-catalyzed approach to 4-chloro-2-methoxypyridine by Merck.

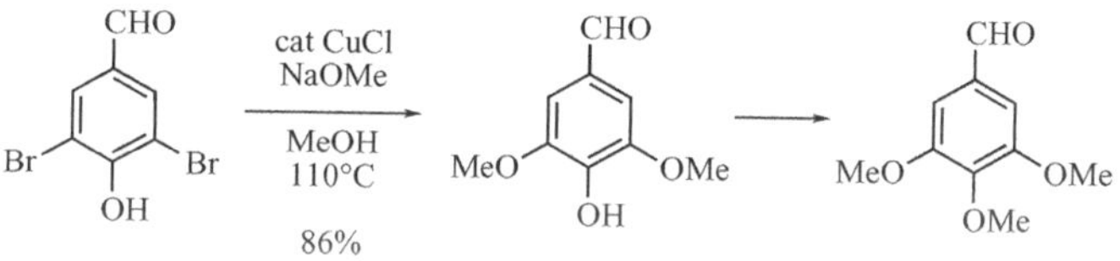

Scheme 19.12 Copper(I) and ligand-free synthesis of 3,4,5-trimethoxybenzaldehyde.

3,4,5-Trimethoxybenzaldehyde is a key intermediate for the manufacture of the antibiotic trimethoprim but also a very interesting building block for the discovery of new biologically active candidates. Gujarat State Fertilizers have developed an improved process including a copper-catalyzed C–O bond coupling without the use of ligands (Scheme 19.12).[26]

Flecainide is an antiarrhythmic agent well established in the market. Providing another example for innovative intermediate chemistry, Par Pharmaceutical described a new process for the synthesis of 2,5-bis(2,2,2-trifluoroethoxy)benzoic acid and related compounds with high yields (Scheme 19.13).[27] In view of the weak nucleophilicity of fluorinated alkoxides, the high copper(I)iodide loading (50 mol %) is not surprising.

Other interesting examples of Ullmann coupling reactions have been described by DuPont[28] (Thymitaq, C–S-coupling, 40 kg scale; Scheme 19.14, Eq. 1), Pfizer[29] (corticotropin-releasing factor antagonist candidate, C–O coupling, 50 kg scale; Scheme 19.14, Eq. 2), Fujisawa Pharmaceuticals[30] (FK3311,

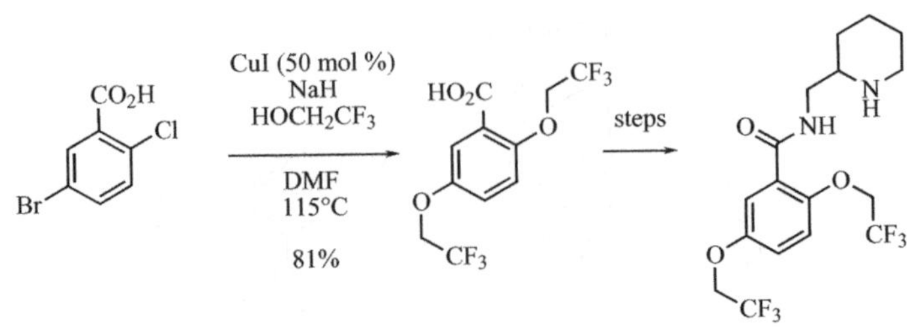

Scheme 19.13 Flecainide synthesis via a double copper(I)-catalyzed alkoxylation by Par.

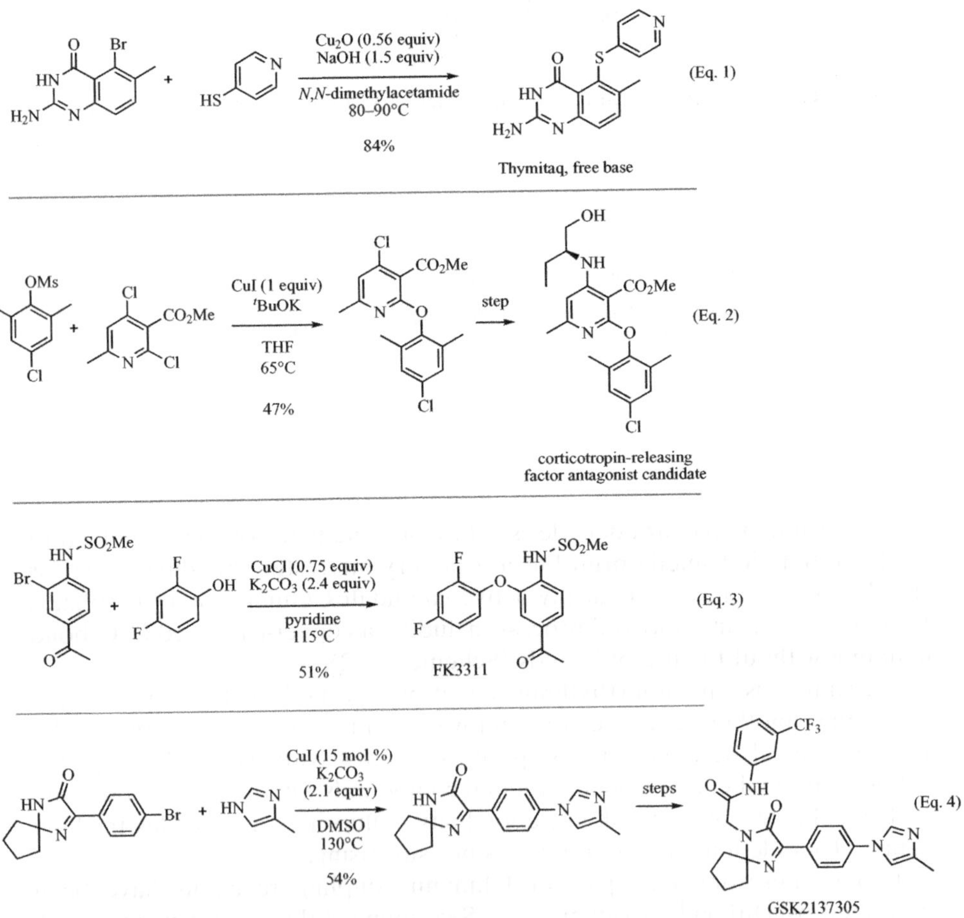

Scheme 19.14 Selected copper-mediated or copper-catalyzed C–S, C–O, and C–N cross couplings.

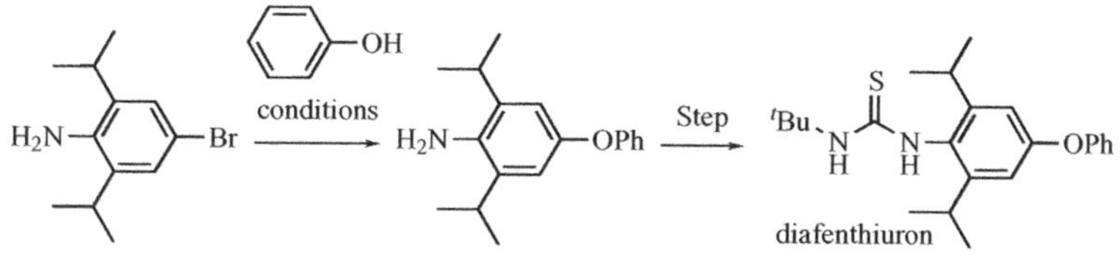

Ciba Geigi (1981)	Ciba Geigi (1991)	Beller–Saltigo (2008)
CuCl (1.5 mol %)	$CuCO_3$ (5 mol %)	CuCl (5 mol %)
–	–	*N*-methylimidazole (50 mol %)
KOH	K_2CO_3	K_2CO_3
o-xylene	toluene/DMF/H_2O	*o*-xylene
150–155°C	140°C	140°C
8 h	10 h	30 h
no yield given	80%	94–100%

Scheme 19.15 Diafenthiuron process by Ciba-Geigy and Beller–Saltigo.

C–O coupling, 45 kg scale; Scheme 19.14, Eq. 3), DSM[31] ((*S*)-2,3-dihydro-1H-indole-carboxylic acid, C–N coupling), and GlaxoSmithKline[32] (GSK2137305, C–N coupling; Scheme 19.14, Eq. 4).

19.3.3 Crop Science Chemistry

The diaryl ether moiety is a common motif in many crop protection products, especially in insecticides and herbicides. A prominent example is the insecticide diafenthiuron, which was first protected in 1981 by a patent application by Ciba-Geigy.[33a] The Ullmann coupling was performed with copper(I) chloride (1.5 mol %) as a catalyst and potassium hydroxide as base under ligand-free conditions (Scheme 19.15); however no yield was specified.

Ten years later, the same company reported on the use of copper carbonate (5 mol %) and potassium carbonate as a base in a three-solvent system at 140°C. A good yield of 80% was then obtained by distillation.[33b] A considerably improved procedure was disclosed by Saltigo GmbH and the academic group of Beller.[34] The use of *N*-methylimidazole (50 mol %) and copper(I) chloride (5 mol %) as a bioinspired catalyst system allowed a smooth elaboration of the central diaryl ether of diafenthiuron in 94% isolated yield.[35]

Inspired by nature, researchers have replaced the lipophilic side chain of insecticidal pyrethrum in many synthetic pyrethroids by a diphenylether moiety (e.g., deltamethrin, cypermethrin, cyfluthrin). A straightforward

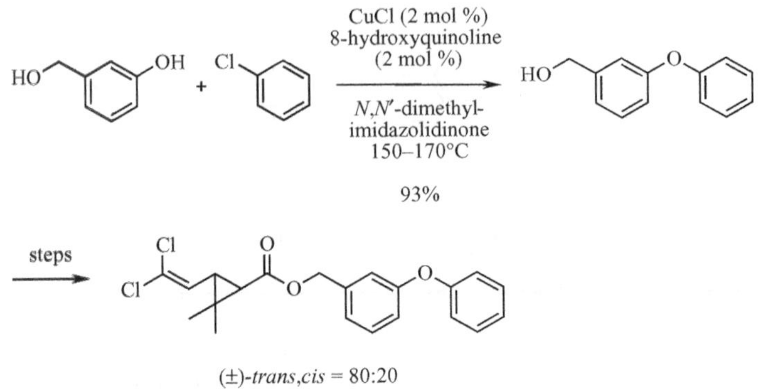

Scheme 19.16 Copper(I)-catalyzed synthesis of permethrin by Mitsui.

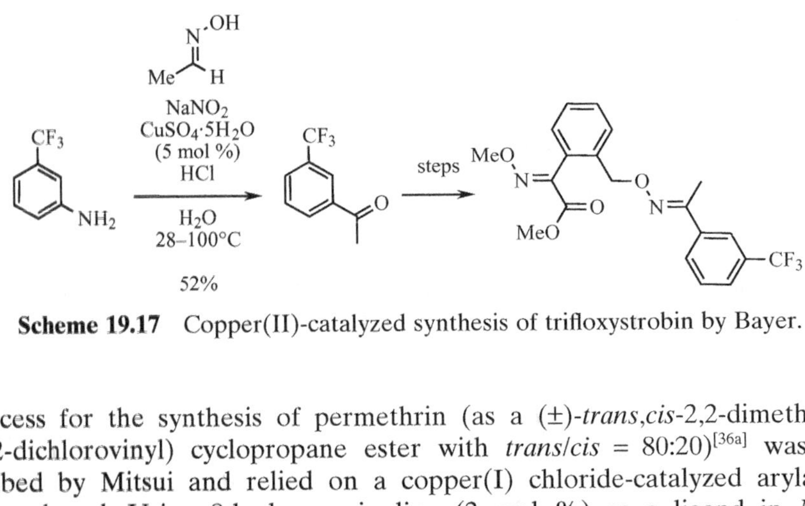

Scheme 19.17 Copper(II)-catalyzed synthesis of trifloxystrobin by Bayer.

process for the synthesis of permethrin (as a (±)-*trans*,*cis*-2,2-dimethyl-3-(2,2-dichlorovinyl) cyclopropane ester with *trans*/*cis* = 80:20)[36a] was described by Mitsui and relied on a copper(I) chloride-catalyzed arylation of a phenol. Using 8-hydroxyquinoline (2 mol %) as a ligand in *N*,*N*′-dimethylimidazolidinone at 150–170°C, an excellent yield was obtained for this chemoselective cross-coupling reaction (Scheme 19.16).[36b]

Going beyond classical Ullmann chemistry, copper-mediated transformations have also proved to be highly valuable for specific C–C coupling reactions such as cyanation reactions. An important illustration of this C–C coupling is a key step in Bayer's synthesis of the strobilurine fungicide trifloxystrobin (Scheme 19.17).[37] The reaction of 3-(trifluoromethyl)phenyldiazonium chloride with acetaldoxime in the presence of copper(II) sulfate (5 mol %) delivered the desired acetophenone in moderate yield. This Beech reaction[38] is a cost-effective approach to diversely substituted acetophenones or benzaldehydes.

2-Amino-5-cyano-*N*,3-dimethylbenzamide is a key building block in the production of Cyazypyr, a new insecticide for the control of lepidopterian pests by DuPont. The importance of patent protection in the area of process

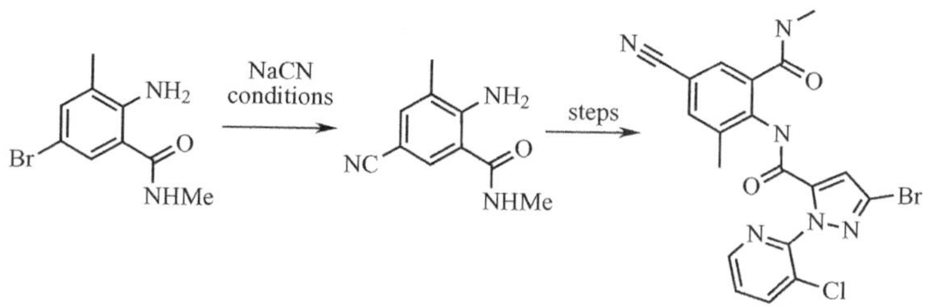

Conditions A[38a]	Conditions B[38b]	Conditions C[38c]
CuI (15 mol %)	CuI (25 mol %)	CuI (15 mol %)
1-butyl-1*H*-imidazole (85 mol %)	3,5-lutidine (60 mol %)	*N,N′*-dimethylethylene-diamine (85 mol %)
mesitylene	anisole	xylene
170°C	155°C	140°C
7 h	12 h	4.5 h
96%	94%	96%

Scheme 19.18 Cyanation reaction in the synthesis of the insecticide Cyazypyr (DuPont).

chemistry is illustrated by the number of applications describing variations of the cyanation reaction. Starting from 2-amino-5-bromo-*N*,3-dimethylbenzamide, copper(I) iodide (15–25 mol %) and sodium cyanide, different ligands were evaluated including 1-butyl-1*H*-imidazole, 3,5-lutidine, and *N,N′*-dimethylethylenediamine (Scheme 19.18). For the elimination of copper waste, ethylenediamine was employed for complexation.[39a]

An interesting copper-catalyzed amidation reaction was described by Syngenta[40] for the scale-up synthesis of the new fungicide isopyrazam, employing copper(I) bromide (20 mol %), *N,N′*-dimethylcyclohexanediamine as a ligand (44 mol %) and ethylene glycol diethyl ether as a solvent (Scheme 19.19). The order of the addition of reactants, catalyst, and ligand on conversion was found to be crucial by the authors. This observation was tentatively assigned to the chelation ability of the amide that could help in maintaining an active catalyst *in situ*.

19.4 CONCLUSION

A closer look at the industrial activities in the last decade in the area of copper-catalyzed C–N, C–O, and C–C couplings reveals that the new

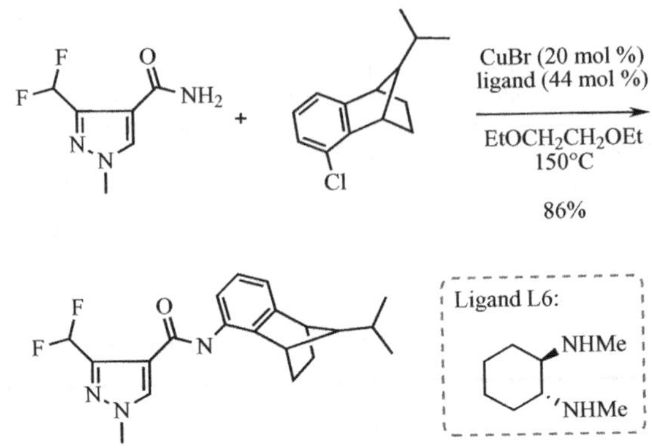

Scheme 19.19 Syngenta's approach to isopyrazam involving a copper(I)-catalyzed amidation reaction.

methodologies and protocols developed in academic laboratories are also being increasingly applied for the design of scale-up processes. As illustrated with some examples, specific efforts are being undertaken to substitute more expensive ligands by cheaper ones in order to maintain the advantage of low catalyst costs. Of particular note, these ligands also play an important role in the purification processes, which are of great importance in modern process chemistry. From our point of view, there is no doubt that Ullmann chemistry will assert its position as a key technology in industrial catalysis over the coming decades.

REFERENCES

[1] For recent reviews, see: (a) Slurry, D. S.; Buchwald, S. L. *Angew. Chem.* **2008**, *120*, 6438–6431; *Angew. Chem. Int. Ed.* **2008**, *47*, 6338–6361. (b) Torborg, C.; Beller, M. *Adv. Synth. Catal.* **2009**, *351*, 3027–3043. (c) Surry, D. S.; Buchwald, S. L. *Chem. Sci.* **2011**, *2*, 27–50.

[2] (a) Ullmann, F.; Bielecki, J. *Ber. Dtsch. Chem. Ges.* **1901**, *34*, 2174–2185. (b) Ullmann, F. *Ber. Dtsch. Chem. Ges.* **1903**, *36*, 2382–2384. (c) Ullmann, F.; Sponagel, P. *Ber. Dtsch. Chem. Ges.* **1905**, *38*, 2211–2212. (d) Goldberg, I. *Ber. Dtsch. Chem. Ges.* **1906**, *39*, 1691–1692.

[3] For reviews, see (a) Ley, S. V.; Thomas, A. W. *Angew. Chem.* **2003**, *115*, 5558–5607; *Angew. Chem. Int. Ed.* **2003**, *42*, 5400–5449. (b) Kunz, K.; Scholz, U.; Ganzer, D. *Synlett* **2003**, 2428–2439. (c) Beletskaya, I. P.; Cheprakov, A. V. *Coord. Chem. Rev.* **2004**, *248*, 2337–2364. (d) Carril, M.; SanMartin, R.; Dominguez, E. *Chem. Soc. Rev.* **2008**, *37*, 639–647. (e) Monnier, F.; Taillefer, M. *Angew. Chem.* **2009**, *121*, 7088–7105; *Angew. Chem. Int. Ed.* **2009**, *48*, 6954–6971. (f) Corbet, J.-P.; Mignani, G. *Chem. Rev.* **2006**, *106*, 2651–2710. (g) Evano, G.; Blanchard, N.; Toumi, M. *Chem. Rev.* **2008**, *108*, 3054–3131. (h) Magano, J.; Dunetz, J. R. *Chem. Rev.* **2011**, *111*, 2177–2250.

[4] Buchwald, S. L.; Klapars, A.; Kwong, F. Y.; Streiter, E.; Zanon, J., inventors; Massachusetts Institute of Technology, assignee. Copper-catalyzed formation of carbon-heteroatom and carbon-carbon bonds. PCT Int. Appl. WO 2004013094 (A2), 2004 February 12.

[5] Cellier, P. P.; Cristau, H.-J.; Spindler, J.-F.; Taillefer, M., inventors; Rhodia Company, assignee. Process for arylating or vinylating or alkynating a nucleophilic compound. U.S. patent 20030236413 (A1), 2003 December 25.

[6] Kunz, K.; Scholz, U.; Gaertzen, O.; Ganzer, D.; Wesener, J., inventors; Bayer AG, assignee. Preparation of copper carbene complexes and their use as coupling reaction catalysts. EP Patent 1437355 (A1), 2004 July 14.

[7] MetalPrices.com. http://www.metalprices.com.

[8] Ellis, G. P. *Chem. Rev.* **1987**, *87*, 779–794.

[9] Ritter, J. C.; Moloy, K. G.; Pollino, J. M.; Mahajan, S., inventors; DuPont De Nemours Company, assignee. Process for the synthesis of fluorinated ethers of aromatic acids. PCT Int. Appl. WO 2011028852 (A2), 2011 March 10.

[10] Buck, E.; Song, Z. J.; Tschaen, D.; Dormer, P. G.; Volante, R. P.; Reider, P. J. *Org. Lett.* **2002**, *4*, 1623–1626.

[11] Ritter, J. C., inventor; DuPont De Nemours Company, assignee. Process for the synthesis of ethers of aromatic acids. PCT Int. Appl. WO 2008079225 A1, 2008 July 3.

[12] (a) Ritter, J. C., inventor; DuPont De Nemours Company, assignee. Process for preparation of 2,6-diaminopyridines from 2-amino-6-chloropyridines and aqueous ammonia in the presence of a copper source. PCT Int. Appl. WO 2009018504 (A1), 2009 February 5. (b) Wibaut, J. P.; Nicolai, J. R. *Rec. Trav. Chim.* **1939**, *58*, 709–721.

[13] Cosimbescu, L., inventor; Eastman Kodak Company, assignee. One step synthesis for quinacridone compounds. U.S. patent 6664396 (B1), 2003 December 16.

[14] Gujadhur, R. K.; Bates, C. G.; Venkataraman, D. *Org. Lett.* **2001**, *3*, 4315–4317.

[15] Haider, J.; Scholz, U.; Sicheneder, H., inventors; Bayer AG, assignee. Process for the preparation of aminodiphenylamines. EP Patent 1437340(A1), 2004 July 14.

[16] Kunz, K.; Haider, J.; Ganzer, D.; Scholz, U.; Sicheneder, A., inventors; Bayer AG, assignee. Process for the preparation of nitrodiphenylamines. EP Patent 1437341(A1), 2004 July 14.

[17] Haider, J.; Kunz, K.; Scholz, U. *Adv. Synth. Catal.* **2004**, *346*, 717–722.

[18] Kubo, S.; Shintou, T.; Aoki, H., inventors; Sankyo Company, assignee. Process for preparation of arylamines. PCT Int. Appl. WO 2004024670, 2004 March 25.

[19] Yoshizaki, H., inventor; Koei Chemical Company, Sumitomo Chemical Company, assignees. Preparation of 4,4′-biscarbazolylarenes via tetrahydrocarbazoles. JP 2008031145 A, 2008 February 14.

[20] Ribecai, A.; Bacchi, S.; Delpogetto, M.; Guelfi, S.; Manzo, A. M.; Perboni, A.; Stabile, P.; Westerduin, P.; Hourdin, M.; Rossi, S.; Provera, S.; Turco, L. *Org. Proc. Res. Dev.* **2010**, *14*, 895–901.

[21] (a) Pu, Y.-M.; Ku, Y.-Y.; Grieme, T.; Black, L. A.; Bhatia, A. V.; Cowart, M. *Org. Proc. Res. Dev.* **2007**, *11*, 1004–1009. (b) For a recent example, see: Kallemeyn, J. M.; Ku, Y.-Y.; Mulhern, M. M.; Bishop, R.; Pal, A.; Jacob, L. *Org. Proc. Res. Dev.* **2013**, doi:10.1021/op3002883.

[22] Pu, Y.-M.; Ku, Y.-Y.; Grieme, T.; Bhatia, A. V. *Tetrahedron Lett.* **2006**, *47*, 149–153.

[23] (a) Ghosh, A.; Sieser, J. E.; Caron, S.; Couturier, M.; Dupont-Gaudet, K.; Girardin, M. *J. Org. Chem.* **2005**, *71*, 1258–1261. (b) Ghosh, A.; Sieser, J. E.; Caron, S.; Watson, T. J. N. *Chem. Commun.* **2002**, 1644–1645.

[24] Enache, L. A.; Kennedy, I.; Sullins, D. W.; Chen, W.; Ristic, D.; Stahl, G. L.; Dzekhtser, S.; Erickson, R. A.; Yan, C.; Muellner, F. W.; Krohn, M. D.; Winger, J.; Sandanayaka, V.; Singh, J.; Zembower, D. E.; Kiselyov, A. S. *Org. Proc. Res. Dev.* **2009**, *13*, 1177–1184.

[25] Campeau, L.-C.; Dolman, S. J.; Gauvreau, D.; Corley, E.; Liu, J.; Guidry, E. N.; Ouellet, S. G.; Steinhuebel, D.; Weisel, M.; O'Shea, P. D. *Org. Proc. Res. Dev.* **2011**, *15*, 1138–1148.

[26] Varshney, A. K.; Vidyasagar, A.; Nalawala, K. J.; Murthy, C. S.; Mehta, M. H., inventors, Gujarat State Fertilizers Company, assignee. A process for the synthesis of 3,4,5-trimethoxybenzaldehyde starting from p-cresol. IN 180997, 1998 April 11.

[27] (a) Gutman, A. L.; Nisnevich, G.; Shkolnik, E.; Zaltzman, I.; Tishin, B., inventors; Par Pharmaceutical Company, assignee. Process and a novel intermediate for the preparation of flecainide. PCT Int. Appl. WO 9902498 (A1), 1999 January 21. (b) Gutman, A. L.; Nisnevich, G.; Shkolnik, E.; Zaltzman, I.; Tishin, B., inventors; Par Pharmaceutical Company, assignee. Process for the preparation of trifluoroethoxyarenecarboxylic acids. PCT Int. Appl. WO 9847853 (A1), 1998 October 29.

[28] Malmgren, H.; Bäckström, B.; Sølver, E.; Wennerberg, J. *Org. Proc. Res. Dev.* **2008**, *12*, 1195–1200.

[29] Caron, S.; Do, N. M.; Sieser, J. E.; Whritenour, D. C.; Hill, P. D. *Org. Proc. Res. Dev.* **2009**, *13*, 324–330.

[30] Zanka, A.; Kubota, A.; Hirabayashi, S.; Nakamura, H. *Org. Proc. Res. Dev.* **1998**, *2*, 71–77.

[31] Poechlauer, P.; Braune, S.; de Vries, A.; May, O. *Chim. Oggi* **2010**, *28*, 14–17.

[32] Graham, J. P.; Langlade, N.; Northall, J. M.; Roberts, A. J.; Whitehead, A. J. *Org. Proc. Res. Dev.* **2011**, *14*, 44–48.

[33] (a) Drabek, J.; Boeger, M., inventors; Ciba-Geigy AG, assignee. Antiparasitic phenoxyphenylisothioureas and intermediates in their production. FR2465720, 1981 March 27. (b) Haessig, R., inventor; Ciba-Geigy AG, assignee. Preparation of 4-phenoxyphenyl isothiocyanates. U.S. Patent 4997967, 1991 March 5.

[34] Schareina, T.; Zapf, A.; Cotté, A.; Müller, N.; Beller, M. *Org. Proc. Res. Dev.* **2008**, *12*, 537–539.

[35] Cotté, A.; Mueller, N.; Gotta, M.; Beller, M.; Schareina, T.; Zapf, A., inventors; Saltigo GmbH, assignee. Improved process for the catalytic synthesis of diaryl ethers. U.S. patent 20090143594 (A1), 2009 June 4.

[36] (a) Elliott, M.; Farnham, A. W.; Janes, N. F.; Needham, P. H.; Pulman, D. A.; Stevenson, J. H. *Nature* **1973**, *246*, 169–170. (b) Takenaka, S.; Oi, R., inventors; Mitsui Toastu Chemicals Inc., assignee. m-Phenoxybenzyl alcohol. EP patent 202838 (B1), 1991 September 18.

[37] (a) Isenring, H. P.; Weiss, B.; Ciba-Geigy AG, assignee. Preparation of [(iminooxy) tolyl]acrylates and analogs as fungicides. EP 0460575(A1), 1991 December 11.

(b) Kühnle, W.; Marhold, A.; Steffan, G.; Bayer AG, assignee. Preparation of fluoroalkyl-substituted acetophenones. U.S. patent 6,184,417 B1, 2001 February 6.

[38] (a) Beech, W. F. *J. Chem. Soc.* **1954**, 1297–1302. (b) Herr, R. J.; Fairfax, D. J.; Meckler, H.; Wilson, J. D. *Org. Proc. Res. Dev.* **2002**, *6*, 677–681.

[39] (a) Dumas, D. J., inventor; DuPont De Nemours Company, assignee. Process for preparing 2-amino-5-cyanobenzoic acid derivatives by cyanation of 2-amino-5-halobenzoic acid derivatives using metal cyanide reagents and a pyridine derivative. PCT Int. Appl. WO 2009111553 (A1), 2009 September 11. (b) Dumas, D. J.; Casalnuovo, A. L., inventors; DuPont De Nemours Company, assignee. Process for preparing 2-amino-5-cyanobenzoic acid derivatives by cyanation of 2-amino-5-halobenzoic acid derivatives using metal cyanide reagents and a pyridine derivative. PCT Int. Appl. WO 2009085816 (A1), 2009 July 9. (c) Annis, G. D.; Bruening, J.; Currie, M. J.; Dumas, D. J.; Shapiro, R., inventors; DuPont De Nemours Company, assignee. Process for preparation of 2-amino-5-cyanobenzamides from the corresponding 5-halo compounds using metal cyanides in the presence of cuprous salts, iodides, and amines. PCT Int. Appl. WO 2008082502 (A2), 2008 July 10.

[40] Hodges, G. R.; Mitchell, L.; Robinson, A. J., inventors; Syngenta Company, assignee. Process for the preparation of N-aryl 4-pyrazolecarboxamides using copper-catalyzed arylation as key step. PCT Int. Appl. WO 2010072632 (A1), 2010 July 1.

20

REUSABLE CATALYSTS FOR COPPER-MEDIATED CROSS-COUPLING REACTIONS UNDER HETEROGENEOUS CONDITIONS

Zhiyong Wang, Changfeng Wan, and Ye Wang

20.1 INTRODUCTION

Many issues have emerged and become more and more serious during the development of society, such as resource and environmental problems, and these problems must be urgently solved for sustainable scientific development and social progress. Therefore, developing efficient, atom-economical, and practical organic reactions is currently one of the most important targets for organic chemists. In the past decades, synthetic organic chemistry has made great progress, and many highly effective and novel reactions were discovered. For example, late-transition metal catalysts, such as palladium,[1] ruthenium,[2] rhodium,[3] gold,[4] and platinum,[5] possess powerful catalytic activity for many important organic reactions. However, their relatively high cost and considerable toxicity limit the scope of their applications and possible improvements. Owing to their own obvious advantages and unique features, the use of first-row transition metals, especially copper and iron, in organic synthesis has been extensively explored by chemists in recent years.[6]

With its unique characteristics, such as low toxicity, low price, and powerful catalytic ability, copper used as a catalyst has already been widely applied in organic synthesis. Among copper-mediated organic reactions, copper-catalyzed

Copper-Mediated Cross-Coupling Reactions, First Edition. Edited by Gwilherm Evano and Nicolas Blanchard.

C–C, C–N, and C–O Ullmann-type coupling reactions are extremely important reactions and were extensively developed in recent years.[7] With the discovery of versatile and efficient new copper/ligand systems for the formation of C–C, C–N, and C–O bonds, it is now possible to catalyze the Ullmann-type coupling reaction under much milder conditions. Many research groups developed highly efficient catalytic systems and made important contributions to the development of these copper-catalyzed transformations.[8] These important discoveries, along with the promise of being able to use only catalytic amounts of copper species in place of the more expensive palladium complexes, have attracted considerable interest in Ullmann-type reactions again, and the main challenge that remains is to devise more effective synthetic procedures to improve reaction efficiency. In addition, the development of heterogeneous copper-based systems is another attractive area of research. Indeed, heterogeneous catalysis is a clean and environmentally friendly technology, and heterogeneous catalysts not only can efficiently catalyze organic reactions but also are repeatedly used without obvious decrease in catalytic activity, at least in ideal cases.

Over the last decade, and due to the development of nanoparticle technology and loading methods, chemists have prepared a variety of transition-metal nanoparticles[9] (including magnetically retrievable ones)[9i–j] and supported metals with different carriers, such as copper nanoparticles,[9a,b] palladium nanoparticles,[9c] mesoporous silica-supported copper oxide, and kaolin-supported copper oxide. In addition, these metallic nanoparticles and nanoparticle composites were successfully applied in organic reactions and gave promising reaction results.[10] Among these heterogeneous catalysts, copper nanoparticles and supported copper were extensively used as catalysts in copper-catalyzed coupling reactions because of their availability, low price, and high catalytic efficiency.[9a,b] Combining the advantages of both copper-catalyzed coupling chemistry and heterogeneous catalysis, the significance of heterogeneous copper-catalyzed coupling reactions has stimulated rapid development up to now. This chapter gives a general analysis and discusses recent advances in heterogeneous copper-catalyzed coupling reactions. In the first section, key bond disconnections (C–C, C–N, C–O, C–S, and C–Se) will be discussed using copper nanoparticle-catalyzed cross couplings, and a second section will focus on the use of supported copper catalysts with an emphasis on the synthesis, recovery, and reuse of these promising catalytic systems.

20.2 COPPER NANOPARTICLE-CATALYZED CROSS-COUPLING REACTIONS

20.2.1 C–C Bond Formation

The Hurtley condensation is a key reaction that can directly construct C–C bonds using activated methylene compounds and aryl halides as substrates, as extensively described in Chapter 8.[11] In recent years, the Hurtley condensation has been further explored, and remarkable advances have been made in the development of efficient catalytic systems for this transformation. Just to mention two key and representative examples, Kwong and coworkers described a mild catalytic system that allowed the arylation of diethylmalonate with a broad range of aryl iodides at 25°C in the presence of 2-picolinic acid,[12] and the first enantioselective Hurtley condensation was reported in 2006 by Ma and coworkers.[13]

Most available methods are based on the use of chelating ligands to promote coupling reactions, and significant efforts have been devoted to the development of heterogeneous systems for this reaction. In this context, Kidwai and coworkers described an efficient catalytic Hurtley coupling reaction by using CuO nanoparticles (10 mol %) as a heterogeneous catalyst under ligand-free conditions, which provided the desired products in good yields (Scheme 20.1).[14] The authors performed recycling experiments, and the CuO-nanoparticle catalyst could be reused up to four cycles without significant loss of activity, with the shape and size of the catalyst being unchanged during the reaction.

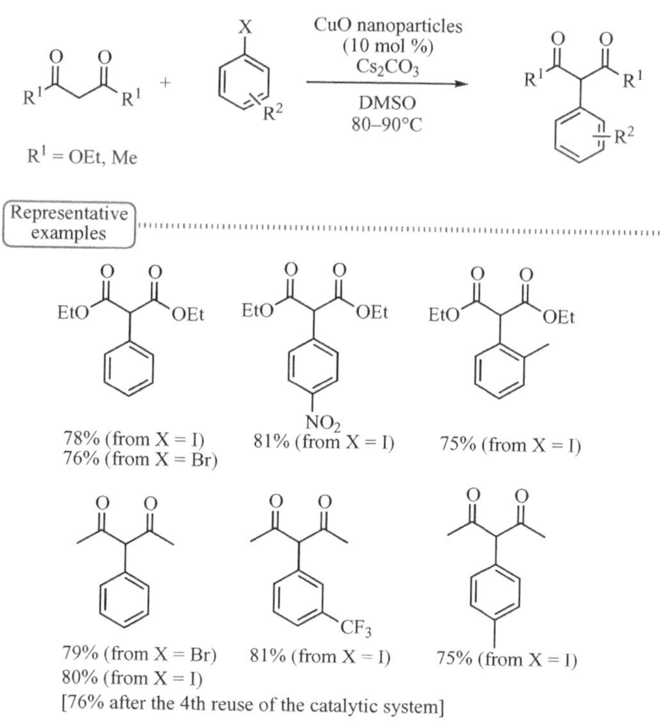

Scheme 20.1 CuO nanoparticle-catalyzed Hurtley reaction.

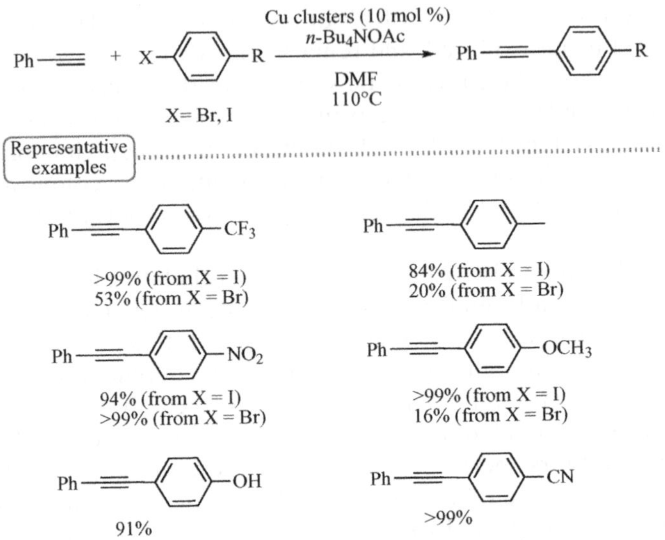

Scheme 20.2 Copper nanocluster-catalyzed Sonogashira reaction.

A useful method to prepare aryl alkynes is the Sonogashira reaction, which corresponds to a palladium/copper cocatalyzed coupling reaction between aryl halides and terminal acetylenes. In 2004, Rothenberg and coworkers reported the first example of copper nanocluster-catalyzed Sonogashira reaction (Scheme 20.2).[15] In their studies, an efficient, inexpensive, and air-stable catalyst was developed, which could be used three times without deactivation in the case of the cross coupling of phenylacetylene and 4-iodobenzotrifluoride, giving a final turnover number of 73. It is noteworthy that tetraoctylammonium formate was required to form an organic corona around the clusters, which prevented their aggregation. In contrast with iodobenzene, bromobenzene as a substrate was less reactive although recent studies have greatly expanded the scope of the Sonogashira reaction of aryl chloride.[15b]

In 2006, Li and coworkers developed a highly efficient copper(I) oxide nanoparticle/tri(*o*-tolyl)phosphine/tetrabutylammonium bromide (TBAB) system to catalyze the coupling reaction of aryl halides with aryl-, vinyl-, and alkynylstannanes (Scheme 20.3).[16] It was found that copper(I) oxide nanoparticles showed a higher catalytic activity than either common Cu_2O or copper(I) halides. This Stille coupling reaction gave the corresponding products in good yields and could tolerate different substituents on the aromatic ring. Moreover, the recycling experiments demonstrated that the copper(I) oxide nanoparticle/tri(*o*-tolyl)phosphine/TBAB catalytic system can be recovered and reused at least three times without loss of activity.

Calò and coworkers reported in 2005 a copper nanocolloid-catalyzed Heck-type reaction of aryl iodides and activated bromides with acrylates by using TBAB as a solvent and tetrabutylammonium acetate as a base (Scheme 20.4).[17] The copper nanoparticles were derived from the reaction of

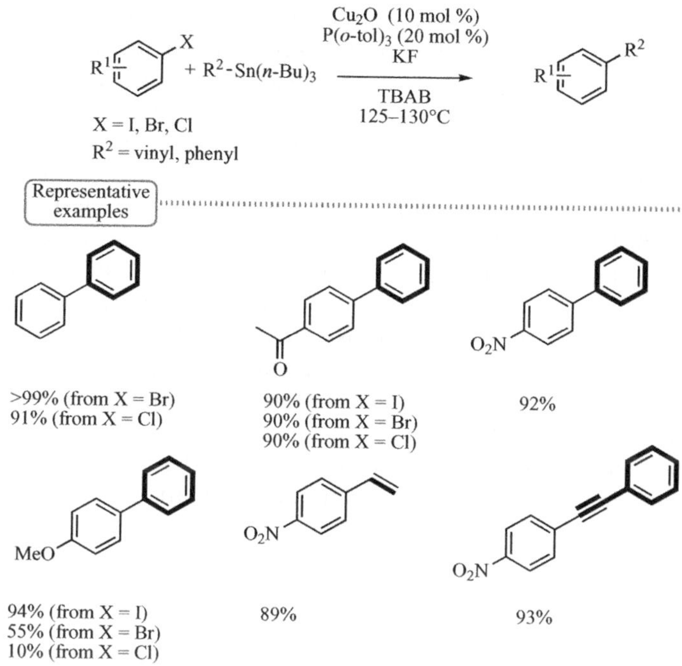

Scheme 20.3 A highly efficient and heterogeneous Stille coupling reaction based on Cu_2O nanoparticle/P(*o*-tol)$_3$/TBAB catalytic system (group originating from the organotin reagent shown in bold).

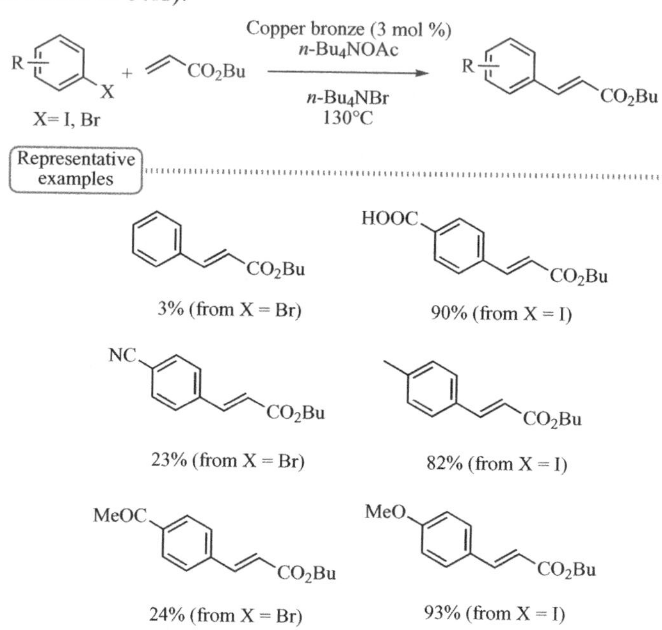

Scheme 20.4 Copper nanocolloid-catalyzed Heck coupling reaction.

iodobenzene with copper bronze and were homogeneously dispersed into TBAB. Aryl iodides as the coupling partners provided higher reaction efficiencies in contrast to aryl bromides. As with previous systems, the catalytic system was readily recovered, still possessed high catalytic activity after 15 runs, and could be stored for months without loss of catalytic efficiency.

As this section demonstrated, copper nanoparticles are efficient catalytic species for the creation of carbon–carbon bonds via the classical Hurtley, Sonogashira, Stille, or Heck cross coupling.[15b] The synthesis of these nanoparticles is usually simple and their catalytic activity could be maintained throughout consecutive recycles, thus increasing considerably interest in their use in the context of sustainable organic transformations. Besides the formation of carbon–carbon bonds, copper nanoparticles have found widespread applications in the formation of diverse carbon–heteroatom bonds, such as C–N, C–O, and C–Se bonds. Selected representative examples of these three types of copper nanoparticle-catalyzed transformations will be presented in the next sections, beginning with carbon–nitrogen bond formation.

20.2.2 C–N Bond Formation

N-Arylated structures are important subunits found in numerous natural products of importance in the fields of biological, pharmaceutical, and materials sciences. A literature survey shows that the arylation reaction of nitrogen nucleophiles was extensively explored and studied as described in Chapter 1. So far, a large number of reports on the modern catalytic Ullmann-type reactions catalyzed by copper catalysts have focused on the formation of C–N bonds, and these studies have been extended to reusable systems.[7,18] In this section, selected examples of copper nanoparticle-catalyzed reactions between aryl iodides (and bromides) and ammonia, anilines, and amides will be presented. A discussion on the use of the less activated aryl chlorides will close this section.

In 2011, Xu and coworkers reported a selective synthesis of anilines by using $NH_3{\cdot}H_2O$ as the nucleophile in the presence of copper(I) iodide nanoparticles (1.5 mol %) in water at room temperature (Scheme 20.5, Eq. 1).[19a] Aryl iodides or aryl bromides react with $NH_3{\cdot}H_2O$ to provide the desired products in good yields. Electronic effects had an obvious influence on the reaction efficiency and aryl halides with electron-withdrawing substituents performed better. Although a slight decrease in catalytic activity was observed after three cycles, the shape and size of the copper(I) iodide nanoparticles almost did not change, which demonstrated that this heterogeneous catalyst was stable in the reaction process. A related reaction based on copper nanobelts as a highly porous catalyst has been reported by Bao, Dai, and coworkers (Scheme 20.5, Eq. 2).[19b] This one-dimensional ribbonlike nanostructure possessing a thickness of 40–50 nm is an efficient catalyst for the synthesis of aniline from phenylboronic acid in methanol at room temperature even after three consecutive reactions.

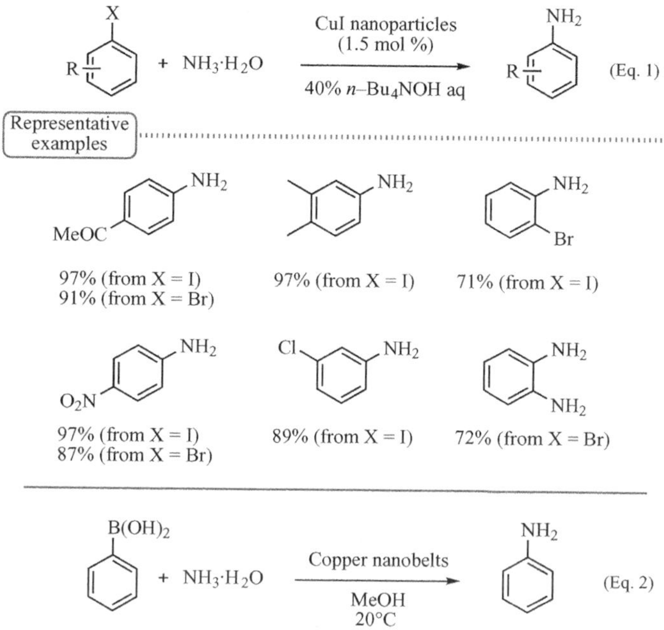

Scheme 20.5 Copper(I) nanoparticle- and copper nanobelt-catalyzed synthesis of anilines from aryl iodides, bromides, and phenylboronic acid using $NH_3{\cdot}H_2O$.

A catalytic system has been reported by Punniyamurthy and coworkers in which recyclable copper oxide nanoparticles are used to catalyze the cross-coupling reaction of anilines with aryl iodides in high yields (Scheme 20.6).[20] This inexpensive catalyst, which operates in air under ligand-free conditions, enabled the condensation of various aryl iodides with N-nucleophiles such as aromatic nitrogen heterocycles and amines. Moreover, the copper oxide nanoparticle catalyst demonstrated high catalytic activity after three runs and still promoted the reaction to generate the desired product in 91% yield.

Besides the above-mentioned report, the Punniyamurthy group also described in 2009 a system for the amidation of aryl iodides catalyzed by reusable copper(I) oxide nanoparticles (7–10 nm in diameter) as a catalyst and PEG_{4000} as a solvent and stabilizer of the catalyst (Scheme 20.7).[21] In the process, aromatic and aliphatic amides were employed as substrates, and the reaction provided the desired coupling products in moderate-to-good yields under ligand-free conditions. Moreover, the recycling experiments demonstrated that the catalyst could be easily recovered by centrifugation and reused three times without any drop in yield, as demonstrated in the synthesis of *p*-benzotoluidide (79% for the first run, then 78% for the next two runs).

Following this report, Wei and coworkers reported the amidation reaction of aryl halides catalyzed by efficient and recyclable copper(I) oxide nanocubes (200–300 nm in diameter) and nanospheres (100–200 nm in diameter)

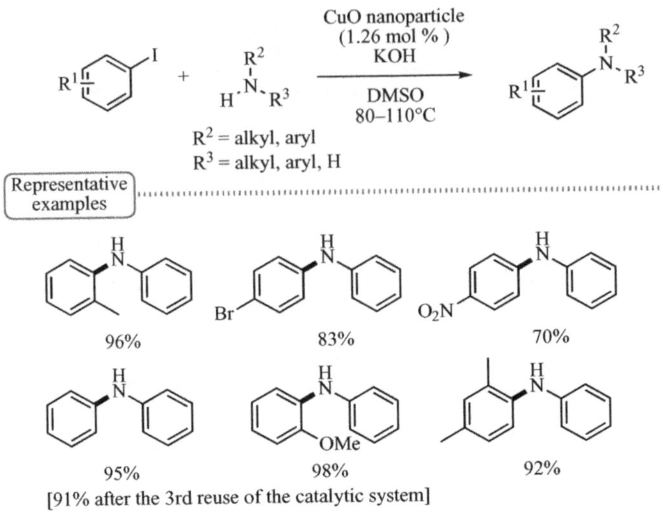

Scheme 20.6 Recyclable CuO nanoparticle-catalyzed C–N cross-coupling reaction of anilines with aryl iodides.

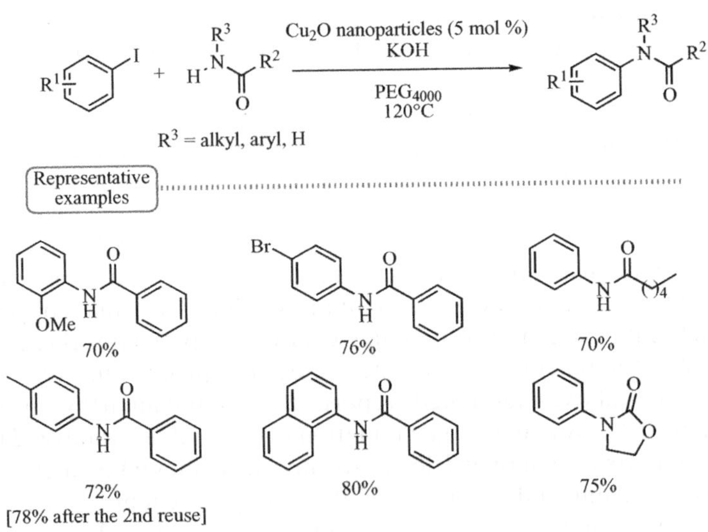

Scheme 20.7 Reusable Cu_2O nanoparticle-catalyzed amidation of aryl iodides.

(Scheme 20.8).[22] In this C–N coupling reaction, a relatively low catalyst loading of 5 mol % was used in the reaction together with 10 mol % dimethylethylenediamine as the ligand and potassium phosphate as the base in toluene. Compared with the 2009 Punniyamurthy report,[21] the yields were generally higher. No obvious loss in the catalytic activity was observed after

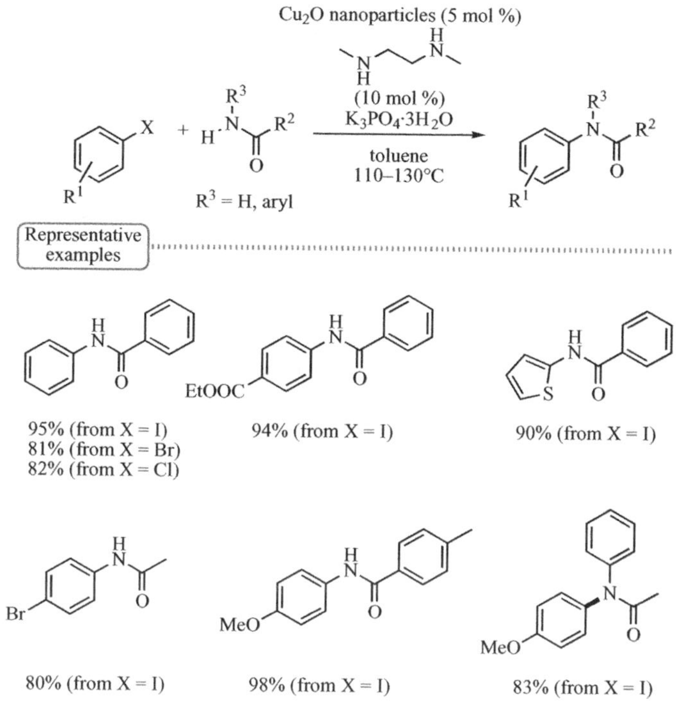

Scheme 20.8 Amidation of aryl halides catalyzed by recyclable Cu_2O nanoparticles.

the catalyst was reused for five times. In addition, the scanning electron microscopy and powder X-ray diffraction analyses showed that copper(I) oxide after five runs was identical with the fresh copper(I) oxide in particle shape, size, and diffraction pattern. These experimental results indicated that the reaction involved a heterogeneous process and that the catalysis may occur on the surface of the Cu_2O nanoparticles. It is worth noting that the recovering of the nanoparticles by simple centrifugation is quite straightforward.

In 2007, Kantam and coworkers employed nanocrystalline copper(I) oxide as an effective heterogeneous catalyst for the N-arylation of heterocycles with activated chloroarenes (Scheme 20.9).[23] This nanocrystalline copper(I) oxide possesses a high surface area (136 $m^2.g^{-1}$ and a crystallite size of 7–9 nm), which makes it more attractive than bulk copper(I) oxide (surface area of 1.9 $m^2.g^{-1}$). In this coupling reaction, electronic effects played an important role in the reaction efficiency and only chloroarene with electron-withdrawing groups provided the desired coupling product in good yields. Recycling experiments demonstrated that the catalyst still possessed a high catalytic activity after five runs as shown in the synthesis of 4-(1*H*-imidazol-1-yl)benzonitrile.

In 2009, the Sreedhar group employed copper(I) iodide nanoparticles as an efficient catalyst to realize the cross coupling of secondary amines and

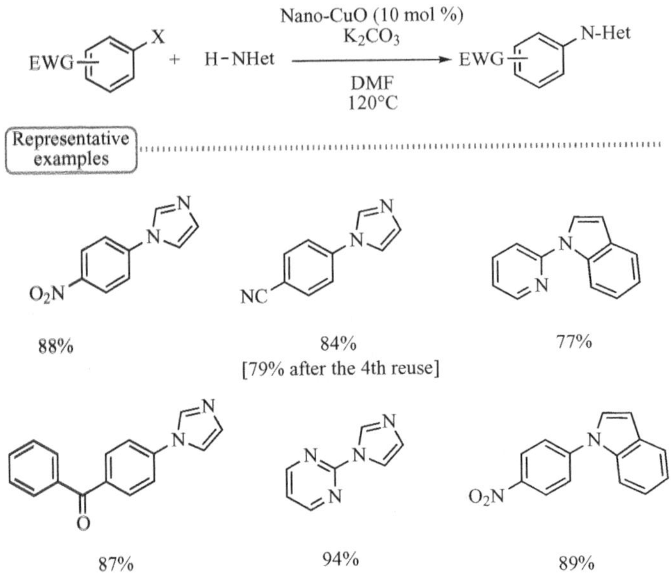

Scheme 20.9 Nanocrystalline copper oxide-catalyzed N-arylation of heterocycles with activated chloroarenes.

π-excessive nitrogen heterocycles with aryl chlorides to afford various arylated nitrogen-containing heterocyclic compounds (Scheme 20.10).[24] In contrast with the traditional methods for the formation of C–N bonds, the procedure featured low loading of catalysts, use of a weak base, and a high catalytic activity. Additionally, the catalyst was easily recoverable and reused without obvious loss in the catalytic activity.

Copper(I) nanoparticles with different morphologies have thus proved to be valuable catalysts for the formation of carbon–nitrogen bonds. The high surface area of the nanoparticles compared with the one of the bulk material clearly makes a difference. The recovery of the catalyst, usually by simple centrifugation followed by washing, also contributes to the attractiveness of these methods. Based on the same concepts, the formation of carbon–sulfur bonds catalyzed by copper nanoparticles has been extensively studied and is discussed in the next section.

20.2.3 C–S Bond Formation

Aryl thioethers represent an important class of compounds in biochemistry, and the transition-metal-catalyzed cross-coupling reaction of thiols with aryl halides now represents the most common strategy for constructing C–S bonds. Among these methods, the copper-catalyzed (Ullmann-type) C–S coupling reaction[25] is the most popular approach due to its low cost in comparison with most of the other metal catalysts.

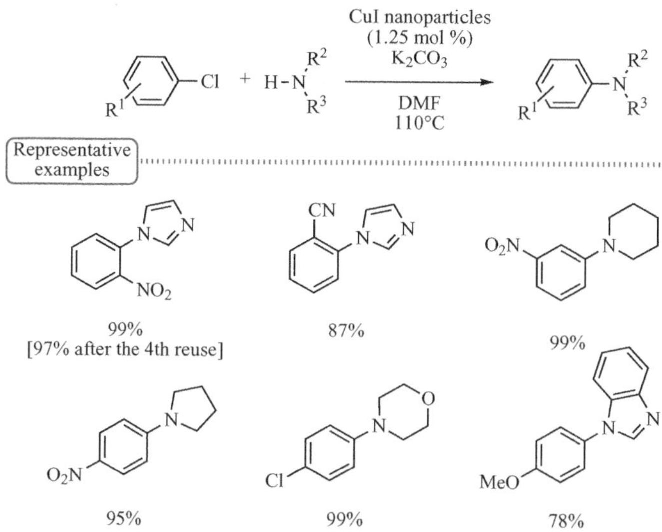

Scheme 20.10 Copper(I) nanoparticle-catalyzed amination of aryl chlorides.

Punniyamurthy and coworkers reported an efficient catalytic system in 2007 to provide the cross-coupling products in moderate-to-good yields, in which aryl iodides and various thiols were treated with copper oxide nanoparticles by using postassium hydroxide as a base in DMSO (Scheme 20.11).[26] This reaction can be carried out under a nitrogen atmosphere without the assistance of any ligand, even if DMSO might actually play this role. The good catalytic activity remained after three sequential coupling reactions of thiols with iodobenzene. When bromobenzene was chosen as the reaction partner, however, the reaction provided the desired product in lower yield under the same conditions.

Ranu and coworkers reported an efficient and convenient methodology for the cross coupling of aryl iodides and thiophenols using copper nanoparticles of 4–6 nm in diameter (Scheme 20.12).[27] In contrast to common copper powder, the copper nanoparticles possessed a higher catalytic efficiency, and investigation of the scope of substrates showed that aryl and benzyl thiols led to the desired products in good yields, while aliphatic thiols displayed a lower reactivity. Additionally, the reactions took place with the assistance of microwave irradiation (120°C, 250 W) without any involvement of ligand. However, no mention of recycling experiments was made in this study.

Nageswar and coworkers developed a green and economical method for the synthesis of symmetrical aryl sulfides by employing recyclable copper oxide nanoparticles as a heterogeneous catalyst (Scheme 20.13).[28a] In this reaction, readily available thiourea was chosen as an effective sulfur surrogate to react with aromatic halides and provided the corresponding coupling

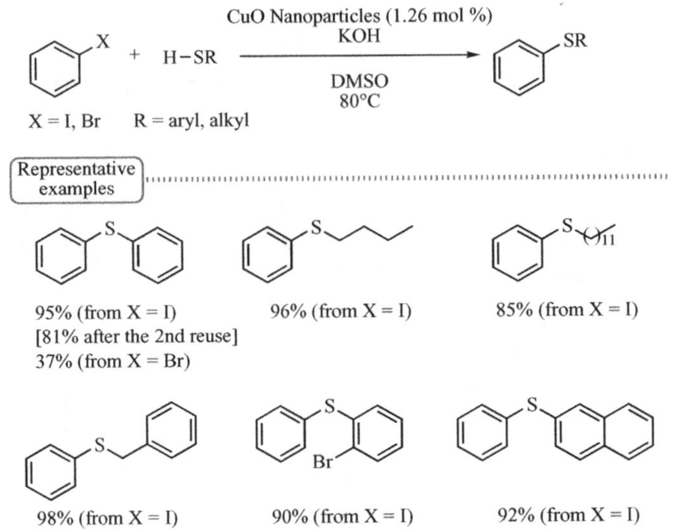

Scheme 20.11 Copper oxide nanoparticle-catalyzed thiolation of aryl iodides.

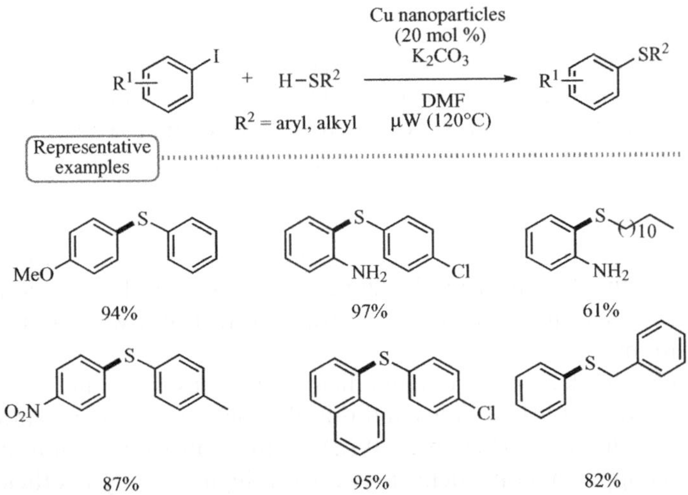

Scheme 20.12 Copper nanoparticle-catalyzed thiolation of aryl iodides under microwave irradiation.

products in moderate-to-good yields. Although no mechanistic details were provided in these studies, it is worth mentioning that thiourea and aminothiourea have already been used for the synthesis of diarylsulfide via photoinduced[28b] or copper-catalyzed reactions, respectively.[28c] Aryl halides with electron-withdrawing as well as electron-donating substituents provided the desired product in good yield (Scheme 20.13). Heterocyclic iodides also gave

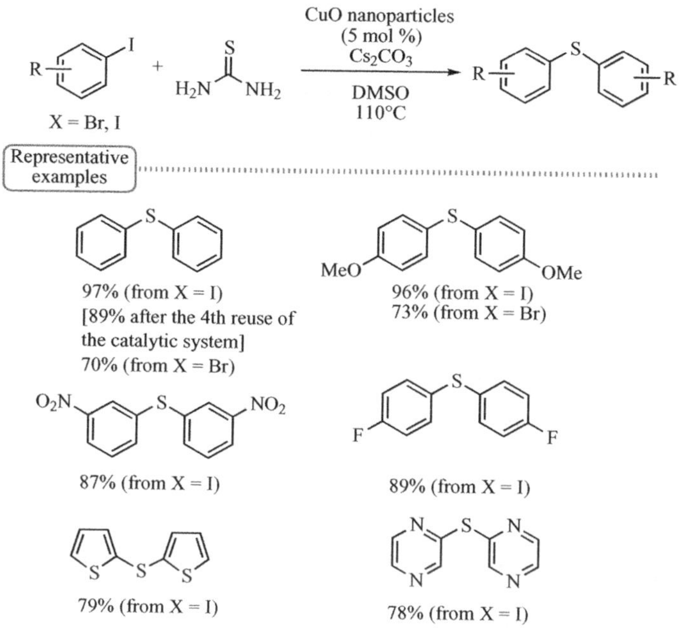

Scheme 20.13 Copper oxide nanoparticle-catalyzed thiolation of aryl halides using thiourea.

the corresponding products in modest yields under the same condition. The catalyst was easily recovered by simple centrifugation after completion of the reaction, and the recovered catalyst was used directly in the next cycle. The catalyst still promoted the coupling reaction to generate the product in 89% yield after the fourth cycle, and no obvious loss of the catalytic activity was observed.

The recent progress that has been made in the copper(I) nanoparticle-catalyzed synthesis of carbon–sulfur bonds illustrated in these examples shows that simple protocols are now available to synthetic chemists. The stability of the catalysts allows for their recycling; microwave irradiation can dramatically increase the rate of cross coupling; and simple non-foul-smelling sources of sulfur have been used. To conclude this section, the impact of copper nanoparticles in the context of the formation of carbon–oxygen bonds will now be discussed.

20.2.4 C–O Bond Formation

The diaryl ether moiety is a common structural unit widely found in biologically active molecules and natural products as well as in monomers for the synthesis of functional polymers. The synthesis of diaryl ethers and their derivatives by classic Ullmann methods has been well known for a long time.

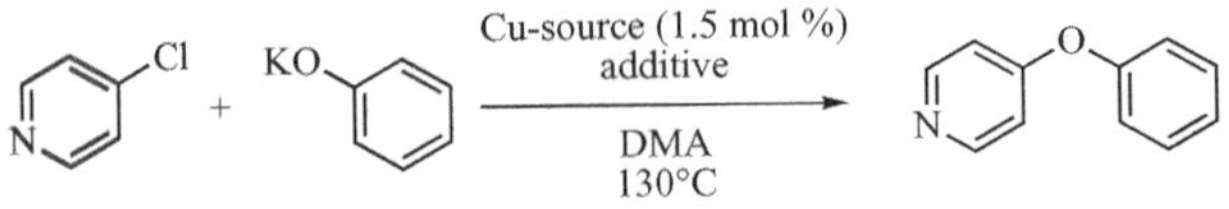

Cu-source	additive	yield (%)
Nano-Cu	–	75
Nano-Cu	18-crown-6	85
Nano-CuSn	–	82
Nano-CuSn	18-crown-6	87
Nano-CuZn	–	87
Nano-CuZn	18-crown-6	90

Scheme 20.14 Copper-based bimetallic nanoparticle-catalyzed coupling of 4-chloropyridine with potassium phenoxide.

Despite the recent renaissance of the catalytic Ullmann method, fewer publications have appeared for the C–O than for the C–N coupling reactions. Only a few efficient and recyclable catalytic systems for the condensation of aryl halides with various phenols have been reported and will be overviewed in this section.[29]

In 2010, Schouten, Wheatley, and coworkers employed bimetallic copper nanoparticles in the Ullmann etherification reaction (Scheme 20.14).[30a] The authors first prepared the oxidatively stable monometallic copper nanoparticles and bimetallic copper nanoparticles (CuSn and CuZn). These catalysts were used in the coupling reaction of 4-chloropyridine with potassium phenolate, thereby avoiding the use of expensive solid base such as cesium carbonate. Experiments showed that these catalysts promoted the desired transformation and provided the coupling product in good yields. Moreover, an increased yield was observed when 18-crown-6 was employed as an alkali metal scavenger.

This research team also published in 2012 a detailed investigation on the same transformation in continuous-flow milli-plant using microwave irradiation.[30b] A copper/titanium oxide fixed-bed reactor and a copper/zinc oxide wall-coated reactor were developed for the continuous Ullmann-type C–O cross-coupling reaction, the latter being more efficient. Leaching of the copper catalyst was observed after 24 hours onstream, and a thermal regeneration at 350°C for a day could partially avoid the drop in yield observed after a few runs.

Wang and coworkers prepared copper(I) oxide nanoparticles having an average size of 28 nm and applied them for the first time to the cross coupling of aryl iodides and aryl bromides with phenols (Scheme 20.15).[31] The reaction can be carried out under mild conditions without any organic ligands or other additives, affording the corresponding diaryl ethers in moderate-to-good

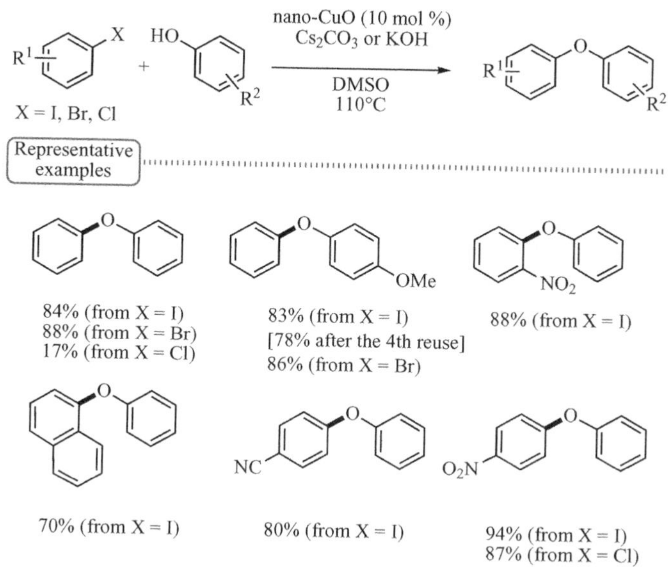

Scheme 20.15 Copper oxide nanoparticle-catalyzed coupling of aryl halides with phenols.

yields. The recyclability of the catalyst was examined in the coupling of iodobenzene and 4-hydroxyanisole, and it was found that the copper(I) oxide nanoparticles could be reused at least five times without obvious loss in catalytic activity.

Most recently, Sreedhar disclosed an efficient and ligand-free cross-coupling reaction of aryl chlorides with phenol derivatives by using a catalytic amount of copper(I) iodide nanoparticles as a recyclable catalyst (Scheme 20.16).[24] The exceptional catalytic activity of these nanoparticles should be emphasized since classical challenging substrates such as *p*-nitrophenol undergo cross coupling with 4-chloroanisole in yields higher than 90%. In addition, the catalytic system can be recovered by simple centrifugation and reused for several cycles without loss of its catalytic activity.

Kidwai and coworkers employed copper nanoparticles as a heterogeneous catalyst to catalyze the Ullmann cross coupling of phenols with aryl halides (Scheme 20.17).[32] The influence of the loading and the size of copper nanoparticles on the reaction efficiency was studied, and it was found that 10 mol % loading of copper nanoparticles with particle size of 20 nm afforded the desired products in good yields. Particle size smaller or larger than 20 nm resulted in the decline of reaction rate. However, recycling experiments demonstrated that the catalytic activity of the catalyst decreased after several runs. Quasielastic light scattering (QELS) data clearly demonstrated that conglomeration of copper nanoparticles occurred after reuse and caused a change in size, which resulted in the decline in catalytic activity.

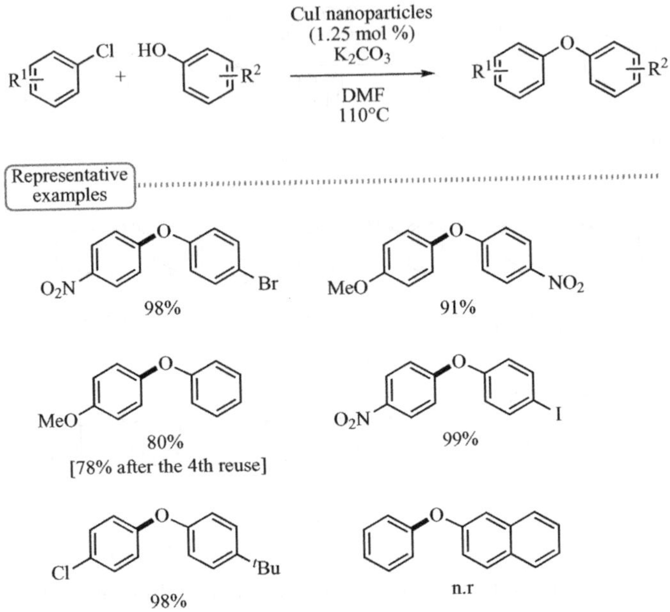

Scheme 20.16 Copper(I) iodide nanoparticle-catalyzed coupling of aryl chlorides with phenols.

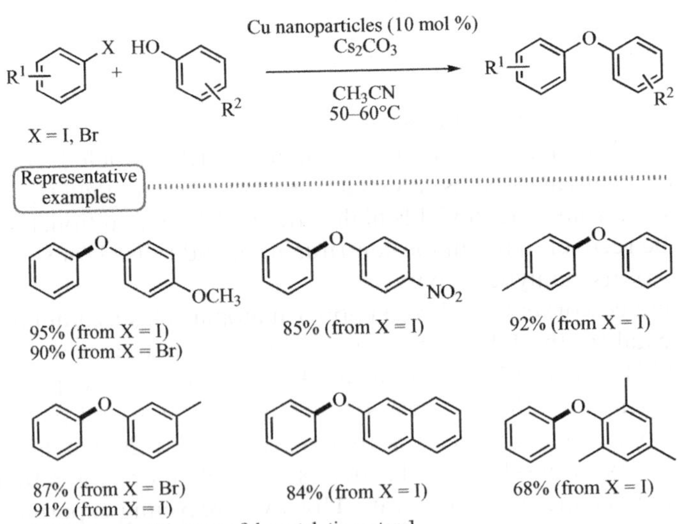

Scheme 20.17 Copper-nanoparticle-catalyzed cross coupling of aryl halides with phenols.

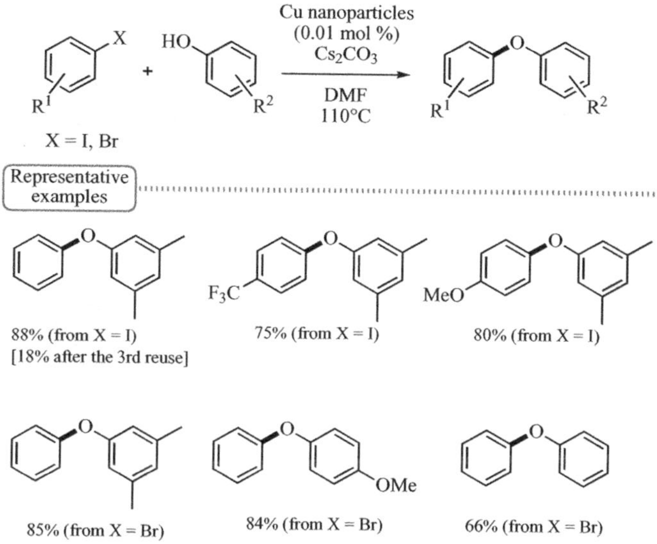

Scheme 20.18 Surfactant-free copper nanoparticles as efficient catalysts of the Ullmann reaction.

Most recently, Obora reported on surfactant-free 2-nm-sized copper nanoparticles as an extremely efficient catalyst in the Ullmann cross-coupling reaction (Scheme 20.18).[33a] Previous studies by this group demonstrated that palladium nanoparticles generated by the dimethylformamide (DMF) reduction method possessed an ultrasmall size (1–1.5 nm) and a surfactant-free active surface and were therefore highly active in Suzuki–Miyaura and Mizoroki–Heck reactions with turnover numbers up to 6×10^8. In the case of the Ullmann cross coupling, it was shown that the reaction of iodobenzene with 3,5-dimethylphenol in the presence of only 0.01 mol % of copper nanoparticles and using cesium carbonate as the base led to an 88% yield in DMF at 110°C.[33b] The reaction proved to be general and aryliodides and bromides could be used, with the exception of 2-iodothiophene, which led to inactivation of the catalytic system by aggregation of the nanoparticles. It is worth mentioning that the recyclability of the single-nanosized colloidal copper nanoparticles is seriously limited since a significant drop in yield was observed after the third cycle of the reaction of iodobenzene with 3,5-dimethylphenol.

Only rare examples of copper nanoparticle-catalyzed cross coupling of aryl halides and aliphatic alcohols were reported. Punniyamurthy demonstrated that pentaerythritol and triethanolamine cross coupling with iodobenzene resulted in excellent yields in the presence of 5 mol % of copper(II) oxide nanoparticles in DMSO at 80°C (Scheme 20.19). However, no recycling studies of the catalytic system were conducted.[34]

The classical Ullmann cross coupling of phenol with aryl halides can therefore be catalyzed by copper nanoparticles with great efficiency and practical

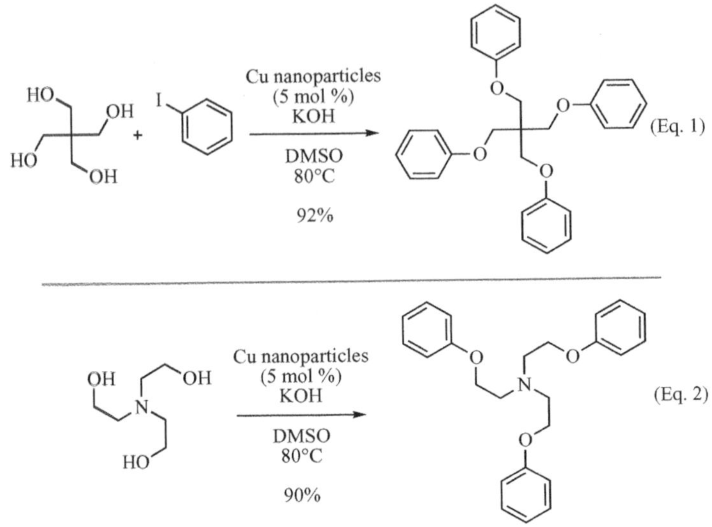

Scheme 20.19 Copper nanoparticle-catalyzed cross coupling of aryl halides and pentaerythritol and triethanolamine.

simplicity, as seen in this section. Astonishingly low catalytic loading of 0.01 mol % could even be used, as shown in the elegant studies by Obora, even though the recycling was not yet optimum. It is worth noting that magnetic copper nanoparticles have also recently attracted a lot of interest.[9h–j] Great progress has also been made in the copper nanoparticle-catalyzed synthesis of carbon–selenium bonds, as described in the next section.

20.2.5 C–Se Bond Formation

If organoselenium compounds are widely used in pharmaceuticals, functional materials, and synthetic chemistry, traditional methods for the formation of C–Se bonds often require harsh reaction conditions. For example, the coupling of selenium reagents with aryl halides takes place in polar solvents, such as hexamethylphosphoramide.[35] To overcome these difficulties, improved methods for the formation of C–Se bonds were reported over the last decades, the most common and powerful one being the transition-metal-catalyzed cross-coupling reaction of selenides and aryl halides. Up to now, different transition-metal-catalyzed systems including palladium, nickel, copper, and iron have been reported, and copper catalysis has been shown to be remarkably efficient.[36] These systems have then been adapted to heterogeneous catalysis, aiming at the development of recyclable catalysts.

In this context, Rao developed in 2009 a nanocrystalline copper(II) oxide-catalyzed cross coupling of aryl halides with diphenyl diselenide, yielding the corresponding diaryl selenide under ligand-free conditions (Scheme 20.20).[37]

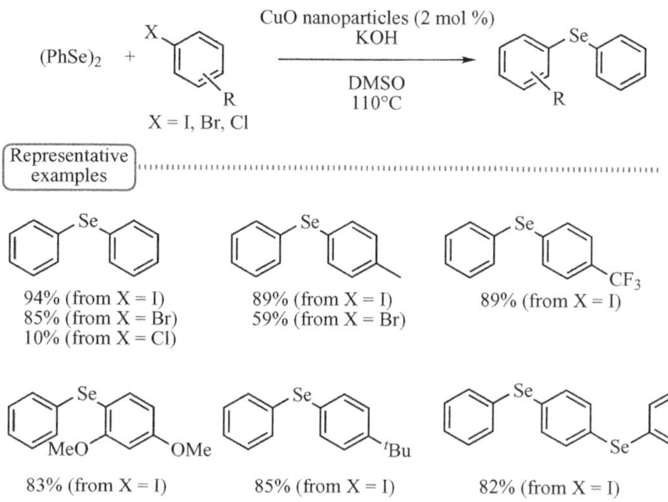

Scheme 20.20 Nanocrystalline CuO-catalyzed coupling of aryl halides with diphenyl diselenide.

The use of nanocrystalline copper(II) oxide as a valuable and inexpensive catalyst for the coupling of carbon–heteroatom is especially appealing since this catalyst can be easily recovered and reused three times without obvious loss of catalytic activity, although no specific example was disclosed in Rao's studies. The reaction was readily performed with aryl iodides, and bromides were also found to be suitable partners. Aryl chlorides gave, not surprisingly, much lower yields.

Subsequently, Rodrigues and coworkers further improved this method, and they performed the coupling reaction by using only 0.5 mol % of copper oxide nanoparticles as catalyst in ionic liquid (Scheme 20.21).[38] In their publication, the authors also reported, quite surprisingly, that alkyl halides could also be used as reaction partners, although the exact role of copper with these substrates is rather unclear and might just serve to reduce the diselenide.

Very recently, Alves also reported a Chan–Lam–Evans-type copper oxide nanoparticle-catalyzed cross coupling of diaryl diselenides with aryl boronic acids (Scheme 20.22).[39] In this process, a variety of substrates were employed in the cross-coupling reaction to afford the corresponding products in good-to-excellent yields. The electronic effect of the aromatic ring attached to the arylboronic acids had little influence on the reaction efficiency and gave the desired product in excellent yields, while diaryl diselenides with electron-donating groups gave lower yields than the diselenides bearing electron-withdrawing groups. When diphenyl ditelluride was chosen as the substrate, diaryl telluride was also obtained in excellent yields. The catalyst can be easily

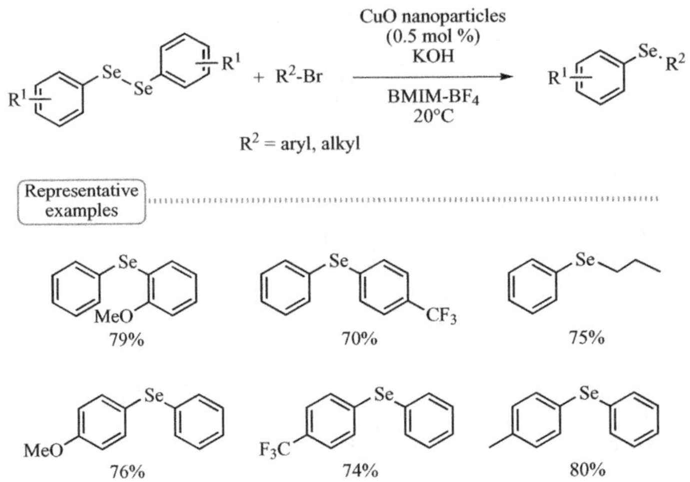

Scheme 20.21 Coupling of aryl and alkyl bromides with diaryl diselenide using nano-CuO in ionic liquid.

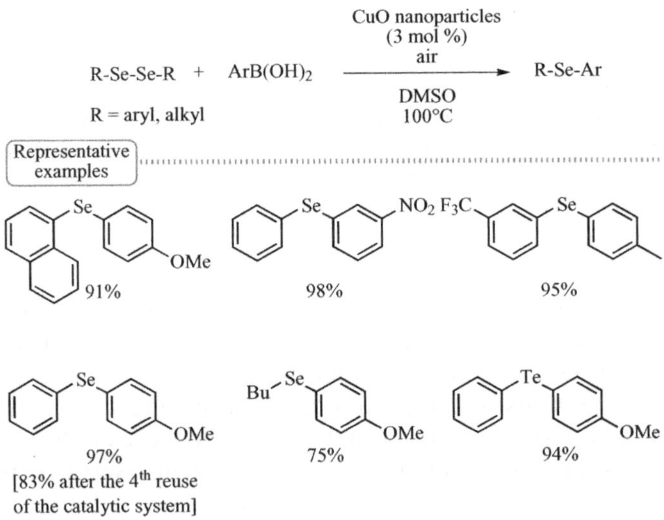

Scheme 20.22 CuO nanoparticle-catalyzed cross coupling of diaryl diselenides with aryl boronic acids.

recovered and utilized for further catalytic reactions, although a slight decrease in yield was observed after the fourth run.

In 2010, Rao and coworkers reported on the use of selenourea as an effective selenium surrogate to prepare symmetric diaryl organoselenium from aryl and benzyl boronic acids (Scheme 20.23).[40] This strategy is quite similar to the one reported by Nageswar for the copper oxide nanoparticle-catalyzed

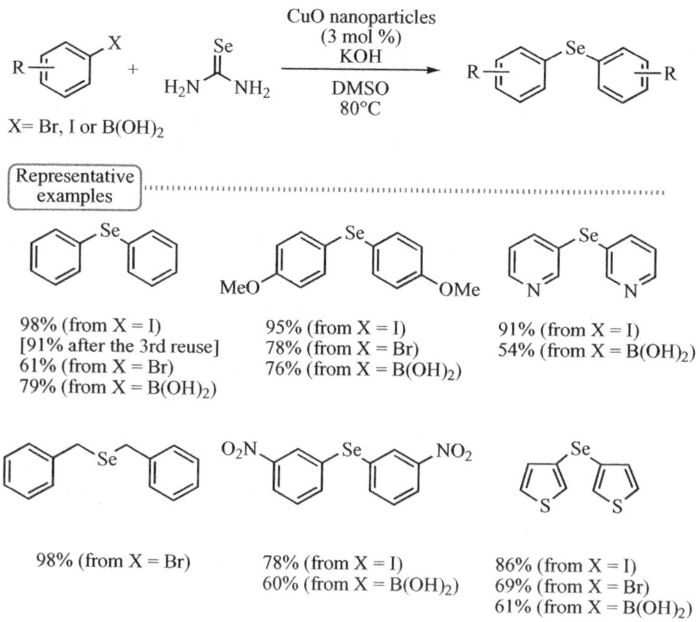

Scheme 20.23 Copper nanoparticle-catalyzed synthesis of symmetric organoselenium derivatives using selenourea as an effective selenium surrogate.

thiolation of aryl halides (Scheme 20.13). The coupling reaction was carried out under ligand-free conditions in the absence of any additive, by using copper oxide nanoparticles as a heterogeneous catalyst. The recovery of the catalyst was easily performed by ultracentrifugation, and, interestingly, no obvious change in the activity and shape of the catalyst was observed after the fourth cycle.

The formation of C–Se bonds is thus efficiently catalyzed by recyclable copper nanoparticles, and selenium surrogates such as selenourea hold great promise for this important transformation.

Overall, copper nanoparticles with defined size emerged as powerful catalysts for a diversity of bond formation, such as C–C, C–N, C–O, and C–Se. One of the most important parameters compared with bulk copper oxide is the greater surface area, which allows for enhanced catalytic activity. These nanoparticles are recycled by centrifugation of the reaction medium followed by washing and drying, and the catalytic activity is usually maintained for five consecutive cross couplings.

The interest in the synthesis, performance, and recycling of copper catalysts is not confined to the nanoparticle area, and supported catalysts have also attracted a great deal of attention. The most relevant advances in this exciting field of research are developed and discussed in the next section, based on representative recent examples.

20.3 SUPPORTED COPPER-CATALYZED CROSS-COUPLING REACTION

Despite the good performance of copper-nanoparticle catalysts in various cross-coupling reactions, there are still several problems associated with these catalysts. One of the shortcomings is that, in many cases, the catalysts are not convenient for handling because of their tendency to aggregate and the small scale of nanomaterial production. Considering the demand for recyclable heterogeneous catalysts, especially for industrial use, researchers have focused on dispersing copper catalysts on different templates to obtain supported systems that would be easier to handle and to recover. Up to now, two kinds of supported copper catalysts have been studied, the first being copper nanoparticles supported on different templates, the second being the anchoring of copper salts or complexes on different heterogeneous materials. There are two advantages for these catalysts: (i) The stability is enhanced due to the statistical dispersion of the catalysts on the templates, and (ii) synthetic access is more convenient and scale-up is in general no longer an issue. Moreover, the interactions between the catalysts and the templates often have an important influence on the catalytic activities, which could impart special properties of these supported copper catalysts. The use of these supported systems in various and representative copper-catalyzed cross-coupling reactions will be overviewed in this section. Reactions have been classified here again according to the nature of the bond formed in the reaction (C–C, C–N, C–S, and C–O).

20.3.1 C–C Bond Formation

The Sonogashira cross coupling could be achieved with both kinds of supported copper catalysts. Biffis and coworkers have reported in 2007 that copper oxide nanoparticles supported on Al_2O_3 (having a surface area of 280 $m^2.g^{-1}$) were able to catalyze the coupling of aryl halides with phenylacetylene without the addition of a palladium cocatalyst, using tetrabutylammonium as a base in DMF at 110°C (Scheme 20.24).[41] Aryl halides bearing electron-withdrawing groups offered higher reaction efficiency in comparison with aryl halides bearing electron-donating groups, and aliphatic alkynes gave the coupling product in modest yields compared with their aromatic analogues, which actually corresponds to the classical reactivity generally noted for the Sonogashira reaction. The authors found that the leaching of copper was too high (62% in the first round) to be reused, and, as a result, the recycling of this catalyst was impossible, which clearly limits the usefulness of this system.

Another kind of silica-anchored proline–copper(I) complex (**Cu-L1**) was also successfully used as catalyst for the Sonogashira reaction (Scheme 20.25).[42] As described in Scheme 20.25, Equation 1, this heterogeneous catalyst was prepared in four steps from a 100–200 mesh silica by functionalization with (4-chloromethyl)phenyltrichlorosilane followed by substitution of the

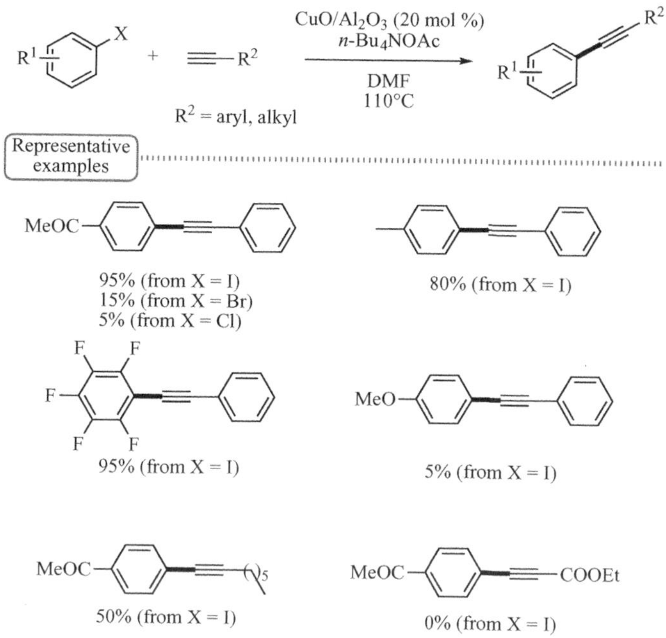

Scheme 20.24 Alumina supported CuO nanoparticle-catalyzed Sonogashira coupling.

benzylic chloride atoms by the disodium salt of 1-(*tert*-butoxycarbonyl)-4-hydroxypyrrolidine-2-carboxylic acid. Subsequent deprotection with trifluoroacetic acid and reaction with copper(I) iodide in DMF afforded the desired catalyst, which proved to consist of 5.48% in weight of copper(I) by atomic absorption spectroscopy. This method has a rather broad substrate scope since unactivated aryl halides and aliphatic alkynes could be successfully used. This catalyst was easily recovered by simple filtration on a sintered glass funnel and could be reused for up to six times when active coupling partners such as phenylacetylene and iodoanisole were employed.

The oxidative homocoupling of terminal alkynes is an important kind of reaction for C–C bond formation in organic synthesis yielding the corresponding symmetrical 1,3-diynes. Examples of such oxidative dimerization using heterogeneous catalysts are quite rare. In 2009, Mizuno and coworkers employed supported copper hydroxide on titanium oxide ($Cu(OH)_x/TiO_2$) in the homocoupling reaction of alkynes (Scheme 20.26).[43] This highly dispersed copper catalyst on titanium dioxide was prepared on a gram scale from calcinated titanium dioxide and copper(II) chloride in basic aqueous solution. The scope of the reaction was quite general as aryl-, alkyl-, and silylacetylenes were successfully dimerized to yield the homocoupling products in good yields. Even conjugated enynes led to high yields of the desired homocoupling products. As usual, the isolation of the catalyst is performed by simple filtration

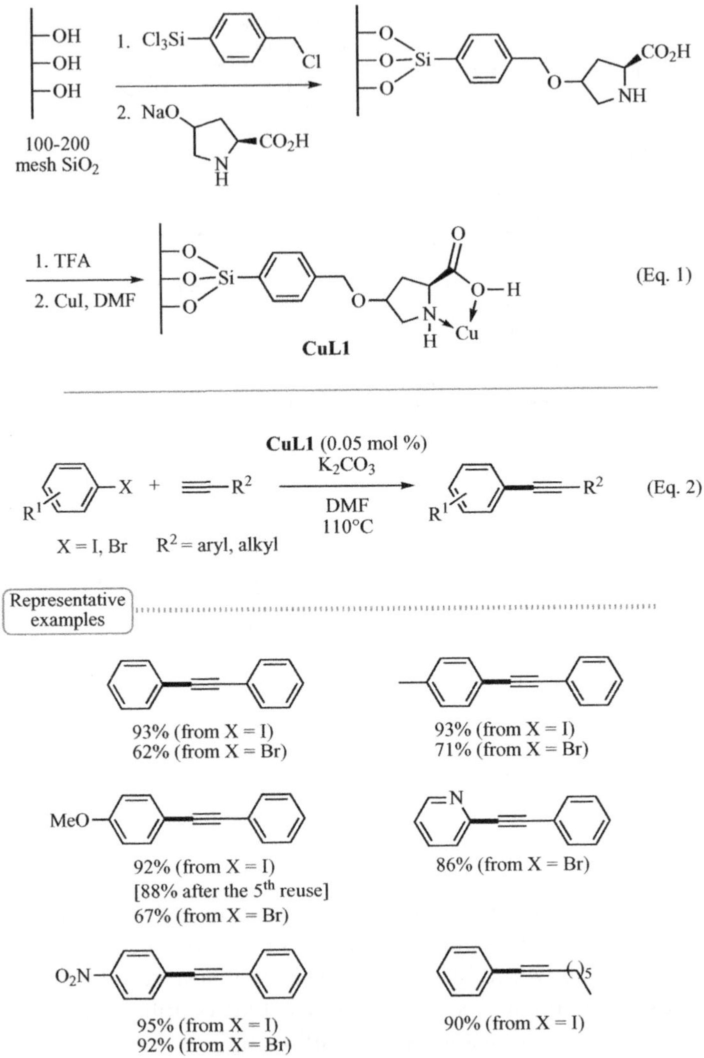

Scheme 20.25 Silica-anchored proline–copper(I) complex-catalyzed Sonogashira reaction.

followed by washing and drying, although the catalytic activity of the recovered heterogeneous copper dramatically decreased (less than 20% yield for the synthesis of 1,4-diphenylbutadiyne after a single reuse of the catalyst).[44] Actually, one of the pitfalls of this approach lies in the deactivation of the copper hydroxide on titanium oxide due to the formation of inactive copper acetylides arising from a slow reoxidation of the copper(I) species.[44]

To circumvent these limitations, Mizuno and coworkers developed in 2011 a very efficient heterogeneous oxidation catalyst that allows a rapid and

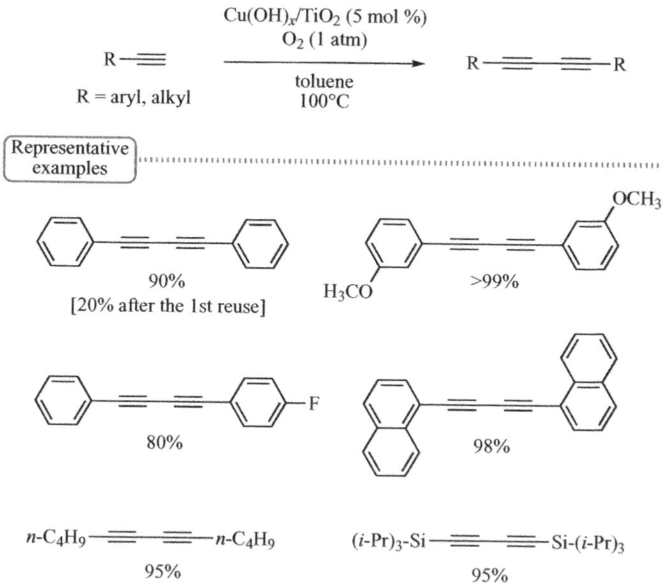

Scheme 20.26 Titanium oxide supported copper hydroxide-catalyzed oxidative homocoupling of alkynes.

thermodynamically favorable electron transfer from the alkyne partner to dioxygen. Manganese oxide-based octahedral molecular sieve (KMn_8O_{16}, OMS-2) as a support proved to be the most efficient due to its large surface area (100 $m^2.g^{-1}$), its electron-conducting properties, and its dioxygen reduction ability. The supported copper hydroxide catalyst on OMS-2 (Cu(OH)*x*/OMS-2) was prepared following the 2009 procedure and was applied in the aerobic oxidative alkyne homocoupling reaction of terminal alkynes (Scheme 20.27).[44] Phenylacetylene derivatives with electron-donating as well as electron-withdrawing substituents were reactive partners leading to the corresponding 1,4-diphenyl-1,3-diyne derivatives, and even the classically less-reactive aliphatic alkynes led to good isolated yield of the desired 1,3-diynes. The authors also performed recycling experiments, and it was found that the heterogeneous catalyst still possessed high catalytic activity after 14 runs—which is quite remarkable—leading to the homocoupling product in 90% yield in the case of the synthesis of 1,4-diphenylbutadiyne.

20.3.2 C–X Bond Formation

20.3.2.1 C–N Bond Formation. Among all studies reporting the use of various copper catalysts for the N-arylation reactions of N-nucleophiles, there are several examples using supported copper catalysts. Compared with homogeneous copper catalysts, these catalysts sometimes show better efficiency, and

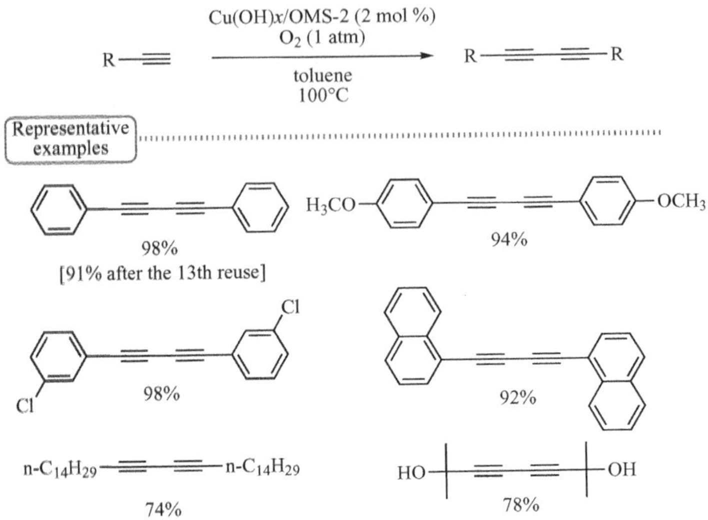

Scheme 20.27 Octahedral molecular sieves (OMSs) supported copper hydroxide-catalyzed oxidative homocoupling of alkynes.

a major advantage is that they are reusable. Moreover, they even exhibit in some cases an original chemoselectivity.

Majoral, Taillefer, and coworkers reported in 2006 that phosphorus dendrimers such as **2-G3** could be used as ligands in the copper-catalyzed N-arylation and N-vinylation of pyrazole (Scheme 20.28).[45] Dendrimer **2-G3** consists of 48 bidentate *N*-[(1*E*)-pyridin-2-ylmethylene]-aniline motifs on the surface and were shown to be the most active ligands in the copper(I) iodide-catalyzed cross coupling between iodobenzene, bromobenzene (Eq. 1), and *E*-β-bromostyrene (Eq. 2), and pyrazole. It is worth noting that if dendrimers are replaced by the monomeric ligand **M**, no reaction was observed, indicating that a strong dendritic effect was operating with ligand **2-G3**.

In this area, Sreedhar and coworkers reported a new kind of supported copper(I) iodide on polyaniline (PANI) nanofibers, which could efficiently catalyze the N-arylation of different amines with aryl iodides and chlorides (Scheme 20.29).[46] The reaction conditions were very mild since the use of aryl iodides enabled the arylation to be performed at ambient temperature. Aryl chlorides, however, required heating at 100°C, which is still reasonable with these substrates. A wide range of amines including aliphatic and aromatic ones as well as π-excessive N-heterocycles could be used as substrates, and the reactions afforded the desired products in good isolated yields. In addition, the copper(I)/PANI nanofibers could be easily reused for at least five times without loss of catalytic activity (98% isolated yield after the fifth cycle in the amination of 2-iodonitrobenzene with pentylamine).

In 2009, Gupte and coworkers investigated the reactivity of heterogenized copper catalysts in the N-arylation of aniline with iodobenzene (Scheme

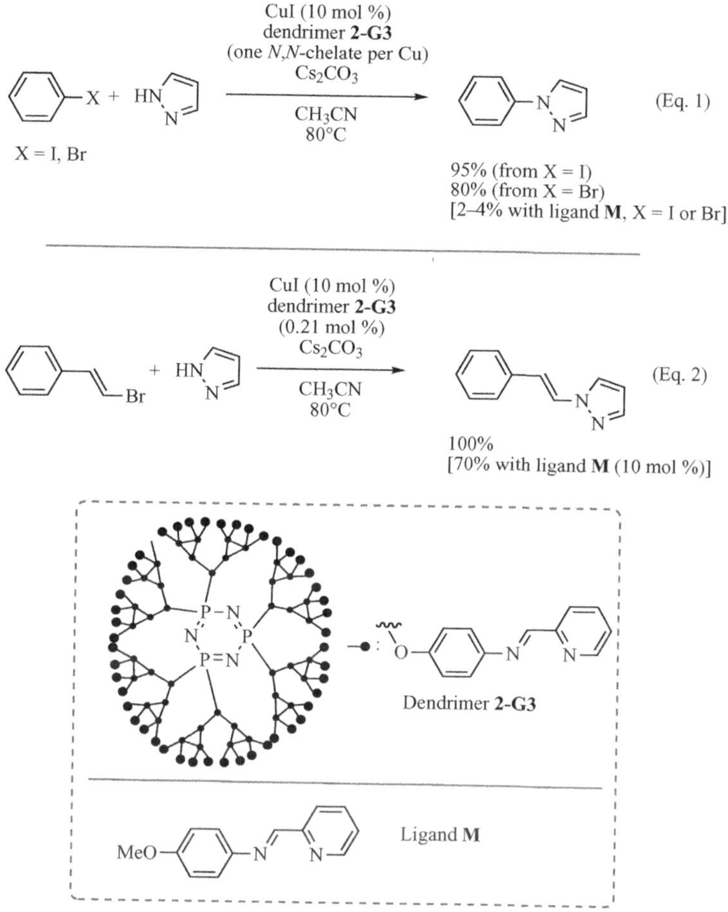

Scheme 20.28 Dendrimers as efficient ligands in the copper(I)-catalyzed N-arylation and N-vinylation of pyrazoles.

20.30).[47] A series of zeolite-supported copper complexes were synthesized such as encapsulated Cu(Bipy)-Y and tethered Cu(phen)(PPh_3)Br on zeolite Y. The former was prepared from copper acetate and Na–Y zeolite followed by heating in the presence of bipyridine, while the latter was prepared from phosphotungstic acid, Na–Y zeolite, and Cu(Phen)(PPh_3)Br in refluxing methanol. Interestingly, the interactions between the zeolite template and the copper complexes had a great influence on the ratio of the two possible products, diphenylamine and triphenylamine. The copper(I) complexes encapsulated in the Na–Y zeolite showed a high selectivity for diphenylamine (up to 87% yield by using Cu(Bipy)/zeolite Y), while the copper complexes tethered on zeolite Y gave a good yield of triphenylamine (up to 72% yield by using Cu(Phen)(PPh_3)Br tethered to phosphotungstic acid on zeolite Y). A basic

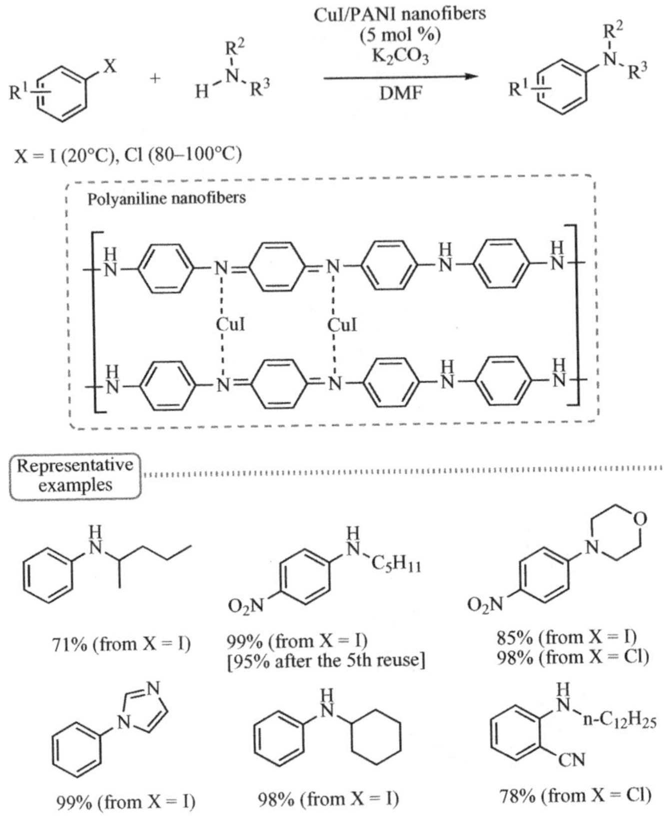

Scheme 20.29 CuI/PANI nanofiber-catalyzed N-arylation of amines and related nucleophiles.

recycling study was performed using the latter catalyst, and after five recycles, the yield of triphenylamine was identical and no leaching of copper was observed in the solvent.

Islam and coworkers prepared a polymer-supported copper(II) catalyst and employed it as a robust heterogeneous catalyst for the arylation of nitrogen-containing heterocycles with aryl halides as well as arylboronic acids (Scheme 20.31).[48] This heterogeneous catalyst was easily obtained in two steps from 4-(2-pyridylazo)resorcinol (PAR) and chloromethylated polystyrene, the intermediate PAR-anchored polymer being further treated with copper(II) acetate in acidic medium. Investigation of the reactivity of various substrates showed that the cross coupling of aryl halides or arylboronic acids with N–H heterocycles generated the corresponding products in good-to-excellent yields, even after five runs. As a final note, Khatri and Jain recently reported a glycerol embedded copper catalyst able to catalyze the union of aryl halides and amines in good yields. Six consecutive reuses were possible without loss of catalytic activity.[48c]

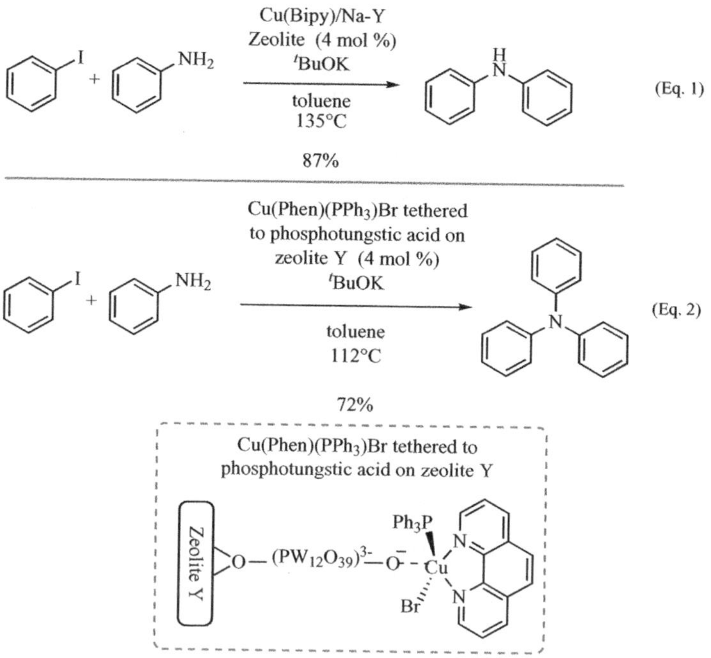

Scheme 20.30 Zeolite-supported copper complex-catalyzed N-arylation of aniline with iodobenzene: influence of the heterogeneous catalyst on the chemoselectivity.

As seen in this section, supported copper catalysts have been developed for the creation of C–N bonds with great efficiency. A diversity of supports has been studied such as nanofibers, zeolites, modified polystyrenes, or glycerol; and in some cases, the interaction between the copper catalyst and the support can modify the chemoselectivity of the cross coupling. Of course, the recycling of the copper catalysts is of fundamental importance in this area, and it was shown that a simple filtration of the catalyst followed by washing still delivered an active species for at least five reuses. Based on the same concepts, the formation of C–S bonds relying on supported copper catalysts has been thoroughly investigated, and relevant examples will be discussed in the next section.

20.3.2.2 C–S Bond Formation. Recently, Ranu and coworkers successfully employed copper(II) sulfate anchored on basic alumina as a new kind of heterogeneous catalyst for the coupling of aryl halides with thiols under aerobic, ligand-free conditions (Scheme 20.32).[49] Basic alumina proved to be superior to neutral alumina, zeolite, and hydrotalcite. The preparation of this catalyst was quite straightforward and involved a single step from an aqueous solution of copper(II) sulfate and basic alumina followed by drying under high vacuum at 120°C. The copper(II) loading was measured to be 0.52 mmol.g^{-1} by inductively coupled plasma mass spectrometry (ICP-MS).

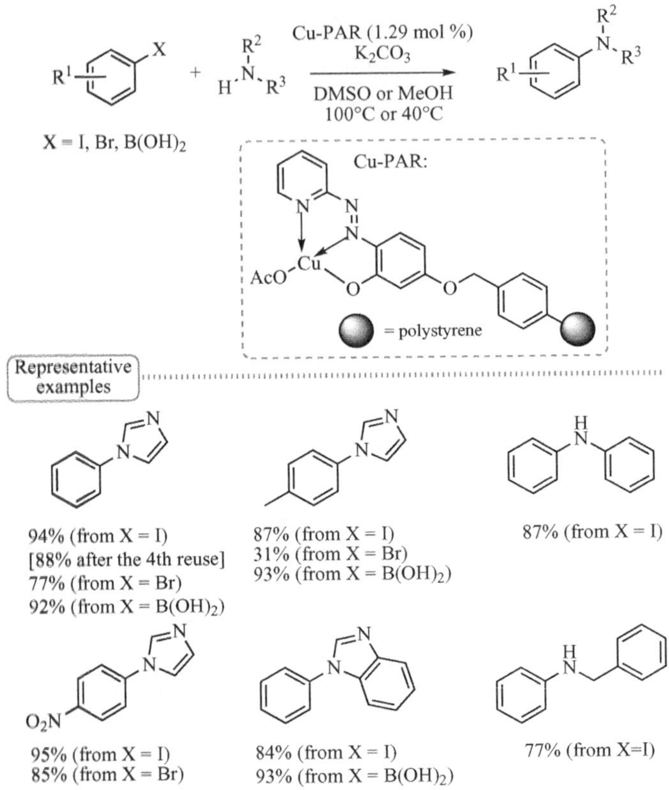

Scheme 20.31 Polymer-supported copper catalyst for the N-arylation of nitrogen-containing heterocycles.

The authors found that potassium carbonate was the most reactive base for the cross coupling of aryl iodides in DMF at 110°C, whereas cesium carbonate was more appropriate for aryl bromides under the same conditions. Based on sequential cross coupling, unsymmetrical bis-thioethers could be efficiently prepared from bromoiodoaryl derivatives and a thiol in two sequential copper(II) on alumina-supported-catalyzed reactions.

Centrifugation of the reaction mixture allows for the recovery of the heterogeneous catalyst that still possesses a copper(II) content of 0.50 $mmol.g^{-1}$ after seven consecutive reuses and was thus able to catalyze the cross coupling between iodobenzene and thiophenol in 77% yield (Scheme 20.33).

In 2010, Lee and coworkers reported the thiolation of aryl iodides catalyzed by a mesoporous silica-supported copper oxide catalyst. Detailed characterization showed that the copper oxide was well dispersed onto the mesoporous silica matrix, which exhibited higher catalytic activity than common copper oxide nanoparticles (Scheme 20.34).[50] A wide range of alkyl-, aryl-, heteroaryl-thiols, and aryl iodides with different substituents was found to be reactive in

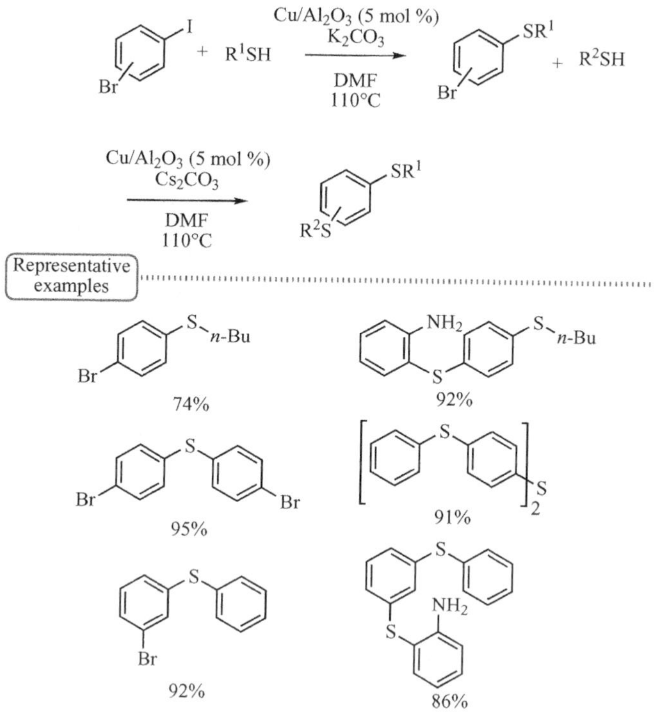

Scheme 20.32 Cu/Al_2O_3-catalyzed coupling of 1-bromo-4-iodobenzene with thiols under aerobic and ligand-free conditions.

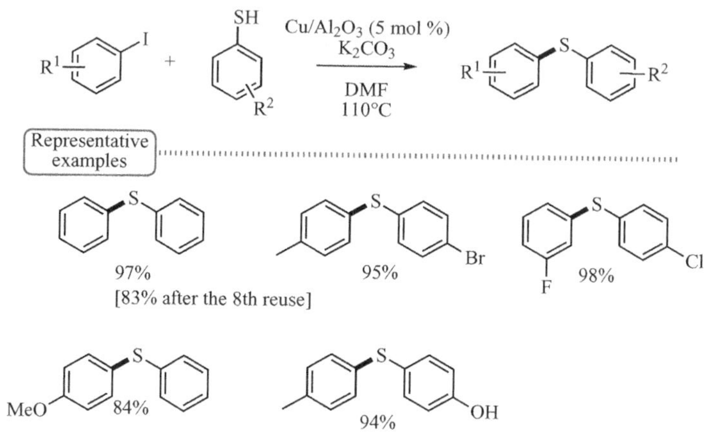

Scheme 20.33 Cu/Al_2O_3-catalyzed cross coupling of aryl halides with thiols under aerobic and ligand-free conditions.

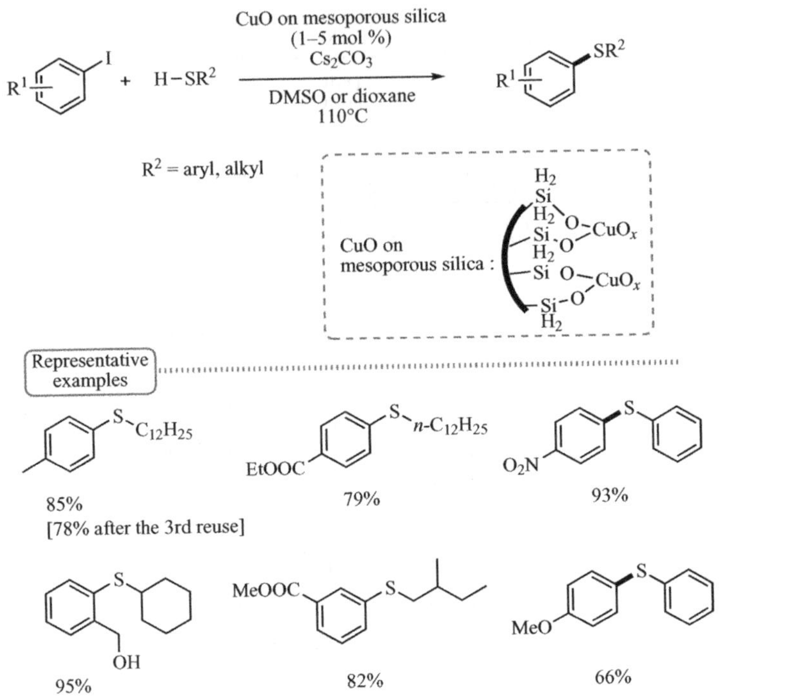

Scheme 20.34 Mesoporous silica-supported CuO-catalyzed thiolation of aryl iodides.

this cross-coupling reaction, leading to the desired thioethers in good yields. Reuse of this catalyst was performed four times and the result showed that there was only a slight loss of catalytic activity.

Wang and coworkers developed a copper(I) iodide–PEG_{1000} catalytic system and employed it as a catalyst in the cross-coupling reaction of aryl iodides with thiophenols or alkanethiols at 110°C (Scheme 20.35).[51] In this reaction, poly(ethylene glycol) served not only as the reaction medium or phase transfer catalyst but also as a ligand. When the coupling reaction was carried out in pure poly(ethylene glycol), various substrates afforded the desired product in good yield. It is worth noting that a chemoselective cross coupling of 4-hydroxythiophenol and iodobenzene was observed since only 4-hydroxy-1-(phenylsulfanyl)benzene was obtained in 96%. When the coupling reaction was carried out in aqueous poly(ethylene glycol) solution, only alkanethiol gave the coupling products and in poor yield. After the cross-coupling reaction, the desired products were extracted directly from the PEG_{1000}–copper(I) iodide solid (PEG_{1000} has a melting point between 38 and 42°C), the latter being then able to cross couple iodobenzene and thiophenol for five consecutive times in excellent yields.

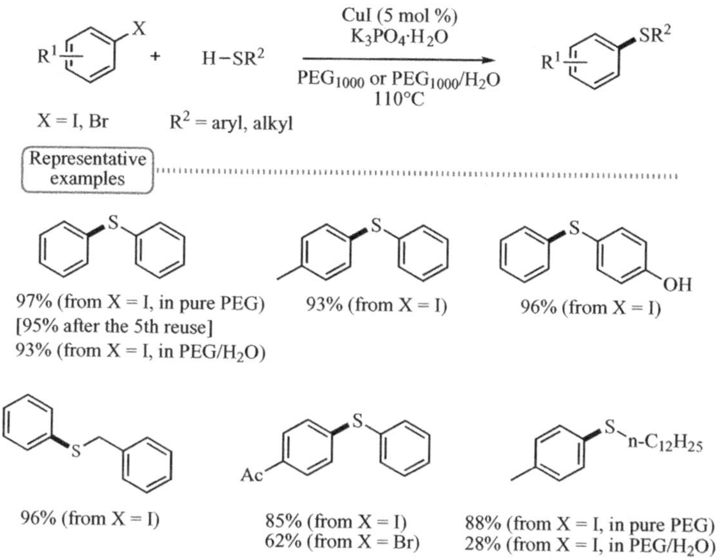

Scheme 20.35 Copper(I) iodide/PEG_{1000}-catalyzed thiolation reaction of aryl halides.

20.3.2.3 C–O Bond Formation. The Ullmann coupling for C–O bond formation catalyzed by supported copper catalysts has also attracted the attention of numerous research groups in the past years. In 2008, Taillefer and coworkers developed a new method to immobilize copper catalyst in O-arylation of phenols with aryl halides (Scheme 20.36).[52] By sol-gel hydrolysis of a silylated bispyridinyl ligand, a new kind of hybrid silica A was obtained. Copper(I) iodide was then added and was very efficiently immobilized by the two nitrogen atoms of the bipyridine motif. This copper-encapsulated bridged silsesquioxane exhibited a higher efficiency than other catalysts involving copper(I) iodide and classical bipyridinic ligands. The recycling study using different substrates showed that this catalyst could be easily reused 11 times with different substrates to give rise to 11 different O-arylated products with excellent yields.

Subsequently, this group prepared β-diamide-based hybrid silica ligand **L2**, which was further used in association with copper(I) iodide to generate *in situ* the corresponding heterogeneous catalyst copper(I) iodide/**L2**. The latter proved to be active in DMF at 110°C; better from an industrial perspective is methylisobutylketone, which is easier to eliminate and is much less toxic (Scheme 20.37).[53] The copper(I) iodide/**L2** catalytic system allowed the cross coupling of phenols with aryl bromides; phenol derivatives bearing electron-withdrawing groups displayed higher reaction efficiency compared with phenol derivatives bearing electron-donating groups. On the other hand, electron-poor aryl bromides led to better results than electron-rich aryl bromides.

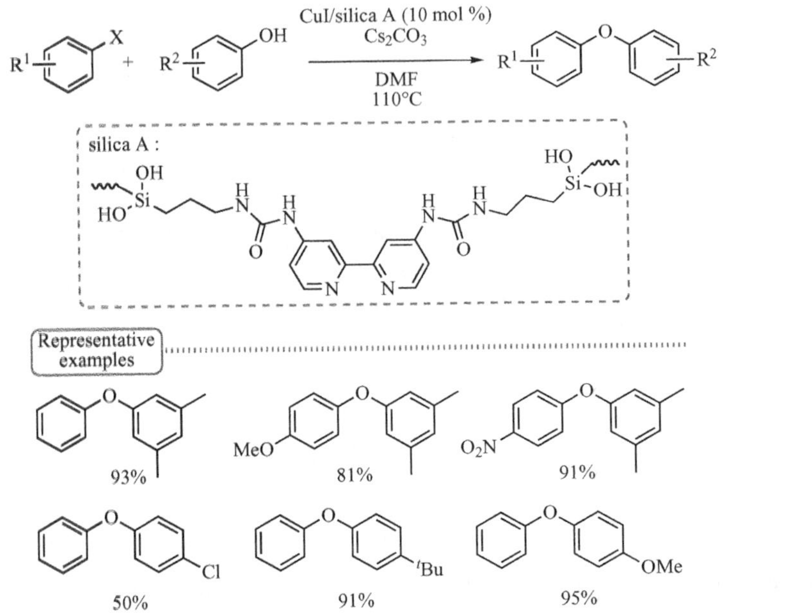

Scheme 20.36 Silica-supported copper(I) iodide nanoparticle-catalyzed coupling reaction of aryl halides with phenols.

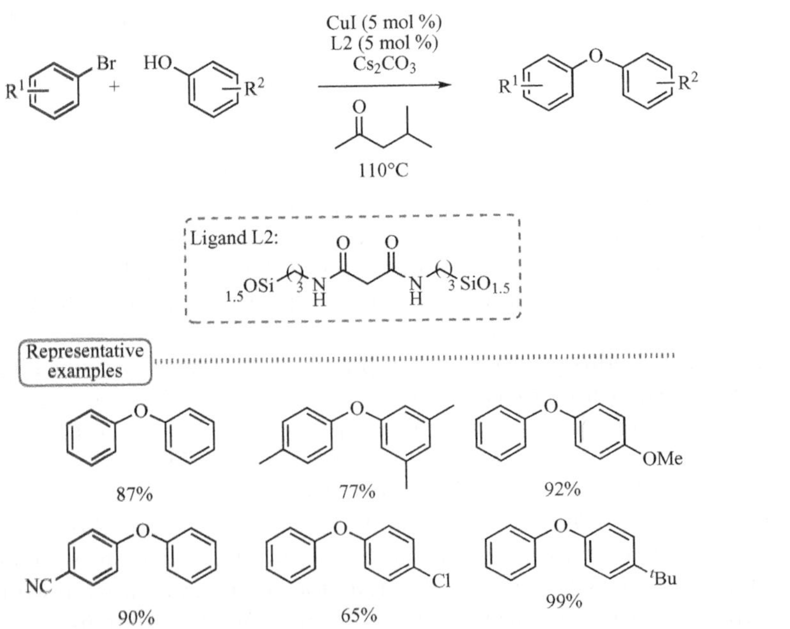

Scheme 20.37 Copper(I) iodide/β-diamide-immobilized hybrid silica-catalyzed arylation of phenols from aryl bromides.

A negligible copper leaching was detected in the reaction process, and the recycling experiments demonstrated that this heterogeneous catalyst possessed an excellent reusability after just a simple filtration of the reaction mixture. As an example, the cross coupling of bromobenzene with six different substituted phenols occurred in excellent yields, the sixth run leading to 99% of the desired substituted biaryl ether.

As described in Section 20.3.2.1, dendrimers are efficient ligands in copper(I)-catalyzed cross coupling between aryl or styryl halides and N-nucleophiles. Logically, dendrimers have also been used in the corresponding O-arylation reaction of phenolic derivatives in the presence of copper(I) iodide (10 mol %) and potassium phosphate in acetonitrile at 55°C. Good yield of the arylated phenol was obtained with only 0.21 mol % of dendrimer.[54]

Very recently, a new system involving polystyrene-anchored copper(II) Schiff base complexes was reported for the synthesis of diaryl ethers by Islam and coworkers (Scheme 20.38).[55]

Using this system, a series of aryl bromides and iodides could be readily coupled to various phenols at 70°C in acetonitrile. The substituents on the

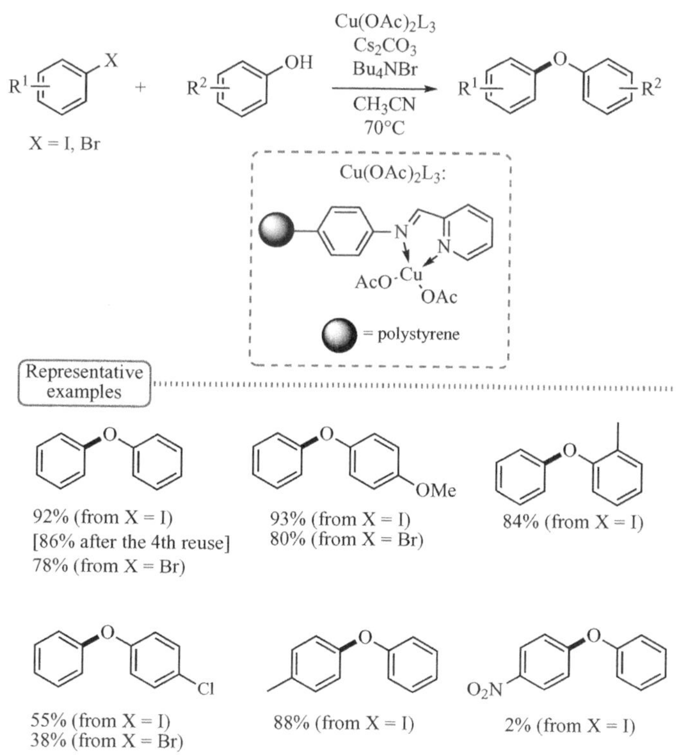

Scheme 20.38 Polymer-anchored copper(II)/Schiff base complex-catalyzed cross-coupling reaction of aryl halides with phenol derivatives.

phenol ring had a clear influence on the reaction efficiency, and phenols substituted with electron-withdrawing groups were found to be much less efficient. The catalyst could be easily recycled by simple filtration and washing. In the cross coupling of iodobenzene and phenol, five consecutive runs were performed with excellent isolated yields. Infrared (IR) and UV–vis spectra were unaffected even after the last run, showing that the catalyst was not modified. Elegant studies confirmed that the reaction was heterogeneous in nature since no leaching was detected by atomic absorption spectroscopic analyses.

20.4 CONCLUSION

Although major achievements were reported for the development of copper-catalyzed Ullmann reactions in the past decade, this area of research is still very active and numerous challenges remain. They include the discovery and preparation of highly efficient catalysts, the optimization of catalytic systems, and the improvement of reaction efficiency.[56] In this regard, the use of heterogeneous copper catalysts has a strong potential due to their inherent advantages, including high efficiency, low waste generation, and use in more sustainable processes. Copper-catalyzed cross-coupling reactions using heterogeneous catalysts should play a key role in organic chemistry and in copper catalysis for the years to come, and efforts will have to be dedicated to the development of more efficient heterogeneous systems that will possess high reactivity, therefore allowing the cross coupling to be performed under milder conditions and using lower catalyst loadings, leading to easier recyclability and extended lifetime. If these challenges are met, these systems should find various applications in chemistry, including in industrial processes.

REFERENCES

[1] Beccalli, E. M.; Broggini, G.; Martinelli, M.; Sottocornola, S. *Chem. Rev.* **2007**, *107*, 5318–5365.

[2] (a) Naota, T.; Takaya, H.; Murahashi, S. I. *Chem. Rev.* **1998**, *98*, 2599–2660. (b) Trost, B. M.; Toste, F. D.; Pinkerton, A. B. *Chem. Rev.* **2001**, *101*, 2067–2096.

[3] (a) Fagnou, K.; Lautens, M. *Chem. Rev.* **2003**, *103*, 169–196. (b) Doyle, M. P.; Duffy, R.; Ratnikov, M.; Zhou, L. *Chem. Rev.* **2010**, *110*, 704–724.

[4] (a) Li, Z.; Brouwer, C.; He, C. *Chem. Rev.* **2008**, *108*, 3239–3265. (b) Stephen, A.; Hashmi, K. *Chem. Rev.* **2007**, *107*, 3180–3211. (c) Jiménez-Núñez, E.; Echavarren, A. M. *Chem. Rev.* **2008**, *108*, 3326–3350.

[5] (a) Chianese, A. R.; Lee, S. J.; Gagné, M. R. *Angew. Chem., Int. Ed.* **2007**, *46*, 4042–4059. (b) Albrecht, M.; van Koten, G. *Angew. Chem., Int. Ed.* **2001**, *40*, 3750–3781.

[6] (a) Do, H. Q.; Daugulis, O. *J. Am. Chem. Soc.* **2008**, *130*, 1128–1129. (b) Viirre, R. D.; Evindar, G.; Batey, R. A. *J. Org. Chem.* **2008**, *73*, 3452–3459. (c) Liu, X. W.; Zhang, Y. M.; Wang, L.; Fu, H.; Jiang, Y. Y.; Zhao, Y. F. *J. Org. Chem.* **2008**, *73*, 6207–6212.

[7] (a) Monnier, F.; Taillefer, M. *Angew. Chem., Int. Ed.* **2009**, *48*, 6954–6971. (b) Monnier, F.; Taillefer, M. *Angew. Chem., Int. Ed.* **2008**, *47*, 3096–3099. (c) Evano, G.; Blanchard, N.; Toumi, M. *Chem. Rev.* **2008**, *108*, 3054–3131.

[8] (a) Cristau, H. J.; Cellier, P. P.; Hamada, S.; Spindler, J. F.; Taillefer, M. *Org. Lett.* **2004**, *6*, 913–916. (b) Cristau, H. J.; Cellier, P. P.; Spindler, J. F.; Taillefer, M. *Chem. Eur. J.* **2004**, *10*, 5607–5622.

[9] (a) Zecchina, A.; Bordiga, S.; Groppo, E. (eds.), *Selective Nanocatalysts and Nanoscience*, Wiley-VCH: Weinheim, Germany, **2011**. (b) Astruc, D. (ed.), *Nanoparticles and Catalysis*, Wiley-VCH: Weinheim, Germany, **2008**. (c) Corain, B.; Schmid, G.; Toshima, N. (eds.), *Metal Nanoclusters in Catalysis and Materials Science: The Issue of Size Control*, Elsevier: Amsterdam, **2007**. (d) Lu, Y.; Wei, W.; Chen, W. *Chin. Sci. Bull.* **2012**, *57*, 41–47. (e) Ranu, B. C.; Dey, R.; Chatterjee, T.; Ahammed, S. *ChemSusChem* **2012**, *5*, 22–44. (f) Cookson, J. *Platin. Met. Rev.* **2012**, *56*, 83–98. (g) Buurmans, I. L. C.; Weckhuysen, B. M. *Nat. Chem.* **2012**, *4*, 873–886. (h) For a review on magnetically retrievable catalysts in organic chemistry, see: Baig, R. B.; Varma, R. S. *Chem. Commun.* **2013**, *49*, 752–770. (i) Yang, S.; Wu, C.; Zhou, H.; Yang, Y.; Zhao, Y.; Wang, C.; Yang, W.; Xu, J. *Adv. Synth. Catal.* **2013**, *355*, 53–58. (j) Yang, S.; Xie, W.; Zhou, H.; Wu, C.; Yang, Y.; Niu, J.; Yang, W.; Xu, J. *Tetrahedron* **2013**, *69*, 3415–3418.

[10] (a) Zhang, Z. H.; Wang, Z. Y. *J. Org. Chem.* **2006**, *71*, 7485–7487. (b) Zhang, Z. H.; Zha, Z. G.; Gan, C. S.; Pan, C. F.; Zhou, Y. Q.; Wang, Z. Y.; Zhou, M. M. *J. Org. Chem.* **2006**, *71*, 4339–4342. (c) Zhang, J. T.; Yu, C. M.; Wang, S. J.; Wan, C. F.; Wang, Z. Y. *Chem. Commun.* **2010**, 5244–5246. (d) Wang, Y.; Ouyang, G. H.; Zhang, J. T.; Wang, Z. Y. *Chem. Commun.* **2010**, 7912–7914.

[11] (a) Hurtley, W. R. H. *J. Chem. Soc.* **1929**, 1870–1873. For additional recent examples, please see: (b) Samim, M.; Kaushik, N. K.; Maitra, A. *Bull. Mater. Sci.* **2007**, *30*, 535–540. (c) Alonso, F.; Melkonian, T.; Moglie, Y.; Yus, M. *Eur. J. Org. Chem.* **2011**, *2011*, 2524–2530. (d) Wei, Y.; Chen, S.; Kowalczyk, B.; Huda, S.; Gray, T. P.; Grzybowski, B. A. *J. Phys. Chem. C* **2010**, *114*, 15612–15616. (e) Thathagar, M. B.; Beckers, J.; Rothenberg, G. *J. Am. Chem. Soc.* **2002**, *124*, 11858–11859.

[12] Yip, S. F.; Cheung, H. Y.; Zhou, Z. Y.; Kwong, F. Y. *Org. Lett.* **2007**, *9*, 3469–3472.

[13] (a) Xie, X. A.; Chen, Y.; Ma, D. *J. Am. Chem. Soc.* **2006**, *128*, 16050–16051. (b) Cai, Q.; Zhou, F. *Synlett* **2013**, 408–411.

[14] Kidwai, M.; Bhardwaj, S.; Poddar, R. *Beilstein J. Org. Chem.* **2010**, *6*(35), 1–6.

[15] (a) Thathagar, M. B.; Beckers, J.; Rothenberg, G. *Green Chem.* **2004**, *6*, 215–218. (b) For a recent report of this reaction from aryl chloride, see: Arundhathi, R.; Damodara, D.; Mohan, K. V.; Kantam, M. L.; Likhar, P. R. *Adv. Synth. Catal.* **2013**, *355*, 751–756.

[16] Li, J. H.; Tang, B. X.; Tao, L. M.; Xie, Y. X.; Liang, Y.; Zhang, M. B. *J. Org. Chem.* **2006**, *71*, 7488–7490.

[17] Calò, V.; Nacci, A.; Monopoli, A.; Leva, E.; Cioffi, N. *Org. Lett.* **2005**, *7*, 617–620.

[18] For reviews, see: (a) Klaus, K.; Ulrich, S.; Dirk, G. *Synlett* **2003**, 2428–2439. (b) Ley, S. V.; Thomas, A. W. *Angew. Chem., Int. Ed.* **2003**, *42*, 5400–5449. (c) Djakovitch,

L.; Batail, N.; Genelot, M. *Molecules* **2011**, *16*, 5241–5267. (d) Beletskaya, I. P.; Cheprakov, A. V. *Coord. Chem. Rev.* **2004**, *248*, 2337–2364. For additional examples, see: (e) Pan, K.; Ming, H.; Yu, H.; Liu, Y.; Kang, Z.; Zhang, H.; Lee, S.-T. *Cryst. Res. Technol.* **2011**, *46*, 1167–1174. (f) Wang, Y.; Biradar, A. V.; Wang, G.; Sharma, K. K.; Duncan, C. T.; Rangan, S.; Asefa, T. *Chem. Eur. J.* **2010**, *16*, 10735–10743. (g) Colacino, E.; Villebrun, L.; Martinez, J.; Lamaty, F. *Tetrahedron* **2010**, *66*, 3730–3735. (h) Saha, P.; Ramana, T.; Purkait, N.; Ali, M. A.; Paul, R.; Punniyamurthy, T. *J. Org. Chem.* **2009**, *74*, 8719–8725.

[19] (a) Xu, H. J.; Liang, Y. F.; Cai, Z. Y.; Qi, H. X.; Yang, C. Y.; Feng, Y. S. *J. Org. Chem.* **2011**, *76*, 2296–2300. (b) Han, M.; Yuan, D.; Liu, S.; Bao, J.; Dai, Z.; Zhu, J. *Mater. Res. Bull.* **2012**, *47*, 4438–4444.

[20] Rout, L.; Jammi, S.; Punniyamurthy, T. *Org. Lett.* **2007**, *9*, 3397–3399.

[21] Jammi, S.; Krishnamoorthy, S.; Saha, P.; Kundu, D. S.; Sakthivel, S.; Ali, M. A.; Paul, R.; Punniyamurthy, T. *Synlett* **2009**, 3323–3327.

[22] Yao, Z. J.; Wei, X. W. *Chin. J. Chem.* **2010**, *28*, 2260–2268.

[23] Kantam, M. L.; Yadav, J.; Laha, S.; Sreedhar, B.; Jha, S. *Adv. Synth. Catal.* **2007**, *349*, 1938–1942.

[24] Sreedhar, B.; Arundhathi, R.; Reddy, P. L.; Kantam, M. L. *J. Org. Chem.* **2009**, *74*, 7951–7954.

[25] (a) Niu, L. F.; Cai, Y.; Liang, C.; Hui, X. P.; Xu, P. F. *Tetrahedron* **2011**, *67*, 2878–2881. (b) Kao, H. L.; Chen, C. K.; Wang, Y. J.; Lee, C. F. *Eur. J. Org. Chem.* **2011**, 1776–1781. (c) Ke, F.; Qu, Y. Y.; Jiang, Z. Q.; Li, Z. K.; Wu, D.; Zhou, X. G. *Org. Lett.* **2011**, *13*, 454–457. (d) Xu, R. S.; Wan, J. P.; Mao, H.; Pan, Y. J. *J. Am. Chem. Soc.* **2010**, *132*, 15531–15533. (e) Feng, Y. S.; Li, Y. Y.; Tang, L.; Wu, W.; Xu, H. J. *Tetrahedron Lett.* **2010**, *51*, 2489–2492. (f) Ramana, T.; Saha, P.; Das, M.; Punniyamurthy, T. *Org. Lett.* **2010**, *12*, 84–87. (g) Reddy, K. H. V.; Reddy, V. P.; Shankar, J.; Madhav, B.; Anil Kumar, B. S. P.; Nageswar, Y. V. D. *Tetrahedron Lett.* **2011**, *52*, 2679–2682. (h) Akkilagunta, V. K.; Kakulapati, R. R. *J. Org. Chem.* **2011**, *76*, 6819–6824. (i) Wang, H.; Jiang, L.; Chen, T.; Li, Y. *Eur. J. Org. Chem.* **2010**, *2010*, 2324–2329. (j) Xu, H.-J.; Zhao, X.-Y.; Deng, J.; Fu, Y.; Feng, Y.-S. *Tetrahedron Lett.* **2009**, *50*, 434–437.

[26] Rout, L.; Sen, T. K.; Punniyamurthy, T. *Angew. Chem., Int. Ed.* **2007**, *46*, 5583–5586.

[27] Ranu, B. C.; Saha, A.; Jana, R. *Adv. Synth. Catal.* **2007**, *349*, 2690–2696.

[28] (a) Reddy, K. H. V.; Reddy, V. P.; Shankar, J.; Madhav, B.; Kumar, B. S. P. A.; Nageswar, Y. V. D. *Tetrahedron Lett.* **2011**, *52*, 2679–2682. (b) Argüello, J. E.; Schmidt, L. C.; Peñéñory, A. B. *Org. Lett.* **2003**, *5*, 4133–4136. (c) Wu, X. M.; Hu, W. Y. *Chin. Chem. Lett.* **2012**, *23*, 391–394.

[29] See also: (a) Benaskar, F.; Engels, V.; Patil, N.; Rebrov, E. V.; Meuldijk, J.; Hessel, V.; Hulshof, L. A.; Jefferson, D. A.; Schouten, J. C.; Wheatley, A. E. H. *Tetrahedron Lett.* **2010**, *51*, 248–251. (b) Kim, J. Y.; Park, J. C.; Kim, A.; Kim, A. Y.; Lee, H. J.; Song, H.; Park, K. H. *Eur. J. Inorg. Chem.* **2009**, 4219–4223. (c) Niu, F.; Jiang, Y.; Song, W. G. *Nano Res.* **2010**, *11*, 757–763. (d) Jammi, S.; Sakthivel, S.; Rout, L.; Mukherjee, T.; Mandal, S.; Mitra, R.; Saha, P.; Punniyamurthy, T. *J. Org. Chem.* **2009**, *74*, 1971–1976.

[30] (a) Engels, V.; Benaskar, F.; Patil, N.; Rebrov, E. V.; Hessel, V.; Hulshof, L. A.; Jefferson, D. A.; Vekemans, J. A. J. M.; Karwal, S.; Schouten, J. C.; Wheatley, A. E. H.

Org. Process Res. Dev. **2010**, *14*, 644–649. (b) Benaskar, F.; Patil, N. G.; Engels, V.; Rebrov, E. V.; Meuldijk, J.; Hulshof, L. A.; Hessel, V.; Wheatley, A. E. H.; Schouten, J. C. *Chem. Eng. J.* **2012**, *207–208*, 426–439.

[31] Zhang, J. T.; Zhang, Z. H.; Wang, Y.; Zheng, X. Q.; Wang, Z. Y. *Eur. J. Org. Chem.* **2008**, 5112–5116.

[32] Kidwai, M.; Mishra, N. K.; Bansal, V.; Kumar, A.; Mozumdar, S. *Tetrahedron Lett.* **2007**, *48*, 8883–8887.

[33] (a) Isomura, Y.; Narushima, T.; Kawasaki, H.; Yonezawa, T.; Obora, Y. *Chem. Commun.* **2012**, *48*, 3784–3786. (b) Hyotanishi, M.; Isomura, Y.; Yamamoto, H.; Kawasaki, H.; Obora, Y. *Chem. Commun.* **2011**, *47*, 5750–5752.

[34] (a) Jammi, S.; Sakthivel, S.; Rout, L.; Mukherjee, T.; Mandal, S.; Mitra, R.; Saha, P.; Punniyamurthy, T. *J. Org. Chem.* **2009**, *74*, 1971–1976. (b) Swapna, K.; Murthy, S. N.; Jyothi, M. T.; Nageswar, Y. V. D. *Org. Biomol. Chem.* **2011**, *9*, 5978–5988.

[35] Suzuki, H.; Abe, H.; Osuka, A. *Chem. Lett.* **1981**, *10*, 151–152.

[36] (a) Wang, M.; Ren, K.; Wang, L. *Adv. Synth. Catal.* **2009**, *351*, 1586–1594. (b) Nishiyama, Y.; Tokunaga, K.; Sonoda, N. *Org. Lett.* **1999**, *1*, 1725–1727.

[37] Reddy, V. P.; Kumar, A. V.; Swapna, K.; Rao, K. R. *Org. Lett.* **2009**, *11*, 951–953.

[38] Singh, D.; Alberto, E. E.; Rodrigues, O. E. D.; Braga, A. L. *Green Chem.* **2009**, *11*, 1521–1524.

[39] Alves, D.; Santos, C. G.; Paixão, M. W.; Soares, L. C.; de Souza, D.; Rodrigues, O. E. D.; Braga, A. L. *Tetrahedron Lett.* **2009**, *50*, 6635–6638.

[40] Reddy, V. P.; Kumar, A. V.; Rao, K. R. *J. Org. Chem.* **2010**, *75*, 8720–8723.

[41] Biffis, A.; Scattolin, E.; Ravasio, N.; Zaccheria, F. *Tetrahedron Lett.* **2007**, *48*, 8761–8764.

[42] Wang, Z. L.; Wang, L.; Li, P. H. *Synthesis* **2008**, 1367–1372.

[43] Oishi, T.; Katayama, T.; Yamaguchi, K.; Mizuno, N. *Chem. Eur. J.* **2009**, *15*, 7539–7542.

[44] Oishi, T.; Yamaguchi, K.; Mizuno, N. *ACS Catal.* **2011**, *1*, 1351–1354.

[45] Ouali, A.; Laurent, R.; Caminade, A.-M.; Majoral, J.-P.; Taillefer, M. *J. Am. Chem. Soc.* **2006**, *128*, 15990–15991.

[46] Arundhathi, R.; Kumar, D.; Sreedhar, B. *Eur. J. Org. Chem.* **2010**, 3621–3630.

[47] Patil, N. M.; Gupte, S. P.; Chaudhari, R. V. *Appl. Catal. A-Gen.* **2010**, *372*, 73–81.

[48] (a) Islam, M.; Mondal, S.; Mondal, P.; Roy, A. S.; Tuhina, K.; Mobarok, M.; Paul, S.; Salam, N.; Hossain, D. *Catal. Lett.* **2011**, *141*, 1171–1181. (b) For a recent report on tandem azidation and C-H amination of biphenyl halides, see: Ou, Y.; Jiao, N. *Chem. Commun.* **2013**, *49*, 3473–3475. (c) Khatri, P. K.; Jain, S. L. *Tetrahedron Lett.* **2013**, *54*, 2740–2743.

[49] Bhadra, S.; Sreedhar, B.; Ranu, B. C. *Adv. Synth. Catal.* **2009**, *351*, 2369–2378.

[50] Chen, C. K.; Chen, Y. W.; Lin, C. H.; Lin, H. P.; Lee, C. F. *Chem. Commun.* **2010**, 282–284.

[51] She, J.; Jiang, Z.; Wang, Y. G. *Tetrahedron Lett.* **2009**, *50*, 593–596.

[52] Benyahya, S.; Monnier, F.; Taillefer, M.; Man, M. W. C.; Bied, C.; Ouazzani, F. *Adv. Synth. Catal.* **2008**, *350*, 2205–2208.

[53] Benyahya, S.; Monnier, F.; Man, M. W. C.; Bied, C.; Ouazzani, F.; Taillefer, M. *Green Chem.* **2009**, *11*, 1121–1123.

[54] Keller, M.; Ianchuk, M.; Ladeira, S.; Taillefer, M.; Caminade, A.-M.; Majoral, J.-P.; Ouali, A. *Eur. J. Org. Chem.* **2012**, *2012*, 1056–1062.

[55] Islam, M.; Mondal, S.; Manir, M.; Singha Roy, A.; Mondal, P.; Hossain, D. *Chin. J. Chem.* **2010**, *28*, 1810–1820.

[56] Zhang, P.; Yuan, J.; Li, H.; Liu, X.; Xu, X.; Antonietti, M.; Wang, Y. *RSC Adv.* **2013**, *3*, 1890–1895.

INDEX

Copper-Mediated Cross-Coupling Reactions, First Edition. Edited by Gwilherm Evano and Nicolas Blanchard.